Bergey's Manual® of
DETERMINATIVE
BACTERIOLOGY

Ninth Edition

John G. Holt
Noel R. Krieg
Peter H. A. Sneath
James T. Staley
Stanley T. Williams

Williams & Wilkins

BALTIMORE • PHILADELPHIA • HONG KONG
LONDON • MUNICH • SYDNEY • TOKYO

A WAVERLY COMPANY

Editor: William R. Hensyl
Copy Editor: Linda E. Forlifer
Illustration Planner: Lorraine Wrzosek
Cover Designer: Wilma E. Rosenberger
Production Coordinator: Barbara J. Felton

Copyright © 1994
Williams & Wilkins
428 East Preston Street
Baltimore, Maryland 21202, USA

Printed in the United States of America

Library of Congress Cataloging-in-Publication Data

Bergey's manual of determinative bacteriology / [edited by] John G. Holt . . . [et al.].—9th ed.
    p. cm.
   Rev. ed. of: Shorter Bergey's manual of determinative bacteriology.
   Includes index.
   ISBN 0-683-00603-7
   1. Bacteriology—Classification. 2. Bacteriology—Terminology. I. Bergey, D. H. (David Hendricks), 1860–1937. II. Holt, John
     G. III. Shorter Bergey's manual of determinative bacteriology.
   QR81.A5 1993
   589.9'0012—dc20

                            CIP
                       93-8438
          93 94 95 96 97
     1 2 3 4 5 6 7 8 9 10

The illustration on the cover is a phase-contrast photomicrograph [Zeiss plan-apochromat, 100x, oil immersion objective] of rumen fluid taken by Frank B. Dazzo, Department of Microbiology and the Center for Microbial Ecology, Michigan State University. The sample was collected by Melvin Yokoyama, Department of Animal Science and the Center for Microbial Ecology, Michigan State University.

Board of Trustees of Bergey's Manual Trust

CONTRIBUTORS

R. C. W. Berkeley
Eberhard Bock
David R. Boone
Don J. Brenner
Richard W. Castenholz
G. Colman
Tom Cross
Cecil Cummins
Michael Goodfellow
William D. Grant
Dorothy Jones
Helmut König
Hans-Peter Koops
J. Gijs Kuenen
John M. Larkin
Hubert Lechevalier
Romano Locci
Alan J. McCarthy
Gloria Maestrojuán
Norbert Pfenning
Hans Reichenbach
Lesley A. Robertson
Karl O. Stetter
Leana V. Vasilyeva
Friedrich Widdel
George A. Zavarzin

and the 263 other authors of sections in
Bergey's Manual of Systematic Bacteriology
from whose contributions material was taken.

PREFACE

The ninth edition of *Bergey's Manual of Determinative Bacteriology* is a departure from past editions that attempted, usually inadequately, to combine systematic and determinative information. Systematic information will continue to be found in *Bergey's Manual of Systematic Bacteriology*, with the *Determinative* manual serving as a reference to aid in the identification of unknown bacteria. This ninth edition is intended solely for the identification of those bacteria that have been described and cultured. They are a small number and represent only a fraction of those existing in nature. For example, the cover of the manual is a photomicrograph of freshly collected rumen fluid, and it is quite possible that many of the bacteria pictured there are not in this book! Much work remains to be done and future *Bergey's Manuals* will only become larger.

Note that the arrangement of the book is strictly phenotypic; no attempt has been made to offer a natural classification. The arrangement chosen is utilitarian and is intended to aid in the identification of bacteria. We have divided the bacteria into 35 groups, which are comparable to the "Parts" in the eighth edition and the "Sections" in the *Systematic* volumes. These groups are not meant to be formal taxonomic ranks, but are a continuation of our tradition of dividing the bacteria into easily recognized phenotypic groups. We feel this arrangement is most useful for diagnostic purposes.

The book was compiled by abstracting the phenotypic information contained in the four volumes of *Bergey's Manual of Systematic Bacteriology*. Introductory material concerning identification and a key to the groups were added. The past decade has seen an explosion in the description of new taxa of bacteria. We have attempted to include as many of them as possible, but, in a manual of this type with its varied production schedule, not all of the new taxa could be included. For inclusion in this manual, we had to set a cut-off date of January 1991 for valid publication. In some cases, we have been able to include more recent taxa and have taken their descriptions directly from the original publications.

The editors wish to express their appreciation to Betty J. Caldwell, Project Secretary at Trust Headquarters, who supervised the word processing and painstakingly typed many pages of text and tables. We are also indebted to Doris Scott of Office Services, Michigan State University, for typing much of the text. The Trust is grateful to the Department of Microbiology at Michigan State University for housing its headquarters and editorial office and for providing a supportive and congenial atmosphere for microbial systematics. We also wish to thank the following people who have kindly and competently reviewed parts of this book: Marvin P. Bryant, Jeffery C. Burnham, Ercole Canale-Parola, Johanna Döbereiner, Claudia Fendrich, James G. Ferry, Howard Gest, Arthur P. Harrison, Barry Holmes, Kari Hovind-Hougen, William E. Inniss, Arnold F. Kaufmann, D. P. Kelly, John M. Larkin, Michael Madigan, Robert A. Mah, E. G. Mulder, Aharon Oren, John L. Penner, John Postgate, Bruno Pot, H. D. Raj, Bernhard Schink, Robert M. Smibert, William R. Strohl, Joseph G. Tully, Anita Van Landschoot, Antonio Ventosa, and Wolfram Zillig.

Comments on this edition are welcomed and should be directed to the Bergey's Manual Trust, Department of Microbiology, Giltner Hall, Michigan State University, East Lansing, MI, USA, 48824-1101. E-mail: bergeys@msu.edu

John G. Holt

CONTENTS

Contents

GROUP 5 FACULTATIVELY ANAEROBIC GRAM-NEGATIVE RODS 175

Contents

Contents

xiv

Contents

Contents

CHAPTER I

USING THIS MANUAL

Although this edition of *Bergey's Manual of Determinative Bacteriology* contains much information derived from the four volumes of *Bergey's Manual of Systematic Bacteriology,* it should not be considered an abridged version of the latter work. Instead, it deals specifically with one aspect of bacterial taxonomy, viz., the **identification of bacteria.** It has been specially designed for those persons who are primarily engaged in the identification of bacteria and who are not concerned with classification (arranging organisms into taxa) or nomenclature (naming of organisms). Moreover, it has been updated with regard to various nomenclatural changes and new taxa that have been described since the *Systematic* volumes were published.

Although this book contains much information that is useful for bacterial identification, it does not contain the wealth of descriptive information on genera and species that is provided by *Bergey's Manual of Systematic Bacteriology.* For example, it does not contain detailed descriptions of the species, information pertaining to the classification and nomenclature of bacteria, or methods for enrichment and isolation of various species. Also, we have not included identification schemes for certain groups of genera or species because either they require very sophisticated methods for cultivation and/or identification (e.g., the rickettsias) or there is a lack of sufficient differential characteristics (e.g., species of *Bacillus*).

A General Approach to Using the Manual

It is assumed that the reader has isolated a microorganism and now wishes to identify it. The following six steps provide a general approach to this task.

Step 1. The Nature of Bacterial Identification Schemes

The reader is urged to read Chapter II (*The Nature of Bacterial Identification Schemes*) to become familiar with the general principles and approaches used in bacterial identification. It is particularly important to understand that the identification schemes presented in this manual are not classification schemes. **The features that are used to differentiate various organisms often have little to do with the fundamental basis for arranging the organisms into taxonomic groups.**

Step 2. Is My Isolate a Procaryotic Microorganism or a Eucaryotic Microorganism?

This manual is used for identification of bacteria (procaryotes). Thus, the microorganism to be identified must be a bacterium rather than a microscopic eucaryote (mold, yeast, alga, or protozoan) in order to use the identification schemes in this manual. Chapter III provides a table of characteristics that differentiate procaryotes from eucaryotes.

Step 3. To Which Major Category of Bacteria Does My Isolate Belong?

For practical purposes, bacteria can be divided into four major categories:

I.   Gram-negative eubacteria that have cell walls,

II.   Gram-positive eubacteria that have cell walls,

III.   eubacteria lacking cell walls,

IV.   the archaeobacteria.

Chapter IV provides characteristics that allow differentiation of these four categories.

1

Step 4. To Which Group Does My Isolate Belong?

After assignment to the proper major bacterial category, the next step is to determine the appropriate section in the manual that discusses it. Chapter V provides listings of the groups within each major category and a brief description of the kinds of bacteria contained within each group.

Step 5. To Which Genus Does My Isolate Belong?

Most groups in the manual provide one or more tables or keys indicating characteristics that can be used to differentiate the genera within the group.

Step 6. To Which Species Does My Isolate Belong?

Most genus descriptions are accompanied by one or more tables that allow differentiation of the species contained in the genus.

CHAPTER II

THE NATURE OF BACTERIAL IDENTIFICATION SCHEMES

Identification schemes are not classification schemes, although there may be a superficial similarity. An identification scheme for a group of organisms can be devised only **after** that group has first been classified (i.e., recognized as being different from other organisms); it is based on one or more characters or on a pattern of characters that all of the members of the group have and that other groups do not have. The characters used are often not those that were involved in classification of the group; for example, classification might be based on a DNA/DNA hybridization study, whereas identification might be based on a phenotypic character that is found to correlate well with the genetic information. In general, the characters chosen for an identification scheme should be **easily determinable,** whereas those used for classification may be quite difficult to determine (such as DNA homology values). The characters should also be **few in number,** whereas classification may involve large numbers of characters, such as in a numerical taxonomy study. These ideal features of an identification scheme may not always be possible, particularly with genera or species that are not susceptible to being characterized by traditional biochemical or physiological tests. In such cases, one may need to resort to relatively difficult procedures to achieve an accurate identification—procedures such as polyacrylamide gel electrophoresis (PAGE) of cellular proteins, cellular lipid composition, or nucleic acid hybridization.

Serological reactions, which generally have only limited value in classification, often have enormous value for identification. Slide agglutination tests, fluorescent antibody techniques, and other serological methods can be performed simply and rapidly and are usually highly specific; therefore, they offer a means for achieving quick, presumptive identification of bacteria. Their specificity is frequently not absolute, however, and confirmation of the identification by additional physiological or biochemical tests is usually required.

With many genera and species, identification may not be based on only a few tests, but rather on the pattern given by applying a whole battery of tests. The members of the family *Enterobacteriaceae* represent one example of this. To alleviate the need for inoculating large numbers of tubed media, a variety of convenient and rapid multitest systems have been devised and are commercially available for use in identifying various taxa, particularly those of medical importance. For a good summary of some of these systems see D'Amato et al., 1991, Substrate profile systems for the identification of bacteria and yeasts by rapid and automated approaches, pp. 128–136, *In* Balows, Hausler, Herrmann, Isenberg, and Shadomy (Eds.), *Manual of Clinical Microbiology,* 5th Ed., American Society for Microbiology, Washington, D.C. New systems are being developed continually. Each manufacturer provides charts, tables, coding systems, computer databases, and characterization profiles for use with the particular multitest system being offered.

Genetic probes represent the newest technology for definitive identification of bacteria. The technique is based on the detection of a specific portion of an organism's genetic material. A probe consists of a small piece of DNA which has a nucleotide sequence unique to a particular species or genus of bacteria. The probe is labeled with a radioisotope, fluorescent dye, or biotin and allowed to react with an organism's denatured DNA. If the organism's DNA has a sequence complementary to that of the probe, the probe DNA will bind to it by complementary base pairing. Fluorescent-labeled DNA probes to ribosomal RNA genes are now available to characterize living organisms under the microscope.

The Need for Standardized Test Methods

One difficulty in devising identification schemes is that the results of characterization tests may vary depending on the size of the inoculum, the incubation temperature, the length of the incubation period, the composition of the medium, the surface-to-volume ratio of the medium, and the criteria used to define a "positive" or "negative" reaction. Therefore, the results of characterization tests obtained by one laboratory often do not match exactly those obtained by another laboratory, although the results within each laboratory may be quite consistent. The blind acceptance of an identification scheme without reference to the particular conditions employed by those who devised the scheme can lead to error (and, unfortunately, such conditions are not always specified). Ideally, it would be desirable to standardize the conditions used for testing various characteristics, but this is easier said than done, especially on an international basis. The use of commercial multitest systems offers some hope of increasing the standardization among various laboratories because of the high degree of quality control exercised over the media and reagents, but no one system has yet been agreed on for universal use for any given taxon. **It is therefore always advisable to include strains whose identity has been firmly established** (type or reference strains, available from national culture collections) **for comparative purposes when making use of an identification scheme,** to make sure that the scheme is valid for the conditions used in one's own laboratory. It is strongly recommended that one compare the characteristics of an isolate suspected of belonging to a particular species to those of the type strain or to an established reference strain of that species. The type strains of bacterial species are given in:

Skerman, V.B.D., V. McGowan, and P.H.A. Sneath. 1989. *Approved Lists of Bacterial Names.* Amended edition. American Society for Microbiology, Washington, D.C.

Moore, W.E.C., and L.V.H. Moore. 1989. Index of the bacterial and yeast nomenclatural changes published in the *International Journal of Systematic Bacteriology* since the 1980 Approved Lists of Bacterial Names (1 January 1980 to 1 January 1989). American Society for Microbiology, Washington, D.C.

Lists of validation of the publication of new names and new combinations previously effectively published outside the *International Journal of Systematic Bacteriology.* These lists are published periodically in the *International Journal of Systematic Bacteriology.*

Articles describing new species, as published in issues of the *International Journal of Systematic Bacteriology.*

A listing of culture collections from which type strains can be obtained is provided in *Bergey's Manual of Systematic Bacteriology.*

The Need For Definitions of "Positive" and "Negative" Reactions

Some tests may be found to be based on plasmid- or phage-mediated characteristics; such characteristics may be highly mutable and therefore unreliable for identification purposes. Even with immutable characteristics, certain tests may not be well suited for use in identification schemes because they may not give highly reproducible results (e.g., the catalase test, oxidase test, Voges-Proskauer test, and gelatin liquefaction are notorious in this regard). Ideally, a test should give reproducible results that are clearly either positive or negative, without equivocal reactions. In fact, no such test may exist. The Gram reaction of an organism may be "Gram variable," the presence of endospores in a strain that makes only a few may be very difficult to determine by staining or by heat-resistance tests, acid production from sugars may be difficult to distinguish from no acid production if only small amounts of acid are produced, and a weak growth response may not be clearly distinguishable from "no growth." A precise, although arbitrary, definition of what constitutes a "positive" and "negative" reaction is often important for a test to be useful in an identification scheme.

The Need For Pure Cultures

Although a few bacteria are so morphologically remarkable as to make them identifiable without isolation, pure cultures are nearly always a necessity before one can attempt identification of an organism using phenotypic characteristics. Newer techniques of identification using gene probes, often with enhancement by a polymerase chain reaction, can be performed on mixed populations. These techniques are presently limited to a small number of clinically important species. **The single selection of a colony from a plate does not assure purity.** This is especially true if selective media are used; live but nongrowing contaminants may often be present in or near a colony and can be subcultured along with the chosen organism. It is for this reason that **nonselective media are preferred for final isolation,** because they allow such contaminants to develop into visible colonies. Even with nonselective media, apparently well-isolated colonies should not be subcultured too soon; some contaminants may be slow-growing and may appear on the plate only after a longer incubation. Another difficulty occurs with bacteria that form extracellular slime or that grow as a network of chains or filaments; contaminants often become firmly embedded or entrapped and are difficult to separate. In the instance of cyanobacteria, contaminants frequently penetrate and live in the gelatinous sheaths that surround the cells, making pure cultures difficult to obtain.

In general, colonies from a pure culture that has been streaked on a solid medium are similar to one another, providing evidence of purity. Although this is usually true, there are exceptions, as in the case of S→R variation, capsular variants, and pigmented or nonpigmented variants, which may be selected by certain media, temperatures, or other growth conditions. Another criterion of purity is morphology: organisms from a pure culture usually exhibit a high degree of morphological similarity in stains or wet mounts. Again, there are exceptions, depending on the age of the culture, the medium used, and other growth conditions. Examples include coccoid body formation, cyst formation, spore formation, and pleomorphism. For instance, examination of a broth culture of a marine spirillum after 2 or 3 days may lead one to believe that the culture is highly contaminated with cocci unless one is previously aware that such spirilla frequently develop into thin-walled coccoid forms after active growth.

General Precautions in Bacterial Identification

The following summary is taken from "The Mechanism of Identification" by S. T. Cowan and J. Liston in the eighth edition of this manual, with some modifications:

1. Make sure that you have a pure culture.

2. Work from broad categories down to a smaller, specific category of organism.

3. Use all information available to you to narrow the range of possibilities.

4. Apply common sense at each step.

5. Use the minimum number of tests to make the identification.

6. Compare your isolate to type or reference strains of the pertinent taxon to make sure that the identification scheme being used actually is valid for the conditions in your particular laboratory.

If, as well may happen, you cannot identify your isolate from the information contained in the manual, neither despair nor immediately assume that you have isolated a new genus or species; many of the problems of microbial classification are the result of people jumping to this conclusion prematurely. When you fail to identify your isolate, check (a) its **purity,** (b) that you have carried out the **appropriate tests,** (c) that your **methods are reliable,** and (d) that you have used correctly the various keys and tables of the manual. It has been said that the most frequent cause of mistaken identity of bacteria is error in the determination of shape, Gram-staining reaction, and motility. In most cases, you should have little difficulty placing your isolate into a genus; allocation to a species or subspecies may need the help of a specialized reference laboratory.

On the other hand, it is always possible that you have actually isolated a new genus or species. A comparison of the present edition of the manual with the previous edition indicates that a number of new genera and species have been added. Some prime examples are genera such as *Oligella, Hydrogenophaga, Acidomonas, Deleya, Chryseomonas,* etc. Undoubtedly, there exist in nature a great number of bacteria that have not yet been classified and therefore cannot yet be identified by existing schemes. Yet, before describing and naming a new taxon, one must be **very sure that it is really a new taxon** and not merely the result of an inadequate identification.

CHAPTER III

PROCARYOTE OR EUCARYOTE?

Although it is usually easy to distinguish bacteria from eucaryotic microorganisms, in some instances it may be difficult, especially with bacteria that exhibit some attribute similar to those of microscopic eucaryotes. For instance, the hyphae formed by actinomycetes might be confused with the hyphae formed by molds; a fascicle of bacterial flagella could give the misleading impression of being a single eucaryotic flagellum; the ability of spirochetes to twist and contort their shape is suggestive of the flexibility exhibited by certain protozoa; some eucaryotic cells are as small as bacteria, and some bacteria are as large as eucaryotic cells. Probably the most reliable is the demonstration of the absence of a nuclear membrane in procaryotes, but this involves electron microscopy of thin sections. Other features range from those that are relatively easy to determine to the molecular biological features that require sophisticated methods. Fluorescent-labeled gene probes that are being developed will easily distinguish between procaryotic and eucaryotic cells.

Bacteria (procaryotes) can be described as follows [1]:

single cells or simple associations of similar cells (0.2–10.0 μm in smallest dimension) forming a group defined by cellular not organismal properties.[2] The nucleoplasm (genophore) is never separated from the cytoplasm by a unit-membrane system (nuclear membrane). Cell division is not accompanied by cyclical changes in the texture or staining properties of either nucleoplasm or cytoplasm; a microtubular (spindle) system is not formed. The plasma membrane (cytoplasmic membrane) is frequently complex in topology and forms vesicular, lamellar, or tubular intrusions into the cytoplasm; vacuoles and replicating cytoplasmic organelles independent of the plasma membrane system (chlorobium vesicles, gas vacuoles) are relatively rare and are enclosed by nonunit membranes. Respiratory and photosynthetic functions are associated with the plasma-membrane system in those members possessing these physiological attributes, although in the cyanobacteria there may be an independence of plasma and thylakoid membranes. Ribosomes of the 70S type (except for one group—the archaeobacteria—with slightly higher S values) are dispersed in the cytoplasm; an endoplasmic reticulum with attached ribosomes is not present. The cytoplasm is immobile; cytoplasmic streaming, pseudopodial movement, endocytosis, and exocytosis are not observed. Nutrients are acquired in molecular form. Enclosure of the cell by a rigid wall is common but not universal. The cell may be nonmotile or may exhibit swimming motility (mediated by flagella of bacterial type) or gliding motility on surfaces.

In organismal terms, these ubiquitous inhabitants of moist environments are predominantly unicellular microorganisms, but filamentous, mycelial, or colonial forms also occur. Differentiation is limited in scope (holdfast structures, resting cells, and modifications in cell shape). Mechanisms of gene transfer and recombination occur, but these processes never involve gametogenesis and zygote formation.

Some characteristics that differentiate between procaryotes and eucaryotes are listed in Table III.1.

---

[1] Terms are defined in the Glossary at the end of Chapter V.

[2] That is, by the structure and components of the cells of an organism rather than by the properties of the organism as a whole.

Table III.1   Some differential characteristics of procaryotes and eucaryotes [a]

Characteristic	Procaryotes	Eucaryotes
CYTOLOGICAL FEATURES		
Nucleoplasm (genophore, nucleoid) separated from the cytoplasm by a unit-membrane system (nuclear membrane)	−	+
Size of smallest dimension of cells (width or diameter):		
Usually 0.2–2.0 μm	+ [b]	−
Usually >2.0 μm	−	+
Mitochondria present	−	+
Chloroplasts present in phototrophs	−	+
Vacuoles, if present, enclosed by unit membranes	−	+
Gas vacuoles present [c]	D	−
Golgi apparatus present	−	D
Lysosomes present	−	D
Microtubular systems present	− [d]	D
Endoplasmic reticulum present	−	+
Ribosome location:		
Dispersed in the cytoplasm	+	−
Attached to an endoplasmic reticulum	−	+
Cytoplasmic streaming, pseudopodial movement, endocytosis, and exocytosis	−	D
Cell division accompanied by cyclical changes in the texture or staining properties of either nucleoplasm or cytoplasm	−	+
Flagella, if present:		
Diameter:		
0.01–0.02 μm	+	−
ca. 0.2 μm	−	+
In cross-section, have a characteristic "9 + 2" arrangement of microtubules	−	+
Endospores present [e]	D	−
ANTIBIOTIC SUSCEPTIBILITY		
Susceptible to:		
Penicillin, streptomycin, or other antibiotics specific for procaryotes	D	−
Cycloheximide or other antibiotics specific for eucaryotes	−	D
FEATURES BASED ON CHEMICAL ANALYSIS		
Poly-β-hydroxybutyrate present (as a storage compound in cytoplasmic inclusions)	D	−
Teichoic acids present (in cell walls)	D	−
Polyunsaturated fatty acids possibly present (in membranes)	Rare	Common
Branched-chain iso- or anteiso-fatty acids and cyclopropane fatty acids present (in membranes)	Common	Rare

Table III.1 *(continued)*

Characteristic	Procaryotes	Eucaryotes
Sterols present (in membranes)	$-^f$	Common
Diaminopimelic acid present (in cell walls)	D^g	−
Muramic acid present (in cell walls)	D^h	−
Peptidoglycan (containing muramic acid) present in cell walls	D^h	−
NUTRITION		
Nutrients acquired by cells as soluble small molecules; to serve as sources of nutrients, particulate matter or large molecules must first be hydrolyzed to small molecules by enzymes external to the plasma membrane.	+	D
METABOLIC FEATURES		
Respiratory and photosynthetic functions and associated pigments and enzymes (e.g., chlorophylls, cytochromes), if present, are associated with the plasma membrane or invaginations thereof.	$+^i$	−
Chemolithotrophic type of metabolism occurs (inorganic compounds can be used as electron donors by organisms that derive energy from chemical compounds).	D	−
Ability to fix N_2	D	−
Ability to dissimilate NO_3^- to N_2O or N_2	D	−
Methanogenesis	D	−
Ability to carry out anoxygenic photosynthesis	D	−
ENZYMIC FEATURES		
Type of superoxide dismutase:		
Cu-Zn type	$-^j$	+
Mn and/or Fe type	+	$-^k$
REPRODUCTIVE FEATURES		
Cell division occurs by mitosis, and a microtubular (spindle) system is present.	−	+
Meiosis occurs.	−	D
Mechanisms of gene transfer and recombination, if they occur, involve gametogenesis and zygote formation.	−	+
MOLECULAR BIOLOGICAL PROPERTIES		
Number of chromosomes present per nucleoid	usually 1	usually > 1
Chromosomes circular	+	−
Chromosomes linear	−	+
Sedimentation constant of ribosomes:		
70S	+	$-^l$
80S	−	+
Sedimentation constants of ribosomal RNA:		
16S, 23S, 5S	+	−
18S, 28S, 5.85S, 5S	−	+

Table III.1 *(continued)*

Characteristic	Procaryotes	Eucaryotes
MOLECULAR BIOLOGICAL PROPERTIES *(continued)*		
First amino acid to initiate a polypeptide chain during protein synthesis:		
Methionine	D	+
N-Formylmethionine	D	–
Messenger-RNA binding site at AUCACCUCC at 3' end of 16S or 18S ribosomal RNA	+	–

[a] Symbols: +, positive; –, negative; D, differs among organisms.

[b] A few bacteria (e.g., certain treponemes, mycoplasmas, *Haemobartonella*) may have a width as small as 0.1 μm; a few bacteria (e.g., *Achromatium, Macromonas*) may have a width greater than 10 μm.

[c] Gas vacuoles are not bounded by a unit membrane. The vesicles composing the vacuoles can be caused to collapse by the sudden application of hydrostatic pressure—a feature essential to their identification.

[d] However, certain intracellular fibrils that may be microtubules have been reported in *Spiroplasma,* certain spirochetes, the cyanobacterium *Anabaena,* and in bacterial L forms.

[e] Bacterial endospores are usually resistant to a heat treatment of 80°C or more for 10 min, however, some types of endospores may be killed by this heat treatment and may require testing at lower temperatures.

[f] Except in membranes of most mycoplasmas.

[g] Present in virtually all Gram-negative eubacteria and in many Gram-positive bacteria.

[h] Present in walled eubacteria except chlamydiae; absent in archaeobacteria.

[i] However, in cyanobacteria there may be an independence of cytoplasmic membrane and thylakoid membranes.

[j] With a few rare exceptions, such as certain photobacteria.

[k] Except in mitochondria, in which the Mn type occurs.

[l] Except in mitochondria and chloroplasts, which have 70S ribosomes.

CHAPTER IV

THE FOUR MAJOR CATEGORIES OF BACTERIA

The four major (on a phenotypic basis) categories of bacteria are described briefly below, followed by tables of characteristics useful for differentiating some of these categories.

I. Gram-negative Eubacteria That Have Cell Walls

These are procaryotes that have a complex (Gram-negative type) cell-wall profile consisting of an outer membrane and an inner, thin peptidoglycan layer (which contains muramic acid and is present in all but a few organisms that have lost this portion of wall) and a variable complement of other components outside or between these layers. They usually stain Gram negative. Cell shapes may be spheres, ovals, straight or curved rods, helices, or filaments; some of these forms may be sheathed or capsulated. Reproduction is by binary fission, but some groups show budding, and a rare group (*Pleurocapsales*) shows multiple fission. Fruiting bodies and myxospores may be formed by the myxobacteria. Swimming motility, gliding motility, and nonmotility are commonly observed. Members of the division may be phototrophic or nonphototrophic (both lithotrophic and heterotrophic) bacteria and include aerobic, anaerobic, facultatively anaerobic, and microaerophilic species; some members are obligate intracellular parasites.

II. Gram-positive Eubacteria That Have Cell Walls

These are procaryotes with a cell-wall profile of the Gram-positive type; reaction with Gram's stain generally, but not always, is positive. Cells may be spheres, rods, or filaments; the rods and filaments may be non-branching, but many show true branching. Cellular reproduction is generally by binary fission; some produce spores as resting forms (endospores or spores on hyphae). They are not photosynthetic; generally, they are chemosynthetic heterotrophs and include aerobic, anaerobic, facultatively anaerobic, and microaerophilic species. The members of this division include simple asporogenous and sporogenous bacteria, as well as the actinomycetes and their relatives.

III. Eubacteria Lacking Cell Walls

These are procaryotes that lack cell walls (commonly called the mycoplasmas and including the class *Mollicutes*) and do not synthesize the precursors of peptidoglycan. They are enclosed by a unit membrane, the plasma membrane. The cells are highly pleomorphic and range in size from large deformable vesicles to very small (0.2 μm), filterable elements. Filamentous forms with branching projections are common. Reproduction may be by budding, fragmentation, and/or binary fission. Some groups show a degree of regularity of form due to the placing of internal structures. Usually, they are nonmotile, but some species show a form of gliding motility. No resting forms are known. Cells stain Gram negative. Most require complex media for growth (high-osmotic-pressure surroundings) and tend to penetrate the surface of solid media forming characteristic "fried egg" colonies. The organisms resemble the naked L-forms that can be generated from many species of bacteria (notably Gram-positive eubacteria) but differ in that the mycoplasmas are unable to revert and make cell wall. Most species are further distinguished by requiring both cholesterol and long-chain fatty acids for growth; unesterified cholesterol is a unique component of the membranes of both sterol-requiring and nonrequiring species if present in the medium. The guanine plus cytosine content of ribosomal RNA is 43–48 mol% (lower than the 50–54 mol% of walled Gram-negative and Gram-positive eubacteria); the guanine plus cytosine content of the DNA is also comparatively low, 23–46 mol%, and the genome size of the mycoplasmas is less than that of other procaryotes at $0.5–1.0 \times 10^9$ daltons. The mycoplasmas may be saprophytic, parasitic, or pathogenic, and the pathogens cause diseases of animals, plants, and tissue cultures.

IV. The Archaeobacteria

The archaeobacteria are predominantly terrestrial and aquatic microbes, occurring in anaerobic or hypersaline or hydrothermally and geothermally heated environments; also, some occur as symbionts in animal digestive tracts. They consist of aerobes, anaerobes, and facultative anaerobes that grow chemolithoautotrophically, organotrophically, or facultatively organotrophically. Archaeobacteria may be mesophiles or thermophiles, with some species growing even above 100°C.

A unique biochemical feature of archaeobacteria is the presence of glycerol isopranyl ether lipids. The lack of murein (peptidoglycan-containing muramic acid) in cell walls makes archaeobacteria insensitive to β-lactam antibiotics. The "common arm" of the tRNAs contains pseudouridine or 1-methylpseudouridine instead of ribothymidine. The sequences of 5S, 16S, and 23S rRNAs are very different from the corresponding ones in eubacteria and eucaryotes.

Archaeobacteria share some molecular features with eucaryotes: (a) the elongation factor 2 (EF-2) contains the amino acid diphthamide and is therefore ADP-ribosylable by diphtheria toxin, (b) amino acid sequences of the ribosomal "A" protein exhibit sequence homologies with the corresponding eucaryotic (L-7/L-12) protein, (c) the methionyl initiator tRNA is not formylated, (d) some tRNA genes contain introns, (e) the aminoacyl stem of the initiator tRNA terminates with the base pair "AU," (f) the DNA-dependent RNA polymerases are multicomponent enzymes and are insensitive to the antibiotics rifampicin and streptolydigin, (g) like the α-DNA polymerases of eucaryotes, the replicating, archaeobacterial DNA polymerases are not inhibited by aphidicolin or butylphenyl-dGTP, and (h) protein synthesis is inhibited by anisomycin but not by chloramphenicol.

Autotrophic archaeobacteria do not assimilate carbon dioxide via the Calvin cycle. In *Methanobacterium*, CO_2 is fixed via an acetyl-CoA pathway, whereas in *Acidianus* and *Thermoproteus*, autotrophic CO_2 is fixed via a reductive tricarboxylic acid pathway. Fixation of N_2 has been demonstrated by some methanogens.

Gram stain results may be positive or negative within the same order because of very different types of cell envelopes. Gram-positive species possess pseudomurein, methanochondroitin, and heteropolysaccharide cell walls; Gram-negative cells have (glyco-) protein surface layers. The cells may have a diversity of shapes, including spherical, spiral, plate- or rod-shaped; unicellular and multicellular forms in filaments or aggregates also occur. The diameter of an individual cell may be 0.1–>15 μm, and the length of the filaments can be up to 200 μm. Multiplication is by binary fission, budding, constriction, fragmentation, or unknown mechanisms. Colors of cell masses may be red, purple, pink, orange-brown, yellow, green, greenish black, gray, and white.

The major groups of archaeobacteria include (a) the methanogenic archaeobacteria, (b) the archaeobacterial sulfate reducers, (c) the extremely halophilic archaeobacteria, (d) the cell wall-less archaeobacteria, and (e) the extremely thermophilic S^0-metabolizers.

Table IV.1 Some characteristics differentiating eubacteria from archaeobacteria [a]

Characteristic	Eubacteria	Archaeobacteria
GENERAL MORPHOLOGIC AND METABOLIC FEATURES		
Strict anaerobes that form methane as the predominant metabolic end product from H_2–CO_2, formate, acetate, methanol, methylamine or H_2–methanol. Cells exhibit a blue-green epifluorescence when excited at 420 nm.	–	D
Strict anaerobes that form H_2S from sulfate by dissimilatory sulfate reduction. Extremely thermophilic (growth up to 92°C). Exhibit blue-green epifluorescence when excited at 420 nm.	–	D
Cells stain Gram negative or Gram positive and are aerobic or facultatively anaerobic chemoorganotrophs. Rods and regular to highly irregular cells occur. Cells require a high concentration of NaCl (1.5 M or above). Neutrophilic or alkaliphilic. Mesophilic or slightly thermophilic (up to 55°C). Some species contain the red-purple photoactive pigment bacteriorhodopsin and are able to use light for ATP synthesis.	–	D
Thermoacidophilic, aerobic, coccoid cells lacking a cell wall.	–	D
Obligately thermophilic, aerobic, facultatively anaerobic, or strictly anaerobic Gram-negative rods, filaments, or cocci. Optimal growth temperature between 70°C and 105°C. Acidophiles and neutrophiles. Autotrophic or heterotrophic. Most species are sulfur metabolizers.	–	D
CELL WALLS (IF PRESENT)		
Contain muramic acid	+	–
ANTIBIOTIC SUSCEPTIBILITY		
Susceptible to penicillin or other antibiotics that inhibit synthesis of muramic acid-containing peptidoglycan	D	–
LIPIDS		
Membrane phospholipids consist of: Long chain alcohols (phytanols) that are ether linked to glycerol to form C_{20} diphytanyl glycerol diethers or C_{40} dibiphytanyl diglycerol tetraethers	–	+
Long chain aliphatic fatty acids that are ester linked to glycerol	+	–
Pathway used in formation of lipids:		
Mevalonate pathway	–	+
Malonate pathway	+	–
MOLECULAR BIOLOGICAL FEATURES		
Ribothymine is present in the "common arm" of the tRNAs.	usually +	–
Pseudouridine or 1-methylpseudouridine is present in the "common arm" of the tRNAs.	–	+

Table IV.1   *(continued)*

Characteristic	Eubacteria	Archaeobacteria
First amino acid to initiate a polypeptide chain during protein synthesis:		
Methionine	−	+
N-Formylmethionine	+	−
Aminoacyl stem of the initiator tRNA terminates with the base pair "AU."	−	+
Protein synthesis by ribosomes inhibited by:		
Anisomycin	−	+
Kanamycin	+	−
Chloramphenicol	+	−
ADP-Ribosylation of the peptide elongation factor EF-2 is inhibited by diphtheria toxin.	−	+
Elongation factor 2 (EF-2) contains the amino acid diphthamide.	−	+
Some tRNA genes contain introns.	−	+
DNA-dependent RNA polymerases are:		
Multicomponent enzymes	−	+
Inhibited by rifampicin and streptolydigin	+	−
Replicating DNA polymerases are inhibited by aphidicolin or butylphenyl-dGTP.	+	−

[a]   Symbols: +, positive; −, negative; D, differs among organisms.

Table IV.2   Some characteristics differentiating eubacterial Major Category I from eubacterial Major Category II [a]

Characteristic	Major Category I	Major Category II
CYTOLOGICAL FEATURES		
Gram staining reaction	−[b]	+[c]
An outer membrane is present (in the cell wall) in addition to the plasma (cytoplasmic) membrane.	+	−
Acid-fast staining	−[d]	D[e]
Endospores present	D	D[f]
Filamentous growth with hyphae that show true branching	−	D
LOCOMOTION		
Gliding motility occurs.	D	−
CHEMICAL FEATURES		
Percentage of the dry weight of the cell wall that is represented by lipid	Usually 11–22%	Usually < 4%[g]
Teichoic or lipoteichoic acids present	−	D

Table IV.2   *(continued)*

Characteristic	Major Category I	Major Category II
Lipopolysaccharide (LPS) [h] occurs (in the outer membrane of the cell wall).	+	–
2-Keto-3-deoxyoctonate (KDO) present [i]	D	–
Percentage of the dry weight of the cell wall is represented by peptidoglycan.	Usually < 10%	Usually > 10%
Mycolic acids present	–	D [j]
Phosphatidylinositol mannosides present	–	D [k]
Phosphosphingolipids present	D [l]	–
METABOLIC FEATURES		
Energy derived by the oxidation of inorganic iron, sulfur, or nitrogen compounds	D	–
ENZYMIC FEATURES		
Citrate synthases:		
Inhibited by reduced nicotinamide adenine dinucleotide (NADH) Molecular weight:	Usually + [m]	Usually –
ca. 250,000	Usually + [n]	Usually –
ca. 100,000	Usually –	Usually +
Succinate thiokinases, molecular weight of:		
70,000–75,000	Usually –	Usually +
140,000–150,000	Usually +	Usually –

[a] Symbols: +, positive; –, negative, D, differs among organisms.

[b] The staining reaction may not be conclusive. A few bacteria that have a Gram-positive type of cell wall profile and whose walls contain teichoic acids may stain Gram negative, probably because of the thinness of the wall. For example, see the genus *Butyrivibrio*.

[c] The staining reaction may not be conclusive. A few bacteria that have a Gram-negative type of cell wall profile and whose walls contain lipopolysaccharides may stain Gram positive. For example, see the genus *Xanthobacter*.

[d] An exception may be the genus *Coxiella*.

[e] Acid-fast staining occurs in the genus *Mycobacterium* and in some *Nocardia* species.

[f] Endospores occur in the genera *Bacillus, Clostridium, Desulfotomaculum, Sporosarcina,* and *Thermoactinomyces*.

[g] Except for *Mycobacterium, Corynebacterium, Nocardia,* and other genera whose walls contain mycolic acids.

[h] LPS consists of Lipid A (a β-linked D-glucosamine disaccharide to which phosphate residues are linked at positions 1 and 4 and fatty acids are linked to both the amino and hydroxyl groups of the glucosamines), a core polysaccharide (a short acidic heteropolysaccharide), and O antigens (side chains that are polysaccharide composed of repeating units).

[i] In many but not all Gram-negative bacteria, the core polysaccharide contains KDO which, if present, can serve as an indicator of the presence of LPS.

[j] Mycolic acids occur in *Corynebacterium, Nocardia, Mycobacterium, Bacterionema, Faenia,* and *Rhodococcus*.

[k] Present in certain actinomycete and coryneform bacteria.

[l] For example, in *Bacteroides*.

[m] Known exceptions include *Acetobacter, Thermus,* and cyanobacteria.

[n] One known exception is *Thermus*.

CHAPTER V

GROUPS WITHIN THE FOUR MAJOR CATEGORIES OF BACTERIA

Each of the four major categories of bacteria described in Chapter IV is divided into various groups. The tables in this chapter provide brief descriptions of these groups.

Table V.1   The groups within Major Category I (Gram-negative eubacteria that have cell walls)

GROUP	NAME	IMPORTANT FEATURES
1 p. 27	THE SPIROCHETES	Gram-negative helically shaped cells that are highly flexible. Motile by periplasmic flagella rather than by flagella that project from the cell into the external medium. Chemoorganotrophic. Anaerobic, microaerophilic, facultatively anaerobic, or aerobic. Occur either free-living or in association with animal, mollusk, arthropod, or human hosts. Some are pathogenic.
2 p. 39	AEROBIC/MICROAEROPHILIC, MOTILE, HELICAL/VIBRIOID GRAM-NEGATIVE BACTERIA	Gram-negative cells that are helical (having one or more complete helical turns) or vibrioid (having less than one complete helical turn). Motile by polar flagella and swim in straight lines with a characteristic corkscrew-like motion. Aerobic or microaerophilic, having a respiratory type of metabolism with oxygen as the normal electron acceptor. Some also may exhibit anaerobic respiration with electron acceptors other than O_2, such as nitrate or fumarate. Chemoorganotrophic, but some can grow autotrophically with H_2 as the electron donor. Occur in soil, freshwater or marine environments, within plant roots, or in the reproductive organs, intestinal tract, and oral cavity of humans and animals. Some are pathogenic for animals or humans. Some are predatory on other microorganisms.
3 p. 65	NONMOTILE (OR RARELY MOTILE), GRAM-NEGATIVE CURVED BACTERIA	Chemoorganotrophic heterotrophs. Saprophytic. Four types of organisms are included in this group, as follows: I. Curved or C-shaped bacteria that may form rings by overlapping of the ends of a cell. Coils and helices may occur. Gas vacuoles may occur. Aerobic. Occur in soil, freshwater, or marine environments. II. Vibrioid or straight rods. Gas vacuoles are formed. Aerotolerant anaerobes, having a strictly fermentative type of metabolism. III. Bow-shaped cells with gas vacuoles. Cells arranged in coenobia of two rings or four rings (cloverleaf appearance). Pretzel-shaped cells may occur. Occur in ponds and lakes where sulfide is present and oxygen is absent. IV. Slender S-shaped cells arranged side-by-side in flat, sigmoid aggregates of four or multiples of four. Aggregates are occasionally motile. Occur in fresh and brackish waters where sulfide is present and oxygen is absent. Note: II above contains straight rods as well as curved rods.
4 p. 71	GRAM-NEGATIVE AEROBIC/ MICROAEROPHILIC RODS AND COCCI	Chemoorganotrophic heterotrophs, but some may grow autotrophically by using H_2 as an electron donor. Do not form prosthecae, stalks, sheaths, or gas vacuoles. Do not possess gliding motility. Do not reproduce by budding. Can grow under an air atmosphere and have a strictly respiratory type of metabolism with O_2 as the terminal electron acceptor. Some are also capable of anaerobic respiration with terminal electron acceptors other than O_2. Occur in soil, freshwater, or marine environments, within plant roots, or in the reproductive organs, intestinal tract, and oral cavity of humans and animals. Some are pathogenic for animals or humans.

Table V.1 *(continued)*

GROUP	NAME	IMPORTANT FEATURES
5 p. 175	**FACULTATIVELY ANAEROBIC GRAM-NEGATIVE RODS**	Chemoorganotrophic heterotrophs, but some may grow autotrophically by using H_2 as an electron donor. Do not form prosthecae, stalks, sheaths, or gas vacuoles. Do not possess gliding motility. Do not reproduce by budding. Capable of growing under an air atmosphere by a respiratory type of metabolism; also capable of growing anaerobically by fermentation. Occur either free-living or in association with animal, human, or plant hosts. Some are pathogenic.
6 p. 291	**GRAM-NEGATIVE, ANAEROBIC, STRAIGHT, CURVED, AND HELICAL RODS**	Chemoorganotrophic heterotrophs. Obtain energy by anaerobic respiration or by fermentation. Do not respire anaerobically with sulfate or other oxidized sulfur compounds or with elemental sulfur as electron acceptors.
7 p. 335	**DISSIMILATORY SULFATE- OR SULFUR-REDUCING BACTERIA**	Anaerobic. Chemoorganotrophic heterotrophs. Respire anaerobically either with sulfate and other oxidized sulfur compounds or with elemental sulfur as terminal electron acceptors. Endospores not formed.
8 p. 347	**ANAEROBIC GRAM-NEGATIVE COCCI**	Chemoorganotrophic heterotrophs. Have a strictly fermentative type of metabolism.
9 p. 351	**THE RICKETTSIAS AND CHLAMYDIAS**	Obligate intracellular parasites of eucaryotic hosts (vertebrates or arthropods). May be rod-shaped, coccoid, or pleomorphic. Many species are pathogenic.
10 p. 353	**ANOXYGENIC PHOTOTROPHIC BACTERIA**	Bacteria that contain bacteriochlorophyll and can use light as an energy source. When growing under illumination the organisms are anaerobic and do not evolve O_2 during photosynthesis. Some are also capable of growing in the dark by respiring with oxygen. Do not contain phycobiliproteins.
11 p. 377	**OXYGENIC PHOTOTROPHIC BACTERIA**	Bacteria that contain chlorophyll *a*, can use light as an energy source, and evolve O_2 in a manner similar to that of green plants. Two subdivisions are included, as follows: I. Contain chlorophyll *a* and have phycobiliproteins (allophyco-cyanin, phycocyanin and sometimes phycoerythrin). These organisms are called *cyanobacteria*. II. Contain chlorophyll *a* and chlorophyll *b* but lack phycobiliproteins.
12 p. 427	**AEROBIC CHEMOLITHOTROPHIC BACTERIA AND ASSOCIATED ORGANISMS**	Nonphototrophic. The following subdivisions can be recognized: I. Nitrifiers. Reduced inorganic nitrogen compounds (ammonia and nitrite) can be used as energy sources for growth. II. Sulfur oxidizers. Reduced inorganic sulfur compounds can be oxidized, and most organisms can utilize this as sole source of energy. III. Obligate hydrogen oxidizers. Hydrogen gas (H_2) is used as the energy source for growth, and organic sources of carbon are not used. IV. Nonprosthecate and nonstalked bacteria that produce or deposit iron and/or manganese oxides on or within the cells. V. Magnetotactic bacteria. Bacteria that exhibit tactic responses to magnetic fields. The cells contain iron-rich, electron-dense intracellular inclusions (magnetosomes).

Table V.1 *(continued)*

GROUP	NAME	IMPORTANT FEATURES

13
p. 457
	BUDDING AND/OR APPENDAGED BACTERIA	Nonphototrophic. The following subdivisions can be recognized:
		I. Bacteria having *prosthecae* (narrow living extensions of the cell and which consist of the cell and which consist of cell wall, cytoplasmic membrane, and cytoplasm).
		   A. Multiply asymetrically by budding. Buds may be produced at the tip of a prostheca or on the cell surface.
		   B. Multiply by binary transverse fission.
		II. Nonprosthecate bacteria.
		   A. Budding bacteria.
		   B. Non-budding bacteria having *stalks* (acellular appendages not containing cytoplasm, mediating attachment to surfaces).
		   C. Other bacteria
		      a. Bacteria that bear tapering filaments encrusted with manganese dioxide.
		      b. Bacteria that bear thin threadlike structures not encrusted with metal oxides
		      c. Bacteria having *stalks* (hollow conical appendages observable by light microscopy and having cross striations when viewed by electron microscopy).

14 p. 477	**SHEATHED BACTERIA**	Nonphototrophic. Aerobic. Do not exhibit gliding motility. Characterized by growing as filaments, the cells of which are enclosed with a *sheath* (a tube or extracellular material). Typically, the sheath is transparent when viewed in wet mounts by phase contrast microscopy and appears much like a microscopic plastic tubule or pipe. Occasionally the sheath is so thin and closely associated with the cells that it cannot be readily discerned by phase microscopy. Addition of 95% ethanol to the wet mount may facilitate visualization. Alternatively, a sheath may be detected within a filament if there are gaps between the cells. Sheaths may appear yellow to dark brown, owing to the deposition of iron and manganese oxides. Single cells may be motile by polar or subpolar flagella, or they may be nonmotile.
15 p. 483	**NONPHOTOSYNTHETIC, NONFRUITING GLIDING BACTERIA**	Nonphototropic rods or filaments that lack flagella but that can glide across solid surfaces. The organisms do not have a complex life cycle in which the cells swarm together in masses and form fruiting bodies. Sheaths may occur in some genera. Resting cells called myxospores are formed by some genera.
16 p. 515	**FRUITING GLIDING BACTERIA: THE MYXOBACTERIA**	Chemoorganotrophic, strictly aerobic bacteria that lack flagella but can glide across solid surfaces. Under conditions of nutrient deprivation, cells aggregate to form fruiting bodies composed of modified slime and cells, which are often brightly colored and of macroscopic dimensions. Fruiting bodies vary in complexity, from simple mounds to complex structures consisting of sporangia of characteristic shape and dimensions, which may be sessile or borne singly or in groups on simple or branched stalks. Within the fruiting bodies, the cells are converted to resting cells called either myxospores or microcysts.

Table V.2   The groups within Major Category II (Gram-positive eubacteria that have cell walls)

GROUP	NAME	IMPORTANT FEATURES
17 p. 527	**GRAM-POSITIVE COCCI**	Chemoorganotrophic, mesophilic, non-spore-forming cocci that stain Gram positive. Three major subdivisions can be recognized: I. Aerobic cocci that occur in pairs, clusters, or tetrads. Catalase positive. Cytochromes present. Teichoic acids not present in cell walls. Acid production from carbohydrates may often be negative or weak. II. Facultatively anaerobic or microaerophilic cocci that occur in pairs, chains, clusters, or tetrads. The presence of catalase, cytochromes, and cell wall teichoic acids varies. Cytochromes may or may not be present. III. Strictly anaerobic cocci that occur in pairs, chains, tetrads, or cuboidal packets. Cytochromes are absent in those genera that have been tested. The catalase reaction is usually negative, although in some instances there is a weak or pseudocatalase reaction.
18 p. 559	**ENDOSPORE-FORMING GRAM-POSITIVE RODS AND COCCI**	Bacteria that produce heat-resistant *endospores*. Endospores are best verified by testing that cultures survive a temperature of 70–80°C for 10 min, followed by cultivation under suitable conditions. Mostly motile rods or filaments; however, one genus contains motile cocci (in tetrads or cuboidal packets). Most stain Gram positive, at least in young cultures; however, one genus stains Gram negative. Strict aerobes, facultative anaerobes, microaerophils, or strict anaerobes. One anaerobic genus respires anaerobically with sulfate.
19 p. 565	**REGULAR, NONSPORING GRAM-POSITIVE RODS**	Rod-shaped cells (coccoid to elongated rods or filaments), Gram positive, nonsporing nonpigmented (one genus has slight yellow pigmentation), mesophilic, chemoorganotrophic, and grow only in complex media. Some are pathogens of animals. Three major physiological subdivisions can be recognized: I. Fermentative, saccharolytic microaerophils that do not possess heme-containing catalase, cytochromes, or menaquinones and which utilize oxygen only via flavin-containing oxidases and peroxidases. II. Aerobes or facultative anaerobes that possess cofactors and enzymes for respiration. These organisms are also able to ferment sugars, mainly to lactic acid, under oxygen-limited or anaerobic conditions. III. Strict aerobes that neither utilize glucose as a carbon or energy source nor ferment sugars to organic acids.
20 p. 571	**IRREGULAR, NONSPORING GRAM-POSITIVE RODS**	The majority are irregular rods that stain Gram positive, grow in the presence of air, and do not produce endospores. Some may exhibit club-shaped forms, branched filamentous elements, or mixtures of rods or filamentous and coccoid forms, and some may have a rod-coccus cycle. One genus stains Gram negative to Gram variable. Oxygen requirements range from strictly aerobic or facultatively anaerobic to microaerophilic or strictly anaerobic. Some are pathogens of animals or plants.
21 p. 597	**THE MYCOBACTERIA**	Aerobic, nonmotile, nonsporing, slow-growing (2–40 days) rod-shaped bacteria that are characteristically *acid-fast* (i.e., after staining they resist decolorization with acidified alcohol or with strong mineral acids). The degree of staining by the Gram method is weak. Branched filaments are formed occasionally. No aerial mycelium is formed.
22–29 p. 605	**ACTINOMYCETES**	Gram-positive bacteria that form branching filaments or hyphae that may persist as a stable mycelium or may break up into rod-shaped or coccoid elements. Motility, when present, is due to flagellation. **See Introduction to Groups 22–29 for further guidance to members of the groups.**

Table V.3 Major Category III (Cell wall-less eubacteria: The mycoplasmas or mollicutes)

GROUP	NAME	IMPORTANT FEATURES
30 p. 705	**MYCOPLASMAS**	Pleomorphic cells devoid of cell walls. Growth on agar shows characteristic "fried egg" appearance. Some require sterols for growth. May show gliding motility. Facultatively anaerobic to obligately anaerobic. Have low mol% G + C of DNA of ~23 to ~46.

Table V.4 The groups within Major Category IV *(Archaeobacteria)*

GROUP	NAME	IMPORTANT FEATURES
31 p. 719	**THE METHANOGENS**	Cells are strict anaerobes that are able to form methane as the dominating metabolic end product. H_2–CO_2, formate, acetate, methanol, methylamines, or H_2–methanol can serve as substrates. S^0 may be reduced to H_2S without gain of energy. Blue-green epifluorescence when excited at 420 nm. Cells possess coenzyme M, factor 420, factor 430, and methanopterin. RNA polymerases are of the AB′B″ type.
32 p. 737	**ARCHAEAL SULFATE REDUCERS**	Cells are strict anaerobes that are able to form H_2S from sulfate by dissimilatory sulfate reduction. Traces of methane are formed in addition. Extremely thermophilic (growth up to 92°C). Exhibit blue-green fluorescence at 420 nm under the UV microscope. Cells possess factor 420 and methanopterin, but no coenzyme M and no factor 430. RNA polymerase is of the (A+C)B′B″ type.
33 p. 739	**EXTREMELY HALOPHILIC ARCHAEOBACTERIA (HALOBACTERIA)**	Cells stain Gram negative or Gram positive, are aerobic or facultatively anaerobic chemoorganotrophs. Rods and regular to highly irregular cells occur. Cells require a high concentration of sodium chloride (1.5 M or above). Neutrophilic or alkaliphilic. Mesophilic or slightly thermophilic (up to 55°C). RNA polymerase is of the AB′B″C type. Some species contain the red-purple photoactive pigment bacteriorhodopsin and are able to use light for ATP synthesis.
34 p. 747	**CELL WALL-LESS ARCHAEOBACTERIA**	Thermoacidophilic, aerobic, coccoid cells lacking a cell envelope. Cytoplasmic membrane contains a mannose-rich glycoprotein and a lipoglycan. RNA polymerase is of the BAC type.
35 p. 749	**EXTREMELY THERMOPHILIC AND HYPERTHERMOPHILIC S^0-METABOLIZERS**	Obligately thermophilic, aerobic, facultatively anaerobic, or strictly anaerobic Gram-negative rods, filaments, or cocci. Optimal growth temperature between 70°C and 105°C. Acidophiles and neutrophiles. Autotrophic or heterotrophic growth. Most species are sulfur metabolizers. RNA polymerase is of the BAC type.

Glossary of Some Terms Used in Chapters III–V

acid-fast The property of resisting decolorization with acidified alcohol or with dilute mineral acids after being stained with a strong dye such as carbol fuchsin.

acidophilic Growing best under acidic conditions.

aerobe An organism that is capable of using oxygen as a terminal electron acceptor, can tolerate a level of oxygen equivalent to or higher than that present in an air atmosphere (21% oxygen), and has a strictly respiratory type of metabolism. Some aerobes may also be capable of growing anaerobically with electron acceptors other than oxygen.

alkaliphilic Growing best under alkaline conditions.

anaerobe An organism that is incapable of oxygen-dependent growth and cannot grow in the presence of an oxygen concentration equivalent to that present in an air atmosphere (21% oxygen). Some anaerobes may have a fermentative type of metabolism; others may carry out anaerobic respiration in which a terminal electron acceptor other than oxygen is used.

anoxygenic Not able to produce oxygen. (Contrast with oxygenic.)

autotroph An organism that uses inorganic compounds as nutrients and carbon dioxide as the sole source of carbon. (Contrast with heterotroph.)

bacteriochlorophyll A magnesium-containing porphyrin that resembles but is not identical to the chlorophyll of cyanobacteria, eucaryotic algae, and green plants. It occurs in anoxygenic phototrophic bacteria.

budding A form of reproduction in which a new cell is formed as an outgrowth from the parent cell.

catalase An enzyme that destroys hydrogen peroxide by catalyzing the reaction $2H_2O_2 \rightarrow 2H_2O + O_2$.

chemolithotroph An organism that relies on chemical compounds for energy and uses inorganic compounds as a source of electrons. (Contrast with chemoorganotroph.)

chemoorganotroph An organism that relies on chemical compounds for energy and uses organic compounds as a source of electrons. (Contrast with chemolithotroph.)

chemotroph An organism that uses chemical compounds for energy. (Contrast with phototroph.)

chlorobium vesicles Intracellular oval bodies that contain photosynthetic pigments and that underlie and are attached to the cytoplasmic membrane. Occur in *Chlorobium,* a genus of photosynthetic bacteria.

coccoid Round, resembling a coccus.

curved Bent without angles. The term does not necessarily imply a vibrioid or helical shape but may also apply to cells curved in only one plane.

cytochrome	One of various iron-porphyrins that can undergo reversible oxidation-reduction and that serves as an electron carrier in an electron transport chain.
endospore	A thick-walled spore formed in a bacterial cell. It is very resistant to being killed by heat and various other chemical and physical agents. Endospores are best verified by demonstrating that cultures survive a temperature of 70–80°C for 10 min.
facultative anaerobe	An organism that can grow well both in the abscence of oxygen and in the presence of a level of oxygen equivalent to that in an air atmosphere (21% oxygen). Some are capable of growing aerobically by respiring with oxygen and of growing anaerobically by fermentation; others have a strictly fermentative type of metabolism and do not respire with oxygen.
fermentation	An energy-yielding metabolic process in which electrons derived from an organic substrate are used ultimately to reduce an organic electron acceptor that is made by the cell itself. Neither an electron transport chain nor an exogenous terminal electron acceptor is involved. (Contrast with respiration.)
flavin	An enzyme that contains as tightly bound prosthetic groups either flavin mononucleotide (FMN) or flavin adenine dinucleotide (FAD).
fruiting body	A specialized spore-producing structure that is composed of slime and bacterial cells, is often brightly colored, and may be visible to the naked eye. Formed by myxobacteria.
gas vacuole	A cavity in the cytoplasm that contains gas of composition similar to that of the surrounding atmosphere. The boundary of the cavity is a membrane consisting of protein subunits. Gas vacuoles appear as bright areas within a cell by phase-contrast microscopy and are identified by their ability to collapse when the cells are subjected to a sudden increase in hydrostatic pressure.
gliding motility	A type of movement across surfaces that is exhibited by some bacteria devoid of flagella and incapable of swimming motility.
Gram variable	Some Gram-positive cells occur in an otherwise Gram-negative pure culture.
halophilic	Growing best at high NaCl concentrations.
helical	Curved in three dimensions and having one or more turns; corkscrew-shaped.
heterotroph	An organism that is unable to use carbon dioxide as its sole source of carbon and requires one or more organic compounds. (Contrast with autotroph.)
hyphae	The individual filaments or threads of a mycelium.
menaquinones	Vitamins K. 2-Methyl-3-*all-trans*-polyprenol-1,4-naphthoquinones possessing side chains varying in length from C_5-C_{65}. Menaquinones function as electron carriers.
mesophilic	Growing best at a moderate temperature range (25–40°C).
methanogenic	Producing methane.
microaerophile	An organism that is capable of oxygen-dependent growth but cannot grow in the presence of a level of oxygen equivalent to that present in an air atmosphere (21% oxygen). Oxygen-dependent growth occurs only at low oxygen levels. In addition to being able to respire with oxygen, some microaerophiles may be capable of respiring anaerobically with electron acceptors other than oxygen.
mutualism	A type of interaction in which two or more organisms living together benefit each other.
mycelium	A mass or network of interwoven filaments (hyphae).

N-acetylmuramic acid	An acetylated amino sugar bearing a lactyl group. Together with N-acetylglucosamine, it forms the polysaccharide backbone of murein, the rigid, cross-linked polymer that forms the rigid portion of a eubacterial cell wall.
neutrophilic	Growing best at pH values near 7.
oxygenic	Able to produce oxygen. (Contrast with anoxygenic.)
parasite	An organism that lives on or in a living animal or plant host and derives nourishment from it. (Contrast with saprophyte.)
pathogenic	Capable of causing disease.
periplasmic	Refers to the periplasmic space, the space between the cytoplasmic membrane and the outer membrane of Gram-negative bacteria.
phototroph	An organism that uses light energy for metabolism. (Contrast with chemotroph.)
phycobiliproteins	Water-soluble red or blue proteins, the colored portion of which is a linear tetrapyrrole. Found in cyanobacteria.
pleomorphic	Having various shapes.
prostheca	A semi-rigid extension of the cell wall, cytoplasmic membrane and cytoplasm and which has a diameter less than that of the cell. (Contrast with stalk.)
respiration	An energy-yielding metabolic process in which electrons from an oxidizable substrate are transferred by a series of oxidation-reduction reactions (i.e., via an electron-transport chain) to an exogenous terminal electron acceptor such as oxygen, nitrate, or fumarate). (Contrast with fermentation.)
RNA polymerase	An enzyme that catalyzes the synthesis of ribonucleic acid (RNA) by using deoxyribonucleic acid (DNA) as a template.
saprophyte	An organism that lives on dead organic matter.
sessile	Attached directly, not by means of some intervening structure.
sheath	A tube of extracellular material resembling a microscopic pipe enclosing the bacterial cells.
spinae	Hollow conical appendages observable by light microscopy and having cross-striations when viewed by electron microscopy.
sporangia	Spore-containing structures.
stalk	A nonliving ribbon-like or tubular appendage, excreted by a bacterial cell, that mediates attachment to a surface. (Contrast with prostheca.)
teichoic acid	A polymer composed of alternating units of ribitol-phosphate or glycerol-phosphate to which may be attached various sugars and amino acids.
thermoacidophilic	Growing best at high temperatures under acidic conditions.
thermophilic	Growing best at high temperatures (above 45°C).
vibrioid	Curved in three dimensions but having less than one complete turn.

GROUP 1

THE SPIROCHETES

The spirochetes are **helically shaped, motile bacteria,** 0.1–3.0 × 5–250 µm. They are unicellular, with one possible exception (*Spirochaeta plicatilis*). They stain Gram negative. The outermost structure of the helical cell is a multilayered **outer membrane,** often referred to as the "outer sheath" or "outer cell envelope" (Figure 1.1).

Figure 1.1. Schematic representation of a spirochete. The broken line indicates the outer sheath (outer cell envelope). The area delimited by the thick, solid line adjacent to the broken line represents the protoplasmic cylinder. The circles near the ends of the protoplasmic cylinder indicate the insertion points of the periplasmic flagella. The solid thin lines wound around the protoplasmic cylinder are the periplasmic flagella. (Reproduced with permission from E. Canale-Parola, Bacteriological Reviews *41*:181-204, 1977, © American Society for Microbiology.)

This outer membrane completely surrounds the **protoplasmic cylinder,** which consists of the cytoplasmic and nuclear regions plus the surrounding cytoplasmic membrane-cell wall complex. Around the helical protoplasmic cylinder are wound **periplasmic flagella.** The periplasmic flagella are enclosed by the outer membrane and, thus, are located between this membrane and the protoplasmic cylinder. One end of each flagellum is inserted near a pole of the protoplasmic cylinder, and the other end is not inserted. The same number of flagella is inserted at each end of the cell (e.g., for spirochetes with two flagella, one flagellum is inserted at each end). The total number of periplasmic flagella per cell ranges from 2 to more than 100, depending on the species. The periplasmic flagella are components of the motility apparatus of the cell and perform a function essential for locomotion and for the other movements typical of spirochetes. Unlike other bacterial flagella, the periplasmic flagella are (a) permanently wound around the cell body and (b) entirely endocellular, being enclosed by the outer sheath. Thus, **the motility of spirochetes differs from that of other bacteria because spirochetes suspended in liquids are able to locomote even though their cells do not have flagella that propel them by rotating in direct contact with the external environment.** Spirochetes have three main types of movements in liquid environments: **locomotion, rotation about their longitudinal axis, and flexing motions.** Cells of spirochetes remain locomotory in environments of relatively high viscosity, whereas other flagellated bacteria usually become immotile in viscous fluids. **Anaerobic, facultatively anaerobic, or aerobic, and chemoheterotrophic.** Carbohydrates, amino acids, long chain fatty acids, or long chain fatty alcohols serve as carbon and energy sources. They are free living or in association with animal and human hosts. Some species are pathogenic.

Differentiation of the genera in **Group 1:** See Table 1.1.

Genus **Borrelia**

Consists of helical cells 0.2–0.5 × 3–20 µm, composed of 3–10 loose coils. They stain well with Giemsa's stain. The cells are surrounded by (a) a surface layer, (b) the outer membrane, and (c) the cytoplasmic membrane/cell wall complex. Seven to 30 periplasmic flagella (axial fibrils, periplasmic fibrils, or endoflagella) originate at each end of the cell and wind about the protoplasmic cylinder to overlap in the middle of the cell. The cells are actively motile with frequent reversal of direction of translational movement. No intracytoplasmic tubules occur. Species that have been grown in vitro are **microaerophilic.** Chemoorganotrophic. Nutritional requirements for in vitro growth are complex. The diamino acid present in the peptidoglycan is L-ornithine. Borrelias are pathogens of humans, other mammals, and birds. They are the causative agents of tick-borne and louse-borne relapsing fever in humans and of tick-borne Lyme disease in humans.

27

Type species: *Borrelia anserina.*

Differentiation of the species of the genus **Borrelia:** See Table 1.2.

Genus **Brachyspira**

Editorial note: The genus *Brachyspira* was not included in *Bergey's Manual of Systematic Bacteriology.* The genus was created in 1982 by Hovind-Hougen et al. (J. Clin. Microbiol. *16:* 1127-1136; Int. J. Syst. Bacteriol. *33:* 896-897, 1983). It includes only one species, *B. aalborgi.*

Cells measure 0.2 × 1.7–6 μm, with a wavelength of about 2 μm. Intracytoplasmic tubules are not present. **Four periplasmic flagella are inserted at each end of the cell. Anaerobic** but will tolerate normal atmosphere for at least 6–8 hours. **They can be cultivated** on solid tryptose soy medium supplemented with 5% calf blood. They are chemoorganotrophic and possess β-galactosidase and traces of esterase (lipase C8), acid phosphatase, and phosphoamidase. Catalase and oxidase negative. Some strains show hemolytic activity. This species is parasitic to humans.

Type (and only) species: *Brachyspira aalborgi.*

Characteristics of the species: As described for the genus. This genus has been isolated from biopsy material from human patients with intestinal spirochetosis.

Genus **Cristispira**

Helical or undulate cells 0.5–3.0 × 30–180 μm, generally displaying 2–10 complete helical turns. Ends of cells are blunt, rounded, or tapered; in fixed and stained preparations a filament or spicule may emanate from one or both ends. Stained preparations reveal a series of **ovoid inclusions** that impart a chambered appearance to the protoplasmic cylinder. **A bundle of 100 or more periplasmic flagella is intertwined with the protoplasmic cylinder and may distend the outer sheath to form a ridge or crest (the so-called "crista") on the protoplasmic cylinder.** The crista is not always obvious on live cells but may be conspicuous when the cells stop moving. It is frequently seen by light microscopy of stained cells. Motility is parallel to the cell's long axis, and individual cells move forward or backward with no anterior-posterior polarity. Translocation may include rotation about the cell's long axis or may take the form of an irrotational traveling helical wave. **Flexing movements are common.** No strains of *Cristispira* have been grown in pure culture. Strains are **widely distributed among marine and freshwater molluscs** (clams, mussels, and oysters) and inhabit the crystalline style (a mucoproteinaceous rod-shaped organ) or fluid of the digestive tract. **They are probably commensals.** They are also found in gastropods and may occur in nonmollusc species as well.

Type (and only) species: *Cristispira pectinis.*

Characteristics of the species: Although cristispires have been observed in many molluscan hosts and although more than one *Cristispira* species might exist, only a single species is presently recognized, *C. pectinis.* The cells of this species are helically coiled, flexible cells, 1.5 μm in diameter. The length of the helix is 36–72 μm. The cells possess no more than four complete turns of the helix. The ends of the cells are round or tapered with no terminal appendages. Stained preparations reveal cross-striations, polar granulation, and multiple inclusions. In fixed and stained preparations a crista extends along the side of the cell. Division is by transverse fission. *C. pectinis* has been found in the crystalline style and intestinal fluid of *Pecten jacobaeus* from the Gulf of Naples, Italy.

Large spirochetes resembling *Cristispira* have been observed in starfish, tunicates, and termites.

Molluscan spirochetes of smaller size than *Cristispira,* and which do not possess the multitude of periplasmic flagella typical of cristispires, have been described in *Anodonta mutabilis, Pachelebra* (probably *Pachylabra*) *moesta* (a snail), *Pinna squamosa, Scrobicularia piperata,* planorbid molluscs, and *Polydora flava* (a marine polychete annelid).

Genus **Leptonema**

Editorial note: The genus *Leptonema* was not included in *Bergey's Manual of Systematic Bacteriology.* The genus was created by Hovind-Hougen in 1979 (Int. J. Syst. Bacteriol. *29:* 245-251) and validly published in 1983 (Int. J. Syst. Bactetriol. *33:* 438-440). It includes one species, *Leptonema illini.*

Morphological characteristics are as described for the genus *Leptospira* except that the **periplasmic flagella are inserted in the cell by a basal body with a single pair of discs** instead of two as in the genus *Leptospira*, and the **cells contain two bundles of intracytoplasmic tubules which originate in the ends of the cell and run parallel to the periplasmic flagella.** Metabolism is respiratory. Fatty acids as a major source of carbon. **Growth occurs on trypticase soy broth without the addition of serum or serum albumin** (unlike leptospires). Cells are nonpathogenic.

Type (and only) species: *Leptonema illini.*

Characteristics of the species: As described for the genus. See also Table 1.3.

Genus **Leptospira**

Flexible helical cells, 0.1 μm in diameter and 6 to 20-24 μm in length. Cells appear faintly stained with aniline dyes. **Unstained cells are not visible by brightfield microscopy** but are visible by darkfield and phase-contrast microscopy. Resting stages are not known. The helical conformation is right-handed (clockwise coiling). One or both ends of the cells are typically hooked. Two periplasmic flagella (axial fibrils or endoflagella) occur per cell, one inserted at each end, and they rarely overlap in the central region of the cell. In liquid environments characteristic movements of the cells appear as an alternating rotation about the long axis and translation in the direction of the unhooked cell end. In environments of higher viscosity flexuous, boring, and serpentine movements also take place. **Intracytoplasmic tubules do not occur. Aerobic.** Diffuse-to-discrete subsurface colonies are formed in 1% agar and turbid-to-clear surface colonies on 2% agar; some strains form both subsurface and surface colonies on 1% or 2% agar. Optimum temperature is 28–30°C. This genus is **chemoorganotrophic, using fatty acids or fatty alcohols having 15 or more carbon atoms as carbon and energy sources.** Do not utilize carbohydrates or amino acids as energy sources. Serum or serum albumin is required in the culture medium for growth. This genus **does not grow in trypticase soy broth without the addition of serum or serum albumin.** Oxidase positive and catalase and/or peroxidase positive. They are resistant to the inhibitory action of 5′-fluorouracil, to which other spirochetes are sensitive. The cellular lipids do not contain glycolipids. The diamino acid present in the peptidoglycan is α,ε-

diaminopimelic acid. Free-living in soil, freshwater or marine habitats or living in association with animal and humans hosts. Nonsaprophytic strains cause leptospirosis, a zoonosis primarily infecting wild and domestic animals, which may act as reservoirs. Humans can contract leptospirosis as accidental hosts. Leptospires and leptospirosis occur throughout the world.

Type species: *Leptospira interrogans.*

Differentiation of the species of the genus **Leptospira:** See Table 1.3.

Serovars of **Leptospira** *species:* The antigenic composition of *Leptospira* strains is used for taxonomic purposes below the species level of classification. The basic taxon is the serovar. The standard method for determining the serovars is the microscopic agglutination test, with cross-agglutinin absorption. For convenience, antigenically related serovars are organized into serogroups. Pathogenic leptospires are divided into 38 serogroups and 65 serovars. Determination of the serogroup and serovar to which an isolate belongs requires a considerable amount of expertise and is done mainly by leptospirosis reference laboratories.

Genus **Serpulina**

Editorial note: The genus *Serpulina* was not included in *Bergey's Manual of Systematic Bacteriology.* It was created in 1991 by Stanton et al. (Int. J. Syst. Bacteriol. *41:* 50–58) and was named *Serpula,* which was illegitimate as a junior homonym of a name used for a genus of fungi. In 1992 Stanton (Int. J. Syst. Bacteriol. *42:* 189–190) changed the name to *Serpulina.*

These are helical cells with loose, regular coils, 7–9 × 0.3–0.4 μm. They are Gram negative. Typical spirochete ultrastructure consisting of outer sheath, protoplasmic cylinder bounded by a cytoplasmic membrane, and endoflagella inserted at each end of a protoplasmic cylinder and lying between the outer sheath and the protoplasmic cylinder. Eight or nine flagella are inserted at each end of a cell. Cells are **motile; flexing and creeping at 22°C; translational movement at 37–42°C.** Host is associated. Cells have been **isolated from intestinal contents and feces of swine and other mammals.** One species, *Serpulina hyodysenteriae,* is enteropathogenic for swine. Genus is chemoorganotrophic and uses soluble sugars for growth. It is weakly fermentative. It produces acetate, butyrate, H_2, and

CO_2 from glucose and does not reduce nitrate. It is negative for lipase and lecithinase and does not grow in media containing 1% glycine. The genus grows anaerobically in Trypticase soy or brain heart infusion broth containing 10% fetal calf, calf, or rabbit serum. Growth occurs at 36–42°C. The genus does not grow at 25 and 30°C. Surface colonies on Trypticase soy blood agar plates are visible after 48–96 hours of incubation at 38°C. Colonies are 0.5–3 mm in diameter, flat, translucent, and, depending on the species, weakly or strongly beta-hemolytic. Selective culture media for isolating *Serpulina hyodysenteriae* and *Serpulina innocens* from intestinal contents contain spectinomycin or a combination of spectinomycin, colistin, vancomycin, rifampin, and spiramycin. The G + C contents of the DNAs of the strains tested are 25–26 mol%.

Type species: *Serpulina hyodysenteriae.*

Differentiation of the species of the genus **Serpulina:** See Table 1.4.

Genus **Spirochaeta**

Helical cells 0.2–0.75 μm in diameter and 5–250 μm in length. Under unfavorable conditions, spherical cells or structures 0.5–2.0 μm (occasionally up to 10 μm) in diameter are formed. All species have two periplasmic flagella, one inserted at each end of the cell, except *Spirochaeta plicatilis*, which has many periplasmic flagella. **Cells locomote when suspended in liquids and crawl or creep when in contact with solid surfaces. Anaerobic or facultatively anaerobic.** Under aerobic growth conditions the **facultatively anaerobic species usually produce carotenoid pigments** that give a yellow, yellow-orange, or red coloration to colonies. Optimum growth temperature is 25–40°C. **Chemoorganotrophic, using a variety of carbohydrates as carbon and energy sources.** The main products of anaerobic carbohydrate metabolism are ethanol, acetate, CO_2, and H_2, except for one species (*Spirochaeta zuelzerae*) that produces succinate and lactate instead of ethanol. Facultatively anaerobic species oxidize carbohydrates aerobically yielding primarily CO_2 and acetate. A marine species (*Spirochaeta isovalerica*) ferments L-valine, L-isoleucine, and L-leucine with ATP generation, but does not utilize these amino acids as sole carbon and energy sources for growth. The diamino acid present in the peptidoglycan is L-ornithine. The genus is indigenous to aquatic environments such as the sediments, mud, and water of ponds, marshes, swamps, lakes, and

rivers. Occur commonly in H_2S-containing environments and are present in freshwater and marine environments. **They are free living.** None is reported to be pathogenic.

Type species: *Spirochaeta plicatilis.*

Differentiation of the species of the genus **Spirochaeta:** See Table 1.5.

Genus **Treponema**

Helical rods 0.1–0.4 μm in diameter and 5–20 μm in length. Most species stain poorly if at all with Gram's or Giemsa's methods and are best observed with darkfield or phase-contrast microscopy. Cells stain well with silver impregnation methods. Cells have tight regular or irregular spirals. Under unfavorable cultural or environmental conditions spherical cells, or spirochetal spheres, are formed. These can also be seen in old cultures. Cells have one or more periplasmic flagella inserted at each end of the protoplasmic cylinder. **Cells have both rotational and translational movement in liquid media.** In a semisolid or solid medium cells exhibit a serpentine-type movement. Intracytoplasmic tubules are present. **Anaerobic or microaerophilic.** Human pathogenic species are now considered to be microaerophiles and have not been cultivated in artificial media, although some have been cultivated (with difficulty and cannot be serially transferred) in tissue cultures. **They are chemoorganotrophs, using a variety of carbohydrates or amino acids as carbon and energy sources.** Cultivated anaerobic species are catalase and oxidase negative. Some require long-chain fatty acids found in serum for growth, while other cultivated species require short-chain volatile fatty acids for growth. The diamino acid present in the peptidoglycan is L-ornithine. **Found in the oral cavity, intestinal tract, and genital areas of humans and animals.** The genus is host-associated. Some species are pathogenic.

Type species: *Treponema pallidum.*

Differentiation of the species of the genus **Treponema:** See Tables 1.6, 1.7, and 1.8.

Genera Of Insect Gut Spirochetes

Editorial note: Four new genera of spirochetes associated with the insect gut (*Pillotina, Diplocalyx, Hol-*

landina, and *Clevelandina)* were described in 1988 by Bermudes et al. (Int. J. Syst. Bacteriol. *38:* 291-302). None have been cultivated, and they are separated from other spirochetes and each other by morphological criteria mainly determined by electron microscopy.

Important Notes for Users of This Manual

Unless otherwise indicated in footnotes to tables, the meanings of symbols are as follows:

+ 90% or more of strains are positive
− 90% or more of strains are negative
d 11-89% of strains are positive
v strain instability (not equivalent to "d")
D Different reactions in different taxa (species of a genus or genera of a family)

All other symbols are defined in footnotes to tables.

Table 1.1 Differential characteristics of the genera of spirochetes [a]

Characteristic	Borrelia	Brachyspira	Cristispira	Leptonema	Leptospira	Serpulina	Spirochaeta	Treponema
Free living	–	–	–	D	D	–	+	–
Host associated	+	+	+	D	D	+	–	+
Obligate aerobes	–	–	–	+	+	–	–	–
Obligate anaerobes	–	+	–	–	–	+	D	D
Facultative anaerobes	–	–	–	–	–	–	D	–
Microaerophiles	+	–	–	–	+	–	D	D
Growth on inanimate laboratory media	D	+	–	+	+	–	D	D
Cell diameter (μm)	0.2–0.5	0.2	0.5–3.0	0.1	0.1	0.3–0.4	0.2–0.75	0.1–0.4
Cell length (μm)	3–20	1.7–6.0	30–180	6–20	6–24	7–9	5–250	5–20
Periplasmic flagella form a bundle that distends the outer sheath to form a ridge (crista)	–	–	+	–	–	–	–	–
Energy and carbon sources:								
Carbohydrates	+			–	–	+	+	D
Amino acids				–	–	D	–[b]	D
Long-chain fatty acids	+			+	Rare	–	–	–
De novo cellular fatty acid synthesis							+	D
Ability to synthesize:								
Monoglycosyl diglyceride				–	–		+	+
Phosphatidyl choline				–	–		–	+
Inhabit the crystalline style or the fluid of the digestive tract of marine and fresh water molluscs	–	–	+	–	–	–	–	–
Strains that grow on tryptose blood agar form visible colonies within 3 days		+				+		–
Number of periplasmic flagella at each end of cell (by electron microscopy)	7–30	4	100 or more	1	1	8–9	1 [c]	1 or more
Intracytoplasmic tubules present as observed by electron microscopy of negatively stained cells	–	–		+	–	+		+
Growth in trypticase soy broth without added serum or serum albumin	–			+	–	–		–

[a] Symbols: see standard definitions.

[b] Spirochaeta isovalerica ferments L-valine, L-isoleucine, and L-leucine with ATP generation but does not utilize these amino acids as sole carbon and energy sources for growth.

[c] S. plicatilis, which has many periplasmic flagella.

Table 1.2 Differential characteristics of the species of the genus _Borrelia_

Species	Vector	Host	Distribution	Disease
B. anserina	Argas miniatus, A. persica, A. reflexus	Numerous birds	Worldwide	Avian borreliosis
B. burgdorferi [a,b]	Ixodes dammini, I. ricinus, I. pacificus	Humans, animal	U.S.A. and Europe	Lyme disease
B. brasiliensis	Ornithodoros brasiliensis	Rodents	Brazil	
B. caucasica	Ornithodoros verrucosus	Rodents, humans	Caucasus	Tick-borne relapsing fever
B. coriaceae [a,c]	Ornithodoros coriaceus	Cattle	Western U.S.A.	Epizootic bovine abortion
B. crocidurae	Ornithodoros erraticus (small variety)	Rodents, humans	Africa, Near East, Central Asia	Tick-borne relapsing fever
B. dugesii	Ornithodoros dugesi	Probably rodents	Mexico	
B. duttonii [a]	Ornithodoros moubata	Humans	Africa	Tick-borne relapsing fever
B. graingeri	Ornithodoros graingeri	Rodents, humans	East Africa	Tick-borne relapsing fever
B. harveyi	Unknown	Monkeys	Africa	
B. hermsii [a]	Ornithodoros hermsi	Rodents, humans	Western U.S.A. and Canada	Tick-borne relapsing fever
B. hispanica [d]	Ornithodoros erraticus (large variety)	Rodents, humans	Spain, Portugal, Morocco, Algeria, Tunisia	Tick-borne relapsing fever
B. latyschewii	Ornithodoros tartakovskyi	Rodents, reptiles, humans	Iran, Central Asia	Tick-borne relapsing fever
B. mazzottii	Ornithodoros talaje	Rodents, armadillos, monkeys, humans	Mexico and Guatemala	Tick-borne relapsing fever
B. parkeri [a]	Ornithodoros parkeri	Rodents, humans	Western U.S.A.	Tick-borne relapsing fever
B. persica	Ornithodoros tholozani	Rodents, bats, humans	Middle East, Central Asia	Tick-borne relapsing fever
B. recurrentis [d]	Pediculus humanus subsp. humanus	Humans	South America, Europe, Africa, Asia	Louse-borne relapsing fever
B. theileri	Rhipicephalus decoloratus, R. evertsi, Boophilus micropus	Ruminants, horses	South Africa, Australia	Cattle and horse borreliosis
B. tillae	Ornithodoros zumpti	Rodents	South Africa	
B. turicatae [a]	Ornithodoros turicatae	Rodents, humans	U.S.A. and Mexico	Tick-borne relapsing fever
B. venezuelensis	Ornithodoros rudis	Rodents, humans	Central and South America	Tick-borne relapsing fever

[a] Grow well in vitro.

[b] The species _B. burgdorferi_ was not included in _Bergey's Manual of Systematic Bacteriology._ See Johnson et al. (Int. J. Syst. Bacteriol. 34:496–497, 1984).

[c] The species _B. coriaceae_ was not included in _Bergey's Manual of Systematic Bacteriology._ See Johnson et al. (Int. J. Syst. Bacteriol. 37: 72–74, 1987).

[d] Grow in vitro, but only low cell numbers are obtained.

Table 1.3   Differential characteristics of the species of the genera *Leptospira* and *Leptonema*[a]

Characteristic	*L. biflexa*	*L. borgpetersenii*[b]	*L. inadai*[b]	*L. interrogans*	*L. meyeri*[b]	*L. noguchii*[b]	*L. parva*[c]	*L. santarosai*[b]	*L. weilii*[b]	*L. wolbachii*[b]	*Leptonema illini*
Saprophytes	+	−	−	−	+	−	+	−	−	+	+
Growth in trypticase broth without serum or serum albumin	−	−	−	−	−	−	−	−	−	−	+
Growth at 37°C	d	−	+	−	+	−	+	−	−	−	+
Growth in presence of:											
8-Azaguanine (225 µg/ml)	d	−	+	−	+	−	+	−	−	+	+
2,6-Diaminopurine (10 µg/ml)	d	d	−	d	+	−	−	d	−	+	−
Copper sulfate (100 µg/ml)	d	−	+	−	+	−	+	−	−	−	−
Lipase (trioleinase) activity	+	−	+	d	+	d	+	−	−	+	+
Presence of intracytoplasmic tubules (electron microscopy of thin sections)	−	−	−	−	−	−	−	−	−	−	+

[a] Symbols: +, all strains positive for characteristic; −, all strains negative; d, differs among strains.

[b] These species were not included in *Bergey's Manual of Systematic Bacteriology.* See Yasuda et al. (Int. J. Syst. Bacteriol. *37:* 407–415, 1987).

[c] *L. parva* was not included in *Bergey's Manual of Systematic Bacteriology.* See Hovind-Hougen et al. (Zentralbl. Bakter. Parasitenkd. Infekstionskr. Hyg. Abt. 1 Orig. Reihe A *250*: 343–354, 1981).

Important Notes for Users of This Manual

Unless otherwise indicated in footnotes to tables, the meanings of symbols are as follows:

    +   90% or more of strains are positive

    −   90% or more of strains are negative

    d   11-89% of strains are positive

    v   strain instability (not equivalent to "d")

    D   Different reactions in different taxa (species of a genus or genera of a family)

All other symbols are defined in footnotes to tables.

Table 1.4   Differential characteristics of the species of the genus _Serpulina_ [a]

Characteristic	S. hyodysenteriae	S. innocens
Fermentation of: [b]		
Fructose	−	+
Maltose	+	−
Indole production	+	−
β-hemolysis	−	+
Enteropathogenic for swine	+	−

[a] Symbols: see standard definitions.

[b] Acid production in weakly buffered medium containing carbohydrates.

Table 1.5   Differential characteristics of the species of the genus _Spirochaeta_ [a]

Characteristic	S. aurantia	S. halophila	S. isovalerica [b]	S. litoralis	S. plicatilis	S. stenostrepta	S. zuelzerae
Cultivation in pure culture	+	+	+	+	−	+	+
Cell diameter (μm)	0.30	0.40	0.4	0.4–.5	0.75	0.2–0.3	0.2–0.4
Number of periplasmic flagella:							
Two flagella/cell	+	+	+	+	−	+	+
Many flagella/cell	−	−	−	−	+	−	−
Relationship to oxygen:							
Obligate anaerobe	−	−	+	+		+	+
Facultative anaerobe	+	+	−	−		−	−
High NaCl concentration required for growth	−	+	+	+		−	−
Ethanol formed	+	+	+	+		+	−
Succinate formed	−	−	−	−		−	+
Optimum growth temperature (°C)	25–30	35–40	15–35	30		30–37	37–39
Pigmentation:							
Yellow-orange	+	−	−	−		−	−
Red	−	+	−	−		−	−
Ferments leucine to form isovalerate			+	−			
Utilization of:							
Raffinose, lactose, inulin			−	+			
Mannitol			+	−			

[a] Symbols: see standard definitions.

[b] _S. isovalerica_ was not included in _Bergey's Manual of Systematic Bacteriology_. See Harwood and Canale-Parola (Int. J. Syst. Bacteriol. _33_: 573-579, 1983).

Table 1.6 Differential characteristics of the noncultivable species of the genus *Treponema*[a]

Characteristic	T. carateum	T. pallidum subsp. endemicum	T. pallidum subsp. pallidum	T. pallidum subsp. pertenue	T. paraluiscuniculi
Natural host:					
Humans	+	+	+	+	–
Rabbits	–	–	–	–	+
Clinical infection:					
Syphilis	–	–	+	–	–
Yaws	–	–	–	+	–
Nonvenereal endemic syphilis	–	+	–	–	–
Pinta	+	–	–	–	–
Rabbit syphilis	–	–	–	–	+
Nature of infection in natural host:					
Systemic; may affect most internal organs	–	–	+	–	–
Usually restricted to cutaneous lesions	+	+	–	+	+
Sexually transmitted	–	–	+	–	+
Geographical distribution:					
Worldwide	–	–	+	–	
Found only in tropical countries in both hemispheres	–	–	–	+	
Found only in tropical countries in the Western hemisphere	+	–	–	–	
Restricted to the Middle East, Africa, southeast Asia, and Yugoslavia	–	+	–	–	
Cutaneous lesions produced in:					
Rabbits	–	+	+	+	
Hamsters	–	+	–	+	
Mice	–	–	–	–	
Guinea pigs	–	+	–[b]	–	

[a] Symbols: see standard definitions.

[b] A slight lesion is occasionally seen at the point of injection of guinea pigs.

Table 1.7 Differential characteristics of the cultivable species of the genus *Treponema* that do not ferment carbohydrates [a,b]

Characteristic	T. denticola	T. minutum	T. refringens	T. scoliodontum	T. vincentii
Cell diameter (μm):					
0.15–0.20	–	+	–	+	–
0.20–0.25	+	–	+	–	+
Require thiamine pyrophosphate	+	–	–	+	+
Esculin hydrolysis	+	+	+	–	d
Indole production	d[c]	+	+	–	Weak
1% Glycine (growth)	d	+	d[d]	–	–
Phosphatase	+	–	–		–
Convert fumarate to succinate	–	–	+		–

[a] Symbols: see standard definitions.

[b] These species do not produce an acidic pH in weakly buffered medium containing carbohydrate.

[c] Distinguishes between biovars of *T. denticola*. Biovar *denticola* is indole positive, whereas biovar *comondonii* is indole negative.

[d] Distinguishes between biovars of *T. refringens*. Biovar *refringens* does not grow with 1% glycine, whereas biovar *calligyrum* does grow in 5–6 days.

Table 1.8 Differential characteristics of the cultivable species of the genus *Treponema* that ferment carboydrates [a]

Characteristic	T. bryantii	T. pectinovorum [b,c]	T. phagedenis	T. saccharophilum [b]	T. socranskii [b,d]	T. succinifaciens
Cell diameter (μm):						
0.15–0.20	–	–	–	–	+	–
0.20–0.25	–	–	+	–	–	–
0.30	+	+	–	–	–	+
0.36–0.38	–	–	–	–	–	–
0.6–0.7	–	–	–	+	–	–
Growth requirements:						
Serum	–	–	+	–	–	–
Volatile fatty acids [e]	+	+	–	+	+	+
Glucose	+	–	–	+	–	+
Fermentation of [f]:						
Glucose	+	–	+	+	d	+
Fructose	–	–	+	+	+	–
Lactose	+	–	+	+	–	+
Maltose	–	–	–	+	+	+
Mannitol	–	–	+	–	–	
Starch	–	–	–	+	+	+
Sucrose	+	–	+	+	+	–
Ribose	–	–	+	–	+	–
Esculin hydrolysis		–	d [g]		–	
Indole production		–	+		–	
1% Glycine (growth)			+			
Phosphatase			+			
Convert fumarate to succinate			+			
β-hemolysis			–		Weak	

[a] Symbols: see standard definitions.

[b] *T. pectinovorum*, *T. socranskii*, and *T. saccharophilum* were not included in *Bergey's Manual of Systematic Bacteriology*. The species *T. pectinovorum* was created in 1983 by Smibert and Burmeister (Int. J. Syst. Bacteriol. *33:* 852–856). The species *T. socranskii* was created in 1984 by Smibert et al. (Int. J. Syst. Bacteriol. *34:* 457–462). The species *T. saccharophilum* was created in 1985 by Paster and Canale-Parola (Appl. Environ. Microbiol. *50:* 212–219).

[c] The major distinguishing characteristic of *T. pectinovorum* is its ability to ferment pectin, polygalacturonic acid, and glucuronic acid. These are the only substrates that are fermented, and at least one of these compounds must be provided in order for growth to occur.

[d] *T. socranskii* has three subspecies. *T. socranskii* subsp. *socranskii* and *T. socranskii* subsp. *buccale* can only be differentiated by serologic tests, while *T. socranskii* subsp. *paredis* can be differentiated from the other two subspecies by its ability to ferment arabinose.

[e] As occur in rumen fluid.

[f] Acid production in weakly buffered medium containing carbohydrate.

[g] Distinguishes between biovars of *T. phagedenis*. Biovar *reiter* does not hydrolyze esculin, whereas biovar *kazan* does.

GROUP 2

AEROBIC/MICROAEROPHILIC, MOTILE, HELICAL/VIBRIOID GRAM-NEGATIVE BACTERIA

This is a diverse assemblage of bacteria whose cells are **helical or vibrioid,** having from less than one complete helical turn to many turns. Many species form intracellular poly-β-hydroxybutyrate granules. **Gas vacuoles, stalks, prosthecae, and endospores are not formed.** Some species have cells that become converted into thin-walled "coccoid bodies" in old cultures. Some species (of *Azospirillum*) produce enlarged, encapsulated, desiccation-resistant cysts. Cells from young cultures stain Gram negative; one genus (*Azospirillum*) may contain forms that stain Gram positive in older cultures. **Cells are motile in broth cultures by polar flagella.** They swim in straight lines with a characteristic corkscrew-like motion. **Aerobic or microaerophilic,** although some species can also grow anaerobically with nitrate, fumarate, or trimethylamine oxide as terminal electron acceptors. Many species are incapable of catabolizing carbohydrates. Some species (of *Azospirillum*) can acidify certain sugar media under anaerobic conditions but grow very poorly and much prefer aerobic conditions. They are typically **oxidase positive. Indole is not produced.** They are **chemoorganotrophic;** however, some species can grow as facultative autotrophs with H_2 gas as the electron donor. None are phototrophic. **Some species can fix N_2, but only under microaerobic conditions.** One species (*Aquaspirillum magnetotacticum*) exhibits tactic responses to magnetic fields. Occur in soil, freshwater, hypersaline or marine environments, within plant roots, or in the reproductive organs, intestinal tract, and oral cavity of humans and animals. Three genera (*Bdellovibrio, Micavibrio,* and *Vampirovibrio*) are predacious on other microorganisms.

Differentiation of the genera in **Group 2:** See Table 2.1.

Genus **Alteromonas: Alteromonas undina**

Curved rods, 0.7–1.5 μm in diameter and 1.8–3.0 μm in length. Do not accumulate **poly-β-hydroxybutyrate** as an intracellular reserve product. Microcysts or endospores are not formed. Cells stain Gram negative. **Motile** by means of a **single polar flagellum.** Chemoorganotrophs having a strictly respiratory type of metabolism with O_2 as the terminal electron acceptor. Some strains grow at 4°C; all grow at 20°C; none grow at 35°C. Colonies are not pigmented. **Nitrate is not reduced to nitrite. Do not denitrify.** Organic growth factors are required. **Require a seawater base for growth.** They produce gelatinase, lipase, and chitinase but not alginase. Some strains produce amylase. Utilize the following carbon sources: sucrose, maltose, acetylglucosamine, succinate, fumarate, and tyrosine. Do not utilize D-mannose, D-galactose, D-fructose, cellobiose, melibiose, lactose, erythritol, mannitol, glycerol, salicin, gluconate, DL-lactate, citrate, D-sorbitol, DL-malate, α-ketoglutarate, and *m*-hydroxybenzoate. They do not possess a constitutive arginine dihydrolase system. **Habitat is coastal waters and the open oceans.**

Alteromonas species whose cells are straight are considered in Group 4 of this manual.

Differentiation of **Altermonas undina** *from* **Marinomonas communis:** Unlike *M. communis, A. undina* (a) has gelatinase, lipase, and chitinase activity, (b) utilizes sucrose, acetylglucosamine, and tyrosine, and (c) does not utilize mannose, fructose, gluconate, DL-lactate, mannitol, glycerol, sorbitol, DL-malate, α-ketoglutarate, and *m*-hydroxybenzoate.

Genus **Aquaspirillum** (*except* **Aquaspirillum fasciculus**)

Rigid, generally helical cells, 0.2–1.4 μm in diameter; however, one species is vibrioid. **Intracellular granules of poly-β-hydroxybutyrate are usually formed.** Some species form thin-walled **coccoid bodies** which predominate in old cultures. Cells are **motile by polar flagella,** usually **bipolar tufts;** one species is monotrichous, others have a single flagellum at each pole. One species exhibits tactic responses to magnetic fields. **Aerobic to microaerophilic.** A few species can grow anaerobically with nitrate as the terminal electron acceptor. The optimum temperature for most species is 30–32°C. Chemoorganotrophic; however, one species is a facultative hydrogen autotroph. **Oxidase positive.** Usually catalase and phosphatase positive and indole and sulfatase negative. Casein, starch, and hippurate are not hydrolyzed. **No growth occurs in the presence of 3% NaCl.** A few species can denitrify. Nitrogenase activity occurs in some species but only under microaerobic conditions. Amino acids or the salts of organic acids serve as carbon sources; **carbohydrates are usually not catabolized,** but a few species can attack a limited variety of carbohydrates. Vitamins are not usually required. Usually occur in **stagnant freshwater environments.**

Type species: *Aquaspirillum serpens.*

Differentiation of the species of the genus **Aquaspirillum:** See Table 2.2.

Genus **Azospirillum**

Plump vibrioid or straight rods, 0.9–1.2 μm in width, often with pointed ends. **Cells stain Gram negative to Gram variable.** Intracellular granules of **poly-β-hydroxybutyrate** are present. Cells are motile with a characteristic corkscrew-like or vibratory motion in liquid media by means of a **single polar flagellum.** Lateral flagella may also be formed in addition to the polar flagellum by some species when cells are cultured on solid media at 30°C. The colonies of some strains form a **light pink or dark pink pigment** on potato agar. Optimum growth temperature is 34–37°C. Some strains grow well at pH 7; others prefer more acidic conditions. They are **nitrogen fixers,** exhibiting N_2-dependent growth under **microaerobic conditions.** Grow well under an air atmosphere in the presence of a source of fixed nitrogen, such as an ammonium salt. They possess mainly a respiratory type of metabolism with O_2 and, with some strains, NO_3^- as terminal electron acceptors. Weak fermentative ability may also occur. Under severe oxygen limitation some strains may dissimilate NO_3^- to NO_2^- or N_2O and N_2. They are **oxidase positive.** They are chemoorganotrophic; however, some strains are facultative hydrogen autotrophs. **Grow well on the salts of organic acids, such as malate, succinate, lactate, and pyruvate;** certain carbohydrates may also serve as carbon sources. Some strains require biotin. Species **occur free-living in soil or in association with the roots of cereal crops, grasses, and tuber plants.** Root nodules are not induced.

Type species: *Azospirillum lipoferum.*

Differentiation of species of the genus **Azospirillum:** See Table 2.3.

Genus **Bdellovibrio**

Consists of bacteria which are **predacious upon other Gram-negative bacteria.** The cells are **comma-shaped rods, 0.2–0.5 μm in diameter** and motile by means of a **single, ensheathed polar flagellum.** Bdellovibrios **exhibit a morphologically and physiological biphasic life cycle, alternating between a nongrowing predatory phase and an intracellular reproductive phase.** The highly motile bdellovibrio appears to locate its prey by means of chance collision; it forcibly strikes and attaches to the generally much larger prey cell, then rapidly penetrates into the prey's periplasmic space. The prey cell containing the invading bdellovibrio usually rounds up and swells into a spherical form (bdelloplast). The bdellovibrio kills the prey cell very early in the attack. The prey cell, soon after attack, is functionally a substrate for bdellovibrio development. The developing bdellovibrio elongates into a snake-like form at the expense of the prey's protoplast. The spiral-shaped nonmotile cell then fragments into motile, unit-sized predacious vibrios which leave the prey (now a "ghost" cell) to begin the cycle anew.

Bacteria suspected of belonging to the genus *Bdellovibrio* must be confirmed to grow in the periplasmic space of bacteria. This is the primary characteristic that binds this diverse group of comma-shaped bacteria. This can be accomplished by phase-contrast light microscopy or transmission electron microscopy. These techniques should confirm the location of the bdellovibrios *within* the host cell. An acceptable procedure is described by Burnham et al. (J. Bacteriol. *96:* 1366-

1381, 1968). Negatively stained specimens examined with the electron microscope will also aid in this identification by confirming the presence of a polar ensheathed flagellum.

Obligately **aerobic,** having a strictly respiratory type of metabolism with oxygen as the terminal electron acceptor. Optimum temperature is generally 28–30°C; growth is poor above 37°C and below 10°C. **All wild-type strains upon initial isolation are dependent on intraperiplasmic growth in susceptible prey.** Growth of isolates in the absence of prey has been achieved in some cases with media supplemented with high concentrations of bacterial cell extracts. Only predacious bdellovibrios have been obtained on initial isolation from nature, but mutants capable of axenic growth ("prey-independent" strains) have been derived from the predacious strains. Some strains are facultative, i.e., capable of growth in the presence or absence of prey cells. Habitats of bdellovibrios include soil, sewage, and freshwater and marine environments.

Type species: *Bdellovibrio bacteriovorus.*

Differentiation of the species of the genus **Bdellovibrio:** See Table 2.4.

In addition to the three species indicated, a number of marine strains have been reported but have not yet been formally assigned to any species. These strains, which are isolated from sea water, are distinguished from strains of the three recognized species by their requirement for Na⁺ (75 mM or higher).

For a more definitive differentiation between *B. bacteriovorus* vs. *B. stolpii* and *B. starrii* than given in Table 2.4, it may be necessary to determine the DNA base composition of isolates. The mol% G + C of *B. bacteriovorus* is 50, whereas that of *B. stolpii* is 42 and of *B. starrii* is 43.5.

Genus **Campylobacter**

Slender, vibrioid cells, 0.2–0.5 μm wide and 0.5–5 μm long. The rods may have one or more helical turns and can be as long as 8 μm. They also appear S-shaped and gull-wing-shaped when two cells form short chains. They are non-sporeforming. Cells in old cultures may form spherical forms (**coccoid bodies**). Stain is Gram negative. Cells are **motile with a characteristic corkscrew-like motion by means of a single, unsheathed,** polar flagellum at one or both ends of the cells. They are typically **microaerophilic** and have a respiratory type of metabolism. They typically require an O₂ concentration of between 3 and 15% and a CO₂ concentration of 3-5%. Occasionally a few strains may grow slightly under aerobic conditions (21% O₂). One species is aerotolerant and can grow under an air atmosphere. **Some species require H₂ or formate for microaerophilic growth. Some species can grow under anaerobic conditions with either fumarate, formate plus fumarate, or H₂ plus fumarate, in the medium. Chemoorganotrophs. Carbohydrates are neither fermented nor oxidized.** No acidic or neutral end products are produced. Neither serum nor blood is required for growth. Energy is obtained from amino acids or tricarboxylic acid cycle intermediates, not carbohydrates. Gelatin is not hydrolyzed. Methyl red and Voges-Proskauer tests are negative. No lipase activity occurs. **Oxidase positive and urease negative,** except for some strains of *C. lari.* Pigments are not produced. Menaquinone-6 and a methyl-substituted menaquinone-6 are the major cellular quinones. Hexadecanoic acid is a major cellular fatty acid in most spcies. Some species are pathogenic for humans and animals. **Found in the reproductive organs, intestinal tract, and oral cavity of humans and animals.** One species is a **nitrogen-fixer** isolated from the roots of plants growing in salt marshes.

Type species: *Campylobacter fetus.*

Differentiation of the species of the genus **Campylobacter:** See Table 2.5.

Genus **Cellvibrio**

Editorial note: The genus *Cellvibrio* was not included in *Bergey's Manual of Systematic Bacteriology.* The genus was created in 1985 by Blackall et al. (J. Appl. Bacteriol. *59:* 81-97; Int. J. Syst. Bacteriol. *36:* 354-356, 1986). It includes only one species, *C. mixtus.*

Slightly curved rods 0.2–0.5 x 1.0–1.3 μm. Cells stain Gram negative and are motile. They have a **single polar flagellum** and up to 11 **lateral flagella** of thinner diameter, especially when the cells are grown on a solid medium. They are aerobic and **oxidase and catalase positive. Glucose is oxidized. Cellulose is hydrolyzed. Organic acids are not utilized as sole carbon sources.** Growth factors are not required.

Type (and only) species: *Cellvibrio mixtus.*

Characteristics of the species: Growth occurs on peptone-yeast extract agar plus 5% sucrose, sucrose-peptone agar, glucose-yeast extract agar, and glucose-ammonium sulfate agar, but is poor or absent on peptone-yeast extract agar. Colonies on sucrose-peptone agar have a diameter of 1–3 mm and are either white to cream, smooth, convex to low convex, translucent, mucoid, and circular with an entire edge, or orange-yellow, dull, flat, translucent, and circular with an irregular edge. All strains on defined mineral salts medium plus cellulose are white, smooth, circular, with an entire edge, and surrounded by a narrow zone of clearing. Acid is produced from L-arabinose, dextrin, galactose, glucose, lactose, maltose, mannose, melibiose, raffinose, sucrose, trehalose, and xylose, but not from adonitol, arabitol, D-arabinose, dulcitol, erythritol, glycerol, inositol, mannitol, ribose, sorbitol, sorbose, or ethanol. Starch, chitin, esculin, and neutral or alkaline pectate gels are hydrolyzed; gelatin, casein, DNA, Tween 80, and acidic pectate gels are not hydrolyzed. α-Glucosidase, β-glucosidase and β-xylosidase are produced; urease, α-fucosidase, arginine dihydrolase, arylsulfatase, acylamidase, and phenylalanine deaminase are not produced. Curdlan polysaccharide is produced from glucose. The following are not produced: 3-ketolactose from lactose, dihydroxyacetone from glycerol, indole from tryptophan, H_2S from cysteine, and gas from nitrate. The methyl red and Voges-Proskauer tests are negative. Gluconate is not oxidized. Growth does not occur on cetrimide agar, in an anaerobic environment, at 42°C, on glucose nitrogen-free medium, in alkylamine mineral salts medium, in phenol mineral salts medium, or in the presence of 0.0075% KCN. *C. mixtus* subsp. *dextranolyticus* hydrolyzes dextran; *C. mixtus* subsp. *mixtus* does not.

Genus **Halovibrio**

Editorial note: The genus *Halovibrio* was not included in *Bergey's Manual of Systematic Bacteriology.* The genus was created in 1988 by Fendrich (Syst. Appl. Microbiol. *11:* 36–43; Int. J. Syst. Bacteriol. *39:* 205–206, 1989). It includes only one species, *H. variabilis.*

Obligately aerobic, vibrioid, Gram-negative rods requiring at least 1.2 M (7.0% w/v) NaCl for growth. Motility occurs by means of a polar flagellum. Chemoorganotrophic. **Carbohydrates are not oxidized.**

Type (and only) species: *Halovibrio variabilis.*

Characteristics of the species: These are vibrio-shaped cells, 0.5–0.8 x 1.0–3.0 μm, occurring singly or in pairs. Coccoid bodies may be found in old cultures. They are halophilic, growing between 1.2 M (7.0%, w/v) and 4.9 M (28.6%, w/v) NaCl, with optimum at 1.6 M (9.4%, w/v) NaCl. pH range: 6.5–8.4; optimum 7.5. Temperature range is 15–37°C; optimum 33°C. Colonies are circular with entire margins, slimy, raised, with light brown color. Growth occurs on acetate, caproate, glutamine, valerate, esculin, glycerol, ethanol, Casamino™ acids, peptone, and yeast extract; catalase and oxidase positive. Gelatin is not liquefied. Urease positive. Starch and cellulose are not hydrolyzed. Indole and H_2S are not produced. Nitrate is not reduced. Negative for arginine dihydrolase and for lysine and ornithine decarboxylases. Habitat is North Arm of Great Salt Lake, Utah, USA.

Genus **Helicobacter**

Editorial note: The genus *Helicobacter* was not included in *Bergey's Manual of Systematic Bacteriology.* The genus was created in 1989 by Goodwin et al. (Int. J. Syst. Bacteriol. *39:* 397-405). It includes two species, *H. pylori* and *H. mustelae.*

Helical, curved, or straight, 0.5–1.0 μm wide x 2.5–5.0 μm long, and have rounded ends. External glycocalyx is produced in vitro in shaken broth cultures. Motility is rapid and darting **by means of multiple sheathed flagella that are unipolar or bipolar and lateral, with terminal bulbs. Microaerophilic.** Variable growth occurs under an air atmosphere enriched with 10% CO_2 and under anaerobic conditions. Slow growth occurs in brain heart infusion (BHI) broth and other liquid media unless shaken; growth occurs in 2–5 days on BHI agar and chocolate agar. Growth also occurs on BHI agar supplemented with charcoal or corn starch. Colonies are nonpigmented, translucent, and 1–2 mm in diameter. Optimum temperature is 37°C; growth occurs at 30°C but not at 25°C; variable growth occurs at 42°C. Growth occurs in the presence of 0.5% glycine and 0.04% triphenyltetrazolium chloride. There is no growth with 3.5% NaCl. **Catalase and oxidase positive. Urea is rapidly hydrolyzed.** The major isoprenoid quinones are MK-6 and an unidentified MK-6 (Un-Mk-6). Hexadecanoic acid is not a major cellular fatty acid. H_2S production is negative on triple sugar iron agar and variable on lead acetate paper.

Nitrate reduction and hippurate hydrolysis are variable. Positive for alkaline phosphatase and γ-glutamyltranspeptidase activities; leucine arylamidase activity is variable. Susceptible to penicillin, ampicillin, amoxicillin, erythromycin, gentamicin, kanamycin, rifampin, and tetracycline. It is resistant to vancomycin, sulfonamides, and trimethoprim. There is variable resistance to nalidixic acid, cephalothin, metronidazole, and polymyxin. **Isolated from the gastric mucosa of primates and ferrets. Some organisms in the genus may be associated with gastritis and peptic ulceration.**

Type species: *Helicobacter pylori.*

Differentiation of the species of the genus **Helicobacter:** See Table 2.6.

Genus **Herbaspirillum**

Editorial note: The genus *Herbaspirillum* was not included in *Bergey's Manual of Systematic Bacteriology.* The genus was created in 1986 by Baldani et al. (Int. J. Syst. Bacteriol. *36:* 86-93). It includes only one species, *H. seropedicae.*

Generally vibrioid, sometimes helical cells, 0.6–0.7 μm in width. Stain is Gram negative. Motility is by means of **1–3 polar flagella at one or both poles. Colonies on potato agar have brownish centers and are raised and smooth.** Optimum growth temperature is 34°C; pH range for good growth is 5.3–8.0. **Nitrogen fixers**, exhibiting N₂-dependent growth under **microaerobic conditions.** Grow well under an air atmosphere in the presence of a source of fixed nitrogen, such as an ammonium salt. Possess a respiratory type of metabolism. **No growth occurs anaerobically with nitrate.** Under severe oxygen limitation NO₃⁻ is dissimilated to NO₂⁻ and, by some strains, to N₂O. **Oxidase positive.** Chemoorganotrophic. **Grow well on the salts of organic acids such as malate, fumarate, succinate, pyruvate, citrate**; certain carbohydrates may also serve as carbon sources. Vitamins are not required. **Occur free-living in soil or in association with the roots of members of the *Gramineae.***

Type (and only) species: *Herbaspirillum seropedicae.*

Characteristics of the species: Cells generally have two polar flagella (occasionally one to three) on one or both poles. On soft nutrient agar at 35°C pronounced swarming occurs, but lateral flagella as described for

some *Azospirillum* species do not occur. The efficiency of nitrogen fixation in semisolid N-deficient media is 12–15 mg of N per g of DL-malate or 13 mg of N per g of mannitol. N₂-dependent growth is slower than that of *Azospirillum* species, but growth in the presence of mineral nitrogen or glutamate is much faster. It does not grow in the presence of 2% NaCl. It can use glucose and α-ketoglutarate as carbon sources; sucrose is not utilized. The optimum temperature for N₂-dependent growth is 34°C; no growth occurs at 22 and 38°C.

Genus **Marinomonas: Marinomonas communis**

Editorial note: The species *Marinomonas communis* was included in *Bergey's Manual of Systematic Bacteriology* under its former name, *Alteromonas communis.* It was reclassified in the genus *Marinomonas* in 1983 by Van Landschoot and De Ley (J. Gen. Microbiol. *129:* 3057-3074; Int. J. Syst. Bacteriol. *34:* 91-92, 1984).

Curved rods, 0.7–1.5 μm in diameter and 1.8–3.0 μm in length. Do not accumulate **poly-β-hydroxybutyrate.** Microcysts or endospores are not formed. Stain is Gram negative. **Motility occurs by means of single polar flagella.** Chemoorganotrophs capable of respiratory but not fermentative metabolism; molecular oxygen is the terminal electron acceptor. **Do not denitrify.** Organic growth factors are not required. **Require a seawater base for growth.** Growth does not occur at 4°C, but does occur at 20 and 40°C. Colonies are not pigmented. Nitrate is not reduced to nitrite. Negative for amylase, gelatinase, lipase, alginase, and chitinase. The following carbon sources are utilized: D-mannose, D-fructose, D-gluconate, saccharate, *meso*-inositol, succinate, fumarate, DL-lactate, citrate, aconitate, D-mannitol, glycerol, γ-aminobutyrate, sarcosine, putrescine, D-sorbitol, DL-malate, α-ketoglutarate, quinate, D-alanine, L-ornithine, betaine, *m*-hydroxybenzoate, and *p*-hydroxybenzoate. The following carbon sources are not utilized: sucrose, cellobiose, melibiose, lactose, salicin, *N*-acetylglucosamine, erythritol, L-threonine, and L-tyrosine. Do not possess a constitutive arginine dihydrolase system. **Habitat is coastal waters and the open oceans.**

The other species of *Marinomonas, M. vaga,* has straight cells and is considered in Group 4.

Differentiation of **Marinomonas communis** *from* **Alteromonas undina:** Unlike *A. undina, M. communis*

(a) does not have gelatinase, lipase, and chitinase activity, (b) does not utilize sucrose, acetylglucosamine, and tyrosine, and (c) does utilize mannose, fructose, gluconate, DL-lactate, mannitol, glycerol, sorbitol, DL-malate, α-ketoglutarate, and *m*-hydroxybenzoate.

Genus **Micavibrio**

Editorial note: The genus *Micavibrio* was not included in *Bergey's Manual of Systematic Bacteriology.* The genus was created in 1982 by Lambina et al. (Mikrobiologiya *51:* 114-117; Int. J. Syst. Bacteriol. *39:* 93-94, 1989). It includes only one species, *M. admirandus.*

Comma-shaped rods, 0.2–0.3 μm. Stain is Gram negative. **Motility occurs by a single flagellum which is unsheathed** (unlike the flagellum of a bdellovibrio). **Predacious upon cells of *Xanthomonas maltophilia*** (so far the only known host) by **attaching to the outside of the host, losing their motility, and developing as exoparasites with eventual lysis of the host.** Unlike bdellovibrios, they do not invade the periplasmic space of the host cell and do not develop into elongated spirillar forms as part of their life cycle. They are resistant to vibriostatic pteridine O/129 and **cannot grow in the absence of host bacteria.**

Type (and only) species: *Micavibrio admirandus.*

Characteristics of the species: As described for the genus.

Genus **Oceanospirillum**

Editorial note: Oceanospirillum kriegii and *Oceanospirillum jannaschii* contain straight rods and thus are considered in Group 4 in this manual. The following description of the genus does not include them.

Rigid helical cells 0.3–1.4 μm in diameter. Intracellular **poly-β-hydroxybutyrate** is formed. Most species form thin-walled **coccoid bodies** which predominate in old cultures. Stain is Gram negative. **Motility occurs by bipolar tufts of flagella or by a single flagellum at each pole. Aerobic,** having a strictly respiratory type of metabolism with oxygen as the terminal electron acceptor. Nitrite respiration does not occur. Optimum temperature is 25–32°C. **Oxidase positive** and indole and arylsulfatase negative. Casein, starch, hippurate, and esculin are not hydrolyzed. **Seawater is required for growth. Carbohydrates are neither oxidized nor fer-**

mented. Amino acids or the salts of organic acids serve as carbon sources. Growth factors are not usually required. Isolated from coastal sea water, decaying seaweed, and from putrid infusions of marine mussels.

Type species: *Oceanospirillum linum.*

Differentiation of the species of the genus **Oceanospirillum:** See Table 2.7.

Genus **Spirillum**

Rigid helical cells, 1.4–1.7 μm in diameter x 14–60 μm in length. Intracellular **poly-β-hydroxybutyrate granules** are present. Coccoid bodies are not formed. Stain is Gram negative. Motility occurs by **large bipolar tufts of flagella** having a long wavelength and about one helical turn; these are easily visible by darkfield or phase-contrast microscopy. Normally, **cells are microaerophilic,** but they can grow aerobically in special liquid media or media containing certain supplements. Colonies on solid media can be obtained only with special media and only under microaerobic conditions. They have a **strictly respiratory type of metabolism** with oxygen as the terminal electron acceptor. Growth does not occur anaerobically with nitrate. Optimum temperature is 30°C. **Oxidase and phosphatase positive. Catalase negative.** Indole and sulfatase negative. Casein, starch, esculin, gelatin, DNA, and RNA are not hydrolyzed. Inhibited by extremely low levels of hydrogen peroxide in the culture medium. NaCl levels above 0.02% and phosphate levels greater than 0.01 M are inhibitory. **Carbohydrates are not catabolized.** The salts of certain organic acids, especially potassium succinate, are used as carbon sources. Vitamins are not required. **Occur in stagnant freshwater environments.**

Type species: *Spirillum volutans.*

Characteristics of **Spirillum volutans:** As described for the genus.

"Spirillum minus"

Rigid cells; usually described as spiral with two or three turns, although the waves have been reported to be planar. The ends of the cell may be blunt or pointed. Cell diameter is ~0.2 μm; cell length is 3–5 μm; wavelength is 0.8–1.0 μm. **Actively motile by one**

or more flagella at each pole. *S. minus* causes "Sodoku," one of the two forms of rat-bite fever in humans. Best observed in blood or exudates from patients by dark-field or phase-contrast microscopy or wet mounts; staining with Giemsa or Wright's stain or by silver impregnation is also useful. **It is questionable whether the organism has ever been cultured successfully on artificial media.** It can be cultured in vivo by intraperitoneal inoculation of patient's blood or exudates from lesions, or blood from naturally infected rats, into spirillum-free mice or guinea pigs.

"Spirillum pleomorphum"

Helical cells, curved rods, crescent-shaped cells 0.7–1.0 x 2.0–4.5 μm. U-form cells and nearly ring-like forms occur on peptone-yeast extract-glucose agar. **Motility occurs by a single polar flagellum.** Colonies on peptone-yeast extract-glucose agar are circular, smooth, convex, entire, opaque, and pale brown. **Optimum temperature is 9°C; maximum 20°C; minimum below 0°C.** Aerobic and **oxidase and catalase positive.** Indole, Voges-Proskauer, and methyl red tests are negative. No growth with 5% NaCl. **Acid but no gas from xylose (aerobically).** No acid or gas from glucose, lactose, sucrose, maltose, arabinose, or glycerol aerobically or anaerobically. No change in litmus milk. H$_2$S is not produced. Starch and cellulose are not degraded. Nitrate is reduced to nitrite. No growth occurs anaerobically with nitrate. Succinate, formate, acetate, fumarate, and propionate are assimilated. Citrate, lactate, protocatechuate, *p*-hydroxybenzoate, and hippurate are not assimilated. **Isolated from antarctic soil.**

"Spirillum pulli"

Rigid spiral cells. By darkfield microscopy the cell diameter is **1 μm,** and the cell length is 5–12 μm. Actively motile by a single flagellum at each pole. Believed to be the cause of a diphtheroid stomatitis in the mouth of chickens. Attempts to culture the organism in artificial media have been unsuccessful. Experimental passage of the disease in chickens has been accomplished by contact and by experimental inoculation.

Genus "Sporospirillum"

Rigid helical bacteria of enormous size, 1.8–4.8 μm in diameter and 40–100 μm in length. **Structures that morphologically resemble endospores occur within the cells.** The sporelike structures have the ability to rotate and to migrate within the cytoplasm of the bacteria. They initially develop near the cell poles and later migrate to the center where they are released after the cell ruptures and disintegrates. **The Gram-staining reaction of the cells has not been reported. The cells are motile,** but no organelles of locomotion are evident. The relationship of the organisms to oxygen is unknown. **Occur in the intestinal contents of tadpoles and have not been isolated.**

Type species: none designated.

Differentiation of the species of the genus "Sporospirillum": Because these organisms have not been isolated, species identification is done entirely on a morphological basis, as indicated in Table 2.8.

Genus Vampirovibrio

Comma-shaped rods, 0.3 μm in diameter. Larger coccoid forms (0.6 μm) have also been described. Stain is Gram negative. **Motility occurs by a single, non-sheathed, polar flagellum.** Reproduction is by binary fission. **Cells attach to strains of the alga *Chlorella* and require viable cells of the alga for development.** Growth occurs outside the host cells, and penetration of the host cells has not been reported. The bacterium does not develop into an elongated spirillar form as part of its life cycle.

Type (and only) species: *Vampirovibrio chlorellavorus.*

Characteristics of the species: As described for the genus.

Genus Wolinella: Wolinella succinogenes *and* Wolinella curva

Helical or curved cells 0.5–1.0 x 2–6 μm, with tapered ends. Nonsporing. **Motility occurs by a single polar flagellum.** Although *Wolinella* species were described as anaerobes in *Bergey's Manual of Systematic Bacteriology*, Volume 1, they are in fact **H$_2$- and formate-requiring microaerophiles, capable of respiring**

with oxygen. In the absence of alternative terminal electron acceptors such as fumarate, they exhibit oxygen-dependent growth with low levels of oxygen and will not grow anaerobically or aerobically (for *W. succinogenes,* see Wolin et al., J. Bacteriol. *81:* 911-917, 1961; for *W. curva,* see Han et al., Int. J. Syst. Bacteriol. *41:* 218–222, 1991; see also the description of *Wolinella recta*—a straight rod—in Groups 4 and 6 of this manual). **Fumarate and nitrate are used as electron acceptors for anaerobic respiration. Oxidase positive and catalase negative.** Contain cytochromes *b* and *c*. **Carbohydrates are not utilized. Isolated from the bovine rumen, the human gingival sulcus, dental root canal infections, and other clinical material.**

Type species: *Wolinella succinogenes.*

Differentiation between **Wolinella succinogenes** *and* **Wolinella curva:** See Table 2.5.

Important Notes for Users of This Manual

Unless otherwise indicated in footnotes to tables, the meanings of symbols are as follows:

+ 90% or more of strains are positive
− 90% or more of strains are negative
d 11-89% of strains are positive
v strain instability (not equivalent to "d")
D Different reactions in different taxa (species of a genus or genera of a family)

All other symbols are defined in footnotes to tables.

GROUP 2 AEROBIC/MICROAEROPHILIC, MOTILE, HELICAL/VIBRIOID GRAM-NEGATIVE BACTERIA

Table 2.1 Differentiation of motile, aerobic, or microaerophilic Gram-negative curved bacteria [a]

Characteristic	*Alteromonas undina* [b]	*Aquaspirillum*	*Azospirillum*	*Bdellovibrio*	*Campylobacter*
Predominant shape:					
Helical	−	+ [d]	−	−	−
Curved in one plane	+	−	−	−	−
Vibrioid	−	−	+ [e]	+	+ [f]
Cell diameter (μm)	0.7–1.5	0.2–1.4	0.9–1.5	0.2–0.5	0.2–0.9
Cultivable on inanimate laboratory media	+	+	+	− [h]	+
Require host or host cells for cultivation	−	−	−	+ [h]	−
Predacious on other Gram-negative bacteria	−	−	−	+	−
Grows in periplasmic space of host cell				+	
Exoparasitic growth				−	
Predacious on eucaryotic algae	−	−	−	−	−
Structures morphologically resembling endospores are present	−	−	−	−	−
Usual arrangement of polar flagella	Monotrichous	Bipolar tufts [i]	Monotrichous	Monotrichous	1 at one or both poles
Sheathed flagella present	−	−	−	+	− [l]
Lateral flagella occur in addition to polar flagella	−	−	D [m]	−	−
Optimum growth temperature 5–9°C; maximum temperature 20°C	−	− [o]	−	−	−
Pathogenic for humans or animals	−	−	−	−	D
Inhibited by 3.5% NaCl	−	+	D	D	D
Sea water or Na⁺ required for growth	+	−	−	D [p]	− [q]
Requires at least 7% NaCl for growth; can grow with 28% NaCl	−	−	−	−	−
Nitrogenase activity under microaerobic conditions	−	− [r]	+	−	− [s]
Relation to oxygen under non-N₂-fixing conditions:					
Aerobic	+	+ [t]	+	+	− [u]
Microaerophilic	−	− [t]	−	−	+ [u]
Grows anaerobically by using H₂ or formate as electron donor and fumarate as electron acceptor	−	−	−	−	D
Some carbohydrates catabolized	+	− [v]	+	−	−
Glucose catabolized	+	− [w]	D		
Cellulose hydrolyzed		−			
Urease					− [x]
Habitat:					
Freshwater	−	+	−	D	−
Marine	+	−	−	D	− [z]
Soil	−	− [aa]	+	D	
Within plant roots	−	−	+	−	− [z]
Humans and/or warm-blooded animals	−	−	−	−	+ [z]
Intestinal contents of tadpoles	−	−	−	−	−

Footnotes are at end of table

Table 2.1 *(continued)*

Characteristic	*Cellvibrio*	*Halovibrio*	*Helicobacter*	*Herbaspirillum*	*Marinomonas communis*[b]
Predominant shape:					
Helical	−	−	−	−	−
Curved in one plane	+	−	−	−	+
Vibrioid	−	+	+	+[g]	−
Cell diameter (μm)	0.2–0.5	0.5–0.8	0.5–1.0	0.6–0.7	0.7–1.5
Cultivable on inanimate laboratory media	+	+	+	+	+
Require host or host cells for cultivation	−	−	−	−	−
Predacious on other Gram-negative bacteria	−	−	−	−	−
Grows in periplasmic space of host cell					
Exoparasitic growth					
Predacious on eucaryotic algae	−	−	−	−	−
Structures morphologically resembling endospores are present	−	−	−	−	−
Usual arrangement of polar flagella	Mono-trichous	Mono-trichous	Multiple at one or both poles	1–3 at one or both poles	Monotrichous
Sheathed flagella present			+		−
Lateral flagella occur in addition to polar flagella	+[m]	−	D[n]	−	−
Optimum growth temperature 5–9°C; maximum temperature 20°C	−	−	−	−	−
Pathogenic for humans or animals	−	−	+	−	−
Inhibited by 3.5% NaCl			+	+	−
Sea water or Na$^+$ required for growth	−	+	−	−	+
Requires at least 7% NaCl for growth; can grow with 28% NaCl	−	+		−	−
Nitrogenase activity under microaerobic conditions	−		−	+	−
Relation to oxygen under non-N$_2$-fixing conditions:					
Aerobic	+	+	−	+	+
Microaerophilic	−	−	+	−	−
Grows anaerobically by using H$_2$ or formate as electron donor and fumarate as electron acceptor	−	−		−	−
Some carbohydrates catabolized	+	−	−	+	+
Glucose catabolized	+		−	+	+
Cellulose hydrolyzed	+	−	−		
Urease	−	+	+		
Habitat:					
Freshwater	−	−	−	−	−
Marine	−	−	−	−	+
Soil	+	−	−	+	−
Within plant roots	−	−	−	+	−
Humans and/or warm-blooded animals	−	−	+	−	−
Intestinal contents of tadpoles	−	−	−	−	−

Footnotes are at end of table

Table 2.1 (continued)

Characteristic	*Micavibrio*	*Oceanospirillum*	*Spirillum*	*"Spirillum minus"*	*"Spirillum pleomorphum"*
Predominant shape:					
Helical	−	+	+	+	−
Curved in one plane	−	−	−	−	+
Vibrioid	+	−	−	−	−
Cell diameter (μm)	0.25–0.35	0.3–1.4	1.4–1.7	0.2	0.7–1.0
Cultivable on inanimate laboratory media	−	+	+	−	+
Require host or host cells for cultivation	+	−	−	+	−
Predacious on other Gram-negative bacteria	+	−	−	−	−
Grows in periplasmic space of host cell	−				
Exoparasitic growth	+				
Predacious on eucaryotic algae	−	−	−	−	−
Structures morphologically resembling endospores are present	−	−	−	−	−
Usual arrangement of polar flagella	Mono-trichous	Bipolar tufts[j]	Bipolar tufts	1 or more at each pole	Monotrichous
Sheathed flagella present	−		−		
Lateral flagella occur in addition to polar flagella	−	−	−	−	−
Optimum growth temperature 5–9°C; maximum temperature 20°C	−	−	−	−	+
Pathogenic for humans or animals	−	−	−	+	−
Inhibited by 3.5% NaCl	−	−	+		
Sea water or Na$^+$ required for growth	−	+	−		−
Requires at least 7% NaCl for growth; can grow with 28% NaCl	−	−			
Nitrogenase activity under microaerobic conditions		−	−		
Relation to oxygen under non-N$_2$-fixing conditions:					
Aerobic	+	+	−		+
Microaerophilic	−	−	+		−
Grows anaerobically by using H$_2$ or formate as electron donor and fumarate as electron acceptor		−	−		
Some carbohydrates catabolized		−	−		
Glucose catabolized					
Cellulose hydrolyzed					
Urease					
Habitat:					
Freshwater	+[y]	−	+	−	−
Marine		+	−	−	−
Soil	−	−	−		+
Within plant roots	−	−	−		−
Humans and/or warm-blooded animals		−	−	+	−
Intestinal contents of tadpoles	−	−	−		−

Footnotes are at end of table

Table 2.1 *(continued)*

Characteristic	"Spirillum pulli"	"Sporospirillum"	Vampirovibrio	Wolinella[c]
Predominant shape:				
Helical	+	+	−	D
Curved in one plane	−	−	−	−
Vibrioid	−	−	+	D
Cell diameter (µm)	1.0	1.4–4.8	0.3	0.5–1.0
Cultivable on inanimate laboratory media	−	−	−	+
Require host or host cells for cultivation	+	+	+	−
Predacious on other Gram-negative bacteria	−	−	−	−
Grows in periplasmic space of host cell				
Exoparasitic growth				
Predacious on eucaryotic algae	−	−	+	−
Structures morphologically resembling endospores are present	−	+	−	−
Usual arrangement of polar flagella	1 at one or both poles	−[k]	Monotrichous	Monotrichous
Sheathed flagella present			−	−
Lateral flagella occur in addition to polar flagella	−	−	−	−
Optimum growth temperature 5–9°C; maximum temperature 20°C	−	−	−	−
Pathogenic for humans or animals	+	−	−	
Inhibited by 3.5% NaCl				
Sea water or Na⁺ required for growth				−
Requires at least 7% NaCl for growth; can grow with 28% NaCl				−
Nitrogenase activity under microaerobic conditions				−
Relation to oxygen under non-N_2-fixing conditions:				
Aerobic				−
Microaerophilic				+[b]
Grows anaerobically by using H_2 or formate as electron donor and fumarate as electron acceptor				+
Some carbohydrates catabolized				−
Glucose catabolized				
Cellulose hydrolyzed				
Urease				−
Habitat:				
Freshwater	−		+	−
Marine	−	−		−
Soil	−	−		−
Within plant roots	−	−	−	−
Humans and/or warm-blooded animals	+	−	−	+
Intestinal contents of tadpoles	−	+	−	−

Footnotes are at end of table

Table 2.1 *(continued)*

[a] Symbols: +, all species positive except where noted; -, all species negative except where noted; D, differs among species.

[b] For differentiation of *Alteromonas undina* from *Marinomonas communis,* see the descriptions of these species.

[c] Only *Wolinella succinogenes* and *Wolinella curva* are covered in Group 2. Although the genus *Wolinella* has been regarded as anaerobic, the organisms are in fact microaerophiles. They are capable of respiring with oxygen when it is provided at low concentrations and cannot grow under an air atmosphere. They can also grow anaerobically by using fumarate as a terminal electron acceptor for anaerobic respiration. For further information see Wolin et al. (J. Bacteriol. *81:* 911-917, 1961) and Han et al. (Int. J. Syst. Bacteriol. *41:* 218–222, 1991). The genus *Wolinella* contains three species, of which only two contain vibrioid or helical organisms: *W. succinogenes* and *W. curva.* The third species, *W. recta,* contains straight rods. See also Group 6, Table 6.23. *W. curva* and *W. recta* were recently moved to the genus *Campylobacter* by Vandamme et al. (Int. J. Syst. Bacteriol. *41:* 88–103, 1991).

[d] *A. delicatum* is mainly vibrioid.

[e] Some cells in *Azospirillum* cultures are straight rods.

[f] Chains of *Campylobacter* cells may have a helical appearance.

[g] Some cells in cultures of *Herbaspirillum* are helical.

[h] All wild-type strains of *Bdellovibrio* upon initial isolation are dependent on intraperiplasmic growth in susceptible bacterial prey. Mutants capable of axenic growth ("prey-independent" strains) have been derived from the predacious strains, and some strains are facultative, i.e., capable of growth in the presence of absence of prey cells.

[i] *A. delicatum* has mainly a single flagellum at one pole; *A. polymorphum* has mainly a single flagellum at each pole, and *"A. arcticum"* has a single polar flagellum.

[j] *O. pusillum* has mainly a single flagellum at each pole.

[k] *Sporospirillum* spp. are motile but no organelles of locomotion have been observed.

[l] *C. cinaedi* and *C. fennelliae* have sheathed flagella.

[m] Especially (or in the case of azospirilla, only) when cells are grown on solid media.

[n] *H. mustelae* is reported to have multiple lateral flagella in addition to polar flagella.

[o] *"A. arcticum"* is positive.

[p] Marine bdellovibrios require Na^+.

[q] *C. nitrofigilis* requires Na^+.

[r] *A. peregrinum* and some strains of *A. itersonii* have nitrogenase activity.

[s] *C. nitrofigilis* has nitrogenase activity.

[t] *A. magnetotacticum* is microaerophilic.

[u] *C. cryaerophila* can grow aerobically (although it may be microaerophilic on primary isolation). *C. nitrofigilis* can grow aerobically on complex media such as Brucella agar.

[v] *A. gracile, A. itersonii, A. peregrinum,* and *"A. arcticum"* can catabolize a very restricted variety of sugars.

[w] *A. gracile* can produce acid from glucose aerobically. *A. itersonii* can produce acid from glucose anaerobically but not aerobically. *"A. arcticum"* can grow on glucose, fructose and ribose but not on other carbohydrates.

[x] A group of urease-positive thermophilic campylobacters (UPTC) have been shown to be variants of *Campylobacter lari* (Mégraud et al., J. Clin. Microbiol. *26:* 1050-1051, 1988).

[y] Occurs in sewage waters.

[z] *C. nitrofigilis* occurs in the roots of salt-marsh grasses.

[aa] *"A. arcticum"* occurs in arctic sediments.

Table 2.2 Differential characteristics of *Aquaspirillum* species, *"Spirillum minus"*, *"Spirillum pleomorphum"*, and *"Spirillum pulli"* [a]

Characteristics	*A. anulus*	*A. aquaticum* [b]	*"A. arcticum"* [c]	*A. autotrophicum*	*A. delicatum*	*A. dispar*	*A. giesbergeri*
Cell diameter (μm) [e]	0.8–1.4	0.5–0.6	~1.0	0.6–0.8	0.3–0.4	0.5–0.7	0.7–1.4
Predominant shape:							
Helical	+	+	–	+	–	+	+
Vibrioid	–	–	+ [g]	–	+	–	–
Type of helix: [h]							
Clockwise	+	+		+		+	+
Counterclockwise	–	–		–		–	–
Poly-β-hydroxybutyrate formed	+	+	+	+	+	+	+
Flagellar arrangement	BT	BT	U(1)	BT	U(1–2)	BT	BT
Coccoid bodies predominant at 3–4 weeks	–	–	+	–	–	–	–
Oxidase test	+	+	+	+	+	+	+
Utilization of sugars or acid production from sugars	–	–	+ [k]	–	–	–	–
Anaerobic growth with NO_3^-	–	–	–	–	–	+	–
NO_3^- reduced only to NO_2^-	–	+	–	–	+	–	–
Denitrification	–	–	–	–	–	+	–
Phosphatase	+	+	+	+	+	+	+
Esculin hydrolysis	–	–	–	–	–	–	–
Urease	–	–	– or W	+	–	–	–
Optimum temperature = 20°C; temperature range = 2–26°C	–	–	–	–	–	–	–
Optimum temperature = 5–9°C; maximum = 20°C	–	–	+	–	–	–	–
Optimum temperature = 41°C	–	–	–	–	–	–	–
Can be cultivated in vitro	+	+	+	+	+	+	+
Pathogenic (humans or animals)	–	–	–	–	–	–	–
Growth factors required [n]	–	+ [n]	–	–	–	–	–
Glutamate used as a sole carbon source	+	+	+	+	–	d	–
Histidine used as a sole carbon source	–	–	–	–	–	–	–
Tryptophan and glycine used as sole carbon sources	–	–	–	+	–	–	–
Nitrogenase activity [o]	–	–	–		–	–	–
Hydrogen autotrophy		–		+		–	
Magnetotactic; magnetosomes present in cells [p]	–	–	–	–	–	–	–
Obligately microaerophilic	–	–	–	–	–	–	–

Footnotes are at end of table

Table 2.2   (continued)

Characteristics	A. gracile	A. itersonii	A. magnetotacticum	A. metamorphum	A. peregrinum	A. polymorphum	A. psychrophilum
Cell diameter (µm) [e]	0.2–0.3	0.4–0.8 [f]	0.2–0.4	0.7–1.3	0.5–0.7	0.3–0.5	0.7–0.9
Predominant shape:							
Helical	+	+	+	+	+	+	+
Vibrioid	–	–	–	–	–	–	–
Type of helix: [h]							
Clockwise	+	–	+	– [i]	–	–	+
Counterclockwise	–	+	–	+ [i]	+	+	–
Poly-β-hydroxybutyrate formed	–	+	+	+	+	+	–
Flagellar arrangement	BT	BT	BS	BT	BT	BS	BT
Coccoid bodies predominant at 3–4 weeks	–	+	+	–	d [j]	+	–
Oxidase test	+	+	–	+	+	+	+
Utilization of sugars or acid production from sugars	+ [k]	+ [k]	–	–	+ [k]	–	–
Anaerobic growth with NO_3^-	–	+	–	–	–	–	+
NO_3^- reduced only to NO_2^-	+	–	–	–	–	+	–
Denitrification	–	+	+	–	–	–	+
Phosphatase	+	+	+	+	+	–	+
Esculin hydrolysis	–	+	–	–	+	+	–
Urease	+	–		+	+	–	
Optimum temperature = 20°C; temperature range = 2–26°C	–	–	–	–	–	–	+
Optimum temperature = 5–9°C; maximum = 20°C	–	–	–	–	–	–	–
Optimum temperature = 41°C	–	–	–	–	–	–	–
Can be cultivated in vitro	+	+	+	+	+	+	+
Pathogenic (humans or animals)	–	–	–	–	–	–	–
Growth factors required [n]	+ [n]	–	–	–	–	–	
Glutamate used as a sole carbon source		+		+	+	+	
Histidine used as a sole carbon source		+		–	–	–	
Tryptophan and glycine used as sole carbon sources		–		–	–	–	–
Nitrogenase activity [o]		d	+	–	+	–	
Hydrogen autotrophy		–		–			
Magnetotactic; magnetosomes present in cells [p]	–	–	+	–	–	–	–
Obligately microaerophilic	–	–	+ [q]	–	–	–	–

Footnotes are at end of table

Table 2.2   *(continued)*

Characteristics	*A. putridiconchylium*	*A. serpens*	*A. sinuosum*	*"S. minus"*	*"S. pleomorphum"*	*"S. pulli"*
Cell diameter (μm) [e]	0.7–1.2	0.6–1.2	0.6–0.9	~0.2	0.7–1.0	~1.0
Predominant shape:						
Helical	+	+	+	+	−	+
Vibrioid	−	−	−	−	+	−
Type of helix: [h]						
Clockwise	+	+	+			
Counterclockwise	−	−	−			
Poly-β-hydroxybutyrate formed	+	+	+			
Flagellar arrangement	BT	BT	BT	BS or BT	U(1)	BS
Coccoid bodies predominant at 3–4 weeks	−	−	−			
Oxidase test	+	+	+		+	
Utilization of sugars or acid production from sugars	−	−	−		+[k]	
Anaerobic growth with NO_3^-	−	−	−		−	
NO_3^- reduced only to NO_2^-	−	d	−		+	
Denitrification	−	−	−			
Phosphatase	+	+	+			
Esculin hydrolysis	−	−	−			
Urease	−	−	+			
Optimum temperature = 20°C; temperature range = 2–26°C	−	−	−		−	
Optimum temperature = 5–9°C; maximum = 20°C	−	−	−		+	
Optimum temperature = 41°C	−	d	−		−	
Can be cultivated in vitro	+	+	+	−	+	−
Pathogenic (humans or animals)	−	−	−	+[l]	−	+[m]
Growth factors required [n]	−	−	−			
Glutamate used as a sole carbon source	+	+	−			
Histidine used as a sole carbon source	+	−	−			
Tryptophan and glycine used as sole carbon sources	−	−	−			
Nitrogenase activity [o]	−	−	−			
Hydrogen autotrophy						
Magnetotactic; magnetosomes present in cells [p]	−	−	−			
Obligately microaerophilic	−	−	−			

Footnotes are at end of table

Table 2.2   *(continued)*

---

[a] Symbols: +, positive for all strains except where noted; negative, negative for all strains except where noted; blank space, not determined; BT, bipolar tufts; U (1-2), 1 or 2 flagella at only one pole; BS, single flagellum at each pole.

[b] *A. aquaticum* was recently shown to be synonomous with *Comamonas terrigena* (Willems et al., Int. J. Syst. Bacteriol. *41:* 427–444, 1991).

[c] *A. arcticum* was not included in *Bergey's Manual of Systematic Bacteriology.* The species was created in 1989 by Butler et al. (Syst. Appl. Microbiol. *12:* 263-266; Int. J. Syst. Bacteriol. *40:* 320–321, 1990). It is distinguished from *"S. pleomorphum"* by its lack of pleomorphism, inability to reduce nitrate, ability to produce acid from glucose but not xylose, temperature optimum of 5°C rather than 9°C, ability to assimilate lactate, and inability to assimilate formate.

[d] Includes the species *A. bengal,* which has been shown to be a subjective synonym of *A. serpens.* See Boivin et al., Int. J. Syst. Bacteriol. *35:* 512-517, 1985. *A. bengal* strains grow optimally at 41°C and are considered to represent a biovar of *A. serpens.*

[e] By phase-contrast microscopy of 24- to 48-h-old broth cultures (for cultivable species).

[f] The range for the subspecies *nipponicum* is 0.5-0.8 μm; for the subspecies *itersonii* the range is 0.4-0.6 μm.

[g] The cells of *"A. arcticum"* are straight to curved, some with an observable twist.

[h] Determined by focusing on the bottom of the cells. The pattern //// indicates a clockwise (right-handed) helix, whereas the pattern \\\\ indicates a counterclockwise (left-handed) helix.

[i] Formerly thought to be clockwise. See Konishi and Yoshii, J. Gen. Microbiol. *132:* 877-881, 1986.

[j] The subspecies *integrum* fails to form coccoid bodies, whereas the subspecies *peregrinum* forms them readily in old cultures.

[k] *A. gracile,* produces acid from glucose, galactose and arabinose (aerobically); growth occurs on glucose, xylose, and arabinose but not fructose. *A. itersonii,* produces acid from fructose (aerobically and anaerobically) and glucose (anaerobically only); growth occurs on fructose but not glucose, xylose or arabinose. *A. peregrinum* produces acid from fructose (aerobically and anaerobically); growth occurs on fructose but not glucose, xylose or arabinose. *"S. pleomorphum"* produces acid from xylose (aerobically) but not from glucose. *"A. arcticum"* produces acid from glucose (aerobically) but not from xylose; growth occurs on glucose, fructose and ribose. For detection of acid production, peptone concentrations must be kept low (0.2% or less) in order to detect change in pH indicator.

[l] Causes rat-bite fever in humans.

[m] Causes a diphtheroid stomatitis in the mouths of adult chickens.

[n] *A. gracile* requires biotin and *A. aquaticum* requires niacin.

[o] By the acetylene reduction test with cultures grown either in nitrogen-deficient semisolid media under aerobic conditions or in liquid media under anaerobic conditions.

[p] An intracellular chain of electron-dense, cuboidal inclusion bodies containing magnetite (Fe_3O_4), which are responsible for tactic responses of the cells to magnetic fields.

[q] Grows best under a 1-3% oxygen atmosphere; does not grow aerobically.

Table 2.3 Characteristics differentiating *Azospirillum* species and *Herbaspirillum seropedicae*[a]

Characteristic	A. amazon-ense[b]	A. brasil-ense	A. halo-praeferens[c]	A. ira-kense[d]	A. lipo-ferum	H. sero-pedicae
Cell width (μm)	0.8–1.0	1.0–1.2	0.7–1.4	0.6–0.9	1.0–1.5	0.6–0.7
Pleomorphic cells in alkaline N-free semisolid medium	−	−	−		+	−
Flagellar arrangement on cells grown in liquid media:						
One to three flagella at one or both poles	−	−	−	−	−	+
Single flagellum at one pole	+	+	+	+	+	−
Lateral flagella formed (in addition to polar flagella) when grown on agar media	−	+		+	+	−
Optimum temperature (°C)	35	37	41	30–33	37	34
Growth at pH 6.0	+	+ or W	−	+	+	+
Biotin requirement	−	−	+	−	+	−
Growth on potato (BMS) agar	+	+	−		+	+
Colony type on potato agar:						
White, flat with raised margin	+	−			−	−
Pink, raised, curled	−	+			+	−
Brownish, small, raised, smooth	−	−			−	+
Growth in presence of 3% NaCl	−	d	+	+	−	
Dissimilation of:						
NO$_3^-$ to NO$_2^-$	±	+	+	d	+	+
NO$_2^-$ to N$_2$O	−	±	+	−	±	±[e]
Capable of NO$_3^-$-dependent anaerobic growth in complex media	−	+		−	+	−
Sole carbon sources for growth in N-deficient semisolid medium:						
Glucose	+	−	−		+	+
Sucrose	+	−	−		−	−
α-Ketoglutarate	−	−	+		+	+
Citrate	−	+			+	+
Carbon source utilization (API gallery method)[f]:						
Glucose	+	d	−	+	+	
Sucrose	+	−	−	+	−	
myo-Inositol	+	−		−	d	
pH range for good growth	5.7–6.5	6.0–7.3	6.8–8.0	5.5–8.5	5.7–6.8	5.3–8.0
Pectin hydrolyzed in 7 days[f]	−	−	−	+	−	

[a] Symbols: see standard definitions. In addition, W, weak; ±, several strains positive but the majority of strains negative.

[b] *A. amazonense* was not included in *Bergey's Manual of Systematic Bacteriology*, Volume 1. The species was created by Falk et al. (Int. J. Syst. Bacteriol. *35:* 117-118, 1985).

[c] *A. halopraeferens* was not included in *Bergey's Manual of Systematic Bacteriology*, Volume 1. The species was created by Reinhold et al. (Int. J. Syst. Bacteriol. *37:* 43-51, 1987).

[d] *A. irakense* was not included in *Bergey's Manual of Systematic Bacteriology*, Volume 1. The species was created by Khammas et al. (Res. Microbiol. *140:* 679–693, 1989; Int. J. Syst. Bacteriol. *41:* 580–581, 1991).

[e] All strains tested have shown very weak N$_2$O production.

[f] Method described by Khammas et al., Res. Microbiol. *140:* 679–693, 1989.

Table 2.4   Differential characteristics of the species of the genus *Bdellovibrio* [a]

Characteristics	*B. bacteriovorus*	*B. starrii*	*B. stolpii*
Facultatively predatory [b]	−	−	+
Sensitive to vibriostatic agent 0/129 [c]	+	−	+
Bacteriophage susceptibility [d]	+	−	+
Protease production [d]	Low	Moderate	High

[a] Symbols: see standard definitions.

[b] Defined to mean that any individual comma-shaped cell may complete its life cycle either prey-dependently or prey-independently.

[c] O/129 = 2,4-diamino-6,7-diisopropylpteridine phosphate.

[d] See Burnham and Conti, *Bergey's Manual of Systematic Bacteriology,* Volume 1, pp. 118-124, 1984.

Important Notes for Users of This Manual

Unless otherwise indicated in footnotes to tables, the meanings of symbols are as follows:

+ 90% or more of strains are positive

− 90% or more of strains are negative

d 11-89% of strains are positive

v strain instability (not equivalent to "d")

D Different reactions in different taxa (species of a genus or genera of a family)

All other symbols are defined in footnotes to tables.

Table 2.5 Differential characteristics of the species of the genus *Campylobacter* and of *Wolinella succinogenes* and *Wolinella curva*[a]

Characteristics	C. cinaedi[b]	C. coli	C. concisus	C. cryaerophila[c]	C. fennelliae[d]	C. fetus subsp. fetus	subsp. venerealis
Catalase	+	+	−	+	+	+	+
Growth at:							
25°C	−	−	−	+	−	+	+
42°C	+	+	+	d	−	d	−
15°C	−	−	−	+	−	−	−
Growth in:							
1% glycine	+	+	+	−	+	+	−
1% oxgall	+	+	+	+	−	+	+
Roop minimal medium[n]		+	−	d		d	d
1.5% NaCl plates		−	+	−		d	−
3.5% NaCl	−	−	−	−	−	−	−
Requires at least 1.5% NaCl for growth	−	−	−	−	−	−	−
Anaerobic growth with 0.1% trimethylamine oxide as the electron acceptor			−				
Sensitive to:							
Nalidixic acid, 30 μg disk	+	+	−	d	+	−	−
Cephalothin, 30 μg disk	−	d	−	d	+	+	+
Hippurate hydrolysis	−	−	−	−	−	−	−
H$_2$S production:							
Sulfide-Indole-Motility medium	−	−	+	−	−	−	−
Triple sugar iron agar slants[o]	−	+	+	−	−	−	−
Alkaline phosphatase		d	−	−		−	−
Indoxyl acetate hydrolysis	−	+	−	+	+	−	−
Aerobic growth on complex solid media	−	−	−	+	−	−	−
Nitrate reduced to nitrite	+	+	+	+	−	+	+
Nitrite reduction	−	−	+		−	−	−
Nitrogenase activity	−	−	−	−	−	−	−
Electron donor required for growth under microaerobic conditions with O$_2$ as the electron acceptor:							
H$_2$ or formate	−	−	+	−	−	−	−
H$_2$ but not formate	+	−			+		
Electron donor required for growth under anaerobic conditions with fumarate as the electron acceptor:							
H$_2$ or formate	−	−	+	−	−	−	−
H$_2$ but not formate	+	−			+		
Fumarate can be used as the electron acceptor for anaerobic growth in the absence of H$_2$ or formate				−			
Sheathed flagella	+	−	−	−	+	−	−
Growth in presence of:							
Crystal violet, 0.00005%			+				
Indulin scarlet, 0.05%							
Sodium fluoride, 0.05%							

Footnotes are at end of table

Table 2.5 *(continued)*

Characteristics	C. hyointestinalis[e]	C. jejuni[f] subsp. jejuni	C. jejuni[f] subsp. doylei	C. lari[g]	C. mucosalis[h]	C. nitrofigillis[j]
Catalase	+	+	+	+	−	+
Growth at:						
25°C	+	−	−	−	+	+
42°C	+	+	W	+	+	−
15°C	−	−	−	−	−	+
Growth in:						
1% glycine	+	+	+	+	+	−
1% oxgall	+	+	+	+	+	−
Roop minimal medium[n]	−	−		−	−	
1.5% NaCl plates	−	−		+	−	
3.5% NaCl	−	−	−	−	−	+
Requires at least 1.5% NaCl for growth	−	−	−	−	−	+
Anaerobic growth with 0.1% trimethylamine oxide as the electron acceptor	+			+		
Sensitive to:						
Nalidixic acid, 30 µg disk	−	+	+	−	d	+
Cephalothin, 30 µg disk	+	−	+	−	+	+
Hippurate hydrolysis	−	+		−	−	−
H$_2$S production:						
Sulfide-Indole-Motility medium	−	−			+	
Triple sugar iron agar slants[o]	+	−	−	−	+	
Alkaline phosphatase	−	+	+	−	−	
Indoxyl acetate hydrolysis	−	+	+	−	−	−
Aerobic growth on complex solid media	−	−	−	−	−	+
Nitrate reduced to nitrite	+	+	−	+	+	
Nitrite reduction	−	−		−	d	
Nitrogenase activity	−	−	−	−	−	+
Electron donor required for growth under microaerobic conditions with O$_2$ as the electron acceptor:						
H$_2$ or formate	−	−	−	−	+	−
H$_2$ but not formate					−	
Electron donor required for growth under anaerobic conditions with fumarate as the electron acceptor:						
H$_2$ or formate	−	−	−	−	+	−
H$_2$ but not formate					−	
Fumarate can be used as the electron acceptor for anaerobic growth in the absence of H$_2$ or formate					−	
Sheathed flagella	−	−	−	−	−	−
Growth in presence of:						
Crystal violet, 0.00005%						
Indulin scarlet, 0.05%						
Sodium fluoride, 0.05%						

Footnotes are at end of table

Table 2.5   *(continued)*

Characteristics	C. sputorum [j] biovar sputorum [e]	biovar bubulus	biovar fecalis	C. upsaliensis [k]	W. curva [l]	W. succinogenes [m]
Catalase	−	−	+	W or −	−	−
Growth at:						
25°C	−	−	−	d	−	+
42°C	+	+	+	+	+	−
15°C	−	−	−	−		−
Growth in:						
1% glycine	+	+	+	d	+	−
1% oxgall	+	−	−	d	+	−
Roop minimal medium [n]	−	d	d	−		
1.5% NaCl plates	d	+	+	−		
3.5% NaCl	−	d	+	−	−	−
Requires at least 1.5% NaCl for growth	−	−	−			
Anaerobic growth with 0.1% trimethylamine oxide as the electron acceptor	d	+	+	−		
Sensitive to:						
Nalidixic acid, 30 μg disk	d	d	−	+	+	+
Cephalothin, 30 μg disk	+	+	+	+	+	+
Hippurate hydrolysis	−	−	−	−	−	−
H$_2$S production:						
Sulfide-Indole-Motility medium	d	+	+	−	+	+
Triple sugar iron agar slants [o]	d	+	+	−	+	+
Alkaline phosphatase	−	−	−	+		
Indoxyl acetate hydrolysis	−	−	−	+	+	−
Aerobic growth on complex solid media	−	−	−	−	−	−
Nitrate reduced to nitrite	+	+	+	+	+	+
Nitrite reduction	d	d	+	−	−	−
Nitrogenase activity	−	−	−		−	−
Electron donor required for growth under microaerobic conditions with O$_2$ as the electron acceptor:						
H$_2$ or formate	−	−	−	−	+	+
H$_2$ but not formate	−	−	−	−	−	−
Electron donor required for growth under anaerobic conditions with fumarate as the electron acceptor:						
H$_2$ or formate	−	−	−		+	+
H$_2$ but not formate	−	−	−		−	−
Fumarate can be used as the electron acceptor for anaerobic growth in the absence of H$_2$ or formate	+	+	+		−	−
Sheathed flagella	−	−	−	−	−	−
Growth in presence of:						
Crystal violet, 0.00005%	−				−	+
Indulin scarlet, 0.05%					+	−
Sodium fluoride, 0.05%					−	+

Footnotes are at end of table

Table 2.5 *(continued)*

---

[a] Symbols: see standard definitions. W = weakly positive.

[b] *C. cinaedi* was added subsequent to publication of *Bergey's Manual of Systematic Bacteriology,* Volume 1. See Totten et al., J. Infect. Dis. *151:* 131-139, 1985; Int. J. Syst. Bacteriol. *38:* 328-329, 1988). In 1991, Vandamme et al. (Int. J. Syst. Bacteriol. *41:* 88-103) assigned the species to the genus *Helicobacter,* as *Helicobacter cinaedi.*

[c] *C. cryaerophila* was added subsequent to publication of *Bergey's Manual of Systematic Bacteriology,* Volume 1. See Neill et al., Int. J. Syst. Bacteriol. *35:* 342-356, 1985. In 1991, Vandamme et al. (Int. J. Syst. Bacteriol. *41:* 88-103) assigned the species to a new genus, *Arcobacter,* as *Arcobacter cryaerophilus.*

[d] *C. fennelliae* was added subsequent to publication of *Bergey's Manual of Systematic Bacteriology,* Volume 1. See Totten, Fennell, Tenover et al., J. Infect. Dis. *151:* 131-139, 1985; (Int. J. Syst. Bacteriol. *38:* 328-329, 1988). In 1991, Vandamme et al. (*Int. J. Syst. Bacteriol. 41:* 88-103) assigned the species to the genus *Helicobacter,* as *Helicobacter fennelliae.*

[e] *C. hyointestinalis* has been added subsequent to publication of *Bergey's Manual of Systematic Bacteriology,* Volume 1. See Gebhart et al. (J. Clin. Microbiol. *21:* 715-720, 1985; Int. J. Syst. Bacteriol. *35:* 535, 1985).

[f] Since publication of *Bergey's Manual of Systematic Bacteriology,* two subspecies of *C. jejuni* were described by Steele and Owen (Int. J. Syst. Bacteriol. *38:* 316-318, 1988).

[g] *C. lari* (previously assigned the grammatically incorrect name *C. laridis*) has been added subsequent to publication of *Bergey's Manual of Systematic Bacteriology,* Volume 1. The strains have sometimes been called "NARTC" strains (nalidixic acid-resistant thermophilic campylobacters). See Benjamin et al. (Curr. Microbiol. *8:* 231-238, 1983; Int. J. Syst. Bacteriol. *34:* 270–271, 1984). Some strains are urease positive.

[h] Previously called *C. sputorum* subsp. *mucosalis.* See Roop et al., Can. J. Microbiol. *31:* 823-831, 1985.

[i] *C. nitrofigilis* was added subsequent to publication of *Bergey's Manual of Systematic Bacteriology,* Volume 1. See McClung et al., Int. J. Syst. Bacteriol. *33:* 605-612, 1983. In 1991, Vandamme et al. (Int. J. Syst. Bacteriol. *41:* 88-103) assigned the species to a new genus, *Arcobacter,* as *Arcobacter nitrofigilis.*

[j] *C. sputorum* subsp. *sputorum,* *C. sputorum* subsp. *bubulus,* and *"C. fecalis"* are biovars of *C. sputorum* (Roop et al., Can. J. Microbiol. *31:* 823-831, 1985, and *36:* 348, 1986).

[k] *C. upsaliensis* was not included in *Bergey's Manual of Systematic Bacteriology.* It was created by Sandstedt and Ursing (Syst. Appl. Microbiol. *14:* 39-45, 1991; Int. J. Syst. Bacteriol. *41:* 331, 1991). It has been isolated from healthy and diarrheic dogs and cats (Gebhart et al., Abts. Annu. Meet. Amer. Soc. Microbiol. C55, 1984; Sandstedt et al., Curr. Microbiol. *8:* 209-213, 1983), blood cultures from pediatric patients (Lastovica et al., J. Clin. Microbiol. *27:* 657-659, 1989), and from human feces (Goosens et al., J. Clin. Microbiol. *28:* 1039-1046, 1990).

[l] *W. curva* was not included in *Bergey's Manual of Systematic Bacteriology.* The species was created in 1984 by Tanner et al. (Int. J. Syst. Bacteriol. *34:* 275-282). Although *W. curva* was initially regarded as an anaerobe, Han et al. (Int. J. Syst. Bacteriol. *41:* 218–222, 1991) have shown that it is a microaerophile capable of growing with O_2 as a terminal electron acceptor (optimum O_2 concentration, approximately 6%). Under anaerobic conditions it uses fumarate as a terminal electron acceptor for anaerobic respiration, thereby forming succinate as an end product. See Group 6, Table 6.23. In 1991, Vandamme et al. (Int. J. Syst. Bacteriol. *41:* 88-103) assigned the species to the genus *Campylobacter,* as *Campylobacter curvus.*

[m] Although *W. succinogenes* was regarded as an anaerobe in *Bergey's Manual of Systematic Bacteriology,* Wolin et al. (J. Bacteriol. *81:* 911-917, 1961) showed that it was oxidase positive and that it was capable of growing with O_2 as a terminal electron acceptor under microaerobic conditions (approximately 2% O_2) but not under atmospheric levels of O_2. Under anaerobic conditions it uses fumarate as a terminal electron acceptor for anaerobic respiration, thereby forming succinate as an end product. See Group 6, Table 6.23.

[n] Roop et al. (Can. J. Microbiol. *30:* 938-951, 1984).

[o] In water of syneresis at base of slant (Roop et al., Can. J. Microbiol. *30:* 938-951, 1984).

Table 2.6   Differential characteristics of the species of the genus *Helicobacter*[a]

Characteristics	H. mustelae[b]	H. pylori[c]
Growth at 42°C	+	−
Multiple lateral flagella in addition to polar flagella	+	−
Growth in the presence of air enriched with 10% CO_2	−	+
Growth on PSD agar[d]	−	+
Growth in presence of 1% glycine	d	−
Nitrate reduction	+	−
Susceptible to:		
Nalidixic acid, 30 µg disk	+	d
Cephalothin, 30 µg disk	−	+
Found in human cases of gastritis and gastric and duodenal ulcers; causative agent of type B gastritis	−	+
Found in the gastric mucosa of ferrets; causes gastritis and peptic ulceration in adult animals	+	−
Cellular fatty acid profile includes 3-OH 18:0 acid	−	+

[a]   Symbols: see standard definitions.

[b]   *H. mustelae* was not included in *Bergey's Manual of Systematic Bacteriology*. It was initially considered to be a subspecies of *Campylobacter pylori* (Fox et al., Int. J. Syst. Bacteriol. *38:* 367-370, 1988) but was later found to be a separate species (Fox et al., Int. J. Syst. Bacteriol. *39:* 301-303, 1989). It was classified in the genus *Helicobacter* by Goodwin et al. (Int. J. Syst. Bacteriol. *39:* 397-405, 1989).

[c]   *H. pylori* (synonym: *Campylobacter pylori*) was not included in *Bergey's Manual of Systematic Bacteriology*. The species was created in 1984 by Marshall et al., Microbios Lett. *25:* 83-88. It was later classified in the genus *Helicobacter* by Goodwin et al. (Int. J. Syst. Bacteriol. *39:* 397-405, 1989).

[d]   Peptone-starch-dextrose agar (Dunkelberg et al., Appl. Microbiol. *19:* 47-52, 1970).

Important Notes for Users of This Manual

Unless otherwise indicated in footnotes to tables, the meanings of symbols are as follows:

    +   90% or more of strains are positive

    −   90% or more of strains are negative

    d   11-89% of strains are positive

    v   strain instability (not equivalent to "d")

    D   Different reactions in different taxa (species of a genus or genera of a family)

All other symbols are defined in footnotes to tables.

Table 2.7 Differential characteristics of the species of the genus *Oceanospirillum*[a]

Characteristic	*O. beijerinckii*	*O. japonicum*	*O. linum*	*O. maris*	*O. minutulum*	*O. multiglobuliferum*	*O. pusillum*
Cell diameter (μm)	0.8–1.4	0.8–1.4	0.4–0.6	0.6–1.2	0.3–0.4	0.5–0.9	0.3–0.5
Type of helix [b]:	C	C	C	C	C	C	CC
Flagellar arrangement:							
Bipolar tuft	+	+	+	+	+	+	–
Bipolar single	–	–	–	–	–	–	+
Coccoid bodies predominant at 3–4 weeks	+	–	+	+	+	+	+
Coccoid bodies predominant at 24–48 h	–	–	–	–	–	+	–
Maximum salt tolerance is low (4% NaCl)	–	–	–	–	–	+	–
Optimum temperature is 25°C rather than 30–32°C	–	–	–	d [c]	–	–	–
Catalase	W or –	W or –	+ or W	d [c]	+	+	W or –
Phosphatase	–	W	+	d [c]	–	+	W
NO$_3^-$ reduced to NO$_2^-$	–	–	–	–	+	–	+
Auxotrophic growth requirement	–	–	+ [d]	d [c]	–	–	–

[a] Symbols: see standard definitions.

[b] Determined by focusing on the bottom of the cells. The pattern //// indicates a clockwise (C) helix, whereas the pattern \\\\ indicates a counterclockwise (CC) helix.

[c] *O. maris* contains three subspecies: *maris, williamsae,* and *hiroshimense* (Pot et al., Int. J. Syst. Bacteriol. *39:* 23-34, 1989). *O. maris* subsp. *maris* has strong catalase activity, is able to grow in the presence of 1% glycine, lacks deoxyribonuclease (DNase) and ribonuclease (RNase) activities, and can grow on oxaloacetate and L-glutamate when cells are tested in a defined vitamin-free medium. *O. maris* subsp. *williamsae* has an unidentified growth factor growth requirement and fails to grow in a defined vitamin-free medium; it also has very weak catalase activity, cannot grow in the presence of 1% glycine, but does exhibit DNase and RNase activities. *O. maris* subsp. *hiroshimense* is distinguished by having phosphatase activity and by having an optimum growth temperature of 25°C instead of 30-32°C. It differs from *O. maris* subsp. *maris* by being able to use the following carbon sources in a defined, vitamin-free medium: succinate, fumarate, pyruvate, lactate, tartrate, acetate, and propionate. The catalase activity is weak or negative.

[d] *O. linum* grows poorly or not at all in defined media with single carbon sources and NH$_4^+$ as the nitrogen source; however, abundant growth occurs in a defined medium containing succinate plus malate as the carbon sources and methionine as the nitrogen source.

Table 2.8 Differentiation of the species of the genus *"Sporospirillum"*

Characteristic	*"S. bisporum"*	*"S. gyrini"*	*"S. praeclarum"*
Cell diameter (μm)	3.5–4.8	1.8–2.6	3.0–4.0
Cell length (μm)	50–90	40–100	50–100
Endospore-like structures:			
Number	2	1	1
Width (μm)	2–4	2	3–4
Length (μm)	10–14	5–7	9–12

NONMOTILE (OR RARELY MOTILE), GRAM-NEGATIVE CURVED BACTERIA

Cells are **curved, S-shaped, bow-shaped, helical, coiled, or ring-shaped. Some species form gas vesicles.** Stalks, prosthecae, and endospores are not produced. Stain is Gram negative; however, the Gram staining reactions for two genera ("*Brachyarcus*" and "*Pelosigma*") have not been reported. Cells are typically **nonmotile;** however, aggregates of "*Pelosigma*" and an occasional strain of *Ancylobacter* may be motile by polar flagella. **Chemoorganotrophic.** None are phototrophic. Cells occur in soil, freshwater, or marine environments or in anaerobic digester sludge. Two genera ("*Brachyarcus*" and "*Pelosigma*") have not been obtained in pure culture.

Differentiation of the genera of **Group 3:** See Table 3.1.

Genus **Ancylobacter**

Editorial note: The genus was included in *Bergey's Manual of Systematic Bacteriology* under the name *Microcyclus,* but this name was found to be illegitimate (see H. D. Raj, Int. J. Syst. Bacteriol. *33:* 397-398, 1983).

Curved rods, 0.3–1.0 μm in diameter and 1.0–3.0 μm in length. Rings (0.9–3.0 μm outer diameter) are formed occasionally prior to cell division. Coils, helical, or filamentous forms are not produced. Cells are encapsulated. Resting stages are not known. **Some strains produce gas vesicles.** Stain is Gram negative. **Cells are generally nonmotile,** but motility occurs in one gas vesicle-containing strain by means of a single polar flagellum. **Obligately aerobic** and possess a strictly respiratory type of metabolism with oxygen as the terminal electron acceptor. Optimum temperature is 22–37°C. Colonies are translucent to opaque and white to cream-colored. Pellicles are produced in liquid media. **Oxidase and catalase positive.** Chemoorganotrophic, using a variety of sugars or salts of organic acids as carbon sources. Chemolithotrophic growth on H_2 has been reported. Strains that have been tested can use methanol and formate (**facultatively methylotrophic**). Occur in soil and freshwater sources.

Type (and only) species: *Ancylobacter aquaticus.*

Characteristics of the species: As described for the genus.

Genus "Brachyarcus"

Consists of rod-shaped cells bent like a bow, 1.0–1.5 × 2.5 μm, usually with several cylindrical **gas vesicles,** which may appear reddish if viewed with a phase-contrast microscope. The cells occasionally contain some minute sulfur **granules. Cells are arranged in groups (coenobia)** as a result of polar growth and median cross-division combined with **tight attachment to a surface by means of a mucoid substance.** Occasionally, cells are merely embedded in the polymer and are thus free-floating as a flat coenobium. After division, **cells form coenobia of two (rings) or four (clover-leaf appearance)** or more cells. Delayed cell division results in the formation of **pretzel-shaped cells.** Division in the coenobia may or may not be synchronous. Coenobia of 2–10 cells may measure 3–6 μm in diameter; secondary families of irregularly humped agglomerates of up to 100 μm or more in size can be found. Often the mucoid capsule is thin and not distinctly delineated. The Gram reaction has not been reported. **Nonmotile,** microaerophilic, or **probably anaerobic.** Has not been obtained in pure culture.

Type (and only) species: *"Brachyarcus thiophilus."*

Characteristics of the species: Originally found in Lake Vuolep Njakajaure, Swedish Lappland, in April, at a depth of 12–13.5 m. They have also been observed between April and October at depths of 0.25 m (pond) to 27 m (lake) in Germany; the temperature ranged from 2–20°C but was usually below 16.5°C; the pH ranged from 7.12–7.8; sulfide was always present; and oxygen was always absent.

Genus **Cyclobacterium**

Editorial note: The genus was not included in *Bergey's Manual of Systematic Bacteriology.* It was created in 1990 by Raj and Maloy (Int. J. Syst. Bacteriol. *40*: 337-347). It contains only a single species, *Cyclobacterium marinus.* (In this manual we have changed *marinus,* the masculine form of the adjective, to the neuter form *marinum*). This species was previously classified in *Bergey's Manual of Systematic Bacteriology,* Volume 1, in the genus *Flectobacillus,* as *Flectobacillus marinus.*

The genus **consists of mainly circle-shaped (ringlike) and horseshoe-shaped cells with an outer diameter of 0.8–1.5 μm and a cell width of 0.3–0.7 μm.** The cells have rounded (never tapered) ends. **Coils, spiral forms, and some straight rods occur less frequently; filamentous or pleomorphic cells are rare.** Stain is Gram negative. Cells are encapsulated. They are **non-flagellated and nonmotile.** No resting or life cycle stages occur. **No gas vesicles occur. Optimum growth occurs at 20–25°C in media containing seawater or 3.0% NaCl.** Convex, mucoid, opaque, smooth, small (<2-mm) colonies grow on modified Zobell marine agar or tryptone-glucose-yeast extract agar containing 3% NaCl. **Aerobic,** having a strictly respiratory type of metabolism. **Acids are produced from most carbohydrates but not from ribose, sorbose, and sugar alcohols.** Malonate and tartrate but not glycerol phosphate can be used as sole carbon sources. **Oxidase and catalase positive.** They are lipase (tributyrin hydrolysis) and urease negative. Habitat is marine environments.

Type (and only) species: *Cyclobacterium marinum.*

Characteristics of the species: Acid is not produced from cellobiose or dextrin. Acetate, citrate, fumarate, and malate can be used as sole carbon sources. Cells are resistant to chlortetracycline, kanamycin, penicillin G, streptomycin, and sulfamethoxazole/trimethoprim.

See Table 3.2 for characteristics that differentiate *C. marinum* from species of *Flectobacillus.*

Genus **Flectobacillus**

These are **rigid, straight to vibrioid, coiled, and helical rods,** the degree of curvature varying among individual cells within a culture in different species. **The most abundantly occurring forms are cells in the shape of the letter C or S, with some forming a closed ringlike structure due to the overlapping of the ends of the cell.** The ends of the cell may be either rounded or tapered (pointed). **Cells measure 0.3–2.0 × 1–31 μm. Rings 1–10 μm in outer diameter may be present.** Long, straight filaments (>50 μm) with bulbous or involuted shapes, short rods, or coccoid forms are seen in older cultures. Stain is Gram negative. Cells are **nonmotile.** Optimum temperature is 20–30°C; some strains of *F. glomeratus* may require temperatures less than 25°C. **Colonies are pale pink, rose-colored, pale yellow, or tan-colored, depending on the species.** They are aerobic, having a strictly respiratory type of metabolism with oxygen as the terminal electron acceptor. **Chemoorganotrophic but not methylotrophic.** One species (*F. major*) produces acid aerobically from a variety of carbohydrates but not from sugar alcohols, whereas the other (*F. glomeratus*) does not oxidize any sugars or sugar alcohols. *F. major* utilizes some organic acids as sole carbon sources but not citrate, fumarate, malate, malonate, or tartrate; *F. glomeratus* does not use organic acids. Gelatin is liquefied, at least weakly. **Oxidase positive and catalase weakly positive. Habitats are fresh and marine waters.**

Type species: *Flectobacillus major.*

Differentiation of the species of the genus **Flectobacillus:** See Table 3.2.

Genus **Meniscus**

These are **curved or straight rods,** 0.7–1.0 × 2.0–3.0 μm. Cultures may show single cells, pairs, tightly coiled spirals, S-shapes (two cells, one inverted), and doughnut-shaped cells, where the ends are overlapping before division by binary fission has occurred. Rings have outer diameters of about 3.0 μm. Stain is Gram negative. Cells are **nonmotile.** Resting stages are not known. **Cells are encapsulated. Gas vesicles are present** and are arranged at random within cells. **Colonies are chalky-white.** Chemoorganotrophic. **They are aerotolerantly anaerobic, possessing a strictly fermentative type of metabolism.** Capable of growth under an air atmosphere, provided at least 1% CO_2 is present. Carbohydrates are fermented with acid but with no gas production. Vitamin B_{12}, thiamine, and CO_2 are required for growth. Optimum temperature is 30°C; there is no growth at 10 or 40°C. **Catalase and oxidase negative.** They are **isolated from anaerobic digester sludge.**

Type (and only) species: *Meniscus glaucopis.*

Characteristics of the species: The following substrates are fermented: agar (weakly), dextrin, melezitose, raffinose, cellobiose, sucrose, lactose, maltose, melibiose, fructose, galactose, glucose, rhamnose (weakly), α-methyl-D-glucoside, esculin, salicin, D-ribose, D-xylose, and arabinose. The following substrates are not fermented: mannose, sorbose, glycerol, lactate, mannitol, sorbitol, adonitol, dulcitol, inositol, or amino acids. There is no hydrolysis of starch, cellulose, DNA, gelatin, casein, pectin, inulin, gum arabic, tributyrin, chitin, xylan, or glycogen. Negative for amino acid deaminases, urease, acetylmethylcarbinol, indole, H$_2$S production, and nitrate reduction.

Genus "Pelosigma"

S-shaped, slender filaments, 0.23–0.35 × 9–30 μm, colorless or pale gray, usually with a slight spiral twist. Cells are **usually arranged side-by-side in sigmoid aggregates of four or multiples of four;** multiple aggregates of considerable thickness have been observed. **The cells may be held together throughout their length by a mucoid substance, or only at one end forming a point, the other end being wider and spread out to reveal the individual filaments.** Multiplication is presumably by synchronous cross-division of the aggregate and sudden separation of the daughter aggregates. The Gram reaction has not been recorded. **The aggregate may be motile** by an organelle visible by light microscopy and located at the pointed end. This organelle may be a tuft of flagella. **Gas vesicles are not present.** The organisms have not yet been obtained in pure culture. **They occur in and on mud in fresh and brackish waters.**

Type species: *"Pelosigma cohnii."*

Differentiation of the species of the genus "Pelosigma": See Table 3.3.

Genus Runella

Rigid straight to curved rods, the degree of curvature varying among individual cells within a culture. The cells measure 0.5–0.9 × 2.0–4.5 μm. **The ends of a cell may overlap, producing a ring-shaped structure** with an outside diameter of 2.0–3.0 μm. Filaments up to 14 μm long may be produced. On rare occasions, a coil of 2–3 turns may be produced. Stain is Gram negative. Cells are **nonmotile.** Resting stages are not known. **Obligately aerobic,** possessing a strictly respiratory type of metabolism with oxygen as the terminal electron acceptor. **Acid is produced aerobically from only a few carbohydrates.** Optimum temperature is 20–30°C. **Colonies contain a pale pink, water-insoluble pigment;** catalase weakly positive. **Oxidase positive** and chemoorganotrophic. **They are isolated from fresh water.**

Type (and only) species: *Runella slithyformis.*

Characteristics of the species: As described for the genus. Acids are produced aerobically from glucose, inulin, maltose, sucrose, and sometimes from galactose, mannose, raffinose, and rhamnose. There is no acid production from arabinose, cellobiose, dextrin, fructose, lactose, melibiose, ribose, salicin, trehalose, xylose, or α-methyl-D-glucoside. Cannot utilize glycerol phosphate, malonate, succinate, and tartrate as sole carbon sources. Gelatin is not hydrolyzed.

Genus Spirosoma

Rigid straight to curved rods, the degree of curvature varying among individual cells within a culture. The cells measure 0.5–1.0 × 1.5–6.0 μm. Long sinuous filaments up to 50 μm long may be present. **Rings 1.5–3.0 μm in outer diameter are formed by overlapping of the ends of a cell. Coils and helices may be present.** Stain is Gram negative. **Cells are nonmotile;** resting stages are not known. **Obligate aerobes,** possessing a strictly respiratory type of metabolism with oxygen as the sole terminal electron acceptor. **Acids are produced aerobically from a variety of carbohydrates.** Optimum temperature is 20–30°C. **Colonies contain a pale to light yellow, water-insoluble pigment. Catalase and oxidase positive.** Chemoorganotrophic. **Isolated from soil and fresh water.**

Type (and only) species: *Spirosoma linguale.*

Characteristics of the species: As described for the genus. Acids are produced aerobically from glucose, inulin, maltose, sucrose, galactose, mannose, raffinose, rhamnose, arabinose, cellobiose, dextrin, fructose, lactose, melibiose, ribose, salicin, trehalose, xylose, and α-methyl-D-glucoside. Utilizes glycerol phosphate, malonate, succinate, and tartrate as sole carbon sources. Gelatin is hydrolyzed.

Table 3.1 Differentiation of the genera of non-motile (or rarely motile), curved bacteria [a]

Characteristic	Ancylobacter	"Brachyarcus"	Cyclobacterium	Flectobacillus	Meniscus	"Pelosigma"	Runella	Spirosoma
Gram reaction	-		-	-	-	-	-	-
Isolated in pure culture	+	-	+	+	+	-	+	+
Respire with oxygen	+		+	+	-	-	+	+
Strictly fermentative type of metabolism	-		-	-	+	+	-	-
Growth in media with seawater or 3% NaCl	-		+	+ or -	-	-	-	-
Rings commonly formed	+ or -	+	+	+ or -	+	+	+	+
Coils and helices formed	-	-	+	+ or -	+	-	-	+
Long filaments or coccoid forms may be present	-	-	-	+ or -	+	-	+	-
Bow-shaped cells, arranged in coenobia of two rings or four rings (clover-leaf appearance); pretzel-shaped cells may occur	-	+	-	-	-	-	-	-
Slender S-shaped cells arranged side-by-side in flat, sigmoid aggregates of four or multiples of four	-	-	-	-	-	+	-	-
Gas vesicles present	+ or -	+	-	-	+	-	-	-
Colonies are yellow or tan	-	-	-	+ or -	-	-	-	+
Colonies are pink	-	-	+	+ or -	-	-	+	-
Formate and methanol utilized	+	-	-	-	-	-	-	-
Sugar alcohols (glycerol, mannitol, sorbitol) oxidized	+		-	-				
Acid production oxidatively from ribose	+	-	-	-	-	-	-	+
Sole carbon sources:								
Citrate	+		+	-			-	-
Fumarate	-		+	-			-	-
Malate	-		+	-			-	+
Malonate	+		+	-			-	+
Tartrate	-		+	-			-	+
Gelatin liquefaction	-		-	Weak or +			-	+
Oxidase	+		+	+	-		+	+
Catalase	+		+	Weak	-		Weak	+
Lipase (tributyrin hydrolysis)	+		-	+ or -			+	-
Urease	+		-	-			-	-
Motility	Rare	-	-	-		Rare[b]	-	-

[a] Symbols: see standard definitions. + or -, results differ either among species or among strains within a species.

[b] Aggregates are occasionally motile, possibly by a tuft of flagella.

Table 3.2 Differential characteristics of *Cyclobacterium marinum* and of the species of *Flectobacillus*[a]

Characteristic	C. marinum	F. glomeratus[b]	F. major
Cell diameter (μm)	0.3–0.7	0.3–0.6	0.6–2.0
Rings formed	+	v	+
Coils and helices formed	+	v	−
Coccoid cells formed in older cultures	−	+	−
Long filaments form in older cultures	−	−	+
Growth in 5% NaCl	+	+	−
Pink	+	−	+
Yellow or tan	−	+	−
Acid production from carbohydrates	+	−	+
Salts of organic acids used as carbon sources	+	−	+

[a] Symbols: see standard definitions.

[b] *F. glomeratus* was not included in *Bergey's Manual of Systematic Bacteriology,* Volume 1. The species was created in 1987 by McGuire et al. (Syst. Appl. Microbiol. *9:* 265–272), based on an examination of strains isolated from antarctic marine environments.

Table 3.3 Differential characteristics of the species of *"Pelosigma"*[a]

Characteristic	"P. cohnii"	"P. palustre"
Aggregates helically curved	+	−
Aggregates flat and band-like	−	+
Aggregate width (μm)	1.2–4.0	8–10
Aggregate length (μm)	9–20	20–25
Occurrence in brackish water	+	−
Found in fresh water	−	+

[a] Symbols: see standard definitions.

Important Notes for Users of This Manual

Unless otherwise indicated in footnotes to tables, the meanings of symbols are as follows:

- + 90% or more of strains are positive
- − 90% or more of strains are negative
- d 11-89% of strains are positive
- v strain instability (not equivalent to "d")
- D Different reactions in different taxa (species of a genus or genera of a family)

All other symbols are defined in footnotes to tables.

GROUP 4

GRAM-NEGATIVE AEROBIC/MICROAEROPHILIC RODS AND COCCI

These Gram-negative bacteria form a very heterogeneous morphological and physiological group. However, they all possess a strictly respiratory type of metabolism and can use oxygen as a terminal electron acceptor. Most can grow under an air atmosphere (21% oxygen), but the microaerophilic members of the group cannot. Some genera can respire anaerobically with nitrate, fumarate, or other terminal electron acceptors. Several genera can fix molecular nitrogen.

Key to the differentiation of the major subgroups in **Group 4:**

I. Aerobic. Rods and cocci that can grow under an air atmosphere (21% oxygen). Some genera are microaerophilic under nitrogen-fixing conditions but, when supplied with a source of fixed nitrogen, they grow as aerobes.

Subgroup 4a

II. Microaerophilic. Straight rods that cannot grow under an air atmosphere (21% oxygen). In the presence of decreased levels of oxygen and in the absence of alternative electron acceptors, they exhibit oxygen-dependent growth. Some can also respire anaerobically with nitrate, fumarate, or other terminal electron acceptors.

Subgroup 4b

SUBGROUP 4A

Differentiation of the Genera in **Subgroup 4A.** Major differential characteristics of the genera are presented in Table 4.1. Some supplementary differential characteristics for certain physiological groups are given in Tables 4.2–4.9.

Genus **Acetobacter**

Ellipsoidal to rod-shaped, straight, or slightly curved, 0.6–0.8 μm × 1.0–1-4.0 μm, occurring singly, in pairs, or in chains. Involution forms are frequent in some strains and may be spherical, elongated, swollen, club-shaped, curved, or filamentous. **Cells may be motile or nonmotile; if motile, the flagella are peritrichous or lateral.** Endospores are not formed. Cells stain Gram negative (in a few cases Gram variable). They are **obligately aerobic;** metabolism is respiratory, never fermentative. Form pale colonies; most strains produce no pigments. A minority of strains produce water-soluble pigments or show pink colonies due to porphyrins. Catalase positive and oxidase negative. There is no gelatin liquefaction, indole production, and H_2S formation. **Oxidize ethanol to acetic acid. Acetate and lactate are oxidized to CO_2 and H_2O.** The best carbon sources for growth are ethanol, glycerol, and lactate. Acid is formed from *n*-propanol, *n*-butanol, and D-glucose. Neither lactose nor starch is hydrolyzed. Chemoorganotrophs. Optimum temperature is 25–30°C. The pH optimum for growth is 5.4–6.3. *Acetobacter* species occur in flowers, fruits, honey bees, **sake, tequila, palm wine, grape wine, cider, beer, South African Bantu beer, kefir,** brewer's yeast, **vinegar, beechwood shavings of vinegar generators, vinegar acetifiers,** sugar cane juice, "tea fungus," vegetable tanning liquors, "**nata**," garden soil, and canal water. Some acetobacters cause **pink disease** in pineapple fruit and **rot** in apples and pears. One species is a **microaerophilic nitrogen fixer** inhabiting the roots and stems of sugarcane.

Type species: *Acetobacter aceti.*

Differentiation of the species of the genus **Acetobacter:** See Table 4.10.

Genus **Acidiphilium**

Editorial note: The genus *Acidiphilium* was not included in *Bergey's Manual of Systematic Bacteriology*. The genus was created in 1981 by Harrison (Int. J. Syst. Bacteriol. *31:* 327-332). It initially contained one species, *A. cryptum,* but the following species have since been added: *A. organovorum* (Lobos et al., Int. J. Syst. Bacteriol. *36:* 139-144, 1986); *A. angustum, A. facilis,* and *A. rubrum* (Wichlacz et al., Int. J. Syst. Bacteriol. *36:* 197-201, 1986).

Cells are straight rods with rounded ends. Strains vary from 0.3 to 1.2 μm in diameter and from 0.6 to 4.2 μm in length. Cells stain Gram negative. **Motility occurs by means of a polar flagellum or by means of two lateral flagella;** a few strains are nonmotile. Endospores are not formed. Aerobic and **acidophilic, growing in the pH range of 2.5–5.9 but not at pH 6.1.** Some strains grow at pH 2.0. **Mesophilic and chemoorganotrophic. Upon initial isolation, some strains may not grow in concentrations of peptone and extracts customarily employed in organic media;** for instance, 0.05% dehydrated Trypticase soy may inhibit growth, whereas 0.01% may stimulate growth. **Acetate is not used as a carbon source and may be inhibitory. Citrate is readily used. All strains grow with glucose,** but the basal medium may need to be supplemented with a trace of yeast extract. Are found in acidic mineral environments such as pyrite mine drainage, pyritic coal refuse, and copper and uranium mine tailings. *Acidiphilium* species also have been isolated as contaminants in enrichment cultures of *Thiobacillus ferrooxidans.*

Type species: *Acidiphilium cryptum.*

Differentiation of the species of the genus **Acidiphilium:** See Table 4.11.

Genus **Acidomonas**

Editorial note: The genus *Acidomonas* was not included in *Bergey's Manual of Systematic Bacteriology.* The genus was created in 1989 by Urakami et al. (Int. J. Syst. Bacteriol. *39:* 50-55). It contains only one species, *A. methanolica* (formerly *Acetobacter methanolica*).

Gram-negative rods, 0.8–1.0 × 1.5–3.0 μm, with rounded ends. Occur singly and rarely in pairs. **Cells are nonmotile.** Colonies on peptone-yeast extract-malt extract agar are shiny, smooth, raised, entire, white to light yellow, and 2–3 mm in diameter after 3 days at 30°C. No water-soluble fluorescent pigment is produced. Aerobic, and metabolism is strictly respiratory and not fermentative. Nitrate is not reduced to nitrite. The Voges-Proskauer test is negative. Indole and H_2S are not produced. There is no hydrolysis of gelatin and starch. Ammonia is produced. Litmus milk is not changed. Dihydroxyacetone is not produced from glycerol. **Acetic acid is produced from ethanol. Acid is produced from D-glucose** oxidatively but not from L-arabinose, D-xylose, D-mannose, D-fructose, D-galactose, maltose, sucrose, lactose, trehalose, D-sorbitol, D-mannitol, inositol, glycerol, or soluble starch. Cells **utilize methanol,** ethanol, **acetic acid,** D-glucose, glycerol, and pectin as sole carbon sources for energy and growth but do not utilize L-arabinose, D-xylose, D-fructose, D-galactose, maltose, sucrose, lactose, trehalose, D-sorbitol, D-mannitol, inositol, soluble starch, citric acid, lactic acid, **methylamine,** methane, or hydrogen. Some strains utilize D-mannose weakly. Calcium pantothenate is required. Ammonia, nitrate, and urea are utilized as nitrogen sources. **Urease, oxidase, and catalase are produced. Growth occurs between pH 2.0 and 5.5.** Good growth occurs between pH 3.0 and 5.0. Growth does not occur above pH 6.0 or below pH 1.5. Good growth occurs at 30°C and 37°C but does not occur at 42°C. Growth does not occur in the presence of 3% NaCl. The cellular fatty acids are composed of a large amount of straight-chain unsaturated $C_{18:1}$ acid and small amounts of straight-chain saturated $C_{15:0}$, $C_{16:0}$, $C_{17:0}$, $C_{18:0}$, and $C_{19:0}$ acids, straight-chain unsaturated $C_{16:1}$ acid, and $C_{19:0}$ cyclopropane acid. The hydroxy acids are composed of large amounts of 3-OH $C_{14:0}$, 3-OH $C_{16:0}$, 2-OH $C_{14:0}$, and 2-OH $C_{16:0}$ acids. The ubiquinone system is Q-10, along with ubiquinone Q-9 and minor ubiquinone Q-11 components.

Type (and only) species: *Acidomonas methanolica.*

Characteristics of the species: As described for the genus.

Genus **Acidothermus**

Editorial note: The genus *Acidothermus* was not included in *Bergey's Manual of Systematic Bacteriology.* The genus was created in 1986 by Mohagheghi et al. (Int. J. Syst. Bacteriol. *36:* 435-443). It includes only one species, *A. cellulolyticus.*

Cells are **slender rods and long slender filaments, 0.4 × 5–20 µm,** with rounded ends. **They do not form endospores, flagella, or rotund bodies.** Cells are **non-motile.** Cells are **Gram variable but usually stain Gram negative.** Thin sections show no outer cell wall membranes, and thus *Acidothermus* cell walls are distinctly different from the typical Gram-negative cell walls of *Thermus* and *Thermomicrobium* species. The principal constituents of purified cell walls are diaminopimelic acid, glucosamine, muramic acid, serine, and alanine. Colonies grown on LPBM mineral salts agar with 0.5 g yeast extract per liter and 5.0 g D-glucose or D-cellobiose per liter are creamy, white, smooth, circular, entire, and 1–3 mm in diameter. Obligate aerobes. They grow on several carbohydrates, including D-glucose and D-cellobiose. **Thermophilic,** growing between 37–70°C. **Acidophilic,** they grow at pH 3.5–7.0. Isolated from acidic hot springs at Yellowstone National Park, pH 4–5.5, temperature 45–65°C.

Type (and only) species: *Acidothermus cellulolyticus.*

Characteristics of the species: Optimum temperature is between 50°C and 60°C. Optimum pH is near 5.0. Carbon and energy sources include D-glucose, D-cellobiose, cellulose, xylan, D-galactose, maltose, sucrose, raffinose, D-mannose, D-mannitol, and D-sorbitol, with ammonium ions or amino acids as a source of nitrogen. Grows on Casamino acids (0.1%) plus tryptone (0.1%); there is no growth on nutrient broth, acetate, lactate, citrate, or pectin. Citrate and acetate are inhibitory at 0.01 M. Catalase positive. It actively digests cellulose.

Genus **Acidovorax**

Editorial note: The genus *Acidovorax* was not included in *Bergey's Manual of Systematic Bacteriology.* The genus was created in 1990 by Willems et al. (Int. J. Syst. Bacteriol. *40:* 384–398). Three species were described: *A. facilis* (formerly *Pseudomonas facilis*), *A. delafieldii* (formerly *Pseudomonas delafieldii*), and *A. temperans.*

Straight to slightly curved rods, 0.2–0.7 × 1.0–5.0 µm, occurring singly or in short chains and **motile by means of a single polar flagellum.** Cells stain Gram negative. They are oxidase positive. Urease activity varies among strains. Some strains grow on Christensen urea agar but lack urease according to API 20NE tests. No pigment is produced on nutrient agar. Aerobic and chemoorganotrophic. *Acidovorax facilis* and several

Acidovorax delafieldii strains are capable of lithoautotrophic growth by using hydrogen as an energy source. **Oxidative carbohydrate metabolism occurs with oxygen as the terminal electron acceptor;** alternatively, some strains of *Acidovorax delafieldii* and *Acidovorax temperans* are capable of heterotrophic denitrification of nitrate. Good growth is obtained on media containing organic acids, amino acids, or peptone, but only a limited number of sugars are used for growth. Two hydroxylated fatty acids, 3-hydroxyoctanoic acid (3-OH-8:0) and 3-hydroxydecanoic acid (3-OH-10:0), are always present, 2-hydroxylated fatty acids are absent, and a cyclopropane-substituted fatty acid (17:cyc) is present in most of the strains. The mean G+C values of the DNAs are 62–66 mol%.

Type species: *Acidovorax facilis.*

Differentiation of the species of the genus **Acidovorax:** See Table 4.11a.

Genus **Acinetobacter**

Rods 0.9–1.6 µm in diameter and 1.5–2.5 µm in length become spherical in the stationary phase of growth. They commonly occur in pairs and also in chains of variable length. Cells do not form spores. Cells stain Gram negative but occasionally are difficult to destain. Swimming motility does not occur, but the cells display "twitching motility," presumably because of the presence of polar fimbriae. Aerobic, having a strictly respiratory type of metabolism with oxygen as the terminal electron acceptor. All strains grow between 20°C and 30°C, with most strains having temperature optima of 33–35°C. They grow well on all common complex media. **They are oxidase negative and catalase positive.** Most strains grow in defined media containing a single carbon and energy source; they use ammonium or nitrate salts as the source of nitrogen and display no growth factor requirements. D-Glucose is the only hexose utilized by some strains. The pentoses D-ribose, D-xylose, and L-arabinose can also be utilized as carbon sources by some strains. Occur naturally in soil, water, and sewage.

Type species: *Acinetobacter calcoaceticus.*

Differentiation of the species of the genus **Acinetobacter:** See Table 4.12.

Genus **Afipia**

Editorial note: The genus *Afipia* was not included in *Bergey's Manual of Systematic Bacteriology.* The genus was created in 1991 by Brenner et al. (J. Clin. Microbiol. *29:* 2450–2460; Int. J. Syst. Bacteriol. *42:* 327, 1992). Three species were described: *A. broomeae, A. clevelandensis,* and *A. felis.*

Rod-shaped cells are motile by means of a single polar, subpolar, or lateral flagellum. Cells stain Gram negative and are oxidase positive. Cells grow on buffered charcoal yeast extract (BCYE) agar and nutrient broth but not on nutrient broth containing 6% NaCl and rarely on MacConkey agar at 25°C and 30°C. Cells grow at least weakly at 35°C, but growth does not occur at 42°. Colonies are gray-white, glistening, convex, and opaque and are 1.5 mm wide with entire edges; they are formed after 72 hours of incubation at 32°C on blood agar or on BCYE agar, except that the size range is 0.5–1.5 mm. They are urease positive and turn litmus milk alkaline. Negative in reactions for hemolysis, gas production from nitrate, indole production, H$_2$S production (triple sugar iron method), gelatin hydrolysis, and esculin hydrolysis and are nonfermentative. Acid is not produced oxidatively from D-glucose, lactose, maltose, or sucrose. Cells contain 11-methyloctadec-12-enoic (C$_{Br19:1}$), *cis*-octadec-11-enoic (C$_{18:1omega7c}$), and, generally, 17- and 19-carbon acids with a cyclopropane ring as major acids, with only trace amounts of hydroxy acids.

A. felis was initially thought to be the etiological agent of cat scratch disease in humans, but recent data indicate that the organism may not be the cause. Other species are presumptively considered to be pathogenic for humans.

Type species: *Afipia felis.*

Differentiation of the species of the genus **Afipia:** See original article for characteristics that differentiate the three species and three unnamed genospecies.

Genus **Agrobacterium**

These rods are 0.6–1.0 × 1.5–3.0 μm and occur singly or in pairs. **They are non-spore-forming** and **Gram negative. Motility occurs** by 1 to 6 **peritrichous flagella. Aerobic,** possessing a **respiratory type of metabolism** with oxygen as the terminal electron acceptor.

Some strains are capable of anaerobic respiration in the presence of nitrate. Most strains are able to grow under reduced oxygen tensions in plant tissues. Optimum temperature is 25–28°C. Colonies are usually convex, circular, smooth, and nonpigmented to light beige. **Growth on carbohydrate-containing media is usually accompanied by copious extracellular polysaccharide slime.** Catalase positive and usually oxidase and urease positive. 3-Ketoglycosides are produced by the majority of strains belonging to *A. tumefaciens* biovar 1 and *A. radiobacter* biovar 1. **Chemoorganotrophs,** utilizing a wide range of carbohydrates, salts of organic acids, and amino acids as carbon sources but not cellulose, starch, agar of glucose, D-galactose, and other carbohydrates. Ammonium salts and nitrates can serve as nitrogen sources for strains of some species and biovars; others require amino acids and additional growth factors. With the exception of *A. radiobacter,* members of this genus **invade the crown, roots, and stems of a great variety of dicotyledonous and some gymnospermous plants via wounds, causing the transformation of the plant cells into autonomously proliferating tumor cells.** The induced plant diseases are commonly known as **crown gall, hairy root, and cane gall.** Some strains possess a wide host range, whereas others (e.g., grapevine isolates) possess a very limited host range. The tumors are self-proliferating and graftable. The tumor induction by *Agrobacterium* is correlated with the presence of a **large tumor-inducing plasmid (Ti-plasmid)** in the bacterial cells. Agrobacteria are **soil inhabitants.** Oncogenic strains occur mainly in soils previously contaminated with diseased plant material. Some nononcogenic *Agrobacterium* strains have been isolated from human clinical specimens.

Type species: *Agrobacterium tumefaciens.*

Differentiation of the species of the genus **Agrobacterium:** See Table 4.13.

Genus **Agromonas**

Editorial note: The genus *Agromonas* was not included in *Bergey's Manual of Systematic Bacteriology.* The genus was created in 1983 by Ohta and Hattori (Antonie van Leeuwenhoek J. Microbiol. Serol. *49:* 429-446; Int. J. Syst. Bacteriol. *35:* 223, 1985) and contains only one species, *A. oligotrophica.*

Bent, branched, and budding cells are 0.6–1.0 × 2–7 μm. Motility occurs by polar flagella. No resting stages are known. Cells stain Gram negative. Colonies are colorless. **Catalase and oxidase positive.** Casein, gelatin, starch, and cellulose are not hydrolyzed. **Atmospheric nitrogen is fixed under low O_2 levels.** Cellular fatty acids mainly consist of a straight-chain unsaturated acid, $C_{18:1}$. Ubiquinone Q-10 is present.

Type (and only) species: *Agromonas oligotrophica.*

Characteristics of the species: Bent, branched, and budding cells are produced on dilute nutrient broth. Several cells adhere to each other and form a rosette. Motility occurs by a polar flagellum. Endospores are not formed. On dilute nutrient broth agar, colonies are punctiform, pulvinate, entire, and colorless. Oligotrophic (i.e., **growth can occur in a medium containing less than 1 mg of an organic carbon source per liter. NaCl, KCl, casamino acids, peptone, and meat extract inhibit growth at 0.5–1.0%.** Several sugars and many organic acids are utilized. Aromatic acids such as ferulic acid, *p*-coumaric acid, and *p*-anisic acid are used, but benzoic acid is not used. Neither acid nor gas is produced from glucose. Not proteolytic to casein and gelatin. Cellulose and starch are not hydrolyzed. **Aerobic.** Cellular fatty acids mainly consist of $C_{18:1}$; a small amount of $C_{16:0}$ and a 19-carbon unsaturated acid with a double bond and possibly with a side chain are found as minor components. Ubiquinone is present.

Genus **Alcaligenes**

Rods, coccal rods, or cocci, 0.5–1.0 × 0.5–2.6, usually occur singly. Resting stages are not known. **Cells stain Gram negative. Motility occurs** with 1 to 8 (occasionally up to 12) **peritrichous** flagella. **Obligately aerobic,** possessing a strictly respiratory type of metabolism with oxygen as the terminal electron acceptor. Some strains are capable of anaerobic respiration in the presence of nitrate or nitrite. Optimum temperature is 20–37°C. Colonies on nutrient agar are **nonpigmented. Oxidase positive** and catalase positive. Indole is not produced. Cellulose, esculin, gelatin, and DNA usually are not hydrolyzed. **Chemoorganotrophic, using a variety of organic acids and amino acids as carbon sources. Alkali is produced** from several organic salts and amides. Carbohydrates are usually not utilized. Some strains produce acid from D-glucose and D-xylose and utilize both carbohydrates as carbon source. **Occur in water and soil.** Some are common, apparently saprophytic, inhabitants of the intestinal tract of vertebrates. Numerous strains have been **iso-**lated from clinical material such as blood, urine, feces, purulent ear discharges, spinal fluid, wounds, etc. Occasionally, cause opportunistic infections in humans.

Type species: *Alcaligenes faecalis.*

Differentiation of the species of the genus **Alcaligenes:** See Table 4.14.

Genus **Alteromonas**

Straight or curved rods, 0.7–1.5 × 1.8–3.0 μm, do not accumulate poly-β-hydroxybutyrate (PHB) as an intracellular reserve product. Microcysts or endospores are not formed. **Cells stain Gram negative. Motility occurs** by means of a **single, unsheathed, polar flagellum;** two species (*A. luteoviolacea* and *A. denitrificans*) have sheathed flagella. **Chemoorganotrophs** capable of **respiratory but not fermentative metabolism.** Molecular oxygen is a universal electron acceptor. **One species is capable of denitrification.** None of the strains has a constitutive arginine dihydrolase system. All **require seawater base for growth;** many strains require organic growth factors. All grow at 20°C. **Common inhabitants of coastal waters and the open oceans.**

Type species: *Alteromonas macleodii.*

Differentiation of the species of the genus **Alteromonas:** See Table 4.15.

Genus **Aminobacter**

Editorial note: The genus *Aminobacter* was not included in *Bergey's Manual of Systematic Bacteriology.* The genus was created in 1992 by Urakami et al. (Int. J. Syst. Bacteriol. *42:* 84–92). Three species were described: *A. aminovorans* (formerly *Pseudomonas aminovorans*) *A. aganoensis,* and *A. niigataensis.*

Cells are rod-shaped with rounded ends, 0.5–0.9 × 1.0–3.0 μm. Stain Gram negative. Motility occurs by means of subpolar flagella. **Reproduce by budding.** Granules of poly-β-hydroxybutyric acid are accumulated in the cells. Colonies are white to light yellow. Grow abundantly in nutrient broth and PYG broth. Water-soluble fluorescent pigment is not produced. Methyl red and Voges-Proskauer negative. Indole and hydrogen sulfide are not produced. Hydrolysis of gelatin and starch are not observed. Ammonia is produced. Denitrification is negative. **Acids are produced from sugars oxidatively but not fermentatively. Monomethylamine, trimethylamine, trimethylamine-N-oxide,** and sugars are utilized, but **methanol, methane, and hydrogen are not utilized.** Growth factors are not required as essential supplements. Ammonia, nitrate, urea, peptone, and methylamine are utilized as nitrogen sources. Oxidase and catalase are produced, but urease is not. Aerobic. Metabolism is strictly respiratory and not fermentative. Good growth occurs between pH 6.0 and 8.0 but not above pH 9.0 or below pH 5.0. Good growth occurs at 30°C and 37°C but not at 42°C. Growth does not occur in the presence of 3% sodium chloride.

The cellular fatty acids include a large amount of straight-chain unsaturated $C_{18:1}$ acid. The hydroxy acids include a large amount of 3-OH $C_{12:0}$ acid. The ubiquinone system is ubiquinone Q-10.

Type species: *Aminobacter aminovorans.*

Differentiation of the species of the genus **Aminobacter:** See Table 4.15a.

Genus **Aquaspirillum: A. fasciculus**

Straight rods, 0.7–0.9 × 3.6–43.0 μm. Curved or S-shaped variants have been reported to occur in one strain after prolonged serial transfer. **Intracellular poly-β-hydroxybutyrate granules are present.** Extensive conversion of the rod-shaped cells to round forms (**"coccoid bodies"**) occurs in older cultures. Cells from broth cultures have **bipolar flagellar fascicles composed of up to 11 flagella.** The fascicles can be seen clearly by darkfield microscopy and show an **unusual and distinctive behavior** when the cells are suspended in ordinary, nonviscous media: helical wave propagation with waves progressing from base to tip, an ability to coil up like springs, and basal bending accompanied by a change in wavelength. **In ordinary media the cells do not swim and, instead, exhibit an ineffectual "floundering about" movement.** When cells from broth cultures are suspended in a medium of high viscosity (10–200 centipoise, obtained by the use of agents such as methylcellulose "400 centipoise") they swim steadily in straight lines. Upon initial isolation, the cells form **highly viscous flocs;** the floc-forming ability is gradually lost during subsequent transfers. When cell flocs are crushed and homogenized in a small quantity of water, free-swimming cells can be seen moving in straight lines; here the tailing flagellar fascicle is extended behind each cell, while the leading fascicle is either coiled into a polar loop or is coiled around the cell. Optimum temperature is 30°C; there is no growth at 20 or 40°C. **Catalase and oxidase positive. Carbohydrates are not catabolized.** Pyruvate and proline are the most effective sole carbon sources. **Nitrogenase activity occurs under microaerobic conditions.** Habitat is pond water.

Genus **Azomonas**

Cells are 2.0 μm in diameter and of various lengths, ranging from rods to ovoid to coccoid in shape. They occur singly, in pairs, or in clumps. Pleomorphism is generally present. **Cells generally stain Gram negative, sometimes Gram variable. They do not produce endospores or cysts. Motility occurs** by peritrichous or polar flagella. **Aerobic,** but can also grow under decreased oxygen tensions. **Water-soluble pigments and fluorescent pigments** are produced by nearly all strains. Chemoorganotrophic, using sugars, alcohols, and the salts of organic acids for growth. **Nitrogen fixers; generally fix nonsymbiotically at least 10 mg of molecular nitrogen/g of carbohydrate (usually glucose) consumed. Molybdenum is required for nitrogen fixation.** Nonproteolytic and can utilize ammonium salts and certain amino acids as sources of nitrogen. **Catalase positive.** The optimum pH for nitrogen fixation is close to neutrality, but certain strains can also fix nitrogen at a pH of 4.6–4.8. They occur in soil and water.

Type species: *Azomonas agilis.*

Differentiation of the species of the genus **Azomonas:** See Table 4.16.

Genus **Azorhizobium**

Editorial note: The genus *Azorhizobium* was not included in *Bergey's Manual of Systematic Bacteriology.* The genus was created in 1988 by Dreyfus et al. (Int. J. Syst. Bacteriol. *38:* 89-98).

Consists of rods 0.5–0.6 × 1.5–2.5 μm. Cells are **motile; they have peritrichous flagella on solid medium and one lateral flagellum in liquid medium.** Colonies on agar are circular and have a creamy color.

Obligately aerobic. **Fix N$_2$ under microaerobic conditions** and grow well on N$_2$ with vitamins present in a nitrogen-free medium. **Oxidase and catalase positive.** Urease-negative. **Among the sugars, only glucose is oxidized.** Organic acids such as lactate or succinate are the favorite carbon substrates for both NH$_4^+$- and N$_2$-dependent growth. Malonate is also used. Starch is not hydrolyzed. Cells can grow on DL-proline. No strain denitrifies. Strains nodulate effectively the roots and stems of *Sesbania rostrata*.

Type (and only) species: *Azorhizobium caulinodans.*

Characteristics of the species: Possess arginine dihydrolase and lysine decarboxylase. Grow in the presence of 8% KNO$_3$. Grow on azelate, maleate, adipate, pimelate, suberate, gluconate, mucate, crotonate, nicotinate, 2-ketogluconate, propionate, butyrate, isobutyrate, valerate, isovalerate, caproate, laurate, 2-ketoglutarate, fumarate, glutarate, sebacate, DL-malate, citrate, pyruvate, aconitate, citraconitate, D-glucuronate, α-D-galacturonate, *m*-hydroxybenzoate, L-aspartate, quinate, L-alanine, L-lysine, L-asparagine, betaine, and sarcosine. Do not oxidize mannitol. Assimilate 1,2-propanediol and 2,3-butanediol. Growth occurs from 12–43°C. Among the vitamins, nicotinic acid is required for N$_2$ fixation under microaerobic conditions. Equally good growth is observed between pH 5.5 and 7.8.

Genus **Azotobacter**

Large ovoid cells 1.5–2.0 μm in diameter. Pleomorphic, ranging from rods to coccoid cells. Occur singly, in pairs, or irregular clumps, and sometimes in chains of varying length. Do not produce endospores, **but form cysts.** Cells stain Gram negative. **Motility occurs by peritrichous flagella, or cells are nonmotile. Aerobic,** but can also grow under decreased oxygen tensions. Water-soluble and water-insoluble pigments are produced by some strains of all species. Chemoorganotrophic, using sugars, alcohols, and salts of organic acids for growth. **Nitrogen fixers; generally fix nonsymbiotically at least 10 mg of N$_2$ per g of carbohydrate (usually glucose) consumed. Molybdenum is required for nitrogen fixation** but may be partially replaced by vanadium. Nonproteolytic. Utilize nitrate and ammonium salts (all but one species) and certain amino acids as sources of nitrogen. **Catalase positive.** The pH range for growth in the presence of combined nitrogen is 4.8–8.5; the optimum pH for growth and nitrogen fixation is 7.0-7.5. Occur in soil and water; one species occurs in association with plant roots.

Type species: *Azotobacter chroococcum.*

Differentiation of the species of the genus **Azotobacter:** See Table 4.17.

Genus **Beijerinckia**

Straight or slightly curved rods, 0.5–1.5 μm × 1.7–4.5 μm, with rounded ends. Occur singly. Sometimes large, misshapen cells 3.0 × 5.0–6.0 μm occur; these are occasionally branched or forked. Large, highly refractile, intracellular granules of **poly-β-hydroxybutyrate** occur, generally one at each pole. **Cysts** (enclosing one cell) and **capsules** (enclosing several cells) occur in some species. Cells stain Gram negative. **Motility occurs by peritrichous flagella, or cells are nonmotile.** They are **aerobic,** having a strictly respiratory type of metabolism with oxygen as the terminal electron acceptor. **Molecular nitrogen is fixed** under aerobic conditions and also under decreased oxygen pressures (microaerobic conditions). Optimum temperature is 20–30°C; no growth occurs at 37°C. **Growth occurs between pH 3.0 and pH 9.5–10.0.** In liquid media **no surface pellicle is formed, but the whole medium becomes a homogeneous, highly viscous, semitransparent mass;** in some species the whole medium becomes opalescent and turbid, and adhering slime is not produced. On agar media, especially under N$_2$-fixing conditions, **copious tenacious and elastic slime is produced and giant colonies develop** with a smooth, folded, or plicated surface; some strains form slime having a more granular consistency similar to that formed by *Azotobacter.* **Catalase positive. Glucose, fructose, and sucrose are utilized** by all strains and are oxidized to CO$_2$ and a small amount of acetic acid. **No growth occurs on peptone agar or in peptone broth. Glutamate is utilized poorly or not at all.** Occur in soils, particularly those of **tropical regions.**

Type species: *Beijerinckia indica.*

Differentiation of the species of the genus **Beijerinckia:** See Table 4.18.

Genus **Bordetella**

Minute coccobacilli, 0.2–0.5 × 0.5–2.0 μm, often bipolar stained, and arranged singly or in pairs, more rarely in chains. Cells stain Gram negative. Cells may be **motile or nonmotile;** if motile, by means of peritrichous flagella. **Strictly aerobic.** Optimum temperature is 35–37°C. Colonies on Bordet-Gengou medium are smooth, convex, pearly, glistening, nearly transparent, and surrounded by a zone of hemolysis without definite periphery. **Metabolism is respiratory, never fermentative.** Chemoorganotrophic, require **nicotinamide, organic sulfur** (e.g., cysteine), and **organic nitrogen** (amino acids). Utilize oxidatively glutamic acid, proline, alanine, aspartic acid, and serine with production of ammonia and CO_2. Litmus milk is made alkaline. **Mammalian parasite and pathogen.** Localize and multiply among the epithelial cilia of the respiratory tract.

Type species: *Bordetella pertussis.*

Differentiation of the species of the genus **Bordetella:** See Table 4.19.

Genus **Bradyrhizobium**

Rods 0.5–0.9 × 1.2–3.0 μm. Commonly pleomorphic under adverse growth conditions, they usually contain granules of poly-β-hydroxybutyrate which are refractile by phase-contrast microscopy. Nonsporeforming. Gram negative. **Motility occurs** by one polar or subpolar flagellum. Fimbriae have not been described. **Aerobic,** possessing a respiratory type of metabolism with oxygen as the terminal electron acceptor. Optimum temperature is 25–30°C. Optimum pH is 6-7, although lower optima may be exhibited by strains from acid soils. Colonies are circular, opaque, rarely translucent, white, and convex, and tend to be granular in texture; **they do not exceed 1 mm in diameter within 5–7 days incubation on yeast-mannitol-mineral salts agar.** Colonies produced by some strains isolated from *Lotononis bainesii* are red because of intracellular pigmentation. Only a moderate turbidity develops after 3–5 days or longer in agitated broth. Faster growing strains are uncommon. Chemoorganotrophic, they utilize a range of carbohydrates; pentoses are preferred as carbon sources. Cellulose and starch are not utilized. They produce an alkaline reaction in mineral salts medium containing mannitol or many other carbohy-drates. **Growth on carbohydrate media is usually accompanied by extracellular polysaccharide slime. Some strains can grow chemolithotrophically in the presence of H_2, CO_2 and low levels of O_2.** Ammonium salts, usually nitrates, and some amino acids, can serve as nitrogen sources. Peptone is poorly utilized (except for strains isolated from *Lotononis*). Casein and agar are not hydrolyzed. There is usually no requirement for vitamins with the rare exception of biotin, which also may be inhibitory to some strains. **3-Ketoglycosides are not produced.** The organisms are characteristically able to invade the root hairs of tropical-zone and some temperate-zone leguminous plants (family *Leguminosae*) and incite the production of root nodules, wherein the bacteria occur as intracellular symbionts. All strains exhibit host range affinities (host "specificity"). **The bacteria are present in root nodules as swollen forms which are normally involved in fixing atmospheric nitrogen into combined forms utilizable by the host plant. Some strains fix nitrogen in the free living state** when examined under special conditions.

Type (and only) species: *Bradyrhizobium japonicum.*

Characteristics of the species: The characteristics are as described for the genus. Bacteroids in root nodules are slightly swollen rods with rare branching, or coccus form (in *Arachis* spp.). Cells of *B. japonicum* have one polar or subpolar flagellum. The species normally causes the formation of root nodules on species of *Glycine* (soybean) and on *Macroptilium atropupureum* (siratro).

Other organisms belonging to **Bradyrhizobium:** Other bradyrhizobia have not yet been classified as species or biovars. These organisms cause nodule production on certain species of *Lotus* (*L. uliginosus* and *L. pedunculatus*), as well as on *Vigna* and species of *Lupinus, Ornithopus, Cicer, Sesbania, Leucaena, Mimosa, Lablab,* and *Acacia,* which are also nodulated by the fast growing *Rhizobium loti.* Some strains fix nitrogen in the free living state under special conditions. It is suggested that until such time as further species or biovars are created within the genus *Bradyrhizobium,* these organisms (other than *B. japonicum*) be designated as *Bradyrhizobium* sp. with the name of the appropriate host plant given in parentheses immediately following: e.g., *Bradyrhizobium* sp. (*Vigna*) or *Bradyrhizobium* sp. (*Lupinus*).

Genus **Brucella**

Cocci, coccobacilli, or short rods, 0.5–0.7 × 0.6–1.5 µm. Arranged singly and, less frequently, in pairs, short chains, or small groups. True capsules are not produced. They do not usually show true bipolar staining. Resting stages are not known. Stain Gram negative. **Cells are nonmotile** and do not produce flagella. **Aerobic,** possessing a respiratory type of metabolism and having a **cytochrome-based electron transport system** with oxygen or nitrate as the terminal electron acceptor. Nitrate reductase is produced. **Many strains require supplementary CO$_2$ for growth,** especially on primary isolation. Colonies on serum-dextrose agar or other clear medium are transparent, raised, convex, with an entire edge and a smooth, shiny surface. They appear a **pale honey color** by transmitted light. Nonsmooth variants of the smooth species occur, but there are also stable nonsmooth species with a distinctive host range. Optimum temperature is 37°C. Growth occurs between 20 and 40°C. Optimum pH is 6.6–7.4. **Catalase positive and usually oxidase positive,** but negative strains occur. Chemoorganotrophic, most strains require complex media containing several amino acids, thiamin, nicotinamide, and magnesium ions; some strains may be induced to grow on minimal media containing an ammonium salt as the sole nitrogen source. Growth is improved by serum or blood, but hemin (X-factor) and nicotinamide adenine dinucleotide (NAD: V-factor) are not essential. **Acid production does not occur from carbohydrates in conventional media,** except for *B. neotomae.* Do not produce indole. **Do not liquefy gelatin** or inspissated serum. Do not lyse erythrocytes and do not produce **acetyl methyl carbinol** (Voges-Proskauer test). **Methyl red negative. Possess characteristic intracellular antigens specific for the genus.** They are intracellular parasites, transmissible to a wide range of animal species including humans.

Type species: *Brucella melitensis.*

Differentiation of the species of the genus **Brucella:** See Tables 4.20 and 4.21.

Genus **Chromohalobacter**

Editorial note: The genus *Chromohalobacter* was not included in *Bergey's Manual of Systematic Bacteriology.* The genus was created in 1989 by Ventosa et al. (Int. J. Syst. Bacteriol. *39:* 382-386). It included only one species, *C. marismortui.*

Rods 0.6–1.0 × 1.5–4.0 µm when grown in the presence of 10% (w/v) NaCl; at higher and lower salt concentrations the cells are longer. Stain is Gram negative. **Motility occurs by means of peritrichous flagella.** Aerobic, having a strictly respiratory type of metabolism. Optimum temperature is 37°C; optimum pH is 7.5. **Colonies are violet blue to brown. Catalase-positive and oxidase-negative.** Chemoorganotrophic. Optimum growth occurs in media containing 10% NaCl. Acid is produced from D-glucose and other sugars. Gelatin, casein, Tween 80, starch, esculin, and tyrosine are not hydrolyzed. Carbohydrates, amino acids, and some polyols can serve as sole carbon sources. Habitat is Dead Sea and marine salterns.

Type (and only) species: *Chromohalobacter marismortui.*

Characteristics of the species: Growth occurs in media containing 1–30% NaCl. On solid, complex media containing 10% NaCl, colonies are circular, convex, smooth, entire, and concentrically ringed with dark brown centers followed by bluish brown, grayish brown, and yellow rings. They produce a yellow pigment and a violet blue pigment which is not violacein. Pigment production is favored by suboptimal growth temperatures, glycerol, and 10% NaCl. pH range is 5–10; temperature range is 5–45°C. Acid but no gas is produced in marine oxidation-fermentation medium containing 10% total salts from D-glucose, D-galactose, maltose, lactose, D-arabinose, D-xylose, sucrose, trehalose, glycerol, and D-mannitol. Generally, NO$_3^-$ is reduced to NO$_2^-$; NO$_2^-$ is not reduced. Simmons citrate is used. H$_2$S is not produced from cysteine. There is no production of phosphatase, indole, urease, acetoin, DNase, β-galactosidase, phenylalanine deaminase, arginine dihydrolase, lysine decarboxylase, or ornithine decarboxylase. The following compounds are used as sole carbon and energy sources: dulcitol, D-fucose, D-galactose, D-gluconate, D-glucose, glutamate, *meso*-inositol, maltose, D-mannitol, D-mannose, pyruvate, D-ribose, sucrose, D-sorbitol, and D-xylose, but not *N*-acetylglucosamine, amygdalin, DL-α-aminobutyrate, butyrate, cellobiose, citrate, esculin, *p*-hydroxybenzoate, hippurate, inulin, malonate, melibiose, oxalate, raffinose, salicylate, salicin, and D-tartrate. The following are used as sole carbon, nitrogen, and energy sources: L-alanine, DL-arginine, L-glutamine, L-ornithine, L-proline, putrescine, and L-serine, but not L-allantoin, betaine, creatine, ethionine, L-isoleucine, L-leucine, phenylalanine, sarcosine, L-threonine, and L-valine.

Genus **Chryseomonas**

Editorial note: The genus *Chryseomonas* was not included in *Bergey's Manual of Systematic Bacteriology*. The genus was created in 1986 by Holmes et al. (Int. J. Syst. Bacteriol. *36:* 161-165). It included only one species, *C. polytricha*. In 1987, Holmes et al. (Int. J. Syst. Bacteriol. *37:* 245-250) found *Pseudomonas luteola* to be a senior subjective synonym of *C. polytricha;* hence the name of the species was changed to *C. luteola*.

Rods with parallel sides and rounded ends. **Intracellular granules of poly-β-hydroxybutyrate are absent.** Do not produce prosthecae or sheaths. No resting stages are known. Cells stain Gram negative. **Motility occurs by multitrichous (10–12) polar flagella.** Aerobic, having a strictly respiratory type of metabolism. Temperature range is 18–42°C. **Growth on solid media is pale to deep yellow.** Colonies are typically circular (1 mm diameter), low convex, smooth (a few strains produce wrinkled colonies) and shiny with entire edges. **Catalase positive; oxidase negative.** Agar is not digested. Chemoorganotrophic and oxidatively saccharolytic. They are not known to be present in the general environment but are apparently saprophytes or commensals of humans and other warm-blooded animals in whom they may occasionally be pathogenic.

Type (and only) species: *Chryseomonas luteola*.

Characteristics of the species: Growth occurs at 42°C but not at 5°C. Hemolysis not present. They produce oxidative reaction in oxidation-fermentation medium. Tributyrin, Tween 20, tyrosine, and gelatin are hydrolyzed. No opalescence is produced on lecithovitellin agar. Nitrate is not reduced. Indole-negative. There is no H_2S production (lead acetate paper and triple sugar iron agar). Grow occurs on β-hydroxybutyrate (without formation of inclusion granules) and on MacConkey agar. Esculin is hydrolyzed but not starch. Alkali is produced on Christensen citrate. Citrate is utilized (Simmons medium). Malonate is utilized. Not fluorescent on King medium B. Gluconate is not oxidized. Arginine dihydrolase is produced but not lysine decarboxylase or ornithine decarboxylase. There is no production of 3-ketolactose. Selenite is not reduced. Phenylalanine deaminase is negative. β-D-Galactosidase is produced but not phosphatase. Acid is produced in ammonium salt medium under aerobic conditions from glucose, arabinose, ethanol, fructose, glycerol, inositol, maltose, mannitol, trehalose, and xylose; no acid from cellobiose, dulcitol, lactose, and raffinose. There is no

production of gas from glucose in peptone-water medium.

Genus **Comamonas**

Editorial note: The genus *Comamonas* was not included in *Bergey's Manual of Systematic Bacteriology*. The genus was created in 1985 by De Vos et al. (Int. J. Syst. Bacteriol. *35:* 443-453) and initially included only one species, *C. terrigena*. In 1987 Tamoaka and Komagata (Int. J. Syst. Bacteriol. *37:* 52-59) reclassified *Pseudomonas acidovorans* and *Pseudomonas testosteroni* as species of *Comamonas*.

Straight or slightly curved rods, 0.5–1.0 × 1–4 μm. Cells occur singly or in pairs. **Motility ocurs by a tuft of polar flagella.** Endospores are not produced. **Poly-β-hydroxybutyrate is accumulated within the cells.** Cells stain Gram negative. **Oxidase and catalase positive.** Strictly aerobic, nonfermentative, and chemoorganotrophic. Good growth is obtained on media containing organic acids, amino acids, or peptone. No fluorescent pigments are produced. Carbohydrates are rarely attacked. The major cellular fatty acids are hexadecanoic acid (16:0), hexadecenoic acid (16:1), and octadecenoic acid (18:1). Cells always contain 3-hydroxydecanoic acid (3-OH 10:0). The major quinone is Q-8. Menaquinone is not produced.

Type species: *Comamonas terrigena*.

Differentiation of the species of the genus **Comamonas:** See Table 4.22.

Genus **Cupriavidus**

Editorial note: The genus *Cupriavidus* was not included in *Bergey's Manual of Systematic Bacteriology*. The genus was created in 1987 by Makkar and Casida (Int. J. Syst. Bacteriol. *37:* 323-326). It includes only one species, *C. necator*.

Coccoid rods 0.7–0.9 × 0.9–1.3 μm. Stain is Gram negative. **Motility ocurs by 2–10 peritrichous flagella.** Chemoheterotrophic. Organic nitrogen source is not required. Glucose is not utilized; fructose is catabolized oxidatively. **Catalase and oxidase positive.** Nitrate reduced. Gelatin, starch, and urea are not hydrolyzed. Indole and H_2S are not produced. Utilizes several amino acids, but not L-lysine or L-methionine, as the only source of carbon and nitrogen for growth. Optimum temperature is 27°C. Optimum pH is 7.0–8.0. NaCl at 3% inhibits growth. **Resistant to copper. Growth initiation is stimulated by copper.** Colonies on nutrient agar after 2 days at 27°C are off-white, glistening, mucoid, smooth, and convex with an entire edge, 2–4 mm in diameter. Isolated from soil. **Nonobligate predator causing lysis in soil of various Gram-positive and Gram-negative bacteria.** Can lyse certain other nonobligate bacterial predators. Growth does not require the presence of prey species.

Type (and only) species: *Cupriavidus necator.*

Characteristics of the species: Characteristics are as described for the genus. During laboratory cultivation, *C. necator* may produce small numbers of a nonmucoid variant. The variant resembles the mucoid form except for a drier, flatter appearance and smaller colonies.

Genus **Deleya**

Editorial note: The genus *Deleya* was not included in *Bergey's Manual of Systematic Bacteriology.* The genus was created in 1983 by Baumann et al. (Int. J. Syst. Bacteriol. *33:* 793-802) and included five species: *D. marina* (formerly *Pseudomonas marina*), *D. aesta* (formerly *Alcaligenes aestus*), *D. pacifica* (formerly *Alcaligenes pacificus*), *D. cupida* (formerly *Alcaligenes cupidus*), and *D. venusta* (formerly *Alcaligenes venustus*). *D. aesta* was designated as the type species. In 1984 a sixth species was added: *D. halophila* (Quesada et al., Int. J. Syst. Bacteriol. *34:* 287-292). In 1989 Akagawa and Yamasato (Int. J. Syst. Bacteriol. *39:* 462-466) reported that *D. aesta, Alcaligenes aquamarinus,* and *Alcaligenes faecalis* subsp. *homari* all belong to a single species and that, because of the priority of the specific epithet *aquamarinus, D. aesta* should be called *D. aquamarina.*

Straight rods 0.8–1.1 × 1.5–3.0 μm. Cells accumulate poly-β-hydroxybutyrate as an intracellular reserve product. Involution forms are usually present in old cultures or under adverse conditions of cultivation. Cells do not form endospores or microcysts. **Most species are motile by 4–12 peritrichous flagella;** one species (*D. marina*) is motile by 2–5 polar flagella. Strict aerobes; molecular oxygen is a universal electron acceptor. All species are chemoorganotrophs that are able to grow on a mineral medium containing a seawater base, D-glucose, and NH₄Cl. **Na⁺ is an absolute requirement for all species;** the minimal concentration necessary for

optimal growth ranges from 75–200 mM. All species utilize D-glucose, acetate, succinate, DL-β-hydroxybutyrate, lactate, glycerol, and L-alanine. All species grow at 35°C. No species has an extracellular lipase, gelatinase, or chitinase. No species is able to utilize H_2 as the sole source of energy and CO_2 as the sole source of carbon. All species lack an arginine dihydrolase system. Found in marine habitats; one species (*D. halophila*) is found in hypersaline soils.

Type species: *Deleya aquamarina.*

Differentiation of the species of the genus **Deleya:** See Table 4.23.

Genus **Derxia**

Rod-shaped cells with rounded ends, 1.0–2.0 × 3.0–6.0 μm, occurring singly or in short chains. Cells are rather pleomorphic, depending on age and the medium. In aging cultures cells often remain together forming long filaments of sometimes locally swollen or distorted cells. Some cells may assume enormous sizes (up to 30 μm). **Young cells have a homogeneous cytoplasm; older cells show typical large refractile bodies throughout the whole cell. Resting stages are not known.** Cells stain Gram negative. **Motility occurs by a short polar flagellum;** motile cells are numerous in liquid glucose media containing combined nitrogen, but rare on nitrogen-deficient solid media. **Aerobic**, having a strictly respiratory type of metabolism with oxygen as the terminal electron acceptor. **Molecular nitrogen is fixed** under aerobic conditions and also under decreased oxygen pressures (microaerobic conditions). Optimum temperature is 25–35°C; growth is slow at 15°C, feeble at 40°C; there is no growth at 50°C. **Growth occurs between pH 5.5 and 9.0;** no growth occurs at pH 4.4. **Broth cultures turn into a gelatinous mass,** but growth near the surface is more luxuriant and forms a thick, tough pellicle. **Colonies on agar media are at first slimy and semitransparent, later massive and opaque, highly raised with a wrinkled surface. Older colonies develop a dark mahogany-brown color. Catalase negative.** A wide range of sugars, alcohols, and organic acids are oxidized mostly to CO_2 and a small amount of acid, probably acetic, when growing in an alkaline medium. **Can grow as a facultative hydrogen autotroph. Growth on methane or methanol as the sole carbon source has been demonstrated.** Found in tropical soils (Asia, Africa, South America).

Type (and only) species: *Derxia gummosa.*

Characteristics of the species: As described for the genus.

Genus **Ensifer**

Rods 0.7–1.1 × 1.0–1.9 μm, occurring singly or in pairs. **Motility occurs by a tuft of 3–5 subterminal flagella. Reproduction is by budding** at one end of the cell, with the bud then elongating to give asymmetric polar growth. Separation of the cells occurs by binary fission. After fission, growth resumes as new buds at the newly formed poles of both the mother and daughter cells. Cells stain Gram negative, but may stain poorly. Aerobic, having a strictly respiratory type of metabolism. Chemoheterotrophic and are not fastidious in nutritional requirements. Colonies 10-15 mm in diameter develop on 0.1× heart infusion agar and are grayish white, convex, slimy, moist, opaque, and circular with undulate margins; they may appear almost translucent due to excessive slime production. **Weakly catalase positive.** Optimum temperature is 27°C. Growth occurs at 20 and 37°C. **Nonobligate predator that attaches end-wise to various living Gram-positive and Gram-negative host bacteria;** no specialized terminal structure for attachment purposes occurs. Depending on the pH (6 is optimal), the nutritive level of the environment, and the species of host cell, the host cells may be killed, and often they are also lysed. **Attachment to and lysis of bacterial cells occur naturally in soil and in vitro with pure cultures.** On agar media a diffusible factor is produced that causes lysis of host cells. Utilize a variety of organic carbon sources, including glucose, galactose, mannose, rhamnose, xylose, mannitol, sorbitol, glycerol, L-glutamic acid, L-alanine, L-asparagine, and L-glutamine. Nitrate and nitrite are reduced. Nitrification is negative with ammonia and nitrite. Nitrogenase activity is absent. Grows on pure gelatin without hydrolysis. Starch is not hydrolyzed. Growth is inhibited by 4% NaCl but not 2% NaCl. Growth occurs on deoxycholate agar; initially the growth is whitish purple and then changes to buff. **Isolated from soil.**

Type (and only) species: *Ensifer adhaerens.*

Characteristics of the species: Characteristics are as described for the genus. Variants have been isolated from cultures and produce less slime, resulting in smaller colonies having a drier, slightly whiter appearance.

Genus **Erythrobacter**

Rod-shaped cells are **motile by subpolar flagella.** Gram negative. **Although the cells contain bacteriochlorophyll *a*, they do not grow anaerobically in the light. Colonies and cell suspensions are orange due to carotenoid pigments. Aerobic chemoorganotrophs,** having a predominantly respiratory type of metabolism. Under microaerobic conditions, small amounts of acid are produced from a wide range of carbohydrates. **Catalase and oxidase positive.** Biotin is required for growth. Methanol is not utilized. Voges-Proskauer and methyl red negative. Gelatin is hydrolyzed. Occur in oxic marine environments, predominantly on seaweeds.

Type (and only) species: *Erythrobacter longus.*

Characteristics of the species: Cells 0.3–0.4 × 2.0–5.0 μm. See also Table 4.45.

Genus **Flavimonas**

Editorial note: The genus *Flavimonas* was not included in *Bergey's Manual of Systematic Bacteriology.* The genus was created in 1987 by Holmes et al. (Int. J. Syst. Bacteriol. *37:* 245-250). It includes only one species, *F. oryzihabitans.*

Rods with parallel sides and rounded ends. **Intracellular granules of poly-β-hydroxybutyrate are generally absent.** Cells do not produce prosthecae or sheaths. No resting stages are known. Cells stain Gram negative. **Aerobic,** having a strictly respiratory type of metabolism. Grow at temperatures from 18–42°C. Growth on solid media is pigmented pale yellow to deeply yellow. Colonies are typically circular (diameter, 1 mm), low convex, smooth (occasional strains produce wrinkled colonies), and shiny, with entire edges. **Catalase-positive; oxidase-negative.** Agar is not digested. Chemoorganotrophic. Saccharolytic. Found in the general environment; they are apparently saprophytes or commensals of humans and other warm-blooded animals, in which they may occasionally prove pathogenic.

Type (and only) species: *Flavimonas oryzihabitans.*

Characteristics of the species: **Cells are motile by a single polar flagellum or nonmotile.** Dark brown, and diffusible melanin-like pigment is produced by some strains on tyrosine agar. Hemolysis is not present.

No fluorescent pigments occur on King Medium B. Grow at room temperature and 37°C but not at 5 or 42°C. Produce oxidation reaction in glucose oxidation-fermentation medium. Hydrolyze tributyrin and Tween 20 but not gelatin. Tween 80 is not hydrolyzed. Casein is not digested. No opalescence is produced on lecithovitellin agar. No reduction of nitrate or nitrite. Indole-negative and H_2S negative (lead acetate paper and triple sugar iron agar). Grow on β-hydroxybutyrate but produce no lipid inclusions. Grow on cetrimide. Grow on MacConkey agar. No hydrolysis of esculin or starch. Alkali is produced on Christensen citrate. Citrate is utilized (Simmons medium). No oxidation of gluconate. No production of arginine desimidase, arginine dihydrolase, lysine decarboxylase, or ornithine decarboxylase and no production of 3-ketolactose. Selenite is not reduced. Phenylalanine deaminase is negative. No production of β-galactosidase. Acid is produced in ammonium salt medium under aerobic conditions from D-glucose, L-arabinose, ethanol, D-fructose, glycerol, mannitol, L-rhamnose, sorbitol, trehalose, and D-xylose. No production of acid in ammonium salt medium under aerobic conditions from adonitol, dulcitol, lactose, raffinose, and salicin. No production of acid or gas from D-glucose in peptone water medium.

Genus **Flavobacterium**

Rods with parallel sides and rounded ends, typically 0.5 × 1.0–3.0 µm. Intracellular granules of poly-β-hydroxybutyrate are absent. Endospores are not formed. Cells stain Gram negative. **Nonmotile.** Not glide or spread. **Aerobic,** having a strictly respiratory type of metabolism. Environmental isolates grow at 37°C. Growth on solid media is typically pigmented (yellow to orange) but nonpigmented strains occur. Colonies are translucent (occasionally opaque), circular (diameter 1–2 mm), convex or low convex, smooth and shiny with entire edges. **Catalase, oxidase, and phosphatase positive.** Agar is not digested. Chemoorganotrophic. Acid, but no gas is produced from carbohydrates in media having a low peptone concentration. Widely distributed in soil and water; also found in raw meats, milk and other foods, and in the hospital environment and in human clinical material.

Type species: *Flavobacterium aquatile.*

Differentiation of the species of the genus **Flavobacterium:** See Table 4.24.

Genus **Francisella**

Rod-shaped cells, 0.2 × 0.2–0.7 µm *(Francisella tularensis)* **or 0.7–1.7 µm** *(Francisella novicida),* when cultured in appropriate media and examined during active growth; pleomorphism occurs subsequently. Cells stain Gram negative, faintly-staining. **Nonmotile and obligately aerobic.** On glucose-cysteine-blood agar smooth gray colonies are formed which reach a maximum size in 2–4 days and are surrounded by a characteristic green zone of discoloration. Weakly catalase-positive. **Oxidase negative.** Catabolism of carbohydrates is characteristically slow with the production of acid but no gas. **Cysteine (or cystine) is either required for growth** *(F. tularensis)* **or is greatly stimulatory for growth** *(F. novicida).* **H_2S is produced.** Unlike other bacteria, the type species contains relatively large amounts of long-chain saturated and monoenoic C_{20}–C_{26} fatty acids as well as 3-hydroxyhexadecanoate, 2-hydroxydecanoate, and 3-hydroxyoctadecanoate. *F. tularensis* **is the causative agent of tularemia in humans and other animals;** *F. novicida* causes experimental infections in laboratory animals.

Type species: *Francisella tularensis.*

Differentiation of the species of the genus **Francisella:** See Table 4.25.

Genus **Frateuria**

Regular straight rods, 0.5–0.7 × 0.7–3.5 µm, occurring singly or in pairs. Cells stain Gram negative. **Motility occurs by polar flagella, or cells are nonmotile. Obligately aerobic.** Optimum temperature for growth is 25–30°C. Colonies on mannitol-yeast extract-peptone (MYP) agar are yellow to orange. On glucose-yeast extract-$CaCO_3$ (GYC) agar most strains produce a typical brown water-soluble pigment. **Oxidase negative. Grows at pH 3.6.** There is no nitrate reduction. Starch and gelatin are not hydrolyzed. **H_2S is produced.** Chemoorganotrophic. Acid is produced from ethanol and a number of carbon sources. On D-glucose and D-xylose, the pH drops below 4.0. From D-glucose, 2-keto- and 2,5-diketogluconic acids are formed, but not 5-ketogluconic acid. **There are no requirements for growth factors.** Isolated from *Lilium auratum* and from the fruit of *Rubus parvifolius* (raspberry) in Japan.

Type (and only) species: *Frateuria aurantia.*

Characteristics of the species: As described for the genus.

Genus **Gluconobacter**

Cells are ellipsoidal to rod shaped, 0.5–1.0 × 2.6–4.2 µm, occurring singly and/or in pairs, rarely in chains. Enlarged, irregular cell forms (involution forms) may occur. Endospores are not formed. Gram negative (in a few cases Gram variable). **Motile or nonmotile; if motile, the cells have 3-8 polar flagella,** rarely a single flagellum. Believed to be obligately aerobic, having a strictly respiratory type of metabolism with oxygen as the terminal electron acceptor, although recent reports show that strains are capable of reducing thiosulfate to form H_2S. H_2S (detected by paper strip method) is formed in sorbitol medium containing thiosulfate. Colonies are pale. Optimum temperature is 25–30°C; there is no growth at 37°C. Optimum pH, 5.5–6.0; most strains will grow at pH 3.6. Strongly catalase positive and oxidase negative. Negative for nitrate reduction, gelatin liquefaction, and indole production. Chemoorganotrophic. **Oxidize ethanol to acetic acid. Do not oxidize acetate or lactate to CO_2 and H_2O. Strong ketogenesis occurs from polyalcohols.** Acid formation from D-glucose and D-xylose is pronounced. All strains produce 2-ketogluconic acid from D-glucose, and the majority of strains also form 5-ketogluconic acid. No acid production or growth occurs on starch. *Gluconobacter* strains occur in **flowers,** garden soil, baker's yeast, **honey bees, fruits, cider,** beer, **wine, wine vinegar, South African Bantu beer, palmsap,** and soft drinks, Some strains are said to cause **pink disease** in pineapples and **rot** in apples and pears; however, the identity of these strains as being members of *Gluconobacter* is in doubt (see Micales et al., Int. J. Syst. Bacteriol. *35:* 79-85, 1985). *Gluconobacter* strains prefer sugar-enriched environments in contrast to *Acetobacter* strains, which prefer alcohol-enriched environments.

Type species: *Gluconobacter oxydans.*

Differentiation of the species of the genus **Gluconobacter:** See Table 4.26.

Genus **Halomonas**

Rod-shaped or pleomorphic. Rods are generally 0.6–0.8 µm × 1.6–1.9 µm. Elongated, flexuous filaments may be formed under certain conditions. Spores are not formed. Cells stain Gram negative. **Motile or nonmotile; motility is by means of several unsheathed lateral or polar flagella. Possess mainly a respiratory type of metabolism with oxygen as the terminal electron acceptor.** Growth of some strains may occur anaerobically with nitrate. Growth on glucose may occur under anaerobic conditions in the absence of nitrate, indicating some fermentative ability; however, anaerobic growth with other carbohydrates or with amino acids requires the presence of nitrate. Colonies are **white to yellow, never red.** Catalase and oxidase positive. Most strains reduce nitrate to nitrite. Chemoorganotrophic. Carbohydrates, amino acids, and some polyols can serve as sole carbon sources in mineral media. Ammonium sulfate can serve as a sole nitrogen source. **Halotolerant, able to grow in NaCl concentrations ranging from 0-0.5–20% (w/v) or more.** Strains have been isolated from a solar salt facility, from a saline lake in Antarctica, and from an estuary in New Hampshire.

Type species: *Halomonas elongata.*

Differentiation of the species of the genus **Halomonas:** See Table 4.27.

Genus **Hydrogenophaga**

Editorial note: The genus *Hydrogenophaga* was not included in *Bergey's Manual of Systematic Bacteriology.* The genus was created in 1989 by Willems et al. (Int. J. Syst. Bacteriol. *39:* 319-333). It includes species previously classified in the genus *Pseudomonas* as *P. flava, P. pseudoflava, P. palleronii, P. taeniospiralis,* and "*P. carboxydoflava*".

Straight to slightly curved rods 0.3–0.6 × 0.6–5.5 µm, occurring singly or in pairs. Motile by means of one, rarely two, polar to subpolar flagella. Cells stain Gram negative. **Oxidase positive.** Catalase reaction varies among species. **A nondiffusible yellow pigment is produced.** Aerobic and facultative H_2 autotrophs. Have an oxidative carbohydrate metabolism with oxygen as the terminal electron acceptor. Some species (*H. pseudoflava* and *H. taeniospiralis*) exhibit anaerobic nitrate respiration, with denitrification. Grow well on media containing organic acids, amino acids, or peptone, but are less versatile in the use of carbohydrates. A cyclopropane-substituted fatty acid (17:cyc) is present; 3-hydroxyoctanoic acid (3-OH-8:0) is present alone or together with 3-hydroxydecanoic acid (3-OH-10:0). 2-

Hydroxy-substituted fatty acids are absent. Ubiquinone Q-8 is present as the main quinone. 2-Hydroxyputrescine and putrescine are present in roughly equimolar concentrations either exclusively or as the dominant polyamine compounds.

Type species: *Hydrogenophaga flava.*

Differentiation of the species of the genus **Hydrogenophaga:** See Table 4.28.

Genus **Janthinobacterium**

Rods 0.8–1.2 × 2.5–6.0 μm with rounded ends, sometimes slightly curved. Occur singly, occasionally with some pairs of short chains. Definite capsules are not evident, although sometimes intercellular slime is formed. No resting stages are known. Cells stain Gram negative, occasionally with barred or bipolar staining and lipid inclusions. **Motility occurs by means of both a single polar flagellum and usually one to four subpolar or lateral flagella.** Strict aerobes. Produce low convex, round, **violet colonies** on solid media; in nutrient broth, a violet ring is formed at the junction of the liquid surface and the container wall. Optimum temperature is 25°C; minimum, 2°C; maximum, 32°C. Optimum pH is 7–8, with no growth below pH 5. No growth occurs in media containing 6% or more of NaCl. Chemoorganotrophs, having a strictly respiratory type of metabolism with oxygen as the terminal electron acceptor. Acid, but no gas is produced from glucose and certain other carbohydrates. Lactate is oxidized to CO_2. Usually oxidase positive by the method of Kovacs (1956), although the violet pigment may interfere with the reading. Catalase positive, indole negative, and Voges-Proskauer-negative. Nitrate and nitrite are reduced, sometimes with visible gas production. Ammonia is formed from peptone. Phosphatase positive. Arysulfatase negative, grow on ordinary peptone media. Utilize citrate and ammonia as sole carbon and nitrogen sources for growth and grow rapidly. Growth factors are not required. Resistant to benzylpenicillin (10 μg/ml) and to vibriostatic agent 0/129 (2,4-diamino-6,7-diisopropylpteridine, 30 μg/disc). Soil and water organisms, common in temperate climates. Occasionally cause food spoilage.

Type (and only) species: *Janthinobacterium lividum.*

Characteristics of the species: As described for the genus.

Genus **Kingella**

Straight rods, 1.0 μm in length with rounded or square ends. Occur in pairs and sometimes short chains. Endospores are not formed. **Cells are Gram negative, but there is a tendency to resist decolorization.** Cells are nonmotile by normal tests but may be fimbriated (piliated) and show "twitching motility." Aerobic or facultatively anaerobic; grow best aerobically but can grow weakly under anaerobic conditions on blood agar. Optimum temperature is 33–37°C. Two types of colonies occur on blood agar: (a) a spreading, corroding type associated with "twitching motility," fimbriation, and transformation competence, and (b) a smooth, convex type not showing twitching, fimbriation, or competence. **Oxidase positive** (when tested with tetramethyl-*p*-phenylenediamine; the dimethyl reagent may give weak or negative reactions). **Catalase negative.** Coagulated serum is not liquefied. Urease negative. Phenylalanine deaminase activity is negative or weak. Chemoorganotrophic. **Glucose and limited number of other carbohydrates are fermented with acid production** but no gas. **Susceptible to penicillin.** Occur in human mucous membranes of the upper respiratory tract.

Type species: *Kingella kingae.*

Differentiation of the species of the genus **Kingella:** See Table 4.29.

Genus **Lampropedia**

Sheets of rounded, almost cubical cells, arranged in square tablets of 16–64 cells, occasionally separated into pairs or tetrads. **Divide synchronously in a sheet and alternately in two planes.** The cells of a tablet are enclosed within a **complex, structured envelope.** Each cell is inclosed in a **Gram-negative type of cell wall.** Intracellular granules of **poly-β-hydroxybutyrate** are prominent. No flagella occur. **Twitching movements** of small groups of cells occur during active growth. **Obligately aerobic,** having a strictly respiratory type of metabolism with oxygen serving as the terminal electron acceptor. **Growth occurs as thin, hydrophobic, extending pellicle** on the surface of both liquid and solid media. Nonpigmented. Optimum temperature is 30°C. Optimum pH is 7.0. Oxidase and catalase positive. Chemoorganotrophic. **Energy sources are limited to intermediates of the tricarboxylic acid cycle.** Carbohydrates, alcohols, glucosides, and fatty acids are not utilized. Ammonium salts of certain amino acids can serve as sole nitrogen sources. **Vitamins may be required for growth.** The ecological niche is unknown, but observations and isolation indicate an environment rich in organic matter.

Type (and only) species: *Lampropedia hyalina.*

Characteristics of the species: Cells are 1.0–1.5 × 1.0–2.5 μm. Morphological characteristics are as described for the genus. Utilizes pyruvate, lactate, butyrate, fumarate, malate, succinate (and acetate in the presence of catalytic levels of pyruvate) as sole energy sources. Utilizes NH$_4$Cl, alanine, arginine, and tyrosine as sole nitrogen sources. Biotin and thiamine are required for growth. Temperature range for growth is 10–35°C; optimum, 30°C. pH range for growth is 6.0–8.6; optimum, 7.0.

Genus **Legionella**

Rods 0.3–0.9 × 2–20 μm or more. Do not form endospores or microcysts and are not encapsulated. Not acid fast. Cells stain Gram negative. **Motile is by one, two, or more straight or curved polar or lateral flagella;** nonmotile strains are occasionally seen. Aerobic. **L-Cysteine hydrochloride and iron salts are required for growth.** The oxidase test is negative or weakly positive. Nitrates are not reduced. Urease-negative. Gelatin is liquefied. **Branched chain fatty acids predominate in the cell wall.** Chemoorganotrophic, using amino acids as carbon and energy sources. **Carbohydrates are neither fermented nor oxidized.** Isolated from surface water, mud, and from thermally polluted lakes and streams. There is no known soil or animal source. Pathogenic for humans, causing pneumonia (Legionnaires' disease) or a mild, febrile disease (Pontiac fever).

Type species: *Legionella pneumophila.*

Differentiation of the species of the genus **Legionella:** There are now 30 species and 47 serologically distinct groups in *Legionella* (Thacker et al., J. Clin. Microbiol. *27:* 1831-1834, 1989; Bornstein et al., Res. Microbiol. *140:* 541-552, 1989). Identification is done serologically, which is satisfactory for the most frequently occurring species and serovars, but antisera for many species are not available commercially. There are tests that will potentially differentiate *Legionella* species, but these

have not yet been tried on all species, or on a sufficient number of strains of many species, and many of these tests are not applicable for routine laboratory use (Vasey et al., J. Appl. Bacteriol. *65:* 339-345, 1988; Fox and Brown, J. Clin. Microbiol. *27:* 1952-1955, 1989). Thus identification of most *Legionella* species must be done by a reference laboratory.

Genus **Marinobacter**

Editorial note: The genus *Marinobacter* was not included in *Bergey's Manual of Systematic Bacteriology.* The genus was created in 1992 by Gauthier et al. (Int. J. Syst. Bacteriol. *42:* 568–576). It includes only one species, *M. hydrocarbonoclasticus.*

Cells are rod-shaped, 0.3–0.6 × 2.0–3.0 μm. Have numerous surface blebs when grown on eicosane in mineral medium. Cells stain Gram negative. Nonsporeforming. **Motility occurs by means of single unsheathed polar flagellum in media containing 0.2–1 M NaCl.** Cells are unflagellated in media with a lower or higher NaCl concentration. Colonies on agar media are white when young and pinky beige after 48 hours of incubation. Grow at temperatures ranging from 10–45°C (mesotrophic), with optimal growth at 32°C. **Exhibit extreme halotolerance and can grow in NaCl concentrations ranging from 0.08–3.5 M. Have an absolute requirement for sodium ion.** Aerobic, with a nonfermentative metabolism. Grow anaerobically with nitrate or on succinate, citrate, or acetate but not on glucose. Denitrify, with N$_2$ production. Oxidase, cytochrome oxidase, catalase, tweenase, and lecithinase positive. Grow on acetate, butyrate, caproate, succinate, fumarate, adipate, DL-lactate, and citrate as sole carbon sources but not on carbohydrates and amino acids (except L-proline and L-glutamate). Degrade a large variety of aliphatic or aromatic hydrocarbons and produce a nondialyzable bioemulsifier when grown on hydrocarbons. Resistant to novobiocin, tetracycline, oleandomycin, staphylomycin, and vibriostatic agent 0/129 and are susceptible to penicillin G, kanamycin, streptomycin, chloramphenicol, erythromycin, cephaloridine, gentamicin, and nalidixic acid.

Type (and only) species: *Marinobacter hydrocarbonoclasticus.*

Characteristics of the species: As described for the genus.

Genus **Marinomonas**

Editorial note: The genus *Marinomonas* was not included in *Bergey's Manual of Systematic Bacteriology.* The genus was created in 1983 by Landschoot and De Ley (J. Gen. Microbiol. *129:* 3057-3074; Int. J. Syst. Bacteriol. *34:* 91-92, 1984). It includes two species: *M. communis* (formerly *Alteromonas communis*) and *M. vaga* (formerly *Alteromonas vaga*).

Straight or curved rods, 0.7–1.5 × 1.8–3.0 µm. Do not accumulate poly-β-hydroxybutyrate. Microcysts or endospores are not formed. Gram negative. **Motility occurs by means of a single polar flagellum.** Chemoorganotrophs are capable of respiratory but not fermentative metabolism. Molecular oxygen is a universal terminal electron acceptor. **Do not reduce nitrate to nitrite and do not denitrify.** None of the strains has a constitutive arginine dihydrolase system. **All require a seawater base for growth.** Utilize D-sorbitol, DL-malate, α-ketoglutarate, D-mannose, D-fructose, succinate, fumarate, glycerol, and *m*-hydroxybenzoate. Negative for extracellular gelatinase and lipase.

Type species: *Marinomonas communis.*

Differentiation of the species of the genus **Marinomonas:** See Table 4.30.

Genus **Mesophilobacter**

Editorial note: The genus *Mesophilobacter* was not included in *Bergey's Manual of Systematic Bacteriology.* The genus was created in 1989 by Nishimura et al. (Int. J. Syst. Bacteriol. *39:* 378-381). It includes only one species, *M. marinus.*

Pleomorphic rods 0.5–0.6 × 1.0–2.0 µm. A **plump form** (0.8–1.0 × 1.5–3.0 µm) and **elongated cells** (15 µm or more) occur in young cultures, and **coccoid cells** (0.6–0.9 µm in diameter) occur in old cultures. Cells stain Gram negative and are nonencapsulated. **Nonmotile. Oxidase and catalase positive. Moderately halophilic.** Optimum temperature is 33–37°C. Growth occurs at 5°C but not at 0°C. Aerobic, having a strictly respiratory type of metabolism. Penicillin-resistant. **Isolated from sea water.**

Type (and only) species: *Mesophilobacter marinus.*

Characteristics of the species: Colonies on nutrient agar are usually circular but sometimes irregular, smooth, slightly convex, entire, glistening, opaque, and pale yellowish brown. There is moderate growth in nutrient broth, with slight sediment with no reaction in litmus milk. Indole production, urease and gelatinase reactions differ among strains. Methyl red is positive; Voges-Proskauer is negative. Phenylalanine deaminase is negative. No hemolysis occurs. Nitrate is reduced to nitrite. Acidity but no gas is produced from D-ribose, D-glucose, D-fructose, D-mannitol, and cellobiose. The following compounds are assimilated: D-ribose, D-glucose, D-fructose, maltose, sucrose, trehalose, acetate, citrate, succinate, fumarate, malate, and gluconate. Growth occurs in the presence of 7% NaCl. The ubiquinone system is Q-8. The major cellular fatty acids are straight-chain $C_{16:0}$, $C_{18:1}$, and $C_{16:0}$ acids.

Genus **Methylobacillus**

Editorial note: The genus *Methylobacillus* was not included in *Bergey's Manual of Systematic Bacteriology.* The genus was created in 1986 by Urakami and Komagata (Int. J. Syst. Bacteriol. *36:* 502-511). It includes only one species, *M. glycogenes.*

Short Gram-negative rods, motile with a single polar flagellum or nonmotile. Obligate methylotrophs which grow on C1 compounds other than methanol and which cannot grow on methane. Utilization of D-fructose varies from strain to strain. Strict aerobes; metabolism is respiratory. The major cellular fatty acids are straight-chain saturated $C_{16:0}$ acid and unsaturated $C_{16:1}$ acid. The major quinone is Q-8, with Q-7 and Q-9 as minor components.

Type (and only) species: *Methylobacillus glycogenes.*

Characteristics of the species: Nonsporeforming rods with rounded ends, 0.3–0.5 × 0.8–2.0 µm. Cells occur singly, rarely in pairs, and are motile by a single polar flagellum; occasionally they are nonmotile. Capsules are not produced. Granules of poly-β-hydroxybutyrate do not accumulate in the cells. No growth occurs in nutrient broth and peptone broth. Colonies on methanol-containing agar are shiny, smooth, raised, entire, white to light yellow, and 1–3 mm in diameter after 3 days at 30°C. Water-soluble pigments are not produced. Nitrate is reduced to nitrite. Methyl red and Voges-Proskauer are negative. Indole is negative. H_2S and ammonia are not produced. Hydrolysis of gelatin and starch are negative. Denitrification is negative. Acid is not produced from D-glucose and D-fructose. Methanol is utilized as a sole carbon source, but methane is not. Obligate methylotrophs. L-arabinose, D-xylose, D-glucose, D-mannose, galactose, maltose, glycerol, soluble starch, succinic acid, citric acid, acetic acid, ethanol, and hydrogen are not utilized. Utilization of D-fructose and methylamine differs among strains. Vitamins and amino acids are not required, but several strains require thiamine for growth. Ammonia, urea, and nitrate are used as sole nitrogen sources. Urease-negative. **Oxidase negative.** Catalase is produced by most strains. Aerobic. Metabolism is strictly respiratory and not fermentative. Good growth occurs between pH 6.0 and 8.0. All strains grow at 30°C, and about 50% of the strains grow at 42°C. Most strains do not grow in media containing 3% NaCl.

Genus **Methylobacterium**

Editorial note: The genus *Methylobacterium* was not included in *Bergey's Manual of Systematic Bacteriology* although the genus was included on the Approved Lists (1980). The type species is *M. organophilum.* The genus was emended in 1983 by Green and Bousfield (Int. J. Syst. Bacteriol. *33:* 875-877), who added three species: *M. rhodinum* (formerly *Pseudomonas rhodos), M. radiotolerans* (formerly *Pseudomonas radiora*), and *M. mesophilicum* (formerly *Pseudomonas mesophilica*). In 1984, Urakami and Komagata (Int. J. Syst. Bacteriol. *34:* 188-201) created the genus *Protomonas,* which included one species, *P. extorquens* (formerly *Pseudomonas extorquens*); however, in 1985 Bousfield and Green (Int. J. Syst. Bacteriol. *35:* 209) reclassified members of the genus *Protomonas* in *Methylobacterium.* Thus *P. extorquens* became *M. extorquens.* In 1988 Green et al. (Int. J. Syst. Bacteriol. *38:* 124-127) added three additional species to *Methylobacterium: M. rhodesianum, M. zatmanii,* and *M. fujisawaense.*

Rods 0.8–1.0 × 1.0–8.0 μm, occurring singly or occasionally in rosettes. Occasionally they are branched and pleomorphic. **Motility occurs by single polar, subpolar, or lateral flagella, although some strains are not vigorously motile.** Cells often contain large sudanophilic inclusions and sometimes volutin granules. Cells stain Gram negative, although many strains are Gram variable; representative strains have the multilayered cell wall structure and the type of citrate synthase characteristic of Gram-negative bacteria. Most strains grow slowly, and some do not grow at all on nutrient agar.

Colonies on glycerol-peptone agar are 1–3 mm in diameter and **pale pink to bright orange-red;** colonies on methanol-salts agar are a more uniform pale pink. The pigment is insoluble and probably carotenoid. In static liquid media strains grow as a pink surface ring or pellicle. Strictly aerobic and catalase and oxidase (often weakly) positive. Chemoorganotrophs, facultative methylotrophs, and occasionally facultative methanotrophs. The ability of some strains to utilize methane as a sole source of carbon and energy is easily lost if strains are not maintained on an inorganic medium in a methane atmosphere. Representative strains have been reported to assimilate C1 compounds via the homoisocitrate pathway (icl-pathway) and to have a complete tricarboxylic acid cycle when they are grown on complex organic substrates. Members of the genus have been isolated from soil, dust, fresh water, lake sediments, leaf surfaces and nodules, rice grains, air, and hospital environments. Optimum temperature is 25–30°C.

Type species: *Methylobacterium organophilum.*

Differentiation of the species of the genus **Methylobacterium:** See Table 4.31.

Genus **Methylococcus**

Cells are **spherical,** usually occurring in pairs. **Non-motile. Resting stage is a cyst.** Cells stain Gram negative. Aerobic, having a strictly respiratory type of metabolism with oxygen as the terminal electron acceptor. **Methane, methanol, and formaldehyde are the only known compounds serving as sole carbon and energy sources.** No organic growth factors are required.

Type species: *Methylococcus capsulatus.*

Characteristics of the species: Cocci are 1.0 μm in diameter. Capsules are formed. Poly-β-hydroxybutyrate is not formed. The cells are not pigmented. Nitrite is superior to ammonium salts or casein hydrolysate as a nitrogen source. No organic growth factors are required, although colony formation on solid media is enhanced by complex extracts such as casein hydrolysate. Optimum temperature is 37°C. Growth occurs between 30 and 50°C but not at 55°C. The major carbon assimilation pathway is the ribulose monophosphate pathway (3-hexulose phosphate synthase-positive). An incomplete tricarboxylic acid cycle is present (2-oxoglutarate dehydrogenase-negative). Isocitrate dehydrogenase is NAD-dependent only. Cells are capable of autotrophic CO_2 fixation (ribulose bisphosphate carboxylase-positive). Nitrogenase activity may or may not occur. The predominant fatty acid chain length is 16.

Genus Methylomonas

Straight, curved, or branched rods, but not helical, 0.5–1.0 × 1.0–4.0 μm. Motility occurs by a single polar flagellum. Sheaths or prosthecae are not known. **One species forms cysts.** Cells stain Gram negative. Aerobic, having a strictly respiratory type of metabolism with oxygen as the terminal electron acceptor. **Methane, methanol, and formaldehyde are the only known sole sources of energy and carbon. Organic growth factors are not required. One species requires NaCl for growth. Temperature range for growth is 20–35°C.**

Type species: *Methylomonas methanica.*

Differentiation of the species of the genus **Methylomonas:** See Table 4.32.

Genus Methylophaga

Editorial note: The genus *Methylophaga* was not included in *Bergey's Manual of Systematic Bacteriology.* The genus was created in 1985 by Janvier et al. (Int. J. Syst. Bacteriol. *35:* 131-139). It includes two species, *M. marina* and *M. thalassica.*

Short straight rods 0.2 μm in diameter. Motility occurs by a single polar flagellum. Cells have a very thick periplasmic space (20–30 nm). Cells can be broken by osmotic shock after washing with 0.5 M NaCl. **Strictly aerobic and moderately halophilic. Auxotrophic for vitamin B$_{12}$.** Strains do not grow on peptone-yeast extract medium containing (or not containing) NaCl. Except fructose, the only growth substrates that are used are C1 compounds, such as methanol and methylamine, which are dissimilated by the RuMP pathway. Do not grow on methane. Isolated from marine environments.

Type species: *Methylophaga marina.*

Differentiation of the species of the genus **Methylophaga:** See Table 4.33.

Genus Methylophilus

Editorial note: The genus *Methylophilus* was not included in *Bergey's Manual of Systematic Bacteriology.* The genus was created in 1987 by Jenkins et al. (Int. J. Syst. Bacteriol. *37:* 446-448). It includes only one species, *M. methylotrophus.*

When grown on methanol-mineral salts agar or in methanol-mineral salts liquid medium, the cells are **straight or slightly curved rods usually 0.3–0.6 × 0.8–1.5 μm** occurring singly or in pairs. Gram negative, but the stain is often not taken up well. **Motility occurs by polar flagella or cells are nonmotile.** Endospores are absent. There are **no cellular inclusions.** No sheath or prosthecae are detected. No capsules are formed, but slime may be produced by some strains. Colonies on methanol-mineral-salts agar plates incubated for 2 days at 30 or 37°C are circular, 1–2 mm in diameter, with entire edge, convex, translucent to opaque. Pyocyanin and fluorescein are not produced. There exists no, or extremely poor, growth on nutrient agar and in nutrient broth incubated at 30 or 37°C for 2 days. There is no, or extremely poor, growth on blood agar; there is no hemolysis. Optimum temperature is 30–37°C; no growth occurs at 4 or 45°C. Optimum pH is 6.5–7.2. Aerobic. Metabolism is respiratory; very little or no acid is produced from glucose. Methanol is oxidized as the sole carbon and energy source by all strains. In addition, a limited range of other carbon compounds such as methylamines, formate, glucose, and fructose may be utilized as the sole carbon and energy source. Nutritionally nonexacting; nitrate and ammonium salts serve as nitrogen sources. Catalase positive and oxidase positive. The fatty acid composition is primarily of the nonhydroxylated straight-chain saturated and monounsaturated types with $C_{16:0}$ and $C_{16:1}$ predominating. The major isoprenoid quinone components are ubiquinones with eight isoprene units (Q8). Isolated from activated sludge, mud, and river and pond water.

Type (and only) species: *Methylophilus methylotrophus.*

Characteristics of the species: Colonies on methanol-mineral-salts agar are grayish white. The cells are motile by single flagella. In addition to growth on methanol as the sole carbon and energy source, good growth occurs on glucose and may or may not occur on methylamines as the sole carbon and energy source. Different strains give different results with fructose as the sole carbon and energy source. Poor growth, which varies between strains, may occur on lactose, sucrose, D-ribose, D-xylose, ethanol, propanol, butanol, acetate, and formate. Acid is not produced from glucose. Acetoin, tested by the Voges-Proskauer method, may or may not be produced. Tween 20, 40, and 60 are hydrolyzed. Tween 80 is not hydrolyzed. Urease is produced. Leucine arylamidase is produced. Phosphatase production is weak and different between different strains. Sulfatase is not produced. H₂S is not produced. Gelatin is not liquefied. Extracellular DNase and RNase are not produced. 2,3,5-Triphenyltetrazolium chloride (0.01%) is reduced. No growth occurs in the presence of 0.01% potassium tellurite, or with 5% NaCl. Resistant to penicillin, oleandomycin; sensitive to nalidixic acid, streptomycin, and a number of other antibiotics.

Genus **Methylovorus**

Editorial note: The genus *Methylovorus* was not included in *Bergey's Manual of Systematic Bacteriology.* The genus was created in 1991 by Govorukhina and Trotsenko (Int. J. Syst. Bacteriol. *41:* 158–162). It includes only one species, *M. glucosotrophus.*

When grown on methanol-mineral salt medium, cells are straight or slightly curved rods, usually 0.5–0.6 × 1.0–1.3 μm occurring singly or in pairs. They stain Gram negative. Endospores are absent. No complex intracellular membranes exist. No sheath or prosthecae are detected. No capsules are formed, but slime may be produced by some strains. Colonies on methanol-mineral salt agar incubated for 2 days at 30°C are circular, 1–2 mm in diameter, with entire edge, convex, and translucent to opaque, pink, creamy or milky in color. Pyocyanin and fluorescein are not produced. Cells multiply by binary fission. No aggregation or pigmentation occurs in liquid medium. No growth or extremely poor growth occurs on nutrient agar and in nutrient broth at 30–37°C; no growth occurs under an atmosphere of CH₄ plus O₂ or H₂ plus CO₂ plus O₂. Optimum pH for growth is 7.0–7.2; temperature, 35–37°C. Obligate aerobes with respiratory metabolism. **Methanol and glucose are utilized as carbon and energy sources.** In addition, some strains are able to grow slowly on methylated amines, inulin, and betaine. Nitrates, ammonium salts, methylated amines, glutamate, and peptone serve as nitrogen sources. Acetoin, indole, H₂S, and NH₃ are not produced in test medium. Milk is not hydrolyzed. Urease, catalase, and oxidase positive. Peroxidase is variable. Arginine dihydrolase is negative. The strains may hydrolyze starch, but not cellulose, gelatin, or Tween 80. They produce acid (but not gases) from glucose.

All the strains assimilate methanol carbon through the RuMP pathway and ammonia via the glutamate cycle (glutamate synthase and glutamine synthetase). Neither α-keto-glutarate dehydrogenase nor the glyoxylate shunt enzymes are present. The fatty acid composition is primarily of the nonhydroxylated straight-chain saturated and monounsaturated types with C₁₆:₀ and C₁₆:₁ω₇. The strains do not contain branched C₁₇ fatty acid. The major phospholipids are phosphatidylethanolamine and phosphatidylglycerol. The strains also possess diphosphatidylglycerol. Isolated from activated sludge, mud, soil, and pond water.

Type (and only) species: *Methylovorus glucosotrophus.*

Characteristics of the species: As described for the genus.

Genus **Moraxella**

Rods (subgenus *Moraxella*) **or cocci** (subgenus *Branhamella*). **The rods are often very short and plump, frequently approaching a coccus shape (1.0–1.5 × 1.5–2.5 μm;** they usually occur in pairs and short chains (one plane of division). Variation in cell size, shape, and filament or chain formation is often seen in cultures, the pleomorphism being enhanced by lack of oxygen and by incubation temperatures above the optimum. The cocci are usually smaller (0.6-1.0 μm in diameter) and occur as single cells or in pairs with the adjacent sides flattened (differing planes of division); division in two planes at right angles to each other sometimes results in the formation of tetrads. Cells may be capsulated. Gram negative, but often with a tendency to resist decolorization. Flagella are absent. Both rod-shaped and coccal species may be fimbriated. Swimming motility is absent, but surface-bound "twitching motility" has been observed in some rod-shaped species. Aerobic, but some strains may grow weakly under anaerobic conditions. Most species (exception: *M. [M.] osloensis*) are nutritionally fastidious, but the specific growth requirements are unknown. Optimum temperature, 33–35°C. Usually catalase positive. Chemoorganotrophic. No acid is produced from carbohydrates. Usually highly sensitive to penicillin. Parasitic on the mucous membranes of humans and other warm-blooded animals.

Type species: *Moraxella (Moraxella) lacunata.*

Differentiation of the species of the genus **Moraxella:** See Table 4.34.

Genus Morococcus

Editorial note: The genus *Morococcus* was not included in *Bergey's Manual of Systematic Bacteriology.* The genus was created in 1981 by Long et al. (Int. J. Syst. Bacteriol. *31:* 294-301). It includes only one species, *M. cerebrosus.*

Cocci, <1 μm in diameter, bound firmly together in tightly packed, mulberry-like aggregates of 10–20 cells. They are nonmotile and nonsporeforming. Poly-β-hydroxybutyrate is not produced. Aerobic. Complex growth factors are not required. Colonies on sucrose agar give a black reaction with iodine. Temperature range for growth, 23–42°C; pH range, 5.5–9.0. **Catalase and oxidase positive.** Nitrate is reduced. H$_2$S is produced. Acid is produced from carbohydrates. Originally isolated from a human brain abscess.

Type (and only) species: *Morococcus cerebrosus.*

Characteristics of the species: May give weak reactions in both tubes of the Hugh-Leifson O/F test. Acid is produced from glucose, fructose, sucrose, and maltose, but not from arabinose, ribose, xylose, rhamnose, galactose, mannose, sorbose, salicin, cellobiose, lactose, melibiose, trehalose, melezitose, raffinose, dextrin, inulin, or starch. Litmus milk is reduced. H$_2$S is produced from cysteine. Positive for DNase, ornithine decarboxylase, and methyl red. Negative for lecithinase, phosphatase, phenylalanine deaminase, and urease. Indole is not produced. Citrate and malonate are not oxidized. There is no hydrolysis of starch, esculin, gelatin, casein, or Tween 80. Pathogenic to mice.

Differentiation of **M. cerebrosus** *from* **Neisseria** *species*: See Table 4.35.

Genus Neisseria

Cocci are 0.6–1.0 μm in diameter, occurring singly but more often in pairs with adjacent sides flattened; one species *(N. elongata)* **is an exception and consists of short rods 0.5 μm wide, often arranged as diplobacilli or in short chains.** Division of the coccal species is in two planes at right angles to each other,

sometimes resulting in tetrads. Capsules and fimbriae (pili) may be present. Endospores are not present. Cells stain Gram negative, but there is a tendency to resist decolorization. Swimming motility does not occur, and flagella are absent. Aerobic. Some species produce a greenish yellow carotenoid pigment. Some species are nutritionally fastidious and hemolytic. Optimum temperature is 35–37°C. **Oxidase positive.** Catalase positive except *N. elongata.* Carbonic anhydrase is produced by all species. All species reduce nitrite except *N. gonorrhoeae* and *N. canis.* Chemoorganotrophic. Some species are saccharolytic. They are the inhabitants of the mucous membranes of mammals. Some species are primary pathogens for humans.

Type species: *Neisseria gonorrhoeae.*

Differentiation of the species of the genus **Neisseria:** See Table 4.35.

Genus Oceanospirillum: Oceanospirillum kriegii *and* Oceanospirillum jannaschii

Editorial note: These species were not included in *Bergey's Manual of Systematic Bacteriology.* The species were created in 1984 by Bowditch et al. (Curr. Microbiol. *10:* 221-230; Int. J. Syst. Bacteriol. *34:* 503, 1984) and differ from all other species of *Oceanospirillum* by consisting of straight rod-shaped cells instead of helical cells (see also Pot et al., Int. J. Syst. Bacteriol. *39:* 23-34, 1989).

Characteristics of **Oceanospirillum kriegii: Straight rods 0.8–1.2 × 2.6–3.6 μm. Motility occurs by a single flagellum at one pole. Accumulate poly-β-hydroxybutyrate as an intracellular reserve product.** Aerobic and possess a strictly respiratory type of metabolism. **Require Na$^+$ or a seawater-based medium to grow.** Growth occurs at 20–35°C but not at 4 or 40°C. Utilize 29–33 organic compounds including D-**glucose, D-fructose, and mannitol** but no other pentose, hexose, or disaccharide. Utilize acetate, succinate, fumarate, DL-malate, pyruvate, α-ketoglutarate, *p*-hydroxybenzoate, L-alanine, D-alanine, L-proline, ethanol, propanol, and quinate as sole sources of carbon and energy. Do not utilize isobutanol, saccharate, glycolate, W-aminovalerate, L-tyrosine, histamine, or sarcosine. **An exocellular lipase is produced.** Nitrate is not reduced to nitrite.

Characteristics of **Oceanospirillum jannaschii:** **Straight rods 1.0–1.4 × 2.4–3.2 μm. Motility occurs by 1–2 flagella at one pole. Accumulate poly-β-hydroxybutyrate as an intracellular reserve product.** Aerobic and possess a strictly respiratory type of metabolism. **Require Na⁺ or a sea water-based medium to grow.** Growth occurs at 20–30°C but not at 4°C or 35°C. Utilize 39–46 organic compounds but **do not utilize carbohydrates.** Utilize acetate, succinate, fumarate, DL-malate, pyruvate, L-alanine, D-alanine, L-proline, W-aminovalerate, histamine, sarcosine, ethanol, propanol, and isobutanol as sole sources of carbon and energy. They do not utilize saccharate, glycolate, quinate, L-tyrosine, and *p*-hydroxybenzoate. **No exocellular lipase is produced.** Nitrate is reduced to nitrate.

Genus **Ochrobactrum**

Editorial note: The genus *Ochrobactrum* was not included in *Bergey's Manual of Systematic Bacteriology.* The genus was created in 1988 by Holmes et al. (Int. J. Syst. Bacteriol. *38:* 406-416). It includes only one species, *O. anthropi.*

Rods with parallel sides and rounded ends, usually occurring singly. Resting stages are not known. Cells stain Gram negative. **Motility occurs by peritrichous flagella.** Obligately aerobic, possessing a strictly respiratory type of metabolism with oxygen as the terminal electron acceptor. Optimum temperature is 20–37°C. Colonies on nutrient agar are nonpigmented. **Oxidase and catalase positive.** Indole negative. Esculin, gelatin, and DNA are not hydrolyzed. Chemoorganotrophic, using a variety of amino acids, organic acids, and carbohydrates as carbon sources. Occur in human clinical specimens. It was previously known as "CDC Group Vd".

Type (and only) species: *Ochrobactrum anthropi.*

Characteristics of the species: Growth occurs on Mac-Conkey agar. Acid produced in ammonium salt medium under aerobic conditions from glucose, arabinose, ethanol, fructose, rhamnose, and xylose. Grows on β-hydroxybutyrate. Negative characteristics are: pigment production; oxidation of gluconate; lysine and ornithine decarboxylase; acid and gas produced in peptone water medium; reduction of selenite; casein digestion; extracellular DNase; production of acid in ammonium salt medium under aerobic conditions from raffinose; arginine desimidase; hydrolysis of Tween 20

and 80; production of brown melanin-like pigment on tyrosine agar; accumulation of lipids after growth on β-hydroxybutyrate; fluorescence on King medium B; growth at 5°C; production of 3-ketolactose and lecithinase; hydrolysis of starch; and production of acid from 10% lactose.

Genus **Oligella**

Editorial note: The genus *Oligella* was not included in *Bergey's Manual of Systematic Bacteriology.* The genus was created in 1987 by Rossau et al. (Int. J. Syst. Bacteriol. *37:* 198-210). It includes two species: *O. urethralis* (formerly *Moraxella urethralis*) and *O. ureolytica* (formerly "CDC Group IVe").

Small rods, mostly not exceeding 1 μm and often occurring in pairs. Cells lack the typical plumpness of moraxellas. Noncapsulated and nonsporeforming. **Mostly nonmotile, but some strains of *O. ureolytica* are peritrichously flagellated. Aerobic.** Moderately fastidious chemoorganotrophs. Grow on nutrient agar, but growth is enhanced by the addition of, e.g., yeast autolysate, serum, or blood. Colonies on blood agar develop rather slowly and are more overtly white than those of all recognized species of *Moraxella.* No pigments and no odor are produced. Nonhemolytic. Biochemically they are rather inert, and only a few organic acids and amino acids are utilized as the sole carbon source. Carbohydrates are neither fermented nor oxidized. **Oxidase positive and usually catalase positive.** Indole and H₂S are not formed. Gelatin is not hydrolyzed. Mainly isolated from the genitourinary tract of humans. Pathogenicity is unknown but probably low.

Type species: *Oligella urethralis.*

Differentiation of the species of the genus **Oligella:** See Table 4.36.

Genus **Paracoccus**

Spherical cells (0.5–0.9 μm in diameter) or short rods (0.9–1.2 μm long). Occur singly, in pairs, or in clusters. **Intracellular granules of poly-β-hydroxybutyrate are present.** No resting stages are known. Cells stain Gram negative. **Nonmotile and aerobic,** having a strictly respiratory type of metabolism; anaerobic growth can occur if nitrate, nitrite or nitrous oxide are available as terminal electron acceptor. **Nitrate is**

reduced to nitrous oxide and molecular nitrogen under anaerobic conditions. One species can grow either **autotrophically with H_2 and CO_2** or heterotrophically with a wide variety of organic compounds as sole carbon sources; this species is not halophilic. **A second species is not capable of autotrophic growth but is halophilic.** Optimum temperature is 25–30°C. **Oxidase and catalase positive.** Occur in soil and presumably in natural and artificial brines.

Type species: *Paracoccus denitrificans.*

Differentiation of the species of the genus **Paracoccus:** See Table 4.37.

Genus **Phenylobacterium**

Editorial note: The genus *Phenylobacterium* was not included in *Bergey's Manual of Systematic Bacteriology.* The genus was created in 1985 by Makkar Lingens et al. (Int. J. Syst. Bacteriol. 35: 26-39). It includes only one species, *P. immobile.*

Rods, coccal rods, or cocci 0.7–1.0 × 1.0–2.0 μm, occurring singly, in pairs, or in short chains. Some strains tend to clump. **In old cultures pleomorphic forms, such as long rods (1.0 × 2.0–4.0 μm), long chains (10–50 μm) connected by filaments, and elliptical forms, may occur. Cells are nonmotile.** They do not produce sheaths or prosthecae. Resting stages are not known. Capsule stain is negative; flexible capsule is present. Cells stain Gram negative. Growth on agar is slow; colonies are small (1-2 mm after 2–3 weeks). Colonies may be smooth, convex, moist with shiny surfaces and entire edges and easily emulsified in saline, or rather rough, dry, and not emulsifiable in saline. **In liquid media there is slight production of greenish yellow nonfluorescent pigment; in L-phenylalanine cultures production of yellowish green fluorescent pigment.** Chemoorganotrophic; metabolism respiratory, never fermentative. Osmotically sensitive. Vitamin B_{12} is is required as a growth factor. **There is high nutritional specialization. Good growth occurs on chloridazon, antipyrin, and L-phenylalanine.** Slow growth occurs on L-glutamate, pyruvate, fumarate, succinate, and malate and on diluted complex media. **Most sugars, alcohols, amino acids, carboxylic acids, and ordinary complex media are not utilized.** Optimum temperature is 29–30°C. No growth occurs at 4 and 37°C. At 37°C cultures die within several days. Optimum pH is 6.8–7.0. Cells are isolated from soil

after enrichment in mineral media containing chloridazon, antipyrin, or pyramidon. **Do not denitrify; do not produce nitrite from nitrate.** NH_4^+ and NO_3^- are used as sole sources of nitrogen. No growth occurs with N_2. **Catalase positive; weakly oxidase positive.** Gelatin, casein, starch, and esculin are not hydrolyzed. Urease negative. Litmus milk is negative. There is weak H_2S production from thiosulfate or cysteine. Methyl red and Voges-Proskauer are negative. Indole negative. No acid and gas are produced from sugars and alcohols. Not pathogenic for rats and rabbits.

Type (and only) species: *Phenylobacterium immobile.*

Characteristics of the species: As described for the genus.

Genus **Phyllobacterium**

Straight rods (in vitro), 0.4–0.8 μm × 0.8–2.0 μm. Gram negative. **Motility occurs by a single polar flagellum or several polar or lateral flagella of long wavelength.** Aerobic, having a strictly respiratory type of metabolism with oxygen as the terminal electron acceptor. Optimum temperature is 28–34°C. Colonies on glucose-yeast extract agar are translucent, colorless, or beige colored, and slimy. **Oxidase positive.** Chemoorganotrophic, using a variety of sugars or salts of organic acids as carbon sources. Do not hydrolyze starch, pectin, or cellulose. **Occur in leaf nodules of higher plants (species of *Myrsinaceae* [myrsine] and *Rubiaceae* [madder]).**

Type species: *Phyllobacterium myrsinacearum.*

Differentiation of the species of the genus **Phyllobacterium:** See Table 4.38.

Genus **Pseudomonas**

Straight or slightly curved rods, but not helical, 0.5–1.0 × 1.5–5.0 μm. Many species accumulate poly-β-hydroxybutyrate as carbon reserve material, which appears as sudanophilic inclusions. They do not produce prosthecae and are not surrounded by sheaths. No resting stages are known. Cells stain Gram negative. **Motility occurs by one or several polar flagella;** they are rarely nonmotile. In some species lateral flagella of shorter wavelength may also be formed. **Aerobic, having a strictly respiratory type of metabolism with oxygen as the terminal electron acceptor; in some**

93

cases nitrate can be used as an alternate electron acceptor, allowing growth to occur anaerobically. Xanthomonadins are not produced. Most, if not all, species fail to grow under acidic conditions (pH 4.5). Most species do not require organic growth factors. Oxidase positive or negative. Catalase positive and chemoorganotrophic; some species are facultative chemolithotrophs, able to use H_2 or CO as energy sources. Widely distributed in nature. Some species are pathogenic for humans, animals, or plants.

Type species: *Pseudomonas aeruginosa.*

Differentiation of the species of the genus **Pseudomonas:** See Tables 4.39 through 4.42.

Genus **Psychrobacter**

Editorial note: The genus *Psychrobacter* was not included in *Bergey's Manual of Systematic Bacteriology.* The genus was created in 1986 by Juni and Heym (Int. J. Syst. Bacteriol. *36:* 388-391). It includes only one species, *P. immobilis.*

Coccobacilli, 0.9–1.3 × 1.5–3.8 μm. The cocci tend to be oval-shaped, whereas the rods can vary in length from extremely short to relatively long. For some strains the rods tend to be somewhat swollen. Cells stain Gram negative. Nonmotile and aerobic and nonpigmented. Nonsporeforming. **Catalase and oxidase positive. Most strains are psychrotrophic; they are able to grow at 5°C and have temperature optima near 20°C, and they are generally unable to grow at 35–37°C. Those strains able to grow well at 35–37°C usually cannot grow at 5°C.** Colonies on heart infusion agar are smooth and opaque. Associated with fish and processed meat and poultry products. Many strains can form acid aerobically from glucose and several other sugars. Strains that grow at 35–37°C have been isolated from pathological specimens derived from humans and animals.

Type (and only) species: *Psychrobacter immobilis.*

Characteristics of the species: Most strains form acid aerobically from glucose and from several other sugars, deaminate phenylalanine and tryptophan, contain urease, reduce nitrate, and hydrolyze Tween 80. Unable to form acid from fructose or sucrose or to hydrolyze starch.

Genus **Rhizobacter**

Editorial note: The genus *Rhizobacter* was not included in *Bergey's Manual of Systematic Bacteriology.* The genus was created in 1988 by Goto and Kuwata (Int. J. Syst. Bacteriol. *38:* 233-239). It includes only one species, *R. daucus.*

Straight to slightly curved, Gram-negative rods. Motility occurs with polar flagella or lateral flagella or both. Poly-β-hydroxybutyrate granules are formed. Chemoorganotrophic. Aerobic, with respiratory metabolism of glucose. Utilize D-glucose as a sole source of carbon and energy. White or yellowish white, plicated, tough or viscid colonies on agar plates. The yellow pigment differs from xanthomonadins. Floccular growth consisting of globular units in liquid media; fingerlike projections never occur. Positive for production of oxidase, catalase, and H_2S from cysteine, nitrate reduction, *o*-nitrophenyl-β-D-galactopyranoside, and growth inhibition by KCN. Negative for denitrification, methyl red, and production of arginine dihydrolase, nitrogenase, fluorescent pigment, indole, and acetoin. Susceptible to 10 μg of vibriostatic agent 0/129 phosphate. Hydrolyzes starch, dextrin, and glycogen. Cannot utilize benzene derivatives as carbon sources. Do not require NaCl and growth factors. Ubiquinone Q8 is present. The type species occurs in soil and is a plant pathogen, causing galls on carrot roots in nature.

Type (and only) species: *Rhizobacter daucus.*

Characteristics of the species: Straight to slightly curved, Gram-negative, encapsulated rods with round ends, 0.9–1.3 × 2.1–2.5 μm. Usually a small number of cells in a population are motile with polar flagella or lateral flagella or both. Pseudooligotrophic. Forms tough, white, plicated colonies on DYPG (potato-peptone-glucose) agar plates, but yellowish white colonies on DYPG (yeast extract-peptone-glucose) agar plates. The species forms copious flocs consisting of globular cell aggregates in liquid media. Positive for urease, lipase (Tween 80 hydrolysis), and lecithinase. Negative for gluconate, casein hydrolysis, nitrogenase, acetoin, and 3-ketolactose. Maximum NaCl tolerance is 0.7%. Utilize as carbon sources D-arabinose, ribose, xylose, lactose, cellobiose, melibiose, sucrose, raffinose, starch, dextrin, glycogen, inulin, adonitol, inositol, mannitol, sorbitol, salicin, galacturonate, lactate, mucate, quinate, tartrate, and methanol. They cannot utilize as carbon sources trehalose, melezitose, dulcitol, erythritol, acetate, citrate, malonate, and benzoate. The tobacco hypersensitivity reaction is negative.

Genus **Rhizobium**

Rods 0.5–0.9 × 1.2–3.0 µm. **Commonly pleomorphic under adverse growth conditions. Usually contain granules of poly-β-hydroxybutyrate** which are refractile by phase-contrast microscopy. Cells stain Gram negative. **Motility occurs by one polar or subpolar flagellum or two to six peritrichous flagella.** Fimbriae have been described on a few strains. Aerobic, possessing a respiratory type of metabolism with oxygen as the terminal electron acceptor. Often able to grow well under oxygen tensions less than 1.0 kPa. Optimum temperature is 25–30°C. Optimum pH is 6–7. Colonies are circular, convex, semitranslucent, raised, and mucilaginous, **usually 2-4 mm in diameter within 3–5 days on yeast-mannitol-mineral salts agar.** Pronounced turbidity develops after 2 or 3 days in agitated broth. Chemoorganotrophic, utilizing a wide range of carbohydrates and salts of organic acids as carbon sources, without gas formation. Cellulose and starch are not utilized. **Produce an acidic reaction in mineral-salts medium containing mannitol or other carbohydrates. Growth on carbohydrate media is usually accompanied by copious extracellular polysaccharide slime.** Ammonium salts, nitrate, nitrite, and most amino acids can serve as nitrogen sources. Some strains will grow in a simple mineral salts medium with vitamin-free casein hydrolysate as the sole source of both carbon and nitrogen. Peptone is poorly utilized. Casein and agar are not hydrolyzed. Some strains require biotin or other water-soluble vitamins. **3-Ketoglycosides are not produced.** The organisms are characteristically **able to invade the root hairs of temperate-zone and some tropical-zone leguminous plants** (family *Leguminosae*) **and incite production of root nodules** wherein the bacteria occur as intracellular symbionts. All strains exhibit host range affinities (host "specificity"). **The bacteria are present in root nodules as pleomorphic forms (bacteroids) which are normally involved in fixing atmospheric nitrogen into a combined form (ammonia) utilizable by the host plant.**

Type species: *Rhizobium leguminosarum.*

Differentiation of the species of the genus **Rhizobium:** See Tables 4.43 and 4.44.

Genus **Rhizomonas**

Editorial note: The genus *Rhizomonas* was not included in *Bergey's Manual of Systematic Bacteriology.* The genus was created in 1990 by van Bruggen et al. (Int. J. Syst. Bacteriol. *40:* 175–188). It includes only one species, *R. suberifaciens.*

Cells straight or slightly curved rods, 0.5 × 1.2 µm. Gram negative. **When motile, have one lateral, subpolar, or polar flagellum.** Resting stages are unknown. Cell division takes place by binary fission. **Cells accumulate PHB granules and are arginine dihydrolase negative.** Colonies are not fluorescent, are white or yellowish, and are smooth or wrinkled. Obligately aerobic and have oxidative metabolism. Optimum growth temperature is ca. 28–33°C; the maximum growth temperature is between 36 and 42°C. Ethanol is not converted to acetic acid. Oxidase and catalase are produced. Denitrification to N_2 gas does not occur. Ubiquinone Q10 is present. Whole-cell fatty acids consist mainly of even-numbered unsaturated (18:1 and 16:1) and saturated (16:0) straight-chain fatty acids, as well as 2-OH-14:0 and 17:1 fatty acids. 18:1 fatty acid represents at least 50% of the total fatty acids. Causes corky root disease on lettuce.

Type (and only) species: *Rhizomonas suberifaciens.*

Characteristics of the species: As described for the genus.

Species **Rochalimaea henselae**

Editorial note: The species *Rochalimaea henselae* was not included in *Bergey's Manual of Systematic Bacteriology.* The species was described in 1992 by Regnery et al. (J. Clin. Microbiol. *30:* 265–274; Int. J. Syst. Bacteriol. *42:* 511, 1992).

Cells are slightly curved rods 0.5–0.6 × 1.0–2.0 µm. Cells stain Gram negative. **Readily stained with Gimenez stain. Flagella are not found. Oxidase, urease, and catalase negative. Optimal growth occurs on enriched media with erythrocytes such as BHIA and TSA supplemented with 5% sheep blood.** Nonreactive in carbohydrate utilization tests. Colonies on initial isolation are durable, invaginated, cauliflower-like and imbedded in the agar. Thought to be a causative agent of bacillary angiomatosis and possibly of cat scratch disease of humans.

Genus **Roseobacter**

Editorial note: The genus *Roseobacter* was not included in *Bergey's Manual of Systematic Bacteriology*. The genus was created in 1991 by Shiba (Syst. Appl. Microbiol. *14:* 140-145; Int. J. Syst. Bacteriol. *41:* 331, 1991). It includes two species, *R. litoralis* and *R. denitrificans.*

Ovoid or rod, 0.6–0.9 × 1.0–2.0 μm. Motile by subpolar flagella. Gram-negative. Cell divides by binary fission. **The bacterium grows heterotrophically under aerobic conditions. Aerobic phototrophic activity is found. Bacteriochlorophyll *a* is present,** and contains phytol as the esterifying alcohol. Bacteriochlorophyll synthesis does not occur under anaerobic conditions. Cell suspension spectra show a large absorption band at 805–7 nm and a smaller band at 868–73 nm in the near infrared region. The major carotenoid is spheroidenone. The major quinone is ubiquinone–10. Menaquinone is not present. The main cellular fatty acid is $C_{18:1}$. Na$^+$ ions, biotin, thiamine, and nicotinic acid are required for growth. The optimum pH for the growth ranges from pH 7.0–8.0, and the optimum temperature ranges from 20–30°C. The bacterium can grow on some organic acids as a sole organic carbon source. Methanol is not utilized. Susceptible to chloramphenicol, penicillin, tetracycline, streptomycin and polymyxin B. Gelatin and Tween 80 are hydrolized. Catalase and oxidase are present. G + C content of DNA ranges from 56–60 mol%.

Type species: *Roseobacter litoralis*

Differentiation of the species of the genus **Roseobacter** *and* **Erythrobacter longus:** See Table 4.45.

Genus **Rugamonas**

Editorial note: The genus *Rugamonas* was not included in *Bergey's Manual of Systematic Bacteriology*. The genus was created in 1986 by Austin and Moss (J. Gen. Microbiol. *132:* 1899-1909; Int. J. Syst. Bacteriol. *37:* 179-180, 1987) It includes only one species, *R. rubra.*

Gram-negative rods with rounded ends. Motility occurs by one or more subpolar or polar flagella. Colonies on agar medium develop a wrinkled elevation and rubbery consistency after incubation for 5 days at 20°C. Flocs are formed in liquid culture after 7 days; cells become pleomorphic after 5 days. Chemoorgano-trophic; metabolism is respiratory, never fermentative. **Catalase and oxidase positive.**

Type (and only) species: *Rugamonas rubra.*

Characteristics of the species: Cells are 0.8–0.9 μm × 2.4–4.0 μm. Intracellular granules of prodigiosin develop in approximately 7 days. Young cells are motile by 1 or more subpolar/polar flagella. Cultures in Bennett's broth form pink/red flocs after 7 days. Colonies on Bennett's agar at 20°C are white, shiny, and circular after 2–3 days, become pink after 4 days, and deep red, wrinkled, and rubbery in consistency after 5 days, with a diameter of about 3 mm. Nitrates are reduced to N_2. Temperature range for growth: 4–30°C, but not 37°C. Optimum temperature is about 25°C. Growth occurs in 0–0.5% NaCl, and at pH 5–9. Arginine decarboxylase negative; lysine and ornithine decarboxylases are produced by some strains. Esculin, chitin, DNA, RNA, gelatin, lecithin, Tween 20, 40, 60, and 80, tyrosine, and urea are degraded but not allantoin, cellulose, elastin, hypoxanthine, starch, and xanthine. β-Galactosidase and phosphatase are positive; H_2S and phenylalanine deaminase are negative. The gluconate and Koser's citrate tests are positive; Voges-Proskauer, methyl red, and malonate tests are negative. The following enzymes are produced: alkaline phosphatase, acid phosphatase, esterase (some strains), esterase-lipase, leucine arylamidase, and phosphoamidase. Utilizes D-alanine, L-arabinose, DL-arginine, *meso*-inositol, sodium DL-lactate, cellobiose, D-fructose, D-galactose, L-histidine, maltose, D-mannose, D-raffinose, D-ribose, DL-serine, sodium acetate, trisodium citrate, D-sorbitol, sucrose, and trehalose as sole carbon sources. Does not utilize adonitol, i-erythritol, sodium malonate, D-melibiose, D-melezitose, D-rhamnose, DL-valine, and D-xylose. Lactose is utilized by some strains. Isolated from river water.

Genus **Serpens**

Rod-shaped cells, 0.3–0.4 × 8–12 μm. Occur singly or in pairs. Cysts or coccoid bodies not formed, but cells in the stationary phase of growth are longer (16–25 μm) and often possess blebs or spherical protuberances. Cells stain Gram negative. **Extremely flexible and capable of serpentine-like motility in agar gels. Cells possess bipolar tufts of 4–10 flagella. Poly-β-hydroxybutyrate or other internal granules are not formed.** Have a strictly respiratory type of metabolism with oxygen as the sole electron acceptor. Grow aerobically but prefer oxygen concentrations less than that of an air atmosphere. **Catalase and oxidase positive.** Chemoorganotrophic. **Lactate is the only effective carbon and energy source,** although very slight growth occurs with acetate or α-ketoglutarate. **Carbohydrates, fatty acids, and sugar alcohols are not catabolized.** Casein hydrolysate, peptone, yeast extract, and, for most strains, ammonium chloride can serve as nitrogen sources; nitrates and nitrites are not used. Vitamins are stimulatory but not required. Optimum temperature is 28–30°C. On media containing 1.8–2.0% agar, colonies are cream colored, round, 3–6 mm in diameter, and have a filamentous edge. On media with less than 1.5% agar, only subsurface spreading colonies occur. Found in the sediments of eutrophic fresh water ponds.

Type species: *Serpens flexibilis.*

Characteristics of the species of the genus **Serpens:** As described for the genus.

Genus **Sinorhizobium**

Editorial note: The genus *Sinorhizobium* was not included in *Bergey's Manual of Systematic Bacteriology.* The genus was created in 1988 by Chen et al. (Int. J. Syst. Bacteriol. *38:* 392-397). It includes two species: *S. fredii* and *S. xinjiangensis.*

Rods 0.5–0.9 × 1.2–3.0 μm. Usually contain granules of poly-β-hydroxybutyrate. Nonsporeforming. Gram negative. **Motility occurs by means of 1 polar flagellum or 1 to more than 3 peritrichous flagella. Aerobic,** possessing a respiratory type of metabolism with oxygen as the terminal electron acceptor. Colonies are circular, convex, semitranslucent, raised, and mucilaginous, usually 2–4 mm in diameter within 3–5 days on YMA medium. Pronounced turbidity develops within 3–5 days in agitated broth. Optimum temperature is 25–30°C, but most strains grow at 35°C and some strains grow at 10°C. Optimum pH is 6–8. Some strains grow at pH 5.0, and other strains grow at pH 10.5. Most strains grow in the presence of 1.0% NaCl, but not 1.5% NaCl, although a few strains grow well on YMA containing 4.5% NaCl. The following substrates are utilized as sole carbon sources: D-arabinose, cellobiose, fructose, D-galactose, glucose, L-glutamine, lactose, D-mannose, mannitol, D-ribose, sodium succinate, xylose, and D-turanose. The following are not utilized:

ammonium tartrate, ammonium oxalate, cellulose, dulcitol, fucose, glycine, sorbose, sodium alginate, and vanillic acid. All strains produce acid on YMA. Ammonium chloride and nitrates rather than amino acids are the preferred sole nitrogen sources, but some strains utilize certain amino acids. Peptone is poorly utilized. All strains require pantothenate and nicotinic acid, but their requirements for other vitamins vary. **Oxidase and catalase positive. Symbiotic nitrogen-fixers.** Do not exhibit a wide host range, but effectively nodulate *Glycine soja, Glycine max* cv. TGm119, *G. max* cv. TGm120, *G. max* cv. TGm344, *G. max* cv. Kai-Yu No. 8, *Vigna unquiculata* cv. VITA-3, and *Cajanus cajan* cv. CITA-1. An ineffective symbiosis is formed on *Vigna radiata, Macroptilium atropurpureum, Macroptilium lathyroides,* and *Sesbania cannabina.*

Type species: *Sinorhizobium fredii.*

Differentiation of the species of the genus **Sinorhizobium:** See Table 4.46.

Genus **Sphingobacterium**

Editorial note: The genus *Sphingobacterium* was not included in *Bergey's Manual of Systematic Bacteriology.* The genus was created in 1983 by Yabuuchi et al. (Int. J. Syst. Bacteriol. *33:* 580-598). It includes three species: *S. spiritivorum* (formerly *Flavobacterium spiritivorum*), *S. multivorum* (formerly *Flavobacterium multivorum*), and *S. mizutae.*

Straight rods. Nonsporeforming. Cells stain Gram negative. **Have no flagella,** but some species may exhibit gliding motility in semisolid media. **Catalase positive.** Chemoorganotrophs without specialized growth factor requirements. **Colonies usually become yellow after several days at room temperature.** Indole and acetylmethylcarbinol are not produced. Nonproteolytic, and gelatin are not hydrolyzed. Acid is produced from carbohydrates oxidatively but not fermentatively. Cellular lipids contain sphingolipids whose ceramide moieties are chiefly branched-chain dihydrosaturated $C_{17:0}$ sphingosin, and the major acid is i-2-OH-$C_{15:0}$.

Type species: *Sphingobacterium spiritivorum.*

Differentiation of the species of the genus **Sphingobacterium:** See Table 4.47.

Genus **Thermoleophilum**

Editorial note: The genus *Thermoleophilum* was not included in *Bergey's Manual of Systematic Bacteriology*. The genus was created in 1984 by Zarilla and Perry (Arch. Microbiol. *137:* 286-290; Int. J. Syst. Bacteriol. *36:* 354-356, 1986). It included only one species, *T. album.* A second species, *T. minutum* was added in 1986 by Zarilla and Perry (Int. J. Syst. Bacteriol. *36:* 13-16).

Short rods, 0.4 × 0.7–1.5 μm. Nonsporeforming. **Aerobic.** Form very small translucent colonies on agar surface. **Grows only at the expense of *n*-alkanes from 13–20 carbons in length in a defined mineral salts medium.** No growth factors are required. **Optimum temperature is 60°C; range is 45–70°C.** Optimum pH is around neutrality; pH range: 5.8–8.0. Can be isolated from mud and water samples, primarily from thermal environments, but also found in nonthermal sources.

Type species: *Thermoleophilum album.*

Differentiation of the species of the genus **Thermoleophilum:** At present there are no routine phenotypic tests available that can differentiate *T. album* from *T. minutum.* The two species can be differentiated by DNA hybridization experiments and by differences in the electrophoretic mobility patterns of proteins released from the cells by sonication.

Genus **Thermomicrobium**

Short, irregularly-shaped rods, 1.3–1.8 × 3.0–6.0 μm. The pleomorphic forms are dumbbell-shaped or appear irregular in diameter and occur singly or in pairs. Neither resting stages (e.g., "rotund bodies") nor endospores are formed. Cells stain Gram negative. **No peptidoglycan diamino acid occurs in the cell walls in significant amounts.** Cells are **nonmotile.** Obligately aerobic. **Optimum temperature for growth is 70–75°C;** maximum, 80°C; minimum, 45°C. Optimal pH is 8.2–8.5, but good growth occurs between 7.5 and 8.7. Chemoorganotrophic, having a strictly respiratory type of metabolism with oxygen as the terminal electron acceptor. **Catalase positive.** Colonies have a rose-pink color. Maximum growth occurs on a medium consisting of yeast extract and peptone (0.5% each). **Growth does not occur on glucose.** *n*-Alkanes are not uti-

lized. **Generation time is 5.5 h.** Isolated from a hot spring in Yellowstone National Park, U.S.A.

Type species: *Thermomicrobium roseum.*

Characteristics of the species: As described for the genus.

Genus **Thermus**

Straight rods, 0.5–0.8 × 5.0–10.0 μm. Filaments from 20 to more than 200 μm may occur under some cultural conditions. Most strains form **rotund bodies**—large spheres 10–20 μm in diameter—derived from the association of individual cells; such bodies are usually seen in old cultures. Cells are **nonmotile;** they do not possess flagella. Endospores are absent. Cells stain Gram negative. Most strains form **yellow, orange, or reddish colonies,** with pigmentation due to carotenoid pigments. **Aerobic,** having a strictly respiratory type of metabolism with oxygen as the terminal electron acceptor. **Oxidase and catalase positive.** Gelatin is usually hydrolyzed. Starch is usually weakly digested. Nitrates are usually reduced to nitrites. **Thermophilic, with an optimum temperature of 70–75°C.** The optimum pH for growth is around neutrality. Found in **hot springs** of neutral to alkaline pH, as well as in **hot water heaters.** Also found in **natural waters subject to thermal pollution.**

Type species: *Thermus aquaticus.*

Differentiation of the species of the genus **Thermus:** See Table 4.48.

Genus **Variovorax**

Editorial note: The genus *Variovorax* was not included in *Bergey's Manual of Systematic Bacteriology*. The genus was created in 1991 by Willems et al. (Int. J. Syst. Bacteriol. *41:* 445–450). It includes only one species, *V. paradoxus* (formerly *Alcaligenes paradoxus*).

Cells are straight to slightly curved rods 0.5–0.6 × 1.2–3.0 μm, **occurring singly or in pairs. Motility occurs by means of degenerate peritrichous flagella.** Cells stain Gram negative. Colonies are yellow because of the presence of carotenoid pigments. Oxidase and catalase positive. Aerobic and chemoorganotrophic. **Some strains are capable of lithoautotrophic growth in which hydrogen is used as an energy source.** Oxidative carbohydrate metabolism occurs with oxygen as the terminal electron acceptor. Good growth is obtained on media containing carbohydrates, organic acids (including amino acids), or peptone.

Type (and only) species: *Variovorax paradoxus.*

Characteristics of the species: As described for the genus. See also Table 4.14.

Genus **Volcaniella**

Editorial note: The genus *Volcaniella* was not included in *Bergey's Manual of Systematic Bacteriology.* The genus was created in 1990 by Quesada et al. (Int. J. Syst. Bacteriol. *40:* 261–267). It includes only one species, *V. eurihalina.*

Cells are capsulated short straight rods 0.8–1.0 × 2.0–2.5 μm, **occurring singly or in pairs. They are nonmotile.** Cells stain Gram negative. Accumulate poly-β-hydroxybutyrate as an intracellular reserve product. Do not form endospores or sheaths. Have a respiratory type of metabolism with oxygen as the terminal acceptor. **Moderate halophiles** capable of growth at salt concentrations between 5 and 20% (wt/vol). Optimal growth occurs at a salt concentration of 7.5% (wt/vol). Chemoorganotrophs. All strains grow well at pH 5–10 and at 15–45°C. Oxidase negative and catalase positive. Isolated from hypersaline habitats (soils, salt ponds) and from sea water.

Type (and only) species: *Volcaniella eurihalina.*

Characteristics of the species: As described for the genus.

Genus **Weeksella**

Editorial note: The genus *Weeksella* was not included in *Bergey's Manual of Systematic Bacteriology.* The genus was created in 1986 by Holmes et al. (Syst. Appl. Microbiol. *8:* 185-190; Int. J. Syst. Bacteriol. *37:* 179-180, 1987). It included one species, *W. virosa* (formerly CDC Group IIF). In 1986 Holmes et al. (Syst. Appl. Microbiol. *8:* 191-196; Int. J. Syst. Bacteriol. *37:* 179, 1987) added a second species, *W. zoohelcum* (formerly CDC Group IIj).

Rods with parallel sides and rounded ends, **typically 0.6 × 2–3 μm. Intracellular granules of poly-β-** hydroxybutyrate are absent. Endospores are not formed. Cells stain Gram negative. **Nonmotile.** Do not glide or spread. Aerobic, having a strictly respiratory type of metabolism. Grow at temperatures from 18–42°C. The growth on solid media is not pigmented. Colonies are circular (diameter 0.5–2 mm), low convex, smooth, and shiny, with entire edges. **Catalase and oxidase positive.** Agar is not digested. Chemoorganotrophic and nonsaccharolytic. **Indolepositive.** Not known from the general environment; apparently parasites, saprophytes, or commensals of the internal surfaces of humans and other warm-blooded animals.

Type species: *Weeksella virosa.*

Differentiation of the species of the genus **Weeksella:** See Table 4.49.

Genus **Xanthobacter**

Rod-shaped cells 0.4–1.0 × 0.8–6.0 μm. **Some species form pleomorphic cells on succinate-containing media or coccoid cells on media containing an alcohol** as the sole carbon source. **Refractile (polyphosphate) and lipid (poly-β-hydroxybutyrate acid) bodies are evenly distributed in the cells.** Resting stages are unknown. **The Gram reaction is positive or variable;** however, the ultrastructure of the cell wall seems to be of the negative Gram type, the peptidoglycan is directly crosslinked by *meso*-diaminopimelic acid, and lipopolysaccharides are present. Cells are **nonmotile or motile (by peritrichous flagella).** Aerobic, having a strictly respiratory type of metabolism with oxygen as the terminal electron acceptor. Optimum temperature is 25–30°C. Optimum pH is 5.8–9.0. Colonies are usually opaque and slimy (although a slime-free species exists) and are **yellow** due to a water-insoluble carotenoid pigment (**zeaxanthin dirhamnoside**). **Catalase positive.** All strains can grow as **chemolithoautotrophs** in mineral media under an atmosphere of H_2, O_2 and CO_2 (7:2:1, v/v) as well as **chemoorganotrophically** on methanol, ethanol, *n*-propanol, *n*-butanol, and various organic acids as sole carbon sources. Carbohydrates utilization is limited, and neither volatile/nonvolatile fatty acids nor gas are produced from some carbohydrates. Some strains require vitamins. **Atmospheric nitrogen is fixed** in nitrogen-deficient media, but by most strains only under a decreased oxygen pressure; the efficiency is more than 10 mg of nitrogen fixed/g of sucrose consumed, and is usually 20 mg/g. The induction of root nodules on plants has so far not been observed. The organisms occur free-living in wet soil containing decaying organic material, and also in water.

Type species: *Xanthobacter autotrophicus.*

Differentiation of the species of the genus **Xanthobacter:** See Table 4.50.

Genus **Xanthomonas**

Straight rods, usually 0.4–0.7 × 0.7–1.8 μm, predominantly single. **Do not produce poly-β-hydroxybutyrate inclusions.** Do not have sheaths or prosthecae. No resting stages are known. Cells stain Gram negative. **Motility occurs by a single polar flagellum** (except *X. maltophilia,* which has multitrichous flagella). **Aerobic,** having a strictly respiratory type of metabolism with oxygen as the terminal electron acceptor. **No denitrification or nitrate reduction occurs** (except *X. maltophilia,* which reduces nitrate to nitrite). Optimum temperature is 25–30°C. Colonies are usually **yellow,** smooth, and butyrous or viscid. **The pigments are highly characteristic brominated aryl polyenes, or "xanthomonadins"** (except for *X. maltophilia,* which does not produce xanthomonadins). **Oxidase negative or weakly positive. Catalase positive.** Chemoorganotrophic, able to use a variety of carbohydrates and salts of organic acids as sole carbon sources. Small amounts of acid are produced from many carbohydrates. Acid is not produced in purple milk or litmus milk. **Grow on calcium lactate but not on glutamine.** Asparagine is not used as a sole source of carbon and nitrogen. Growth is inhibited by 0.1% (and usually by 0.02%) triphenyltetrazolium chloride. **Growth factors usually required include methionine, glutamic acid, nicotinic acid, or a combination of these. Plant pathogens;** occur in association with plants (except for *X. maltophilia,* which is an opportunistic pathogen of humans).

Type species: *Xanthomonas campestris.*

Differentiation of the species of the genus **Xanthomonas:** See Table 4.51.

Genus **Xylella**

Editorial note: The genus *Xylella* was not included in *Bergey's Manual of Systematic Bacteriology.* The genus was created in 1987 by Wells et al. (Int. J. Syst. Bacteriol. *37:* 136–143). It includes only one species, *X. fastidiosa.*

Single straight rods, 0.25–0.35 × 0.9–3.5 μm, with long filamentous strands under some cultural conditions. Colonies are of two types: convex to pulvinate smooth opalescent with entire margins and umbonate rough with finely undulated margins. Cells stain Gram negative. **Nonmotile. Oxidase negative and catalase positive.** Strictly aerobic, nonfermentative, nonhalophilic, nonpigmented. **Nutritionally fastidious,** requiring a specialized medium such as BCYE containing charcoal or glutamine-peptone medium (PW) containing serum albumin. Optimal temperature for growth is 26–28°C. Optimum pH is 6.5–6.9. **Habitat is the xylem of plant tissue.**

Type (and only) species: *Xylella fastidiosa.*

Characteristics of the species. Hydrolyzes gelatin and utilizes hippurate. Most strains produce β-lactamase. Glucose is not fermented. Negative in tests for indole, H_2S, β-galactosidase, lipase, amylase, coagulase, and phosphatase. The species has been isolated as a phytopathogen from tissues of a number of host plants. The type strain was isolated from grapevine with Pierce's disease.

Genus **Xylophilus**

Editorial note: The genus *Xylophilus* was not included in *Bergey's Manual of Systematic Bacteriology.* The genus was created in 1987 by Willems et al. (Int. J. Syst. Bacteriol. *37:* 422–430). It includes only one species, *X. ampelinus* (formerly *Xanthomonas ampelina*).

Straight to slightly curved rods, 0.4–0.8 × 0.6–3.3 μm. Filamentous cells (length, 30 μm or more) may occur in older cultures. Cells occur singly, in pairs, or in short chains. **Motility occurs by a single polar flagellum.** Cells stain Gram negative. **Oxidase negative, catalase positive.** Aerobic, chemoorganotrophic. Oxidative carbohydrate metabolism. Even at the optimal temperature of 24°C, growth is generally very slow and poor. **Growth occurs on L-glutamine but not on**

calcium lactate (in contrast to *Xanthomonas* strains). Plant pathogens, causing bacterial necrosis and canker.

Type (and only) species: *Xylophilus ampelinus.*

Characteristics of the species: On nutrient agar, colonies are circular, semitranslucent, slightly raised, glistening, and pale yellow with entire margins. After 6 days the colonies are 0.2–0.3 mm in diameter; after 15 days the colonies are 0.6–0.8 mm in diameter. Better growth is obtained on GYAC medium (Willems et al., 1987), and the best growth is obtained on a medium containing 1% yeast extract, 2% D-galactose, 2% $CaCO_3$, and 2% agar at 24°C. On this medium, colonies are yellow, and a brown diffusible pigment is produced. Some strains may produce two colony types on GYCA medium. Average colony diameters of 1.4 × 0.6 and 0.6 × 0.3 mm are attained after 15 days at 24°C by the fast- and slow-growing types, respectively. Minimal and maximal growth temperatures are 6 and 30°C, respectively. The maximal NaCl concentration tolerated is 1%. Growth is very slow in the presence of 0.01-0.02% 2,3,5-triphenyltetrazolium chloride. L-Glutamate (0.1%) is required for growth: L-methionine is not required. Acid is produced from L-arabinose and D-galactose but not from D-xylose, D-ribose, L-rhamnose, D-glucose, D-mannose, D-fructose, sucrose, trehalose, cellobiose, lactose, maltose, D-melibiose, raffinose, glycogen, inulin, dextrin, adonitol, dulcitol, D-mannitol, D-sorbitol, L-sorbose, salicin, α-methylglycoside, arbutin, and *meso*-inositol. Grows on acetate (0.2% but not 0.5%), citrate, DL-malate, succinate, DL-tartrate, and fumarate; does not grow on formate, propionate, malonate, maleate, oxalate, benzoate, or calcium gluconate. Grows on Dye asparagine medium. No hydrolysis of gelatin, esculin, starch, casein, arbutin, and sodium hippurate; lipolysis of Tween 80. Potato soft rot test is negative. H_2S is formed from cysteine and weakly from thiosulfate. Urease-positive. The following features are lacking in all strains that have been studied: nitrate reduction, production of indole and ammonia, Voges-Proskauer, arginine dihydrolase, lysine and ornithine decarboxylase, lecithinase, and acid from glucose. Has been isolated from *Vitis vinifera,* on which it causes bacterial necrosis and canker, mainly of the woody parts, in the Mediterranean region and South Africa. Similar symptoms have been reported in Switzerland, Austria, Bulgaria, the Canary Islands, and Argentina.

Genus **Zoogloea**

Straight to slightly curved, plump rods, 1.0–1.3 × 2.1–3.6 μm, with rounded ends; sometimes tapered to a blunt point at one or both poles. Nonsporeforming and noncystforming. Cells in older cultures are demonstrably encapsulated. Cells stain Gram negative. **Actively motile,** especially in young cultures, by means of a **single polar flagellum.** Intracellular granules of poly-β-hydroxybutyrate are formed on media containing the salts of organic acids. Cultures enter into formation of **flocs and films** in liquid media at late growth stages; **the cells become embedded in gelatinous matrices to form zoogloeae, which are distinguished by a "tree-like" or "finger-like" morphology.** Young colonies on solid media under a normal air atmosphere are translucent and punctiform but may increase to 1 or 2 mm in diameter and exhibit opaque centers. Cells are nonpigmented. **Aerobic,** having a strictly respiratory type of metabolism with oxygen as the terminal electron acceptor; **growth can also occur anaerobically in the presence of nitrate** (nitrate respiration). **Denitrification occurs with formation of N_2.** Optimum temperature for growth, 28–37°C. Optimum pH is 7.0–7.5. **Oxidase positive.** Weakly catalase positive. Chemoorganotrophic. **Acid is not formed from carbohydrates** except xylose, glycerol, and ethanol, which are attacked by a few strains. **Proteolytic on gelatin.** Most strains are **urease positive.** Litmus milk is unchanged. H_2S is not usually produced from cysteine. Major carbon sources include salts of several organic acids (e.g., lactate, pyruvate, and fumarate), dicarboxylic amino acids (e.g., aspartate, glutamate, and asparagine), alcohols, and salts of certain aromatic acids (e.g., benzoate and *m*-toluate). **Benzene derivatives are attacked by *meta* cleavage of the ring structure.** Organic nitrogen compounds (e.g., dicarboxylic amino acids) and ammonia serve as nitrogen sources; nitrate is unsuitable. Specific growth factor requirements, if any, are unknown. **Occur free-living in organically polluted fresh water and in wastewater at all stages of treatment.**

Type (and only) species: *Zoogloea ramigera.*

Characteristics of the species: No growth occurs on Koser's citrate. H_2S is not produced from cysteine or on Kligler iron agar. Indole is not produced. Ammonia is produced from asparagine. Lipolytic activity occurs. Growth on benzoate, *m*-toluate, and *o*-cresol occurs with *meta* cleavage of the aromatic ring. Methanol, formate, and formaldehyde are not utilized as sole carbon sources.

SUBGROUP 4B

Differentiation of the genera and species in **Subgroup 4B:** The major differential characteristics are presented in Table 4.52.

Genus **Taylorella**

Editorial note: The genus *Taylorella* was not included in *Bergey's Manual of Systematic Bacteriology.* The genus was created in 1983 by Sugimoto et al. (Curr. Microbiol. *9:* 155-162; Int. J. Syst. Bacteriol. *34:* 503-504, 1984). It included only one species, *T. equigenitalis.* This species was formerly included in *Bergey's Manual of Systematic Bacteriology* as a *species incertae sedis* in the genus *Haemophilus.*

Rods, often approaching a spherical shape, 0.7 × 0.7–1.8 μm, with occasional filaments 5-6 μm long. Cells stain Gram negative. **Nonmotile. Microaerophilic:** very little growth occurs in air or anaerobically. **Optimum growth occurs under 5–10% CO₂.** Metabolism is strictly respiratory with oxygen as the terminal electron acceptor. Optimum temperature is 35-37°C; range, 30–42°C. **Catalase and oxidase positive. Nitrate and nitrite are not reduced.** Chemoorganotrophic. Grows best on chocolate agar, but X- and V-factors do not stimulate growth. No growth occurs on ordinary media, and very poor, if any, growth occurs on blood agar. **Acid is not produced from carbohydrates.** Cells are positive for phosphatase and phosphoamidase. They are negative in tests for indole, H₂S, lysine and ornithine decarboxylase, arginine dihydrolase, urease, gelatinase, lipase, and DNase. **Parasites of horses,** occurring as commensals on the external genitals of healthy stallions but may survive for prolonged periods in the clitoridal region and vulva of mares following infection. **Pathogenic for mares,** causing endometritis and cervicitis as a result of venereal exposure to carrier stallions.

Type (and only) species: *Taylorella equigenitalis.*

Characteristics of the species: As described for the genus. See also Table 4.52.

Genus **Wolinella: Wolinella recta**

Predominantly straight rods 0.5–1.0 × 2–6 μm, with rounded ends. Cells are nonsporing. **Motility occurs by a single polar flagellum.** Although *Wolinella recta* was described as an anaerobe in *Bergey's Manual of Systematic Bacteriology,* Volume 1, it is in fact a **H₂- and formate-requiring microaerophile capable of respiring with oxygen.** In the absence of alternative terminal electron acceptors such as fumarate, it exhibits oxygen-dependent growth with low levels of oxygen and will not grow anaerobically or aerobically (Ohta and Gottschal, FEMS Microbiol. Ecol. *53:* 79-86, 1988; Ohta and Gottschal, FEMS Microbiol. Lett. *50:* 163-168, 1988). **Fumarate and nitrate are used as electron acceptors for anaerobic respiration** (see also Group 6 in this manual). **Oxidase positive and catalase negative.** Contain cytochromes *b* and *c*. **Carbohydrates are not utilized. Isolated from cases of human periodontitis.**

Differentiation of **Wolinella recta** *from other Gram-negative microaerophilic rods:* See Table 4.51.

Genus **Bacteroides: Bacteroides ureolyticus** *and* **Bacteroides gracilis**

Rods 0.4–0.5 × 1.4–6.0 μm. Cells are **nonmotile.** Although the two species were described as anaerobes in *Bergey's Manual of Systematic Bacteriology,* Volume 1, they are in fact **H₂-and formate-requiring microaerophiles capable of respiring with oxygen.** In the absence of alternative terminal electron acceptors such as fumarate, they exhibit oxygen-dependent growth with low levels of oxygen and will not grow anaerobically or aerobically (Han et al., Int. J. Syst. Bacteriol. *41:* 218-222, 1991). **Fumarate and nitrate are used as electron acceptors for anaerobic respiration** (see also Group 6 in this manual). Contain cytochromes *b* and *c*. They grow at 25 and 37°C but not at 42°C. **Catalase negative.** One species (*B. ureolyticus*) is oxidase positive. **Carbohydrates are not utilized.** H₂S is produced in Sulfide-Indole-Motility (SIM) medium and in the water of syneresis at the base of fresh triple-sugar iron agar slants. *B. ureolyticus* is isolated from infections of the respiratory and intestinal tracts, from the buccal cavity, intestinal tract, and urogenital tract, from blood drawn after dental extraction. *B. gracilis* is isolated from the gingival crevice of humans and from visceral or head or neck infections and anaerobic pleuropulmonary infections.

Differentiation of **Bacteroides ureolyticus** *and* **Bacteroides gracilis** *from other Gram-negative microaerophilic rods:* See Table 4.52.

Table 4.1 General characteristics of the genera of Gram-negative, aerobic eubacterial rods and cocci [a]

Characteristic	Aceto-bacter	Acidi-philium	Acido-monas	Acido-ther-mus	Acido-vorax	Acine-tobacter	Afipia
Gram reaction	−	−	−	− or v	−	−	−
Cell shape:							
Rods or coccobacilli, straight or curved	+	+	+	+	+	+	+
Cocci	−	−	−	−	−	−	−
Cell diameter, μm	0.6–0.8	0.3–0.8	0.8–1.0	0.4	0.2–0.7	0.9–1.6	0.2–0.5
Cysts or cyst-like forms occur	−	−	−	−	−	−	−
Cells arranged in square tablets of 16–64 cells	−	−	−	−	−	−	−
Motility in liquid media	+ or −	+ or −		−	+	−	+
Flagellar arrangement:							
Polar or subpolar	−	+ or −			+		+
Lateral	+	+ or −			−		+
Poly-β-hydroxybutyrate accumulated	−		−		+	−	
Cells embedded into gelatinous matrices to form tree-like or finger-like zoogloeas	−	−	−	−	−	−	−
Grow at 60°C or higher [c]	−	−	−	+	−	−	−
Grow at 5°C but not 37°C					−		
Fluorescent pigment	−		−				
Yellow colonies	−	−	+ or −	−	−	−	−
Red or orange colonies	−	+ or −	−	−	−	−	−
Violet colonies	−	−	−	−	−	−	−
Oxidase	−	− or W	+		+	−	+
Acid from glucose	+		+			+ or −	−
Indole from tryptophan	−		−				−
Oxidize ethanol to acetic acid at pH 4.5 [e]	+	−	+	−	−		
Fix N_2 in vitro aerobically [f]	−		−	−	−	−	−
Fix N_2 in vitro under microaerobic conditions [g]	+ or −	−	−	−	−	−	−
Fix N_2 in root nodules of plants [h]	−	−	−	−	−	−	−
Denitrification (to N_2)	−		−		D	−	−
Grow in lean organic media between pH 2.0 and 5.9 but not 6.1 or higher	−	+		−		−	
Marine or brine bacteria that require at least 0.1% NaCl or more [i]	−	−	−		−	−	−
Grow with 20% or more NaCl	−	−	−		−	−	−
Growth greatly stimulated by or dependent upon cysteine							−
Can use 1-C compounds as sole carbon sources [j]	+ or −	−	+	−		−	
Grow on L-glutamine but not on calcium lactate							
Presence of carbonic anhydrase							
Facultative H_2 autotrophs [k]	−	−	−	−	D	−	
Grow well on chloridazon, antipyrin, and phenylalanine							
Grow on n-alkanes having 13–20 carbon atoms							
Genus includes pathogens, frank or opportunistic:							
Humans and/or animals	−	−	−	−	−	+	+
Plants	+	−	−	−	−	−	−
Cause transformation of plant cells into autonomously proliferating tumor cells	−						−
Cause nodules on leaves of myrsine and madder plants	−	−	−	−	−	−	−
Cause lysis in soil of various Gram-positive and Gram-negative bacteria	−	−	−	−	−	−	−
Cellular lipids contain sphingolipids							
Contain bacteriochlorophyll a	−	−	−	−	−	−	−

Footnotes are at end of table

Table 4.1 (continued)

Characteristic	Agro-bacte-rium	Agro-monas	Alca-ligenes	Altero-monas	Amino-bacter	Aqua-spi-rillum[b]	Azo-monas
Gram reaction	−	−	−	−	−	−	− or v
Cell shape:							
Rods or coccobacilli, straight or curved	+	+	+	+	+	+	Ovoid
Cocci	−	−	−	−	−	−	−
Cell diameter, µm	0.6–1.0	0.6–1.0	0.5–1.0	0.7–1.5	0.5–0.9	0.7–0.9	≥2.0
Cysts or cyst-like forms occur	−	−	−	−	−	−	−
Cells arranged in square tablets of 16–64 cells	−	−	−	−	−	−	−
Motility in liquid media	+	+	+	+	+	+[b]	+ or −
Flagellar arrangement:							
Polar or subpolar	−	+	−	+	+	+	−
Lateral	+	−	+	−	−	−	+
Poly-β-hydroxybutyrate accumulated			−	−	+	+	+
Cells embedded into gelatinous matrices to form tree-like or finger-like zoogloeas	−	−	−	−	−	−	−
Grow at 60°C or higher[c]	−	−	−	−	−	−	−
Grow at 5°C but not 37°C					−		
Fluorescent pigment	−	−	−	−	−	W	+ or −
Yellow colonies	−	−	+ or −	+ or −	d	−	−
Red or orange colonies	−	−	−	+ or −	−	−	−
Violet colonies	−	−	−	+ or −	−	−	−
Oxidase	+ or −	+	+	+	+	+	+ or −
Acid from glucose	+	−			+	−	
Indole from tryptophan			−		−	−	
Oxidize ethanol to acetic acid at pH 4.5[e]	−	−	−	−	−	−	−
Fix N₂ in vitro aerobically[f]	−	−	−	−	−	−	+
Fix N₂ in vitro under microaerobic conditions[g]	−	+	−	−	−	+	
Fix N₂ in root nodules of plants[h]	−	−	−	−	−	−	−
Denitrification (to N₂)	+ or −		+ or −	+ or −	−	−	−
Grow in lean organic media between pH 2.0 and 5.9 but not 6.1 or higher	−	−	−	−	−	−	−
Marine or brine bacteria that require at least 0.1% NaCl or more[i]	−	−	−	+	−	−	−
Grow with 20% or more NaCl	−	−	−	−	−	−	−
Growth greatly stimulated by or dependent upon cysteine					−		
Can use 1-C compounds as sole carbon sources[j]	−	−	−	−	−	−	−
Grow on L-glutamine but not on calcium lactate							
Presence of carbonic anhydrase							
Facultative H₂ autotrophs[k]	−	−	+ or −	−	−	−	−
Grow well on chloridazon, antipyrin, and phenylalanine							
Grow on n-alkanes having 13–20 carbon atoms							
Genus includes pathogens, frank or opportunistic:							
Humans and/or animals	−	−	+	−	−	−	−
Plants	+	−	−	−	−	−	−
Cause transformation of plant cells into autonomously proliferating tumor cells	+ or −	−	−	−	−	−	−
Cause nodules on leaves of myrsine and madder plants	−	−	−	−	−	−	−
Cause lysis in soil of various Gram-positive and Gram-negative bacteria	−	−	−	−	−	−	−
Cellular lipids contain sphingolipids						−	
Contain bacteriochlorophyll a	−	−	−	−	−	−	−

Footnotes are at end of table

Table 4.1 *(continued)*

Characteristic	Azo-rhizo-bium	Azoto-bacter	Beije-rinckia	Borde-tella	Brady-rhizo-bium	Brucella
Gram reaction	−	−	−	−	−	−
Cell shape:						
Rods or coccobacilli, straight or curved	+	Ovoid	+	+	+	+
Cocci	−	−	−	−	−	−
Cell diameter, μm	0.5–0.6	≥1.5	0.5–1.5	0.2–0.5	0.5–0.9	0.5–0.7
Cysts or cyst-like forms occur	−	+	+	−	−	−
Cells arranged in square tablets of 16–64 cells	−	−	−	−	−	−
Motility in liquid media	+	+ or −	+ or −	+ or −	+	−
Flagellar arrangement:						
Polar or subpolar	−	−	−	−	+	
Lateral	+	+	+	+	−	
Poly-β-hydroxybutyrate accumulated		+	+	−	+	−
Cells embedded into gelatinous matrices to form tree-like or finger-like zoogloeas	−	−		−	−	−
Grow at 60°C or higher [c]	−	−	−	−	−	−
Grow at 5°C but not 37°C						
Fluorescent pigment		+ or −	+ or −	−	−	−
Yellow colonies		+ or −	−	−	−	−
Red or orange colonies		−		−	+ or −	−
Violet colonies		−	−	−	−	−
Oxidase	+	+ or −		+ or −		+ or −
Acid from glucose				−		−
Indole from tryptophan				−		
Oxidize ethanol to acetic acid at pH 4.5 [e]	−	−	−	−	−	−
Fix N$_2$ in vitro aerobically [f]	−	+	+	−	−	−
Fix N$_2$ in vitro under microaerobic conditions [g]	+			−	+ or −	−
Fix N$_2$ in root nodules of plants [h]	+	−	−	−	−	−
Denitrification (to N$_2$)	−	−	−	−	+ or −	
Grow in lean organic media between pH 2.0 and 5.9 but not 6.1 or higher	−	−	−	−	−	−
Marine or brine bacteria that require at least 0.1% NaCl or more [i]	−	−	−	−	−	−
Grow with 20% or more NaCl	−	−	−	−	−	−
Growth greatly stimulated by or dependent upon cysteine						−
Can use 1-C compounds as sole carbon sources [j]	−	−	−	−	−	−
Grow on L-glutamine but not on calcium lactate						
Presence of carbonic anhydrase						
Facultative H$_2$ autotrophs [k]	−	−	−	−	+ or −	−
Grow well on chloridazon, antipyrin, and phenylalanine						
Grow on n-alkanes having 13–20 carbon atoms						
Genus includes pathogens, frank or opportunistic:						
Humans and/or animals	−	−	−	+	−	+
Plants	−	−	−	−	−	−
Cause transformation of plant cells into autonomously proliferating tumor cells	−	−	−	−	−	−
Cause nodules on leaves of myrsine and madder plants	−	−	−	−	−	−
Cause lysis in soil of various Gram-positive and Gram-negative bacteria	−	−	−	−	−	−
Cellular lipids contain sphingolipids						
Contain bacteriochlorophyll a	−	−	−	−	−	−

Footnotes are at end of table

Table 4.1 *(continued)*

Characteristic	Chromohalo-bacter	Chry-seo-monas	Coma-monas	Cupri-avidus	Deleya	Derxia
Gram reaction	−	−	−	−	−	−
Cell shape:						
Rods or coccobacilli, straight or curved	+	+	+	+	+	+
Cocci	−	−	−	−	−	−
Cell diameter, μm	0.6–1.0		0.5–1.0	0.7–0.9	0.8–1.1	1.0–2.0
Cysts or cyst-like forms occur	−	−	−	−	−	−
Cells arranged in square tablets of 16–64 cells	−	−	−	−	−	−
Motility in liquid media	+	+	+	+	+	+
Flagellar arrangement:						
Polar or subpolar	−	+	+	−	+ or −	+
Lateral	(peritrichous)	−	−	+	+ or −	−
Poly-β-hydroxybutyrate accumulated		−	+		+	+
Cells embedded into gelatinous matrices to form tree-like or finger-like zoogloeas	−					−
Grow at 60°C or higher [c]	−	−	−	−	−	−
Grow at 5°C but not 37°C	−					
Fluorescent pigment	−	−	−	−	−	−
Yellow colonies	−	+	−	−	−	−
Red or orange colonies	−	−	−	−	−	−
Violet colonies	+[d]	−	−	−	−	−
Oxidase	−	−	+	+	+ or −	
Acid from glucose	+	+	−	−		−
Indole from tryptophan	−	−	−	−	−	−
Oxidize ethanol to acetic acid at pH 4.5 [e]	−	−	−	−	−	−
Fix N$_2$ in vitro aerobically [f]	−	−	−	−	−	+
Fix N$_2$ in vitro under microaerobic conditions [g]	−		−	−	−	
Fix N$_2$ in root nodules of plants [h]	−	−	−	−	−	−
Denitrification (to N$_2$)	−		−			−
Grow in lean organic media between pH 2.0 and 5.9 but not 6.1 or higher	−		−		−	−
Marine or brine bacteria that require at least 0.1% NaCl or more [i]	+	−		−	+	
Grow with 20% or more NaCl	+	−		−	+ or −	−
Growth greatly stimulated by or dependent upon cysteine	−					
Can use 1-C compounds as sole carbon sources [j]	−		−		−	+
Grow on L-glutamine but not on calcium lactate						
Presence of carbonic anhydrase						
Facultative H$_2$ autotrophs [k]	−	−	−	−	−	+
Grow well on chloridazon, antipyrin, and phenylalanine						
Grow on n-alkanes having 13–20 carbon atoms						
Genus includes pathogens, frank or opportunistic:						
Humans and/or animals	−	+	−	−	−	−
Plants	−	−			−	−
Cause transformation of plant cells into autonomously proliferating tumor cells						
Cause nodules on leaves of myrsine and madder plants	−	−	−	−	−	−
Cause lysis in soil of various Gram-positive and Gram-negative bacteria	−	−	−	+	−	−
Cellular lipids contain sphingolipids	−					
Contain bacteriochlorophyll a	−	−	−	−	−	−

Footnotes are at end of table

Table 4.1 *(continued)*

Characteristic	Ensifer	Erythrobacter	Flavimonas	Flavobacterium	Francisella	Frateuria	Gluconobacter
Gram reaction	–	–	–	–	–	–	–
Cell shape:							
Rods or coccobacilli, straight or curved	+	+	+	+	+	+	+
Cocci	–	–	–	–	–	–	–
Cell diameter, µm	0.7–1.1	0.5–0.7		0.5	0.2–1.7	0.5–0.7	0.5–0.9
Cysts or cyst-like forms occur	–	–	–	–	–	–	
Cells arranged in square tablets of 16–64 cells	–	–	–	–	–	–	
Motility in liquid media	+	+	+ or –	–	–	+ or –	+ or –
Flagellar arrangement:							
Polar or subpolar	+	+	+			+	+
Lateral	–	–	–			–	–
Poly-β-hydroxybutyrate accumulated			–	–	–	–	–
Cells embedded into gelatinous matrices to form tree-like or finger-like zoogloeas	–	–	–	–	–	–	–
Grow at 60°C or higher [c]	–	–	–	–	–	–	–
Grow at 5°C but not 37°C							
Fluorescent pigment	–		–	–	–	–	–
Yellow colonies	–	–	+	+	–	+	–
Red or orange colonies	–	+	–	–	–	–	–
Violet colonies	–	–	–	–	–	–	–
Oxidase		+	–	+	–	–	–
Acid from glucose	trace		+	+ or –	+	+	+
Indole from tryptophan		+	–	+ or –	–	–	–
Oxidize ethanol to acetic acid at pH 4.5 [e]	–	–	–	–	–	+	+
Fix N_2 in vitro aerobically [f]	–	–	–	–	–	–	–
Fix N_2 in vitro under microaerobic conditions [g]	–	–	–	–	–	–	–
Fix N_2 in root nodules of plants [h]	–	–	–	–	–	–	–
Denitrification (to N_2)		–	–	–			
Grow in lean organic media between pH 2.0 and 5.9 but not 6.1 or higher	–	–	–	–	–	–	–
Marine or brine bacteria that require at least 0.1% NaCl or more [i]	–	+	–	–	–	–	–
Grow with 20% or more NaCl	–	–	–	–	–	–	–
Growth greatly stimulated by or dependent upon cysteine	–				+ or –		
Can use 1-C compounds as sole carbon sources [j]	–	–	–	–	–	–	–
Grow on L-glutamine but not on calcium lactate							
Presence of carbonic anhydrase							
Facultative H_2 autotrophs [k]					–	–	–
Grow well on chloridazon, antipyrin, and phenylalanine							
Grow on n-alkanes having 13–20 carbon atoms							
Genus includes pathogens, frank or opportunistic:							
Humans and/or animals	–	–	+	+ or –	+	–	–
Plants	–	–	–	–	–	+	+ or –
Cause transformation of plant cells into autonomously proliferating tumor cells	–	–	–	–	–	–	–
Cause nodules on leaves of myrsine and madder plants	–	–	–	–	–	–	–
Cause lysis in soil of various Gram-positive and Gram-negative bacteria	+	–	–	–	–	–	–
Cellular lipids contain sphingolipids					–		
Contain bacteriochlorophyll *a*	–	+	–	–	–	–	–

Footnotes are at end of table

Table 4.1 *(continued)*

Characteristic	Halomonas	Hydrogenophaga	Janthinobacterium	Kingella	Lampropedia	Legionella	Marinobacter
Gram reaction	–	–	–	–	–	–	–
Cell shape:							
Rods or coccobacilli, straight or curved	+	+	+	+	–	+	+
Cocci	–	–	–	–	cubical	–	–
Cell diameter, μm	0.6–0.8	0.3–0.6	0.8–1.2	~1.0		0.3–0.9	0.3–0.6
Cysts or cyst-like forms occur	–	–	–	–	–	–	–
Cells arranged in square tablets of 16–64 cells	–	–	–	–	+	–	–
Motility in liquid media	+	+	+	–	–	+	+
Flagellar arrangement:							
Polar or subpolar	+ or –	+	+			+ or –	+
Lateral	+ or –	–	+			+ or –	–
Poly-β-hydroxybutyrate accumulated		+		–	+	–	–
Cells embedded into gelatinous matrices to form tree-like or finger-like zoogloeas	–	–	–	–	–	–	–
Grow at 60°C or higher [c]	–	–	–	–	–	–	–
Grow at 5°C but not 37°C							–
Fluorescent pigment	–	–	–	–	–	–	–
Yellow colonies	+ or –	+	–	–	–	–	–
Red or orange colonies	–	–	–	–	–	–	–
Violet colonies	–	–	+	–	–	–	–
Oxidase	+	+	+	+	+	– or W	+
Acid from glucose	+	+ or –	+	+	–	–	
Indole from tryptophan	+ or –	–	–	+ or –	–		
Oxidize ethanol to acetic acid at pH 4.5 [e]	–	–	–	–	–	–	–
Fix N_2 in vitro aerobically [f]	–	–	–	–	–	–	–
Fix N_2 in vitro under microaerobic conditions [g]	–	+ or –	–	–	–	–	–
Fix N_2 in root nodules of plants [h]	–	–	–	–	–	–	–
Denitrification (to N_2)		+ or –	+ or –	+ or –		–	+
Grow in lean organic media between pH 2.0 and 5.9 but not 6.1 or higher	–	–	–	–	–	–	–
Marine or brine bacteria that require at least 0.1% NaCl or more [i]	+ or –	–	–	–	–	–	+
Grow with 20% or more NaCl	+	–	–	–	–	–	+
Growth greatly stimulated by or dependent upon cysteine						+	
Can use 1-C compounds as sole carbon sources [j]	–	–	–	–	–	–	–
Grow on L-glutamine but not on calcium lactate							
Presence of carbonic anhydrase							
Facultative H_2 autotrophs [k]	–	+	–	–	–	–	
Grow well on chloridazon, antipyrin, and phenylalanine							
Grow on n-alkanes having 13–20 carbon atoms							
Genus includes pathogens, frank or opportunistic:							
Humans and/or animals	–	–	–	–	–	+	–
Plants	–	–	–	–	–	–	–
Cause transformation of plant cells into autonomously proliferating tumor cells	–	–	–	–	–	–	–
Cause nodules on leaves of myrsine and madder plants	–	–	–	–	–	–	–
Cause lysis in soil of various Gram-positive and Gram-negative bacteria	–	–	–	–	–	–	–
Cellular lipids contain sphingolipids							–
Contain bacteriochlorophyll a	–	–	–	–	–	–	–

Footnotes are at end of table

Table 4.1 *(continued)*

Characteristic	Marino-monas	Meso-philo-bacter	Methyl-obacillus	Methyl-obacterium	Methyl-ococcus	Methyl-omonas
Gram reaction	−	−	−	− or v	−	−
Cell shape:						
Rods or coccobacilli, straight or curved	+	+	+	+	−	+
Cocci	−	−	−	−	+	−
Cell diameter, μm	0.7–1.5	0.5–0.6	0.3–0.5	0.8–1.0	1.0	0.5–1.0
Cysts or cyst-like forms occur	−	−			+	+
Cells arranged in square tablets of 16–64 cells	−	−	−	−	−	−
Motility in liquid media	+	−	+ or −	+	−	+
Flagellar arrangement:						
Polar or subpolar	+		+	+ or −		+
Lateral	−		−	+ or −		−
Poly-β-hydroxybutyrate accumulated	−	−	−	+	−	−
Cells embedded into gelatinous matrices to form tree-like or finger-like zoogloeas	−	−	−	−	−	−
Grow at 60°C or higher [c]	−	−	−	−	−	−
Grow at 5°C but not 37°C						
Fluorescent pigment		−				
Yellow colonies	−	−	+ or −	−	−	−
Red or orange colonies	−	−	−	+	−	−
Violet colonies	−	−	−	−	−	−
Oxidase	+ or −	+	−	+ or W		+
Acid from glucose		−	−		−	
Indole from tryptophan		+ or −	−			
Oxidize ethanol to acetic acid at pH 4.5 [e]	−	−		−		
Fix N$_2$ in vitro aerobically [f]	−	−				
Fix N$_2$ in vitro under microaerobic conditions [g]	−	−				
Fix N$_2$ in root nodules of plants [h]	−	−	−	−	−	−
Denitrification (to N$_2$)	−	−				
Grow in lean organic media between pH 2.0 and 5.9 but not 6.1 or higher	−	−				−
Marine or brine bacteria that require at least 0.1% NaCl or more [i]	+	+	−	−	−	−
Grow with 20% or more NaCl	−		−	−	−	−
Growth greatly stimulated by or dependent upon cysteine						
Can use 1-C compounds as sole carbon sources [j]	−	−	+	+	+	+
Grow on L-glutamine but not on calcium lactate						
Presence of carbonic anhydrase						
Facultative H$_2$ autotrophs [k]	−	−		−	−	−
Grow well on chloridazon, antipyrin, and phenylalanine						
Grow on *n*-alkanes having 13–20 carbon atoms						
Genus includes pathogens, frank or opportunistic:						
Humans and/or animals	−	−	−	−	−	−
Plants	−	−	−	−	−	−
Cause transformation of plant cells into autonomously proliferating tumor cells	−	−				
Cause nodules on leaves of myrsine and madder plants	−	−	−	−	−	−
Cause lysis in soil of various Gram-positive and Gram-negative bacteria	−	−	−	−	−	−
Cellular lipids contain sphingolipids						
Contain bacteriochlorophyll *a*	−	−	−	−	−	−

Footnotes are at end of table

Table 4.1   (continued)

Characteristic	Methyl-ophaga	Methyl-ophilus	Methyl-ovorus	Moraxella subgenus Branha-mella	Moraxella subgenus Moraxella
Gram reaction	−	−	−	−	−
Cell shape:					
Rods or coccobacilli, straight or curved	+	+	+	−	+
Cocci	−	−	−	+	−
Cell diameter, µm	0.2	0.3–0.6	0.5–0.6	0.6–1.0	1.0–1.5
Cysts or cyst-like forms occur	−		−	−	−
Cells arranged in square tablets of 16–64 cells	−	−	−	−	−
Motility in liquid media	+	+ or −	+	−	−
Flagellar arrangement:					
Polar or subpolar	+	+	+		
Lateral	−				
Poly-β-hydroxybutyrate accumulated		−			
Cells embedded into gelatinous matrices to form tree-like or finger-like zoogloeas	−	−	−	−	−
Grow at 60°C or higher [c]	−	−	−	−	−
Grow at 5°C but not 37°C			−		
Fluorescent pigment			−	−	−
Yellow colonies	−	−	−		−
Red or orange colonies	−	−	−	−	−
Violet colonies	−	−	−	−	−
Oxidase	+	+	+	+	+
Acid from glucose	−	− or W		−	−
Indole from tryptophan			−	−	−
Oxidize ethanol to acetic acid at pH 4.5 [e]	−	−	−	−	−
Fix N_2 in vitro aerobically [f]			−	−	−
Fix N_2 in vitro under microaerobic conditions [g]			−		
Fix N_2 in root nodules of plants [h]	−	−	−	−	−
Denitrification (to N_2)					
Grow in lean organic media between pH 2.0 and 5.9 but not 6.1 or higher	−	−	−	−	−
Marine or brine bacteria that require at least 0.1% NaCl or more [i]	+	−	−	−	−
Grow with 20% or more NaCl	−	−	−	−	−
Growth greatly stimulated by or dependent upon cysteine			−		
Can use 1-C compounds as sole carbon sources [j]	+	+	+	−	−
Grow on L-glutamine but not on calcium lactate					
Presence of carbonic anhydrase				−	−
Facultative H_2 autotrophs [k]	−	−	−	−	−
Grow well on chloridazon, antipyrin, and phenylalanine				−	−
Grow on n-alkanes having 13–20 carbon atoms					
Genus includes pathogens, frank or opportunistic:					
Humans and/or animals	−	−	−	+	+
Plants	−	−	−	−	−
Cause transformation of plant cells into autonomously proliferating tumor cells					
Cause nodules on leaves of myrsine and madder plants	−	−	−	−	−
Cause lysis in soil of various Gram-positive and Gram-negative bacteria	−	−	−	−	−
Cellular lipids contain sphingolipids			−		
Contain bacteriochlorophyll a	−	−	−	−	−

Footnotes are at end of table

Table 4.1 *(continued)*

Characteristic	Moro-coccus	Neis-seria	Oceano-spir-rillum[b]	Ochro-bactrum	Oligella	Para-coccus
Gram reaction	−	−	−	−	−	−
Cell shape:						
Rods or coccobacilli, straight or curved	−	+ or −	+	+	+	+ or −
Cocci	+	+ or −	−	−	−	+ or −
Cell diameter, μm	<1.0	0.6–1.0	0.8–1.4		~0.6	0.5–1.2
Cysts or cyst-like forms occur	−	−	−	−	−	−
Cells arranged in square tablets of 16–64 cells	−	−	−	−	−	−
Motility in liquid media	−	−	+	+	+ or −	−
Flagellar arrangement:						
Polar or subpolar			+	−	−	
Lateral			−	+	+	
Poly-β-hydroxybutyrate accumulated	−		+		+	+
Cells embedded into gelatinous matrices to form tree-like or finger-like zoogloeas	−	−	−	−	−	−
Grow at 60°C or higher [c]	−	−	−	−	−	−
Grow at 5°C but not 37°C	−					
Fluorescent pigment	−	−		−	−	−
Yellow colonies	−	+ or −	−	−	−	−
Red or orange colonies	−	−	−	−	−	−
Violet colonies	−	−	−	−	−	−
Oxidase	+	+	+	+	+	+
Acid from glucose	+	+ or −		+	−	−
Indole from tryptophan	−			−	−	
Oxidize ethanol to acetic acid at pH 4.5 [e]	−	−	−	−	−	−
Fix N_2 in vitro aerobically [f]	−	−	−	−	−	−
Fix N_2 in vitro under microaerobic conditions [g]	−	−	−	−	−	−
Fix N_2 in root nodules of plants [h]	−	−	−	−	−	−
Denitrification (to N_2)	v					+ or −
Grow in lean organic media between pH 2.0 and 5.9 but not 6.1 or higher	−	−				−
Marine or brine bacteria that require at least 0.1% NaCl or more [i]	−	−	+	−	−	+ or −
Grow with 20% or more NaCl	−	−	−	−	−	+ or −
Growth greatly stimulated by or dependent upon cysteine	−					
Can use 1-C compounds as sole carbon sources [j]	−	−	−	−	−	+ or −
Grow on L-glutamine but not on calcium lactate						
Presence of carbonic anhydrase		+				
Facultative H_2 autotrophs [k]	−	−	−	−	−	+ or −
Grow well on chloridazon, antipyrin, and phenylalanine	−	−	−	−	−	−
Grow on n-alkanes having 13–20 carbon atoms						
Genus includes pathogens, frank or opportunistic:						
Humans and/or animals	+	+	−			−
Plants	−	−	−	−	−	−
Cause transformation of plant cells into autonomously proliferating tumor cells	−	−	−	−	−	−
Cause nodules on leaves of myrsine and madder plants	−	−	−	−	−	−
Cause lysis in soil of various Gram-positive and Gram-negative bacteria	−	−	−	−	−	−
Cellular lipids contain sphingolipids						
Contain bacteriochlorophyll a	−	−	−	−	−	−

Footnotes are at end of table

Table 4.1   *(continued)*

Characteristic	Phenylobacterium	Phyllobacterium	Pseudomonas	Psychrobacter	Rhizobacter	Rhizobium
Gram reaction	–	–	–	–	–	–
Cell shape:						
Rods or coccobacilli, straight or curved	+ or –	+	+	+	+	+
Cocci	+ or –	–	–	–	–	–
Cell diameter, μm	0.7–1.0	0.4–0.8	0.5–1.0	0.9–1.3	0.9–1.3	0.5–0.9
Cysts or cyst-like forms occur	–	–	–	–	–	–
Cells arranged in square tablets of 16–64 cells	–	–	–	–	–	–
Motility in liquid media	–	+	+ or –	–	+	+
Flagellar arrangement:						
Polar or subpolar		+ or –	+		+ or –	+ or –
Lateral		+ or –	–		+ or –	+ or –
Poly-β-hydroxybutyrate accumulated			+ or –		+	+
Cells embedded into gelatinous matrices to form tree-like or finger-like zoogloeas	–	–	–	–	–	–
Grow at 60°C or higher [c]	–	–	–	–	–	–
Grow at 5°C but not 37°C				+ or –		
Fluorescent pigment		–	+ or –	–	–	–
Yellow colonies	+	–	+ or –	–	+	–
Red or orange colonies	–	–	+ or –	–	–	–
Violet colonies	–	–	–	–	–	–
Oxidase	W	+	+ or –	+	+	
Acid from glucose	–	+	+ or –	+ or –	+	+
Indole from tryptophan	–					–
Oxidize ethanol to acetic acid at pH 4.5 [e]	–	–	–	–	–	
Fix N$_2$ in vitro aerobically [f]	–	–	–	–	–	
Fix N$_2$ in vitro under microaerobic conditions [g]	–		–	–	–	+ or –
Fix N$_2$ in root nodules of plants [h]	–	–	–	–	–	+
Denitrification (to N$_2$)			+ or –		–	–
Grow in lean organic media between pH 2.0 and 5.9 but not 6.1 or higher	–	–	–	–	–	–
Marine or brine bacteria that require at least 0.1% NaCl or more [i]	–	–	+ or –	–	–	–
Grow with 20% or more NaCl	–	–	–	–	–	–
Growth greatly stimulated by or dependent upon cysteine			+ or –		–	
Can use 1-C compounds as sole carbon sources [j]	–	–	–	–	+	–
Grow on L-glutamine but not on calcium lactate						
Presence of carbonic anhydrase						
Facultative H$_2$ autotrophs [k]	–	–	+ or –	–	–	–
Grow well on chloridazon, antipyrin, and phenylalanine	+	–	–	–	–	–
Grow on *n*-alkanes having 13–20 carbon atoms						
Genus includes pathogens, frank or opportunistic:						
Humans and/or animals	–	–	+ or –	–	–	–
Plants	–	+	+ or –	–	+	–
Cause transformation of plant cells into autonomously proliferating tumor cells	–	–	–	–	–	–
Cause nodules on leaves of myrsine and madder plants	–	+	–	–	–	–
Cause lysis in soil of various Gram-positive and Gram-negative bacteria	–	–	–	–	–	–
Cellular lipids contain sphingolipids						
Contain bacteriochlorophyll *a*	–	–	–	–	–	–

Footnotes are at end of table

Table 4.1 *(continued)*

Characteristic	Rhizo-monas	Roseo-bacter	Ruga-monas	Serpens	Sino-rhizo-bium	Sphin-gobac-terium
Gram reaction	–	–	–	–	–	–
Cell shape:						
Rods or coccobacilli, straight or curved	+	+	+	+	+	+
Cocci	–	–	–	–	–	–
Cell diameter, μm	0.4–0.5	0.6–0.9	0.8–0.9	0.3–0.4	0.5–0.9	0.5
Cysts or cyst-like forms occur	–	–	–	–	–	–
Cells arranged in square tablets of 16–64 cells	–	–	–	–	–	–
Motility in liquid media	+	+	+	+	+	–
Flagellar arrangement:						
Polar or subpolar	+	+	+	+	+ or –	
Lateral	–	–	–	–	+ or –	
Poly-β-hydroxybutyrate accumulated	+			–		–
Cells embedded into gelatinous matrices to form tree-like or finger-like zoogloeas	–	–	–	–	–	–
Grow at 60°C or higher[c]	–	–	–	–	–	–
Grow at 5°C but not 37°C	–					
Fluorescent pigment	–		–	–	–	–
Yellow colonies	+	–	–	–	–	+
Red or orange colonies	–	+	+	–	–	–
Violet colonies	–	–	–	–	–	–
Oxidase	+	+	+	+	+	+
Acid from glucose	+			–		+
Indole from tryptophan						–
Oxidize ethanol to acetic acid at pH 4.5[e]	–	–	–	–	–	–
Fix N_2 in vitro aerobically[f]	–		–	–	–	–
Fix N_2 in vitro under microaerobic conditions[g]	–		–			–
Fix N_2 in root nodules of plants[h]	–	–	–	–	+	–
Denitrification (to N_2)	–		+		–	
Grow in lean organic media between pH 2.0 and 5.9 but not 6.1 or higher	–	–	–	–	–	–
Marine or brine bacteria that require at least 0.1% NaCl or more[i]	–	+	–	–	–	–
Grow with 20% or more NaCl	–	–	–	–	–	–
Growth greatly stimulated by or dependent upon cysteine						
Can use 1-C compounds as sole carbon sources[j]		–	–	–	–	–
Grow on L-glutamine but not on calcium lactate						
Presence of carbonic anhydrase						
Facultative H_2 autotrophs[k]				–	–	–
Grow well on chloridazon, antipyrin, and phenylalanine				–		
Grow on n-alkanes having 13–20 carbon atoms						
Genus includes pathogens, frank or opportunistic:						
Humans and/or animals	–	–	–	–	–	
Plants	+	–	–	–	–	–
Cause transformation of plant cells into autonomously proliferating tumor cells	–	–	–	–	–	–
Cause nodules on leaves of myrsine and madder plants	–	–	–	–	–	–
Cause lysis in soil of various Gram-positive and Gram-negative bacteria	–	–	–	–	–	–
Cellular lipids contain sphingolipids	–					+
Contain bacteriochlorophyll *a*	–	+	–	–	–	–

Footnotes are at end of table

Table 4.1 *(continued)*

Characteristic	Thermo-leo-philum	Thermo-micro-bium	Thermus	Variovorax	Volcan-iella	Weeks-ella
Gram reaction	−	−	−	−	−	−
Cell shape:						
Rods or coccobacilli, straight or curved	+	+	+	+	+	+
Cocci	−	−	−	−	−	−
Cell diameter, μm	0.4	1.3–1.8	0.5–0.8	0.5–0.6	0.8–1.0	0.6
Cysts or cyst-like forms occur	−	−	−	−	−	−
Cells arranged in square tablets of 16–64 cells	−		−	−	−	−
Motility in liquid media			−	+	−	−
Flagellar arrangement:						
Polar or subpolar				−	−	
Lateral				(peritrichous)	−	
Poly-β-hydroxybutyrate accumulated				+	+	
Cells embedded into gelatinous matrices to form tree-like or finger-like zoogloeas	−	−	−	−	−	−
Grow at 60°C or higher [c]	+	+	+	−	−	−
Grow at 5°C but not 37°C				−	−	
Fluorescent pigment	−	−	−	−	−	−
Yellow colonies	−	−	+ or −	+	−	−
Red or orange colonies	−	+	+ or −	−	−	−
Violet colonies	−	−	−	−	−	−
Oxidase			+	+	−	+
Acid from glucose	−	−				−
Indole from tryptophan					−	+
Oxidize ethanol to acetic acid at pH 4.5 [e]	−	−	−	−	−	−
Fix N_2 in vitro aerobically [f]	−	−	−	−	−	−
Fix N_2 in vitro under microaerobic conditions [g]	−	−	−	−	−	−
Fix N_2 in root nodules of plants [h]	−	−	−	−	−	−
Denitrification (to N_2)				−	−	
Grow in lean organic media between pH 2.0 and 5.9 but not 6.1 or higher	−	−	−	−	−	−
Marine or brine bacteria that require at least 0.1% NaCl or more [i]	−	−	−	−	+	−
Grow with 20% or more NaCl	−	−	−	−	+	−
Growth greatly stimulated by or dependent upon cysteine						
Can use 1-C compounds as sole carbon sources [j]	−	−	−			−
Grow on L-glutamine but not on calcium lactate						
Presence of carbonic anhydrase						
Facultative H_2 autotrophs [k]	−	−	−	d		−
Grow well on chloridazon, antipyrin, and phenylalanine						
Grow on n-alkanes having 13–20 carbon atoms	+	−	−			
Genus includes pathogens, frank or opportunistic:						
Humans and/or animals	−	−	−	−	−	
Plants	−	−	−	−	−	−
Cause transformation of plant cells into autonomously proliferating tumor cells	−	−	−	−	−	−
Cause nodules on leaves of myrsine and madder plants	−	−	−	−	−	−
Cause lysis in soil of various Gram-positive and Gram-negative bacteria	−	−	−	−	−	−
Cellular lipids contain sphingolipids					−	−
Contain bacteriochlorophyll a	−	−	−	−	−	−

Footnotes are at end of table

114

Table 4.1 *(continued)*

Characteristic	Xantho-bacter	Xantho-monas	Xylella	Xylophilus	Zoogloea
Gram reaction	+ or −	−	−	−	−
Cell shape:					
Rods or coccobacilli, straight or curved	+	+	+	+	+
Cocci	−	−	−	−	−
Cell diameter, μm	0.4–1.0	0.4–0.7	0.2–0.4	0.4–0.8	1.0–1.3
Cysts or cyst-like forms occur	−	−	−	−	−
Cells arranged in square tablets of 16–64 cells	−	−	−	−	−
Motility in liquid media	+ or −	+	−	+	+
Flagellar arrangement:					
Polar or subpolar	−	+		+	+
Lateral	+	−		−	−
Poly-β-hydroxybutyrate accumulated	+			−	+
Cells embedded into gelatinous matrices to form tree-like or finger-like zoogloeas	−	−	−	−	+
Grow at 60°C or higher [c]	−	−	−	−	−
Grow at 5°C but not 37°C					
Fluorescent pigment	−	−	−	−	−
Yellow colonies	+	+	−	+	−
Red or orange colonies	−	−	−	−	−
Violet colonies	−	−	−	−	−
Oxidase	+	−	−	−	+
Acid from glucose		+ or −		−	−
Indole from tryptophan	−	−	−	−	−
Oxidize ethanol to acetic acid at pH 4.5 [e]	−	−	−	−	−
Fix N_2 in vitro aerobically [f]	−	−	−	−	−
Fix N_2 in vitro under microaerobic conditions [g]	+	−	−	−	−
Fix N_2 in root nodules of plants [h]	−	−	−	−	−
Denitrification (to N_2)	−		−	+	
Grow in lean organic media between pH 2.0 and 5.9 but not 6.1 or higher	−	−	−	−	−
Marine or brine bacteria that require at least 0.1% NaCl or more [i]	−	−	−	−	−
Grow with 20% or more NaCl	−	−	−	−	−
Growth greatly stimulated by or dependent upon cysteine					
Can use 1-C compounds as sole carbon sources [j]	+	−	−	−	−
Grow on L-glutamine but not on calcium lactate		−		+	
Presence of carbonic anhydrase					
Facultative H_2 autotrophs [k]	+	−	−	−	−
Grow well on chloridazon, antipyrin, and phenylalanine					
Grow on n-alkanes having 13–20 carbon atoms					
Genus includes pathogens, frank or opportunistic:					
Humans and/or animals	−	−	−	−	−
Plants	−	+	+	+	−
Cause transformation of plant cells into autonomously proliferating tumor cells	−	−	−	−	−
Cause nodules on leaves of myrsine and madder plants	−	−	−	−	−
Cause lysis in soil of various Gram-positive and Gram-negative bacteria	−	−	−	−	−
Cellular lipids contain sphingolipids					
Contain bacteriochlorophyll a	−	−	−	−	−

Footnotes are at end of table

Table 4.1 *(continued)*

[a] Symbols: see standard definitions: + or –, reaction differs among species or among strains within species

[b] For the genus *Aquaspirillum*, only *Aquaspirillum fasciculus*, a straight rod, is considered here (see Table 4.5 for further information). For the genus *Oceanospirillum*, only *Oceanospirillum kriegii* and *Oceanospirillum jannaschii*, which are straight rods, are considered here (see Table 4.7 for further information).

[c] See Table 4.2 for further information.

[d] Violet-blue to brown. On complex media containing 10% NaCl, colonies are concentrically ringed with dark brown centers followed by bluish-brown, grayish-brown, and yellow rings. They produce a yellow pigment and a violet blue pigment.

[e] See Table 4.3 for further information.

[f] See Table 4.4 for further information.

[g] See Table 4.5 for further information.

[h] See Table 4.6 for further information.

[i] See Table 4.7 for further information.

[j] See Table 4.8 for further information.

[k] See Table 4.9 for further information.

Table 4.2 **Differentiation of Gram-negative, aerobic, heterotrophic, rod-shaped eubacteria that grow at 60°C or higher** [a]

Characteristic	*Acidothermus*	*Thermoleophilum*	*Thermomicrobium*	*Thermus*
Cell shape	Slender rods and long, slender filaments	Short rods	Short, irregularly shaped rods	Straight rods
Cell diameter, μm	0.4	0.4	1.3–1.8	0.5–0.8
Growth on 0.1% tryptone + 0.1% yeast extract in mineral salts broth		–	+	+
Growth on glucose	+	–	–	+
Growth on *n*-alkanes 13–20 carbons in length		+	–	–
Optimum temperature, °C	50–60	60	70–75	70–75
Temperature range, °C	37–70	45–70	45–80	40–79
Growth at pH 4.5	+	–	–	–
Form rotund bodies (large spheres 10–20 μm in diameter) in old cultures	–	–	–	+
Pigmentation of colonies	White	None or white	Rose or pink	White, yellow, orange, pink, red

[a] Symbols: see standard definitions.

Table 4.3   Differentiation of aerobic Gram-negative bacteria that oxidize ethanol to acetic acid in neutral or acidic (pH 4.5) media. [a]

Characteristic	Acetobacter	Acidomonas	Frateuria	Gluconobacter
Motility	+ or −	−	+ or −	+ or −
Flagellar arrangement:				
Polar	−		+	+
Peritrichous	+		−	−
Overoxidation of ethanol	+		−	−
Oxidation of DL-lactate to CO_2 and H_2O	+	−	+	−
Oxidation of acetate to CO_2 and H_2O	+	+	−	−
Ketogenesis	+ or −		+	+
Formation of brown water-soluble pigments on GYC agar	−		+[b]	−
Growth factors required	+ or −		−	+
Products formed from D-glucose:				
2-Ketogluconic acid	+ or −		+	+
5-Ketogluconic acid	+ or −		−	+
2,5-Diketogluconic acid	+ or −		−	+ or −
Acetylmethylcarbinol (Voges-Proskauer)	+ or −	−	−	+ or −
Growth in presence of 30% glucose	−		+[b]	−
Growth on Frateur's Hoyer mannitol medium	−		+	−
Acid produced from:				
D-Arabinose	−		+[b]	+
i-Inositol	−	−	+	+ or −
Maltose	−	−	−	+ or −
Carbon sources for growth:				
L-Arabinose	−	−	+	−
D-Mannose, D-lyxose, L-lyxose	−		+	−
n-Propanol	+ or −		−	−
Acetate	+ or −	+	+	−
Glycerate	+ or −		+	−
Lactate	+ or −	−	+	−
D-Mannitol	+	−	+	+
Methanol	−	+	−	−
Oxidase test	−	+	−	−

[a] Symbols: see standard definitions. + or −, reaction differs among species or among strains within species.

[b] Some strains are negative.

Important Notes for Users of This Manual

Unless otherwise indicated in footnotes to tables, the meanings of symbols are as follows:

    + 90% or more of strains are positive

    − 90% or more of strains are negative

    d 11-89% of strains are positive

    v strain instability (not equivalent to "d")

    D Different reactions in different taxa (species of a genus or genera of a family)

All other symbols are defined in footnotes to tables.

Table 4.4   Differentiation of Gram-negative eubacteria that fix N$_2$ under aerobic conditions [a]

Characteristic	*Azomonas*	*Azotobacter*	*Beijerinckia*	*Derxia*
Cysts (*sensu stricto*) formed	–	+	–	–
Cell morphology	Rods with rounded ends	Blunt rods to oval cells	Rods to ovoid to coccoid	Rods with rounded ends
Cell diameter, μm	2.0 or more	1.5–2.0 or more	0.5–1.5	1.0–1.2
Motility	+	D	D	+
Flagellar arrangement:				
Polar	D	–	–	+
Peritrichous	D	+	+	–
Growth at pH 4–5	–	–	+	–
Pellicle formed on surface of liquid media			–	+
Growth on plain peptone agar	–	–	+	
Intracellular lipoid bodies:				
Bipolar	–	–	+	–
Numerous throughout whole cell			–	+
Dark mahogany-brown colonies produced with aging			–	+
Catalase	+	+	+	–
Autotrophic use of H$_2$ to fix N$_2$	–	–	–	+
Root-associated N$_2$ fixation	–	D	–	–

[a] Symbols: see standard definitions.

Table 4.5 Aerobic Gram-negative rod-shaped eubacteria that fix N$_2$ in vitro under microaerobic conditions (as in semi-solid media) and do not incite nodules on the roots or stems of plants [a]

Characteristic	Acetobacter diazotrophicus	Agromonas	Aquaspirillum fasciculus	Xanthobacter
Gram reaction	−	−	−	+ or variable
Shape	Rod	Bent, branched, and budding rods	Rod	Rods; pleomorphic on succinate-containing media; coccoid on media containing an alcohol
Cell diameter, μm	0.7–0.9	0.6–1.0	0.7–0.9	0.4–1.0
Motility	+	+	+, but unusual [b]	− or +
Flagella (if present):				
Peritrichous	+	−	−	+
Polar	−	+	+, but unusual [c]	−
Colony color	Dark brown [d]	−	−	Yellow
Grows and fixes N$_2$ at pH values less than 3.0	+	−	−	−
Methanol used as a sole carbon source	−	−	−	+
Some carbohydrates are catabolized	+	+	−	D
Habitat	Roots and stems of sugarcane	Soil	Water	Soil, water

[a] Symbols: see standard definitions.

[b] In ordinary media the cells do not swim but instead exhibit an ineffectual floundering about movement. When cells are suspended in media of high viscosity (10–200 centipoises) produced by agents such as methylcellulose or DNA, they swim steadily in straight lines.

[c] Cells have bipolar flagellar fascicles which are composed of up to 11 flagella and which can be seen clearly by dark-field microscopy. The fascicles exhibit a distinctive behavior when cells are suspended in ordinary media. This behavior includes helical wave propagation with waves progressing from base to tip, an ability to coil up into a loop like a spring, and basal bending accompanied by a change in wavelength. In viscous media the aft fascicle is extended while the fore fascicle is either coiled into a loop or is coiled around the body of the cell.

[d] On potato agar supplemented with 10% sucrose.

119

Table 4.6 Differentiation of genera of N$_2$-fixing, Gram-negative eubacteria that incite the development of nodules on the roots and/or stems of plants [a]

Characteristic	Azorhizobium	Bradyrhizobium	Rhizobium	Sinorhizobium
Flagellar arrangement	One or more lateral	One subpolar	Peritrichous [b]	One polar or 1–3 lateral
Colony diameter on yeast extract-mannitol agar		−	+	+
within 3 days at 28°C	1–2 mm	<1 mm	2–4 mm	2–4 mm
Growth at 44°C	+	−	−	
Growth at pH 8	+	−	+	+
Lysine decarboxylase	+	−	−	
Arginine dihydrolase	+	−	D	
Acid production on yeast extract-mannitol agar	−	−	+	+
Carbon source utilization:				
D-ribose, D-xylose, D-mannose, D-galactose, D-arabinose	−	+	+	+
D-cellobiose	−	−	+	+
D-maltose, lactose	−	−	+	
Inositol, raffinose	−	−	+	D
Azelate, maleate, adipate, pimelate, suberate	+	+	−	
L-tartrate, meso-tartrate	−	+	−	
1,2-Propanediol, 2-3-butanediol, itaconate	+	−	−	
Dulcitol		−	+	−
Nitrogen source utilization:				
DL-asparagine		−	+	−
L-histidine		+	+	−
Tolerance to 0.1% brilliant cresyl blue		+	+	−
Requirement for pantothenate		D	−	+
Phenylalanine deaminase		−	−	+

[a] Symbols: see standard definitions.
[b] R. loti has a single subpolar flagellum.

Important Notes for Users of This Manual

Unless otherwise indicated in footnotes to tables, the meanings of symbols are as follows:
+ 90% or more of strains are positive
− 90% or more of strains are negative
d 11-89% of strains are positive
v strain instability (not equivalent to "d")
D Different reactions in different taxa (species of a genus or genera of a family)
All other symbols are defined in footnotes to tables.

Table 4.7   Differential characteristics of some aerobic, rod-shaped marine bacteria that require NaCl to grow [a]

Characteristic	Alteromonas[b]	Deleya marina	Deleya other species	Marino-monas[c]	Meso-philo-bacter	Methylo-monas pelagica	Methylo-phaga
Cell shape:							
Straight rod	+	+	+	D	+[g]	+	+
Curved rod	−[h]	−	−	D[i]	−	−	+
Motility	+	+	+	+	−	+	+
Flagellar arrangement:							
Polar	+	+	−	+		+	+
Lateral	−	−	+	−		−	−
Cell diameter, µm	0.5–1.5	0.8–1.1	0.8–1.1	0.7–1.5	0.5–0.6	0.8–1.2	0.2
Poly-β-hydroxybutyrate accumulation	−	+	+	−	−	−	
Utilization of:							
DL-malate	−	+	D	−	+	−	−
α-Ketoglutarate	−[k]	+	D	+		−	−
β-Hydroxybutyrate	−			+		−	−
Methane	−	−	−	−		+	+
Methanol	−					+	+
Methylamine	−	−	−	−		−	D
NO₃⁻ reduced to NO₂⁻	−[l]	−	D	−	+		
Oxidase	+	−	D	D	+		+
Lipase	+			−			
Gelatinase	+	−	−	−			
Utilization of:							
D-Glucose	+[m]	+	+	+	+	−	−
D-Fructose	D	+	D	+	+	−	+
Saccharate		−	D	+			
Glycolate		−	D	−			
Mannitol		+	D	+	+		
Isobutanol		−	D	−			
Quinate		−	D	+			
δ-Aminovalerate		−	D	−			
L-Tyrosine	D	+	D	−			
Sarcosine		d	D	+			

[a] Symbols: see standard definitions.

[b] The species considered are *A. macleodii, A. haloplanktis, A. espejiana, A. undina, A. hanedai, A. rubra, A. luteoviolacea, A. citrea, A. aurantia, A. nigrifaciens, A. denitrificans,* and *A. colwelliana.*

[c] The species considered are *M. communis* and *M. vaga,* which were formerly assigned to the genus *Alteromonas* in *Bergey's Manual of Systematic Bacteriology,* Vol. 1.

[d] The two species considered here are straight rods. All other species of *Oceanospirillum* are helical.

[e] Marine pseudomonads differ from terrestrial species by requiring over 75 mM NaCl for optimum growth rate and cell yield.

[f] *P. perfectomarina* may be synonymous with *P. stutzeri,* based on DNA homology experiments (Döhler et al., Int. J. Syst. Bacteriol. *37:* 1-3, 1987).

Footnotes continued on next page.

Table 4.7 *(continued)*

Characteristic	Oceanospirillum[d]		Pseudomonas[e]			
	kriegii	*jannaschii*	*stanieri*	*perfectomarina*[f]	*doudoroffii*	*nautica*
Cell shape:						
Straight rod	+	+	+	+	+	+
Curved rod	−	−	−	−[j]	−	−
Motility	+	+	+	+	+	+
Flagellar arrangement:						
Polar	+	+	+	−	+	+
Lateral	−	−	−	−	−	−
Cell diameter, μm	0.8–1.2	1.0–1.4	0.6–0.8	0.5–0.7	0.7–1.2	0.3–0.5
Poly-β-hydroxybutyrate accumulation	+	+	+	−	+	−
Utilization of:						
DL-malate	+	d	d	+	+	d
α-Ketoglutarate	+		+	+	+	−
β-Hydroxybutyrate	+	−	+	+	+	d
Methane	−	−	−	−	−	−
Methanol	−	−	−	−	−	−
Methylamine	−	−	−	−	−	−
NO_3^- reduced to NO_2^-	−	+	−	−	d	d
Oxidase	+	+	+	+	+	+
Lipase	+	−	−	−	−	+
Gelatinase	−	−	−	−	−	d
Utilization of:						
D-Glucose	+	−	−	+	−	−
D-Fructose	+	−	−	−	+	−
Saccharate	−	−	−	+	−	
Glycolate	−	−	−	+	+	−
Mannitol	+	−	−	−	−	−
Isobutanol	−	+	+	+	−	d
Quinate	+	−	d	−	−	−
δ-Aminovalerate	−	+	d	−	+	−
L-Tyrosine	−	−	+	+	−	−
Sarcosine	−	+	−	−	+	−

[g] Coccoid cells occur in old cultures.

[h] Only one species, *A. undina*, is a curved rod. Long filamentous helices often exceeding 20 μm occur when *A. colwelliana* grows on surfaces, in nutrient-poor media, or during late growth phases.

[i] *M. communis* is curved; *M. vaga* is straight.

[j] Some cells of *P. perfectomarina* may be slightly curved.

[k] *A. colwelliana* utilizes α-ketoglutarate.

[l] *A. hanedai*, *A. colwelliana*, and some strains of *A. haloplanktis* are positive. *A. denitrificans* carries out denitrification with gas formation.

[m] *A. colwelliana* and some strains of *A. hanedai* are unable to use D-glucose.

Table 4.8 Differential characteristics of some aerobic Gram-negative eubacteria that can use 1-carbon compounds as sole carbon sources [a]

Characteristic	Acetobacter methanolicus	Acido-monas	Derxia	Methylo-bacillus	Methylo-bacterium	Methylo-coccus	Methylo-monas	Methylo-phaga	Methylo-philus	Paracoccus denitrificans/ alcaliphilus	Rhizo-bacter	Xantho-bacter
Gram reaction	–	–	–	–	– or v	–	–	–	–	–	–	+ or v
Cell shape:												
Rod	+	+	+	+	+	+	+	+	+	+ or –	+	+
Coccus	–	–	+	–	–	+	–	–	–	+ or –	+	–[b]
Cell diameter, μm	0.6–0.8	0.8–1.0	1.0–1.2	0.3–0.5	0.8–1.0	1.0	0.5–1.2	0.2	0.3–0.6	0.5–0.9	0.9–1.3	0.4–1.0
Cysts formed	–	–	–	D	–	+	D	–	D	–	–	D
Motility	+	–	+	D	+	–	+	+	D	–	+	D
Flagellar arrangement:												
Polar	–	+	+	+	D	+	+	+	+	–	d	–
Lateral	+	–	–	–	D	+	–	–	–		d	+
Subpolar	–	–	–	–	D	–	–	–	–	–	–	–
NaCl or seawater required for growth	–	–	–	–	–	–	D	+	–	–	–	–
Vitamin B$_{12}$ required	–	–	+	–	–	–	–	+	–	–	–	+
Thiamine required	–	–	–	d	–	–	–	–	d	–	–	–
Sole carbon sources:												
Methane	–	–	+	–	D	+	+	–	–	–	–	–
Methanol	+	+	+	+	+	+	+	+	+	+	+	+
Methylamine	+	+	+	D	D	+	+	+	d	d	+	+
Formaldehyde	–					+	D					
Organic compounds other than 1-C compounds	+	+	+	D	+	–	–	+	d	+	+	+
Major 1-C pathway:												
Ribulose mono-phosphate				+	–	+	+	+	+			
Serine				–	+	–	–	–	–			
Oxidase	–	+	+	–	+ or W	+	+	+	+	+	+	+
Catalase	+	+	+	D	+	+	+	+	+	+	+	+
Poly-β-hydroxybutyrate accumulation	–	–	+	–	+	–	–	–	–	+	+	+
Pigmentation	White to pale yellow, rarely pink	White to light yellow	Dark mahogany brown[c]	White to light yellow	Pink to orange-red	None	Pink or white	Pink	None	White or light yellow	Yellow[d]	Yellow
Autotrophic CO$_2$ fixation	–	–	–	–	–	–	–	–	–	D	–	+
Growth at 45°C	–	–	+	–	–	+	–	–	–	–	–	D
Grows only below pH 5.5	–	+	–	–	–	–	–	–	–	–	–	–

Table 4.8 *(continued)*

Characteristic	Acetobacter methanolicus	Acido-monas	Derxia	Methylo-bacillus	Methylo-bacterium	Methylo-coccus	Methylo-monas	Methylo-phaga	Methylo-philus	Paracoccus denitrificans/ alcaliphilus	Rhizo-bacter	Xantho-bacter
Oxidize ethanol to acetic acid at pH 4.5	+	+	−		−	−	−	−	−	−	−	−
Fatty acid composition: Major fatty acids:												
16:0				+	−		+	+	+			
16:1				+	−		+	+	+			
18:1				−	+		−	−	−			
Major quinones:												
Q-8	−			+	−		+	+	+		+	
Q-10	+			−	+		−	−	−		−	
Major phospholipid: Diphosphati-dylglycerol				+					−			
Plant pathogen	−	−	−	−	−	−	−	−	−	−	+	−

[a] Symbols: see standard definitions.

[b] Coccoid cells produced on alcohol-containing media.

[c] At first the colonies of *Derxia* are slimy and semi-transparent, later massive and opaque, highly raised with a wrinkled surface. Older colonies develop a dark mahogany-brown color.

[d] On yeast extract-peptone-glucose medium. On other media the colonies may be white.

Table 4.9 Some aerobic, Gram-negative eubacteria that are facultative hydrogen autotrophs [a]

Characteristic	Alcaligenes[b]	Bradyrhizobium[c]	Derxia	Hydrogenophaga[d]	Methylococcus	Paracoccus denitrificans	Pseudomonas[e]	Xanthobacter
Gram reaction	−	−	−	−	−	−	−	+ or v
Cell shape	Short rods to coccoid	Rods	Rods	Rods	Cocci	Cocci or short rods	Rods	Rods or coc-coid cells
Cell diameter, μm	0.5–1.0	0.5–0.9	1.0–1.2	0.3–0.6		0.5–0.9	0.3–0.6	0.4–1.0
Motility	+	+	+	+	−	−	+	D
Flagellar arrangement	Lateral or subpolar	1 polar	1 polar	1 or more polar or subpolar			1 or more polar or subpolar	Lateral
Yellow pigmentation	D	−	−	+	−	−	D[f]	+
Utilization of:								
$\quad$D-Glucose				+	−	+	D	
$\quad$Malonate				−			D	
$\quad$p-Hydroxybenzoate				D	−	+	−	
$\quad$β-Hydroxybutyrate	+			+				−
Hydrolysis of poly-β-hydroxybutyrate				−			D	
N₂-fixers that can incite formation of nodules on the roots of plants	−	+	−	−	−	−	−	−
N₂ fixed aerobically in vitro	−	−	+	−			−	−
N₂ fixed in vitro but only under microaerobic conditions	−	D	−	D		−	D	+
Can use methanol as a carbon source	−	−	−	−	+	+	−	+

[a] Symbols: see standard definitions.

[b] Refers to the following species: *A. paradoxus* biovar 1, *A. eutrophus*, *A. latus*, and some strains of *A. denitrificans* subspecies *xylosoxidans*.

[c] Some strains are facultative hydrogen autotrophs.

[d] Refers to the following species: *H. flava (Pseudomonas flava), H. palleronii (Pseudomonas palleronii), H. pseudoflava (Pseudomonas pseudoflava; Pseudomonas carboxydoflava),* and *H. taeniospiralis (Pseudomonas taeniospiralis).*

[e] Refers to the following species: *P. saccharophila, P. facilis, "P. hydrogenovora", "P. hydrogenothermophila", "P. carboxydohydrogena", "P. compransoris",* and *"P. carboxydovorans".*

[f] Only *"P. hydrogenothermophila"* has yellow pigmentation, whereas all species of *Hydrogenophaga* have yellow pigmentation. *"P. hydrogenothermophila"* is distinguished by having an optimum growth temperature of 52°C.

Table 4.10   Differential characteristics of the species of the genus *Acetobacter*[a,b]

Characteristic	A. aceti	A. diazo-trophicus[c]	A. hansenii	A. liquefaciens	A. methano-licus[d]	A. pasteur-ianus	A. xylinum[e]
Formation of:							
Water-soluble brown pigments on GYC[f]	–	+	–	+	–	–	–
Gamma-pyrones from D-glucose	–	+	–	d	–	–	–
Gamma-pyrones from D-fructose	–	+	–	+	–	–	–
5-Ketogluconic acid from D-glucose	+	–	d	d	–	–	+
2,5-Diketogluconic acid from D-glucose	–	+	–	+	–	–	–
Ketogenesis from glycerol	+	d	+	+	Weak	–	+
Growth on carbon sources:							
Ethanol	+	+	–	+	Weak	d	–
Dulcitol	–	–	d	–	–	–	–
Sodium acetate	+	+	–	d	Weak	d	–
Methanol	–	–	–	–	+	–	–
Growth on L-amino acids in the presence of D-mannitol as the carbon source:							
L-glycine, L-threonine, L-tryptophan	–	–	–	d	–	–	–
L-glutamine	d	–	+	+	–	–	–
L-asparagine	d	+	+	+	–	–	–
Growth in the presence of 10% ethanol	–	–	–	–	–	d	–
Growth in the presence of 30% D-glucose	–	+	–	–	–	–	–
N₂ fixation and growth on N₂	–	+	–	–	–	–	–

[a] Symbols: see standard definitions.

[b] Data taken from Gillis et al. (Int. J. Syst. Bacteriol. *39*: 361-364, 1989).

[c] *A. diazotrophicus* was not included in *Bergey's Manual of Systematic Bacteriology*, Vol. 1. 364). It was created in 1989 by Gillis et al. (Int. J. Syst. Bacteriol. *39*: 361-364).

[d] *A. methanolicus* was not included in *Bergey's Manual of Systematic Bacteriology*, Vol. 1. It was created in 1986 by Uhlig et al. (Int. J. Syst. Bacteriol. *36*: 317-322).

[e] *A. xylinum* was not included in *Bergey's Manual of Systematic Bacteriology*, Vol. 1. The name was revived in 1983 by Yamada (J. Gen. Appl. Microbiol. *29*: 417-420).

[f] GYC, 5% D-glucose + 1% yeast extract + 3% CaCO₃ + 2.5% agar.

Table 4.11 Differential characteristics of the species of the genus *Acidiphilium*[a,b]

Characteristic	*A. angustum*[c]	*A. cryptum*	*A. facilis*[c]	*A. organovorum*	*A. rubrum*[c]
Cell diameter, μm	0.8	0.3–0.5	0.7	0.6	0.6
Cell length, μm	2.9	0.6–1.5	1.8	0.8–1.0	2.2
Population doubling time, hours	11	6[d]	3.3[e]	2.5[f]	14[e]
Colony pigmentation:					
Red-violet	–	–	–	–	+
Pink	+	d	–	–[g]	–
White or light brown	–	d	+	+	–
Optimum pH				3.0	
pH range for growth	[h]	2.5–5.9[i]	[h]	2.0–5.5	[h]
Inhibition by acetate	+	+	–	+	+
Urease activity	–		+		–
H₂S from cysteine	–	d			–
Growth on sole carbon sources:					
Glucose	–[j]	– or Weak[k]	+	+	d[j]
Lactose	–		+	–	–
Glycerol	+		+	+	–
Ethanol	+		+		–
ʟ-Malate	–		+	+	+
α-Ketoglutarate	–		+		+
cis-Aconitate	–		+		–
Succinate	–		+	+	d
Fumarate	–		+	+	–
Glutamate	–	–	+	+	d
Pyruvate	–		d	–	–

[a] Symbols: +, 80% or more of strains positive; d, 20–79% of strains positive; –, less than 20% of strains positive.

[b] The genus *Acidiphilium* was created in 1981 by Harrison (Int. J. Syst. Bacteriol. *31*: 327-332). It initially contained a single species *A. cryptum*, which was described in *Bergey's Manual of Systematic Bacteriology*, Vol. 3, p. 1863-1868. The following species have since been added: *A. organovorum* (Lobos et al., Int. J. Syst. Bacteriol. *36*: 139-144, 1986); *A. angustum, A. facilis*, and *A. rubrum* (Wichlacz et al., Int. J. Syst. Bacteriol. *36*: 197-201, 1986).

[c] Data from Wichlacz et al., Int. J. Syst. Bacteriol. *36*: 197-201, 1986.

[d] In shaking cultures in basal salts-0.1% glucose-0.01% dehydrated Trypticase soy medium at 25°C (Harrison, Int. J. Syst. Bacteriol. *31*: 327-332, 1981).

[e] In shaking cultures at 27 to 30°C in 0.1% (w/v) glucose-0.01% (w/v) yeast autolysate-basal salts medium, pH 3.0 (Wichlacz et al., Int. J. Syst. Bacteriol. *36*: 197-201, 1986).

[f] In citrate-phosphate buffered medium at 37°C, pH 3.0 (Lobos et al., Int. J. Syst. Bacteriol. *36*: 139-144, 1986).

[g] Initially white, but turn pink in 1–2 weeks.

[h] Growth occurs at pH 2.5 but not at 6.0; the exact range for growth has not been established.

[i] Some strains can grow at pH 2.0.

[j] All strains can grow with glucose in basal medium in the presence of 0.01% yeast autolysate.

[k] Growth is poor (<10⁸ cells/ml) unless medium is supplemented with peptone or yeast extract. Some strains are inhibited by 0.2% glucose when they are first isolated, but later, after "adaptation," grow well in laboratory media. At a concentration of 0.01%, Trypticase soy stimulates growth, but 0.05% inhibits growth (Harrison, Int. J. Syst. Bacteriol. *31*: 327-332, 1981).

Table 4.11a   Differential characteristics of the species of the genus *Acidovorax*[a]

Characteristics	A. delafieldii	A. facilis	A. temperans
Growth on:			
L-arabinose, D-galactose, D-ribose, D-mannose	+	+	−
Adipate	+	−	+
2-Ketogluconate, 2-ketoglutarate, citraconate	+	−	d
NO$_2^-$ reduction	d	−	+
Gelatinase	d	+	−
Autotrophic growth with hydrogen	d	+	−

[a]   See standard definitions.

Important Notes for Users of This Manual

Unless otherwise indicated in footnotes to tables, the meanings of symbols are as follows:

+   90% or more of strains are positive

−   90% or more of strains are negative

d   11-89% of strains are positive

v   strain instability (not equivalent to "d")

D   Different reactions in different taxa (species of a genus or genera of a family)

All other symbols are defined in footnotes to tables.

Table 4.12 Differentiation of the species of the genus *Acinetobacter*[a,b]

Characteristic	A. baumanni	A. calcoaceticus	A. haemolyticus	A. johnsonii	A. junii	A. lwoffii	Unnamed species "3"	"6"	"10"	"11"	"12"
Growth at:											
44°C	+	−	−	−	−	−	−	−	−	−	−
41°C	+	−	−	−	90	−	+	−	−	−	−
37°C	+	+	+	−	+	+	+	+	+	+	+
Gelatin hydrolysis	−	−	96	−	−	−	−	+	−	−	−
Hemolysis	−	−	+	−	−	−	−	+	−	−	−
Glutamyltransferase	99	+	4	−	−	−	+	66	−	−	−
Citrate (Simmons)	+[c]	+	91	+	82	−	+	+[c]	+	+	−
Acid from glucose	95	+	52	−	−	6	+	66	+	−	33
β-Xylosidase	95	−	52	−	−	6	+	66	−	−	−
Utilization of:											
DL-Lactate	+	+	−	+	+	+	+	−	+	+	+
Glutarate	+	+	−	−	−	−	+	−	+	+	+
L-Phenylalanine	87	+	−	−	−	−	+	−	−	−	+
Phenylacetate	87	+	−	−	−	94	66	−	25	50	+
Malonate	98	+	−	13	−	−	87	−	−	−	+
L-Histidine	98	+	96	−	+	−	94	+	+	+	−
Azelate	90	+	−	−	−	+	+	−	50	25	+
D-Malate	98	−	96	22	+	76	+	66	+	+	−
L-Aspartate	+	+	64	61	40	−	+	66	+	75	−
L-Leucine	97	38	96	−	11	−	94	+	−	−	+
Histamine	−	−	−	−	−	−	−	−	75	+	−
L-Tyrosine	+	+	5	70	60	3	+	66	+	75	+
β-Alanine	95	+	−	−	−	−	94	−	+	+	−
Ethanol	+[c]	+	96	+	+	97	+	+	+	+	+
2,3-Butanediol	+	+	−	35	−	−	+	−	+	+	+
trans-Aconitate	99	+	52	−	−	−	+	−	−	−	−
L-Arginine	98	+	96	35	95	−	+	+	−	−	+
L-Ornithine	93	+	−	4	−	2	+	−	−	−	
DL-4-Amino-butyrate	+	+	+	35	88	40	+	−	+	+	+

[a] Data from Bouvet and Grimont, Int. J. Syst. Bacteriol. *36*: 228-240, 1986.

[b] Symbols: +, all strains positive; −, all strains negative. The numbers are percentages of positive strains.

[c] All strains except one or two auxotrophic strains.

Table 4.13 Differential characteristics of the species of the genus *Agrobacterium*[a]

Characteristic	A. radiobacter		A. rhizogenes[b]	A. rubi[c]	A. tumefaciens		
	biovar 1	biovar 2	biovar 2		biovar 1	biovar 2	biovar 3
Growth:							
at 35°C:	+	−	−	d	+	−	d
On selective medium of Scroth et al.[d]	+	−	−		+	−	−
On selective medium of New and Kerr[e]	−	+	+		−	+	−
In presence of 2% NaCl	+	−[f]	−		+	−[f]	+
3-Ketolactose produced	+	−	−	−	+	−	−[f]
Acidic reaction produced from:							
meso-Erythritol	−	+	+	+	−	+	−[f]
Melezitose					+	−	−[f]
Ethanol	+	−	−	−	+	−	−[f]
Alkaline reaction produced from:							
Sodium malonate	−	+	+	+	−	+	+
Sodium L-tartrate					d	+	+
Sodium propionate					d	−	
Simmons' citrate with 0.0005% yeast extract	−	+	+	−	−	+	
Reaction in litmus milk:[g]							
Alkaline	+	−	−	+	+	−	+
Acidic	−	+	+	−	−	+	
Formation of pellicle in ferric ammonium citrate solution	+	−	−	−	+	−	d
Growth factor requirements:							
Biotin and/or glutamic acid	−	+[f]	+[f]		−	+[f]	
L-Glutamic acid and yeast extract	−	−	−	+	−	−	
Phytopathogenicity:							
Tumors produced on wounded stems of e.g., tomato plants, *Helianthus annuis, Nicotiana tabacum* and/or on discs of *Daucus carota*	−	−	+	+	+	+	d[h]
Roots produced on discs of *Daucus carota*	−	−	+	−	−	−	−

[a] Symbols: see standard definitions.

[b] The majority of the investigated strains belong to biovar 2. See *Bergey's Manual of Systematic Bacteriology*, Vol. 1, 248-252, for further information.

[c] Only the following three strains are considered to belong to *A. rubi*: ATCC 13334, 13335, and Braun EU6. Thus only these strains should be used as reference strains for comparison with an unidentified isolate.

[d] Scroth et al., Phytopathology *55*: 645-647, 1965.

[e] New and Kerr, J. Appl. Bacteriol. *34*: 233-236, 1971.

[f] Some strains have been reported to give reactions different from those indicated. See *Bergey's Manual of Systematic Bacteriology*, Vol. 1, p. 252, for further information.

[g] An alkaline reaction in litmus milk is frequently accompanied by a brown discoloration; and acid reaction (pink color) is frequently accompanied by a clot formation.

[h] Biovar 3 strains have been isolated mainly from grapevines. The majority of these isolates display a very limited host range. For such strains phytopathogenicity can only be demonstrated on young shoots of grapevines.

Table 4.14 Differential characteristics of the species of the genus *Alcaligenes*[a,b]

Characteristic	A. eutrophus	A. faecalis	A. latus	A. paradoxus	A. piechaudii[c]	A. xylosoxidans[d] subsp. denitrificans	subsp. xylosoxidans
Cell width >1.2 μm	−	−	+	−	−	−	−
Yellow carotenoid cellular pigments	−	−	−[e]	+	−	−	−
NO_3^- reduced to NO_2^-	+	−	+	d[f]	+	[+]	+
Anaerobic growth with NO_3^-	+	−	−	−	−	[+]	+
Anaerobic growth with NO_2^-	−	+	−	−	−	[+]	+
H_2 chemolithotrophy	+	−[g]		d[f]		−[g]	−[h]
Gelatin hydrolysis	−	−	+	d	−	−	−
Acid from D-xylose in O-F medium	−				−	−	+
Carbon sources for growth:							
D-Glucose	M	−	+	+	−	−	+
L-Arabinose	−	−	−	+	−	−	−
D-Xylose	−	−[i]	−	+	−[i]	−[i]	+[i]
D-Fructose	+	−	+	+	−	−	d
D-Mannitol	−	−	−	+	−	−	−
D-Mannose	−	−	−	+	−	−	d
D-Gluconate	+	−	+	+	+	[+]	+
Acetate	+	+	−	+		+	+
Adipate	+	−[i]	−	[+]	+[i]	+[i]	+[i]
Pimelate	+	−[i]	−	[+]	+[i]	+[i]	+[i]
Sebacate	+	−	d	+		+	+
Suberate	+	−	+	d		+	+
meso-Tartrate	+	−	−	d		+	+
Itaconate	+	−	+	d	+	+	+
Mesaconate		−			+	d	[+]
Isovalerate		+			+	d	−
n-Valerate		+			+	d	d
Isolated from clinical specimens	−	+	−	−	+	+	+

[a] Symbols: +, 90% or more of strains are positive; [+], 80% or more of strains are positive; d, 11–79% of strains are positive; −, 10% or less of strains are positive; M, mutant growth.

[b] The species *A. aestus*, *A. cupidus*, *A. pacificus*, *A. venustus*, and *A. aquamarinus*, previously included in this genus, have been reclassified in the genus *Deleya* (Baumann et al., Int. J. Syst. Bacteriol. *33*: 793-802, 1983).

[c] *A. piechaudii* was not included in *Bergey's Manual of Systematic Bacteriology*, Vol. 1. The species is described by Kiredjian et al., Int. J. Syst. Bacteriol. *36*: 282-287, 1986.

[d] The former names *A. denitrificans*, *A. denitrificans* subsp. *denitrificans* and *A. denitrificans* subsp. *xylosoxidans* are illegitimate and have been corrected to *A. xylosoxidans*, *A. xylosoxidans* subsp. *denitrificans* and *A. xylosoxidans* subsp. *xylosoxidans*, respectively (Kiredjian et al., Int. J. Syst. Bacteriol. *36*: 282-287, 1986).

[e] The colonies of *A. latus* are grayish pink or yellowish.

[f] Two biovars of *A. paradoxus* are recognized: biovar I can grow chemolithotrophically with H_2 and reduces NO_3^- to NO_2^-; biovar II does not possess these properties.

[g] Only a limited number of strains have been tested.

[h] Strain ATCC 15749 (previously named *A. ruhlandii*) can grow chemolithotrophically with H_2.

[i] Tested by using API 50AO and API 50CH galleries.

Table 4.15 Differential characteristics of the species of the genus *Alteromonas*[a,b]

Characteristic	A. aurantia	A. citrea	A. colwelliana[c]	A. denitrificans[d]	A. espejiana	A. haloplanktis	A. hanedai	A. luteoviolacea	A. macleodii	A. nigrifaciens[e]	A. rubra	A. undina
Cell shape:												
Straight	+	+	+[f]	+	+	+	+	d[g]	+	+	+	−
Curved	−	−	−	−	−	−	−	d[g]	−	−	−	+
Sheathed flagellum	−	−	−	+	−	−	−	d[g]	−	−	−	−
Growth at:												
4°C	+	−	−	+	−	−	+	d	−	+	−	d
35°C	−	d	−	−	d	d	−	+	+	−	+	−
40°C	−	d	−	−	−	−	−	−	d	−	−	−
NO_3^- reduced to NO_2^-	−	−	+	−	−	d	+	−	−	−	−	−
Denitrification with gas formation	−	−	−	+	−	−	−	−	−	−	−	−
Bioluminescence	−	−	−	−	−	−	+	−	−	−	−	−
Organic growth factors required	+	+	+	+	+	d	+	+	−	+	+	+
Production of:												
Amylase	+	+	+	+	+	d	−	+	+	−	+	d
Alginase	−		+	+	−	−			d	−		−
Chitinase	−	d	−	+	−	d	+	−	−	−	−	+
Utilization of:												
D-Glucose	+	+	−	+	+	+	d	+	+		+	+
D-Mannose	+	+	−		d	+	−	d[g]	−	−	+	−
D-Galactose	−	−	−	−	+	d	d	−	+	+	−	−
D-Fructose	+	+	−		d	d	−	d[g]	+	+	−	−
Sucrose	−	−	−	−	+	+	−	−	+	−	−	+
Maltose	d	−	−	+	+	+	−	+	+		−	+
Cellobiose	−	−	−		d	−	−	−	+	−	−	−
Melibiose	−				+	−	−	−	+	+		−
Lactose	−	−	−	−	+	−	−	−	+	+	−	−
Salicin	−	−	−		−	−	−	−	+	−	−	−
D-Gluconate	−	−	+	−	−	−	d	−	+	+	−	−
N-Acetylglucosamine	+	−	−	−	+	+	+	+	d	−		+
Succinate	−	−	−	−	+	−	−	−	−	+	−	+
Fumarate	−	−	+	−	−	+	−	−	−	+	−	+
DL-Lactate	−	−	−	−	−	−	−	d	d	+	−	−
DL-Glycerate	−		−		−	−	+	−	+	−		−
α-Ketoglutarate	−	−	+		−	−	−	−	−	−	−	−
Citrate	−	−	−		+	+	−	−	−	+	−	−
Aconitate	−				+	+		−	−	+		
D-Mannitol	−	−	−	−	+	d	−	−	d	−	−	−
Glycerol	−	−	−	−	−	−	−	−	+	+	−	−
L-Threonine	d	−	−		+	−	−	+	−	−		
L-Tyrosine	+	−	+	+	+	d	d	+	+			+
L-Glutamate	−	−	+	−	d	+	−	d	+			
L-Arginine	+	−	+	d	d	−	d	d	−			d
Putrescine		−	−	−	−	+		−	−		−

Table 4.15 *(continued)*

Characteristic	A. aurantia	A. citrea	A. colwelliana [c]	A. denitrificans [d]	A. espejiana	A. haloplanktis	A. hanedai	A. luteoviolacea	A. macleodii	A. nigrifaciens [e]	A. rubra	A. undina
Colony pigmentation:												
Brown, water-soluble	–	–	+	–	–	–	d	–	–	–	–	–
Red, water-insoluble	–	–	–	+[h]	–	–	–	–	–	–	+	–
Violet, water-insoluble	–	–	–	–	–	–	–	+	–	–	–	–
Yellow, water-insoluble	–	+	–	–	–	–	–	+[i]	–	–	–	–
Orange, water-insoluble	+	–	–	–	–	–	–	–	–	–	–	–
Black, water-insoluble	–	–	–	–	–	–	–	–	–	+	–	–

[a] Symbols: see standard definitions.

[b] Two species previously included in this genus, *A. communis* and *A. vaga*, have been reclassified in the genus *Marinomonas* (Van Landschoot and DeLey, J. Gen. Microbiol. *129:* 3057-3074, 1983).

[c] *A. colwelliana* was not included in *Bergey's Manual of Systematic Bacteriology*, Vol. 1. The species was created in Weiner et al., Int. J. Syst. Bacteriol. *38:* 240-244, 1988.

[d] *A. denitrificans* was not included in *Bergey's Manual of Systematic Bacteriology*, Vol. 1. The species was created by Enger et al., Int. J. Syst. Bacteriol. *37:* 416-421, 1987.

[e] *A. nigrifaciens* was not included in *Bergey's Manual of Systematic Bacteriology*, Vol. 1. The species was created by Baumann et al., Int. J. Syst. Bacteriol. *34:* 145-149, 1984.

[f] Long filamentous helices often exceeding 20 μm in length occur when *A. colwelliana* grows on surfaces, in nutrient-poor media, or during late growth phases.

[g] See Novick and Tyler, Int. J. Syst. Bacteriol. *35:* 111-113, 1985.

[h] Pigmentation of cells grown in liquid medium varies from bright blue to purple to red, depending on the growth medium (Enger et al., Int. J. Syst. Bacteriol. *37:* 416-421, 1987).

[i] In addition to violacein, a yellow choroform-extractable pigment having an absorption peak at 432 nm is produced by *A. luteoviolacea.*

Table 4.15a Differential characteristics of the species of the genus _Aminobacter_[a]

Characteristic	A. aminovorans	A. aganoensis	A. niigataensis
NO_3^- reduction	–	w	w
Growth in peptone water	+	w or –	w or –
Production of acid from L-arabinose oxidatively	+	–	–
Utilization of:			
Dimethylamine	–	+	+
Formamide	–	–	w
_N_-Methylformamide	–	–	+
_N,N_-Dimethylformamide	–	–	+
Tetramethylammonium hydroxide	–	+	–

[a] Symbols: see standard definitions; w, weak.

Important Notes for Users of This Manual
Unless otherwise indicated in footnotes to tables, the meanings of symbols are as follows:
+ 90% or more of strains are positive
– 90% or more of strains are negative
d 11-89% of strains are positive
v strain instability (not equivalent to "d")
D Different reactions in different taxa (species of a genus or genera of a family)
All other symbols are defined in footnotes to tables.

Table 4.16 Differential characteristics of the species of the genus *Azomonas*[a]

Characteristic	A. agilis	A. insignis	A. macrocytogenes
Presence of enlarged cells in media with ethanol	–	–	+
Flagellar arrangement:			
Peritrichous	+	–	–
Lophotrichous	–	+	–
Monotrichous	–	–	+[b]
Gram staining reaction	–	–	v
Formation of colony-retained homopolysaccharides	–	–	+
Diffusible pigments:			
Brown-black, on benzoate medium	–	d	–
Blue-white fluorescent, on iron-deficient medium[c]	+	–	d
Utilization as carbon sources:			
Mannitol	–	–	+
Maltose	d	–	+
Malonate	+	+	–

[a] Symbols: see standard definitions.

[b] Rarely, two flagella may occur at one pole.

[c] See Pigment Production, *Bergey's Manual of Systematic Bacteriology*, Vol. 1, page 225.

Table 4.17 Differential characteristics of the species of the genus *Azotobacter*[a]

Characteristic	A. armeniacus	A. beijerinckii	A. chroococcum	A. nigricans	A. paspali	A. vinelandii
Motility	+	–	+	–	+	+
Long filaments in young cultures	–[b]	–[b]	–[b]	–[b]	+	–[b]
Water soluble pigments:						
Yellow-green fluorescent[c]	–	–	–	–	+	+
Green	–	–	–	–	–	d
Brown-black	–	–	–	d	–	–
Brown-black to red-violet	+	–	–	+	–	–
Red-violet	+	–	–	d	+	d
Utilization as carbon source:						
Rhamnose	–	–	–	–	–	+
Caproate	–	–	+	–	–	+
Caprylate	+	–	–	–	–	+
meso-Inositol	d	d	–	–	–	+
Mannitol	+	d	+	d	–	+
Malonate	–	+	d	d	–	+

[a] Symbols: see standard definitions.

[b] These species may sporadically produce filamentous forms of different lengths.

[c] On iron-deficient medium: see *Bergey's Manual of Systematic Bacteriology*, Vol. 1, p. 225.

Table 4.18   Differential characteristics of the species of the genus *Beijerinckia*[a]

Characteristic	B. derxii	B. fluminensis	B. indica	B. mobilis
Water-soluble green fluorescent pigment	+	−	−	−
Colony color after aging	Buff	Fulvous or pink	Fulvous or pink	Amber-brown
Motility	−	[−][b]	[−][b]	+
Growth on casein agar	+	[−](15)	[−]	+
Growth with NO_3^- as nitrogen source	[−][c]	[−][c]	d(40–50)	+
Growth on asparagine as the sole carbon and nitrogen source	[−][c]	[−][c]	[−](6)	d(60)
Growth on carbon sources:				
Lactose	+	[−](20)	d(70)	−
Erythritol	−	−	−	d(45)
Propanol	−	−	d(50)	+
Acetate, butyrate, fumarate, lactate, malate	[−][c]	d(20–80)	d(50–95)	+
Benzoate	−	−	[−](6)	+

[a] Symbols: +, all strains positive; d, differs among strains; [−], most strains negative; ( ), % of strains giving positive reaction.

[b] If positive, motility occurs mostly in young stages and the cells are usually only weakly motile.

[c] If positive, the reactions are weak.

Table 4.19   Differential characteristics of the species of the genus *Bordetella*[a]

Characteristic	B. avium[b]	B. bronchiseptica	B. parapertussis	B. pertussis
Motility	+	+	−	−
Number of days required for colonies to appear on Bordet-Gengou medium	1–2	1–2	2–3	3–6
Growth on MacConkey agar	+	+	+	−
Browning on peptone agar	d	−	+	−
NO_3^- reduced to NO_2^-	−	+	−	−
Urease	−	+[c]	+	−
Oxidase	+	+	−	+
Carbon sources for growth[d]:				
Acetate, adipate	d	+	−	
meso-Tartrate	−	d	−	
Itaconate	−	+	−	
Succinate	+	+	−	
Alkalinization of[e]:				
Acetamide or formamide	+	−	−	
Malonamide	−	+	−	

[a] Symbols: see standard definitions.

[b] B. avium was not included in *Bergey's Manual of Systematic Bacteriology*, Vol. 1. The species was created by Kersters et al., Int. J. Syst. Bacteriol. *34*: 56-70, 1984.

[c] Positive within 4 hours.

[d] Tested by using the API auxanographic system (see Kersters et al., Int. J. Syst. Bacteriol. *34*: 56-70, 1984).

[e] Tested on Greenwood low-peptone medium (see Pickett, M.J., *Nonfermentative Gram-negative bacilli: a syllabus for detection and identification,* Scientific Development Press, Los Angeles, 1980).

Table 4.20 Differentiation of the species of the genus *Brucella*[a,b]

Characteristic	B. abortus	B. canis	B. melitensis	B. neotomae	B. ovis	B. suis biovar 1	biovar 2	biovar 3	biovar 4
Lysis by phage at RTD[c]:									
Td	L	NL	NL	PL	NL	NL	NL	NL	NL
Wb	L	NL	NL	L	NL	L	L	L	L
Fi	L	NL	NL	L	NL	PL	PL	PL	L
Bk$_2$	L	NL	L	L	NL	L	L	L	L
R/O	PL	NL	NL	NL	L	NL	NL	NL	NL
R/C	NL	L	NL	NL	L	NL	NL	NL	NL
Oxidation of substrates:									
L-Alanine	+	d	+	d	d	d	−	d	−
L-Asparagine	+	−	+	+	+	−	d	−	−
L-Glutamic acid	+	+	+	+	+	−	d	d	d
L-Arabinose	+	d	−	+	−	+	+	−	−
D-Galactose	+	d	−	+	−	d	d	−	−
D-Ribose	+	+	−	d	−	+	+	+	+
D-Glucose	+	+	+	+	−	+	+	+	+
D-Xylose	d	−	−	−	−	+	+	+	+
L-Arginine, DL-Citrulline, DL-Ornithine	−	+	−	−	−	+	+	+	+
L-Lysine	−	+	−	−	−	+	−	+	
meso-Erythritol	+	d	+	+	−	+	+	+	+
Preferred host	Cattle	Dogs	Sheep, goats	Desert wood rat	Sheep	Swine	Swine, hares	Swine	Reindeer

[a] Symbols: NL, no lysis; L, lysis; PL, partial lysis; +, Q$_{O_2}$N is 50 or greater; −, Q$_{O_2}$N is less than 50; d, Q$_{O_2}$N values differ among strains (may be higher or lower than 50). See *Bergey's Manual of Systematic Bacteriology*, Vol. 1, p. 383.

[b] Results apply to smooth strains only, except for *B. canis* and *B. ovis* (which occur only in the non-smooth phase).

[c] RTD, routine test dilution (see *Bergey's Manual of Systematic Bacteriology*, Vol. 1, p. 384).

Table 4.21 Differentiation of the species and biovars of the genus *Brucella*[a]

Characteristic	B. abortus biovars								B. canis	B. melitensis biovars			B. neotomae	B. ovis	B. suis biovars			
	1	2	3	4	5	6	7	9		1	2	3			1	2	3	4
CO_2 requirement	[+]	[+]	[+]	[+]	–	–	–	–	–	–	–	–	–	+	–	–	–	–
H_2S produced	+	+	+	+	–	[–]	[+]	+	–	–	–	–	+	–	+	–	–	–
Growth on media containing:																		
Thionine[b]	–	–	+[c]	–	+	+[c]	+	+	+	+	+	+	–[d]	+	+	+	+	+
Basic fuchsin	+	–	+	+	+	+	+	+	[–]	+	+	+	–	[–]	[–]	–	+	[–]
Agglutination with monospecific antisera against antigens:																		
A	+	+	+	–	–	+	+	–	–	–	+	+	+	–	+	+	+	+
M	–	–	–	+	+	–	+	+	–	+	–	+	–	–	–	–	–	+
R	–	–	–	–	–	–	–	–	+	–	–	–	–	+	–	–	–	–

[a] Symbols: +, positive for all strains; [+], positive for most strains; [–], negative for most strains; –, negative for all strains.

[b] Dye concentration, 1:50,000 (w/v).

[c] For more certain differentiation of biovars 3 and 6, thionine at 1:25,000 (w/v) is used; biovar 3 gives a positive growth response, biovar 6 is negative.

[d] Growth will occur in the presence of thionine at a concentration of 1:150,000 (w/v).

Table 4.22 Differentiation of the species of the genus *Comamonas*[a,b]

Characteristic	C. acidovorans	C. terrigena	C. testosteroni
Requirement for methionine and nicotinamide[c]	–	+	–
Assimilation of:[c,d]			
D-Fructose, D-mannitol, DL-tartrate, ethylene glycol, propylene glycol	+	–	–
DL-β-Hydroxybutyrate, testosterone	–	–	+
Fatty acid composition (%):[c]			
14:0	<1	>3	<1
2-OH 16:0	<1	<1	>2
Phosphoamidase[e]	+	+	–
Carbon sources utilized for growth:[e,f]			
D-Fructose, D-mannitol, phenylacetate, maleate	–	–	+
Glycolate, L-histidine, L-phenylalanine	+	–	+
Benzoate	+	–	–
β-Alanine	–	+	+

[a] Symbols: see standard definitions.

[b] The genus *Comamonas* was not included in *Bergey's Manual of Systematic Bacteriology*, Vol. 1. The genus was revived in 1985 by De Vos et al. (Int. J. Syst. Bacteriol. *35*: 443-453 and initially included only one species, *Comamonas terrigena*. However, two additional species were added in 1987 by Tamaoka et al. (Int. J. Syst. Bacteriol. *37*: 52-59): *Comomonas acidovorans* (previously *Pseudomonas acidovorans*) and *Comamonas testosteroni* (previously *Pseudomonas testosteroni*).

[c] Data from Tamaoka et al. (Int. J. Syst. Bacteriol. *37*: 52-59, 1987).

[d] Determined on the basal medium of Stanier et al. (J. Gen. Microbiol. *43*: 159-271, 1966).

[e] Data from De Vos et al. (Int. J. Syst. Bacteriol. *35*: 443-453, 1985).

[f] Tested with the API auxanographic system.

Table 4.23   Differential characteristics of the species of the genus _Deleya_[a]

Characteristic	_D. aquamarina_[b]	_D. cupida_[c]	_D. halophila_[d]	_D. marina_[e]	_D. pacifica_[f]	_D. venusta_[g]
Flagellar arrangement:						
Peritrichous	+	+	+	−	+	+
Polar (2–5 flagella)	−	−	−	+	−	−
Oxidase	+	−	+	−	+	+
Amylase	d	−	−	−	−	−
NO$_3^-$ reduced to NO$_2^-$	+	d	+	−	−	+
Utilization of sole carbon and energy sources:						
D-Ribose	d	d		+	−	−
L-Arabinose	−	+	+	−	−	−
D-Galactose	d	+	+	+	−	−
D-Mannose	−	+	d	−	−	−
Cellobiose	−	d	+	−	−	−
Maltose	d	d	+	−	−	d
Salicin	−	d	d	−	−	−
D-Trehalose	d	d	+	−	−	d
D-Xylose	−	d	d	−	d	−
Sucrose	d	d	+	−	−	d
Mannitol	+	+	d	+	−	d
Sorbitol	+	d	d	−	−	d
Adonitol	−	d	−	−	d	−
meso-Inositol	d	d	−	d	−	+
Saccharate		+		−	−	−
Glycolate	−	+		−	−	−
D-Gluconate	+	+	+	+	+	+
Suberate	+	−		−	−	−
Aconitate	−	+		+	d	+
δ-Aminovalerate	−	d		−	+	+
L-Histidine	−	d		−	+	d
L-Tyrosine	−	+		+	+	
DL-Kynurenine	−	−		−	+	−
Ethanolamine	−	−		−	−	+
Benzylamine	−	−		−	d	−
Putrescine		+		d	+	+
Sarcosine	−	+		d	+	+
Allantoin		+		−	d	+
Benzoate	−	+	−	−	+	d
Hippurate		d	−	−	−	−
Caprylate	d	+	−	d	d	+

[a] Symbols: see standard definitions.

[b] Formerly _Alcaligenes aestus_ in _Bergey's Manual of Systematic Bacteriology_, Vol. 1. The species was placed in the genus _Deleya_, as _D. aesta_, by Baumann et al. (Int. J. Syst. Bacteriol. _33_: 793-802, 1983). The name was later changed to _D. aquamarina_ by Akagawa and Yamasato (Int. J. Syst. Bacteriol. _39_: 462-466, 1989), who found that _D. aestus_ and _Alcaligenes aquamarinus_ were synonymous.

[c] Formerly _Alcaligenes cupidus_ in _Bergey's Manual of Systematic Bacteriology_, Vol. 1. The species was placed in the genus _Deleya_ by Baumann et al. (Int. J. Syst. Bacteriol. _33_: 793-802, 1983).

[d] _D. halophila_ was not included in _Bergey's Manual of Systematic Bacteriology_. It was created by Quesada et al., Int. J. Syst. Bacteriol. _34_: 287-292, 1984.

[e] Formerly _Pseudomonas marina_ in _Bergey's Manual of Systematic Bacteriology_, Vol. 1. The species was placed in the genus _Deleya_ by Baumann et al. (Int. J. Syst. Bacteriol. _33_: 793-802, 1983).

[f] Formerly _Alcaligenes pacificus_ in _Bergey's Manual of Systematic Bacteriology_, Vol. 1. The species was placed in the genus _Deleya_ by Baumann et al. (Int. J. Syst. Bacteriol. _33_: 793-802, 1983).

[g] Formerly _Alcaligenes venustus_ in _Bergey's Manual of Systematic Bacteriology_, Vol. 1. The species was placed in the genus _Deleya_ by Baumann et al. (Int. J. Syst. Bacteriol. _33_: 793-802, 1983).

Table 4.24   Differential characteristics of the species of the genus *Flavobacterium* [a,b]

Characteristic	*F. aquatile*	*F. balustinum*	*F. branchiophila* [c]	*F. breve*	*F. gleum* [d]	*F. indologenes* [e]	*F. meningosepticum*	*F. odoratum*	*F. thalpophilum* [f]
Acid produced aerobically from: [g]									
Glucose	+ [h]	+	+	d	+	+	d	–	+
Arabinose	–	–	–	–	d	–	–	–	+
Cellobiose	–	–	Weak	–	–	–	–	–	+
Ethanol	–	+		–	–	d	d	–	–
Lactose	+ [h]	–	–	–	–	–	d	–	+
Mannitol	–	–	–	–	–	d	d	–	–
Maltose	+ [h]	–	+	d	+	+	+	–	+
Raffinose	–	–	Weak	–	–	–	–	–	+
Salicin	–	–	–	–	–	–	–	–	+
Sucrose	+ [h]	–	+	–	–	–	–	–	+
Rhamnose	–	–	–	–	–	–	–	–	+
Trehalose	–	–	+	–	+	+	d	–	+
Xylose	–	–	–	–	d	–	–	–	+
Adonitol	–	–	–	–	–	–	–	–	+
Glycerol	–	–		–	d		d	–	+
Casein digestion	+	+	+	+	–		+	+	–
Esculin hydrolysis	–	+	–	–	+	+	+	–	+
Indole production (Ehrlich reagent)	–	+	–	+	+	+	d	–	
NO_3^- reduction	NG	+	–	–		d	–	–	+
NO_2^- reduction	NG	–		–	d	d	d	+	–
Starch hydrolysis	NG	–	+	–		+	–	–	–
Urease production	–	–		–	d	–	d	+	+
β-Galactosidase production (ONPG hydrolysis)	–	–	–	–	d		+	–	+
Gelatinase production (plate method)	–	+	+	+		+	+	+	d
Growth at 42°C	–	–	–	–			d	–	+

[a] Symbols: see standard definitions; NG, no growth on test medium.

[b] Two species *(F. spiritivorum* and *F. multivorum)* previously included in the genus *Flavobacterium* in *Bergey's Manual of Systematic Bacteriology*, Vol. 1, have been reclassified in the genus *Sphingobacterium* (Yabuuchi et al., Int. J. Syst. Bacteriol. *33*: 580-598, 1983).

[c] *F. branchiophila* (sic) was not included in *Bergey's Manual of Systematic Bacteriology*, Vol. 1. The species was created by Wakabayashi et al. (Int. J. Syst. Bacteriol. *39*: 213-216, 1989). The species causes bacterial gill disease of freshwater fishes.

[d] *F. gleum* was not included in *Bergey's Manual of Systematic Bacteriology*, Vol. 1. The species was created by Holmes et al. (Int. J. Syst. Bacteriol. *34*: 21-25, 1984).

[e] *F. indologenes* was not included in *Bergey's Manual of Systematic Bacteriology*, Vol. 1. The species was created by Yabuuchi et al. (Int. J. Syst. Bacteriol. *33*: 580-598, 1983).

[f] *F. thalpophilum* was not included in *Bergey's Manual of Systematic Bacteriology*, Vol. 1. The species was created by Holmes et al. (Int. J. Syst. Bacteriol. *33*: 677-682, 1983).

[g] Tested in ammonium salts medium.

[h] Delayed reaction.

Table 4.25   Differential characteristics of the species of the genus *Francisella*[a]

Characteristic	F. novicida	F. tularensis biovar *tularensis*[b]	F. tularensis biovar *palaearctica*[c]
Capsule present	–	v[d]	v[d]
Growth requirement for cysteine or cystine	–	+	+
Growth on ordinary media (blood agar, gelatin, peptone broth)	+	–	–
Colonies on glucose-cystine-blood agar >5 mm in diameter	+	–	–
Acid produced from:			
Maltose	–	+	+[e]
Sucrose	+	–	–
Glycerol	+	+	–[e]
Citrulline ureidase	+	+	–[e]
Serum agglutination:			
F. tularensis	–	+	+
F. novicida	+	–	–
Vaccine efficacy:			
Killed vaccine			
F. tularensis	–	+[f]	+[f]
F. novicida	+	–	–
Live vaccine			
F. tularensis	+	+	+
F. novicida	+	+[f]	+[f]

[a] Symbols: see standard definitions.

[b] This biovar has been proposed as a subspecies, *Francisella tularensis* subsp. *tularensis*, by Olsufjev and Meshcheryakova (Int. J. Syst. Bacteriol. *33*: 872-874, 1983).

[c] This biovar has been proposed as a subspecies, *Francisella tularensis* subsp. *holarctica*, by Olsufjev and Meshcheryakova (Int. J. Syst. Bacteriol. *33*: 872-874, 1983).

[d] Capsule associated with virulence of the strain.

[e] Central Asian strains (proposed biovar *mediaasiatica*) reported to give only a weak fermentation of maltose (which is revealed by the Rodionova method based on a determination of dehydrogenases) but a positive fermentation of glycerol and citrulline. Biovar *mediaasiatica* has been proposed as a subspecies, *Francisella tularensis* subsp. *mediaasiatica*, by Olsufjev and Meshcheryakova (Int. J. Syst. Bacteriol. *33*: 872-874, 1983).

[f] Protects fully susceptible hosts only against strains of partially reduced virulence.

Table 4.26 Differential characteristics of three species of the genus *Gluconobacter*[a]

Characteristic	G. asaii[b]	G. frateurii[c]	G. oxydans[d]
Growth without nicotinate[e]	d[f]	+	−[g]
Growth on ribitol (adonitol)[h]	−	+	−
Growth on arabitol[h]	−	+	−

[a] Symbols: see standard definitions.

[b] The species *G. asaii* was not included in *Bergey's Manual of Systematic Bacteriology*, Vol. 1. It was created by Mason and Claus (Int. J. Syst. Bacteriol. *39*: 174-184, 1989).

[c] The species *G. frateurii* was not included in *Bergey's Manual of Systematic Bacteriology*, Vol. 1. It was created in 1989 by Mason and Claus (Int. J. Syst. Bacteriol. *39*: 174-184). On the basis of DNA homology experiments, many of the strains in this species were removed from another species, *G. cerinus*, which had been created in 1984 by Yamada and Akita (J. Gen. Appl. Microbiol. *30*: 115-126). The two species apparently are phenotypically similar but genetically different, since the type strain of *G. cerinus* has only a low degree of DNA homology with the type strain of *G. frateurii*. Further studies are required to indicate what phenotypic characteristics can be used to differentiate *G. cerinus* reliably from *G. frateurii*; meanwhile, only DNA hybridization experiments can provide reliable differentiation.

[d] Data based on Mason and Claus (Int. J. Syst. Bacteriol. *39*: 174-184, 1989).

[e] Tested with standardized inocula in the Casamino acids-mannitol-vitamin medium described by Mason and Claus (Int. J. Syst. Bacteriol. *39*: 174-184, 1989). Positive growth response = an optical density (O.D.) of 1.0 or more after three passages (24 h each) in nicotinate-deficient media. Negative growth response = an O.D. of 0.5 or less more after three passages (24 h each) in nicotinamide-deficient media.

[f] Most strains are positive. The only negative strain so far encountered had the unique ability to cause an alkalinization of ribitol or arabitol medium, thereby differentiating it from *G. oxydans*.

[g] Ten percent of strains tested grow to an O.D. between 0.5 and 1.0.

[h] Tested with standardized inocula in a medium containing 0.5% vitamin-free Casamino acids, 0.5% yeast extract, and 0.5% of the test polyol (Mason and Claus, Int. J. Syst. Bacteriol. *39*: 174-184, 1989). Positive growth response = growth to an O.D. of 1.0 or more within 24 h; negative growth response = O.D. of 0.5 or less within 24 h.

Table 4.27 Differential characteristics of the species of the genus *Halomonas*[a]

Characteristic	H. elongata	H. halodurans[b]	H. subglaciescola[c]
Motility	+	+	d
Lysine decarboxylase	+	+	−
β-Galactosidase	+	−	−
Utilization of:[d]			
Glucose, sucrose, cellobiose, lactose, citrate, gluconate, valine	+		−
Glycerol	+	+	−
DL-serine, L-tyrosine	+	+	−
L-cysteine, dihydrocholesterol, heptadecanoic acid	−	+	
Inulin	−	+	

[a] Symbols: see standard definitions

[b] *H. halodurans*, previously named *Pseudomonas halodurans*, was not included in *Bergey's Manual of Systematic Bacteriology*, Vol. 1. It was included in the genus *Halomonas* by Hebert and Vreeland, Int. J. Syst. Bacteriol. *37*: 347-350, 1987.

[c] *H. subglaciescola* was not included in *Bergey's Manual of Systematic Bacteriology*, Vol. 1. It was created in 1987 by Franzmann et al., Int. J. Syst. Bacteriol. *37*: 27-34.

[d] Tested in media containing 8% NaCl.

Table 4.28 Differential characteristics of the species of the genus *Hydrogenophaga*[a,b]

Characteristic	H. flava[c]	H. palleronii[d]	H. pseudoflava[e]	H. taeniospiralis[f]
Growth on:				
Glycolate	+	+	+	−
L-arabinose, sucrose, D-galactose, D-fructose, D-mannose, mannitol, D-arabitol, sorbitol, D-cellobiose, butylamine, ethanolamine	+	−	+	+
Maltose, D-turanose, L-histidine	+	−	+	−
L-Fucose	+	−	−	+
Azelate	−	+	+	+
D-Xylose, DL-5-aminovalerate	−	−	+	+
Lactose	−	−	+	−
DL-3-Aminobutyrate	−	−	−	+
Denitrification	−	−	+	+
Reduction of NO_3^-	+	−	+	+
Reduction of NO_2^-	−	−	+	+
Presence of 3-hydroxy-decanoic acid	+	−	+	−

[a] Symbols: see standard definitions.

[b] Data from Willems et al. (Int. J. Syst. Bacteriol. *39*: 319-333, 1989).

[c] Previously classified as *Pseudomonas flava* in *Bergey's Manual of Systematic Bacteriology*, Vol. 1. This species grows slowly and can easily lose its chemolithoautotrophic ability.

[d] Previously classified as *Pseudomonas palleronii* in *Bergey's Manual of Systematic Bacteriology*, Vol. 1.

[e] Previously classified as *Pseudomonas pseudoflava* and "*Pseudomonas carboxydoflava*" in *Bergey's Manual of Systematic Bacteriology*, Vol. 1.

[f] *H. taeniospiralis* was not included in *Bergey's Manual of Systematic Bacteriology*, Vol. 1. It was originally described by Lalucat et al. (Int. J. Syst. Bacteriol. *32*: 332-338, 1982) as *Pseudomonas taeniospiralis* and was classified in the genus *Hydrogenophaga* by Willems et al. (Int. J. Syst. Bacteriol. *39:* 319-333, 1989).

Table 4.29 Differential characteristics of the species of the genus *Kingella*[a]

Characteristic	K. denitrificans	K. indologenes	K. kingae
β-Hemolytic	−	−	+
Growth in presence of 4% NaCl	−	+	−
NO_3^- reduction	+	−	−
NO_2^- reduction [b]	+	−	−
Gas produced from NO_2^-	+	−	−
Phosphatase activity	−	+	+
Casein digestion	−	+	+
Indole production	−	+	−
Tween 40 hydrolysis	−	+	−

[a] Symbols: see standard definitions.

[b] NO_2^- reduction is positive in all three species when a rich basal medium such as Mueller-Hinton broth with yeast extract is used. Using less rich media, *K. kingae* and *K. indologenes* give negative reactions.

Table 4.30 Differentiation of the species of the genus *Marinomonas*[a,b]

Characteristic	M. communis[c]	M. vaga[d]
Cell shape:		
Curved	+	−
Straight	−	+
Growth at 40°C	+	−
Utilization of:		
Erythritol, N-Acetylglucosamine, heptanoate	−	+
Ethanol, *n*-propanol	+	−
Cellobiose, ᴅ-ribose, ᴅ-arabinose, caproate, caprylate, pelargonate, adonitol, trigonelline	−	d
ᴅʟ-Glycerate, ethanolamine	d	−
ʟ-Arabinose, ʟ-rhamnose	d	+
ᴅ-Alanine, ʟ-serine, ʟ-lysine, ʟ-arginine, spermine	+	d

[a] Symbols: see standard definitions.

[b] The genus *Marinomonas* was not included in *Bergey's Manual of Systematic Bacteriology*, Vol. 1. The genus was created in 1983 by Landschoot and De Ley (J. Gen. Microbiol. *129*: 3057-3074).

[c] Previously classified as *Alteromonas communis* in *Bergey's Manual of Systematic Bacteriology*, Vol. 1.

[d] Previously classified as *Alteromonas vaga* in *Bergey's Manual of Systematic Bacteriology*, Vol. 1.

Important Notes for Users of This Manual

Unless otherwise indicated in footnotes to tables, the meanings of symbols are as follows:

+ 90% or more of strains are positive

− 90% or more of strains are negative

d 11-89% of strains are positive

v strain instability (not equivalent to "d")

D Different reactions in different taxa (species of a genus or genera of a family)

All other symbols are defined in footnotes to tables.

Table 4.31 Differential characteristics of the species of the genus *Methylobacterium*[a,b]

Characteristic	M. extor-quens[c]	M. fujisa-waense[d]	M. meso-phili-cum[e]	M. organo-philum	M. radio-tole-rans[e]	M. rhodesi-anum[d]	M. rhodi-num[e]	M. zat-manii[d]
Substrates utilized as sole carbon sources:[f]								
D-Glucose	−	+	+	+	+	−	weak	−
D-Fucose, D-xylose, L-arabinose	−	+	+	−	+	−	−	−
Fructose	−	d	−	+	−	+	+	+
L-Aspartate, L-glutamate	d	+	+	−	+	d	+	−
Citrate	−	+	+	−	+	−	+	−
Sebacate	−	+	d	−	+	−	−	−
Acetate	+	+	−	+	+	+	+	+
Betaine	+	−	−	−	+	+	+	−
Tartrate	d	d	−	−	−	−	−	d
Ethanol	+	d	+	+	d	+	+	+
Methylamine	+	−	−	+	−	+	+	+
Trimethylamine	−	−	−	+	−	−	−	d
Methane	−	−	−	d	−	−	−	−
Growth on peptone-rich nutrient agar (Oxoid CM55)	d	+	−	+	+	+	+	+

[a] Symbols: see standard definitions.

[b] Although included on the 1980 Approved Lists, the genus *Methylobacterium* was not described in *Bergey's Manual of Systematic Bacteriology*, Vol. 1.

[c] Urakami and Komagata (Int. J. Syst. Bacteriol. *34*: 188-201, 1984) created a genus called *Protamonas*, which included one species, *P. extorquens* (formerly *Pseudomonas extorquens*). However, in 1985 Bousfield and Green (Int. J. Syst. Bacteriol. *35*: 209) reclassified this species in the genus *Methylobacterium*.

[d] *M. rhodesianum, M. zatmanii*, and *M. fujisawaense* were created by Green et al. (Int. J. Syst. Bacteriol. *38*: 124-127, 1988).

[e] *M. rhodinum* (formerly *Pseudomonas rhodos*), *M. radiotolerans* (formerly *Pseudomonas radiora*), and *M. mesophilicum* (formerly *Pseudomonas mesophilica*) were included in the genus *Methylobacterium* by Green and Bousfield (Int. J. Syst. Bacteriol. *33*: 875-877, 1983).

[f] Data from Green et al. (Int. J. Syst. Bacteriol. *38*: 124-127, 1988). The tests are read after 14 days of incubation at 30°C. Doubtful results are checked by twice subculturing in liquid medium.

Table 4.32 Differential characteristics of the species of the genus *Methylomonas*[a]

Characteristic	M. methanica	M. pelagica[b]
NaCl or sea water required for growth	−	+
Cysts formed	+	−
Formaldehyde utilized	+	−
Colony pigmentation	Pink	White

[a] Symbols: see standard definitions.

[b] The species *Methylomonas pelagica* was not included in *Bergey's Manual of Systematic Bacteriology*, Vol. 1. It was created in 1987 by Sieburth et al. (Curr. Microbiol. *14*: 285-293; Int. J. Syst. Bacteriol. *38*: 136-137, 1988).

Table 4.33   Differential characteristics of the species of the genus _Methylophaga_ [a,b,c]

Characteristic	M. marina	M. thalassica
Utilization of:		
Fructose	+(100)	d (86)
Dimethylamine	d (14)	d (86)
Trimethylamine	– (0)	d (71)
Enzymes of monomethylamine oxidation:		
Methylamine dehydrogenase	d (71)	d (14)
N-Methylglutamate dehydrogenase	d (29)	d (86)

[a] Symbols: see standard definitions. Numbers in parentheses indicate the percentage of positive strains.

[b] The genus _Methylophaga_ was not included in _Bergey's Manual of Systematic Bacteriology_, Vol. 1. The genus was created in 1985 by Janvier et al. (Int. J. Syst. Bacteriol. _35_: 131-139) and included two species, _M. marina_ and _M. thalassica_.

[c] Unequivocal differentiation between the species is not yet achievable by routine phenotypic characterization tests. However, it can be achieved by DNA hybridization experiments (see Janvier et al., Int. J. Syst. Bacteriol. _35_: 131-139, 1985).

Important Notes for Users of This Manual

Unless otherwise indicated in footnotes to tables, the meanings of symbols are as follows:

+   90% or more of strains are positive

−   90% or more of strains are negative

d   11-89% of strains are positive

v   strain instability (not equivalent to "d")

D   Different reactions in different taxa (species of a genus or genera of a family)

All other symbols are defined in footnotes to tables.

Table 4.34 Differential characteristics of the species of the genus *Moraxella*[a]

Characteristic	Subgenus *Branhamella*				Subgenus *Moraxella*					
	M. (B.) catarrhalis	*M. (B.) caviae*	*M. (B.) cuniculi*	*M. (B.) ovis*	*M. (M.) atlantae*	*M. (M.) bovis*	*M. (M.) lacunata*	*M. (M.) nonliquefaciens*	*M. (M.) osloensis*	*M. (M.) phenylpyruvica*
Cell shape:										
Rods	−	−	−	−	+	+	+	+	+	+
Cocci	+	+	+	+	−	−	−	−	−	−
Hemolysis (human blood)	−	W	−	[+]	−	[+]	−	−	−	−
Colony consistency:										
Friable	+	−								
Butyrous	−	+								
Colony size:										
Large					−			+	−	−
Small					+			−	−	−
Gelatin liquefaction	−	−	−	−	−	[+]		−	−	−
Serum liquefaction	−	−	−	−	−	[+]	+	−	−	−
Growth on mineral salts + NH$_4^+$ + acetate	−	−	−	−	−	−	−	−	[+]	[−][b]
Growth at 5°C	−	−	−	−	−	[−]	−	−	−	[+]
Growth in presence of 6% NaCl	−	−	−	−	+	−	−	−	−	[+]
Growth stimulated by bile salts	−	−	−	−	−	−	−	−	−	+
Phenylalanine deaminase	−	−	−	−	−	−	−	−	−	[+]
Urease	−	−	−	[+]	−	[−]	−	−	[−][c]	d
NO$_3^-$ reduction	[+]	+	−	−	−	[−]	+	+	d	[+]
NO$_2^-$ reduction	[+]	[+]	−	−	−	−	−	−	−	−
Sensitive to penicillin:										
1.0 U/ml	[+][d]	+	+	+	+	+	+	+	[+][d]	[+][d]
0.1 U/ml (β-lactamase positive strains omitted)	+	−	+	+	+	+	+	+	d	+
In complex media, utilization of:										
Butyrate	d	+	−	+	+	+	+	−	[+]	−
Caproate	−	−	−	+	+	+	+	−	[+]	−
Ethanol	−	−	−	−	d	+	d	−	+	−
Propionate	−	−	−	−	−	−	−	+	+	−
Acetate	d	d	−	d	+	+	+	[−]	+	+
Presence of thymidine phosphorylase, nucleoside deoxyribosyltransferase, and thymidine kinase	−	−	−	−		−	−	[−][e]	−	−
True waxes present in cell wall	+	+	+	+	+	+	+	+	+	−

Table 4.34 (continued)

Characteristic	Subgenus *Branhamella*				*M. (M.) atlantae*	*M. (M.) bovis*	*M. (M.) lacu-nata*	Subgenus *Moraxella*		
	M. (B.) catar-rhalis	*M. (B.) caviae*	*M. (B.) cuniculi*	*M. (B.) ovis*				*M. (M.) nonliquefaciens*	*M. (M.) osloensis*	*M. (M.) phenylpyruvica*
Has been isolated from:[f]										
Humans	+				+		+	+	+	+
Cattle				+		+				+
Sheep				+						
Goats										+
Horses				+		+				
Pigs										+
Rabbits			+							
Guinea pigs		+					+			

[a] Symbols: +, all tested strains positive (but see footnote *f*); [+], most strains positive; [−], most strains negative; −, all tested strains negative; W, weakly positive.

[b] Some strains of *M. (M.) phenylpyruvica* may grow in the minimal medium; these strains are distinguished from *M. (M.) osloensis* by the phenylalanine deaminase reaction and/or a strong urease reaction. The ability of *M. (M.) phenylpyruvica* to grow weakly at 4–5°C will further help to distinguish it from *M. (M.) osloensis*.

[c] A few strains are weakly urease-positive.

[d] Some strains have been found to be penicillin-resistant on the basis of β-lactamase production (no β-lactamase-negative strain grows in the presence of 1.0 U/ml penicillin).

[e] The type strain is positive for thymidine kinase only.

[f] Mainly based on strains with genetically or gas-chromatographically confirmed identity; at least one isolate from a host is indicated by the symbol +.

Table 4.35   Differential characteristics of the species of the genus *Neisseria* and of *Morococcus cerebrosus*[a,b]

Characteristic	*N. canis*	*N. cinerea*	*N. denitrificans*	*N. elongata*	*N. flavescens*	*N. gonorrhoeae*	*N. lactamica*	*N. macacae*[c]	*N. meningitidis*	*N. mucosa*	*N. polysaccharea*[d]	*N. sicca*	*N. subflava*	*Morococcus cerebrosus*
Cell shape:														
Cocci	+	+	+	−	+	+	+	+	+	+	+	+	+	+
Short rods	−	−	−	+	−	−	−	−	−	−	−	−	−	−
Cell arrangement:														
Pairs	+	+	+	+	+	+	+	+	+	+	+	+	+	+
Tetrads	−	−	−	−	+	−	−	+	+	−	+	+	+	−
Short chains	−	−	−	+	−	−	−	−	−	−	−	−	−	+
Mulberry-like aggregates of 10–20 cells	−	−	−	−	−	−	−	−	−	−	−	−	−	−
Yellowish pigment	−	d	d	Weak	+	−	+	+	−	d	+	d	+	−
Hemolysis on blood agar:														
Sheep	−	−	−	−	−	−	−	−	−	−		d	−	
Horse	−	−	−	−	−	−	d	+	−	−		d	−	+
Rabbit	d	−	−	−	−	−	−	+	−	−		d	−	
Human	−	−	−	−	−	−	−	−	−	−		d	−	
Acid produced from:														
Glucose	−	−	+	−[e]	−	+	+	+	+	+	+	+	+	+
Maltose	−	−	−	−	−	−	+	+	+	+	+	+	+	+
Fructose	−	−	+	−	−	−	−	+	−	+	−	+	d	+
Sucrose	−	−	+	−	−	−	−	+	−	+	Rare	+	d	+
Mannose	−	−	+	−	−	−	−	−	−	−	−	−	−	−
Lactose	−	−	−	−	−	−	+	−	−	−	−	−	−	−
NO₃⁻ reduction	+	−	−	−	−	−[f]	−	−	−[f]	+	+	−	−	+
NO₂⁻ reduction	−	+	+	+	+	−	+	+	−	+	−	+	+	v
Gas from NO₂⁻	−	+	+	+	+	−	+	+	−	+	−	+	+	v
Polysaccharide synthesized from sucrose (iodine test)	−	−	+	−	+	−	−	+	−	+	+	+	d[g]	+

[a] Symbols: see standard definitions.

[b] The "false neisserias" (*N. caviae*, *N. ovis*, and *N. cuniculi*) are now classified in the genus *Moraxella*, subgenus *Branhamella*.

[c] *N. macacae* was not included in *Bergey's Manual of Systematic Bacteriology*, Vol. 1. It was created in 1983 by Vedros et al. (Int. J. Syst. Bacteriol. *33*: 515–520).

[d] *N. polysaccharea* was not included in *Bergey's Manual of Systematic Bacteriology*, Vol. 1. It was created in 1987 by Riou and Guibourdenche (Int. J. Syst. Bacteriol. *37*: 163–165).

[e] A few strains may form a small amount of acid from glucose but most strains are negative.

[f] Potassium nitrite in low concentrations (0.01%) can be reduced by *N. gonorrhoeae* and by serogroups A, D, and Y of *N. meningitidis*.

[g] Biovar *perflava* is positive; the reaction differs among strains of biovar *subflava*.

149

Table 4.36 Differential characteristics of the species of the genus *Oligella*[a,b]

Characteristic	O. ureolytica[c]	O. urethralis[d]
Motility	d[e]	−
Growth at 42°C	−	+
Urease	+	−
NO$_3^-$ reduction	d	−
Denitrification	d[f]	+[f]
Utilization of *p*-hydroxybenzoate as a carbon source for growth[g]	+	−
Presence of 19:cyc cellular fatty acid (>1%)	+	−

[a] Symbols: see standard definitions.
[b] Data from Rossau et al. (Int. J. Syst. Bacteriol. *37:* 198-210, 1987).
[c] Previously classified as CDC group IVe strains.
[d] Previously classified as *Moraxella urethralis*.
[e] Motility, when it occurs, is by means of long peritrichous flagella.
[f] Demonstration of gas may be difficult.
[g] Tested with API ZYM galleries.

Table 4.37 Differential characteristics of the species of the genus *Paracoccus*[a]

Characteristic	P. alcaliphilus[b]	P. denitrificans	P. halodenitrificans
Halophilic, requiring at least 3% NaCl for growth	−	−	+
Facultative hydrogen autotroph	−	+	−
Thiamin required for growth	−	−	+
Biotin required for growth	+	−	−
Denitrification (to N$_2$)	−	+	+
Growth at pH 6	−	+	
Growth at pH 9	+	−	
Utilization of:			
D-Xylose	+	−	
Trehalose	−	+	

[a] Symbols: see standard definitions
[b] *Paracoccus alcaliphilus* was not included in *Bergey's Manual of Systematic Bacteriology*, Vol. 1. The species was created in 1987 by Rossau et al. (Int. J. Syst. Bacteriol. *37:* 198-210).

Table 4.38 Differential characteristics of the species of the genus *Phyllobacterium*[a]

Characteristic	P. myrsinacearum	P. rubiacearum
Flagellar arrangement:		
One to several flagella with polar or lateral attachment	+	−
Single polar flagellum only	−	+
Occur in leaf nodules of	*Myrsinaceae*	*Rubiaceae*
NO$_3^-$ reduced to NO$_2^-$	+	−

[a] Symbols: see standard definitions.

Table 4.39 Differentiation of some well studied species of the genus *Pseudomonas*[a,b]

Characteristic	P. aeruginosa	P. alcaligenes	P. caryophylli	P. cepacia	P. chlororaphis[c]
Number of flagella	1	1	>1	>1	>1
Fluorescent, diffusible pigments	+	−	−	−	+
Diffusible non-fluorescent pigments	+ (blue-green[e])	−	+ (yellow-green)	+ (various colors)	−
Non-diffusible nonfluorescent pigments	−	d (yellow-orange)	−	−	+ (green or orange)
Poly-β-hydroxybutyrate accumulation	−	−	+	+	−
Extracellular poly-β-hydroxybutyrate hydrolysis		−	−	−	
Autotrophic growth with H_2	−	−	−	−	−
Growth at 41°C	+	+	+	d	−
Growth at 4°C	−		−	−	+
Organic growth factors required (pantothenate, biotin, cyanocobalimin, methionine, or cystine)	−	−	d	−	−
Levan formation from sucrose	−	−			+
Arginine dihydrolase	+	+	+	−	+
Oxidase reaction	+	+	+	+	+
Denitrification	+	+	+	−	+
Gelatin hydrolysis	+	d	−	d	+
Starch hydrolysis	−	−	−	−	−
Utilization of:					
Glucose	+	−	+	+	+
Trehalose	−	−	d	+	+
2-Ketogluconate	+	−	+	+	+
meso-Inositol	−	−	+	+	+
Geraniol	+	−	−	−	−
L-Valine	d	−	d	d	+
β-Alanine	+	d	+	+	+
L-Arginine	+	+	+	+	+
D-Xylose			+	d	
D-Ribose			+	+	
L-Rhamnose			+	d	
Saccharate			+	+	
Levulinate			−	+	
Citraconate			−	+	
Mesaconate			−	−	
D(-)-Tartrate			−	−	
meso-Tartrate			+	+	
Erythritol			−	−	
Adonitol			−	+	
2,3-Butylene glycol			+	+	
m-Hydroxybenzoate			−	+	
Tryptamine			−	+	
α-Amylamine			−	+	
Sucrose					
Malonate					
Nitrate used as a nitrogen source	+	+	+	+	+
Lecithinase (egg yolk reaction)	−	−	d	d	d
Catechol, *ortho* cleavage	+				+
Protocatechuate:					
ortho cleavage	+		+	+	+
meta cleavage	−		−	−	−
Saprophytic or opportunistic animal pathogens	+	+	−	+	+
Phytopathogens	−	−	+	−	−
Parasite of animals and humans	−	−	−	−	−

Footnotes are at end of table

Table 4.39 *(continued)*

Characteristic	P. cichorii	P. delafieldii	P. diminuta	P. facilis
Number of flagella	>1	1	1	1
Fluorescent, diffusible pigments	+	−	−	−
Diffusible non-fluorescent pigments	−	−	−	−
Non-diffusible nonfluorescent pigments	−	−	−	−
Poly-β-hydroxybutyrate accumulation	−	+	+	+
Extracellular poly-β-hydroxybutyrate hydrolysis		+	−	+
Autotrophic growth with H_2	−	−	−	+
Growth at 41°C	−	−	d	−
Growth at 4°C	−		−	
Organic growth factors required (pantothenate, biotin, cyanocobalimin, methionine, or cystine)	−	−	+	−
Levan formation from sucrose	−			
Arginine dihydrolase	−	−		−
Oxidase reaction	+	+	+	+
Denitrification	−	−	−	−
Gelatin hydrolysis	−	−	−	+
Starch hydrolysis	−	−	−	−
Utilization of:				
Glucose	+	+	−	+
Trehalose	−	−	−	−
2-Ketogluconate	−	+	−	−
meso-Inositol	d	−	−	−
Geraniol	−		−	
L-Valine	−	−	−	−
β-Alanine	−	+	−	+
L-Arginine	+		−	
D-Xylose		+	−	d
D-Ribose		+	−	+
L-Rhamnose		−	−	−
Saccharate		+	−	−
Levulinate		−	−	−
Citraconate		+	−	−
Mesaconate		−	−	−
D(-)-Tartrate		−	−	−
meso-Tartrate			−	
Erythritol		−	−	−
Adonitol		−	−	−
2,3-Butylene glycol		−	−	−
m-Hydroxybenzoate		−	−	−
Tryptamine		−	−	−
α-Amylamine		−	−	−
Sucrose		−	−	−
Malonate		+	−	+
Nitrate used as a nitrogen source	+	+	−	+
Lecithinase (egg yolk reaction)	+	−		
Catechol, *ortho* cleavage	+			
Protocatechuate:				
ortho cleavage		−	−	−
meta cleavage		+	+	+
Saprophytic or opportunistic animal pathogens	−	+	+	+
Phytopathogens	+	−	−	−
Parasite of animals and humans	−	−	−	−

Footnotes are at end of table

Table 4.39 (continued)

Characteristic	P. fluorescens				
	biovar I	biovar II	biovar III	biovar IV	biovar V
Number of flagella	>1	>1	>1	>1	>1
Fluorescent, diffusible pigments	+	+	+	+	+
Diffusible non-fluorescent pigments	–	–	–	–	–
Non-diffusible nonfluorescent pigments	–	–	–	+ (blue)	–
Poly-β-hydroxybutyrate accumulation	–	–	–	–	–
Extracellular poly-β-hydroxybutyrate hydrolysis					
Autotrophic growth with H_2	–	–	–	–	–
Growth at 41°C	–	–	–	–	–
Growth at 4°C	+	+	+	+	d
Organic growth factors required (pantothenate, biotin, cyanocobalimin, methionine, or cystine)	–	–	–	–	–
Levan formation from sucrose	+	+	–	+	–
Arginine dihydrolase	+	+	+	+	+
Oxidase reaction	+	+	+	+	+
Denitrification	–	+	+	+	–
Gelatin hydrolysis	+	+	+	+	+
Starch hydrolysis	–	–	–	–	–
Utilization of:					
Glucose	+	+	+	+	+
Trehalose	+	+	+	+	+
2-Ketogluconate	+	+	+	+	+
meso-Inositol	+	+	+	+	+
Geraniol	–	–	–	–	–
L-Valine	+	+	+	+	+
β-Alanine	+	+	+	+	+
L-Arginine	+	+	+	+	+
D-Xylose					
D-Ribose					
L-Rhamnose					
Saccharate					
Levulinate					
Citraconate					
Mesaconate					
D(-)-Tartrate					
meso-Tartrate					
Erythritol					
Adonitol					
2,3-Butylene glycol					
m-Hydroxybenzoate					
Tryptamine					
α-Amylamine					
Sucrose					
Malonate					
Nitrate used as a nitrogen source	+	+	+	+	+
Lecithinase (egg yolk reaction)	+	Weak	+	+	d
Catechol, ortho cleavage	+	+	+	+	+
Protocatechuate:					
ortho cleavage	+	+	+	+	+
meta cleavage	–	–	–	–	–
Saprophytic or opportunistic animal pathogens	+	+	+	+	+
Phytopathogens	–	–	–	–	–
Parasite of animals and humans	–	–	–	–	–

Footnotes are at end of table

153

Table 4.39 *(continued)*

Characteristic	P. gladioli	P. mallei	P. mendocina	P. pickettii	P. pseudo-alcaligenes
Number of flagella	>1	0	1 [d]	1	1
Fluorescent, diffusible pigments	–	–	–	–	–
Diffusible non-fluorescent pigments	+ (yellow-green)	–	–	–	–
Non-diffusible nonfluorescent pigments	–	–	+ (yellow-orange)	–	–
Poly-β-hydroxybutyrate accumulation	+	+		+	–
Extracellular poly-β-hydroxybutyrate hydrolysis	–	d	–	–	
Autotrophic growth with H_2	–	–	–	–	
Growth at 41°C	+	+	+	+	+
Growth at 4°C	–			–	
Organic growth factors required (pantothenate, biotin, cyanocobalimin, methionine, or cystine)	–	–	–	–	–
Levan formation from sucrose			–	–	–
Arginine dihydrolase	–	+	+	–	d
Oxidase reaction	+	+	+	+	+
Denitrification	–	+	+	+	+
Gelatin hydrolysis	+	+	–	–	d
Starch hydrolysis	–	d	–	–	–
Utilization of:					
Glucose	+	+	+	+	–
Trehalose	+	+	–	–	–
2-Ketogluconate	+	d	–	+	–
meso-Inositol	+	+	–	–	–
Geraniol	–	–	+	–	–
L-Valine	+	d	+	+	–
β-Alanine	+	+	+	+	d
L-Arginine	+	+	+	–	+
D-Xylose	+	+		+	
D-Ribose	+	–		–	
L-Rhamnose	–	–		–	
Saccharate	+	–		+	
Levulinate	–	–		+	
Citraconate	+	–		+	
Mesaconate	+	–		–	
D(-)-Tartrate	+	–		–	
meso-Tartrate	+	–		+	
Erythritol	–	–			
Adonitol	+	–		–	
2,3-Butylene glycol	–	–		–	
m-Hydroxybenzoate	–	–		–	
Tryptamine	–	–		–	
α-Amylamine	–	+		–	
Sucrose					
Malonate					
Nitrate used as a nitrogen source		+	+		+
Lecithinase (egg yolk reaction)	+		–	–	
Catechol, *ortho* cleavage		+	+		
Protocatechuate:					
ortho cleavage	+	+		+	
meta cleavage	–	–		–	
Saprophytic or opportunistic animal pathogens	–	–	+	+	+
Phytopathogens	+	–	–	–	d [h]
Parasite of animals and humans	–	+ [i]	–	–	–

Footnotes are at end of table

154

Table 4.39 *(continued)*

Characteristic	P. pseudo-mallei	P. putida biovar A	P. putida biovar B	P. saccharophila	P. solana-cearum
Number of flagella	>1	>1	>1	1	>1
Fluorescent, diffusible pigments	–	+	+	–	–
Diffusible non-fluorescent pigments	–	–	–	–	d (brown)
Non-diffusible nonfluorescent pigments	d (orange)	–	–	–	–
Poly-β-hydroxybutyrate accumulation	+	–	–	+	+
Extracellular poly-β-hydroxybutyrate hydrolysis	+	–	–	–	–
Autotrophic growth with H$_2$	–	–	–	+	–
Growth at 41°C	+	–	–	–	–
Growth at 4°C	–	d	+		
Organic growth factors required (pantothenate, biotin, cyanocobalimin, methionine, or cystine)	–	–	–	–	–
Levan formation from sucrose		–	–		–
Arginine dihydrolase	+	+	+	–	–
Oxidase reaction	+	+	+	+	+
Denitrification	+	–	–	–	+
Gelatin hydrolysis	+	–	–	+	–
Starch hydrolysis	+	–	–	+	–
Utilization of:					
Glucose	+	+	+	+[f]	+
Trehalose	+	–	–	+	+
2-Ketogluconate	+	+	+	–	–
meso-Inositol	+	–	–	–	d
Geraniol	–	–	–	–	–
L-Valine	+	+	+	–	–
β-Alanine	+	+	+	–	d
L-Arginine	+	+	+		–
D-Xylose	–			+	–
D-Ribose	+			+	d
L-Rhamnose	–			–	–
Saccharate	–			–	+
Levulinate	+			+	d
Citraconate	–			–	–
Mesaconate	–			–	–
D(-)-Tartrate	–			–	–
meso-Tartrate	–				d
Erythritol	+			–	–
Adonitol	d			–	–
2,3-Butylene glycol	–			–	–
m-Hydroxybenzoate	–			–	–
Tryptamine	–			–	–
α-Amylamine	+			–	–
Sucrose				+	
Malonate				–	
Nitrate used as a nitrogen source	+	+	+	+	
Lecithinase (egg yolk reaction)		–	–		–
Catechol, *ortho* cleavage	+	+	+		
Protocatechuate:					
ortho cleavage	+	+	+	–	+
meta cleavage	–	–	–	+	–
Saprophytic or opportunistic animal pathogens	+[g]	+	+	+	–
Phytopathogens	–	–	–	–	+
Parasite of animals and humans	–	–	–	–	–

Footnotes are at end of table

Table 4.39 *(continued)*

Characteristic	*P. stutzeri*	*P. syringae* pathovars	*P. vesicularis*	*P. viridiflava*
Number of flagella	1 [d]	>1	1	1–2
Fluorescent, diffusible pigments	–	+	–	+
Diffusible non-fluorescent pigments	–	–	–	–
Non-diffusible nonfluorescent pigments	–	–	+ (yellow-orange)	d (blue-green)
Poly-β-hydroxybutyrate accumulation	–	–	+	–
Extracellular poly-β-hydroxybutyrate hydrolysis	–			
Autotrophic growth with H_2	–	–	–	–
Growth at 41°C	d	–	–	–
Growth at 4°C		d	–	
Organic growth factors required (pantothenate, biotin, cyanocobalimin, methionine, or cystine)	–	–	+	–
Levan formation from sucrose	–	d		
Arginine dihydrolase	–	–		–
Oxidase reaction	+	–	Weak	–
Denitrification	+	–	–	–
Gelatin hydrolysis	–	d	–	+
Starch hydrolysis	+	–	–	–
Utilization of:				
Glucose	+	+	+	+
Trehalose	–	–	–	–
2-Ketogluconate	–	–	–	–
meso-Inositol	–	d	–	+
Geraniol	–	–	–	–
L-Valine	+	–	–	–
β-Alanine	–	–	–	–
L-Arginine	–	d	–	+
D-Xylose			–	
D-Ribose			–	
L-Rhamnose			–	
Saccharate			–	
Levulinate			–	
Citraconate			–	
Mesaconate			–	
D(-)-Tartrate			–	
meso-Tartrate			–	
Erythritol			–	
Adonitol			–	
2,3-Butylene glycol			–	
m-Hydroxybenzoate			–	
Tryptamine			–	
α-Amylamine			–	
Sucrose			–	
Malonate			–	
Nitrate used as a nitrogen source	+	+	–	+
Lecithinase (egg yolk reaction)	–	d	–	d
Catechol, *ortho* cleavage	+			+
Protocatechuate:				
ortho cleavage		+	–	
meta cleavage		–	+	
Saprophytic or opportunistic animal pathogens	+	–	+	–
Phytopathogens	–	+	–	+
Parasite of animals and humans	–	–	–	–

Footnotes are at end of table

Table 4.39 *(continued)*

[a] Symbols: see standard definitions.

[b] These species include those in rRNA groups I–IV as described in N. J. Palleroni in *Bergey's Manual of Systematic Bacteriology*, Vol. 1. However, some species have been reclassified into other genera since *Bergey's Manual of Systematic Bacteriology* was published. These include *P. acidovorans* and *P. testosteroni* (now classified in the genus *Comamonas*), and *P. flava*, *P. pseudoflava*, and *P. palleronii* (now classified in the genus *Hydrogenophaga*). Additional *Pseudomonas* species are listed in Table 4.40, 4.41, and 4.42.

[c] The species *Pseudomonas chlororaphis* now includes *Pseudomonas aureofaciens*, which was formerly considered to be a separate species. See Johnson and Palleroni, Int. J. Syst. Bacteriol. *39:* 230-235, 1989.

[d] Lateral flagella of short wavelength may also be produced under certain conditions.

[e] May be obscured in a few strains by production of reddish or brown pigments.

[f] Positive for all strains studied, but may require mutation in strains isolated from nature.

[g] Probably a soil organism and accidental pathogen, causing melioidosis in animals and humans.

[h] Two phytopathogenic subspecies of *P. pseudoalcaligenes* have been proposed. *P. pseudoalcaligenes* subsp. *citrulli*, pathogenic for watermelon (*Citrullus lanatus*; see Schaad et al., Int. J. Syst. Bacteriol. *28:* 117-125, 1978) and *P. pseudoalcaligenes* subsp. *konjaci*, pathogenic for the leaves of konjac (*Amorphophalus konjac*; see Goto, Int. J. Syst. Bacteriol. *33:* 539-545, 1983).

[i] Parasitic on horses and donkeys, in which it causes glanders and farcy. The infection is transmissible to humans and other animal species.

Table 4.40 Sources and characteristics of additional *Pseudomonas* species [a]

SOURCE	SPECIES	CHARACTERISTICS AND REFERENCES
ISOLATED FROM DISEASED PLANTS AND MUSHROOMS		
Cultivated mushrooms	*P. agarici*	One, rarely two, flagella. Fluorescent pigment produced. Acid is produced from glucose and various other sugars. Oxidase positive. **Causes drippy gill of mushrooms.** One of the main differences with another mushroom pathogen, *P. tolaasii*, is in the utilization of benzoate. See *Bergey's Manual of Systematic Bacteriology*, Vol. 1, p. 188.
	P. tolaasii	One to five flagella. Fluorescent pigment produced. Acid is produced from glucose. Gelatin is liquefied. **Isolated from brown spot of cultivated mushrooms.** See *Bergey's Manual of Determinative Bacteriology*, 7th ed., p. 136.
Birds-nest fern	*P. asplenii*	One to three flagella. Fluorescent pigment produced. Temperature range: 1–34°C. Acid is produced from glucose and various other sugars. Gelatin is liquefied. **Isolated from lesions of the bird's nest fern (*Asplenium nidus*).** See *Bergey's Manual of Determinative Bacteriology*, 7th ed., p. 124.
Pawpaw	*P. caricapapayae*	Three to six flagella. Fluorescent pigment produced. Temperature range: 7–45°C. Acid is produced from glucose and various other sugars. Gelatin is liquefied. **Isolated from water-soaked, angular spots on leaves of pawpaw.** See *Bergey's Manual of Systematic Bacteriology*, Vol. 1, p. 188.
Almond tree	*P. amygdali*	One to six flagella. No fluorescent pigment produced. Temperature range: 3–32°C. Acid is produced from glucose and various other sugars. Gelatin is not hydrolyzed. **Produces a hyperplastic bacterial canker in the almond tree (*Prunus dulcis*, fam. *Rosaceae*).** Not pathogenic for other fruit trees. See *Bergey's Manual of Systematic Bacteriology*, Vol. 1, p. 188–189.
Ficus erecta tree	*P. ficuserectae*	Motile by 1–5 flagella. Poly-β-hydroxybutyrate is accumulated. Pigments are not produced. Similar to *P. amygdali* in many properties but differs by forming larger colonies on nutrient agar, utilizing raffinose and glycerol, and failing to hydrolyze Tween 80, to produce H₂S, and to utilize ribose, mannitol, and sorbitol. **Causes dark brown, water-soaked spots on the leaves and stems of *Ficus erecta*** Thumb., resulting either in defoliation or shoot blight on severely infected plants. For other characteristics see Goto, Int. J. Syst. Bacteriol. *33*: 546–550, 1983.
Sorghum, corn, clover, and velvet bean	*P. andropogonis*	One, rarely two, flagella. Sheathed flagella have been reported in some strains. No fluorescent pigment produced. Poly-β-hydroxybutyrate is accumulated. Most strains are oxidase negative. Glucose and various other sugars are utilized. Gelatin is not hydrolyzed. The species may be divided into two specialized pathovars, namely, pv. *andropogonis*, the agent of a **stripe disease of sorghum**, and pv. *stizolobii*, which has been described as the cause of **leaf spot of velvet bean** (*Stizolobium deeringianum*). See *Bergey's Manual of Systematic Bacteriology*, Vol. 1, p. 189.
Oats (*Avena sativa*) and foxtail (*Chaetochloa lutescens*)	*P. avenae*	No fluorescent pigment produced. Poly-β-hydroxybutyrate is probably accumulated. Oxidase negative. Acid is produced from glucose and various other sugars. Gelatin liquefaction is variable. **Pathogenic for oats (*Avena sativa*) and foxtail** (*Chaetochloa lutescens*). See *Bergey's Manual of Systematic Bacteriology*, Vol. 1, p. 189.

Table 4.40 *(continued)*

SOURCE	SPECIES	CHARACTERISTICS AND REFERENCES
Orchids	*P. cattleyae*	One or two bipolar flagella. No fluorescent pigment produced. Acid is produced from glucose and various other sugars. Gelatin is not liquefied. **Pathogenic for *Cattleya* sp. and *Phalaenopsis* sp.** (fam. *Orchidaceae*). See *Bergey's Manual of Determinative Bacteriology*, 7th ed., p. 148.
Cissus plants	*P. cissicola*	Non-motile immediately after isolation, but motile clones appear after subculturing, the cells of which have a polar flagellum. Poly-β-hydroxybutyrate is accumulated. No fluorescent pigment is produced. No growth below 5°C or above 37°C. Acid is produced from glucose and various other sugars by most isolates. Gelatin is liquefied. Starch is hydrolyzed. Arginine dihydrolase is negative. **Pathogenic for *Cissus japonica*** (fam. *Vitaceae*). See *Bergey's Manual of Systematic Bacteriology*, Vol. 1, p. 189.
Tomato plants	*P. corrugata*	Mulitrichous polar flagella. Poly-β-hydroxybutyrate is accumulated. No fluorescent pigment is produced. Yellow-green diffusible, non-fluorescent pigment is produced. Colonies are wrinkled, yellowish, sometimes with green center. Growth occurs at 37°C but not at 41°C. Gelatin is hydrolyzed. Among the characters that differentiate this species from *P. cepacia* and *P. gladioli* are the absence of pectate hydrolysis and rot of onion slices, and the lack of utilization of D-arabinose, cellobiose, adipate, *meso*-tartrate and citraconate. **Isolated from tomato pith necrosis.** See *Bergey's Manual of Systematic Bacteriology*, Vol. 1, p. 189.
Rice	*P. glumae*	Two to four flagella. Fluorescent pigment produced on potato agar. Temperature range: 11–40°C. Acid is produced from glucose and various other sugars. **Pathogenic for rice plants** (*Oryza sativa* fam. *Gramineae*). See *Bergey's Manual of Systematic Bacteriology*, Vol. 1, p. 189–190.
	P. fuscovaginae	Motile by 1–4 polar flagella. Produce a green, fluorescent, diffusible pigment. Oxidase positive. No growth at 37°C. Unable to denitrify. Acid is produced from glucose and various other carbohydrates. Arginine dihydrolase positive. **Distinguished from other arginine dihydrolase positive fluorescent pseudomonads by its inability to utilize 2-ketogluconate or inositol. Pathogenic for *Oryza sativa*, *Hordeum vulgare*, *Triticum aestivum*, *Avena sativa*, *Zea mays*, *Lolium perenne*, *Bromus marginatus*, *Phleum pratense*, and *Phalaris arundinacea*.** For other characteristics see Miyajima et al., Int. J. Syst. Bacteriol. *33*: 656-657, 1983.
	P. plantarii	Motile by 1–3 polar flagella. Poly-β-hydroxybutyrate is accumulated. Colonies have a slight yellow tint and weakly produce a diffusible, reddish brown pigment under certain conditions. Fluorescent pigments not produced. Oxidase positive. Organic growth factors not required. Able to denitrify. Arginine dihydrolase negative. Gelatin is liquefied. Temperature range: 4–10°C to 38°C. **Tobacco hypersensitivity reaction negative.** Acid is produced from glucose and various other carbohydrates. **Causes seedling blight of rice.** For other characteristics see Azegami et al., Int. J. Syst. Bacteriol. *37*: 144-152, 1987.
Sugarcane	*P. rubrilineans*	Motile by a single flagellum. Poly-β-hydroxybutyrate is accumulated. No pigments are produced. Oxidase positive. Gelatin liquefaction is weak. Capable of growth at 40°C. Acid is produced from glucose and various other sugars. **The agent of red stripe of sugarcane.** See *Bergey's Manual of Systematic Bacteriology*, Vol. 1, p. 190.

Table 4.40 Sources and characteristics of additional *Pseudomonas* species [a]

SOURCE	SPECIES	CHARACTERISTICS AND REFERENCES
	P. rubrisubalbicans	Slightly curved rods motile by several polar flagella. Poly-β-hydroxybutyrate is accumulated. No pigments are produced. Oxidase positive. Gelatin is not hydrolyzed. Capable of growth at 40°C. Acid is produced from glucose and various other sugars. **The agent of mottled stripe of sugarcane.** See *Bergey's Manual of Systematic Bacteriology*, Vol. 1, p. 190.
Carnation	*P. woodsii*	Motile by a single polar flagellum. Gelatin is not liquefied. Acid is produced from glucose and various other sugars. Isolated from water-soaked lesions on carnation leaves. **Pathogenic for carnation** (*Dianthus caryophyllus*, fam. *Caryophyllaceae*). See *Bergey's Manual of Determinative Bacteriology*, 7th ed., p. 150-151.
ISOLATED FROM ANIMALS		
Eels	*P. anguilliseptica*	Motile by a single polar flagellum. No fluorescent pigment. Gelatin is liquefied. No acid is produced from glucose or other sugars that have been tested. **Isolated from diseased pond-cultured eels** (*Anguilla japonica*). See *Bergey's Manual of Systematic Bacteriology*, Vol. 1, p. 190.
ISOLATED FROM SOIL		
By enrichment in an atmosphere of H₂, O₂ and CO₂	"*P. hydrogenovora*"	Motile by several polar flagella. Poly-β-hydroxybutyrate is accumulated. **Can grow autotrophically with H₂. Aside from *P. saccharophila*, it is the only species of hydrogen pseudomonad that can use starch for heterotrophic growth.** For other characteristics see Table 4.41 and *Bergey's Manual of Systematic Bacteriology*, Vol. 1, p. 191.
	"*P. hydrogenothermophila*"	Motile by polar flagella. **Optimum growth temperature, 52°C. Can grow autotrophically with H₂.** For other characteristics see Table 4.41 and *Bergey's Manual of Systematic Bacteriology*, Vol. 1, p. 191.
By enrichment in an atmosphere of CO and O₂	*P. carboxydohydrogena* "*P. compransoris*" "*P. carboxydovorans*"	**These three species can oxidize CO.** For other characteristics see Table 4.41 and *Bergey's Manual of Systematic Bacteriology*, Vol. 1, p. 191.
	"*P. gasotropha*"	**Can oxidize CO.** Motile by a single subpolar flagellum. Slime is produced. Colonies are very small. Little or no growth is observed on meat peptone agar. Cyanocobalamin is required as a growth factor. **Only organic acids and alcohols (including methanol) are utilized heterotrophically.** Nitrate is reduced to nitrite. Starch, cellulose and gelatin are not hydrolyzed. For other information see *Bergey's Manual of Systematic Bacteriology*, Vol. 1, p. 192.
By direct isolation on agar containing poly-β-hydroxybutyrate as the sole carbon source	*P. lemoignei*	Motile by a single polar flagellum. Poly-β-hydroxybutyrate is accumulated within the cells. No cellular or fluorescent pigments are produced. Nitrate is not reduced to nitrite of N₂. Organic growth factors not required. **Extremely limited nutritional spectrum.** Carbohydrates, polyols and amino acids are not utilized. Growth occurs at 41°C. For other characteristics see *Bergey's Manual of Determinative Bacteriology*, 8th ed., p. 231-232.

Table 4.40 *(continued)*

SOURCE	SPECIES	CHARACTERISTICS AND REFERENCES
By denitrification enrichment with calcium citrate	*P. indigofera*	Motile by a single polar flagellum. Poly-β-hydroxybutyrate is accumulated within the cells. **Colonies are blue with a metallic sheen;** margin thin, of yellow color. The blue color is abundantly produced in media with sucrose. Gelatin is not liquefied. Glucose and various other carbohydrates are utilized. Optimum temperature, 18–20°C. No pigmentation occurs at 37°C. For other characteristics see *Bergey's Manual of Systematic Bacteriology*, Vol. 1, p. 192-193.
By enrichment with dextran as the sole carbon source	*P. mixta*	In liquid medium, motile by a single polar flagellum. On solid media **lateral flagella of shorter wavelength are also formed.** Poly-β-hydroxybutyrate is accumulated within the cells. Colonies are pale buff tan to rusty red. Oxidase positive. Arginine dihydrolase negative. Unable to denitrify. Grows at 42°C but not at 4°C. **Hydrolyzes dextran, starch, pectate,** gelatin, casein, Tweens 20–80, DNA, and esculin, but not cellulose, chitin, tributyrin, or lecithin. **Some strains hydrolyze xylan, xanthan gum, alginate, and poly-β-hydroxybutyrate.** *Ortho*-cleavage of protocatechuate. See Bowman et al., Syst. Appl. Microbiol. *11*: 53-59, 1988.
By enrichment with amines	*P. aminovorans*	**Strains have the ability to use certain amines as sole carbon and energy sources, including methylamine and trimethylamine.** Other amines such as dimethylamine, tetramethylamine, ethylamine, and diethylamine can be used by some strains. However, some amines, such as triethylamine, tetraethylamine, tetrapropylamine, and hexylamine are not used by any strain. Glucose and various other carbohydrates can be utilized. For other characteristics see *Bergey's Manual of Systematic Bacteriology*, Vol. 1, p. 193.
By enrichment with isoprenoid compounds	*P. citronellolis*	Motile by a single polar flagellum. **Can grow on citronellol.** Glucose can be utilized. Gelatin not liquefied. Nitrates are reduced to nitrites. For other characteristics see *Bergey's Manual of Systematic Bacteriology*, Vol. 1, p. 193.
By enrichment with natural resins	*P. resinovorans*	Motile by a single polar flagellum. Fluorescent pigment is produced. **Can grow on colophony, Canada balsam or abietic acid, phenol, phenanthrene, salicylic acid, *m*-cresol, and naphthalene as carbon and energy sources.** Gelatin not liquefied. Unable to denitrify. No acid from glucose or other carbohydrates tested. For other characteristics see *Bergey's Manual of Systematic Bacteriology*, Vol. 1, p. 193.
By enrichment with aromatic compounds	*P. boreopolis*	Motile. Fluorescent pigment is produced. Some strains produce a red pigment in gelatin medium. **Naphthalene is used by all strains.** Acid is produced from glucose by most strains. For other characteristics see *Bergey's Manual of Systematic Bacteriology*, Vol. 1, p. 193.
	P. pictorum	Motile by a single polar flagellum. Colonies are yellow. **Phenol is utilized.** Acid is produced from glucose. Gelatin is not hydrolyzed. For other characteristics see *Bergey's Manual of Systematic Bacteriology*, Vol. 1, p. 193.
In antibiotic screenings	*"P. acidophila"*	Motile by one to several polar flagella. Poly-β-hydroxybutyrate is not accumulated. No diffusible pigment is produced. Oxidase negative. Temperature range: 2–37°C. Growth factors not required. Weak acid production from glucose and various other carbohydrates. Gelatin is not liquefied. Unable to denitrify. **Produces the β-lactam antibiotic sulfazecin.** For other characteristics see *Bergey's Manual of Systematic Bacteriology*, Vol. 1, p. 194.

161

Table 4.40 (continued)

SOURCE	SPECIES	CHARACTERISTICS AND REFERENCES
	"P. mesoacidophila"	Motile by one to several polar flagella. Poly-β-hydroxybutyrate is not accumulated. No pigments are produced. Oxidase negative. Temperature range: 2–37°C. Weak acid production from glucose and various other carbohydrates. Gelatin is liquefied. Unable to denitrify. Arginine dihydrolase positive. **Produces the β-lactam antibiotic isosulfazecin.** For other characteristics see *Bergey's Manual of Systematic Bacteriology*, Vol. 1, p. 194.
	P. pyrrocinia	Motile with polar flagellum. No pigment is produced. No growth at 42°C. Oxidase negative. Unable to denitrify. Acid is produced from glucose and various other carbohydrates. **Produces the antibiotic pyrrolnitrin** (which is also produced by some strains of *P. chlororaphis, P. aureofaciens,* and *P. cepacia*). For other characteristics see *Bergey's Manual of Systematic Bacteriology*, Vol. 1, p. 194.
By the use of acid, low nitrogen media	*P. glathei*	Motile by a polar flagellum. Oxidase positive. Arginine dihydrolase negative. Unable to denitrify. Growth factors not required. Glucose and various other carbohydrates are utilized. **Differs from other species of the genus by its ability to grow in nitrogen-deficient media, its acid tolerance (pH 4.5), and the utilization of oxalate as a substrate for growth.** For other characteristics see *Bergey's Manual of Systematic Bacteriology*, Vol. 1, p. 194.
By direct isolation	*P. aurantiaca*	Lophotrichous, with 4–6 flagella. **Produces two main diffusible pigments, green and orange.** The colony may remain orange, while the medium is stained green and fluoresces. Gelatin is liquefied. Capable of growth and acid production from glucose and various other carbohydrates. For other characteristics see *Bergey's Manual of Systematic Bacteriology*, Vol. 1, p. 194.

ISOLATED FROM WATER

SOURCE	SPECIES	CHARACTERISTICS AND REFERENCES
Fresh and distilled water	*P. huttiensis*	One to three polar flagella. Growth factors not required. Acid is produced from glucose and various other carbohydrates. Gelatin is not hydrolyzed. **Isolated from distilled water.** For other characteristics see *Bergey's Manual of Systematic Bacteriology*, Vol. 1, p. 194-195.
	P. lanceolata	Small or large oval cells, motile by a polar flagellum. Gelatin not hydrolyzed. Slight amount of acid is produced from glucose. No acid produced from other sugars that have been tested. **Isolated from distilled water.** For other characteristics see *Bergey's Manual of Systematic Bacteriology*, Vol. 1, p. 195.
	P. spinosa	One, two, or occasionally three polar flagella. Under unfavorable conditions the rods may be curved. **In their natural habitat (river water) the curvature is more marked, and no flagella are observed. Instead, numerous "spines" (fimbriae?) may be seen in the stained preparations.** No growth at 37°C. Weak acidity is produced from glucose, sucrose, and maltose; no acid is produced from other sugars that have been tested. Gelatin is not hydrolyzed. For other characteristics see *Bergey's Manual of Systematic Bacteriology*, Vol. 1, p. 195.

Table 4.40 (continued)

SOURCE	SPECIES	CHARACTERISTICS AND REFERENCES
Sea water	P. doudoroffii	One to three polar flagella. **Poly-β-hydroxybutyrate is accumulated.** No pigment production. **Sodium ions are required for growth.** No levan produced from sucrose. Gelatin and starch not hydrolyzed. **Unable to denitrify.** Arginine dihydrolase negative. No autotrophic growth with H₂. Grows at 41°C but not at 4°C. **Aromatic ring cleavage is of the ortho type.** For other characteristics see Table 4.7, Baumann et al. (Int. J. Syst. Bacteriol. 33: 857-865, 1983), and Bergey's Manual of Systematic Bacteriology, Vol. 1, p. 195-196.
	P. nautica	Motile by a single polar flagellum. Lateral flagella may also be produced by some strains. **Poly-β-hydroxybutyrate is not accumulated.** No pigment production. **Sodium ions are required for growth.** No levan produced from sucrose. Some strains hydrolyze gelatin and/or starch. **Aromatic ring cleavage is of the ortho type. Some strains can denitrify.** Arginine dihydrolase negative. No autotrophic growth with H₂. Some strains grow at 41°C; no growth at 4°C. For other characteristics see Table 4.7, Baumann et al. (Int. J. Syst. Bacteriol. 33: 857-865, 1983), and Bergey's Manual of Systematic Bacteriology, Vol. 1, p. 195-196.
	P. perfectomarina	**Non-motile. Poly-β-hydroxybutyrate is not accumulated. Sodium ions are required for growth. Vigorous denitrifier.** Utilizes glucose and maltose but no other pentose, hexose, or disaccharide. Does not utilize aromatic compounds or carry out aromatic ring cleavage. For other characteristics see Table 4.7 and Baumann et al. (Int. J. Syst. Bacteriol. 33: 857-865, 1983).
	P. stanieri	Motile by a single polar flagellum. **Poly-β-hydroxybutyrate is accumulated. Sodium ions are required for growth. Does not denitrify. Carbohydrates not utilized. Aromatic ring cleavage is of the meta type.** For other characteristics see Table 4.7 and Baumann et al. (Int. J. Syst. Bacteriol. 33: 857-865, 1983).
	P. elongata	Colonies in agar are sunken due to **agar liquefaction.** Yellowish-brown pigment produced from peptone. **Sodium ions are required for growth.** Glucose and various other carbohydrates are utilized. Gelatin is slowly liquefied. Starch is hydrolyzed. Isolated from intertidal sand, sea water and bottom sediments. For other characteristics see Bergey's Manual of Systematic Bacteriology, Vol. 1, p. 195.
	P. gelidicola	Motile by a single polar flagellum. Colonies are yellow. **Sodium ions are required for growth. Hydrolyzes agar** but not gelatin. Utilizes glucose and various other carbohydrates. Isolated from sea water, algae and rotted straw submerged in sea water. For other characteristics see Bergey's Manual of Systematic Bacteriology, Vol. 1, p. 192-193.
Hypersaline lakes	P. halophila	Single polar flagellum. Reddish-brown colonies. **Halotolerant: growth occurs between 0.02 and 3.3 M (0.012 and 19.3%, w/v) NaCl; optimum = 0.8 M (4.7%, w/v).** Poly-β-hydroxybutyrate not accumulated. Temperature range: 4-37°C. Oxidase positive. Arginine dihydrolase negative. Gelatin is hydrolyzed. Glucose and various other sugars are utilized. Isolated from the Great Salt Lake, Utah, USA. (See Fendrich, Syst. Appl. Microbiol. 11: 36-43, 1988).

Table 4.40 (continued)

SOURCE	SPECIES	CHARACTERISTICS AND REFERENCES
ISOLATED FROM FOODS		
Milk and dairy products	*P. fragi*	Motile by a polar flagellum. No pigments reported. Temperature range: 10–30°C. Gelatin is liquefied. Acid is produced from glucose and various other carbohydrates. **Fat is generally hydrolyzed. Isolated from milk and other dairy products, dairy utensils, meat, water, etc.** For other characteristics see *Bergey's Manual of Determinative Bacteriology*, 7th ed., p. 110.
	P. mephitica	Motile by a polar flagellum. No pigments produced. No growth at 37°C. **After 1 or 2 days a skunk-like odor develops.** Gelatin is liquefied slowly. Acid is produced slowly from glucose and various other carbohydrates. **Presumably isolated from rinse water.** For other characteristics see *Bergey's Manual of Determinative Bacteriology*, 7th ed., p. 111-112.
	P. synxantha	Motile by polar flagella. Large spreading transparent colonies having a bluish cast by reflected light. Colonies may show flesh color. **Produces an intense, diffusible, yellow to orange color in cream or in the cream layer of milk.** Acid is produced from glucose and various other carbohydrates. Gelatin is liquefied. For other characteristics see *Bergey's Manual of Determinative Bacteriology*, 7th ed., p. 104-105.
Eggs	*P. mucidolens*	**Motile. Fluorescent pigment is produced.** Slight growth at 10°C and 37°C. Acid is produced from glucose and various other carbohydrates. Gelatin is liquefied. **Causes mustiness in eggs.** For other characteristics see *Bergey's Manual of Systematic Bacteriology*, Vol. 1, p. 197.
	P. taetrolens	Motile by 1–5 polar flagella. Gelatin not liquefied. **Produces a strong musty odor.** Isolated from musty eggs and also from milk. For other characteristics see *Bergey's Manual of Determinative Bacteriology*, 7th ed., p. 108.
Meats	*P. fragi*	See entry above for this species.
Grains	*P. azotoformans*	Motile with polar flagella. Fluorescent pigment is produced. No growth at 37°C. Oxidase positive. Able to denitrify. Gelatin is liquefied. Glucose, and various other sugars are assimilated. Protocatechuate is assimilated. **Isolated from Japanese rice paddies and oil brines.** For other characteristics see *Bergey's Manual of Systematic Bacteriology*, Vol. 1, p. 197.
	P. fulva	Motile by 1–3 polar flagella. Fluorescent pigment is produced. Poor growth at 37°C and no growth at 42°C. Oxidase reaction feebly positive. Gelatin is not liquefied. Unable to denitrify. Glucose and various other compounds including protocatechuate are assimilated. **Isolated from Japanese rice paddies.** For other characteristics see *Bergey's Manual of Systematic Bacteriology*, Vol. 1, p. 197.
	P. straminea	Motile by a polar flagellum. No growth at 37°C. Colonies are yellow. Fluorescent pigment is produced. Oxidase positive. Gelatin is liquefied. Unable to denitrify. Glucose and various other compounds are assimilated. Protocatechuate is not assimilated. **Isolated from Japanese rice paddies.** For other characteristics see *Bergey's Manual of Systematic Bacteriology*, Vol. 1, p. 197.

Table 4.40 *(continued)*

SOURCE	SPECIES	CHARACTERISTICS AND REFERENCES
Salted beans	*P. beijerinckii*	Motile by polar flagella. **An insoluble purple pigment is produced**, but not in all media; it is favored by reduced oxygen tension, a pH of 8.0, and when the organisms are grown in extracts of beans or some other vegetable. Gelatin is not liquefied. Acid is produced from glucose. Can grow in the presence of 12% NaCl. Isolated from beans preserved with salt. For other characteristics see *Bergey's Manual of Determinative Bacteriology*, 7th ed., p. 121.
Copra (coconut)	*P. cocovenenans*	See *Bergey's Manual of Systematic Bacteriology*, Vol. 1, p. 197.
ISOLATED FROM SEWAGE		
	"P. butanovora"	Motile by a single polar flagellum. Poly-β-hydroxybutyrate is accumulated. Colonies are pale yellow to brownish yellow. Growth factors are not required. Maximum growth temperature, 42.5°C. Oxidase positive. Able to denitrify. No hydrolysis of gelatin. **Able to assimilate C_2 to C_9 normal hydrocarbons.** Glucose and other carbohydrates are not assimilated. Resembles *P. pseudoalcaligenes* in many properties. For other characteristics see *Bergey's Manual of Systematic Bacteriology*, Vol. 1, p. 198.
ISOLATED FROM OIL SUSPENSIONS		
	P. oleovorans	The cells are almost coccoid but increase in length during the exponential phase of growth in broth. Colonies have a fluorescence that is not imparted to the medium. Gelatin is not liquefied. Starch is hydrolyzed. **Isolated from oil-water emulsions used as lubricants and cooling agents in the cutting and grinding of metals. Apparently the organism lives on some normal component of the cutting compound, probably the naphthalenic acids which act as emulsifying agents.** For other characteristics see *Bergey's Manual of Systematic Bacteriology*, Vol. 1, p. 198.
ISOLATED FROM OIL BRINE		
	P. iners	Motile by a single polar flagellum. No pigments are produced. No growth at 42°C; only scanty growth at 37°C. Oxidase negative. Gelatin is not liquefied. Unable to denitrify. Growth fails to occur in synthetic media, suggesting growth factor(s) requirement. No acid from glucose or other carbohydrates tested. **Isolated from oil brines in Japan.** For other characteristics see *Bergey's Manual of Systematic Bacteriology*, Vol. 1, p. 192-193.
	P. nitroreducens	Motile by polar flagella. Fluorescent pigment produced. Oxidase positive. Differs from most other fluorescent pseudomonads by having both the ability to denitrify plus an inability to liquefy gelatin. Glucose is utilized. Protocatechuate is utilized. **The fresh isolate could use kerosene, but the activity was lost after subcultivation in the laboratory.** For other characteristics see *Bergey's Manual of Systematic Bacteriology*, Vol. 1, p. 198.
	P. azotoformans	See entry above for this species.

165

Table 4.40 *(continued)*

SOURCE	SPECIES	CHARACTERISTICS AND REFERENCES
ISOLATED FROM CLINICAL SPECIMENS	*P. paucimobilis*	Motile by a single polar flagellum. **Motility occurs at 18–22° but not at 37°C.** Cells have lipid inclusions, probably poly-β-hydroxybutyrate. Yellow, insoluble, nonfluorescent carotenoid pigment is produced. No growth at 5 or 42°C. Oxidase positive. Gelatin is not liquefied. Starch is hydrolyzed. Arginine dihydrolase negative. Acid is produced from glucose and various other carbohydrates. **Isolated from a respirator.** For other characteristics see *Bergey's Manual of Systematic Bacteriology,* Vol. 1, p. 198.
ISOLATED AS LABORATORY CONTAMINANTS	*P. echinoides*	Rods, slightly curved, with sharp ends. Monotrichous flagella. Yellow intracellular pigments (carotenoids) are produced. **Cell aggregates (rosettes) form in broth and also in agar media.** In liquid media they can be observed as small flocs. Gelatin is not liquefied. Starch is slowly hydrolyzed. For other characteristics see *Bergey's Manual of Systematic Bacteriology,* Vol. 1, p. 192-193.
	P. pertucinogena	Motile by single polar flagella. Poly-β-hydroxybutyrate is not accumulated. No pigments are produced. Oxidase positive. No hydrolysis of gelatin or starch. Arginine dihydrolase negative. Acid is produced from glucose and a few other sugars, but in general acid reactions from most sugars are weak or negative. **Produce pertucin, a bacteriocin active against *Bordetella pertussis.*** For other characteristics see *Bergey's Manual of Systematic Bacteriology,* Vol. 1, p. 192-193.

---

[a] Some species previously included in *Bergey's Manual of Systematic Bacteriology,* Vol. 1, have been reclassified in other genera: *Methylobacterium rhodos* (formerly *Pseudomonas rhodos*), *Methylobacterium mesophilicum* (formerly *Pseudomonas mesophilica*), *Methylobacterium radiora* (formerly *Pseudomonas radiora*), *Hydrogenophaga pseudoflava* (formerly *Pseudomonas pseudoflava* and also the organism formerly called "*Pseudomonas carboxydoflava*"), and *Deleya marina* (formerly *Pseudomonas marina*).

Table 4.41 Characteristics differentiating various hydrogen and CO pseudomonads [a]

Characteristic	Hydrogen bacteria				Carboxydobacteria		
	P. facilis	*"P. hydrogeno-thermophila"*	*"P. hydrogeno-vora"*	*P. saccharo-phila*	*P. carboxy-dohydrogena*	*"P. carboxy-dovorans"*	*"P. compran-soris"*
Yellow cellular pigment	−	+	−	−	−	−	−
Hydrolysis of:							
Gelatin	+	−	−	−	−	−	−
Starch	−	−	+	+	−	−	−
Poly-β-hydroxybutyrate	+			−	−	−	−
Tween 80	−		+	−	−	−	−
Utilization of:							
Glucose	+	−	+	+[b]	−	−	−
Fructose	+	−	+	+[b]	+	−	−
Oxalate	−	−	−	−	+	+	−
Succinate	−	+	+	+	+	−	−
Malonate	+	−	−	−	−	−	−
β-hydroxybutyrate	+		+	+	−	−	+
L-tyrosine	+	+		−	−	−	−
L-proline	+			+	−	−	−

[a] Symbols: see standard definitions.

[b] Positive for all strains studied, but may require mutation in strains isolated from nature.

Important Notes for Users of This Manual

Unless otherwise indicated in footnotes to tables, the meanings of symbols are as follows:

+ 90% or more of strains are positive

− 90% or more of strains are negative

d 11-89% of strains are positive

v strain instability (not equivalent to "d")

D Different reactions in different taxa (species of a genus or genera of a family)

All other symbols are defined in footnotes to tables.

Table 4.42 Characteristics useful for differentiation of various denitrifying pseudomonads [a]

Characteristic	P. aeruginosa	P. alcali-genes	P. caryo-phylli	P. fluorescens II, III P. aureofaciens	P. mallei	P. mendocina	P. pickettii	P. pseudo-alcali-genes [b]	P. pseudo-mallei	P. solana-cearum	P. stutzeri
Number of flagella	1	1	>1	>1	0	1	1	1	>1	>1	1
Poly-β-hydroxybutyratic acid accumulation	-	-	+	-	+	-	+	d	+	+	-
Growth at 40°C	+	+	+	-	+	+	+	+	+	-	+
Pyoverdin production	+	-	-	+	-	-	-	-	-	-	-
Pyocyanin production	+	-	-	-	-	-	-	-	-	-	-
Yellow cellular pigment	-	d	-	-	-	+	-	-	d	-	-
Arginine dihydrolase	+	+	+	+	+	+	-	d	+	-	+
Starch hydrolysis	-	-	-	-	d	-	-	-	+	-	+
Poly-β-hydroxybutyrate hydrolysis	-	-	-	-	d	-	-	-	+	-	-
Growth on:											
D-Xylose	-	-	+	d	+	-	+	-	-	-	-
Maltose	-	-	-	-	d	-	-	-	+	-	+
Saccharate	-	-	+	d	-	+	+	-	-	+	d
Mannitol	+	-	+	+	+	-	-	-	+	d	d
Ethylene glycol	-	-	-	-	-	+	d	d	-	-	+
2,3-Butylene glycol	+	-	+	d	-	-	d	-	-	-	d
Geraniol	+	-	-	-	-	+	-	-	-	-	-
Azelate	+	-	-	d	-	+	+	-	+	d	-
Levulinate	+	-	-	d	-	+	+	-	+	d	-
Glycolate	-	-	+	-	-	+	+	-	-	d	+
L-Serine	d	-	+	d	d	+	+	d	+	d	d
L-Arginine	+	+	+	+	+	+	-	+	+	-	-
L-Histidine	+	d	+	+	+	+	d	d	+	d	-
Betaine	+	-	+	+	+	+	-	+	+	+	-
Sarcosine	+	-	-	+	d	+	-	d	d	d	-

[a] Symbols: see standard definitions.

[b] Not all strains of *P. pseudoalcaligenes* are denitrifiers.

Table 4.43 Differential characteristics of the species of the genus *Rhizobium*[a]

Characteristics	*R. galegae*[b]	*R. leguminosarum*	*R. loti*	*R. meliloti*
Flagellar arrangement:				
Single polar or subpolar flagellum	d	–	d	–
1–2 polar or subpolar flagella	d	–	–	–
2–6 peritrichous flagella	–	+	–	+
Agglutination with antiserum against *R. meliloti*[c]		–	–	+
Growth in presence of 2% NaCl	–	–	–	d
Growth at 39–40°C	–	–	–	d
Thiamine required	–	d		–
Pantothenate required	+	+		–
H₂S produced		–	–	d
Precipitate on Ca glycerophosphate medium[d]	–	–	–	d
Acid reaction in litmus milk	–	–	–	d
Susceptibility to phage gal 1/R	+	–	–	–

[a] Symbols: see standard definitions.

[b] The species *Rhizobium galegae* was not included in *Bergey's Manual of Systematic Bacteriology*, Vol. 1. It was created in 1989 by Lindstrom (Int. J. Syst. Bacteriol. *39:* 365-367).

[c] Antisera for somatic and/or flagellar antigens.

[d] See Hofer, J. Bacteriol. *41:* 193-224, 1941.

Table 4.44 Nodulation characteristics of *Rhizobium* species[a]

Host plant	*R. galegae*[b]	*R. leguminosarum* biovar viceae	*R. leguminosarum* biovar trifolii	*R. leguminosarum* biovar phaseoli	*R. loti*	*R. meliloti*
Pisum sativum, Vicia hirsuta and *Vicia sativa*	–	+[b]	±	±	–	–
Phaseolus vulgaris	–	(±)	(±)	+	±	–
Trifolium repens	–	±	+[c]	(±)	–	–
Lotus corniculatus	–	–	–	–	+	–
Medicago sativa	–	–	–	–	(±)	+[d]
Macroptilium atropurpureum	–	–	–	±	±	–
Galega orientalis, Galega officinalis	+[e]	–	–	–	–	–

[a] Symbols: +, generally nodulates; ±, sometimes nodulates, nodules commonly ineffective; (±), rarely nodulates, nodules commonly ineffective; –, does not nodulate.

[b] Some negatives have been reported with *V. hirsuta* and *P. sativum*.

[c] Some negatives have been reported with isolates from *T. ambiguum* and African clovers.

[d] Other species of *Medicago* are likely to be more strain-specific.

[e] Effective nodules are formed by strains isolated from *G. orientalis* on that plant and by strains isolated from *G. officinalis* on their original host. Strains isolated from *G. orientalis* form ineffective nodules on *G. officinalis* and vice versa. *R. galagae* strains do not infect other leguminous plants that have been tested. Plants of *Galega* sp. are only occasionally infected by other rhizobia, and such infection does not result in nitrogen fixation.

Table 4.45 Differential characteristics of *Roseobacter* species and *Erythrobacter longus*[a]

Characteristic	R. denitrificans	R. litoralis	E. longus
Cell shape	Ovoid or rod	Ovoid or rod	Rod
Cell width, μm	0.6–0.9	0.6–0.9	0.3–0.4
Cell length, μm	1.0–2.0	1.2–2.0	2.0–5.0
Colony color:			
Orange	–	–	+
Pink	+	+	–
Thiamine and niacin required for growth	+	+	–
NO_3^- dissimilatorily reduced to N_2O	+	–	–
Susceptible to streptomycin (50 μg disk), polymyxin B (100 U disk)	+	+	–
Absorption spectra of cell suspensions show:			
Small peak at 802–804 nm; large peak at 863–867 nm	–	–	+
Large peak at 805–807 nm; smaller peak at 868–873 nm	+	+	–

[a] Symbols: see standard definitions.

Table 4.46 Differential characteristics of the species of the genus *Sinorhizobium*[a,b]

Characteristic	S. fredii	S. xianjinensis
Carbon sources:		
Raffinose	+	d
Rhamnose	+	d
D-Amygdalin	d	–
Acetate	+	d
Malate	+	d
DL-Asparagine	d	–
Nitrogen sources:		
L-Valine	d	–
L-Lysine	d	–
L-Arginine	+	d
Growth at pH 5.5	d	–
Growth at pH 8.5	d	–
Acid production in litmus milk	d	+
Growth at 10°C	d	–
Nitrate reduction	d	–
Resistance to vancomycin (25 μg/ml), chloramphenicol (125 μg/ml), penicillin (25 μg/ml), gentamicin, (25 μg/ml), streptomycin (5 μg/ml)	d	–

[a] Symbols: 95% or more of strains are positive; d, between 5 and 94% of strains are positive; –, less than 5% of strains are positive.
[b] Data taken from Chen et al. (Int. J. Syst. Bacteriol. *38*: 392–397, 1988).

Table 4.47 Differential characteristics of the species of the genus *Sphingobacterium* [a,b]

Characteristic	S. mizutae	S. multivorum [c]	S. spiritivorum [d]
Oxidative acidity from:			
Ethanol (3%)	−	−	+
Mannitol	−	−	+
Rhamnose	+	−	+
Glycogen	−	+	−
Deoxyribonuclease	−	+	+

[a] Symbols: see standard definitions.

[b] Data taken from Yabuuchi et al. (Int. J. Syst. Bacteriol. *33*: 580-598, 1983).

[c] Previously classified as *Flavobacterium multivorum* in *Bergey's Manual of Systematic Bacteriology*, Vol. 1.

[d] Previously classified as *Flavobacterium spiritivorum* in *Bergey's Manual of Systematic Bacteriology*, Vol. 1.

Table 4.48 Differential characteristics of the species of the genus *Thermus* [a,b]

Characteristic	T. aquaticus	T. filiformis [c]	T. ruber [d]
Nitrate reduction	−	−	+
Hydrolysis of:			
Arbutin	−	+	+
Hide powder azure, elastin	+	−	+
Casein	+	−	−
Deoxyribonuclease	+	−	−
α-Galactosidase	−	+	−
β-Galactosidase	+	+	−
Carbon sources:			
Acetate, glutamate	−	+	+
Succinate, glucose, mannitol, melibiose, sucrose, galactose, trehalose	−	+	−
Sorbitol, pyruvate	−	−	+
Proline	+	+	−
Maltose, salicin	+	−	−

[a] Symbols: see standard definitions.

[b] Data taken from Hudson et al. (Int. J. Syst. Bacteriol. *37*: 431-436, 1987). Tests performed at 70°C for *T. aquaticus* and *T. filiformis* and at 60°C for *T. ruber*.

[c] *Thermus filiformis* was not included in *Bergey's Manual of Systematic Bacteriology*, Vol. 1. It was created in 1987 by Hudson et al. (Int. J. Syst. Bacteriol. *37*: 431-436).

[d] *Thermus ruber* was not included in *Bergey's Manual of Systematic Bacteriology*, Vol. 1. It was created in 1984 by Loginova et al. (Int. J. Syst. Bacteriol. *34:* 498-499).

Table 4.49 Differential characteristics of the species of the genus *Weeksella*[a,b]

Characteristic	W. virosa	W. zoohelcum
Urease	−	+
Growth on MacConkey agar	+	−
Growth at 42°C	+	−
Growth on β-hydroxybutyrate	+	−

[a] Symbols: see standard definitions.
[b] Data taken from Holmes et al. (Syst. Appl. Microbiol. *8:* 191-196, 1986).

Table 4.50 Differential characteristics of the species of the genus *Xanthobacter*[a,b]

Characteristic	X. agilis	X. autotrophicus	X. flavus
Pleomorphism	−	+	+
Slime production	−	+	+
Motility	+	−	−
Color of colonies	Pale yellow	Yellow	Yellow
Growth in nutrient broth	−	+	+
Autotrophic growth at 37°C	−	+	+
Sensitive to chloramphenicol and erythromycin	+	−	−
Biotin requirement	−	−	+
Sole carbon sources:			
Glycolate, glyoxylate, oxalate, D-gluconate, L(+)-ascorbate	−	+	+
meso-Tartrate, D-glutamine	−	+	−
DL-Alanine	+	−	+
Malonate	−	−	+

[a] Symbols: see standard definitions.
[b] Data taken from Jenni and Aragno (Syst. Appl. Microbiol. *9:* 254-257, 1987).

Table 4.51 Differential characteristics of the species of the genus *Xanthomonas*[a]

Characteristic	X. albilineans	X. axonopodis	X. campestris[b]	X. citri[c]	X. fragariae	X. maltophilia[b]	X. phaseoli[c]	"X. populi"
Number of flagella	1	1	1	1	1	>1	1	1
Reduction of NO_3^- to NO_2^-	−	−	−	−	−	+	−	−
Lysine decarboxylase	−	−	−		−	+		−
Methionine or cystine required for growth	+		d	−		+	−	
Mucoid growth on nutrient agar + 5% glucose	−	−	+		+			+
Xanthomonadins produced	+	+	+	+	+	−	+	
Hydrolysis of:								
Gelatin	d	−	d		+	+		−
Esculin	+	+	+		−			
Starch	−	+	d	+	+	−	+	
Milk proteolysis	−	−	+		−	+	+	Slow
H_2S from peptone	−	+	+		−			−
Pectinase activity in culture[d]			d	+			−	
Maximum growth temperature, °C	37	35–37	35–39	38	33		38	27.5
Maximum NaCl tolerance, %	0.5	1.0	2.0–5.0		0.5–1.0			0.4–0.6
Acid production within 21 days on Dye's medium C[e] from:								
Arabinose	−	−	+		−			−
Mannose	+	−	+		+			+
Galactose	d	−	+		−			+
Trehalose	−	+	+		−			+
Cellobiose	−	−	+		−			−
Fructose	−	−	+		+			+
Opportunistic pathogen of humans	−	−	−	−	−	+	−	−

[a] Symbols: see standard definitions.

[b] Previously classified in *Bergey's Manual of Systematic Bacteriology*, Vol. 1, as *Pseudomonas maltophilia*. It was reclassified in the genus *Xanthomonas* by Swings et al. (Int. J. Syst. Bacteriol. *33*: 409-413,1983).

[c] *Xanthomonas citri* and *X. phaseoli* were not included in *Bergey's Manual of Systematic Bacteriology*, Vol. 1. The two species were revived in 1989 by Gabriel et al. (Int. J. Syst. Bacteriol. *39*: 14-22).

[d] On minimal sodium polypectate medium (Hildebrandt, Phytopathology *62*: 1430-1436, 1971).

[e] See *Bergey's Manual of Systematic Bacteriology*, Vol. 1, p. 206.

Table 4.52 Differential characteristics of *Taylorella equigenitalis*, *Wolinella recta*, *Bacteroides ureolyticus*, and *Bacteroides gracilis*

Characteristic	Taylorella equigenitalis	Wolinella recta	Bacteroides ureolyticus	Bacteroides gracilis
Motility	–	+	–	–
Urease	–	–	+	–
Oxidase	+	+	+	–
Nitrate reduced	–	+	+	+
Nitrite reduced	–	+	–	+
H$_2$ or formate is required as an electron donor	–	+	+	+
Habitat:				
Horses	+	–	–	–
Humans	–	+	+	+

[a] Symbols: see standard definitions.

Important Notes for Users of This Manual

Unless otherwise indicated in footnotes to tables, the meanings of symbols are as follows:

+ 90% or more of strains are positive

– 90% or more of strains are negative

d 11-89% of strains are positive

v strain instability (not equivalent to "d")

D Different reactions in different taxa (species of a genus or genera of a family)

All other symbols are defined in footnotes to tables.

GROUP 5

FACULTATIVELY ANAEROBIC GRAM-NEGATIVE RODS

These Gram-negative bacteria include the traditional families of *Enterobacteriaceae, Vibrionaceae,* and *Pasteurellaceae,* as well as several other genera that are facultatively anaerobic rods. Cells are 0.1–1.5 μm in diameter, although longer filamentous forms occur in many genera. They are straight rods, except for *Vibrio,* which has curved or vibrioid cells. They are found free-living or in association with animal or plant (*Erwinia*) hosts. Many species are pathogenic for humans and animals, and some are pathogenic for insects and plants.

Differentiation of the genera and families in **Group 5:** See Table 5.1.

SUBGROUP 1: Family Enterobacteriaceae

Straight rods, usually 0.3–1.8 μm. Cells stain Gram negative. **Motile by peritrichous flagella, except for** *Tatumella,* **or cells are nonmotile. Facultatively anaerobic** and chemoorganotrophic, **having both a respiratory and a fermentative type of metabolism.** Most species grow well at 37°C; however, some species grow better at 25–30°C and are often more active metabolically at these temperatures. D-Glucose and other carbohydrates are catabolized with the production of acid and, in many species, gas. **Oxidase negative and catalase positive,** except for *Shigella dysenteriae* O Group 1 and *Xenorhabdus* species other than *X. luminescens.* **Reduce nitrates,** except *Arsenophonus,* a number of *Erwinia* species, most *Xenorhabdus* species, and some strains of *Klebsiella pneumoniae* subsp. *ozaenae, Pantoea,* and *Yersinia.* Distributed worldwide. Found in soil, water, fruits, vegetables, grains, flowering plants and trees, and animals from worms and insects to humans. There is substantial heterogeneity in ecology, host range, and pathogenic potential for humans and animals, insects, and plants. A number of species cause diarrheal diseases including typhoid fever and bacillary dysentery. Many species not normally associated with diarrheal disease are often referred to as opportunistic pathogens. Most of these, as well as the species causing diarrheal disease, can cause a variety of extraintestinal infections including bacteremia, meningitis, and urinary tract, respiratory, and wound infections. *Enterobacteriaceae* are responsible for about 50% of nosocomial infections, most frequently caused by *Escherichia coli, Klebsiella, Enterobacter, Proteus, Providencia,* and *Serratia marcescens.*

Type genus: *Escherichia.*

Differentiation of **Enterobacteriaceae** *from other families and genera in Group 5:* Characteristics that differentiate *Enterobacteriaceae* from the other families and genera are presented in Table 5.1. The genera in the family *Vibrionaceae* are oxidase positive and have polar flagella when grown in liquid media—characteristics that distinguish them from *Enterobacteriaceae.* However, at least two *Vibrio* species (*V. metschnikovii* and *V. gazogenes*) are oxidase negative, and some strains of other species may be oxidase negative or weakly positive.

Differentiation of the genera in **Enterobacteriaceae:** It is virtually impossible to provide genus definitions and a single, meaningful biochemical table that differentiate the 30 genera in *Enterobacteriaceae.* In diagnostic and other laboratories, one attempts to identify members of this family directly to the species level, using essentially the same battery of tests for all of the more than 115 named species and subspecies. If an organism is not identified with certainty, additional tests are available to differentiate among subspecies, species within a given genus, or genera. Biochemical characteristics of the species of *Enterobacteriaceae* are given in Table 5.2.

Editorial note: Table 5.2 has been updated from *Bergey's Manual of Systematic Bacteriology;* in some cases additional data have changed the reaction percentage so that it is now in a different category (e.g., from "d" to [+]; from [+] to +, etc.). Almost all reactions are reported for 48 h of incubation at 37°C. A number of species have delayed positive (3 days or longer) reactions in one or more tests (gelatin is one good example). A number of species, including many yersiniae, erwiniae, and *Xenorhabdus* are metabolically more active at 25–30°C than at 37°C (motility in yersiniae is an example).

Genus **Arsenophonus**

Editorial note: The genus *Arsenophonus* was not included in *Bergey's Manual of Systematic Bacteriology.* The genus was created by Werren et al. (Int. J. Syst. Bacteriol. *41:* 563–565, 1991). It includes one species, *A. nasoniae.*

Rods, occasionally filamentous in young cultures, are 0.4–0.6 μm × 7–10 μm. Gram negative. **Nonmotile and facultatively anaerobic.** Chemoorganotrophic, having both a respiratory and a fermentative type of metabolism. Optimal temperature is 30°C. D-Glucose and other carbohydrates are catabolized with production of acid but no gas. **Oxidase negative; catalase positive; and indole, methyl red, and Voges-Proskauer negative. Tests are negative (no growth) for lysine and ornithine decarboxylase and for arginine dihydrolase. Nitrates are not reduced; H₂S negative. There is no growth on MacConkey agar.** Occur in female wasps. *Arsenophonus* is pathogenic for male egg lethality in wasps.

Type (and only) species: *Arsenophonus nasoniae.*

Characteristics of the species: See Table 5.2.

Editorial note: The organism is fastidious. Determination of its biochemical characteristics requires unusual media.

Colonies are mucoid, gray-white, round, and convex with entire edges. One percent proteose-peptone is required for growth in characterization media (Werren et al., Int. J. Syst. Bacteriol., *41:* 563–565, 1991). Negative reactions or no growth is obtained in tests for deoxyribonuclease, esculin hydrolysis, growth on malonate, Tween 20 and Tween 80 hydrolysis, and urease. Gelatin is liquefied. Acid is produced from fructose and sucrose but not from adonitol, L-arabinose, cellobiose, dulcitol, glycerol, *myo*-inositol, lactose, maltose, D-mannitol, raffinose, trehalose, and D-xylose. Genus grows on media containing D-glucose, fructose, and sucrose and weakly on cellobiose, maltose, trehalose, and D-xylose but not on adonitol, L-arabinose, dulcitol, glycerol, *myo*-inositol, lactose, D-mannitol, and raffinose.

Genus **Budvicia**

Editorial note: The genus *Budvicia* was not included in *Bergey's Manual of Systematic Bacteriology.* The genus was created in 1983 by Aldova et al. (Zentralbl. Bakteriol. Parasitenkd. Infektionskr. Hyg. Abt. 1 Orig. Reihe A *254:* 95–108) and validated by Bouvet et al. in 1985 (Int. J. Syst. Bacteriol. *35:* 60–64). It includes one species, *Budvicia aquatica.*

Straight rods stain Gram negative. **Motile with peritrichous flagella when grown at 22°C.** Nonmotile strains occur at 22°C, and less than one-half of strains are motile at 37°C. **Facultatively anaerobic.** Chemoorganotrophic, having **both a respiratory and a fermentative type of metabolism.** Optimal temperature is 30–37°C. D-Glucose and some other carbohydrates, including L-arabinose (most strains), L-rhamnose, and D-xylose, are catabolized with the production of acid and little or no gas. **Oxidase negative and catalase positive. Indole is not produced.** Methyl red positive and **Voges-Proskauer negative. Citrate is not utilized. Lysine decarboxylase, arginine dihydrolase, and ornithine decarboxylase tests are negative. Most strains produce H₂S.** Reduces nitrates. Isolated predominantly from freshwater, with rare human stool and animal isolates. There is no known clinical significance.

Type (and only) species: *Budvicia aquatica.*

Budvicia strains were originally thought to be atypical strains of *Citrobacter,* but they are easily differentiated from *Citrobacter* species (Table 5.6). Biochemical differentiation of *Budvicia* from the H₂S-positive genera *Leminorella* and *Pragia* is given in Tables 5.22 and 5.26. Differentiation from genera with negative reactions in tests for lysine, arginine, and ornithine is shown in Table 5.2.

Characteristics of the species: As described for the genus and as shown in Table 5.2.

Genus **Buttiauxella**

Editorial note: The genus *Buttiauxella* was mentioned as an "Other Organism" under the genus *Kluyvera,* but it was not included as a separate genus in *Bergey's Manual of Systematic Bacteriology.* The genus was created by Ferragut et al. in 1981 (Zentralbl. Bakteriol. Parasitenkd. Infektionskr. Hyg. Abt. 1 Orig. *C 2:*

33–44; Int. J. Syst. Bacteriol. *32*: 266–268, 1982). It includes one species, *B. agrestis*.

Small, **rod-shaped cells are motile by peritrichous flagella.** Gram negative. **Facultatively anaerobic.** Chemoorganotrophic, having **both a respiratory and a fermentative type of metabolism.** Optimal temperature is 30–37°C. D-Glucose and other carbohydrates are catabolized with the production of acid and gas. **Oxidase negative and catalase positive. Indole and Voges-Proskauer negative; methyl red and Simmons citrate positive. Lysine decarboxylase and arginine dihydrolase negative; ornithine decarboxylase positive.** Reduces nitrates. Carbohydrates fermented include L-arabinose, cellobiose, lactose, maltose, D-mannitol, D-mannose, melibiose, raffinose, L-rhamnose, salicin, trehalose, and D-xylose. Isolated from fresh water; there are no human or animal isolates.

Type (and only) species: *Buttiauxella agrestis*.

Characteristics of the species: It is phenotypically similar to the genus *Kluyvera*. Tests that differentiate these genera are listed in Table 5.3.

Differentiation of the species of the genus **Buttiauxella:** Strains representing genospecies closely related to *B. agrestis*, designated Enteric Group 63 and Enteric Group 64 (Farmer et al., J. Clin. Microbiol. *21*: 46–76, 1985), are differentiated from *B. agrestis* in Table 5.3. Characteristics of *B. agrestis* are given in Table 5.2.

Genus **Cedecea**

Rod-shaped cells are 0.5–0.6 × 1–2 μm. Gram negative. **Most strains are motile. Facultatively anaerobic.** Chemoorganotrophic, having **both a respiratory and a fermentative type of metabolism.** Optimal temperature is 37°C. D-Glucose and other carbohydrates are catabolized with the production of acid and usually gas. **Oxidase negative and catalase positive, indole negative, usually methyl-red positive, and Simmons citrate positive, with a variable Voges-Proskauer reaction. Lysine decarboxylase negative and lipase (corn oil) positive.** Reduce nitrates. The carbohydrates fermented include D-arabitol, cellobiose, maltose, D-mannitol, D-mannose, salicin, and trehalose. Isolated from human clinical specimens, more than 50% from the respiratory tract. This is an infrequent opportunistic pathogen.

Type species: *Cedecea davisae*.

Cedecea species were first thought to be intermediate between typical *Serratia* species and *S. fonticola*. They are also somewhat similar to *Ewingella*. Reactions differentiating *Cedecea* from these genera are listed in Table 5.5.

Differentiation of the species of the genus **Cedecea:** *Cedecea* species, including the unnamed "Species 3 and 5," are differentiated by the reactions listed in Table 5.4. Characteristics of *Cedecea* species are given in Table 5.2.

Genus **Citrobacter**

Straight rods, about 1 μm in diameter and 2–6 μm in length, occur singly and in pairs. Gram negative. **Usually motile by peritrichous flagella. Facultatively anaerobic.** Chemoorganotrophic, having **both a respiratory and a fermentative type of metabolism.** Optimal temperature is 37°C. D-Glucose and other carbohydrates are catabolized with the production of acid and gas. **Oxidase negative and catalase positive, methyl red positive, usually citrate positive, Voges-Proskauer negative, and lysine decarboxylase negative.** Reduce nitrates. Carbohydrates usually fermented include L-arabinose, cellobiose, glycerol, maltose, D-mannitol, L-rhamnose, D-sorbitol, trehalose, and D-xylose. Occur in the feces of humans and animals, probably as normal intestinal inhabitants. Often isolated from clinical specimens as opportunistic pathogens. *C. diversus* can cause neonatal meningitis. Also found in soil, water, sewage, and food.

Type species: *Citrobacter freundii*.

Characteristics that separate *Citrobacter* from biochemically similar genera are shown in Table 5.6.

Differentiation of the species of the genus **Citrobacter:** See Table 5.7. Complete characteristics are given in Table 5.2.

Editorial note: The names *Citrobacter koseri* and *Levinea malonatica* are synonyms for *C. diversus*, and *Levinea amalonatica* is a synonym for *C. amalonaticus*. These names have and continue to appear in the literature, and the reader should recognize them as synonymous.

Genus **Edwardsiella**

Small, straight rods, about 1 µm in diameter × 2–3 µm, Gram negative. **Motile by peritrichous flagella** (*E. ictaluri* is motile at 25°C but not at 37°C). **Facultatively anaerobic** and chemoorganotrophic, having **both a respiratory and a fermentative type of metabolism.** Optimal temperature is 37°C, except for *E. ictaluri,* which prefers a lower temperature. D-Glucose and some other carbohydrates are catabolized with the production of acid and often gas, but they are **inactive compared to most genera in** *Enterobacteriaceae.* **Oxidase negative, catalase positive, and Voges-Proskauer and Simmons citrate negative. Lysine decarboxylase and usually ornithine decarboxylase positive.** Reduce nitrates. Carbohydrates fermented by all species are maltose and D-mannose. Most frequently occur in the intestine of cold-blooded animals and their environment, particularly fresh water, but also occur in warmblooded animals and humans. Pathogenic for eels, catfish, and other animals; they are a rare opportunistic pathogen for humans.

Type species: *Edwardsiella tarda.*

Edwardsiella is biochemically somewhat similar to *Escherichia coli, Shigella,* and *Salmonella,* but it is easily differentiated on the basis of a complete set of biochemical test results (Table 5.8).

Differentiation of the species of the genus **Edwardsiella:** Differential characteristics of *Edwardsiella* species are given in Table 5.9. Additional biochemical results are shown in Table 5.2.

Genus **Enterobacter**

Straight rods, 0.6–1.0 µm wide × 1.2–3.0 µm long, Gram negative. **Motile by peritrichous flagella (except** *E. asburiae).* **Facultatively anaerobic** and chemoorganotrophic, having **both a respiratory and a fermentative type of metabolism.** Optimal temperature is 30–37°C. D-Glucose and other carbohydrates are catabolized with the production of acid and gas. Indole negative. Most strains are **Voges-Proskauer positive and Simmons citrate positive.** Methyl red reaction varies. **Lysine negative** (except *E. gergoviae*) and **ornithine positive** (except *E. agglomerans*). **Malonate is usually utilized, and gelatin is slowly liquified** (3–14 days) by most strains. **H₂S, deoxyribonuclease, and lipase are not produced.** Carbo-

hydrates fermented by all or most strains include L-arabinose, cellobiose, maltose, D-mannitol, D-mannose, salicin, and trehalose. Widely distributed in nature, occurring in fresh water, soil, sewage, plants, vegetables, and animal and human feces. Several species, most notably *E. cloacae, E. sakazakii, E. aerogenes, E. agglomerans,* and *E. gergoviae,* are opportunistic pathogens, causing burn, wound, and urinary tract infections and occasionally septicemia and meningitis.

Type species: *Enterobacter cloacae.*

Characteristics that distinguish *Enterobacter* species from klebsiellae are given in Tables 5.10 and 5.11. Table 5.12 lists characteristics that can be used to differentiate *Enterobacter* from *Klebsiella, Hafnia,* and *Serratia.*

Differentiation of the species of the genus **Enterobacter:** See Table 5.13.

Editorial note: The species listed below were not included in *Bergey's Manual of Systematic Bacteriology* or were briefly mentioned under "Other organisms belonging to the genus *Enterobacter*." *Enterobacter amnigenus* was created by Izard et al. in 1981 (Int. J. Syst. Bacteriol. *31:* 35–42); *Enterobacter asburiae* was created by Brenner et al. in 1986 (J. Clin. Microbiol. *23:* 1114–1120; Int. J. Syst. Bacteriol. *38:* 220–222, 1988); *Enterobacter cancerogenus* was transferred to *Enterobacter* from the genus *Erwinia* by Dickey and Zumoff in 1988 (Int. J. Syst. Bacteriol. *38:* 371–374); *Enterobacter dissolvens* was transferred to *Enterobacter* from the genus *Erwinia* by Brenner et al. in 1986 (J. Clin. Microbiol. *23:* 1114–1120; Int. J. Syst. Bacteriol. *38:* 220–222, 1988); *Enterobacter hormaechei* was created by O'Hara et al. in 1989 (J. Clin. Microbiol. *27:* 2046–2049; Int. J. Syst. Bacteriol. *40:* 105–106, 1990); *Enterobacter intermedium* was created by Izard et al. in 1980 (Zentralbl. Bakteriol. Parasitenkd. Infektionskr. Hyg. Abt. 1 Orig. Reihe C: *1:* 51–60; Int. J. Syst. Bacteriol. *30:* 601, 1980). The specific epithet *"intermedium"* did not agree in gender with the genus name and therefore has been changed to *intermedius* (von Graevenitz, Int. J. Syst. Bacteriol. *40:* 211, 1990). *Enterobacter nimipressuralis* was transferred to *Enterobacter* from the genus *Erwinia* by Brenner et al. in 1986 (J. Clin. Microbiol. *23:* 1114–1120; Int. J. Syst. Bacteriol. *38:* 220–222, 1988); and *Enterobacter taylorae* was created by Farmer et al. in 1985 (J. Clin. Microbiol. *21:* 77–81; Int. J. Syst. Bacteriol. *35:* 223–225, 1985). *Enterobacter agglomerans* is a senior subjective synonym of *Erwinia herbicola* and *Erwinia milletiae* (Beji et al., Int. J. Syst.

Bacteriol. *38*: 77–88, 1988). There is a proposal to transfer *Enterobacter agglomerans* to the new genus *Pantoea* (see below). *Enterobacter cancerogenus* has been proposed as a senior subjective synonym of *Enterobacter taylorae* (Grimont and Ageron, Res. Microbiol. *140*: 459–465, 1989).

Characteristics that differentiate *Enterobacter* species are given in Table 5.13. Additional reactions are given in Table 5.2.

Genus **Erwinia**

Editorial note: The genus *Erwinia* has been studied mainly by phytopathologists. The specialist should use the tests shown in Tables 5.13 through 5.15, which are favored by phytopathologists. Many of the tests and test conditions are markedly different from those listed for other *Enterobacteriaceae* (Table 5.2). The biochemical characteristics of erwiniae in Table 5.2 are presented for comparison with other genera. Several species were sometimes placed in the genus *Pectobacterium*. Although valid, *Pectobacterium* is no longer used.

Straight rods, 0.5–1.0 × 1–3 μm, occur singly, in pairs, and sometimes in short chains. Gram negative. **Motile by peritrichous flagella** (except *E. stewartii*). **Facultatively anaerobic.** Chemoorganotrophic, having **both a respiratory and a fermentative type of metabolism.** Optimum temperature is 27–30°C. D-Glucose and other carbohydrates are catabolized with the production of acid; most species do not produce gas. **Oxidase negative and catalase positive; they are lysine decarboxylase, arginine dihydrolase, and ornithine decarboxylase negative. Nitrates are not reduced by many species. Ferment fructose, galactose, β-methylglucoside, and sucrose; usually D-mannitol, D-mannose, ribose, and D-sorbitol; but rarely adonitol, dextrin, dulcitol, and melezitose. Utilize acetate, fumarate, gluconate, malate, and succinate but not benzoate, oxalate, or propionate as carbon and energy yielding sources.** Associated with plants as pathogens, saprophytes, or constituents of the epiphytic flora. Very rarely isolated from humans.

Type species: *Erwinia amylovora.*

A meaningful comparison of *Erwinia* with other genera in *Enterobacteriaceae* has not been adequately done. This is because few nonerwiniae have been tested by the methods used for erwiniae and few erwiniae have been subjected to the routine test battery used for other *Enterobacteriaceae.* Thus, the data presented in Table 5.2 to differentiate erwiniae from other genera may not be comparable because of the different methods used.

Differentiation of the species of the genus **Erwinia:** See Tables 5.14, 5.15, and 5.16, and 5.2.

The species listed below were not included in *Bergey's Manual of Systematic Bacteriology. Erwinia cacticida* was created by Alcorn et al. in 1991 (Int. J. Syst. Bacteriol. *41*: 197–212); *Erwinia carotovora* subsp. *wasabiae* was created by Goto and Matsumoto in 1987 (Int. J. Syst. Bacteriol. *37*: 130–135); *Erwinia persicinus* was created by Hao et al. in 1990 (Int. J. Syst. Bacteriol. *40*: 379–383); and *Erwinia psidii* was created by Neto et al. in 1987 (Fitopatol. Bras. *12*: 345–350; Int. J. Syst. Bacteriol. *38*: 328, 1988).

Genus **Escherichia**

Straight rods, 1.1–1.5 μm × 2.0–6.0 μm, occur singly or in pairs. Capsules or microcapsules occur in many strains. Gram negative. **Motile by peritrichous flagella or are nonmotile. Facultatively anaerobic.** Chemoorganotrophic, having **both a respiratory and a fermentative type of metabolism.** Optimal temperature is 37°C. D-Glucose and other carbohydrates are catabolized with the formation of acid and gas. **Oxidase negative, catalase positive, methyl red positive, Voges-Proskauer negative, and usually citrate negative. Negative for H_2S, urea hydrolysis, and lipase.** *Escherichia* species reduce nitrates. All or most strains ferment L-arabinose, maltose, D-mannitol, D-mannose, L-rhamnose, trehalose, and D-xylose. *O*-Nitrophenyl-β-D-galactopyranoside positive. Occur as normal flora in the lower part of the intestine of warm-blooded animals and, in the case of *E. blattae,* of cockroaches. *E. coli* strains that contain enterotoxins and/or other virulence factors, including invasiveness and colonization factors, cause diarrheal disease. *E. coli* is also a major cause of urinary tract infections and nosocomial infections including septicemia and meningitis. Other species, except for *E. blattae,* are rarely occurring opportunistic pathogens, usually associated with wound infections.

Type species: *Escherichia coli.*

Editorial note: E. coli is often subdivided serologically or by the presence of virulence factors to identify and characterize epidemiologically pathogenic strains. Complete serotyping includes somatic (O), capsular (K), and flagellar (H) antigens.

Biochemically typical strains of *Escherichia* species are not difficult to differentiate from other genera (Table 5.17). *E. coli*, however, presents special problems. *E. coli* and all shigellae represent a single genetic species in which shigellae are metabolically inactive, nonmotile biogroups. It is difficult to differentiate metabolically inactive *E. coli* strains from shigellae. This is especially true for a biogroup of *E. coli* strains that are atypically lactose negative, nonmotile, and anaerogenic (do not produce gas in the fermentation of carbohydrates). Another biogroup of *E. coli* is negative in reactions for lysine decarboxylase, arginine dihydrolase, and ornithine decarboxylase, which makes them similar to *Enterobacter (Pantoea) agglomerans* and other species that are negative in these tests. Differential reactions for separating *E. coli* and other *Escherichia* species from these genera are shown in Table 5.17. Finally, *E. coli* strains sometimes exhibit atypical reactions in a variety of tests including H_2S, citrate, urease, KCN, adonitol, inositol, and indole, making it essential to consider the overall biochemical profile rather than specific "key" reactions before eliminating *E. coli* from consideration.

Editorial note: The species *Escherichia fergusonii*, *Escherichia hermannii*, and *Escherichia vulneris* were not included in *Bergey's Manual of Systematic Bacteriology*. *Escherichia fergusonii* was created by Farmer et al. in 1985 (J. Clin. Microbiol. *21:* 77–81; Int. J. Syst. Bacteriol. *35:* 223–225, 1985). *Escherichia hermannii* was created by Brenner et al. in 1982 (J. Clin. Microbiol. *15:* 703–713; Int. J. Syst. Bacteriol. *33:* 438–440, 1983). *Escherichia vulneris* was created by Brenner et al. in 1982 (J. Clin. Microbiol. *15:* 1133–1140; Int. J. Syst. Bacteriol. *33:* 438–440, 1983).

Differentiation of the species of the genus **Escherichia:** Characteristics useful in the differentiation of *Escherichia* species are given in Table 5.18. *Obesumbacterium proteus* and *Leclercia adecarboxylata* are included for comparison as the former species is most closely related to *Escherichia* by DNA hybridization and the latter has been included in the genus *Escherichia*. Complete reactions are shown in Table 5.2.

Genus **Ewingella**

Editorial note: The genus *Ewingella* was not included in *Bergey's Manual of Systematic Bacteriology*. The genus

was created in 1983 by Grimont et al. (Ann. Microbiol. Inst. Pasteur *134 A:* 39–52; Int. J. Syst. Bacteriol. *34:* 91–92, 1984). It includes one species, *Ewingella americana*.

Straight rods, 0.6–0.7 µm × 1–1.8 µm, stain Gram negative. **Motile with peritrichous flagella. Facultatively anaerobic** and chemoorganotrophic, having **both a respiratory and a fermentative type of metabolism.** Optimal temperature is 37°C. D-Glucose and some other carbohydrates are catabolized with the production of acid. **Oxidase negative, catalase positive, and indole negative. Most strains are methyl red positive. Voges-Proskauer and Simmons citrate positive. Negative in tests for lysine and ornithine decarboxylase, arginine dihydrolase, H_2S, urease, and malonate.** Reduce nitrates. Carbohydrates fermented include D-mannitol, D-mannose, and trehalose. Isolated from human clinical sources, most often sputum, wounds, and blood. Is an infrequent opportunistic pathogen.

Type (and only) species: *Ewingella americana.*

Characteristics of the species: *Ewingella* is most similar to *Cedecea* species and is differentiated from these species by tests listed in Table 5.5. *Ewingella* must also be differentiated from other organisms that are negative in tests for lysine and ornithine decarboxylases and for arginine dihydrolase. This is easily accomplished by comparing complete biochemical profiles, as given in Table 5.2.

Genus **Hafnia**

Straight rods, ~1 µm in diameter and 2–5 µm in length, stain Gram negative. **Motile by peritrichous flagella; nonmotile strains occur. Facultatively anaerobic.** Chemoorganotrophic, having **both a respiratory and a fermentative type of metabolism.** Optimal temperature is 30–37°C. D-Glucose and other carbohydrates are catabolized with the production of acid and gas. **Oxidase negative, catalase positive, and indole and Simmons citrate negative; the majority of strains are methyl red and Voges-Proskauer positive. Lysine and ornithine decarboxylase positive and are arginine dihydrolase negative. H_2S and urease negative and KCN positive.** Reduce nitrates. Ferments L-arabinose, glycerol, maltose, D-mannitol, D-mannose, L-rhamnose, trehalose, and D-xylose. Occur in the feces of humans and other animals, including birds, and in sewage, soil, water, and dairy products. Opportunistic pathogen for humans, usually in blood, urine, or wound infections in patients with underlying illness or predisposing factors.

Type (and only) species: *Hafnia alvei.*

Editorial note: The organism designated Biogroup 1 of *Obesumbacterium proteus* is actually a metabolically inactive biogroup of *H. alvei* that has become adapted to a brewery environment (Table 5.24).

Characteristics of the species: *H. alvei* is differentiated from *Enterobacter* and *Serratia* by the reactions given in Table 5.19. Reactions to differentiate *H. alvei* from specific species in these genera are shown in Table 5.2.

Genus **Klebsiella**

Straight rods, 0.3–1.0 μm in diameter and 0.6–6.0 μm in length, arranged singly, in pairs, or in short chains. Cells are **capsulated.** Gram negative. **Nonmotile. Facultatively anaerobic.** Chemoorganotrophic, having **both a respiratory and a fermentative type of metabolism.** Optimal temperature is 37°C. D-Glucose and other carbohydrates are catabolized with the production of acid and gas, but anaerogenic strains occur. **Oxidase negative and catalase positive.** Indole, methyl red, Voges-Proskauer, and Simmons citrate reactions vary among species. **Usually lysine decarboxylase positive and ornithine decarboxylase negative. Arginine dihydrolase negative. Several species hydrolyze urea. Grow on KCN. H$_2$S is not produced.** Reduce nitrates. **Most species ferment all commonly tested carbohydrates except dulcitol and erythritol.** Occur in human feces and clinical specimens, soil, water, grain, fruits, and vegetables. *K. pneumoniae, K. oxytoca,* and occasionally other species are opportunistic pathogens that can cause bacteremia, pneumonia, and urinary tract and other human infections. Frequently cause nosocomial infections in urological, neonatal, intensive care, and geriatric patients.

Type species: *Klebsiella pneumoniae.*

K. pneumoniae and *K. oxytoca* contain a large polysaccharide capsule, which gives rise to large mucoid colonies, especially on a carbohydrate-rich medium. This characteristic distinguishes them from other *Enterobacteriaceae,* except for some strains of *Enterobacter aerogenes* and *Escherichia coli.* There are more than 80 capsular (K) antigens that can be used to serotype klebsiellae.

Characteristics to differentiate *Klebsiella* species from *Enterobacter* species are given in Tables 5.10 and 5.12. Additional reactions are given in Table 5.11. The greatest problem is to distinguish *K. pneumoniae* strains from nonmotile *Enterobacter aerogenes* strains, which liquefy gelatin very slowly. The urease test may be of decisive importance in such cases. (*K. pneumoniae* is urease positive.)

Differentiation of the species of the genus **Klebsiella:** Reactions that differentiate the species and subspecies of *Klebsiella* are given in Table 5.20. Complete reactions are given in Table 5.2.

Editorial note: The species *Klebsiella ornithinolytica* and *Klebsiella terrigena* were not included in *Bergey's Manual of Systematic Bacteriology. K. ornithinolytica* was created in 1989 by Sakazaki et al. (Curr. Microbiol. *18:* 201–206; Int. J. Syst. Bacteriol. *39:* 495–497, 1989) for strains previously referred to as ornithine positive *K. oxytoca,* NIH (Japan) Group 12 (Sakazaki et al., Curr. Microbiol. *18:* 201–206, 1989); CDC *Klebsiella* Group 47 (indole positive, ornithine positive); and indole positive, ornithine positive biogroup of *K. planticola* (Farmer et al., J. Clin. Microbiol. *21:* 46–76, 1985). Both laboratories agree on the phenotypic properties of this organism, but they disagree on whether DNA relatedness data indicate that it should be retained as a biogroup of *K. planticola* or that it should be a separate species. *K. terrigena* was created in 1981 by Izard et al. (Int. J. Syst. Bacteriol. *31:* 116–127). The former species *Klebsiella ozaenae* and *Klebsiella rhinoscleromatis* were proposed as subspecies of *Klebsiella pneumoniae* in 1984 by Orskov (*Bergey's Manual of Systematic Bacteriology,* Volume 1, p. 461–465; Int. J. Syst. Bacteriol. *34:* 355–357, 1984). There is no doubt that *K. ozaenae* and *K. rhinoscleromatis* are the same genospecies as *K. pneumoniae,* but most laboratories still treat and report them as species.

Genus **Kluyvera**

Straight rods, 0.5–0.7 × 2–3 μm, stain Gram negative. **Motile by scant peritrichous flagella. Facultatively anaerobic** and chemoorganotrophic, having **both a respiratory and a fermentative type of metabolism.** Optimal temperature is 37°C. D-Glucose and other carbohydrates are catabolized with the production of acid and gas. **Oxidase negative; catalase positive; indole, methyl red, and usually Simmons citrate positive; Voges-Proskauer negative; ornithine decarboxylase positive; and arginine dihydrolase negative. Utilize malonate and grow on KCN. H_2S is not produced, and urea is not hydrolyzed.** Reduce nitrate. Ferment most carbohydrates but generally do not ferment polyhydroxyl alcohols. Occur in food, soil, sewage, and human clinical specimens, most often in respiratory tract, urinary tract, and feces and occasionally in blood. *Kluyvera* species are infrequent opportunistic pathogens.

Type species: *Kluyvera ascorbata.*

Kluyvera species most closely resemble *Buttiauxella agrestis,* which has not been isolated from humans, and they superficially resemble *E. coli, Citrobacter,* and *Enterobacter.* Tests to differentiate *Kluyvera* from these organisms are given in Tables 5.3 and 5.2.

Differentiation of the species of the genus **Kluyvera:** See Table 5.21. In addition, *K. cryocrescens* usually has big zones of inhibition for carbenicillin and cephalothin, whereas *K. ascorbata* has much smaller zones of inhibition.

Genus **Leclercia**

Editorial note: The genus *Leclercia* was not included in *Bergey's Manual of Systematic Bacteriology.* The genus was created in 1986 by Tamura et al. (Curr. Microbiol. *13:* 179–184; Int. J. Syst. Bacteriol. *37:* 179–180, 1987). It includes one species, *Leclercia adecarboxylata.*

Straight rods stain Gram negative and are **motile by peritrichous flagella. Facultatively anaerobic.** Chemoorganotrophic, having **both a respiratory and a fermentative type of metabolism.** Optimal temperature is 37°C. D-Glucose and other carbohydrates are catabolized with the production of acid and gas. **Oxidase negative, catalase positive, indole and methyl red positive, and Voges-Proskauer and Simmons citrate negative. Negative reactions for lysine and ornithine decarboxylase and arginine dihydrolase occur. Utilize malonate and grow on KCN. H_2S and urease negative.** Some strains produce yellow pigment. Reduce nitrates. Ferment most carbohydrates except *myo*-inositol and D-sorbitol. Occur in human clinical specimens, especially sputum and blood, and in food,

water, and environmental sources. An occasional opportunistic human pathogen.

Type (and only) species: *Leclercia adecarboxylata.*

Editorial note: L. adecarboxylata was formerly named *Escherichia adecarboxylata* and was frequently considered to be a synonym for *Enterobacter agglomerans.* It is included as a *species incertae sedis* within *Escherichia* in *Bergey's Manual of Systematic Bacteriology.*

Characteristics of the species: Leclercia must be differentiated from other genera that give negative reactions in tests for lysine and ornithine decarboxylase and arginine dihydrolase (Table 5.2). It is differentiated from lysine-, arginine-, and ornithine-negative *E. coli* strains by reactions in tests for KCN, malonate, and the fermentation of D-adonitol, cellobiose, and D-sorbitol (Table 5.18).

Genus **Leminorella**

Editorial note: The genus *Leminorella* was not included in *Bergey's Manual of Systematic Bacteriology.* The genus was created in 1985 by Hickman-Brenner et al., (J. Clin. Microbiol. *21:* 234–239; Int. J. Syst. Bacteriol. *35:* 375–376, 1985). It contained two species, *Leminorella grimontii* and *Leminorella richardii.*

Straight rods. Gram negative. **Nonmotile and facultatively anaerobic.** Chemoorganotrophic, having **both a respiratory and a fermentative type of metabolism.** Optimal temperature is 37°C. D-Glucose and other carbohydrates are catabolized with production of acid. Gas production from D-glucose is negative in one species and variable or delayed in the other species. **Oxidase negative; catalase positive. Indole and Voges-Proskauer negative; one species is positive and the other is negative in methyl red and Simmons citrate tests. Negative in tests for lysine and ornithine decarboxylase, and arginine dihydrolase. Produce H_2S. Negative for urease, KCN, and malonate tests.** Reduce nitrates. **L-arabinose and D-xylose are fermented; almost all other commonly tested carbohydrates are not fermented.** Isolated from human stools and urines. Clinical significance is unknown.

Type species: *Leminorella grimontii.*

Like *Proteus* species, *Leminorella* species produce H_2S, are positive for tyrosine clearing, and do not ferment D-mannose, but unlike *Proteus* they ferment L-arabinose and are negative in tests for urea hydrolysis and phenyl-alanine deaminase. Tests that differentiate *Leminorella* from *Proteus* and *Providencia,* as well as *Budvicia,* are given in Table 5.22. Differentiation from *Pragia* is shown in Table 5.26. Tests to differentiate *Leminorella* from other lysine, arginine, ornithine negative species are given in Table 5.2.

Differentiation of the species of the genus **Leminorella:** See Table 5.23. Other biochemical characteristics are given in Table 5.2.

Genus **Moellerella**

Editorial note: The genus *Moellerella* was not included in *Bergey's Manual of Systematic Bacteriology.* The genus was created in 1984 by Hickman-Brenner et al., (J. Clin. Microbiol. *19:* 460–463; Int. J. Syst. Bacteriol. *34:* 355–357, 1984). It included one species, *Moellerella wisconsensis.*

Straight rods. Gram negative. **Nonmotile and facultatively anaerobic.** Chemoorganotrophic, having **both a respiratory and a fermentative type of metabolism.** Optimal temperature is 37°C. D-Glucose and other carbohydrates are catabolized with production of acid, but no gas. **Oxidase negative; catalase positive. Indole and Voges-Proskauer negative and methyl red and Simmons citrate positive. Lysine and ornithine decarboxylase and arginine dihydrolase negative. H₂S is not produced; urease in not hydrolyzed; malonate is not utilized. A majority of strains grow in KCN.** Reduce nitrates. Carbohydrates fermented include adonitol, D-arabitol, D-galactose, lactose, D-mannose, melibiose, raffinose, and sucrose. Glycerol and D-mannitol are fermented within 7 days. Occur in human stool and in water. Associated with, but not a proven cause of, human diarrhea.

Type (and only) species: *Moellerella wisconsensis.*

Characteristics of the species: Moellerella must be differentiated from other genera with negative reactions for lysine and ornithine decarboxylase, and arginine dihydrolase, including the biogroup of *E. coli* strains that are negative in these tests. This is done on the basis of reactions for indole, Simmons citrate, motility, gas production from D-glucose, and fermentation of a number of carbohydrates (Table 5.2).

Genus **Morganella**

Straight rods, 0.6–0.7 μm in diameter and 1.0–1.7 μm in length. Gram negative. **Motile by peritrichous flagella. Swarming does not occur. Facultatively anaerobic.** Chemoorganotrophic having **both a respiratory and a fermentative type of metabolism.** Optimal temperature is 37°C. **D-Glucose and D-mannose are the only common carbohydrates catabolized** with production of acid and usually gas (which may be delayed). **Oxidase negative, catalase positive, indole and methyl red positive, Voges-Proskauer and Simmons citrate negative, lysine decarboxylase and arginine dihydrolase negative, and ornithine decarboxylase positive. Phenylalanine and tryptophan are deaminated oxidatively. Urease is hydrolyzed. There is growth on KCN. H₂S is not produced. Decompose tyrosine to produce a clearing on media containing the insoluble amino acid.** Reduce nitrates. Occur in feces of humans, dogs, other mammals, and reptiles. Are opportunistic secondary invaders, isolated from bacteremia, respiratory tract, wound, and urinary tract infections.

Type (and only) species: *Morganella morganii.*

Characteristics of the species: Characteristics that differentiate *Morganella* from *Proteus* and *Providencia* are given in Table 5.27.

Genus **Obesumbacterium**

Pleomorphic rods 0.8–2.0 μm in diameter, 1.5–100 μm in length (short, "fat" rods predominate when grown in beer wort with live yeasts; long pleomorphic rods usually predominate on bacteriological media). Gram negative. **Nonmotile and facultatively anaerobic.** Chemoorganotrophic, having **both a respiratory and a fermentative type of metabolism.** Optimum temperature is ~32°C. D-Glucose and a few other carbohydrates are catabolized with the production of acid. **Oxidase negative, catalase positive, and indole, Voges-Proskauer, and Simmons citrate negative; most strains are methyl red negative. Lysine and ornithine decarboxylase positive and arginine dihydrolase negative. H₂S is not produced, urea is not hydrolyzed, there is no growth in KCN, and malonate is not utilized.** Reduce nitrates. **Most strains ferment D-mannose and trehalose, but rarely any other carbohydrates. Many biochemical tests normally used for differentiation of** *Enterobacteriaceae* are negative or delayed. **Grows slowly, forming colonies less than 0.5 mm at 24 h. Occurs as a brewery contaminant** which can survive and grow in the presence of live yeasts during beer production.

Type (and only) species: *Obesumbacterium proteus.*

Characteristics of the species: Biochemical reactions for the differentiation of *Obesumbacterium* from other genera are listed in Table 5.24.

Editorial note: There are two biogroups in *O. proteus.* Within the brewery industry, both are designated as *"Hafnia protea,"* a name that is not valid. Biogroup 2 represents *O. proteus sensu stricto.* Biogroup 1 is actually an unusual (metabolically inactive) biogroup of *Hafnia alvei* that has become adapted to the brewery environment (Table 5.24).

Genus **Pantoea**

Editorial note: The genus *Pantoea* was not included in *Bergey's Manual of Systematic Bacteriology.* The genus was created in 1989 by Gavini et al., (Int. J. Syst. Bacteriol. *39:* 337–345). It included two species, *Pantoea agglomerans* (synonyms: *Enterobacter agglomerans, Erwinia herbicola, Erwinia milletiae*) and *Pantoea dispersa.*

Straight rods, 0.5–1.0 × 1–3 μm. Gram negative. **Motile by peritrichous flagella. Most strains produce a yellow pigment. Facultatively anaerobic.** Chemoorganotrophic, having **both a respiratory and a fermentative type of metabolism.** Optimal temperature is 30°C. D-Glucose and other carbohydrates are catabolized with production of acid, but not gas. **Oxidase negative, catalase positive, indole negative, Voges-Proskauer and Simmons citrate positive,** and methyl red variable. **Negative for lysine and ornithine decarboxylase and for arginine dihydrolase** (Gavini et al. found 30% of *P. agglomerans* **to be ornithine decarboxylase positive;** previous studies reported all strains negative). **No production of H$_2$S; urea is not hydrolyzed. Most strains grow on KCN. Malonate utilization varies among species.** Reduce nitrates. Carbohydrates fermented by all or most strains include L-arabinose, D-galactose, maltose, D-mannitol, D-mannose, L-rhamnose, sucrose, trehalose, and D-xylose. Isolated from plant surfaces, seeds, soil, and water, as well as from animals and human wounds, blood, and urine. An opportunistic human pathogen.

Type species: *Pantoea agglomerans.*

*Differentiation of the species of the genus **Pantoea:*** See Table 5.25.

Characteristics that differentiate *Pantoea* species from other genera are given in Table 5.2.

Editorial note: *P. agglomerans* is frequently referred to by its synonyms, especially *Enterobacter agglomerans.* In this volume it is included in both *Enterobacter* and *Pantoea.*

Genus **Pragia**

Editorial note: The genus *Pragia* was not included in *Bergey's Manual of Systematic Bacteriology.* The genus was created in 1988 by Aldova et al., (Int. J. Syst. Bacteriol. *38:* 183–189). It included one species, *Pragia fontium.*

Straight rods. Gram negative. **Motile by peritrichous flagella. Facultatively anaerobic.** Chemoorganotrophic, having **both a respiratory and a fermentative type of metabolism.** Optimal temperature is 37°C. D-Glucose and a few other carbohydrates are catabolized with the production of acid but not gas. **Oxidase negative, catalase positive, indole and Voges-Proskauer negative, and methyl red and usually Simmons citrate positive. Negative for lysine and ornithine decarboxylase and arginine dihydrolase. Produce H$_2$S. Do not hydrolyze urea, grow in KCN, or utilize malonate.** Reduce nitrates. Ferment D-galactose and one-half to three-quarters of strains ferment glycerol (delayed), *myo*-inositol (delayed), and salicin; **other commonly tested carbohydrates are not fermented.** Occur in drinking water; there is one human stool isolate. There is no evidence for pathogenicity.

Type (and only) species: *Pragia fontium.*

Characteristics of the species: Characteristics that differentiate *Pragia* from other lysine-, arginine-, and ornithine-negative genera are given in Table 5.26. Complete biochemical reactions are given in Table 5.2.

Genus **Proteus**

Straight rods, 0.4–0.8 μm in diameter × 1–3 μm in length. Gram negative. **Motile by peritrichous fla-**

gella. **Most strains swarm with periodic cycles of migration producing concentric zones, or spread in a uniform film, over moist surfaces** of nutrient media solidified with agar or gelatin. **Facultatively anaerobic** and chemoorganotrophic, having **both a respiratory and a fermentative type of metabolism.** Optimal temperature is 37°C. D-Glucose and a few other carbohydrates are catabolized with production of acid and usually gas. **Oxidase negative, catalase positive, and methyl red positive; species vary in indole, Voges-Proskauer, and Simmons citrate tests. Lysine decarboxylase and arginine dihydrolase negative; only *P. mirabilis* decarboxylates ornithine. Phenylalanine and tryptophan are oxidatively deaminated, and urea is hydrolyzed. Decompose tyrosine to produce a clearing on agar media in which the insoluble amino acid is incorporated. Grow on KCN. H$_2$S is usually produced. Malonate is not utilized.** Reduces nitrates. One or more species ferment glycerol, maltose, sucrose, trehalose, and D-xylose. Occur in intestines of humans and a wide variety of animals; also occur in manure, soil, and polluted waters. *P. myxofaciens* has been isolated only from gypsy moth larvae. Human pathogens, causing urinary tract infections; also are secondary invaders, causing septic lesions, often in burn patients.

Type species: *Proteus vulgaris.*

Phenylalanine (or tryptophan) deaminase is useful in differentiating *Proteus, Providencia,* and *Morganella* from other *Enterobacteriaceae.* Characteristics for differentiating *Proteus* from *Providencia* and *Morganella* are given in Table 5.27.

Differentiation of the species of the genus **Proteus:** See Table 5.28. Additional biochemical characteristics are given in Table 5.2.

Editorial note: Proteus penneri was not included in *Bergey's Manual of Systematic Bacteriology.* The species was created in 1982 by Hickman et al., (J. Clin. Microbiol. *15:* 1097–1102; Int. J. Syst. Bacteriol. *33:* 438–440, 1983).

Genus **Providencia**

Straight rods, 0.6–0.8 × 1.5–2.5 µm. Gram negative. **Motile by peritrichous flagella. Swarming does not occur. Facultatively anaerobic.** Chemoorganotrophic, having **both a respiratory and a fermentative type of**

metabolism. Optimal temperature is 37°C. D-Glucose and other carbohydrates are catabolized with the production of acid; some strains produce gas. **Oxidase negative, catalase positive, indole positive (except *P. heimbachae*), usually methyl red positive, and Voges-Proskauer negative. Simmons citrate is utilized by some species. Lysine and ornithine decarboxylase and arginine dihydrolase test negative. Oxidatively deaminate phenylalanine and tryptophan. Decompose tyrosine to produce a clearing on agar media in which the insoluble amino acid is incorporated. H$_2$S is not produced; urea is not hydrolyzed except for *P. rettgeri*. Grown on KCN except for *P. heimbachae*. Malonate is not utilized. Ferment mannose and one or more of the following polyhydric alcohols:** adonitol, D-arabitol, erythritol, *myo*-inositol, D-mannitol. Isolated from human diarrhetic stools, urinary tract infections, wounds, burns and bacteremias, and from penguins. A human pathogen.

Type species: *Providencia alcalifaciens.*

Characteristics that differentiate *Providencia* species from *Proteus* and *Morganella* are given in Tables 5.27.

Differentiation of the species of the genus **Providencia.** See Table 5.29. Other characteristics of these species are given in Table 5.2.

Editorial note: Providencia heimbachae and *Providencia rustigianii* were not included in *Bergey's Manual of Systematic Bacteriology. Providencia heimbachae* was created in 1986 by Muller et al., (Int. J. Syst. Bacteriol. *36:* 252–256) and *Providencia rustigianii* (formerly known as biogroup 3 of *Providencia alcalifaciens*) was created in 1983 by Hickman-Brenner et al., (J. Clin. Microbiol. *17:* 1057–1060; Int. J. Syst. Bacteriol. *33:* 672–674, 1983).

Genus **Rahnella**

Small rod-shaped cells 0.5–0.7 × 2–3 µm. Gram negative. **Nonmotile at 36°C and motile by peritrichous flagella when grown at 25°C. Facultatively anaerobic.** Chemoorganotrophic, having **both a respiratory and a fermentative type of metabolism.** Grow well from 25–37°C. D-Glucose and other carbohydrates are catabolized with the production of acid and, in a majority of strains, gas. **Oxidase negative, catalase positive, and indole negative. Most strains are methyl red positive. Voges-Proskauer and Simmons citrate posi-**

tive and lysine and ornithine decarboxylase and arginine dihydrolase negative. **H$_2$S is not produced, urea is not hydrolyzed; malonate is utilized. There is no growth on KCN. Weakly positive for phenylalanine deaminase.** Reduce nitrates. Carbohydrates fermented include L-arabinose, cellobiose, dulcitol, lactose, maltose, D-mannitol, D-mannose, melibiose, raffinose, L-rhamnose, salicin, D-sorbitol, sucrose, trehalose, and D-xylose. Occurs in freshwater. Occasionally isolated from human clinical specimens; there is no known clinical significance.

Type (and only) species: *Rahnella aquatilis.*

Characteristics of the species: Characteristics that differentiate *Rahnella* from other lysine, arginine, ornithine negative genera are given in Table 5.2.

Genus **Salmonella**

Straight rods, 0.7–1.5 × 2–5 μm. Gram negative. **Usually motile by peritrichous flagella. Facultatively anaerobic.** Chemoorganotrophic, having **both a respiratory and a fermentative type of metabolism.** Optimal temperature is 37°C. D-Glucose and other carbohydrates are catablized with the production of acid and usually gas. **Oxidase negative, catalase positive, indole and Voges-Proskauer negative, and methyl red and Simmons citrate positive. Lysine and ornithine decarboxylase positive; there is a variable arginine dihydrolase reaction. H$_2$S is produced; urea is not hydrolyzed; growth on KCN and utilization of malonate are variable.** Reduce nitrates. Carbohydrates usually fermented include L-arabinose, maltose, D-mannitol, D-mannose, L-rhamnose, D-sorbitol, trehalose, and D-xylose. Occur in humans, warm and cold blooded animals, foods, and the environment. Pathogenic for humans and many animal species. Causative agents of typhoid fever, enteric fevers, gastroenteritis, and septicemia.

Type species: *Salmonella choleraesuis.*

Differentiation of the species of the genus **Salmonella:** Most salmonellae are aerogenic; however, *Salmonella choleraesuis* serovar typhi (*Salmonella typhi*; *Salmonella* serovar typhi), an important exception, and some other serovars never produce gas. Characteristics useful for differentiating the genus *Salmonella* from other *Enterobacteriaceae* are given in Table 5.30.

Division into serovars: In the Kauffmann-White scheme, serovars are represented by the numbers and letters given to the different O (somatic), Vi (capsular) and H (flagellar) antigens. Antigenic formulae (for example, 6,7:r:1,7) represent the O antigens (6,7); the phase 1 H antigen(s) (r); and the phase 2 H antigens (1,7), respectively. Those formulae with particular O antigens in common are collected into an O group and arranged alphabetically by H antigens within the group. The scheme serves both *Salmonella* species and all subspecies (see below). Serovars may be further divided on the basis of biochemical tests which may be of epidemiological interest (for example, the xylose-positive and xylose-negative-strains of *Salmonella* serovar typhi).

Phage typing: Phage typing is done for epidemiologic surveillance of certain serovars, most notably *Salmonella* serovars typhi and enteritidis. The 01 bacteriophage of Felix and Callow is highly specific for salmonellae, lysing more than 98% of strains studied in routine *Salmonella* diagnosis (Kallings, Acta Pathol. Microbiol. Scand. *70:* 446–454, 1967).

Differentiation of the species and subspecies of the genus **Salmonella:** Characteristics differentiating the species and subspecies of *Salmonella* are given in Table 5.31. Complete biochemical reactions are given in Table 5.2.

Editorial note: *Salmonella bongori* and subspecies of *Salmonella choleraesuis* were not included in *Bergey's Manual of Systematic Bacteriology.* *Salmonella bongori* (formerly *Salmonella choleraesuis* subsp. *bongori*) was created in 1989 by Reeves et al., (J. Clin. Microbiol. *27:* 313–320; Int. J. Syst. Bacteriol. *39:* 371, 1989). *Salmonella chloeraesius* subsp. *arizonae, choleraesuis, diarizonae, houtenae,* and *salamae* were created in 1982 by Le Minor et al., (Ann. Microbiol. *133B:* 245–254; Int. J. Syst. Bacteriol. *35:* 375–376, 1985). *Salmonella choleraesuis* subsp. *indica* was created in 1986 by Le Minor et al., (Ann. Inst. Pasteur *137B:* 211–217; Int. J. Syst. Bacteriol. *37:* 179–180, 1987).

Nomenclature: All of the *Salmonella* (including "Arizona") serovars belong to two species. *Salmonella bongori* contains less than 10 serovars that are extremely rare. The more than 2500 remaining serovars are all part of *Salmonella choleraesuis,* which is divisible, both phenotypically and genetically, into 6 subspecies. All serovars in subspecies *choleraesuis* are named, whereas serovars in other subspecies (except for some in subspecies *salamae* and *houtenae*) are not named. It has been recommended that diagnostic laboratories report named *Salmonella* serovars by name and unnamed servovars by antigenic formula and subspecies. Some examples are: *Salmonella typhi* or *Salmonella* serovar typhi; *S. typhimurium* or *Salmonella* serovar typhimurium; *Salmonella* subsp. *arizonae* serovar 50:z_4,z_{24}:-(formerly *"Arizona hinshawii"*), *Salmonella* subsp. *salamae* serovar 56:z_{10}:e,n,x. It must also be noted that the Arizona group (formerly classified as *"Arizona hinshawii"* in the U.S. and now subspecies *arizonae* [monophasic strains] and subspecies *diarizonae* [for diphasic strains] were serotyped using an *"Arizona"* scheme in the U.S. and are now serotyped by the *Salmonella* serotyping scheme. For this reason (at least in the U.S.) it is useful to give both antigenic formulae; for example *Salmonella* serovar 50:z_4,z_{24}:-(formerly *"Arizona hinshawii"* 9a,9b:1,3, 11:-). Further details of the complex nomenclatural problems in *Salmonella* are found in the publications of Le Minor and colleagues (*Bergey's Manual of Systematic Bacteriology*, vol. 1, p. 427–448, 1984; Ann. Microbiol. *133B*: 223–243, 1982; Ann. Microbiol. *133B*: 245–254, 1982) and of Farmer et al., (Clin. Microbiol. Newsl. *6*: 63–66, 1984).

Genus **Serratia**

Straight rods, 0.5–0.8 μm in diameter and 0.9–2.0 μm in length. Gram negative. **Usually motile by peritrichous flagella. Facultatively anaerobic.** Chemoorganotrophic, having **both a respiratory and a fermentative type of metabolism.** They grow well at 30–37°C. D-Glucose and other carbohydrates are catabolized with the production of acid and often gas. **Indole negative except for some *S. odorifera* strains. The methyl red test varies. Simmons citrate test is positive; Voges-Proskauer test is usually positive except for *S. fonticola*. Most species are lysine decarboxylase positive, arginine dihydrolase negative, and ornithine decarboxylase positive. H$_2$S is not produced; urea is not hydrolyzed; malonate is usually not utilized. Most strains produce DNase and hydrolyze corn oil. Gelatin is usually hydrolyzed.** Reduce nitrates. Carbohydrates fermented by all or most strains include maltose, D-mannitol, D-mannose, salicin, sucrose, and trehalose. Occur in human clinical specimens, soil, water, plant surfaces, and other environmental sites, digestive tract of rodents, and insects. *S. marcescens* is a prominent opportunistic pathogen for hospitalized humans, causing septicemia and urinary tract infections. Several other species can be involved in bacteremia or can be isolated from sputum without clinical significance. Cause of mastitis in cows and other animal infections.

Type species: *Serratia marcescens.*

Characteristics that differentiate *Serratia* from biochemically similar taxa are given in Tables 5.32 and 5.33. Characteristics that differentiate each *Serratia* species from other taxa are given in Table 5.34.

Differentiation of the species of the genus **Serratia:** See Tables 5.35 and 5.36. Since it is difficult to separate the *S. liquefaciens, S. grimesii, S. proteamaculans* complex using routine biochemical tests, most laboratories identify strains as *S. liquefaciens* or *S. liquefaciens* group.

Editorial note: Serratia entomophila, Serratia grimesii, Serratia proteamaculans, Serratia proteamaculans subsp. *proteamaculans, Serratia proteamaculans* subsp. *quinovora, "Serratia rubidaea* subsp. *burdigalensis," "Serratia rubidaea* subsp. *colindalensis,"* and *"Serratia rubidaea* subsp. *rubidaea"* were either mentioned without description or not included in *Bergey's Manual of Systematic Bacteriology. Serratia entomophila* was created in 1988 by Grimont et al., (Int. J. Syst. Bacteriol. *38:* 1–6). *Serratia grimesii* and *Serratia proteamaculans* with two subspecies, *proteamaculans* and *quinovora* were created in 1982 by Grimont et al., (Curr. Microbiol. *7:* 69–74; Int. J. Syst. Bacteriol. *33:* 438–440, 1983). Subspecies *burdigalensis, colindalensis,* and *rubidaea* of *Serratia rubidaea* were created in 1990 by Grimont et al., (P.A.D. Grimont, personal communication).

Genus **Shigella**

Straight rods. Gram negative. **Nonmotile. Facultatively anaerobic.** Chemoorganotrophic, having **both a respiratory and a fermentative type of metabolism.** Optimal temperature is 37°C. D-Glucose and other carbohydrates are catabolized with the production of acid; a few strains form gas. **Oxidase negative and catalase positive. Production of indole varies. Methyl red positive, Voges-Proskauer and Simmons citrate negative, and lysine decarboxylase and arginine dihydrolase negative; the ornithine reaction varies. H$_2$S is not produced; urea is not hydrolyzed; malonate is not utilized, and there is no growth on KCN.** Reduce nitrates. Carbohydrates fermented include D-mannitol, D-mannose, and usually maltose and trehalose. Intestinal pathogens of humans and other primates, causing bacillary dysentery.

Type species: *Shigella dysenteriae.*

It is often difficult to differentiate shigellae from non-motile, lactose negative, anaerogenic strains and other inactive *E. coli* strains. Identification is further complicated by the fact that some *E. coli* strains share with *Shigella* the ability to cause bacillary dysentery. Biochemical characteristics of use in differentiating shigellae from *E. coli* are given in Table 5.17.

Differentiation of the species of the genus **Shigella:** *Shigella* contains four species, *S. dysenteriae, S. flexneri, S. boydii,* and *S. sonnei.* These are often referred to as subgroups A, B, C, and D, respectively. It is often possible to separate *S. sonnei* from other shigellae solely on the basis of biochemical reactions (Table 5.37), but this cannot be done to distinguish between *S. boydii, S. dysenteriae,* and *S. flexneri.* Somatic (O) antigen typing is necessary to confirm species identification of shigellae.

Genus **Tatumella**

Small rod-shaped cells 0.6–0.8 × 0.9–3 μm. Gram negative. **Nonmotile at 36°C; over half the strains are motile.** Facultatively anaerobic. Chemoorganotrophic, having **both a respiratory and a fermentative type of metabolism.** Grow well at 25–36°C, but are **more active metabolically at 25°C.** D-Glucose and other carbohydrates are catabolized with the production of acid but not gas. **Oxidase negative; catalase positive. Indole, methyl red, Voges-Proskauer, Simmons citrate, lysine and ornithine decarboxylase, arginine dihydrolase, H₂S production, urea hydrolysis, growth on KCN, and malonate utilization are negative. Phenylalanine deaminase is positive.** Reduces nitrates. Carbohydrates fermented include D-mannose, sucrose, and trehalose. Isolated from human clinical specimens, mainly respiratory tract, but some blood isolates, and from animals. Probable rare opportunistic human pathogen.

Type (and only) species: *Tatumella ptyseos.*

Characteristics of the species: Its relative metabolic inactivity at 36°C allows simple differentiation from other *Enterobacteriaceae.* It must be distinguished from phenylalanine deaminase-positive species as shown in Table 5.38. Complete biochemical reactions are given in Table 5.2.

Genus **Xenorhabdus**

Rod-shaped cells 0.3–2 × 2–10 μm; occasional filaments 15–50 μm. In older cultures the cells contain crystalline inclusions (not poly-β-hydroxybutyrate). **Spheroplasts or coccoid bodies,** resulting from disintegration of the cell wall, occur in the last third of exponential growth. Gram negative. **Motile by peritrichous flagella. Facultatively anaerobic.** Chemoorganotrophic, having **both a respiratory and a fermentative type of metabolism. Optimal temperature is 25°C; they grow poorly or not at all at 36°C.** D-Glucose **and a few other carbohydrates are catabolized with the production of acid but not gas; the reactions are usually weak or delayed with small amounts of acid. Oxidase negative; catalase negative except for** *X. luminescens.* **Most strains are indole negative. Methyl red, Voges-Proskauer, and Simmons citrate negative. Lysine and ornithine decarboxylase and arginine dihydrolase negative. H₂S is not produced; most strains do not hydrolyze urea; no growth on KCN; malonate is not utilized. Nitrates are usually not reduced.** D-Mannose **is the only carbohydrate fermented by almost all strains. All but one species produce various color pigments. Phase variants (different colony forms) occur in all strains.** Natural habitat is the intestinal lumen of entomopathogenic nematodes, and the body cavities of insects infected by these nematodes. One genospecies has been isolated from human wounds and blood (Farmer et al., J. Clin. Microbiol. *27:* 1594–1600, 1989).

Type species: *Xenorhabdus nematophilus.*

Editorial note: Different laboratories have reported substantial differences in biochemical reactions for *Xenorhabdus* (Akhurst and Boemare, J. Gen. Microbiol. *134:* 1835–1845, 1988; Farmer et al., J. Clin. Microbiol. *27:* 1594–1600, 1989; Grimont et al., Int. J. Syst. Bacteriol. *34:* 378–388, 1984). Many of the differences are probably due to slow and weak reactions, and to differences in media. Complete biochemical reactions have not been reported for species other than *X. luminescens* and *X. nematophilus.* No genera in *Enterobacteriaceae* are phenotypically similar to *Xenorhabdus.* The source of isolation, poor or absent growth at 37°C, and the generally inactive biochemical profile of *Xenorhabdus* make it a phenotypically unique genus.

Differentiation of the species of the genus **Xenorhabdus:** See Table 5.39. Additional reactions are given in Table 5.2.

Genus **Yersinia**

Straight rods, sometimes approaching a spherical shape, 0.5–0.8 μm in diameter and 1–3 μm in length. Gram negative. **Nonmotile at 37°C, but motile by peritrichous flagella when grown below 30°C, except for some** *Y. ruckeri* **strains and** *Y. pestis* **which is always nonmotile. Facultatively anaerobic.** Chemoorganotrophic, having **both a respiratory and a fermentative type of metabolism. Optimal temperature is 28–30°C.** D-Glucose and other carbohydrates are catabolized with the production of acid but little or no gas. **Oxidase negative and catalase positive, and indole production varies among species. Usually methyl red positive. Voges-Proskauer and citrate test negative at 37°C but variable at 25–28°C. Usually cells are lysine decarboxylase negative, arginine dihydrolase negative, and ornithine decarboxylase positive except** *Y. pestis, Y. pseudotuberculosis,* **and** *Y. rohdei.* **H₂S is not produced. Urea is generally hydrolyzed except for** *Y. bercovieri, Y. pestis,* **and** *Y. ruckeri.* **Few strains grow on KCN, and malonate is not utilized.** Reduce nitrates. Carbohydrates fermented by all or most species include L-arabinose, maltose, D-mannitol, D-mannose, and trehalose. Occur in a broad spectrum of habitats including humans, animals, especially rodents and birds, soil, water, dairy products and other foods. *Y. pestis* is the causative agent of plague, which is primarily a disease of wild rodents. *Y. pestis* is transmitted among wild rodents by fleas, in which the bacteria multiply and block the esophagus and the pharynx. The fleas regurgitate the bacteria when they take their next blood meal and transmit the disease to humans if no other hosts are available. Infective flea bites produce the typical bubonic form of plague in humans. A secondary pneumonia can develop and primary pneumonic plague can spread by droplet infection. *Y. pseudotuberculosis* is pathogenic for many animal species and occasionally humans, causing mesenteric adenitis, chronic diarrhea, and severe septicemia.

Y. enterocolitica causes similar infections in animals and humans. *Y. ruckeri* causes red mouth disease in fish. Other species are occasional opportunistic human pathogens or are nonpathogenic.

Type species: *Yersinia pestis.*

Characteristics that differentiate *Yersinia* from other genera are given in Table 5.40. Table 5.41, which gives key *Yersinia* reactions at both 28°C and 36°C, should also be consulted since many *Yersinia* species are more metabolically active at the lower incubation temperature.

Differentiation of the species of the genus **Yersinia.** See Table 5.42. Different biogroups of *Y. enterocolitica* are differentiated in Table 5.43. Complete biochemical reactions at 37°C are given in Table 5.2.

Genus **Yokenella**

investigators agreed that *Yokenella* should be the genus name.

Straight rods. Gram negative. **Motile by peritrichous flagella. Facultatively anaerobic and** chemoorganotrophic, having **both a respiratory and a fermentative type of metabolism.** Optimal temperature is 37°C. D-Glucose and other carbohydrates are catabolized with the production of acid and gas. **Oxidase negative, catalase positive, indole and Voges-Proskauer negative, methyl red and Simmons citrate positive, lysine and ornithine decarboxylase positive, and arginine dihydrolase negative. H₂S is not produced; urease is not hydrolyzed; there is growth on KCN; malonate is not utilized.** Reduce nitrates. Carbohydrates fermented include L-arabinose, cellobiose, maltose, D-mannitol, D-mannose, melibiose, L-rhamnose, trehalose, and D-xylose. Isolated from human wounds, urine, sputum, and stool, and insect intestine. Clinical significance is unknown.

Type (and only) species: *Yokenella regensburgei.*

Characteristics of the species: Both human and insect strains of *Yokenella* were first thought to be *Hafnia alvei* or a *Hafnia*-like species. Characteristics to differentiate *Yokenella* from *Hafnia* are given in Table 5.44. Further biochemical reactions are given in Table 5.2. *Yokenella* is also somewhat similar to species in the genera *Citrobacter* and *Escherichia. Yokenella* strains frequently give delayed (3–7 days) positive reactions for several biochemical tests (Table 5.45).

SUBGROUP 2: Family Vibrionaceae

Straight or curved rods, 0.3–1.3 μm in width and 1.4–5.0 μm in length. Gram negative. **Motile by polar flagella or nonmotile** (genus *Enhydrobacter*). Additional lateral flagella may be synthesized when grown on solid media. **Facultatively anaerobic.** Chemoorganotrophic, having **both a respiratory and a fermentative type of metabolism.** Most species grow well at 37°C, but photobacteria and several marine vibrios require temperatures of 25°C or less for growth. D-Glucose and other carbohydrates are catabolized with production of acid and, in some species, gas. **Most are oxidase positive** (exceptions are *Vibrio metschnikovii, V. gazogenes, Photobacterium phosphoreum,* most *P. angustum,* and some *P. leiognathi*). Photobacteria and a majority of *Vibrio* species require 2–3% NaCl for optimum growth. Distributed worldwide, primarily **aquatic inhabitants** found in seawater and fresh water

and in association with aquatic animals. Several species are pathogenic for humans, fish, eels, and frogs, as well as other vertebrates and invertebrates. *Vibrio cholerae* is the causative agent of cholera. A number of *Vibrio* and *Aeromonas* species, and *Plesiomonas* cause diarrheal, wound, and septicemic infections in humans.

Type genus: *Vibrio.*

Editorial note: Since the publication of *Bergey's Manual of Systematic Bacteriology,* it has been proposed that the genera *Aeromonas* and *Plesiomonas* be removed from *Vibrionaceae. Aeromonas* would be placed in the new family *Aeromonadaceae* and *Plesiomonas* would be transferred to the family *Enterobacteriaceae* (Colwell et al., Int. J. Syst. Bacteriol. *36:* 473–477, 1986; MacDonnell and Colwell, Syst. Appl. Microbiol. *6:* 171–182, 1985). For the purpose of this diagnostic volume, both genera are retained in the family *Vibrionaceae,* without prejudice to the validity of these proposals.

Differentiation of **Vibrionaceae** *from other families and genera in Group 5:* Characteristics that differentiate *Vibrionaceae* from the other families and genera are presented in Table 5.1. The genera in the family *Enterobacteriaceae* are oxidase negative and, when motile, have peritrichous flagella—characteristics that distinguish them from *Vibrionaceae.* However, *Tatumella ptyseos* (in the family *Enterobacteriaceae*) has lateral flagella; at least two *Vibrio* species (*V. metschnikovii* and *V. gazogenes*) are oxidase negative, and some strains of other species may be oxidase negative or weakly positive.

Differentiation of the genera in **Vibrionaceae:** See Table 5.46.

Editorial note: Two new genera, *Listonella* (MacDonell and Colwell, Syst. App. Microbiol. *6:* 171–182, 1985; Int. J. Syst. Bacteriol. *36:* 354, 1986) and *Colwellia* (Deming et al., Syst. Appl. Microbiol. *10:* 152–160, 1988; Int. J. Syst. Bacteriol. *38:* 328, 1988) have been proposed for species formerly considered members of the genus *Vibrio.* For the purpose of this diagnostic volume, all of these species are retained in the genus *Vibrio* without prejudice. The new genus names are given in parentheses to eliminate any confusion.

Genus **Aeromonas**

Straight rods with rounded ends to cells approaching a spherical shape, 0.3–1.0 μm in diameter and 1.0–3.5

μm in length. Occur singly, in pairs, or short chains. Gram negative. **Usually motile by a single polar flagellum** (peritrichous flagella may be formed on solid media in young cultures). **Facultatively anaerobic.** Chemoorganotrophic, having **both a respiratory and a fermentative type of metabolism. Optimal temperature is 22–28°C; most species grow well at 37°C, but some strains do not.** D-Glucose and other carbohydrates are catabolized with the production of acid and often gas. **Oxidase positive and catalase positive. Usually they are arginine dihydrolase positive and ornithine decarboxylase negative. Urease and phenylalanine deaminase test negative. Gelatinase and DNase are positive.** Reduce nitrates. Carbohydrates fermented by almost all strains include maltose, D-galactose, and trehalose. **Resistant to the vibriostatic agent** 2,4-diamino-6,7-diisopropylpteridine **(0/129).** Occur in fresh water and sewage. Some species are pathogenic to frogs, fish, and humans. Human disease is usually diarrhea or bacteremia.

Type species: *Aeromonas hydrophila.*

Characteristics that differentiate *Aeromonas* from other genera are given in Tables 5.46 and 5.47.

Differentiation of the species of the genus **Aeromonas:** It is difficult to differentiate the new *Aeromonas* species from the better known species *A. caviae, A. hydrophila, A. salmonicida,* and *A. sobria.* For this reason most clinical laboratories will report isolates as members of the *A. caviae* group, which includes *A. caviae, A. eucrenophila,* and *A. media;* the *A. hydrophila* group, which includes *A. hydrophila* and a motile biogroup of *A. salmonicida;* the *A. sobria* group, which includes *A. sobria* and *A. veronii;* or *A. schubertii.* Biochemical characteristics of *Aeromonas* species are given in Table 5.48. It must also be noted that essentially all clinical isolates identified as *A. sobria* are actually *A. veronii* (Hickman-Brenner et al., J. Clin. Microbiol. *25:* 900–906, 1987). *A. sobria* sensu stricto is almost never encountered in clinical samples.

Editorial note: Aeromonas eucrenophila, Aeromonas media, Aeromonas salmonicida subsp. *smithia, Aeromonas schubertii,* and *Aeromonas veronii* were not included in *Bergey's Manual of Systematic Bacteriology. Aeromonas eucrenophila* was created in 1988 by Schubert and Hegazi (Zentralbl. Bakteriol. Parasitenkd. Infektionskr. Hyg. Abt. 1 Orig. Reihe A *268:* 34–39; Int. J. Syst. Bacteriol. *38:* 449, 1988). *Aeromonas media* was created in 1983 by Allen et al., (Int. J. Syst. Bacteriol. *33:* 599–604). *Aeromonas salmonicida* subsp. *smithia* was

created in 1989 by Austin et al., (Syst. Appl. Microbiol. *11:* 277–290; Int. J. Syst. Bacteriol. *39:* 495–497, 1989). *Aeromonas schubertii* was created in 1988 by Hickman-Brenner et al., (J. Clin. Microbiol. *26:* 1561–1564; Int. J. Syst. Bacteriol. *39:* 205–206, 1989). *Aeromonas veronii* was created in 1987 by Hickman-Brenner et al., (J. Clin. Microbiol. *25:* 900–906; Int. J. Syst. Bacteriol. *38:* 220–222, 1988).

Genus **Enhydrobacter**

Editorial note: The genus *Enhydrobacter* was not included in *Bergey's Manual of Systematic Bacteriology.* The genus was created in 1987 by Staley et al., (Int. J. Syst. Bacteriol. *37:* 289–291). It included one species, *Enhydrobacter aerosaccus.*

Very short straight rods, 0.5–0.7 μm in diameter by 1.0–5.0 μm in length, approaching a spherical shape. Contain gas vacuoles. Single cells, pairs, and short chains occur. Gram negative. **Nonmotile and facultatively anaerobic.** Chemoorganotrophic, having **both a respiratory and a fermentative type of metabolism.** Optimal temperature is 37–39°C. **There is very slow growth; biochemical tests read for up to 60 days.** D-Glucose and other carbohydrates are catabolized with production of acid (gas production has not been reported). **Oxidase and catalase positive and indole negative. No growth in methyl red-Voges-Proskauer medium. Positive for lysine and ornithine decarboxylases and arginine dihydrolase.** Reduces nitrates. **Resistant to vibriostatic agent 0/129.** Isolated from the oxygen depleted zone of a eutrophic lake.

Type (and only) species: *Enhydrobacter aerosaccus.*

Characteristics of the species:

Editorial note: The organism is fastidious and very slow-growing. Determination of its biochemical characteristics requires unusual media and 30–60 days. It has only been isolated once and is not recommended for study by students.

Colonies on RM-2 agar (Staley et al., Int. J. Syst. Bacteriol. *37:* 289–291, 1987) are colorless, convex with an entire margin, and adhere to agar surface. Folic acid and biotin are required for growth on defined mineral medium. Best growth occurs microaerophilically. No growth occurs on deoxyribonuclease medium; no hydrolysis of *O*-nitrophenyl-β-D-galactopyranoside.

Tween 80 is hydrolyzed. The following carbon sources are utilized: acetate, L-arabinose, citrate, ethanol, D-fructose, formate, L-fucose, fumerate, D-galactose, D-glucose, glycerol, inulin, lactate, *meso*-malate, maltose, D-mannose, oxaloacetate, pyruvate, L-rhamnose, D-ribose, L-sorbose, succinate, sucrose, *meso*-tartrate, and D-xylose. The following carbon sources are not utilized: adonitol, agar, amygdalin, benzoate, butyrate, cellobiose, cellulose, chitin, dextrin, dulcitol, erythritol, formate, gelatin, *myo*-inositol, lactose, malonate, D-mannitol, melezitose, melibiose, methanol, pectin, phthalate, salicin, D-sorbitol, starch, raffinose, and trehalose.

Reactions that differentiate *E. aerosaccus* from other genera in *Vibrionaceae* are given in Table 5.46.

Genus **Photobacterium**

Plump, straight rods, 0.8–1.3 µm in diameter and 1.8–2.4 µm in length. Involution forms usually seen in old cultures or under adverse conditions of cultivation. Gram negative. **Motile by 1–3 unsheathed polar flagella; some nonmotile. Facultatively anaerobic.** Chemoorganotrophic, having **both a respiratory and a fermentative type of metabolism.** Optimal temperature appears to be 18–25°C. **Sodium ions are required for growth.** D-Glucose and D-mannose are catabolized with the production of acid and, in *P. phosphoreum*, gas. **Oxidase reaction is variable.** A majority of strains are lysine decarboxylase and arginine dihydrolase positive. Ornithine decarboxylase negative. **Accumulate poly-β-hydroxybutyrate** under certain conditions of cultivation; **do not use the exogenous monomer β-hydroxybutyrate.** Most strains **grow in a mineral medium containing a seawater base, D-glucose, and NH₄Cl;** other strains also require L-methionine. In addition to D-glucose, utilize D-fructose, glycerol, and D-mannose. Two species are bioluminescent. Isolated from the marine environment and on the surfaces and in the intestinal contents of marine animals; some are found as symbionts in specialized luminous organs of marine fish.

Type species: *Photobacterium phosphoreum.*

Characteristics that differentiate *Photobacterium* from other genera are given in Table 5.46.

Differentiation of the species of the genus **Photobacterium:** See Table 5.49A. Other characteristics of these species are given in Table 5.49B.

Genus **Plesiomonas**

Editorial note: MacDonell and Colwell (Syst. Appl. Microbiol. *6:* 171182, 1985) recommended that *Plesiomonas* be transferred to the genus *Proteus* in the family *Enterobacteriaceae* because its 5S rRNA is closely related to that of *Proteus mirabilis.* Such a change would cause problems in the phenotypic definition of the genus *Proteus.*

Roundended, straight, rod-shaped cells, 0.8–1.0 × 3.0 µm. Gram negative. **Motile by polar flagella,** generally lophotrichous. **Facultatively anaerobic** and chemoorganotrophic, having **both a respiratory and a fermentative type of metabolism.** Optimal temperature is 37°C. D-Glucose and other carbohydrates are catabolized with production of acid but no gas. **Oxidase and catalase positive. Produce indole. Voges-Proskauer negative. Positive for lysine, ornithine decarboxylase, and arginine dihydrolase. Lipase negative.** Reduces nitrates. Most strains are **sensitive to vibriostatic agent 0/129. Possesses the enterobacterial common antigen.** Occurs in fish and other aquatic animals and in a variety of mammals. It is associated with diarrhea and occasional opportunistic infection in humans.

Type (and only) species: *Plesiomonas shigelloides.*

Characteristics that differentiate *Plesiomonas* from other genera are given in Table 5.46 and 5.47.

Characteristics of **Plesiomonas shigelloides:** See Table 5.50.

Colonies on nutrient agar or blood agar are 1.0–1.5 mm in diameter grayish, shiny, and opaque, with a smooth surface and an entire edge after 24 h at 37°C. The methyl red test is variable. Negative reactions in tests for citrate utilization, deoxyribonuclease, esculin hydrolysis, gelatin liquefaction, H₂S, growth in KCN, malonate, *O*-nitrophenyl-β-D-galactopyranoside, phenylalanine deaminase, and urease occur. Produce acid from glycerol, *myo*-inositol, maltose, and trehalose.

Genus **Vibrio**

Straight or curved rods, 0.5–0.8 µm in width and 1.4–2.6 µm in length. Gram negative. **Motile by one or more polar flagella which are enclosed in a sheath continuous with the outer membrane of the cell wall.**

Facultatively anaerobic. Chemoorganotrophic, having **both a respiratory and a fermentative type of metabolism.** Optimal temperature varies considerably; all grow at 20°C, most grow at 30°C. D-Glucose and other carbohydrates are catabolized with production of acid, but not gas (except for *V. furnissii, V. gazogenes,* and some strains of *V. (Listonella) damsela).* **Oxidase positive** (except *V. gazogenes* and *V. metschnikovii).* Reduce nitrates (except *V. gazogenes, V. metchnikovii,* and *V. ordalii).* Carbohydrates fermented by most species include maltose, D-mannose, and trehalose. **Most species are sensitive to the vibriostatic agent 0/129. Sodium ions stimulate the growth of all species and are an absolute requirement for most species.** Found in aquatic habitats with a wide range of salinities. Very common in marine and estuarine environments and on the surfaces and in the intestinal contents of marine animals. Some species are also found in fresh water. Some dozen species are pathogenic for humans, and several species are pathogenic for marine vertebrates and invertebrates. The most notable human pathogens are *V. cholerae,* the causative agent of cholera; *V. parahaemolyticus,* a major cause of food poisoning caused by contaminated fish or shellfish; and *V. vulnificus* which causes a highly fatal septicemia. These and other species are associated with wound infections, diarrheal disease, and a variety of extraintestinal infections.

Type species: *Vibrio cholerae.*

Characteristics that differentiate *Vibrio* from other genera are given in Table 5.46. Although the number of readily determinable diagnostic traits between these genera is limited, there is little difficulty in their differentiation since the individual species are generally well defined.

Differentiation of the species of the genus **Vibrio:**

Characteristics that distinguish the biovars of *V. cholerae* are given in Table 5.51. Characteristics helpful in differentiating *Vibrio* species, especially the marine vibrios, are given in Table 5.52A. Characteristics using tests more likely to be done in a clinical laboratory are given in Table 5.52B for those species likely to be encountered in a clinical laboratory.

A microcomputer program with a complete data matrix for all *Vibrionaceae* and a separate one for the marine vibrios (tests done at 25°C with marine cations) is available. Information can be obtained by writing J.J.

Farmer III, Centers for Disease Control, Building 1, Room B310, Mailstop C03, Atlanta, Georgia 30333.

Editorial note: The genus *Listonella* was created in 1985 by MacDonell and Colwell (Syst. Appl. Microbiol. *6:* 171–182; Int. J. Syst. Bacteriol. *36:* 354–356, 1986) to house the species *Listonella (Vibrio) anguillarum, Listonella (Vibrio) damsela,* and *Listonella (Vibrio pelagius) pelagia,* all of which had been in the genus *Vibrio.* The genus *Colwellia* was created in 1988 by Deming et al., (Syst. Appl. Microbiol. *10:* 152–160; Int. J. Syst. Bacteriol. *38:* 328–329, 1988) to house the species *Colwellia (Vibrio) psychroerythrus* and *Colwellia (Vibrio) hadaliensis. C. psychroerythrus* had been included in the genus *Vibrio,* although the name had never been validated. Since it is clear that the genus *Vibrio* still requires detailed study to determine its evolutionary heterogeneity, we chose to maintain all of these species in *Vibrio,* without prejudice, for the purpose of this diagnostic volume. The genus names *Listonella* and *Colwellia* are used in parentheses to prevent confusion.

Vibrio aestuarianus, Vibrio (Listonella) damsela, Vibrio diazotrophicus, Vibrio hollisae, Vibrio mimicus, Vibrio ordalii, and *Vibrio orientalis* were mentioned, but not described under "Other organisms" in the "Genus Vibrio" in *Bergey's Manual of Systematic Bacteriology. Vibrio cincinnatiensis, Vibrio furnissii, Colwellia (Vibrio) hadaliensis, Vibrio mediterranei, Colwellia (Vibrio) psychroerythrus, Vibrio salmonicida,* and *Vibrio tubiashii* were not included in *Bergey's Manual of Systematic Bacteriology. Vibrio aestuarianus* was created in 1983 by Tison and Seidler (Int. J. Syst. Bacteriol. *33:* 699–702). *Vibrio cincinnatiensis* was created in 1986 by Brayton et al., (J. Clin. Microbiol. *23:* 104–108; Int. J. Syst. Bacteriol. *36:* 354–356, 1986). *Vibrio (Listonella) damsela* was created in 1981 by Love et al., (Science *214:* 1139–1140; Int. J. Syst. Bacteriol. *32:* 266–268, 1982). *Vibrio diazotrophicus* was created in 1982 by Guerinot et al., (Int. J. Syst. Bacteriol. *32:* 350–357). *Vibrio furnissii* was created in 1983 by Brenner et al., (J. Clin. Microbiol. *18:* 816–824; Int. J. Syst. Bacteriol. *34:* 91–92, 1984). *Colwellia (Vibrio) hadaliensis* was created in 1988 by Deming et al., (Syst. Appl. Microbiol. *10:* 152–160; Int. J. Syst. Bacteriol. *38:* 328–329, 1988). *Vibrio hollisae* was created in 1982 by Hickman et al., (J. Clin. Microbiol. *15:* 395–401; Int. J. Syst. Bacteriol. *32:* 384–385, 1982). *Vibrio mediterranei* was created in 1986 by Pujalte and Garay (Int. J. Syst. Bacteriol. *36:* 278–281). *Vibrio mimicus* was created in 1981 by Davis et al., (J. Clin. Microbiol. *14:* 631–639; Int. J. Syst. Bacteriol. *32:* 266–268, 1982). *Vibrio ordalii* was created in 1981 by Schiewe et al., (Curr. Microbiol. *6:*

343–348; Int. J. Syst. Bacteriol. *32:* 384–385, 1982). *Vibrio orientalis* was created in 1983 by Yang et al., (Curr. Microbiol. *8:* 95–100; Int. J. Syst. Bacteriol. *33:* 672–674, 1983). "*Vibrio psychroerythrus*" was created in 1972 by D'Aoust and Kushner (J. Bacteriol. *111:* 340–342). It was redescribed in 1988 by Deming et al., as *Colwellia (Vibrio) psychroerythrus* (Syst. Appl. Microbiol. *10:* 152–160; Int. J. Syst. Bacteriol. *38:* 328–329, 1988). *Vibrio salmonicida* was created in 1986 by Wiik and Egidium (Int. J., Syst. Bacteriol. *36:* 521–523). *Vibrio tubiashi* was created in 1984 by Hada et al., (Int. J. Syst. Bacteriol. *34:* 1–4).

Vibrio carchariae, created in 1984 by Grimes et al., (Microb. Ecol. *10:* 271–282; Int. J. Syst. Bacteriol. *35:* 223–225, 1985) has been shown by DNA hybridization to be a junior subjective synonym for *Vibrio harveyi* (F.W. Hickman-Brenner, J.J. Farmer III, D.J. Brenner, G.R. Fanning, unpublished observations).

SUBGROUP 3: Family Pasteurellaceae

Coccoid to straight rod-shaped cells, usually 0.2–0.4 × 0.4–2.0 µm. Pleomorphism with cell swelling and formation of filaments may occur. Gram negative. **Contain demethylmenaquinones;** ubiquinones may or may not be produced. **Nonmotile. Aerobic with varying degrees of microaerophilia, facultatively anaerobic.** Chemoorganotrophic, having **both a respiratory and a fermentative type of metabolism.** Optimal growth temperature is 37°C. D-Glucose and other carbohydrates are catabolized with production of acid; however, conventional fermentation test media may fail to detect the accumulation of acid fermentation products with some of the most fastidious species. Usually anaerogenic, but gas producing species do occur. **Oxidase, catalase, and alkaline phosphatase reactions are characteristically positive,** but negative reactions occur in some species. **Nitrates are reduced to nitrites.** Complex media supplemented with yeast extract and serum or whole blood lysate are used for primary isolation. Require organic nitrogen sources. **Varying patterns of nutritional requirements may include several amino acids, B vitamins, β-nicotinamide, adenine nucleotides, and haematin or protoporphyrin. Parasitic in vertebrates,** particularly mammals and/or birds.

Type genus: *Pasteurella.*

Differentiation of **Pasteurellaceae** *from other families and genera in Group 5:* Characteristics that differentiate

Pasteurellaceae from the other families and genera are presented in Table 5.1.

Differentiation of the genera in **Pasteurellaceae:**

Editorial note: In *Bergey's Manual of Systematic Bacteriology* it is indicated that each of the genera in *Pasteurellaceae* is heterogenous and that each must be redefined on the basis of genetic and biochemical criteria. Since the publication of *Bergey's Manual of Systematic Bacteriology* new studies have confirmed this heterogeneity, transferred species between existing genera, and added new species to each of the existing genera (Mutters et al., Int. J. Syst. Bacteriol. *35:* 309–322, 1985; Schlater et al., J. Clin. Microbiol. *27:* 2169–2174, 1989; De Ley et al., Int. J. Syst. Bacteriol. *40:* 126–137, 1990; Sneath and Stevens, Int. J. Syst. Bacteriol. *40:*148–153, 1990). The data from these studies indicate that there may be a need for at least 4 additional genera, but that phenotypic definitions for the new genera as well as the established genera are difficult to establish.

Pending further studies to establish and define genera in the family *Pasteurellaceae,* differentiation among the genera will not be attempted. A biochemical profile of each species in the family is given in Table 5.53. Where proposals to transfer species to another genus have been made, the new genus name is used, without prejudice, with the old genus name given in parentheses. It must be noted that both names are valid.

Genus **Actinobacillus**

Spherical, oval, or rod-shaped cells, 0.4 × 1.0 µm. Cells are mostly bacillary but are interspersed with coccal elements which often lie at the pole of a bacillus giving a characteristic "Morse code" form. Occasional longer forms up to 6 µm, especially on media containing glucose or maltose. Cells are arranged singly, in pairs, or, more rarely, in chains. Gram negative. Exhibit irregular staining. Small amounts of extracellular slime may be demonstrated in wet India ink preparations. **Nonmotile.** Cultures are very sticky on primary isolation; colonies may be difficult to remove completely from the agar surface. Surface cultures have low viability and die in 5–7 days. **Facultatively anaerobic** and chemoorganotrophic, having **both a respiratory and a fermentative type of metabolism. Optimal temperature is 37°C.** D-Glucose and fructose are catabolized with the production of acid but no gas. **Positive for β-galactosidase. Methyl red and indole**

are negative. Parasitic or commensal in humans, sheep, cattle, horses, pigs, other mammals, and birds.

Type species: *Actinobacillus lignieresii.*

Differentiation of the species of the genus **Actinobacillus.** See Table 5.54. Other characteristics of these species are given in Table 5.53.

The close similarities between actinobacilli and the genera *Haemophilus* and *Pasteurella* pose problems of differentiation at the genus level, usually necessitating a comparison of biochemical characteristics at the species level (Table 5.53).

Editorial note: Actinobacillus hominis, Actinobacillus muris, and *Actinobacillus rossii* were not included in *Bergey's Manual of Systematic Bacteriology; Actinobacillus (Haemophilus) pleuropneumoniae* was listed in the genus *Haemophilus; Actinobacillus seminus* was listed as a species *incertae sedis* whose name had not been validated; and *Actinobacillus (Pasteurella) ureae* was listed in the genus *Pasteurella. Actinobacillus hominis* was created in 1981 by Friis-Møller (In Kilian, Frederiksen, and Biberstein (ed), *Haemophilus, Pasteurella,* and *Actinobacillus,* Academic Press, London, P. 151–157; Int. J. Syst. Bacteriol. *35:* 375–376, 1985); *Actinobacillus muris* was created in 1986 by Bisgaard (Acta Pathol. Microbiol. Immunol. Scand. Sect. B *94:* 1–18; Int. J. Syst. Bacteriol. *38:* 220–222, 1988); *Actinobacillus pleuropneumoniae* was transferred from *Haemophilus* to *Actinobacillus* in 1983 by Pohl et al., (Int. J. Syst. Bacteriol. *33:* 510–514); *Actinobacillus rossii* and *Actinobacillus seminis* were created in 1990 by Sneath and Stevens (Int. J. Syst. Bacteriol. *40:* 148–153); *Actinobacillus ureae* was transferred from *Pasteurella* to *Actinobacillus* in 1986 by Mutters et al. (Int. J. Syst. Bacteriol. *36:* 343–344).

Genus **Haemophilus**

Minute to medium-sized spherical, oval, or rod-shaped cells; generally less than 1 μm in width and variable in length, sometimes forming threads of filaments and showing **marked pleomorphism.** Gram negative. **Nonmotile and facultatively anaerobic.** Almost all species **require preformed growth factors present in the blood, particularly X factor (protoporphyrin IX or protoheme) and/or V factor (nicotinamide adenine dinucleotide (NAD) or NAD phosphate (NADP).** Even after specific growth factors

have been provided, growth is best on complex media. Chemoorganotrophic, having **both a respiratory and a fermentative type of metabolism. Optimal temperature is 35–37°C.** D-Glucose and other carbohydrates are catabolized with the production of acid; a few species produce gas. **Nitrates are reduced to, or beyond, nitrites.** Oxidase and catalase reactions vary among species. Occur **as obligate parasites or commensals on the mucous membranes of humans and a variety of animals.** *H. influenzae* is the leading cause of meningitis in children. It also causes other septicemic conditions, otitis media, sinusitis, and chronic bronchitis. *H. influenzae* biovar aegyptius is mainly responsible for conjunctivitis, and certain of its strains are responsible for the newly described disease, Brazilian purpuric fever (Brenner et al., J. Clin. Microbiol. *26:* 1524–1534, 1988). *H. ducreyi* the causative agent of the venereal disease soft chancre or chancroid.

Type species: *Haemophilus influenzae.*

Since not all *Haemophilus* species require X and/or V factors and some species of *Actinobacillus* and *Pasteurella* do require NAD, the requirement for these factors is not definitive for *Haemophilus.* The problems in differentiating *Haemophilus* from the other genera in *Pasteurellaceae* usually necessitate a comparison of biochemical characteristics at the species level (Table 5.53).

Differentiation of the species of the genus **Haemophilus:** See Table 5.53. Differentiation of *H. influenzae* biovars and of *H. parainfluenzae* biovars is given in Table 5.55.

Editorial note: Nomenclatural changes have occurred in a number of species classified in *Haemophilus* in *Bergey's Manual of Systematic Bacteriology. Haemophilus aegyptius* was shown to be the same species as *Haemophilus influenzae* (Casin et al., Ann. Microbiol. (Paris) *137B:* 155–163, 1986). Here, we consider it as a biovar of *H. influenzae. Haemophilus equigenitalis* was transferred to the new genus *Taylorella* (Sugimoto et al., Curr. Microbiol. *9:* 155–162, 1983; Int. J. Syst. Bacteriol. *34:* 503–504, 1984). *Haemophilus pleuropneumoniae* was transferred to the genus *Actinobacillus* (Pohl et al., Int. J. Syst. Bacteriol. *33:* 510–514, 1983). *"Haemophilus agni"* and *"Haemophilus somnus"* have not been validated. Both are listed as species *incertae sedis* in *Bergey's Manual of Systematic Bacteriology.* De Ley et al., (Int. J. Syst. Bacteriol. *40:* 126–137, 1990) state that *"Histophilus ovis,"* another invalid name, is a senior synonym for *"H. agni"* and suggest that both *"H. agni"* and *"H. somnus"* be transferred to the genus *"Histophilus."*

Genus **Pasteurella**

Spherical, ovoid, or rod-shaped cells, 0.3–1.0 μm in diameter and 1.0–2.0 μm in length. Occur singly or less frequently in pairs or short chains. **Bipolar staining is common,** especially in preparations made from infected animal tissues. Gram negative. **Nonmotile and facultatively anaerobic.** Chemoorganotrophic, having **both a respiratory and a fermentative type of metabolism. Optimal temperature is 37°C.** D-Glucose and other carbohydrates are catabolized with the production of acid but usually no gas. Most species are oxidase and catalase positive. **Nitrates are reduced to nitrites** (except *P. lymphangitidis*). Negative in reactions for methyl red, acetoin production (Voges-Proskauer), lysine decarboxylase, arginine dihydrolase, and gelatinase. **Parasitic or commensal on the mucous membranes of the upper respiratory and digestive tracts of mammals (rarely humans) and birds.** *P. multocida* causes hemorrhagic septicemia of cattle, fowl cholera, and pneumonia in farm animals. *P. haemolytica* causes pneumonia and "shipping fever" in cattle, sheep, and goats.

Type species: *Pasteurella multocida.*

The overlap and exceptions to the phenotypic descriptions of *Pasteurella*, *Haemophilus*, and *Actinobacillus* make it quite difficult to differentiate these genera. Differentiation is best done directly at the species level (Table 5.53).

Differentiation of the species of the genus **Pasteurella:** See Table 5.53.

Editorial note: In *Bergey's Manual of Systematic Bacteriology*, *Actinobacillus (Pasteurella) ureae* was included in the genus *Pasteurella* and *Pasteurella (Haemophilus) avium* was included in the genus *Haemophilus*. Mutters et al., (Int. J. Syst. Bacteriol. *35:* 5–9, 1985) proposed the new combination *Pasteurella avium;* Mutters et al., (Int. J. Syst. Bacteriol. *36:* 343–344, 1986) proposed the new combination *Actinobacillus ureae. Pasteurella anatis, Pasteurella bettyae, Pasteurella caballi, Pasteurella canis, Pasteurella dagmatis, Pasteurella granulomatis, Pasteurella langaa, Pasteurella lymphangitidis, Pasteurella mairii, Pasteurella multocida* subsp. *gallicida, Pasteurella multocida* subsp. *multocida, Pasteurella multocida* subsp. *septica, Pasteurella stomatis, Pasteurella testudinis, Pasteurella trehalosi,* and *Pasteurella volantium* were not included in *Bergey's Manual of Systematic Bacteriology. Pasteurella anatis, Pasteurella canis, Pasteurella dagmatis,*

Pasteurella langaa, Pasteurella multocida subsp. *gallicida, Pasteurella multocida* subsp. *multocida, Pasteurella multocida* subsp. *septica,* and *Pasteurella stomatis* were created in 1985 by Mutters et al., (Int. J. Syst. Bacteriol. *35:* 309–322). *Pasteurella bettyae, Pasteurella lymphangitidis, Pasteurella mairii,* and *Pasteurella trehalosi* were created in 1990 by Sneath and Stevens (Int. J. Syst. Bacteriol. *40:* 148–153; see Sneath, Int. J. Syst. Bacteriol. *42:* 658–659, 1992 for correction of orthography of the epithets *bettyae* and *mairii*). *Pasteurella caballi* was created in 1989 by Schlater et al., (J. Clin. Microbiol. *27:* 2169–2174; Int. J. Syst. Bacteriol. *40:* 320–321, 1990). *Pasteurella granulomatis* was created in 1989 by Ribeiro et al., (J. Clin. Microbiol. *27:* 1401–1402; Int. J. Syst. Bacteriol. *32:* 105–106, 1982). *Pasteurella testudinis* was created in 1982 by Snipes and Biberstein (Int. J. Syst. Bacteriol. *32:* 201–210). *Pasteurella volantium* was created in 1985 by Mutters et al., (Int. J. Syst. Bacteriol. *35:* 5–9).

SUBGROUP 4: Other Genera

The 7 genera included in Subgroup 4 are no more similar to one another than to the other Subgroups in Group 5. They are placed in a single Subgroup for convenience. Phenotypic characteristics that differentiate the genera in Subgroup 4 from each other and from Subgroups 1–3 are given in Table 5.1.

Differentiation of the genera included in **Subgroup 4:** See Table 5.1.

Genus **Calymmatobacterium**

Pleomorphic rods, 0.5–1.5 μm wide by 1.0–2.0 μm in length, with rounded ends. Occur singly or in clusters. **Exhibit single or bipolar condensation of chromatin. Capsules are present.** Gram negative. Nonmotile. The exudate from infected tissues, when stained by Wright's or Giemsa stain, demonstrates **characteristic intracellular organisms in the cytoplasm of large mononuclear phagocytes. Quite fastidious;** can be cultivated in vivo in the yolk sack of embryonated chicken eggs or in vitro on special egg yolk-containing media, but cultivation is rarely successful, and **no strains are available in culture collections.** Optimum temperature is 37°C. Pathogenic for humans, causing **donovanosis (granuloma inguinale). Diagnosis is clinical and microscopic** (Dienst and Brownell, *Bergey's Manual of Systematic Bacteriology*, vol. 1,

p. 585–587, 1984; Albritton et al., Manual Clin. Microbiol., 4th ed., p. 869–873, 1985).

Type (and only) species: *Calymmatobacterium granulomatis.*

Differentiation of the genus **Calymmatobacterium** *from other genera:* The examination of diseased tissue smears stained by Wright's blood stain or Giemsa stain is the simplest procedure for identification of *C. granulomatis.* The characteristic appearance of the intracellular organisms (Figure 5.1) is specific for the diagnosis of donovanosis. Differentiation from other genera is shown in Table 5.1.

Genus Cardiobacterium

Straight rods 0.5–0.75 μm in diameter and 1–3 μm in length, with rounded ends. Occasional long filaments may occur. **Pleomorphic.** Arranged singly, in pairs, in short chains, and in rosette clusters. **Gram negative, but retention of crystal violet may occur in the swollen ends or central portions of cells. Nonmotile. Facultatively anaerobic.** CO_2 is required by some strains on isolation. Aerobic growth is scant unless humidity is elevated. Growth in candle extinction jars or under anaerobic conditions is not dependent on elevated humidity. Optimum temperature is 30–37°C. Colonies on blood agar are smooth, convex, and opaque, reaching 1–2 mm after 48 hours. Chemoorganotrophic, having a fermentative type of metabolism. **Oxidase positive and catalase negative. Small amounts of indole are formed. Nitrates are not reduced.** Acid but not gas is produced from glucose and from fructose, maltose, mannitol, mannose, sorbitol, and sucrose (within 7 days). No growth occurs on MacConkey agar. Urease negative and ornithine decarboxylase negative. Occur in nasal flora of humans. Pathogenic for humans, causing endocarditis.

Type (and only) species: *Cardiobacterium hominis.*

Differentiation of **Cardiobacterium** *from other genera:* One or both ends of cells are frequently enlarged, and the crystal violet of the Gram stain tends to be retained in these areas. Indole formation is an important characteristic for identification. Indole production, however, may be weak and may not be detected by procedures which do not concentrate the indole by xylene extraction (Weaver, *Bergey's Manual Systematic Bacteriology,* vol. 1, p. 583–585, 1984). Negative reactions occur in

tests for esculin hydrolysis, gelatin liquefaction, H_2S production, Tween 20 and Tween 40 hydrolysis, and fermentation of adonitol, arabinose, cellobiose, dulcitol, erythritol, galactose, inositol, lactose, melezitose, melibiose, rhamnose, salicin, trehalose, and xylose.

Characteristics that differentiate *Cardiobacterium* from other genera are given in Tables 5.1 and 5.56.

Genus Chromobacterium

Rods 0.6–0.9 × 1.5–3.5 μm with rounded ends, sometimes slightly curved. Occur singly; occasionally pairs, elongated forms or short chains occur. No capsules and **no resting stages.** Gram negative, often with barred or bipolar staining and lipid inclusions. **Motile by means of both a single polar flagellum and usually 1–4 subpolar or lateral flagella. Facultative anaerobes.** Produce butyrous, **violet colonies** on solid media; in nutrient broth; a **violet ring** is formed at the junction of the liquid surface and the container wall. Growth occurs on ordinary media at 25°C, but species differ in their optimum temperature. Chemoorganotrophs, having mainly a **fermentative metabolism.** Acid, but no gas, is produced from glucose, fructose, trehalose, and usually mannose, but not from dulcitol, inositol, inulin, lactose, mannitol, and xylose. Usually oxidase positive (Kovacs' reagent), although the violet pigment may interfere with the reading. Catalase positive. Indole negative. Voges-Proskauer negative. Nitrate is reduced. Liquefies gelatin. Negative reactions occur in tests for lysine and ornithine decarboxylase, phenylalanine deaminase, and esculin hydrolysis. Where tested (only *C. violaceum*), negative test for acid production from adonitol, melezitose, melibiose, and raffinose, and negative reactions in tests for *O*-nitrophenyl-β-D-galactopyranoside, deoxyribonuclease, and pectate digestion, and a positive reaction occurs for growth in KCN. **Resistant to benzylpenicillin** (10 μg/ml) **and to vibriostatic agent 0/129.** Soil and water organisms. *C. violaceum* occasionally causes serious pyogenic or septicemic infections of mammals, including humans.

Type species: *Chromobacterium violaceum.*

Differentiation of **Chromobacterium** *from other genera:* Characteristics that differentiate *Chromobacterium* from other genera in Subgroup 4 are given in Table 5.1. Some species previously classified in *Chromobacterium* are now classified in the genus *Janthinobacterium.*

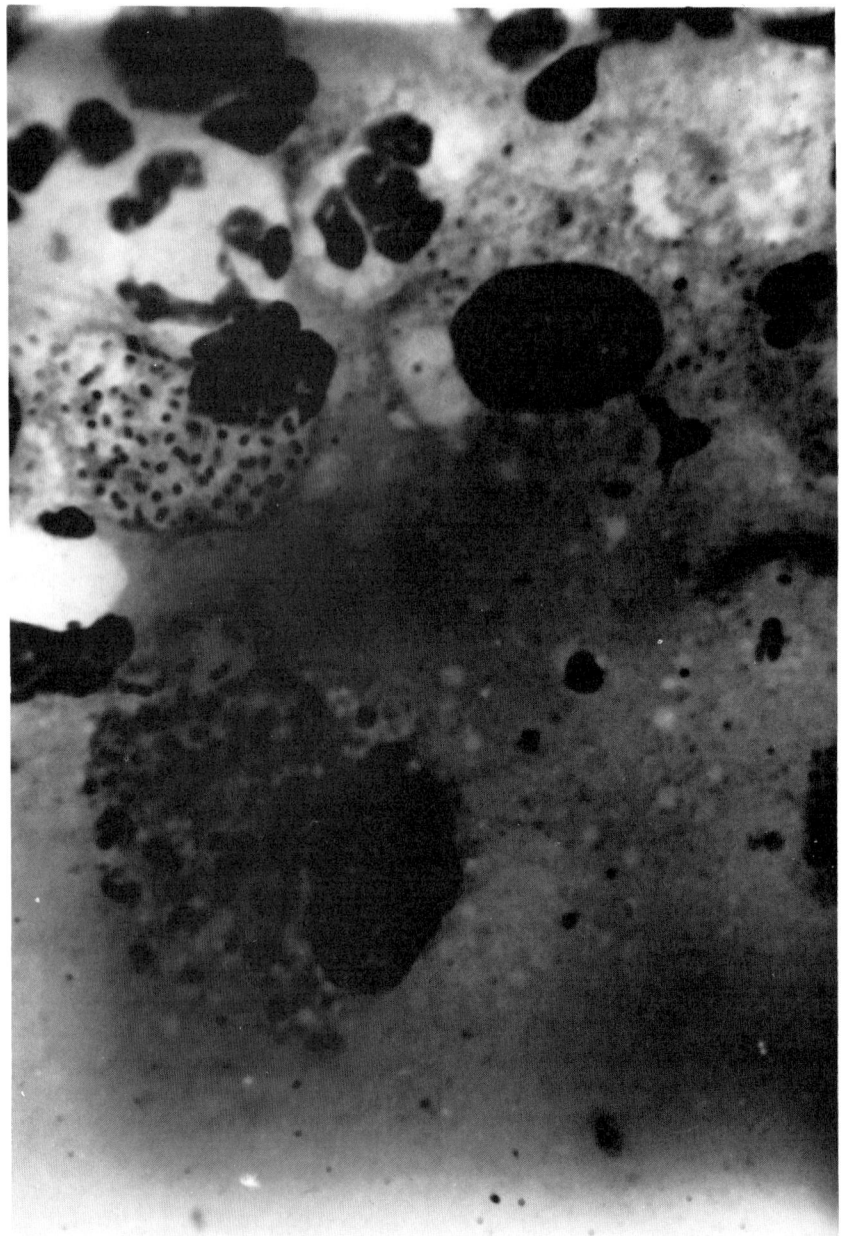

Figure 5.1. Large mononuclear phagocytes filled with *Calymmatobacterium granulomatis.* Wright's stain (X 900). (Reproduced with permission from R. B. Dienst and G. H. Brownell in M. P. Starr et al., (editors) *The Prokaryotes: a Handbook on Habitats, Isolation and Identification of Bacteria.* p. 1410, 1981, Springer Verlag, New York.)

Characteristics differentiating *Chromobacterium* from *Janthinobacterium* are presented in Table 5.57.

Differentiation of the species of the genus **Chromobacterium:** See Table 5.58.

Genus **Eikenella**

Straight rods, 0.3–0.4 × 1.5–4.0 μm. Short filaments are occasionally formed. Nonsporeforming. Gram negative. **Nonmotile,** possessing no flagella; however, a "twitching motility" may occur on agar surfaces. **Facultatively anaerobic.** Optimum temperature is 35–37°C. **Colonies may appear to corrode the surface of the**

agar. Nonhemolytic; a slight greening of blood media around colonies may occur. **Oxidase positive** (Kovacs' method). **Negative for catalase, urease, arginine dihydrolase, and indole. Lysine decarboxylase positive. Nitrates are reduced to nitrites. No acid is formed from glucose or other carbohydrates. Hemin is usually required for growth under aerobic conditions.** Occur in the human mouth and intestine; can be opportunistic pathogens.

Type (and only) species: *Eikenella corrodens.*

Differentiation of **Eikenella** *from other genera:* Characteristics that differentiate *Eikenella* from other genera in Subgroup 4 are given in Table 5.1. Differentiation of *Eikenella* from other genera and species is shown in Table 5.59.

Characteristics of the species: Colonies are 0.2–0.5 mm at 24 hours; 0.5–1.0 mm at 48 hours. Colonies appear as if in small pits in the surface of the medium (Figure 5.2). Plate cultures have an odor described as "bleach-like" or as resembling that of *Haemophilus* and *Pasteurella* species. Growth in liquid media is usually poor, but may be improved by addition of cholesterol (Henricksen, Int. J. Syst. Bacteriol. *19:* 165–166, 1969) or 3% blood serum. The addition of 0.2% agar improves growth (Jackson et al., J. Med. Microbiol. *4:* 171–184, 1971). Hemin (5 to 25 μg/ml) is required for aerobic, but not anaerobic growth of fresh isolates. Aerobic and anaerobic growth of fresh isolates is enhanced by 5–10% CO_2. Negative reaction occurs for gelatin hydrolysis; variable reaction occurs for ornithine decarboxylase.

Genus **Gardnerella**

Pleomorphic rods about 0.5 μm in diameter and 1.5–2.5 μm in length. No capsules or endospores are formed. **Gram negative to Gram variable. Nonmotile. Facultatively anaerobic.** Fastidious in growth requirements but do not need X factor or V factor. **Catalase and oxidase negative.** Chemoorganotrophic, having a fermentative type of metabolism. Optimum growth temperature is 35–37°C. **Acid but no gas** is produced from glucose and some other carbohydrates. Does not reduce nitrates. **Hippurate is hydrolyzed. Human blood, but not sheep blood, is hemolyzed.** Found in the human genital/urinary tract. **Considered to be a major cause of bacterial "nonspecific" vaginitis.**

Type (and only) species: *Gardnerella vaginalis.*

Differentiation of the genus **Gardnerella** *from other genera:* Characteristics that differentiate *Gardnerella* from other genera in Subgroup 4 are given in Table 5.1. Differentiation of *Gardnerella* from other genera is given in Table 5.60.

Characteristics of the species: Little or no growth on nutrient agar or on most common selective media. On Vaginalis agar colonies are 0.5 mm in diameter after 48 hours of incubation in a candle extinction jar or in a CO_2 incubator. Special procedures are used to test for acid production from carbohydrates, oxidase, catalase, hemolysis, and hippurate hydrolysis (Greenwood and Pickett, *Bergey's Manual of Systematic Bacteriology,* vol. 1, p. 587–591, 1984). Negative reactions in tests for arginine dihydrolase, esculin hydrolysis, gelatin hydrolysis, H_2S, indole, lysine decarboxylase, phenylalanine deaminase, ornithine decarboxylase, Tween 80 hydrolysis, urease, and acetoin (Voges-Proskauer). There is no acid production from arbutin, cellobiose, glycerol, inositol, mannitol, melibiose, raffinose, rhamnose, and salicin. Variable reactions in tests for lipase, *O*-nitrophenyl-β-D-galactopyranoside, and acid production from L-arabinose, fructose, galactose, inulin, lactose, mannose, sucrose, and xylose. Positive reactions occur in tests for H_2O_2 inhibition, methyl red, starch hydrolysis, and acid production from dextrin, maltose, ribose, and starch.

Genus **Streptobacillus**

Rods, 0.1–0.7 × 1–5 μm long, with rounded or pointed ends. Occur singly or in long, wavy chains or filaments 10–150 μm long. **May be highly pleomorphic.** Single rods may show central swelling; chains or filaments may have a series of swellings resulting in a "string of beads" appearance. Gram negative. Nonmotile. Facultatively anaerobic. **Serum, ascitic fluid, or blood is required for growth.** A moist environment or soft agar may enhance growth. **Conversion to L-phase or transitional-phase variants may occur spontaneously during cultivation.** Optimum temperature is 35–37°C. Chemoorganotrophic. **Glucose and other carbohydrates are fermented with acid production** but no gas. **Catalase and oxidase negative. Indole is not produced. Nitrate is not reduced to nitrite. Inhabitants of the throat and nasopharynx of wild and laboratory rats. Causes one form of rat bite fever in humans.**

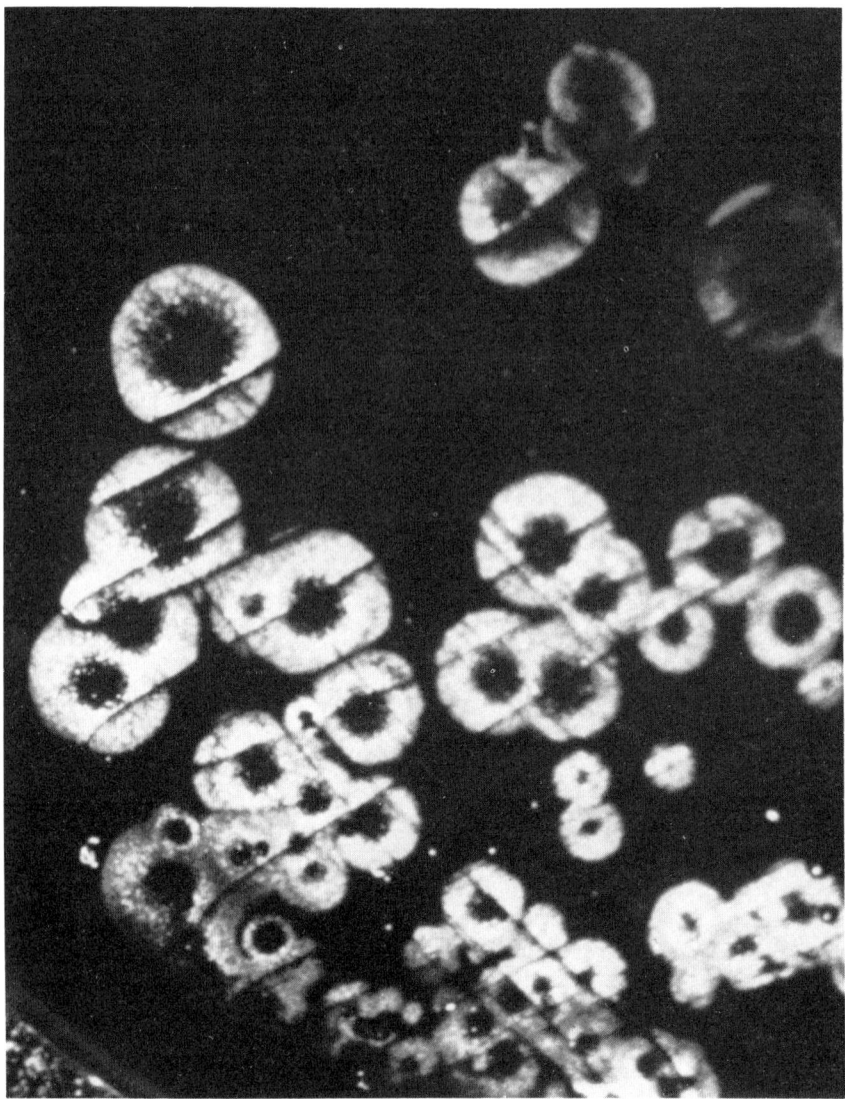

Figure 5.2. Colonies of *Eikenella corrodens* on sheep blood agar at 48 h. Colonies are 0.5–1.0 mm. (Reproduced with permission from E. J. Bottone, J. Kittick, and S. S. Schneierson, American Journal of Clinical Pathology *59:* 560–566, 1973, American Society of Clinical Pathologists.)

Type (and only) species: *Streptobacillus moniliformis.*

Differentiation of **Streptobacillus** *from other genera:* Characteristics that differentiate *Streptobacillus* from other genera are given in Tables 5.1 and 5.61.

Characteristics of the species: Colonies on serum agar are 1–2 mm in diameter after 3 days. Colonies of the L-phase variant are much smaller and exhibit a typical "fried egg" appearance with a dense center that penetrates the agar. Colonies are nonhemolytic on horse or sheep blood agar. Serum broth cultures form a white, flocculent sediment or whitish granules at the bottom and sides of tubes; the supernatant medium is usually clear. Isolation from most clinical lesions is difficult because of the requirement for serum, a humid environment, the slow growth of the organism, and the probability of overgrowth by contaminating organisms. The viability of cultures is brief. Transfers should be made soon after detection of growth. For transfer of L-phase colonies, small agar blocks containing heavy growth should be streaked on agar. **All media for biochemical tests require serum, blood, or ascites supplementation.** Biochemical reactions, particularly for the L-phase variant, may best be determined using agar plates containing the basal medium supplemented with horse serum and filter sterilized substrates and/or indicators (Cohen et al., Appl. Microbiol. *16:* 1655–1662, 1968). All tests should include uninoculated controls containing substrate and inoculated controls in which distilled water is substituted for the substrate. It is negative in tests for gelatin liquefaction, methyl red, phenylalanine deaminase, urease, and acetoin (Voges-Proskauer). It is variable for esculin hydrolysis and is positive in tests for arginine dihydrolase and H$_2$S production. Acid is produced from fermentation of dextrin, fructose, galactose, glycogen, inulin, maltose, mannose, salicin, and starch. Carbohydrates not fermented are adonitol, arabinose, cellobiose, dulcitol, glycerol, inositol, lactose, mannitol, melezitose, melibiose, raffinose, rhamnose, sorbitol, sorbose, sucrose, trehalose, and xylose.

Genus **Zymomonas**

Rod-shaped cells with rounded ends, usually in pairs, 2–6 μm long and 1.0–1.4 μm wide. Gram negative. Usually nonmotile; if motile, the cells possess **1–4 polar flagella.** Facultative anaerobic; some strains are obligately anaerobic. Chemoorganotrophic, **growing on and fermenting 1 mol of glucose or fructose to almost 2 mol of ethanol, 2 mol of CO$_2$ and some lactic acid. Some strains may also utilize sucrose, but other carbon sources are not used.** Optimum temperature is 25–30°C. Colonies on the standard medium (D-glucose, 20 g; yeast extract, 5 g, per liter of distilled water) are glistening, regularly edged, white to cream colored, 1–2 mm in diameter after 2 days, and produce **a fruity odor. Oxidase negative and catalase positive.** Gelatinase negative. **Nitrates are not reduced, and indole is not produced.** A mixture of amino acids is required for good growth, but no one amino acid is essential. **Require biotin and pantothenate.** Occurs as a spoiler in **beers, ciders, and perries;** as fermenting agents in **Agave sap, palm sap and sugarcane juice; and on honey bees and in ripening honey.**

Type species: *Zymomonas mobilis.*

Differentiation of the genus **Zymomonas** *from other genera:* Zymomonas is phenotypically well defined and is easily identified. Its most outstanding feature is the quantitative fermentation of glucose, fructose, or sucrose—but no other sugars—to equimolar amounts of ethanol and CO$_2$. Characteristics that differentiate *Zymomonas* from other genera are given in Tables 5.1 and 5.62.

Characteristics of **Zymomonas mobilis:** It does not form spores or capsules. No growth occurs in nutrient or 1% peptone broth. The medium of Dadds (see Swings and De Ley, *Bergey's Manual of Systematic Bacteriology,* vol. 1, p. 576–580, 1984) is used for enrichment. Weakly positive Voges-Proskauer reaction and variable reactions in tests for H$_2$S, lysine and ornithine decarboxylases, arginine dihydrolase, urease. The following substrates are not fermented or used as carbon sources: adonitol, D-arabinose, L-arabinose, cellobiose, citrate, dextrin, dulcitol, erythritol, ethanol, D-galacturonate, glycerol, DL-lactate, lactose, DL-malate, maltose, D-mannitol, D-mannose, pyruvate, raffinose, L-rhamnose, D-ribose, salicin, D-sorbitol, L-sorbose, starch, succinate, tartrate, trehalose, D-xylose. Growth occurs at 36°C and colony diameter of 1.5 mm after aerobic growth for 7 days at 30°C separate *Z. mobilis* subsp. *mobilis* from *Z. mobilis* subsp. *pomacii* (no growth and colony size of <1 mm).

Table 5.1 Differential characteristics of genera and families in Group 5 [a]

Characteristic	Zymomonas	Chromobacterium	Cardiobacterium	Calymmatobacterium	Gardnerella	Eikenella	Streptobacillus	Family Enterobacteriaceae	Family Vibrionaceae	Family Pasteurellaceae
Cell diameter, μm	1.0–1.4	0.6–0.9	0.5–0.75	0.5–0.15	ca. 0.5	0.3–0.4	0.1–0.7	0.3–1.5	0.3–1.3	0.2–0.4
Main shape of cells:										
Straight rods	+	+	+	+	+	+	+	+	D	+
Curved or vibrioid	−	−	−	−	−	−	−	−	D	−
May stain Gram variable	−	−	+	−	+	−	−	−	−	−
Acid from D-glucose	+	+[b]	+	+	+	−	+	+	+	+
Major product of sugar fermentation:										
Ethanol	+	−	−	−	−			−	−	−
Lactic acid	−	+	+		−					
Acetic acid	−	−	−		+					
Colonies are violet	−	+[c]	−	−	−	−	−	−	−	−
Motility (swimming)	−[d]	+	−	−	−	−	−	D	+[e]	−
Flagellar arrangement:										
Polar	+[d]	−						+[f]	+	
Lateral	+[d]	−						+	−	
Mixed [g]	−	+						−	D	
Oxidase	−	+	+		−	+	−	−	D	D
Catalase	+	+[h]	−		−	−	−	+[e]	D	D
NO₃⁻ reduced to NO₂⁻	−	+[i]	−	−	−	+	−	+[e]	+[e]	+[e]
Indole produced	−	−[j]	+[j]	−	−	+	−	D	D	D
Hemin required for aerobic growth	−	−	+		−	−	−	−	−	D
Many strains appear to corrode surface of agar media	−	−	−	−	−	+	−	−	−	−
Na⁺ required or stimulatory for growth	−	−	−	−	−	−	+	−[e]	D	−
Organic nitrogen sources required	−	−	−	−	−	−	−	−[e]	−[e]	+
Plant pathogenicity	−	−	−	−	+	−	−	D	−	−
Causes donovanosis (granuloma inguinale) in humans	−	−	−	+	−	−	−	−	−	−
Causes one form of rat-bite fever in humans	−	−	−	−	−	−	+	−	−	−
Causes vaginitis in humans	−	−	−	−	+	−	−	−	−	−

[a] Symbols: see standard definitions.

[b] Most strains show a fermentative attack on glucose, but approximately 20% show an oxidative attack.

[c] White variants occur that are difficult to identify and may be mistaken for *Aeromonas* or *Vibrio* species.

[d] Most strains are nonmotile but a few are motile by means of 1–4 subpolar or lateral flagella.

[e] A few exceptions occur.

[f] Except *Tatumella*, which may have polar, subpolar, or lateral flagella.

[g] A single polar flagellum plus one or more lateral flagella.

[h] Most strains also reduce NO₂⁻.

[i] Negative by usual testing methods, but under some conditions compounds that give positive reactions with indole test reagent may accumulate.

[j] Only small amounts of indole are formed.

Table 5.2 Biochemical differentiation of the species of the family *Enterobacteriaceae*

Test	*Arsenophonus nasoniae*	*Budvicia aquatica*	*Buttiauxella agrestis*	*Cedecea davisae*	*Cedecea lapagei*	*Cedecea neteri*	*Citrobacter amalonaticus*
Gram stain (24h)	−	−	−	−	−	−	−
Oxidase (24h)	−	−	−	−	−	−	−
Indole production	−	−	−	−	−	−	+
Methyl red	−	+	+	+	d	+	+
Voges-Proskauer	−	−	−	d	[+]	d	−
Citrate (Simmons)	a	−	+	+	+	+	[+]
Hydrogen sulfide production	−	[+]	−	−	−	−	−
Urea hydrolysis	−[b]	d	−	−	−	−	[+]
Phenylalanine deaminase (24h)	−	−	−	−	−	−	−
Lysine decarboxylase	−[b]	−	−	−	−	−	−
Arginine dihydrolase	−[b]	−	−	d	[+]	+	[+]
Ornithine decarboxylase	−[b]	−	+	+	−	−	+
Motility	−	d	+	+	[+]	+	+
Gelatin hydrolysis, 22°C	+	−	−	−	−	−	−
KCN, growth		−	[+]	[+]	+	d	+
Malonate utilization	−	−	d	+	+	+	−
D-Glucose, acid production	+	+	+	+	+	+	+
D-Glucose, gas production	−	d	+	d	+	+	+
Acid production:							
D-Adonitol	−	−	−	−	−	−	−
L-Arabinose	−	[+]	+	−	−	−	+
Cellobiose	−	−	+	+	+	+	+
Dulcitol	−	−	−	−	−	−	−
Glycerol	−	−	d	−	−	−	d
myo-Inositol	−	−	−	−	−	−	−
Lactose	−	[+]	+	[−]	d	d	d
Maltose	−	−	+	+	+	+	+
D-Mannitol	−	d	+	+	+	+	+
D-Mannose		−	+	+	+	+	+
Melibiose	−	−	+	−	−	−	−
α-Methyl-D-glucoside	−	−	−	−	−	−	−
Raffinose	−	−	+	−	−	−	−
L-Rhamnose	−	+	+	−	−	−	+
Salicin		−	+	+	+	+	d
D-Sorbitol		−	−	−	−	+	+
Sucrose	+	−	−	+	−	+	[−]
Trehalose	−	−	+	+	+	+	+
D-Xylose	−	+	+	+	−	+	+
Mucate		[−]	+	−	−	−	+
Tartrate, Jordans		d	d	−	−	−	[+]
Esculin hydrolysis	−	−	+	d	+	+	−
Acetate utilization		−	−	−	d	−	[+]
Nitrate reduction	−	+	+	+	+	+	+
Deoxyribonuclease, 25°C	−[b]	−	−	−	−	−	−
Lipase		−	−	+	+	+	−
ONPG [c]		+	+	+	+	+	+
Pigment [d]	−	−	−	−	−	−	−
Flagella arrangement [e]	−	P	P	P	P	P	P
Catalase production (24h)	+	+	+	+	+	+	+
Oxidation-fermentation [f]	F	F	F	F	F	F	F

Footnotes are at end of table

Table 5.2 *(continued)*

Test	Citrobacter diversus	Citrobacter freundii	Edwardsiella hoshinae	Edwardsiella ictaluri	Edwardsiella tarda	Enterobacter aerogenes
Gram stain (24h)	−	−	−	−	−	−
Oxidase (24h)	−	−	−	−	−	−
Indole production	+	−	[−]	−	+	−
Methyl red	+	+	+	−	+	−
Voges-Proskauer	−	−	−	−	−	+
Citrate (Simmons)	+	+	−	−	−	+
Hydrogen sulfide production	−	[+]	−	−	+	−
Urea hydrolysis	d	d	−	−	−	−
Phenylalanine deaminase (24h)	−	−	−	−	−	−
Lysine decarboxylase	−	−	+	+	+	+
Arginine dihydrolase	d	d	−	−	−	−
Ornithine decarboxylase	+	[−]	+	d	+	+
Motility	+	+	+	−	+	+
Gelatin hydrolysis, 22°C	−	−	−	−	−	−
KCN, growth	−	+	−	−	−	+
Malonate utilization	+	[−]	+	−	−	+
D-Glucose, acid production	+	+	+	+	+	+
D-Glucose, gas production	+	+	d	d	+	+
Acid production:						
D-Adonitol	+	−	−	−	−	+
L-Arabinose	+	+	[−]	−	−	+
Cellobiose	+	d	−	−	−	+
Dulcitol	d	d	−	−	−	−
Glycerol	+	+	d	−	d	+
myo-Inositol	−	−	−	−	−	+
Lactose	d	d	−	−	−	+
Maltose	+	+	+	+	+	+
D-Mannitol	+	+	+	−	−	+
D-Mannose	+	+	+	+	+	+
Melibiose	−	d	−	−	−	+
α-Methyl-D-glucoside	d	−	−	−	−	+
Raffinose	−	d	−	−	−	+
L-Rhamnose	+	+	−	−	−	+
Salicin	[−]	−	d	−	−	+
D-Sorbitol	+	+	−	−	−	+
Sucrose	d	d	+	−	−	+
Trehalose	+	+	+	−	−	+
D-Xylose	+	+	−	−	−	+
Mucate	+	+	−	−	−	+
Tartrate, Jordans	[+]	+	−	−	[−]	+
Esculin hydrolysis	−	−	−	−	−	+
Acetate utilization	[+]	[+]	−	−	−	d
Nitrate reduction	+	+	+	+	+	+
Deoxyribonuclease, 25°C	−	−	−	−	−	−
Lipase	−	−	−	−	−	−
ONPG [c]	+	+	−	−	−	+
Pigment [d]	−	−	−	−	−	−
Flagella arrangement [e]	P	P	P	−	P	P
Catalase production (24h)	+	+	+	+	+	+
Oxidation-fermentation [f]	F	F	F	F	F	F

Footnotes are at end of table

Table 5.2 (continued)

Test	Enterobacter agglomerans	Enterobacter amnigenus biogroup 1	Enterobacter amnigenus biogroup 2	Enterobacter asburiae	Enterobacter cloacae	Enterobacter (Erwinia) dissolvens
Gram stain (24h)	−	−	−	−	−	−
Oxidase (24h)	−	−	−	−	−	−
Indole production	[−]	−	−	−	−	−
Methyl red	d	−	d	+	−	−
Voges-Proskauer	d	+	+	−	+	+
Citrate (Simmons)	d	d	+	+	+	+
Hydrogen sulfide production	−	−	−	−	−	−
Urea hydrolysis	[−]	−	−	d	d	+
Phenylalanine deaminase (24h)	[−]	−	−	−	−	−
Lysine decarboxylase	−	−	−	−	−	−
Arginine dihydrolase	−	−	d	[−]	+	+
Ornithine decarboxylase	−	d	+	+	+	+
Motility	[+]	+	+	−	+	+
Gelatin hydrolysis, 22°C	−	−	−	−	−	−
KCN, growth	d	+	+	+	+	+
Malonate utilization	d	+	+	−	[+]	+
D-Glucose, acid production	+	+	+	+	+	+
D-Glucose, gas production	[−]	+	+	+	+	+
Acid production:						
D-Adonitol	−	−	−	−	[−]	−
L-Arabinose	+	+	+	+	+	+
Cellobiose	d	+	+	+	+	+
Dulcitol	[−]	−	−	−	[−]	−
Glycerol	d	−	−	[−]	d	−
myo-Inositol	[−]	−	−	−	[−]	d
Lactose	d	d	d	[+]	+	d
Maltose	+	+	+	+	+	+
D-Mannitol	+	+	+	+	+	+
D-Mannose	+	+	+	+	+	+
Melibiose	d	+	+	−	+	+
α-Methyl-D-glucoside	−	d	+	+	[+]	+
Raffinose	d	+	−	d	+	+
L-Rhamnose	[+]	+	+	−	+	+
Salicin	d	+	+	+	[+]	+
D-Sorbitol	d	−	+	+	+	+
Sucrose	[+]	+	−	+	+	+
Trehalose	+	+	+	+	+	+
D-Xylose	+	+	+	+	+	+
Mucate	d	d	+	[−]	[+]	+
Tartrate, Jordans	[−]	−	−	d	d	−
Esculin hydrolysis	d	+	+	+	d	+
Acetate utilization	d	−	−	[+]	[+]	+
Nitrate reduction	[+]	+	+	+	+	+
Deoxyribonuclease, 25°C	−	−	−	−	−	−
Lipase	−	−	−	−	−	d
ONPG [c]	+	+	+	+	+	+
Pigment [d]	[+Y]	−	−	−	−	−
Flagella arrangement [e]	P	P	P	−	P	P
Catalase production (24h)	+	+	+	+	+	+
Oxidation-fermentation [f]	F	F	F	F	F	F

Footnotes are at end of table

Table 5.2 *(continued)*

Test	Enterobacter gergoviae	Enterobacter hormaechei	Enterobacter intermedius	Enterobacter (Erwinia) nimipressuralis	Enterobacter sakazakii
Gram stain (24h)	–	–	–	–	–
Oxidase (24h)	–	–	–	–	–
Indole production	–	–	–	–	[–]
Methyl red	–	d	+	–	–
Voges-Proskauer	+	+	+	+	+
Citrate (Simmons)	+	+	d	+	+
Hydrogen sulfide production	–	–	–	–	–
Urea hydrolysis	+	[+]	–	–	–
Phenylalanine deaminase (24h)	–	–	–	–	d
Lysine decarboxylase	+	–	–	–	–
Arginine dihydrolase	–	[+]	–	+	+
Ornithine decarboxylase	+	+	+	+	+
Motility	+	d	+	+	+
Gelatin hydrolysis, 22°C	–	–	–	+	–
KCN, growth	–	+	d	+	+
Malonate utilization	+	+	+	+	[–]
D-Glucose, acid production	+	+	+	+	+
D-Glucose, gas production	+	[+]	+	+	+
Acid production:					
D-Adonitol	–	–	–	–	–
L-Arabinose	+	+	+	+	+
Cellobiose	+	+	+	+	+
Dulcitol	–	[+]	+	–	–
Glycerol	+	–	+	+	[–]
myo-Inositol	–	–	–	–	[+]
Lactose	d	–	+	+	+
Maltose	+	+	+	+	+
D-Mannitol	+	+	+	+	+
D-Mannose	+	+	+	+	+
Melibiose	+	–	+	+	+
α-Methyl-D-glucoside	–	[+]	+	+	+
Raffinose	+	–	+	–	+
L-Rhamnose	+	+	+	+	+
Salicin	+	d	+	+	+
D-Sorbitol	–	–	+	+	–
Sucrose	+	+	d	–	+
Trehalose	+	+	+	+	+
D-Xylose	+	+	+	+	+
Mucate	–	+	+	+	–
Tartrate, Jordans	+	[–]	+	–	–
Esculin hydrolysis	+	–	+	+	+
Acetate utilization	+	d	–	–	+
Nitrate reduction	+	+	+	+	+
Deoxyribonuclease, 25°C	–	–	–	–	–
Lipase	–	–	–	–	–
ONPG[c]	+	+	+	+	+
Pigment[d]	–	–	–	–	Y
Flagella arrangement[e]	P	P	P	P	P
Catalase production (24h)	+	+	+	+	+
Oxidation-fermentation[f]	F	F	F	F	F

Footnotes are at end of table

Table 5.2 *(continued)*

Test	Enterobacter taylorae (Enterobacter cancerogenus; Erwinia cancerogena)	Erwinia amylovora	Erwinia ananas	Erwinia cacticida	Erwinia carotovora	Erwinia chrysanthemi
Gram stain (24h)	–	–	–	–	–	–
Oxidase (24h)	–	–	–	–	–	–
Indole production	–		–	+	–	+
Methyl red	–					
Voges-Proskauer	+	+	+	+	+	+
Citrate (Simmons)	+			+		
Hydrogen sulfide production	–		d		+	+
Urea hydrolysis	–	–	–	–	–	–
Phenylalanine deaminase (24h)	–	–	–	–	–	–
Lysine decarboxylase	–		–	–		
Arginine dihydrolase	+		–			
Ornithine decarboxylase	+		–	–		
Motility	+	+	+	+	+	+
Gelatin hydrolysis, 22°C	–	+	+	–	+	+
KCN, growth	+	–	–	–	d	d
Malonate utilization	+			+		
D-Glucose, acid production	+	+	+	+	+	+
D-Glucose, gas production	+	–	–	–	d	+
Acid production:						
D-Adonitol	–	–	–	–	–	–
L-Arabinose	+	d	+	–	+	+
Cellobiose	+	–	+	–	+	+
Dulcitol	–	–	–	–	–	–
Glycerol	–	–	+	[+]	d	+
myo-Inositol	–	–	+	–	d	d
Lactose	–	–	+	[–]	+	d
Maltose	+	–	+	–	d	–
D-Mannitol	+	–	+	+	+	+
D-Mannose	+	–	+	+	+	+
Melibiose	–	–	+	–	+	+
α-Methyl-D-glucoside	–	–	–	–	d	–
Raffinose	–	–	+	–	+	+
L-Rhamnose	+	–	d	+	+	+
Salicin	+	–	+	+	+	+
D-Sorbitol	–	d	+	–	+	+
Sucrose	–			+		
Trehalose	+	+	+	[+]	+	–
D-Xylose	+	–	+	d	+	+
Mucate	[+]					
Tartrate, Jordans	–			–		
Esculin hydrolysis	+			+		
Acetate utilization	d					
Nitrate reduction	+	–	–	+	+	+
Deoxyribonuclease, 25°C	–	–	–		–	–
Lipase	–				–	+
ONPG [c]	+			+		
Pigment [d]	–	–	Y	–	–	–
Flagella arrangement [e]	P	P	P	P	P	P
Catalase production (24h)	+	+	+	+	+	+
Oxidation-fermentation [f]	F	F	F	F	F	F

Footnotes are at end of table

Table 5.2   *(continued)*

Test	*Erwinia cypripedii*	*Erwinia mallotivora*	*Erwinia nigrifluens*	*Erwinia persicinus*	*Erwinia psidii*	*Erwinia quercina*	*Erwinia rhapontici*
Gram stain (24h)	–	–	–	–	–	–	–
Oxidase (24h)	–	–	–	–	–	–	–
Indole production	–	–	–	–	–	–	–
Methyl red				–			
Voges-Proskauer	–	+	+	+	+	+	+
Citrate (Simmons)				+			
Hydrogen sulfide production	+	–	+	–	+	+	+
Urea hydrolysis	–	–	+	–	–	–	–
Phenylalanine deaminase (24h)	+	–	–	–		–	–
Lysine decarboxylase				–			
Arginine dihydrolase				–			
Ornithine decarboxylase				–			
Motility	+	+	+	+	+	+	+
Gelatin hydrolysis, 22°C	–	–	–	–	–	–	–
KCN, growth	+	–	–	–		–	+
Malonate utilization				+	–		
D-Glucose, acid production	+	+	+	+	+	+	+
D-Glucose, gas production	+	–	–	–	–	–	–
Acid production:							
D-Adonitol	–	–	–	–		–	–
L-Arabinose	+	–	+	+		–	+
Cellobiose	+	–	–	+	–	–	+
Dulcitol	–	–	–	–	+	–	d
Glycerol	d	–	+	+		+	+
myo-Inositol	+	–	+	+	+	–	+
Lactose	–	–	–	+	–	–	+
Maltose	+	–	–	+		–	+
D-Mannitol	+	+	+	+	+	+	+
D-Mannose	+	+	+	+	+	+	+
Melibiose	+	–	+	+		–	+
α-Methyl-D-glucoside	–	–	–	–	+	+	d
Raffinose	–	–	+	+	–	–	+
L-Rhamnose	+	–	+	+	+	–	+
Salicin	+	–	+	+	+	+	+
D-Sorbitol	+	–	+	+	+	+	+
Sucrose				+			
Trehalose	+	+	+	+	–	–	+
D-Xylose	+	+	+	–	–		d
Mucate				–			
Tartrate, Jordans				–			
Esculin hydrolysis				+			
Acetate utilization				–	–		
Nitrate reduction	+	–	–	+	–	–	+
Deoxyribonuclease, 25°C	–		–	–		–	–
Lipase	–			–			–
ONPG [c]				+			
Pigment [d]	–	–	–	R	–	–	–
Flagella arrangement [e]	P	P	P	P	P	P	P
Catalase production (24h)	+	+	+	+	+	+	+
Oxidation-fermentation [f]	F	F	F	F	F	F	F

Footnotes are at end of table

Table 5.2 *(continued)*

Test	*Erwinia rubrifaciens*	*Erwinia salicis*	*Erwinia stewartii*	*Erwinia tracheiphila*	*Erwinia uredovora*	*Escherichia blattae*	*Escherichia coli*
Gram stain (24h)	–	–	–	–	–	–	–
Oxidase (24h)	–	–	–	–	–	–	–
Indole production	–	–	–	–	+	–	+
Methyl red						+	+
Voges-Proskauer	–	+	–	d	+	–	–
Citrate (Simmons)						d	–
Hydrogen sulfide production	+	+	–	+	–	–	–
Urea hydrolysis	–	–	–	–	–	–	–
Phenylalanine deaminase (24h)	–	–	–	–	–	–	–
Lysine decarboxylase			–		–	+	+
Arginine dihydrolase			–		–	–	[–]
Ornithine decarboxylase			–		–	+	d
Motility	+	+	–	+	+	–	+
Gelatin hydrolysis, 22°C	–	–	–	–	+	–	–
KCN, growth	–	–	–	–	–	–	–
Malonate utilization						+	–
D-Glucose, acid production	+	+	+	+	+	+	+
D-Glucose, gas production	–	–	–	–	–	+	+
Acid production:							
D-Adonitol	–	–	–	–	+	–	–
L-Arabinose	+	–	+	–	+	+	+
Cellobiose	–	–	–	–	+	–	–
Dulcitol	–	–	–	–	–	–	d
Glycerol	d	d	–	–	+	+	d
myo-Inositol	–	+	–	–	+	–	–
Lactose	–	–	+	–	+	–	+
Maltose	–	–	–	–	+	+	+
D-Mannitol	+	+	+	–	+	–	+
D-Mannose	+	+	+	–	+	+	+
Melibiose	–	+	+	–	+	–	[+]
α-Methyl-D-glucoside	+	–	–	–	–	–	–
Raffinose	–	+	+	–	+	–	d
L-Rhamnose	–	–	–	–	+	+	[+]
Salicin	–	+	–	–	d	–	d
D-Sorbitol	+	+	+	–	+	–	+
Sucrose						–	d
Trehalose	–	–	+	–	+	[+]	+
D-Xylose	–	–	+	–	+	+	+
Mucate						d	+
Tartrate, Jordans						d	+
Esculin hydrolysis						–	d
Acetate utilization						–	+
Nitrate reduction	–	–	–	–	+	+	+
Deoxyribonuclease, 25°C	–	–	–	–	+	–	–
Lipase						–	–
ONPG [c]						–	+
Pigment [d]	–	–	+Y	–	+Y	–	–
Flagella arrangement [e]	P	P	–	P	P	–	P
Catalase production (24h)	+	+	+	+	+	+	+
Oxidation-fermentation [f]	F	F	F	F	F	F	F

Footnotes are at end of table

Table 5.2 (continued)

Test	Escherichia coli inactive	Escherichia fergusonii	Escherichia hermannii	Escherichia vulneris	Ewingella americana	Hafnia alvei	Klebsiella oxytoca
Gram stain (24h)	−	−	−	−	−	−	−
Oxidase (24h)	−	−	−	−	−	−	−
Indole production	[+]	+	+	−	−	−	+
Methyl red	+	+	+	+	[+]	d	[−]
Voges-Proskauer	−	−	−	−	+	[+]	+
Citrate (Simmons)	−	[−]	−	−	+	−	+
Hydrogen sulfide production	−	−	−	−	−	−	−
Urea hydrolysis	−	−	−	−	−	−	+
Phenylalanine deaminase (24h)	−	−	−	−	−	−	−
Lysine decarboxylase	d	+	−	[+]	−	+	+
Arginine dihydrolase	−	−	−	d	−	−	−
Ornithine decarboxylase	[−]	+	+	−	−	+	−
Motility	−	+	+	+	d	[+]	−
Gelatin hydrolysis, 22°C	−	−	−	−	−	−	−
KCN, growth	−	−	+	[−]	−	+	+
Malonate utilization	−	d	−	[+]	−	d	+
D-Glucose, acid production	+	+	+	+	+	+	+
D-Glucose, gas production	−	+	+	+	−	+	+
Acid production:							
D-Adonitol	−	+	−	−	−	−	+
L-Arabinose	[+]	+	+	+	−	+	+
Cellobiose	−	+	+	+	−	[−]	+
Dulcitol	d	d	[−]	−	−	−	d
Glycerol	d	[−]	−	[−]	[−]	+	+
myo-Inositol	−	−	−	−	−	−	+
Lactose	[−]	−	d	[−]	d	−	+
Maltose	[+]	+	+	+	[−]	+	+
D-Mannitol	+	+	+	+	+	+	+
D-Mannose	+	+	+	+	+	+	+
Melibiose	d	−		+	−	−	+
α-Methyl-D-glucoside	−	−	−	[−]	−		+
Raffinose	[−]	−	d	+	−	−	+
L-Rhamnose	d	+	+	+	[−]	+	+
Salicin	−	d	d	d	[+]	[−]	+
D-Sorbitol	[+]	−	−	−	−	−	+
Sucrose	[−]	−	d	−	−	−	+
Trehalose	+	+	+	+	+	+	+
D-Xylose	d	+	+	+	[−]	+	+
Mucate	d	−	+	[+]	−	−	+
Tartrate, Jordans	[+]	+	d	−	d	d	+
Esculin hydrolysis	−	d	d	[−]	d	−	+
Acetate utilization	d	+	[+]	d	−	[−]	+
Nitrate reduction	+	+	+	+	+	+	+
Deoxyribonuclease, 25°C	−	−	−	−	−	−	−
Lipase	−	−	−	−	−	−	−
ONPG [c]	d	[+]	+	+	[+]	+	+
Pigment [d]	−	−	+Y	dY	−	−	−
Flagella arrangement [e]	P	P	P	P	P	P	−
Catalase production (24h)	+	+	+	+	+	+	+
Oxidation-fermentation [f]	F	F	F	F	F	F	F

Footnotes are at end of table

Table 5.2 *(continued)*

Test	*Klebsiella planticola*	*Klebsiella pneumoniae* subsp. *ozaenae*	*Klebsiella pneumoniae* subsp. *pneumoniae*	*Klebsiella pneumoniae* subsp. *rhinoscleromatis*	*Klebsiella terrigena*	*Kluyvera ascorbata*
Gram stain (24h)	–	–	–	–	–	–
Oxidase (24h)	–	–	–	–	–	–
Indole production	[–]	–	–	–	–	+
Methyl red	+	+	[–]	+	d	+
Voges-Proskauer	+	–	+	–	+	–
Citrate (Simmons)	+	d	+	–	d	+
Hydrogen sulfide production	–	–	–	–	–	–
Urea hydrolysis	+	–	+	–	–	–
Phenylalanine deaminase (24h)	–	–	–	–	–	–
Lysine decarboxylase	+	d	+	–	+	+
Arginine dihydrolase	–	–	–	–	–	–
Ornithine decarboxylase	–	–	–	–	[–]	+
Motility	–	–	–	–	–	+
Gelatin hydrolysis, 22°C	–	–	–	–	–	–
KCN, growth	+	[+]	+	[+]	+	+
Malonate utilization	+	–	+	+	+	+
D-Glucose, acid production	+	+	+	+	+	+
D-Glucose, gas production	+	d	+	–	[+]	+
Acid production:						
D-Adonitol	+	+	+	+	+	+
L-Arabinose	+	+	+	+	+	+
Cellobiose	+	+	+	+	+	+
Dulcitol	[–]	–	d	–	[–]	[–]
Glycerol	+	d	+	d	+	d
myo-Inositol	+	d	+	+	[+]	–
Lactose	+	d	+	–	+	+
Maltose	+	+	+	+	+	+
D-Mannitol	+	+	+	+	+	+
D-Mannose	+	+	+	+	+	+
Melibiose	+	+	+	+	+	+
α-Methyl-D-glucoside	+	d	+	–	+	+
Raffinose	+	+	+	+	+	+
L-Rhamnose	+	d	+	+	+	+
Salicin	+	+	+	+	+	+
D-Sorbitol	+	d	+	+	+	d
Sucrose	+	[–]	+	[+]	+	+
Trehalose	+	+	+	+	+	+
D-Xylose	+	+	+	+	+	+
Mucate	+	[–]	+	–	+	+
Tartrate, Jordans	+	d	+	d	+	d
Esculin hydrolysis	+	[+]	+	d	+	+
Acetate utilization	d	–	[+]	–	[–]	d
Nitrate reduction	+	[+]	+	+	+	+
Deoxyribonuclease, 25°C	–	–	–	–	–	–
Lipase	–	–	–	–	–	–
ONPG [c]	+	[+]	+	–	+	+
Pigment [d]	–	–	–	–	–	–
Flagella arrangement [e]	–	–	–	–	–	P
Catalase production (24h)	+	+	+	+	+	+
Oxidation-fermentation [f]	F	F	F	F	F	F

Footnotes are at end of table

211

Table 5.2 *(continued)*

Test	Kluyvera cryocrescens	Leclercia (Escherichia) adecarboxylata	Leminorella grimontii	Leminorella richardii	Moellerella wisconsensis	Morganella morganii
Gram stain (24h)	–	–	–	–	–	–
Oxidase (24h)	–	–	–	–	–	–
Indole production	+	+	–	–	–	+
Methyl red	+	+	+	–	+	+
Voges-Proskauer	–	–	–	–	–	–
Citrate (Simmons)	[+]	–	+	–	[+]	–
Hydrogen sulfide production	–	–	+	+	–	–
Urea hydrolysis	–	[–]	–	–	–	+
Phenylalanine deaminase (24h)	–	–	–	–	–	+
Lysine decarboxylase	[–]	–	–	–	–	–
Arginine dihydrolase	–	–	–	–	–	–
Ornithine decarboxylase	+	–	–	–	–	+
Motility	+	[+]	–	–	–	+
Gelatin hydrolysis, 22°C	–	–	–	–	–	–
KCN, growth	[+]	+	–	–	d	+
Malonate utilization	[+]	+	–	–	–	–
D-Glucose, acid production	+	+	+	+	+	+
D-Glucose, gas production	+	+	d	–	–	+
Acid production:						
D-Adonitol	–	+	–	–	+	–
L-Arabinose	+	+	+	+	–	–
Cellobiose	+	+	–	–	–	–
Dulcitol	–	[+]	[+]	–	–	–
Glycerol	–	–	[–]	–	–	–
myo-Inositol	–	–	–	–	–	–
Lactose	+	+	–	–	+	–
Maltose	+	+	–	–	d	–
D-Mannitol	+	+	–	–	d	–
D-Mannose	+	+	–	–	+	+
Melibiose	+	+	–	–	+	–
α-Methyl-D-glucoside	+	–	–	–	–	–
Raffinose	+	d	–	–	+	–
L-Rhamnose	+	+	–	–	–	–
Salicin	+	+	–	–	–	–
D-Sorbitol	d	–	–	–	–	–
Sucrose	[+]	d	–	–	+	–
Trehalose	+	+	–	–	–	–
D-Xylose	+	+	[+]	+	–	–
Mucate	[+]	d	+	d	–	–
Tartrate, Jordans	[–]	[+]	+	+	d	+
Esculin hydrolysis	+	+	–	–	–	–
Acetate utilization	[+]	[–]	–	–	–	–
Nitrate reduction	+	+	+	+	+	+
Deoxyribonuclease, 25°C	–	–	–	–	–	–
Lipase	–	–	–	–	–	–
ONPG [c]	+	+	–	–	+	–
Pigment [d]	–	[–Y]	–	–	–	–
Flagella arrangement [e]	P	P	–	–	–	P
Catalase production (24h)	+	+	+	+	+	+
Oxidation-fermentation [f]	F	F	F	F	F	F

Footnotes are at end of table

212

Table 5.2 *(continued)*

Test	Obesumbacterium proteus	Pantoea agglomerans	Pantoea dispersa	Pragia fontium	Proteus mirabilis	Proteus myxofaciens
Gram stain (24h)	–	–	–	–	–	–
Oxidase (24h)	–	–	–	–	–	–
Indole production	–	–	–	–	–	–
Methyl red	[–]	d	d	+	+	+
Voges-Proskauer	–	+	+	–	d	+
Citrate (Simmons)	–	+	+	[+]	d	+
Hydrogen sulfide production	–	–	–	+	+	–
Urea hydrolysis	–	–	–	–	+	+
Phenylalanine deaminase (24h)	–	[+]	–	[–]	+	+
Lysine decarboxylase	+	–	–	–	–	–
Arginine dihydrolase	–	–	–	–	–	–
Ornithine decarboxylase	+	d	–	–	+	–
Motility	–	+	+	+	+	+
Gelatin hydrolysis, 22°C	–	–	–	–	+	+
KCN, growth	–	[+]	+	–	+	+
Malonate utilization	–	+	–	–	–	–
D-Glucose, acid production	+	+	+	+	+	+
D-Glucose, gas production	–	–	–	–	+	+
Acid production:						
D-Adonitol	–	–	–	–	–	–
L-Arabinose	–	+	d	–	–	–
Cellobiose	–	d	d	–	–	–
Dulcitol	–	–	–	–	–	–
Glycerol	–	–	[–]	–	d	+
myo-Inositol	–	–	d	–	–	–
Lactose	–	[–]	–	–	–	–
Maltose	d	+	+	–	–	+
D-Mannitol	–	+	+	–	–	–
D-Mannose	[+]	+	+	–	–	–
Melibiose	–	–	–	–	–	–
α-Methyl-D-glucoside	–	–	–	–	–	+
Raffinose	–	–	–	–	–	–
L-Rhamnose	[–]	+	+	–	–	–
Salicin	–	+	–	[+]	–	–
D-Sorbitol	–	–	–	–	–	–
Sucrose	–	+	+	–	[–]	+
Trehalose	[+]	+	+	–	+	+
D-Xylose	[–]	+	+	–	+	–
Mucate	–	–	d	–	–	–
Tartrate, Jordans	[–]	–	d	–	[+]	+
Esculin hydrolysis	–	+	–	[+]	–	–
Acetate utilization	–			–	[–]	–
Nitrate reduction	+	+	d	+	+	+
Deoxyribonuclease, 25°C	–	–	–	–	d	+
Lipase	–			–	+	+
ONPG [c]	–	+	+	–	–	–
Pigment [d]	–	+Y	dY	–	–	–
Flagella arrangement [e]	–	P	P	P	P	P
Catalase production (24h)	+	+	+	+	+	+
Oxidation-fermentation [f]	F	F	F	F	F	F

Footnotes are at end of table

213

Table 5.2 *(continued)*

Test	*Proteus penneri*	*Proteus vulgaris*	*Providencia alcalifaciens*	*Providencia heimbachae*	*Providencia rettgeri*	*Providencia rustigianii*
Gram stain (24h)	−	−	−	−	−	−
Oxidase (24h)	−	−	−	−	−	−
Indole production	−	+	+	−	+	+
Methyl red	+	+	+	[+]	+	d
Voges-Proskauer	−	−	−	−	−	−
Citrate (Simmons)	−	[−]	+	−	+	[−]
Hydrogen sulfide production	d	+	−	−	−	−
Urea hydrolysis	+	+	−	−	+	−
Phenylalanine deaminase (24h)	+	+	+	+	+	+
Lysine decarboxylase	−	−	−	−	−	−
Arginine dihydrolase	−	−	−	−	−	−
Ornithine decarboxylase	−	−	−	−	−	−
Motility	[+]	+	+	d	+	d
Gelatin hydrolysis, 22°C	d	+	−	−	−	−
KCN, growth	+	+	+	−	+	+
Malonate utilization	−	−	−	−	−	−
D-Glucose, acid production	+	+	+	+	+	+
D-Glucose, gas production	d	[+]	[+]	−	−	d
Acid production:						
D-Adonitol	−	−	+	+	+	−
L-Arabinose	−	−	−	−	−	−
Cellobiose	−	−	−	−	−	−
Dulcitol	−	−	−	−	−	−
Glycerol	d	d	[−]	−	d	−
myo-Inositol	−	−	−	d	+	−
Lactose	−	−	−	−	−	−
Maltose	+	+	−	d	−	−
D-Mannitol	−	−	−	−	+	−
D-Mannose	−	−	+	+	+	+
Melibiose	−	−	−	−	−	−
α-Methyl-D-glucoside	[+]	d	−	−	−	−
Raffinose	−	−	−	−	−	−
L-Rhamnose	−	−	−	+	d	−
Salicin	−	d	−	−	d	−
D-Sorbitol	−	−	−	−	−	−
Sucrose	+	+	[−]	−	[−]	d
Trehalose	d	d	−	−	−	−
D-Xylose	+	+	−	−	−	−
Mucate	−	−	−	−	−	−
Tartrate, Jordans	[+]	[+]	+	d	+	d
Esculin hydrolysis	−	d	−	−	d	−
Acetate utilization	−	[−]	d	−	d	[−]
Nitrate reduction	+	+	+	+	+	+
Deoxyribonuclease, 25°C	d	[+]	−	−	−	−
Lipase	d	[+]	−	−	−	−
ONPG [c]	−	−	−	−	−	−
Pigment [d]	−	−	−	−	−	−
Flagella arrangement [e]	P	P	P		P	P
Catalase production (24h)	+	+	+	+	+	+
Oxidation-fermentation [f]	F	F	F	F	F	F

Footnotes are at end of table

Table 5.2 *(continued)*

Test	Providencia stuartii	Rahnella aquatilis	Salmonella bongori	Salmonella choleraesuis subsp. arizonae	Salmonella choleraesuis subsp. choleraesuis	Salmonella choleraesuis subsp. diarizonae
Gram stain (24h)	−	−	−	−	−	−
Oxidase (24h)	−	−	−	−	−	−
Indole production	+	−	−	−	−	−
Methyl red	+	[+]	+	+	+	+
Voges-Proskauer	−	+	−	−	−	−
Citrate (Simmons)	+	+	+	+	+	+
Hydrogen sulfide production	−	−	+	+	+	+
Urea hydrolysis	d	−	−	−	−	−
Phenylalanine deaminase (24h)	+	+	−	−	−	−
Lysine decarboxylase	−	−	+	+	+	+
Arginine dihydrolase	−	−	+	d	d	d
Ornithine decarboxylase	−	−	+	+	+	+
Motility	[+]	−	+	+	+	+
Gelatin hydrolysis, 22°C	−	−	−	−	−	−
KCN, growth	+	−	+	−	−	−
Malonate utilization	−	+	−	+	−	+
D-Glucose, acid production	+	+	+	+	+	+
D-Glucose, gas production	−	+	[+]	+	+	+
Acid production:						
D-Adonitol	−	−	−	−	−	−
L-Arabinose	−	+	+	+	+	+
Cellobiose	−	+	−	−	−	−
Dulcitol	−	[+]	+	−	+	−
Glycerol	d	[−]	−	−	−	−
myo-Inositol	+	−	−	−	d	−
Lactose	−	+	−	[−]	−	[+]
Maltose	−	+	+	+	+	+
D-Mannitol	−	+	+	+	+	+
D-Mannose	+	+	+	+	+	+
Melibiose	−	+	[+]	+	+	+
α-Methyl-D-glucoside	−	−	−	−	−	−
Raffinose	−	+	−	−	−	−
L-Rhamnose	−	+	+	+	+	+
Salicin	−	+	−	−	−	−
D-Sorbitol	−	+	+	+	+	+
Sucrose	d	+	−	−	−	−
Trehalose	+	+	+	+	+	+
D-Xylose	−	+	+	+	+	+
Mucate	−	d	+	+	+	d
Tartrate, Jordans	+	−	−	−	+	[−]
Esculin hydrolysis	−	+	−	−	−	−
Acetate utilization	[+]	−	+	+	+	[+]
Nitrate reduction	+	+	+	+	+	+
Deoxyribonuclease, 25°C	−	−	−	−	−	−
Lipase	−	−	−	−	−	−
ONPG [c]	−	+	+	+	−	+
Pigment [d]	−	−	−	−	−	−
Flagella arrangement [e]	P	P	P	P	P	P
Catalase production (24h)	+	+	+	+	+	+
Oxidation-fermentation [f]	F	F	F	F	F	F

Footnotes are at end of table

Table 5.2 *(continued)*

Test	Salmonella choleraesuis subsp. houtenae	Salmonella choleraesuis subsp. indica	Salmonella choleraesuis subsp. salamae	Serratia entomophila	Serratia ficaria	Serratia fonticola
Gram stain (24h)	–	–	–	–	–	–
Oxidase (24h)	–	–	–	–	–	–
Indole production	–	–	–	–	–	–
Methyl red	+	+	+	[–]	d	+
Voges-Proskauer	–	–	–	+	d	–
Citrate (Simmons)	+	[+]	+	+	+	+
Hydrogen sulfide production	+	+	+	–	–	–
Urea hydrolysis	–	–	–	–	–	[–]
Phenylalanine deaminase (24h)	–	–	–	–	–	–
Lysine decarboxylase	+	+	+	–	–	+
Arginine dihydrolase	d	d	+	–	–	–
Ornithine decarboxylase	+	+	+	–	–	+
Motility	+	+	+	+	+	+
Gelatin hydrolysis, 22°C	–	–	–	+	+	–
KCN, growth	+	–	–	+	d	d
Malonate utilization	–	–	+	–	–	[+]
D-Glucose, acid production	+	+	+	+	+	+
D-Glucose, gas production	+	+	+	–	–	[+]
Acid production:						
D-Adonitol	–	–	–	–	–	+
L-Arabinose	+	+	+	–	+	+
Cellobiose	d	–	–	–	+	–
Dulcitol	–	d	+	–	–	+
Glycerol	–	d	[–]	–	–	[+]
myo-Inositol	–	–	–	–	d	d
Lactose	–	[–]	–	–	[–]	+
Maltose	+	+	+	+	+	+
D-Mannitol	+	+	+	+	+	+
D-Mannose	+	+	+	+	+	+
Melibiose	+	[+]	–	–	d	+
α-Methyl-D-glucoside	–	–	–	–	–	+
Raffinose	–	–	–	–	d	+
L-Rhamnose	+	+	+	–	d	[+]
Salicin	d	–	–	+	+	+
D-Sorbitol	+	–	+	–	+	+
Sucrose	–	–	–	+	+	[–]
Trehalose	+	+	+	+	+	+
D-Xylose	+	+	+	d	+	[+]
Mucate	–	+	+	–	–	–
Tartrate, Jordans	d	+	d	+	[–]	d
Esculin hydrolysis	–	–	[–]	+	+	+
Acetate utilization	d	[+]	+	[+]	d	[–]
Nitrate reduction	+	+	+	+	+	+
Deoxyribonuclease, 25°C	–	–	–	+	+	–
Lipase	–	–	–	[–]	[+]	–
ONPG [c]	–	d	[–]	+	+	+
Pigment [d]	–	–	–	–	–	–
Flagella arrangement [e]	P	P	P	P	P	P
Catalase production (24h)	+	+	+	+	+	+
Oxidation-fermentation [f]	F	F	F	F	F	F

Footnotes are at end of table

216

Table 5.2 *(continued)*

Test	*Serratia grimesii*	*Serratia liquefaciens*	*Serratia marcescens*	*Serratia odorifera* biogroup 1	*Serratia odorifera* biogroup 2	*Serratia plymuthica*
Gram stain (24h)	–	–	–	–	–	–
Oxidase (24h)	–	–	–	–	–	–
Indole production	–	–	–	d	d	–
Methyl red	[+]	+	[–]	+	d	+
Voges-Proskauer	d	+	+	d	+	[+]
Citrate (Simmons)	+	+	+	+	+	d
Hydrogen sulfide production	–	–	–	–	–	–
Urea hydrolysis	–	–	[–]	–	–	–
Phenylalanine deaminase (24h)	–	–	–	–	–	–
Lysine decarboxylase	+	+	+	+	+	–
Arginine dihydrolase	+	–	–	–	–	–
Ornithine decarboxylase	+	+	+	+	–	–
Motility	+	+	+	+	+	d
Gelatin hydrolysis, 22°C	+	+	+	+	+	d
KCN, growth	+	+	+	d	[–]	d
Malonate utilization	–	–	–	–	–	–
D-Glucose, acid production	+	+	+	+	+	+
D-Glucose, gas production	+	[+]	d	–	[–]	d
Acid production:						
D-Adonitol	–	–	d	d	d	–
L-Arabinose	+	+	–	+	+	+
Cellobiose	–	–	–	+	+	[+]
Dulcitol	–	–	–	–	–	–
Glycerol		+	+	d	d	d
myo-Inositol		d	[+]	+	+	d
Lactose	–	–	–	d	+	[+]
Maltose	+	+	+	+	+	+
D-Mannitol	+	+	+	+	+	+
D-Mannose	+	+	+	+	+	+
Melibiose	+	[+]	–	+	+	+
α-Methyl-D-glucoside	–	–	–	–	–	d
Raffinose	+	[+]	–	+	–	+
L-Rhamnose	–	[–]	–	+	+	–
Salicin	+	+	+	+	d	+
D-Sorbitol	+	+	+	+	+	d
Sucrose	+	+	+	+	–	+
Trehalose	+	+	+	+	+	+
D-Xylose	+	+	–	+	+	+
Mucate	–	–	–	–	–	–
Tartrate, Jordans		[+]	[+]	+	+	+
Esculin hydrolysis	+	+	+	+	d	[+]
Acetate utilization		d	d	d	d	d
Nitrate reduction	+	+	+	+	+	+
Deoxyribonuclease, 25°C	+	[+]	+	+	+	+
Lipase	+	[+]	+	d	d	d
ONPG [c]	+	+	+	+	+	d
Pigment [d]	–	–	dR	–	–	dR
Flagella arrangement [e]	P	P	P	P	P	P
Catalase production (24h)	+	+	+	+	+	+
Oxidation-fermentation [f]	F	F	F	F	F	F

Footnotes are at end of table

Table 5.2 *(continued)*

Test	*Serratia proteamaculans*	*Serratia rubidaea*	*Shigella boydii, S. dysenteriae, and S. flexneri*	*Shigella sonnei*	*Tatumella ptyseos*	*"Xenorhabdus beddingii"*
Gram stain (24h)	–	–	–	–	–	–
Oxidase (24h)	–	–	–	–	–	–
Indole production	–	–	d	–	–	–
Methyl red	d	[–]	+	+	–	–
Voges-Proskauer	[+]	+	–	–	–	–
Citrate (Simmons)	+	+	–	–	–	–
Hydrogen sulfide production	–	–	–	–	–	–
Urea hydrolysis	–	–	–	–	–	–
Phenylalanine deaminase (24h)	–	–	–	–	+	–
Lysine decarboxylase	+	d	–	–	–	–
Arginine dihydrolase	–	–	–	–	–	–
Ornithine decarboxylase	+	–	–	+	–	–
Motility	+	[+]	–	–	–	+
Gelatin hydrolysis, 22°C	+	+	–	–	–	+
KCN, growth	+	[–]	–	–	–	+
Malonate utilization	–	+	–	–	–	–
D-Glucose, acid production	+	+	+	+	+	
D-Glucose, gas production	+	d	–	–	–	
Acid production:						
D-Adonitol	–	+	–	–	–	
L-Arabinose	+	+	d	+	–	
Cellobiose	–	+	–	–	–	
Dulcitol	–	–	–	–	–	
Glycerol		[–]	–	[–]	–	d
myo-Inositol		[–]	–	–	–	
Lactose	–	+	–	–	–	
Maltose	+	+	d	+	–	+
D-Mannitol	+	+	+	+	–	
D-Mannose	+	+	+	+	+	+
Melibiose	+	+	d	[–]	d	
α-Methyl-D-glucoside	–	–	–	–	–	
Raffinose	+	+	d	–	[–]	
L-Rhamnose	d	–	–	[+]	–	
Salicin	d	+	–	–	d	[+]
D-Sorbitol	[+]	–	d	–	–	
Sucrose	+	+	–	–	+	
Trehalose	+	+	[+]	+	+	+
D-Xylose	+	+	–	–	–	
Mucate	–	–	–	–	–	
Tartrate, Jordans		d	d	+	–	
Esculin hydrolysis	d	+	–	–	–	+
Acetate utilization		[+]	–	–	–	
Nitrate reduction	+	+	+	+	+	–
Deoxyribonuclease, 25°C	+	+	–	–	–	+
Lipase	+	+	–	–	–	–
ONPG[c]	+	+	–	+	–	
Pigment[d]	–	+R	–	–	–	+B
Flagella arrangement[e]	P	P	–	–	L	P
Catalase production (24h)	+	+	+	+	+	–
Oxidation-fermentation[f]	F	F	F	F	F	F

Footnotes are at end of table

Table 5.2 *(continued)*

Test	"Xenorhabdus bovienii"	Xenorhabdus luminescens	Xenorhabdus nematophilus	"Xenorhabdus poinarii"	Yersinia aldovae	Yersinia bercovieri
Gram stain (24h)	–	–	–	–	–	–
Oxidase (24h)	–	–	–	–	–	–
Indole production	d	d	d		–	–
Methyl red		–	–		+	+
Voges-Proskauer		–	–		–	–
Citrate (Simmons)		d	–		–	–
Hydrogen sulfide production		–	–		–	–
Urea hydrolysis	–	[–]	–	–	[+]	d
Phenylalanine deaminase (24h)	–	–	–	–	–	–
Lysine decarboxylase		–	–		–	–
Arginine dihydrolase		–	–		–	–
Ornithine decarboxylase		–	–		d	[+]
Motility	+	+	+	+	–	–
Gelatin hydrolysis, 22°C		d	[+]		–	–
KCN, growth		–	–		–	–
Malonate utilization		–	–		–	–
D-Glucose, acid production	+	[+]	[+]		+	+
D-Glucose, gas production	–	–	–		–	–
Acid production:						
D-Adonitol		–	–		–	–
L-Arabinose		–	–		d	+
Cellobiose		–	–		–	+
Dulcitol		–	–		–	–
Glycerol	–	–	–	–	–	–
myo-Inositol		–	–		–	–
Lactose		–	–		–	[–]
Maltose		[–]	–		–	+
D-Mannitol		–	–		[+]	+
D-Mannose		+	[+]		+	+
Melibiose		–	–		–	–
α-Methyl-D-glucoside		–	–		–	–
Raffinose		–	–		–	–
L-Rhamnose		–	–		–	–
Salicin	–	–	–	–	–	[–]
D-Sorbitol		–	–		d	+
Sucrose		–	–		[–]	+
Trehalose		–	–		[+]	+
D-Xylose		–	–		d	+
Mucate		–	–		–	–
Tartrate, Jordans		d	d		+	+
Esculin hydrolysis	–	–	–	–	–	[–]
Acetate utilization		–	–		–	–
Nitrate reduction	d	–	[–]		+	+
Deoxyribonuclease, 25°C		–	[–]		–	–
Lipase		–	–		–	–
ONPG [c]		–	–		–	[+]
Pigment [d]	+Y	+Y,O,R	dy	+B	–	–
Flagella arrangement [e]	P	P	P	P	P	P
Catalase production (24h)	–	+	–	–	+	+
Oxidation-fermentation [f]		F	F		F	F

Footnotes are at end of table

219

Table 5.2 *(continued)*

Test	Yersinia enterocolitica	Yersinia frederiksenii	Yersinia intermedia	Yersinia kristensenii	Yersinia mollaretii
Gram stain (24h)	–	–	–	–	–
Oxidase (24h)	–	–	–	–	–
Indole production	d	+	+	d	–
Methyl red	+	+	+	+	+
Voges-Proskauer	–	–	–	–	–
Citrate (Simmons)	–	[–]	–	–	–
Hydrogen sulfide production	–	–	–	–	–
Urea hydrolysis	[+]	[+]	[+]	[+]	[–]
Phenylalanine deaminase (24h)	–	–	–	–	–
Lysine decarboxylase	–	–	–	–	–
Arginine dihydrolase	–	–	–	–	–
Ornithine decarboxylase	+	+	+	+	[+]
Motility	–	–	–	–	–
Gelatin hydrolysis, 22°C	–	–	–	–	–
KCN, growth	–	–	–	–	–
Malonate utilization	–	–	–	–	–
D-Glucose, acid production	+	+	+	+	+
D-Glucose, gas production	–	d	[–]	[–]	–
Acid production:					
D-Adonitol	–	–	–	–	–
L-Arabinose	+	+	+	[+]	+
Cellobiose	[+]	+	+	+	+
Dulcitol	–	–	–	–	–
Glycerol	+	[+]	d	d	[–]
myo-Inositol	d	[–]	[–]	[–]	–
Lactose	–	d	d	–	d
Maltose	[+]	+	+	+	d
D-Mannitol	+	+	+	+	+
D-Mannose	+	+	+	+	+
Melibiose	–	–	[+]	–	–
α-Methyl-D-glucoside	–	–	[+]	–	–
Raffinose	–	d	d	–	–
L-Rhamnose	–	+	+	–	–
Salicin	[–]	+	+	[–]	[–]
D-Sorbitol	+	+	+	+	+
Sucrose	+	+	+	–	+
Trehalose	+	+	+	+	+
D-Xylose	d	+	+	[+]	d
Mucate	–	–	–	–	–
Tartrate, Jordans	[+]	d	[+]	d	+
Esculin hydrolysis	[–]	[+]	+	–	–
Acetate utilization	[–]	[–]	[–]	–	–
Nitrate reduction	+	+	+	+	+
Deoxyribonuclease, 25°C	–	–	–	–	–
Lipase	d	d	[–]	–	–
ONPG [c]	+	+	+	d	[–]
Pigment [d]	–	–	–	–	–
Flagella arrangement [e]	P	P	P	P	P
Catalase production (24h)	+	+	+	+	+
Oxidation-fermentation [f]	F	F	F	F	F

Footnotes are at end of table

Table 5.2 *(continued)*

Test	Yersinia pestis	Yersinia pseudotuberculosis	Yersinia rohdei	Yersinia ruckeri	Yokenella (Koserella trabulsii) regensburgei
Gram stain (24h)	–	–	–	–	–
Oxidase (24h)	–	–	–	–	–
Indole production	–	–	–	–	–
Methyl red	[+]	+	d	+	+
Voges-Proskauer	–	–	–	–	–
Citrate (Simmons)	–	–	–	–	+
Hydrogen sulfide production	–	–	–	–	–
Urea hydrolysis	–	+	d	–	–
Phenylalanine deaminase (24h)	–	–	–	–	–
Lysine decarboxylase	–	–	–	d	+
Arginine dihydrolase	–	–	–	–	–
Ornithine decarboxylase	–	–	[–]	+	+
Motility	–	–	–	–	+
Gelatin hydrolysis, 22°C	–	–	–	d	–
KCN, growth	–	–	–	[–]	+
Malonate utilization	–	–	–	–	–
D-Glucose, acid production	+	+	+	+	+
D-Glucose, gas production	–	–	–	–	+
Acid production:					
D-Adonitol	–	–	–	–	–
L-Arabinose	+	d	+	–	+
Cellobiose	–	–	[–]	–	+
Dulcitol	–	–	–	–	–
Glycerol	d	d	d	d	–
myo-Inositol	–	–	–	–	–
Lactose	–	–	–	–	–
Maltose	[+]	+	–	+	+
D-Mannitol	+	+	+	+	+
D-Mannose	+	+	+	+	+
Melibiose	[–]	d	d	–	+
α-Methyl-D-glucoside	–	–	–	–	–
Raffinose	–	[–]	d	–	[–]
L-Rhamnose	–	d	–	–	+
Salicin	d	[–]	–	–	–
D-Sorbitol	d	–	+	d	–
Sucrose	–	–	+	–	–
Trehalose	+	+	+	+	+
D-Xylose	+	+	d	–	+
Mucate	–	–	–	–	–
Tartrate, Jordans	–	d	+	d	–
Esculin hydrolysis	d	+	–	–	d
Acetate utilization	–	–	–	–	[–]
Nitrate reduction	[+]	+	[+]	[+]	+
Deoxyribonuclease, 25°C	–	–	–	–	–
Lipase	–	–	–	d	–
ONPG [c]	d	d	d	d	+
Pigment [d]	–	–	–	–	–
Flagella arrangement [e]	–	P	P	P	P
Catalase production (24h)	+	+	+	+	+
Oxidation-fermentation [f]	F	F	F	F	F

Footnotes are at end of table

Table 5.2 *(continued)*

Symbols: −, 0–10% positive; [−], 11–25% positive; d, 26–75% positive; [+], 76–89% positive; +, 90–100% positive. Results are for 48 h incubation. Tests done at 36 ± 1° except as indicated in the "Test" column and except for *Xenorhabdus* species and *Erwinia* species where tests otherwise done at 36°C were done at 25°C and 27°C respectively. Tests for *Erwinia* species were done as described in Tables 5.14 through 5.16.

[a] Data not available.

[b] No growth.

[c] ONPG, *O*-nitrophenyl-β-ᴅ-galactopyranoside.

[d] B, brown; O, orange; R, red or pink; Y, yellow.

[e] P, peritrichous; L, lateral.

[f] F, fermentative.

Important Notes for Users of This Manual

Unless otherwise indicated in footnotes to tables, the meanings of symbols are as follows:

+   90% or more of strains are positive

−   90% or more of strains are negative

d   11-89% of strains are positive

v   strain instability (not equivalent to "d")

D   Different reactions in different taxa (species of a genus or genera of a family)

All other symbols are defined in footnotes to tables.

Table 5.3 Differentiation of *Buttiauxella agrestis* from Enteric groups 63 and 64 and from the genus *Kluyvera*

Test	B. agrestis	Enteric group 63	Enteric group 64	K. ascorbata	K. cryocrescens
Sucrose, acid	− (60)	−	−	+	[+]
Ascorbate (7 days)	−	+	−	+	−
D-arabitol, acid	−	−	+	−	−
Melibiose, acid	+	− (67)	−	+	+
Raffinose, acid	+	−	−	+	+
α-methyl-D-glucoside, acid	−	d (100)	− (100)	+	+
Citrate, Simmons	+	− (67)	d (100)	+	[+]
Lysine decarboxylase	−	+	−	+	[−]
Lactose, acid	+	−	+	+	+

Symbols: see Table 5.2 on p. 222. Numbers in parentheses indicate the percentage of positive strains in delayed reactions (3-7 days).

Table 5.4 Differentiation of *Cedecea* species

Test	C. davisae	C. lapagei	C. neteri	Cedecea species 3	Cedecea species 5
Ornithine decarboxylase	+	−	−	−	d
Sucrose, acid	+	−	+	d	+
D-Sorbitol, acid	−	−	+	−	+
Raffinose, acid	−	−	−	+	+
D-Xylose, acid	+	−	+	+	+
Melibiose, acid	−	−	−	+	+
Malonate	+	+	+	−	−

Symbols: see Table 5.2 on p. 222.

Table 5.5 Differentiation of *Cededea* from *Ewingella* and *Serratia*

Test	Cedecea	Ewingella	Serratia	S. fonticola
Lysine decarboxylase	−	−	+ [a]	+
Arginine dihydrolase	[+]	−	− [b]	−
Gelatin hydrolysis	−	−	+	−
KCN	[+]	−	[+] [c]	d
Malonate utilization	+	−	− [d]	[+]
D-Glucose, gas production	[+]	−	d	[+]
L-Arabinose	−	−	+ [e]	+
Cellobiose	+	−	d	−
Maltose	+	[−]	+	+
Deoxyribonuclease	−	−	+	−
Lipase	+	−	d	−

Symbols: see Table 5.2 on p. 222.
[a] Except *S. entomophila*, *S. ficaria*, *S. plymuthica*.
[b] Except *S. grimesii*.
[c] Most strains positive except *S. odorifera* biogroup 2, *S. rubidaea*.
[d] Except *S. rubidaea*.
[e] Except *S. entomophila*, *S. marcescens*.

Table 5.6   Differentiation of the genus *Citrobacter* from other genera

Test	*C. amalonaticus*	*C. diversus*	*C. freundii*	*Budvicia*	*Enterobacter*	*E. coli*[a]	*Salmonella*
Indole production	+	+	−	−	−	+	−
Voges-Proskauer	−	−	−	−	+[b]	−	−
Citrate, Simmons	[+]	+	+	−	+	−	+
H$_2$S production	−	−	[+]	[+]	−	−	+
Urea hydrolysis	[+]	d	d	d	d	−	−
Lysine decarboxylase	−	−	−	−	d	+	+
KCN growth	+	−	+	−	d	−	−[g]
Malonate utilization	−	+	[−]	−	+[c]	−	d
D-Adonitol, acid production	−	+	−	−	−[d]	−	−
Cellobiose, acid production	+	+	d	−	+	−	−
Melibiose, acid production	−	−	d	−	+[e]	[+]	+
Esculin hydrolysis	−	−	−	−	+[f]	d	−

Symbols: see Table 5.2 on p. 222.

[a] See Table 5.2 for reactions of other *Escherichia* species.
[b] *Enterobacter asburiae* is negative.
[c] *Enterobacter asburiae* and most *Enterobacter sakazakii* strains are negative.
[d] *Enterobacter aerogenes* is positive.
[e] *Enterobacter asburiae*, *Enterobacter hormaechei*, and *Enterobacter taylorae* are negative.
[f] *Enterobacter hormaechei* is negative or delayed positive.
[g] *S. bongori* and *S. choleraesuis* subsp. *salamae* are positive.

Table 5.7   Biochemical differentiation of *Citrobacter* species

Test	*C. amalonaticus*	*C. amalonaticus* BG 1	*C. diversus*	*C. freundii*
Indole production	+	+	+	−
Citrate, Simmons	[+]	−	+	+
H$_2$S production	−	−	−	[+]
Ornithine decarboxylase	+	+	+	[−]
KCN, growth	+	+	−	+
Malonate utilization	−	−	+	[−]
D-Adonitol, acid production	−	−	+	−
Melibiose, acid production	−	+	−	d
Raffinose, acid production	−	+	−	d
Sucrose, acid production	[−]	+	d	d

Symbols: see Table 5.2 on p. 222.

Table 5.8 Differentiation of *Edwardsiella* species from other genera

Test	E. hoshinae	E. ictaluri	E. tarda	E. coli	Salmonella	Shigella boydii, dysenteriae, flexneri	sonnei
Indole production	[−]	−	+	+	−	d	−
Methyl red	+	−	+	+	+	+	+
Citrate, Simmons	−	−	−	−	+	−	−
H₂S production	−	−	+	−	+	−	−
Lysine decarboxylase	+	+	+	+	+	−	−
Ornithine decarboxylase	+	d	+	d	+	−	+
Motility	+	−	+	+	+	−	−
Malonate utilization	+	−	−	−	[−]	−	−
D-Glucose, gas production	d	d	+	+	+	−	−
Acid production from:							
L-Arabinose	[−]	−	−	+	+	d	+
Lactose	−	−	−	+	−	−	−
D-Mannitol	+	−	−	+	+	+	+
Melibiose	−	−	−	[+]	+	d	[−]
L-Rhamnose	−	−	−	[+]	+	−	[+]
D-Sorbitol	−	−	−	+	+	d	−
Trehalose	+	−	−	+	+	[+]	+
D-Xylose	−	−	−	+	+	−	−
ONPG	−	−	−	+	[−]	−	+

Symbols: see Table 5.2 on p. 222.

Table 5.9 Biochemical differentiation of *Edwardsiella* species

Test	E. hoshinae	E. ictaluri	E. tarda	E. tarda biogroup 1
Indole production	[−]	−	+	+
Methyl red	+	−	+	+
H₂S production	−	−	+	−
Motility	+	−	+	+
Malonate utilization	+	−	−	−
Acid production from:				
L-Arabinose	[−]	−	−	+
D-Mannitol	+	−	−	+
Sucrose	+	−	−	+
Trehalose	+	−	−	−

Symbols: see Table 5.2 on p. 222.

225

Table 5.10 Differentiation of *Enterobacter* species from klebsiellae

Test	K. oxytoca	K. planticola	K. pneumoniae subsp. ozaenae	K. pneumoniae subsp. pneumoniae	K. pneumoniae subsp. rhinoscleromatis	K. terrigena	E. aerogenes	E. agglomerans	E. amnigenus
Indole production	+	[-]	-	-	-	-	-	[-]	-
Methyl red	[-]	+	+	[-]	+	d	-	d	[+]
Urea hydrolysis	+	+	-	+	-	-	-	[-]	-
Lysine decarboxylase	+	+	d	+	-	+	+	-	-
Arginine dihydrolase	-	-	-	-	-	-	-	-	d
Ornithine decarboxylase	-	-	-	-	-	[-]	+	-	[+]
Motility	-	-	-	-	-	-	+	[+]	+
KCN, growth	+	+	[+]	+	[+]	+	+	d	+
Malonate utilization	+	+	-	+	+	+	+	d	+
D-Glucose, gas production	+	+	d	+	-	[+]	+	[-]	+
D-Adonitol, acid production	+	+	+	+	+	+	+	-	-
myo-Inositol, acid production	+	+	d	+	+	[+]	+	[-]	-
Melibiose, acid production	+	+	+	+	+	+	+	d	+

Table 5.10 *(continued)*

Test	E. asburiae	E. cloacae	E. dissolvens	E. gergoviae	E. hormaechei	E. intermedius	E. nimipressuralis	E. sakazakii	E. taylorae
Indole production	-	-	-	-	-	-	-	[-]	-
Methyl red	+	-	-	-	d	+	-	-	-
Urea hydrolysis	d	d	+	+	[+]	-	-	-	-
Lysine decarboxylase	-	-	-	+	-	-	-	-	+
Arginine dihydrolase	[-]	+	+	-	[+]	-	+	+	+
Ornithine decarboxylase	+	+	+	-	+	+	+	+	+
Motility	-	+	+	+	d	+	+	+	+
KCN, growth	+	+	+	-	+	d	+	+	+
Malonate utilization	-	[+]	+	+	+	+	+	[-]	+
D-Glucose, gas production	+	+	+	+	[+]	+	+	+	+
D-Adonitol, acid production	-	[-]	-	-	-	-	-	-	-
myo-Inositol, acid production	-	[-]	d	-	-	-	-	[+]	-
Melibiose, acid production	-	+	+	+	-	+	+	+	-

Symbols: see Table 5.2 on p. 222.

Table 5.11 Additional tests for the separation of *Enterobacter* species and klebsiellae [a]

Test [b]	*K. oxytoca*	*K. planticola*	*K. pneumoniae* subsp. *ozaenae*	*K. pneumoniae* subsp. *pneumoniae*	*K. pneumoniae* subsp. *rhinoscleromatis*	*K. terrigena*	*E. aerogenes*
Acid production from:							
L-Fucose	+	+	+	+	+	+	d
5-Ketogluconate	+	+	−	−	−	+	−
D-Lyxose	−	+	−	−	−	d	−
D-Sorbose	+	+	−	−	−	[−]	−

Table 5.11 *(continued)*

Test [b]	*E. agglomerans*	*E. amnigenus*	*E. cloacae*	*E. gergoviae*	*E. intermedius*	*E. sakazakii*	*E. taylorae*
Acid production from:							
L-Fucose	[−]	−	d	−	−	−	+
5-Ketogluconate	−	−	−	+	+	−	−
D-Lyxose	−	−	−	−	−	−	−
D-Sorbose	−	−	−	−	−	−	−

Symbols: see Table 5.2 on p. 222.

[a] These tests have not been reported for *E. asburiae*, *E. dissolvens*, *E. hormaechei*, and *E. nimipressuralis*.

[b] In most cases positive reactions occur within 48 h, but tests should be held for 4 d.

Important Notes for Users of This Manual

Unless otherwise indicated in footnotes to tables, the meanings of symbols are as follows:

+ 90% or more of strains are positive

− 90% or more of strains are negative

d 11-89% of strains are positive

v strain instability (not equivalent to "d")

D Different reactions in different taxa (species of a genus or genera of a family)

All other symbols are defined in footnotes to tables.

Table 5.12 Differentiation of *Enterobacter* and related genera

Test	Enterobacter	Klebsiella	Hafnia	Serratia
Motility	+ [a]	−	[+]	+
Ornithine decarboxylase	+ [b]	−	+	+ [c]
Arginine dihydrolase	d	−	−	− [d]
Deoxyribonuclease	−	−	−	+
Citrate utilization	+	+ [e]	−	+
Susceptible to *Hafnia* phage	−	−	+	−

Symbols: see Table 5.2 on p. 222.

[a] *E. asburiae* and almost one-half of *E. hormaechei* strains are nonmotile.
[b] *E. agglomerans* is negative.
[c] *S. odorifera* biogroup 2, *S. plymuthica*, and *S. rubidaea* are negative.
[d] *S. grimesii* is positive.
[e] *K. pneumoniae* subsp. *rhinoscleromatis* and some strains of *K. pneumoniae* subsp. *ozaenae* are negative.

Table 5.13 Biochemical differentiation of *Enterobacter* species

Test	E. aerogenes	E. agglomerans	E. amnigenus biogroup 1	E. amnigenus biogroup 2	E. asburiae	E. cloacae	E. dissolvens
Methyl red	−	d	−	d	+	−	−
Voges-Proskauer	+	d	+	+	−	+	+
Urea hydrolysis	−	[−]	−	−	d	d	+
Lysine decarboxylase	+	−	−	−	−	−	−
Arginine dihydrolase	−	−	−	d	[−]	+	+
Ornithine decarboxylase	+	−	d	+	+	+	+
Motility	+	[+]	+	+	−	+	+
Gelatin hydrolysis 22°C	−	−	−	−	−	−	−
KCN, growth	+	d	+	+	+	+	+
Malonate utilization	+	d	+	+	−	[+]	+
Acid production from:							
D-Adonitol	+	−	−	−	−	[−]	−
Dulcitol	−	[−]	−	−	−	[−]	−
Glycerol	+	d	−	−	[−]	d	−
myo-Inositol	+	[−]	−	−	−	[−]	d
Melibiose	+	d	+	+	−	+	+
α-CH₂-D-glucoside	+	−	d	+	+	[+]	+
Raffinose	+	d	+	−	d	+	+
L-Rhamnose	+	[+]	+	+	−	+	+
D-Sorbitol	+	d	−	+	+	+	+
Sucrose	+	[+]	+	−	+	+	+
Yellow pigment 25°C	−	[+]	−	−	−	−	−

Table 5.13 *(continued)*

Test	E. gergoviae	E. hormaechei	E. intermedius	E. nimipressuralis	E. sakazakii	E. taylorae
Methyl red	−	d	+	−	−	−
Voges-Proskauer	+	+	+	+	+	+
Urea hydrolysis	+	[+]	−	−	−	−
Lysine decarboxylase	+	−	−	−	−	−
Arginine dihydrolase	−	[+]	−	+	+	+
Ornithine decarboxylase	+	+	+	+	+	+
Motility	+	d	+	+	+	+
Gelatin hydrolysis 22°C	−	−	−	+	−	−
KCN, growth	−	+	d	+	+	+
Malonate utilization	+	+	+	+	[−]	+
Acid production from:						
D-Adonitol	−	−	−	−	−	−
Dulcitol	−	[+]	+	−	−	−
Glycerol	+	−	+	+	[−]	−
myo-Inositol	−	−	−	−	[+]	−
Melibiose	+	−	+	+	+	−
α-CH₂-D-glucoside	−	[+]	+	+	+	−
Raffinose	+	−	+	−	+	−
L-Rhamnose	+	+	+	+	+	+
D-Sorbitol	−	−	+	+	−	−
Sucrose	+	+	d	−	+	−
Yellow pigment 25°C	−	−	−	−	+	−

Symbols: see Table 5.2 on p. 222.

Important Notes for Users of This Manual

Unless otherwise indicated in footnotes to tables, the meanings of symbols are as follows:

+ 90% or more of strains are positive
− 90% or more of strains are negative
d 11-89% of strains are positive
v strain instability (not equivalent to "d")
D Different reactions in different taxa (species of a genus or genera of a family)

All other symbols are defined in footnotes to tables.

Table 5.14   Cultural, physiological and biochemical characteristics of *Erwinia* species [a]

Test	*E. amylovora*	*E. ananas*	*E. cacticida*	*E. carotovora*	*E. chrysanthemi*	*E. cypripedii*	*E. mallotivora*	*E. nigrifluens*	*E. persicinus*	*E. psidii*	*E. quercina*	*E. rhapontici*	*E. rubrifaciens*	*E. salicis*	*E. stewartii*	*E. tracheiphila*	*E. uredovora*
Motility	+	+	+	+	+	+	+	+	+	+	+	+	+	+	−	+	+
Anaerobic growth	W	+	+	+	+	+	+	+	+	+	+	+	+	W	+	W	+
Growth factors required	+	−	−	−	−	+	−		−	+	−	−	−	−	−	+	−
Pink diffusible pigment	−	−	−	−	−	−	−	−	+	−	−	+	+	−	−	−	−
Blue pigment	−	−	−	−	d	−	−	−	−	−	−	−	−	−	−	−	−
Yellow pigment	−	+	−	−	−	−	−	−	−	−	−	−	−	−	+	−	+
Mucoid growth	+	+	−	d	d	d	+	−	−	+	+	+	+	+	+	−	−
Symplasmata		−													−		d
Growth at 36°C	−	+	+	d	+	+	−	+	+	−	+	d	+	−	d	−	+
H₂S from cysteine	−	d	+	+	+		+	−	+	+	+	+	+	+	−	+	−
Reducing substances from sucrose	+	+	−	d	−	−	+	−		+	+	d	−	+	d	d	+
Acetoin	+	+	+	+	+	−	+	+	+	+	+	+	−	+	−	d	+
Urease	−	−	−	−	−	−	−	+	−	−	−	−	−	−	−	−	−
Pectate degradation	−	−	+	+	+	−	−	−	−	−	−	−	+	+	−	−	−
Gluconate oxidation	−	−	−	−	−	+	−	−	−	−	−	d	−	−	−	−	−
Gas from D-glucose	−	−	−	d	+	+	−	−	−	−	−	−	−	−	−	−	−
Casein hydrolysis	−	−		d	d												
Growth in KCN broth	−	−	−	d	d	+					−	+	−	−	−	−	−
Cotton seed oil hydrolysis	−	−		d	+	−	−					d					
Gelatin liquefaction	+	+	−	+	+	−	−	−	−	−	−	−	−	−	−	−	+
Phenylalanine deaminase	−	−	−	−	−	+	−	−	−	−	−	−	−	−	−	−	−
Indole	−	+	−	−	+	−	−	−	−	−	−	−	−	−	−	−	+
Nitrate reduction	−	−	+	+	+	+	−	−	+	−	−	+	−	−	−	−	+
Growth in 5% NaCl		+	+	+	d	+	−					+			+	−	+
Deoxyribonuclease	−	−		−					−			−	−	−	−	−	+
Phosphatase		d	−	+	d	−			+			d					
Lecithinase				−	+	−						−					
Sensitivity to erythromycin (15 μg/disk)		−	−	+	+				+			+					

Symbols: +, 80% or more positive; −, 2-% or less positive; d, 21-79% positive; W, weak growth; blank space, insufficient or no data.

[a] See *Bergey's Manual of Systematic Bacteriology* for methods (except for *E. cacticida* [Alcorn et al., Int. J. Syst. Bacteriol. *41:* 197-212, 1991]; *E. persicinus* Hao et al., Int. J. Syst. Bacteriol. *40:* 379-383, 1990], and *E. psidii* [Neto et al., Fitopatol. Bras. *12:* 345-350, 1987).

Table 5.15 Acid production from organic compounds by *Erwinia* species [a]

Compound	*E. amylovora*	*E. ananas*	*E. cacticida*	*E. carotovora*	*E. chrysanthemi*	*E. cypripedii*	*E. mallotivora*	*E. nigrifluens*	*E. persicinus*	*E. psidii*	*E. quercina*	*E. rhapontici*	*E. rubrifaciens*	*E. salicis*	*E. stewartii*	*E. tracheiphila*	*E. uredovora*
D-Adonitol	−	−	−	−	−	−	−	−	−		−	−	−	−	−	−	+
L-Arabinose	d	+	−	+	+	+	−	+	+	+	−	+	+	−	+	−	+
Cellobiose	−	+	−	+	+	+	(+)	−	+	−	−	+	−	−	−	−	+
Dextrin	−	−		−	−	−	−	−	−		−	−	−	−	−	−	+
Dulcitol	−	−	−	−	−	−	−	−	−	+	−	d	−	−	−	−	−
Esculin	−	d	+	+	+	+	−	+	+		+	+	−	+	−	−	d
Fructose	+	+	+	+	+	+	+	+	+	+	+	+	+	+	+	+	+
D-Galactose	+	+	+	+	+	+	+	+	+	+	+	+	+	+	+	+	+
D-Glucose	+	+	+	+	+	+	+	+	+	+	+	+	+	+	+	+	+
Glycerol	−	+	+	d	+	d	(+)	+		−	+	+	d	d	−	−	+
myo-Inositol	−	+	−	d	d	+	−	+	+	−	+	−	+	−	+	−	+
Inulin	−	d	−	−	d	−	−	−		−	+	−	−	d	−	+	
Lactose	−	+	−	+	d	−	−	−	+	−	−	+	−	−	+	−	+
Maltose	−	+	−	d	−	+	−	−	+	−	−	+	−	−	−	−	+
D-Mannitol	−	+	+	+	+	+	+	+	+	+	+	+	+	+	+	−	+
D-Mannose	−	+	+	+	+	+	+	+	+	+	+	+	+	+	+	−	+
Melezitose	−	−	−	−	−	−	−	−		−	−	d	−	−	−	−	+
Melibiose	−	+		+	+	+	−	+	+		−	+		+	+	−	+
α-CH₂-D-glucoside	−	−	−	d	−	−	−	−	−	+	+	d	+	−	−	−	−
Raffinose	−	+	−	+	+	−	−	+	+	−	−	+	−	+	+	−	+
L-Rhamnose	−	d	+	+	+	+	−	+	+	+	−	+	−	−	−	−	+
Ribose	+	+	+	+	+	+	+	+	+	+	+	+	+	+	+	−	+
Salicin	−	+	+	+	+	+	−	+	+	+	+	+	−	+	−	−	d
D-Sorbitol	d	+	−	+	+	+	−	+	+	+	+	+	+	+	+	−	+
Starch	−	+	−	−	−	−	−	−	−		−	+	−	−	−	−	+
Sucrose	+	+	+	+	+	+	+	+	+	+	+	+	+	+	+	+	+
Trehalose	+	+	d	+	−	+	+	+	+	−	−	+	−	−	+	−	+
D-Xylose	−	+	d	+	+	+	+	+	−	−	−	d	−	−	+	−	+

Symbols: see Table 5.14; (+), delayed positive reaction.

[a] After 7-days growth at 27°C in unshaken aqueous solution of 1% organic compound, 1% peptone with bromcresol purple as an indicator, except for *E. cacticida*, *E. persicinus*, and *E. psidii* (see references given in Table 5.14). *E. tracheiphila* grows very slowly in this medium.

Table 5.16   Utilization of some organic compounds as a source of carbon and energy for *Erwinia* species [a]

Species	Citrate	Formate	Lactate	Tartrate	Galacturonate	Malonate
E. amylovora	+	+	+	−	−	−
E. ananas	+	+	+	+	d	−
E. cacticida	+	+	−	b	b	+
E. carotovora	+	+	+	−	d	−
E. chrysanthemi	+	+	+	d	d	+
E. cypripedii	+	+	+	+	+	d
E. mallotivora	+	−	−	−	−	−
E. nigrifluens	−	+	+	+	−	−
E. persicinus	+	b	+	b	b	+
E. psidii	b	b	b	−	b	−
E. quercina	+	+	+	−	−	−
E. rhapontici	+	+	+	d	d	+
E. rubrifaciens	+	+	+	+	−	−
E. salicis	−	−	−	−	−	−
E. stewartii	+	+	+	+	−	−
E. tracheiphila	d	d	−	−	−	−
E. uredovora	+	+	+	+	−	−

Symbols: +, 80% or more positive; −, 20% or less positive; d, 21–89% positive.
[a] In 21 days at 27°C on OY medium.
[b] Data not available.

Important Notes for Users of This Manual

Unless otherwise indicated in footnotes to tables, the meanings of symbols are as follows:

  + 90% or more of strains are positive

  − 90% or more of strains are negative

  d 11-89% of strains are positive

  v strain instability (not equivalent to "d")

  D Different reactions in different taxa (species of a genus or genera of a family)

All other symbols are defined in footnotes to tables.

Table 5.17 Differentiation of *Escherichia* species from other species

Test	*E. blattae*	*E. coli*	*E. coli* (metabolically inactive strains)	*E. fergusonii*	*E. hermannii*	*E. vulneris*	*Shigella sonnei*	other shigellae	*Enterobacter agglomerans*	*Edwardsiella tarda*	*Leclercia adecarboxylata*
Indole	-	+	[+]	+	+	-	-	d	[-]	+	+
Voges-Proskauer	-	-	-	-	-	-	-	-	d	-	-
Citrate, Simmons	d	-	-	[-]	-	-	-	-	d	-	-
H₂S production	-	-	-	-	-	-	-	-	-	+	-
Urea hydrolysis	-	-	-	-	-	-	-	-	[-]	-	[-]
Lysine decarboxylase	+	+	d	+	-	[+]	-	-	-	+	-
Ornithine decarboxylase	+	d	[-]	+	+	-	+	-	-	+	-
Motility	-	+	-	+	+	+	-	-	[+]	+	[+]
KCN, growth	-	-	-	-	+	[-]	-	-	d	-	+
Malonate utilization	+	-	-	d	-	[+]	-	-	d	-	+
D-Glucose, gas production	+	+	-	+	+	+	-	-	[-]	+	+
Acid production from:											
D-Adonitol	-	-	-	+	-	-	-	-	-	-	+
Cellobiose	-	-	-	+	+	+	-	-	d	-	+
Lactose	-	+	[-]	-	d	[-]	-	-	d	-	+
D-Mannitol	-	+	+	+	+	+	+	+	+	-	+
Melibiose	-	[+]	d	-	-	+	[-]	d	d	-	+
L-Rhamnose	+	[+]	d	+	+	+	[+]	-	[+]	-	+
D-Sorbitol	-	+	[+]	-	-	-	-	d	d	-	-
D-Xylose	+	+	d	+	+	+	-	-	+	-	+
Mucate	d	+	d	-	+	[+]	-	-	d	-	d
Acetate utilization	-	+	d	+	[+]	d	-	-	d	-	[-]
Yellow pigment	-	-	-	-	+	d	-	-	[+]	-	[-Y]

Symbols: see Table 5.2 on p. 222.

233

Table 5.18   Biochemical differentiation of *Escherichia* species and related species

Test	E. blattae	E. coli	E. fergusonii	E. hermannii	E. vulneris	Leclercia adecarboxylata	Obesumbacterium proteus
Indole production	−	+	+	+	−	+	−
Lysine decarboxylase	+	+	+	−	[+]	−	+
Ornithine decarboxylase	+	d	+	+	−	−	+
Motility	−	+	+	+	+	[+]	−
KCN, growth	−	−	−	+	[−]	+	−
Malonate utilization	+	−	d	−	[+]	+	−
D-Glucose, gas production	+	+	+	+	+	+	−
Acid production from:							
D-Adonitol	−	−	+	−	−	+	−
L-Arabinose	+	+	+	+	+	+	−
D-Arabitol	−	−	+	−	−	+	ND
Cellobiose	−	−	+	+	+	+	−
Lactose	−	+	−	d	[−]	+	−
D-Mannitol	−	+	+	+	+	+	−
Melibiose	−	[+]	−	−	+	+	−
D-Sorbitol	−	+	−	−	−	−	−
Mucate	d	+	−	+	[+]	d	−
Acetate utilization	−	+	+	[+]	d	[−]	−
Yellow pigment	−	−	−	+	d	[−]	−

Symbols: see Table 5.2 on p. 222. ND, not done.

Table 5.19   Differential characteristics of the genus *Hafnia* and biochemically similar genera

Test	Hafnia	Enterobacter	Serratia
Citrate, Simmons	−[a]	+	+
Gelatin hydrolysis	−	−[b]	+[c]
Lysine decarboxylase	+	−[d]	D
Arginine dihydrolase	−	D	−[e]
Lipase (Tween 80)	−	−	+[f]
Deoxyribonuclease	−	−	+[g]
Acid production from:			
Raffinose	−	D	D
Sucrose	−	+[h]	+[i]
Lactose, D-Adonitol, *myo*-Inositol, D-Sorbitol	−	D	D
Susceptible to *Hafnia*-specific bacteriophage	+	−	−

Symbols: +, 90–100% of strains are positive; −, 90–100% of strains are negative; D, different reaction given by different species of a genus.

[a] Late positive reactions are given by ~50% of *Hafnia* strains.
[b] Except *E. nimipressuralis*.
[c] Except *S. fonticola*.
[d] Except *E. gergoviae*.
[e] Except *S. grimesii*.
[f] Most strains positive except *S. entomophila, S. fonticola*.
[g] Except *S. fonticola*.
[h] Except *E. amnigenus* biogroup 2, *E. nimipressuralis, E. taylorae*.
[i] Except *S. fonticola, S. odorifera* biogroup 2.

Table 5.20 Differentiation of species and subspecies of the genus *Klebsiella*

Test	*K. oxytoca*	*K. planticola*	*K. pneumoniae* subsp. *ozaenae*	*K. pneumoniae* subsp. *pneumoniae*	*K. pneumoniae* subsp. *rhinoscleromatis*	*K. terrigena*
Indole production	+	[−]	−	−	−	−
Methyl red	[−]	+	+	[−]	+	d
Voges-Proskauer	+	+	−	+	−	+
Citrate, Simmons	+	+	d	+	−	d
Urea hydrolysis	+	+	−	+	−	−
Lysine decarboxylase	+	+	d	+	−	+
Malonate utilization	+	+	−	+	+	+
Dulcitol, acid production	d	[−]	−	d	−	[−]
β-Gentibiose, acid production	+	+	+	+	−	+
Lactose, acid production	+	+	d	+	−	+
D-Melezitose, acid production	d	−	−	−	−	+
Mucate, acid production	+	+	[−]	+	−	+
Gentisate utilization	+	−	−	−	−	+
m-Hydroxybenzoate utilization	+	−	−	−	−	+
Growth 10°C	+	+	−	−	−	+
Lactose, gas production 44°C	−	−	+	+	+	−
Pectate hydrolysis	+	−	−	−	−	−

Symbols: see Table 5.2 on p. 222.

Table 5.21 Differentiation of *Kluyvera* species

Test	*K. ascorbata*	*K. cryocrescens*
Ascorbate	+	−
D-Glucose fermentation at 5°C (21 days)	−	+
Irgasan susceptibility (disk)	R	S
Lysine decarboxylase	+	[−]

Symbols: see Table 5.2 on p. 222. R, resistant; S, sensitive

235

Table 5.22   Differentiation of *Leminorella* from *Budvicia*, *Proteus*, and *Providencia*

Test	Leminorella	Budvicia	Proteus	Providencia
Indole production	−	−	−[a]	+[b]
H$_2$S production	+	[+]	d	−
Urease	−	d	+	d
Phenylalanine deaminase	−	−	+	+
Motility	−	d	+	[+]
Gelatin hydrolysis	−	−	+[c]	−
KCN, growth	−	−	+	+[b]
L-Arabinose, acid	+	[+]	−	−
Lactose, acid	−	[+]	−	−
D-Mannitol, acid	−	d	−	−[d]
D-Mannose, acid	−	+	−	+
L-Rhamnose, acid	−	+	−	d
D-Xylose, acid	[+]	+	+[e]	−
ONPG	−	+	−	−

Symbols: see Table 5.2 on p. 222.

[a] Except *P. vulgaris*.

[b] Except *P. heimbachae*.

[c] Except some strains of *P. penneri*.

[d] Except *P. rettgeri*.

[e] Except *P. myxofaciens*.

Table 5.23   Differentiation of *Leminorella* species

Test	L. grimontii 2 days	L. grimontii 7 days	L. richardii 2 days and 7 days
Methyl red	+	ND	ND
Citrate, Simmons	+	+	−
D-Glucose, gas	d	+	−
Dulcitol, acid	d	[+]	−

Symbols: see Table 5.2 on p. 222.   ND, not done.

Table 5.24 Biochemical differentiation of *Hafnia alvei*, *Obesumbacterium proteus*, and *Escherichia blattae*

Test	H. alvei	O. proteus biogroup 1	O. proteus biogroup 2	E. blattae
Susceptibility to *Hafnia*-specific bacteriophage	+	+	−	−
Voges-Proskauer	[+]	+	−	−
Motility	[+]	−	−	−
KCN, growth	+	d	−	−
Malonate utilization	d	d	−	+
D-Glucose, gas	+	−	−	+
L-Arabinose, acid	+	d	−	+
D-Mannitol, acid	+	+	−	−
L-Rhamnose, acid	+	d	[−]	+
Salicin, acid	[−]	+	−	−
D-Xylose, acid	+	−	[−]	+
Esculin hydrolysis	−	+	−	−
Tartrate, Jordans	d	d	[−]	d
ONPG	+	d	−	−

Symbols: see Table 5.2 on p. 222.

Table 5.25 Differentiation of *Pantoea* species

Test	P. agglomerans	P. dispersa
Phenylalanine deaminase	[+]	−
Malonate utilization	+	−
L-Arabinose, acid	+	d
Glycerol, acid	−	[−]
myo-Inositol, acid	−	d
Lactose, acid	[−]	−
Salicin, acid	+	−
Mucate, acid	−	d
Tartrate, Jordans	−	d
Esculin hydrolysis	+	−
Nitrate reduction	+	d

Symbols: see Table 5.2 on p. 222.

Table 5.26 Differentiation of *Pragia* from *Budvicia* and *Leminorella*

Test	P. fontium	B. aquatica	L. grimontii	L. richardii
Growth, 4°C	+	+	−	−
Growth, 42°C	−	−	+	+
Methyl red	+	+	+	−
Citrate, Simmons	[+]	−	+	−
Urea hydrolysis	−	d	−	−
Motility	+	d	−	−
L-Arabinose, acid	−	[+]	+	+
Dulcitol, acid	−	−	[+]	−
D-Galactose, acid	+	+	−	−
Lactose, acid	−	[+]	−	−
L-Rhamnose, acid	−	+	−	−
Salicin, acid	[+]	−	−	−
D-Xylose, acid	−	+	[+]	+
Esculin hydrolysis	[+]	−	−	−
Tartrate, Jordans	−	d	+	+
ONPG	−	+	−	−
Tyrosine clearing	−	−	+	+

Symbols: see Table 5.2 on p. 222.

Important Notes for Users of This Manual

Unless otherwise indicated in footnotes to tables, the meanings of symbols are as follows:

+ 90% or more of strains are positive
− 90% or more of strains are negative
d 11-89% of strains are positive
v strain instability (not equivalent to "d")
D Different reactions in different taxa (species of a genus or genera of a family)

All other symbols are defined in footnotes to tables.

Table 5.27 Biochemical differentiation of *Proteus*, *Providencia*, and *Morganella morganii*

Test	Proteus				Morganella morganii	Providencia				
	mirabilis	*myxofaciens*	*penneri*	*vulgaris*	*morganii*	*alcalifaciens*	*heimbachae*	*rettgeri*	*rustigianii*	*stuartii*
Indole production	–	–	–	+	+	+	–	+	+	+
Voges-Proskauer	d	+	–	–	–	–	–	–	–	–
Citrate, Simmons	d	+	–	[–]	–	+	–	+	[–]	+
H₂S production	+	–	d	+	–	–	–	–	–	–
Urea hydrolysis	+	+	+	+	+	–	–	+	–	d
Ornithine decarboxylase	+	–	–	–	+	–	–	–	–	–
Gelatin hydrolysis	+	+	d	+	–	–	–	–	–	–
D-Glucose, gas production	+	+	d	[+]	+	[+]	–	–	d	–
Acid production from:										
D-Adonitol	–	–	–	–	–	+	+	+	–	–
D-Arabitol	–	–	–	–	–	–	+	+	–	–
Glycerol	d	+	d	d	–	[–]	–	d	–	d
myo-Inositol	–	–	–	–	–	–	d	+	–	+
Maltose	–	+	+	+	–	–	d	–	–	–
D-Mannose	–	–	–	–	+	+	+	+	+	+
α-CH₂-D-glucoside	–	+	[+]	d	–	–	–	–	–	–
L-Rhamnose	–	–	–	–	–	–	+	d	–	–
Sucrose	[–]	+	+	+	–	[–]	–	[–]	d	d
Trehalose	+	+	d	d	–	–	–	+	–	+
D-Xylose	+	–	+	+	–	–	–	+	–	+
Deoxyribonuclease	d	+	d	[+]	–	–	–	–	–	–
Lipase	+	+	d	[+]	–	–	–	–	–	–
Tyrosine clearing	+	–	+	+	+	+	+	+	+	+
Swarming	+	+	d	+	–	–	–	–	–	–

Symbols: see Table 5.2 on p. 222.

239

Table 5.28   Differential characteristics of *Proteus* species

Test	*P. mirabilis*	*P. myxofaciens*	*P. penneri*	*P. vulgaris*
Indole production	−	−	−	+
Voges-Proskauer	d	+	−	−
Citrate, Simmons	d	+	−	[−]
H₂S production	+	−	d	+
Ornithine decarboxylase	+	−	−	−
Acid production from:				
Maltose	−	+	+	+
α-CH₂-D-glucoside	−	+	[+]	d
Sucrose	[−]	+	+	+
D-Xylose	+	−	+	+
Tyrosine clearing	+	−	+	+

Symbols: see Table 5.2 on p. 222.

Table 5.29   Differentiation of *Providencia* species

Test	*P. alcalifaciens*	*P. heimbachae*	*P. rettgeri*	*P. rustigianii*	*P. stuartii*
Indole production	+	−	+	+	+
Citrate, Simmons	+	−	+	[−]	+
Urea hydrolysis	−	−	+	−	d
KCN, growth	+	−	+	+	+
D-Glucose, gas	[+]	−	−	d	−
D-Adonitol, acid	+	+	+	−	−
D-Arabitol, acid	−	+	+	−	−
D-Galactose, acid	−	+	+	+	+
myo-Inositol, acid	−	d	+	−	+
D-Mannitol, acid	−	−	+	−	−
L-Rhamnose, acid	−	+	d	−	−
Trehalose, acid	−	−	−	−	+

Symbols: see Table 5.2 on p. 222.

Table 5.30   Differentiation of the genus *Salmonella* from other genera

Test	Salmonella	Citrobacter			Edwardsiella
		amalonaticus	diversus	freundii	
Indole production	–	+	+	–	d
Citrate, Simmons	+	[+]	+	+	–
H₂S production	+	–	–	[+]	d
Urea hydrolysis	–	[+]	d	d	–
Lysine decarboxylase	+	–	–	–	+
Ornithine decarboxylase	+	+	+	[–]	[+]
D-Adonitol, acid production	–	–	+	–	–
L-Arabinose, acid production	+	+	+	+	–
L-Rhamnose, acid production	+	+	+	+	–
D-Sorbitol, acid production	+	+	+	+	–
D-Xylose, acid production	+	+	+	+	–
Acetate utilization	+	[+]	[+]	[+]	–

Symbols: see Table 5.2 on p. 222.

Table 5.31   Differentiation of *Salmonella* species, subspecies, and some serovars

Test	Salmonella bongori	Salmonella choleraesuis				
		subsp. arizonae	subsp. choleraesuis	subsp. diarizonae	subsp. houtenae	subsp. indica
Citrate, Simmons	+	+	+	+	+	[+]
H₂S production	+	+	+	+	+	+
Lysine decarboxylase	+	+	+	+	+	+
Ornithine decarboxylase	+	+	+	+	+	+
Motility	+	+	+	+	+	+
KCN, growth	+	–	–	–	+	–
Malonate utilization	–	+	–	+	–	–
D-Glucose, gas	[+]	+	+	+	+	+
L-Arabinose, acid	+	+	+	+	+	+
Dulcitol, acid	+	–	+	–	–	d
Lactose, acid	–	[–]	–	[+]	–	[–]
Maltose, acid	+	+	+	+	+	+
Melibiose, acid	[+]	+	+	+	+	[+]
L-Rhamnose, acid	+	+	+	+	+	+
D-Sorbitol, acid	+	+	+	+	+	–
Trehalose, acid	+	+	+	+	+	+
D-Xylose, acid	+	+	+	+	+	+
Mucate, acid	+	+	+	d	–	+
Tartrate, Jordans	–	–	+	[–]	d	+
ONPG	+	+	–	+	–	d

Footnote is at end of table.

Table 5.31 *(continued)*

Test	Salmonella choleraesuis subsp. salamae	Salmonella choleraesuis subsp. choleraesuis				
		serovar choleraesuis	serovar gallinarum	serovar paratyphi A	serovar pullorum	serovar typhi
Citrate, Simmons	+	[−]	−	−	−	−
H₂S production	+	d	+	−	+	+
Lysine decarboxylase	+	+	+	−	+	+
Ornithine decarboxylase	+	+	−	+	+	−
Motility	+	+	−	+	−	+
KCN, growth	−	−	−	−	−	−
Malonate utilization	+	−	−	−	−	−
D-Glucose, gas	+	+	−	+	+	−
L-Arabinose, acid	+	−	[+]	+	+	−
Dulcitol, acid	+	−	+	+	−	−
Lactose, acid	−	−	−	−	−	−
Maltose, acid	+	+	+	+	−	+
Melibiose, acid	−	d	−	+	−	+
L-Rhamnose, acid	+	+	−	+	+	−
D-Sorbitol, acid	+	[+]	−	+	[−]	+
Trehalose, acid	+	−	d	+	[+]	+
D-Xylose, acid	+	+	d	−	[+]	[+]
Mucate, acid	+	−	d	−	−	−
Tartrate, Jordans	d	[+]	+	−	−	+
ONPG	[−]	−	−	−	−	−

Symbols: see Table 5.2 on p. 222.

GROUP 5 FACULTATIVELY ANAEROBIC GRAM-NEGATIVE RODS

Table 5.32 Differential characteristics of the genus *Serratia* and biochemically similar taxa

| Test | *Serratia* species | *Enterobacter* | | | pectinolytic erwiniae (includes *E. carotovora* and *E. chrysanthemi*) | *Klebsiella pneumoniae* subsp. *pneumoniae* | *Serratia fonticola* |
		aerogenes	*agglomerans*	*cloacae*			
Voges-Proskauer	+	+	d	+	+	+	–
Urea hydrolysis	–	–	[–]	d	–	+	[–]
Gelatin hydrolysis	+	–	–	–	D	–	–
Malonate utilization	–	+	d	[+]	[a]	+	[+]
α-CH₂-D-glucoside, acid production	–	+	–	[+]	d	+	+
Deoxyribonuclease	+	–	–	–	–	–	–
Lipase (corn oil)	[+]	–	–	–	d	–	–
D-Glucose, acid production in presence of 0.001 M iodoacetate	+	+	+	+	–	+	+

Symbols: see Table 5.2 on p. 222. D, reaction varies among species.
[a] Data not available.

243

Table 5.33 Additional tests to differentiate the genus *Serratia* from biochemically similar taxa

Test	Serratia species	Enterobacter aerogenes	Enterobacter agglomerans	Enterobacter cloacae	pectinolytic erwiniae (includes *E. carotovora* and *E. chrysanthemi*)	Klebsiella pneumoniae subsp. pneumoniae	Serratia fonticola
Carbon source utilization:							
4-Aminobutyrate	+	D	+	d	–	D	–
5-Aminovalerate	–	D	–	–	–	D	–
Arginine	–	+	–	–	–	+	–
Caprate, Caproate, and Caprylate	+	–	–	–	–	–	–
D-Dulcitol	–	–	–	–	–	–	+
L-Fucose	+	+	–	–	–	+	–
Pelargonate	D	–	–	–	–	–	–
Tagatose	–	D	–	–	–	D	+
Tyrosine	+	D	–	–	–	D	a
Tributyrin hydrolysis	+	–	–	–	–	–	a
Gluconate, reducing substances from	+	+	–	+	–	+	+
Pectate hydrolysis	–	–	–	–	+	–	–

Symbols: see Table 5.2 on p. 222. D, reaction varies among species.

Table 5.34 Differentiation of *Serratia* species

Test	S. entomophila	S. ficaria	S. fonticola	S. grimesii	S. liquefaciens	S. marcescens	S. odorifera biogroup 1	S. odorifera biogroup 2	S. plymuthica	S. proteamaculans	S. rubidaea
Voges-Proskauer	+	d	–	d	+	+	d	+	[+]	[+]	+
Lysine decarboxylase	–	–	+	+	+	+	+	+	–	+	d
Arginine dihydrolase	–	–	–	+	–	–	–	–	–	–	–
Ornithine decarboxylase	–	–	+	+	+	+	+	–	–	+	–
Gelatin hydrolysis	+	+	+	+	+	+	+	+	d	+	+
Malonate utilization	–	–	[+]	–	–	–	–	–	–	–	+
D-Glucose, gas	–	–	[+]	+	[+]	d	–	[–]	d	+	d
D-Adonitol, acid	–	–	+	–	–	d	d	d	–	–	+
L-Arabinose, acid	–	+	+	+	+	–	+	+	+	+	+
Cellobiose, acid	–	+	+	–	–	–	+	+	[+]	–	+
Dulcitol, acid	–	–	+	–	–	–	–	–	–	–	–
Lactose, acid	–	[–]	+	–	–	–	d	+	[+]	–	+
Melibiose, acid	–	d	+	+	[+]	–	+	+	+	+	+
α-CH$_2$-D-glucoside	–	–	+	–	–	–	+	+	d	–	–
Raffinose, acid	–	d	+	+	[+]	–	+	–	+	+	+
L-Rhamnose, acid	–	d	[+]	–	[–]	–	+	+	–	d	–
D-Sorbitol, acid	–	+	+	+	+	+	+	+	d	[+]	+
Sucrose, acid	+	+	[–]	+	+	+	+	–	+	+	+
D-Xylose, acid	d	+	[+]	+	+	–	+	+	+	+	+
Deoxyribonuclease	+	+	–	+	[+]	+	+	+	+	+	+
Lipase	[–]	[+]	–	+	[+]	+	d	d	d	+	+
Pigment (red, pink, orange)	–	–	–	–	–	d	–	–	d	–	+

Symbols: see Table 5.2 on p. 222.

Table 5.35 Differentiation of *Serratia* species *liquefaciens*, *proteamaculans*, and *grimesii*

Test	S. liquefaciens	S. proteamaculans subsp. proteamaculans			S. proteamaculans subsp. quinovora	S. grimesii	
		biotype Clc	biotype EB	biotype RB		biotype Cld	biotype ADC
trans-Aconitate, growth	–	+	+	–	+	–	+
Adonitol, growth	–	–	(+)	–	–	–	–
Benzoate, growth	–	–	(+)	d	–	+	–
m-Erythritol, growth	–	–	+	–	–	–	–
Gentisate, growth	–	–	–	+	–	–	–
D-Malate, growth	+	–	–	–	d	(d)	(d)
L-Rhamnose, growth	–	–	–	+	d	–	–
m-Tartrate, growth	+	–	–	–	d	–	–
Arginine dihydrolase	–	–	–	–	–	+	+
Esculin hydrolysis	+	+	+	+	–	+	+

Symbols: +, 90% or more positive; –, 10% or less positive; d, 11–89% positive; ( ), positive in 3-7 days.

Table 5.36 Differentiation of *Serratia rubidaea* subspecies

Test	*S. rubidaea* subspecies		
	"burdigalensis"	*"colindalensis"*	*"rubidaea"*
Histamine, growth	d	−	+
D-Melezitose, growth	−	+	+
D-Tartrate, growth	+	−	d
Voges-Proskauer	+	−	d
Lysine decarboxylase	+	+	−
Malonate utilization	+	+	−

Symbols: see Table 5.35.

Table 5.37 Biochemical differentiation of *Shigella sonnei* from other shigellae

Test	*S. sonnei*	other shigellae
Indole production	−	d
Ornithine decarboxylase	+	−
Raffinose, acid production	−	d
L-Rhamnose, acid production	[+]	−
ONPG	+	−

Symbols: see Table 5.2 on p. 222.

Important Notes for Users of This Manual

Unless otherwise indicated in footnotes to tables, the meanings of symbols are as follows:

- \+   90% or more of strains are positive
- −   90% or more of strains are negative
- d   11-89% of strains are positive
- v   strain instability (not equivalent to "d")
- D   Different reactions in different taxa (species of a genus or genera of a family)

All other symbols are defined in footnotes to tables.

Table 5.38 *Differentiation of Tatumella* from other phenylalanine deaminase-positive genera and species

Test	Tatumella ptyseos	Enterobacter (Pantoea) agglomerans	Enterobacter sakazakii	Erwinia cypripedii	Morganella morganii	Proteus	Providencia	Rahnella aquatilis
Methyl red	−	d	−	a	+	+	+[b]	[+]
Voges-Proskauer	−	d	+	−	−	d	−	+
H₂S production	−	−	−	+	−	d	−	−
Urease	−	[−]	−	−	+	+	d	−
Ornithine decarboxylase	−	−	+		+	−[d]	−	−
Motility	−	[+]	+	+	+	+	[+]	−
KCN, growth	−	d	+	+	+	+	+[c]	−
Malonate utilization	−	d	[−]		−	−	−	+
D-Glucose, gas	−	[−]	+	+	+	[+]	[−]	+
L-Arabinose, acid	−	+	+	+	−	−	−	+
Cellobiose, acid	−	d	+	+	−	−	−	+
Lactose, acid	−	d	+	−	−	−	−	+
Maltose, acid	−	+	+	+	−	+[d]	−	+
D-Mannitol, acid	−	+	+	+	−	−	−[e]	+
L-Rhamnose, acid	−	[+]	+	+	−	−	[−]	+
D-Xylose, acid	−	+	+	+	−	+[f]	−	+
ONPG	−	+	+		−	−	−	+

Symbols: see Table 5.2 on p. 222.

[a] Data not available.
[b] Except some strains of *P. rustigianii* and *P. heimbachae*.
[c] Except *P. heimbachae*.
[d] Except *P. mirabilis*.
[e] Except *P. rettgeri*.
[f] Except *P. myxofaciens*.

Table 5.39 Differentiation of *Xenorhabdus* species [a]

Test	"X. beddingii"	"X. bovienii"	X. luminescens	X. nematophilus	"X. poinarii"
Catalase	−	−	+	−	−
Bioluminescence	−	−	+	−	−
Pigment on Loeffler's blood serum	b		O	B	
Pigment on nutrient agar	B	Y	Y,O,R,P	U	B
Colony color on MacConkey agar [c]	B	B	R	B	R
Adsorption of bromthymol blue	+	+	+	+	−
Lipase (Tween 80) [c]	−	+		−	+
Phenylalanine deaminase [d]	−	−		+	−
Urease	−	−	d	−	−
Phosphatase	+	−	+	−	−
Esculin hydrolysis	+	−	+	−	d
Glycerol, acid	d	−		−	−
Salicin, acid	+	−		−	−

Symbols: see Table 5.2 on p. 222. O, orange; B, brown; Y, yellow; R, red; P, pink; U, buff.

[a] Data from Akhurst, Int. J. Syst. Bacteriol. *36*: 454-457, 1986 and Akhurst and Boemare, J. Gen. Microbiol. *134*: 1835-1845, 1988.
[b] Data not available.
[c] Colony form 1 only.
[d] Colony form 2 only.

Table 5.40   Differentiation of *Yersinia* from other genera

Test	*Yersinia*	*Citrobacter*	*Enterobacter*	*Escherichia*	*Hafnia*	*Klebsiella*	*Proteus*	*Salmonella*
Voges-Proskauer	−	−	+[c]	−	[+]	d	d	−
Citrate, Simmons	−	+	+	[−]	−	d	d	+
H_2S production	−	d	−	−	−	−	d	+
Phenylalanine deaminase	−	−	−	−	−	−	+	−
Lysine decarboxylase	−[a]	−	[−][b]	[+]	+	[+]	−	+
Motility, 37°C	−	+	+[c]	[+]	[+]	−	+	+
Motility, 25°C	+[d]	+	+[c]	[+]	[+]	−	+	+
KCN, growth	−[a]	d	+[e]	−[f]	+	+	+	−[g]
Malonate utilization	−	d	+[h]	d	d	+[i]	−	d
D-Glucose, gas	− or W	+	+[j]	+	+	[+]	[+]	+
L-Arabinose, acid	+[k]	+	+	+	+	+	−	+
D-Mannitol, acid	+[l]	+	+	+[m]	+	+	−	+
Mucate, acid	−	+	d	d	−	+[n]	−	+[o]

Symbols: see Table 5.2 on p. 222. W, weak.

[a] Except some strains of *Y. ruckeri*.

[b] *E. aerogenes* and *E. gergoviae* are positive.

[c] Except *E. asburiae*.

[d] Except *Y. pestis* and some strains of *Y. ruckeri*.

[e] Except *E. gergoviae* and some strains of *E. agglomerans*.

[f] Except *E. hermannii* and a few strains of *E. vulneris*.

[g] Except *S. bongori* and some strains of *S. choleraesuis* subsp. *houtenae*.

[h] Except *E. asburiae* and *E. sakazakii*.

[i] Except *K. ozaenae*.

[j] Except *E. agglomerans*.

[k] Except *Y. ruckeri* and some strains of *Y. aldovae, Y. kristensenii*, and *Y. pseudotuberculosis*.

[l] Except some strains of *Y. aldovae*.

[m] Except *E. blattae*.

[n] Except *K. ozaenae* and *K. rhinoscleromatis*.

[o] Except *S. choleraesuis* subspecies *diarizonae* and *houtenae*.

Table 5.41   Reactions of yersiniae at 25–28°C and 37°C .

Test	Y. aldovae	Y. bercovieri	Y. enterocolitica	Y. frederiksenii	Y. intermedia	Y. kristensenii	Y. mollaretii	Y. pestis	Y. pseudotuberculosis	Y. rohdei	Y. ruckeri
Voges-Proskauer											
37°C	–	–	–	–	–	–	–	–	–	–	–
25–28°C	+	–	+	+	+	–	–	–	–	–	–
Citrate, Simmons											
37°C	–	–	–	[–]	–	–	–	–	–	–	–
25–28°C	d	–	–	d	–	–	–	–	–	[+]	–
Urease											
37°C	[+]	d	[+]	[+]	[+]	[+]	[–]	–	+	d	–
25–28°C	+	+	+	+	+	+	+	–	+	d	–
Ornithine decarboxylase											
37°C	d	[+]	+	+	+	+	[+]	–	–	[–]	+
25–28°C	+	+	+	+	+	+	+	–	–	[+]	+
Motility											
37°C	–	–	–	–	–	–	–	–	–	–	–
25–28°C	+	+	+	+	+	+	+	–	+	+	[+]
Glycerol, acid											
37°C	–	–	+	[+]	d	d	[–]	d	d	d	d
25–28°C	+	[+]	+	+	+	+	+	d	+	[+]	d
myo-Inositol, acid											
37°C	–	–	d	[–]	[–]	[–]	–	–	–	–	–
25–28°C	+	d	d	+	[+]	d	d	–	–	–	–
Melibiose, acid											
37°C	–	–	–	–	[+]	–	–	[–]	d	d	–
25–28°C	–	–	–	–	+	–	–	d	+	d	–
Raffinose, acid											
37°C	–	–	–	d	d	–	–	–	[–]	d	–
25–28°C	–	–	–	–	+	–	–	–	[–]	d	–
L-Rhamnose, acid											
37°C	–	–	–	+	+	–	–	–	d	–	–
25–28°C	+	–	–	+	+	–	–	–	+	–	–
Salicin, acid											
37°C	–	[–]	[–]	+	+	[–]	[–]	d	[–]	–	–
25–28°C	+	–	[–]	+	+	–	[–]	[+]	d	–	–
D-Xylose, acid											
37°C	d	+	d	+	+	[+]	d	+	+	d	–
25–28°C	+	+	d	+	+	+	+	+	+	+	–
Mucate, acid											
37°C	–	–	–	–	–	–	–	–	–	–	–
25–28°C	d	+	–	[–]	d	–	+	–	–	–	–
Esculin hydrolysis											
37°C	–	[–]	[–]	[+]	[–]	–	–	d	+	–	–
25–28°C	+	[+]	d	+	+	–	[–]	+	+	–	–

Symbols: see Table 5.2 on p. 222.

Table 5.42 Differentiation of *Yersinia* species

Test	*Y. aldovae*	*Y. bercovieri*	*Y. enterocolitica*	*Y. frederiksenii*	*Y. intermedia*	*Y. kristensenii*	*Y. mollaretii*	*Y. pestis*	*Y. pseudotuberculosis*	*Y. rohdei*	*Y. ruckeri*
Indole production	–	–	d	+	+	d	–	–	–	–	–
Voges-Proskauer	+	–	[+]	[+]	+	–	–	–	–	–	d
Citrate, Simmons	d	–	–	d	+	–	–	–	–	d	+
Urease	+	+	+	+	+	+	+	–	+	d	–
Lysine decarboxylase	–	–	–	–	–	–	–	–	–	–	+
Ornithine decarboxylase	+	+	+	+	+	+	+	–	–	[+]	+
Motility	+	+	+	+	+	+	+	–	+	+	[+]
Gelatin hydrolysis	–	–	–	–	–	–	–	–	–	–	+
Cellobiose, acid	–	+	+	+	+	+	+	–	–	+	–
Melibiose, acid	–	–	–	–	+	–	–	d	+	d	–
α-CH₂-D-glucoside, acid	–	–	–	–	+	–	–	–	–	–	–
Raffinose, acid	–	–	–	–	+	–	–	–	d	d	–
L-Rhamnose, acid	+	–	–	+	+	–	–	+	+	–	–
D-Sorbitol, acid	+	+	+	+	+	+	+	–	–	+	–
Sucrose, acid	–	+	+	+	+	–	+	–	–	+	–
Mucate, acid	d	+	–	d	d	–	+	[a]	–	–	–
L-Fucose, acid	d	+	d	+	d	d	–		–		
L-Sorbose, acid	–	–	d	+	+	+	+		–		
Pyrazinamidase	+	+	d	+	+	+	+	–	–	+	
β-Xylosidase	–	–	–	d	–	–	–	+	+		
τ-Glutamyl transferase	+		+	+	+	+		–	d	+	+

Symbols: see Table 5.2 on p. 222. Incubation at 25–28°C.
[a] Data not available.

251

Table 5.43   Differentiation of the biotypes of *Yersinia enterocolitica*

Test	Biotype				
	1	2	3	4	5
Indole production	+	+	−	−	−
Sucrose, acid	+	+	+	+	d
Trehalose, acid	+	+	+	+	−
ᴅ-Xylose, acid	+	+	+	−	−
Deoxyribonuclease	−	−	−	+	+
Lipase (Tween 80)	+	−	−	−	−
Nitrate reduction	+	+	+	+	−

Symbols: see Table 5.2 on p. 222. Tests done at 28°C.

Table 5.44   Differentiation of *Yokenella* from *Hafnia*

Test	*Yokenella*	*Hafnia*
Methyl red	+	d
Voges-Proskauer	−	[+]
Citrate, Simmons	+	−
Glycerol, acid	−	+
Cellobiose, acid	+	[−]
Melibiose, acid	+	−
Susceptibility to *Hafnia*-specific bacteriophage	−	+
Catalase reaction	W	S

Symbols: see Table 5.2 on p. 222. W, weak reaction (15 seconds); S, strong reaction (1 second).

Table 5.45   Delayed positive reactions in *Yokenella regensburgei*

Test	Percent positive at	
	2 days	7 days
Arginine dihydrolase	8	100
Glycerol	0	17
myo-Inositol, acid	0	83
Salicin, acid	8	92
Esculin hydrolysis	67	100
Acetate utilization	25	50

Table 5.46   Differentiation of the genera in *Vibrionaceae* [a]

Characteristic	Vibrio	Photobacterium	Aeromonas	Plesiomonas	Enhydrobacter
Sheathed polar flagella	+	−	−	−	−
Motility	+	+	[+] [e]	+	−
Accumulation of poly-β-hydroxybutyrate coupled with the inability to utilize β-hydroxybutyrate	−	+			
Na⁺ required for or stimulates growth	+	+	−	−	−
Lipase	[+] [b]	D	[+]	−	+
D-Mannitol, utilization	[+] [c]	−	[+]	−	−
Sensitivity to vibriostatic compound 0/129 [d]	+	+	−	+	−
Mol% G+C of DNA	38–51	40–44	57–63	51	66

[a] Symbols: +, all species positive; [+], most species positive; −, all species negative; D, some species positive, some species negative.
[b] *V. nereis*, *V. anguillarum* biovar II, and *V. costicola* are negative.
[c] *V. nereis*, *V. anguillarum* biovar II, and *V. marinus* are negative.
[d] There are some exceptions.
[e] Except *A. media* and *A. salmonicida*.

Table 5.47   Differentiation of *Aeromonas* from *Plesiomonas*

Characteristic	A. salmonicida	Other aeromonads	Plesiomonas
Motility	−	+	+
Monotrichous flagella in liquid medium	−	+	−
Lophotrichous flagella in liquid medium	−	−	+
Coccobacilli in pairs, chains and clumps	+	−	−
Single and paired rods	−	+	+
Brown water-soluble pigment	d	−	−
Growth in nutrient broth at 37°C	−	+	+
Indole production in 1% peptone water	d	+	+
KCN, growth	−	d	−
L-Arabinose, utilization	d	d	−
Salicin, acid	d	d	−
Sucrose, acid	d	+	−
D-Mannitol, acid	d	+	−
myo-Inositol, breakdown	−	−	+
Voges-Proskauer	d	d	−
D-Glucose, gas	d	d	−

Symbols: +, typically positive; −, typically negative; d, differs among species or strains.

Table 5.48 Differentiation of *Aeromonas* species [a]

Characteristic	A. caviae	A. eucrenophila	A. hydrophila	A. media	A. salmonicida subsp. achromogenes	subsp. masoucida	subsp. salmonicida	subsp. smithia	A. schubertii	A. sobria	A. veronii
Indole production	+	+	+	d	+	+	−	−	−	+	+
Methyl red	+	d	+	+	+	+	+	−	+	−	+
Voges-Proskauer	−	−	+	−	−	+	−	−	−	d	+
Citrate, Simmons	d	−	d	d	−	−	−	−	d	−	+
H$_2$S production	−	−	+	−	−	+	−	+	−	−	−
Urea hydrolysis	−	−	−	−	−	−	−	−	−	−	−
Phenylalanine deaminase	−	d	−	d	−	−	−		d	+	[+]
Lysine decarboxylase	−	−	d	−	d	d	d	−	+	+	+
Arginine dihydrolase	+	+	+	+	+	+	+	[−]	+	+	−
Ornithine decarboxylase	−	−	−	−	−	−	−	−	−	−	+
Motility	+	+	+	−	−	−	−	−	+	+	+
Gelatin hydrolysis	+	+	+	+	+	+	+	+	+	+	[+]
KCN, growth	+	+	+	+	−	−	−		−	−	d
Malonate utilization	−	−	−	−	−	−	−		−	−	−
D-Glucose, acid	+	+	+	+	+	+	+	[+]	+	+	+
D-Glucose, gas	−	+	+	−	−	+	+	[+]	−	−	+
Adonitol, acid	−	−	−	−	−	−	−		−	−	−
L-Arabinose, acid	+	+	+	+	−	+	+	+	+	+	−
D-Arabitol, acid	−	−	−	−	−	−	−		−	−	−
Cellobiose, acid	[+]	+	−	+	−	−	−	−	−	d	[+]
Dulcitol, acid	−	−	−	−	−	−	−		−	−	−
Erythritol, acid	−	−	−	−	−	−	−		−	−	−
D-Galactose, acid	+	+	+	+	+	+	+	−	+	+	+
Glycerol, acid	d	+	+	d	d	d	d	[−]	d	d	+
myo-Inositol, acid	−	−	−	−	−	−	−		−	−	−
Lactose, acid	d	−	d	d	−	−	−	−	+	−	−
Maltose, acid	+	+	+	+	+	+	+	−	−	+	+
D-Mannitol, acid	+	+	+	+	−	+	+	−	+	d	+
D-Mannose, acid	d	+	[+]	+	+	+	+		−	+	+
Melibiose, acid	−	−	−	−	−	−	−		−	−	−
α-CH$_2$-D-glucoside, acid	−	−	d	−	−	d	−	−	−	d	[+]
Raffinose, acid	−	−	−	−	−	−	−		−	−	−
L-Rhamnose, acid	−	−	−	−	−	−	−		−	−	−
Salicin, acid	+	+	+	d	d	d	d	[−]	−	d	+
D-Sorbitol, acid	−	−	−	−	−	−	−		−	d	−

Table 5.48 *(continued)*

Characteristic	A. caviae	A. eucrenophila	A. hydrophila	A. media	A. salmonicida subsp. achromogenes	A. salmonicida subsp. masoucida	A. salmonicida subsp. salmonicida	A. salmonicida subsp. smithia	A. schubertii	A. sobria	A. veronii
Sucrose, acid	+	d	+	+	+	+	−	d	−	d	+
Trehalose, acid	+	+	+	+	+	+	+	−	+	d	+
D-Xylose, acid	−	−	−	−	−	−	−	−	−	−	−
Esculin hydrolysis	+	+	+	d	−	+	+	−	−	−	+
Mucate, acid	−	−	−	−	−	−	−	−	−	−	−
Tartrate, Jordan	−	−	−	d	−	−	−	−	−	−	+
Acetate utilization	d	d	d	d	d	+	+	+	d	d	[+]
Lipase (corn oil)	+	−	d	d	+	+	+	−	+	−	+
DNase	+	+	+	+	+	+	+	+	+	+	+
Nitrate reduction	+	+	+	+	+	+	+	+	+	+	+
Oxidase	+	+	+	+	+	+	+	+	+	+	+
ONPG	+	d	+	[+]	d	d	d	+	+	−	[+]
Citrate, Christensen	d	−	−	−	−	−			+	−	+
Tyrosine clearing	d	d	+	d	−	d			+	d	[+]
String test											
0% NaCl, growth	+	+	+	+	+	+	+	+	+	+	+
1% NaCl, growth	+	+	+	+	+	+	+	+	+	+	+
6% NaCl, growth				−				−	−	−	−
0/129 sensitivity	−	−	−			−		−	−	−	−
Brown soluble pigment	−	−	−	+	−	+	+	−	−	−	−

Symbols: see Table 5.2 on p. 222.

a Data are from the publications listed below and were obtained using a number of different methods; incubation temperatures of 20–37°C, and times of incubation from 1–7 d. All of these factors can influence test results. Most *A. salmonicida* strains are nonmotile and do not grow at 37°C, however, there is a motile biogroup that does grow at this temperature. All other species grow at 37°C, although the optimal growth temperature may be from 22–30°C. Some strains of all species are more metabolically active when 1% NaCl is incorporated into the biochemical test medium. Further details are given in *Bergey's Manual of Systematic Bacteriology* and in Allen et al., Int. J. Syst. Bacteriol. *33*: 599–604, 1983; Hickman-Brenner et al., J. Clin. Microbiol. *25*: 900–906, 1987; Schubert and Hegazi, Zbl. Bakt. Hyg. A *268*: 34–39, 1988; Hickman-Brenner et al., J. Clin. Microbiol. *26*: 1561–1564, 1988; Janda and Duffey, Rev. Inf. Dis. *10*: 980–997, 1988; Arduino, M.J. Phenotypic characterization of the genus *Aeromonas*, Ph.D. dissertation, Department of Parasitology and Laboratory Practice, School of Public Health, University of North Carolina, 1989; Austin et al., Syst. Appl. Microbiol. *11*: 277–290, 1989; Kuijper et al., J. Clin. Microbiol. *27*: 132–139, 1989; Altwegg et al., J. Clin. Microbiol. *28*: 258–264, 1990.

Table 5.49A Differentiation of *Photobacterium* species [a]

Characteristic	P. angustum	P. leiognathi	P. phosphoreum
D-Glucose, gas	−	−	+
Luminescence	−	+	+
Growth, 4°C	d [b]	−	+
Growth, 35°C	+	+	−
Gelatinase	d [b]	−	−
Lipase	d	+	−
Acetate utilization	+	+	−
DL-Glycerate utilization	−	d	+
Maltose utilization	d	−	+
L-Proline utilization	−	+	−
Pyruvate utilization	+	+	−
D-Xylose utilization	+	−	−

Symbols: see standard definitions.

[a] Incubated at 15–20°C, unless otherwise indicated.

[b] Positive in 80% of strains.

Important Notes for Users of This Manual

Unless otherwise indicated in footnotes to tables, the meanings of symbols are as follows:

- \+ 90% or more of strains are positive
- − 90% or more of strains are negative
- d 11-89% of strains are positive
- v strain instability (not equivalent to "d")
- D Different reactions in different taxa (species of a genus or genera of a family)

All other symbols are defined in footnotes to tables.

Table 5.49B   Other biochemical characteristics of *Photobacterium* species

Characteristic	P. angustum	P. leiognathi	P. phosphoreum
Nitrate reduction	d	+	+
20°C, growth	+	+	+
30°C, growth	+	+	d
40°C, growth	–	–	–
L-Methionine requirement	–	–	d
Chitinase	d	+	+
Amylase	–	–	–
Alginase	–	–	–
2,3-Butanediol production	–	–	d
N-Acetylglucosamine utilization	+	+	+
L-Alanine utilization	+	d	d
L-Aspartate utilization	d	+	d
Caprate utilization	–	d	–
D-Fructose utilization	+	+	+
Fumarate utilization	+	+	d
D-Galactose utilization	+	+	+
D-Gluconate utilization	+	+	d
D-Glucose utilization	+	+	+
D-Glucuronate utilization	–	–	d
L-Glutamate utilization	–	d	d
DL-Glycerate utilization	–	d	d
Glycerol utilization	+	+	+
DL-Lactate utilization	+	+	d
DL-Malate utilization	–	d	d
D-Mannose utilization	+	+	+
D-Ribose utilization	+	+	d
L-Serine utilization	+	d	d
Succinate utilization	+	+	d
Sucrose utilization	d	–	–
Trehalose utilization	d	–	–
L-Threonine utilization	–	d	d

Symbols: see Table 5.2 on p. 222.

Table 5.50 Biochemical reactions of *Plesiomonas shigelloides*

Characteristic	Reaction
Indole production	+
Methyl red	[+]
Voges-Proskauer	−
Citrate, Simmons	−
H$_2$S production	−
Urea hydrolysis	−
Phenylalanine deaminase	−
Lysine decarboxylase	+
Arginine dihydrolase	+
Ornithine decarboxylase	+
Motility	+
Gelatin hydrolysis	−
KCN, growth	−
Malonate utilization	−
D-Glucose, acid	+
D-Glucose, gas	−
Adonitol, acid	−
L-Arabinose, acid	−
D-Arabitol, acid	−
Cellobiose, acid	−
Dulcitol, acid	−
Erythritol, acid	−
D-Galactose, acid	+
Glycerol, acid	d
myo-Inositol, acid	+
Lactose, acid	+
Maltose, acid	+
D-Mannitol, acid	−
D-Mannose, acid	[−]
Melibiose, acid	−
α-CH$_2$-D-glucoside, acid	−
Raffinose, acid	−
L-Rhamnose, acid	−
Salicin, acid	−
D-Sorbitol, acid	−
Sucrose, acid	−
Trehalose, acid	+
D-Xylose, acid	−
Esculin hydrolysis	−
Mucate, acid	−
Tartrate, Jordan	d
Acetate utilization	[−]
Lipase (corn oil)	−
DNase	−
Nitrate reduction	+
Oxidase	+
ONPG	+
Citrate, Christensen	−

Table 5.50 *(continued)*

Characteristic	Reaction
Tyrosine clearing	−
String test	−
0% NaCl, growth	+
1% NaCl, growth	+
6% NaCl, growth	−
0/129 sensitivity	[+]
Brown soluble pigment	−

Symbols: see Table 5.2 on p. 222.

Table 5.51 **Differentiation of the biovars of *V. cholerae***

Characteristic	Classica l	Eltor
Hemolysis [a]	−	d [b]
Voges-Proskauer [c]	−	+
Hemagglutination [d]	−	+
Sensitivity to:		
Polymyxin B 50 IU [e]	+	−
Classical phage IV [f]	+	−
Eltor phage V [g]	−	+

Symbols: see standard definitions.

[a] Sheep erythrocytes, brain-heart-thioglycolate-cystine agar (Sakazaki et al., Jpn. J. Med. Sci. Biol. *24:* 83-91, 1971).

[b] The first strains of the current pandemic were strongly hemolytic but later strains have been nonhemolytic.

[c] Method 3 of Cowan (*Cowan and Steel's Manual for the Identification of Medical Bacteria*, 2nd Ed., Cambridge University Press, London, 1965.

[d] Chicken erythrocytes, slide test (Finkelstein and Mukerjee, Proc. Soc. Exp. Biol. Med. *112:* 355-359, 1963).

[e] Disk test (Gan and Tija, Amer. J. Hyg. *77:* 184-186, 1963).

[f] Mukerjee et al., Ann. Biochem. Exp. Med. *17:* 161-176, 1957.

[g] Basu and Mukerjee, Experientia *24:* 299-300, 1968.

Table 5.52A Differentiation of *Vibrio* species and biovars

Characteristic	V. aestuarianus	V. alginolyticus	V. (Listonella) anguillarum biovar 1	V. campbellii	V. cholerae	V. cincinnatiensis
3–12 polar flagella	–	–	–	–	–	–
Lateral flagella on solid media		+	–	+	–	–
Swarming	–	+	–	–	–	–
Straight rods [a]	–	+	–	+	d	+
PHB accumulation [b]		–	–	–	–	
Pigment	–	–	–	–	–	–
Arginine dihydrolase [d]	+	–	+	–	–	–
Oxidase	+	+	+	+	+	+
Nitrate reduction		+	+	+	+	+
Luminescence	–	–	–	–	–[e]	–
D-Glucose, gas	–	–	–	–	–	–
Acetoin and/or diacetyl production	d	+	+	–	+	+
Na⁺ required for growth	+	+	+	+	–	+
Organic growth factor requirement		–	–	–	d	
4°C, growth	+	–	–	–	–	–
30°C, growth	+	+	+	+	+	+
35°C, growth	+	+	+	+	+	+
40°C, growth		+	–	–	+	
Amylase	–	+	+	+	+	+
Gelatinase	–	+	+	+	+	–
Lipase		+	+	+	+	–
Alginase	–	–	–	–	–	–
Chitinase		+	+	+	+	+
Utilization of:						
Acetate		+	+	d	+	+
Aconitate		+	+	d	+	
β-Alanine		–	–	–	–	
D-Alanine		+	+	+	d	+
L-Alanine		+	d	d	–	+
γ-Aminobutyrate		–	–	–	–	+
δ-Aminovalerate		–	–	–	–	
L-Arabinose		–	+	–	–	+
L-Arginine		+	–	–	+	+
L-Aspartate		d	+	–	d	+
Butyrate		+	–	–	–	
Caprate		+	d	d	d	
Caproate		–	+	–	+	
Caprylate		+	+	d	+	
Cellobiose		–	+	d	–	+
Citrate	+	+	+	d	d	+
Citrulline		–	–	–	–	+

Footnotes are at end of table

Table 5.52A *(continued)*

Characteristic	V. aestuarianus	V. alginolyticus	V. (Listonella) anguillarum biovar 1	V. campbellii	V. cholerae	V. cincinnatiensis
Ethanol		d	−	−	−	−
D-Galactose		d	d	−	+	
D-Galacturonate		−	−	−	−	
D-Gluconate		+	+	−	+	+
L-Glutamate		+	+	−	+	+
Glutarate		−	−	−	−	
D-Glucuronate		−	−	−	−	−
DL-Glycerate		+	+	d	d	
Glycine		+	−	−	−	
Heptanoate		+	−	−	−	
L-Histidine		+	+	−	d	
p-Hydroxybenzoate		−	−	−	−	
β-Hydroxybutyrate		−	−	−	−	
myo-Inositol		−	d	−	−	
Isobutyrate		d	−	−	−	
α-Ketoglutarate		+	+	+	+	+
DL-Lactate		+	+	+	+	
Lactose		−	−	−	−	−
L-Leucine		+	−	−	−	−
DL-Malate		+	+	d	+	
D-Mannitol		+	+	d	+	+
D-Mannose		d	+	d	d	
Melibiose		−	−	−	−	
L-Ornithine		−	−	−	+	−
Pelargonate		+	−	d	d	
L-Proline		+	+	+	+	+
Propanol		d	−	−	−	
Propionate		+	+	+	+	
Putrescine		d	−	−	−	−
Pyruvate		+	+	+	+	
Quinate		−	−	−	−	
L-Rhamnose		−	−	−	−	
D-Ribose		+	d	+	+	
Salicin		−	−	−	−	+
L-Serine		+	+	d	d	
D-Sorbitol		−	+	−	−	−
Sucrose		+	+	−	+	+
Trehalose		+	+	+	+	+
L-Tyrosine		+	−	d	−	−
Valerate		+	−	−	−	
D-Xylose		−	−	−	−	+

Footnotes are at end of table

Bergey's Manual of Determinative Bacteriology-9

Table 5.52A   *(continued)*

Characteristic	V. costicola	V. (Listonella) damsela	V. diazotrophicus	V. fischeri	V. fluvialis biovar 1	V. fluvialis biovar 2	V. furnissii
3–12 polar flagella	−		−	+	−	−	
Lateral flagella on solid media	−		+	−	d	d	
Swarming	−	−	−	−	−	−	−
Straight rods [a]	−	+	−	+	d	−	−
PHB accumulation [b]	−			−	−	d	
Pigment	−	−	−	Y	−	−	−
Arginine dihydrolase [d]	+	+	+	−	+	+	+
Oxidase	+	+	+	+	+	+	+
Nitrate reduction	d	+	+	d	+	+	+
Luminescence	−	−	−	+	−	−	−
D-Glucose, gas	−	+	−	−	−	+	+
Acetoin and/or diacetyl production	−	+	−	−	−	−	−
Na⁺ required for growth	+	+	+	+	+	+	−
Organic growth factor requirement	d			−	−	−	
4°C, growth	−	−	+	−	−	−	−
30°C, growth	+	+	+	+	+	+	+
35°C, growth	+	+	+	d	+	+	+
40°C, growth	−		d	−	+	+	
Amylase	−		+	−	+	d	
Gelatinase	−	−	−	−	+	d	+
Lipase	−	−	−	+	+	+	d
Alginase	−	−		−	−	−	−
Chitinase	−	+	−	d	+	+	+
Utilization of:							
Acetate	+	−	+	−	+	+	d
Aconitate	−			d	+	+	
β-Alanine	−			−	−	−	
D-Alanine	+		+	−	+	+	
L-Alanine	+		+	−	+	+	
γ-Aminobutyrate	−	−	+		+	+	+
δ-Aminovalerate	−	−	−	−	−	d	d
L-Arabinose	−	−	+	−	+	+	+
L-Arginine	−		+	−	+	+	
L-Aspartate	−			d	+	+	
Butyrate	−	−	−	−	−	−	−
Caprate	−			−	+	+	
Caproate	−			−	+	+	
Caprylate	−			−	+	+	
Cellobiose	−	−	+	+	d	d	−
Citrate	−	−	+	d	+	+	+
Citrulline	−	−	−	−	−	−	−

Footnotes are at end of table

Table 5.52A *(continued)*

Characteristic	V. costicola	V. (Listonella) damsela	V. diazotrophicus	V. fischeri	V. fluvialis biovar 1	V. fluvialis biovar 2	V. furnissii
Ethanol	−		−	−	+	+	d
D-Galactose	−	+	+	+	+	+	+
D-Galacturonate	−	−	+	−	+	+	+
D-Gluconate	+	−	+	−	+	+	+
L-Glutamate	−		+	d	+	+	
Glutarate	−	−		−	−	+	+
D-Glucuronate	−	−	+	−	+	−	−
DL-Glycerate	−			d	+	+	
Glycine	−		−	−	d	d	
Heptanoate	−			−	+	+	
L-Histidine	+			−	+	d	
p-Hydroxybenzoate	−			−	+	d	
β-Hydroxybutyrate	−	−		−	+	d	+
myo-Inositol	−	−	−	−	−	−	−
Isobutyrate	−			−	−	−	
α-Ketoglutarate	−		+	−	+	+	+
DL-Lactate	+			−	+	+	
Lactose	−	−	+	−	−	−	−
L-Leucine	−	−	−	−	−	−	−
DL-Malate	−			−	+	+	
D-Mannitol	+	−	+	+	+	+	+
D-Mannose	−	+	−	+	+	d	d
Melibiose	−	−	−	−	−	−	−
L-Ornithine	−			−	+	+	
Pelargonate	−		−	−	+	+	−
L-Proline	−		+	+	+	+	
Propanol	−	−	−	−	+	+	+
Propionate	+	−	d	−	+	+	+
Putrescine	−			−	d	+	
Pyruvate	+		+	−	+	+	+
Quinate	−			−	+	d	
L-Rhamnose	−		−	−	−	−	−
D-Ribose	+			+	+	+	
Salicin	−		+	d	+	−	−
L-Serine	+		d	−	+	+	
D-Sorbitol	−	−	−	−	−	−	−
Sucrose	+	−	+	−	+	+	+
Trehalose	+	−	+	d	+	+	+
L-Tyrosine	+			−	+	+	
Valerate	−	−	−	−	−	−	−
D-Xylose	−	−	+	−	−	−	−

Footnotes are at end of table

Table 5.52A *(continued)*

Characteristic	V. gazogenes	V. (Coliver) hadaliensis [g]	V. harveyi	V. hollisae	V. logei	V. marinus	V. mediterranei
3–12 polar flagella	−		−	−	+	−	−
Lateral flagella on solid media	−		+	−	−	−	
Swarming	−		−	−	−	−	−
Straight rods [a]	−	−	+	−	+	+	+
PHB accumulation [b]	−		−		−	−	
Pigment	R	−	−	−	Y	−	−
Arginine dihydrolase [d]	−		−	−	−	−	−
Oxidase	−	+	+	+	+	+	+
Nitrate reduction	−		+	+	+	+	+
Luminescence	−		d		+	−	− [e]
D-Glucose, gas	+		−	−	−	−	−
Acetoin and/or diacetyl production	−		−	−	−	−	−
Na [+] required for growth	+		+	+	+	+	+
Organic growth factor requirement	−		−		d	+	
4°C, growth	−	+	−		+	+	−
30°C, growth	+	−	+	+	−	−	+
35°C, growth	+	−	+	+	−	−	
40°C, growth	+	−	d		−	−	−
Amylase	+		+		−	−	
Gelatinase	+		+	−	−	+	−
Lipase	+		+	−	d	+	+
Alginase	−		d		−	−	
Chitinase	−	+	+		+	+	−
Utilization of:							
Acetate	+		+	−	−	+	
Aconitate			+		−	−	
β-Alanine	−		−		−	−	
D-Alanine	−		+		−	+	
L-Alanine			d		−	−	+
γ-Aminobutyrate	−		−		−	−	−
δ-Aminovalerate	−		−		−	−	
L-Arabinose	+		d		−	−	
L-Arginine	−		d		−	−	
L-Aspartate	+		d		+	+	+
Butyrate	+		−		−	−	
Caprate			d		−	+	
Caproate	−		d		−	−	
Caprylate			d		−	−	
Cellobiose	+		+		+	−	+
Citrate	+		+	−	−	−	
Citrulline	−		d		−	−	

Footnotes are at end of table

Table 5.52A   *(continued)*

Characteristic	V. gazogenes	V. (Coliver) hadaliensis[g]	V. harveyi	V. hollisae	V. logei	V. marinus	V. mediterranei
Ethanol	–		–		–	–	
D-Galactose	+		d		+	+	+
D-Galacturonate	–		–		–	–	
D-Gluconate	–		+		+	+	–
L-Glutamate	+		+		+	+	+
Glutarate			–		–	–	
D-Glucuronate	–		+		–	–	
DL-Glycerate			+		–	–	–
Glycine	–		d		–	–	–
Heptanoate	–		+		–	–	
L-Histidine			d		–	–	
p-Hydroxybenzoate	–		–		–	–	
β-Hydroxybutyrate	–		–		–	–	
myo-Inositol	–		–		–	–	+
Isobutyrate	–		–		–	–	
α-Ketoglutarate	+		+		–	+	
DL-Lactate	+		+		–	+	+
Lactose	–		–[f]		–	–	+
L-Leucine	–		–		–	–	
DL-Malate	+		d		–	+	+
D-Mannitol	+		+		+	–	
D-Mannose	+		+		d	–	
Melibiose			–		–	–	
L-Ornithine	–		–		–	–	–
Pelargonate			d		–	–	
L-Proline	+		+		+	+	+
Propanol	–		d		–	–	
Propionate	+		+		–	–	
Putrescine	–		–		–	–	+
Pyruvate	+		+		–	–	
Quinate			–		–	–	
L-Rhamnose	–		–		–	–	
D-Ribose	+		+		d	+	
Salicin	+		d		–	–	+
L-Serine	+		+		d	+	+
D-Sorbitol	+		–		–	–	+
Sucrose	+		d		–	–	+
Trehalose			+		–	–	
L-Tyrosine	–		+		–	–	
Valerate	–		–		–	–	
D-Xylose	+		–		–	–	

Footnotes are at end of table

265

Table 5.52A *(continued)*

Characteristic	V. metschnikovii	V. mimicus	V. natriegens	V. nereis	V. nigripulchritudo	V. ordalii	V. orientalis
3–12 polar flagella	–	–	–	–	–	–	–
Lateral flagella on solid media	–		–	–	–	–	–
Swarming	–	–	–	–	–	–	–
Straight rods [a]	d	–	+	+	+ [c]	d	–
PHB accumulation [b]	–		+	+	d	–	+
Pigment	–	–	–	–	B	–	–
Arginine dihydrolase [d]	–	–	–	+	–	–	+
Oxidase	–	+	+	+	+	+	+
Nitrate reduction	–	+	+	+	+	–	+
Luminescence	–	–	–	–	–	–	+
D-Glucose, gas	–	–	–	–	–	–	
Acetoin and/or diacetyl production	+	–	–	–	–	–	–
Na$^+$ required for growth	d	–	+	+	+	+	+
Organic growth factor requirement	d		–	–	–	+	d
4°C, growth	–		–	d	–	–	+
30°C, growth	+	+	+	+	+	–	+
35°C, growth	+	+	+	+	–	–	+
40°C, growth	+		+	d	–	–	–
Amylase	+		d	–	+	–	+
Gelatinase	+		d	d	+	+	+
Lipase	+	–	+	–	+	–	+
Alginase	–		–	–	–	–	–
Chitinase	+		–	d	+	d	+
Utilization of:							
Acetate	+	d	+	d	+	+	+
Aconitate	d		+	+	+	d	+
β-Alanine	–		d	–	d	–	
D-Alanine	d		+	+	+	–	+
L-Alanine	+		+	+	+	–	+
γ-Aminobutyrate	–		+	+	–	–	
δ-Aminovalerate	–		+	+	–	–	
L-Arabinose	–		+	–	–	–	–
L-Arginine	+		+	+	–	–	
L-Aspartate	+		d	d	+	+	+
Butyrate	–		+	+	–	–	
Caprate	d		+	+	–	–	d
Caproate	+		d	d	+	–	–
Caprylate	d		+	d	d	–	–
Cellobiose	–		d	–	+	–	+
Citrate	d	+	+	+	+	+	+
Citrulline	–		+	+	–	–	

Footnotes are at end of table

Table 5.52A *(continued)*

Characteristic	V. metschnikovii	V. mimicus	V. natriegens	V. nereis	V. nigripulchritudo	V. ordalii	V. orientalis
Ethanol	−		+	+	d	−	
D-Galactose	d		+	−	+	−	+
D-Galacturonate	−		−	−	−	−	
D-Gluconate	+		+	+	d	−	+
L-Glutamate	+		+	+	+	+	+
Glutarate	−		+	+	−		−
D-Glucuronate	−		d	−	+	−	−
DL-Glycerate	−		+	−	+	d	+
Glycine	−		+	+	+	−	d
Heptanoate	d		+	d	+	−	−
L-Histidine	d		+	d	+	d	+
p-Hydroxybenzoate	−		+	−	−	−	
β-Hydroxybutyrate	−		+	+	+	−	+
myo-Inositol	d		d	−	+		
Isobutyrate	−		+	+	−	−	
α-Ketoglutarate	−		+	+	+		−
DL-Lactate	+		+	+	+	−	+
Lactose	d		−	−	+	−	
L-Leucine	−		d	+	−	−	−
DL-Malate	+		+	+	+	−	+
D-Mannitol	+		+	−	+	−	+
D-Mannose	d		d	−	+	−	+
Melibiose	−		d	−	+	−	
L-Ornithine	+		+	+	−		−
Pelargonate	−		+	d	−	−	−
L-Proline	+		+	+	+	+	+
Propanol	−		+	+	d	−	
Propionate	−		+	+	+	−	−
Putrescine	−		+	+	−	−	+
Pyruvate	+		+	+	+	+	+
Quinate	−		+	−	−	−	−
L-Rhamnose	−		+	−	−	−	
D-Ribose	+		+	+	−	+	+
Salicin	−		+	−	−	−	−
L-Serine	+		+	d	+	−	d
D-Sorbitol	d		−	−	+	−	
Sucrose	+		+	+	−	+	+
Trehalose	+		+	+	+	−	+
L-Tyrosine	−		d	d	−	−	−
Valerate	−		+	+	−		−
D-Xylose	−		−	−	−	−	

Footnotes are at end of table

Table 5.52A   *(continued)*

Characteristic	V. parahaemolyticus	V. pelagius (Listonella pelagia)		V. proteolyticus	V. (Colwellia) psychroerythrus
		biovar 1	biovar 2		
3–12 polar flagella	−	−	−	−	−
Lateral flagella on solid media	+	−	−	+	
Swarming	−	−	−	+	
Straight rods [a]	+	+	−	+	−
PHB accumulation [b]	−	−	−	−	
Pigment	−	−	−	−	R
Arginine dihydrolase [d]	−	−	−	+	
Oxidase	+	+	+	+	+
Nitrate reduction	+	+	+	+	+
Luminescence	−	−	−	−	
D-Glucose, gas	−	−	−	−	−
Acetoin and/or diacetyl production	−	−	−	+	
Na+ required for growth	+	+	+	+	+
Organic growth factor requirement	−	−	−	−	
4°C, growth	−	d	d	−	+
30°C, growth	+	+	+	+	−
35°C, growth	+	+	+	+	−
40°C, growth	+	−	−	+	−
Amylase	+	−	+	+	+
Gelatinase	+	−	+	+	
Lipase	+	+	+	+	
Alginase	−	+	+	−	
Chitinase	+	d	+	+	
Utilization of:					
Acetate	+	+	+	−	
Aconitate	+	+	+	+	
β-Alanine	−	−	d	+	
D-Alanine	+	+	+	+	
L-Alanine	d	+	+	+	
γ-Aminobutyrate	−	−	+	−	
δ-Aminovalerate	−	−	+	−	
L-Arabinose	d	−	−	−	
L-Arginine	+	+	+	+	
L-Aspartate	+	d	+	+	
Butyrate	d	−	−	−	
Caprate	+	+	+	+	
Caproate	d	−	−	−	
Caprylate	+	−	−	+	
Cellobiose	−	−	−	−	
Citrate	+	+	+	+	−
Citrulline	−	+	+	−	

Footnotes are at end of table

Table 5.52A *(continued)*

Characteristic	V. parahaemolyticus	V. pelagius (Listonella pelagia)		V. proteolyticus	V. (Colwellia) psychroerythrus
		biovar 1	biovar 2		
Ethanol	+	d	–	–	
D-Galactose	+	+	+	–	
D-Galacturonate	–	–	–	–	
D-Gluconate	+	+	+	+	
L-Glutamate	+	+	+	+	
Glutarate	–	–	d	–	
D-Glucuronate	d	–	–	–	
DL-Glycerate	+	–	–	+	
Glycine	d	+	+	+	
Heptanoate	+	–	–	–	
L-Histidine	+	d	d	+	
p-Hydroxybenzoate	–	–	–	–	
β-Hydroxybutyrate	–	–	–	–	
myo-Inositol	–	–	–	–	
Isobutyrate	–	–	–	–	
α-Ketoglutarate	+	–	–	+	
DL-Lactate	+	+	+	+	
Lactose	–	d [f]	d [f]	–	
L-Leucine	+	–	–	–	
DL-Malate	+	d	–	+	
D-Mannitol	+	+	+	+	
D-Mannose	+	d	+	+	
Melibiose	–	–	–	–	
L-Ornithine	–	d	+	+	
Pelargonate	+	–	–	+	
L-Proline	+	+	+	+	
Propanol	+	d	d	–	
Propionate	+	+	+	–	
Putrescine	+	+	+	+	
Pyruvate	+	+	+	+	
Quinate	–	–	d	–	
L-Rhamnose	–	–	–	–	
D-Ribose	+	+	+	+	
Salicin	–	–	–	–	
L-Serine	+	+	+	+	
D-Sorbitol	–	–	–	+	
Sucrose	–	+	d	–	
Trehalose	+	+	+	+	
L-Tyrosine	+	d	–	–	
Valerate	–	–	–	–	
D-Xylose	–	–	–	–	

Footnotes are at end of table

Table 5.52A *(continued)*

Characteristic	*V. salmonicida*	*V. splendidus* biovar 1	*V. splendidus* biovar 2	*V. tubiashii*	*V. vulnificus*
3–12 polar flagella	+	–	–	–	–
Lateral flagella on solid media	–	–	d	+	–
Swarming	–	–	–	–	–
Straight rods [a]	–	–	d	–	–
PHB accumulation [b]		–	–		–
Pigment	–	–	–	–	–
Arginine dihydrolase [d]	–	+	–	d	–
Oxidase	+	+	+	+	+
Nitrate reduction	–	+	+	+	+
Luminescence	–	+	–	–	–
D-Glucose, gas	–	–	–	–	–
Acetoin and/or diacetyl production	–	–	–	–	–
Na$^+$ required for growth	+	+	+	+	+
Organic growth factor requirement		–	–	–	–
4°C, growth	+	d	–	–	–
30°C, growth	–	+	+	+	+
35°C, growth	–	d	–	d	+
40°C, growth	–	–	–		+
Amylase		+	+	+	+
Gelatinase	–	+	+	+	+
Lipase	–	+	+	+	+
Alginase		d	–	–	–
Chitinase	–	+	+	+	+
Utilization of:					
Acetate		d	+		+
Aconitate		+	+		+
β-Alanine		–	–		–
D-Alanine		+	+	d	+
L-Alanine		+	d		+
γ-Aminobutyrate		–	–	d	–
δ-Aminovalerate		–	–	d	–
L-Arabinose	–	–	–	–	–
L-Arginine		+	–		+
L-Aspartate		+	d		+
Butyrate		–	–	–	–
Caprate		d	+		+
Caproate		–	–		–
Caprylate		–	–		+
Cellobiose	–	+	d	+	+
Citrate		+	+	+	+
Citrulline		+	–	d	–

Footnotes are at end of table

Table 5.52A *(continued)*

Characteristic	V. salmonicida	V. splendidus biovar 1	V. splendidus biovar 2	V. tubiashii	V. vulnificus
Ethanol		−	−	−	−
D-Galactose	d	+	−	+	+
D-Galacturonate		−	−	−	−
D-Gluconate	+	d	−	+	+
L-Glutamate		+	d		+
Glutarate		−	−	−	−
D-Glucuronate		+	−	−	+
DL-Glycerate		d	d		+
Glycine		+	d	+	−
Heptanoate		+	−		−
L-Histidine		d	d		d
p-Hydroxybenzoate		−	−		−
β-Hydroxybutyrate		−	−	d	−
myo-Inositol	−	−	−	−	−
Isobutyrate		−	−		−
α-Ketoglutarate		+	+	d	+
DL-Lactate		+	+		+
Lactose	−	−	−	d	d[f]
L-Leucine		−	−	−	−
DL-Malate		+	d		d
D-Mannitol	d	+	+	+	d
D-Mannose	d	+	−	+	+
Melibiose	−	−	−	d	−
L-Ornithine		−	−		−
Pelargonate		−	−	−	−
L-Proline		+	+		+
Propanol		−	−	−	−
Propionate		+	+	−	+
Putrescine		−	−	d	−
Pyruvate		+	+	+	+
Quinate		d	−		−
L-Rhamnose	−	−	−	−	−
D-Ribose	d	+	+		+
Salicin	−	−	−		−
L-Serine		+	d		−
D-Sorbitol	−	−	−		−
Sucrose	−	d	−	+	
Trehalose	+	+	+	+	+
L-Tyrosine		d	d		+
Valerate		−	−	−	−
D-Xylose	−	−	−	−	−

Footnotes are at end of table

271

Table 5.52A   *(continued)*

Symbols: see standard definitions (except for footnote [a] below). Incubation temperature and duration varies among species. Y, yellow-orange; B, blue-black; R, red.

[a]   +, straight rods; −, curved rods.

[b]   Polyhydroxybutyrate.

[c]   Straight rods in exponential phase of growth becoming curved in stationary phase.

[d]   Determined by the anaerobic production of ornithine from arginine.

[e]   Luminous strains of this species have been found.

[f]   Wild-type strains are unable to utilize lactose; many strains may readily acquire this property by mutation.

[g]   This organism has been isolated once. It requires hydrostatic pressure of at least 300 atmospheres for growth.

Important Notes for Users of This Manual

Unless otherwise indicated in footnotes to tables, the meanings of symbols are as follows:

+   90% or more of strains are positive

−   90% or more of strains are negative

d   11-89% of strains are positive

v   strain instability (not equivalent to "d")

D   Different reactions in different taxa (species of a genus or genera of a family)

All other symbols are defined in footnotes to tables.

Table 5.52B   Differentiation of *Vibrio* species found in clinical specimens [a]

Characteristic	V. alginolyticus	V. cholerae	V. cincinnatiensis	V. damsela	V. fluvialis	V. furnissii	V. harveyi [c]	V. hollisae	V. metschnikovii	V. mimicus	V. parahaemolyticus	V. vulnificus
Indole production [b]	[+]	+	−	−	[−]	[−]	+	+	[−]	+	+	+
Methyl red [b]	d	+	+	+	+	+	+	−	+	+	[+]	[+]
Voges-Proskauer [b]	+	d	−	+	−	−	d	−	+	−	−	−
Citrate, Simmons	−	+	[−]	−	+	+	−	−	d	+	−	d
H₂S production	−	−	−	−	−	−	−	−	−	−	−	−
Urea hydrolysis	−	−	−	−	−	−	−	−	−	−	[−]	−
Phenylalanine deaminase	−	−	−	−	−	−	d	−	−	−	−	d
Lysine decarboxylase [b]	+	+	+	d	−	−	+	−	d	+	+	+
Arginine dihydrolase [b]	−	−	−	+	+	+	−	−	d	−	−	−
Ornithine decarboxylase [b]	d	+	−	−	−	−	−	−	−	+	+	d
Motility	+	+	+	d	d	[+]	−	−	d	+	+	+
Gelatin hydrolysis [b] (22°C)	+	+	−	−	[+]	[+]	−	−	d	d	+	d
KCN, growth	[−]	−	−	−	d	[+]	−	−	−	−	[−]	−
Malonate utilization	−	−	−	−	−	[−]	−	−	−	−	−	−
D-Glucose, acid	+	+	+	+	+	+	d	+	+	+	+	+
D-Glucose, gas	−	−	−	−	−	+	−	−	−	−	−	−
Adonitol, acid	−	−	−	−	−	−	−	−	−	−	−	−
L-Arabinose, acid	−	−	+	−	+	+	−	+	−	−	[+]	−
D-Arabitol, acid	−	−	−	−	d	[+]	−	−	−	−	−	−
Cellobiose, acid	−	−	+	−	d	[−]	d	−	−	−	−	+
Dulcitol, acid	−	−	−	−	−	−	−	−	−	−	−	−
Erythritol, acid	−	−	−	−	−	−	−	−	−	−	−	−
D-Galactose, acid	[−]	+	+	+	+	+	−	+	d	[+]	+	+
Glycerol, acid	[+]	d	+	−	−	d	−	−	+	[−]	d	−
myo-Inositol, acid	−	−	+	−	−	−	−	−	d	−	−	−
Lactose, acid	−	−	−	−	−	−	−	−	d	[−]	−	[+]
Maltose, acid	+	+	+	+	+	+	+	−	+	+	+	+
D-Mannitol, acid	+	+	+	−	+	+	d	−	+	+	+	d
D-Mannose, acid	+	[+]	+	+	+	+	d	+	+	+	+	+
Melibiose, acid	−	−	−	−	[−]	−	−	−	−	−	−	d
α-CH₂-D-glucoside, acid	−	−	d	−	−	−	−	−	[−]	−	−	−
Raffinose, acid	−	−	−	−	[−]	−	−	−	−	−	−	−
L-Rhamnose, acid	−	−	−	−	−	d	−	−	−	−	−	−
Salicin, acid	−	−	+	−	−	−	−	−	−	−	−	+
D-Sorbitol, acid	−	−	−	−	−	−	−	−	d	−	−	−
Sucrose, acid	+	+	+	−	+	+	d	−	+	−	−	[−]
Trehalose, acid	+	+	+	[+]	+	+	d	−	+	+	+	+
D-Xylose, acid	−	−	+	−	−	−	−	−	−	−	−	−
Mucate, acid	−	−	−	−	−	−	−	−	−	−	−	−
Tartrate, Jordans	+	[+]	+	−	d	[−]	d	d	d	[−]	+	[+]
Esculin hydrolysis [b]	−	−	−	−	−	−	−	−	d	−	−	d
Acetate utilization	−	+	[−]	−	d	d	−	−	[−]	[+]	−	−
Nitrate reduction [b]	+	+	+	+	+	+	+	+	−	+	+	+

273

Table 5.52B   *(continued)*

Characteristic	*V. alginolyticus*	*V. cholerae*	*V. cincinnatiensis*	*V. damsela*	*V. fluvialis*	*V. furnissii*	*V. harveyi* [c]	*V. hollisae*	*V. metschnikovii*	*V. mimicus*	*V. parahaemolyticus*	*V. vulnificus*
Oxidase	+	+	+	+	+	+	+	+	−	+	+	+
DNase (25°C)	+	+	[+]	d	+	+	+	−	d	d	+	d
Lipase (corn oil)	[+]	+	d	−	+	[+]	−	−	+	[−]	+	+
ONPG	−	+	[+]	−	d	d	−	−	d	+	−	d
Yellow pigment (25°C)	−	−	−	−	−	−	−	−	−	−	−	−
Tyrosine clearing	d	[−]	−	−	d	d	−	−	−	d	[+]	d
0% NaCl, growth	−	+	−	−	−	−	−	−	−	+	−	−
1% NaCl, growth	+	+	+	+	+	+	+	+	+	+	+	+
6% NaCl, growth	+	d	+	+	+	+	+	[+]	[+]	d	+	d
8% NaCl, growth	+	−	+	−	d	[+]	−	−	d	−	[+]	−
Swarming	+	−	−	−	−	[+]	−	−	−	−	[+]	−
String test	+	+	[+]	[+]	+	+	+	+	+	+	d	+
0/129 sensitivity	[−]	+	[−]	+	d	−	+	d	+	+	[−]	+
Polymyxin B, sensitivity	d	[−]	+	[+]	+	[+]	+	+	+	[+]	d	−

Symbols: see Table 5.2 on p. 222. Results after 48 h at 36°C unless otherwise stated.

[a] Data from Farmer et al., *In* E. H. Lennette, A. Balows, W. J. Hausler, Jr., H. J. Shadomy (Eds.), *Manual of Clinical Microbiology,* 4th ed., American Society of Microbiology, Washington, D.C. 1985, pp. 282-301.

[b] NaCl was added to these media at a final concentration of 1%.

[c] Human isolates may have a somewhat different biotype.

[d] No growth.

Important Notes for Users of This Manual

Unless otherwise indicated in footnotes to tables, the meanings of symbols are as follows:

- +   90% or more of strains are positive
- −   90% or more of strains are negative
- d   11-89% of strains are positive
- v   strain instability (not equivalent to "d")
- D   Different reactions in different taxa (species of a genus or genera of a family)

All other symbols are defined in footnotes to tables.

Table 5.53 Biochemical characteristics of species in the family *Pasteurellaceae*[a]

Characteristic	Actinobacillus					
	(Haemophilus) actinomycetemcomitans	capsulatus	equuli	hominis	lignieresii	muris
Catalase	+	+	d	–	d	+
Oxidase	+	+	(+)	+	(+)	+
Nitrate reduction	+	+	+	+	+	+
ONPG	–	+	d	+	d	–
Phosphatase	+	+	+	+	+	–
Gelatinase	–	–	d	–	–	–
H₂S production	–	–	d	N	+	–
Ornithine decarboxylase	–	–	–	–	–	–
Indole production	–	–	–	–	–	–
Urease	–	+	+	+	+	+
Esculin hydrolysis	–	+L	–	d	–	+L
NAD requirement	–	–	–	–	–	–
X-factor requirement	–	–	–	–	–	N
MacConkey agar, growth[c]	d	+	+	–	+	–
β-Hemolysis, sheep cells	–	–	–	–	–	–
Methyl red	–	–	–	N	–	–
Voges-Proskauer	–	–	–	–	d	–
Lysine decarboxylase	–	–	–	–	–	–
Arginine dihydrolase	–	–	–	–	–	–
Cyclic AMP reaction	–	–	–	–	–	N
α-Fucosidase	N	N	N	N	N	+
D-Glucose, gas production	(+)	–	–	–	–	–
D-Adonitol, acid	–	–	–	–	–	–
L-Arabinose, acid	–	–	(–)	–	(–)	–
Arbutin, acid	–	N	–	N	–	+
Cellobiose, acid	–	+	–	–	–	+L
Dextrin, acid	d	–	+	N	+	–
Dulcitol, acid	–	–	–	–	–	–
meso-Erythritol	N	N	N	N	N	–
Fructose, acid	+	+	+	N	+	+
D-Galactose, acid	+	+	d	+	+	–
D-Glucose, acid	+	+	+	+	+	+
Glycerol, acid	–	N	d	N	d	(–)
myo-Inositol, acid	–	–	–	–	–	dw
Inulin, acid	–	–	–	N	–	–
Lactose, acid	–	+	+	+	dL	–
Maltose, acid	+	+	+	+	+	+
D-Mannitol, acid	(+)	+	+	+	+	+
D-Mannose, acid	+	+	+	–	+	+
Melezitose, acid	N	N	N	N	N	–
Melibiose, acid	–	+	+	+L	–	+L
Raffinose, acid	–	+	+	+	d	+
L-Rhamnose, acid	–	–	–	–	–	–
D-Ribose, acid	–	N	+	N	+	+
Salicin, acid	–	+	–	d	–	+

Footnotes are at end of table

Table 5.53 *(continued)*

| Characteristic | Actinobacillus | | | | | |
	(Haemophilus) *actinomycetemcomitans*	*capsulatus*	*equuli*	*hominis*	*lignieresii*	*muris*
D-Sorbitol, acid	(−)[d]	+	d	−	(−)	−
L-Sorbose, acid	−	−	−	−	−	−
Starch, acid	d	−	d	N	−	N
Sucrose, acid	−	+	+	+	+	+
Trehalose, acid	−	+	+	+	−	+
D-Xylose, acid	d	+	+	+	+	−
Ubiquinones present	−	+	d	N	+	N
Naphthoquinones present	+	+	+	N	+	N

Table 5.53 *(continued)*

| Characteristic | Actinobacillus | | | | | Haemophilus | |
	(Haemophilus) *pleuropneumoniae*	*rossii*	*seminis*	*suis*	*(Pasteurella)* *ureae*	*"agni"*	*aphrophilus*
Catalase	d	+	+	+	d	−	−
Oxidase	d	+	d	(+)	(+)	+	−
Nitrate reduction	+	+	+	+	+	+	+
ONPG	+	(+)	−[b]	(+)	−	N	+
Phosphatase	+	+	−	+	+	−	+
Gelatinase	N	N	N	−	−	N	N
H₂S production	+	N	N	−	N	N	−
Ornithine decarboxylase	−	−	d	−	−	+	−
Indole production	−	−	−	−	−	+	−
Urease	+	+	−	+	+	d	−
Esculin hydrolysis	d	−	d	+	−	−	−
NAD requirement	+	−	−	−	−	−	−
X-factor requirement	−	N	N	−	−	−	−
MacConkey agar, growth[c]	−	(+)	−	+	d	−	(−)
β-Hemolysis, sheep cells	+	d	−	+	−	−	−
Methyl red	N	N	N	−	−	N	N
Voges-Proskauer	−	N	−	−	−	N	N
Lysine decarboxylase	−	N	−	−	−	−	−
Arginine dihydrolase	−	N	−	−	−	−	−
Cyclic AMP reaction	+	N	N	−	N	−	−
α-Fucosidase	−	N	N	N	−	N	−
D-Glucose, gas production	−	d	−	−	−	−	+
D-Adonitol, acid	−	N	−	−	−	−	−
L-Arabinose, acid	−	+	dL	(+)	−	+w	−
Arbutin, acid	N	−	−	+	−	N	N
Cellobiose, acid	−	−	−	+	−	−	−
Dextrin, acid	N	N	−	+	+	N	N
Dulcitol, acid	−	−	−	−	−	−	−
meso-Erythritol	−	N	N	N	−	N	−

Footnotes are at end of table

Table 5.53 *(continued)*

Characteristic	Actinobacillus					Haemophilus	
	(Haemophilus) pleuropneumoniae	rossii	seminis	suis	(Pasteurella) ureae	"agni"	aphrophilus
Fructose, acid	+	N	−	+	+	+	+
D-Galactose, acid	+w	+	dL	(+)	−	+w	+
D-Glucose, acid	+	+	wL	+	+	+	+
Glycerol, acid	−	N	−	+	−	N	d
myo-Inositol, acid	−	+	d	−	−	−	d
Inulin, acid	−	N	−	−	−	N	−
Lactose, acid	d	d	−	+	−	−	+
Maltose, acid	+	(−)	dL	+	(+)	+w	+
D-Mannitol, acid	+	+	dL	−	+	+w	−
D-Mannose, acid	+	d	−	+	d	−	+
Melezitose, acid	+	N	N	N	−	N	−
Melibiose, acid	−	−	−	+	−	N	−
Raffinose, acid	d	(−)	−	+	−	−	+
L-Rhamnose, acid	−	N	N	−	−	−	−
D-Ribose, acid	N	N	N	N	+	N	+
Salicin, acid	−	−	−	+	−	−	−
D-Sorbitol, acid	−	+	−	−	(−)	−	−
L-Sorbose, acid	−	N	N	−	−	−	−
Starch, acid	N	N	−	d	N	N	N
Sucrose, acid	+	−	−	+	+	−	+
Trehalose, acid	−	−	−	+	−	−	+
D-Xylose, acid	+	+	−	+	−	+w	−
Ubiquinones present	N	N	+	+	+	−	−
Naphthoquinones present	N	N	+	+	+	+	+

Table 5.53 *(continued)*

Characteristic	Haemophilus					
	ducreyi	haemoglobinophilus	haemolyticus	influenzae	paracuniculus	paragallinarum
Catalase	−	+	+	+	+	−
Oxidase	+	+	+	+	+	−
Nitrate reduction	+	+	+	+	+	+
ONPG	−	d	−	−	+	+
Phosphatase	+	−	+	+	+	+
Gelatinase	N	−	N	N	N	N
H₂S production	−	d	+	−	−	N
Ornithine decarboxylase	−	−	−	(+)	+	−
Indole production	−	+	d	d	+	−
Urease	−	−	+	(+)	+	−
Esculin hydrolysis	−	−	−	−	−	−
NAD requirement	−	−	+	+	+	+
X-factor requirement	+	+	+	+	−	−

Footnotes are at end of table

Table 5.53 *(continued)*

Characteristic	Haemophilus					
	ducreyi	*haemoglobinophilus*	*haemolyticus*	*influenzae*	*paracuniculus*	*paragallinarum*
MacConkey agar, growth [c]	–	–	–	–	–	–
β-Hemolysis, sheep cells	dL	–	+	–	–	–
Methyl red	N	N	N	N	N	N
Voges-Proskauer	N	N	N	N	N	N
Lysine decarboxylase	–	–	–	d	N	–
Arginine dihydrolase	–	–	–	–	+	–
Cyclic AMP reaction	–	–	–	–	–	–
α-Fucosidase	–	–	–	–	N	–
D-Glucose, gas production	–	–	dL	–	–	–
D-Adonitol, acid	–	–	–	–	–	–
L-Arabinose, acid	–	–	–	–	N	–
Arbutin, acid	N	N	N	N	N	N
Cellobiose, acid	–	–	–	–	–	–
Dextrin, acid	N	N	N	N	N	N
Dulcitol, acid	–	–	–	–	–	–
meso-Erythritol	–	–	–	–	–	–
Fructose, acid	–	–	+w	–	+	+
D-Galactose, acid	–	–	+	+	–	–
D-Glucose, acid	–	+	+	+	+	+
Glycerol, acid	–	–	–	–	–	–
myo-Inositol, acid	–	–	–	–	–	–
Inulin, acid	–	–	–	–	–	–
Lactose, acid	–	–	–	–	–	–
Maltose, acid	–	+	+	+	+	+
D-Mannitol, acid	–	+	–	–	–	+
D-Mannose, acid	–	+	–	–	N	+
Melezitose, acid	–	–	–	–	–	–
Melibiose, acid	–	–	–	–	–	–
Raffinose, acid	–	–	–	–	–	–
L-Rhamnose, acid	–	–	–	–	N	–
D-Ribose, acid	–	d	+	+	N	+
Salicin, acid	–	–	–	–	–	–
D-Sorbitol, acid	–	–	–	–	–	+
L-Sorbose, acid	–	–	–	–	–	–
Starch, acid	N	N	N	N	N	N
Sucrose, acid	–	+	–	–	+	+
Trehalose, acid	–	–	–	–	–	–
D-Xylose, acid	–	+	d	+	–	d
Ubiquinones present	–	+	+	–	+	+
Naphthoquinones present	+	+	+	+	+	+

Footnotes are at end of table

Table 5.53 *(continued)*

| Characteristic | Haemophilus | | | | |
	parahaemolyticus	*parainfluenzae*	*paraphrohaemolyticus*	*paraphrophilus*	*parasuis*
Catalase	d	d	+	−	+
Oxidase	+	+	+	+	−
Nitrate reduction	+	+	+	+	+
ONPG	d	d	d	+	d
Phosphatase	+	+	+	+	+
Gelatinase	N	N	N	N	N
H$_2$S production	+	+	+	+	d
Ornithine decarboxylase	d	d	−	−	−
Indole production	−	−	−	−	−
Urease	+	d	+	−	−
Esculin hydrolysis	−	−	−	−	−
NAD requirement	+	+	+	+	+
X-factor requirement	−	−	−	−	−
MacConkey agar, growth [c]	−	−	−	−	−
β-Hemolysis, sheep cells	+	−	+	−	−
Methyl red	N	N	N	N	N
Voges-Proskauer	N	N	N	N	N
Lysine decarboxylase	−	d	−	−	−
Arginine dihydrolase	−	−	−	−	−
Cyclic AMP reaction	−	−	−	−	−
α-Fucosidase	−	−	d	−	+
D-Glucose, gas production	dL	[d]	−	+	−
D-Adonitol, acid	−	−	−	−	−
L-Arabinose, acid	−	−	−	−	−
Arbutin, acid	N	N	N	N	N
Cellobiose, acid	−	−	−	−	−
Dextrin, acid	N	N	N	N	N
Dulcitol, acid	−	−	−	−	−
meso-Erythritol	−	−	−	−	−
Fructose, acid	+	+	+	+	+
D-Galactose, acid	d	+	d	−	+
D-Glucose, acid	+	+	+	+	+
Glycerol, acid	−	−	−	−	−
myo-Inositol, acid	−	−	−	−	d
Inulin, acid	−	−	−	−	+
Lactose, acid	−	−	−	+	d
Maltose, acid	+	+	+	+	+
D-Mannitol, acid	−	−	−	−	−
D-Mannose, acid	−	+	−	+	+
Melezitose, acid	−	−	−	+	−
Melibiose, acid	−	−	−	+	−
Raffinose, acid	−	−	−	−	−
L-Rhamnose, acid	−	−	−	−	−
D-Ribose, acid	−	−	−	+	+
Salicin, acid	−	−	−	−	−

Footnotes are at end of table

Table 5.53 *(continued)*

| Characteristic | Haemophilus | | | | |
	parahaemolyticus	*parainfluenzae*	*paraphrohaemolyticus*	*paraphrophilus*	*parasuis*
D-Sorbitol, acid	−	−	−	−	−
L-Sorbose, acid	−	−	−	+	−
Starch, acid	N	N	N	N	N
Sucrose, acid	+	+	+	+	+
Trehalose, acid	−	−	−	+	−
D-Xylose, acid	−	−	−	−	−
Ubiquinones present	−	−	−	−	+
Naphthoquinones present	+	+	+	+	+

Table 5.53 *(continued)*

| Characteristic | Haemophilus | | Pasteurella | | | | | | |
	segnis	*"somnus"*	*aerogenes*	*anatis*	*avium*	*bettyae*	*caballi*[e]	*canis*	*dagmatis*
Catalase	d	−	+	+	+w	d	−	+	+
Oxidase	−	+	d	+w	+	−	+	+	+
Nitrate reduction	+	+	+	+	+	+	+	+	+
ONPG	d	d	d	+	−	−	+	−	−
Phosphatase	+	N	+	+	+	+	+	+	+
Gelatinase	N	N	−	N	−	N	−	−	−
H$_2$S production	−	N	+	N	−	N	N	N	N
Ornithine decarboxylase	−	−	d	−	−	−	d	+	−
Indole production	−	+	(−)	−	−	−	−	dw	+
Urease	−	−	+	−	−	−	−	−	+
Esculin hydrolysis	−	−	−	−	−	−	−	−	−
NAD requirement	+	−	−	−	d	−	−	−	−
X-factor requirement	−	−	−	−	−	N	−	N	N
MacConkey agar, growth[c]	−	−	+	+w	−	d	−	−	−
β-Hemolysis, sheep cells	−	−	−	−	−	−	−	−	−
Methyl red	N	N	−	N	N	N	−	N	N
Voges-Proskauer	N	N	−	N	N	N	−	N	N
Lysine decarboxylase	−	−	−	N	−	N	−	−	−
Arginine dihydrolase	−	−	−	N	−	N	−	−	−
Cyclic AMP reaction	−	−	−	N	N	N	N	N	N
α-Fucosidase	−	N	N	N	−	N	−	N	N
D-Glucose, gas production	−	−	(+)	−	−	dw	+	−	+w
D-Adonitol, acid	−	−	−	N	−	N	−	−	−
L-Arabinose, acid	−	d	(+)	−	−	−	−	−	−
Arbutin, acid	N	N	−	−	−	−	N	−	−
Cellobiose, acid	−	−	−	−	−	−	−	−	−
Dextrin, acid	N	N	N	N	N	N	N	N	N
Dulcitol, acid	−	d	−	−	−	−	−	−	−
meso-Erythritol	−	−	N	N	N	N	N	N	N

Footnotes are at end of table

Table 5.53 *(continued)*

Characteristic	*Haemophilus*		*Pasteurella*						
	segnis	*"somnus"*	*aerogenes*	*anatis*	*avium*	*bettyae*	*caballi*[e]	*canis*	*dagmatis*
Fructose, acid	+w	+	N	N	+	N	+	+	+
D-Galactose, acid	+w	d	+	+	+L	–	+	+	+
D-Glucose, acid	+w	+	+	+	+	+	+	+	+
Glycerol, acid	+w	–	+	N	N	N	N	N	N
myo-Inositol, acid	–	d	d	–	–	–	(–)	–	–
Inulin, acid	–	–	N	N	N	N	–	N	N
Lactose, acid	–	–	(–)	+L	–	–	+L	–	–
Maltose, acid	+w	+	+	–	–	d	(+)	–	+
D-Mannitol, acid	–	+	d	+	–	–	+	–	–
D-Mannose, acid	–	+	+	+	+L	d	+	+	+
Melezitose, acid	–	–	–	N	N	N	N	N	N
Melibiose, acid	–	–	d	–	–	–	N	–	–
Raffinose, acid	–	–	d	+w	–	–	dL	–	+w
L-Rhamnose, acid	–	–	d	N	–	N	(–)	–	–
D-Ribose, acid	–	N	N	N	N	N	N	N	N
Salicin, acid	–	–	–	–	–	–	–	–	–
D-Sorbitol, acid	–	+	(–)	–	–	–	(–)	–	–
L-Sorbose, acid	–	N	N	N	–	N	N	–	–
Starch, acid	N	N	N	N	–	N	–	–	–
Sucrose, acid	+w	–	+	+	–	–	+	+	+
Trehalose, acid	–	+	–	+	+	–	–	d	+
D-Xylose, acid	–	+	+	+	d	–	d	(–)	–
Ubiquinones present	–	–	+	N	+	–	N	+	+
Naphthoquinones present	+	+	+	N	+	+	N	+	+

Table 5.53 *(continued)*

Characteristic	*Pasteurella*						
	gallinarum	*granulomatis*	*haemolytica*	*langaa*	*lymphangitidis*	*mairii*	*multocida* subsp. *gallicida*
Catalase	+	+w	+	–	+	(+)	+
Oxidase	+	+	+	+w	–	+	+
Nitrate reduction	+	+	+	+	–	+	+
ONPG	–	+	d	+	–	d	–
Phosphatase	+	N	+	+	+	+	+
Gelatinase	–	–	(–)	N	N	N	–
H$_2$S production	(+)	–	N	N	N	N	+
Ornithine decarboxylase	–	–	–	–	–	(+)	+
Indole production	–	–	–	–	–	–	+
Urease	–	–	–	–	+	+	–
Esculin hydrolysis	–	+	d	–	+	d	–
NAD requirement	–	–	–	–	–	–	–
X-factor requirement	–	N	–	N	N	N	–

Footnotes are at end of table

Table 5.53   (continued)

Characteristic	Pasteurella						
	gallinarum	granulomatis	haemolytica	langaa	lymphangitidis	mairii	multocida subsp. gallicida
MacConkey agar, growth [c]	–	+	+	–	d	d	d
β-Hemolysis, sheep cells	–	+	(+)	–	–	d	–
Methyl red	–	N	–	N	N	N	–
Voges-Proskauer	–	N	–	N	N	N	–
Lysine decarboxylase	–	–	–	–	N	N	–
Arginine dihydrolase	–	–	–	–	N	N	–
Cyclic AMP reaction	–	N	d	N	N	N	–
α-Fucosidase	N	N	N	N	N	N	N
D-Glucose, gas production	–	–	–	–	–	–	–
D-Adonitol, acid	–	–	–	–	N	N	(–)
L-Arabinose, acid	–	–	+	–	+	+	d
Arbutin, acid	–	N	(–)	N	+	–	–
Cellobiose, acid	–	+L	(–)	–	–	–	–
Dextrin, acid	+	N	+	N	N	N	d
Dulcitol, acid	–	–	(–)	–	–	–	–
meso-Erythritol	N	N	–	N	N	N	(–)
Fructose, acid	+	N	+	+	N	+	+
D-Galactose, acid	+	+	+	+	+	+	+
D-Glucose, acid	+	+	+	+	+	+	+
Glycerol, acid	–	N	d	N	N	N	d
myo-Inositol, acid	–	–	d	–	–	d	–
Inulin, acid	–	N	–	N	N	N	(–)
Lactose, acid	–	+L	d	+L	–	(–)	–
Maltose, acid	+	+L	+	–	d	d	–
D-Mannitol, acid	–	+L	+	+	+	(+)	+
D-Mannose, acid	+	–	(–)	+	+	+	+
Melezitose, acid	N	N	N	N	N	N	(–)
Melibiose, acid	–	–	–[d]	–	+	–	N
Raffinose, acid	(+)	N	d	–	–	–	–
L-Rhamnose, acid	–	N	d	–	N	N	(–)
D-Ribose, acid	N	N	N	N	N	N	N
Salicin, acid	–	+	(–)	–	d	–	–
D-Sorbitol, acid	(–)	+	+	–	d	(+)	+
L-Sorbose, acid	–	N	–	–	N	N	(–)
Starch, acid	–	N	+	–	N	N	–
Sucrose, acid	+	+	+	+	d	+	+
Trehalose, acid	+	–	–	–	+	(–)	–
D-Xylose, acid	d	–	+	–	–	+	+
Ubiquinones present	+	N	+	N	N	N	+
Naphthoquinones present	+	N	+	N	N	N	+

Footnotes are at end of table

Table 5.53 *(continued)*

	Pasteurella						
Characteristic	*multocida* subsp. *multocida*	*multocida* subsp. *septica*	*pneumotropica*	*stomatis*	*testudinis*	*trehalosi*	*volantium*
Catalase	+	+	+	+	+	−	+
Oxidase	+	+	+	+	+	+	+
Nitrate reduction	+	+	+	+	+	+	+
ONPG	−	−	+	−	(+)	−	+
Phosphatase	+	+	+	+	−	+	+
Gelatinase	−	−	d	−	−	N	−
H₂S production	+	+	(+)	N	−	N	N
Ornithine decarboxylase	+	d	+	−	−	−	d
Indole production	+	(+)	+	+w	+	−	−
Urease	−	−	+	−	−	−	−
Esculin hydrolysis	−	−	−	−	+	d	−
NAD requirement	−	−	−	−	−	−	+
X-factor requirement	−	−	−	N	N	N	N
MacConkey agar, growth ᶜ	d	d	d	−	d	+	−
β-Hemolysis, sheep cells	−	−	−	−	+	(+)	−
Methyl red	−	−	−	N	N	N	N
Voges-Proskauer	−	−	−	N	−	N	N
Lysine decarboxylase	−	−	−	−	−	N	−
Arginine dihydrolase	−	−	−	−	−	N	−
Cyclic AMP reaction	−	−	−	N	N	N	N
α-Fucosidase	N	N	−	N	N	N	N
D-Glucose, gas production	−	−	d	−	−	−	−
D-Adonitol, acid	(−)	(−)	(−)	−	N	N	−
L-Arabinose, acid	−	−	(−)	−	d	−	−
Arbutin, acid	−	−	−	−	+L	d	N
Cellobiose, acid	−	−	−	−	−	d	N
Dextrin, acid	d	d	d	N	d	N	N
Dulcitol, acid	−	−	−	−	−	−	−
meso-Erythritol	(−)	(−)	(−)	N	N	N	N
Fructose, acid	+	+	+	+	N	N	+
D-Galactose, acid	+	+	+	+	(+)	−	N
D-Glucose, acid	+	+	+	+	+	+L	+
Glycerol, acid	d	d	(+)	N	−	N	N
myo-Inositol, acid	−	−	d	−	+L	d	N
Inulin, acid	(−)	(−)	−	N	−	N	N
Lactose, acid	−	−	d	−	(−)	−	dw
Maltose, acid	−	(−)	d	−	(+)	+	+
D-Mannitol, acid	+	(+)	−	−	d	+	+
D-Mannose, acid	+	+	+	+	−	+	N
Melezitose, acid	(−)	(−)	−	N	N	N	N
Melibiose, acid	N	(−)	d	−	+	−ᵈ	N
Raffinose, acid	−	−	d	−	d	−	N
L-Rhamnose, acid	(−)	(−)	(−)	−	+	N	−
D-Ribose, acid	N	N	+	N	N	N	N
Salicin, acid	−	−	(−)	−	(−)	d	N

Footnotes are at end of table

Table 5.53 *(continued)*

Characteristic	*Pasteurella*						
	multocida subsp. multocida	*multocida subsp. septica*	*pneumotropica*	*stomatis*	*testudinis*	*trehalosi*	*volantium*
D-Sorbitol, acid	–	+	–	–	d	+	d
L-Sorbose, acid	(–)	(–)	–	–	N	N	–
Starch, acid	–	–	(+)	–	d	N	–
Sucrose, acid	+	+	+	+	+	+	N
Trehalose, acid	+	d	(+)	+	d	+	+
D-Xylose, acid	+	d	d	–	+	–	d
Ubiquinones present	+	+	–	+	N	N	+
Naphthoquinones present	+	+	+	+	N	N	+

Symbols: +, 90% or more positive; (+), 80–89% positive; d, 21–79% positive; (–), 11–20% positive; –, 0–10% positive; w, weak reaction; L, delayed reaction; N, not tested.

[a] In addition to *Bergey's Manual of Systematic Bacteriology*, data taken from Mutters et al., (Int. J. Syst. Bacteriol. *35:* 309-322, 1985), Ribeiro et al., (J. Clin. Microbiol. *27:* 1401-1402, 1989), and Sneath and Stevens (Int. J. Syst. Bacteriol. *40:* 148-153, 1990).

[b] This test was reported positive in another study.

[c] Different formulations of this medium may explain different test results; results for *A. pleuropneumoniae, P. avium,* and *P. volantium* are for medium supplemented with NAD.

[d] Positive and negative results have been reported for this test.

[e] 76% of strains produce a yellow pigment.

Important Notes for Users of This Manual

Unless otherwise indicated in footnotes to tables, the meanings of symbols are as follows:

    +   90% or more of strains are positive

    –   90% or more of strains are negative

    d   11-89% of strains are positive

    v   strain instability (not equivalent to "d")

    D   Different reactions in different taxa (species of a genus or genera of a family)

All other symbols are defined in footnotes to tables.

Table 5.54   Differentiation of *Actinobacillus (Haemophilus/Pasturella)* species [a]

Characteristic	A. actinomycetemcomitans	A. capsulatus	A. equuli	A. hominis	A. lignieresii	A. muris	A. (H.) pleuropneumoniae	A. rossii	A. seminis	A. suis	A. (P.) ureae
Catalase	+	+	d	−	d	+	d	+	+	+	d
Oxidase	+	+	(+)	+	(+)	+	d	+	d	(+)	(+)
ONPG	−	+	d	+	d	−	+	(+)	−[b]	(+)	−
Phosphatase	+	+	+	+	+	−	+	+	−	+	+
Ornithine decarboxylase	−	−	−	−	−	−	−	−	d	−	−
Urease	−	+	+	+	+	+	+	+	−	+	+
Hemolysis on sheep blood agar	−	−	−	−	−	−	+	d	−	+	−
MacConkey agar, growth[c]	d	+	+	−	+	−	−	(+)	−	+	d
NAD requirement	−	−	−	−	−	−	+	−	−	−	−
Esculin hydrolysis	−	+L	−	d	−	+L	d	−	d	+	−
D-Glucose, gas production	(+)	−	−	−	−	−	−	d	−	−	−
L-Arabinose, acid	−	−	(−)	−	(−)	−	−	+	dL	(+)	−
Arbutin	−	N	−	N	−	+	N	−	−	+	−
Cellobiose, acid	−	+	−	−	−	+L	−	−	−	+	−
D-Galactose	+	+	d	+	+	−	+w	+	dL	(+)	−
myo-Inositol	−	−	−	−	−	dw	−	+	d	−	−
Lactose, acid	−	+	+	+	dL	−	d	d	−	+	−
Maltose	+	+	+	+	+	+	+	(−)	dL	+	(+)
D-Mannitol, acid	(+)	+	+	+	+	+	+	+	dL	−	+
D-Mannose	+	+	+	−	+	+	+	d	−	+	d
Melibiose, acid	−	+	+	+L	−	+L	−	−	−	+	−
Raffinose	−	+	+	+	d	+	d	(−)	−	+	−
Salicin, acid	−	+	−	d	−	+	−	−	−	+	−
D-Sorbitol	(−)[d]	+	d	−	(−)	−	−	+	−	−	(−)
Sucrose	−	+	+	+	+	+	+	−	−	+	+
Trehalose, acid	−	+	+	+	−	+	−	−	−	+	−
D-Xylose	d	+	+	+	+	−	+	+	−	+	−

Symbols: +, 90% or more positive; (+), 80–89% positive; d, 21–79% positive; (−), 11–20% positive; −, 0–10% positive; w, weak reaction; L, delayed positive; N, not tested.

[a] Except for *A. ureae*, data from Sneath and Stevens (Int. J. Syst. Bacteriol. *40:* 148-153, 1990).

[b] This test was reported as positive in another study.

[c] Different formulations of this medium may explain discrepancies in the reported results; results for *A. pleuropneumoniae* are for medium supplemented with NAD.

[d] Positive and negative results have been reported for this test.

Table 5.55   Differentiation of the biovars of *Haemophilus influenzae* and of *H. parainfluenzae*

Characteristic	*H. influenzae* biovar									*H. parainfluenzae* biovar					
	I	II	III	IV	V	VI	VII	VIII	aegyptius	I	II	III	IV	V	VI
Indole production	+	+	−	−	+	−	+	−	−	−	−	−	+	+	+
Urease	+	+	+	+	−	−	−	−	+	−	+	+	+	−	+
Ornithine decarboxylase	+	−	−	+	+	+	−	−	−	+	+	−	+	+	−
D-Glucose, gas production										(d)	(d)	−	NT	−	+
D-Xylose [a]	+	+	+	+	+	+	+	+	−	−	−	−	v	−	−

Symbols: see standard definitions; (d), 11–89% give a delayed positive reaction; NT, not tested.

[a] Biovar aegyptius is always negative; occasional negative strains occur in biovars I-VIII. References to *H. influenzae* biovars VI, VII, and VIII are given in J. Clin. Microbiol *20:* 815-816, 1984. *H. parainfluenzae* biovar IV was described by Bruun et al., (Acta Path. Microbiol. Immunol. Scand. Sect. B. *92:* 135-138, 1984). *H. parainfluenzae* biovars V and VI were described by Sturm (J. Med. Microbiol. *21:* 349-352, 1986). Sturm designates these as VI and VII because he includes *H. segnis* as biogroup IV.

Table 5.56   Differentiation of *Cardiobacterium hominis* and phenotypically similar genera and species

Characteristics	Cardio-bacterium hominis	Haemophilus aphrophilus	Kingella	Actinobacillus actinomycetem-comitans	Pasteurella	Eikenella corrodens	Capnocy-tophaga
Oxidase	+	d	+	d	+	+	−
Catalase	−	−	−	+	+	−	−
Indole	+	−	D [a]	−	D	−	−
Fermentative	+ [b]	+	+ [b]	+	+	−	+
Nitrate reduction	−	+	D [c]	+	+	+	D

Symbols: see standard definitions.

[a] Produced only by *K. indologenes.*

[b] *C. hominis* acidifies sorbitol and usually mannitol; *K. indologenes* does not acidify mannitol or sorbitol.

[c] Not reduced by *K. indologenes.*

Table 5.57   Characteristics differentiating the genus *Chromobacterium* from the genus *Janthinobacterium*

Characteristics	Chromobacterium	Janthinobacterium
Glucose is catabolized by:		
Fermentation	D (80)	−
Oxidation	D (20)	+ (95)
Turbid zone on egg yolk agar (lecithinase activity)	+	−
Acid from:		
Trehalose	+	−
L-Arabinose	−	+ (95)
D-Xylose	−	+ (95)
Casein hydrolysis	+	− (10 [w])
Esculin hydrolysis	−	+ (95)

Symbols: see standard definitions. Numbers in parentheses indicate the % of strains giving a positive reaction.

[w] Weak.

Table 5.58 Characteristics of the species of the genus *Chromobacterium*[a]

Characteristics	*C. violaceum*	*C. fluviatile*
Growth at 4°C, 7 d	–	+
Growth at 37°C, 7 d	+	–
Colonies flat, spreading	–	+
HCN production	[+] (95)	–
H₂S production	– or W	–
Arginine dihydrolase	d (50)	–
Turbid zone on egg-yolk agar (lecithinase)	+	Weak, only under colony
Hemolytic on horse blood	[+] (95)	W or –
Urease	W or –	–
Acid from:		
Maltose	d (50)	+
Mannose	[+] (80)	[+] (90)
Sorbitol	d (60)	–
Sucrose	[–] (25)	[–] (10)
Glycerol	[–] (10)	[–] (30)
Salicin	[–] (10)	–
L-Arabinose	–	[–] (10)
Galactose	–	[–] (10)
Cellobiose	–	[–] (30)

[a] Symbols: +, all strains positive; [+], positive for 80% or more strains; d, positive for 31–79% of strains; [–], positive for 30% or fewer strains; –, negative for all strains; W, weak. Numbers in parentheses indicate the % of strains giving a positive reaction.

Table 5.59 Differentiation of *Eikenella* from superficially similar organisms

Characteristics	*Eikenella*	*Actino-bacillus*	*Bordetella* (excluding *B. bronchiseptica*)	*Brucella*	*Cardio-bacterium*	*Haemophilus*	*Moraxella*	*Kingella*	*Yersinia*	*Bacteroides ureolyticus*
Oxidase (dimethyl-*p*-phenylenediamine)	+	–					+	+	–	d
Catalase	–	+	+	+			+	–	+	–
Indole	–				+					
Acid from glucose	–	+			+	+		+	+	
Urease	–			+						+
Grow only anaerobically	–									+
Proteolytic	–									+

Symbols: see standard definitions.

Table 5.60 Differential characteristics of the genus *Gardnerella* and other morphologically or physiologically similar genera

Characteristics	*Gardnerella*	*Capnocytophaga*	*Actinobacillus*	*Cardiobacterium*	*Haemophilus*
Oxidase	−	−	−[a]	+	D
Catalase	−	+	+	−	D
Cell morphology:					
Pleomorphic	+	−	−	+	−
Short to coccoid	−	−	+	−	+
Fusiform	−	+	−	−	−
Yellow-orange colonies	−	+	−	−	−
Nitrate reduced to nitrite	−	D	+	−	+
Indole production	−	−	−	+[w]	D
Growth on:					
Blood agar, 35°C, in air	+	−	−[a]	d	−
Blood agar, 35°C, in air + 5% CO_2	+	+	+	+	−[b]
Kligler iron agar reactions:					
Slant	NG	K or A	K or A	K or NG	NG
Butt	NG	N or A	A	A or NG	NG
β Hemolysis, 5% human blood agar	+	−	−	−	−
Acid production from:					
Lactose	d	D	D	−	D
D-Mannitol	−	−	D	+	D

Symbols: +, typically positive; D, differs among species; d, differs among strains of a genus containing only a single species; −, typically negative; A, acid; K, alkaline; N, neutral; NG, no growth; w, weakly positive.

[a] An occasional strain may be weakly positive.

[b] One species, *H. aphrophilus,* can grow on blood agar.

Important Notes for Users of This Manual

Unless otherwise indicated in footnotes to tables, the meanings of symbols are as follows:

    + 90% or more of strains are positive

    − 90% or more of strains are negative

    d 11-89% of strains are positive

    v strain instability (not equivalent to "d")

    D Different reactions in different taxa (species of a genus or genera of a family)

All other symbols are defined in footnotes to tables.

Table 5.61   Differentiation of *Streptobacillus* from other genera

Characteristics	Streptobacillus	Actinobacillus	Cardiobacterium	Haemophilus
Serum requirement	+	−	−	−
Flocculent growth	+	−	−	−
L-phase variant	+	−	−	−
Catalase	−	d	−	d
Oxidase	−	d	+	d
Indole	−	−	+	d
Nitrate reduction	−	+	−	+

Symbols: see standard definitions.

Table 5.62   Differential characteristics of the genus *Zymomonas* and other genera [a]

Characteristics	Zymomonas	Acetobacter	Gluconobacter	Pseudomonas[b]	Vibrio	Aeromonas
Flagellar arrangement:						
Polar	+	−	+	+	+	+
Peritrichous	−	+	−	−	−	−
Facultatively anaerobic	+	−	−	−	+	+
Oxidase	−	−	−	+[c]	+[c]	+
Sensitive to vibriostatic compound 0/129	−			−	+	−
Fermentative	+	−	−	−	+	+
Gas from D-glucose	+	−	−	−	−	D
Nitrate reduction	−	−	−	D	+	+

[a] Symbols: +, typically positive; −, typically negative; D, differs among species.
[b] *Pseudomonas* groups I, II and III as defined in this manual.
[c] Some species exhibit a negative or weak oxidase reaction.

GROUP 6

GRAM-NEGATIVE, ANAEROBIC, STRAIGHT, CURVED, AND HELICAL BACTERIA

This group corresponds very broadly to the organisms allocated to the family *Bacteroidaceae* in *Bergey's Manual of Systemic Bacteriology,* Volume 1. However, the group has become so diverse that essentially all they now have in common is that they have been considered to be anaerobes and that they stain Gram negative. In fact, they are not all anaerobes; some, i.e., *Wolinella* species, *Bacteroides gracilis,* and *Bacteroides ureolyticus,* have now been shown to be microaerophiles, capable of respiring with oxygen; however, they are included here for convenience. Although the organisms in Group 6 do stain Gram negative, some (e.g., *Butyrivibrio*) have a Gram-positive type of cell wall as seen by electron microscopy.

Some members of the group have unusual features. For instance, one genus, *Sporomusa,* forms heat-resistant endospores. Eight genera are thermophilic, and two genera are extreme halophiles.

Members of the group have various habitats, but most have been isolated from the mouth or intestinal tract of humans and animals or from anoxic muds and sewage sludge.

Differentiation of the genera in **Group 6:** See Table 6.1.

Genus **Acetivibrio**

Slightly curved rods, 0.4–0.8 x 4–10 μm, occurring in pairs or short chains. **Motile by a single flagellum attached one-third of the way along the concave side, or by a bundle of flagella attached linearly on the concave side.** Mesophilic, growing best at 35°C. Chemoorganotrophic, with **acetate as the major acid product** of carbohydrate fermentation. Other products are ethanol, CO_2 and H_2. Propionate and lactate are not formed. **Isolated from sewage sludge and pig feces.**

Type species: *Acetivibrio cellulolyticus.*

Differentiation of the species of the genus **Acetivibrio:** See Table 6.2.

Genus **Acetoanaerobium**

Editorial note: The genus *Acetoanaerobium* was not included in *Bergey's Manual of Systematic Bacteriology.* The genus was created in 1985 by Sleat et al. (Int. J. Syst. Bacteriol. *35:* 10–15). It includes a single species, *A. noterae.*

Straight rods, 0.8 μm × 1.0–5.0 μm. Gram negative, but have an atypical wall structure and are not lysed by KOH. Nonsporeforming. **Motile by 3–4 peritrichous flagella.** Ferment a limited range of substrates including yeast extract, glucose, and maltose. **With glucose and maltose only acetate is produced during growth: with yeast extract, propionate, butyrate, isobutyrate, and isovalerate are formed as well. With 80% H_2 and 20% CO_2 as the gas phase and 0.2% yeast extract, acetate is synthesized from H_2 and CO_2.** Optimum temperature is 37°C; optimum pH, 7.6. **Isolated from swamp sediment** near Lake Hula, Galilee, Israel.

Type (and only) species: *Acetoanaerobium noterae.*

For differentiation of *Acetoanaerobium* from other H_2-oxidizing acetogenic bacteria, see Sleat et al. (Int. J. Syst. Bacteriol. *35:* 10–15, 1985, Table 2).

Characteristics of the species: As described for the genus.

Genus **Acetofilamentum**

Editorial note: The genus *Acetofilamentum* was not included in *Bergey's Manual of Systematic Bacteriology.*

The genus was created in 1988 by Dietrich et al. (Syst. Appl. Microbiol. *10:* 273–278; Int. J. Syst. Bacteriol. *39:* 93, 1989). It includes a single species, *A. rigidum.*

Long, thin rigid filaments, 0.18–0.30 wide and 100 μm or more in length. Gram negative, but the cell wall profile is not typical of either Gram-positive or Gram-negative cells. Nonsporing. **Filaments can be disaggregated by SDS, giving single cells 2–7 μm long,** indicating that the filaments are surrounded by a protein sheath. **Nonmotile.** Optimum temperature is 35–38°C. Chemoorganotrophic, utilizing several different hexoses, pentoses, or amino acids: yeast extract is required for growth. Glucose is fermented to acetate, CO_2, and H2, increasing concentrations of H_2 in the gas phase (5–20%) are increasingly inhibitory. **Isolated from municipal sewage sludge.**

Type (and only) species: *Acetofilamentum rigidum.*

Characteristics of the species: As described for the genus.

Genus **Acetogenium**

Editorial note: The genus *Acetogenium* was not included in *Bergey's Manual of Systematic Bacteriology.* The genus was created in 1983 by Leigh and Wolfe (Int. J. Syst. Bacteriol. *33:* 886, 1983). The genus includes a single species, *A. kivui.*

Straight rods measure 0.7–0.8 × 2.0–7.5 μm. Gram negative, but have a cell wall of the Gram-positive type. Nonsporing and **nonmotile. Thermophilic:** optimum temperature is 66°C; temperature range is 50–72°C. Optimum pH is 6.4. **Ferment** glucose, mannose, fructose, and pyruvate, **producing almost entirely acetate. Acetate is also formed from H_2 and CO_2** (80% H_2 and 20% CO_2, two atmospheres pressure). **Growth requires sulfide and/or cysteine. Isolated from lake mud.**

Type (and only) species: *Acetogenium kivui.*

Characteristics of the species: As described for the genus. The species was isolated from the mud of Lake Kivu, Africa.

Genus **Acetomicrobium**

Editorial note: The genus *Acetomicrobium* was not included in *Bergey's Manual of Systematic Bacteriology.* The genus was created in 1984 by Soutschek et al. (Syst. Appl. Microbiol. *5:* 377–390; Int. J. Syst. Bacteriol. *35:* 223, 1985) and included a single species, *A. flavidum.* A second species, *A. faecalis (sic),* was added in 1987 by Winter et al. (Syst. Appl. Microbiol. *9:* 71–76; Int. J. Syst. Bacteriol. *38:* 136, 1988).

Curved rods 0.6–0.8 × 2.0–7.0 μm. Nonsporing. **Motile by a single subpolar flagellum or a few lateral flagella. Thermophilic:** optimum temperature is 58–73°C. Chemoorganotrophic. Ferment a variety of hexoses and pentoses. **Yeast extract is required for growth. Glucose is fermented to acetate, CO_2 and H_2, or to acetate, lactate, ethanol, CO_2 and H_2.** Isolated from sewage sludge.

Type species: *Acetomicrobium flavidum.*

Differentiation of the species of the genus **Acetomicrobium:** See Table 6.3.

Genus **Acetothermus**

Editorial note: The genus *Acetothermus* was not included in *Bergey's Manual of Systematic Bacteriology.* It was created in 1988 by Dietrich et al. (Syst. Appl. Microbiol. *10:* 174–179; Int. J. Syst. Bacteriol. *38:* 328, 1988). It includes a single species, *A. paucivorans.*

Straight rods, 0.5 × 5-8 μm when adequate levels of vitamin B_{12} are present. Long filaments (>50 μm) are formed in the absence of this vitamin. Nonsporing, **nonmotile. Thermophilic:** optimum temperature is 58°C; temperature range is 50–60°C. Chemoorganotrophic. **Requires yeast extract and, especially, vitamin B_{12} for growth. Glucose and fructose are fermented to acetate, CO_2, and H_2. No other sugars are attacked. Isolated from sewage digested at 60°C.**

Type (and only) species: *Acetothermus paucivorans.*

Acetomicrobium is similar to *Acetothermus* in that it is found in sewage and produces acetate, CO_2, and H_2 from glucose. *Acetothermus* differs from *Acetomicrobium* by (a) being nonmotile; (b) requiring vitamin B_{12}; (c) by fermenting only glucose and fructose; and (d) by having a higher mol% G + C content of the DNA (62 vs. 48).

Characteristics of the species: As described for the genus.

Genus **Acidaminobacter**

Editorial note: The genus *Acidaminobacter* was not included in *Bergey's Manual of Systematic Bacteriology.* The genus was created in 1984 by Stams and Hansen (Arch. Microbiol. *137:* 329–337; Int. J. Syst. Bacteriol. *35:* 223, 1985). It includes a single species, *A. hydrogenoformans.*

Straight rods 0.5–0.6 × 1.5–3.7 μm, with pointed ends. Occur singly or in pairs. **Nonmotile.** Optimum temperature is 30°C; temperature range, 15–42°C. Chemoorganotrophic. **Ferment amino acids, especially glutamic acid. Acetate is the major end product,** with lesser amounts of NH$_3$, formate, CO$_2$, H$_2$, propionate, and sometimes other fatty acids, depending on the amino acid utilized. Lactate, succinate, butyrate, ethanol are not formed. **Indole is produced from tryptophan.** Growth on many substrates is stimulated by and may be dependent on the presence of H$_2$-utilizing bacteria such as *Methanospirillum* or *Desulfovibrio.* **Isolated from black estuarine mud.**

Type (and only) species: *Acidaminobacter hydrogenoformans.*

Acidaminobacter can be differentiated from *Pelobacter,* a genus of Gram-negative anaerobic mesophilic rods isolated from anoxic muds, by the following features: (a) *Acidaminobacter* cells have pointed rather than rounded ends; (b) they attack various amino acids, whereas *Pelobacter* strains attack only unusual substrates such as gallic acid, ethylene glycol, or polyethylene glycol; and (c) *Acidaminobacter* strains produce indole from tryptophan, whereas *Pelobacter* does not produce indole. *Acidaminobacter* can be differentiated from *Syntrophobacter,* a genus of Gram-negative nonmotile rods isolated from sewage, because *Syntrophobacter* grows only in co-culture with *Desulfovibrio* and attacks only propionate.

Characteristics of the species: As described for the genus.

Genus **Anaerobiospirillum**

Spiral-shaped cells with rounded ends, 0.6–0.8 μm wide and 3–15 μm long, with a wavelength of ca. 1.5–2.0 μm. Occasionally cells up to 32 μm in length occur. Exhibit a **corkscrew-like motility by means of bipolar tufts of flagella.** No axial fibrils are present. Nonsporing. Chemoorganotrophic. **Carbohydrates are fermented, mainly to acetic and succinic acids.** Isolated from the **throats and colons of beagle dogs** and from cases of **human diarrhea.**

Type (and only) species: *Anaerobiospirillum succiniciproducens.*

For differentiation of *Anaerobiospirillum* from similar genera of Gram-negative motile organisms, see Table 6.4.

Characteristics of the species: Most strains ferment glucose, lactose, sucrose, and raffinose; some may also produce acid from fructose, maltose, trehalose, melibiose, and other sugars. Very similar organisms have been reported from cases of septicemia in humans (Park et al., Am. J. Clin. Pathol. *85:* 73–76, 1986) and from patients with diarrhea (Malnick et al., J. Clin. Microbiol. *28:* 1380–1384, 1990). Strains of *A. succiniciproducens* have been isolated from the feces of healthy dogs and cats, but not from the feces of healthy humans. Strains from human cases of diarrhea differ in fermentative pattern from the original isolates from beagle dogs, and may represent a distinct species. Malnick et al. (J. Clin. Microbiol. *28:* 1380–1384, 1990) devised a selective medium for *Anaerobiospirillum.* Of various common laboratory media, chocolate agar provides the best growth (Park et al., Am. J. Clin. Pathol. *85:* 73–76, 1986).

Genus **Anaerorhabdus**

Editorial note: the genus *Anaerorhabdus* was not included in *Bergey's Manual of Systematic Bacteriology.* It was created in 1986 by Shah and Collins (Syst. Appl. Microbiol. *8:* 86–88; Int. J. Syst. Bacteriol. *36:* 573, 1986). It includes a single species, *A. furcosus* (formerly *Bacteroides furcosus*).

Pleomorphic short rods 0.3–1.5 μm wide and 1–3 μm long, occuring singly, in pairs, or short chains: some cells are forked or Y-shaped. Nonsporing and nonmotile. **Generally asaccharolytic,** although a few carbohydrates are fermented weakly. **Acetic and lactic acids are the major metabolic end products.** Sphingolipids and menaquinones not produced, and the non-hydroxylated cellular fatty acids are primarily straight chain saturated and unsaturated; methyl-branched acids are either absent or present only in traces. Isolated from appendix abscesses, lung and abdominal abscesses, and occasionally from human and pig feces.

Type (and only) species: *Anaerorhabdus furcosus.*

Anaerorhabdus differs from *Bacteroides fragilis* and similar species by lacking both sphingolipids and menaquinones, and its non-hydroxylated cellular fatty acids are predominantly of the straight-chain type.

Characteristics of the species: Glucose, fructose, and occasionally sucrose are weakly fermented. Esculin is hydrolyzed. Optimum temperature is 30–37°C. Growth is stimulated by rumen fluid and generally inhibited by bile.

Genus **Anaerovibrio**

Curved or spiral-shaped rods 0.5 μm wide and 1.2–10.0 μm long. Nonsporing. **Motility occurs by a single polar flagellum.** Optimum temperature is 30–37°C. Chemoorganotrophic, attacking a very limited range of sugars. Propionate, acetate, and succinate are produced as end products. **Some strains attack only glycerol or diolein, producing only propionate. Others are characteristically lipolytic, hydrolyzing triglycerides to glycerol and fatty acids and producing mostly propionate and succinate from glycerol.** Found in the rumen of cattle and sheep or in anoxic freshwater muds and sewage sludge.

Type species: *Anaerovibrio lipolytica.*

Differentiation of the species of the genus **Anaerovibrio:** See Table 6.5. For differentiation of *A. lipolytica* from morphologically similar motile anaerobic organisms from the rumen, see Table 6.6.

Genus **Bacteroides**

Rod-shaped organisms of variable size. Many species are pleomorphic and show terminal or central swellings, vacuoles, or filaments. Generally nonmotile (two species are motile, and others may show "twitching motility"). Anaerobic (but see note below), chemoorgano-trophic, metabolizing carbohydrates, peptone, or metabolic intermediates. **Especially with strongly saccharolytic species, fermentation products include acetate, succinate, lactate, formate, or propionate. Butyrate is not usually a major product, but when it is formed it is accompanied by isobutyrate and isovalerate.** Many species contain high levels of branched chain fatty acids, generally anteiso-C_{15} acids, and also sphingolipids (see Shah and Collins, J. Appl. Bacteriol. *55:* 403–416, 1983). **Hemin and Vitamin K are highly stimulatory for the growth of many species** and are generally added to media for growth of bacteroides. Isolated from a wide range of anaerobic habitats: gingival crevice, intestinal tract (cecum and rumen), sewage sludge, and infective and purulent conditions in humans and animals.

Type species: *Bacteroides fragilis.*

Although *B. ureolyticus* and *B. gracilis* were described as anaerobes in *Bergey's Manual of Systematic Bacteriology,* vol. 1, they are in fact H_2- and formate-requiring microaerophiles capable of respiring with oxygen. In the absence of alternative terminal electron acceptors such as fumarate, they exhibit oxygen-dependent growth with low levels of oxygen and will not grow anaerobically or aerobically (Wolin et al., J. Bacteriol. *81:* 911–917, 1961; Han et al., Int. J. Syst. Bacteriol. *41:* 218–222, 1991).

Two described species, *B. polypragmatus* and *B. xylanolyticus,* are motile by peritrichous flagella; *B. galacturonicus* also has peritrichous flagella, but is apparently nonmotile. Two other species, *B. ureolyticus* and *B. gracilis,* are described as showing "twitching" movements but no true translocational motility. For a description of twitching motility, see Henrichsen, Acta. Pathol. Scand. B. *83:* 171–178, 1975.

It was thought that *Bacteroides* differed from most other Gram-negative organisms (in particular *Fusobacterium*) by not having keto-deoxyoctonate (KDO) in their lipopolysaccharide. However, KDO is present in *Bacteroides,* but in a phosphorylated form, so that it cannot easily be recognized until after hydrolysis in strong acid (Kumada et al., FEMS Microbiol. Lett. *51:* 77–80, 1988).

Differentiation the species of the genus **Bacteroides:** See Table 6.7.

Genus **Butyrivibrio** *(See also Group 20)*

Curved rods 0.4–0.6 × 2–5 μm. Gram negative, but cells have a Gram-positive type cell wall with an unusually thin peptidoglycan layer (Cheng and Costerton, J. Bacteriol. *129:* 1506–1512, 1977: Cheng et al., Can. J. Microbiol *35:* 274–282, 1989). **Motile by one or several polar or subpolar flagella. Have a fermentative metabolism, attacking a range of carbohydrates with butyric acid as the major end product.** Most strains produce extracellular polysaccharides, and some are capsulated (Stack et al., Appl. Environ. Microbiol. *54:* 878–883, 1988).

Type species: *Butyrivibrio fibrisolvens.*

Differentiation of the species of the genus **Butyrivibrio:** See Table 6.8.

Genus **Centipeda**

Editorial note: The genus *Centipeda* was not included in *Bergey's Manual of Systematic Bacteriology.* It was created by Lai et al. in 1983 (Int. J. Syst. Bacteriol. *33:* 628–635). It includes a single species, *C. periodontii.*

Serpentine rods 0.65 × 4–17 μm. Nonsporing. **Motility is by means of flagella inserted in a spiral line along the cell body** (Males et al., J. Gen. Microbiol. *130:* 185–191, 1984). Movement occurs by flexion of the entire cell and rotation around the long axis. Temperature range for growth is 32–37°C. Chemoorganotrophic. Saccharolytic and may ferment a wide range of sugars giving acid but no gas. **Propionic acid is the major product of sugar breakdown,** with variable smaller amounts of acetic, lactic, or succinic acids. **Isolated from subgingival lesions in patients with periodontitis.**

Type (and only) species: *Centipeda periodontii.*

Characteristics of the species: As described for the genus. *C. periodontii* is most likely to be confused with oral strains of *Selenomonas* species isolated from the mouth. The main points of distinction are given in Table 6.9.

Genus **Fervidobacterium**

Editorial note: The genus *Fervidobacterium* was not included in *Bergey's Manual of Systematic Bacteriology.* It was created in 1985 by Patel et al. (Arch. Microbiol. *141:* 63–69; Int. J. Syst. Bacteriol. *35:* 535, 1985). It includes two species, *F. nodosum* and *F. islandicum.*

Predominantly straight rods, 0.50–0.55 × 1.0–2.5 μm, occuring singly, in pairs, or short chains. Many cells produce terminal swellings (spheroids). Round bodies containing several cytoplasmic units also occur. Nonsporing and **motile** (flagellar arrangement has not been described). **Thermophilic.** Optimum temperature is 70°C; range for growth is 40–80°C. **Chemoorganotrophic, fermenting a variety of sugars to acetate, lactate, CO_2, H_2, and ethanol.** Isolated from hot springs at Rotorua and Waimangu in New Zealand and in Iceland.

Type species: *Fervidobacterium nodosum.*

Differentiation of the species of the genus **Fervidobacterium:** The two species, *F. nodosum* and *F. islandicum,* differ in DNA base ratio, DNA/DNA homology, and ability to utilize cellulose (*F. islandicum* is positive).

Genus **Fibrobacter**

Editorial note: The genus *Fibrobacter* was not included in *Bergey's Manual of Systematic Bacteriology.* It was created in 1988 by Montgomery et al. (Int. J. Syst. Bacteriol. *38:* 430–435). It included two species, *F. succinogenes* (formerly *Bacteroides succinogenes*) and *F. intestinalis.*

Rods or pleomorphic ovoid cells, 0.4–0.8 × 0.8–2.0 μm. Nonsporing. Nonmotile. **Characteristically ferment cellulose and cellobiose but attack few other sugars. Major fermentation products are acetic and succinic acids:** sometimes a small amount of formic acid is produced. Require CO_2, volatile fatty acids, ammonia, and one or more vitamins for growth. Isolated from the mammalian gastrointestinal tract.

Type species: *Fibrobacter succinogenes.*

Differentiation of the species of the genus **Fibrobacter:** See Table 6.10.

Genus **Fusobacterium**

Rods that may or may not be spindle-shaped, and are often very pleomorphic. Nonsporing. **Nonmotile.** Chemoorganotrophic, metabolizing peptone or carbohydrates, but in general only weakly fermentative. **Major end product of metabolism is butyrate,** often with acetate and lactate and lesser amounts of propionate, succinate, and formate. **Isobutyrate and isovalerate not produced.** Found principally in the gingival sulcus and in the intestinal and genital tracts; also isolated from blood cultures and various purulent lesions in humans and animals, and from tropical ulcers.

Type species: *Fusobacterium nucleatum.*

Differentiation of **Fusobacterium** *from other nonmotile genera of anaerobes:* The classic description of members of the genus *Fusobacterium* is that they are Gram-negative obligately anaerobic spindle-shaped rods which produce large amounts of butyric acid. However, the typical fusiform or spindle-shaped appearance is not shown by all fusobacteria, and it may be difficult to distinguish them from strains of *Bacteroides,* or from strains of *Clostridium* or *Eubacterium* where the cells are easily decolorized. Cultures should be examined at all stages of growth, tested for heat-resistance, and stained for spores.

In general, members of *Bacteroides* do not produce butyrate as an end product, or if they do it is accompanied by isobutyrate and isovalerate. *Leptotrichia* produces large amounts of lactic acid but no butyrate.

Bennett and Duerden (J. Appl. Bacteriol. *59:* 171–181, 1985) found that all the strains of *Fusobacterium* that they tested were sensitive to 300 µg/ml phosphomycin, whereas almost all strains of *Bacteroides* were resistant. (See also Rodriguez et al., J. Appl. Bact. *41:* 251–254, 1974; and Essers, J. Appl. Bacteriol. *52:* 319–323, 1982.)

Differentiation of the species of the genus **Fusobacterium:** See Table 6.11. As is evident from the table, one of the problems of identification in the group is the general lack of reactivity in conventional tests. Determination of fatty acid profiles using gas chromatography (Jantzen and Hofstad, J. Gen. Microbiol. *123:* 163–171, 1981), electrophoretic patterns of glutamate dehydrogenase (GDH) (Gharbia and Shah, J. Gen. Microbiol. *134:* 327–332, 1988), and pyrolysis-mass spectrometry (Magee et al., J. Med. Microbiol. 28: 2274, 33–236,

1989) have all identified groups which correspond well to already recognized species. GDH electrophoresis and pyrolysis also showed that *F. nucleatum* strains formed a heterogeneous group, and this has been confirmed in more detail by Dzink et al. (Int. J. Syst. Bacteriol. *40:* 74–78, 1990) using DNA hybridization and electrophoresis of soluble proteins. They proposed three subspecies of *F. nucleatum:* subsp. *nucleatum,* subsp. *polymorphum,* and subsp. *vincentii.* These subspecies could not be separated on the basis of commonly used phenotypic tests.

Genus **Haloanaerobium**

Editorial note: The genus *Haloanaerobium* was not included in *Bergey's Manual of Systematic Bacteriology.* It was created in 1983 by Zeikus et al. (Curr. Microbiol. *9:* 225–234; Int. J. Syst. Bacteriol. *34:* 503, 1984). It includes a single species, *H. praevalens.*

Straight rods, 0.9–1.1 × 2.0–2.6 µm. Nonsporing and nonmotile. Halophilic, growing optimally in about 13% NaCl. Optimum temperature is 37°C. Chemoorganotrophic, utilizing carbohydrates, pectin, amino acids, and amino sugars. **The major products of fermentation are acetate, butyrate, propionate, H_2, and CO_2.** Isolated from anoxic sediments in the Great Salt Lake, Utah, USA.

Type (and only) species: *Haloanaerobium praevalens.*

Characteristics of the species: As described for the genus. For differentiation of *H. praevalens* from *Halobacteroides halobius,* see Table 6.12.

Genus **Halobacteroides**

Editorial note: The genus *Halobacteroides* was not included in *Bergey's Manual of Systematic Bacteriology.* It was created in 1984 by Oren et al. (Syst. Appl. Microbiol. *5:* 58–70; Int. J. Syst. Bacteriol. *34:* 355, 1984).

Straight or curved rods, 0.5 × 10–20 µm long. Nonsporing. **Motile by means of peritrichous flagella. Halophilic,** growing optimally in 8.4–14% NaCl. Optimum temperature is 37–42°C. Chemoorganotrophic, fermenting a number of sugars and some organic acids. **End products from glucose fermenta-**

tion are acetate, ethanol, H_2, and CO_2. Isolated from Dead Sea sediments.

Type (and only) species: *Halobacteroides halobius.*

Characteristics of the species: Cells appear flexible and may be coiled up at one end. In older cultures cells become spherical. They are sensitive to chloramphenicol (30 µg/ml) and penicillin (20 U/ml). Requires biotin and *p*-aminobenzoic acid for growth. Other properties are as described for the genus. For differentiation of *Halobacteroides halobius* from *Haloanaerobium praevalens,* see Table 6.12.

Genus **Ilyobacter**

Editorial note: The genus *Ilyobacter* was not included in *Bergey's Manual of Systematic Bacteriology.* It was created in 1984 by Stieb and Schink (Arch. Microbiol. *140:* 139–146; Int. J. Syst. Bacteriol. *35:* 375, 1985) and included a single species, *I. polytropus.* A second species, *I. tartaricus,* was added in 1984 by Schink (Arch. Microbiol. *139:* 409–414; Int. J. Syst. Bacteriol. *35:* 375, 1985).

Generally short, straight rods 0.7–1.0 µm wide and 1.3–3.0 µm long, occurring in pairs or chains. Nonsporing. May be **motile or nonmotile.** Generally require 1% NaCl in medium for good growth. Optimum temperature is 28–34°C. Chemoorganotrophic, fermenting a variety of sugars and organic acids, such as 3-hydroxybutyrate, crotonate, L-tartarate, citrate, pyruvate, malate, fumarate, glucose, fructose, and glycerol, and giving a variety of end products depending on the substrate. **Glucose is fermented to acetate, formate, and ethanol. Isolated from anoxic marine muds.**

Type species: *Ilyobacter polytropus.*

Differentiation of the species of the genus **Ilyobacter:** See Table 6.13.

Genus **Lachnospira** *(See also Group 20)*

Curved rods, 0.4–0.6 × 2.0–4.0 µm. Gram negative but have a Gram-positive type cell wall. **Young cultures stain Gram positive.** Nonsporing. **Motile by monotrichous lateral to subpolar flagella.** Chemoorgano-

trophic, having a fermentative type of metabolism. **End products from glucose fermentation are acetate, formate, lactate, ethanol, H_2, and CO_2.** Pectin fermentation results in methanol being produced in addition, due to pectinmethylesterase. Isolated from the bovine rumen.

Type (and only) species: *Lachnospira multiparus.*

Characteristics of the species: The characteristics are as described for the genus. For differentiation from *Anaerovibrio lipolytica* and other morphologically similar rumen bacteria see Table 6.6.

Genus **Leptotrichia**

Straight or slightly curved rods 0.8–1.5 × 5–15 µm. Generally stain Gram negative, although very young cultures may stain Gram positive; have an atypical Gram-negative type wall structure. Nonsporing. Nonmotile. Grows best under an atmosphere containing 5–10% CO_2; on repeated transfer some strains may grow aerobically with CO_2. Optimum temperature is 35–37°C. No growth occurs below 25°C. Chemoorganotrophic, fermenting carbohydrates. **The major end product of glucose metabolism is lactic acid.** Found chiefly in dental plaque, but similar organisms have been reported from the female genital tract.

Type (and only) species: *Leptotrichia buccalis.*

Characteristics of the species: Fusiform rods, often in pairs with adjacent ends flattened. Cells may also also occur as long filaments. For good growth the organism requires nutrient broth supplemented with yeast extract (0.3%), glucose (0.5%), cysteine hydrochloride (0.1%), and serum (5%). The medium of Kasai may also be used (Kasai, J. Dent. Res. *40:* 800–811, 1961). CO_2 (5–10%) is needed for growth. Colonies on blood agar are "medusa-head," 1–2 mm in diameter at 48 hours.

Although resembling some *Fusobacterium* species morphologically, *L. buccalis* differs in being strongly saccharolytic and producing large amounts of lactic acid (see Table 6.11). Another filamentous organism from dental material, *Corynebacterium matruchotii (Bacterionema matruchotii),* differs from *L. buccalis* by being aerobic, Gram positive, and showing a "whip-handle" morphology.

Genus **Malonomonas**

Editorial note: The genus *Malonomonas* was not included in *Bergey's Manual of Systematic Bacteriology*. It was created in 1989 by Dehning and Schink (Arch. Microbiol. *151:* 427–433; Int. J. Syst. Bacteriol. *40:* 320, 1990). It includes a single species, *M. rubra.*

Straight to slightly curved rods, 0.4 × 3.1–4.0 μm, with rounded ends, occurring singly, in pairs, in short chains, or in large aggregates. Nonsporing. **Motile by 1–2 polar flagella.** Optimum temperature is 28–30°C; range, 22–45°C. Chemoorganotroph, growing best anaerobically but able to tolerate up to 5% of oxygen. Able to grow with malonate as the sole source of carbon and energy. **Malonate is decarboxylated to acetate; fumarate and malate are fermented to succinate and CO_2. No other organic acids, sugars, or alcohols are utilized.** Contain large amounts of cytochrome *c.* Isolated from anoxic marine muds.

Type (and only) species: *Malomonas rubra.*

Characteristics of the species: Requires at least 150 mM (0.88%) NaCl for growth. Red disc-shaped colonies are formed in solid media, the color being due to the large amount of cytochrome. Other properties are as described for the genus.

Genus **Megamonas**

Editorial note: The genus *Megamonas* was not included in *Bergey's Manual of Systematic Bacteriology*. It was created in 1982 by Shah and Collins (Zentralbl. Bakteriol. Hyg. *C3:* 394–398; Int. J. Syst. Bacteriol. *33:* 439, 1983). It includes a single species, *M. hypermegas* (formerly *Bacteroides hypermegas*).

Large rods, 0.8–3.0 × 3–20 μm, with rounded ends, usually granular in appearance due to volutin. Nonsporing. **Nonmotile.** Chemoorganotrophic, fermenting a variety of carbohydrates. **The end products are acetic, propionic, and lactic acids.** Do not contain sphingolipids or menaquinones, and non-hydroxylated fatty acids are predominantly straight-chain. **Isolated from the intestinal tracts of humans, animals, and poultry.**

Type species: *Megamonas hypermegas.*

Characteristics of the species: Actively saccharolytic but not proteolytic. Esculin is hydrolyzed. Other characteristics are as described for the genus.

Genus **Mitsuokella**

Editorial note: The genus *Mitsuokella* was not included in *Bergey's Manual of Systematic Bacteriology*. It was created in 1982 by Shah and Collins (Zentralbl. Bakteriol. Hyg. *C3:* 491–494; Int. J. Syst. Bacteriol. *33:* 439, 1983). It included a single species, *M. multiacidus* (formerly *Bacteroides multiacidus*.) A second species, *M. dentalis*, was added in 1986 by Haapasalo et al. (Int. J. Syst. Bacteriol. *36:* 566–568).

Regular or ovoid rods 0.7–1.5 × 1.2–1.5 μm. Nonsporing and **nonmotile.** Fimbriae may be present, and some strains are capsulated. Carbohydrates are fermented, sometimes vigorously. **The end products from glucose fermentation are acetic and succinic acids with a variable amount of lactic acid.** Do not contain sphingolipids or menaquinones, and branched-chain fatty acids are only minor components. **Isolated from the feces of humans and pigs and from dental root canal infections in humans.**

Type species: *Mitsuokella multiacidus.*

Differentiation of the species of the genus **Mitsuokella:** See Table 6.14.

Genus **Oxalobacter**

Editorial note: The genus *Oxalobacter* was not included in *Bergey's Manual of Systematic Bacteriology*. It was created in 1985 by Allison et al. (Arch. Microbiol. *141:* 1-7; Int. J. Syst. Bacteriol. *35:* 375, 1985). It includes a single species, *O. formigenes.*

Straight or curved rods 0.4–0.6 × 1.2–1.5 μm, with rounded ends. Nonsporing. **Nonmotile. Oxalate is the only source of carbon and energy, and formate and CO_2 are produced as end products. Isolated from the rumen or large bowel of humans and animals, and from lake sediments.**

Type (and only) species: *Oxalobacter formigenes.*

Characteristics of the species: Produces zones of clearing around colonies when grown on agar medium containing calcium oxalate. Other properties are as described for the genus.

Genus **Pectinatus**

Slightly curved rods 0.7–0.9 × 3–30 µm or more, with rounded ends. Nonsporing. **Motile by means of flagella arranged in a comb-like fashion on the concave side of the cell.** Optimum temperature is 30°C. **Sugars are fermented to acetic and propionic acids with smaller amounts of succinic and lactic. Isolated from spoiled beer and pitching yeast.**

Type species: *Pectinatus cerevisiiphilus.*

Differentiation of the species of the genus **Pectinatus:** See Table 6.15.

Genus **Pelobacter**

Editorial note: The genus *Pelobacter* was not included in *Bergey's Manual of Systematic Bacteriology.* It was created in 1982 by Schink and Pfennig (Arch. Microbiol. *133:* 195–201; Int. J. Syst. Bacteriol. *33:* 896, 1983). It included a single species, *P. acidigallici.* A second species, *P. venetianus,* was added in 1983 by Schink and Stieb (Appl. Environ. Microbiol. *45:* 1905–1913; Int. J. Syst. Bacteriol. *34:* 91, 1984). Two more species, *P. carbinolicus* and *P. propionicus,* were added in 1984 by Schink (Arch. Microbiol. *137:* 33–41; Int. J. Syst. Bacteriol. *34:* 356, 1984). A fifth species, *P. acetylenicus,* was added in 1985 by Schink (Arch. Microbiol. *142:* 295–301; Int. J. Syst. Bacteriol. *36:* 355, 1986).

Straight or slightly curved rods 0.5–0.8 µm wide and 1.2–6.0 µm long, with rounded or slightly pointed ends, occuring singly, in pairs or short chains. Nonsporing. **Some species are motile.** Optimum temperature is 33–35°C. **Utilize a limited range of substrates such as pyrogallol, 2,3-butanediol, ethylene glycol, polyethylene glycol, acetylene, acetoin, pyruvate, and lactate. Found in anoxic marine or freshwater muds.**

Type species: *Pelobacter acidigallici.*

Differentiation of the species of the genus **Pelobacter:** See Table 6.16.

Genus **Porphyromonas**

Editorial note: The genus *Porphyromonas* was not included in *Bergey's Manual of Systematic Bacteriology.* It was created in 1988 by Shah and Collins (Int. J. Syst. Bacteriol. *38:* 128–131). It includes three species: *P. asaccharolytica* (formerly *Bacteroides asaccharolyticus*), *P. gingivalis* (formerly *Bacteroides gingivalis*), and *P. endodontalis* (formerly *Bacteroides endodontalis*).

Short rods 0.5–0.8 × 1.0–3.0 µm. Nonsporing and nonmotile. **Generally cells form brown to black colonies on blood agar due to protoheme production. Asaccharolytic:** growth is not significantly affected by carbohydrates but is greatly enhanced by protein hydrolysates such as peptone or by yeast extract. **Major fermentation products are *n*-butyric and acetic acids:** minor products are propionic, isobutyric, isovaleric, and sometimes phenylacetic acids. Nonhydroxylated cellular fatty acids are predominantly isomethyl branched (iso-C_{15}) acids. **Isolated from oral infections and root canal infections.**

Type species: *Porphyromonas asaccharolyticus.*

Differentiation of the species of the genus **Porphyromonas:** See Table 6.17.

Genus **Prevotella**

Editorial note: The genus *Prevotella* was not included in *Bergey's Manual of Systematic Bacteriology.* The genus was created in 1990 by Shah and Collins (Int. J. Syst. Bacteriol. *40:* 205–208) with the type species *P. melaninogenica* (formerly *Bacteroides melaninogenicus*). Fifteen other species were transferred from the genus *Bacteroides* to *Prevotella.*

Pleomorphic rods, nonmotile, nonsporing. Obligately anaerobic. Chemoorganotrophic. **Moderately saccharolytic;** major fermentation products are acetate and succinate, with lower levels of isobutyrate, isovalerate, and lactate. Growth is inhibited by 20% (w/v) bile.

Differentiation of the species of the genus **Prevotella:** See Table 6.7 (Species of *Bacteroides* now included in *Prevotella* are marked as footnote *b*).

Genus **Propionigenium**

Editorial note: The genus *Propionigenium* was not included in *Bergey's Manual of Systematic Bacteriology*. It was created in 1982 by Schink and Pfennig (Arch. Microbiol. *133:* 209–216; Int. J. Syst. Bacteriol. *33:* 896, 1983). It includes a single species, *P. modestum*.

Short rods 0.5–0.6 × 0.5–2.0 μm, with rounded ends, occuring singly, in pairs, or short chains. Nonsporing and **nonmotile.** Optimum temperature is 33°C. **Utilize short-chain organic acids for growth and producing propionate as the principal end product. Isolated from marine and freshwater muds and from human saliva.**

Type (and only) species: *Propionigenium modestum.*

Characteristics of the species: Ferments succinate, fumarate, malate, aspartate, oxalacetate, and pyruvate, producing propionate, acetate, and CO_2. In pure culture, 1% NaCl is necessary in the growth medium.

As a Gram-negative propionate-forming anaerobe, *Propionigenium* may be confused with *Propionispira*. However, *Propionigenium* is a short, nonmotile rod that ferments only a limited variety of organic acids and lacks cytochromes, whereas *Propionispira* forms long, curved, motile cells, ferments carbohydrates, and contains cytochromes.

Genus **Propionispira**

Editorial note: The genus *Propionispira* was not included in *Bergey's Manual of Systematic Bacteriology*. It was created in 1982 by Schink et al. (J. Gen. Microbiol. *128:* 2771–2779; Int. J. Syst. Bacteriol. *33:* 673, 1983). It includes a single species, *P. arboris*.

Curved to helical rods, 1.0 × 7.0 μm. May produce long spiral cell filaments. Nonsporing. **Motile by peritrichous flagella.** Optimum temperature, 30–33°C. **Ferments a wide range of carbohydrates, producing mainly propionate, acetate, and CO_2. Can grow with N_2 as sole nitrogen source and reduce acetylene to ethylene.** Cytochrome *b* is present. **Isolated from alkaline wetwoods of poplar trees.**

Type (and only) species: *Propionispira arboris.*

Characteristics of the species: As a Gram-negative propionate-forming anaerobe, *Propionispira* may be confused with *Propionigenium*. However, *Propionispira* is a curved motile organism that ferments a wide range of sugars and contains cytochromes, whereas *Propionigenium* is a short, nonmotile rod that lacks cytochromes and ferments only a limited variety of organic acids.

Genus **Rikenella**

Editorial note: The genus *Rikenella* was not included in *Bergey's Manual of Systematic Bacteriology*. It was created in 1985 by Collins et al. (Syst. Appl. Microbiol. *6:* 79–81; Int. J. Syst. Bacteriol. *35:* 375, 1985). It includes a single species, *R. microfusus* (formerly *Bacteroides microfusus*).

Small rods with pointed ends. Generally 0.15–0.3 × 0.3–1.5 μm but may be up to 5.0 μm long. Nonsporing. **Nonmotile.** Optimum temperature is 37°C. **Ferment glucose and a few other sugars, producing principally propionic and succinic acids.** Fatty acids are principally isomethyl branched-chain acids, and menaquinones are present. Optimum temperature is 37°C. Isolated from fecal or cecal samples from calves, chickens, and Japanese quail.

Type (and only) species: *Rikenella microfusus.*

Characterisitcs of the species: Ferments glucose, mannose, lactose, and melibiose, and sometimes galactose: growth is improved by the presence of carbohydrate. Strains are β-hemolytic, liquefy gelatin, and grow well in the presence of 20% bile. Produces propionic and succinic acids, but not butyric or lactic acids.

Genus **Roseburia**

Editorial note: The genus *Roseburia* was not included in *Bergey's Manual of Systematic Bacteriology*. It was created in 1983 by Stanton and Savage (Int. J. Syst. Bacteriol. *33:* 618–627). It includes a single species, *R. cecicola*.

Slightly curved rods, 0.5 × 2.0–5.0 μm. Nonsporing. **Actively motile by a bundle of flagella inserted subterminally on the concave side of the cell, or sometimes at one end. Ferment several carbohydrates with butyrate as the principal or the only product of metabolism. Isolated from the cecum of the mouse.**

Type (and only) species: *Roseburia cecicola.*

Characteristics of the species: Cells have 20–35 flagella in a bundle which may appear as a single subterminal fasicle by phase-contrast microscopy. Good growth is obtained in a medium containing volatile fatty acids, yeast extract, trypticase, inorganic salts, hemin, and glucose. Other end products formed from glucose in addition to butyrate are CO_2, H_2, and traces of ethanol.

The only other curved motile Gram-negative anaerobic rod from the intestine that produces large amounts of butyric acid is *Butyrivibrio.* Unlike *Roseburia,* however, *Butyrivibrio* cells have predominantly a single polar or subpolar flagellum and a Gram-positive type of cell wall (although the cells stain Gram negative). *Roseburia* may also be confused with *Fusobacterium,* which produces butyrate as a principal end product; however, in addition to a difference in cell shape, *Roseburia* is motile, whereas *Fusobacterium* is nonmotile.

Genus **Ruminobacter**

Editorial note: The genus *Ruminobacter* was not included in *Bergey's Manual of Systematic Bacteriology.* It was created in 1986 by Stackebrandt and Hippe (Syst. Appl. Microbiol. *8:* 204–207; Int. J. Syst. Bacteriol. *37:* 175, 1987). It includes a single species, *R. amylophilus* (formerly *Bacteroides amylophilus).*

Pleomorphic ovals to short rods 0.9–1.2 × 1.0–3.0 μm, with round or tapered ends, and sometimes swollen or irregular forms. Nonsporing. **Nonmotile.** Optimum temperature is 37–39°C. **Carbohydrates are metabolized to acetate, formate, and succinate.** Sphingolipids and menaquinones are absent; predominant cell fatty acids are straight chain saturated and unsaturated. **Found in the rumen of cattle and sheep.**

Type (and only) species: *Ruminobacter amylophilus.*

Characteristics of the species: Carbon dioxide, ammonia, and a fermentable carbohydrate are required for growth. Acid is produced from dextrin, glycogen, starch, and maltose; cellulose and cellobiose are not fermented.

Genus **Sebaldella**

Editorial note: The genus *Sebaldella* was not included in *Bergey's Manual of Systematic Bacteriology.* It was created in 1986 by Collins and Shah (Int. J. Syst. Bacteriol. *36:* 349–350). It includes a single species, *S. termitidis* (formerly *Bacteroides termitidis).*

Rod-shaped cells, 0.3–0.5 × 2.0–12.0 μm, with central swellings. Occur singly, in pairs, and in filaments. Nonsporing. **Nonmotile. Ferment sugars to acetic and lactic acids and sometimes also formic.** Menaquinones are absent, and the major cellular fatty acids are straight chain saturated and unsaturated. **Found in the posterior intestinal contents of termites.**

Type (and only) species: *Sebaldella termitidis.*

Characteristics of the species: Ferments glucose, fructose, maltose, mannitol, mannose, rhamnose, sucrose, trehalose, and xylose. Degrades uric acid to CO_2, acetate, and ammonia. Most strains produce H_2S. Other properties are as described for the genus.

Genus **Selenomonas**

Curved, usually crescent-shaped rods, 0.9–1.1 × 3.0–6.0 μm, with ends often tapered. Occur singly, in pairs, or short chains. Capsules are not present. Nonsporing. **Have an active tumbling motility due to flagella (up to 16) arranged as a tuft or short line near the center of the concave side of the cell.** Have a fermentative type of metabolism. **Fermentation of glucose gives mainly acetic and propionic acids with CO_2 and/or lactate. Found mainly in the human buccal cavity, the rumen of herbivores, and the cecum of pigs and several rodents;** however, one species has been isolated from a commercial anaerobic digestor, and another from brewery yeast.

Type species: *Selenomonas sputigena.*

Differentiation of the species of the genus **Selenomonas:** See Table 6.18. In addition, the following key is helpful in differentiating *Selenomonas* species isolated from the mouth.

Key to the species of **Selenomonas** *isolated from the mouth*

I. Acid from sucrose

    A. Esculin hydrolyzed

        1. Acid from trehalose

            S. dianae

        2. No acid from trehalose

            S. infelix

    B. Esculin not hydrolyzed

        1. Acid from lactose

            a. Acid from mannitol

                S. flueggei

            aa. No acid from mannitol

                S. sputigena

        2. No acid from lactose

            S. artemidis

II. No acid from sucrose

        S. noxia

Genus **Sporomusa**

Editorial note: The genus *Sporomusa* was not included in *Bergey's Manual of Systematic Bacteriology.* It was created in 1984 by Moller et al. (Arch. Microbiol. *139:* 388–396; Int. J. Syst. Bacteriol. *35:* 224, 1985) and included two species, *S. sphaeroides* and *S. ovata.* Four other species were subsequently added: *S. acidovorans* (Olliver et al., Arch. Microbiol. *142:* 307–310,1985; Int. J. Syst. Bacteriol. *40:* 105, 1990); *S. paucivorans* (Hermann et al., Int. J. Syst. Bacteriol. *37:* 93-101, 1987); *S. termitida* (Breznak et al., Arch. Microbiol. *150:* 282–288, 1988; Int. J. Syst. Bacteriol. *40:* 212, 1990); and *S. malonica* (Dehning et al., Arch. Microbiol. *151:* 421–426, 1989; Int. J. Syst. Bacteriol. *40:* 321, 1990).

Curved rods 0.5–1.0 × 1–8 µm, with tapered ends. Most strains form round or oval, terminal or subterminal, heat-resistant spores. The spores withstand exposure to 80°C for at least 10 min. **Motile by up to 15 flagella arising from the concave side of the cell.** Optimum temperature is 30–39°C. **Can utilize certain N-methyl compounds, amino acids, alcohols, hydroxy fatty acids, or other organic acids, but very few sugars or sugar alcohols can be used as an energy source. All species can use H_2 plus CO_2 as growth substrates. The major fermentation end product is acetate,** but butyrate or propionate may be produced additionally from some substrates. Sulfate is not reduced. Isolated from river mud, factory waste, and the gut of termites.

Type species: *Sporomusa sphaeroides.*

Additional note: Morphologically, the genus *Sporomusa* resembles the genus *Selenomonas* in consisting of curved, motile, Gram-negative rods with lateral flagellation on the concave side of the cell. However, in addition to the failure to form spores, *Selenomonas* cannot grow with H_2 and CO_2 as substrates, and the pattern of substrate utilization is very different (see Table 6.19). For similar reasons, *Sporomusa* can be differentiated from other curved, motile, Gram-negative genera such as *Roseburia, Centipeda, Pectinatus,* and *Acetivibrio.*

Differentiation of the species of the genus **Sporomusa:** See Table 6.19.

Genus **Succinimonas**

Short, straight rods, 1.0–1.5 × 1.2–3.0 μm, with rounded ends. Nonsporing. **Motile by a single polar flagellum.** Optimum temperature is 30–37°C. **Ferment glucose, maltose, dextrin, or starch, but not other carbohydrates, and they produce large amounts of succinic and smaller amounts of acetic acid. Found in the rumen of cattle** fed hay-grain diets.

Type (and only) species: *Succinimonas amylolytica.*

Characteristics of the species: *S. amylolytica* is composed of short straight rods, sometimes almost coccal, unlike other Gram-negative motile rods from the rumen, which are curved or helical. It is also distinguished by producing large amounts of succinic acid. Strains grow well in a synthetic medium containing glucose, mineral salts, ammonia, B vitamins, and CO_2/HCO_3^- buffer. Acetate is stimulatory, and CO_2 is consumed during fermentation of glucose.

Genus **Succinivibrio**

Curved or helical rods, 0.4–0.6 × 1.0–7.0 μm, with pointed ends. Nonsporing. **Motile by means of a single polar flagellum. Ferment a variety of sugars, with acetic and succinic acids as principal end products** and smaller amounts of formate and lactate. A large net uptake of CO_2 may occur during growth. **Isolated from the rumen of cattle and sheep, rarely reported from human infections.**

Type (and only) species: *Succinivibrio dextrinosolvens.*

Characteristics of the species: Species is found in the rumen of cattle and sheep, especially if fed high grain diets. Many strains are reported to have urease activity. The species has been reported from cases of septicemia in humans (Porschen and Chan, J. Clin. Microbiol. *5:* 444–447, 1977; Southern, Am. J. Clin. Pathol. *64:* 540–543, 1975). In many ways, *S. dextrinosolvens* resembles *Anaerobiospirillum succiniciproducens;* however, the latter has tufts of flagella at each pole (see Table 6.4).

Genus **Syntrophobacter**

Editorial note: The genus *Syntrophobacter* was not included in *Bergey's Manual of Systematic Bacteriology.* It was created in 1980 by Boone and Bryant (Appl. Environ. Microbiol. *40:* 626–632; Int. J. Syst. Bacteriol. *34:* 356, 1984). It includes a single species, *S. wolinii.*

Rods and long filaments, 0.6–1.0 × 1.0–35.0 μm. Nonsporing. **Nonmotile.** Optimum temperature is 35°C. **Grow only in co-culture with strains of *Desulfovibrio.*** There is no growth in the absence of H_2-utilizing organisms. **Oxidize propionate, but not other fatty acids, to acetate, H_2, and CO_2. Isolated from an anaerobic sewage digestor.**

Type (and only) species: *Syntrophobacter wolinii.*

Characteristics of the species: The properties are as described for the genus. For differentiation from *Syntrophomonas* spp., see Table 6.20.

Genus **Syntrophomonas**

Editorial note: The genus *Syntrophomonas* was not included in *Bergey's Manual of Systematic Bacteriology.* It was created in 1981 by McInerney et al. (Appl. Environ. Microbiol. *41:* 1029–1039; Int. J. Syst. Bacteriol. *32:* 267, 1982), who included a single species, *S. wolfei.* A second species, *S. sapovorans,* was added in 1986 by Roy et al. (Arch. Microbiol. *145:* 142–147; Int. J. Syst. Bacteriol. *37:* 179, 1987).

Slightly helical rods 0.5–1.0 × 2–7 μm. Nonsporing. **Motile by 2–8 flagella inserted linearly on the concave side.** Under most conditions cells are only sluggishly motile. May contain granules of poly-β-hydroxybutyrate. Optimum temperature is 30–37°C. **Obtain energy by β-oxidation of fatty acids in co-**

culture with organisms that utilize H$_2$, such as *Desulfovibrio* spp. or *Methanospirillum hungatei.* Most strains can be adapted to grow in pure culture on crotonate, although growth is slow. **Fatty acids are degraded primarily to acetate and H$_2$,** but propionate or isovalerate may be formed depending on the fatty acid attacked. Acetate and butyrate are produced from crotonate. **Isolated from anoxic muds, sewage digestors, and the bovine rumen.**

Type species: *Syntrophomonas wolfei.*

Differentiation of the species of the genus **Syntrophomonas:** See Table 6.20.

Genus **Thermobacteroides**

Editorial note: The genus *Thermobacteroides* was not included in *Bergey's Manual of Systematic Bacteriology.* It was created in 1981 by Ben-Bassat and Zeikus (Arch. Microbiol. *128*: 365–370; Int. J. Syst. Bacteriol. *33*: 673, 1983) and the genus description was emended in 1985 by Ollivier et al. (Int. J. Syst. Bacteriol. *35*: 425–428). The genus originally included a single species, *T. acetoethylicus,* but two other species were added subsequently: *T. proteolyticus* (Ollivier et al., Int. J. Syst. Bacteriol. *35*: 425–428, 1985) and *T. leptospartum* (Toda et al., Agr. Biol. Chem. *52*: 1339–1344, 1988; Int. J. Syst. Bacteriol. *39*: 93, 1989).

Straight rods, 0.5–0.6 × 1.0–6.0 μm, occurring singly or in pairs. Some strains have filamentous forms. Nonsporing. **Motile by peritrichous flagella or cells are nonmotile. Thermophilic; optimum temperature is 60–65°C; range is 35–80°C. Metabolize carbohydrates or peptones; the end products of metabolism may include acetate, butyrate, isobutyrate, isovalerate, propionate, ethanol, CO$_2$, and H$_2$. Found in thermophilic digestors, cattle compost, and natural thermal environments where organic matter is decomposing.**

Type species: *Thermobacteroides acetoethylicus.*

Differentiation of the species of the genus **Thermobacteroides:** See Table 6.21.

Genus **Thermosipho**

Editorial note: The genus *Thermosipho* was not included in *Bergey's Manual of Systematic Bacteriology.* It was created in 1989 by Huber et al. (Syst. Appl. Microbiol. *12*: 32–37; Int. J. Syst. Bacteriol. *39*: 496, 1989). It includes a single species, *T. africanus.*

Rods 0.5 × 3.0–4.0 μm surrounded by a sheath-like structure ballooning over the ends of the cell: up to 12 cells can be enclosed on the same sheath. Nonmotile. Thermophilic; optimum temperature is 75°C, but growth can occur at temperatues as low as 35°C. Carbohydrates are not attacked; complex organic material such as yeast extract, peptone, or tryptone is required for growth. Hydrogen is inhibitory, and inhibition is overcome by the addition of sulfur, with formation of H$_2$S. Will grow in NaCl concentrations as high as 3.6%. **Isolated from geothermally heated tidal springs at Djibouti, Africa.**

Type (and only) species: *Thermosipho africanus.*

Characteristics of the species: Characteristics differentiating *T. africanus* from other thermophilic, Gram-negative, sheathed bacteria are listed in Table 6.22.

Genus **Thermotoga**

Editorial note: The genus *Thermotoga* was not included in *Bergey's Manual of Systematic Bacteriology.* It was created in 1986 by Huber et al. (Arch. Microbiol. *144*: 324–333; Int. J. Syst. Bacteriol. *36*: 575, 1986). It initially contained a single species, *T. maritima.* Two other species were added subsequently: *T. neopolitana* (Jannasch et al., Arch. Microbiol. *150*: 103–104, 1986; Int. J. Syst. Bacteriol. *39*: 93, 1989) and *T. thermarum* (Windberger et al., Arch. Microbiol. *151*: 506–512, 1989; Int. J. Syst. Bacteriol. *42*: 327, 1992).

Rods 0.6 × 1.5–11.0 μm, surrounded by a sheath-like structure which protrudes balloon-like beyond the ends of the cell. Nonsporing. May be motile by polar, lateral, or peritrichous flagella, or nonmotile. **Thermophilic; optimum temperature is 70–80°C; growth occurs at temperatures as high as 90°C. Metabolize carbohydrates, tryptone, or yeast extract; fermentation products being principally lactate, acetate, CO$_2$, and H$_2$.** Hydrogen is inhibitory to growth, but with most strains the inhibition can be overcome by adding sulfur, with formation of H$_2$S. Marine strains will grow at NaCl concentrations up to 3.75%. **Isolated from geothermally heated marine sediments or hot springs.**

Type species: *Thermotoga maritima.*

Differentiation of the species of the genus **Thermotoga:** See Table 6.22.

Genus **Tissierella**

Editorial note: The genus *Tissierella* was not included in *Bergey's Manual of Systematic Bacteriology.* It was created in 1986 by Collins and Shah (Int. J. Syst. Bacteriol. *36:* 461–463). It includes a single species, *T. praeacuta* (formerly *Bacteroides praeacutus*).

Rods 0.6–0.9 × 2.0–20.0 µm, with rounded or pointed ends, sometimes filamentous. Nonsporing. **Motile by peritrichous flagella.** Optimum temperature is 37°C; range, 25–45°C. **Nonfermentative or weakly fermentative: end products in glucose broth are acetic, butyric, and isovaleric acids.** Menaquinones are absent; cell fatty acids are primarily isomethyl branched acids. Malate and glutamate dehydrogenase are absent. **Isolated from infant and adult feces, and occasionally from clinical specimens.**

Type (and only) species: *Tissierella praeacuta.*

Characteristics of the species: As described for the genus.

Genus **Wolinella**

Helical, curved, or straight cells 0.5–1.0 × 2–6 µm, with rounded or tapered ends. Nonsporing. **Motile by a single polar flagellum.** Although *Wolinella* species were described as anaerobes in *Bergey's Manual of Systematic Bacteriology,* vol. 1, they are in fact **H$_2$- and formate-requiring microaerophiles capable of respiring with oxygen.** In the absence of alternative terminal electron acceptors such as fumarate, they exhibit oxygen-dependent growth with low levels of oxygen and will not grow anaerobically or aerobically (For *W. succinogenes,* see Wolin et al., J. Bacteriol. *81:* 911–917, 1961; for *W. recta,* see Ohta and Gottschal, FEMS Microbiol. Ecol. *53:* 79–86, 1988, and FEMS

Microbiol. Lett. *50:* 163–168, 1988; for *W. curva,* see Han et al., Int. J. Syst. Bacteriol. *41:* 218–222, 1991). See also Groups 2 and 4 in this manual. **Oxidase positive** and catalase negative. Contain cytochromes. **Under anaerobic conditions, growth is dependent upon the presence of H$_2$ and fumarate, or formate and fumarate:** formate or H$_2$ is oxidized to CO$_2$, and fumarate is reduced to succinate. **Carbohydrates are not utilized. Isolated from the bovine rumen, the human gingival sulcus, dental root canal infections, and other clinical material.**

Type species: *Wolinella succinogenes.*

Additional note: Although wolinellas are not anaerobes, they will normally be isolated by methods used for strict anaerobes, and from a practical point of view they are included here with the other members of Group 6.

Differentiation of the species of the genus **Wolinella:** See Table 6.23.

Genus **Zymophilus**

Editorial note: The genus *Zymophilus* was not included in *Bergey's Manual of Systematic Bacteriology.* It was created in 1990 by Schleifer et al. (Int. J. Syst. Bacteriol. *40:* 19–27). It includes two species, *Z. raffinosivorans* and *Z. paucivorans.*

Straight to slightly curved or helical rods, 0.7–1.0 × 3–30 µm, occurring singly, in pairs, or in short chains. Motile (but motility may be lost after several subcultivations). The flagellar arrangement has not been described. Have a fermentative type of metabolism. **Glucose is fermented to acetic and propionic acids; trace amounts of lactic acid may sometimes occur. Isolated from pitching yeast or brewery wastes.**

Type species: *Zymophilus raffinosivorans.*

Differentiation of the species of the genus **Zymophilus:** *Z. raffinosivorans* can be differentiated from *Z. paucivorans* by its ability to utilize xylose, raffinose, inositol, and melibiose. Also, cells of *Z. raffinosivorans* are straight to slightly curved, while cells of *Z. paucivorans* are curved to helical.

Table 6.1 Differential characteristics of Gram-negative, anaerobic, straight, curved, and helical bacteria[a]

Characteristic	Acetivibrio	Acetoanaerobium	Acetofilamentum	Acetogenium	Acetomicrobium	Acetothermus	Acidaminobacter	Anaerobiospirillum
Predominant shape	Curved	Straight	Long rigid filaments[c]	Straight	Curved	Straight	Straight	Coiled spirals
Cell dimensions (µm):								
Width	0.4–0.8	0.8	0.18–0.3	0.7–0.8	0.6–0.8	0.5	0.5–0.6	0.6–0.8
Length	4.0–10.0	1.0–5.0	100 or more	2.0–7.5	2.0–7.0	5.0–8.0	1.5–3.7	3.0–15.0
Motility	+	+	–	–	+		–	+
Flagellar arrangement:								
Single polar	+	–			+			–
Polar tufts	–	–	+		–			+
Subpolar	–	–	–		–			–
Lateral (one side)	–	–	–		+			–
Peritrichous	–	+	–		–			–
Cell wall type:								
Gram negative	+	–	–	–	+	+	+	+
Atypical Gram negative	–	+	+	+	–	–	–	–
Gram positive	–	–	–	–	–	–	–	–
Temperature relationships:								
Thermophilic	–	–	–	+	–	+	–	–
Range for growth (°C)	20–40		35–38	5–72	35–75	50–60	15–42	18–50
Optimum (°C)	35	37		66	35–73	58	30	37
Grow only in co-culture[h]	–	–	–	–	–	–	–	–
Fermentative type of metabolism	+	+	+	+	–	+	+	+
Principal end products[n]	A,Eth, CO$_2$,H$_2$	A,(p,b, ib,iv)	A,CO$_2$,H$_2$	A	A,L,Eth, CO$_2$,H$_2$	A	A,(p)	A,S
NaCl requirements					Inhibited by 4%		Growth up to 3.2%	
Pathogenicity	–	–	–	–		–	–	+
Principal habitat	Sewage, pig colon	Hula Swamp, Israel	Sewage sludge	Lake Kivu, Africa	Sewage sludge	Sewage sludge	Estuarine mud	Dogs; cases of diarrhea in humans

Footnotes are at end of table

Table 6.1 (continued)

Characteristic	Anaero-rhabdus	Anaero-vibrio	Bacter-oides	Butyrivibrio	Centipeda	Fervido-bacterium	Fibro-bacter	Fusobacte-rium	Haloanaero-bium
Predominant shape	Pleomorphic short rods	Curved	Variable	Curved	Serpentine rods	Straight	Short or coccoid rods	Variable	Straight
Cell dimensions (μm):									
Width	0.3–1.5	0.5	Variable	0.4–0.6	0.65	0.50–0.55	0.4–0.8	Variable	0.9–1.1
Length	1.0–3.0	1.2–10.0	Variable	2.0–5.0	4.0–17.0	1.0–2.5	0.8–2.0	Variable	2.0–2.6
Motility	–	+	–[d]	+	+	+[e]	–	–	–
Flagellar arrangement:									
Single polar		+	–	+	–				
Polar tufts		–	–		–				
Subpolar		–	–		–				
Lateral (one side)		–			+				
Peritrichous		–	+[d]	–					
Cell wall type:									
Gram negative	+	+	+	+	+	+	+	+	+
Atypical Gram negative	–	–	–	–	–	–	–	–	–
Gram positive	–	–	–	–	–	–	–	–	–
Temperature relationships:									
Thermophilic	–	–	–	–	–	+	–	–	–
Range for growth (°C)	25–44	30–37		30–45	32–37	40–80			>5 – <60
Optimum (°C)	30–37	30–37	37	37	37	65–70	37	37	37
Grow only in co-culture[h]	–	–	+ or –	–	–	–	–	–	
Fermentative type of metabolism	–	+	+ or –	+	+	+	+	–	+
Principal end products[n]	A,L	A,P, CO₂,s	A,F,L, P,S°	B,(A,l,s)	P,a,l,s	L,A,H₂, CO₂,(Eth)	A,S,f	B,a, f,l,p	A,B,P, CO₂,H₂
NaCl requirements		–		–					Opt., 13%
Pathogenicity	+		+ or –	–	+		–	+	
Principal habitat	Pyogenic lesions; feces	Rumen; river mud	Wide range of anaerobic habitats	Rumen; human and animal intestine	Human periodontal disease	Hot springs, New Zealand	Rumen and cecum	Gingival sulcus; intestine, genital tract	Great Salt Lake, Utah, USA

Footnotes are at end of table

Table 6.1 (continued)

Characteristic	Halobacteroides	Ilyobacter	Lachnospira	Leptotrichia	Malonomonas	Megamonas	Mitsuokella	Oxalobacter	Pectinatus	Pelobacter
Predominant shape	Straight or curved	Straight	Curved	Straight or slightly curved	Straight or slightly curved	Large rods	Straight	Straight or curved	Slightly curved	Straight or slightly curved
Cell dimensions (µm):										
Width	0.5	0.7–1.0	0.4–0.6	0.8–1.5	0.4	0.8–3.0	0.7–1.5	0.4–0.6	0.7–0.9	0.5–0.8
Length	10.0–20.0	1.3–3.0	2.0–4.0	5.0–15.0	3.0–4.0	3.0–20.0	1.2–1.5	1.2–1.5	3.0–20.0	1.2–6.0
Motility	+	+ or – [f]	+	–	+	–	–	–	+	+ or –
Flagellar arrangement:										
Single polar	–	–	–	–	+	–	–	–	–	+
Polar tufts	–	–	–	–	–	–	–	–	–	–
Subpolar	–	–	–	–	–	–	–	–	–	–
Lateral (one side)	–	–	+	–	–	–	–	–	+	–
Peritrichous	+	–	–	–	–	–	–	–	–	+
Cell wall type:										
Gram negative	+	+	–	–	+	+	+	+	+	+
Atypical Gram negative	–	–	–	+	–	–	–	–	–	–
Gram positive	–	–	+	–	–	–	–	–	–	–
Temperature relationships:										
Thermophilic	–	–	–	–	–	–	–	–	–	–
Range for growth (C°)	37–45	10–35	>22– <50	35–37	22–45	–	–	14–45	33–35	33–35
Optimum (°C)	–	28–34	38	–	28–30	37	37	37	–	–
Grow only in co-culture [h]	–	–	–	–	–	–	–	–	–	–
Fermentative type of metabolism	+	+	+	+	+ [i]	+	+	+ [j]	+	+ [k]
Principal end products [n]	A,Eth, CO₂,H₂	A,B,F, Eth [p]	A,F,L, Eth,CO₂,H₂ [q]	L,(a,s)	A,S,CO₂	A,L,S	A,L,S	F,CO₂	A,P,s, Acetoin	A,P,Eth,CO₂ [r]
NaCl requirements	Opt., 8.4–14%	Opt., 1%			At least 0.87%			–		Marine spp. need 1%
Pathogenicity	–	–	–	–	–	+	+	–	–	–
Principal habitat	Dead Sea, Israel	Marine mud	Rumen	Gingival crevice	Marine mud	Intestinal tract of humans and animals	Intestinal tract; root canal sepsis	Colon; rumen; lake mud	Spoiled beer	Freshwater or marine mud

Footnotes are at end of table

308

Table 6.1 (continued)

Characteristic	Porphyromonas	Propionigenium	Propionispira	Rikenella	Roseburia	Ruminobacter	Sebaldella	Selenomonas	Sporomusa
Predominant shape	Short rods	Short rods or coccoid	Curved or helical; sometimes long filaments	Small rods with pointed ends	Slightly curved	Short rods	Rods with central swellings	Curved	Spore-forming rods
Cell dimensions (µm):									
Width	0.5–0.8	0.5–0.6	1.0	0.15–0.30	0.5	0.9–1.2	0.3–0.5	0.9–1.1	0.5–1.0
Length	1.0–3.0	0.5–2.0	7.0	0.3–5.0	2.0–5.0	1.0–3.0	2.0–12.0	3.0–6.0	1.0–8.0
Motility	–	–	+	–	+	–	–	+	+
Flagellar arrangement:									
Single polar			–		–			–	–
Polar tufts			–					–	–
Subpolar			–		+[g]			–	–
Lateral (one side)			–		–			+	+
Peritrichous			+		–			–	–
Cell wall type:									
Gram negative	+	+	+	+	+	+	+	+	–
Atypical Gram negative	–	–	–	–	–	–	–	–	+
Gram positive	–	–	–	–	–	–	–	–	–
Temperature relationships:									
Thermophilic	–	–	–	–	–	–	–	–	–
Range for growth (°C)	37	15–40	4–46	25–45	>22 – <45	37–39		30–45	30–39
Optimum (°C)	37	33	30–35	37	37	37–39	37	35–40	
Grow only in co-culture[h]	–	–	–	–	–	–	–		–
Fermentative type of metabolism	–	+[l]	+	+	+	+	+	+	+[m]
Principal end products[n]	A,B, pA[s],S,N	a,P,CO$_2$	A,P,CO$_2$	a,P,S	B,CO$_2$	A,F,S	A,L,(f)	A,P,CO$_2$(L)	A,(B,P)
NaCl requirements	0.5–0.8%	At least 1%	–	–	–	–	–		
Pathogenicity	+	–	–	–	–	–	–	+ or –	–
Principal habitat	Oral infections; root canals	Freshwater or marine mud; saliva	Alkaline wetwoods of poplar trees	Feces of birds and animals	Mouse cecum	Rumen	Termite gut	Mouth; cecum; rumen[w]	River mud; factory waste; termite gut

Footnotes are at end of table

Table 6.1 *(continued)*

Characteristic	Succini-vibrio	Succini-monas	Syntropho-bacter	Syntropho-monas	Thermo-bacter-oides	Thermo-sipho	Thermo-toga	Tissierella	Wolinella[b]	Zymophilus
Predominant shape	Curved	Short rods or coccoid	Rods and long filaments	Slightly curved	Straight	Sheathed rods	Sheathed rods	Straight, filaments	Curved or straight	Slightly curved to helical
Cell dimensions (µm):										
Width	0.4–0.6	1.0–1.5	0.6–1.0	0.5–1.0	0.5–0.6	0.5	0.6	0.6–0.9	0.5–1.0	0.7–1.0
Length	1.0–7.0	1.2–3.0	1.0–35.0	2.0–7.0	1.0–6.0	3.0–4.0	1.5–11.0	2.0–20.0	2.0–6.0	3.0–30.0
Motility	+	+	−	+	+ or −	−	+ or −	+	+	+[e]
Flagellar arrangement:										
Single polar	+	+		−	−		−	−	+	
Polar tufts	−	−	+	−	−	+	−	−	−	+
Subpolar	−	−	−	−	−	−	+	−	−	−
Lateral (one side)	−	−	−	+	−	−	+	−	−	−
Peritrichous	−	−	−	−	+		−	+	−	−
Cell wall type:										
Gram negative	+	+		−	−	+	+	+	+	−
Atypical Gram negative	−	−	−	+	+	−	−	−	−	+
Gram positive	−	−	−	−	−	−	−	−	−	−
Temperature relationships:										
Thermophilic	−	−	−	+	+	+	+	−	−	−
Range for growth (°C)	>22 − <45	>22 − <45			>40 − <80	33–77	55–90	25–45		
Optimum (°C)	30–39	30–37	35	35	65	75	70–80	37	35	30
Grow only in co-culture[h]	−	−	+	+	−	−	−	−	−	−
Fermentative type of metabolism	+	+	+	+	+	−	+	−	−	+
Principal end products[n]	A,S,f,l	a,S	A,CO2,H2	A,P, iB,iV	A,B,iB, iV,Eth,H2	t	A,L,CO2,H2	A,B,iV	S,CO2[u]	A,P
NaCl requirements	−	−	−	−	−	0.11–3.6%	0.25–6.0%[v]			
Pathogenicity	−	−	−	−	−			+		−
Principal habitat	Rumen	Rumen	Sewage digestor	Sewage digestor; rumen	Tannery waste; hot springs, Yellowstone N.P., USA	Hot springs, Africa	Hot springs, Italy, Azores, Africa	Feces; clinical samples	Rumen; human mouth	Pitching yeast; brewery waste

Footnotes are at end of table

Table 6.1 *(continued)*

^{a} Symbols: see standard definitions unless otherwise noted. Also: + or –, different reactions occur but a positive reaction is more common; – or +, different reactions occur, but a negative reaction is more common.

^{b} Although *Wolinella* species were described as anaerobes in *Bergey's Manual of Systematic Bacteriology*, Vol. 1, they are in fact H_2- and formate-requiring microaerophiles capable of respiring with oxygen. In the absence of alternative terminal electron acceptors such as fumarate, they exhibit oxygen-dependent growth with low levels of oxygen and will not grow anaerobically or aerobically (Wolin et al., J. Bacteriol. *81*: 911-917, 1961; Han et al., Int. J. Syst. Bacteriol. *41*: 218–222, 1991). See also Group 2 in this manual.

^{c} The filaments in *Acetofilamentum* are disaggregated by sodium dodecyl sulfate, giving individual cells 2-7 μm long; this is probably due to breakdown of a protein sheath.

^{d} Only 2 of the 47 species presently assigned to *Bacteroides* are motile, both by peritrichous flagella.

^{e} The flagellar arrangement has not been described.

^{f} The flagellar arrangement in motile strains of *Ilyobacter* has not been described.

^{g} *Roseburia* has a bundle of 20-35 flagella arising subterminally from the concave side of the cell.

^{h} With H_2-utilizing organisms such as *Desulfovibrio* or *Methanospirillum*.

^{i} *Malonomonas* uses malonate as the sole source of carbon and energy.

^{j} *Oxalobacter* uses oxalate as the sole source of carbon and energy.

^{k} *Pelobacter* uses only a limited range of substrates, such as pyrogallol, 2,3-butanediol, ethylene glycol, acetylene, lactate, and pyruvate.

^{l} *Propionigenium* uses short-chain organic acids for growth and produces mainly propionic acid as an end product.

^{m} *Sporomusa* uses few sugars but can utilize N-methyl compounds, amino acids, alcohols, hydroxy fatty acids, and other organic acids. All strains can use H_2 plus CO_2 as growth substrate.

^{n} Capital letters indicate >1 meq of acid per 100 ml of broth; lower case letters, <1 meq/100 ml. A or a, acetic; B or b, butyric; F, formic; iB or ib, isobutyric; iV or iv, isovaleric; L or l, lactic; P or p, propionic; S or s, succinic; Eth, ethanol. Products in parentheses may or may not be detected.

^{o} Butyrate is very seldom a product of metabolism by *Bacteroides* strains. Only two species, *B. salivosus* and *B. levii*, are reported to produce major amounts.

^{p} Glucose is fermented to acetate, formate, and ethanol; fermentation of other substrates yields a variety of end products.

^{q} Methanol is produced during digestion of pectin due to the action of pectinmethylesterase.

^{r} Other end products may be formed depending on the substrate (see Table 6.16).

^{s} Phenylacetic acid (pA) is formed by strains of *P. gingivalis*.

^{t} The end products have not been described.

^{u} *Wolinella* species require H_2 or formate as electron donors. Anaerobic growth requires fumarate or nitrate as a terminal electron acceptor. Formate is oxidized to CO_2, while fumarate is reduced to succinate.

^{v} Marine strains require higher concentrations of NaCl for growth and will tolerate up to 6%.

^{w} One species, *S. lacticifex*, was isolated from brewery waste (Schleifer et al., Int. J. Syst. Bacteriol. *40*: 19-27, 1990).

Table 6.2   Differential characteristics of the species of the genus *Acetivibrio*[a]

Characteristic	A. cellulolyticus[b]	A. ethanolgignens
Substrates utilized:		
Cellobiose, cellulose	+[c]	−
Fructose, galactose, lactose, maltose, mannitol, mannose,	−	+
Glucose	−[d]	+
Reduction of nitrate to nitrite	−	+
H_2S production	−	+
Voges-Proskauer test for acetylmethylcarbinol	−	+
Ammonia production	−	+
Flagellar arrangement:		
Single, subpolar, from concave side of cell	+	−
Multiple, often fasciculated, from concave side of cell	−	+
Yellow pigment produced in cellulose broth	+[c]	−

[a] Symbols: see standard definitions.

[b] Growth of *A. cellulolyticus* is improved by addition of Na_2S or cysteine-HCl to the medium (Patel et al., Can. J. Microbiol. *28:* 772–777, 1982).

[c] Other anaerobic organisms that hydrolyze cellulose and produce a yellow pigment are *Bacteroides cellulosolvens* (Murray et al., Int. J. Syst. Bacteriol. *34:* 185–187, 1984) and some cellulolytic marine bacteria described by Miyoshi (Bull. Jpn. Soc. Sci. Fish. *44:* 197–202, 1978). However, *B. cellulosolvens* is nonmotile, and the marine strains have a growth temperature optimum of 30°C instead of 37°C, liquefy gelatin, and produce isobutyric acid as well as acetic acid.

[d] *A. cellulolyticus* can be adapted to grow on glucose.

Table 6.3   Differential characteristics of the species of the genus *Acetomicrobium*[a]

Characteristic	A. faecalis (sic)	A. flavidum
Source	Sewage sludge incubated at 72°C	Sewage sludge incubated at 60°C
Morphology:		
Cell width (μm)	0.6	0.8
Cell length (μm)	3–7	2–3
Flagellar arrangement	1–2 subpolar	Single polar, or a few lateral
Optimum temperature (°C)	70–73	58
Temperature range (°C)	60–75	35–65
Yellow pigmented colonies	−	+
Fermentation of arabinose, ribose, xylose	+	−
End products from glucose fermentation	Acetic, Lactic, Ethanol, CO_2, H_2	Acetic, CO_2, H_2

[a] Symbols: see standard definitions.

Table 6.4 Differentiation of *Anaerobiospirillum* from similar Gram-negative motile organisms [a]

Characteristic	Anaerobio-spirillum	Campylo-bacter	Desulfo-vibrio	Mobi-luncus [b]	Succini-vibrio	Succini-monas	Wolinella
Cell shape	Spiral	Vibrioid to spiral	Curved rods	Curved rods	Curved rods	Short straight rods	Curved or straight rods
Flagellar arrangement	Bipolar tufts	Single polar or bipolar	Single polar	Multiple subpolar	Single polar	Single polar	Single polar
Utilization of carbohydrates	+	−	−	+	+	+	−
Fermentation of galactose and xylose	−				+		
Stimulation by formate/fumarate [c]	−	D	−	−	−	−	+
H_2S from SO_4	−	−	+	−	−	−	−
Relation to oxygen:							
Anaerobic	+	−	+	+	+	+	−
Microaerophilic	−	+	−	−	−	−	+ [d]
Oxidase test	−	+	−	−	−	−	+

[a] Symbols: see standard definitions.

[b] *Mobiluncus* has a Gram-positive-type cell wall.

[c] Formate/fumarate, 0.2% sodium formate + 0.3% sodium fumarate.

[d] Although *Wolinella* species were described as anaerobes in *Bergey's Manual of Systematic Bacteriology,* Vol. 1, they are in fact H_2- and formate-requiring microaerophiles capable of respiring with oxygen. In the absence of alternative electron terminal electron acceptors such as fumarate, they exhibit oxygen-dependent growth with low levels of oxygen and will not grow anaerobically or aerobically (Wolin et al., J. Bacteriol. *81:* 911–917, 1961; Han et al., Int. J. Syst. Bacteriol. *41:* 218–222, 1991). See also Group 2 in this manual.

Table 6.5 Differential characteristics of the species of the genus *Anaerovibrio* [a]

Characteristic	"A. glycerini" [b]	A. lipolytica
Morphology	Curved or spiral-shaped	Slightly curved
Cell length (μm)	2.0–10.0	1.2–1.8
Only substrates fermented	Glycerol or diolein (glycerol dioleic ester)	Glycerol, ribose, fructose, DL-lactate
End products formed from glycerol	Propionate only	Propionate, succinate
Source	Anoxic freshwater muds and sewage sludge	Rumen of cattle and sheep

[a] Symbols: see standard definitions.

[b] "A. glycerini" was not included in *Bergey's Manual of Systematic Bacteriology*. It was created in 1989 by Schauder and Schink (Arch. Microbiol. *152:* 473–478).

Table 6.6 Differentiation of *Anaerovibrio lipolytica* from morphologically similar motile anaerobic bacteria from the rumen [a]

Characteristic	Anaerovibrio lipolytica	Butyrivibrio fibrisolvens	Lachnospira multiparus	Selenomonas ruminantium	Succinivibrio dextrinosolvens	Succinimonas amylolytica	Wolinella succinogenes [b]
Cell shape	Curved rods	Curved rods or helical rods	Curved rods	Crescent-shaped	Curved rods	Short straight rods	Curved rods
Ends of cell:							
Rounded ends	+	-	-	v	-	+	v
Pointed ends	-	+	+	v	+	-	v
Flagellar arrangement:							
Single polar	+	+	-	-	+	+	+
Single lateral or subterminal	-	-	+	-	-	-	-
Tuft on concave side	-	-	-	+	-	-	-
Gram staining	-	-	weakly +	-	-	-	-
Cell wall type:							
Gram positive	-	-	+	-	-	-	-
Gram negative	+	+	-	+	+	+	+
Fermentative type of metabolism	+	+	+	+	+	+	-
Fermentation of:							
Dextrin	-	+	-	+ or -	+	+	-
Glycerol	+	-	-	d	-	-	-
Arabinose	-	+	-	+	+	-	-
Growth stimulation by formate/fumarate [c]	-	- or +	-	-	-	-	+
H$_2$S production	+	- or +	-	+ or -	-	-	+
Lipolytic activity for long chain triglycerides	+	-	-	-	-	-	-
Clearing around colonies in linseed oil agar	+	-	-	-	-	-	-
"Woolly ball" colonies in solid media	-	-	+	-	-	-	-
End products of metabolism [d]	A,P,s, CO$_2$	B,a,f,l	A,F,L,Eth, CO$_2$,H$_2$	A,P	A,S	A,S	s,CO$_2$,H$_2$

[a] Symbols: see standard definitions. Also: + or –, different reactions occur but a positive reaction is more common; – or +, different reactions occur, but a negative reaction is more common.

[b] Although *Wolinella* species were described as anaerobes in *Bergey's Manual of Systematic Bacteriology*, Vol. 1, they are in fact H$_2$- and formate-requiring microaerophiles capable of respiring with oxygen. In the absence of alternative terminal electron acceptors such as fumarate, they exhibit oxygen-dependent growth with low levels of oxygen and will not grow anaerobically or aerobically (Wolin et al., J. Bacteriol. *81*: 911–917, 1961; Han et al., Int. J. Syst. Bacteriol. *41*: 218–222, 1991). See also Group 2 in this manual.

[c] Formate/fumarate, 0.2% sodium formate + 0.3% sodium fumarate.

[d] Capital letters indicate >1 meq of acid per 100 ml of broth; lower case letters, <1 meq/100 ml. A or a, acetic; B or b, butyric; F, formic; L or l, lactic; P or p, propionic; S or s, succinic; Eth, ethanol.

Table 6.7   Differential characteristics of the species of the genus *Bacteroides* [a]

Characteristic	B. buccae [b]	B. buccalis [b,c]	B. denticola [b]	B. forsythus [c]	B. gracilis [b,d]	B. heparinolyticus [b,c]
Principal habitat [f]	1	1	1	1	1	1
Isolation from clinical samples [h]	+H		+H	+H	+H	+H,A
Growth in bile [i]	–	–	–		– or +	–
Indole produced	–	–	–		–	+
Catalase produced [j]	–	–	–		–	–
Gelatin digested	+ or w	–	+		–	–
Meat digested	–	–	–		–	–
Esculin hydrolyzed	+	+	+	+	–	+
Starch hydrolyzed	+	–	+		–	+
H_2S produced					+	–
H_2 produced [k]	–	–	–		1	
Hemolysis	–	– or beta	–			–
Pigment produced on laked blood agar [l]	–	–	B or –	–	–	–
Acid produced from: [m]						
Arabinose	+	–	–	–	–	
Ribose	+ or –	–	+ or –	–	–	
Xylose	+	–	–	–	–	+
Rhamnose	+ or –	–	–	–	–	
Glycerol		–	+ or w	–	–	
Glucose	+	+	+	–	–	+
Cellobiose	+	+	–	–	–	+
Lactose	+	+	+	–	–	+
Maltose	+	+	+	–	–	+
Melibiose	+	+	–	–	–	
Melezitose	–	–	–	–	–	
Sucrose	+	+	+	–	–	+
Trehalose	–	–	–	–	–	
Raffinose	+	+	+ or –	–	–	–
Salicin	+	–	–	–	–	+
Urease	–				–	
Products from peptone-yeast extract-glucose (PYG) broth [n]	A,S	A,S, *i*v,l	A,s,l	a,p,s	a,S	a,S,p, *i*v
Phenylacetic acid produced						

Footnotes are at end of table

315

Table 6.7 *(continued)*

Characteristic	B. intermedius[b]	B. loeschei[b]	B. melaninogenicus[b]	B. oralis[b]	B. oris[b]	B. oulorum[b,c]
Principal habitat[f]	1	1	1	1	1	1
Isolation from clinical samples[h]	+H	+H	+H	+H	+H	
Growth in bile[i]	−	−	−	−	−	−
Indole produced	−	−	−	−	−	−
Catalase produced[j]	−	−	−	−	−	+
Gelatin digested	+	+	+	+	+ or w	−
Meat digested	+ or −	−	−	−	−	−
Esculin hydrolyzed	−	+	− or +	+	+	+
Starch hydrolyzed	+ or −	+	+	+	+	−
H$_2$S produced	d		−	−		
H$_2$ produced[k]	−	−	−	−	−	
Hemolysis	beta or −	beta or −	beta or alpha	− or beta	−	beta
Pigment produced on laked blood agar[l]	B	B or −	B	−	−	−
Acid produced from:[m]						
Arabinose	−	−	−	−	+ or −	−
Ribose	−	−	−	− or w	+ or −	
Xylose	−	−	−	−	+	−
Rhamnose	−	−	−	+ or −	+ or −	−
Glycerol				−		
Glucose	+	+	+	+	+	+
Cellobiose	−	+	−	+	+ or w	−
Lactose	−	+	+	+	+	+
Maltose	+	+	+	+	+	+
Melibiose	−	+ or −	− or +	+	+ or −	
Melezitose	−	−	−	−		−
Sucrose	+	+	+	+	+	+
Trehalose	−	−	−	−	−	−
Raffinose	+ or −	+	+	+	+	+
Salicin	−	−	−	+	+	−
Urease				−	−	
Products from peptone-yeast extract-glucose (PYG) broth[n]	A,S	a,S	A,f,S	A,S,f,l	A,S,p, *i*v,*i*b	A,S
Phenylacetic acid produced	−	−				

Footnotes are at end of table

Table 6.7 *(continued)*

Characteristic	B. pneumosintes	B. veroralis [b,c]	B. zoogleoformans [b]	B. macacae	B. salivosus [c]	B. tectum [c]
Principal habitat [f]	1	1	1	1	1	1
Isolation from clinical samples [h]	+H		+H,A	+A	+A	+A
Growth in bile [i]	–	–	–	–	–	+
Indole produced	–	–	+ or –	+	+	–
Catalase produced [j]	–		–	+	+	–
Gelatin digested	–	–	+ or –	+	+	+
Meat digested	–	–	–	+		–
Esculin hydrolyzed	–	+	+	–	–	+
Starch hydrolyzed	–	+	+ or –	–		
H_2S produced		–	+ or –	+		
H_2 produced [k]	–		–			
Hemolysis			alpha	–		
Pigment produced on laked blood agar [l]	–	–	–	B	B	–
Acid produced from: [m]						
Arabinose	–	–	+ or –	–		–
Ribose	–	–	–	–		–
Xylose	–	–	+ or –	–	–	–
Rhamnose	–	–	–	–		–
Glycerol						
Glucose	–	+	+ or w	+	–	w–
Cellobiose	–	+	+ or w	–	–	
Lactose	–	+	+ or w	+	–	–
Maltose	–	+	+ or w	–	–	– or w
Melibiose	–	w	+ or –	–		–
Melezitose	–	–	–	–		
Sucrose	–	+	+	–		–
Trehalose	–	–	–	–	–	–
Raffinose	–	+	+	–		–
Salicin	–	–	+ or –	–	–	+ or –
Urease	–	–	–	–	–	
Products from peptone-yeast extract-glucose (PYG) broth [n]	a,s	A,S	A,S, p,*i*v	A,P,s, *i*v,*i*b	A,B	A,P,S
Phenylacetic acid produced				+	+(strongly)	+

Footnotes are at end of table

Table 6.7 *(continued)*

Characteristic	B. levii	B. ruminicola[b,e]	B. cellulosolvens[c]	B. polypragmatus[c]	B. xylanolyticus[c]	B. caccae[c]
Principal habitat[f]	2R	2R	4	4	6	2C
Isolation from clinical samples[h]	+A					(+H)
Growth in bile[i]	−	−	−	w		+
Indole produced	−	−	−	+	−	−
Catalase produced[j]	−		−	−	−	−
Gelatin digested	+	+	−	−	+	− or w
Meat digested	+			−		−
Esculin hydrolyzed	−	+		+		+
Starch hydrolyzed	−			−		−
H_2S produced	+	− or +		+	+	−
H_2 produced[k]	1,3	−			+	trace
Hemolysis					−	− or beta
Pigment produced on laked blood agar[l]	B	−	Y	−	−	−
Acid produced from:[m]						
Arabinose	−	+	−	+	+	+
Ribose	−	−	−	+		+
Xylose	−	+ or −	−	+	+	+
Rhamnose	−	−	−	+	+	+ or −
Glycerol	−	− or +	−	+	−	−
Glucose	+ or w	+	−	+	+	+
Cellobiose	−	+	+	+	+	+ or −
Lactose	+ or w	+	−	+	+	+
Maltose	−	+	−	+	+	+
Melibiose	−	+	−	+		+
Melezitose	−	− or w	−	+		+
Sucrose	−	+	−	−	−	+
Trehalose	−	−	−	+	+	+
Raffinose	−	+ or w	−	+		+
Salicin	−	+ or −	−	+	+	+ or −
Urease			−	−	−	−
Products from peptone-yeast extract-glucose (PYG) broth[n]	A,B,p	A,F,S	A,L,Eth	A,Eth, CO_2,H_2	A,Eth, CO_2,H_2	A,S,p, *iv*
Phenylacetic acid produced	−	trace				−

Footnotes are at end of table

Table 6.7 (continued)

Characteristic	B. distasonis	B. eggerthii	B. fragilis	B. galacturonicus[c]	B. merdae[c]	B. ovatus	B. pectin-ophilus[c]
Principal habitat[f]	2C	2C	2C	2C	2C	2C	2C
Isolation from clinical samples[h]	(+H)	(+H)	+H	–	–	(+H)	–
Growth in bile[i]	+	+	+	+	+	+	+
Indole produced	–	+	–		–	+	
Catalase produced[j]	+	+	+	–	– or +	+ or –	–
Gelatin digested	– or w	–	– or w	–	– or w	+ or –	+
Meat digested	–	–	–		–	–	
Esculin hydrolyzed	+	+	+	–	+	+	–
Starch hydrolyzed	+ or –	+	+ or –	–	–	+ or –	–
H$_2$S produced	+		+		+	+	
H$_2$ produced[k]	–	–	2,3		– or trace	2,1	
Hemolysis	– or alpha		–		– or beta	–	
Pigment produced on laked blood agar[l]	–	–	–	–	–	–	–
Acid produced from:[m]							
Arabinose	–	+	–	–	– or +	+	–
Ribose	+	–	–	–	+ or –	+	–
Xylose	+	+	+	–	+	+	–
Rhamnose	v	+	–	–	+	+	–
Glycerol	–	–	–	–	–	–	–
Glucose	+	+	+	–	+	+	
Cellobiose	+	–	–	–	w	+	
Lactose	+	+	+	–	+	+	–
Maltose	+	+	+	–	+	+	–
Melibiose	+	–	+		+	+	
Melezitose	+	–	–		+	+	
Sucrose	+	–	+	–	+	+	–
Trehalose	+	–	–		+	+	
Raffinose	+	–	+		+	+	
Salicin	+	–	–		+	+	
Urease	–	–	–		–	–	
Products from peptone-yeast extract-glucose (PYG) broth[n]	A,S,p	A,S,p	A,S,p	A,F,l	A,S,p, ib,iv	A,S,p, l,ib,iv	A,F,l, Eth
Phenylacetic acid produced	+	–	+	–	–	+	–

Footnotes are at end of table

Table 6.7 *(continued)*

Characteristic	B. splanch-nicus	B. stercoris[c]	B. thetaiota-omicron	B. uniformis	B. vulgatus	B. helco-genes[c]	B. pyogenes[c]	B. suis[c]
Principal habitat[f]	2C,3	2C	2C	2C	2C	2C	2C	2C
Isolation from clinical samples[h]	(+H)	−	+H	+H	(+H)	+A	+A	+A
Growth in bile[i]	+	+	+	+	+	−	−	−
Indole produced	+	+	+	+	−	−	−	−
Catalase produced[j]	−	−	+ or −	−	+ or −	−	−	−
Gelatin digested	+ or −	−	− or w	−	+	−	−	−
Meat digested	−	−	−	−	−	w	−	−
Esculin hydrolyzed	+	+	+	+	+	+	+ or −	+
Starch hydrolyzed	−	+	+	+ or −	+ or −	+ or −	+ or −	+
H$_2$S produced			+		+	−	+ or −	−
H$_2$ produced[k]	4	− or trace	1,3	trace,2	1,3			
Hemolysis		beta	−		−	− or beta	−	− or beta
Pigment produced on laked blood agar[l]	−	−	−	−	−	−	−	−
Acid produced from:[m]								
Arabinose	+	− or +	+	+	+	−	−	+
Ribose	−	+	+	−	−	−	w	w
Xylose	−	+	+	+	+	+	−	+
Rhamnose	−	+	+	−	+	−	−	−
Glycerol	−	−	−	−	−	−	+	−
Glucose	+	+	+	+	+	+	+	+
Cellobiose	−	− or +	+	+	−	+	+ or −	+
Lactose	+	+	+	+	+	+	+	+
Maltose	−	+	+	+	+	+	+ or −	+
Melibiose	−	− or +	+	+	+	+	− or +	+
Melezitose	−	−	d	−	−	−	−	−
Sucrose	−	+	+	+	+	+	+ or −	+
Trehalose	−	−	+	−	−	−	−	−
Raffinose	−	+	+	+	+	+	+ or −	+
Salicin	−	−	−	+	−	+	− or +	+
Urease	−	−	−	−	−	−	−	−
Products from peptone-yeast extract-glucose (PYG) broth[n]	A,S	A,S,f, p,*i*b,*i*v	A,S,p, l,*i*b,*i*v	a,S,p, l,*i*b,*i*v	A,S,p, l,*i*b,*i*v	A,S,p, *i*b	A,S,p, *i*b	A,S,p, *i*b
Phenylacetic acid produced	+	−	+	−	−	−	−	−

Footnotes are at end of table

Table 6.7 (continued)

Characteristic	B. bivius[b]	B. capil- losus	B. coagulans	B. corporis[b]	B. disiens[b]	B. putredinis	B. nodosus	B. ureolyt- icus[d]
Principal habitat[f]	3[g]	2C,3	2C,3		3[g]	2C	5	3
Isolation from clinical samples[h]	+HA	+HA	+H	H	+H	+HA	+A	+H
Growth in bile[i]	−	−	−	−	−	−	− or w	−
Indole produced	−	−	+	−	−	+	−	−
Catalase produced[j]	−	−	−	−	−	− or w		−
Gelatin digested	+	− or w	+	+	+	+	+	+ or w
Meat digested	+	−	−	− or +	+	+ or −	+	− or +
Esculin hydrolyzed	−	+	−	−	−	−	−	−
Starch hydrolyzed	+	− or +	−	+	+	−	−	−
H$_2$S produced		− or +	+	−		+	+	+
H$_2$ produced[k]	−	−	−	−	−	2,4	−	1,2
Hemolysis		−	−			−	−	− or beta
Pigment produced on laked blood agar[l]	−	−	−	B	−	−	−	−
Acid produced from:[m]								
Arabinose	−	−	−	−	−	−	−	−
Ribose	−	− or +	−	−	−	−	−	−
Xylose	−	− or +	−	−	−	−	−	−
Rhamnose	−	−	−	−	−	−	−	−
Glycerol	+ or −	−	−	−	−	−	−	−
Glucose	+	+ or −	−	−	+	−	−	−
Cellobiose	−	−	−	−	−	−	−	−
Lactose	+	−	−	−	−	−	−	−
Maltose	+	−	−	−	+	−	−	−
Melibiose	−	−	−	−	−	−	−	−
Melezitose	−	−	−	−	−	−	−	−
Sucrose	−	−	−	−	−	−	−	−
Trehalose	−	−	−	−	−	−	−	−
Raffinose	−	−	−	−	−	−	−	−
Salicin	−	−		−	−		−	−
Urease			−				−	+
Products from peptone-yeast extract-glucose (PYG) broth[n]	A,S,l, ib,iv	A,S,f, l,p	A,l,p,s	A,S, ib, iv	A,S,ib, iv,f,l,p	s,l,ib, iv	A,P,s	A,S,l,f
Phenylacetic acid produced	−	−		−	−	+		−

Footnotes are at end of table

Table 6.7 *(continued)*

[a] Symbols (unless otherwise noted): +, 90–100% of strains positive; –, 10% or less of strains are positive; v, strains give variable results on repeated testing; d, 11–89% of strains positive; w, 90–100% of strains are weakly positive; blank space, reaction not reported. When two symbols are given and separated by "or" (e.g, + or –, – or +, etc.), the first reaction is the more common.

[b] A new genus, *Prevotella,* has been proposed for these species by Shah and Collins (Int. J. Syst. Bacteriol. *40:* 205-208, 1990).

[c] These species were not included in *Bergey's Manual of Systematic Bacteriology* but have been validly published. See the following references: *B. buccalis* (Shah and Collins, Zentralbl. Bakteriol. Parasitenkd. Infektionskr. Hyg., I Abt. Orig. *C2:* 235-241, 1981; Int. J. Syst. Bacteriol. *32:* 266, 1982); *B. caccae* (Johnson et al., Int. J. Syst. Bacteriol. *36:* 499-501, 1986); *B. cellulosolvens* (Murray et al., Int. J. Syst. Bacteriol. *34:* 185-187, 1984); *B. forsythus* (Tanner et al., Int. J. Syst. Bacteriol. *36:* 213-221, 1986); *B. galacturonicus* (Jensen and Canale-Parola, Appl. Environ. Microbiol. *52:* 880-887, 1986; Int. J. Syst. Bacteriol. *37:* 179, 1987); *B. heparinolyticus* (Okuda et al., Int. J. Syst. Bacteriol. *35:* 438-442, 1985); *B. helcogenes* (Benno et al., Syst. Appl. Microbiol. *4:* 396-407, 1983; Int. J. Syst. Bacteriol. *33:* 896, 1983); *B. merdae* (Johnson et al., Int. J. Syst. Bacteriol. *36:* 499-501, 1986); *B. oulorum* (Shah et al., Int. J. Syst. Bacteriol. *35:* 193-197, 1985); *B. pectinophilus* (Jensen and Canale-Parola, Appl. Environ. Microbiol. *52:* 880-887, 1986; Int. J. Syst. Bacteriol. *37:* 179, 1987); *B. polypragmatus* (Patel and Breuil, pp. 291-296, *In* Moo-Young and Robinson (eds.), Adv. Biotechnol., Pergamon Press, Toronto, 1981; Int. J. Syst. Bacteriol. *32:* 266, 1982); *B. pyogenes* (Benno et al., Syst. Appl. Microbiol. *4:* 396-407, 1983; Int. J. Syst. Bacteriol. *33:* 896, 1983); *B. suis* (Benno et al., Syst. Appl. Microbiol. *4:* 396-407, 1983; Int. J. Syst. Bacteriol. *33:* 896, 1983); *B. salivosus* (Love et al., Int. J. Syst. Bacteriol. *37:* 307-309, 1987); *B. stercoris* (Johnson et al., Int. J. Syst. Bacteriol. *36:* 499-501, 1986); *B. tectum* (Love et al., Int. J. Syst. Bacteriol. *36:* 123-128, 1986); *B. veroralis* (Watabe et al., Int. J. Syst. Bacteriol. *33:* 57-64, 1983); *B. xylanolyticus* (Scholten-Koersehman et al., Antonie van Leeuwenhoek J. Microbiol. Serol. *52:* 543–554, 1986; Int. J. Syst. Bacteriol. *38:* 136, 1988).

[d] Although *B. gracilis* and *B. ureolyticus* were described as anaerobes in *Bergey's Manual of Systematic Bacteriology,* Vol. 1, they are in fact H_2- and formate-requiring microaerophiles capable of respiring with oxygen. In the absence of alternative electron terminal electron acceptors such as fumarate, they exhibit oxygen-dependent growth with low levels of oxygen and will not grow anaerobically or aerobically (Han et al., Int. J. Syst. Bacteriol. *41:* 218–222, 1991). See also Group 2 in this manual.

[e] Two subspecies of *B. ruminicola* have been designated, *ruminicola* and *brevis,* but there are several biovars of each, and DNA hybridization studies indicate considerable diversity among strains (Churn and Johnson, Abstr. Annu. Meet. Am. Soc. Microbiol., p. 281, 1980; Hudman and Gregg, Curr. Microbiol. *19:* 313, 1989). The results given here are the commoner findings; for reactions of different biovars and subspecies, see *Bergey's Manual of Systematic Bacteriology,* Vol. 1, p. 623. Human strains from the mouth originally thought to be *B. ruminicola* are now designated *B. oris* or *B. buccae* (Holdeman et al., Int. J. Syst. Bacteriol. *32:* 125, 1982).

[f] 1, mouth; 2C, cecum; 2R, rumen; 3, genitourinary tract; 4, sewage sludge; 5, foot rot in sheep; 6, fermenting cow manure.

[g] *B. bivius* and *B. disiens* have been isolated from clinical specimens mostly connected with the genitourinary tract; their normal habitat is not known with certainty.

[h] +H, isolated from human samples; +A, isolated from animal samples; +HA, isolated from human and animal samples; -, not reported from clinical samples.

[i] The concentration generally used is 20% ox bile or 2% Bacto oxgall (Difco), which is a 10X concentrate of ox bile.

[j] For accurate and reproducible results, the medium should not contain fermentable carbohydrate and hemin should be added (Wilkins et al., J. Clin. Microbiol. *8:* 533, 1978).

[k] H_2 in headspace: 1, 0.5%; 2, 1.0%; 3, 2.0%; 4, 3.0% or more; +, present, amount not reported.

[l] –, colonies not pigmented (i.e., they are white, cream-colored, or pale tan); B, brown to black pigment; Y, yellow pigment. The brown to black pigment develops best on media that contain laked blood and may take up to 14 days to develop. The pigment is not melanin but protoheme or protoporphyrin or a mixture of the two (Shah et al., Biochem. J. *180:* 45, 1979). The yellow pigment of *B. cellulosolvens* develops on cellobiose agar.

[m] +, pH change of more than 1.0 pH unit; w, pH change of between 0.5 and 1.0 pH unit; –, pH change of less than 0.5 pH unit.

[n] Capital letters indicate >1 meq of acid per 100 ml of broth; lower case letters, <1 meq/100 ml. A or a, acetic; B or b, butyric; F, formic; *i* b, isobutyric; *i* v, isovaleric; L or l, lactic; P or p, propionic; S or s, succinic; Eth, ethanol. Products in parentheses may or may not be detected.

Table 6.8 Differential characteristics of the species of the genus *Butyrivibrio*[a]

Characteristic	B. crossotus[b]	B. fibrisolvens[c]
Source	Human intestine	Rumen of sheep, cattle, and other herbivores, and intestine of humans, rabbits, and horses
Flagellar arrangement	Lophotrichous, polar or subpolar	Usually monotrichous and polar
Fermentation of:		
Glucose and fructose	Weak	+
Arabinose, xylose, galactose, cellobiose, sucrose, inulin, esculin	–	+
Production of H_2 from glucose and maltose	–	+

[a] Symbols: see standard definitions.

[b] *B. crossotus* requires carbohydrate for growth, but the materials fermented seem to be limited to starch, glycogen, dextrin, and maltose. The fermentation pattern indicates that it is quite distinct from the strains isolated from human feces by Brown and Moore (J. Dairy Sci. *43:* 1570–1574, 1960).

[c] Much phenotypic diversity has been noted in strains assigned to *B. fibrisolvens,* both in fermentation patterns and in end products. DNA hybridization experiments have shown the existence of at least five distinct groups among such strains, with many strains still ungrouped (Mannarelli, Int. J. Syst. Bacteriol. *38:* 340–347, 1988; Hudman and Gregg, Curr. Microbiol. *19:* 313–318, 1989), and some strains seem to have a Gram-negative-type cell wall rather than an unusually thin Gram-positive one (Dibbayawan et al., J. Ultrastr. Res. *90:* 286–293, 1985). There are also similarities between *B. fibrisolvens* and *Eubacterium rectale,* which is Gram positive but very easily decolorized (Moore and Holdeman, Appl. Microbiol. *27:* 961–979, 1974).

Table 6.9 Differentiation of *Centipeda periodontii* from *Selenomonas* species isolated from the mouth[a]

Characteristic	Centipeda periodontii	Selenomonas species
Cell size	0.65 x 4–17 μm	1.0–1.3 x 3–14 μm
Cell shape	Serpentine rods with occasional branching	Curved rods
Flagellar arrangement	Arise in a linear zone which spirals around the cell	Flagellar tufts on concave side
Number of flagella per cell	50–250	Up to 16
Lactate utilization	+	–

[a] Symbols: see standard definitions.

Table 6.10 Differential characteristics of the species of the genus *Fibrobacter*[a]

Characteristic	F. intestinalis	F. succinogenes subsp. elongata	F. succinogenes subsp. succinogenes
Source	Cecum	Rumen	Rumen
Cell shape	Rods	Rods	Coccoid
Cell width (μm)	0.3–0.4	0.4	0.7–0.8
Vitamin requirements:			
Biotin	−	+	+
p-Aminobenzoic acid	+	+	− or +
B$_{12}$ (cyanocobalamine)	+	d	−
Thiamine	d	−	−
Fermentation of lactose	−	−	+[b]

[a] Symbols: see standard definitions. − or +, both positive and negative reactions occur, but a negative reaction is the more common.

[b] Fermentation of lactose may be slow.

Important Notes for Users of This Manual

Unless otherwise indicated in footnotes to tables, the meanings of symbols are as follows:

+ 90% or more of strains are positive

− 90% or more of strains are negative

d 11-89% of strains are positive

v strain instability (not equivalent to "d")

D Different reactions in different taxa (species of a genus or genera of a family)

All other symbols are defined in footnotes to tables.

Table 6.11 Differential characteristics of *Fusobacterium* species and *Leptotrichia buccalis*[a]

Characteristic	F. alocis[d]	F. gonidiaformans	F. mortiferum	F. naviforme	F. necrogenes	F. necrophorum	F. nucleatum	F. perfoetans
Principal habitat[f]	1[h]	2C	2C	1,2C,2R	2C	2C,3	1	2C
Isolation from clinical samples[j]	H	HA	H	H	A	HA	H	H
Morphological features:								
Regular rods	+	-	-	+	-	-	+	-
Long filaments common	-	+	+		+	+	-	-
Swellings and globular forms common	-	+[l]	+	-	+	+	+	-
Cells generally short or coccoid	-	-	-	-	-	-	-	+
Rounded ends	v	v	+	-	v	v	-	+
Pointed ends	v	v	-	+	v	v	+	-
Capsulated	-	-		-	-	-	-	
Propionate produced from:								
Threonine	-[m]	+	+	-	+	+	+	+
Lactate	-[m]	-	-	-	-	+	-	-
Hippurate hydrolysis	-	v	-	- or +	-	-	- or +	-
Esculin hydrolyzed	-	-	+	-	+	-	-	-
Growth in bile[n]	-[m]	-	+	-	- or +	- or +	-	-
Indole produced	-[m]	+	-	+	-	+	+	-
H₂S produced	-	+	+	+	+	+	+ or -	+
H₂ produced[o]	4	4	4	+	4	4	-	4
Hemolysis	-	alpha or -	-	-	beta or -	beta or alpha	- or beta	-
Acid produced from:[p]								
Glucose	-	-	+ or w	w or -	w or +	- or w	- or w	w
Galactose	-	-	-	-	-	-	-	-
Mannose	-	-	+ or w	-	w or +	-	- or w	-
Fructose	-	-	+ or w	-	w or +	- or w	- or w	w
Cellobiose	-	-	w or -	-	- or w	-	-	-
Lactose	-	-	+ or w	-	-	-	-	-
Maltose	-	-	w or -	-	- or w	-	-	-
Melibiose	-	-	w or -	-	-	-	-	-
Sucrose	-	-	+ or w	-	- or w	-	-	w
Trehalose	-	-	- or +	-	w or -	-	-	-
Raffinose	-	-	+ or w	-	- or w	-	-	-
Starch	-	-	- or w	-	-	-	-	-
Salicin	-	-	- or w	-	- or w	-	-	-
Products from peptone-yeast extract-glucose (PYG) broth[q]	a,b	A,B,p,(f,l,s)	A,B,p,(f,l,s,v)	a,B,L,(f,p,s)	a,B,p,(f,l,s)	a,B,p,(l,s)	a,B,p,(F,L,s)	a,B,p,(L)

Footnotes are at end of table

Table 6.11 (continued)

Characteristic	F. periodonticum[c]	F. prausnitzii	F. pseudonecrophorum[e]	F. russii	F. simiae	F. sulci[d]	F. ulcerans[b]	F. varium[g]	L. buccalis
Principal habitat[f]	1	2C		2C	1	1[h]	4	2C[g]	1
Isolation from clinical samples[i]	H	[j]	B	HA	A	H	H	H	H
Morphological features:									
Regular rods	+	+[k]	+	+	+	+	+	-	+
Long filaments common	+	- or w	-	+	-	+	-	-	+
Swellings and globular forms common	-	-	-	-	-	-	-	-	-
Cells generally short or coccoid	-	-	+	-	-	-	-	+	-
Rounded ends	v	v		+	+	+	-	+	v
Pointed ends	v	v		v	+	-	+	-	v
Capsulated	-	+	-	-		-	-	-	-
Propionate produced from:									
Threonine	+	-	+	-	+	-	+	+	-
Lactate	-	-	+	-	+	-	-	-	-
Hippurate hydrolysis	+	-	-	- or +	+	-	-	- or +	+
Esculin hydrolyzed	-	+	-	-	-	-	-	-	-
Growth in bile[n]	-	+ or -	+	-	+	-	-	+	-
Indole produced	+	-	+	+	+	-	-	d	-
H₂S produced	+	+	+	- or +	+	-	-	+	+
H₂ produced[o]	-	-	4	-	-	- or trace	+	4	-
Hemolysis	-	-	-	- or beta	-	-	-	-	-
Acid produced from:[p]									
Glucose	+	w or -	w	-	+	-	+	w or +	+
Galactose	+	- or w	-	-	-	-	-	- or w	+
Mannose	-	- or w	-	-	-	-	- or +	w or +	+
Fructose	+	w or -	w	-	+	-	-	w or +	+
Cellobiose	-	- or w	-	-	-	-	-	-	+
Lactose	-	w or -	-	-	-	-	-	-	+
Maltose	-	w or -	-	-	-	-	-	-	+
Melibiose	-	-	-	-	-	-	-	-	-
Sucrose	-	- or w	-	-	-	-	-	-	+
Trehalose	-	- or w	-	-	-	-	-	-	+
Raffinose	-	-	-	-	-	-	-	-	- or +
Starch	-	- or w	-	-	-	-	-	-	- or +
Salicin	-	- or w	-	-	-	-	-	-	+
Products from peptone-yeast extract-glucose (PYG) broth[q]	A,B,l, (f,p,s)	B,F,L,s	B	a,B,L, (f)	A,B, l,p,s	a,B,(s)	a,B, l,p,s	A,B,L, p,(s)	L,(a,s)

Footnotes are at end of table

Table 6.11 *(continued)*

a   Symbols (unless otherwise noted): +, 90–100% of strains positive; –, 10% or less of strains are positive; w, 90–100% of strains are weakly positive; v, varies within strains or within a single culture; blank space, reaction not reported. When two symbols are given and separated by "or" (e.g, + or –, – or w, etc.), the first reaction is the more common.

b   *F. ulcerans* was not included in *Bergey's Manual of Systematic Bacteriology.* See Adriaans and Shah (Int. J. Syst. Bacteriol. *38:* 447-448, 1988). The organism has been isolated from tropical ulcers and from mud samples in tropical areas.

c   *F. periodonticum* was not included in *Bergey's Manual of Systematic Bacteriology.* See Slots et al. (J. Dent. Res. *62:* 960-963, 1983; Int. J. Syst. Bacteriol. *34:* 270, 1984).

d   *F. alocis* and *F. sulci* were not included in *Bergey's Manual of Systematic Bacteriology.* See Cato et al. (Int. J. Syst. Bacteriol. *35:* 475-477, 1985).

e   *F. pseudonecrophorum* was not included in *Bergey's Manual of Systematic Bacteriology.* See Shinjo et al. (Int. J. Syst. Bacteriol. *40:* 71-73, 1990). Of the three strains examined, two were isolated from ovine abscesses and one from bovine feces. In addition to being nonhemolytic and lipase-negative, they differ from *F. necrophorum* in being resistant to 500 U/ml penicillin and in not producing liver abscesses in mice.

f   1, mouth; 2R, rumen; 2C, cecum; 3, genitourinary tract; 4, tropical ulcers and mud samples in tropical areas.

g   *F. varium* has been isolated from the intestines of humans, mice, insects, and fish.

h   *F. alocis* and *F. sulci* have been isolated from subgingival areas associated with gingivitis or periodontitis.

i   H, human samples; A, animal samples.

j   *F. prausnitzii* is rarely isolated from clinical samples, although it is common in human feces.

k   Rods may be up to 14 µm long, and the longer forms are curved.

l   The spheroidal or gonidial forms implied by the species name are most often seen in old cultures or in media that are not highly reduced.

m   Strains of *F. alocis* isolated from cats differ from human strains in being indole positive, producing propionate from threonine and lactate, and growing in 20% bile (Love et al., Int. J. Syst. Bacteriol. *37:* 23-26, 1987).

n   The concentration generally used is 20% ox bile or 2% Bacto oxgall (Difco), which is a 10X concentrate of ox bile.

o   H₂ in headspace: 1, 0.5%; 2, 1.0%; 3, 2.0%; 4, 3.0% or more; +, present, amount not reported.

p   +, pH change of more than 1.0 pH unit; w, pH change of between 0.5 and 1.0 pH unit; –, pH change of less than 0.5 pH unit.

q   Capital letters indicate >1 meq of acid per 100 ml of broth; lower case letters, <1 meq/100 ml. A or a, acetic; B or b, butyric; F or f, formic; V or v, valeric; L or l, lactic; P or p, propionic; S or s, succinic. Products in parentheses may or may not be detected.

Table 6.12 Differentiation of *Halobacteroides halobius* from *Haloanaerobium praevalens* [a]

Characteristic	Haloanaerobium praevalens	Halobacteroides halobius
Cell shape	Short straight rods	Long rods, some with coiled ends
Cell length (μm)	2.0–2.6	10–20
Motility	−	+
End products from glucose fermentation	Acetic, butyric, propionic, CO_2, H_2	Acetic, Ethanol, CO_2, H_2
Source	Great Salt Lake, Utah, USA	Dead Sea, Palestine

[a] Symbols: see standard definitions.

Table 6.13 Differential characteristics of the species of the genus *Ilyobacter* [a]

Characteristic	I. polytropus	I. tartaricus
Cell width (μm)	0.7	1.0
Cell length (μm)	1.5–3.0	1.2–2.0
Motility	+ on initial isolation	−
Slime formation	−	+
Fermentation of:		
3–hydroxybutyrate, malate, fumarate	+	−
L-tartrate	−	+

[a] Symbols: see standard definitions.

Table 6.14 Differential characteristics of the species of the genus *Mitsuokella*

Characteristic	M. dentalis	M. multiacidus
Cell shape	Ovoid rods	Regular rods
Cell width (μm)	0.7	0.8–1.5
Cell length (μm)	1–2	3–20
Source	Dental root canal infection in humans	Feces of humans and pigs
Major end products from glucose fermentation	Acetic, succinic	Acetic, lactic, succinic
Colonies on blood agar	Water-drop appearance; 1–2 mm diameter at 2 days	Greyish white; 3–8 mm diameter at 2 days

Table 6.15 Differential characteristics of the species of the genus *Pectinatus*[a]

Characteristic	*P. cerevisiiphilus*	*P. frisingensis*[b]
Acid from fermentation of:		
Xylose, melibiose	+	−
Cellobiose, inositol	−	d[c]
N-acetylglucosamine	−	+[c]

[a] Symbols: see standard definitions.

[b] *P. frisingensis* was not included in *Bergey's Manual of Systematic Bacteriology*. The species was created in 1990 by Schleifer et al., (Int. J. Syst. Bacteriol. *40:* 19–27).

[c] Some strains give only a weakly positive reaction.

Table 6.16 Differential characteristics of the species of the genus *Pelobacter*[a,b]

Characteristic	*P. acetylenicus*	*P. acidigallici*	*P. carbinolicus*	*P. propionicus*	*P. venetianus*
Cell width (μm)	0.6–0.8	0.5–0.8	0.5–0.7	0.5–0.7	0.5
Cell length (μm)	1.5–4.0	1.5–3.5	1.2–3.0	1.2–6.0	1.0–2.5
Shape of end of cell	Rounded or pointed[c]	Rounded	Rounded	Rounded	Rounded
Motility	+	+	−	−	+
Flagellar arrangement	Not described	Subpolar or peritrichous			Polar or subpolar
Principal substrates attacked	Acetylene, acetoin, choline, ethanolamine, 1,2–propanediol, glycerol	Gallic acid, pyrogallol, phloroglucinol	2,3–Butanediol, acetoin, ethylene glycol	2,3–Butanediol, acetoin, ethanol, pyruvate, lactate	Polyethylene glycol (PEG) and PEG-containing compounds, acetoin
End products	Acetate, ethanol, propanol, 1,3-propanediol, triethanolamine, depending on substrate	Acetate, CO_2	Acetate, ethanol	Acetate, propionate	Acetate, ethanol, propionate, propanol butyrate, butanol, depending on substrate
Source	Freshwater or marine muds	Marine or freshwater mud	Marine muds	Freshwater mud, sewage sludge	Marine mud, sewage sludge

[a] Symbols: see standard definitions.

[b] *Pelobacter* strains have generally been isolated by enrichment of the original material with highly specific substrates (e.g., 20 mM gallic acid in the case of *P. acidigallici*). The original articles should be consulted for details. Marine strains usually require media containing 2% NaCl for growth.

[c] Some marine strains have rounded ends; these strains also differ from the other marine strains (which have pointed ends) and from freshwater strains (which have pointed ends) in giving yellowish colonies in agar shake cultures (Schink et al., Arch. Microbiol. *142:* 295-301, 1985).

Table 6.17 Differential characteristics of the species of the genus *Porphyromonas*[a]

Characteristic	P. asaccharolytica	P. endodontalis	P. gingivalis
α-Fucosidase	+	−	−
Agglutination of sheep red blood cells	−	−	+
Trypsin-like activity	−	−	+
Production of phenylacetic acid [b]	−	−	+
Electrophoretic mobility of malate dehydrogenase [c]	Medium	Fast	Slow
Principal source	Human feces	Root canal infection; periodontal pockets	Gingival flora in periodontal disease

[a] Symbols: see standard definitions.
[b] For detection of phenylacetic acid, see van Assche, J. Clin. Microbiol. *8:* 614–615, 1978.
[c] Compared to that of *Bacteroides intermedius* strain T588 (Shah and Williams, J. Gen. Microbiol. *128:* 2955–2965, 1982).

Important Notes for Users of This Manual

Unless otherwise indicated in footnotes to tables, the meanings of symbols are as follows:

+ 90% or more of strains are positive

− 90% or more of strains are negative

d 11-89% of strains are positive

v strain instability (not equivalent to "d")

D Different reactions in different taxa (species of a genus or genera of a family)

All other symbols are defined in footnotes to tables.

Table 6.18 Differential characteristics of the species of the genus *Selenomonas*[a]

Characteristic	"S. acidaminophila"[b]	S. artemidis[b]	S. dianae[b]	S. flueggei[b]	S. infelix[b]	S. lacticifex[b]	S. noxia[b]	S. ruminantium[c]	S. sputigena
Principal habitat[d]	3	1	1	1	1	4	1	2R,2C	1
Isolation from clinical samples[e]		H	H	H	H		H		H
Growth in bile[f]	+	–	–	–	–		–	+	–
Esculin hydrolysis	–	–	+	–	+	+	–	+	–
H$_2$S produced		–	–	–	–		–	+	–
Gelatinase	+	–	–	–	–		–	–	–
Acid produced from:[g]									
Arabinose	–	–	–	–	–	+	–[h]	+	–
Xylose		–	–	–	–	+	–		–
Dulcitol		–	–	–	–	–	–	+	–
Mannitol	–	+ or –	+	+	+	–	–	+	–
Sorbitol	–	– or +	–	+	+ or –	–	–	+	+
Glucose	+	+	+	+	+	+	– or w	+	+
Mannose	–	+	+	+	+	+	+	+	+
Cellobiose	–	–	–	+	–	– or +	–	+	–
Lactose	–	–	+ or –	+	+		–	+	+
Melibiose		+ or –	+ or –	+	+	+	–	+	+
Sucrose	–	+	+	+	+	+	–	+	+
Trehalose	–	–	+	–	–	–	–	+ or –	–
Raffinose	–	–	+ or –	+	+	+ or –	–		+
Salicin	–	–	–	–	–		–	+	–
Products from peptone-yeast extract-glucose (PYG) broth[i]	A,P,s	A,P	A,P,L,pyr	A,P,l	A,P,l	A,P,L	A,P	A,P,L,CO$_2$,s	A,P

[a] Symbols (unless otherwise noted): +, 90–100% of strains positive; –, 10% or less of strains are positive; w, 90–100% of strains are weakly positive; blank space, reaction not reported. When two symbols are given and separated by "or" (e.g., + or –, – or +, etc.), the first reaction is the more common.

[b] These species were not included in *Bergey's Manual of Systematic Bacteriology* but have been validly published. See the following references: *S. noxia, S. dianae, S. infelix, S. flueggii, S. artemidis* (Moore et al., Int. J. Syst. Bacteriol. *37*: 271–280, 1987); "*S. acidaminophila*" (Nanninga et al., Arch. Microbiol. *147*: 152–157, 1987); *S. lacticifex* (Schleifer et al., Int. J. Syst. Bacteriol. *40*: 19–27, 1990).

[c] Two subspecies of *S. ruminantium* have been defined, *ruminantium* and *lactilytica*. The latter differs from the former by fermenting lactate and glycerol; also, some strains of *S. ruminantium* subsp. *ruminantium* produce urease. See Robinson et al. (Appl. Environ. Microbiol. *41*: 950–955, 1981) and John et al. (J. Dairy Sci. *57*: 1003–1014, 1974).

[d] 1, mouth; 2R, rumen; 2C, pig cecum; 3, anaerobic fermentor; 4, brewery yeast.

[e] H, isolated from human samples.

[f] The concentration generally used is 20% ox bile or 2% Bacto oxgall (Difco), which is a 10X concentrate of ox bile.

[g] +, pH change of more than 1.0 pH unit; w, pH change of between 0.5 and 1.0 pH unit; –, pH change of less than 0.5 pH unit.

[h] Strains of *S. noxia* seem to ferment none of the carbohydrates tested, but growth in PYG is abundant and the strains produce more acetic and propionic acids in peptone-yeast extract-glucose (PYG) broth than in peptone-yeast extract (PY) broth, even though the pH of PYG cultures is not lower than that of cultures grown in PY broth (Moore et al., Int. J. Syst. Bacteriol. *37*: 271–280, 1987).

[i] Capital letters indicate >1 meq of acid per 100 ml of broth; lower case letters, <1 meq/100 ml. A or a, acetic; L or l, lactic; P or p, propionic; S or s, succinic; Pyr or pyr, pyruvic.

Table 6.19   Differential characteristics of the species of the genus *Sporomusa*[a]

Characteristic	S. acidovorans	S. malonica	S. ovata	S. pauci- vorans	S. sphaeroides	S. termitida
Cell width (μm)	0.7–1.0	0.7	0.7–1.0	0.4–0.7	0.5–0.8	0.5–0.8
Cell length (μm)	2.0–8.0	2.6–4.8	1.0–5.0	2.0–3.0	2.0–4.0	2.0–8.0
Spore shape	Round	Oval	Oval	No spores found[b]	Round	Round or oval
Catalase	ND	–	– or weak	–	+	+
Utilization of:						
Fructose	+	+	+ or –	–	–	–
Ribose	+	–	–	–	–	–
Mannitol	ND	–	+	+	+	+
Glycerol	+	–	–	+	+	–
n-Propanol	ND	+	+	+	+	–
N,N-dimethylglycine	ND	ND	+	–	+	–
Citrate	–	+	–	–	–	+
Succinate	+	+	–	–	–	+
Fumarate	+	+	–	–	–	–
Malonate	–	+	ND	ND	ND	+
Isolated from	Alcohol factory waste	Freshwater mud	River mud; sugar beet fields	Lake mud	River mud; sugar beet fields	Intestinal tract of termites

[a] Symbols: see standard definitions. Also, – or weak, negative or weakly positive, but a negative reaction is the more common; + or –, positive or negative, but a positive reaction is the more common; ND, not determined.

[b] Spores have not been detected in strains of *S. paucivorans*, but the type strain shows 38% DNA sequence similarity to *S. sphaeroides*. Moreover, one strain of *S. ovata* never produced spores; thus, nonsporing strains are known to occur in the group.

Table 6.20   Differential characteristics of the species of the genus *Syntrophomonas* and of *Syntrophobacter wolinii*[a]

Characteristic	Syntrophomonas sapovorans	Syntrophomonas wolfei		Syntrophobacter wolinii
		subsp. saponavida[b]	subsp. wolfei	
Motility	+	+	+	–
Fatty acids principally attacked:				
C₄–C₈ straight chain, saturated	+	+	+	–
C₈–C₁₈ straight chain, saturated	+	+	–	–
Attacks *only* propionate	–	–	–	+
Attacks unsaturated acids such as oleate, elaidate, and linoleate	+	–	–	–
Requires Ca²⁺ for attack of fatty acids longer than C₈	+	–	–	NA[c]
Growth in pure culture will occur with crotonate	+	+	+	–

[a] Symbols: see standard definitions.

[b] The distinction between *S. wolfei* subsp. *saponavida* and *S. sapovorans* is not clear-cut, and further investigation may well show that the strains are identical.

[c] NA, not applicable; attacks only propionate.

Table 6.21 Differential characteristics of the species of the genus *Thermobacteroides*[a]

Characteristic	*T. acetoethylicus*	*T. leptospartum*	*T. proteolyticus*
Cell morphology	Short rods	Long thin rods	Short rods, pleomorphic in old cultures
Cell width (μm)	0.6	0.25–0.45	0.5
Cell length (μm)	2.0	4.5–15.0	1–6
Motility	+ (peritrichous flagella)	−	−
Temperature optimum (°C)	65	60	63
Temperature optimum (°C)	>40 – <80	>45 – <71	35–75
Acid produced from: [b]			
Xylose	−	+	−[c]
Galactose		+	−
Fructose		−	+
Cellobiose, lactose	+	+	−
Sucrose	+	−	+
End products from glucose [d]	A, Eth, CO_2 H_2, b, *i*b	A, Eth	A, CO_2, H_2
Habitat	Octopus spring, Yellowstone N.P., USA	Cattle compost	Tannery waste, cattle manure

[a] Symbols: see standard definitions.

[b] All species produce acid from glucose, maltose, and mannose.

[c] Strains of *T. proteolyticus* need the addition of yeast extract (0.1%) and rumen fluid (2.0%) to the medium for good results in fermentation tests.

[d] Capital letters indicate >1 meq of acid per 100 ml of broth; lower case letters, <1 meq/100 ml. A or a, acetic; B or b, butyric; F, formic; *i*B or *i*b, isobutyric; *i*V or *i*v, isovaleric; L or l, lactic; P or p, propionic; S or s, succinic; Eth, ethanol. Products in parentheses may or may not be detected.

Table 6.22 Characteristics differentiating the species of the genus *Thermotoga* and *Thermosipho africanus*[a]

Characteristic	*Thermotoga maritima*	*Thermotoga neopolitana*	*Thermotoga thermarum*	*Thermosipho africanus*
Number of cells enclosed within one sheath	1–4	1–4	1–4	Up to 12
Motility	+	+ or −	+	−
Flagellar arrangement:				
Single	+	−	−	
Peritrichous	−	+	−	
Lateral	−	−	+	
Range of NaCl concentrations that allow growth, %	0.25–3.75	0.25–6.0	0.2–0.55	0.11–3.5
Ability to utilize sugars	+	+	+	−
Susceptibility to rifampicin (μg/ml)	>100	>100	<1.0	>10, <100
Inhibition by H_2 reversed by sulfur	+	+	−[b]	+
Site of isolation	Geothermal marine sediments in Italy and Azores	Hot springs, Bay of Naples; Hot springs, Djibouti, Africa	Hot springs, Djibouti, Africa	Hot springs in intertidal zone, Djibouti, Africa

[a] Symbols: see standard definitions.

[b] Sulfur by itself is inhibitory to *Thermotoga thermarum*.

Table 6.23 Differential characteristics of the species of the genus *Wolinella*[a,b]

Characteristic	W. curva[c]	W. recta	W. succinogenes
Cell morphology:			
Helical or curved cells dominant	+	−	+
Straight cells dominant	−	+	−
Cells with tapered ends	+	−	+
Growth in the presence of:			
Janus green (0.01%), basic fuchsin (0.0032%), oxgall (1%), sodium deoxycholate (0.1%), methyl orange (0.032%), penicillin (16 μg/ml), polymyxin B (4 μg/ml)	+	−	+
Glycine (1%)	+	+	−
Crystal violet (0.0005%), sodium fluoride (0.05%)	−	−	+
Indulin scarlet (0.05%)	+	−	−
Source	Lesions in human oral cavities; blood cultures	Human periodontitis	Bovine rumen

[a] Symbols: see standard definitions.

[b] The results in this table are for cultures incubated at 35°C under anaerobic conditions (atmosphere = 80% N_2, 10% H_2, 10% CO_2) in media containing 0.2% sodium formate as the electron donor (Tanner et al., Int. J. Syst. Bacteriol. *34:* 275–282, 1984).

[c] *W. curva* was not included in *Bergey's Manual of Systematic Bacteriology*. The species was created in 1984 by Tanner et al. (Int. J. Syst. Bacteriol. *34:* 275–282).

Important Notes for Users of This Manual

Unless otherwise indicated in footnotes to tables, the meanings of symbols are as follows:

    + 90% or more of strains are positive

    − 90% or more of strains are negative

    d 11-89% of strains are positive

    v strain instability (not equivalent to "d")

    D Different reactions in different taxa (species of a genus or genera of a family)

All other symbols are defined in footnotes to tables.

GROUP 7

DISSIMILATORY SULFATE- OR SULFUR-REDUCING BACTERIA

Cells are spherical, ovoid, rod-shaped, spiral, or vibrioid-shaped; 0.4–3.0 µm in diameter; occurring singly, in pairs, or sometimes in aggregates. **Uniseriately multicellular filamentous forms also occur.** Most genera stain Gram negative; filamentous and spore-forming types may stain Gram positive. A few species contain gas vacuoles. These bacteria are **motile and nonmotile.** Their motility is mainly due to flagella; filamentous forms exhibit gliding motility. Flagella are polar or peritrichous. Colonies in agar are usually yellowish brown, pink, or reddish; whitish colonies sometimes occur. **Strictly anaerobic. Sulfate-reducing bacteria reduce sulfate and in a few cases also sulfur to H$_2$S; sulfur-reducing bacteria are unable to reduce sulfate or other oxoanions of sulfur. H$_2$ or organic compounds serve as electron donors; oxidation of organic compounds is either incomplete, leading to acetate as an end product, or complete, leading to CO$_2$.** Electron donors utilized by many species are H$_2$, lactate, fatty acids, ethanol, or dicarboxylic acids. Autotrophic growth on H$_2$, CO$_2$, and sulfate may occur. Ammonium salts are generally used as the nitrogen source. All species examined thus far can fix dinitrogen. Typical habitats are anoxic sediments or bottom waters of freshwater, marine, or hypersaline aquatic environments; thermophilic species occur in hot springs and submarine hydrothermal vents.

The thermophilic sulfate-reducing archaeobacteria and the extremely thermophilic sulfur-reducing archaeobacteria are described in Groups 32 and 35, respectively, of this manual.

Differentiation of the four subgroups in **Group 7:** See Table 7.1.

SUBGROUP 1

Subgroup 1 consists of the one genus *Desulfotomaculum,* which comprises all spore-forming sulfate-reducing bacteria.

Genus **Desulfotomaculum**

Cells are straight to slightly curved rods, 0.5–2 × 2–9 µm, often with pointed ends; drop-shaped forms may also occur. **Central to terminal, heat- and desiccation-resistant endospores are formed,** sometimes with adjacent gas vacuoles. Cells may stain Gram positive. Motility by peritrichous or polar flagella is common. Desulfotomacula are **strictly anaerobic. Sulfate and in some cases also sulfite or thiosulfate are reduced to H$_2$S.** Sulfur is usually not reduced. **Many but not all species use H$_2$, lactate, and monocarboxylic acids C$_1$ through C$_{18}$; fewer species use sugars or amino acids. Organic substrates are either incompletely oxidized to acetate or completely oxidized to CO$_2$.** Species utilizing H$_2$ may be heterotrophic, requiring acetate as the carbon source, or autotrophic. All species are able to grow in simple, defined media; however, yeast extract may stimulate growth significantly. Many *Desulfotomaculum* species are more sensitive to H$_2$S produced from sulfate or to Na$_2$S added as reductant than are non-spore-forming sulfate- or sulfur-reducing bacteria. Vitamins may be required. The optimum pH range 6.6–7.4. The optimum temperature range for mesophilic species is 25–40°C, and that for thermophilic species is 40–65°C. *Desulfotomaculum* species may be the predominant sulfate reducers in temporarily flooded soils such as rice paddies. They also occur in anoxic freshwater and marine sediments and in the intestines of animals.

Type species: *Desulfotomaculum nigrificans.*

Differentiation of the species of the genus **Desulfotomaculum:** See Table 7.2.

SUBGROUP 2

Subgroup 2 comprises non-spore-forming sulfate-reducing bacteria that oxidize organic substrates incompletely to acetate. A few species are able to use sulfur as electron acceptor. Cells of all genera stain Gram negative.

Differentiation of the genera in **Subgroup 2:** See Table 7.3.

Genus **Desulfobulbus**

Cells are ovoid to rod-shaped or lemon-shaped, 0.6–1.3 × 1.5–2.5 µm, and occur singly or in pairs. They are motile by single polar flagella or are nonmotile. *Desulfobulbus* species are **strictly anaerobic. Sulfate and often also sulfite or thiosulfate are reduced to H₂S.** Sulfur is never reduced. **The characteristic organic electron donor is propionate. H₂, lactate, or ethanol is also oxidized. Organic substrates are incompletely oxidized to acetate.** In the absence of an external electron acceptor, lactate or ethanol and CO₂ may be fermented to propionate and acetate. All species are able to grow in simple, defined media. *p*-Aminobenzoate is required for growth. The optimum pH range is 6.6–7.5. The optimum temperature range is 25–40°C. Thermophilic species are not known. Habitats are anoxic freshwater and marine sediments.

Type species: *Desulfobulbus propionicus.*

Differentiation of the species of the genus **Desulfobulbus:**

1. Rod-shaped cells 0.6–0.7 × 1.5–2.5 µm.

    Desulfobulbus elongatus

2. Ovoid to lemon-shaped cells 1.0–1.3 × 1.8–2.0 µm.

    Desulfobulbus propionicus

The species *D. elongatus* was not included in *Bergey's Manual of Systematic Bacteriology.* The species was created in 1984 by Samain et al. (Syst. Appl. Microbiol. *5:* 394–401; Int. J. Syst. Bacteriol. *35:* 224–225, 1985).

Genus **Desulfomicrobium**

Editorial note: The genus *Desulfomicrobium* was not included in *Bergey's Manual of Systematic Bacteriology.* The genus was created in 1988 by Rozanova et al. (Mikrobiologiya *57:* 634–641).

Ovoid to rod-shaped cells are 0.6 × 1.3 µm. They are motile by single polar flagella or are nonmotile. **Strictly anaerobic. Sulfate and often also sulfite or thiosulfate are reduced to H₂S.** Sulfur may also be reduced. **H₂, lactate, or malate is oxidized as the electron donor; utilization of ethanol has not been observed. Organic substrates are incompletely oxidized to acetate.** *Desulfomicrobium* species share many physiological properties with *Desulfovibrio* species; however, desulfoviridin is absent from *Desulfomicrobium.* All species are able to grow in simple, defined media; however, yeast extract may stimulate growth. The vitamin requirement is unknown. The optimum pH range is 6.6–7.5. The optimum temperature range is 25–40°C. Thermophilic species have not been described. Habitats are anoxic freshwater and marine sediments and oil field waters.

Type species: *Desulfomicrobium baculatum.*

Differentiation of the species of the genus **Desulfomicrobium:** Species cannot readily be distinguished. The following species were described: *Desulfomicrobium apsheronum* and *D. baculatum.* *Desulfomicrobium baculatum* is the former species *Desulfovibrio baculatus.*

Characteristics of the species: As described for the genus.

Genus **Desulfomonas**

Editorial note: The genus *Desulfomonas* will be reclassified with *Desulfovibrio.*

Rod-shaped, somewhat irregular cells are 0.8–1.3 × 1.2–5 µm. Cells are nonmotile and strictly anaerobic. Sulfate is reduced to H₂S. Other electron acceptors have not been tested. **H₂, lactate, or ethanol serves as the electron donor. Organic substrates are incompletely oxidized to acetate.** Except for morphology, *Desulfomonas* resembles the genus *Desulfovibrio.* Able to grow in simple, defined media; however, yeast extract may stimulate growth. *p*-Aminobenzoate is required for growth. The optimum pH range is 6.6–7.5. The optimum temperature range is 30–40°C. Thermophilic species have not been described. *Desulfomonas* has been isolated from human feces.

Type (and only) species: *Desulfomonas pigra.* It will be reclassified as *Desulfovibrio piger.*

Characteristics of the species: As described for the genus.

Genus **Desulfovibrio**

Spiral to vibrioid-shaped cells are 0.5–1.3 × 0.8–5 µm. They are motile by single polar flagella or tufts of polar flagella; *Desulfovibrio carbinolicus* is nonmotile. **Strictly anaerobic. Sulfate and often also sulfite or thiosulfate are reduced to H₂S.** Sulfur may be also reduced. **H₂, lactate, ethanol, and often also malate or fumarate serve as electron donors; some species may utilize sugars, glycerol, choline, or a number of amino acids. Organic substrates are incompletely oxidized to acetate.** Desulfoviridin is present. All species are able to grow in simple, defined media; however, yeast extract may stimulate growth. A few species require biotin and/or other vitamins. The optimum pH range is 6.6–7.5. The optimum temperature range is 25–40°C. Thermophilic species have not been described. Habitats are anoxic freshwater and marine sediments, oil fields and other industrial water systems, and intestines of animals.

Type species: *Desulfovibrio desulfuricans.*

Differentiation of the species of the genus **Desulfovibrio:** See Table 7.4.

Genus **Thermodesulfobacterium**

Editorial note: The genus *Thermodesulfobacterium* was not included in *Bergey's Manual of Systematic Bacteriology.* The genus was created in 1983 by Zeikus et al. (J. Gen. Microbiol. *129:* 1159–1169).

Cells are ovoid to rod-shaped and 0.3 × 0.9–2.5 µm. They are motile by single polar flagella or are nonmotile. **Strictly anaerobic. Sulfate is reduced to H₂S. H₂ or lactate is oxidized as the electron donor; ethanol is not utilized. Organic substrates are incompletely oxidized to acetate.** Desulfoviridin is absent. All species are able to grow in simple, defined media; however, yeast extract may stimulate growth. The vitamin requirement is unknown. The optimum pH range is 6.6–7.5. The optimum temperature range is 65–70°C. *Thermodesulfobacterium* species have been isolated from hot springs and oil field water systems.

Type species: *Thermodesulfobacterium commune.*

Differentiation of the species of the genus **Thermodesulfobacterium:**

1. Rod-shaped, sometimes oval cells are 0.3 × 0.9 µm and nonmotile. The optimal growth temperature is about 65°C.

   T. commune

2. Rod-shaped, sometimes oval cells are 0.5 × 1.2–2.5 µm and motile by single polar flagella. The optimal growth temperature is about 70°C.

   T. mobile

The species was created in 1988 by Rozanova and Pivovarova (Mikrobiologiya *57:* 102–106; Int. J. Syst. Bacteriol. *41:* 178–179, 1990). *T. mobile* is the former *Desulfovibrio thermophilus.*

SUBGROUP 3

These are non-spore-forming, sulfate-reducing bacteria that oxidize organic substrates completely to CO₂. Growth with sulfur as an electron acceptor has never been observed. **Thermophilic sulfate-reducing archaeobacteria** that oxidize organic substrates completely are included in **Group 32** of this manual.

Differentiation of the genera in **Subgroup 3:** See Table 7.5.

Genus **Desulfobacter**

Cells are oval to rod-shaped or slightly curved to vibrioid-shaped and 0.5–2.4 × 1.7–7 µm. They stain Gram negative. Cells are motile by single polar flagella or are nonmotile and are **strictly anaerobic. Sulfate and usually also sulfite and thiosulfate are reduced to H₂S. The preferred, generally utilized electron donor is acetate.** In addition, some species use ethanol and a few species use H₂ or lactate. H₂-utilizing *Desulfobacter hydrogenophilus* grows autotrophically with CO₂ as the sole carbon source. Desulfoviridin is absent. Growth occurs in simple, defined media. Most species require vitamins. The addition of >7 g NaCl and >1 g MgCl₂·6H₂O is usually stimulatory or required for growth. The optimum pH range is 6.5–7.4. The optimum temperature range is 20–33°C. Thermophilic species have not been described. *Desulfobacter* species are most common in anoxic marine or brackish sediments. Some types may be found in anoxic freshwater sediments. Occurrence in oil field waters has been reported.

Types species: *Desulfobacter postgatei.*

Differentiation of the species of the genus **Desulfobacter:** See Table 7.6.

Genus **Desulfobacterium**

Editorial note: The genus *Desulfobacterium* was not included in *Bergey's Manual of Systematic Bacteriology.* The genus was created in 1986 by Bak and Widdel (Arch. Microbiol. *146:* 170–176; Int. J. Syst. Bacteriol. *38:* 136–137, 1988).

Cells are oval to rod-shaped or slightly curved to vibrioid-shaped and 0.7–3 × 1.5–2.8 μm. They stain Gram negative. Cells are motile by single polar flagella or are nonmotile and are **strictly anaerobic. Sulfate and in some cases also sulfite and/or thiosulfate are reduced to H₂S.** Nutritionally, *Desulfobacterium* species are very diverse and versatile. **Many species use H₂, formate and higher monocarboxylic acids up to C₁₆, lactate, or ethanol as electron donors. Several species may use aromatic compounds such as phenyl-substituted organic acids, phenolic compounds, or N-heterocyclic compounds.** Growth on acetate or propionate is usually very poor. H₂-utilizing species are autotrophic. Desulfoviridin is absent. Growth occurs in simple, defined media. Yeast extract usually has no stimulatory effect and may even inhibit growth. Most species require vitamins. The optimum pH range is 6.6–7.6. The optimum temperature range in most cases is 20–30°C and is seldom 30–35°C. Thermophilic species have not been described. *Desulfobacterium* species are most common in anoxic marine or brackish sediments. A number of types have been found in anoxic freshwater sediments.

Type species: *Desulfobacterium autotrophicum.*

Differentiation of the species of the genus **Desulfobacterium:** See Table 7.7.

Genus **Desulfococcus**

Cells are spherical or lemon-shaped and 1.4–2.3 μm in diameter. They stain Gram negative. The cells are motile by single polar flagella or are nonmotile and are **strictly anaerobic. Sulfate and often also sulfite and thiosulfate are reduced to H₂S.** Nutritionally versatile, **many species use formate and higher monocarboxylic acids up to C₁₆, lactate, or ethanol as electron donors.** Acetone, phenyl-substituted organic acids, or nicotinate may be also utilized. Desulfoviridin is present in all species but *Desulfococcus niacini.* Growth occurs in simple, defined media. Vitamins are required. The optimum pH range is 6.7–7.6. The optimum temperature range is 28–35°C. Thermophilic species have not been described. *Desulfococcus* species are found in freshwater, marine, or brackish sediments.

Type species: *Desulfococcus multivorans.*

Differentiation of the species of the genus **Desulfococcus:** See Table 7.8.

Genus **Desulfomonile**

Editorial note: The genus *Desulfomonile* was not included in *Bergey's Manual of Systematic Bacteriology.* The genus was created in 1990 by DeWeerd et al. (Arch. Microbiol. *154:* 23–30; Int. J. Syst. Bacteriol. *41:* 178–179, 1991).

Rod-shaped cells, 0.8–1 × 5–10 μm, with rounded ends stain Gram negative. Cells are nonmotile and **strictly anaerobic. Sulfate, sulfite, and thiosulfate are reduced to H₂S. Best growth occurs on pyruvate. H₂, formate, or benzoate is also utilized.** Poor growth occurs on lactate, acetate, or butyrate. Desulfoviridin is present. Growth occurs in simple, defined media. Vitamins are required; in addition, 1,4-naphthoquinone serves as growth factor. The optimum pH range is 6.8–7.0. The optimum temperature is 37°C. Thermophilic species have not been described. *Desulfomonile* has been isolated from sewage sludge in an enrichment with **3-chlorobenzoate that is reductively dehalogenated** by *Desulfomonile* to benzoate.

Type (and only) species: *Desulfomonile tiedjei.*

Characteristics of the species: As described for the genus.

Genus **Desulfonema** *(See also Group 15)*

Filaments of uniseriately arranged cells, 2.5–8 μm in diameter, may be up to 2 mm long. Gram stain is variable. **Gliding motility** occurs. **Strictly anaerobic. Sulfate is reduced to H$_2$S; sulfite and thiosulfate may or may not be used.** Sulfur does not serve as electron acceptor for growth but is inhibitory. **Acetate and higher monocarboxylic acids up to at least C$_{10}$, fumarate, and succinate serve as electron donors. H$_2$, lactate, or benzoate may or may not be utilized.** Desulfoviridin is present or absent. Growth occurs in simple, defined media. Vitamins are required. The optimum pH range is 7.0–7.6. The addition of >15 g NaCl and >2 g MgCl$_2$ · 6H$_2$O is required for growth. The optimum temperature range is 28–32°C. Thermophilic species have not been described. *Desulfonema* species occur in organic compound-rich, anoxic marine sediments.

Type species: *Desulfonema limicola.*

Differentiation of the species of the genus **Desulfonema:**

1. Multicellular filaments are 2.5–3 μm in diameter. H$_2$ or lactate is utilized with sulfate. Benzoate is not utilized. Desulfoviridin is present.

     D. limicola

2. Multicellular filaments are 6–8 μm in diameter. H$_2$ or lactate is not utilized. Benzoate is utilized. Desulfoviridin is absent. In addition to other minerals, 1 g CaCl$_2$ · 2H$_2$O/L is required for growth.

     D. magnum

Genus **Desulfosarcina**

Cells are rod-shaped, oval, or almost coccoid and are 1–1.5 × 1.5–2.5 μm. Cells grow partly in dense, **sarcina-like cell packets** that may adhere to surfaces. They stain Gram negative. Single cells may be motile by single polar flagella. **Strictly anaerobic. Sulfate, sulfite, or thiosulfate is reduced to H$_2$S. H$_2$, formate and higher monocarboxylic acids up to C$_{14}$, lactate, ethanol, fumarate, and benzoate serve as electron donors.** Growth on acetate is very poor. Are able to grow autotrophically. Desulfoviridin is absent. Growth occurs in simple, defined media. Vitamins are not required. The optimum pH range is 7.2–7.6. The optimum temperature is 33°C. Thermophilic species have not been described. Cells occur in marine sediments.

Type (and only) species: *Desulfosarcina variabilis.*

Characteristics of the species: As described for the genus.

SUBGROUP 4

Non-spore-forming, sulfur-reducing bacteria oxidize organic substrates completely to CO$_2$. Growth with sulfate or other oxoanions of sulfur has never been observed. **Extremely thermophilic sulfur-reducing archaeobacteria** are described in **Group 35** of this manual.

Differentiation of the genera in **Subgroup 4:** See Table 7.9.

Genus **Desulfurella**

Editorial note: The genus *Desulfurella* was not included in *Bergey's Manual of Systematic Bacteriology.* The genus was created in 1990 by Bonch-Osmolovskaya et al. (Arch. Microbiol. *153*: 151–155).

Rod-shaped to oval, sometimes curved cells are 0.4–0.8 × 1–4 μm. They stain Gram negative. Single, laterally inserted flagella have propeller-like movement. **Strictly anaerobic. Sulfur is reduced to H$_2$S.** Sulfate or other oxoanions of sulfur are never reduced. **Acetate serves as the electron donor and is completely oxidized to CO$_2$.** Agar colonies are whitish to grayish; cytochromes are absent. Growth occurs in simple, defined media; however, yeast extract is stimulatory. The growth factor requirement is unknown. The optimum pH range is 6.8–7.0. Moderately thermophilic, with an optimum temperature of 52–57°C. Extremely thermophilic species are not known. Habitats are organic compound-rich geothermal springs.

Type (and only) species: *Desulfurella acetivorans.*

Characteristics of the species: As described for the genus.

Genus **Desulfuromonas**

Rod-shaped to oval, sometimes curved cells are 0.4–0.8 × 1–4 μm. Cells stain Gram negative. Single, laterally inserted flagella have propeller-like movement. **Strictly anaerobic. Sulfur is reduced to H₂S.** Sulfate or other oxoanions of sulfur are never reduced. **Acetate, propionate, and ethanol serve as electron donors. Organic substrates are completely oxidized to CO₂.** Agar colonies are pink because of a high content of cytochromes or other pigments. Growth occurs in simple, defined media. Biotin is required. The optimum pH range is 6.8–7.5. Extremely thermophilic species are not known. The optimum temperature is about 30°C. Habitats are anoxic marine sediments.

Type (and only) species: *Desulfuromonas acetoxidans.*

Characteristics of the species: As described for the genus.

Important Notes for Users of This Manual

Unless otherwise indicated in footnotes to tables, the meanings of symbols are as follows:

+ 90% or more of strains are positive

− 90% or more of strains are negative

d 11-89% of strains are positive

v strain instability (not equivalent to "d")

D Different reactions in different taxa (species of a genus or genera of a family)

All other symbols are defined in footnotes to tables.

Table 7.1 Differentiation of Subgroups 1 through 4 of the dissimilatory sulfate- or sulfur-reducing bacteria [a]

Characteristic	1. Spore-forming sulfate-reducing bacteria	2. Non-spore-forming sulfate-reducing bacteria; incomplete substrate oxidation	3. Non-spore-forming sulfate-reducing bacteria; complete substrate oxidation	4. Sulfur-reducing bacteria
Sulfate reduced to sulfide	+	+	+	−
Sulfur reduced to sulfide	−	+ or −	−	+
Organic substrates completely oxidized to CO_2	+ or −	− [b]	+	+

[a] Symbols: see standard definitions.

[b] *Desulfovibrio baarsii* oxidizes substrates completely to CO_2; it will be reclassified as *Desulfoarculus baarsii*.

Table 7.2 Differential characteristics of the species of the genus *Desulfotomaculum* [a]

Characteristic	D. acetoxidans	D. antarcticum	D. geothermicum [b]	D. guttoideum [c]	D. kuznetsovii [d]	D. nigrificans	D. orientis	D. ruminis	D. sapomandens [e]	D. thermoacetoxidans [f]
Organic substrates completely oxidized to CO_2	+	−	−	−	+	−	−	−	+	+
Utilized with sulfate:										
$H_2 + CO_2$, + or − acetate as carbon source	−	−	+	−	+	+	+	+	−	+
Lactate	−	+	+	+	−	+	+	+	−	+
Acetate	+	−	−	−	+	−	−	−	+	+
Butyrate	+	−	+	−	+	−	−	−	+	+
Fatty acids (C_6–C_{16})	−	−	+	−	+	−	−	−	+	−
Optimum temperature range:										
25–40°C	+	+	−	+	−	−	+	+	+	−
50–60°C	−	−	+	−	+	+	−	−	−	+
60–65°C	−	−	−	−	+	−	−	−	−	−
Gas vacuoles formed during sporulation	+	−	+	−	−	−	−	−	+	−

[a] Symbols: see standard definitions.

[b-f] These five species were not included in *Bergey's Manual of Systematic Bacteriology*.

[b] *D. geothermicum* was created in 1988 by Daumas et al. (Antonie van Leeuwenhoek J. Microbiol. Serol. *54:* 165–178; Int. J. Syst. Bacteriol. *40:* 105–106, 1990).

[c] *D. guttoideum* was created in 1983 by Gogotova and Vainshtein (Mikrobiologiya *52:* 789–793; Int. J. Syst. Bacteriol. *36:* 573–576, 1986).

[d] *D. kuznetsovii* was created in 1988 by Nazina et al. (Mikrobiologiya *57:* 823–827; Int. J. Syst. Bacteriol. *40:* 470–471, 1990).

[e] *D. sapomandens* was created in 1985 by Cord-Ruwisch and Garcia (FEMS Microbiol. Lett. *29:* 325–330; Int. J. Syst. Bacteriol. *40:* 105–106, 1990).

[f] *D. thermoacetoxidans* was created in 1990 by Min and Zinder (Arch. Microbiol. 153: 399–404).

Table 7.3 **Differential characteristics of the genera of Subgroup 2** [a]

Characteristic	*Desulfobulbus*	*Desulfomicrobium*	*Desulfomonas* [b]	*Desulfovibrio*	*Thermodesulfobacterium*
Cells spiral or vibrioid-shaped	−	−	−	+	−
Cells ovoid to rod-shaped	+	+	+	−	+
Motile by polar flagella	+ or −	+	−	+ or −	+ or −
Optimum temperature range:					
25–40°C	+	+	+	+	−
65–70°C	−	−	−	−	+
Utilized with sulfate:					
H₂ + CO₂ + acetate as carbon source	+	+	+	+[c,d]	+
Lactate	+	+	+	+[c]	+
Propionate	+	−	−	−	−
Desulfoviridin present	−	−	+	+[e]	−

[a] Symbols: see standard definitions.

[b] *Desulfomonas* will be reclassified with *Desulfovibrio*.

[c] Not utilized by *D. baarsii*.

[d] Not utilized by *D. sapovorans*.

[e] Not present in *D. baarsii* and *D. sapovorans*.

Important Notes for Users of This Manual

Unless otherwise indicated in footnotes to tables, the meanings of symbols are as follows:

+ 90% or more of strains are positive

− 90% or more of strains are negative

d 11-89% of strains are positive

v strain instability (not equivalent to "d")

D Different reactions in different taxa (species of a genus or genera of a family)

All other symbols are defined in footnotes to tables.

Table 7.4 Differential characteristics of the species of the genus *Desulfovibrio*[a]

Characteristic	D. africanus, D. carbinolicus,[b] D. desulfuricans, D. fructosovorans,[c] D. furfuralis,[d] D. simplex,[e] D. sulfodismutans,[f] D. termitidis,[g] D. vulgaris	D. baarsii[i]	D. giganteus,[h] D. gigas	D. salexigens	D. sapovorans[j]
Cell size (μm):					
0.5–0.8 × 1.5–4	+	+	–	+	–
0.8–1.0 × 6–11	–	–	+	–	+
Utilized with sulfate:					
H_2 + CO_2 + acetate as carbon source	+	–	+	+	–
Lactate	+	–	+	+	+
Fatty acids (C_4–C_{16})	–	+	–	–	+
Obligate requirement for 20 g NaCl/L	–	–	–	+	–
Desulfoviridin	+	–	+	+	–

[a] Symbols: see standard definitions.

[b-h] These species were not included in *Bergey's Manual of Systematic Bacteriology*.

[b] *D. carbinolicus* was created in 1987 by Nanninga and Gottschal (Appl. Environ. Microbiol. *53:* 802–809).

[c] *D. fructosovorans* was created in 1988 by Ollivier et al. (Arch. Microbiol. *149:* 447–450; Int. J. Syst. Bacteriol. *40:* 105–106, 1990).

[d] *D. furfuralis* was created in 1989 by Folkerts et al. (Syst. Appl. Microbiol. *11:* 161–169; Int. J. Syst. Bacteriol. *39:* 495–497, 1989).

[e] *D. simplex* was created in 1989 by Zellner et al. (Arch. Microbiol. *152:* 329–334; Int. J. Syst. Bacteriol. *40:* 470–471, 1990).

[f] *D. sulfodismutans* was created in 1987 by Bak and Pfennig (Arch. Microbiol. *147:* 184–189; Int. J. Syst. Bacteriol. *38:* 136–137, 1988).

[g] *D. termitidis* was created in 1990 by Trinkerl et al. (Syst. Appl. Microbiol. *13:* 372–377; Int. J. Syst. Bacteriol. *41:* 178–179, 1991).

[h] *D. giganteus* was created in 1988 by Esnault et al. (Syst. Appl. Microbiol. *10:* 147–151; Int. J. Syst. Bacteriol. *38:* 328–329, 1988).

[i] Will be reclassified as *Desulfoarculus baarsii.*

[j] Will be reclassified as *Desulfobotulus sapovorans.*

Important Notes for Users of This Manual

Unless otherwise indicated in footnotes to tables, the meanings of symbols are as follows:

+ 90% or more of strains are positive

– 90% or more of strains are negative

d 11-89% of strains are positive

v strain instability (not equivalent to "d")

D Different reactions in different taxa (species of a genus or genera of a family)

All other symbols are defined in footnotes to tables.

Table 7.5 Differential characteristics of the genera of Subgroup 3 [a]

Characteristic	Desulfobacter	Desulfobacterium	Desulfococcus	Desulfomonile	Desulfonema	Desulfosarcina
Desulfoviridin present	−	−	+	+	+ or −	−
Cells spiral or vibrioid-shaped	+ or −	−	−	−	−	−
Cells ovoid to rod-shaped	+ or −	+	−	+	−	+
Cells spherical	−	+ or −	+	−	−	+ or −
Multicellular filaments	−	−	−	−	+	−
Motility:						
Flagella	+ or −	+ or −	−	−	−	+ or −
Gliding	−	−	−	−	+	−
Utilized with sulfate:						
Acetate:						
Doubling time 15–30 h	+	−	−	−	−	−
Doubling time ≥35 h	−	+	+	+	+	+
Fatty acids (C_4–C_8)	−	+ or −	+	−	+	+
Benzoate	−	+ or −	+ or −	+	+ or −	+

[a] Symbols: see standard definitions.

Table 7.6 Differential characteristics of the species of the genus *Desulfobacter* [a]

Characteristics	D. curvatus [b]	D. hydrogenophilus [b]	D. latus [b]	D. postgatei
Cells curved rods or vibrioid-shaped	+	−	−	−
Cells oval to rod-shaped	−	+	+	+
Cell size (μm):				
0.5–1.0 × 1.7–3.5	+	−	−	−
1.0–1.5 × 1.7–3.0	−	+	−	+
1.6–2.4 × 5.0–7.0	−	−	+	−
Motile by flagella	+	−	+ or −	+ or −
Utilized with sulfate:				
H_2 + CO_2	−	+	−	−
Ethanol	+	+	−	−

[a] Symbols: see standard definitions.

[b] *D. curvatus, D. hydrogenophilus,* and *D. latus* were not included in *Bergey's Manual of Systematic Bacteriology.* These species were created in 1987 by Widdel (Arch. Microbiol. *148:* 286–291; Int. J. Syst. Bacteriol. *38:* 328–329, 1988).

Table 7.7 Differential characteristics of the species of the genus *Desulfobacterium*[a]

Characteristic	D. anilini[b]	D. autotrophicum[c]	D. catecholicum[d]	D. indolicum[e]	D. macestii[f]	D. phenolicum[g]
Motile by flagella	–	+	–	+	+	+
Utilized with sulfate:						
H$_2$ + CO$_2$	+	+	+	–	+	–
Lactate	–	+	+	–	+	–
Propionate	+	+	+	+	–	–
Butyrate	+	+	+	–	–	+
Fatty acids C$_6$–C$_{16}$	+	+	+	–	–	–
Benzoate	+	–	+	–	–	+
Phenol	+	–	–	–	ND	+
Catechol	+	–	+	–	ND	–
Indole	–	–	–	+	–	+
Aniline	+	–	–	–	–	–

[a] Symbols: see standard definitions; ND, not determined.

[b–g] No species of the genus *Desulfobacterium* were included in *Bergey's Manual of Systematic Bacteriology*.

[b] *D. anilini* was created in 1989 by Schnell et al. (Arch. Microbiol. *152:* 556–563; Int. J. Syst. Bacteriol. *40:* 320–321, 1990).

[c] *D. autotrophicum* was created in 1987 by Brysch et al. (Arch. Microbiol. *148:* 264–274; Int. J. Syst. Bacteriol. *38:* 328–329, 1988).

[d] *D. catecholicum* was created in 1987 by Szewzyk et al. (Arch. Microbiol. *147:* 163–168; Int. J. Syst. Bacteriol. *38:* 136–137, 1988).

[e] *D. indolicum* was created in 1986 by Bak and Widdel (Arch. Microbiol. *146:* 170–176; Int. J. Syst. Bacteriol. *38:* 136–137, 1988).

[f] *D. macestii* was created in 1989 by Gogotova and Vainshtein (Mikrobiologiya *58:* 76–80; Int. J. Syst. Bacteriol. *39:* 495–497, 1989).

[g] *D. phenolicum* was created in 1986 by Bak and Widdel (Arch. Microbiol. *146:* 177–180; Int. J. Syst. Bacteriol. *38:* 136–137, 1988).

Table 7.8 Differential characteristics of the species of the genus *Desulfococcus*[a]

Characteristic	D. biacutus[b]	D. multivorans	"D. niacini"[b,c]
Cells spherical	–	+	+
Cells lemon-shaped	+	–	–
Motile by flagella	–	–	+
Utilized with sulfate:			
H$_2$ + CO$_2$	ND	–	+
Lactate	–	+	–
Acetone	+	–	ND
Benzoate	–	+	–
Nicotinate	–	–	+
Desulfoviridin	+	+	–

[a] Symbols: see standard definitions; ND, not determined.

[b] The species *D. biacutus* and *"D. niacini"* were not included in *Bergey's Manual of Systematic Bacteriology*. The species *D. biacutus* was created in 1990 by Platen et al. (Arch. Microbiol. *154:* 355–361). The species *"D. niacini"* was created in 1983 by Imhoff-Stuckle and Pfennig (Arch. Microbiol. *136:* 194–198).

[c] Will be reclassified as *Desulfobacterium niacini*.

Table 7.9 Differential characteristics of the genera of Subgroup 4 [a]

Characteristic	*Desulfurella*	*Desulfuromonas*
Electron acceptor reduce to sulfide:		
Sulfate	−	−
Sulfur	+	+
Optimum temperature range:		
25–40°C	−	+
50–57°C	+	−
Color of colonies in agar:		
Pink to reddish brown	−	+
Whitish	+	−
Growth in the absence of sulfur on fumarate or malate with or without acetate as additional substrate	−	+ or −

[a] Symbols: see standard definitions.

Important Notes for Users of This Manual

Unless otherwise indicated in footnotes to tables, the meanings of symbols are as follows:

- \+ 90% or more of strains are positive
- − 90% or more of strains are negative
- d 11-89% of strains are positive
- v strain instability (not equivalent to "d")
- D Different reactions in different taxa (species of a genus or genera of a family)

All other symbols are defined in footnotes to tables.

ANAEROBIC GRAM-NEGATIVE COCCI

Cocci, ca. 0.3–0.5 – ca. 2.5 µm in diameter, occur characteristically in pairs. Single cells, masses, or chains may also occur, although the chains may show gaps, illustrating the basic diplococcal arrangement. Adjacent sides of cell pairs may be flattened. **Gram negative, but cells tend to resist decolorization. Nonmotile;** flagella do not occur. **Anaerobic** and **oxidase negative.** Catalase negative, but some strains decompose peroxide by a non-heme-containing pseudocatalase. **Chemoorganotrophic.** Possess complex nutritional requirements. **The genera *Acidaminococcus, Megasphaera,* and *Veillonella* have a fermentative type of metabolism.** Gas may be produced (sometimes abundantly); carbohydrates may or may not be fermented; and lactic acid may not be produced. If lactic acid is present it is not a major product of fermentation. Lactate is fermented by some genera with the production of CO_2, H_2, and various volatile fatty acids containing 2–6 carbon atoms. **The genus *Syntrophococcus* has a respiratory type of metabolism** and requires a sugar as an electron donor and formate, methoxybenzenenoids, acrylate side chains of benzenoids, or a hydrogenotrophic bacterium as an exogenous electron acceptor system for effective growth. These are **mainly parasites of homothermic animals** such as humans, ruminants, rodents, and pigs, but one species (*Megasphaera cerevisiae*) occurs in spoiled, bottled beer.

Differentiation of the genera included in **Group 8:** See Table 8.1.

Genus **Acidaminococcus**

Cocci, 0.6–1.0 µm in diameter, often occur as oval or kidney-shaped diplococci. Optimum temperature is 30–37°C. Optimum pH is 7.0. **Possess a fermentative type of metabolism. Amino acids,** especially glutamate, **are the main energy sources.** Pyruvate, lactate, fumarate, malate, succinate, and citrate **are not used as energy sources.** Only ca. 40% of strains catabolize glucose, and the reaction is weak. In amino acid-containing media, **acetic and butyric acids** accumulate in a molar ratio of 2:1; CO_2 is also formed, **but H_2 and**

propionate are not detectable. Isolated from the intestinal tract of pigs and humans.

Type (and only) species: *Acidaminococcus fermentans.*

Characteristics of the species: As described for the genus.

Genus **Megasphaera**

Cocci, 1.3–2.0 µm or more in diameter, occur in pairs or occasionally in chains. Growth occurs from 15–40°C, but generally not at 45°C. **Possess a fermentative type of metabolism. Fructose and lactate are fermented; glucose may or may not be fermented. Found in the rumen of cattle and sheep, in the feces and intestine of humans, and in spoiled bottled beer.**

Type species: *Megasphaera elsdenii.*

Differentiation of the species of the genus **Megasphaera:** See Table 8.2.

Genus **Syntrophococcus**

Editorial note: The genus *Syntrophococcus* was not included in *Bergey's Manual of Systematic Bacteriology.* The genus was created in 1986 by Krumholz and Bryant (Arch. Microbiol. *143:* 313-318; Int. J. Syst. Bacteriol. *36:* 489, 1986). See also Doré and Bryant (Appl. Environ. Microbiol. *55:* 927-933, 1989; *56:* 984-989, 1990). The genus includes only one species, *S. sucromutans.*

Coccus-shaped cells 1.0–1.3 µm in diameter. Possess a respiratory type of metabolism and require a sugar as an electron donor and one of a limited variety of electron acceptors to grow effectively in pure culture. Sugars are used as electron donors with acetate as the only organic product. H_2/CO_2-utilizing methanogens can serve as an electron acceptor system. The genus differs from *Megasphaera* in that **lactate and glucose are not fermented,** and 4-carbon straight- and branched-chain fatty acids, propionate, valerate and isovalerate, and caproate are not produced. It differs from *Acidaminococcus, Megasphaera,* and *Veillonella* in that **growth requires a lipid supplement such as 150 µm oleate. Found in the rumen of hay-fed cattle.**

Type (and only) species: *Syntrophococcus sucromutans.*

Characteristics of the species: Cells have a rounded to flattened plane of division and occur in short chains. An electron donor and an electron acceptor are required for energy and growth. Electron donors include pyruvate, glucose, fructose, galactose, maltose, cellobiose, lactose, arabinose, maltose, ribose, xylose, salicin, and esculin. The major organic product formed from these substrates is acetate. Compounds not used as electron donors include hydrocaffeate, gallate, malate, aspartate, glutamate, glycine, 3-hydroxybutyrate, valine, serine, isoleucine, butyrate, pectin, and starch. Electron acceptors (followed by the major organic products in parentheses) are as follows: formate (acetate); caffeate (hydrocaffeate), ferulate (caffeate, hydrocaffeate, acetate), syringate and 3,4,5-trimethoxybenzoate (gallate and acetate), vanillin (protocatechuic aldehyde, protocatechuate and acetate), and vanillate (protocatechuate and acetate). The following compounds are not used as electron acceptors or as single energy sources: methanol, formaldehyde, CO, ethanol, fumarate, malate, crotonate, lactate, aspartate, glutamate, glycine, valine, serine, isoleucine, 1-methoxybenzene, 4-methoxybenzoate, cinnamate, 4-hydroxycinnamate, 4:1 $H_2:CO_2$, and 4:1 $H_2:CO_2$ plus fumarate, crotonate, or caffeate. Growth occurs with an electron donor such as cellobiose or xylose when an H_2-CO_2-using methanogen is the electron acceptor system. Good growth occurs in an anaerobic medium containing 4:1 $N_2:CO_2$ gas phase, minerals, 40 mM $NaHCO_3$, 2 mM cysteine, 6 mM sodium formate, 6 mM cellobiose or xylose, 0.5% Casitone™, and a lipid supplement such as 150 µm oleate. pH range is 6.0–7.6; optimum, pH 6.4. Temperature range is 30.5–44.0°C; optimum, 35–42°C. Habitat is a numerous microbe catabolizing ferulate and syringate in the rumen of hay-fed cattle.

Genus **Veillonella**

Cocci 0.3–0.5 µm in diameter, appearing by light microscopy as diplococci, masses, and short chains. Optimum temperature is 30–37°C. Optimum pH is 6.5–8.0. Catalase negative, but some species produce an atypical catalase lacking porphyrin. **Possess a fermentative type of metabolism. Pyruvate, lactate,** malate, fumarate, and oxaloacetate **are fermented. Carbohydrates and polyols are not fermented,** except for one species where fructose fermentation has been detected. **Acetate, propionate, CO_2, and H_2 are produced from lactate.** CO_2 is required for growth. **Parasitic in the mouths and in the intestinal and respiratory tracts of humans and other animals.**

Type species: *Veillonella parvula.*

Differentiation of the species of the genus **Veillonella:** See Table 8.3.

Important Notes for Users of This Manual

Unless otherwise indicated in footnotes to tables, the meanings of symbols are as follows:

+ 90% or more of strains are positive
− 90% or more of strains are negative
d 11-89% of strains are positive
v strain instability (not equivalent to "d")
D Different reactions in different taxa (species of a genus or genera of a family)

All other symbols are defined in footnotes to tables.

Table 8.1 Differential characteristics of the genera of anaerobic Gram-negative cocci [a]

Characteristic	*Acidaminococcus*	*Megasphaera*	*Syntrophococcus*	*Veillonella*
Cell diameter (μm)	0.6–1.0	1.3–2.0 or more	1.0–1.3	0.3–0.5
Red fluorescence of colonies under ultraviolet light (360 nm)	–	–	+	
Lipid supplement required for growth	–	–	+	–
Effective growth requires a sugar as an electron donor and formate, methoxybenzenenoids, acrylate side chains of benzenoids, or a hydrogenotrophic bacterium as an exogenous electron acceptor system	–	–	+	–
Ability to ferment carbohydrates	–[b]	+	–[c]	–[b]
Lactate fermented	–	+	–	+
Amino acids are the main energy source	+	–	–	–
Pyruvate utilized	–	+		+
Succinate decarboxylated	–	–		+
Products in growth media:				
Gases:				
CO_2	+	+		+
H_2	–	Weak	+[d]	+
Volatile fatty acids:				
2-carbon	+	+	+	+
3-carbon	–	+	–	+
4-carbon	+	+	–	–
5-carbon	–	+	–	–
6-carbon	–	+	–	–

[a] Symbols: see standard definitions.

[b] Generally negative; slight or variable reactions may occur.

[c] Some carbohydrates including cellobiose can be used as an electron donor provided that an exogenous electron acceptor is present.

[d] H_2 is produced but is inhibitory in very low concentration.

Table 8.2 Differential characteristics of the species of the genus *Megasphaera* [a]

Characteristic	*M. cerevisiae* [b]	*M. elsdenii*
Growth at 40°C	–	+
Fermentation of glucose and maltose	–	+
Propionic acid and valeric acid formed when grown on fructose	+	–

[a] Symbols: see standard definitions.

[b] *M. cerevisiae* was not included in *Bergey's Manual of Systematic Bacteriology*, Volume 1. The species was created by Engelmann and Weiss (Syst. Appl. Microbiol. *6:* 287-290, 1985; Int. J. Syst. Bacteriol. *36:* 354-346, 1986).

Table 8.3 Differential characteristics of the species of the genus *Veillonella*[a]

Characteristic	V. atypica	V. caviae	V. criceti	V. dispar	V. parvula	V. ratti	V. rodentium
Fructose fermented	–	–	+	–	–	–	–
Catalase[b]	–	–	+	+	–	+	–
Serological group[c]	V[d] or VI[d]	[e]	I	VII[d]	II, IV[d], or VI[d]	III	II
Putrescine or cadaverine requirement	–	78% +	Most +	+	Most –	–	ca. 35% +
Sources from which strains have so far been isolated	Human, rarely rodent; buccal	Guinea pig; mouth	Hamster; mouth	Human; mouth and respiratory tract	Human, rat, and rabbit; buccal or intestinal	Rat; mouth and intestine	Hamster, rat, and rabbit; buccal or intestinal

[a] Symbols: standard definitions except as noted.

[b] Non-heme-containing pseudocatalase.

[c] As defined by Rogosa (J. Bacteriol. *90:* 704-709, 1965 and *Bergey's Manual of Determinative Bacteriology,* 8th ed., pp. 446-447, 1974).

[d] Serogroups from which human strains have been isolated.

[e] The serogrouping of *V. caviae* strains has not yet been reported.

Important Notes for Users of This Manual

Unless otherwise indicated in footnotes to tables, the meanings of symbols are as follows:

+ 90% or more of strains are positive

– 90% or more of strains are negative

d 11-89% of strains are positive

v strain instability (not equivalent to "d")

D Different reactions in different taxa (species of a genus or genera of a family)

All other symbols are defined in footnotes to tables.

THE RICKETTSIAS AND CHLAMYDIAS

This group includes two subgroups, as follows:

I. **Rickettsias.** This subgroup multiplies by binary fission, not by a complex developmental cycle. Most live in an obligate intracellular association, parasitic or mutualistic, with eucaryotic hosts (vertebrates or arthropods); a few can be grown on moderately complex bacteriological media containing blood. The cell walls contain muramic acid. Glutamate is oxidized with generation of ATP. Rickettsias are mainly rod-shaped, coccoid, and often pleomorphic bacteria that stain Gram negative and lack flagella; exceptions may occur for each of these features, as follows:

> Some appear ring-shaped in stained preparations.

> Some have a flagellum.

> Some may stain Gram positive.

The parasitic species are associated with the reticuloendothelial and vascular endothelial cells or erythrocytes of vertebrates and often with various organs of arthropods, which may act as vectors or primary hosts. Some cause disease in humans or other vertebrate and invertebrate hosts. The mutualistic species occur in insects and are regarded as essential for development and reproduction of the host.

There are three genera that are able to grow on bacteriological media, *Bartonella*, *Grahamella*, and *Rochalimaea*, but which have been traditionally classified with the rickettsias because of their intracellular parasitic growth habit. Phylogenetically, *Bartonella* and *Rochalimaea* fall into the α-2 subgroup of the *Proteobacteria* and are discussed in Group 4 of this manual.

II. **Chlamydias.** Nonmotile, obligately parasitic, coccoid bacteria that multiply only within membrane-bounded vacuoles in the cytoplasm of the cells of humans, other mammals, and birds. Arthropods do not serve as hosts. Multiplication occurs by means of a unique developmental cycle characterized by change of small elementary bodies into larger reticulate bodies that divide by fission. They are pathogenic. Cell walls contain no muramic acid or only a trace. Glucose is not oxidized with net generation of ATP; the parasites are dependent upon the ATP of the host cell.

The bacteria in this group require very specialized cultivation techniques for study. Only certain laboratories are equipped for their identification and, for that reason, they will not be considered further in this manual.

Important Notes for Users of This Manual

Unless otherwise indicated in footnotes to tables, the meanings of symbols are as follows:

+ 90% or more of strains are positive
− 90% or more of strains are negative
d 11-89% of strains are positive
v strain instability (not equivalent to "d")
D Different reactions in different taxa (species of a genus or genera of a family)

All other symbols are defined in footnotes to tables.

GROUP 10

ANOXYGENIC PHOTOTROPHIC BACTERIA

Cells are spherical, spiral, or rod- or vibrioid-shaped and are 0.3–6.0 µm in diameter. They occur singly or in regular or irregular aggregates; unicellular or uniseriately multicellular filamentous forms also occur. Gram negative. In most cases, multiplication is by binary fission; some species multiply by budding. With or without gas vacuoles. Motile or nonmotile; motility is by flagella or by gliding. Flagella are either monotrichous or multitrichous. Photosynthetic pigments are located in the cytoplasmic membrane (Subgroup 4), in various types of intracytoplasmic membrane systems (Subgroups 1–3), or in chlorosomes (Subgroups 5 and 6). Colors of cell suspensions are purple-violet to purple-red, rose-red, yellowish brown, brown, and green. Common to all species is the presence of bacteriochlorophylls (see Table 10.1) and of carotenoid pigments (see Table 10.2). Photoautotrophic or photoorganotrophic under anaerobic or microaerobic conditions. In contrast to oxygenic photosynthesis of cyanobacteria, anoxygenic photosynthesis is dependent on external electron donors, such as reduced sulfur compounds, molecular hydrogen, or organic compounds. During sulfide oxidation, highly refractile globules of sulfur are transiently stored either inside the cells (Subgroup 1) or outside the cells (Subgroups 2 and 5). Storage materials are polysaccharides, poly-β-hydroxybutyrate, and polyphosphate. Carbon dioxide is assimilated through the reductive pentose phosphate cycle or the reductive citric acid cycle (Subgroup 5). Ammonium salts are generally used as the nitrogen source. The fixation of dinitrogen has been demonstrated in most representatives of all subgroups except Subgroup 6. With the exception of Subgroup 5, many species are capable of growing as chemoautotrophs or chemoorganotrophs under aerobic or microaerobic conditions. Fatty acids, organic acids, or alcohols serve as electron donors and carbon sources. Habitats are the anoxic parts of moist soils and aquatic environments including fresh water, brackish water, and marine and hypersaline environments.

The anoxygenic phototrophic bacteria are presented in seven subgroups that can be differentiated by the characteristics given in Table 10.3.

SUBGROUP 1

Cells are able to grow with sulfide and sulfur as the sole photosynthetic electron donor for CO_2 assimilation; grow well under photoautotrophic conditions. In the presence of both sulfide and light, globules of sulfur appear inside the cells and may be further oxidized to sulfate. Contain bacteriochlorophyll a or b and carotenoids of Subgroups 1–4 (Table 10.2). Vitamin B_{12} may be required for growth.

Differentiation of the genera in Subgroup 1: See Table 10.4.

Genus Amoebobacter

Cells are spherical and 1.5–3 µm in diameter; occurring singly or in irregular or platelet-shaped aggregates. Cells are surrounded by slime. Nonmotile. Cells contain gas vacuoles centrally or peripherally. Cultures are rose-red; only one species is purple-red. Intracytoplasmic membrane systems are of the vesicular type. The photosynthetic pigments are bacteriochlorophyll a and carotenoids of the spirilloxanthin or okenone series. Phototrophic under anaerobic conditions; may be chemotrophic under microaerobic or aerobic conditions. Photoautotrophic growth with sulfide, sulfur, thiosulfate, or molecular hydrogen. During sulfide oxidation, globules of sulfur are transiently stored within the cells. The final oxidation product is sulfate. Acetate, propionate, pyruvate, or lactate is photoassimilated. Vitamin B_{12} may be required for growth. The pH range is 6.8–7.5; the growth temperature is 20–35°C. *Amoebobacter* species may occur as bloom-forming bacteria together with *Thiocapsa* in wastewater lagoons rich in organic substances. *Amoebobacter purpureus* occurs mainly in sulfide-containing water of stratified freshwater lakes.

Type species: *Amoebobacter roseus.*

Differentiation of the species of the genus **Amoebobacter:** See Table 10.5.

Genus **Chromatium**

Cells are straight to slightly curved rod-shaped or ovoid, with rounded ends, 1.0–6.0 µm in diameter and 1.5–16.0 µm long (Fig. 10.1). **Motile by monotrichous or multitrichous polar flagella.** In the presence of sulfide and light, globules of sulfur appear evenly distributed within the cells, which do not contain gas vacuoles. Intracytoplasmic membrane systems are of the vesicular type. Photosynthetic pigments bacteriochlorophyll *a* and carotenoids of Subgroups 1, 3, and 4. Under anaerobic conditions, there is **photoautotrophic growth with sulfide or sulfur as the electron donor** for CO_2 assimilation. The final oxidation product is sulfate; molecular hydrogen or thiosulfate may also be used by certain species. All species photoassimilate a number of simple organic compounds in the presence of sulfide and bicarbonate; acetate or pyruvate is most widely used. Vitamin B_{12} may be required for growth. The optimum growth temperatures are 20–35°C; only one species, *Chromatium tepidum,* grows at 50°C. The pH range is 6.8–7.5. *Chromatium* species occur in sulfide-containing parts of freshwater, estuarine, marine, or hypersaline environments.

Type species: *Chromatium okenii.*

Differentiation of the species of the genus **Chromatium:** See Table 10.6.

Genus **Lamprobacter**

Cells are rod-shaped or ovoid, 2.0–2.5 µm wide and 4–5 µm long. Cells with flagellar motility are devoid of gas vacuoles; when cells lose motility, they form **gas vacuoles** and slime capsules. In the presence of sulfide or thiosulfate and light, globules of sulfur are evenly distributed within the cells. Cultures are purple-red. There is an intracytoplasmic membrane system of the vesicular type. Cells contain bacteriochlorophyll *a* and carotenoids of the okenone series. **Under anaerobic conditions, there is photoautotrophic growth with reduced sulfur compounds** as electron donors for CO_2 assimilation. May be chemoautotrophic under microaerobic conditions. Photoassimilate a number of

simple organic compounds in the presence of sulfide and bicarbonate; acetate, pyruvate, lactate, ethanol, and glycerol are used. The optimum growth temperature is 23–27°C; the optimum pH is 7.4–7.6. The optimum salinity is 1–2%, and there is no growth in the absence of NaCl. *Lamprobacter* occurs in hydrogen sulfide-containing mud and water of saline water bodies with 1–4% salt in southern areas of the Commonwealth of Independent States (CIS).

Type (and only) species: *Lamprobacter modestohalophilus.*

Characteristics of the species: As described for the genus.

Genus **Lamprocystis**

Cells are spherical to ovoid, 2–3.5 µm in diameter. They occur singly, in small irregular aggregates, or in long and branching aggregates embedded in slime (Fig. 10.2). **Gas vacuoles are found in the central part of the cells.** Under favorable growth conditions, aggregates may break up into smaller clusters and spherical cell colonies, which become motile; **individual motile cells** are also liberated. Cells are motile by a single flagellum. Cultures are purple-violet. There is an intracytoplasmic membrane system of the vesicular type. Bacteriochlorophyll *a* and carotenoids of the rhodopinal series are present. **Under anaerobic conditions, there is photoautotrophic growth with sulfide or sulfur** as the electron donor for CO_2 assimilation; the final oxidation product is sulfate. During sulfide oxidation, globules of sulfur are transiently stored in the peripheral part of the cells. In the presence of sulfide and bicarbonate, acetate and pyruvate are photoassimilated. May require vitamin B_{12}. The growth temperature is 20–30°C, and the pH range is 6.8–7.5. Occurs in mud and stagnant water of ponds and lakes containing hydrogen sulfide; common planktonic bacterium in the top layer of the sulfide-containing hypolimnion of stratified freshwater lakes.

Type (and only) species: *Lamprocystis roseopersicina.*

Characteristics of the species: As described for the genus.

Genus **Thiocapsa**

Cells are spherical to slightly ovoid, 1.0–3 µm in diameter; Occurring singly or in aggregates of two,

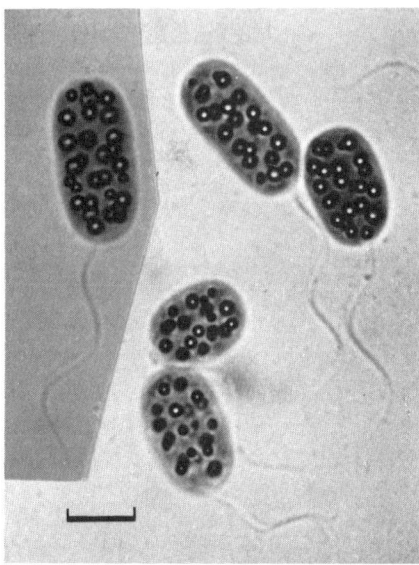

Figure 10.1. *Chromatium okenii* DSM 169 cultured photoautotrophically with sulfide. Note the evenly distributed intracellular globules of elemental sulfur and the spiral tufts of polar flagella. Brightfield micrograph. *Bar*, 5 μm.

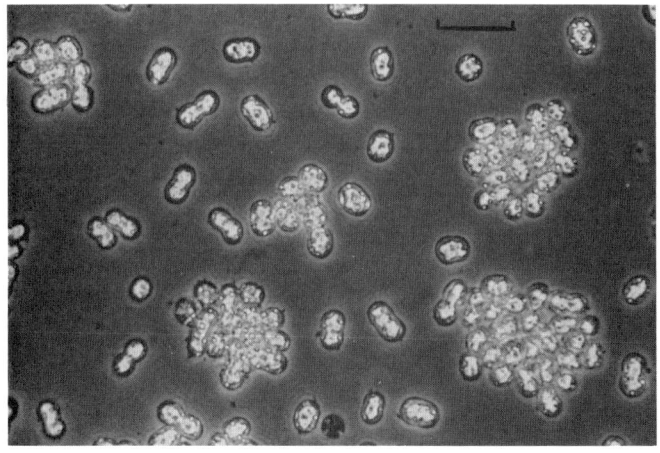

Figure 10.2. *Lamprocystis roseopersicina* DSM 229 cultured photoautotrophically with sulfide. Single cells and cell colonies are motile in liquid medium. The *irregular whitish areas* within the cells are the gas vacuoles. The *small spherical bodies* in the cells are the sulfur globules. Phase-contrast micrograph. *Bar*, 10 μm.

four, or more cells; cells are usually surrounded by slime (Fig. 10.3). **Cells nonmotile** and **without gas vacuoles.** Cultures are rose-red or yellowish brown. The intracytoplasmic membrane system is of the vesicular or tubular type. Bacteriochlorophyll *a* or *b* and carotenoids of the spirilloxanthin series are present. **Under anaerobic conditions, there is photoautotrophic growth with sulfide or sulfur** as the electron donor for CO_2 assimilation; the final oxidation product is sulfate. During sulfide oxidation, globules of sulfur are transiently stored inside the cells. Acetate, pyruvate, and

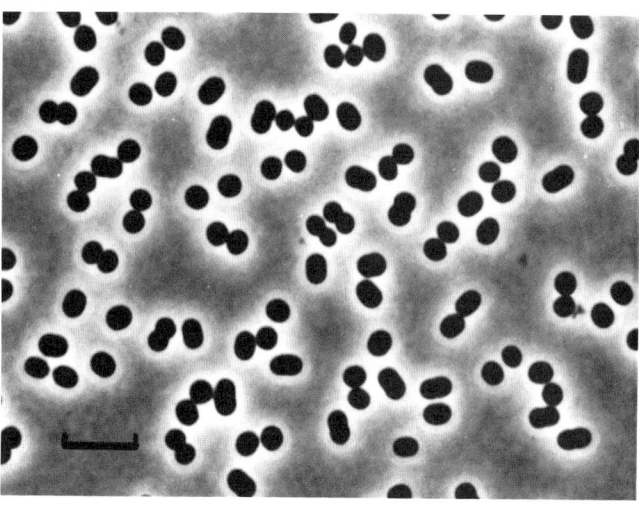

Figure 10.3. *Thiocapsa roseopersicina* DSM 219 cultured photoautotrophically with sulfide. Phase-contrast micrograph *Bar*, 5 μm.

other simple organic compounds are photoassimilated. Obligately phototrophic or facultatively chemoautotrophic under aerobic conditions in the dark. The growth temperature is 20–35°C, and the pH range is 6.8–7.5. Found in the stagnant water and mud of ponds, pools, or wastewater lagoons with organic substances and hydrogen sulfide; they are common also in estuaries, sandy marine beaches, and microbial mats of salt marshes.

Type species: *Thiocapsa roseopersicina.*

Differentiation of the species of the genus **Thiocapsa:** See Table 10.7.

Genus **Thiocystis**

Cells are spherical or slightly ovoid and 2.5–3 μm in diameter. Larger individual cells may occur, and cells may be diplococcus-shaped before division (Fig. 10.4). Motile by means of a single flagellum. Irregular aggregates of nonmotile cells surrounded by slime are formed in the presence of sulfide and bright light. Do not contain gas vacuoles. Cultures are purple-violet or purple-red. Photosynthetic membrane system of the vesicular type. Bacteriochlorophyll *a* and carotenoids of Subgroups 3 or 4 are present. **Under anaerobic conditions, photoautotrophic with sulfide or sulfur as the electron donor** for CO_2 assimilation; during sulfide oxidation, globules of sulfur are transiently formed inside the cells. The final oxidation product is sulfate. In the presence of sulfide and bicarbonate, acetate and pyruvate are photoassimilated. **Facultatively chemoautotrophic** under microaerobic or aerobic conditions. The pH range is 6.8–7.6. The growth temperature is 25–35°C. Marine isolates require 1–2% NaCl. Occur in sulfide-containing water and mud of freshwater and brackish water or seawater environments: sewage lagoons, estuaries, salt marshes, and sulfur springs.

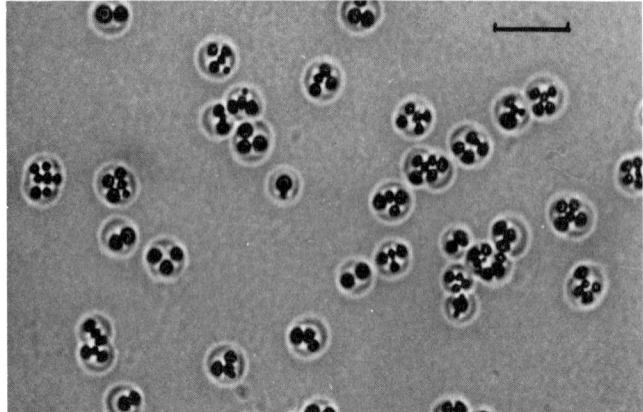

Figure 10.4. *Thiocystis violacea* DSM 207 cultured photoautotrophically with sulfide. The cells contain sulfur globules. Brightfield micrograph. *Bar,* 5 μm.

Type species: *Thiocystis violacea.*

Differentiation of the species of the genus **Thiocystis:**

1. Purple-violet cultures; carotenoids of the rhodopinal series; sulfur globules in the center of the cell.

   T. violacea

2. Purple-red cultures; carotenoids of the okenone series; sulfur globules only at the inner periphery of the cell.

   T. gelatinosa

Genus **Thiodictyon**

Cells are rod-shaped with rounded ends, 2.0 μm wide and 4–8 μm long; nonmotile under all conditions. May form aggregates in which they are arranged end to end in an irregular netlike structure, the shape of which is not constant (Fig. 10.5). May also form more compact clumps or break up into individual cells. **Contain**

large gas vacuoles in their central parts. Cultures are purple-violet. The photosynthetic membrane systems are of the vesicular type. Bacteriochlorophyll *a* and carotenoids of the rhodopinal series are present. **Obligately anaerobic and phototrophic.** Photoautotrophic growth with sulfide or sulfur as the electron donor for CO_2 assimilation. During sulfide oxidation, globules of sulfur are transiently stored in the gas vacuole-free peripheral part of the cell. The final oxidation product is sulfate. In the presence of sulfide and bicarbonate, acetate or pyruvate is photoassimilated. The pH optimum is 7.3; the growth temperature is 20–30°C. Occur in sulfide-containing water and mud of ponds and lakes; common planktonic bacterium in the top layer of the sulfide-containing hypolimnion of stratified freshwater lakes.

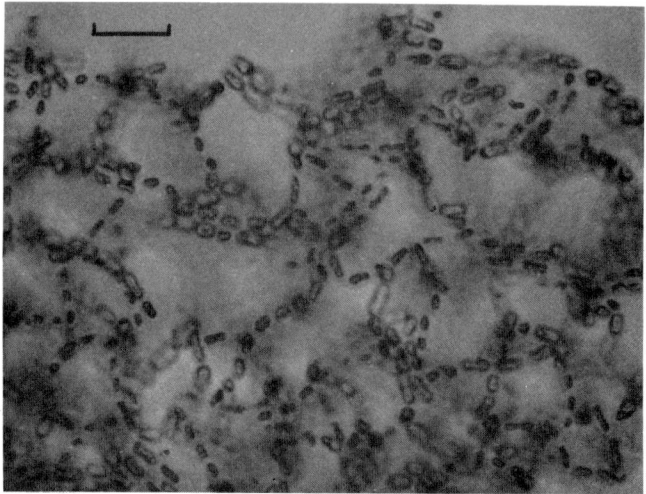

Figure 10.5. *Thiodictyon elegans* DSM 232 showing the typical, somewhat irregular netlike arrangement of the cells. Brightfield micrograph. *Bar,* 12 μm.

Type species: *Thiodictyon elegans.*

Differentiation of the species of the genus **Thiodictyon:**

1. Cells able to grow in the form of typical netlike aggregates in the presence of sulfide and bright light.

   T. elegans

2. Cells occurring singly, not forming netlike aggregates.

   T. bacillosum

Genus **Thiopedia**

Cells are spherical to ovoid or elongated ovoid, 1.4–1.8 μm wide and 1.5–2.5 μm long; regularly arranged in rectangular platelets with 4, 8, 16, 32, or more cells. Platelet formation may be lost upon prolonged cultivation in the laboratory. Nonmotile. **Cells contain irregularly shaped gas vacuoles in the central part.** Cultures are bright purple-red. Photosynthetic membrane system of the vesicular type. Bacteriochlorophyll *a* and carotenoids of the okenone series are present. **Obligately anaerobic and phototrophic. Photoautotrophic growth is possible with sulfide concentrations up to 0.6 mM;** higher concentrations inhibit growth completely. During sulfide oxidation, globules of sulfur are transiently stored in the peripheral part of the cell. Sulfate is the final oxidation product. In the presence of low sulfide concentrations (0.1–0.5 mM) and bicarbonate, the following substrates are photoassimilated: **acetate, butyrate,** or **valerate** (not more than 3 mM to be added). Use of the latter two fatty acids is characteristic of *Thiopedia* in comparison to *Amoebobacter purpureus*. Growth is enhanced by the addition of 100 μM dithionite. Vitamin B_{12} is required for growth. The optimum pH is 7.3; the optimum growth temperature is 20°C. Light intensities above 200–300 lux (tungsten lamp) inhibit growth. Inhabit the mud and stagnant water of ponds and lakes containing hydrogen sulfide; common planktonic bacterium in the top layer of the sulfide-containing hypolimnion of stratified freshwater lakes.

Type (and only) species: *Thiopedia rosea.*

Characteristics of the species: As described for the genus.

Genus **Thiospirillum**

Cells are curved rod- or vibrioid-shaped, sigmoid or spiral with rounded ends, 2.5–4.0 μm in diameter. Sigmoid cells are usually 30–40 μm long; spiral cells are up to 100 μm long. Coil depth is 3–7 μm, and nearly straight rod-shaped cells may occur. Motile by means of a (multitrichous) polar flagellar tuft, 10–12 μm long. Cells are rarely tufted at both ends. In the presence of sulfide and light, globules of sulfur are evenly distributed within the cells. Do not contain gas vacuoles. Intracytoplasmic membrane system of the vesicular type. Cell suspensions are yellowish brown. Bacteriochlorophyll *a* and the carotenoids lycopene and rhodopin are present. **Obligately anaerobic and**

phototrophic. Photoautotrophic growth with sulfide or sulfur as the electron donor for CO_2 assimilation. During sulfide oxidation, globules of sulfur are transiently stored in the cells; the final oxidation product is sulfate. In the presence of sulfide and bicarbonate, acetate is photoassimilated. Vitamin B_{12} is required for growth. The pH range is 6.8–7.5; The optimum pH is 7.3; growth temperature is 20–28°C. Found in the mud and stagnant water of ditches and freshwater ponds containing hydrogen sulfide.

Type (and only) species: *Thiospirillum jenense.*

Characteristics of the species: As described for the genus.

SUBGROUP 2

Cells are able to grow with sulfide and sulfur as the sole photosynthetic electron donors for CO_2 assimilation. In the presence of sulfide and light, **globules of sulfur appear outside the cells** and may be further oxidized to sulfate. Ammonia and dinitrogen are used as the nitrogen source. Growth is dependent on saline and alkaline conditions. Contain bacteriochlorophyll *a* or *b* and carotenoids of carotenoid group 1 (Table 10.2).

Subgroup 2 consists of only one genus.

Genus **Ectothiorhodospira**

Cells are vibrioid or rod-shaped, 0.5–1.5 μm in diameter, and motile by polar flagella; they multiply by binary fission. **Internal photosynthetic membranes are present as lamellar stacks** that are continuous with the cytoplasmic membrane. Photosynthetic pigments are bacteriochlorophyll *a* or *b* and carotenoids. Growth occurs **photoautotrophically under anaerobic conditions with reduced sulfur compounds** or hydrogen as electron donors or photoheterotrophically with a limited number of simple organic compounds. Sulfide is oxidized to **sulfur,** which is **deposited outside the cells,** and may be **further oxidized to sulfate.** Some species are able to grow under microaerobic to aerobic conditions in the dark. **Growth is dependent on saline and alkaline conditions.** Growth factors are not required, but vitamin B_{12} enhances the growth of some strains. The optimum growth temperature is 25–44°C; the optimum pH is 7.6–9.5. Occur in marine to extremely saline environments containing sulfide, with neutral to extremely alkaline pH, such as estuaries, salt

357

flats, salt lakes, soda lakes, and others; occasionally they may be found in soil.

Type species: *Ectothiorhodospira mobilis.*

Differentiation of the species of the genus **Ectothiorhodospira:** See Table 10.8.

SUBGROUP 3

Cells preferably grow by photoassimilation of simple organic substances; some species are capable of using sulfide or thiosulfate as the electron donor for CO_2 assimilation. In the presence of sulfide and light, **globules of sulfur may appear only outside the cells,** never inside. Sulfur is rarely oxidized further to sulfate. Most genera are able to grow as chemoheterotrophs under microaerobic or aerobic conditions. Ammonia or dinitrogen is used as the nitrogen source. Most genera **depend on one or more growth factors;** the most commonly required are biotin, thiamine, niacin, and *p*-aminobenzoic acid.

Differentiation of the genera in **Subgroup 3:** See Table 10.9.

Genus **Rhodobacter**

Cells are ovoid or rod-shaped, 0.5–1.2 μm in diameter, and motile or nonmotile; motile forms have polar flagella. **Cells divide by binary fission,** may produce capsules and slime, and may form chains of cells. **Internal photosynthetic membranes are vesicles.** Photosynthetic pigments are bacteriochlorophyll *a* and carotenoids of the spheroidene series. Cultures are yellowish brown under anaerobic conditions and reddish brown in the presence of air. Photoautotrophic growth is possible in the presence of sulfide as an electron donor and, in some species, with thiosulfate and molecular hydrogen. **Growth occurs photoheterotrophically under anaerobic conditions in the light** with a variety of organic compounds as carbon and electron sources. **Most species grow chemoheterotrophically under aerobic conditions.** Vitamins are required for growth. The optimum growth temperature is 25–35°C; the optimum pH is 6.5–7.5.

Type species: *Rhodobacter capsulatus.*

Differentiation of the species of the genus **Rhodobacter:** See Table 10.10.

Genus **Rhodocyclus**

Cells are slender, curved or straight thin rods, 0.3–1.0 μm in diameter, nonmotile or motile by means of polar flagella; **multiply by binary fission. Internal photosynthetic membranes are present in the form of small, single finger-like intrusions** of the cytoplasmic membrane. Photosynthetic pigments are bacteriochlorophyll *a* and carotenoids. Photoautotrophic growth with molecular hydrogen is possible. Preferably **grow photoheterotrophically** under anaerobic conditions in the light with different organic substrates as the carbon and electron source. Growth is also possible under microaerobic to aerobic conditions in the dark. Reduced sulfur compounds are not used as photosynthetic electron donors. Vitamins are required for growth. The optimum growth temperature is 30°C; the optimum pH is 6.5–7.5.

Type species: *Rhodocyclus purpureus.*

Differentiation of the species of the genus **Rhodocyclus:** See Table 10.11.

Genus **Rhodomicrobium**

Ovoid to elongate-ovoid bacteria, 1.0–1.2 μm in diameter, showing **polar growth and a characteristic vegetative growth cycle** (Fig. 10.6). This cycle includes the formation of **peritrichously flagellated swarmer cells** and nonmotile mother cells that form filaments from one to several times the length of the mother cell. Daughter cells originate as spherical buds at the end of the filaments and may undergo differentiation in various ways. **Dry-resistant polyhedral exospores may be formed.** Cells have **intracytoplasmic membranes of the lamellar type** and contain bacteriochlorophyll and carotenoids. Cells **grow preferably photoheterotrophically** under anaerobic conditions in the light with various organic substrates as carbon and electron sources; molecular hydrogen and sulfide at low concentrations may be used as photosynthetic electron donors. Growth factors are not required. Cells are able to grow under microaerobic to aerobic conditions in the dark. The optimum growth temperature is 25–35°C; the optimum pH is 6.0.

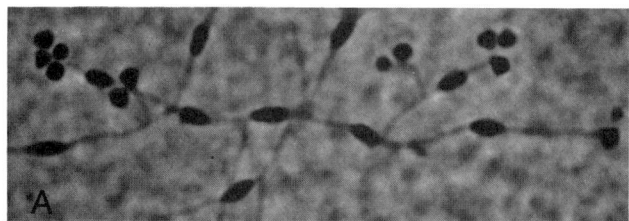

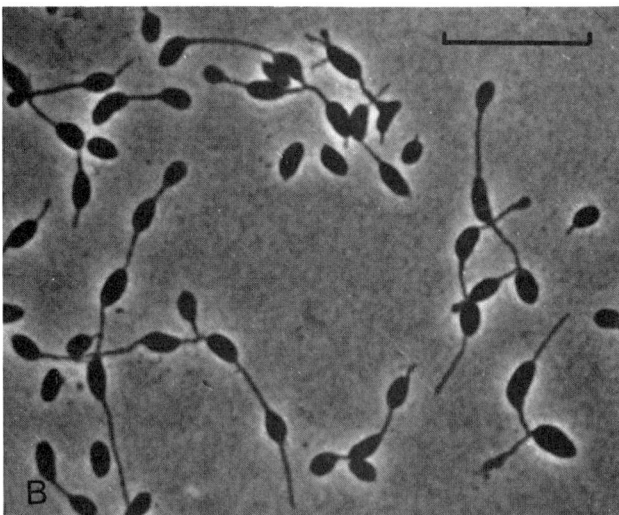

Figure 10.6. *Rhodomicrobium vannielii* strain DSM 163. *A*, polyhedral exospores as buds at the ends of short filaments with common branching points. *B*, cells with filaments and buds of various sizes. Phase-contrast micrographs. *Bar*, 10 μm. (Courtesy of N. Pfennig.)

Type (and only) species: *Rhodomicrobium vannielii.*

Characteristics of the species: As described for the genus.

Genus **Rhodopila**

Cells are spherical to ovoid, 1.6–1.8 μm in diameter under optimal growth conditions, and motile by means of polar flagella. **Divide by binary fission.** Cultures are purple-red. The **internal membrane system is of the vesicular type,** and photosynthetic pigments are bacteriochlorophyll *a* and carotenoids. **Cells grow preferably photoheterotrophically** under anaerobic conditions in the light. Cells are sensitive to oxygen but grow under microaerobic conditions in the dark. The optimum growth temperature is 30–35°C; the optimum pH is 4.8–5.0. The only isolates were obtained from warm acidic sulfur springs in Yellowstone National Park (Wyoming, USA).

Type (and only) species: *Rhodopila globiformis.*

Characteristics of the species: As described for the genus.

Genus **Rhodopseudomonas**

Cells are rod-shaped, 0.6–2.5 μm wide and 0.6–5.0 μm long, and **motile by means of flagella or nonmotile.** They show **polar growth, budding, and asymmetric cell division. Internal photosynthetic membranes are present as lamellae** underlying and parallel to the cytoplasmic membrane. Photosynthetic pigments are bacteriochlorophyll *a* or *b* and various types of carotenoids. **Photoheterotrophic growth occurs** with a number of organic compounds as carbon source and electron donor. Photoautotrophic growth is possible under anaerobic conditions with hydrogen, thiosulfate, or sulfide as the electron donor. Some species grow chemotrophically under microaerobic to aerobic conditions. Vitamins are required for growth. The optimum growth temperature is 25–35°C.

Type species: *Rhodopseudomonas palustris.*

Differentiation of the species of the genus **Rhodopseudomonas:** See Table 10.12.

Genus **Rhodospirillum**

Cells are spiral, 0.7–1.5 μm wide, and motile by means of polar flagella; they divide by binary fission. Internal photosynthetic membranes are present as vesicles or lamellae lying parallel to the cytoplasmic membrane or forming a sharp angle to it. Photosynthetic pigments are bacteriochlorophyll *a* and carotenoids of the spirilloxanthin series. **Growth occurs preferably photoheterotrophically** under anaerobic conditions in the light but also under microaerobic to aerobic conditions in the dark. Growth factors are required. Some species are very sensitive to oxygen. Molecular hydrogen may be used as a photosynthetic electron donor. The growth temperature is 25–35°C; the optimum pH is 6.8–7.2.

Type species: *Rhodospirillum rubrum.*

Differentiation of the species of the genus **Rhodospirillum:** See Table 10.13.

SUBGROUP 4

Cells grow by photoassimilation of simple organic substances; **strictly anaerobic and photoheterotrophic.** Reduced sulfur compounds are not utilized. Ammonia and dinitrogen are used as nitrogen sources. Vitamins are required for growth. **Internal membrane systems or chlorosomes are absent. Cells contain bacteriochlorophyll g** and carotenoids. Cell walls lack lipopolysaccharide.

Differentiation of the genera of **Subgroup 4:** Subgroup 4 comprises two genera.

  1. Motile by peritrichous flagella.

      Heliobacillus

  2. Nonflagellated; motile by gliding.

      Heliobacterium

Genus **Heliobacillus**

Editorial note: The genus *Heliobacillus* was not included in *Bergey's Manual of Systematic Bacteriology.* The genus was created in 1987 by Beer-Romero and Gest (FEMS Microbiol. Lett. *41:* 109–114). It includes only one species, *H. mobilis.*

Cells are rod-shaped and 1.0 μm wide and 7–10 μm long, frequently longer. **Motile by peritrichous flagella.** Cultures are green. Contain bacteriochlorophyll g and carotenoids. Heliobacilli are **obligately anaerobic and phototrophic. Photoheterotrophic** growth occurs with acetate, pyruvate, lactate, and butyrate. Vitamins are required for growth. The optimum growth temperature is 40–42°C; cells grow at pH 7.0–7.2. Occur in paddy fields and temporarily flooded soils.

Type (and only) species: *Heliobacillus mobilis.*

Characteristics of the species: As described for the genus.

Genus **Heliobacterium**

Cells are rod-shaped and frequently bent, 1.0 μm wide and 4–10 μm long or longer. **Motile by gliding.** Cultures are green and contain bacteriochlorophyll g

and carotenoids. **Obligately anaerobic and photoheterotrophic;** grow preferentially with organic acids or complex organic carbon sources. Vitamins are required for growth. The optimum growth temperature is 35–42°C; cells grow at pH 7.0–7.2. Isolated from surface soil.

Type (and only) species: *Heliobacterium chlorum.*

Characteristics of the species: As described for the genus.

SUBGROUP 5

Cells are able to grow with sulfide or sulfur as the sole photosynthetic electron donor for CO_2 assimilation. In the presence of both sulfide and light, **globules of sulfur appear outside the cells,** never inside. All species are obligately anaerobic and phototrophic; they grow well under photoautotrophic conditions. Simple organic substrates are photoassimilated only in the presence of sulfide and bicarbonate. Vitamin B_{12} may be required for growth. Cultures are green (bacteriochlorophyll c or d) or brown (bacteriochlorophyll e; see Table 10.1). Antenna bacteriochlorophylls are located in **chlorosomes** that underlie and are attached to the cytoplasmic membrane.

Differentiation of the genera in **Subgroup 5:** See Table 10.14.

Genus **Ancalochloris**

Cells have an irregular, starlike shape, 0.5–1.0 μm in diameter without the prosthecae. One cell may have up to six prosthecae (Fig. 10.7). Prosthecae vary in length up to 2 μm; they are 0.5–0.7 μm wide at the base and 0.1 μm wide at the tapering ends. Cells multiply by unequal fission and may remain attached in irregular chains that give rise to microcolonies of up to 30 cells. **Cells contain gas vacuoles** and are nonmotile. Bacteriochlorophylls and carotenoids are located in the cytoplasmic membrane and chlorosomes. **Obligately anaerobic and phototrophic. Photoautotrophic growth with sulfide and polysulfide-sulfur as electron donor for CO_2 assimilation; globules of sulfur appear outside the cells.** The optimum pH is 6.8–7.0; the optimum temperature is 20°C. Vitamin B_{12} is required for growth. Habitat is the anoxic hypolimnion of stratified freshwater lakes or meromictic lakes with low sulfide content (0.1–1.0 mM).

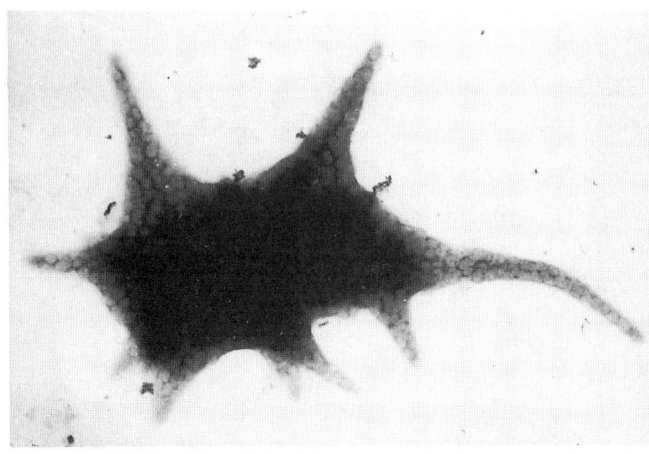

Figure 10.7. *Ancalochloris perfilievii* from a water sample of the hypolimnion of meromictic Lake Bol'shoi, Kichier, Russia. Electron micrograph of a single cell stained with phosphotungstic acid (x30,000).

Type (and only) species: *Ancalochloris perfilievii.*

Characteristics of the species: As described for the genus.

Genus **Chlorobium**

Straight or curved rod-shaped, ovoid, or spherical cells are 0.3–1.1 μm wide and 0.4–3.0 μm long, although much longer cells may occur. **Cells often remain attached to each other in filaments or in streptococcus-like chains;** curved rod-shaped cells may form **tightly or loosely wound spirals. Nonmotile.** Pure cultures are either **green** (grass-green) or **brown** (chocolate-brown); these colors can also be recognized in the light microscope with brightfield illumination. Photosynthetic pigments are located in the cytoplasmic membrane and in **chlorosomes** that are attached to the cytoplasmic membrane. Bacteriochlorophylls *c, d,* or *e* occur as major pigments, and there are also small amounts of bacteriochlorophyll *a*. Chlorobactene (green strains) or isorenieratene (brown strains) are the major carotenoids. **Obligately anaerobic and phototrophic. There is photoautotrophic growth with sulfide or polysulfide-sulfur as electron donor for CO_2 assimilation.** During sulfide oxidation, **globules of sulfur appear outside the cells;** the final oxidation product is sulfate. In sulfide-reduced media, thiosulfate or molecular hydrogen may be used as the electron donor. In the presence of sulfide and bicarbonate, a number of simple organic substrates are photoassimilated; acetate or propionate are most widely used. Vitamin B_{12} may

be required for growth. Ammonia serves as the nitrogen source; dinitrogen is fixed by many strains. The growth temperature is 20–35°C. The optimum pH is 6.8–7.0. Habitat is hydrogen sulfide-containing mud and water of freshwater, brackish water, and marine environments.

Type species: *Chlorobium limicola.*

Differentiation of the species of the genus **Chlorobium:** See Table 10.15.

Genus **Chloroherpeton**

Long rod-shaped cells, 0.6–1.0 μm wide and 8–20 μm long, have rounded ends. Cells are flexible, straight, or bent in various ways; cells flex through up to 180°. After division, cells separate promptly, and no septa are seen in each unit (unicellular). Cells may contain **gas vacuoles. Cells are motile by gliding at about 10 μm·min⁻¹ at 20°C; they grow preferably on solid surfaces or in semisolid media.** If grown in liquid media, cells clump and form extracellular slime. Cultures are green. Photosynthetic pigments are located in the cytoplasmic membrane and chlorosomes; the main pigments are bacteriochlorophyll *c* and γ-carotene. **Obligately anaerobic and phototrophic, show photoautotrophic growth with sulfide and sulfur as the electron donor for CO_2 assimilation.** During sulfide oxidation, **globules of sulfur appear outside the cells** and are only slowly oxidized further to sulfate. Acetate, propionate, malate, succinate, or glutamate is photoassimilated in the presence of sulfide and bicarbonate. Vitamin B_{12} is required for growth. The nitrogen source is ammonia. The optimum temperature is 25°C; the optimum pH is 6.8–7.2. Found in sulfide-containing marine littoral sediments, salt marshes, and tidal inlets rich in rotting plant material.

Type (and only) species: *Chloroherpeton thalassium.*

Characteristics of the species: As described for the genus.

Genus **Pelodictyon**

Cells are rod-shaped to ovoid, 0.6–1.2 μm wide and 1.2–2.5 μm long. In two species cells occur singly or in irregular netlike aggregates; branching is due to ternary fission. In the other two species cells occur singly or in spherical or irregular aggregates. **Cells contain gas vacuoles and are nonmotile.** Cultures are green or brown. Photosynthetic pigments are located in the cytoplasmic membrane and chlorosomes. Bacteriochlorophyll *c, d,* or *e* is present, in addition to small amounts of bacteriochlorophyll *a.* Chlorobactene or isorenieratene is the major carotenoid. **Obligately anaerobic and phototrophic. Photoautotrophic growth occurs with sulfide and sulfur as the electron donor for CO_2 assimilation.** During sulfide oxidation, **globules of sulfur appear outside the cells.** In the presence of sulfide and bicarbonate, a few simple organic substrates are photoassimilated; acetate or propionate is most widely used. The nitrogen source is ammonia; dinitrogen may be assimilated. Vitamin B_{12} is required for growth. The optimum temperature is 15–25°C; the optimum pH is 6.7–7.1. Inhabit hydrogen sulfide-containing water and mud of freshwater, brackish water, and marine environments. Gas-vacuolated species may occur as planktonic and bloom-forming cells in sulfide-containing stagnant water of lakes.

Type species: *Pelodictyon clathratiforme.*

Differentiation of the species of the genus **Pelodictyon:** See Table 10.16.

Genus **Prosthecochloris**

Cells are spherical or short rod-shaped before division, 0.5–0.7 μm wide and 0.5–1.2 μm long; each cell has 10–20 prosthecae, 0.1–0.17 μm wide and 0.07–0.3 μm long with rounded ends. Cells singly or in groups and chains. Do not contain gas vacuoles. **Nonmotile.** Cultures green or brown. Bacteriochlorophyll *c* or *e* and the carotenoids chlorobactene or isorenieratene are major pigments, and they are located in the cytoplasmic membrane and in chlorosomes. **Obligately anaerobic and phototrophic. Photoautotrophic growth occurs with sulfide and sulfur as the electron donor for CO_2 assimilation.** During sulfide oxidation, **globules of sulfur appear outside the cells** and may be incompletely oxidized further to sulfate. In the presence of sulfide and bicarbonate, acetate, pyruvate, and some other substrates are photoassimilated. The nitrogen source is ammonia, and dinitrogen may be fixed. Vitamin B_{12} is required for growth. The temperature range is 20–30°C, and the optimum pH is 6.7–7.0. The salinity range is 0.2–7.0% NaCl, with the optimum about 2% NaCl. **Most common marine green sulfur bacteria.** Habitat is the hydrogen sulfide-containing mud and water of marine ponds and lakes with up to 18% NaCl.

Type species: *Prosthecochloris aestuarii.*

Differentiation of the species of the genus **Prosthecochloris:**

1. Cultures and cell material green; major photosynthetic pigments bacteriochlorophyll *c* and chlorobactene.

    P. aestuarii

2. Cultures and cell material brown; major photosynthetic pigments bacteriochlorophyll *e* and isorenieratene.

    P. phaeoasteroidea

Consortia, symbiotic aggregates

Consortia of **Subgroup 5** bacteria (green sulfur bacteria) with colorless bacteria occur together with other phototrophic bacteria of **Subgroups 1 and 5** in mud and sulfide-containing water of stagnant freshwater ponds or stratified freshwater lakes. None of the consortia has thus far been isolated in pure culture. The generic and species designations existing in the literature are illegitimate and are given in quotation marks as laboratory names in Table 10.17.

SUBGROUP 6

Cells are uniseriately arranged in multicellular filaments that are capable of gliding motility. All genera are facultatively aerobic and preferentially utilize organic substances in their phototrophic or chemotrophic metabolism. They contain bacteriochlorophylls and various carotenoids.

Differentiation of the genera in **Subgroup 6:** See Table 10.18.

Genus **Chloroflexus**

Multicellular filaments of indefinite length, transverse septa hardly visible in the light microscope; cells are 0.5–1.0 μm wide and 2–6 μm long. A thin sheath is sometimes present. **Motile by gliding** (0.01–0.04 μm·s⁻¹). Anaerobic phototrophic cultures are dirty yellowish green, and aerobic cultures are orange-reddish. Anaerobically light-grown cells contain **chlorosomes** and bacteriochlorophylls *c* and *a*, as well as the carotenoids β- and γ-carotene. **Under anaerobic conditions cells are primarily photoheterotrophic,** and under aerobic conditions they are chemohetero-trophic. Under both conditions the following organic substrates are used: yeast extract, casamino acids, ace-tate, glycerol, glucose, pyruvate, or glutamate. Folic acid and thiamine may be required for growth. The nitrogen source is ammonia. The optimum pH is 7.6–8.4. The temperature range of thermophilic strains is 40–66°C, and the optimum temperature is 52–60°C. For mesophilic strains the optimum temperature is 20–25°C and the optimum pH is 7.0–7.2. Thermo-philic strains occur worldwide in neutral to alkaline hot springs at 40–68°C; they form gel-like orange-reddish mats, generally under a thin surface layer of cya-nobacteria. Mesophilic strains form shallow, marine microbial mats in salt marshes or in freshwater lake sediments exposed to light.

Type (and only) species: *Chloroflexus aurantiacus.*

Characteristics of the species: As described for the genus.

Genus **Chloronema**

Multicellular, clearly septate filaments of varying lengths are made up of cells 2.0–2.5 μm wide and 3.5–4.5 μm long. Filaments are surrounded by a sheath. **Filaments are straight or spiral and are motile by gliding** (~10 μm·s⁻¹). **Cells contain large, centrally located gas vacuoles.** Filaments in suspen-sion and in the brightfield microscope are yellowish green. Cells contain chlorosomes, bacteriochlorophyll *c* or *d,* and carotenoids. Cells are capable of anoxygenic photosynthesis and of aerobic chemotrophic metabo-lism. Not in pure culture. Occur at and below the chemocline of stratified freshwater lakes together with other species of phototrophic bacteria of Subgroups 1 and 5.

Type species: *Chloronema giganteum.* In addition to the type species, only one other species is incompletely described: *C. spiroideum.*

Characteristics of both species: As described for the genus.

Genus **Heliothrix**

Multicellular, clearly septate filaments of indefinite length contain cells about 1.5 μm wide and much longer than they are wide. A thin sheath may or may not be present. Motile by gliding. Phototrophic cul-tures are bright orange. Cells contain only bacteri-ochlorophyll *a* and abundant carotenoids; chlorosomes and accessory bacteriochlorophylls are absent. **Aer-otolerant; metabolism is mainly photoheterotrophic,** and acetate is utilized. Not in pure culture; co-culture with the aerobic chemoheterotrophic bacterium *Isosphaera pallida* is possible. The temperature range is 35–56°C, with an optimum of 40–55°C. Occurs in some hot springs in Oregon (USA), forming an orange mat a few millimeters thick; this mat is usually covered by a greenish layer of cyanobacteria.

Type (and only) species: *Heliothrix oregonensis.*

Characteristics of the species: As described for the genus.

Genus **Oscillochloris**

Multicellular filaments of indefinite length show transverse septa in the light microscope; cells are either 1.0–1.4 μm or 4.5–5.5 μm wide. Cells have no sheath but do contain **gas vacuoles.** Gram stain posi-tive or negative. **Filaments are flexible and are motile by gliding.** Filaments are yellowish green or green. Under anaerobic conditions cells contain chlorosomes, bacteriochlorophyll *c,* and carotenoids. **Phototrophic under anaerobic conditions** and may be capable of chemoheterotrophic growth under aerobic conditions. Growth temperature is 10–20°C; the optimum pH is 7.5. Not in pure culture. Occur in mats on the surface of sulfide-containing mud of freshwater ponds and lakes. They thrive together with other species of photo-trophic bacteria of Subgroups 1 and 5.

Type species: *Oscillochloris chrysea.*

Differentiation of the species of the genus **Oscillochloris:**

1. Filaments 4.5–5.5 μm wide and 3.5–7.0 μm long; yellowish green under anaerobic and aero-bic conditions.

    O. chrysea

2. Filaments 1.0–1.4 μm wide and 2.3–3.8 μm long; cell suspensions dark green under anaerobic conditions in the light and orange under aerobic conditions.

O. trichoides

SUBGROUP 7

Cells **grow chemoheterotrophically under aerobic conditions; no growth occurs under anaerobic conditions in the light.** Metabolism is predominantly respiratory. Cells **contain bacteriochlorophyll *a* and carotenoids.**

Subgroup 7 consists of only one genus.

Genus **Erythrobacter**

Cells are ovoid to rod-shaped, 0.5 μm wide and 1.0–5.0 μm long. Gram-negative cells multiply by binary fission and are motile by subpolar flagella. Cultures and colonies are orange or pink. **Cells contain bacteriochlorophyll *a* and carotenoids. Growth is aerobic and chemoheterotrophic.** There is no growth under anaerobic conditions in the light. The carbon sources utilized are acetate, pyruvate, butyrate, glutamate, and glucose. Vitamins are required for growth. The optimum growth temperature is 25–30°C; the pH range is 7.0–8.0; the optimum salinity is 1.7–3.5% NaCl. Occur in oxic marine environments, predominantly on seaweeds.

Type (and only) species: *Erythrobacter longus.*

Characteristics of the species: As described for the genus.

Important Notes for Users of This Manual

Unless otherwise indicated in footnotes to tables, the meanings of symbols are as follows:

+   90% or more of strains are positive

−   90% or more of strains are negative

d   11-89% of strains are positive

v   strain instability (not equivalent to "d")

D   Different reactions in different taxa (species of a genus or genera of a family)

All other symbols are defined in footnotes to tables.

Table 10.1   Characteristic absorption maxima of bacteriochlorophylls in living cells

Bacteriochlorophyll	nm
a	375, 590, 800–810, 830–890
b	400, 605, 835–850, 1015–1035
c	Long wavelength abs. max. 745–760
d	Long wavelength abs. max. 725–745
e	Long wavelength abs. max. 715–725
g	370, 419, 575, 670,[a] 780–790

[a] Breakdown product of bacteriochlorophyll g. The intensity of this peak depends on the degree of exposure to O_2.

Table 10.2   Carotenoid groups of anoxygenic phototrophic bacteria

Group	Name	Major components
1	Normal spirilloxanthin series	Lycopene, rhodopin, spirilloxanthin
2	Alternative spirilloxanthin series	Chloroxanthin, spheroidene, spheroidenone, (spirilloxanthin)
3	Okenone series	Okenone
4	Rhodopinal series (variation of Group 1)	Lycopene, lycopenal, lycopenol, rhodopin, rhodopinal, rhodopinol, (spirilloxanthin)
5	Chlorobactene series	Chlorobactene, isorenieratene, β-carotene, γ-carotene

---

Important Notes for Users of This Manual

Unless otherwise indicated in footnotes to tables, the meanings of symbols are as follows:

  +   90% or more of strains are positive

  −   90% or more of strains are negative

  d   11-89% of strains are positive

  v   strain instability (not equivalent to "d")

  D   Different reactions in different taxa (species of a genus or genera of a family)

All other symbols are defined in footnotes to tables.

Table 10.3 Differentiation of Subgroups 1–7 of the anoxygenic phototrophic bacteria[a]

Characteristic	1. Purple sulfur bacteria: sulfur globules inside cells	2. Purple sulfur bacteria: sulfur globules outside cells	3. Purple nonsulfur bacteria	4. Bacteria with bchl *g*	5. Green sulfur bacteria	6. Multicellular filamentous green bacteria	7. Anaerobic chemotrophic bacteria with bchl
Cells occur:							
Singly or in aggregates	+	+	+	+	+	+	+
Multicellular filaments	–	–	–	–	–	+	–
Motility:							
Flagellar	+ or –	+	+ or –	+ or –	–	–	+
Gliding	–	–	–	+ or –	+ or –	+	–
Major absorption maxima in vivo:							
790–1030 nm	+	+	+	–	–	–[b]	+
780–790 nm	–	–	–	+	–	+	–
700–755 nm	–	–	–	–	+	+[b]	–
Main bacteriochlorophylls:							
Bchl *a* or *b*	+	+	+	–	–	–[b]	+
Bchl *g*	–	–	–	+	–	–	–
Bchl *c*, *d*, or *e*	–	–	–	–	+	+[b]	–
Carotenoids of group:							
Group 1–4	+	+	+	–[c]	–	–	+
Group 5	–	–	–	–	+	+	–
Special photosynthetic structures:							
Internal membranes	+	+	+	+	–	–	ND
Chlorosomes	–	–	–	–	+	+[b]	–
Sulfur globules occur:							
Inside the cells	+	–	–	–	–	–	–
Outside the cells	–	+	+ or –	–	+	+ or –	–
Final oxidation product of sulfide:							
Sulfate	+	+	+ or –	–	+	–	–
Sulfur (or no oxidation)	–	–	+ or –	–	–	+ or –	–
Phototrophic growth:							
Anaerobic conditions	+	+	+	+	+	+	–
Chemotrophic growth:							
Microaerobic conditions	+ or –	+ or –	+	–	–	+	+
Aerobic conditions	+ or –	+ or –	+ or –	–	–	+ or –	+

[a] Symbols: see standard definitions; ND, not determined.

[b] *Heliothrix oregonensis* lacks chlorosomes and bchl *c*, *d*, and *e*. It contains bchl *a* in low amounts.

[c] A neurosporene-like carotenoid is present in *Heliobacterium chlorum*.

Table 10.4 Differential characteristics of the genera of Subgroup 1 [a]

Characteristic	Amoebobacter	Chromatium	Lamprobacter	Lamprocystis	Thiocapsa	Thiocystis	Thiodictyon	Thiopedia	Thiospirillum
Motile by polar flagella	−	+	−	+	−	+	−	−	+
Gas vacuoles	+	−	+	+	−	−	+	+	−
Cells spherical	+	−	−	+	+	+	−	−	−
Cells spherical to ovoid, arranged in flat sheets	+ or −	−	−	−	−	−	−	+	−
Cells ovoid to rod-shaped	−	+	+	−	−	−	+	−	−
Cells spiral or vibrioid-shaped	−	−	−	−	−	−	−	−	+
Cells always with slime capsules	+	−	−	−	+	−	−	−	−

[a] Symbols: see standard definitions

Table 10.5 Differential characteristics of the species of the genus *Amoebobacter* [a]

Characteristic	A. pedioformis [b]	A. pendens	A. purpureus [b]	A. roseus
Cell diameter (µm):				
1.5–2.0	−	+	−	−
2.0–3.0	+	−	+	+
3.0–4.0	−	−	+	−
Shape of cell aggregates:				
Cells singly	+	+	+	+
Irregular aggregates	−	+	+	+
Rectangular platelets of 4–32 cells	+	−	−	−
Gas vacuoles located:				
In central part	−	+	+	+
In peripheral part	+	−	−	−
Color of culture, main carotenoid:				
Pink, spirilloxanthin	+	+	−	+
Purple–red, okenone	−	−	+	−
Optimum temperature range (°C):				
23–25	−	−	+	−
30–37	+	+	−	+

[a] Symbols: see standard definitions.

[b] *A. pedioformis* and *A. purpureus* were not included in *Bergey's Manual of Systematic Bacteriology*. The species *A. pedioformis* was created in 1986 by Eichler and Pfennig (Arch. Microbiol. *146:* 295–300; Int. J. Syst. Bacteriol. *37:* 179–180, 1987). The species *A. purpureus* was created in 1988 by Eichler and Pfennig (Arch. Microbiol. 149: 395-400; Int. J. Syst. Bacteriol. *39:* 205–206, 1989).

Table 10.6 Differential characteristics of the species of the genus *Chromatium*[a]

Characteristic	C. buderi	C. gracile	C. minutissimum	C. minus	C. okenii	C. purpuratum	C. salexigens[b]	C. tepidum[b]	C. vinosum	C. violascens	C. warmingii	C. weissei
Cell diameter (μm):												
4.5–6.0	−	−	−	−	+	−	−	−	−	−	−	−
3.5–4.5	+	−	−	−	−	−	−	−	−	−	+	+
2.0–2.5	−	−	−	+	−	−	+	−	+	+	−	−
1.0–2.0	−	−	−	−	−	+	−	+	−	−	−	−
1.0	−	+	+	−	−	−	−	−	−	−	−	−
Color of culture, carotenoid series:												
Purple-red, okenone	−	−	−	+	+	+	−	−	−	−	−	+
Purple–violet, rhodopinal	+	−	−	−	−	−	−	−	−	+	+	−
Brown-red, normal spirilloxanthin	−	+	+	−	−	−	+	+	+	−	−	−
Vitamin B$_{12}$ requirement	+	−	−	−	+	−	+	−	−	−	+	+
Optimum salinity (%NaCl):												
0–0.5	−	−	+	+	+	−	−	+	+	+	+	+
1–5	+	+	−	−	−	−	+	−	−	−	−	−
10	−	−	−	−	−	−	+	−	−	−	−	−
Optimum temperature range (°C):												
20–28	+	−	−	+	+	+	+	−	−	+	+	+
25–35	−	+	+	−	−	−	−	−	+	−	−	−
48–50	−	−	−	−	−	−	−	+	−	−	−	−

[a] Symbols: see standard definitions.

[b] *C. salexigens* and *C. tepidum* were not included in *Bergey's Manual of Systematic Bacteriology*. The species *C. salexigens* was created in 1988 by Caumette et al. (Syst. Appl. Microbiol. *10:* 288–292; Int. J. Syst. Bacteriol. *39:* 93–94, 1989). The species *C. tepidum* was created in 1986 by Madigan (Int. J. Syst. Bacteriol. *36:* 222–227).

Table 10.7 Differential characteristics of the species of the genus *Thiocapsa*[a]

Characteristic	T. pfennigii	T. roseopersicina
Cell diameter (μm):		
0.8–1.5	+	−
1.2–3.0	−	+
Bacteriochlorophyll *a* or *b*, in vivo absorption maximum:		
a, 800–900 nm	−	+
b, 1015–1035 nm	+	−
Color of culture, carotenoid:		
Pink to rose-red, spirilloxanthin	−	+
Yellow to yellowish brown, tetrahydrospirilloxanthin	+	−
Obligately phototrophic	+	−
Facultatively aerobic and chemotrophic	−	+
Utilization of thiosulfate	−	+

[a] Symbols: see standard definitions.

Table 10.8 Differential characteristics of the species of the genus _Ectothiorhodospira_ [a]

Characteristic	E. abdelmalekii	E. halochloris	E. halophila	E. marismortui [b]	E. mobilis	E. shaposhnikovii	E. vacuolata
Cell diameter (µm):							
0.5–0.6	–	+	–	–	–	–	–
0.6–0.9	–	–	+	–	+	+	–
0.9–1.3	+	–	–	+	–	–	–
1.3–1.5	–	–	–	–	–	–	+
Type of flagellation:							
Polar tuft	–	–	–	+	+	+	+
Bipolar	+	+	+	–	–	–	–
Color of cultures, carotenoid:							
Red, spirilloxanthin	–	–	+	+	+	+	+
Green, rhodopin	+	+	–	–	–	–	–
Bacteriochlorophyll:							
Bchl _a_	–	–	+	+	+	+	+
Bchl _b_	+	+	–	–	–	–	–
Gas vacuoles	–	–	–	–	–	–	+
Nitrate utilization	–	–	–	ND	–	+	–
Optimum salinity:							
1–6%	–	–	–	+	+	+	+
15%	+	+	–	–	–	–	–
15–25%	–	–	+	–	–	–	–

[a] Symbols: see standard definitions; ND, not determined.

[b] _E. marismortui_ was not included in _Bergey's Manual of Systematic Bacteriology_. It was created in 1989 by Oren et al. (Arch. Microbiol. _151:_ 524–529; Int. J. Syst. Bacteriol. _40:_ 105–106, 1990).

Table 10.9 Differential characteristics of the genera of Subgroup 3 [a]

Characteristic	*Rhodobacter*	*Rhodocyclus*	*Rhodomicrobium*	*Rhodopila*	*Rhodopseudomonas*	*Rhodospirillum*
Cell shape:						
Spiral or vibrioid-shaped	−	+	−	−	−	+
Spherical, ovoid, or rod-shaped	+	−	+	+	+	−
Cell division:						
Binary fission	+	+	−	+	−	+
Budding	−	−	+	−	+	−
Internal membranes:						
Vesicles	+	−	−	+	−	+ or −
Lamellae	−	−	+	−	+	+ or −
Finger-like intrusions	−	+	−	−	−	−
Motility:						
Polar flagella	+ or −	+ or −	−	−	+ or −	+
Peritrichous flagella	−	−	+	−	−	−
Exospores	−	−	+	−	−	−

[a] Symbols: see standard definitions.

Important Notes for Users of This Manual

Unless otherwise indicated in footnotes to tables, the meanings of symbols are as follows:

    +   90% or more of strains are positive

    −   90% or more of strains are negative

    d   11-89% of strains are positive

    v   strain instability (not equivalent to "d")

    D   Different reactions in different taxa (species of a genus or genera of a family)

All other symbols are defined in footnotes to tables.

Table 10.10 Differential characteristics of the species of the genus Rhodobacter[a]

Characteristic	R. adriaticus	R. capsulatus	R. euryhalinus[b]	R. sphaeroides	R. sulfidophilus	R. veldkampii
Cell diameter (μm):						
0.5–1.0	+	+	+	−	+	+
1.0–2.0	−	−	−	+	−	−
Motility	−	+	+	+	+	−
Slime formation	+	d	−	d	d	−
NaCl requirement	+	−	+	−	+	−
Sulfate assimilated	−	+	d	+	+	−
Final oxidation product of sulfide:						
Sulfur	−	+	−	+	−	−
Sulfate	+	−	+	−	+	+
Thiosulfate oxidized to sulfate	+	−	ND	−	+	+
Growth on H_2/CO_2	−	+	+	+	+	−
Aerobic dark growth	−	+	−	+	+	+
Nitrate assimilated	−	d	ND	d	−	+
Utilization of:						
Tartrate	−	−	−	d	−	−
Citrate	−	−	−	+	−	−
Mannitol	−	−	d	+	−	−
Glycerol	+	−	+	+	+	−
Ethanol	+	−	d	+	d	−
Gluconate	ND	−	+	+	+	ND

[a] Symbols: see standard definitions; ND, not determined.
[b] R. euryhalinus was not included in Bergey's Manual of Systematic Bacteriology and was created in 1985 by Kompantseva (Mikrobiologiya 54: 974–982; Int. J. Syst. Bacteriol. 39: 205–206, 1989).

Table 10.11 Differential characteristics of the species of the genus Rhodocyclus[a]

Characteristic	R. gelatinosus	R. purpureus	R. tenuis
Cell diameter (μm):			
0.3–0.6	+	−	+
0.6–1.0	−	+	−
Cell shape:			
Half-circle to circle	−	+	−
Curved rod	+	−	+
Motility	+	−	+
Slime production	+	−	+
Carotenoids of:			
Group 2	+	−	−
Group 4	−	+	+
Vitamins required	+	+	−
N_2 fixation	+	−	+
Gelatin utilized	+	−	−
Benzoate utilized	−	+	−
Citrate utilized	+	−	−

[a] Symbols: see standard definitions.

Table 10.12   Differential characteristics of the species of the genus *Rhodopseudomonas*[a]

Characteristic	R. acidophila	R. blastica	R. marina	R. palustris	R. rutila	R. sulfoviridis	R. viridis
Cell diameter (μm):							
0.6–0.9	–	+	+	+	+	+	+
1.0–1.3	+	–	–	–	–	–	–
Motility	+	–	+	+	+	+	+
Color of cultures:							
Red	+	–	+	+	+	–	–
Green	–	–	–	–	–	+	+
Yellowish brown	–	+	–	–	–	–	–
Bacteriochlorophyll:							
Bchl *a*	+	+	+	+	+	–	–
Bchl *b*	–	–	–	–	–	+	+
Vitamins required	–	+	ND	+	–	+	+
Final oxidation products of sulfide:							
Sulfate	–	–	–	+	–	+	–
Sulfur and thiosulfate	–	–	+	–	–	–	–
Sulfate assimilated	+	+	+	+	+	–	+
Growth on H_2/CO_2	+	+	ND	+	ND	ND	–
Optimum pH:							
5.5–6.0	+	–	–	–	–	–	–
6.5–7.5	–	+	+	+	+	+	+

[a] Symbols: see standard definitions; ND, not determined.

Important Notes for Users of This Manual

Unless otherwise indicated in footnotes to tables, the meanings of symbols are as follows:

+   90% or more of strains are positive

–   90% or more of strains are negative

d   11-89% of strains are positive

v   strain instability (not equivalent to "d")

D   Different reactions in different taxa (species of a genus or genera of a family)

All other symbols are defined in footnotes to tables.

Table 10.13 Differential characteristics of the species of the genus *Rhodospirillum*[a]

Characteristics	R. centenum	R. fulvum	R. mediosalinum[b]	R. molischianum	R. photometricum	R. rubrum	R. salexigens	R. salinarum
Cell diameter (μm):								
0.5–0.7	–	+	–	–	–	–	+	–
0.7–1.0	–	–	+	+	–	+	–	+
1.0–1.5	+	–	–	–	+	–	–	–
Color of cultures:								
Red	+	–	+	–	–	+	+	+
Brown	–	+	–	+	+	–	–	–
Internal membranes:								
Vesicles	–	–	+	–	–	+	–	+
Lamellae	+	+	–	+	+	–	+	–
Aerobic growth	+	–	+	–	–	+	+	+
Optimum salinity:								
0–0.5%	+	+	–	+	+	+	–	–
4–7%	–	–	+	–	–	–	–	–
6–20%	–	–	–	–	–	–	+	+
Optimum temperature:								
25–35°C	–	+	+	+	+	+	–	–
35–42°C	+	–	–	–	–	–	+	+

[a] Symbols: See standard definitions.

[b] *R. centenum* was not included in *Bergey's Manual of Systematic Bacteriology* and was created in 1989 by Favinger et al. (Antonie van Leeuwenhoek J. Microbiol. Serol. *55:* 291–296).

[c] *R. mediosalinum* was not included in *Bergey's Manual of Systematic Bacteriology* and was created in 1984 by Kompantseva and Gorlenko (Mikrobiologiya *53:* 954–961).

373

Table 10.14   Differential characteristics of the genera of Subgroup 5 [a]

Characteristic	*Ancalochloris*	*Chlorobium*	*Chloroherpeton*	*Pelodictyon*	*Prosthecochloris*	Consortia, symbiotic aggregates
Cultures and cells:						
Green only	+	−	+	−	−	−
Green or brown	−	+	−	+	+	+
Gas vacuoles	+	−	+	+	−	+ or −
Motility:						
Flagellar	−	−	−	−	−	+ or −
Gliding	−	−	+	−	−	−
Flexible unicellular filaments	−	−	+	−	−	−
Cells starlike, with prosthecae	+	−	−	−	+	−
Cells spherical, ovoid, or rod-shaped	−	+ or −	−	+	−	−
Cells curved, vibrioid, or in spirals	−	+ or −	−	−	−	−
Cell aggregates barrel-shaped, green or brown, motile	−	−	−	−	−	+
Cell aggregates platelet-shaped, green, nonmotile	−	−	−	−	−	+

[a]   Symbols: see standard definitions.

Important Notes for Users of This Manual

Unless otherwise indicated in footnotes to tables, the meanings of symbols are as follows:

+   90% or more of strains are positive

−   90% or more of strains are negative

d   11-89% of strains are positive

v   strain instability (not equivalent to "d")

D   Different reactions in different taxa (species of a genus or genera of a family)

All other symbols are defined in footnotes to tables.

Table 10.15 Differential characteristics of the species of the genus *Chlorobium*[a]

Characteristic	*C. chlorovibrioides*	*C. limicola*	*C. limicola* f. sp. *thiosulfatophilum*	*C. phaeobacteroides*	*C. phaeovibrioides*	*C. vibrioforme*	*C. vibrioforme* f. sp. *thiosulfatophilum*
Color of cultures:							
Green	+	+	+	−	−	+	+
Brown	−	−	−	+	+	−	−
Cells rod-shaped to spherical	−	+	+	+	(+)	−	−
Cells curved rod- and vibrioid-shaped to ringlike or spiral	+	−	−	−	+	+	+
Cell diameter (µm):							
0.7–1.1	−	+	+	−	−	−	−
0.5–0.8	−	−	−	+	−	+	+
0.3–0.4	+	−	−	−	+	−	−
Thiosulfate used	−	−	+	−	−	−	+
NaCl requirement:							
None	−	+	+	+	−	−	−
1–3%	+	−	−	−	+	+	+

[a] Symbols: see standard definitions.

Table 10.16 Differential characteristics of the species of the genus *Pelodictyon*[a]

Characteristic	*P. clathratiforme*	*P. luteolum*	*P. phaeoclathratiforme*	*P. phaeum*
Color of cultures:				
Green	+	+	−	−
Brown	−	−	+	+
Cells rod-shaped, in netlike aggregates	+	−	+	−
Cells rod-shaped ovoid, singly or in aggregates	−	+	−	+

[a] Symbols: see standard definitions.

375

Table 10.17 Differential characteristics of the consortia, symbiotic associations of green sulfur bacteria with colorless bacteria (laboratory names are in quotation marks)

Characteristic	"Chlorochromatium aggregatum" consortium	"Chlorochromatium glebulum" consortium	"Chloroplana vacuolata" consortium	"Pelochromatium roseo-viride" consortium	"Pelochromatium roseum" consortium
Consortia flat sheets (platelets of variable size) of alternating green and colorless bacteria, both with gas vacuoles, nonmotile	-	-	+	-	-
Consortia barrel-shaped, phototrophic bacteria surround central bacterium in 5–6 rows	+	+	-	+	+
Size of consortium:					
2.5–4.0 µm wide and 4.0–10.0 µm long	+	+	-	+	+
Central bacterium rod-shaped, motile, polar flagellum	+	+	-	+	+
Surrounding bacteria resemble:					
Chlorobium limicola	+	+	-	-	-
Chlorobium phaeobacteroides	-	-	-	+	+
Pelodictyon luteolum	-	-	+	+	-

a Symbols: see standard definitions.

Table 10.18 Differential characteristics of the genera of Subgroup 6 [a]

Characteristic	*Chloroflexus*	*Chloronema*	*Heliothrix*	*Oscillochloris*
Diameter of filaments (µm):				
0.5–1.0	+	-	-	-
1.0–1.5	-	-	+	+
1.5–2.5	-	+	-	-
4.5–5.5	-	-	-	+
Gas vacuoles	-	+	-	+
Chlorosomes	+	+	-	+
Bacteriochlorophyll:				
a	+	+	+	+
c or *d*	+	+	-	+
Optimum temperature range (°C):				
10–20	-	+	-	+
20–25	+	-	-	-
20–35	-	-	-	+
40–55	-	-	+	-
50–60	+	-	-	-

a Symbols: see standard definitions.

OXYGENIC PHOTOTROPHIC BACTERIA[*]

The oxygenic photosynthetic procaryotes comprise two separate groups. The most thoroughly studied and understood group is the cyanobacteria. This is a widely distributed and diverse collection of unicellular to multicellular photosynthetic bacteria that possess chlorophyll *a* and carry-out oxygenic photosynthesis. A more recently discovered group of oxygenic photosynthetic procaryotes is placed in the order *Prochlorales*. These bacteria share many features with the cyanobacteria, but they also contain chlorophyll *b* as well as chlorophyll *a*, lack phycobilin pigments, and differ in some other features. This group appears to be heterogeneous phylogenetically and related to diverse groups of the cyanobacteria on the basis of 16S rRNA data, but their exact relationships have not yet been determined.

Key to the Oxygenic Photosynthetic Bacteria

I. Contain chlorophyll *a* and have phycobilins

> I. The Cyanobacteria

II. Contain both chlorophylls *b* and *a* and lack phycobilins

> II. The Prochlorophytes (Order *Prochlorales*)

I. THE CYANOBACTERIA

Diverse and divergent group of **unicellular, colonial,** and **filamentous oxygenic and photosynthetic** bacteria. Filamentous types **simple** or with **false** or **true branches** (sometimes with both); **some filamentous types multiseriate,** forming more complex multicellular thalli. Cell breadth or diameter ranges from less than 0.5 μm to over 100 μm. **Stain Gram negative; outer membrane present; peptidoglycan** layer 2–200 nm thick (T.E.M.); most filamentous forms have minute pores between cells (microplasmodesmata) (T.E.M). May possess an **extracellular sheath, glycocalyx (capsule), or merely mucilage or slime. Pili** present in many (E.M.). **Flagella are not present,** but **swimming** capability exists in strains of one unicellular genus. **Gliding motility** exists in many genera and species, particularly filamentous forms with speeds of < 1–6 (10) μm·sec⁻¹. Propulsion mechanism is uncertain, but circumstantial evidence implicates bands of proteinaceous microfibrils in periplasm, probably between peptidoglycan wall and outer membrane.

Unicellular, colonial, and some filamentous cyanobacteria undergo **binary fission by a constrictive type of growth while many filamentous forms divide by invagination of only the cell membrane and peptidoglycan layer** (T.E.M.). A **budding** division in a few unicellular types. In one subgroup reproduction entirely or partially by **multiple internal fissions.**

Light-harvesting pigments include **phycobiliproteins** (i.e., phycobilins): **phycoerythrin** (PE, in some), **phycoerythrocyanin** (PEC, in some), **phycocyanin** (PC, in all), and **allophycocyanin** (APC, in all), **chlorophyll *a*** (chl *a*), and carotenoids (**especially β-carotene, zeaxanthin, echinenone, myxoxanthophyll, and oscilloxanthin**). PE may contain only phycoerythrobilin as chromophore (C-PE) or a mixture of **phycoerythrobilin and phycourobilin. All known cyanobacteria contain some PC and APC.**

Color of cells (depending on species, light intensity history, or wavelength maxima) may be red, blue-green, purple, green, brown, or seemingly black. **Sheaths or glycocalyx may also contain yellow pigment (scytonemin) or red-blue pigment (gloeocapsin) which may mask cellular pigmentation.**

Chromatic adaptation (in the classical sense) occurs in some PE- or PEC-containing cyanobacteria. In

---

[*]See glossary of terms used to describe oxygenic phototrophic bacteria at the end of Group 11.

general, predominantly low intensity greenish light results in continued synthesis of PE, whereas a greater predominance of red light results in partial or complete shut-down of PE synthesis, sometimes with enhanced synthesis of PC. Phycobiliprotein pigments form **phycobilisomes** on both stromal surfaces of double unit internal membranes (**thylakoids**). Chlorophyll *a* and carotenoids are part of thylakoid membranes (except in one genus which has photosystems in cytoplasmic membrane only, with no internal thylakoids).

Photosynthesis is **oxygenic** and **autotrophic** in all species, but some are capable of sulfide-dependent, anaerobic, **anoxygenic photoautotrophy** when sulfide inhibits photosystem II and consequently oxygenic photosynthesis. **Photoheterotrophy** and **aerobic** (respiratory) and **anaerobic** (fermentative) **chemoheterotrophy** occur, but these modes are generally only capable of maintenance or of sustaining slow growth. In one case, **light-activated heterotrophy** is known and may apply to other so-called cases of heterotrophy (see Anderson and McIntosh, J. Bacteriol. *173*: 2761–2767, 1991).

Photosynthate is stored as **glycogen;** rarely species contain poly-β-hydroxybutyrate. Other cellular inclusions include multi-L-arginyl-poly-[L-aspartic acid] (**cyanophycin** or structured granules), **polyphosphate** granules, **carboxysomes** (polyhedral bodies), and **gas vesicles** (the last mainly in planktonic species).

Species of some groups are capable of differentiating specialized cells. **Heterocysts** have modified thylakoids, generally lack phycobilisomes, lack oxygenic and autotrophic capacity, have additional wall layers, and function primarily in nitrogen fixation. **Akinetes** are thick walled, resistant storage cells and occur in most species that are also capable of heterocyst formation.

Most species are free-living; freshwater, marine, or terrestrial; planktonic or benthic; sometimes species are involved in deposition of marl; they are major components of microbial mats. Habitats are from high to extremely low light intensity. Temperatures are from < 2°C (Antarctic saline ponds) to 74°C (hot springs). A few species are endosymbionts of eukaryotes (e.g., fungi, diatoms, dinoflagellates); some are N-fixing exosymbionts (e.g., in "tissue" of lichens, liverworts, hornworts, aquatic ferns, and a few seed plants). Some endosymbionts lack typical cyanobacterial wall and are termed **cyanelles.** Some cyanobacteria have been shown to be closely related to chloroplasts of eukaryotic algae.

Culture information will be provided, when relevant, with the description of each genus. Information on selected culture media is in Tables 11.1, 11.2, and 11.3.

In order to identify a microorganism microscopically as a cyanobacterium, the following procedures are recommended:

1. It should appear procaryotic, that is, possess no discernible chloroplasts or nucleus (the latter are often difficult to distinguish except under phase contrast optics).

2. Verify that they are **not** swimming (i.e., if motile, only by slow gliding in contact with substrate). One exception: see some strains of genus *Synechococcus.*

3. Color will be variable (see above), but presence of phycobilins or chlorophyll can be confirmed by visible orange to red fluorescence under epifluorescent microscopy. Filamentous and unicellular anoxygenic bacteria (which contain bacteriochlorophylls and no phycobilins) may also vary in color from reddish to greenish, but will fluoresce only in the near infrared, not in the visible range.

4. Confusion with the *Prochlorales,* procaryotes that possess chlorophyll *a* and *b* and **no** phycobilins, can be easily avoided by extracting the specimens in question with methanol or acetone to check if residual color (blue to red) caused by the non-organic soluble phycobilins remains in the cells.

The Classification of Cyanobacteria

The classification of cyanobacteria is in transition and essentially all of the genera included here are provisional and subject to great modification. At this time it would be possible or even likely for a scientist to use a combination of any of the 5 systems listed below. It is important, in such cases, to include the alternative names when designating the organism to be worked with.

The newer systems (i.e., Nr. 1 and 5) are still largely **phenotypic** classifications. However, System Nr. 1, especially, has moved in part to a **genotypic** classification. Both this system and Nr. 5 are systems in **transition,** and much of the classification is provisional. A genotypic classification (indicative of phylogeny and genetic relatedness) may never completely replace a phenotypic system for ease of identifying specimens, particularly collections from natural habitats. In all of the following systems cyanobacteria that look alike and reproduce in a similar manner will probably be grouped together, yet some may be genetically unrelated. Unfortunately, lack of many physiological differences among cyanobacteria (e.g., all are photoautotrophs with about the same nutritional requirements) make traditional phenotypic bacteriological criteria of little use.

Systems of Cyanobacterial Classification

1. The simplest system, based almost entirely on the limited number of cyanobacteria in culture, is used here (see Rippka et al., J. Gen. Microbiol. *111:* 1–61, 1979 and Rippka, Methods Enzymol. *167:* 28–67, 1988). Criteria for classification include: morphology, mode of reproduction, ultrastructure, physiology, chemistry, and sometimes genetics (*Bergey's Manual of Systematic Bacteriology,* Volume 3, pp. 1710–1727). Isolation data and histories of most strains included in *Bergey's Manual* are to be found in Rippka et al., J. Gen. Microbiol. *111:* 1–61, 1979.

2. The simple, but inappropriate system of Drouet (Beih. Nova Hedwigia *66:* 135–209, 1981) based mainly on morphology of herbarium specimens; seldom used today.

3. The "Geitlerian" system is complex and based almost entirely on morphological characteristics of collected specimens (Geitler, Rabenhorst's Kryptogamenflora *14:* 1–1196, 1932).

4. A critical reevaluation of the "Geitlerian" genera still based primarily on morphological and reproductive characteristics (Bourrelly, Les algues d'eau douce III, 2nd ed., 606 pp. N. Boubée, Paris, 1985).

5. A recent and extensive, complex modification of the "Geitlerian" system using morphology, ultrastructure, modes of reproduction, variance, and other criteria from collected and cultured specimens by Anagnostidis and Komárek (Arch. Hydrobiol. Suppl. *71:* 291–302, 1985; Suppl. *73:* 157–226, 1986; Suppl. *80:* 327–472, 1988; Suppl. *82:* 247–345, 1989; Suppl. *86:* 1–73, 1990). More narrowly defined genera than System 1. **Essential** reading for a more **in-depth** examination of cyanobacterial classification.

Key to the Five Subgroups of Cyanobacteria

I. Unicellular or nonfilamentous aggregates of cells held together by outer walls or a gel-like matrix (colonies)

    A. Binary fission in one, two, or three planes, symmetric or asymmetric; or by budding.

        Subgroup 1

    B. Reproduction by internal multiple fissions with production of daughter cells smaller than 1/2 the parent; or by multiple **and** binary fission.

        Subgroup 2

II. Filamentous; trichome of cells branched or unbranched, uniseriate or multiseriate

    A. Binary fission in one plane only giving rise to uniseriate, unbranched trichomes, although false branching may occur.

        1. Trichomes composed of cells which do not differentiate into heterocysts or akinetes.

            Subgroup 3

2. One or a few cells of each trichome differentiate into heterocysts, at least when concentration of external combined nitrogen is low; some also produce akinetes.

Subgroup 4

B. Binary fission periodically or commonly in more than one plane, giving rise to multiseriate trichomes or trichomes with true branches or both.

Subgroup 5

SUBGROUP 1 (=Order: Chroococcales)

Single cells or aggregates or regular colonies of cells. Cell division occurs by binary fission or budding, spherical to elongate cells (<0.5–30 μm in diameter or breadth). Binary fission is in 1, 2, or 3 planes or irregular; cells of aggregates or colonies are held together by firm sheaths to poorly defined mucilage. Accurate identification with regard to true relatedness (and generic designation) requires nucleotide sequence data. Physiological and morphological comparisons generally allow only tentative differentiation of genera. Some of the genera described here are referred to as "culture groups" or "super genera," and various clusters within each may eventually warrant separate generic status. Greater numbers of culture isolations and more genetic studies in the future will certainly result in recognition of additional genera and species. **Since few diagnostic phenotypic properties are apparent in many members of this subgroup, a largely genotypic classification is required.** See Figure 11.1, 11.2, and Waterbury and Rippka (*Bergey's Manual of Systematic Bacteriology*, Volume 3, pp. 1729–1731); and Rippka (Methods Enzymol. *167:* 28–67, 1988).

Komárek and Anagnostidis (Arch. Hydrobiol. Suppl. *73:* 157–226, 1986) have recently revised extensively the traditional orders *Chroococcales* (Subgroup 1), *Pleurocapsales* (Subgroup 2) and *Chamaesiphonales* under one order (*Chroococcales*) with seven definable and distinguishable families. These authors recognize 43 genera among the groups that would correspond to Subgroup 1 in the present treatment. Many unique unicellular and colonial cyanobacteria collected from natural habitats will not be classifiable here, since the great majority of existing genera are not represented in culture or have not been studied extensively in culture. One main criterion for classification in Subgroup 1 is whether cells undergo binary fission in one, two, or three or more planes of division (see Key below). That this is a reliable characteristic is supported by an extensive study by Kovácik (Arch. Hydrobiol. Suppl. *80:* 149–190, 1988) of 20 culture strains of 6 genera.

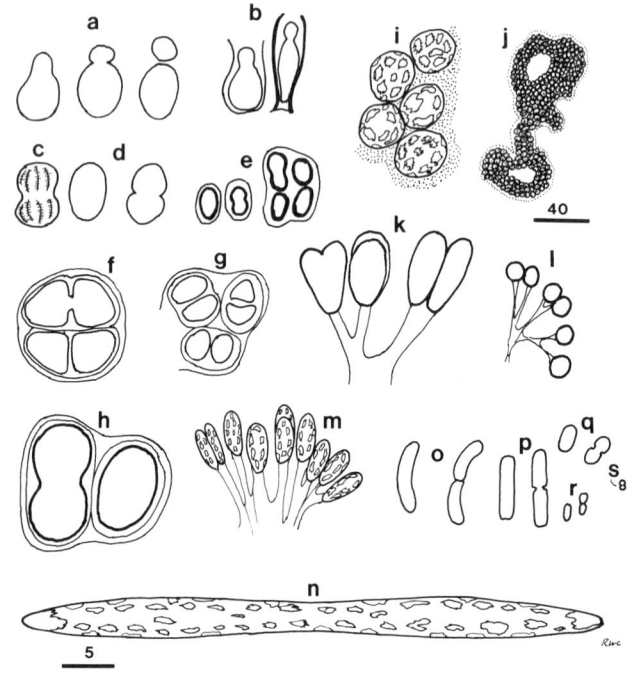

Figure 11.1. Schematic representations of genera of Subgroup 1. *a. Chamaesiphon* sp. in three phases of division. *b. Chamaesiphon:* two "species" with sheath (pseudovagina). *c. Cyanothece* sp. (cf. *Synechococcus minervae*): nondividing cell from hot springs. *d. Cyanothece* sp.: nondividing and dividing cell. *e. Gloeobacter (G. violaceus)* with sheath indicated by *solid line. f.* and *g. Gloeocapsa* spp. with laminated sheaths. *h. Gloeothece* sp. with laminated sheath. *i. Microcystis* sp.: five cells of planktonic colony embedded in gel matrix; gas vacuoles depicted within cells. *j. Microcystis* sp.: planktonic colony; *marker* indicates 40 μm. *k. Gomphosphaeria* showing part of colony with cells forming gel-like stalks. *l. Gomphosphaeria*-like organism (cf. *Snowella* sp.) showing part of colony with cells forming thread-like stalks. *m. Coelosphaerium*-like planktonic organism (cf. *Woronichinia* sp.) showing part of colony with cells forming gel-like stalks. Gas vacuoles depicted in cells. *n. Myxobactron* sp. (=*Dactylococcopsis salina*): single planktonic nondividing cell with gas vacuoles. *o. Synechococcus* sp. (cf. *S. lividus*): nondividing and dividing cell from hot springs. *p. Synechococcus* sp.: before and during division. *q. Synechococcus* sp.: before and during division. *r. Synechococcus* sp.: before and during division; pico-planktonic marine form. *s. Synechococcus* sp.: pico-planktonic form. *Scale marker* (5 μm) for all except *j*, where other scale is indicated.

Key to the genera of **Subgroup 1**

[T.E.M.= requires transmission electron microscopy]

I. Binary fission (cell in equal halves)

A. Cells with thylakoids [T.E.M.]

    1. Fission in one plane only

        a. Cell diameter or breadth > 3 μm

            i. Very elongate cell with both ends pointed

                Genus *Myxobaktron*

            ii. Rod-shaped cell with sheath

                Genus *Gloeothece*

            iii. Rod-shaped cell without sheath

                Cyanothece "culture group"

        b. Cell diameter or breadth < 3 μm

            Synechococcus "culture group"

    2. Fission in two or three planes

        a. Cells bound in multicellular aggregates or pairs by **multilaminated** sheath, coccoid to hemispherical cells

            Gloeocapsa "culture group"

        b. Cells single, in pairs, or in aggregates bound by **amorphous** mucilage or sheath, coccoid

            Synechocystis "culture group" and *Microcystis*

B. Cells without thylakoids [T.E.M.]; cells of known species small (~1.5 μm wide), rod-shaped with sheath, fission in one plane [only genus known to lack thylakoids]

        Genus *Gloeobacter*

II. Budding: asymmetric cell division, larger "mother" cell continues to divide and form smaller "buds"; cells ovoid; thylakoids present [T.E.M.]

        Genus *Chamaesiphon*

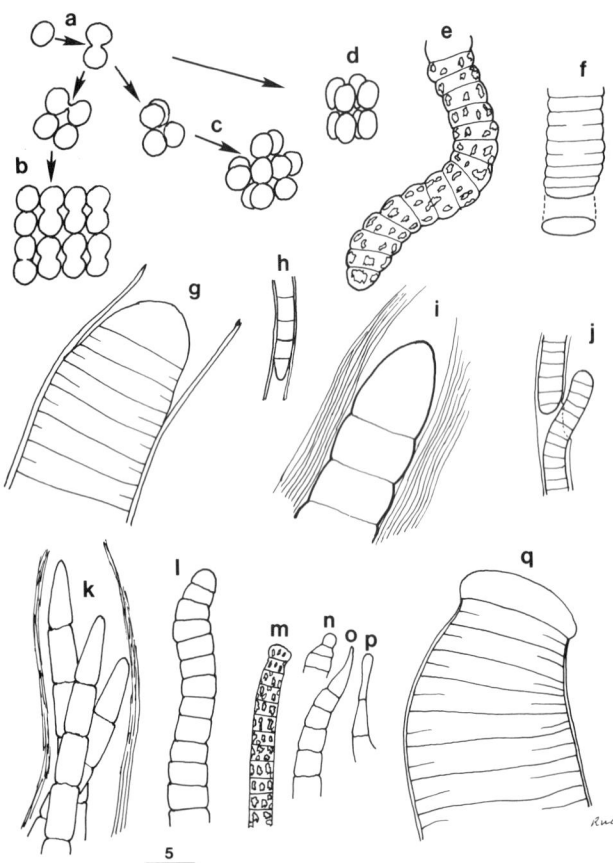

Figure 11.2. Schematic representations of genera of Subgroups 1 and 3.

Subgroup 1 (continued). *a. Synechocystis* sp. before and during division. *b. Synechocystis* sp. with typical divisions in two planes and with synchronous divisions forming plate-like colony of "*Merismopedia.*" *c. Synechocystis* sp. with irregular divisions in three planes. *d. Synechocystis* sp. with synchronous divisions in three planes forming cube-like colony of "*Eucapsis.*"

Subgroup 3. *e. Arthrospira* sp.: planktonic form with gas vacuoles depicted. *f. Crinalium (C. epipsammum):* lowermost cell shown turned 90°. *g.* and *h. Lyngbya* spp. *i. Lyngbya sp.* with multi-laminated sheath (cf. *Porphyrosiphon* sp.) *j. Lyngbya* sp. with false branch (cf. *Plectonema* sp.) *k. Microcoleus* sp. with three trichomes in common sheath. *l.–q. Oscillatoria* spp. showing diversity in trichome breadth, sinuosity, taper, and terminal cell shape. *m.* is planktonic and contains gas vesicles (cf. *O. rubescens* or *Planktothrix rubescens*). *q.* shows centripetally developing septa.

Scale marker (5 μm) applies to all drawings.

Genus **Chamaesiphon**

Spherical to oval-oblong cells; **cells are heteropolar, elongate narrowly at one pole; divide asymmetrically towards narrower pole, releasing smaller of two daughter cells ("bud", exospore, or exocyte).** Larger **basal** or **mother cell** is usually attached in nature to substrate (e.g., aquatic plants, algae, rocks) at opposite pole; it continues to bud (Figure 11.1*a,b*). There is one plane of division. Sheath (**pseudovagina**) may envelope basal cell only. A few spherical buds may remain attached, forming chain.

Description is based primarily on concept of *Chamaesiphon* by Waterbury and Rippka (see *Bergey's Manual of Systematic Bacteriology*, Volume 3, pp. 1729–1731). Several other *Chamaesiphon*-like cyanobacteria genera are recognized by Komárek and Anagnostidis (Arch. Hydrobiol. Suppl. *73:* 193–199, 1986).

Two culture strains or species range in cell diameter from 3–5 μm with mother cell length from 5–7 μm. Both are from fresh water, have a high content of polyunsaturated fatty acids, have a maximum temperature for growth of 27–30°C; use BG-11 or similar culture medium. One strain is a facultative photoheterotroph (using glucose, fructose, or sucrose). The other contains C-PE and shows chromatic adaptation. Neither are motile and neither show nitrogenase activity under anaerobiosis. Mol% G + C of DNA 46.7–46.9 (2 strains). See Table 11.4. Komárek and Anagnostidis (Arch. Hydrobiol. Suppl. *73:* 195, 1986) recognize 24 species of *Chamaesiphon.*

Differentiation of **Chamaesiphon** *from other genera:* This is the only elongate unicell with asymmetric fission and production of exocytes (smaller of 2 cell products of fission). Generally prior to fission, cell is elongate or oval with one pole narrower than the other. Other unicellular genera with or without sheaths (*Synechococcus, Gloeothece, Cyanothece, Gloeobacter*) that divide in a single plane are not asymmetric in shape (across transverse axis) and do not undergo fission in an unequal fashion. However, see *Pleurocapsa* in Subgroup 2.

Genus **Cyanothece**

The genus *Cyanothece* (in the sense of Waterbury and Rippka, *Bergey's Manual of Systematic Bacteriology*, Volume 3, pp. 1739–1741) is currently regarded as a "super genus" or a "culture group" and therefore may be separated into additional genera, but including *Cyanothece*.

Unicellular rod-shaped to coccoid cells that divide by **central transverse binary fission, in a single plane. Known strains are 3 μm or larger in diameter. Contain intracellular thylakoids; no defined sheath** (Figure 11.1*c,d*; Table 11.4). All seven strains are obligate phototrophs; all are photoautotrophic. Medium varies. Some strains occur in fresh water, some marine, and two occur in hypersaline solar evaporation ponds. There are halophilic strains (also known as *Aphanothece halophytica*; Yopp et al., Phycologia *17:* 172–177, 1978). Halophilic strains grow optimally between 0.7–2.0 M NaCl and require elevated concentrations of Mg^{2+} and Ca^{2+}. Nitrogenase activity, sometimes even under aerobic conditions, occurs in all but one halophilic strain. One strain contains C-PE. Maximum temperature was reported in two strains (e.g., 43 and 35°C). Collected from rice fields, mangrove "mats," and solar evaporation ponds. *Cyanothece* (i.e., *Aphanothece halophytica*) often dominant cyanobacterium found on surface of microbial mats in ponds up to 2–3 M NaCl. Mol% G + C of DNA (known in three strains) is 41.2–48.6.

Differentiation of **Cyanothece** *from other genera:* See Table 11.4.

Genus **Gloeobacter**

Oval to rod-shaped, unicellular, or irregular aggregates held together by common multilayered sheath; cells are 2.0–3.0 × 1.5 μm. Divide by **central transverse binary fission, in single plane. Lack intracellular thylakoids; photosystems and light-harvesting phycobiliproteins are associated with cytoplasmic membrane only.** *G. violaceus* is only species described (Rippka et al., Arch. Microbiol. *100:* 419–436, 1974) (Figure 11.1*e*).

G. violaceus occurs terrestrially on calcareous rocks. It is violet in color *en masse,* PE with phycourobilins. Phycocyanin to phycoerythrin ratio is 3:2. No chromatic adaptation is known: obligate photoautotroph with upper temperature limit between 25 and 37°C. Nonmotile, and there is no nitrogenase activity under anaerobiosis. Mol% G + C of the DNA is 64 (one strain) (See Table 11.4).

Differentiation of **Gloeobacter** *from other genera:* It is like *Gloeothece* in all morphological respects except for lack of thylakoids in *Gloeobacter* (a characteristic requiring T.E.M. for verification). Although present cultured strains of *Gloeothece* are larger than *Gloeobacter* in cell diameter, this difference will probably disappear with additional culture strains. Also, other small, unicellular members of Subgroup 1 may demonstrate lack of thylakoids when examined with T.E.M. *G. violaceus* contains large refractile granules (polyphosphate) especially near the cell poles.

Genus **Gloeocapsa**

The genus *Gloeocapsa* (in the sense of Waterbury and Rippka, *Bergey's Manual of Systematic Bacteriology,* Volume 3, pp. 1741–1742) is here regarded as a provisional "super genus" or "culture group." It includes the species of both *Gloeocapsa* and *Chroococcus* of the traditional botanical literature. In Komárek and Anagnostidis (Arch. Hydrobiol. Suppl. *73:* 212–214; 188–191, 1986) the two genera are classified in separate "families."

Unicellular or in aggregates or packets; **spherical to hemispherical shaped cells;** binary fission central to form equal size daughter cells; **fission occurs in 2–3 planes, usually in succession and at right angles to the prior plane.** Each cell or cell pair is surrounded by **distinct sheath which expands with new cell division and new sheaths, hence, multilaminated sheaths** (Figure 11.1*f,g;* Table 11.5).

Cells of the six strains studied range in diameter from 3 to about 30 μm. Culture medium type depends on source of cells: ranging from marine to fresh water (including hot springs and somewhat acid bogs/and terrestrial). Some strains show slow gliding motility. Some are facultatively heterotrophic (positive substrates were glucose, fructose, ribose, and sucrose). There are some strains with PE. PE of one strain contains phycourobilin. There is no chromatic adaptation. Mol% G + C of DNA ranges from 39.8–48.9

Differentiation of **Gloeocapsa** *from other genera: Gloeocapsa* is distinguished from *Synechocystis* by the presence of laminated sheath, and from *Microcystis* by the presence of gas vesicles and amorphous sheath in the latter. See Table 11.5.

Genus **Gloeothece**

Unicellular, rod-shaped cell; binary fission central and transverse to form two equal size daughter cells; in 1 plane only. Distinct sheath is present, holding many cells together in aggregates. The cell diameter of six culture strains is 5–6 µm. Thylakoids present (T.E.M. characteristic). All strains contain PE with phycourobilin and phycoerythrobilin, but PC is dominant. No chromatic adaptation is known. Obligate photoautotrophs. All strains fix nitrogen under aerobic conditions. The mol% G + C of DNA from five strains is 40.8–42.7 (Figure 11.1*h*).

Differentiation of **Gloeothece** *from other genera:* Well-defined sheath distinguishes *Gloeothece* from the *Synechococcus* and *Cyanothece* "culture groups." Presence of thylakoids separates *Gloeothece* from *Gloeobacter*. Division in 1 plane only distinguishes *Gloeothece* from *Gloeocapsa* (also with distinctive sheath). See Table 11.4.

Genus **Microcystis**

Seven strains are referred to by Waterbury and Rippka (*Bergey's Manual of Systematic Bacteriology,* Volume 3, pp. 1743–1746) as the "*Microcystis*-cluster" of the "*Synechocystis* group." Additional information by Fahrenkrug et al. (Int. J. Syst. Bacteriol. *42:* 182–184, 1992) has defined this group more discretely, and it now seems appropriate to refer a total of 11 strains to the genus *Microcystis.*

Cells are oval to spherical, 3–8 µm in diameter. Original collected samples are **amorphous aggregates (colonies) of gas vesiculate cells held together by amorphous "mucilaginous" sheath.** In some cultures, aggregates, sheath, or gas vesicles may be lost, but this may be dependent on conditions (Doers and Parker, J. Phycol. *24:* 502–508, 1988; Parker, J. Phycol. *18:* 471–477, 1982). **Binary fission is in 3 planes, but cells are irregularly oriented in aggregates.** Distinguished as **planktonic colonies, buoyant by virtue of gas vesicles; usually produce and excrete the toxic peptides microcystin or cyanoginosin** (see Codd and Poon, in *Biochemistry of the Algae and Cyanobacteria,* Rogers and Gallon, eds., pp. 283–296, 1988). **β-cyclocitral is also released by** *Microcystis,* and is one of the odor contributing compounds of old *Microcystis*

blooms in lakes and ponds (Jüttner, Z. Naturforsch. C. *39:* 867–871, 1984). **Mol% G + C of the DNA is 39.0–45.4 (4 strains)** (see Table 11.5, Figure 11.1*i,j*).

Differentiation of **Microcystis** *from other genera:* Differentiated provisionally from *Synechocystis* "culture group" by having **gas vesicles, the tendency to form large 3-dimensional aggregates, toxin, and β-cyclocitral.** Also by the **mol% G + C range of about 38–45,** which is in somewhat intermediate position between the "low GC" and "high GC" clusters of fresh water *Synechocystis.* Another genus that may form 3-dimensional aggregates by dividing in 2–3 planes is *Gloeocapsa*, which has well defined sheaths or capsules.

Although not presently represented by analyzed cultures, four distinct and recognizable freshwater and usually planktonic genera (*Coelosphaerium, Snowella, Woronichinia,* and *Gomphosphaeria*) may be confused with *Microcystis.* These genera form more or less spherical colonies with most cells located radially near the surface layer of the gel matrix and usually joined to mucilaginous stalks radiating from the colony center (Figure 11.1*l,m*). See Komárek and Hindák (Arch. Hydrobiol. Suppl. *80:* 203–225, 1988).

Genus **Myxobaktron** (=Dactylococcopsis)

Elongate, single cells with pointed ends (Figure 11.1*n*), single species (strain) in culture, 4–8 µm in width, 35–80 µm in length (see Walsby et al., Proc. Roy. Soc. Lond. B. Biol. Sci. *217:* 417–447, 1983; Komárek and Anagnostidis, Arch. Hydrobiol. Suppl. *73:* 157–226, 1986). Transverse binary fission into two equal daughter cells occurs, each with one hemispherical end until elongation and differentiation occurs. It continues in one plane only. Single strain in culture is marine. *Myxobaktron* (*Dactylococcopsis*) *salina,* in culture, contains extensive gas vesicles (peripheral and terminal) and does not form "colonies." Contains PC, no PE. Grows well at 34°C in medium at or above salinity of seawater. Planktonic in Solar Lake, Sinai, Egypt.

Differentiation of **Myxobaktron** *from other genera:* This cyanobacterium, because of the elongate cell with pointed ends, cannot be confused with other cyanobacteria. However, no genetic information is available, and it is placed with Subgroup 1 simply because of its being unicellular.

Genus Synechococcus

The genus *Synechococcus* (in the sense of Waterbury and Rippka, *Bergey's Manual of Systematic Bacteriology*, Volume 3, pp. 1731–1738) is currently regarded as a provisional "super genus" or a "culture group" which will later be separated into at least a few genera, including the genus *Synechococcus*. It is separated into six clusters based primarily on the range of mol% G + C of the DNA. Over the whole *Synechococcus* "culture group," the range is 39–71%, almost as broad a range as for all procaryotes.

At present, *Synechococcus* "culture group" is defined as follows: **unicellular, rod-shaped to coccoid, less than 3 μm in cell diameter, binary fission central transverse in one plane only, thylakoids present—peripheral and concentric.**

For details of each of the "strain clusters" see *Bergey's Manual of Systematic Bacteriology*, Volume 3, pp. 1731–1738, and Table 11.4, Figure 11.1*o–s*.

a. *"Cyanobacterium-cluster."* Mol% G + C = 39–56.0 (2 strains). Cell diameter 1.7–2.3, non-motile, PC, no PE. Obligate photoautotrophs. From fresh water.

b. *"Synechococcus-cluster."* Mol% G + C = 47.5–55.6 (10 strains). Cell diameter 1–2–(3) μm, some motile by gliding, PC, no PE. Obligate photoautotrophs. Fresh water, including hot springs. High temperature strains (i.e., growth at 55–73°C) probably fit in this cluster.

c. *"Cyanobium-cluster."* Mol% G + C = 65.7–71.4 (8 strains). Cell diameter 0.8–1.4 μm, nonmotile, PC, no PE. Obligate photoautotrophs. Fresh water (one strain from brackish water).

d. *"Marine cluster C."* Mol% G + C = 47.5–49.5 (5 strains). Cell diameter 1.2–2.0 μm, nonmotile, PC, C-PE in one strain which shows chromatic adaptation. Same strain synthesizes nitrogenase under anaerobic conditions. This and three other strains capable of photoheterotrophic growth (DCMU-mediated). Two strains halophilic, others halotolerant. All from coastal marine or brackish waters.

e. *"Marine cluster A."* Mol% G + C = 55–62 (15 strains). Cell diameter 0.6–1.7 μm, most strains nonmotile, four capable of **swimming motility, unique to this cluster.** All strains with PE with or without phycourobilin. No chromatic adaptation. All strains obligate photoautotrophs. All have requirements of Na^+, Cl^-, Mg^{2+}, and Ca^{2+} similar to concentrations in seawater. None synthesize nitrogenase under anaerobiosis. From coastal or oceanic waters.

f. *"Marine cluster B."* Mol% G + C = 63–69.5 (4 strains). Cell diameter 0.8–1.4 μm, nonmotile, PC but no PE. Obligate photoautotrophs. No nitrogenase activity. One strain halophilic; others halotolerant. From coastal waters.

Differentiation of **Synechococcus** *from other genera:* *Synechococcus*, a unicellular cyanobacterium that divides into two equal daughter cells, in one plane only, is distinguished from *Gloeothece* and *Gloeobacter* by lacking sheath and from *Cyanothece* by being less than 3 μm in cell diameter.

Genus Synechocystis

The genus *Synechocystis* (somewhat modified, but in the sense of Waterbury and Rippka, *Bergey's Manual of Systematic Bacteriology*, Volume 3, pp. 1742–1746) is regarded as a provisional "super genus" or "culture group" which will later most certainly warrant division. It is here separated into three clusters which have non-overlapping mol% G + C values of the DNA. An additional "cluster" based mainly on culture characteristics is here regarded as the genus *Microcystis* (see Table 11.5).

Synechocystis ("culture cluster") is **unicellular, coccoid, or spherical, with binary fission central but in 2 or 3 successive planes** at right angles to each other. **Aggregates sometimes occur with amorphous slime holding cells together.** Cells are 2–7 μm in diameter, most strains in the 2–4 μm range; thylakoids is present (Figure 11.2*a–d*).

Details of each "cluster" of strains are in Waterbury and Rippka (*Bergey's Manual of Systematic Bacteriology*, Volume 3, pp. 1742–1746) and Table 11.4.

a. "Marine cluster." Mol% G + C = 30.5–31.7 (2 strains). Cell diameter 2.5–4 μm (spherical), 2 division planes, remain in pairs or single, nonmotile, no slime sheaths or aggregation. Obligate photoautotrophs, aerobic nitrogenase activity, all contain PE with high phycourobilin content. Growth between 26–32°C. Tropical. Elevated requirements for Na⁺, Cl⁻, Mg^{2+}, and Ca^{2+}.

b. "Low GC cluster." Mol% G + C = 35–37 (5 strains). Cell diameter 3–7 μm. Cells spherical to oval; two planes of division (or more) (two planes determined for one strain only). May form flat sheet of cells if restricted to two planes (see genus *Merismopedia* in Geitler, Rabenhorst's Kryptogamenflora *14:* 258–266, 1932). (see Figure 11.2*b*) Amorphous sheath (two strains). Slow gliding motility (three strains); all obligate photoautotrophs; C-PE in three strains. Type II chromatic adaptation in two strains (see Tandeau de Marsac, J. Bacteriol. *130:* 82–91, 1977). Maximum growth temperature 37–39°C (for 5 strains). All from fresh water.

c. "High GC cluster." Mol% G + C = 42.1–48.0 (11 strains). Cell diameter 2–3 μm. Cells spherical to oval; planes of division presumably 2–3 (three planes determined for one strain; see genus *Eucapsis* in Geitler, Rabenhorst's Kryptogamenflora *14:* 257–258, 1932). (see Figure 11.2 *d)* Amorphous sheath material known in one strain. Slow gliding motility known in four strains. Photoheterotrophic or light-activated heterotrophic growth known in nine strains (see Anderson and McIntosh, J. Bacteriol. *173:* 2761–2767, 1991). No PE. Maximum growth temperature 37–39°C (known in five strains). All from fresh water.

Differentiation of **Synechocystis** *from other genera: Synechocystis* is generally unicellular and therefore should not be confused with other unicellular cyanobacteria as long as fission in more than one plane can be confirmed. *Synechococcus* and *Cyanothece,* both without sheaths, divide in one plane only. The early single celled stages of several genera of Subgroup 2 may be confused with *Synechocystis.* Therefore culture isolates must be followed carefully. In addition, early packet or aggregate formation by some genera of Subgroup 2 may also be confused with *Synechocystis* (also see *Myxosarcina* and *Chroococcidiopsis* of Subgroup 2). *Microcystis* forms small to large aggregates of **gas-vesiculate** cells which also divide in 2–3 places and are held together in an amorphous sheath (see genus *Microcystis*). *Synechocystis* may sometime form aggregates of cells held together by an amorphous slime or mucilaginous sheath, unlike the distinct, layered sheaths of *Gloeocapsa* (also two or more planes of division) or distinct sheaths of *Gloeothece* (one plane of division only).

SUBGROUP 2 (=Order: Pleurocapsales)

Single cells to aggregates, sometimes with filament-like outgrowths (pseudofilaments). Reproduction (at least in part) occurs by internal multiple fissions to produce small spherical cells (baeocytes = nanocytes = "endospores") which are released after rupture of outer fibrous wall of parent cell. Unicellular genera reproduce exclusively by multiple internal fissions. Aggregates and pseudofilaments in other genera produced by binary fission, some cells of which then reproduce by multiple internal fissions, releasing baeocytes.

In this subgroup **each round of internal fissions is not followed by increase in cell volume of the daughter cells or baeocytes** (until after release); this differs from series of binary fissions in which cell growth occurs after each round (e.g., in Subgroup 1). Also, **fibrous outer wall of parent cell persists during internal baeocyte formation. Baeocytes are initially surrounded only by typical Gram-negative wall; fibrous outer wall may form before or after release. In some cases baeocytes show gliding motility and phototaxis.** In these, motility ceases when fibrous outer wall is synthesized.

Subgroup 2, although most often considered distinct (as the order *Pleurocapsales*), is grouped with members of Subgroup 1 in the treatment by Komárek and Anagnostidis (Arch. Hydrobiol. Suppl. *73:* 157–226, 1986) who recognize over 20 genera that would fit into the present Subgroup 2. In the present treatment, the system of Waterbury and Rippka (*Bergey's Manual of Systematic Bacteriology,* Volume 3, pp. 1746–1770) is followed.

Some members of Subgroup 2 that divide in two or more planes by binary fission may be easily confused with some members of Subgroup 1 that divide in two or more planes (i.e., *Gloeocapsa, Synechocystis, Microcystis*) if the potential for baeocyte formation is not realized in Subgroup 2 cells.

Key to the genera of **Subgroup 2**

[T.E.M. = requires transmission electron microscopy]

I. Internal multiple fissions only (unicellular)

    A. Baeocytes with gliding motility

        Genus *Dermocarpa*

    B. Baeocytes nonmotile

        Genus *Xenococcus*

II. Internal multiple fissions, **following binary fissions** (combination of both modes)

    A. Vegetative cell from enlarged baeocyte undergoes 1–3 **binary fissions,** with resulting single "apical" cell undergoing multiple internal fissions

        Genus *Dermocarpella*

    B. Vegetative cell from enlarged baeocyte undergoes repeated **binary fissions** resulting in cell aggregates, **some taxa** with projecting filaments or "pseudofilaments" of various size.

        1. Binary fissions in 3 regular successive planes resulting in **cubical aggregate (without filaments or pseudofilaments),** all cells of which usually undergo baeocyte formation.

            a. Baeocytes show gliding motility upon release

            Genus *Myxosarcina*

            b. Baeocytes nonmotile

            Genus *Chroococcidiopsis*

        2. Binary fissions in several planes forming irregular to regular (cubical) aggregates or thalli, with filamentous or semi-filamentous outgrowths. Some or all cells of multicellular aggregate may undergo baeocyte formation.

            Pleurocapsa "culture group."

Genus **Chroococcidiopsis**

Binary fission occurs in three regular planes, often forming irregular to nearly true cubical aggregates. Usually cells of aggregate each form four small non-motile baeocytes by internal multiple fissions. **Baeocytes are surrounded by thin fibrous outer wall at time of release** (requires T.E.M. for verification) (Figure 11.3, Table 11.6).

Of eight strains examined: vegetative cell diameter at time of binary fission is 5–6.3 μm; baeocyte diameter is 3–4 μm. Facultative photoheterotrophy is known in all strains, using glucose, sucrose, or fructose (in most). Nitrogenase activity occurs in all strains under anaerobiosis. Phycoerythrocyanin (PEC) is present in six strains and PE containing phycourobilin in one strain which also shows chromatic adaptation. Maximum growth temperature is 39–44°C (seven strains). All are from fresh water. Mol% G + C of DNA = 40.2–46.4 (Table 11.6, Figure 11.3).

Differentiation of **Chroococcidiopsis** *from other genera:* In the sense of Waterbury and Rippka (*Bergey's Manual of Systematic Bacteriology*, Volume 3, pp. 1758–1762) *Chroococcidiopsis* is distinguished from the genus *Myx-*

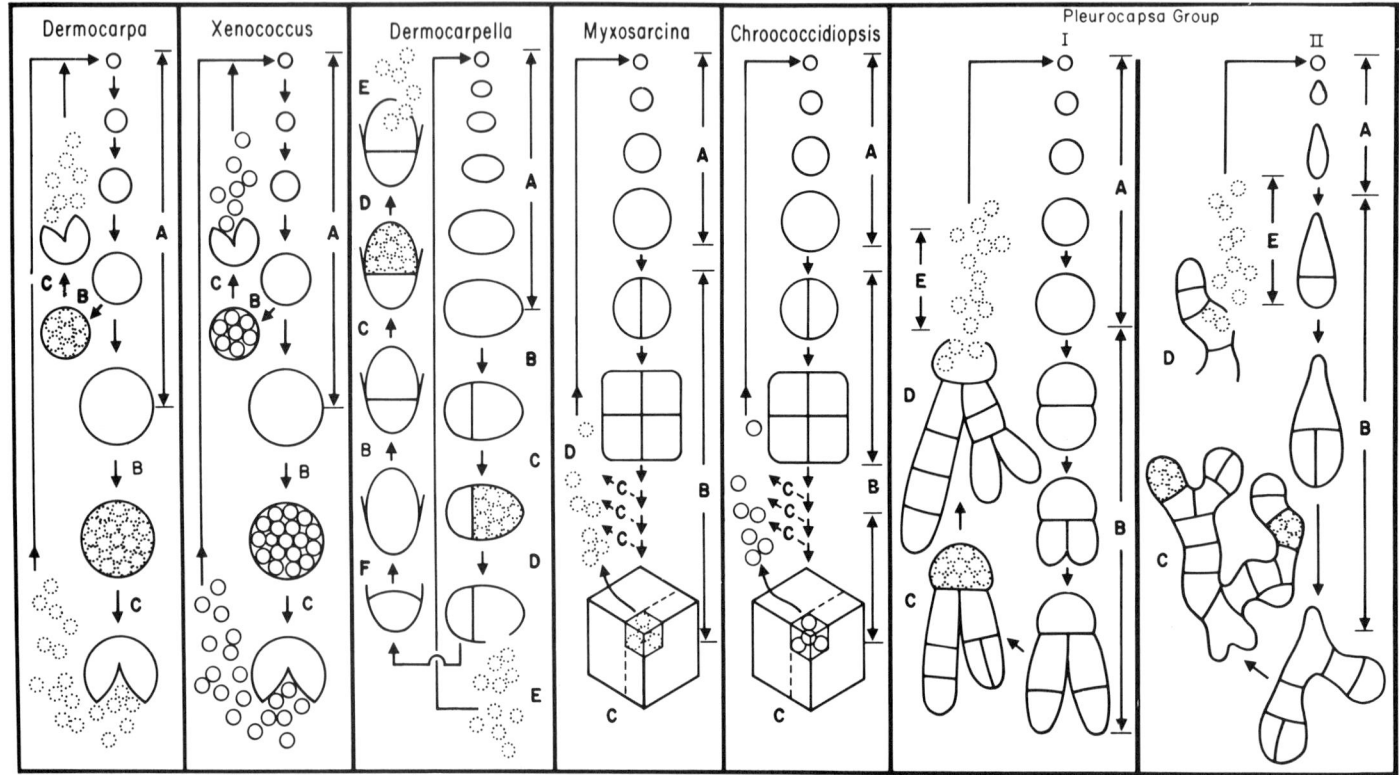

Figure 11.3. Schematic representation of genera of Subgroup 2. Baeocytes that are not surrounded by a fibrous layer at the time of release and are, consequently, motile are symbolized by *dotted circles* to distinguish them from baeocytes that are surrounded by a fibrous layer (*solid circles*). *Dermocarpa:* **A**, symmetric baeocyte enlargement; **B**, multiple fission, leading to baeocyte formation; **C**, baeocyte release followed by a brief period of baeocyte motility. *Xenococcus:* **A**, asymmetric baeocyte enlargement; **B**, multiple fission, leading to baeocyte formation; **C**, release of immotile baeocytes. *Dermocarpella:* **A**, asymmetric baeocyte enlargement; **B**, binary fission, giving rise to a small basal cell and a larger apical cell; **C**, multiple fission of the apical cell, leading to baeocyte formation; **D**, baeocyte release; **E**, period of baeocyte motility; **F**, enlargement of basal cell. *Myxosarcina:* **A**, symmetric baeocyte enlargement to a predetermined size; **B**, repeated binary fission in three regular planes; **C**, multiple fission of almost all the cells in the aggregate, followed by baeocyte release; **D**, period of baeocyte motility. *Chroococcidiopsis:* **A**, symmetric baeocyte enlargement to a predetermined size; **B**, repeated binary fissions in three regular planes; **C**, multiple fission of almost all the cells in the aggregate, followed by the release of immotile baeocytes. *Pleurocapsa* Type I: **A**, symmetric baeocyte enlargement; **B**, binary fissions in many irregular planes; **C**, multiple fission of some vegetative cells; **D**, baeocyte release; **E**, period of baeocyte motility. *Pleurocapsa* Type II: **A**, asymmetric baeocyte enlargement; B, binary fission in many irregular planes; **C**, multiple fission of some vegetative cells; **D**, baeocyte release; **E**, period of baeocyte motility. (Reprinted with permission from Waterbury, J., in *The Prokaryotes*, 2nd ed., Vol. II, Balows, A., et al., eds., pp. 2058–2078, 1992).

osarcina by the nature of the baeocytes which are non-motile in *Chroococcidiopsis* and which show gliding motility upon release in *Myxosarcina*. As cubical aggregates are enlarging (i.e., before baeocyte formation) *Chroococcidiopsis* and *Myxosarcina* can easily be confused with the "high GC cluster" of *Synechocystis* "culture group" of Subgroup 1 which may also, by regular three-plane binary fissions, form cubical aggregates, **(but which never undergo multiple internal fissions to produce baeocytes)**. Also, *Gloeocapsa* (Subgroup 1) may form cubical packets but has well defined sheaths surrounding each cell.

Genus **Dermocarpa**

In the sense of Waterbury and Rippka (*Bergey's Manual of Systematic Bacteriology,* Volume 3, pp. 1751–1754) *Dermocarpa* is **unicellular, spherical, incapable of binary fission, and reproduces exclusively by internal multiple fissions, resulting in 10–1000 baeocytes which are motile by gliding upon release.**

The baeocytes range in diameter from about 1.5–4 μm (five strains). Parental cell may reach 20–30 μm in diameter before forming baeocytes. Enlarged "parental cell" may become pyriform or clavate when developing in packed mass. Facultative photoheterotrophy occurs in three of five strains with glucose in all three, and sucrose and fructose in two. There is nitrogenase activity with anaerobiosis (two strains). C-PE is in four strains; three show chromatic adaptation; PE with phycourobilin occurs in one strain, also with chromatic adaptation. Vitamin B$_{12}$ stimulates growth in two strains and is required in another. Source is marine (three strains); fresh water (two strains). Mol% G + C of DNA = 40.7–44.0 (four strains). See Table 11.6, Figure 11.3.

Differentiation of **Dermocarpa** *from other genera:* *Dermocarpa* is easily confused with *Xenococcus* which, by definition, differs only in that the latter has baeocytes that are nonmotile, having a thin fibrous outer wall layer, in addition to the Gram-negative wall at the time of release (T.E.M. characteristic). Both genera are distinguished from single detached mature parental cells of *Chroococcidiopsis* and *Myxosarcina* (which result from binary fission prior to "maturation") by producing ten or more baeocytes per mother cell rather than the typical four of the last two mentioned genera. Both *Dermocarpa* and *Xenococcus* are usually spherical during enlargement, while the oval *Dermocarpella* shows polar-

ity in having at least one basal vegetative cell in addition to the baeocyte parental cell at the opposite pole (see below). However, a single, non-dividing cell of *Dermocarpa* (or *Xenococcus*), prior to baeocyte formation, is virtually indistinguishable from many non-dividing spherical cells of Subgroup 1 (e.g., some strains of *Cyanothece, Synechococcus, Synechocystis*).

Genus **Dermocarpella**

Ovoid aggregate of cells with larger "apical" cell and 1–3 smaller "basal" cells. As vegetative initial cell develops from liberated baeocyte, 1–3 asymmetric **binary fissions** occur, resulting in the polar ovoid aggregate or thallus. **At maturity "apical" cell undergoes internal multiple fissions to produce usually 60–120 baeocytes.** Intrinsic polarity with one reproductive cell (baeocyte producer) is unique among Subgroup 2 cyanobacteria. Information is based on one culture strain from marine source with mol% G + C of DNA = 45.1. See Table 11.6, Figure 11.3.

Differentiation of **Dermocarpella** *from other genera:* Polarity of aggregate and "specialized" baeocyte-forming "apical" cell is unique, but can be confused with some early stages of development of Type II of the *Pleurocapsa* "culture group" which also begins with an asymmetric elongation of the enlarging baeocyte and an asymmetric binary fission (see Figure 11.3). Earlier unicellular stages may be confused with numerous spherical to ovoid cyanobacteria. See Waterbury and Rippka (*Bergey's Manual of Systematic Bacteriology,* Volume 3, pp. 1754–58) for taxonomic comments, particularly in relation to the system of Komárek and Anagnostidis (Arch. Hydrobiol. Suppl. *73:* 196–204, 1986).

Genus **Myxosarcina**

See genus *Chroococcidiopsis* for a more detailed description. In the sense of Waterbury and Rippka (*Bergey's Manual of Systematic Bacteriology,* Volume 3, pp. 1758–1762) *Myxosarcina* differs from *Chroococcidiopsis* only by the release of **motile baeocytes** in *Myxosarcina* which also means that structurally the motile baeocytes have not yet acquired the fibrous outer wall which surrounds the baeocytes of *Chroococcidiopsis*. Thus, *Myxosarcina* also produces cube-like aggregates of cells by binary fissions in three successive planes and releases at least four baeocytes per vegetative cell. See Table 11.6, Figure 11.3 for more details and comparisons.

Genus **Pleurocapsa**

The genus *Pleurocapsa,* in the sense of Waterbury and Rippka (*Bergey's Manual of Systematic Bacteriology,* Volume 3, pp. 1762–1769), is regarded as a provisional "super genus" or "culture group" from which other genera will eventually be separated.

At present it is defined as a cyanobacterium with **binary fission in several planes, producing cell aggregates ranging from compact irregular masses of cells to masses from which uniseriate to irregular multiseriate filaments arise. Formation of baeocytes by multiple internal fissions occurs in some cells of aggregate or of filamentous portion of structure.** Baeocytes are motile. Two types of the *Pleurocapsa* group are recognized. **Type I:** baeocyte enlarges into spherical cell before first equal binary fission. **Type II:** baeocyte enlarges asymmetrically into elongated or pyriform cell before first binary fission which may be skewed towards wider pole. See Figure 11.3.

Number of binary fissions preceding baeocyte formation not fixed. "Good" growth conditions favor earlier baeocyte production. Multiple fission may occur after only a few binary fissions even in strains that are capable of complex vegetative growth which includes filaments or pseudofilaments. Other factors may also modify aggregate morphology. Photoheterotrophic growth may result in cubical aggregates in some strains, whereas photoautotrophic growth results in aggregates with pseudofilamentous outgrowths. In another strain addition of sucrose results in the apical, uniseriate portion of filaments becoming multiseriate.

Pleurocapsa and *Hyella* (following Geitler, Rabenhorst's Kryptogamenflora *14:* 344–375, 1932) are separated mainly on the basis of growth habit in nature. *Pleurocapsa* is an epiphyte or epilith and *Hyella* is an endolith, penetrating calcareous substrates such as mollusc shells. At this time *Hyella* will not be described separately (see Al-Thukair and Golubic, Algol. Stud. *64:* 167–197, 1991). Among the eight marine culture strains of *Pleurocapsa* included here, only one grew as an endolith in oyster shell chips, and in this case, acid-producing heterotrophic bacteria present in the culture may have been required.

Of the eight strains, seven are capable of photoheterotrophy; six showed nitrogenase activity under anaerobiosis. All eight strains possess PE, three of which contain phycourobilin; four show chromatic adaptation (two of

these have phycourobilin). The maximum temperature for growth ranged from 30–39°C. All eight strains were marine, but members of the *Pleurocapsa* group are also known from fresh water (Rippka et al., J. Gen. Microbiol. *111:* 22–23, 1979) and hot spring isolates. Will grow at temperatures up to 55°C (Wickstrom and Castenholz, J. Phycol. *14:* 84–88, 1978). The mol% of G + C of DNA of eight strains is 43.0–45.4. See Table 11.6; Figure 11.3.

Differentiation of **Pleurocapsa** *from other genera:* This is the only baeocyte-producing cyanobacterium that forms a large amorphous aggregate with filament-like outgrowths. However, regular cubical aggregates may occur under some culture condition and result in *Chroococcidiopsis-* or *Myxosarcina*-like aggregates. In the latter two genera, usually all cells produce baeocytes (usually four in number), whereas fewer cells undergo multiple fission in *Pleurocapsa,* and usually several more baeocytes are produced by each parental cell.

Some of the amorphous and pseudofilamentous habits of the *Pleurocapsa* group resemble the diverse forms of *Chlorogloeopsis* (Subgroup 5). *Chlorogloeopsis,* however, under combined nitrogen limitation, is capable of producing heterocysts.

Genus **Xenococcus**

Unicellular, spherical cells that reproduce exclusively by multiple fissions, and are incapable of binary fission: baeocytes 10–1000 per parental cell. Non-motile. Baeocytes form fibrous outer wall prior to release from mother cell (T.E.M. characteristic). In three strains, baeocyte diameter is 2–3 μm, from parental cells 10–25 μm in diameter. Photoheterotrophy (glucose and sucrose) occurs in one strain, anaerobic nitrogenase activity occurs in one other strain. All three strains contain PE with phycourobilin, and all show chromatic adaptation. Maximum growth temperature is 30–35°C. All are from marine sources. Mol% G + C of DNA = 43.3–44.2 (two strains). See Table 11.6 and Figure 11.3.

Differentiation of **Xenococcus** *from other genera:* *Xenococcus* is differentiated in the sense of Waterbury and Rippka (*Bergey's Manual of Systematic Bacteriology,* Volume 3, pp. 1751–1754) from *Dermocarpa* simply by the **absence of motile baeocytes** in *Xenococcus.* Confusion with other genera during non-reproductive stage is discussed under genus *Dermocarpa.*

SUBGROUP 3 (=Order: Oscillatoriales)

Filamentous cyanobacteria that undergo **binary fission in a single plane and that produce "vegetative" cells only. No heterocysts or akinetes.** False branching may occur in some genera. Trichome diameters range from about 0.4 to over 100 µm (rarely). Terminal cell of some species may be distinctly shaped and often less pigmented. Terminal cell may be differentiated (tapered, and sometimes with calyptra); in some species the trichome taper may include several subterminal cells as well. Generally within 1 strain diameter of central part of trichome varies little (<10%) in contrast to some of the other subgroups of cyanobacteria. Generally all cells appear to retain the ability to divide, but certain portions of a trichome may be more active meristematically than terminal regions. The terminal cell in some strains may never divide, once differentiated. Trichomes may be flexible or semi-rigid. Entire trichome may be wound into a loose or tight spiral; in some only terminal portions of the trichome may be openly spiraled.

Sheath may be present, but even species without an easily visible sheath leave behind at least a very thin elastic sheath or slime trail when gliding. Those species with thickened (sometimes laminated) sheath may move only slightly within this confinement. When short fragments of a few cells separate from the remainder of the trichome near the free, open end of a sheath, these free trichomes (hormogonia) may glide out and disperse, eventually forming new sheaths. Although trichomes without apparent sheaths also fragment, a separable, migrating, hormogonial phase is difficult to distinguish, since all lengths of trichome are generally motile. Gliding motility occurs on solid to semi-solid substrates. Movement forward or back is usually or always accompanied by a right- or left-handed rotation of the trichome. Fragmentation of trichomes occurs in some forms

at the site of a cell which loses much of its contents and dies. In some cases, there appears to be an orderly sacrificial death of these cells (**necridial cells**) which determine the sites of trichome breakage.

There is a large range in DNA base composition (40–67 mol% G + C). The range of genome sizes is also great (2.14–5.19×10^9 daltons) but almost all sizes are less than in Subgroups 4 or 5.

The range in DNA base composition and the genome size indicates a very great artificiality in classification of the assemblage of Subgroup 3, a conclusion supported by the large degree of disparity in nucleotide sequences of 16S rRNA from several strains now classified within this Subgroup (Giovannoni et al., J. Bacteriol. *170:* 3584–3592, 1988; Wilmotte and Golubic, Algol. Stud. *64:* 1–24, 1991). **The triviality and provisional nature of some generic distinctions used here is emphasized.** Often only one (but convenient) character is used, a character that may be the result of only a slight difference in genotype.

Members of Subgroup 3 (the order *Oscillatoriales*) occur in an enormous diversity of habitats: freshwater and marine, both as plankton, mats, or periphyton. Terrestrial crusts, mats, or turfs are also common. Hot spring mats of some cyanobacteria of this subgroup develop up to temperatures of about 62°C. Intimate symbiotic relationships are rare. For more detailed analyses of this subgroup see Castenholz (*Bergey's Manual of Systematic Bacteriology,* Volume 3, pp. 1771–1780), Rippka (Methods Enzymol. *167:* 41–48, 1988), and for an extensive revision see Anagnostidis and Komárek (Arch. Hydrobiol. Suppl. *80:* 327–472, 1988). In the latter treatment 43 genera of this subgroup were recognized, 9 of which were newly erected.

Key to **Subgroup 3**

I. Trichomes not cylindrical (flattened or triradiate in cross-section)

 A. Trichomes flattened, i.e., elliptical in cross-section

 Genus *Crinalium*

 B. Trichomes triradiate in cross-section

 Genus *Starria*

II. Trichomes cylindrical

 A. Trichomes helically coiled (open or closed), usually for entire length

 1. Trichome helix usually nearly closed (i.e., spring-like coil); **cross walls thin and usually invisible with light microscope.** With T.E.M.: pores near cross walls in semi-circular patches on concave side of coil

 Genus *Spirulina*

 2. Trichome helix usually open (i.e., as stretched spring); **cross walls visible with light microscope.** With T.E.M.: circular, multiple rows of junctional pores near cross walls

 Genus *Arthrospira*

 B. Trichomes straight or sinuous for only a portion of length

 1. Trichomes usually immobile within a persistent sheath (hormogonia may migrate out from sheaths at times)

 a. Common sheath regularly harbors two or more trichomes

 Genus *Microcoleus*

 b. Single trichome per sheath (in some, sheath may be confluent or mucilaginous resulting in a mat or fabric.

 Genus *Lyngbya* (includes *Porphyrosiphon* and *Phormidium*)

 2. Trichomes usually motile; no persistent sheath (trichomes when gliding may leave nearly transparent, thin, sheath-like or mucous trail)

 a. Constrictions absent at cross walls, or feeble, total constriction **not** exceeding 1/8 of trichome diameter

 Genera *Oscillatoria* and *Trichodesmium*

 b. Constrictions at cross walls strong to moderate, total constriction exceeding 1/8 of trichome diameter

 Genus *Pseudanabaena*

Genus **Arthrospira**

Filamentous cyanobacteria that divide exclusively by binary fission and in one plane. **The entire trichome is arranged as an open helix in which transverse walls may be seen by light microscopy. Cells are generally shorter than broad or quadrate, but are occasionally elongate** (see Komárek and Lund, Arch. Hydrobiol. Suppl. *85:* 1–13, 1990). Constrictions at cross-walls may be present or absent. With T.E.M. **a single complete circle of junctional pores may be seen.** Persistent sheaths are not produced. Gliding motility is evident in most srains. Trichome widths vary from about 3–12 µm in a variety of forms. The helix is an open spiral with widths ranging from about 35–60 µm. On solid medium, the helix undergoes a transition to a "flat spiral." Considerable variation occurs in degree of helix pitch within some strains, and culture variants occur that are nearly straight. The mol% G + C content of the reference strain is 44.3.

One strain is a gas-vacuolate marine cyanobacterium with 16 µm wide trichomes. It is an obligate photoautotroph, as are all other known strains, and it is unable to synthesize nitrogenase anaerobically. It contains PC, APC, and PE which lacks phycourobilin. Many strains, however, also lack PE. *Arthrospira (Spirulina) maxima* and *A. (Spirulina) platensis* are used as sources of human protein supplement.

The "life cycle" of *Arthrospira* in culture involves the breaking up of trichomes at the sites of necridia (lysing cells) at intervals of every 4–6 cells. **The resulting short and uncoiled hormogonia form a short-lived migratory phase.** Each hormogonium then undergoes cell division, growing into a new helical trichome.

This group with spirally formed trichomes and visible cross-walls is found in marine, brackish water, and saline lake environments of tropical and semi-tropical regions. Many culture isolates have been used in aquaculture. Some forms are planktonic and gas vesiculate, others are benthic and without gas vacuoles. *Arthrospira* often dominates the plankton of warm lakes high in carbonate/bicarbonate with pH levels as high as 11. See Figure 11.2*e*.

Differentiation of **Arthrospira** *from other genera:* *Arthrospira* has commonly been submerged in the genus *Spirulina*. It appears that the more tightly coiled *Spirulina* strains that lack easily discernible cross-walls may constitute a natural taxonomic unit. Since *Oscillatoria* species show various degrees of spiraling near the trichome termini, this genus may also be confused or even grade into the present definition of *Arthrospira*.

Genus **Crinalium**

An *Oscillatoria*-like cyanobacterium in which the **trichome is elliptical in cross-section** rather than circular (or triradiate as in *Starria*) (Figure 11.2*f*). In the one species in culture, the cells are 2–2.5 µm in width and 5–7 µm in breadth; cells are short (1–1.5 µm). Terminal cells are non-differentiated. Binary fission occurs as in *Oscillatoria*. **Cells are non-motile;** thylakoids are mainly parallel and close to cell periphery (T.E.M. characteristic). No PE; β-carotene and echinenone are major carotenoids. There is no nitrogen fixation under anaerobic or aerobic conditions. "Freshwater"-terrestrial, surface layers of sand; drought resistant; optimum growth temperature is 25°C in BG-11 medium. Mol% G + C of DNA is 33.9. See de Winder et al. (J. Gen. Microbiol. *136:* 1645–1653, 1990) for further details.

Differentiation of **Crinalium** *from other genera:* This is the only filamentous cyanobacterium with a flattened (elliptical in cross-section) trichome.

Genus **Lyngbya**

Filamentous organisms that share the entire range of cellular types with *Oscillatoria* but which produce **a distinct and persistent sheath. The sheath may be thin, but can be seen with phase contrast optics, particularly where it extends beyond the terminal cell of the trichome** (Figure 11.2*g–i*). The trichome diameters range from about 1 µm to about 80 µm. **Cell features are as for the description of the genus *Oscillatoria*, including coloration.**

Trichomes are usually immotile within the sheath, but short sections of trichome (hormogonia) sometimes move slowly out of the sheath when placed on new agar-solidified medium. Some strains produce many hormogonia which glide free of the sheaths and appear to be an *Oscillatoria* until new sheath production again immobilizes them. In some cases rapid growth extends trichomes out of old sheaths, and terminal portions appear sheathless.

The mol% G + C is 43.4 and the genome size is about 4.58×10^9 daltons in one reference strain. However, using the present concept of *Lyngbya,* several strains of the "LPP group" (Rippka et al., J. Gen. Microbiol. *111:* 1–61, 1979) would also fall within this generic boundary, since most have at least thin sheaths (see Table 15 of Rippka et al.). Stam (Arch. Hydrobiol. Suppl. *56:* 351–374, 1980) included several sheathed "Oscillatorian" forms in his DNA-DNA hybridization studies. The range of mol% G + C in these ranged from about 42–49.

The sheaths of some strains are quite prominent and strong so that an entire entangled mass of filaments in liquid culture will hold together if attempts are made to remove only a small part with forceps. Laminated sheaths occur commonly in the large diameter species, both in freshwater and marine forms. A yellow, near UV-absorbing pigment (scytonemin) commonly occurs in the sheaths of some marine mat-forming species, giving the whole filament a brownish coloration (see Garcia-Pichel and Castenholz, J. Phycol. *27:* 395–409, 1991). The broad absorption maximum of this pigment lies in the near ultraviolet, violet, and blue regions of the spectrum. In some species, intense purple to red pigments occur in multi-layered sheaths (e.g., the genus *Porphyrosiphon,* Figure 11.2*i*).

In contrast to *Oscillatoria,* few species of *Lyngbya* are planktonic. Gas vesiculate forms are uncommon. *Lyngbya* (including *Symploca* and *Porphyrosiphon*) commonly forms turf-like terrestrial mats. *Lyngbya* (including *Phormidium*) also forms fabric-like benthic mats in shallow marine (including intertidal) and freshwater habitats.

In general, the thicker sheathed species of *Lyngbya* are more difficult to isolate in culture than *Oscillatoria* spp. On agar the production of motile hormogonia is not assured, and the older sheaths sometimes carry a heavy burden of attached, contaminating bacteria. Growth rates of the thicker-sheathed species in culture may also be low (< 1 doubling/24 hours).

Since *Lyngbya* has been little studied in culture, little can be said regarding physiology. In one reference strain, a marine mat-forming type, nitrogenase was synthesized under sustained anaerobic conditions.

Differentiation of **Lyngbya** *from other genera:* The provisional genus *Lyngbya* is here differentiated from all other "oscillatorian" (Subgroup 3) genera by producing a persistent sheath encompassing the individual trichome. The migratory phase of free trichomes (hormogonia) which have moved out of the sheaths may

easily be mistaken for the genus *Oscillatoria.* Therefore, a longer period of growth in culture is necessary.

The uncertainty will still remain in cases where sheaths are very thin and transparent. Some of these thin-sheathed forms and those with more gelatinous, diffluent sheaths would have fallen into the "botanical" genus *Phormidium* or *Symploca.* Bourrelly (Les Algues d'eau douce—algues blues et rouges III, N. Boubée et Cie, pp. 428–453, 1985) also considers the boundaries between these genera too slight to demarcate, and therefore recognizes only *Lyngbya* by right of nomenclatural priority. *Plectonema,* also in the "LPP" group, with more frequent false branching than other sheathed types, may also fit into *Lyngbya,* if the value of the false branching is discounted at the generic level (see Figure 11.2*j*). *Porphyrosiphon* is another genus similar to *Lyngbya* but separated from it by having a multi-laminated sheath with a red or brown sheath pigment (see Figure 11.2*i*). However, a greater problem comes with distinguishing the several "botanical" genera that have multiple, unbranched trichomes within a common sheath. These are *Microcoleus, Hydrocoleum, Sirocoleus, Schizothrix,* and others less well known (see Geitler, Rabenhorst's Kryptogamenflora *14:* 1068–1160, 1932; Anagnostidis and Komárek, Arch. Hydrobiol. Suppl. *80:* 327–472, 1988).

Genus **Microcoleus**

Several *Oscillatoria*-like trichomes within a common, homogeneous sheath. Usually several parallel trichomes are often spirally and tightly interwoven (Figure 11.2*k*). Species from marine and fresh water have been described (Geitler, Rabenhorst's Kryptogamenflora *14:* 1131–1146, 1932). Trichome diameter of several species is ~2–10 µm. One species in particular has been studied extensively, but mainly as collected material from intertidal mats (*M. chthonoplastes*). Although cultures (including axenic) have been used at times, little characterization of *Microcoleus* has come from work with cultures.

In axenic culture, some workers have found that the common sheath disappears. In other cultures fascicles of trichomes were maintained with thin common sheaths, particularly under unidirectional light.

Differentiation of **Microcoleus** *from other genera:* There are several other genera with a habit fairly similar to that of *Microcoleus.* These generic names (e.g., *Hydrocoleum,*

Schizothrix, Sirocoleus) are used in the botanical literature, but there has as yet been little characterization of these from culture strains. They are, therefore, not included in the present treatment.

Genus **Oscillatoria**

Filamentous cyanobacteria that divide exclusively by binary fission and in one plane. **The trichomes tend to be straight to loosely sinuous, particularly near apices; they are flexible or semi-rigid. Transverse septa are generally visible using light microscopy. Constrictions may or may not occur at cross-walls, but the total indentation does not exceed 1/8 of the trichome diameter.** Generally, the transverse septum (cross-wall) is thinner than the longitudinal wall. During fission the cytoplasmic membrane invaginates with a thinner peptidoglycan layer separating the new membranes of the daughter cells (T.E.M.). This characteristic applies to the genera *Spirulina, Arthrospira,* and *Lyngbya* in addition to *Oscillatoria.*

Cells may be much shorter in length than broad (appearing as stacked discs) to a few times longer than broad. The trichome diameters range from about 1 μm to occasionally over 100 μm. **Invariably in broader trichomes (>15 μm diam.) the cells are shorter than broad.** The trichome is usually motile and rotates consistently either in a left- or right-handed manner with respect to the direction of movement. If terminal regions are not in contact with substrate, the free end may oscillate as the trichome rotates, particularly if the free end is curved. Rates of movement range from less than 1 μm·sec⁻¹ to about 11 μm·sec⁻¹. Persistent sheaths are usually absent, but nearly invisible, gossamer thin tubes are shed as trails when the trichome moves on solid substrates or simply slime is shed. Occasionally a more visible sheath may build up around some trichomes, particularly during periods of immobility in old liquid cultures. Trichomes are solitary, but if clustered or in fabric-like mats they are not surrounded by a morphologically distinct common sheath. Copious amounts of gel-like matter, however, may be produced in liquid culture, particularly in axenic culture.

The terminal cell of many species of *Oscillatoria* is differentiated to the extent of developing a shape distinct from the simple bulging of an unattended cross-wall such as that which occurs immediately after trichome fragmentation. Species-consistent shapes of terminal cells include rounded, blunt, truncate, conical, prolonged-attenuate, and capitate. In addition, some terminal cells acquire an outside thickening of the outer cell wall termed the calyptra. Trichomes may be attenuated as well, but only for the length of a few to several cells near the terminus. The terminus of recently fragmented trichomes will generally appear different from those allowed to differentiate. The terminus of many species may be bent (whether tapered or not), and this may extend for the length of several cells. Further cell divisions may not be possible in the differentiated terminal cell.

The color is variable, ranging from bright blue-green to deep red. Several species show "chromatic adaptation" (see Tandeau de Marsac, J. Bacteriol. *130:* 82–91, 1977). Some species, almost black in color, contain abundant C-PE and phycocyanin. Phycourobilin-containing PE with a large 493–495 nm absorption maximum, in addition to the 543–546 nm maximum, occurs in some shade-inhabiting species of *Oscillatoria* (R.W. Castenholz, unpublished) and in marine planktonic *Trichodesmium* (=*Oscillatoria* of some authors).

In the Pasteur Culture Collection the mol% G + C of DNA of 10 strains included in *Oscillatoria* ranged from about 40–50, and the genome size ranged from 2.50–4.38 × 10⁹ daltons in 6 strains examined.

The species of *Oscillatoria* are distributed in fresh marine, and brackish waters. They also occur in inland saline lakes, and a few species tolerate temperatures as high as 56–60°C in some hot springs. Some species are known as mat-formers in streams. A number of species are planktonic in fresh water (e.g., *O. agardhii, O. rubescens, O. borneti*) and warmer marine waters [e.g., *Trichodesmium* (*Oscillatoria*) spp.]. These almost invariably contain gas vesicles (exception: *O. borneti*). Several species are known as motile (gliding) components of microbial mats which are often sulfidic habitats. Species of *Oscillatoria* are also known commonly in other anaerobic, sulfide-containing habitats such as sediments of stratified eutrophic lakes. A few species occur in terrestrial habitats subjected to severe drying or in shallow freshwater ponds in polar regions which are frozen solid for over nine months each year.

All *Oscillatoria* species are capable of photoautotrophic growth; some are obligate photoautotrophs. None of the strains in the Pasteur Collection were chemoheterotrophs, but some were photoheterotrophic. A few species (e.g., *O. terebriformis*) are capable of a very slow fermentative growth under dark anaerobic conditions with exogenous sugars. Sulfide-dependent anoxygenic photosynthesis is known in several species of oscillatorian cyanobacteria. A few synthesized nitrogenase under anaerobiosis. Marine planktonic forms [*Trichodesmium* (*Oscillatoria*) spp.] fix nitrogen when growing as clusters of trichomes in nature. In one thermophilic species only reduced N compounds can be used.

Isolation can usually be done by "self-isolation" on agar using gliding motility which is sometimes rapid (Castenholz, *In:* Parker and Glazer (eds.), Methods Enzymol. *167:* 68–93, 1988). However, many species are extremely refractory to present culture techniques. These have included a large number of apparently oligotrophic marine forms, including the species of *Trichodesmium* which have been difficult to sustain in the laboratory.

Differentiation of **Oscillatoria** *from other genera:* *Lyngbya, Phormidium, Microcoleus, Symploca, Plectonema, Porphyrosiphon, Borzia, Schizothrix,* and others are common, long recognized "botanical" genera nearly identical to *Oscillatoria* in the sense that a similar range of cell and trichome types have been described within each genus (see Rippka, Methods Enzymol. 167, 41–48, 1988; Anagnostidis and Komárek, Arch. Hydrobiol. Suppl. *80:* 327–472, 1988). The boundaries of these and some additional genera have been based on what may be variable characteristics, such as the presence or absence of a firm or diffluent sheath, the number of trichomes within each sheath, the frequency of false branching, and length of trichome. Moreover, *Oscillatoria,* when gliding, also produces a very thin, sheath-like casing. Thus, there appears to be a cline from extremely thin to thicker and even laminated sheaths, but with a similar range of oscillatorian cell and trichome types within each type of sheath.

The genus *Oscillatoria* is not considered here in such restrictive terms as by Rippka et al. (J. Gen. Microbiol. *111:* 1–61, 1979). Nevertheless, 16 S rRNA sequence data indicate a great disparity among some species of this provisional genus (Giovannoni et al., J. Bacteriol. *170:* 3584–3592, 1988). *Oscillatoria* here includes strains with isodiametric or elongated cells (not just those with shortened disc-like cells) and also those which show slight to moderate constrictions at the cross walls, but less than about 1/8 the diameter of the trichome (Figure 11.2*l–q,* 11.4*a–c*). The latter break-off point is arbitrary. The T.E.M. criteria of Guglielmi and Cohen-Bazire (Protistologica *18:* 151–165, 167–177, 1982) on pore patterns and arrangement of fimbriae may help to define the taxonomic unit when more strains are examined.

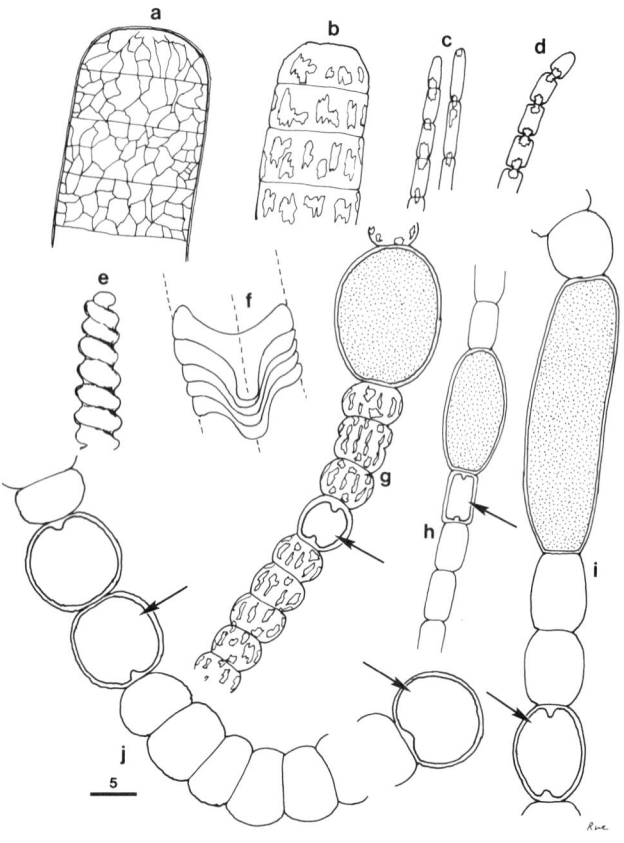

Figure 11.4. Schematic representation of genera of Subgroups 3 and 4.

Subgroup 3 (continued). *a. Oscillatoria* sp.: a large diameter type with "sap vacuole" and intracellular "cytoplasmic strands" (depicted as *lines*). *b. Trichodesmium* (*Oscillatoria*) sp.: marine planktonic form with gas vacuoles shown. Trichomes usually in large fascicles or star-like clusters. *c. Oscillatoria* sp.: freshwater planktonic form with polar gas vacuoles (cf. *O. redekei* or *Limnothrix redekei*). *d. Pseudanabaena* sp. with polar gas vacuoles. *e. Spirulina* sp. with tight coil; transverse septa not shown. *f. Starria* sp. (three-dimensional view showing combined lateral and face views of trichome.

Subgroup 4. g.–j. Anabaena spp.; *g.* planktonic sp. with gas vacuoles depicted. *j. Anabaena* (=*Anabaenopsis*) sp. with intercalary adjacent heterocysts which separate, thus becoming terminal.

In all: *arrows* indicate heterocysts; *stippled cells* are akinetes. *Scale marker* (5 μm) applies to all drawings.

The species of the genus *Trichodesmium* are regarded as species of *Oscillatoria* by some cyanobacteriologists, but this easily identified cluster of marine, planktonic organisms shows distinctive cytological and other features which warrant separation from *Oscillatoria* (see genus *Trichodesmium*, Figure 11.4*b*).

Recently, the genus *Limnothrix* was recognized by some workers, separating well-known polarly gas vesiculate species with peripheral thylakoids and isodiametric to elongate cells from *Oscillatoria* (e.g., *O. redekei*). This is merely one of many "clusters" within the artificial genus *Oscillatoria* that may earn separate generic status after greater analysis in culture (see Anagnostidis and Komárek, Arch. Hydrobiol. Suppl. *80:* 327–342, 1988).

Genus **Pseudanabaena**

Filamentous organisms that divide exclusively by binary fission in one plane and that **have conspicuous constrictions at the cross-walls; in most strains the constriction indents into the trichome 1/2 or more of the diameter** (Figure 11.4*d*). **In a few strains, the constriction is less but still more than 1/8 the diameter (see *Oscillatoria*). Cells are longer than broad to isodiametric and are often barrel-shaped.** The diameter of trichomes of strains presently characterized in culture ranges from about 1–3 μm. **The transverse septum involves a partial centripetal ingrowth of all wall layers.** In some cases the remaining connection appears quite narrow as if the cells were strung as beads. **The structural (peptidoglycan) layer of the cross-wall is 3–6 times thicker than that layer surrounding the rest of the cell** (T.E.M. characteristic). The trichomes are usually straight and quite frequently short, sometimes consisting of only a few cells. Single, detached cells are frequent in most culture populations. Gliding motility occurs in trichomes and unicells, probably without rotation. Gliding motility has apparently been lost in a few culture strains.

In one species, gliding, comet-shaped aggregates of many trichomes develop and usually move at rates greater than individual trichomes (Castenholz, *In:* Biology of Cyanobacteria, Carr and Whitton, eds., Blackwell, pp. 413–439, 1982).

Many strains have clusters of polar gas vesicles, but this characteristic is commonly not expressed under all culture conditions and in some clones seems to have been lost permanently. Although similar in appearance to some forms of *Anabaena* in Subgroup 4, **Pseudanabaena as here defined, is incapable of producing specialized cells, such as heterocysts or akinetes.**

Terminal cells of trichomes are not differentiated with respect to shape as in many species of *Oscillatoria* and *Lyngbya,* and they do not acquire a calyptra. *Pseudanabaena* strains and some strains that had the appearance of *Pseudanabaena* possessed rings (400–500 nm wide) of multiple pores (300–500) near the cell poles, in contrast to the single ring of pores found in strains of *Oscillatoria* (T.E.M. characteristic) (see Guglielmi and Cohen-Bazire, Protistologica *18:* 167–177, 1982).

The color of *Pseudanabaena* may be blue-green to reddish, depending on the presence or absence of PE. The mol% G + C content of 11 strains ranges from about 42–48, and the genome size of 10 strains is 2.14–5.19×10^9 daltons.

Pseudanabaena occurs in hot springs (probably not above 55°C) and on marine and freshwater muds. It is particularly common in anaerobic, sulfide-containing sediments. Some forms are known in freshwater plankton and are often included in the mucilage of other planktonic cyanobacteria.

All strains known are obligate photoautotrophs. Some are capable of synthesizing nitrogenase under anaerobiosis. Culture isolation can usually be done using gliding self-isolation.

Differentiation of **Pseudanabaena** *from other genera:* The genus *Pseudanabaena,* as here defined, is not as restrictive at that proposed by Rippka et al. (J. Gen. Microbiol *111:* 1061, 1979), but corresponds closely to the redefinition of this genus by Guglielmi and Cohen-Bazire (Protistologica *20:* 377–391, 393–413, 1984). Gliding motility of *Pseudanabaena* was used by Rippka et al. to distinguish *Pseudanabaena* (including unicells) from certain strains or species of the unicellular genus *Synechococcus* (Subgroup 1) which sometimes forms short chains in addition to unicells. The criterion is imperfect, since some strains of *Pseudanabaena* have lost gliding ability and some strains of *Synechococcus* do glide. In contrast to the criterion applied by Rippka et al., the possession of polar gas vesicles is not considered a valid character since they are lost in some cultures. On the other hand, populations of apparent unicellular

Synechococcus from hot springs sometime possess polar gas vacuoles.

The comparative sequences of nucleotides of 16S rRNA of two strains of *Pseudanabaena,* as here defined, place these quite distantly from all other members of all subgroups, including Subgroup 3 (Giovannoni et al., J. Bacteriol. *170:* 3584–3592, 1988). This finding again emphasizes the artificial groupings of taxa presented here in Subgroup 3.

Genus **Spirulina**

Filamentous organisms that divide exclusively by binary fission and in one plane but that **grow in the form of a tightly closed to nearly tightly closed right-handed or left-handed helix. The cross walls are thin and are invisible or nearly so with light microscopy** (Figure 11.4e). No sheath is visible under light microscopy and "healthy" trichomes are in constant motion. Gliding motility consists of a "turning of the screw," thus with great transverse movement and little forward motion. The trichome does not truly rotate but moves along the outer surface of the helix. Free ends not in contact with substrate may oscillate greatly as the coil turns. The apex of the trichome is either blunt or pointed. In different species the width of the trichome may be from less than 1 μm to about 5 μm. In the latter case the width of the tight helix may be as great as 12 μm. Color is variable, blue-green to red; some marine strains are extremely red, containing PE as the major light-harvesting pigment. Variations in the tightness of the trichome helix occur, and several culture strains have lost much of their coiled structure.

Using T.E.M., patches of pores at cross-walls occur on only the inner concave surfaces of the helix, at least in 4 strains examined. The mol% G + C of the reference strain is 54, and the genome size is 2.53×10^9 daltons.

The members of this genus occur in fresh, marine, and brackish waters. Species are also seen in inland saline lakes and in some hot springs at temperatures as high as 50°C. They are tolerant of free sulfide in many habitats. They are aquatic and uncommon in terrestrial habitats that are subjected to periodic drying. They are also unknown as intimate endosymbionts or exosymbionts.

Little is known of the physiology of *Spirulina.* One strain examined is a strict photoautotroph and is able to synthesize nitrogenase anaerobically. Some red (shade-adapted) marine strains require "oligotrophic" medium and grow very slowly. A thermophilic strain (cf. *S. labyrinthiformis*) is capable of sulfide-dependent anoxygenic photosynthesis.

The best isolation procedures involve self-isolation by gliding motility, but because of poor forward progress on agar, isolation on glass or plastic surfaces in liquid medium are often more successful.

Differentiation of **Spirulina** *from other genera: Spirulina* has, in the opinion of various authors, included the genus *Arthrospira* (e.g., Geitler, Rabenhorst's Kryptogamenflora *14:* 916–931, 1932). *Spirulina* (here) has a continuous helical coil with thin (invisible) cross-walls and a possibly diagnostic pore pattern. The most easily confused genus, *Arthrospira* has a spirally arranged trichome in which cross-walls are clearly visible with light microscopy. The species of *Arthrospira* described on the basis of morphology have open coils. Some coils are so open that they could easily be regarded as species of *Oscillatoria,* many of which are sinuous or with a very open helix near the terminus of the trichome.

Genus **Starria**

This non-branching, filamentous cyanobacterium is unique in that the **short trichomes in cross-section are triradiate.** The trichomes are straight to helically twisted, about 15 μm in diameter. The cells are short (1–2 μm). The triradiate form usually has broad, arm-like projections 120° apart separated by U-shaped depressions (Figure 11.4f). The pigmentation is concentrated in the arms. A thin sheath (3 μm thick) covers the trichome. Cross-walls are as in the genus *Oscillatoria* with an invagination of the peptidoglycan layer but not of the outer membrane. The relatively thick peptidoglycan layer of longitudinal walls is characterized by evenly distributed pits 70 nm in diameter, as in some species of *Oscillatoria* (e.g., *O. princeps*).

This organism is known from a single strain isolated from a soil sample in Zimbabwe (see Lang, J. Phycol. *13:* 288–296, 1977). Growth occurred on common cyanobacterial medium with soil extract added. Low light intensity was recommended.

Little is known about the physiology or biochemistry of this organism, but its very unusual structure has been documented by light and electron microscopy.

Differentiation of **Starria** *from other genera:* Although morphological variants were derived from the wild-type clone, all possessed a distinct semblance to the triradiate feature. The only genus slightly similar is the ribbon-like *Crinalium* (also in Subgroup 3) and the "botanical" *Gomontiella* which is similar to a U in cross-section with nearly closed arms.

Genus **Trichodesmium**

A genus not easily distinguished from *Oscillatoria* on morphological grounds. **A filamentous *Oscillatoria*-like cyanobacterium that forms parallel (fascicle-like) or radiate clusters in warm water, tropical to semi-tropical planktonic marine habitats.** Trichomes are 6–22 μm in diameter (Figure 11.4*b*). Thylakoids are sparsely distributed in trichomes of natural populations. There is some slime production. Difficult to maintain in culture (see Castenholz, Methods in Enzymology *167*, p. 89, 1988 for references). A few species are recognized (see Anagnostidis and Komárek, Arch. Hydrobiol. Suppl. *80*: 416–419, 1988). All contain **abundant gas vesicles resistant to collapse even at high hydrostatic pressure** (12–37 bars). **All are reddish in color, containing PE with phycourobilin.** Cells are capable of slow gliding motility on solid medium. Nitrogenase activity is confirmed in one species or strain, suspected in all. Cells form large populations over extensive oceanic areas. Two freshwater species are described.

Differentiation of **Trichodesmium** *from other genera:* This genus may be confused with several common species of *Oscillatoria* that are gas vesiculate and planktonic. However, these species of *Oscillatoria* are freshwater, whereas *Trichodesmium* in abundance is marine. The main feature of *Trichodesmium*, however, that distinguishes it from *Oscillatoria*, **is the formation of radiate or fasciculate** bundles. Radiate or fasciculate bundles also occur in freshwater planktonic *Gloeotrichia* and *Aphanizomenon*, respectively, but these produce heterocysts and periodically akinetes.

SUBGROUP 4 (=Order: Nostocales)

This subgroup is distinguished from all other cyanobacteria as **filamentous organisms dividing exclusively by binary fission in one plane only (some with the possibility of "false branching") and having the potential to produce heterocysts.** Some possess heter-

ocysts under almost all environmental conditions. In others, heterocyst differentiation occurs only when ammonium (or nitrate) nitrogen concentration is low in the medium. False branching may occur frequently in same genera but this does not involve division in more than one plane. See Figure 11.4, 11.5, 11.6, and 11.7.

Trichome diameter may range from 1–30 μm in various species with untapered trichomes. In tapered types the basal heterocyst may be as large as 25 μm, whereas the terminal hair cells may be only 1 μm in diameter. Cell divisions are diffuse or limited to intercalary or basal portions of trichome.

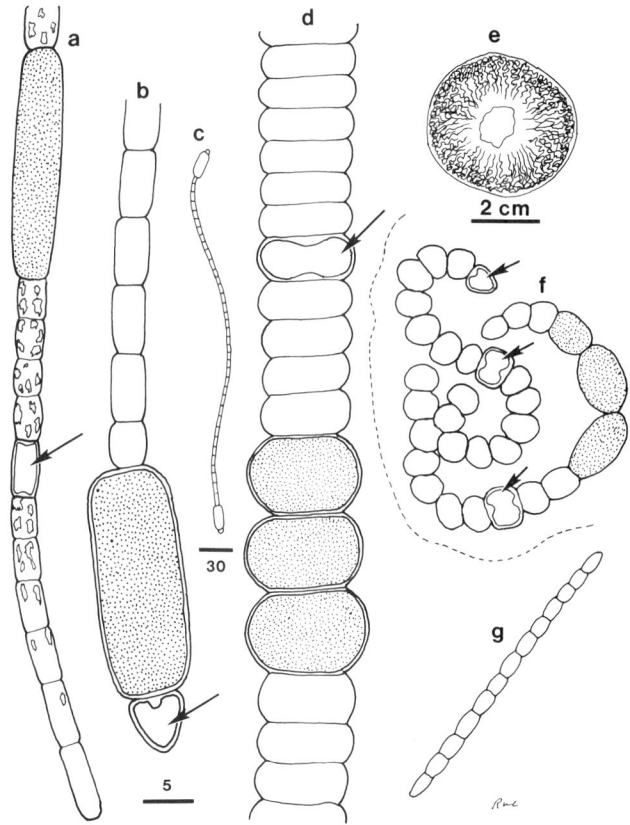

Figure 11.5. Schematic representation of genera of Subgroup 4 (continued). *a. Aphanizomenon* sp.: planktonic form with gas vacuoles depicted. *b. Cylindrospermum* sp. *c. Cylindrospermum* sp. at smaller scale (*marker* = 30 μm). Heterocyst and akinete at each terminus. *d. Nodularia* sp. *e. Nostoc* sp.: gelatinous "colony" at smaller scale (*marker* = 2 cm). *f. Nostoc* sp.: trichomes within gelatinous "colony." Border of "colony" shown by *dashed line. g. Nostoc* sp.: motile hormogonium.

Arrows indicate heterocysts; *stippled cells* are akinetes. *Scale marker* (5 μm) for all except *c.* and *e.*, where other scales are indicated.

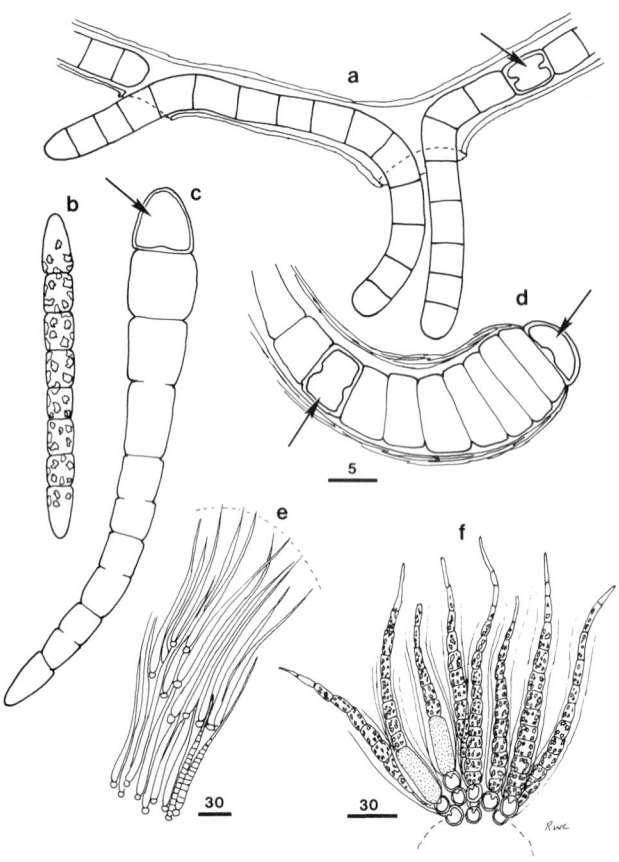

The mol% G + C of many strains (35) of Subgroup 4 are similar, ranging only from about 35–47. The genome sizes, however, are quite different, ranging from 3.17–8.58×10^9 daltons. The results of sequencing nucleotides of 16S rRNA in single strains of 5 different genera covering the three sections of Subgroup 4 show that there is a great degree of genetic relatedness among these representatives and with Subgroup 5 genera than with any genera of other subgroups of cyanobacteria (Giovannoni, J. Bacteriol. *170:* 3584–3592, 1988).

Members of Subgroup 4 may occur in a greater diversity of habitats than members of the other subgroups and commonly enter into symbiotic associations. Several species of *Anabaena, Nodularia, Aphanizomenon,* and *Gloeotrichia* are planktonic and form major blooms in lakes, particularly in perennially warm waters or during summer and fall in temperate lakes. *Nodularia* and *Anabaena* may form major blooms in inland saline lakes and in brackish waters such as the Baltic Sea. However, essentially no free-living members of Subgroup 4 are major contributors to the plankton of oceans. All of the planktonic members produce gas vesicles. Certain species of *Anabaena* and *Aphanizomenon* are known worldwide for their blooms in eutrophic lakes, and some strains produce toxins that act against animals. Many species of most genera are also commonly attached to substrate in freshwater and marine habitats—species of *Calothrix,* and the separable genera *Gloeotrichia, Rivularia, Dichothrix,* and *Gardnerula,* form firm to gelatinous cushions on solid substrates; some species are commonly encrusted by $CaCO_3$. *Nostoc* in fresh water forms firm spherical to amorphous gelatinous colonies, ranging in size from microscopic to the size of bowling balls. *Calothrix, Scytonema,* and related genera are better known than others for subaerial habitats, though particularly in moist or tropical climates. Many, however, are capable of seemingly complete desiccation. *Nostoc* is well known to form terrestrial mats which dry seasonally. No members of this section occur at very high temperatures in hot springs; 52–54°C is the maximum for the most thermophilic strain known (*Calothrix* sp.).

Members of this Subgroup are very well known as intra- and extra-cellular symbionts where, in some cases, they serve as the source of fixed carbon and nitrogen. In other associations they appear to be utilized by the host strictly as a source of fixed nitrogen. Symbiotic associations are mainly restricted to species of *Nostoc* and *Anabaena,* although *Richelia* occurs in planktonic cells of some marine centric diatoms. The most common associations occur in lichens, but also in green plants

Figure 11.6. Schematic representations of genera of Subgroup 4 (continued).

a. Scytonema sp. showing single and geminate false branching. Sheath is present. *b. Calothrix* sp.: hormogonium with gas vacuoles depicted. *c. Calothrix* sp.: young trichome having developed from hormogonium. *d. Calothrix* sp. with multilaminate sheath. *e. Rivularia* sp.: portion of multitiered colony; trichomes with polar heterocysts (*scale marker* = 30 μm). *f. Gloeotrichia* sp.: portion of planktonic, gelatinous colony; gas vacuoles depicted; all trichomes with polar heterocysts; some with subterminal akinete (*scale marker* = 30 μm).

Heterocysts marked by *arrows;* akinetes are *stippled. Scale marker* (5 μm) for all except *e.* and *f.*

Many members of Subgroup 4 produce gel-like or mucilage-like sheaths or extracellular matrices which may range in consistency from firm and leathery to very soft and slimy. The most conspicuous sheaths, however, are thick and laminated. The "branching" sheaths of such genera as *Dichothrix* and *Gardnerula* form some of the most orderly and morphologically complex thallus structures in the cyanobacteria.

such as liverworts, hornworts, ferns, cycads, and in a genus of angiosperms, *Gunnera*.

The visual, microscopic identification of members of Subgroup 4 is often easier than for the preceding subgroups. The heterocyst is the key; although its presence is indicative, its absence is not conclusive. Heterocysts are rare in some natural populations and, in culture, may have to be induced. Also, some genera possibly related to tapered members of Subgroup 4 lack heterocysts under all known conditions.

There are 3 apparent sections within Subgroup 4 into which all but a few genera can be placed. Komárek and Anagnostidis (Arch. Hydrobiol. Suppl. *82:* 247–345, 1989) recognize 4 sections (families) and 32 genera.

A. Within section A (family: *Nostocaceae*) **all vegetative cells normally divide; the position (intercalary or terminal) of heterocysts and of akinetes** is often used to distinguish genera; trichomes are composed of **vegetative cells of uniform diameter (i.e., the trichomes do not taper)**; and **false branching is lacking.** Trichomes have gliding motility or not; motility may be restricted to **hormogonial** (dispersal) phase; trichomes probably do not rotate when gliding.

B. Within section B (family: *Scytonemataceae*) **false branching occurs in all genera; trichomes are untapered** (except **slightly** near terminus in one uncertain genus); **cell divisions usually occur in subterminal region of trichome or "branch."** Heterocysts are frequently located near site of false branch; akinetes are present.

C. Within section C (family: *Rivulariaceae*) **trichomes are slightly to strongly tapered towards apex (i.e., trichomes polar with base and apex). Apex often ends in long, colorless, multicellular "hair." Cell division is generally restricted to basal or intercalary region of trichome. Heterocyst is usually the basal cell of trichome,** but may be intercalary as well. Akinetes may or may not develop; false branching is present or absent.

Key to the genera of **Subgroup 4**

(Heterocysts are potentially present in all genera included here.)

I. Reproduction by diffuse cell divisions and trichome breakage; trichomes are untapered or barely so, trichomes usually **with gliding motility** (often very slow)

    A. Heterocysts intercalary and often terminal as well; position of akinetes (if present) variable

    B. Heterocysts exclusively terminal, usually on one end but sometimes on both; akinete or akinetes form adjacent to heterocyst

        Genus *Cylindrospermum* (Section A)

        1. Vegetative cells spherical, ovoid, or cylindrical-elongate

            a. Vegetative cells cylindrical; (usually in bundles of parallel trichomes) cells more elongate and mainly colorless near ends of trichome

            Genus *Aphanizomenon* (Section A)

            b. Vegetative cells cylindrical, ovoid, or spherical

            Genus *Anabaena* (Section A) (includes *Anabaenopsis* and *Cyanospira*)

2. Vegetative cells shorter than broad, may be disc shaped

Genus *Nodularia* (Section A)

II. Reproduction by diffuse or localized cell divisions; trichomes tapered or untapered; **sheathed or many trichomes in common matrix; nonmotile** except for migrating short chains of cells that lack heterocysts (i.e., hormogonia)

A. Trichomes **tapered** and with sheaths; heterocyst (when present) is the basal cell at wide end of trichome; untapered motile hormogonium (composed of cells **narrower** than vegetative cells) forms a basal heterocyst when motility ceases

Genus *Calothrix* (Section C) (includes *Rivularia, Gloeotrichia,* and others)

B. Trichomes **untapered** and with distinct sheaths; **false branches** occur, often at site of intercalary heterocyst; hormogonia composed of cells of **same** diameter as vegetative cells; hormogonium also forms a basal heterocyst when motility ceases

Genus *Scytonema* (Section B) (includes *Tolypothrix*)

C. Trichomes **untapered; without** false branching. **Common gel** surrounds assemblage of trichomes, at least in nature; hormogonium of same diameter as vegetative trichome **or** narrower diameter, forms a terminal heterocyst, usually at both ends when motility ceases

Genus *Nostoc* (Section A)

SUBGROUP 4, SECTION A
(Family: Nostocaceae)

In culture this section is known primarily for strains that fit under the generic names *Anabaena, Nostoc, Nodularia, Cylindrospermum,* and *Aphanizomenon.* All are genera that lack polarity in terms of tapered trichomes and also lack false or true branching (= family *Nostocaceae*).

There are other genera, not completely described here, that have been included in Section A or the botanical family *Nostocaceae* by Geitler, (Rabenhorst's Kryptogamenflora *14:* 801–905, 1932), Desikachary, (Cyanophyta, New Delhi, pp. 349–433, 1959), and Bourrelly (Les Algues d'eau Douce III, Paris, pp. 414–428, 1985). Some of these are:

Richelia: an unbranched and untapered short trichome, heterocyst polar when present. Occurs intracellularly in planktonic marine diatoms, mainly *Rhizosolenia* spp.; not in culture.

Wollea: a genus differentiated from *Nostoc* by sac-like gelatinous thalli in which the trichomes are arranged in parallel fashion; one species described.

Anabaenopsis: a genus possibly distinct from *Anabaena* or *Cylindrospermum* (Figure 11.4*j*). The principal criteria are that heterocysts are terminal (intercalary and in pairs during differentiation just before separation) and that akinetes are formed as distally as possible from heterocysts (thus, unlike *Cylindrospermum*). Since species of *Anabaenopsis* sometimes have single intercalary heterocysts, there may be insufficient reason at this point for the separation of *Anabaenopsis* from *Anabaena,* although future information from cultures may prove otherwise. The genus *Cyanospira* (Florenzano et al., Arch. Microbiol. *140:* 301–306, 1985) probably fits into this group (see genus *Anabaena*).

Hormothamnion: a genus well-known from marine specimens which form bush-like tufts of false-branching filaments; heterocysts intercalary only.

Aulosira: a genus with several common species in tropical fresh waters. Similar to *Anabaena* but with persistent sheath; heterocysts intercalary; akinetes often in series. Bourrelly (Les Algues d'eau Douce III, Paris, p. 416, 581, 1985) includes *Aulosira* in *Nodularia.*

Genus **Anabaena**

Trichomes are **untapered with conspicuous constrictions at cross-walls.** Trichomes may be straight, curved, or helically (spirally) formed. The cells are cylindrical, spherical, or ovoid (barrel-shaped)—not shorter than broad (or only slightly so) usually ranging in width from about 2–10 μm, but in some species to over 20 μm. The mature terminal cells may be rounded, tapered, or conical in shape. **Heterocysts are intercalary or terminal or both** (Figure 11.4*g–j*). Intercalary heterocysts range from nearly spherical to cylindrical with rounded ends; terminal heterocysts are similar or sometimes conical. Akinetes are usually formed late in growth season and their position in trichomes differs with species. **A distinct individual sheath is absent,** but a slime covering is often present.

Trichomes, when free of adhesive mucilage, are normally motile (usually less than 1 μm/s), and **colonies are not formed. Reproduction is by fragmentation of "parental" trichomes into shorter trichomes indistinguishable in cell dimensions from the former trichome.** Gas vesicles occur in many species, however, mainly those that are planktonic. Many species are known worldwide as major components of the freshwater plankton and also of many saline lakes. Others occur as tychoplankton or periphyton. Symbiotic *Anabaena* (e.g., *A. azollae* of the aquatic fern, *Azolla*) should probably be considered a species of *Nostoc* (see Rippka, Methods Enzymol. *167:* 28–67, 1988). *Anabaena* is not known as a major component of marine plankton. There are at least 15 gas-vesiculate species in freshwater plankton and *A. spiroides, A. circinalis,* and *A. flos-aquae* are among the most common.

The physiology and biochemistry of *Anabaena* strains have been studied in some detail. The most extensive studies of cyanobacterial nitrogen-fixation and of heterocyst structure and differentiation have used *Anabaena.*

Of seven strains examined by Rippka et al., all were obligate phototrophs. All contained phycoerythrocyanin (PEC) in addition to PC, but species do exist that are reddish (i.e., contain PE). Some culture strains of *Anabaena* are facultative heterotrophs. The DNA base composition of about 20 strains ranged from about 35–47 mol% G + C. The genome size of 6 strains ranged from 3.17–3.89 × 10^9 daltons.

The isolation of *Anabaena,* using washing or self-isolation-by-gliding methods works as it does with many other filamentous, gliding cyanobacteria, but prior enrichment with medium free of combined nitrogen has positive selective value. Heterocysts will be produced (if not constitutive), and nitrogen-fixation will occur within a few days, often following an initial yellowing (N-free) period. For the maintenance of cultures, N-free or low N medium should be used for "normal" development and appearance of trichomes.

Differentiation of the genus **Anabaena** *from other genera:* Rippka et al. (J. Gen. Microbiol. *111:* 34–46, 1979; Methods Enzymol. *167:* 48–61, 1988), on the basis of culture strains alone, differentiate *Anabaena* from *Nostoc* primarily on the basis of motility. In *Anabaena* all vegetative trichomes are normally motile, and any fragmented shorter trichomes appear no different from the "parental" trichome. In *Nostoc,* on the other hand, vegetative trichomes are nonmotile. *Nostoc* appears identical to *Anabaena* in essentially all morphological aspects, but motility is restricted to hormogonia, short chains of cells (lacking heterocysts) often are of lesser width than normal vegetative trichomes. However, in culture, motility is commonly lost, in which case *Nostoc* is still distinguishable by the production of short, heterocyst-free trichomes (hormogonia) which are often narrower than typical trichomes and often possess gas vesicles, even when the parental trichome does not.

Nostoc is traditionally distinguished from *Anabaena* by forming large or small gelatinous colonies, often of distinctive shape. However, in some cultures colony growth did not occur. However, colonies are known to form in some axenic cultures, and this feature in nature is well correlated with the immotility stage, followed by the dispersal of motile hormogonia.

Anabaena is here differentiated from *Nodularia,* although the basis for this may be trivial. *Nodularia* has cells which are easily seen to be shorter than broad and are usually disc-like. Like *Anabaena,* no distinguishable hormogonia are formed and the vegetative trichomes are motile. *Anabaena* is also distinguished from the well-known genus, *Aphanizomenon,* by one main characteristic, although other features are also associated with the few species known in the latter genus. In *Aphanizomenon* the cells are longer than broad, but the last few to several cells at the extremities of the trichomes become gradually more elongate, narrower, and essentially hyaline (Figure 11.5*a*). Other specific features of *Aphanizomenon* are given under the description of that genus.

Anabaena is distinguished from *Cylindrospermum* which is characterized by (initially) a single terminal heterocyst which eventually subtends a subterminal, adjacent akinete (but see genus *Trichormus* in Komárek and Anagnostidis, Arch. Hydrobiol. Suppl. *82:* 247–345, 1989). Again, no specialized hormogonia are formed, and the vegetative trichomes are normally motile. Commonly, when trichomes of *Cylindrospermum* elongate enough to fragment medially another terminal heterocyst (and sometimes adjacent akinete) will form on the opposite terminus, either before or after trichome fragmentation (see Figure 11.5*b,c*).

Anabaenopsis is a genus which, for now, may best be included in *Anabaena*. Most species are curved or helically arranged, but heterocysts occur only terminally except when trichomes have elongated and double heterocysts form at midlength; the trichome finally fragments between the 2 heterocysts which results again in terminal heterocysts. Although both species of *Cyanospira*, described by Florenzano et al. (Arch. Microbiol. *140:* 301–306, 1985), had intercalary as well as terminal heterocysts, they appear very similar in other respects to two traditionally described species of *Anabaenopsis,* also known from alkaline African lakes (Figure 11.4*j*).

Genus **Aphanizomenon**

Trichomes are straight, usually **slightly tapered at both ends; cells near the termini are more elongate and hyaline (low pigment content), appearing vacuolate. Flake-like bundles of parallel trichomes; typically planktonic.** Trichomes glide against others of the fascicle. Heterocysts are intercalary, often infrequent and more elongate than the vegetative cells but not much broader. Akinetes are distant from heterocysts and extremely elongate, somewhat wider than vegetative cells. **Gas vesicles are usually abundant in vegetative cells** (Figure 11.5*a*).

This genus is best known worldwide for the abundant planktonic species *A. flos-aquae* and *A. gracile* which characterize many mesotrophic and eutrophic lakes in temperate climates, particularly in the late summer and fall. *Aphanizomenon* also occurs in brackish waters of the Baltic Sea along with *Nodularia spumigena* and *Anabaena* sp.

As yet little work of a taxonomic nature has been done with pure cultures of *Aphanizomenon.* There are problems of isolation and maintenance of this impor-

tant cyanobacterium in culture (see O'Flaherty and Phinney, J. Phycol. *6:* 95–97, 1970; and Heaney and Jaworski, Brit. Phycol. J. *12:* 171–174, 1977). In unicyanobacterial cultures normal flake bundle formation was easily maintained if sufficient iron was available.

Genus **Cylindrospermum**

Trichomes are untapered with a single terminal heterocyst (Figure 11.5*b,c*). **However, when trichome length increases beyond a certain point, a heterocyst will form at the other end, followed eventually by mid-trichome breakage. A single akinete or series of akinetes form adjacent to the heterocyst.** Most species possess a trichome diameter 3–6 μm; akinetes are much wider, up to 20 μm. *Cylindrospermum* is slowly motile and **does not produce specialized hormogonia.** Individual sheaths are not produced, but a confluent mucilage holding many trichomes together may be present. The mean DNA base composition of three strains ranges from 42.1–46.7 mol% G + C with genome sizes of $5.71–6.15 \times 10^9$ daltons for two strains. All 3 strains examined produce PEC, two are obligate photoautotrophs, and one strain is a facultative aerobic chemoheterotroph that grows on fructose or sucrose.

Cylindrospermum is best known as non-planktonic, i.e., as a part of the tychoplankton or periphyton of fresh waters. Some species also occur in moist subaerial (terrestrial) habitats. Cells will grow in BG-11 medium or modifications free of combined N.

Differentiation of the genus **Cylindrospermum** *from other genera:* **Cylindrospermum**, because of the polarity of its trichomes with respect to heterocysts, may be most easily confused with *Calothrix* or some other member of Section C of Subgroup 4. *Calothrix,* when grown in a nitrogen-rich medium, often lacks taper as well as the terminal heterocyst, and it may be confused with *Cylindrospermum* or other genera under similar conditions. However, cells of "young" *Cylindrospermum* trichomes are more rectangular than those of *Calothrix*. It is always best to grow material under restricted nutrient conditions in order to distinguish genera that have trichome polarity.

Genus **Nodularia**

Vegetative cells, akinetes, and heterocysts that are shorter than broad and often discoid or disc-like

(Figure 11.5*d*). **Cell breadth is constant the entire length of the trichome; ranges are from 4–18 μm in the few species described. The compressed heterocysts are intercalary with equally compressed akinetes forming usually distant from heterocysts, and often in series.** Slow gliding motility occurs under most conditions. No PE or PEC exist in cultured strains.

The mean DNA base composition of 1 strain is 40.5 mol% G + C, and the genomic size is 3.89×10^9 daltons. The same strain is a facultative aerobic chemoheterotroph, using glucose, fructose, or sucrose.

This genus, and particularly *N. spumigena,* is known worldwide as a planktonic, gas-vesiculate organism which is often dominant in inland saline lakes and in brackish marine waters such as the Baltic Sea. Salinity appears to be the key factor in the distribution of *Nodularia.* Isolates have generally been made from waters of 3–67 $^0/_{00}$ salinity. Optimum growth occurred with salinity of 5– 20 $^0/_{00}$ NaCl. *Nodularia* is also characteristic of waters with pH values of 8.2–10.0.

Strains have been cultured in BG-11 based medium and may also be cultured in the culture medium described by Booker and Walsby (see Walsby, in *The Prokaryotes,* Vol. 1, pp. 224–235, 1981). Aerobic photosynthetic growth at the expense of N_2 was slow. The range of 25–30°C is usually optimal for growth.

Differentiation of the genus **Nodularia** *from other genera:* Like *Aphanizomenon, Nodularia* is distinguished from *Anabaena* by essentially one characteristic. *Nodularia* has compressed cells (i.e., shorter than broad). Cells of *Anabaena* are longer than broad or more or less spherical. The use of this characteristic may or may not be sufficient to maintain *Nodularia* as a separate genus unless future research reveals additional differences.

Genus **Nostoc**

The trichomes are untapered with conspicuous constrictions at cross-walls. The cells are cylindrical, spherical, or ovoid (barrel-shaped) and are not shorter than broad. Heterocysts are intercalary under most circumstances. Trichome diameter varies with species: ~2–8 μm. In general, the description of individual trichomes of *Nostoc* is covered by the description of *Anabaena.*

Nostoc **is characterized by a confluent gel holding masses of trichomes together, often in the form of a massive thallus which may be spherical, ovoid, or of a less discernable shape.** Some colonies or thalli take the form of flattened discs or large sheets, or may be soft and amorphous. In many cases the outer layer of the colony is firm and leathery and contains most of the trichomes, while the interior is a soft gel with few trichomes but which may be radially arranged (Figure 11.5*e–g*). The size of colonies ranges from microscopic (originating from a single hormogonium come to rest) to over 20 cm in diameter.

Rippka et al. (J. Gen. Microbiol. *111:* 41–46, 1979; Methods Enzymol. *167:* 48–54, 1988) consider the formation of a gelatinous colony to be a secondary character, since some strains do not form a gel and thallus in culture. *Nostoc* **is distinguished (in the absence of colony formation) by the presence of a developmental cycle. Vegetative trichomes are not capable of gliding motility. However, short chains of cells (hormogonia) are formed and released. These, usually motile, trichomes are initially located adjacent to a heterocyst,** or between two heterocysts, but lack heterocysts themselves. After a period of "migration," the hormogonia cease movement, at which time two terminal heterocysts are differentiated. Cell division and growth then resumes accompanied by gel formation under natural and some culture conditions. **The width of hormogonia is usually less than that of vegetative trichomes;** hormogonia are often gas–vesiculate and buoyant when the vegetative trichomes are not. These characteristics may be sufficient to distinguish hormogonia even when motility is not apparent (as in some culture strains).

The DNA base composition of 13 strains of *Nostoc* ranges from 39–45 mol% G + C. The genome size of 11 of these strains ranges from $4.00–6.42 \times 10^9$ daltons.

Of the 13 strains examined by Rippka et al. (J. Gen. Microbiol. *111:* 41–46, 1979) only 3 were unable to grow as photoheterotrophs. Several species and strains of *Nostoc* grown by others are known to grow slowly as dark chemoheterotrophs. Of the 13 strains referred to above, four synthesized C-PE in addition to phycocyanin. Seven others synthesized PEC instead of PE. A few have apparently lost the ability to fix nitrogen aerobically despite the presence of heterocysts. In 10 strains, akinetes were capable of being induced. In two strains, trichomes of old cultures commonly break up into unicells. Detached unicells commonly occur in masses in developing colonies and in symbiotic associations as well.

Gas vesicles occur in the vegetative cells of several species of *Nostoc,* although few colonies are buoyant enough to be planktonic. Various species of *Nostoc* are known from benthic habitats in freshwater lakes. In many cases the semi-spherical colonies are not attached but rest lightly on firm or unconsolidated sediments. Some species are attached to solid substrate in lakes or streams. Many species of *Nostoc* occur as amorphous sheets or masses of gel-bound trichomes in fresh water or in moist terrestrial locations. As such, *Nostoc* is known to form thick, soft mats in polar regions where presumably it is a major contributor of fixed nitrogen. As a symbiont, *Nostoc* is the major phycobiont in cyanolichens, but also occurs as the N_2-fixing symbiont in tripartite, cephalodiate lichens, and in several embryophytes such as bryophytes, cycads, and in the angiosperm genus *Gunnera.*

Many cultures are grown in medium BG-11 lacking NH_4^+ and NO_3^-, the medium for nitrogen-fixing freshwater cyanobacteria (Table 11.1). However, many species and strains of *Nostoc* exist and many cyanobacterial culture media are suitable. Since motile or buoyant, (gas-vesiculate), hormogonia are produced by *Nostoc,* the isolation of new strains usually employs the recovery of these on agar or liquid surfaces after freshly collected material (or aged crude cultures) is placed on or in nutrient-replete medium.

Differentiation of the genus **Nostoc** *from other genera:* *Anabaena,* the genus most closely resembling *Nostoc,* is distinguished by generally being motile in the vegetative state and lacking a developmental cycle which involves differentiated hormogonia. Hormogonia of some species of *Nostoc* are very similar microscopically to trichomes of *Pseudanabaena* (Subgroup 3), including the presence of polar gas vesicles. However, hormogonia will eventually develop into the immotile, thickened, and heterocystous trichomes of *Nostoc* with or without gel formation.

The formation of a gelatinous colony or thallus was considered earlier as the principal criterion distinguishing this genus. This criterion certainly still applies to natural populations and also to many culture strains.

SUBGROUP 4, SECTION B
(Family: Scytonemataceae)

This section is characterized by **uniseriate** trichomes, usually **untapered,** and **sheathed, false branching single or double (geminate)**—frequent or infrequent. **Regions of cell division usually develop near the apex of the trichome** instead of throughout the trichome. **Heterocysts are predominantly intercalary and are often associated with false branches.** Akinetes are absent or rare. **Non-heterocyst containing hormogonia** may be formed (usually terminally) and released. Although a distinctive short chain of cells, **hormogonia are the same cell width as the parent trichome.** Hormogonia, after coming to rest, differentiate a single terminal heterocyst. Cells are usually attached to substrate; they are aquatic (freshwater or marine) or terrestrial—often forming macroscopic tufts, cushions, or entangled masses.

Geitler (Rabenhorst's Kryptogamenflora *14:* 677–801, 1932) considered that there were 12 genera within this group. Bourrelly (Les Algues d'eau Douce III, Paris, pp. 390–402, 1985) recognized 6 or 7. Few strains of any genus within this group have been studied in detail in culture for the purposes of taxonomic characterization. Although several genera will be described briefly for the purposes of general recognition, only *Scytonema* will be characterized. Although *Plectonema* has sometimes been included within this group, this genus has been mentioned earlier as a part of Subgroup 3. The lack of heterocysts under any conditions should exclude it from Subgroup 4 as here defined. *Plectonema* is essentially a sheathed (i.e., *Lyngbya*-like) trichome that exhibits frequent to infrequent false branching (Figure 11.2*j*).

Some additional genera of Section B are briefly described below:

Tolypothrix: single trichome per sheath and false branches single and generally associated with a single or double heterocyst.

Hydrocoryne: a few to several trichomes within one sheath, false branches single, long, and lying close to the main filament.

Coleodesmium (=*Desmonema*): a few to several trichomes within one sheath, false branches single, forming a thallus with pseudodichotomous branching pattern.

Petalonema: a *Scytonema*-like trichome within elaborately laminated thick sheaths.

Scytonematopsis: single trichome per sheath, false branches usually single, sometimes double (geminate). The single characteristic separating this genus from *Scytonema* is that trichomes are somewhat tapered at the termini.

Diplocolon: several **short contorted** trichomes within a common tegument surrounding gel; individual trichomes sheathed and with false branching; forming crusts on moist or intermittently dry terrestrial substrates.

Genus **Scytonema**

Trichomes are **uniseriate and sheathed, with false branches double (geminate), sometimes single, with the region of cell division near the apex of trichome or false branch** (Figure 11.6*a*). Cells may be longer or shorter than broad. Trichome diameter varies with species (~2– >20 μm). However, sheath may add considerable thickness. If the genus *Tolypothrix* is included in *Scytonema* or excluded, **geminate** false branching is absent or rare in *Tolypothrix* by definition.

In the principal strain studied, **hormogonia are formed which have the same lateral dimensions as vegetative trichomes. When the period of motility ends, a single terminal heterocyst is formed before cell division** and growth resumes. The mol% G + C content of 1 strain is 44.4. The genome size is 7.40×10^9, the largest of all cyanobacteria tested except for some strains of *Calothrix* (See Section C).

One strain (cultured in BG-11 medium, Table 11.1) possesses PEC in addition to PC. It is a facultative aerobic chemoheterotroph, growing in the dark on glucose, fructose, or sucrose. On agar-solidified medium, upright, aerial growth is conspicuous. Studies of another species of *Scytonema* in non-axenic, unicyanobacterial culture showed that the degree of geminate, multiple, or single false branching was positively related to heterocyst frequency and, therefore, inversely to nutrient concentration (i.e., combined nitrogen) of the medium (Jeeji-Bai, Schweiz. Z. Hydrol. *38:* 55–62, 1976). The most typical *Scytonema*-like appearance occurred in the low nutrient medium. False branching often occurred when immobile heterocysts appeared to restrict further growth of the trichome, thus resulting in the bulging and breaking out of trichomes through the sheath.

There may be reasons to use the genus *Tolypothrix* and possibly other genera as separate taxa if the slight degree of trichome polarity as well as **single** false branching develop into reliable characters substantiated by **genetic** criteria (see Rippka, Methods Enzymol. *167:* 48–58, 1988; and Komárek and Anagnostidis, Arch. Hydrobiol. Suppl. *82:* 284–289, 1989). *Tolypothrix tenuis* is well known for axenic culture studies of chromatic adaptation (contains C-PE), but few taxonomic evaluations have been made.

Differentiation of **Scytonema** *from other genera:* *Scytonema* may be easily confused with *Lyngbya* (Subgroup 3) if no heterocysts or false branches are evident at the time. If false branches occur (even infrequently) but no heterocysts, this genus may be confused with the botanical genus, *Plectonema*, which in this treatment has been subsumed into the genus *Lyngbya* (see Subgroup 3). In all cases, a sheath is quite evident.

SUBGROUP 4, SECTION C (Family: Rivulariaceae)

The general distinguishing characteristics are: **trichome tapered, at least under low combined nitrogen, uniseriate, no true branching. Heterocysts present terminally under low combined nitrogen** (intercalary heterocysts may also be present). **Cell divisions are intercalary and often localized.** Tapered end of trichome sometimes extends into long, **pale multicellular hair** (particularly with deficiency of phosphorus).

Hormogonia are smaller in diameter than the larger cells of the vegetative trichome; they are formed in culture in response to phosphate repletion, following depletion. After a period of gliding, hormogonia come to rest, **produce a basal heterocyst** and begin tapered growth. Trichomes are usually attached to substrates individually or in groups forming hemispherical or subspherical "cushions." Colonial planktonic forms with trichomes radially arranged, forming spheres; these also contain gas vesicles as do hormogonia of many sessile members of this group.

A large number of "traditional" or botanical genera have been included in Section C. *Calothrix,* and to a lesser extent *Gloeotrichia* and *Rivularia*, has been studied extensively in culture. Only *Calothrix* will be described in detail here. Several species of *Gloeotrichia* and *Rivularia,* however, are well known as field populations.

Many of the "traditional" genera of the family *Rivulariaceae* are noted below. Others are described by Geitler (Rabenhorst's Kryptogamenflora *14:* 581–673, 1932), T.V. Desikachary (Cyanophyta, New Delhi, pp. 507–565, 1959) and Bourrelly (Les Algues d'eau Douce III, Paris, pp. 402–412, 1985).

With Heterocysts:

Calothrix: basal heterocyst, multicellular terminal hair sometimes present, akinetes rare—if present, form adjacent to heterocyst, sheath or gelatinous mucilage may be present, aggregates of filaments may occur.

Gloeotrichia: similar to *Calothrix,* but radially arranged, sheathed trichomes forming spherical (planktonic) or hemispherical (attached) colony; **akinete** formed in the later growth phase adjacent to heterocyst.

Rivularia: similar to *Gloeotrichia;* gelatinous or crustose hemispherical colonies of *Calothrix*-like filaments attached, but **no akinetes** are formed.

Sacconema: colonial, sheath is soft, thick and confluent between *Calothrix*-like trichomes.

Isactis: crustose expanse of *Calothrix*-like filaments.

Dichothrix: *Calothrix*-like trichomes with numerous false branches, a few to many trichomes in a common sheath often forming brush-like macrothallus.

Gardnerula (=*Polythrix*): an elaboration of *Dichothrix* in which the attached thallus branches sub-dichotomously or dichotomously, reaching heights of 1–3 cm.

Microchaete (=*Fremyella*): differs from *Calothrix* by having less or essentially no taper in the trichome, no terminal hair, and a greater number of intercalary heterocysts. Placed in separate family (*Microchaetaceae*) by Komárek and Anagnostidis, (Arch. Hydrobiol. Suppl. *82:* 284–289, 1989) and others.

Without Heterocysts:

Several genera have been described in which trichomes consistently taper, as in *Calothrix;* however, no heterocyst is seen. Since the induction of heterocyst differentiation is usually dependent on external concentrations of ammonium or nitrate, these genera can only be evaluated when they have been critically studied in culture. Examples are the genera *Homoeothrix, Leptochaete, Ammotoidea,* and *Raphidiopsis.* These genera, lacking heterocysts, perhaps should be classified with Subgroup 3.

Genus Calothrix

Trichome, when mature, with taper and basal, 1-pored heterocyst; sheath is usually present, open at narrowed end of trichome, often laminated (Figure 11.6*b*–*d,* 11.7). Basal (wide) end of trichome from approximately 2.5–18 μm in diameter, with the filament (trichome and sheath) as thick as 30 μm in some species. For more details on *Calothrix* see Whitton (*In:* The Cyanobacteria, Fay and Van Baalen, eds., Elsevier, pp. 513–534, 1987).

Although the many species of *Calothrix* are not described in this account, the following morphological characteristics are used for differentiating species: trichome width, degree of trichome swelling at base, presence or absence of elongate, colorless, multicellular "hair" extending narrow end of trichome, degree of aggregation into tufts, whether akinetes occur (rare). Environmental conditions influence several of these characters, so culture conditions are critical. In most strains, substantial taper and differentiation of a basal heterocyst occur only when the medium is deficient in combined nitrogen. Under phosphorus-rich conditions the apical region of the trichome gives rise to hormogonia, usually some 8–10 cells long. **These hormogonia are often gas-vacuolate, buoyant, and sometimes very narrow—perhaps only one-quarter of the width of a mature trichome base and can easily be mistaken for *Oscillatoria* or *Pseudanabaena*** (Subgroup 3). The motile hormogonia often aggregate into clumps. In related "traditional" botanical genera these aggregations of hormogonia often persist and give rise to macroscopic colonies.

Each hormogonium differentiates a single-pored heterocyst at one end. Usually only one heterocyst develops at the base. However, when iron and molybdenum are limiting, in addition to N, a new 1-pored heterocyst may develop from the basal vegetative cell as the previous heterocyst collapses.

The trichome elongates by cell elongation and by cell division. Cell division is restricted to the basal part of the trichome. When the phosphate concentration

and subterminal cells. These cells become unpigmented and lose most of their cytoplasm.

After inorganic or organic phosphate is added to trichomes with hairs, the most apical region of vegetative cells, below the hair, differentiates into a hormogonium (without any cell division) and the hair falls off. Hairs may sometimes form with iron-deficiency as well.

Many strains of PE-containing *Calothrix* show "chromatic adaptation." Others contain PEC, or PC only as the major phycobilin. Facultative photoheterotrophy occurs in most strains using glucose, fructose, sucrose, ribose, or 2–3 of these sugars.

There are differences in the arrangement and structure of the sheath of *Calothrix*. The sheath sometimes covers the basal heterocyst or may reach only to the most basal vegetative cell. Sheaths often contain scytonemin and range in color from golden-brown to very dark brown.

Eleven *Calothrix* strains showed 39.8–44.4 mol% G + C content of the DNA. The genome sizes of 10 of these fell into 2 groups: 7 in the range 5.07×10^9 daltons, and 3 in the range $7.75–8.58 \times 10^9$ daltons. However, a division of the genus on this basis is not consistent with any known phenotypic properties.

Most species of *Calothrix* are easily isolated, particularly those that commonly grow in alternating wet-dry environments. Hormogonia may be produced on agar medium high in phosphate when filaments from phosphate-deficient material are added, at least with suitable temperature and light. The plate should be observed with a dissecting microscope every few hours in order to remove a hormogonium as soon as possible after it has moved out of the parent sheath. Isolation can also be carried out in liquid culture, where gas-vesiculate hormogonia will tend to float to the surface.

Although most *Calothrix* strains grow in a range of laboratory media, care is needed if the organism is to develop a morphology similar to that usual in nature. Use a medium with low phosphate and low combined nitrogen. Strains that have been subcultured in a high-phosphate medium for a long time often develop strange morphologies, even if rapid growth occurs.

Differentiation of **Calothrix** *from other genera:* A number of other genera resemble *Calothrix* by having a tapered trichome. The genus *Homoeothrix*, which includes tapered forms without heterocysts, has been the

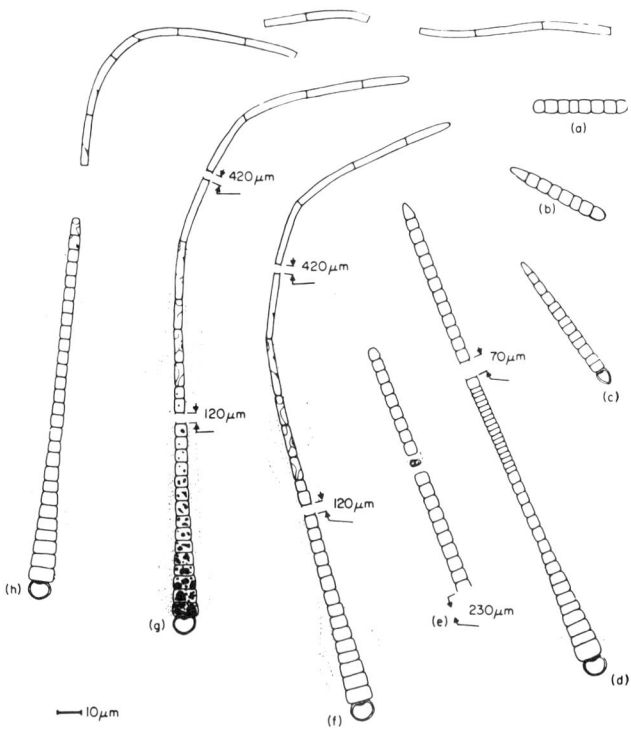

Figure 11.7. Morphological changes in batch culture of *Calothrix parietina* grown to phosphorous deficiency (**a–f**), followed by subsequent changes after addition of further phosphate (**g–h**). Granular inclusions are illustrated only in **g**; *dotted lines* indicate sheath. **a.** Stage I, hormogonium. **b.** Stage I, asymmetric hormogonium. **c.** Stage II, heterocyst formation. **d.** Stage II, trichome before hair formation. **e.** Stage II hormogonium release from Stage II trichome. **f.** Stage III, trichome with hair. **g.** Stage IV, formation of polyphosphate bodies after the addition of phosphate. **h.** Stage IV, hair cells breaking away from trichome, prior to release of hormogonia (Stage I). (Reprinted with permission from D. Livingstone and B. A. Whitton, Brit. Phycol. J. *18*: 29–38, 1983.)

is high a hormogonium develops at the trichome apex, and in some species, formation of new hormogonia takes place repeatedly. In some cases a hormogonium is separated from the rest of the trichome by cell death (i.e., a necridial cell).

Hormogonium formation and release ceases with depletion of phosphorus. In some strains the tapered end of the trichome then starts to differentiate into a multicellular hair. No cell divisions occur, merely a narrowing and elongation of several terminal

subject of a detailed study using classical botanical approaches (Komárek and Kann, Arch. Protistenk. *115:* 173–283, 1973). Some species look somewhat like a tapered *Lyngbya* (Subgroup 3), others look like species of *Calothrix* without a heterocyst. Growth of *Calothrix* strains in the presence of combined nitrogen usually leads to profound morphological changes. In *Calothrix*, heterocyst formation is suppressed in the presence of high NH_4^+ or NO_3^-, and in most cases the trichome lacks taper. If tapering does occur under these conditions, the trichome will resemble the genus *Homoeothrix*, or if taper occurs in both directions it will resemble the genus *Ammotoidea*.

Rivularia and *Gloeotrichia* are "classical" botanical genera that resemble *Calothrix*, but form macroscopic colonies, usually hemispherical or spherical. These genera have tapered trichomes so that the basal heterocyst ends are towards the inside of the colony (Figure 11.6*e,f*). Colonies apparently develop in most cases from an aggregation of many hormogonia. Almost all species in these genera form long hairs in nature. *Rivularia* lacks akinetes, *Gloeotrichia* forms them, usually one subtending the terminal heterocyst.

There are advantages in using the generic names *Gloeotrichia* and *Rivularia* for descriptive purposes in natural habitats. They should be reserved as genera, at least until the organisms have been studied in detail, using additional criteria.

Earlier in the description of Subgroup 4, Section C, a list of other "traditional" genera was provided with brief descriptions of each. These (*Sacconema, Isactis, Dichothrix, Gardnerula, Microchaete,* and others) can only be evaluated well as separate taxa at the generic level with more critical studies of cultures.

SUBGROUP 5 (=Order: Stigonematales)

This subgroup of potentially filamentous cyanobacteria exhibits a high degree of morphological complexity and differentiation. **Longitudinal and/or oblique cell divisions occur in addition to transverse cell divisions. This results in periodic true branching in all genera and often, in some species, in multiseriate trichomes (two or more rows of cells)** (See Figure 11.8, 11.9). False branching also occurs in addition to true branching in some genera. **Pit-like synapses or pore-channels occur between cells of trichomes in some genera** (probably a result of incomplete closure of septum). Heterocysts are intercalary and/or terminal and some-

times lateral. Although reproduction occurs by random breakage of trichomes, hormogonia are also formed in most strains. Akinete-like cells (or pseudoakinetes) are produced in some genera. Sheaths are present or absent.

In most cases the width of the trichomes varies greatly even within a single clone, since narrower, secondary trichomes, arising as branches of thickened primary trichomes, occur as a regular phenomenon, even in uniseriate species. Although dimensions of filaments or thalli may be great, even macroscopic, cell dimensions are similar to those of common cyanobacteria (e.g., 5–30 μm).

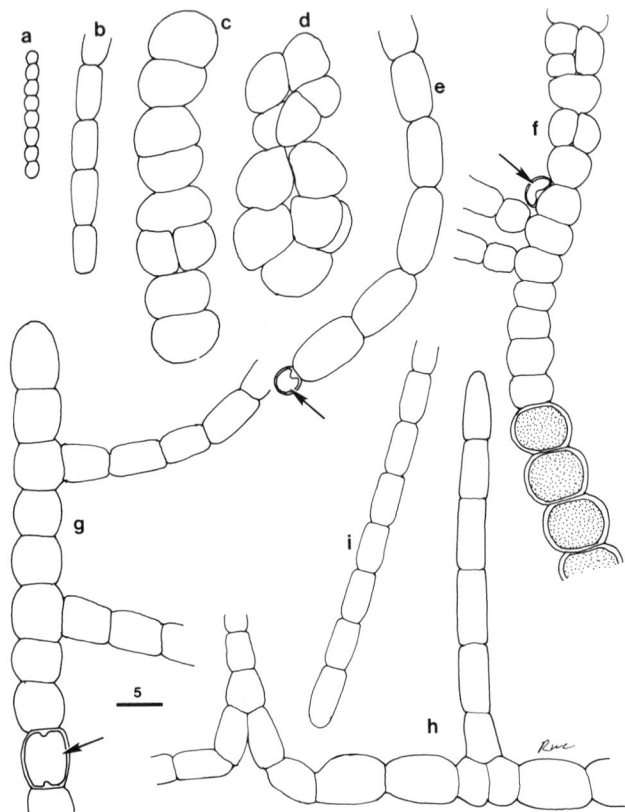

Figure 11.8. Schematic representations of genera of Subgroup 5.

a.–e. Chlorogloeopsis sp. *a.* Motile hormogonium. *b.* Stationary trichome developing from hormogonium. *c.* Trichome developing into amorphous stage (with one division in second plane). *d.* Amorphous stage resulting from divisions in three or more planes. *e.* Trichome in medium depleted of combined nitrogen. *f. Fischerella* sp. with true branches and "akinetes" (pseudoakinetes). *g.–i. Fischerella* sp. (= *Mastigocladus laminosus* from hot springs). *g.* With true branching. *h.* With reverse Y-branching. *i.* Motile hormogonium.

Arrows indicate heterocysts. *Stippled cells* are "akinetes" (pseudoakinetes). *Scale marker* (5 μm) for all drawings.

410

The mol% G + C content of the DNA ranges only from 41.9–46.3 within nine strains examined. However, these represent only two genera and probably two species as well. The genome size of the six strains ranges from 3.62–5.24×10^9 daltons. These two genera, on the basis of similarity of nucleotide sequences of 16S rRNA, lie relatively close to each other within the main branch that includes the genera of Subgroup 4 (Giovannoni, J. Bacteriol. *170:* 3584–3592, 1988).

The best known aquatic forms in culture are the various isolates of *Fischerella* (i.e., *Mastigocladus laminosus*) which typify flowing water of many hot springs below temperatures of 57–58°C. However, a large number of uncultured or poorly studied stigonematalean cyanobacteria are known from somewhat acidic, oligotrophic lakes and from fast flowing streams. An equally large number occur in moist terrestrial environments, particularly in the tropics. A few occur as marine endoliths and as lichen symbionts. None are planktonic.

Only two genera (*Fischerella* and *Chlorogloeopsis*) have been extensively characterized from axenic culture strains, but see Aziz (Nova Hedwigia *49:* 447–454, 1989) on *Stigonema*. It is certain from field observations that other very distinctive genera exist, although most have not yet been cultured. Bourrelly (Les algues d'eau douce III, 2 ed., pp. 350–387, Boudée, Paris, 1985) recognizes 35 genera within the order *Stigonematales*. Geitler (Rabenhorst's Kryptogamenflora *14:* 459–561, 1932), not recognizing the order, nevertheless, includes a similar number of genera in 5 families. Anagnostidis and Komárek (Arch. Hydrobiol. Suppl. *86:* 1–73, 1990) have made an extensive revision of Subgroup 5 (Order: *Stigonematales*) in which they include eight families and 48 genera (several not clearly defined). Because few of the tentatively large number of genera and species are known in culture, only four genera will be described here. The following simplified key includes a few of the additional distinctive genera.

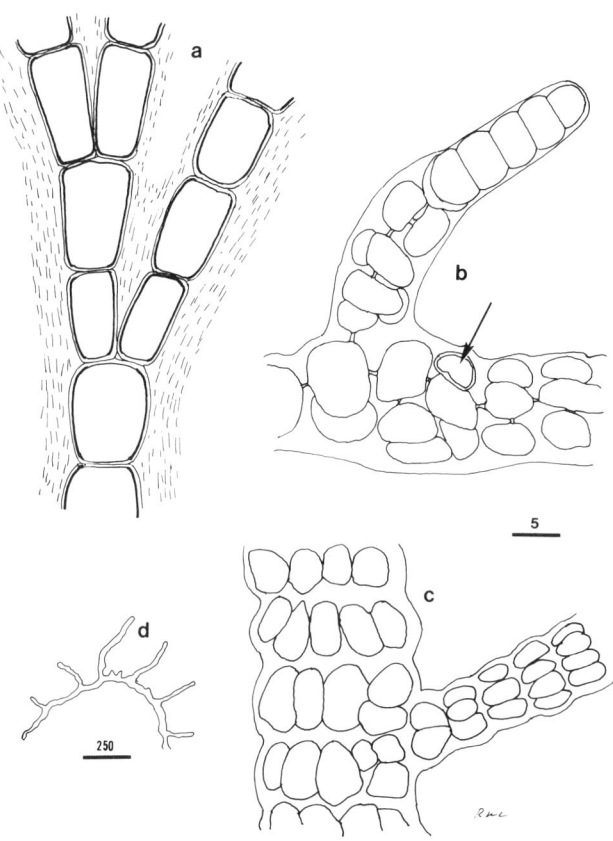

Figure 11.9. Schematic representations of genera of Subgroup 5 (continued).

a. Geitleria sp. with CaCO₃ deposition (*dashed lines*). *b. Stigonema* sp. with "pore channels" between cells derived by divisions. Enclosed hormogonium at tip of branch. *c. Stigonema* sp. *d. Stigonema* sp.: thallus at smaller scale (*marker* = 250 μm).

Heterocyst indicated by *arrow. Scale marker* (5 μm) for all except *d.*

Key to selected genera of **Subgroup 5**

I.   True branching: dichotomous or subdichotomous (calcareous deposits usually surrounding filaments)

    A.   Lacks heterocysts

        Genus *Geitleria*

    B.   With heterocysts

        Genus *Loriella*

II. True branching lateral, irregular, or multiseriate; filamentous habit often indistinct

    A.   Heterocysts as distinct lateral single cell, or terminal on short lateral branches, but not intercalary

        Genus *Nostochopsis* and *Mastigocladopsis*

      Or if marine and endolithic in carbonates

        Genus *Mastigocoleus*

    B.   Heterocysts intercalary, also sometimes terminal

      1.   Produces **hormocysts** and hormogonia, trichomes uniseriate

        Genus *Westiella*

      2.   Hormogonia may be produced but not hormocysts; trichomes uniseriate, multiseriate, or irregular

        a.   True branches, distinct or irregular, but cells not different from those of main axis

            i.   Trichome uniseriate throughout

            Genus *Hapalosiphon*

            ii. Trichome multiseriate, at least in older part

            Genus *Stigonema*

            iii. Trichome usually multiseriate with lateral branches that liberate unicells (phragmocytes)

            Genus *Cyanobotrys*

            iv. Mature trichomes divide in more than one plane, but fragment into irregular *Gloeocapsa*-like aggregates (see Subgroup 1)

            Genus *Chlorogloeopsis*

        b.   True branches with cells different dimensions from those of main axis

        Genus *Fischerella* (includes *Mastigocladus*)

Genus **Chlorogloeopsis**

The filamentous nature of this organism is often unclear, except when hormogonia are produced. Hormogonia are composed of short chains of cylindrical or barrel-shaped cells which, after ceasing motility, enlarge or swell to semi-spherical cells (Figure 11.8*a–e*). When NH_4^+ and NO_3^- are low, both intercalary and terminal heterocysts occur. Growth continues with cell divisions in more than one plane (Figure 11.8*c,d*) so that multiseriate trichomes or irregular "aggregates" develop. The filamentous nature of the organism is usually lost, however, since the growing mass of cells commonly fragments into clusters or amorphous aggregates of cells, often within a mucilaginous sheath (Figure 11.8*d*). Hormogonia arise from such aggregates. Masses of vegetative cells may also enlarge to form thicker-walled cells (pseudoakinetes). Germination of these cells takes place with division in several planes and the shedding of the extra wall layers. **Synaptic "pore channels," common in many genera of Subgroup 5, are not present.**

The mol% G + C contents of 2 strains are 42.1 and 42.9. The genome sizes are 4.20 and 5.24×10^9 daltons. Both strains included by R. Rippka et al. (J. Gen. Microbiol. *111*: 51–58, 1979) are facultative, aerobic chemoheterotrophs utilizing sucrose best, but also glucose, fructose, or ribose. PEC is synthesized by both strains, but no PE.

Chlorogloeopsis is one of the cyanobacteria most easily grown as a chemoheterotroph. A common photoautotrophic growth medium used is that of Kratz and Myers supplemented with $NaHCO_3$ (1 g·l⁻¹) (Table 11.1). Sucrose (10 mM) is included when desired. Under photoautotrophic conditions there is a progression (after inoculation) from short trichomes or hormogonia (Figure 11.8*a*) to multiseriate trichomes, to aseriate clusters or aggregates. Growth in the light in the presence of sucrose, however, did not allow the development of trichomes. Unicells or small groups of cells gave rise to larger spherical clusters of cells and eventually to larger aseriate aggregates (Figure 11.8*c,d*). Dark chemoheterotrophic growth on sucrose maintained aseriate clusters of cells and again no trichome formation (Evans et al., J. Gen. Microbiol. *92*: 147–155, 1976).

Differentiation of the genus **Chlorogloeopsis** *from other genera:* There is a close resemblance of *Chlorogloeopsis* to phases of the developmental cycle of some species of *Nostoc* (Subgroup 4, Section A). The aseriate stages of *Nostoc* are hardly distinguishable from similar phases of *Chlorogloeopsis.* The motile hormogonia are also similar. However, in **Chlorogloeopsis** divisions in more than one plane clearly occur, i.e., in hormogonia that have come to rest and in the "unicellular" and cluster phases. In **Nostoc** during aseriate stages, **cells become detached and disoriented, but two or more planes of division are not discernible.**

Although there are few criteria besides morphological characters to use at present, the thermophilic cyanobacterium referred to recently as "High temperature form (HTF) Mastigocladus" appears to resemble *Chlorogloeopsis* in most respects (Castenholz, Mitt. Int. Ver. Theor. Angew. Limnol. *21*: 296–315, 1978). It occurs in hot springs worldwide up to temperatures of 63–64°C. Hormogonia similar to those of *Chlorogloeopsis* are formed by both aseriate and unicellular aggregate cultures as well (Figure 11.8*a*). The mol% G + C content of the DNA of three strains of "HTF Mastigocladus" are 43.2, 43.5, and 44.8.

Earlier authors observing forms in nature corresponding to "HTF Mastigocladus" have simply referred to these as forms or varieties or forms of *Mastigocladus* (=*Fischerella*) *laminosus* or *Hapalosiphon laminosus. Mastigocladus laminosus,* in the strict sense, displays true branches and inverted "V" branches and is here included in the genus *Fischerella* (see Castenholz, *Bergey's Manual of Systematic Bacteriology,* Volume 3, pp. 1794–1797).

Genus **Fischerella**

The true branches of this filamentous cyanobacteria are uniseriate and composed of cells that are generally longer than broad, particularly those distal from the base. The axis (primary trichome) from which they arise is mainly uniseriate as well but may become multiseriate in part, with divisions in more than one plane. The older axes, however, are seldom more than 2–3 cells in thickness. In addition, the cells of the axes become enlarged, often semi-spherical in shape. The older cells of a main axis may become separated from each other by sheath material and may act as akinetes (pseudoakinetes). Most of the widened cells of the axial trichome, however, possess a true filamentous nature with only a peptidoglycan and cell membrane septum separating cells. **The primary axis forms when a hormogonium comes to rest, cells**

enlarge, and some cell divisions begin that are parallel to the long axis or diagonal (oblique); some of the resulting cells elongate and continue division in 1 plane to form branches (secondary trichomes) (Figure 11.8*f–i*). In some, reverse Y-branching also occurs (see Figure 11.8*h* and glossary at the end of Group 11).

The narrow **secondary trichomes** (which may taper somewhat) become progressively longer with cell elongation and continued transverse divisions. In field populations of *Fischerella* (*Mastigocladus*) *laminosus* in flowing hot springs, almost the entire mass is composed of tufts or streamers of secondary trichomes, several centimeters in length, and no branching is seen except in the prostrate attached mass of primary trichomes (main axes). In culture, at least, secondary trichomes of thermophilic *Fischerella* may eventually differentiate series of spherical, thick-walled cells that are akinete-like (pseudoakinetes) (Figure 11.8*f*). Typical cyanobacterial types of akinetes (as in Subgroup 4) are not easily recognizable.

The hormogonium of *Fischerella* is a gliding trichome composed of a few (ca. 11–16), narrow, morphologically uniform cells that are cylindrical or slightly barrel-shaped (Figure 11.8*i*). They are formed at the distal portion of branches. No heterocyst differentiation occurs until after hormogonia have ceased motility. Heterocysts of *Fischerella* (*Mastigocladus*) *laminosus* are elongate, spherical, or even compressed (shorter than broad) and are lateral, terminal, or intercalary (Figure 11.8*f,g*). Heterocysts of thermophilic strains, at least, are different from typical heterocysts of Subgroup 4 in that they possess only one additional wall layer (homogeneous type) and have densely stacked lamellar membranes (T.E.M. characteristic; see Nierzwicki-Bauer et al., Arch. Microbiol. *137:* 97–103, 1984).

Most of the information on *Fischerella* from culture has been from the thermophilic species, which is common in neutral pH and alkaline hot springs throughout the world (see Castenholz, Mitt. Int. Ver. Theor. Angew. Limnol. *21:* 296–315, 1978). However, Martin and Wyatt (J. Phycol. *10:* 57–65, 1974) have compared the physiology of six strains of stigonematalean cyanobacteria, including five strains of three species of *Fischerella*, none of which were thermophilic. Besides the true branching habit, the great diversity of form also applies to these non-thermophilic representatives. In another non-thermophilic species of *Fischerella* synaptic connections by microplasmodesmata were shown joining adjacent cells of the secondary trichomes (Thurston and Ingram, J. Phycol. *7:* 203–210, 1971).

All seven strains of *Fischerella* examined by Rippka et al. (J. Gen. Microbiol. *111:* 1–61, 1979) synthesize PEC. All strains were capable of photoheterotrophic and dark chemoheterotrophic growth, utilizing glucose, fructose, or sucrose—and in two cases, ribose.

The mol% G + C content of the DNA of the seven strains ranged from 41.9–46.3. The genome size of four strains ranged from 3.62–4.75 × 10⁹ daltons.

Although *Fischerella* (*Mastigocladus*) *laminosus* forms dominant, almost monotypic, populations in many hot springs, species of non-thermophilic *Fischerella* are generally not as conspicuous and often occur in moist subaerial habitats such as rocks and tree trunks. Marine forms appear to be rare or lacking.

The several strains used for the description of *Fischerella* by Rippka et al. (J. Gen. Microbiol. *111:* 1–61, 1979) were all thermophiles formerly ascribed to the genus *Mastigocladus*. Cultures are easily maintained in a variety of media with or without combined nitrogen (e.g., BG–11, D medium, see Table 11.1). When mature cultures are inoculated on agar-solidified medium, motile hormogonia are readily formed. Although thermophilic forms are the cosmopolitan "weeds" of hot springs, medium with a high content of ammonium (i.e., >2 mM) is usually inhibitory. Non-thermophilic species of *Fischerella* have been grown in a variety of freshwater media.

Differentiation of the genus **Fischerella** *from other genera:* The complex variety of forms or developmental stages of *Fischerella*, which may include primary and secondary trichomes, hormogonia, unicells and amorphous cell aggregates, makes identifying any single stage difficult. *Fischerella*, however, is most easily confused with the genera, *Stigonema* and *Hapalosiphon*. In *Stigonema* **the older main axis and the branches become conspicuously multiseriate;** however, the growing tips are usually uniseriate, and this condition may extend for some distance. *Fischerella,* **on the other hand, has secondary trichomes (branches) with more elongate and narrower cells than those of the primary trichome (main axis).** The older main axes may be multiseriate in part, but not more than 2–3 cells thick. *Hapalosiphon* (not described here) is defined as being uniseriate throughout with branches similar to the main axis. The limits separating these three genera are not clear-cut.

Genus **Geitleria**

Branching is pseudodichotomous or dichotomous and lateral (Figure 11.9a). **The sheath becomes heavily calcified and only the cells near the tips of trichomes are able to give rise to a lateral branch or to a false dichotomy when an apical cell undergoes an oblique division followed by further divisions of both cells.** The ultrastructure (T.E.M.) of decalcified specimens, including that of pore-channels (i.e., "pit connections") has been studied. No cultures of this genus appear to have been established. *Geitleria calcarea* is a calcified cyanobacterium common in limestone caves.

Genus **Stigonema**

Many species of this genus are extreme examples of multicellular complexity in the cyanobacteria. **Main axes with numerous branches may reach thicknesses of 1 mm** (Figure 11.9b–d). **Pore-channels (possibly incomplete closures of invaginating septa) usually occur between all cells derived by divisions (T.E.M. mainly). Branches, although initially uniseriate, may eventually become as complex as main axis.** Tips of branches may be uniseriate for the distance of several cells.

A few species of *Stigonema* have been isolated in culture (unicyanobacterial), but extensive taxonomic studies have not been made using these (see Aziz, Nova Hedwigia *49:* 447–454, 1989).

Stigonema is a freshwater or subaerial (terrestrial) genus, found commonly on moist rocks or soil (often forming turfs), or in oligotrophic, slightly acidic lakes and in some streams. Conspicuous tufted benthic mats of *Stigonema* occur in some oligotrophic lakes.

II. THE PROCHLOROPHYTES
(Order *Prochlorales*)

Unicellular or filamentous, branched or unbranched procaryotes resembling cyanophytes, i.e., **cyanobacteria,** from which they differ in that they **form chlorophylls** *a* **and** *b* **and lack accessory red or blue bilin pigments.**

There are currently two valid genera: *Prochloron,* which is unicellular and *Prochlorothrix,* which is filamentous. Further differentiation is given in Table 11.7.

Genus **Prochloron**

Unicellular, spherical, without evident mucilaginous **sheath.** Only form of reproduction so far observed is binary division, by equatorial constriction. So far, the genus is found **almost exclusively** associated as **extracellular symbionts of colonial ascidians** (chiefly didemnids) on subtropical or tropical marine shores.

Type (and only) species: *Prochloron didemni.*

Characteristics of the species: As described for the genus.

Genus **Prochlorothrix**

Unbranched trichomes of indefinite length. No cell differentiation. Gram negative. **Photosynthetic membranes lack phycobilisomes.** Cells contain **chlorophylls** *a* **and** *b.* **Phycobilin pigments are absent.** Oxygen evolving photoautotrophs with respiratory oxygen uptake in the dark. The mol% G + C of the DNA is 53 (one species, one strain).

Type (and only) species: *Prochlorothrix hollandica.*

Characteristics of the species: As described for the genus.

OXYGENIC PHOTOTROPHIC BACTERIA

GLOSSARY

Akinete: a resistant "resting" cell of some cyanobacteria that differentiates from a vegetative cell by developing a thick outer wall, enlarging, and accumulating cyanophycin, glycogen, lipids, and carotenoids.

False branch: a branch formed by slipping to one side of a section or loop of trichome through the sheath; a branch not formed by lateral division of a cell (see Figure 11.2*j*, 11.6*a*).

Filament: a chain of cells together with an investing sheath (in cyanobacteria).

Gas vacuole: an area in cells that refracts light because of its gaseous nature, often appearing red; composed of numerous gas vesicles.

Gas vesicle: a cylindrical structure with conical ends (as seen with E.M.), composed of a single protein envelope (ca. 2 nm thick), impervious to water and containing gas.

Geminate: to become doubled or paired; arranged in pairs.

Heterocyst: a specialized cell in some cyanobacteria that functions in nitrogen-fixation. It differentiates from vegetative cells by adding a new outer multilayered wall, reorganizing the thylakoid structure, and by developing polar granules (cyanophycin) adjacent to pores which connect to other cells in trichome. Appear clearer, paler, and less granular than vegetative cells.

Hormocyst: a hormogonial-like short chain of cells enveloped or encapsulated by a thick outer envelope or sheath.

Hormogonium: a fragmented, short differentiated segment of trichome that is usually motile, and apparently functions as a disseminule.

Multiseriate: arranged with more than one row of cells, as a result of cells in trichome dividing in more than one plane.

Necridial cell: a localized dead or dead appearing cell of a trichome, the possible result of regulated differentiation and lysis; the cell at which trichome fragmentation occurs.

Periphyton: biota attached to submerged surfaces; community of sessile organisms on lake and stream substrate.

Reverse Y branch: a type of branch in certain members of Subgroup 5 in which a false branch loop develops an "apical cell" and results in a branch (see Figure 11.8*h*).

Thallus: a plant-like body composed of many cells or filaments; the thallus may have a definite shape but lacks internal differentiation.

Trichome: a chain of cells without an investing sheath (in cyanobacteria).

True branch: a branch formed by lateral or oblique division of a cell in a trichome.

Tychoplankton: floating or free-living organisms in shallow water of a lake, intermingled with attached vegetation and periphyton, usually near shore.

Uniseriate: arranged in a single row or series of cells, such as in a branched or unbranched trichome.

Important Notes for Users of This Manual

Unless otherwise indicated in footnotes to tables, the meanings of symbols are as follows:

+ 90% or more of strains are positive
− 90% or more of strains are negative
d 11-89% of strains are positive
v strain instability (not equivalent to "d")
D Different reactions in different taxa (species of a genus or genera of a family)

All other symbols are defined in footnotes to tables.

Table 11.1 Composition of Freshwater Media for Cyanobacteria [a]

Ingredient	Concentration					
	Chu No. 10 (modified)	Gerloff et al.	BG-11 [b]	D Medium [c]	Allen and Arnon	Kratz and Myers
Disodium EDTA	–	–	1 [e]	–	4 [j]	–
Nitrilotriacetic acid (NTA)	–	–	–	100	–	–
Citric acid	3	3	6	–	–	165 [k]
NaNO$_3$	–	41	1500 [f]	700	–	–
KNO$_3$	–	–	–	100	2020	1000
Ca(NO$_3$)$_2$ · 4H$_2$O	40–60	–	–	–	–	25
K$_2$HPO$_4$ · 3H$_2$O	13	–	40	–	456	1000
KH$_2$PO$_4$	–	–	–	–	–	–
Na$_2$HPO$_4$	–	8	–	110	–	–
MgSO$_4$ · 7H$_2$O	25	15	75	100	246	250
CaSO$_4$ · 2H$_2$O	–	–	–	60	–	–
MgCl$_2$ · 6H$_2$O	–	21	–	–	–	–
CaCl$_2$ · 2H$_2$O	–	36	36	–	74	–
KCl	–	9	–	–	–	–
NaCl	–	–	–	8	232	–
Na$_2$CO$_3$ (H$_2$O)	20	20	20	–	–	20 (opt.)
Ferric ammonium citrate	–	–	6	–	–	–
Ferric citrate	3 or	3	–	–	–	–
FeCl$_3$	3	–	–	0.3 [h]	–	–
Fe$_2$(SO$_4$)$_3$ · 6H$_2$O	–	–	–	–	–	4
Micronutrients	– [d]	– [d]	1 ml [g]	0.5 ml [i]	– [d]	1 ml [g]
Vitamin mix	– [d]	– [d]	– [d]	– [d]	– [d]	– [d]

[a] Unless indicated, concentrations are in mg liter^{-1} of double-distilled or deionized water; see Castenholz, Methods Enzymol. *167:* 84–85, 1988 for details.

[b] pH 7.4 after cooling.

[c] Prepared as a 20-fold concentrated stock, stored at 4°C. Micronutrients and FeCl$_3$ included in stock. pH adjusted to 8.2 with NaOH before autoclaving. After cooling and clearing, pH is about 7.5. Several variations of this medium are described in Castenholz, Methods Enzymol. *167:* 84–85, 1988.

[d] Micronutrients and vitamins optional. If used, 0.5–1.0 ml of any mixture in Table 11.3.

[e] Disodium-magnesium EDTA is generally used.

[f] The nitrate concentration is often lowered.

[g] The medium generally uses A$_5$ + Co (Table 11.3).

[h] Sometimes 2–4 times this amount is used. A stock solution of 0.29 g liter^{-1} is kept at 4°C.

[i] D micro (Table 11.3).

[j] 13% ferric-sodium EDTA.

[k] Trisodium citrate dihydrate.

Table 11.2 Composition of Marine and Hypersaline Media for Cyanobacteria[a]

Ingredient	Grund[b]	F/2[b]	MN[c]	Ong et al.[d]	ASP-M[e]	Aquil[f]	Erdschreiber's[b]	Yopp et al.[g]
Disodium EDTA	2	10[j]	0.5	5	0.8[j]	—	10[j]	5[j]
Citric acid	—	—	3	—	—	—	—	—
NaNO$_3$	40	90	750	750	40–70	8.5	150	—
Ca(NO$_3$)$_2 \cdot$ 4H$_2$O	—	—	—	—	—	—	—	1,000
K$_2$HPO$_4 \cdot$ 3H$_2$O	—	—	—	—	—	—	—	—
Na$_2$HPO$_4$	4	—	—	—	—	—	40	—
NaH$_2$PO$_4 \cdot$ H$_2$O	—	5–20	20	15	7–14	0.5	—	65
MgSO$_4 \cdot$ 7H$_2$O	—	—	38	—	4,920	—	—	10,000
MgCl$_2 \cdot$ 6H$_2$O	—	—	—	—	4,040	11,030	—	10,680
CaCl$_2 \cdot$ 2H$_2$O	10	—	18	—	1,270	1,000–1,350	—	—
Na$_2$CO$_3$($\cdot$ H$_2$O)	—	—	20	—	—	—	—	—
NaHCO$_3$	—	—	—	—	168	200	—	—
Na$_2$SO$_4$	—	—	—	—	—	4,090	—	—
NaCl	—	—	—	—	23,200	24,360	—	117,000
KCl	—	—	—	—	740	695	—	2,000
KBr	—	—	—	—	—	10	—	—
NaF	—	—	—	—	—	3	—	—
Fe$_2$(SO$_4$)$_3 \cdot$ 6H$_2$O	0.2	—	—	—	—	—	—	—
Na$_2$SeO$_4$ (0.01 mM stock)	1 ml	—	—	—	—	—	—	—
NiSO$_4$(NH$_4$)$_2$SO$_4 \cdot$ 6H$_2$O (0.1 mM stock)	1 ml	—	—	—	—	—	—	—
Micronutrients	0.2 ml[h]	1 ml[k]	1 ml[l]	1 ml	1 ml[m]	0.5 ml[o]	—	1 ml[r]
Ferric ammonium citrate	—	—	—	—	—	—	—	—
Vitamin mix	0.5 ml[i]	1 ml[i]	3	—	1 ml[n]	0.5–1.0 ml[p]	0.5 ml[p]	—
Natural or artificial seawater	1000 ml	1000 ml	750 ml	877 ml	—	—	950 ml	—
Distilled/deionized water	—	—	250 ml	120 ml	1000 ml	1000 ml	1000 ml	1000 ml
Soil extract	—	—	—	—	—	—	50 ml[q]	—

[a] Unless indicated, concentrations are in mg liter^{-1} of natural or artificial seawater or distilled water; see Castenholz, Methods Enzymol. *167*: 84-85, 1988 for details.

[b] McLachlan, *In* Stein (Ed.), Handbook of Phycological Methods, Culture Methods and Growth Measurements, *p. 25.* Cambridge University Press, Cambridge, England, 1973.

[c] Rippka et al., J. Gen. Microbiol. *111*: 1, 1979.

[d] Ong et al., Science *224*: 80, 1984.

[e] See reference b; in addition to ingredients listed, 660–1320 mg liter^{-1} glycylglycine may be added as buffer for axenic cultures.

Table 11.2 *(continued)*

f See Castenholz, Methods Enzymol. *167*: 84–85, 1988 for preparation details; 30 mg liter^{-1} H$_3$BO$_3$ and 17 mg liter^{-1} SrCl$_2$ · 6H$_2$O is also added.

g See Waterbury and Stanier, *In* Starr, Stolp, Trüper, Balows, and Schlegel (Eds.), *The Prokaryotes*, Vol. 1, p. 221. Springer-Verlag, Berlin, 1981; 500 mg liter^{-1} glycylglycine is used as buffer.

h A$_5$ + Co or D Micro generally used (Table 11.3).

i Optional: mix of Ong et al. would generally be adequate (Table 11.3).

j 13% ferric-sodium EDTA.

k Optional, but if used eliminate ferric EDTA of original formula.

l A$_5$ + Co (Table 11.3).

m See reference *b* for suggested micronutrient addition; however, F/2 should be adequate (Table 11.3).

n S-3 mix (Table 11.3).

o Use PIV or F/2 (Table 11.3).

p Use Ong et al. (Table 11.3).

q See Castenholz, Methods Enzymol. *167*: 84–85, 1988 for preparation.

r Sheridan and Castenholz solution (Table 11.3).

Table 11.3 Composition of Micronutrient Solutions and Vitamin Mixes

Ingredient	Micronutrients (g liter⁻¹)					Ingredient	Vitamins (mg ml⁻¹)[f]		
	A₅ + Co[a]	D Micro[b]	Sheridan and Castenholz[c]	PIV[d]	F/2[e]		DN[g]	S-3[e]	Ong et al.[h]
H_2SO_4 (conc)	–	0.5 ml	–	–	–	Nicotinic acid	0.100	0.1	–
HCl (conc)	–	–	3 ml	–	–	PABA	0.010	0.10	–
H_3BO_3	2.86	0.5	0.5	–	–	Biotin	0.001	0.001	0.001
$MnSO_4 \cdot H_2O$	–	2.28	–	–	–	Thiamin	0.200	0.5	2.0
$MnCl_2 \cdot 4H_2O$	1.81	–	2.0	0.041	0.177	Cyanocobalamin	0.001	0.001	0.001
$ZnNO_3 \cdot 6H_2O$	–	–	0.5	–	–	Folic acid	0.001	0.002	–
$ZnSO_4 \cdot 7H_2O$	0.22	0.5	–	–	0.018	myo-Inositol	0.001	5.0	–
$ZnCl_2$	–	–	–	0.005	–	Thymine	–	3.0	–
$CuCl_2 \cdot 2H_2O$	–	–	0.025	–	–	Calcium pantothenate	0.100	0.10	–
$CuSO_4 \cdot 5H_2O$	0.08	0.025	–	0.004	0.010				
$Na_2MoO_4 \cdot 2H_2O$	0.39	0.025	0.025	0.004	0.007				
$Co(NO_3)_2 \cdot 6H_2O$	0.049	–	0.025	–	–				
$CoCl_2 \cdot 6H_2O$	–	0.045	–	0.002	0.011				
$VOSO_4 \cdot 6H_2O$	–	–	0.025	–	–				
$FeCl_3 \cdot 6H_2O$	–	–	–	0.097	1.90				
$NiSO_4(NH_4)_2SO_4 \cdot 6H_2O$	–	0.019	–	–	–				
Na_2SeO_4	–	0.004	–	–	–				
Disodium EDTA	–	–	–	0.75 (add first)	4.35				

[a] Rippka et al., J. Gen. Microbial. *111*:1, 1979.

[b] Castenholz, In Starr, Stolp, Trüper, Balows, and Schlegel (Eds.), *The Prokaryotes*, Vol. 1, p. 236. Springer-Verlag, Berlin, 1981.

[c] Waterbury and Stanier, In Starr, Stolp, Trüper, Balows, and Schlegel (Eds.), *The Prokaryotes*, Vol. 1, p. 221. Springer-Verlag, Berlin, 1981. Ni and Se have recently been added by Castenholz.

[d] Starr, J. Phycol. Suppl. *14*:1978.

[e] McLachlan, In Stein (Ed.), Handbook of Phycological Methods, Culture Methods and Growth Measurements, p. 25. Cambridge University Press, Cambridge, England, 1973.

[f] Concentrations are designed for additions of usually 1 ml liter⁻¹. Vitamin mixes are generally filter sterilized and added after the medium is autoclaved. No cyanobacteria have been shown to have a complex vitamin requirement; most have none at all.

[g] Nelson et al., Arch. Microbiol. *133*:172, 1982.

[h] Ong et al., Science *224*:80, 1984.

Table 11.4 Summary of characteristics of genera in Subgroup 1 that divide in 1 plane only (excluding *Myxobaktron*) [a]

Characteristic	Chamaesiphon (2)	Gloeobacter (2)	Gloeothece (6)	Cyanothece (7)
Mol% G + C of DNA	46.7–46.9(2)	64(1)	40.4–42.7(5)	44.2–48.6(3)
Cell diameter (μm)	3–5(m.c.)	~1.5	5–6	>3
Budding (asymmetric fission)	+	−	−	−
Motility:				
Gliding	−	−	−	−
Swimming	−	−	−	−
Defined sheath	−(+ i.n.)	+	+	−
Facultative heterotrophy:				
Dark	−	−	−	−
Photo	+(1)	−	−	−
Nitrogen fixation:				
Aerobic	ND	ND	+	+(3)
Anaerobic	−	−	+	+
Pigments:				
Phycoerythrin (PE)	+(1)	+	+	+(1)
PE w/phycourobilin	−	+	+	−(1)
Chromatic adaptation	+(1)	−	−	−(1)
Polyunsat. fatty acid content:				
High	+			
Low				
Maxiumum growth temp. (°C)	27–30	<37	35–39(2)	35–43(4)
Salt requirements for growth (i.e. higher than for fresh water)	−	−	−	+(4)
Sources:				
Fresh water	+	+	+	+(3)
"Terrestrial"		+		
Marine (or brackish)				+(2)
Hypersaline (brines)				+(2)
Hot springs (growth >45°C)				

Footnotes are at end of table

421

Table 11.4 *(continued)*

Characteristic	Synechococcus group					
	"Cyanobacterium" cluster (2)	"Cyanobium" cluster (8)	"Synechococcus" cluster (10)	marine A cluster (15)	marine B cluster (4)	marine C cluster (5)
Mol% G + C of DNA	39–40.5	65.7–71.4	47.5–56	55–62	63–69.5	47.4–49.5
Cell diameter (μm)	1.7–2.3	0.8–1.4	1–1.5(9)3(1)	0.6–1.7	0.8–1.4	1.2–2.1(5)
Budding (asymmetric fission)	–	–	–	–	–	–
Motility:						
Gliding	–	–(?)	+(1)(+/– i.n.)	–(?)	–(?)	–(?)
Swimming	–	–	–	+(4)	–	–
Defined sheath	–	–	–	–	–	–
Facultative heterotrophy:						
Dark	–	–	–	–	–	–
Photo	–	–	–	–	–	+(4)
Nitrogen fixation:						
Aerobic	ND	ND	ND	ND	ND	ND
Anaerobic	–	–	–(5)	–	–	+(1)
Pigments:						
Phycoerythrin (PE)	–	–	–	+	–	+(1)
PE w/phycourobilin	NA	NA	NA	+(11)	NA	–
Chromatic adaptation	NA	NA	NA	–	NA	+(1)
Polyunsat. fatty acid content:						
High	+(1)	+(4)	+(1)	ND	+(1)	+(3)
Low	ND		+(6)	ND		
Max. growth temp. (°C)	ND	35–37(6)	37–43(5), >53(3)	~30(15)	41(1)	39–43(2)
Salt Requirements for growth (i.e. higher than for fresh water)	–	–	–	+	+(1)	+(2)
Sources:						
Fresh water	+	+	+			
"Terrestrial"						
Marine (or brackish)				+	+	+
Hypersaline (brines)						
Hot springs (growth >45°C)			+(4)			

[a] Symbols: numerals in parentheses ( ) indicate number of strains, does not imply that all strains have been tested, if no number is given in parentheses, all strains conform; (+) in some of the culture strains only or probable; i.n. = in nature (probable occurrence of feature in natural populations); ND = not determined; – indicates negative result; NA = not applicable.

Table 11.5 Summary of characteristics of genera in Subgroup 1 that divide in 2 or 3 planes [a]

Characteristic	Gloeocapsa (6)	Microcystis (8)	Synechocystis "Low GC cluster" (5)	Synechocystis "High GC cluster" (11)	Synechocystis "Marine cluster" (2)
Mol% G + C of DNA	39.8–48.9(6)	44.2–45.4	35–37(5)	42.1–48.0(11)	30.5–31.7(2)
Cell diameter (µm)	3–30	3–8	3–7(5)	2–3(5)	2.5–4(2)
Sheath:					
Defined	+	–	–	–	–
Amorphous	–	+	+(3)	–	–
Aggregates or colonies	+	+	+(1)	+(2)	–
Motility (gliding)	+(4)	–(1)	+(3)	+(4)	–
Facultative heterotrophy:					
Dark	+(4)	–	–	+(9)	–
Photo	ND	ND	ND	ND	ND
Nitrogen fixation:					
Aerobic	ND	–	–	–	+(2)
Anaerobic	+(2)	–	–	–	+(2)
Pigments:					
Phycoerythrin (PE)	+(3)	–	+(3)	–	+(2)
PE w/phycourobilin	+(1)	NA	–(3)	NA	+(2)
Chromatic adaptation	–(1)	NA	+(2)	NA	–
Polyunsat. fatty acids:					
High	ND	+(1)	+(5)	+(5)	ND
Low	ND		+(5)		ND
Maximum growth temp. (°C)	ND	35(1)	37–39(5)	ND	32(2)
Sources:					
Fresh water	+(i.n.)	+	+(5)	+(6)	
"Terrestrial"	+(3)			+(2)	
Marine (or brackish)	+(1)			+(2)	+(2)
Hypersaline (or saline lake)				+(1)	
Hot springs (growth >45°C)	+?(1)				
"Acid" bogs	+(1)				
Toxins produced	ND	+(2)	ND	ND	ND
β-cyclocitral	ND	+(2)	ND	ND	ND
Gas vesicles	ND	+	–	–	–

[a] Symbols: numerals in parentheses ( ) indicate number of strains; does not imply that all strains have been tested; if no number is given in parentheses, all strains conform; (+) in some of the culture strains only or probable; i.n. = in nature (probable occurrence of feature in natural populations); ND = not determined; – indicates negative result; NA = not applicable.

Table 11.6 Summary of characteristics of genera of Subgroup 2 [a]

Characteristics	Chroococcidiopsis (8)	Dermocarpa (5)	Dermocarpella (1)	Myxosarcina (2)	Pleurocapsa Type I (5)	Pleurocapsa Type II (3)	Xenococcus (3)
Mol% G + C of DNA	40.2–46.4(8)	40.7–44.0(4)	45.1	42.7–44.0(2)	43.0–45.4(5)	43.0–43.3(2)	43.4–44.2(2)
Baeocyte diameter (μm)	3–4(8)	1.5–4(5)	2	2–3	2–3	2.5–3(3)	2–3(3)
Approx. number of baeocytes per mother cell	4(8)	10–~1000	60–120	4–>4	4(1)–30(3)	8–16(2)	10–~1000
Baeocytes motile	–	+	+	+	+	+	–
Vegetative cell diameter (μm):							
At time of 1st binary fission	5–6.3(8)	ND	ND	5–10	Symmetric fission 5–8(5)	Asymmetric fission <6–8(1)	ND
At time of baeocyte formation	ND	20–30(max)	10–13	ND	ND	ND	10–25(3)
Filamentous outgrowths	–	–	–	–	+(3)	+(3)	–
Facultative photoheterotrophy	+(8)	+(3)	+	+	+(4)	+(3)	+(1)
Nitrogen fixation (anaerobic)	+(8)	+(2)	–	+(1)	+(3)	+(3)	+(1)
Pigments:							
Phycoerythrin (PE)	+(7)	+(5)	+	+(2)	+(5)	+(3)	+(3)
PE with phycourobilin	+(1)	+(1)	–	+(1)	+(2)	+(1)	+(3)
Chromatic adaptation	+(1)	+(4)	+	+(2)	+(3)	+(1)	+(3)
Maximum growth temp. (°C)	30–44(7)	35–44(5)	35	35–39(2)	30–39(5)	37(2)	30–35(3)
Source:							
Fresh water	+	+(2)	?	(+)	(+)	(+)	(+)
Marine	(+)	+(3)	+	+	+	+	+(3)

[a] Symbols: numerals in parentheses ( ) indicate number of strains, does not imply that all strains have been tested, if no number is given in parentheses, all strains conform; (+) in some of the culture strains only or probable; i.n. = in nature (probable occurrence of feature in natural populations); ND = not determined; – indicates negative result.

Table 11.7 Characteristics differentiating *Prochloron* and *Prochlorothrix*[a]

Characteristic	*Prochloron*	*Prochlorothrix*
Pigments:		
Chlorophyl *a*/*b* ratio	>4	>7
Echinenon	+	−
Morphology:		
Unicellular	+	−
Filamentous	−	+
Spherical cells	+	−
Cylindrical cells	−	+
Habitat:		
Fresh water	−	+
Marine	+	−
Free living	−	+
Symbiotic	+	−
Growth conditions:		
Mineral medium	−	+
Host needed	+	−
N_2 fixation	+	−
Temperature (°C)	>25	20–30
Mol% G + C of DNA	31–41	53

[a] Symbols: −, 90% or more of strains are negative; +, 90% or more of strains are positive.

Important Notes for Users of This Manual

Unless otherwise indicated in footnotes to tables, the meanings of symbols are as follows:

+ 90% or more of strains are positive
− 90% or more of strains are negative
d 11-89% of strains are positive
v strain instability (not equivalent to "d")
D Different reactions in different taxa (species of a genus or genera of a family)

All other symbols are defined in footnotes to tables.

GROUP 12

AEROBIC CHEMOLITHOTROPHIC BACTERIA AND ASSOCIATED ORGANISMS

This group is divided into three subgroups: 1, the colorless sulfuroxidizing bacteria; 2, the iron- and manganese-oxidizing and/or -depositing bacteria; 3, the nitrifying bacteria.

SUBGROUP 1: COLORLESS SULFUR-OXIDIZING BACTERIA

The colorless sulfur bacteria comprise a very heterogeneous group of organisms which share the ability to oxidize reduced or partially oxidized inorganic sulfur compounds. The use of this ability as a taxonomic criterion has linked many genera that have no taxonomic relationship. The position has been further complicated by the results of 5S and 16S RNA analysis carried out in the last few years, which have indicated that species currently classified within some of the larger genera (e.g., *Thiobacillus*) may be only distantly related to each other, although they are (eco)physiologically similar (Figure 12.1).

Virtually all morphological forms and types of motility occur among the colorless sulfur bacteria, and representatives can be found growing over most of the pH range (pH 1.0–10.5). The main factors linking the genera known as the "colorless sulfur bacteria" are that they are Gram-negative aerobes (some may denitrify) and chemolithotrophic. They do not contain bacteriochlorophyll. Obligate and facultative chemolithoautotrophs occur, as well as chemolithoheterotrophs (Table 12.1). A few organisms classified as colorless sulfur bacteria (e.g., some *Beggiatoa* spp.) may not be true chemolithoheterotrophs, as they do not gain energy from the oxidation of sulfur compounds. They do, however, profit from the oxidation in other ways. Colorless sulfur bacteria can be found wherever reduced sulfur compounds are available (e.g., in sediments, soil, at aerobic/anaerobic interfaces in water, and at volcanic sources such as the hydrothermal vents). Natural enrichments with sometimes 10^8 cells/ml or more may be found at sulfide/oxygen interfaces and where sulfur compounds and oxygen mix. The presence of colorless sulfur bacteria may often be evident to the naked eye by the appearance of copious white deposits of sulfur, streamers, veils, rosettes, or films. Samples taken from sites exposed to light may also contain (at least below 55°C) photolithotrophic (purple) sulfur-oxidizing bacteria, many of which can also be grown as chemolithotrophs. Pellets or concentrated suspensions of the bacteria should be examined for color (red/purple, brown), and the presence of a phototroph can be confirmed by scanning the in vivo light absorption of the suspension in a spectrophotometer and checking for the characteristic (bacterio)chlorophyll peaks of purple bacteria between 850 and 1040 nm. This is especially important with pinkish or brownish pellets that may owe their color, instead, to cytochromes.

Care should be taken during the isolation and identification of the colorless sulfur bacteria because of the toxicity of potential substrates (e.g., S^{2-}, SO_3^{2-}). In addition, the obligately autotrophic species do not always grow well in the presence of organic solidifying agents such as agar, and better results may be obtained with silica gel or inorganic filters floating on mineral medium (de Bruyn et al., Appl. Env. Microbiol. *56:* 2891–2894, 1990). Supposedly pure cultures must be rechecked for contaminating phototrophs by incubating cultures anaerobically in the light under appropriate conditions (see Group 10). The obligate chemolithoautotrophs are notorious for supporting satellite populations of heterotrophs growing on excretion products and must therefore be screened for purity using dilute organic media. A further complication is that some species are not capable of denitrifying growth under anoxic conditions in pure culture because they can only reduce nitrate to nitrite, which is toxic if it accumulates, but can do so, perhaps poorly, in a mixed culture with a nitrite-reducing species.

Once the presence of phototrophs has been eliminated, preliminary separation into genera can be made on the basis of the simple key shown below. Some of the more conspicuous species can be recognized under phase contrast microscopy, but pure cultures are necessary for a sure identification of the rest, especially those in II.

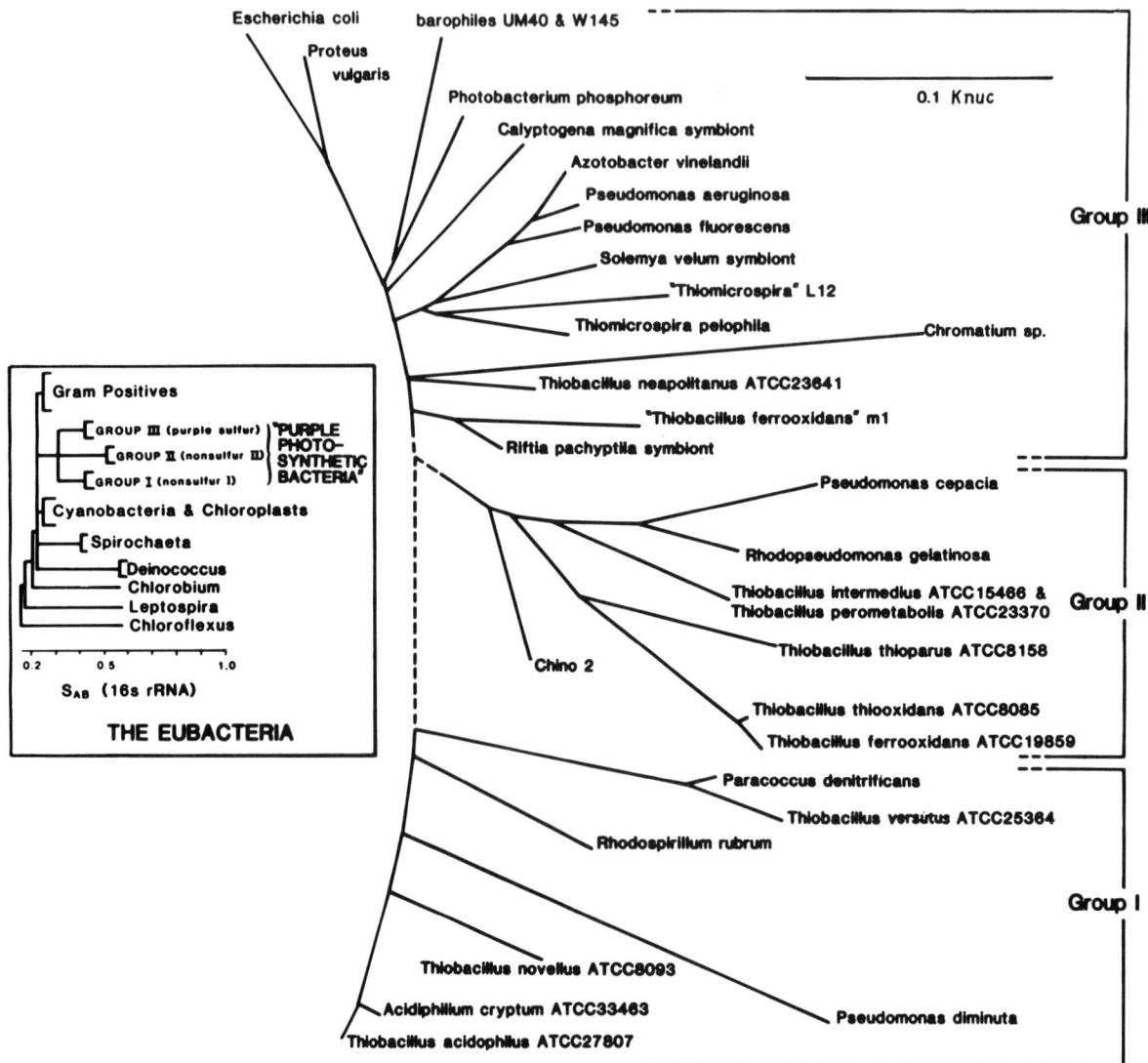

Figure 12.1. Phylogenetic relationships among various bacteria and *Acidiphilium cryptum*. The length of the branches of this "tree" (described, in part, by Stahl et al., Science *224*: 409-411, 1984) are in proportion to "evolutionary distances," i.e., the degree of dissimilarity between species based on the number of nucleotide differences in 5S RNA. The *insert diagram* summarizes the lines of eubacteria descent so far defined by partial 16S rRNA sequence characterization (Fox et al., Science *209*: 457-463. 1980). (Reproduced with permission from D.J. Lane, D.A. Stahl, G.J. Olsen, D.J. Heller and N.R. Pace, Journal of Bacteriology *183*: 75-81, 1985, ©American Society for Microbiology, Washington, D.C.)

Key to the genera of the Colorless Sulfur-Oxidizing Bacteria

The genera marked * are described in full elsewhere. Consult the pages shown in parentheses.

I. These contain intracellular sulfur, visible under the phase contrast microscope, after incubation with sulfide and oxygen.

    A. Filaments (with or without sheaths) or gonidia may move by gliding.

        Beggiatoa * (p. 491) or
        Thiothrix * (p. 493) or
        Thioploca * (p. 492) or
        "Thiospirillopsis" * (p. 493)

    B. Very large cells (5–20 μm), spherical to ovoid, may contain calcium carbonate particles. Has not been isolated in pure culture.

        Achromatium * (p. 491)

    C. Nonmotile rods embedded in gelatinous mass. Has not been isolated in pure culture.

        Thiobacterium

    D. Cells not embedded in gelatinous mass, motile.

        1. Cells cylindrical to bean-shaped with polar flagella.

            Macromonas

        2. Cells spiral with polar flagella.

            Thiospira

        3. Cells round to oval with peritrichous flagella.

            Thiovulum

    E. Cells magnetotactic.

        "Bilophococcus"

II. No intracellular sulfur is visible under the light microscope, but these may contain finely dispersed sulfur, visible under the electron microscope after staining with silver salts. Many, if not all, may deposit sulfur in the colonies after prolonged incubation in the presence of sulfide, making the colonies (yellowish) white and opaque.

    A. Chemolithoautotrophic (obligate or facultative), growth below 55°C.

        1. Rod-shaped, motile with polar flagella, or nonmotile.

            Thiobacillus

2. Cells spiral or vibroid, motile with polar flagella, or nonmotile, no stalks.

   Thiomicrospira

3. Cells vibroid, both ends tapered, stalks (0.15–0.25 μm) on either or both poles.

   "Thiodendron"

4. Cells spherical, sometimes in chains.

   Thiosphaera or
   Paracoccus * (p. 92)

B. Chemolithoheterotrophic, growth below 55°C.

   1. Rod shaped, motile with polar flagella, or nonmotile.

      Thiobacillus or
      Pseudomonas * (p. 93)

   2. Motile rods with daughter cells attached by stalks.

      Hyphomicrobium * (p. 461)

C. Chemoorganotrophs, do not gain metabolically useful energy from the oxidation of reduced sulfur compounds, but may benefit in other ways. Growth below 55°C.

   1. Neutrophilic rods, motile with polar flagella.

      Pseudomonas * (p. 93)

   2. Acidophilic rods, motile with polar or lateral flagella. Growth may be stimulated by Fe^{2+}.

      Acidiphilium

D. Chemolithoautotrophs or chemolithoheterotrophs. Growth above 55°C.

   1. Rods that may form filaments.

      Thermothrix

   2. Cells spherical, irregular, with lobes. Cannot reduce S^0.

      Sulfolobus * (p. 750)

   3. Cells spherical, irregular, with lobes. Under anaerobic conditions can reduce S^0 to H_2S with H_2.

      Acidianus * (p. 749)

   4. Nonmotile rods, also able to grow chemolithoautotrophically on hydrogen.

      Hydrogenobacter

As the subgroup is so diverse, it has been split, for simplicity, into three sections. Within each section, the genera are listed alphabetically. The sections are:

A. Morphologically conspicuous bacteria.
B. Less conspicuous sulfur-oxidizing bacteria.
C. Related bacteria.

SECTION A. MORPHOLOGICALLY CONSPICUOUS BACTERIA

The genera in this section have two characteristics in common: they are all relatively large, and they all generally contain intracellular sulfur, which is readily seen under unfiltered phase contrast as yellow globules that appear to turn pink as the focus of the phase contrast microscope is changed. A reader confronted with large, mesophilic colorless sulfur bacteria should also consult the genera *Beggiatoa* and *Thiothrix* (Group 15).

The main subdivisions between the various species in this section are shown in Table 12.2. It should be remembered that many of these species have not been isolated in pure culture, and their classification is predominantly based on morphological features.

Genus Macromonas

Cylindrical to bean-shaped cells (Figure 12.2), **motile by one polar flagellum, consisting of a tuft of flagella. Several large inclusions,** possibly of calcium carbonate, sometimes accompanied by sulfur globules. Multiplication occurs by constriction, followed by fission. No resting stages are known. The genus comprises two species; the type species has not been grown in pure culture.

Type species: *Macromonas mobilis.*

Differentiation of the species of the genus **Macromonas:** See Table 12.2.

Genus Thiobacterium

Rod-shaped cells, each containing one or more **sulfur globules.** Cells are embedded in **gelatinous masses** that are **spherical** when free-floating or are **dendroid**

when attached to a solid substrate. **Nonmotile.** No resting stages are known. Has not been grown in pure culture.

Type species: *Thiobacterium bovista.*

Genus "Thiodendron"

Vibrio-shaped cells spirally twisted and with both ends somewhat tapered, $0.4–1.0 \times 3–11$ μm; cells **bear thin threads ("stalks"** with a diameter of 0.15–0.25 μm) on either one or both cell poles (Fig. 12.3). The **stalks may be prosthecae;** they are straight or more or less flexuous, often are of considerable length, and occasionally appear branched dichotomously; they are arranged radially from a common center. The vibrio-shaped cells **may be motile** with flagella. Budding of the thin threads has been mentioned; coccoid buds appear to grow out into vibrio-shaped cells, which then develop stalks. Colonies are concentrically layered, sometimes globular and grayish to bluish white with alternating lighter and darker zonation. The appearance is similar to that of thalli of the alga *Padina pavonia.* The layering is assumed to reflect rhythmical external **deposition of granular or colloidal sulfur** as a result of H_2S oxidation, but S^0 deposition is not always observed. In the water of natural sulfur springs, colonies may grow to a size of 4 cm in 3–4 days. Gram reaction is not recorded. This organism is presumably aerobic to microaerophilic. H_2S is required for growth. This organism was originally isolated in pure culture from a sulfur spring near Chokrakskoye in Russia. It is also observed in peat mud, sand, or various freshwater samples.

Type and only species: *"Thiodendron latens."*

Genus Thiospira

Spirilla, usually with pointed ends, with **sulfur inclusions** (Fig. 12.4). **Motile** by monotrichous or polytrichous **polar flagella.** No resting stages are known. Has not been cultivated in pure culture.

Type species: *Thiospira winogradskyi.*

Differentiation of the species of the genus **Thiospira:** See Table 12.2.

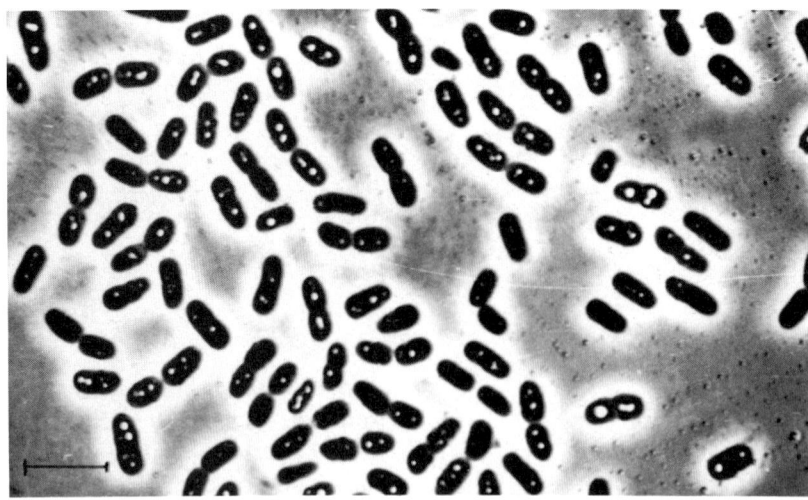

Figure 12.2. *Macromonas bipunctata* cells grown on acetate medium. *Bar,* 10 μm. (Reproduced with permission from G.A. Dubinina and M.Y. Grabovich, Mikrobiologiya *53:* 748-755, 1984.)

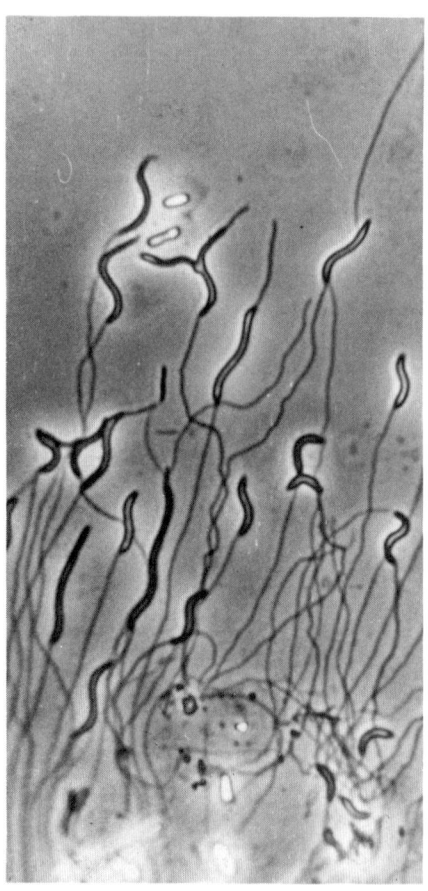

Figure 12.3. *"Thiodendron"* species observed in mud and water sample from a pond. Stalks are ~0.2 μm in diameter. (Courtesy of Dr. Hans Hippe.)

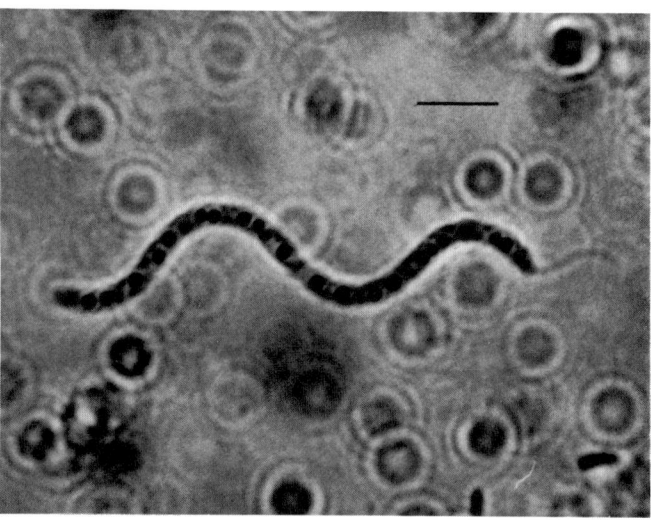

Figure 12.4. *Thiospira* series dividing cell with tuft of flagella. *Bar,* 5 μm. (Photographed by J. Klein, Delft, The Netherlands. Reproduced with permission from J.W.M. la Rivière and K. Schmidt, 1981. Morphologically conspicuous sulfur-oxidizing bacteria. *In* Starr, Stolp, Trüper, Balows and Schlegel (Editors), *The Prokaryotes. A Handbook on Habitats, Isolation, and Identification of Bacteria.* Springer-Verlag, Berlin, chap. 82).

Genus **Thiovulum**

Cells are round to ovoid, 5–25 μm in diameter (Figure 12.5). Cytoplasm is often concentrated at one end of the cell, with the remaining space being occupied by a large vacuole. Cytoplasm normally contains **ortho-rhombic sulfur inclusions,** sometimes concentrated at one end and sometimes filling cells almost completely. Cells are **strongly motile by peritrichous flagella;** forward movement is accompanied by rotation around the long axis. Cells are characterized by the presence of one polar fibrillar organelle, visible in thin sections by electron microscopy; its function is not known. No resting stages are known. **Multiplication is by constriction followed by fission.** Gram negative. Microaerophilic and catalase negative. No pure cultures are available.

Type species: *Thiovulum majus.*

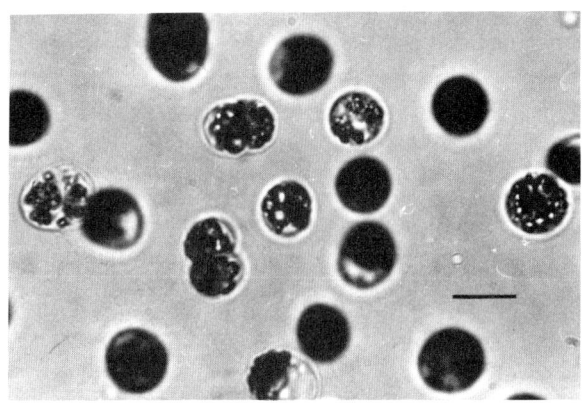

Figure 12.5. *Thiovulum majus* swarming cells showing sulfur inclusions. Note the pair of dividing cells. *Bar,* 10 μm.

subdivided according to morphology, environmental requirements, and obligate or facultative autotrophy. Some of the key characteristics of the species within these genera are summarized in Tables 12.3 and 12.4, but the species are most readily differentiated by means of a key. Most of the facultative autotrophs can grow on a wide range of organic compounds, but only the few that may be of diagnostic use will be listed here. A wider view of the heterotrophic capabilities of the facultative autotrophs can be found in the relevant chapters in *Bergey's Manual of Systematic Bacteriology.*

SECTION B. LESS CONSPICUOUS SULFUR-OXIDIZING BACTERIA

This section contains the better-known colorless sulfur bacteria, all of which have been isolated and studied in pure culture. The section is quite large but is readily

Key to the species of the genera **Thermothrix, Thiobacillus, Thiomicrospira,** *and* **Thiosphaera:**

For growth temperatures and mol% G + C, see Tables 12.2 and 12.3.

1. Rod-shaped cells.

   A. Neutrophilic; can begin growth at pH 6-8 but not at pH 3.

      1. Optimum growth temperature 73°C; facultatively chemolithoautotrophic; forms long filaments under oxygen limitation.

         Thermothrix thiopara (type species).

      2. Obligate chemolithoautotrophs, being dependent on CO_2 and reduced sulfur compounds for biosynthetic carbon and energy.

         a. Mesophilic, no growth above 42°C.

            i. Oxidizes thiocyanate; final pH in liquid thiosulfate medium 3.5-4.5; can reduce nitrate to nitrite (but no further) under anaerobic conditions.

            Thiobacillus thioparus (type species)

ii. Does not oxidize thiocyanate; final pH in liquid thiosulfate medium > 3.0; strictly aerobic, does not reduce nitrate under anaerobic conditions; the mol% G + C of the DNA is 52-57.

Thiobacillus neapolitanus

b. Can use oxygen or, under anoxic conditions, nitrate, nitrite, or N_2O as terminal electron acceptor.

Thiobacillus denitrificans

c. Growth above 42°C.

i. Final pH in liquid thiosulfate medium 4.5-5; produces tetrathionate from thiosulfate before growing on the former; can reduce nitrate to nitrite (but no further) under anoxic conditions.

Thiobacillus tepidarius

3. Facultatively chemolithoautotrophic, can grow with CO_2 and reduced sulfur compounds or organic compounds as carbon and energy sources.

a. Grow efficiently as chemolithotrophs or chemoorganotrophs, can grow autotrophically on formate.

i. Strictly aerobic; can grow on tetrathionate; some strains require biotin; nonmotile.

Thiobacillus novellus

ii. Able to denitrify under anaerobic conditions if provided with organic compounds or formate, does not grow on reduced sulfur compounds in the absence of oxygen; nonmotile or motile by means of polar flagella.

Thiobacillus versutus

b. Grow slowly and poorly on reduced sulfur compounds; grow best as mixotrophs.

i. Strict aerobe; can give a pH of 1.9–2.2 in liquid thiosulfate medium; growth on thiosulfate stimulated by glucose but not by pentoses; motile by single polar flagellum.

Thiobacillus intermedius

ii. Strict aerobe; can reduce the pH of thiosulfate medium supplemented with yeast extract or fructose to 2.8; pentoses stimulate growth on thiosulfate; motile by single polar flagellum.

Thiobacillus perometabolis

iii. Aerobic or can reduce nitrate to nitrite under anoxic conditions; requires thiosulfate for growth on single carbon compounds; produces tetrathionate during growth on thiosulfate and yeast extract.

Thiobacillus delicatus

iv. Aerobic or can reduce nitrate to nitrite under anoxic conditions. Heterotrophic growth on complex organic compounds only, does not grow on sugars, organic acids, or methylamine.

Thiobacillus aquaesulis

     v. Aerobic and can also grow heterotrophically in the absence of oxygen by means of denitrification; moderately halophilic with optimum growth at 0.25 M NaCl; extremely halotolerant with growth occurring at 3.9 M NaCl; may produce extracellular slime.

     "Thiobacillus thyasiris"

B. Acidophilic; can grow at pH values below 3.0; cannot grow at pH 7.0.

   1. Obligately chemolithoautotrophic.

     a. Can grow autotrophically using reduced sulfur compounds or ferrous iron as the source of energy. Does not grow at NaCl concentrations higher than 1%.

     Thiobacillus ferrooxidans

     b. Can grow autotrophically using reduced sulfur compounds, especially ores, as the source of energy. Growth on elemental sulfur or ferrous iron poor. Halotolerant, can grow in concentrations up to 3.5% NaCl.

     Thiobacillus prosperus

     c. Cannot oxidize ferrous iron.

       i. May generate pH values below 1.0 while growing on sulfur; motile with a single polar flagellum.

       Thiobacillus thiooxidans

       ii. pH can fall to 2.0 during growth on sulfur; motile by a tuft of polar flagella; adhesion to surfaces by means of a glycocalyx.

       Thiobacillus albertis

   2. Facultatively chemolithoautotrophic, capable of growth on reduced sulfur compounds and organic compounds.

     a. Able to grow autotrophically on elemental sulfur or tetrathionate but not on H_2S, metal sulfides, thiosulfate, or ferrous iron.

     Thiobacillus acidophilus

     b. Can grow autotrophically on H_2S, metal sulfides, and elemental sulfur but not on ferrous iron.

     Thiobacillus cuprinus

II. Cells vibroid to spiral.

A. Obligately chemolithoautotrophic, requiring reduced sulfur compounds and CO_2.

1. Strictly aerobic; maximum specific growth rate on thiosulfate 0.3 h^{-1}; cells motile by means of a single polar flagellum; diameter of cells 0.2–0.3 μm; may require vitamin B$_{12}$; requires 1.5–3.0% NaCl for growth.

   Thiomicrospira pelophila (type species)

2. Microaerophilic, inhibited by oxygen concentrations greater than 0.5%; denitrification enzymes constitutive; nonmotile; has no requirement for NaCl.

   Thiomicrospira denitrificans

3. Strictly aerobic; maximum specific growth rate on thiosulfate 0.8 h^{-1}; cells motile by means of a single polar flagellum; diameter of cells 0.5 μm; requires 0.5% NaCl for growth.

   Thiomicrospira crunogena

III. Cells coccoid, as single cells, pairs, or chains.

   A. Facultatively chemolithotrophic, able to grow autotrophically on reduced sulfur compounds, heterotrophically and mixotrophically.

   1. Able to grow aerobically or while denitrifying on reduced sulfur compounds, hydrogen, and organic compounds. Sulfur may be produced during denitrifying growth on sulfide or thiosulfate. Denitrifying enzymes constitutive. Cells pleomorphic during rapid heterotrophic growth.

   Thiosphaera pantotropha (type species)

Genus Thermothrix

Rod-shaped cells are usually 0.5–1.0 × 3–5 μm (Fig. 12.6). **Filamentous cells are produced when the oxygen concentration limits the rate of growth** (Fig. 12.7). Rod-shaped cells are motile by a single polar flagellum. **Facultatively anaerobic and facultatively chemolithotrophic,** using an oxidative type of metabolism. The temperature range is 40–80°C, and the pH range is 6.0–8.0. Inorganic sulfur compounds and organic compounds can be used as electron donors. Either oxygen or nitrate can be used as the terminal electron acceptor. **Gram negative.** No resting stages are known. **The rod-shaped growth form is easily confused with *Thiobacillus* spp.**

Type species: *Thermothrix thiopara.*

Genus Thiobacillus

Small, **Gram-negative, rod-shaped cells** (~0.5 × 1.0–4.0 μm) with **some** species **motile by** means of **polar flagella.** No resting stages are known. **Energy** is derived **from the oxidation of** one or more **reduced sulfur compounds,** including sulfides, sulfur, thiosulfate, polythionates, and thiocyanate. **Sulfate is the end product** of sulfur compound oxidation, but sulfur, sulfite, or polythionates may be accumulated, sometimes transiently, by most species. One species also derives energy from oxidizing ferrous iron to ferric iron. **All** species can **fix carbon dioxide by** means of the **Benson-Calvin cycle and are capable of autotrophic growth; some** species are **obligately chemolithotrophic,** while **others** are **also** able to **grow chemoorganotrophically.** The genus includes **obligate aerobes and facultative denitrifying types,** and its species exhibit pH optima of 2–8 with temperature optima of 20–43°C. Different species exhibit mol% G + C of the DNA of 50–68%. Distribution is seemingly ubiquitous in marine, freshwater, and soil environments, especially where oxidizable sulfur is abundant (e.g., sulfur springs, sulfide minerals, sulfur deposits, sewage treatment areas, and sources of sulfur gases, such as H$_2$S from sediments or anaerobic soils).

Type species: *Thiobacillus thioparus.*

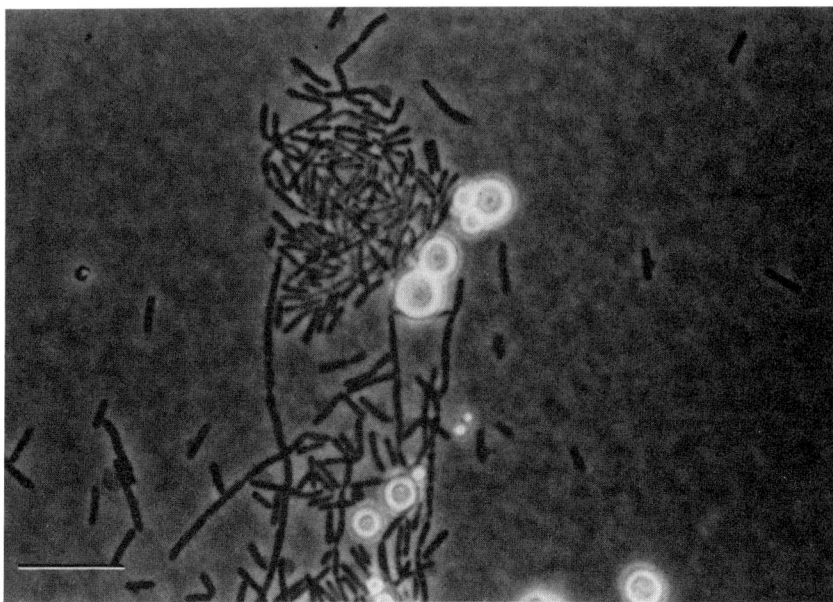

Figure 12.6. *Thermothrix thiopara* during oxidation of thiosulfate, with extracellular deposition of elemental sulfur (spherical granules) shown. The culture was highly aerated, thus producing rod-shaped cells and cell chains but no cell filaments.

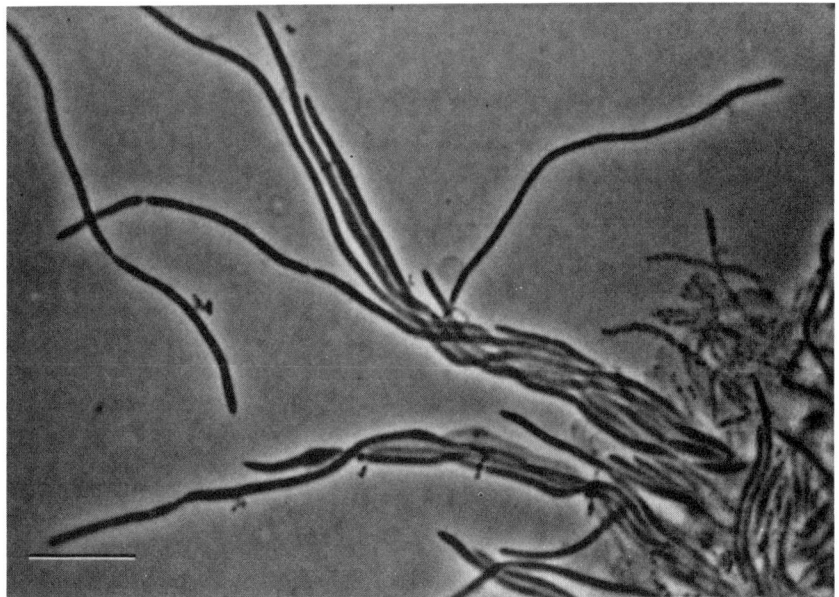

Figure 12.7. *Thermothrix thiopara* cell filaments produced under oxygen-limited growth conditions in nutrient broth. In well-aerated cultures, filaments are not produced. Phase-contrast micrograph. *Bar,* 10 μm.

Differentiation of the species of the genus **Thiobacillus:** See Table 12.3.

Genus **Thiomicrospira**

Small, **spiral-shaped cells** forming long screws or portions of a turn. Cells are 0.2–0.3 μm in diameter and 1–2 μm long; they are sometimes seen in spirals of individual cells up to 30 μm long. Gram negative. Motile by means of a polar flagellum or are nonmotile. No resting stages are known. **Metabolism is respiratory. Energy is derived from** the **oxidation of** one or more reduced or partially **reduced inorganic sulfur compounds,** including sulfide, elemental sulfur, and thiosulfate. The final oxidation product is sulfate, but elemental sulfur may accumulate in the medium. **Chemolithotrophs** able to use carbon dioxide as their major or only carbon source. At least one strain requires vitamin B$_{12}$. Metabolic properties are very similar to those of the genus *Thiobacillus*. The mol% G + C of the DNA varies from 36 to 44 (T_m).

Type species: *Thiomicrospira pelophila*.

Differentiation of the species of the genus **Thiomicrospira:** See Table 12.3.

Genus **Thiosphaera**

Cells are **coccoid** (but on rich media may fail to divide) and **occur singly, in pairs, and as a chain** (Figure 12.8). Gram negative. No resting stages are known. **Metabolism is respiratory,** with cells able to use oxygen and/or nitrate, nitrite, or a nitrogen oxide as the terminal electron acceptor. **Facultative chemolithoautotrophs** able to **utilize reduced sulfur compounds as the energy source** for growth. CO$_2$ can serve as the source of carbon. Oxidase and catalase positive. The mol% G + C of the DNA is 66 (T_m).

Type species: *Thiosphaera pantotropha*.

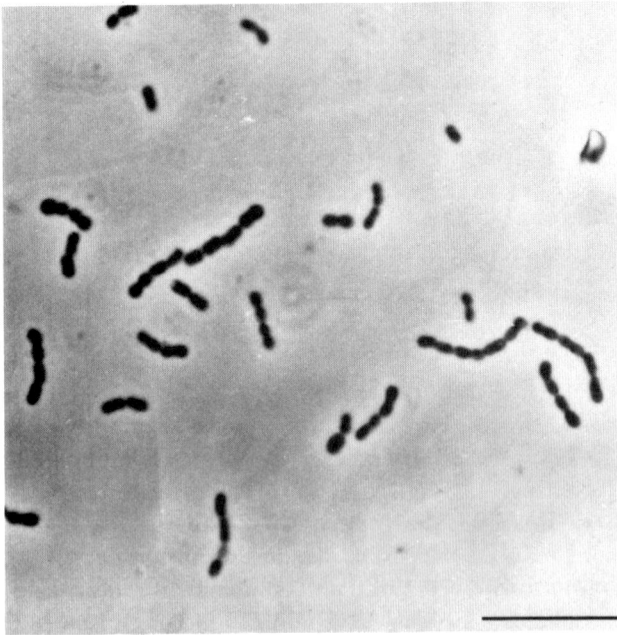

Figure 12.8. *Thiosphaera pantotropha* cells grown on a rich medium to promote chain formation. Phase-contrast micrograph. *Bar,* 10 μm.

SECTION C. RELATED BACTERIA

The members of the genus *Acidiphilium* are chemoorganotrophic, straight rods with rounded ends, which do not oxidize reduced sulfur compounds. They are mesophilic and grow in the pH range between 2.0–5.6. The differentiation of the various species is outlined in Table 12.5. *Acidiphilium* species have frequently been found in close association with *Thiobacillus ferrooxidans,* and their presence in supposedly axenic cultures is now believed to have given rise to some of the reports of facultatively autotrophic *Thiobacillus ferrooxidans* strains, although some of these may have been *Thiobacillus acidophilus.*

Genus **Acidiphilium**

Straight rods with rounded ends. Strains vary from 0.3 to 1.2 μm in diameter and from 0.6 to 4.2 μm in length. **Gram negative.** Motile by means of a polar flagellum or by means of two lateral flagella. A few strains are nonmotile. **Endospores are not formed. Aerobic,** weakly catalase positive, and **acidophilic.** Grows in the pH range of 2.5–5.9 but not at pH 6.1. Some strains grow at pH 2.0. **Mesophilic.** Growth is slow below 20°C, is fastest at 31–41°C, and is absent at 47°C. The cells die rapidly at 67°C. The mol% G + C of the DNA is 63–70 (T_m, Bd, Ez). **Chemoorganotrophic.** Upon initial isolation, however, most strains **will not grow in concentrations of peptones and extracts customarily employed in organic media.** This organism grows in 0.01% trypticase or 0.005–0.05% yeast extract, especially if 0.1% glucose is present, and will not grow with elemental sulfur, inorganic sulfur compounds, or Fe^{2+} as the source of energy, but weak oxidation of elemental sulfur may occur gratuitously in some strains as cells grow as chemoorganotrophs, and Fe^{2+} may stimulate growth. In inorganic media, growth is due to organic impurities in reagents and glassware and from the atmosphere. In agar media with dilute organic substrates (pH 3) after 2 weeks at 30°C, surface colonies are smooth, opaque, and white, pink, or light brown and are 0.5–2 mm in diameter. In ferrous sulfate-agarose medium used to cultivate *Thiobacillus ferrooxidans* and lacking added organic material, subsurface colonies of *Acidiphilium* are barely visible to the unaided eye and are smooth, especially when near a colony of *T. ferrooxidans,* but surface colonies may be lobate and 0.3 mm or less in diameter. *Acidiphilium* is common in acidic mineral environments such as pyritic mine drainage, pyritic coal refuse, and copper and uranium mine tailings. It may be isolated, unwittingly, with *Thiobacillus ferrooxidans* in Fe^{2+} enrichment cultures; thus, it is a common contaminant in these cultures.

Type species: *Acidiphilium cryptum*.

Differentiation of the species of the genus **Acidiphilium:** See Table 12.5.

SUBGROUP 2: IRON- AND MANGANESE-OXIDIZING AND/OR -DEPOSITING BACTERIA

Bacteria that deposit iron and/or manganese oxides. Organisms of diverse morphology mostly in characteristic aggregates forming easily recognizable structures stained by ferric or manganese oxides. This is an environmentally important group of mainly noncultivable bacteria classified mainly on morphology. In addition to iron bacteria that form microscopically obvious deposits, there are many bacteria capable of oxidizing Mn (II) to Mn (IV), which belong to well-known heterotrophic taxa. They are not included in the key below. Bacteria that are involved in the reduction of oxidized iron and manganese compounds are not treated specifically in this chapter. These bacteria are Gram positive, Gram negative, or wall-less. **Ferric and manganese oxides are precipitated within extracellular structures: capsules, sheaths, stalks, and holdfasts.** Oxides are easily recognizable under the light microscope and serve as an ultimate character to assign organisms to the group. Minerals formed might serve as an aid in the identification of deposits. Iron bacteria produce ferrihydrite $(2,5Fe_2O_3 \cdot 4,5H_2O)$. Manganese-oxidizing bacteria form laminated manganese oxides as vernadite or associations of Fe-vernadite with ferroxhite. *"Metallogenium"* produces ε-MnO_2, akhtenskite. Magnetotactic bacteria form magnetite within the cells in the form of magnetosomes—crystallites enveloped by a membrane, which are aligned in chains. In addition to magnetite in sulfide-rich sediments, magnetic iron sulfides such as greigite (Fe_3S_4) are formed. Most of the iron bacteria are nonmotile when covered by oxides (an exception is *Ochrobium*) but possess motile stages. Most of the iron bacteria are microaerophiles and develop at very low O_2 levels, but some grow at atmospheric levels of O_2. Growth is dependent on the E_h - pH domain of the product stability. Metabolic groups that are recognized are: (a) acidophilic chemolithotrophs that oxidize sulfides and produce Fe^{3+}, which is partly hydrolyzed and precipitates; (b) microaerophilic chemolithotrophs at neutral pH (e.g., *Gallionella*) and chemoorganotrophs that precipitate iron oxides in stalks or sheaths; (c) chemoorganotrophs that utilize ferrous and manganese compounds to scavenge toxic oxygen products; (d) chemoorganotrophs that precipitate oxides within the glycocalyx by physicochemical reaction independent from metabolism. Some iron bacteria are chemolithotrophic, but most are oligotrophs utilizing simple organic compounds at low concentrations. At high organic carbon concentration, oxide formation usually does not occur. Iron bacteria inhabit three different biotopes according to E_h - pH domains of Fe^{2+} stability: (a) acidiphilic chemolithotrophs at high E_h and low pH, oxidizing sulfide minerals in mine waters or ore bodies; (b) microaerophilic chemolithotrophs and organotrophs at moderate E_h and close to neutral pH, which are represented mainly by aquatic bacteria developing at the chemocline region or on the surface of sediments and in iron-bearing waters; (c) chemoorganotrophs decomposing organic compounds of iron at high E_h and neutral pH (*"Siderocapsaceae"*). Manganese-oxidizing bacteria develop at high E_h and around pH 8 depending on the upper limit of the $MnCO_3$ domain of stability. Iron bacteria are quite common in continental bodies of water, wetlands, and wells in the ecotone between reductive and oxidative environments. Magnetotactic bacteria develop at low E_h and neutral pH.

Key to the identification of iron- and manganese-oxidizing and/or -depositing bacteria

The genera marked * are described in full elsewhere. Consult the pages shown in parentheses.

I.   Bacteria that deposit iron or manganese oxides externally to the cell. Unicellular bacteria.

   A. Cells without appendages, free from oxides in acidic environment, might oxidize sulfides.

1. Straight rods of pseudomonad type, chemolithotrophic.

   Genus *Thiobacillus*

2. Curved rods

   Genus *"Leptospirillum"*

3. Irregular cocci, extremely thermophilic

   Genus *Sulfolobus* (p. 750) *
   Genus *Acidianus* (p. 749) *
   Genus *Metallosphaera* (p. 749) *

B. Cells with appendages

   1. Prosthecate bacteria with buds on the end of prosthecae

      Genus *Hyphomicrobium* (p. 461) *
      Genus *Pedomicrobium* (p. 462) *

   2. Nonprosthecate stalked bacteria, iron oxides deposited within the stalks

      a. Stalks twisted

         Genus *Gallionella*

      b. Stalks in rosettes

         Genus *Planctomyces* (p. 465) *

C. Cells without appendages, which deposit iron and/or manganese oxides within glycocalyx or on surface of cell.

   Family *"Siderocapsaceae"*

II. Organisms without recognizable cell wall, which deposit manganese oxides in form of "spiders," "leaves," etc.

   A. Manganese oxides are deposited in wavy filaments radiating from the center.

      Genus *"Metallogenium"*

   B. Manganese oxides are deposited in flat sheets

      Genus *"Caulococcus"*
      (not included in this manual)

   C. Manganese oxides are deposited as "lily of the valley" leaves

      Genus *"Kusnezovia"*
      (not included in this manual)

III. Filamentous bacteria with the sheaths impregnated by oxides

A. Filaments immotile, but flagellated unicells may be produced

   Genus *Leptothrix* (p. 478) *

B. Filaments motile by gliding, iron oxide-impregnated thin slime threads are produced

   Genus *Toxothrix* (p. 498) *

C. Spiral filaments formed by the chain of cells, rapidly disintegrating

   Genus *Lieskeela* (p. 479) *

IV. Bacteria that deposit iron minerals in the cells as magnetosomes

   Group Magnetotactic bacteria

Genus **Gallionella**

Cells are kidney shaped, 0.5–0.7 × 0.8–1.8 μm. Cells **secrete colloidal ferric hydroxide** (ferrihydrite) from the concave side without any organic matrix, **forming twisted,** when sessile, or **nontwisted stalks** 0.3–0.5 μm in width and up to 400 μm in length. **Stalks consist of a bundle of numerous fibers about 2 nm in diameter. Stalks dissolve completely in reducing agents and weak acids.** The shape and structure of the stalks are the main diagnostic features of *Gallionella*. **Cells always lie apically at the end of the stalk,** perpendicular to its axis. **Multiplication occurs by transverse binary fission causing dichotomous branching of the stalk** (Figures 12.9 and 12.10). Gram negative. **Motility by means of single polar flagellum. Strictly aerobic and microaerophilic. Only Fe^{2+} serves as electron donor.** Chemolithoautotrophic. **CO_2 is a carbon source** assimilated via Calvin cycle. Cell yield is 1 g dry weight per 150 g ferrous iron oxidized. Manganese is not oxidized. Typically found in oligotrophic ferrous iron-bearing waters; optimally E_h +200 to +300 mV, O_2 content about 1%, temperature 17° and lower, Fe^{2+} 5–25 mg/l, CO_2 more than 150 mg/l. Depend on the stability of Fe^{2+} at pH 6. This is one of the most important iron bacteria, forming large masses of ferrihydrite in bodies of water and water supply systems.

Type (and only) species: *Gallionella ferruginea.*

Characteristics of the species: As described for the genus.

Genus **"Leptospirillum"**

Vibrios or spiral-shaped cells. Gram negative. **Motile by means of single polar flagellum.** Cells multiply by fission and are 0.2–0.4 × 0.9–1.1 μm. **Obligate chemolithotrophic cells use Fe^{2+} as energy source. Utilize sulfides syntrophically with thiobacilli and are aerobic.** Acidophilic pH range is 1.5–4.0 (optimum 2.5–3.0). Some strains are moderately thermophilic. Widely distributed in sulfide ore deposits where it might predominate over *Thiobacillus* because of higher affinity to ferrous iron.

Type (and only) species: *"Leptospirillum ferrooxidans."*

Characteristics of the species: As described for the genus.

The Magnetotactic Bacteria

Magnetotactic bacteria are defined as cells whose swimming direction is influenced by the direction of the (geo)magnetic field. This phenetic assemblage includes morphologically diverse procaryotes. They all appear to be Gram negative, motile by means of flagella, and microaerophilic. They are morphologically and metabolically diverse. The permanent magnetic character of magnetotactic bacteria is caused by magnetosomes, enveloped single crystals of the iron oxide magnetite of crystal size 400–1000 Å. In unstained cells they are easily visualized by transmission electron microscopy. In magnetotactic bacteria from sulfide-bearing waters, iron sulfide greigite is identified. Nonmagnetic ferrihydrite appears in cells blocked in magnetite formation. Size and shape of magnetosomes are genetically

441

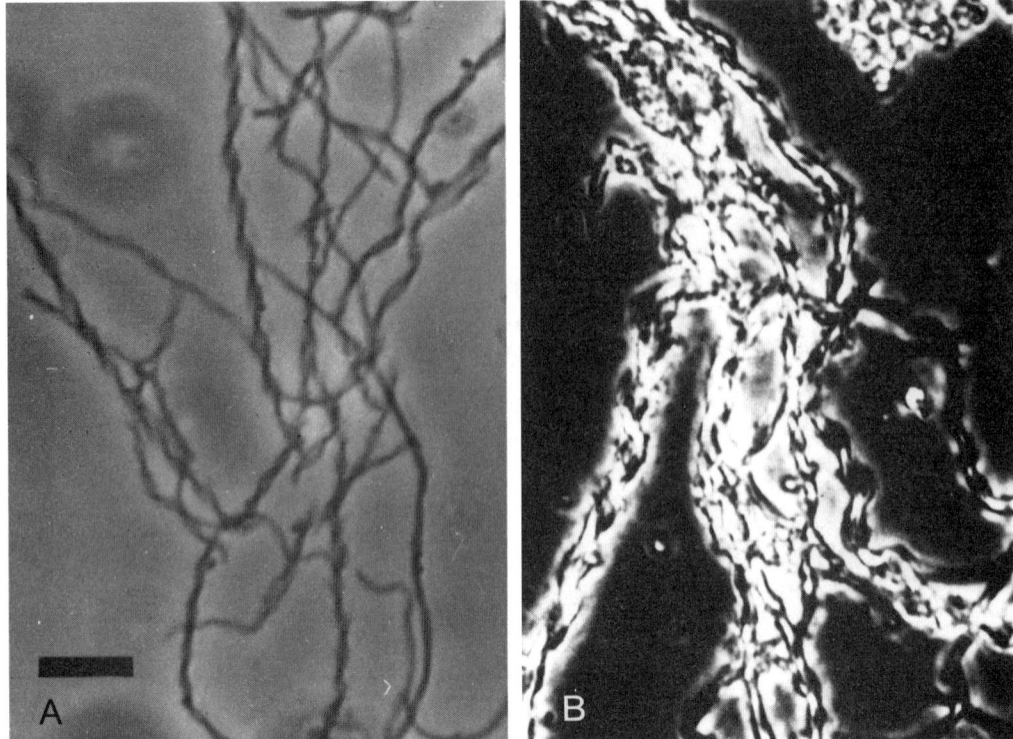

Figure 12.9. Stalks of *Gallionella ferruginea* showing *"primary stalks"* (A, Fe(III) content of 4 X 10^{-8}μg Fe (III) per μm stalk length (*bar,* 10 μm)) and *"secondary stalks"* (B, Fe (III) content of 25 x 10^{-8}μg Fe (III) per μm stalk length)

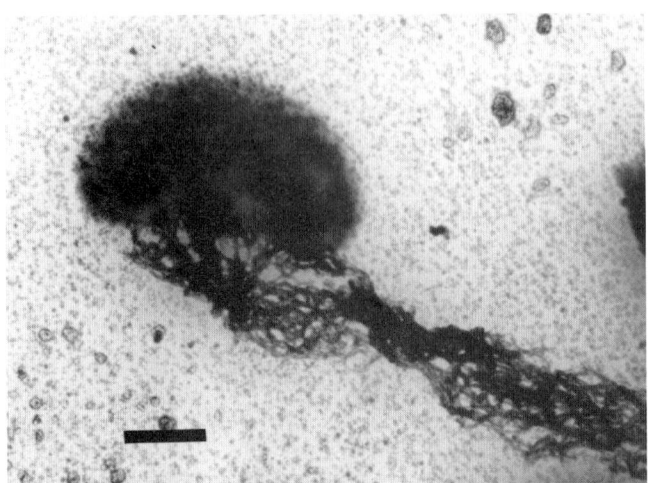

Figure 12.10. Apical cell, region of stalk secretion, and ultrastructure of the stalk of ongrowing *Gallionella ferruginea* from the drain pipe from which strain BD was isolated. Four electron-dense regions indicate PHB granules. (Reproduced with permission from H.H. Hanert, Archives of Microbiology *60:* 348-376, 1968, ©Springer-Verlag, Berlin.)

determined, and they are uniform in shape and arrangement within the cell. Only one morphological type of magnetosome is found within the culture (Figure 12.11); however, the number may vary with culture conditions, iron, and oxygen supply. Magnetotactic behavior of the cells directs their downward swimming to the north in the Northern Hemisphere and to the south in the Southern Hemisphere.

Representatives of the group: *Aquaspirillum magnetotacticum* (helical cells), *"Bilophococcus magnetotacticus"* (spherical cells).

Genus *"Metallogenium"*

Cells are coccoid, 0.2–1.5 μm, and usually in clusters. **Sprout with tapering filaments, 0.2–0.02 μm in diameter** and about 10 μm in length, **heavily encrusted by manganese dioxide** (Figures 12.12 and 12.13). Life cycle includes cocci and cocci with tapering filaments radiating from the center and encrusted by oxides. The "spider" appearance of *"Metallogenium"* colonies makes them easily recognizable in this stage. Cells multiply by budding. Cocci may also be found at the end of filaments. Filaments are straight in liquid and irregularly bent in viscous medium. Old microcolonies are heavily encrusted by MnO_2, and in this stage the organism is not recognizable. As microfossil *Eoastrion, "Metallogenium"* is known in deposits more than 2 Ga. old.

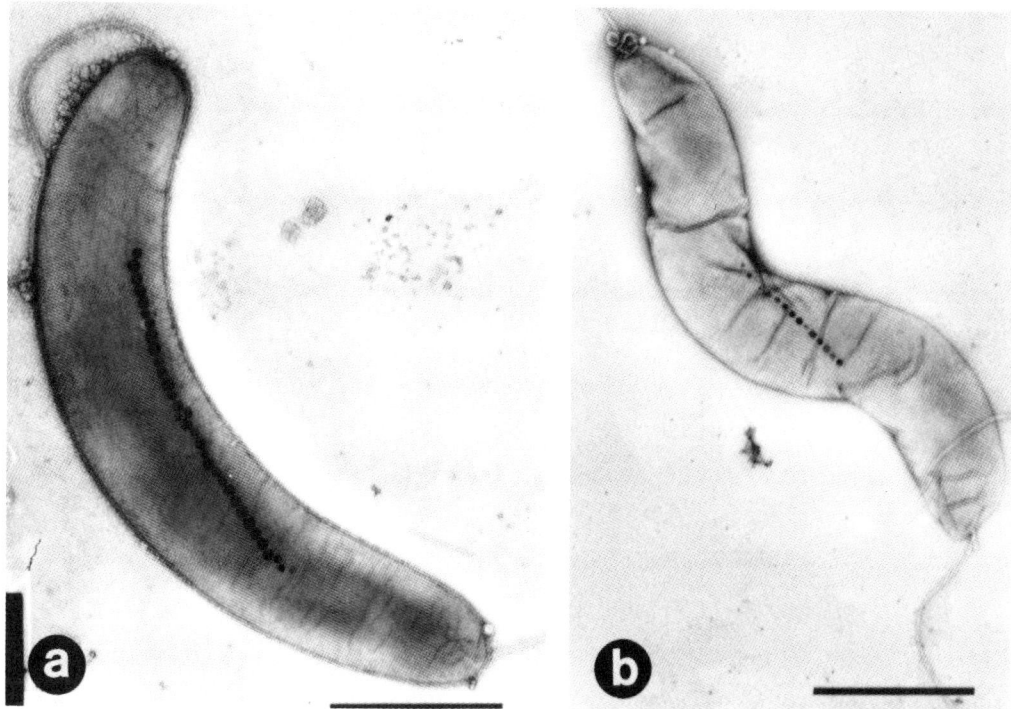

Figure 12.11. Magnetotactic bacteria containing chains of cuboidal or octahedral magnetosomes. *a*, cell is of undescribed species magnetically separated from sediments collected from a Durham, New Hampshire water treatment plant; *b, Aquaspirillum magnetotacticum* strain MS-1 ATCC 31632. Transmission electron micrographs of negatively stained cells. *Bars*, each 1 μm. *b* is reproduced with permission from D. Maratea and R.P. Blakemore, Int. J. of Syst. Bacteriol. *31*: 425-455, 1981, © American Society for Microbiology, Washington, D.C.

Cells are wall-less. Type of locomotion remains unknown. Free-living mycoplasmas able to parasitize a variety of microorganisms. **Chemoorganotrophic or parasitic on fungal mycelia.** Aerobic. Oxidizes manganous compounds leached from various Mn (II)-containing minerals. Is easily cultivated in symbiotic cultures with fungi on solid medium (agar 1.5%, manganese acetate 0.01% or $MnCO_3$, no nitrogen or phosphorus sources added), where it appears as brown microcolonies between hyphae. Many authors doubt that *"Metallogenium"* is a bacterium and regard it as a phenomenon caused by microbial activity in manganese-containing media. It is widely distributed in continental bodies of water, where it causes blooms; in soils; and on the surfaces of rocks. It is the most usual cause of large-scale manganic deposits. The pH range is 6–8, with high E_h. Two other genera of manganese-oxidizing bacteria of somewhat similar morphology have been described: *"Caulococcus"* and *"Kusnezovia."* The latter has "lily of the valley" microcolonies. *"Caulococcus"* contains minute cells within manganic film. They have not been reported since their first description. Differentiation of the species in the genus *"Metallogenium"* is arbitrary. Type species *"Metallogenium personatum"* was reported to contain minute cells within the filaments (Figure 12.12, an herbarium specimen VKM1341).

"Metallogenium symbioticum" contains only cells sprouting by tapering filaments with organic matrix covered by MnO_2 precipitates (Figure 12.13).

Family *"Siderocapsaceae"*

Cells display much variation in morphology and capsulation. The spherical, ellipsoidal, and rod-shaped cells are typically encrusted with oxides of Fe (III) and/or Mn (IV). The cells become nuclei of metal precipitation, resulting in the formation of insoluble metal oxides, which may completely encrust and envelop the cell. Capsular material characteristically serves as the core of iron and manganese deposition, which does not serve as a source of energy. These bacteria are aerobic or microaerophilic, although anaerobiocity was reported for *"Ochrobium."* These bacteria are supposed to be organotrophic, but because of the lack of pure cultures this cannot be proven. In a few instances organotrophic bacteria, *Arthrobacter,* closely resembled *"Siderocapsa"* under laboratory conditions. In aquatic environment, these bacteria are found as either planktonic forms or attached to plants, algae, or other submerged surfaces. Under microscopic examination they appear as yellow to dark brown, rusty masses.

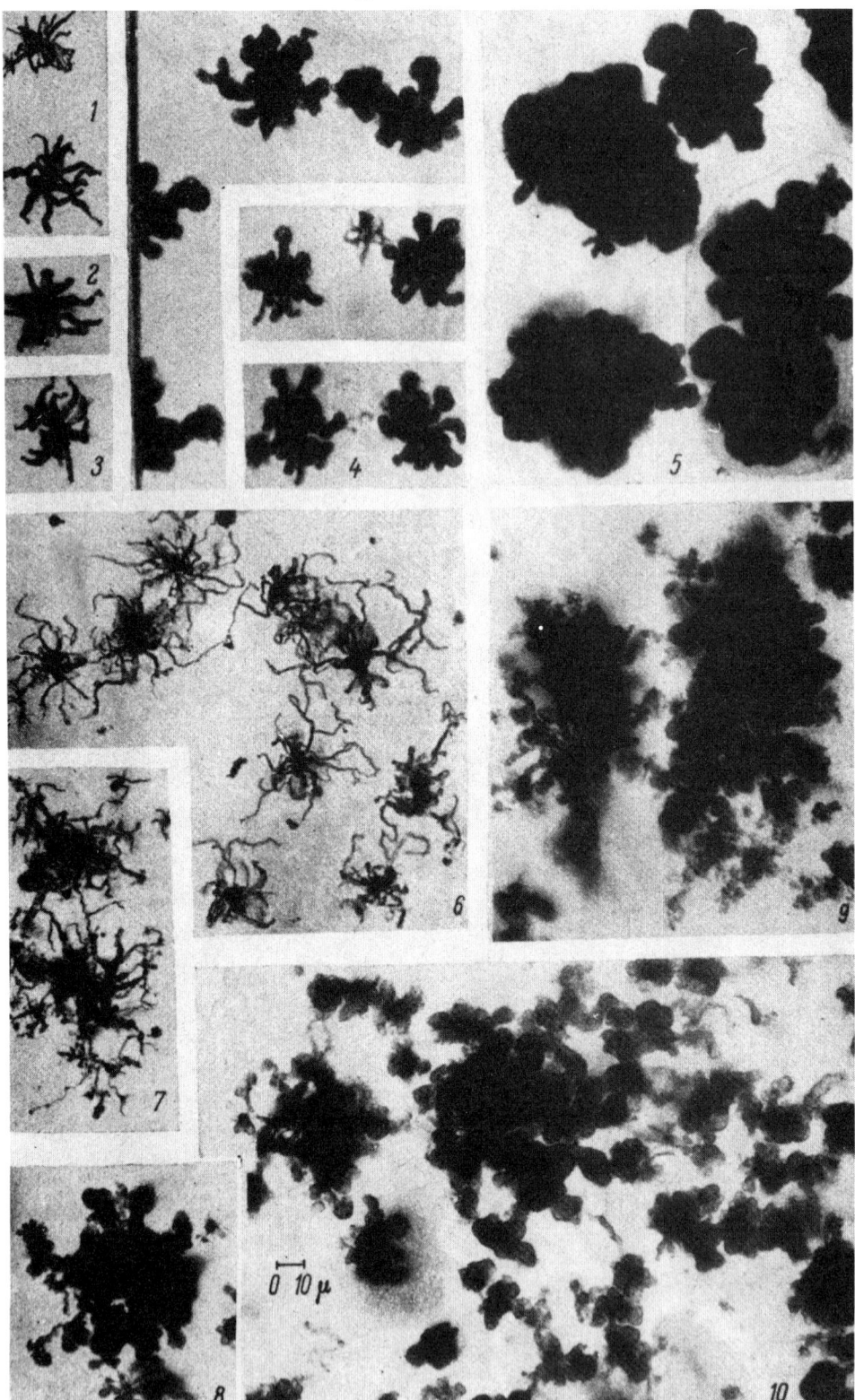

Figure 12.12. *"Metallogenium personatum"* microcolonies in different stages of manganese encrustation. Brightfield light microscopy. *1–3,* beginning of encrustation of trichosphaeric microcolonies; *4,* lobate microcolony; *5,* tuberculate microcolony; and *6–10,* stages of encrustation. *Bar,* 10 μm. (After Perfil'ev and Gabe, *In* Gurevich (Ed), *The Role of Microorganisms in the Formation of Iron Manganese Lake Ores.* Nauka, Moskow, 1964, pp. 16–53).

Key to the genera of the family "Siderocapsaceae"

I. Iron or manganese oxides deposited within the capsules.

   A. Oxides deposited in/on capsular material that surrounds a group of spherical or ovoid cells; capsule appears friable.

   Genus *"Siderocapsa"*

   B. Oxides envelop the cell in very regular fashion as thin sheath; cells resemble minute diatoms.

   Genus *"Naumanniella"*

   C. Oxides coat the cell with an opening on one end; resembles a horseshoe.

   Genus *"Ochrobium"*

II. Iron oxides are deposited on the surface of minute cells. Encrustations bright orange-yellow.

   Genus *"Siderococcus"*

Genus *"Siderocapsa"*

One to several spherical to ovoid cells surrounded by a common capsule impregnated on the surface with oxides of iron or manganese (Figure 12.14); the inner part may be free from oxides. The more or less loose **capsular material may be rust brown, owning to presence of iron oxides, or olive, owing to the presence of manganese oxides.** Cell morphology is often obscure due to iron or manganese incrustation, which can be removed by dilute solutions of either oxalic acid, HCl, or EDTA. Cells are oligocarbophilic; enrichments are in low concentration of organic substances with insoluble iron or $MnCO_3$ compounds. Common freshwater habitats are submerged surfaces or planktonic.

Type species: *"Siderocapsa treubii."*

Differentiation of the species of the genus "Siderocapsa": See Table 12.6.

Genus *"Naumanniella"*

Rod-shaped or ellipsoidal cells. Each cell is **surrounded by a delicate, regular capsule that becomes encrusted with iron oxide or manganese oxide** (Figure 12.15). **Capsular encrustation** emphasizes the cell margin and **gives the appearance of a chain link. Multiply by fission,** sometimes unequally, **or by bud-**

ding. Oligocarbophilic and may be enriched with Fe-citrate. Aerobic. Found in cold iron springs and deep wells or in brown forest soil and crusts of podzols.

Type species: *"Naumanniella neustonica."*

Differentiation of the species of the genus "Naumanniella": See Table 12.7.

Genus *"Siderococcus"*

Minute cells 0.2–0.5 μm diameter are found singly or in small aggregates (Figure 12.16). **Cells are coccoid, bearing short, thin appendages.** Capsule, if present, is very thin. **Multiplication occurs by budding; the cells become pear-shaped.** Cells attach to one another or to other surfaces, making small aggregates. The organism usually **develops as large accretions.** Iron depositions around the cells may be considerable; the **deposits contain only ferric hydroxide** and form distinctly colored horizons on stratified sediments or sediment/water interfaces. Found in cold water about 10°C, pH 6.2–7.0, 1–15 mg/L Fe^{2+}, 2–4 mg/L O_2. Not observed in culture, only in microcosms.

Type (and only) species: *"Siderococcus limoniticus."*

Characteristics of the species: As described for the genus.

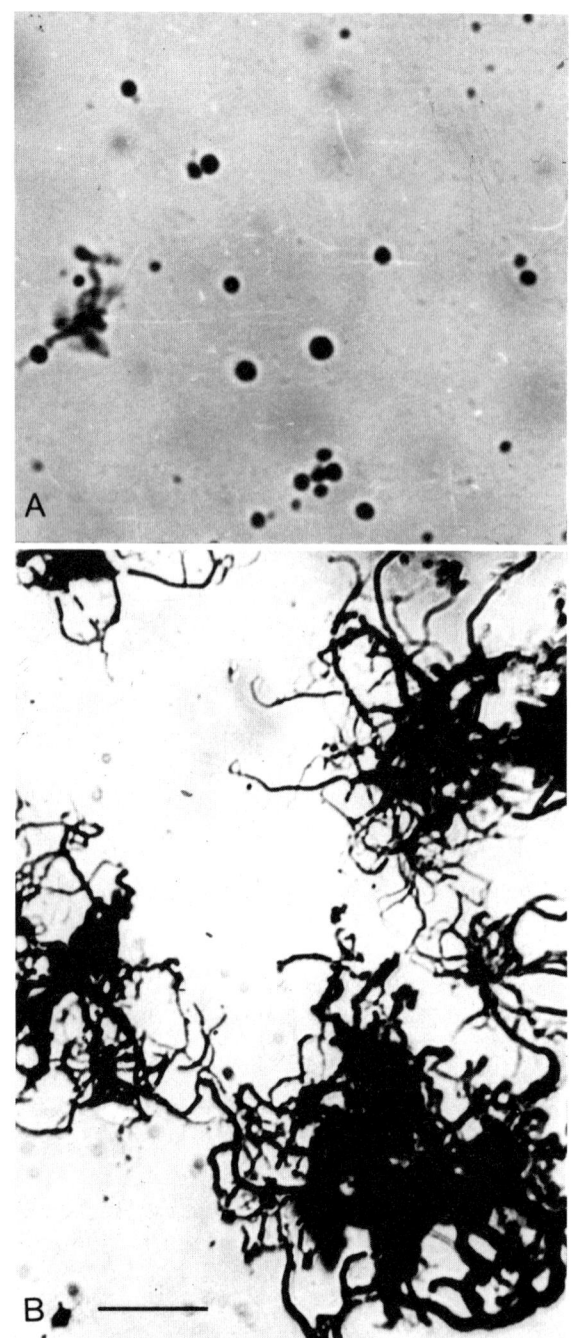

Figure 12.13. *"Metallogenium symbioticum"* in the binary culture with the fungus. Brightfield light microscopy. *A*, coccoid cells in the very beginning of encrustation. *B*, filamentous stage of growth. *Bar*, 2 μm.

Genus *"Ochrobium"*

Cells ellipsoidal to rod-shaped, 0.5–0.7 × 0.7–1.5 μm, **partially surrounded by a marginal thickening (capsule) that may be impregnated by iron oxides.** It is **open at one end, and through this end the flagellum emerges;** this results in a **horseshoe appearance. Cells**

may unite to form aggregates up to 8 cells. Actively motile. Widely distributed in fresh water at low oxygen content. May be capable of anaerobic growth.

Type (and only) species: *"Ochrobium tectum."*

Characteristics of the genus: As described for the genus.

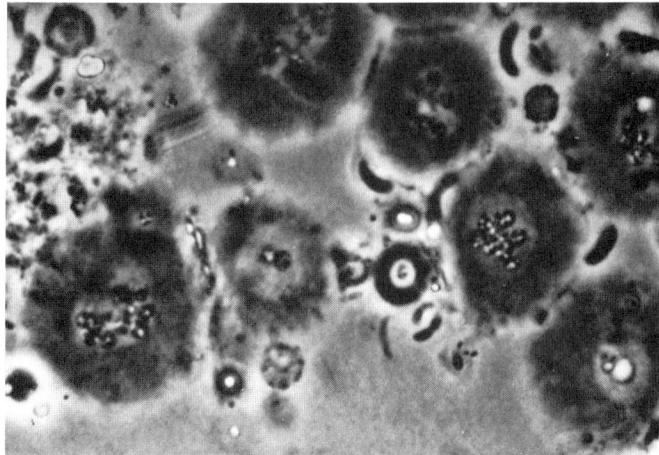

Figure 12.14. *"Siderocapsa eusphaera"* from the metalimnion of Knaack Lake (Wisconsin, U.S.A.) at a 2-m depth. This sample was collected in May.

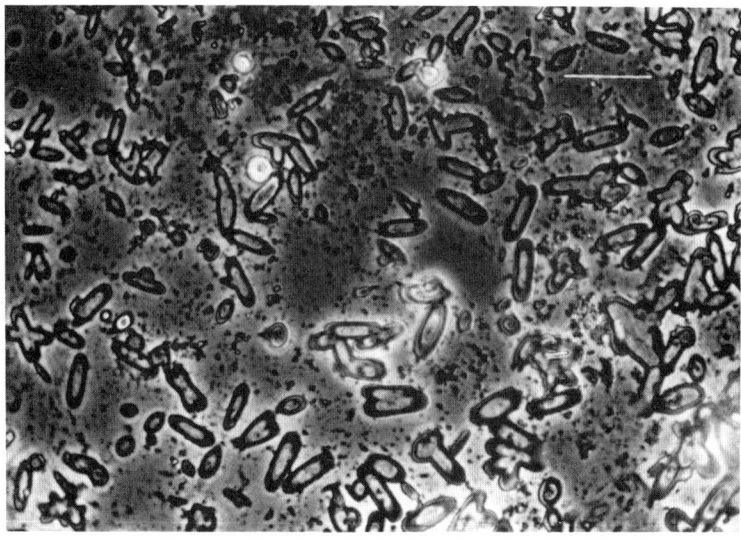

Figure 12.15. *"Naumanniella neustonica"* from an acid bog iron spring in Michigan (U.S.A.). A clean glass slide was exposed in October for 5 days. Some of the cells show lateral branching or exhibit a Y shape, while others share a joint capsule. The encrustations are yellowish olive. *Bar,* 10 μm.

Genus **Sulfobacillus**

Rods with rounded or tapered ends. Nonmotile and occur in pairs or short chains. **Gram positive. Spore-forming. Spores are round or oval, paracentral or terminal with swollen sporangia** and are 0.6–0.8 × 1.0–3.0 μm. **Strictly aerobic and facultatively chemolithoautotrophic.** Can grow heterotrophically with 0.1% glucose or saccharose and 0.2% yeast extract; pH range is 1.1–5.0 (optimum pH 2). **Moderately thermophilic** (range, 20–60°C; **optimum, 50°C). Oxidize Fe²⁺, sulfides, or sulfur.** Occur widely in thermal springs, sulfide ore deposits, and corrosion spots in municipal heat-supply systems.

Type (and only) species: *Sulfobacillus thermosulfidoox-idans.*

Characteristics of the species: As described for the genus.

SUBGROUP 3: NITRIFYING BACTERIA

This subgroup is further subdivided into two sections based upon the nitrogen compound being oxidized. The sections are:

A. Nitrite-oxidizing bacteria.
B. Ammonia-oxidizing bacteria.

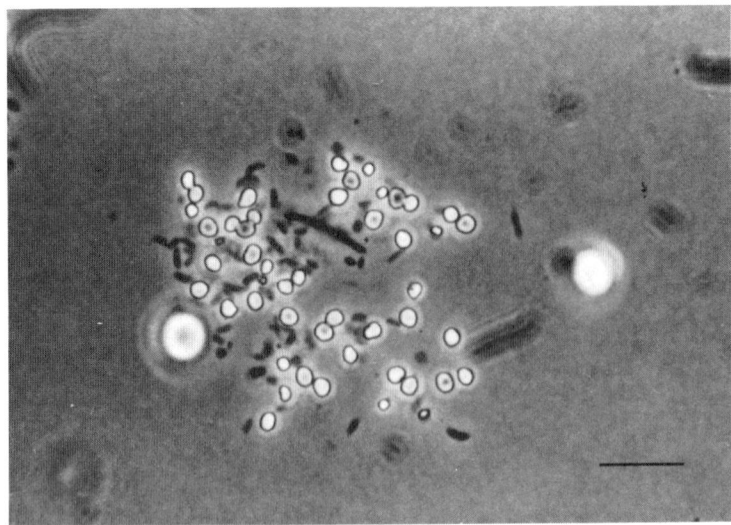

Figure 12.16. *Siderococcus limoniticus* from the hypolimnion of Knaack Lake (Wisconsin, U.S.A.) This sample was drawn from a 10-m depth in May. The original micrograph shows bright yellowish-orange iron deposits on the cell surfaces. Note budding cells. Bar, 10 μm.

SECTION A. NITRITE-OXIDIZING BACTERIA

Diverse group of rod, ellipsoidal, spherical, and spiral-shaped cells. Gram negative. Cells are motile or non-motile. When motile, flagella are polar or lateral. Nitrite-oxidizers are aerobes. As an exception *Nitrobacter* can grow by anaerobic respiration. NO and N_2O are formed if organic substances and nitrate are available. Investigations have shown that one strain of *Nitrobacter* was able to oxidize NO to nitrate. Chemolithoautotrophs obtain energy for cell growth from the oxidation of nitrite to nitrate. CO_2 is fixed via the Calvin cycle. Members of the genus *Nitrobacter* possess more than one type of the CO_2-fixing enzyme ribulose-1,5-bisphosphate carboxylase. They are mesophiles, with a temperature optimum of 28°C and a pH range of 5.8–8.5, with an optimum between 7.6 and 7.8. Unicellular, with the ability to form microcolonies that consist of loosely associated cells embedded in a polysaccharide layer or closely packed cells surrounded by a firm slime layer. The biofilm is referred to as zoogloeae or cysts. Nitrite-oxidizing bacteria are found in aerobic, but occasionally also in anaerobic, environments where organic matter is mineralized. They are widely distributed in soils, fresh water, brackish water, seawater, mudlayers, sewage disposal systems, and inside stones of historical buildings and rocks. They are also found inside corroded bricks and on concrete surfaces such as in cooling towers and highway-automobile tunnels.

Differentiation of the genera of **nitrite-oxidizing bacteria:** See Table 12.8.

Genus **Nitrobacter**

Rod-shaped, pear-shaped, or pleomorphic cells. Cells normally reproduce by budding. **Motile or nonmotile.** Motile by means of a single subterminal or lateral flagellum. **Intracytoplasmic membranes occur as polar flattened vesicles in the peripheral region of the protoplasm.** Cell wall differs from that found in most Gram-negative bacteria when thin sections are observed by electron microscopy. It is triple layer (electron dense, electron transparent, electron dense), with the inner layer more electron dense than the outer layer. Carboxysomes are always found in lithoautotrophically and mixotrophically grown cells. Additional inclusions are PHB-granules and polyphosphates. **Facultative lithoautotrophs.** The major source of energy is nitrite. **Mixotrophic growth is possible.** When both nitrite and organic substances are present, cells often show biphasic growth: first, nitrite is used, and after a lag phase, organic matter is oxidized. Chemoorganotrophic growth is slow and unbalanced, resulting in large amounts of poly-β-hydroxybutyrate granules that distort the shape and size of the cells. One strain was shown to gain energy from the oxidation of nitric oxide to nitrate (Freitag and Bock, FEMS Microbiol. Lett. *66:* 157–162, 1989). Among the nitrite oxidizing bacteria, terrestrial forms exclusively belong to the genus *Nitrobacter.* Species when distributed in oceans often are obligately halophilic.

Type species: *Nitrobacter winogradskyi.*

Differentiation of the species of the genus **Nitrobacter:** This species cannot be distinguished from the species

"*N. hamburgensis*" and "*N. vulgaris*" (Bock, Koops, Möller, and Rudert, Arch. Microbiol. *153:* 105–110, 1990) without genotypic investigations e.g., mol% G + C of the DNA, DNA homology studies or comparison of the protein pattern of whole cells; therefore, it might be preferable to leave species identification to specialized reference laboratories.

Genus **Nitrospina**

Slender rod-shaped cells. In the stationary growth phase and in old cultures, often spherical forms 1.4–1.5 μm in diameter are found. Nonmotile. **Cells lack an intracytoplasmic membrane system,** but occasionally invaginations of the cytoplasmic membrane occur when observed in thin sections by electron microscopy. **Aerobic and obligate chemolithoautotrophs. The oxidation of nitrite to nitrate is the only mode of energy generation.** CO_2 is the main carbon source. **Cells are obligately halophilic,** requiring 70–100% seawater. Thus far they are only found in marine environments.

Type (and only) species: *Nitrospina gracilis.*

Characteristics of the species: As described for the genus.

Genus **Nitrococcus**

Spherical cells. Single cells are elongated shortly before binary fission. After fission they occur in pairs. Motile by means of one or two flagella. **Intracytoplasmic membranes are** randomly arranged throughout the cytoplasm **in form of tubules.** Carboxysomes are present. **Aerobic obligate chemolithoautotrophs** Mesophilic. **The major source for cell growth is the oxidation of nitrite to nitrate.** The main carbon source is CO_2. **Obligate halophilic.** Optimal growth occurs in 70–100% seawater. Exclusively isolated from marine environments.

Type (and only) species: *Nitrococcus mobilis.*

Characteristics of the species: As described for the genus.

Genus **Nitrospira**

Loosely coiled spirals to vibrioid-shaped cells 0.3–0.4 μm in width and a spherical amplitude of 0.8–1.0 μm. Motility is not observed. Intracytoplas- mic membranes are lacking. Unusual thick periplasmic space whose width is twice that found in other Gram-negative bacteria. **Chemolithoautotrophs. Nitrite is the only energy source,** but mixotrophic growth is better than lithoautotrophic growth. In the presence of nitrite and organic substances the generation time decreases from 90 hours to 23 hours and the cell yield increases. In cell-free preparations no nitrite-oxidoreductase activity could be detected. **Seawater is essential for growth.** Optimal growth occurs in a medium containing 70–100% seawater enriched with nitrite, pyruvate, glycerol, yeast extract, or peptone. Isolated from diverse marine environments as ocean waters and marine sediments. One described enrichment culture originated from soil samples collected in Namibia.

Type (and only) species: *Nitrospira marina.*

Characteristics of the species: As described for the genus.

SECTION B. AMMONIA-OXIDIZING BACTERIA

Rod-shaped, spherical, spirillar, or lobular Gram-negative cells. Flagellation of motile cells is polar to subpolar or peritrichous. All species are aerobic but can grow at reduced oxygen partial pressure. The organisms are chemoautotrophs, growing with ammonia as the energy substrate and CO_2 as the main carbon source. Generally, the optimum growth temperature is 30°C and the optimum pH is between 7.5 and 8.0. In enrichment cultures, but not in pure cultures, cells often form aggregations called zoogloeae or cysts. Species are distributed in a great variety of soils, oceans, brackish environments, rivers, lakes, and sewage disposal systems.

Differentiation of the genera of **ammonia-oxidizing bacteria:** See Table 12.9.

Genus **Nitrosomonas**

Generally cells are **straight rods,** but **coccoid forms are often observed. Some species are motile by polar flagellation.** The cell wall is typically Gram negative, but marine species reveal an additional outer protein layer. **Intracytoplasmic membranes are arranged as flattened vesicles in the peripheral cytoplasm.** Some but not all strains possess carboxysomes. Urea can be used as an ammonia source via urease by some but not all species. Organic carbon compounds can be assimilated at a limited extent and can stimulate lithotrophic cell growth. Species are distributed in oceans, brackish environments, rivers, lakes, sewage plants and soils. Isolates from marine environments generally are obligately halophilic.

Type species: *Nitrosomonas europaea.*

Differentiation of the species of **Nitrosomonas:** See Table 12.10.

The presence of further species has been indicated by DNA homology experiments. Suitable distinguishing characteristics for these species are the G + C content of DNA, shape and size of the cells, salt requirement, urea utilization, presence of carboxysomes, and cell protein patterns. At this time species identification can be carried out only by specialized laboratories.

Genus **Nitrosococcus**

Cells are spherical to ellipsoidal. The cell envelope is typically Gram negative, but marine species reveal an additional outer cell wall layer. Two types of arrangement of intracytoplasmic membranes are described. Type I resembles that of *Nitrosomonas,* showing flattened vesicles in the peripheral cytoplasm. Type II is characterized by a more centrally located stack of flattened vesicles. Some of the species can use urea as ammonia source. Organic compounds are assimilated at a limited extent. With exception of the type species, which was isolated from soil, the *Nitrosococcus* species are restricted to marine and brackish environments and are obligately halophilic.

Type species: *Nitrosococcus nitrosus.* (Not available in culture)

Differentiation of the species of the genus **Nitrosococcus:** See Table 12.11.

Genus **Nitrosospira**

Cells are tightly coiled spirals, 0.3–0.4 μm in width, with 3–20 turns. The flagellation of motile cells is peritrichous. **Extensive intracytoplasmic membrane systems are missing,** but invaginations of the cytoplasmic membrane into the protoplasm are often observed. Most strains can use urea as ammonia source and assim-

ilate organic compounds at a limited extent. *Nitrosospira* is distributed in soils and freshwater environments.

Type species: *Nitrosospira briensis.*

Characteristics of the species: N. briensis was first isolated from the soils of Brie (France); the neotype strain originates from the soils of Crete. The characteristics are as described for the genus. (The presence of other species of *Nitrosospira* has been indicated by DNA hybridization experiments.)

Genus **Nitrosolobus**

Cells are pleomorphic lobate. The **protoplasm is compartmentalized by the cytomembrane.** Motile cells are peritrichous flagellated. Most strains utilize urea as ammonia source and can assimilate limited amounts of organic compounds. *Nitrosolobus* was isolated exclusively from soil.

Type species: *Nitrosolobus multiformis.*

Characteristics of the species: N. multiformis was isolated from the soils of Paramaribo (Surinam). The characteristics are as described for the genus. (Another species of *Nitrosolobus* indicated by DNA homology analysis has not been described until now.)

Genus **"Nitrosovibrio"**

Cells are curved slender rods. In the early stationary phase of growth spherical forms are observed. Motile cells have 1–4 polar to subpolar flagella. **Extensive intracytoplasmic membranes are lacking;** however, membrane invaginations into the protoplasm are observed. Most strains can utilize urea as ammonia source. Organic compounds are assimilated to a limited extent. All isolates of *"Nitrosovibrio"* originate from soil or building stones.

Type species: *"Nitrosovibrio tenuis."*

Characteristics of the species: "N. tenuis" has been isolated from a soil of Hawaii. The characteristics are as described for the genus. (Another species indicated by DNA homology analyses has not been described until now.)

GROUP 12 AEROBIC CHEMOLITHOTROPHIC BACTERIA AND ASSOCIATED ORGANISMS

Table 12.1 Definition of the physiological types of bacteria able to oxidize reduced sulphur compounds.

Physiological type	Carbon source		Energy Source	
	Inorganic	Organic	Inorganic	Organic
Obligate chemolithoautotroph [A]	+	−	+	−
Facultative chemolithoautotroph [B]	+	+	+	+
Chemolithoheterotroph	−	+	+	+
Heterotroph	−	+	−	+

Synonyms: [A] obligate autotrophs; [B] facultative autotrophs, mixotrophs.

Table 12.2 Physical characteristics of the morphologically conspicuous colorless sulfur bacteria. [a]
If filaments are observed, the reader should also consult *Beggiatoa*, *Thiothrix*, *Thioploca*, and *Thiospirillopsis* (Group 15).

Species	Size (μm)	Shape	Motility	$CaCO_3$
Achromatium oxaliferum [b]	>5 × <100	Round to cylindrical	Peritrichous or gliding	+
Achromatium volutans [b]	5 × <40	Spherical to ovoid	Slow, jerky gliding	−
Macromonas mobilis	6–14 × 10–30	Cylindrical to oval	Polar tuft	+
Macromonas bipunctata	2–4 × 3–7	Cylindrical to pear shaped	Polar tuft	+
Thiobacterium bovista	0.4– >1 × <3–9	Rods	Nonmotile	−
Thiospira winogradskyi	2–2.5 × <50	Spirilla, pointed ends	Polar or polar tuft	−
Thiospira bipunctata	1.7 × >2 × 6–14	Spirilla, pointed ends	Polar or polar tuft	−
Thiovulum majus	5–25	Round to ovoid	Peritrichous	−

[a] + indicates $CaCO_3$ particles may be present; − indicates $CaCO_3$ never observed. < indicates can be less than..; > indicates can be greater than...
[b] See Group 15.

Table 12.3 Basic characteristics of the obligately autotrophic colorless sulfur bacteria.
Carboxysomes indicates the possession of carboxysomes under some, if not all growth conditions; pH and temp. indicate the most favorable ranges for growth. ND, not determined.

Species	% G + C	Motility	Carboxysomes	pH	NO_3^- reduction to:		Opt. temp.	Ubiquinone
					NO_2^-	N_2		
Thiobacillus thioparus	61-66	+	+	6-8	+	−	25–30	Q-8
Thiobacillus neapolitanus	52-56	+	+	6-8	−	−	25–30	Q-8
Thiobacillus capsulatus	54.5	+	+	5-7	−	−	25–30	ND
Thiobacillus tepidarius	66.6	+	ND	6-8	+	−	40–45	Q-8
Thiobacillus denitrificans	63–68	+	−	6–8	+	+	25–30	Q-8
Thiobacillus ferrooxidans	55-65	+	+[a]	2-4	−	−	30–35	Q-8
Thiobacillus thiooxidans	51-53	+	+	2-4	−	−	25–30	Q-8
Thiobacillus albertis	61.5	+	+	2-4	−	−	25–30	ND
Thiobacillus prosperus	61-64	+	+	1-4	ND	ND	30–35	Q-8
Thiomicrospira pelophila	44	+	−	6-8	−	−	25–30	ND
Thiomicrospira denitrificans	36	−	−	7	+	+	20–25	ND
Thiomicrospira crunogena	42-43	+	−	7-8	−	−	28–32	ND

[a] Under CO_2 limitation

451

Table 12.4   Basic characteristics of the facultatively autotrophic colorless sulfur bacteria. [a]

Although not strictly a member of this group, *Paracoccus denitrificans* has been included because of its ability to grow autotrophically on thiosulfate in aerobic cultures.   "NO_3^- reduction" generally indicates heterotrophic denitrifying growth, *Thiosphaera pantotropha* is the only member of this group known to be able to denitrify on reduced sulfur compounds. Both *T. pantotropha* and *P. denitrificans* can denitrify with hydrogen as the electron donor.   Carboxysomes indicates the possession of carboxysomes under some, if not all, growth conditions; pH and temp. indicate the most favorable ranges for growth.

Species	% G + C	Motility	Carboxysomes	pH	NO_3^- reduction to: NO_2^-	N_2	Opt. temp.
Thiobacillus novellus	66-68	−	−	6-8	−	−	25-30
Thiobacillus versutus	65-68	+	−	6-8	+	+	30-35
Thiobacillus intermedius	65-67	+	+	5-7	−	−	30-35
Thiobacillus perometabolis	65-68	+	−	5-7	−	−	30-35
Thiobacillus delicatus	66-67	−	ND	5-7	+	−	30-35
Thiobacillus aquaesulis	65-66	+	ND	7-9	+	−	40-50
Thiobacillus thyasiris	52	−	+	7-8	+	+	35-40
Thiobacillus acidophilus	61-64	+	+	2-4	−	−	25-30
Thiobacillus cuprinus	66–69	+	ND	3–4	ND	ND	30–36
Thiosphaera pantotropha	66	−	−	7-9	+	+	30-40
Paracoccus denitrificans	64-67	−	−	7-9	+	+	25-35

[a] Symbols: see standard definitions; ND, not determined.

Table 12.5   Differentiation of the species of the genus *Acidiphilium* [a]

Characteristic	*A. angustum*	*A. cryptum*	*A. facilis*	*A. organovorum*	*A. rubrum*
Size (μm)	0.8 × 2.9	0.3 × 1.5	0.7 × 1.8	0.6 × 1.0	0.6 × 2.2
% G + C	67	66–70	65	64	63
Colony color	Pink	White or pink	White-brown	White	Red-violet
Acetate inhibits	+	+	−	+	+
Growth on:					
Citrate	+	+	+	+	+
Glucose	−	+	+	+	+
L-Malate	−	ND	+	ND	+
α-Ketoglutarate	−	ND	+	ND	+
cis-Aconitate	−	ND	+	ND	−
Glycerol	+	+	+	ND	−
Lactose	−	+	+	−	−
Fumarate	−	ND	+	ND	−
Pyruvate	−	ND	ND	−	−
Ethanol	+	ND	+	ND	−
Glutamate	−	−	+	ND	ND
Succinate	−	ND	+	ND	ND

[a] Symbols: see standard definitions; ND, not determined.

Table 12.6 Differential characteristics of the species of the genus "Siderocapsa"[a]

Characteristic	"S. anulata"	"S. arlbergensis"	"S. coronata"	"S. eusphaera"	"S. geminata"	"S. hexagonata"	"S. major"	"S. monoica"	"S. quadrata"	"S. treubii"
Cell form	Cocci	Cocci to rods	Cocci	Cocci to rods	Ovoid	Cocci to ovoid cells	Cocci	Cocci	Cocci to ovoid cells	Cocci
Diameter (μm)	0.2–0.5	0.4–1.0		1–2	0.5–0.6	0.6–0.8	0.7–1.8	0.5–0.8	0.5–1.0	0.4–0.6
Growth attached to surfaces	–	–	–	–		–	+	+	–	+
Free-floating cells	+	+	+	+		+	–	–	+	–
Number of cells/capsule	1	1–4	Several	Up to 60	(1)–2	Several	Several	1	1?	Several
Capsule size and structure	Small	Granular		Stratified: 2–3 concentric layers	Large radially structured or granular and layered	Hexagonal (cross-section lenticular)	Large; loose	Small	Square (cross-section lenticular)	Large
Neustonic	–	–	+	–	–	–	–	–	–	–
Aggregating capsules	–	(+)	+	–	–	–	–	–	–	–
Remarks				Arthrobacter siderocapsulatus	May be an Arthrobacter sp.	An obligate autotroph?				Type species

[a] Symbols: +, 90% or more of strains are positive; –, 90% or more of strains are negative; and (+) rarely positive.

453

Table 12.7 Differential characteristics of the species of the genus *"Naumanniella"*

Characteristic	*"N. catenata"*	*"N. elliptica"*	*"N. minor"*	*"N. neustonica"*	*"N. polymorpha"*	*"N. pygmaea"*
Cell shape	Rods, slightly curved	Ellipsoidal	Curved or spiral	Straight rods	Ellipsoidal to coccoid	Straight rods with rounded ends
Cell size (with capsule) (μm)	1.0–1.2 × 4.9–5.5	2.0 × 2.5–3.0	1.2–1.5 × 3.1–3.6	1.8–3.3 × 4.9–10.0	0.7–1.0 × 1.0–2.0 [a]	1.0 × 2.0
Cell aggregation	In chains	Single	Single	Single	Single	Single
Multiplication	Binary fission?		Binary fission	Binary fission	Budding	Budding?
Habitat			Sediment from an iron well	Neuston		
Growth conditions				Ferrous citrate solution	Growth in manganese carbonate agar or manganese acetate agar; no growth in ordinary organic media	Soil extract agar
Remarks	Psychrophilic		Psychorophilic	Type species psychrophilic	Buds motile; oxidizes only manganese	Psychrophilic

[a] Measurements without capsule

Table 12.8 Differential characteristics of the genera of the nitrite-oxidizing bacteria

Characteristics	Nitrobacter	Nitrococcus	Nitrospina	Nitrospira
Cell shape	Pear-shaped/pleomorphic rods	Spherical	Slender straight rods	Loosely-coiled spirals
Cells size (μm)	0.5–0.8 × 1.0–2.0	1.5	0.3–0.4 × 1.7–6.6	0.3–0.4 × 0.8–1.0
Flagellation of motile cells	Polar to lateral	Polar	Not observed	Not observed
Arrangement of intracytoplasmic membranes	Polar flattened vesicles	Randomly arranged tubules	None	Invaginations
Capability of using organic substances	Heterotrophic growth	None	None	Mixotrophic growth

Table 12.9 Differential characteristics of the genera of the ammonia-oxidizing bacteria

Characteristics	*Nitrosococcus*	*Nitrosolobus*	*Nitrosomonas*	*Nitrosospira*	*"Nitrosovibrio"*
Cell shape	Spherical to ellipsoidal	Pleomorphic lobate	Straight rods	Tightly coiled spirals	Slender curved rods
Cell size (μm)	1.5–1.8 × 1.7–2.5	1.0–1.5 × 1.0–2.5	0.7–1.5 × 1.0–2.4	0.3–0.8 × 1.0–8.0	0.3–0.4 × 1.1–3.0
Flagellation of motile cells	Tuft of flagella	Peritrichous	Polar to subpolar	Peritrichous	Polar to subpolar
Arrangement of intracytoplasmic membranes	Peripheral or central stacks of vesicles	Compartmentalizing	Peripheral flattened vesicles	Invaginations	Invaginations

Table 12.10 Differential characteristics of the species of the genus *Nitrosomonas*

Characteristics	*"N. cryotolerans"*[a]	*N. europaea*
Cell size (μm)	1.2–2.2 × 2.0–4.0	0.8–1.1 × 1.0–1.7
Motility	Not observed	Not observed
Utilization of urea	Yes	No
Salt requirement	Yes (optimum around 2% NaCl)	No
Carboxysomes	Not observed	Not observed
Minimum growth temperature	–5°C	+5°C
Habitats	Marine environments	Soils and freshwater environments

[a] The species *"N. cryotolerans"* was not included in *Bergey's Manual of Systematic Bacteriology*. See Jones et al. (Can. J. Microbiol. *34:* 1122–1128, 1988).

Table 12.11 Differential characteristics of the species of the genus *Nitrosococcus*

Characteristics	*"N. halophilus"*[a]	*"N. mobilis"*	*N. nitrosus*	*N. oceanus*
Cell size (diameter μm)	1.8–2.5	1.5–1.7	1.5–1.7	1.8–2.2
Arrangement of intracytoplasmic membranes	Central stack	Peripheral	Unknown	Central stack
Motility	Tuft of flagella	Tuft of flagella	Unknown	Tuft of flagella
Optimum NaCl⁻ concentration	700 mM	200 mM	Unknown	500 mM
Utilization of urea	Negative	Positive	Unknown	Positive
Habitats	Salt lagoons, salt lakes	Brackish environments	Soils	Marine environments

[a] The species *"N. halophilus"* was not included in *Bergey's Manual of Systematic Bacteriology*. See Koops et al. (*Arch. Microbiol. 154:* 244–248, 1990).

GROUP 13

BUDDING AND/OR APPENDAGED BACTERIA

This is a diverse and heterogeneous group of bacteria noted for their unusual cell shapes, appendages, and complex life cycles. The simplest organisms are rod-shaped or coccoidal bacteria that lack appendages and multiply by budding. The more elaborate organisms produce appendages and have complex life cycles in which cell morphogenesis occurs.

Appendages. Not all bacteria in this group have appendages. The appendages may be either **prosthecae** or **stalks**. The *prostheca* (pl. *-ae*) is a **cellular appendage** (i.e., it contains cytoplasm and is bound by the cell membrane and cell wall of the organism, as illustrated by the prosthecate bacterium, *Ancalomicrobium adetum*) (Figure 13.1).

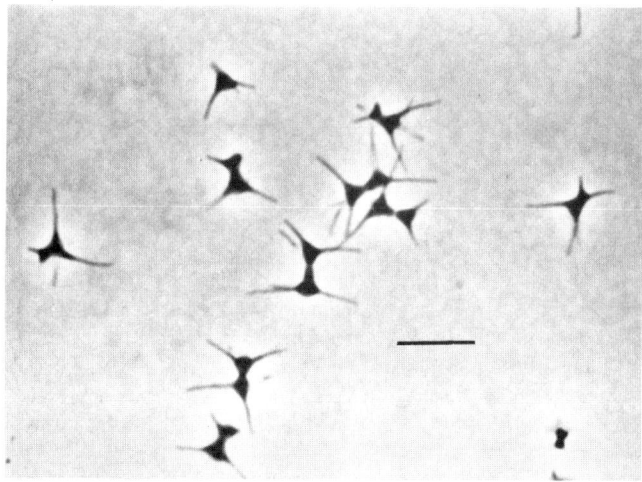

Figure 13.1. *Ancalomicrobium adetum.* **A.** Scanning electron micrograph showing the appearance of prosthecae. *Bar,* 5 µm. (Courtesy of A. R. W. van Neerven) **B.** Phase photomicrograph showing several cells with prosthecae. *Bar,* 5.0 µm. (Reproduced with permission from J. T. Staley, J. Bacteriol. *95:* 1929, 1968, ©American Society for Microbiology, Washington, DC.)

The *stalk* (pl. *-s*) refers to an **acellular appendage** that does not contain cytoplasm and is not bound by the cell membrane. This type of appendage is illustrated by members of the genus *Planctomyces* (Figure 13.2).

Cell division. Many, but not all, bacteria in this group multiply by **budding.** The remainder divide by binary transverse fission. Budding division in bacteria resembles that of the yeast *Saccharomyces cerevisiae.* A small protuberance, or *bud*, is produced at a specific site on the cell (Figures 13.3 and 13.15). This bud, or daughter cell, enlarges as growth proceeds and eventually separates from the mother cell. As in binary transverse fission, this process results in the formation of two cells, both of which can subsequently divide.

Differentiation of groups within the budding and/or appendaged bacteria. This group is subdivided into three separate subgroups. The prosthecate bacteria are treated as Subgroup 1. The remaining nonprosthecate bacteria are treated in Subgroups 2 and 3. The order *Planctomycetales* comprises Subgroup 2, and Subgroup 3 contains the remaining genera. Table 13.1 differentiates among the genera of the prosthecate bacteria, and Table 13.2 differentiates among the genera of nonprosthecate budding and/or stalked bacteria.

SUBGROUP 1: PROSTHECATE BACTERIA

Unicellular bacteria that produce one or more prosthecae per cell. The number, location, size (0.1–0.3 × 0.1– >10 µm), and shape of the prosthecae are distinctive for a genus or group of genera. **Divide by budding or binary transverse fission.** Buds are produced at the tips of prosthecae in some genera. Cells are rod-shaped, vibrioid, coccoid, tetrahedral, or cone-shaped without prosthecae. The prosthecae may make cells appear as miniature clubs, stars, cockleburs, etc., depending upon the species. Holdfasts are produced by some genera, often at the tip of a prostheca.

Heterotrophic. Most are aerobic; some are facultatively anaerobic or denitrifying. Most grow by aerobic respiration of soluble organic compounds including sugars, organic acids, and amino acids. Some are methylotrophic and may denitrify. Rarely fermentative, but two

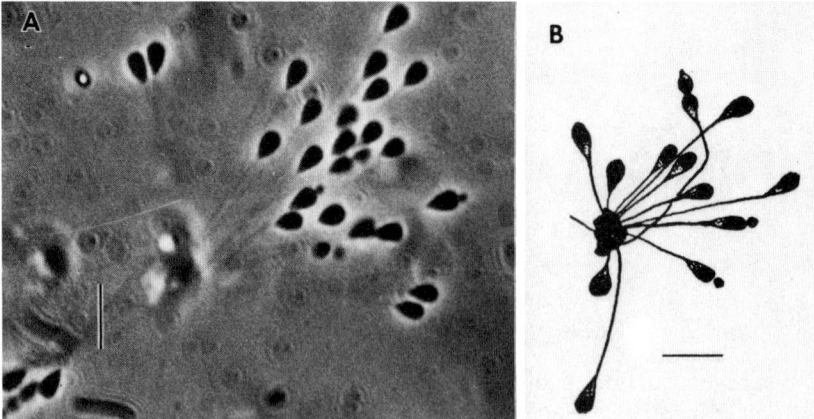

Figure 13.2. A rosette cluster of *Planctomyces* cells as observed by phase microscopy (**A**) and by illustration (**B**). The thin filaments extending from the narrow tip of the ovoid cells are stalks. The stalks have a holdfast at their distal tips, which permits attachment. Note that some cells are producing buds at the opposite (i.e., nonstalked) poles.

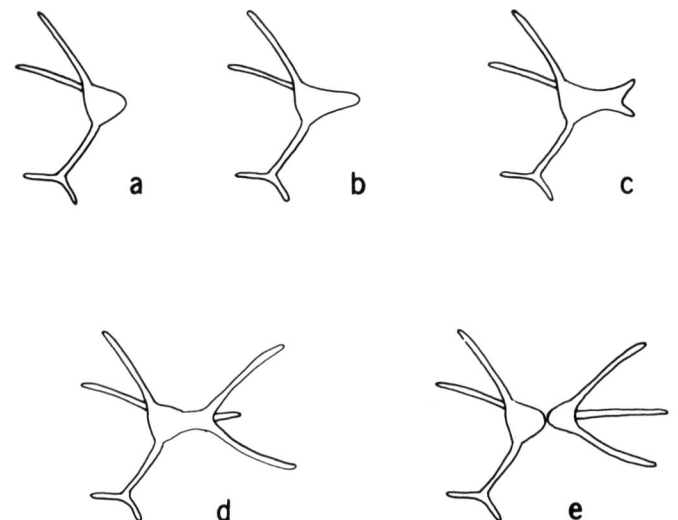

Figure 13.3. A diagram of *Ancalomicrobium adetum* illustrating the life cycle of this budding bacterium. The mother cell has three appendages, one of which is bifurcated (*a*). Bud formation occurs at the apex of the conical cell (*b*). The bud begins to differentiate (*c* and *d*) and ultimately separates from the mother cell (*e*). After division, the mother cell, whose appearance after reproduction remains essentially identical with its initial appearance in *a*, will produce another daughter cell from the same location. This process will be repeated again and again as long as conditions are favorable for growth. (Reproduced with permission from J. T. Staley, J. Bacteriol. *95*: 1921–1942, 1968, ©American Society of Microbiology, Washington, DC.)

genera, *Ancalomicrobium* and *Verrucomicrobium,* ferment glucose.

Oligocarbophilic and grow well on dilute media, typically containing 0.01–0.1% carbon source. Many species grow best if yeast extract is available. May require vitamins. Found ubiquitously in soils and fresh and saltwater habitats.

Differentiation of genera of the **prosthecate bacteria:** See Table 13.1.

Genus **Ancalomicrobium**

Unicellular bacterium with conical cells about 1.0 μm in diameter. Cells produce **two to eight or more prosthecae** (Figure 13.1). Prosthecae are cylindrical without crossbands and taper gradually from the cell to a distal diameter of about 0.2 μm and a length of 2–4 μm when fully differentiated. Prosthecae may be bifurcated. **Budding bacterium.** Buds are formed directly from the mother cell, never from tips of prosthecae. Cells occur singly or in pairs before division; rarely form aggregates. **Gram negative.** Flagella and holdfasts are not produced. **Gas vacuoles** are produced by the type strain. **Facultatively anaerobic** and **chemoorganotrophic,** *Ancalomicrobium* cells use sugars anaerobically and aerobically. Ferments sugars by mixed acid fermentation; some organic acids are utilized aerobically but are not fermented. Ammonium can be used as the sole source of nitrogen. Vitamins are required for growth. Oxidase and catalase positive. Temperature range for the type strain is 9–39°C, and the pH optimum is 7.0. Found in freshwater habitat and pulp mill oxidation ponds.

Type (and only) species: *Ancalomicrobium adetum.*

Characteristics of the species: As described for the genus.

Genus **Asticcacaulis**

Rod-shaped, 0.5–0.7 × 1–3 µm. Some cells have **one subpolar or one or two lateral prosthecae** (Figure 13.4). Prostheca diameter 0.10–0.15 µm. Crossbands are formed in prosthecae. Other cells in the same population bear a **single, subpolar flagellum.** Each type of cell bears a **holdfast** at one pole; the holdfast site is not coincident with the site of the flagellum or of the prostheca(e). Binary fission results in the production of a longer, prosthecate cell and a shorter, flagellated cell. Fission may occur in cells lacking prosthecae. In both instances, **cell division** is **unequal.** Strictly **respiratory** and **aerobic** but may be somewhat O_2 sensitive; only O_2 serves as the terminal electron acceptor for growth, although nitrate may be reduced to nitrite. **Heterotrophic** and **oligotrophic;** they grow readily in media such as peptone-yeast extract below 0.1% (w/v) organic material. Temperature range for growth is 15–35°C. Optimal pH is near neutrality. All isolates require biotin. Glucose, fructose, maltose, or lactose is utilized as the sole carbon source. Colonies are circular and convex, glistening, and with a smooth margin. The habitat is aquatic.

Type species: *Asticcacaulis excentricus.*

Differentiation of the species of the genus **Asticcacaulis:** See Table 13.3.

Genus **Caulobacter**

Cells are **rod-shaped, vibrioid, or fusiform** and 0.4–0.6 × 1–2 µm (rarely larger). Nonbudding bacteria. The younger pole of the cell bears a **single flagellum,** and the older pole bears a prostheca (the "stalk") derived from the cell envelope (Figure 13.5). Stalk diameter is constant along its length, about 0.1–0.12 µm, depending on strain. Crossbands are formed in prosthecae. A **holdfast** is produced at the pole of the cell or the tip of the stalk. Binary fission is **constrictive.** The stalk-bearing progeny cell grows and eventually repeats the **asymmetric cell division process.** The flagellum-bearing progeny cell, after a period of motility, releases the flagellum and develops its stalk at the previously flagellated site as it grows and proceeds to its **asymmetric cell division. Gram negative.** There is a single, polar flagellum in the motile stage. **Heterotrophic** and **oligotrophic,** grow readily in media such as peptone-yeast extract below 0.1% (w/v) organic material. Strictly **respiratory** and **aerobic;** only O_2 serves as the terminal electron acceptor for growth, although nitrate may be reduced to nitrite. No acid or gas is formed from sugars. Temperature range for growth is 10–35°C. Typically, require B vitamins, amino acids, or other unidentified substances. Glucose and glutamic acid are the most widely used carbon sources. Colonies are circular, convex, and glistening, with a smooth margin. Found ubiquitously in water bodies and soils.

Type species: *Caulobacter vibrioides.*

Differentiation of the species of the genus **Caulobacter:** See Table 13.4.

Genus **Dichotomicrobium**

Editorial note: The genus *Dichotomicrobium* was not included in *Bergey's Manual of Systematic Bacteriology.* The genus was created in 1989 by Hirsch and Hoffman (Int. J. Syst. Bacteriol. *39:* 495; effective publication: Syst. Appl. Microbiol. *11:* 291–301). It includes only one species, *D. thermohalophilum.*

Spherical to tetrahedral cells are 0.8–1.8 × 0.8–2.0 µm, **with up to four prosthecae. Buds formed at tips of prosthecae are nonmotile. Prosthecae may branch and may have pili. Cells remain aggregated to form clusters. PHB may be stored. Aerobic, chemoorganotrophic, moderately halophilic (8–222 $^o/_{oo}$ salinity),**

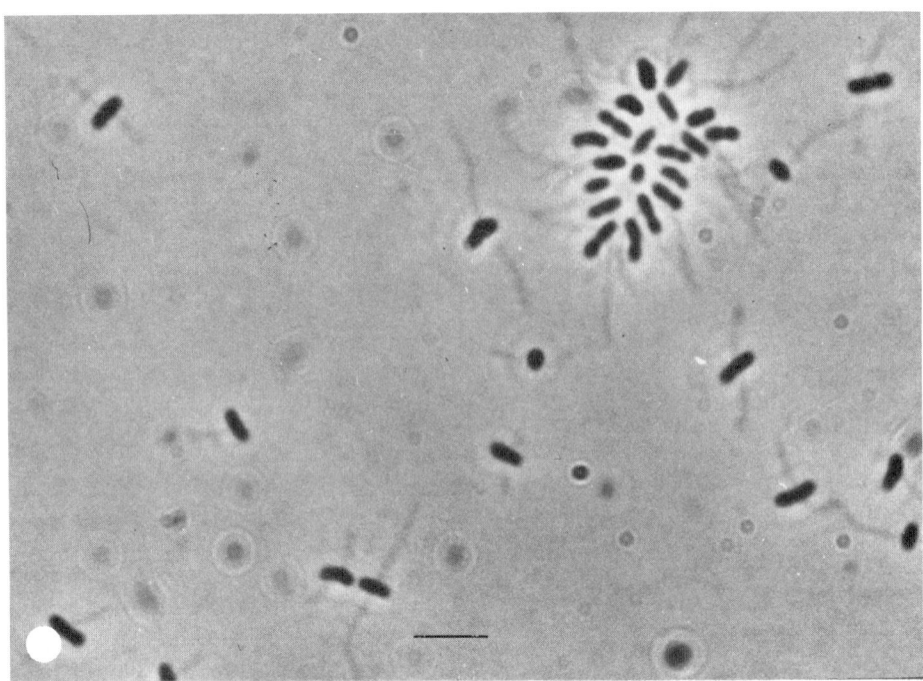

Figure 13.4. *Asticcacaulis biprosthecum.* Note that prosthecae radiate away from cell rosettes because they lack apical holdfasts. Phase-contrast microscopy, wet mount. *Bar,* 5 µm.

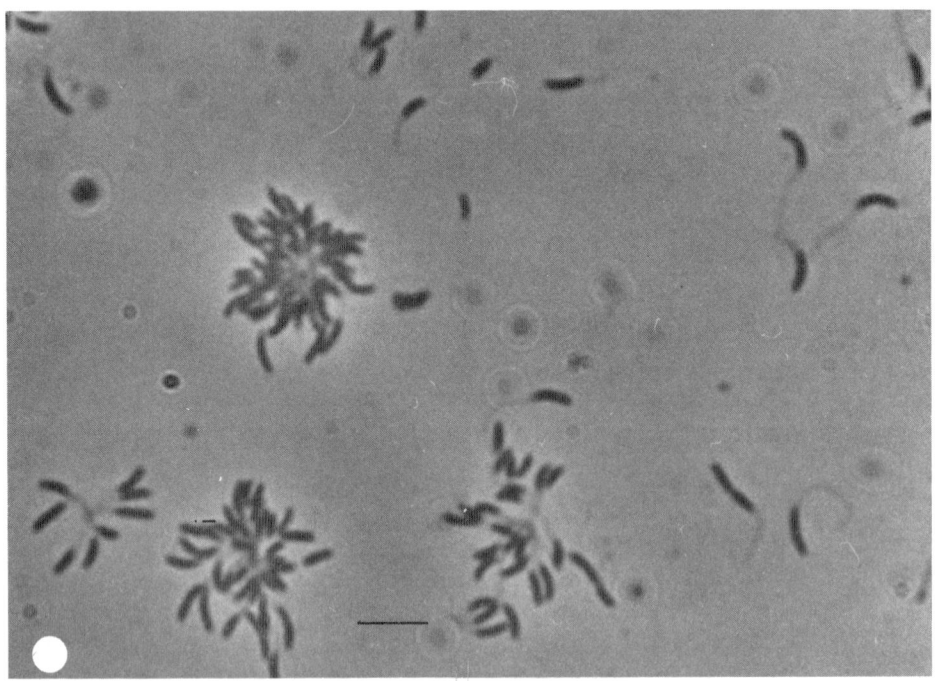

Figure 13.5. *Caulobacter subvibrioides* as observed by phase-contrast microscopy, wet mount. Note that some cells are clustered together in rosettes because prosthecae have apical holdfasts. *Bar,* 5 µm.

and **moderately thermophilic** (optimum temperature, 35–55°C). Yeast extract is required for growth. Use acetate and other organic acids as carbon source and amino acids or yeast extract as nitrogen sources (do not use ammonia, nitrate, or urea). Habitat is saline ponds and lakes.

Type (and only) species: *Dichotomicrobium thermohalophilum.*

Characteristics of the species: As described for the genus.

Genus **Filomicrobium**

Editorial note: The genus *Filomicrobium* was not included in *Bergey's Manual of Systematic Bacteriology.* The genus was created by Schlesner (Int. J. Syst. Bacteriol. *38:* 220, 1988; effective publication: Syst. Appl. Microbiol. *10:* 63–67, 1987). It includes only one species, *F. fusiforme.*

Fusiform cells, 0.5–0.7 × 1.5–4.0 μm, **have two or three polar prosthecae. Buds are formed at tips of prosthecae,** which are about 0.2 μm in diameter and **up to 40 μm in length. Nonmotile,** aerobic, and chemoorganotrophic. Glucose is used as carbon source with ammonia, nitrate, or urea as nitrogen source. Do not denitrify. **Require eighth- to full-strength sea water** for growth. Red pigment in type strain. Habitat is brackish water.

Type (and only) species: *Filomicrobium fusiforme.*

Characteristics of the species: As described for the genus.

Genus **Hirschia**

Editorial note: The genus *Hirschia* was not included in *Bergey's Manual of Systematic Bacteriology.* The genus was created in 1990 by Schlesner et al. (Int. J. Syst. Bacteriol. *40:* 443–451). It includes only one species, *H. baltica.*

Rod-shaped or oval cells, 0.5–1.0 × 0.5–6.0 μm, **have one or two polar prosthecae. Buds are formed at tips of prosthecae,** about 0.2 μm in diameter. **Buds are motile by a single polar flagellum. Aerobic and chemoorganotrophic. PHB is not stored.** Various sugars and organic acids serve as carbon sources, and ammonia and amino acids are nitrogen sources for growth. **Methanol is not used as a carbon source.** Habitat is brackish water.

Type (and only) species: *Hirschia baltica.*

Characteristics of the species: As described for the genus.

Genus **Hyphomicrobium**

Rod-shaped, oval, or bean-shaped cells, 0.3–1.2 × 1–3 μm, have polar prosthecae of varying lengths and 0.2–0.3 μm in diameter. Prosthecae may be branched and may extend from both poles of the cell. Cells reproduce by **budding at the tip of a prostheca.** Mature buds, which separate from the mother cell, are motile by one to three polar to subpolar flagella. Buds may produce holdfasts on the cell surface and aggregate to form rosettes. Motility is rare in older cultures. PHB is stored in cells, usually at one pole. **Aerobic,** chemoorganotrophic, oligocarbophilic, and **methylotrophic (i.e., grow best with one-carbon compounds, e.g., methanol or methylamine as carbon source** and ammonia or amino acids as nitrogen source). May denitrify. Carbon dioxide is required for growth. Habitat is soil and fresh water.

Type species: *Hyphomicrobium vulgare.*

Differentiation of the genus **Hyphomicrobium** *from closely related genera:* The genera, *Hyphomicrobium, Hyphomonas,* and *Hirschia* all share the same general morphological features (Figure 13.6). However, they differ physiologically (see Table 13.1).

Differentiation of the species of the genus **Hyphomicrobium:** See Table 13.5.

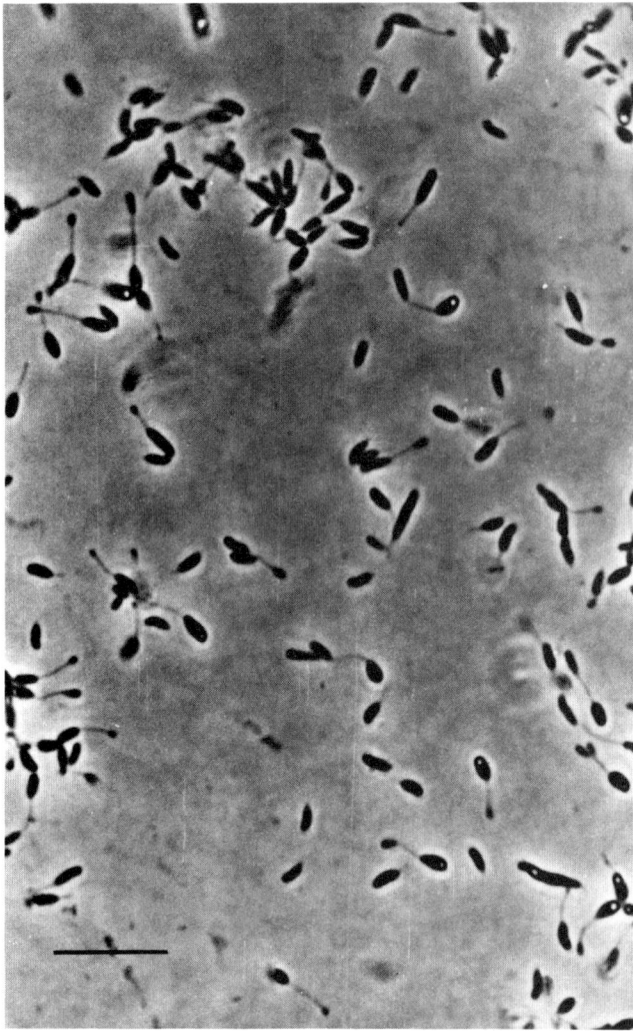

Figure 13.6. *Hyphomicrobium zavarzinii* ZV-580. Phase-contrast light micrograph of young growing culture showing all stages of development as well as rosettes. Some of the mother cells carry large granules of poly-β-hydroxybutyric acid (bright granules). *Bar,* 10 μm.

Genus **Hyphomonas**

Rod-shaped or oval cells, 0.5–1.0 × 1–3 μm, **have polar prosthecae. Buds are produced at tips of prosthecae,** which are 0.2–0.3 μm in diameter and may exceed 1.0 μm in length. Buds are motile with single polar to lateral flagellum. **Aerobic and chemoorganotrophic. Amino acids are required** for growth. **Do not use methanol** as a carbon source.

Type species: *Hyphomonas polymorpha.*

Differentiation of the species of the genus **Hyphomonas:** See Table 13.6.

Genus **Labrys**

Unicellular, flat bacterium, 1.1–1.3 × 1.3–1.5 μm. Cells are flat and **triangular. Two or three tapering short prosthecae** (0.6 μm) protrude from two corners of the triangle; the third remains free and is associated with multiplication. **Cells multiply by budding.** Buds are produced directly from the mother cell at the tip of the triangle lacking prosthecae. In this stage, cells resemble a double-headed ax or labrys. **Gram negative, are nonmotile,** and do not possess fimbriae. **Obligately aerobic, chemoorganotrophic,** and oligocarbophilic. Utilize carbohydrates and some organic acids as sole carbon and energy sources. The type strain requires B vitamins for growth. Found in freshwater lakes.

Type (and only) species: *Labrys monachus.*

Characteristics of the species: As described for the genus.

Genus **Pedomicrobium**

Oval or spherical cells, 0.4–2.0 × 0.4–2.5 μm, **have up to five or more prosthecae,** 0.15–0.3 μm in diameter, **extending from all sites of the cell surface. Buds are formed at tips of prosthecae** (Figure 13.7). Mature buds may be motile with single flagellum. **Oxides of iron or manganese or both are deposited on mother cells and hyphae.** Aerobic and chemoorganotrophic. **PHB is produced.** Acetate, pyruvate, or caproate is used as the carbon source for growth; organic sources of nitrogen, such as peptone, casamino acids, or yeast extract, serve as nitrogen sources. Slow growth occurs with 0.1–1% fulvic acid iron sesquioxide as sole carbon and nitrogen sources. Occur in soil and aquatic habitats. Upon isolation, colonies exhibit encrustations of iron and/or manganese oxides that may be so heavy that individual cells are difficult to visualize in the microscope. Oligocarbophilic. Grow slowly to form small colonies on most media. Habitat is brackish water.

Type species: *Pedomicrobium ferrugineum.*

Differentiation of the species of the genus **Pedomicrobium:** See Table 13.7.

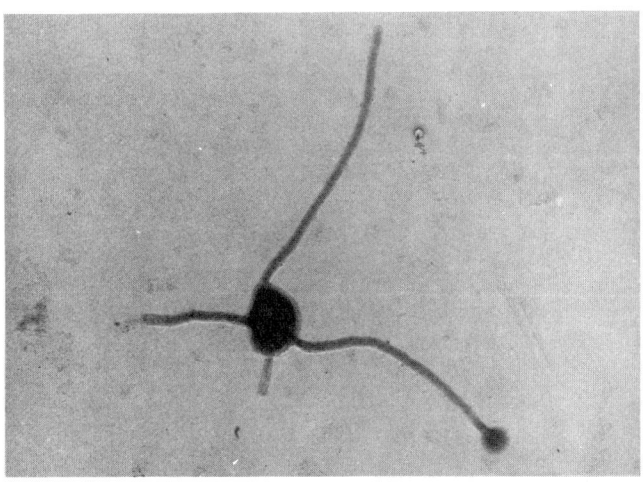

Figure 13.7. *Pedomicrobium manganicum* E-1129. Transmission electron micrograph showing mother cell with hyphae and young bud. *Bar*, 2.5 µm.

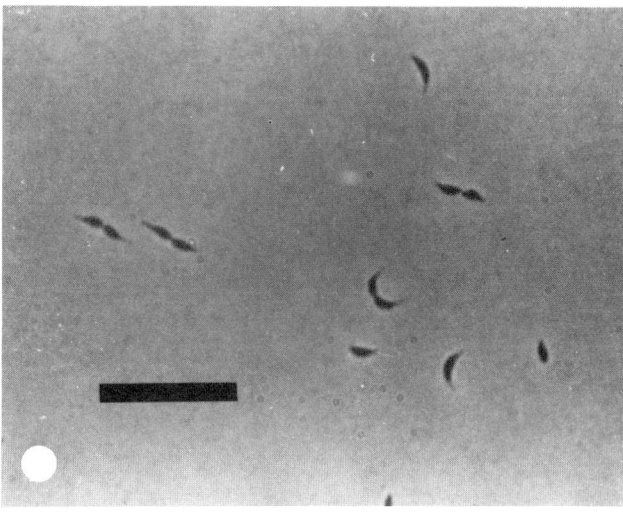

Figure 13.8. Phase-contrast photomicrograph illustrating *Prosthecobacter fusiformis*. *Bar*, 10 µm. (Reproduced with permission from J. T. Staley, J. A. M. de Bont, and K. de Jonge, Antonie van Leeuwenhoek J. Microbiol. Serol. *42:* 333–343, 1976, ©H. Veenman en Zonen, Wageningen, The Netherlands.)

Genus **Prosthecobacter**

Cells are **fusiform** or **vibrio-shaped**, 0.5–0.9 × 2–5 µm, exclusive of appendages. Each cell bears at least **one polar prostheca** of a type unique to this genus (Figure 13.8). The prosthecal diameter is 0.1–0.2 µm, tapering gently away from the cell pole but typically with a bulbous distal tip, which has a **holdfast** and is **adhesive**. Crossbands are not formed in prosthecae. Binary fission is **constrictive** and is completed without formation of a septum. **Nonmotile.** Prostheca development occurs at the younger pole as the cell grows; at the time of fission, each progeny cell bears a single polar prostheca, so that cell division is morphologically **symmetric. Gram negative,** strictly **aerobic, heterotrophic,** and **oligotrophic,** growing readily in complex media containing < 0.1% (w/v) organic material. Temperature for growth is 1–10°C to 35–40°C. Vitamins are not required. Sugars are the only generally utilized carbon sources. Colonies are circular, convex, and glistening, with a smooth margin. Found in aquatic habitats, soil, and sewage.

Type (and only) species: *Prosthecobacter fusiformis.*

Characteristics of the species: As described for the genus.

Genus **Prosthecomicrobium**

Unicellular bacterium has coccobacillary to rod-shaped cells ranging in diameter from 0.8 to 1.2 µm and con-taining **numerous prosthecae** extending from all locations on the cell surface (Figure 13.9). **Prosthecae,** which may number 10– >30 per cell, **are typically short** (i.e., less than 1.0 µm in length); however, some species also produce longer prosthecae (in excess of 2.0 µm). Cells multiply by **budding.** Buds are produced directly from the mother cell, never from tips of prosthecae. Motile and nonmotile species exist. **Motile organisms produce single polar to subpolar flagella;** one species forms gas vacuoles but not flagella. **Obligately aerobic,** nonfermentative, and **heterotrophic.** A variety of sugars and organic acids are used as energy sources for growth; few species utilize C-1 compounds. All strains tested require one or more B vitamins for growth. **Oxidase and catalase positive.** Found in soils and fresh and marine waters.

Type species: *Prosthecomicrobium pneumaticum.*

Differentiation of the species of the genus **Prosthecomicrobium:** See Table 13.8.

Genus **Stella**

Cells are flat, six-pointed stars, radially symmetric, 0.7–3.0 µm (Figure 13.10). Occur singly or in pairs. Divide by binary transverse fission. Spores are not formed. **Gram negative** and aerobic. One species produces gas vesicles. Chemoorganotrophic, using a variety of amino acids or organic acids. Oxidative and **oligocarbophilic.** Found in soil, fresh water, brackish water, and artificial ecosystems where complete decomposition of organic matter occurs.

Type species: *Stella humosa.*

Differentiation of the species of the genus **Stella:** See Table 13.9.

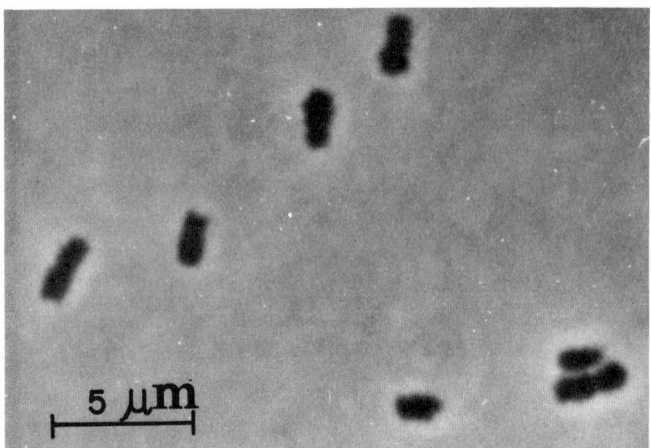

Figure 13.9. *Prosthecomicrobium enhydrum.* Phase photomicrograph showing several cells. Note the irregular surface of the cells due to the presence of numerous short prosthecae.

Genus **Verrucomicrobium**

Editorial note: The genus *Verrucomicrobium* was not included in *Bergey's Manual of Systematic Bacteriology.* The genus was created by Schlesner (Int. J. Syst. Bacteriol. *38:* 221, 1988; effective publication: Syst. Appl. Microbiol. *10:* 54–56, 1987). It includes only one species, *V. spinosum.*

Unicellular rod-shaped bacterium, 0.8–1.0 × 1.0–3.8 μm, **has numerous prosthecae,** about 0.5 μm in length, **emanating from all locations on the cell surface. Fimbriae extend from prosthecal tips. Nonmotile. Chemoheterotrophic using some sugars, and facultatively anaerobic, fermenting glucose with acid production and no gas.** Found in soils and the surface water of an eutrophic lake.

Type (and only) species: *Verrucomicrobium spinosum.*

Differentiation of the species: This bacterium resembles *Prosthecomicrobium* spp. except for the prosthecal fimbriae, sugar fermentation, and lower mol% G + C.

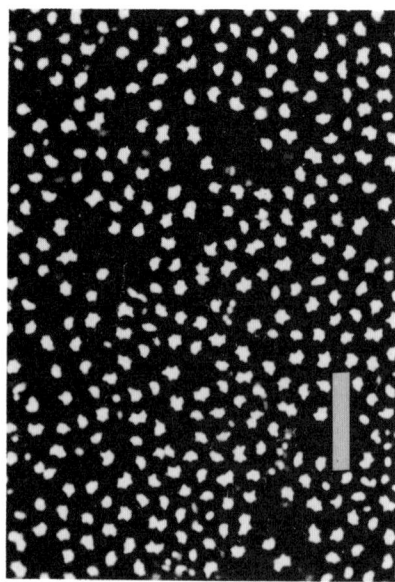

Figure 13.10. Cells of *Stella humosa* showing their typical six-pointed star shape. *Bar,* 10 μm. (Courtesy of L. V. Vasilyeva)

SUBGROUP 2: ORDER PLANCTOMYCETALES

Unicellular, rosette-forming, or filamentous bacteria **Coccoid, ovoid, or pear-shaped cells. Lack peptidoglycan.** Those tested are resistant to penicillin. Cells produce characteristic **crateriform structures** on the cell surface (Figure 13.11). Divide by **budding and may produce** nonprosthecate appendages called **stalks.** Unicellular and rosette-forming species may produce motile, flagellated cells. Rosette-forming species produce microcolonies in natural environment. *"Isosphaera pallida"* motile by gliding. Found in freshwater, marine, and other saline habitats. *"Isosphaera"* found in neutral to alkaline hot springs (35–55°C).

Genus **Gemmata**

Unicellular coccoid to ovoid bacteria (1.4–3.0 x 1.4–3.0 μm) with cell indentation. Motile by **a polar tuft of flagella.** Crateriform structures are uniformly distributed on the cell surface. Cells do not produce stalks. Contain phase-dark cell inclusions. Aerobic and use glucose but not fructose and pyruvate as carbon sources. Catalase positive and oxidase negative.

Type (and only) species: *Gemmata obscuriglobus.*

Characteristics of the species: As described for the genus.

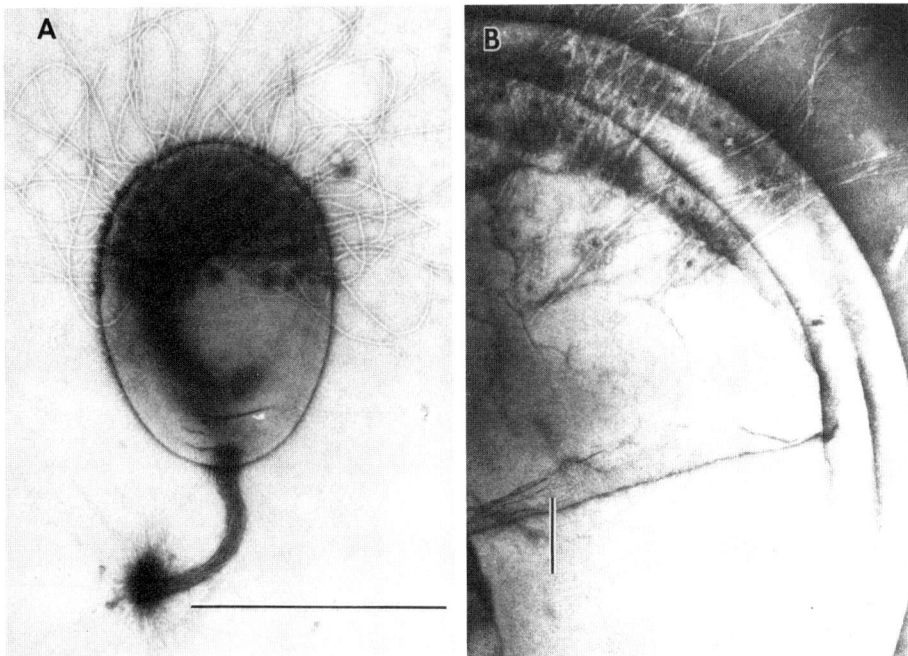

Figure 13.11. A. An ovoid *Planctomyces* cell showing the typical braided, multifibrillar stalk with a terminal holdfast. Note the numerous pili and crateriform surface structures. **B.** Portion of a cell of *Pirellula staleyi* showing crateriform surface structures and pili. *Bars,* 1.0 μm for **A** and 0.1 μm for **B.**

Genus "Isosphaera"

Spherical cells (2.0–2.5 μm in diameter) **in chains,** often more than 100 cells long (Figure 13.12). Divide by budding. **Gas vesicles** may be produced. Cells move by **gliding motility. Obligate aerobes and heterotrophic,** using glucose or lactate as carbon sources at low concentrations (glucose < 0.025%). Temperature range is 40–55°C. Habitat is hot springs.

Type (and only) species: *"Isosphaera pallida."*

Characteristics of the species: As described for the genus.

Genus **Pirellula**

Editorial note: The genus *Pirellula* was not included in *Bergey's Manual of Systematic Bacteriology.* The genus was created in 1987 by Schlesner and Hirsch (Int. J. Syst. Bacteriol. *37:* 441). It was previously named *Pirella* by Schlesner and Hirsch (Int. J. Syst. Bacteriol. *34:* 492–495, 1984), but that name is illegitimate as a later homonym of a genus of fungi.

Unicellular ovoid to pear-shaped bacteria (0.5–3.0 × 1.0–5.0 μm) **produce a holdfast at the narrow tip of** the cell (Figure 13.13). **Do not** commonly **produce stalks.** Buds are motile by **subpolar flagellum.** Holdfast allows attachment to other particulate material in habitat, including sheathed organisms as well as detritus, and permits rosette formation in pure culture. Crateriform structures are located at the reproductive pole. Aerobic and use glucose, fructose, and pyruvate as carbon sources. Grow well at < 0.1% concentration of organic carbon sources on growth media. Catalase and oxidase positive. Common in freshwater lakes and ponds.

Type species: *Pirellula staleyi.*

Differentiation of the species of the genus **Pirellula:** See Table 13.10.

Genus **Planctomyces**

Coccoid or ovoid bacteria produce stalks or bulb-shaped bacteria without stalks. A holdfast is borne at the tip of each stalk or bulb. Organism may be unicellular (Figure 13.11), or cells may attach to one another to form microcolonial rosettes (Figures 13.2 and 13.14). **Divide by budding** to form motile daughter cells from the nonstalked (i.e., reproductive) pole.

Motile cells have a single subpolar flagellum. Iron and manganese oxides may be deposited on stalks. Found in freshwater, marine, and other saline habitats.

Type species: *Planctomyces bekefii.*

Differentiation of the species of the genus **Planctomyces:** See Table 13.11.

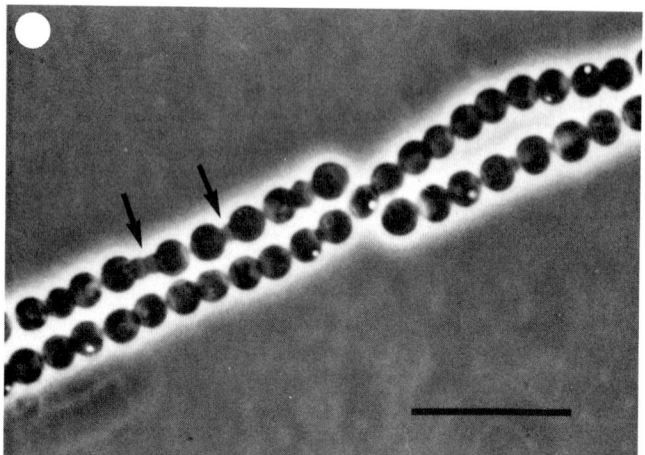

Figure 13.12. Phase-contrast photomicrograph of *"Isosphaera pallida".* *Arrows* indicate pairs of buds forming between adjacent cells. The light spots showing in some cells are gas vacuoles. *Bar,* 10 μm. (Reproduced with permission from S. J. Giovannoni, E. Schabtach, and R. W. Castenholz, Arch. Microbiol. *147:* 276–284, 1987, ©Springer-Verlag.)

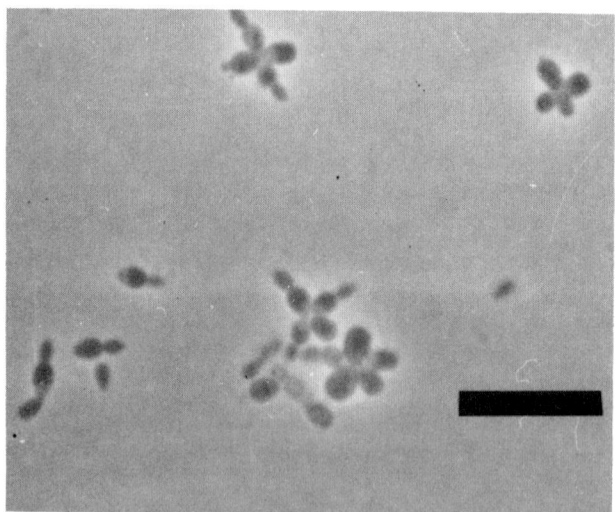

Figure 13.13. Phase photomicrograph of *Pirellula staleyi* showing bud formation and rosettes. (Reprinted by permission from J. T. Staley, Can. J. Microbiol. *19:* 609–614, 1973.)

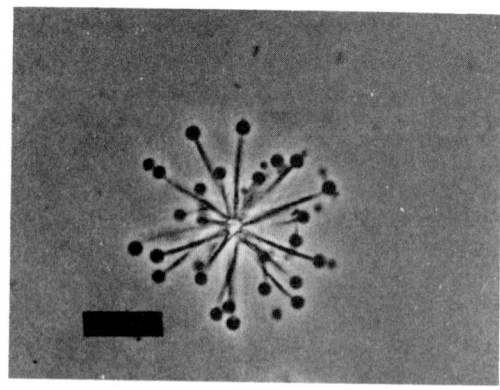

Figure 13.14. A rosette-forming species of *Plantomyces.* Each cell is borne at the tip of a stalk. Note that some cells are producing a bud from their nonstalked pole. Phase photomicrograph. *Bar,* 10 μm.

SUBGROUP 3: OTHER BUDDING AND/OR APPENDAGED BACTERIA

This is a miscellaneous collection of bacteria that do not seem to be related to one another or to the other budding and/or appendaged bacteria. See Table 13.2 for characteristics useful in differentiating them from one another.

Genus **Angulomicrobium**

Unicellular **tetrahedron-shaped cells,** 1.1–1.5 μm on each side, **form buds at the conical pole** of the cell, giving the appearance of a mushroom fruiting structure (Figure 13.15). Nonmotile. **Obligately aerobic heterotrophs** use many organic acids, sugars, and amino acids. **Methanol and formate used** if yeast extract is present. Catalase and oxidase positive. Found in swamps and pulp mill aeration lagoons.

Type (and only) species: *Angulomicrobium tetraedrale.*

Characteristics of the species: As described for the genus.

Genus **Blastobacter**

Irregular ovoid pleomorphic rods, 0.5–1.0 × 1.0–5 μm, show **occasional branching.** **Budding** occurs at the cell pole or laterally. Newly separated buds may become motile. Aerobic. May use methanol or ethanol, sugars, organic acids, or amino acids as carbon source and ammonia, nitrate, or amino acids for nitrogen source. CO_2 may be fixed. Found in fresh water sources and activated sludge.

Type species: *Blastobacter henricii.*

Differentiation of the species of the genus **Blastobacter:** See Table 13.12.

Genus **Ensifer**

Unicellular rods, 0.7–1.1 × 1.0–1.9 μm, **divide by budding. Attach to Gram-positive and Gram-negative host bacteria and may cause lysis of host cells. Host cells are not required** for growth. Motile by **a subpolar tuft of flagella.** Aerobic and heterotrophic; glucose is respired. Grow well on many ordinary media. Reduce nitrate and nitrite. Mol% G + C is 63–67. Found in soil and on the surface of soil bacteria.

Type (and only) species: *Ensifer adhaerans.*

Characteristics of the species: As described for the genus.

Genus **Gallionella**

This genus is described with the iron- and manganese-depositing bacteria in Group 12, Subgroup 2.

Genus **Gemmiger**

Ovoid to hourglass-shaped bacterium, 1.0 × 0.9–2.5 μm, divides by **budding** (Figure 13.16). Cells **may form chains. Nonmotile. Obligately anaerobic heterotroph** uses carbohydrates as the only energy source. **Glucose fermentation produces butyrate** and usually lactate and formate. Catalase negative. Found in ruminal fluid, fecal matter, humans, and chickens.

Type (and only) species: *Gemmiger formicilis.*

Characteristics of the species: As described for the genus.

Genus **Nevskia**

Long, rod-shaped cells, 1–6 × 3–12 μm, **have transparent "stalks."** Stalks consist of capsular material excreted from one side of the cell. After **cells divide by binary transverse fission,** the stalks hold daughter cells together to form Y-shaped branches. **Cauliflower-like aggregates** are found in the natural environment (Figure 13.17). Found in the neuston layer of fresh water habitats. The original type strain was never isolated, and a neotype strain has not yet been formally described.

Type (and only) species: *Nevskia ramosa.*

Characteristics of the species: As described for the genus.

Genus **Seliberia**

Helically sculptured rods (0.5–0.8 × 1–12 μm) **have blunt or rounded ends** (Figure 13.18). Polar holdfasts allow **rosette formation** in pure culture. Motile flagellated cells are produced. **Polar growth** of elongated cells **results in the formation of short swarmer cells and long immotile cells** by asymmetric cell division. A single **subpolar sheathed flagellum** is formed, but several unsheathed lateral flagella may also be produced. Strict aerobe. Heterotrophic with oxidative metabolism of carbohydrates such as glucose, fumarate, or threonine. Catalase and oxidase positive. Temperature range is 15–37°C. Found in oligotrophic water and soil environments.

Type (and only) species: *Seliberia stellata.*

Characteristics of the species: As described for the genus.

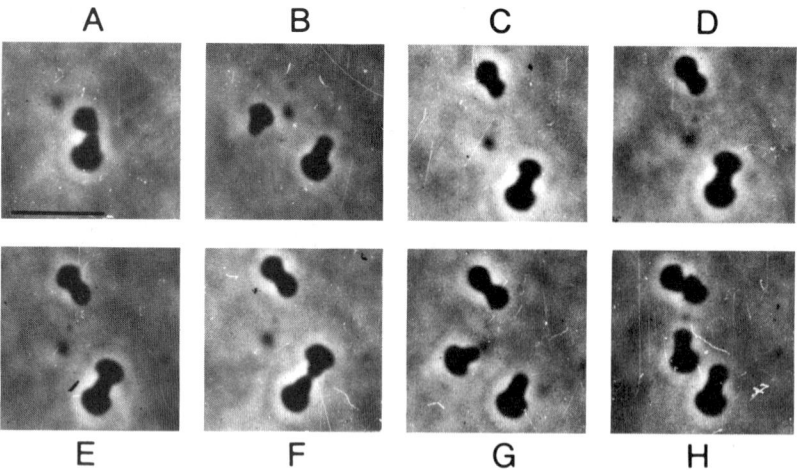

Figure 13.15. *Angulomicrobium tetraedrale.* Sequential series of phase photomicrographs illustrating budding. *Bar,* 5 μm. (Reproduced with permission from L. V. Vasilyeva, T. N. Lafitskaya, and B. B. Namsaraev, Microbiologiya *48:* 1033–1039, 1979.)

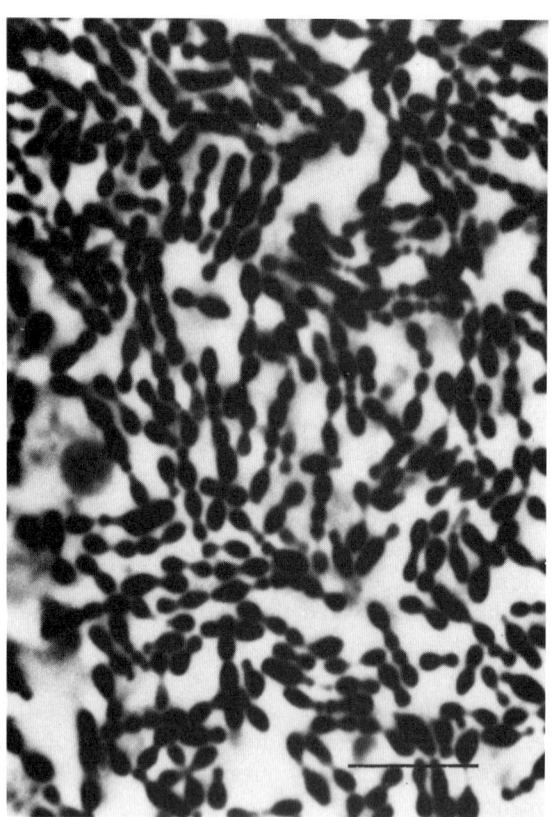

Figure 13.16. *Gemmiger formicilis.* Phase-contrast photomicrograph. *Bar,* 5 μm. (Reproduced with permission from S. C. Croucher and E. M. Barnes, Rev. Inst. Pasteur Lyon *14:* 95–102, 1981, ©Institut Pasteur de Lyon.)

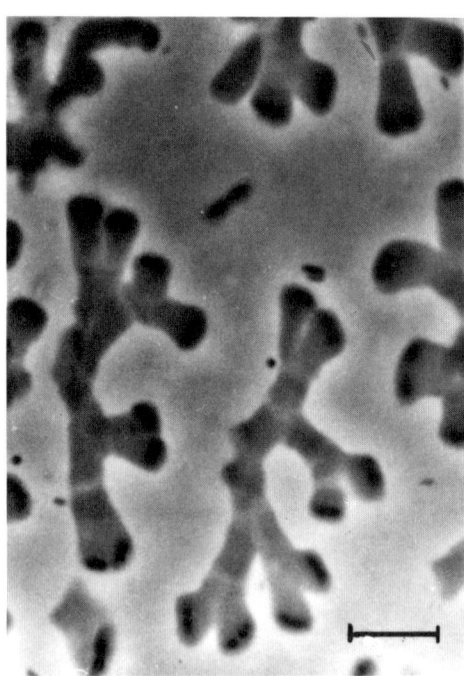

Figure 13.17. *Nevskia ramosa* showing typical morphology. Phase-contrast micrograph. *Bar,* 10 μm.

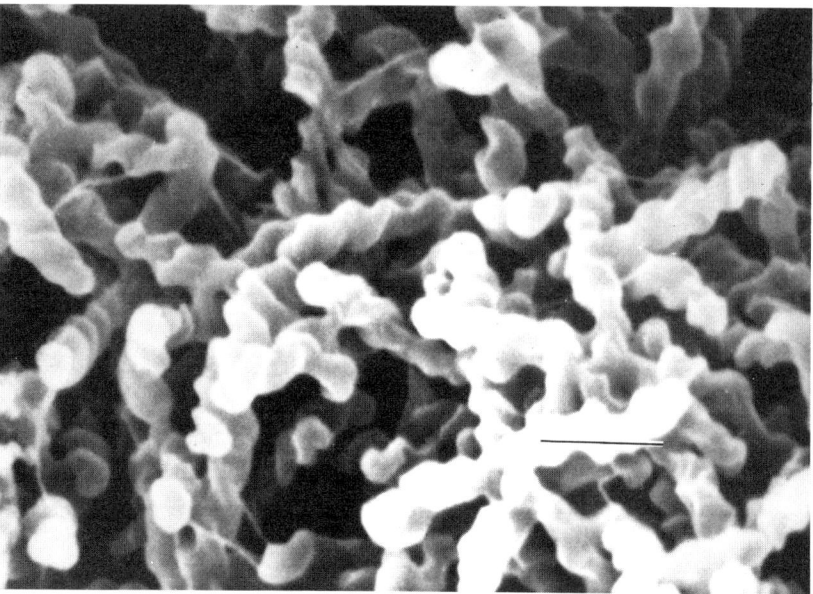

Figure 13.18. Scanning electron micrograph of *Seliberia stellata* showing the helically sculptured topography of the cells. *Bar*, 1 μm. (Courtesy of J. R. Swafford)

Important Notes for Users of This Manual

Unless otherwise indicated in footnotes to tables, the meanings of symbols are as follows:

+ 90% or more of strains are positive

− 90% or more of strains are negative

d 11-89% of strains are positive

v strain instability (not equivalent to "d")

D Different reactions in different taxa (species of a genus or genera of a family)

All other symbols are defined in footnotes to tables.

Table 13.1 Differentiation among genera of *heterotrophic prosthecate bacteria* (Subgroup 1) [a,b]

Characteristic	*Ancalomicrobium*	*Asticcacaulis*	*Caulobacter*	*Dichotomicrobium*	*Filomicrobium*	*Hirschia*	*Hyphomicrobium*	*Hyphomonas*	*Labrys*	*Pedomicrobium*	*Prosthecobacter*	*Prosthecomicrobium*	*Stella*	*Verrucomicrobium*
Polar prosthecae only	−	−	+	−	+	+	+	+	−	−	+	−	−	−
Subpolar/lateral prosthecae only	−	+	−	−	−	−	−	−	−	−	−	−	−	−
Multiple prosthecae, all sites on cell	+	−	−	2–4	−	−	−	−	−	+	−	+	−	+
Several prosthecae in one plane	−	−	−	−	−	−	−	−	+	−	−	−	+	−
Multiply by budding	+	−	−	+	+	+	+	+	+	+	−	+	−	?
Multiply by fission	−	+	+	−	−	−	−	−	−	−	+	−	+	?
Prosthecal bud formation	−	−	−	+	+	+	+	+	−	+	−	−	−	−
Prosthecal pili	−	−	−	±	−	−	−	−	−	−	−	−	−	+
Prosthecae may exceed 2 μm in length	+	+	+	+	+	+	+	+	−	+	+	D	−	−
Flagella	−	+	+	−	−	+	+	+	−	+	−	D	−	−
Gas vacuoles	+	−	−	−	−	−	−	−	−	−	−	D	D	−
Preferred carbon source for growth:														
CH_3OH	−	−	−	−	−	−	+	−	−	−	−	D	−	−
Carbohydrates	+	+	D	+	+	+	−	−	+	+	+	+	−	+
Amino acids	−	−	D	−	−	−	−	+	−	−	−	D	+	−
Metal oxide deposition on cells:														
Fe(III)	−	−	−	−	−	−	D	−	−	D	−	−	−	−
Mn(IV)	−	−	−	−	−	−	D	−	−	D	−	−	−	−
Obligate aerobes	−	+	+	+	+	+	D	+	+	+	+	+	+	−
Denitrifiers	−	−	−	−	−	−	D	−	−	−	−	−	−	−
Fermenters	+	−	−	−	−	−	−	−	−	−	−	−	−	+
Mol% G + C	70–71	55–61	62–67	62–64	62	45–47	50–65	57–62	68	62–67	54–60	64–70	69–74	58–59

[a] Phototrophic prosthecate bacteria are treated in Group 10.
[b] Symbols: see standard definitions; ?, unknown.

Table 13.2 Differentiation among genera of nonprosthecate budding and/or stalked bacteria[a]

Characteristic	Planctomycetales (Subgroup 2)				Other Genera (Subgroup 3)						
	Gemmata	*"Isosphaera"*	*Pirellula*	*Planctomyces*	*Angulomicrobium*	*Blastobacter*	*Ensifer*	*Gallionella*	*Gemmiger*	*Nevskia*	*Seliberia*
Penicillin resistance (10–100 µg/ml)	+	+	+	+	–	–	–	–	–	–	–
Peptidoglycan in cell walls[b]	–	–	–	–	+	+	+	+	+	+	+
Stalks (acellular)	–	–	–	+	–	–	–	+	–	+	–
Flagella	+	–	+	+	–	D	+	+	–	?	+
Gliding motility	–	+	–	–	–	–	–	–	–	?	–
Crateriform structures	+	+	+	+	–	–	–	–	–	–	–
Budding	+	+	+	+	+	+	+	–	+	–	+
Helical cells	–	–	–	–	–	–	–	–	–	–	+
Rosette formation	–	–	D	D	–	–	–	–	–	+	+
Obligate aerobes/microaerophiles	–	+	–	–	+	D	+	+	–	?	D
Obligate anaerobes	–	–	–	–	–	–	–	–	+	?	–
Facultative anaerobes	+	–	+	+	+	D	–	–	–	?	D
Denitrifiers	–	–	–	–	+	D	?	–	–	?	D
Methanol utilizers		?			+	D	?	–	–	?	–
Metal oxide deposition:											
Fe(III)	–	–	–	D	–	–	–	+	–	–	–
Mn(IV)	–	–	–	D							
Mol% G + C	54–57	62	64	50–58	64–69	59–69	63–67	55	59	?	63–66

[a] Symbols: see standard definitions; ?, unknown.

[b] True for species that have been grown in pure culture.

Table 13.3 Differential characteristics of the species of the genus *Asticcacaulis*[a]

Characteristic	*A. biprosthecum*	*A. excentricus*
Number of prosthecae:		
1	–	+
2	+	–
Position of prostheca:		
Subpolar	–	+
Lateral	+	–

[a] Symbols: +, 90% or more of strains are positive; –, 90% or more of strains are negative.

Table 13.4 Differential characteristics of the species of the genus *Caulobacter*[a]

Characteristic	*C. bacteroides*	*C. crescentus*	*C. fusiformis*	*"C. glutinosus"*	*C. halobacteroides*	*C. henricii*	*C. intermedius*	*"C. kusnezovii"*	*C. leidyi*	*C. maris*	*C. subvibrioides*	*C. variabilis*	*C. vibrioides*
Morphology:[b]													
Long axis of cell curved	–	+	–	–	–	+	+	–	–	–	±	–	+
Nonstalked pole tapered	–	+	+	–	–	+	+	+	+	–	±	–	+
Stalk invariably central	+	+	+	+	+	+	+	+	+	+	+	–	+
NaCl:													
0.05% required for growth	–	–	–	–	+	–	–	–	–	+	–	–	–
4% tolerated	–	–	–	–	–	–	–	–	–	+	–	–	–
Carbon sources generally used:													
Carbohydrates	+	+	–	+	+	+	+	+	+	+	+	+	+
Amino acids	d	+	+	+	+	+	+	+	+	–	+	+	d
Other organic acids	+	+	–	–	+	d	+	d	d	–	d	d	d
Primary alcohols	–	+	–	–	–	–	–	–	–	–	–	–	–
Organic growth factors needed:													
Riboflavin	–	–	–	–		–	–	–	–		–	–	+
Biotin	–	–	–	d		–	+	–	–		–	–	–
Vitamin B$_{12}$	–	–	–	–		d	–	–	–		–	–	d
Amino acids	d	–	–	+		d	–	+	–		–	–	–
Other, unidentified	+	–	+	+		–	+	+	–		+	+	+
Colonies distinctly pigmented at all ages	d	–	+	–	–	+	–	+	–	–	d	d	d

[a] Symbols: +, 90% or more of strains are positive; ±, varies among cells within a clone; –, 90% or more of strains are negative; d, 11–89% of strains are positive.

[b] Morphology as observed during growth in 0.1% (w/v) peptone, 0.05% (w/v) yeast extract, 0.01% (w/v) MgSO$_4$·7H$_2$0 broth prepared in tap water or, for marine strains, in 0.05% (w/v) peptone, 0.05% (w/v) casamino acids prepared in 80% seawater; agitated cultures, 30°C.

Important Notes for Users of This Manual

Unless otherwise indicated in footnotes to tables, the meanings of symbols are as follows:

+ 90% or more of strains are positive

– 90% or more of strains are negative

d 11-89% of strains are positive

v strain instability (not equivalent to "d")

D Different reactions in different taxa (species of a genus or genera of a family)

All other symbols are defined in footnotes to tables.

Table 13.5   Characteristics differentiating among the species of the genus _Hyphomicrobium_[a]

Characteristic	_H. aestuarii_	_H. coagulans_	_H. facilis_	_H. hollandicum_	_H. methylovorum_	_H. vulgare_	_H. zavarzinii_
Mother cell ovoid	−	+	+	+	+	+	+
Mother cell bean-shaped	+	−	−	−	−	−	−
Mother cell width:							
0.5–1.2 μm	+	+	+	+	−	+	+
0.3–0.65 μm	−	−	−	−	+	−	−
Flagella polar	−	+	−	−	−	−	−
Flagella lateral	−	−	−	−	+	−	−
Flagella (1–3) subpolar	+	−	+	+	−	+	+
Holdfast, mother cell	−	+	−	−	−	−	−
Carbon sources:							
Acetate	+	−	(+)	−	−	+	(+)
Lactate	+	−	D	+	−	−	−
Succinate	+	NT	−	+	−	+	−
Aspartate	−	+	−	+	+	−	+
Nitrate reduction	+	−	−	−	−	+	+
Growth with 2.5% NaCl	+	NT	+	−	NT	+	+
Gelatin liquefaction	−	+	−	−	−	−	−
H₂S production	−	+	−	−	−	−	−
Mol% G + C	64	NT	59–60	62	61	61	65

[a] Symbols: see standard definitions; NT, not tested.

Table 13.6   Characteristics differentiating the species of the genus _Hyphomonas_[a]

Characteristic	_H. hirschiana_	_H. jannaschiana_	_H. neptunium_	_H. oceanitis_	_H. polymorpha_
Growth at 37°C	+	+	+	−	+
Require methionine	+	−	+	+	+
Require biotin	−	+	−	+	−
Reduce nitrate	+	+	+	+	−
Hemolysis	γ	α	α	γ	α/γ
Brown colony pigment (31–37°C)	−	+	−	−	−
Tween 80 (0.1–1%) inhibition	−	−	−	+	−
Ampicillin (10 μg) resistance	−	−	−	+	−
Streptomycin (10 μg) resistance	±	−	−	+	±

[a] Symbols: see standard definitions.

Table 13.7 Characteristics differentiating the species of the genus *Pedomicrobium*[a]

Characteristic	*P. americanum*	*P. australicum*	*P. ferrugineum*	*P. manganicum*
Iron [Fe(III)] deposition	+	+	+	−
Manganese [Mn(IV)] deposition	+	+	−	+
Intracytoplasmic membranes	+	+	−	+
Spindle-shaped cells	+	+	−	−
Spherical cells	−	−	−	+
Source	North America	Australia	Europe[b]	Europe[b]
Mol% G + C	64–65	63–65	65–67	65–67

[a] Symbols: see standard definitions.

[b] *P. ferrugineum* strains were isolated from podzolic soils in Germany; *P. manganicum* (one strain) was isolated from a European pond, but they presumably have a global distribution.

Table 13.8 Characteristics differentiating the species of the genus *Prosthecomicrobium*[a]

Characteristic	"*P. consociatum*"	*P. enhydrum*	*P. hirschii*	*P. litoralum*	"*P. mishustinii*"	*P. pneumaticum*	"*P. polyspheroidum*"
Short prosthecae (<1.0 mm)	+	+	+	+	+	+	+
Long prosthecae (>2.0 mm)	−	−	+	−[b]	−	−[b]	−
Lateral buds	−	−	−	−	−	−	+[c]
Flagella	−	+	+	−	−	−	+
Gas vacuoles	−	−	−	−	−	+	−
Sodium ion requirement; optimum salinity of 25 °/oo	−	−	−	+	−	−	−
Carbon source utilization:							
Maltose, cellobiose, lactose	−	+	−	+	+	+	+
Melibiose	−	−	−	+	−	+	ND
Rhamnose	+/−	+	−	+	+/−	+	−
Sorbitol	−	−	−	+	−	+	+
Pyruvate	−	+	+	+	+/−	−	−
Propionate	−	−	+	−	−	−	+
Agar digestion	−	−	−	+	−	−	−
Methanol	+/−	ND	+	ND	−	ND	−
Mol% G + C	66–68	66	68–70	66–67	64–65	69–70	64–67

[a] Symbols: +, 90% or more of strains are positive; −, 90% or more of strains are negative; ND, not determined; +/−, indefinite.

[b] Strains of both species rarely produce long appendages.

[c] These are formed rarely.

Table 13.9 Differentiation between species of the genus *Stella*[a]

Characteristic	*S. humosa*	*S. vacuolata*
Gas vacuoles	−	+
Growth in 1% NaCl	+	−
Mol% G + C	69–73	70–74

[a] Symbols: see standard definitions.

Table 13.10 Differentiation of the species of the genus *Pirellula*[a]

Characteristic	*P. marina*	*P. staleyi*
Cell size (μm)	0.7–1.5 × 1.0–2.0	0.9–1.0 × 1.0–1.5
Cell shape:		
Mature cell	Ovoid	Ovoid
Bud	Bean	Ovoid
Salinity requirement (^{0}/oo)	13–175	0
Carbon source utilization (0.1%)		
Glucose	+	+
Fructose	+	−
Glycerol	+	−
Mannitol	+	−
Source	Estuarine	Fresh water

[a] Symbols: see standard definitions.

Table 13.11 Differentiation among species of the genus *Planctomyces*[a]

Characteristic	*P. bekefii*	*P. brasiliensis*	*P. guttaeformis*	*P. limicola*	*P. maris*	*P. stranskae*
Cell shape	Coccus	Spheroid	Bulb	Ovoid	Ovoid	Bulb
Stalk	+	+	−	+	+	−
Other appendages:						
Fimbriae	+	+	?	+	+	+
Distal spike	2 or more	−	1 long	−	−	−
Pure culture	No	Yes	No	Yes	Yes	No
Colony pigmentation		Yellow		Red	Light rose	
Source	Fresh water	Salt pit	Fresh water	Fresh water	Marine water	Fresh water
Salinity growth range		0.7–10.0%		<1.0%	1.5–4.0%	
Mol% G + C		55–58		53	51	

[a] Symbols: see standard definitions; ?, unknown.

475

Table 13.12 Differential characteristics of the species of the genus *Blastobacter*[a]

Characteristics	B. aggregatus	"B. aminooxidans"	B. capsulatus	B. denitrificans	B. henricii	B. natatorius	"B. viscosus"
Initial bud shape:							
Rod	+	–	–	+	–	–	–
Ovoid or spherical	–	+	+	–	+	+	+
Exopolymer and capsule	–	+[b]	+	–	–	–	+
Rosette formation	+	–	–	–	+	+	–
Motility	+	–	–	+	–	+	–
Colony pigmentation yellow	–	+	–	–	ND	+	+
Utilization of methanol (0.5%, w/v)	–	+	–	–	ND	+	+
Utilization of ethanol (0.4%, w/v)	+	+	–	+	ND	–	+
Acid from D(+)-glucose (aerobically)	+	–	+	–	ND	+	+
NO$_3^-$ reduction (assimilatory)	–	+	+	–	ND	–	+

[a] Symbols: –, 90% or more of strains are negative; +, 90% or more of strains are positive; ND, not determined.
[b] Fibrillar capsule.

Important Notes for Users of This Manual

Unless otherwise indicated in footnotes to tables, the meanings of symbols are as follows:

    + 90% or more of strains are positive

    – 90% or more of strains are negative

    d 11-89% of strains are positive

    v strain instability (not equivalent to "d")

    D Different reactions in different taxa (species of a genus or genera of a family)

All other symbols are defined in footnotes to tables.

GROUP 14

SHEATHED BACTERIA

Sheathed bacteria grow as chains of cells in filaments, 0.4–7 μm in width. Gram negative. **Filaments grow in a tube of extracellular material referred to as a sheath.** Typically, the sheath is transparent when viewed in wet mounts by phase microscopy, appearing like a microscopic plastic tubule or pipe, usually but not always containing cells. Occasionally, the sheath is so thin and closely associated with the cells that it cannot be readily discerned by phase microscopy. The addition of 95% ethanol to the wet mount may help enable its visualization. Alternatively, it may be detected where there are gaps between cells of the filament (although lysed cells within a filament of ordinary chain-forming bacteria may give the false impression that a sheath is present). Sheaths may appear yellow to dark brown because of the deposition of iron and manganese oxides. Sheathed bacteria are found in aquatic habitats. The motile species in pure culture treated in this group have flagella.

Sheathed bacteria studied in pure culture are **aerobic chemoheterotrophs** that **use organic acids and sugars** as carbon sources. **Some deposit iron and manganese oxides in or on their sheaths** and may carry out the oxidation of Fe(II) and Mn(II); however, they are not known to obtain energy from this process. The sheathed bacteria are found in aquatic habitats including lakes, streams, and springs. Some are also found in wastewater treatment systems.

Other sheathed procaryotic organisms are treated in other groups. For example, the genera *Herpetosiphon, Thioploca,* and *Thiothrix* are included in Group 15, the "Nonphotosynthetic, Nonfruiting Gliding Bacteria," and certain of the cyanobacteria (e.g., the genus *Lyngbya,* Group 11) have sheaths.

Differentiation of the genera in **Group 14:** See Table 14.1.

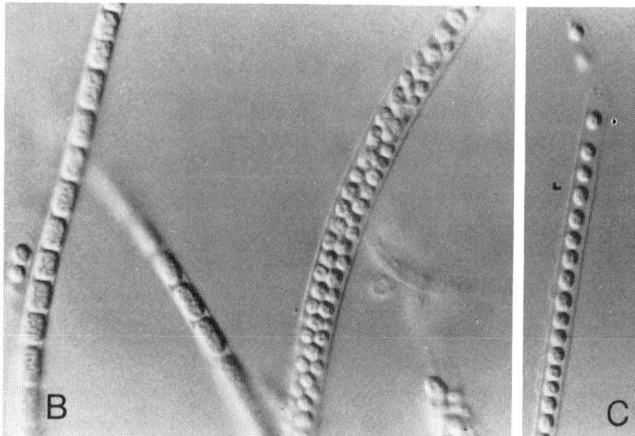

Figure 14.1 *Crenothrix polyspora* from a freshwater source. **A.** Filaments with normal cross-septation and the beginning of longitudinal septation. **B.** Normal filaments and the ensheathed and inflated part of a *Crenothrix* filament with microgonidia. **C.** Macrogonidia released from the tip of an open filament. Nomarski interference contrast light micrographs. (× 620) (Courtesy of P. Hirsch)

Genus "Clonothrix"

Filaments 3–7 μm wide and up to 1.5 cm long, **attached or free,** surrounded by a more or less distinct sheath that **may be encrusted with iron or manganese oxides**, giving a yellowish-brown color. **Filaments taper** and may occur singly or show false branches. Cells are cylindrical, colorless, or bluish, 2–2.5 × 12–18 μm, being larger at the base and smaller at the tip of the filament. Reproduction is by separation of individual cells, followed by breakage of the sheath and release to the outside; such cells frequently attach themselves parallel to older cells in the sheath and produce new filaments, resulting in false branches. These organisms have not been grown in pure culture.

Type (and only) species: *"Clonothrix fusca."*

Characteristics of the species: As described for the genus.

Genus **Crenothrix**

Cells are cylindrical to disk-shaped, 0.6–5.0 μm in diameter, dividing by cross-septation to **form sheathed filaments** up to 1 cm long and **often attached to a firm substrate.** The **very thin sheaths** surrounding the filaments **may be colorless at the tip or encrusted with iron** (or manganese) **oxides at the base.** Filaments are unbranched, but false branching may occur. Filaments may show increased cross-septation on one or both free ends to **form spherical propagation cells, "macrogonidia,"** of the same diameter as the rod-shaped cells. Presumably, these macrogonidia are liberated from the sheath ends and may form new sheathed filaments. Some **filaments widen at the end(s) where the cells undergo cross-septation and longitudinal septation** (Figure 14.1). Numerous, small, **cubical cells arising in this fashion ("microgonidia") are rounded and released from the filament.** Gram negative. Some cell chains show a gliding motility. Found in stagnant and running waters containing low concentrations of organic matter, Fe^{2+}, and traces of methane. These organisms have not been grown on artificial media in pure culture.

Type (and only) species: *Crenothrix polyspora.*

Characteristics of the species: As described for the genus.

Genus **Haliscomenobacter**

Thin rods, 0.4–0.5 × 3–5 μm, usually **in chains, enclosed by a** narrow, barely visible **hyaline sheath. No ferric or manganic oxides are deposited** in or on the sheaths. **Branching of the filaments may** incidentally **occur** in stationary culture (Figure 14.2). Branching cells disrupt the sheath and form a new sheath outside the envelope. Lateral branches are usually short. Cells outside the sheaths are rare; **flagellation** and the motility of these cells have **not** been **observed.** Gram negative. Colonies on poor agar media are filamentous and very small (i.e., < 0.5 mm in diameter). On a sucrose-peptone-yeast extract medium enriched with vitamin B_{12} and thiamine, pinkish, smooth, slightly filamentous colonies 1–3 mm in diameter develop. Liquid cultures turn pink because of the formation of carotenoid pigments. Chemoorganotrophic and **aerobic. Glucose, glucosamine, lactose, sucrose, starch, and,** to a lesser extent, **mannitol serve as carbon and energy sources;** acetate, lactate, succinate, β-hydroxybutyrate, glycerol, and sorbitol are not utilized. Inorganic nitrogen compounds (nitrate, ammonium salts) are moderately good nitrogen sources; amino acids and peptone are preferred. Thiamine and vitamin B_{12} are required. Temperature range for growth is 8–30°C; the optimum is about 26°C. Optimum pH is 7.5.

Type (and only) species: *Haliscomenobacter hydrossis.*

Characteristics of the species: As described for the genus.

Genus **Leptothrix**

Straight rods, 0.6–1.4 × 1–12 μm, **occurring in chains within a sheath** or free-swimming as single cells, in pairs, or, in some species, as motile short chains containing up to eight cells. One species has well-developed holdfasts (Table 14.2). Free cells are **motile by** means of **a single polar flagellum** (one species has a subpolar tuft of several flagella). Most species may contain globules of poly-β-hydroxybutyrate (PHB) as reserve material. **Sheaths** have a pronounced tendency to become **impregnated or covered with ferric and manganic oxides** (Figure 14.3). Encrusted sheaths are often empty because cells have left (Figure 14.4). Resting stages are not known. Gram negative. Chemoorganotrophs; **metabolism is respiratory,** never fermentative. Growth and manganese oxidation may proceed at low oxygen tensions. Temperature range is 10–35°C, with optimum around 25°C. Optimum pH is 6.5–7.5. **A number of sugars** (including glucose, fructose, and sucrose), **organic acids** (including lactic, malic, and β-hydroxybutyric acids), **and glycerol are utilized** by most species **as carbon and energy sources.** Complex nitrogen compounds such as casamino acids

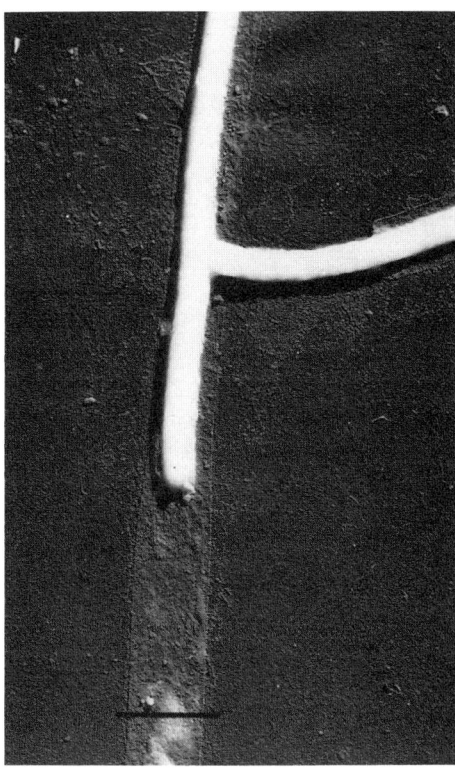

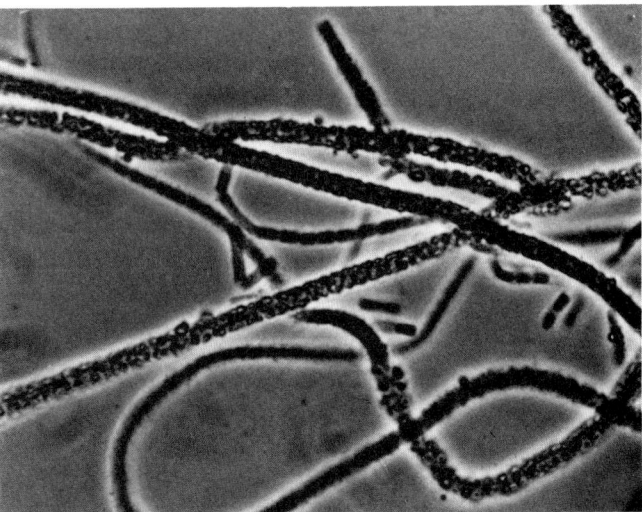

Figure 14.3 *Leptothrix cholodnii* sheaths encrusted with manganese oxide (× 1280). (Reproduced with permission from E. G. Mulder and W. L. van Veen, Antonie van Leeuwenhoek J. Microbiol. Serol. *29:* 121-153, 1963)

Figure 14.2 *Haliscomenobacter hydrossis* branched filament with thin hyaline sheath. Electron micrograph. *Bar,* 1 µm. (Reproduced with permission from E. G. Mulder and M. H. Deinema. 1981. *In* Starr, Stolp, Trüper, Balows, and Schlegel (Eds.), *The Prokaryotes: A Handbook on Habitats, Isolation, and Identification of Bacteria,* ©Springer-Verlag, Berlin, pp. 425-440)

or peptone or mixtures of aspartic and glutamic acids are preferred nitrogen sources. Some species use NH_4^+ and NO_3^-, but most poorly growing stains do not. Unless methionine is supplied, vitamin B_{12} should be added to the nutrient medium. For some strains, a requirement for biotin and thiamine has been reported. Adenine and guanine are also recorded in the literature as essential growth factors from some strains.

Type species: *Leptothrix ochracea.*

Differentiation of the species of the genus **Leptothrix:** See Table 14.2.

Genus "Lieskeella"

Cells are rod-shaped with rounded ends, 0.6 × 2–3 µm, **in chains.** Usually **two filaments are wound**

around one another to give a **double spiral that is surrounded by a yellowish, slimy capsule,** often with heavy deposits of small ferric hydroxide granules. When the deposits are dissolved with dilute HCl, the chains fragment and the individual rods appear as a double zig-zag band. **Cells show bipolar straining when treated with methylene blue.** Cell chains may aggregate in distinct layers or solid skeins. **Filaments disaggregate rapidly upon removal from the natural environment.** There is a **slow but incessant motion** similar to that of filamentous cyanobacteria. The Gram reaction has not been recorded. Presumably aerobic and chemoorgano-trophic but have not been studied in pure culture.

Type (and only) species: *"Lieskeella bifida."*

Characteristics of the species: As described for the genus.

Genus "Phragmidiothrix"

Cells of variable size, usually small and **disk-shaped,** with a diameter 4–6 times the thickness of the cell. Arranged **in colorless, articulate, and unbranched filaments,** 3–6 µm in diameter and over 100 µm in length. **Filaments are attached, forming** grayish white **tufts.** The free end may be of larger diameter than the base, but there is no tapering in either direction. Cells are surrounded by a very **thin, delicate, gelatinous,**

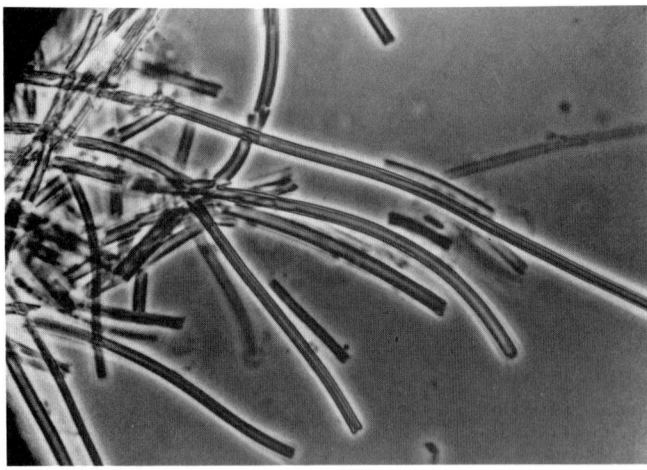

Figure 14.4 Empty sheaths of *Leptothrix ochracea* encrusted with iron and manganese oxides as observed by phase microscopy (× 1186). (Reproduced with permission from E. G. Mulder and W. L. van Veen, Antonie van Leeuwenhoek J. Microbiol. Serol. *29:* 121-153, 1963.)

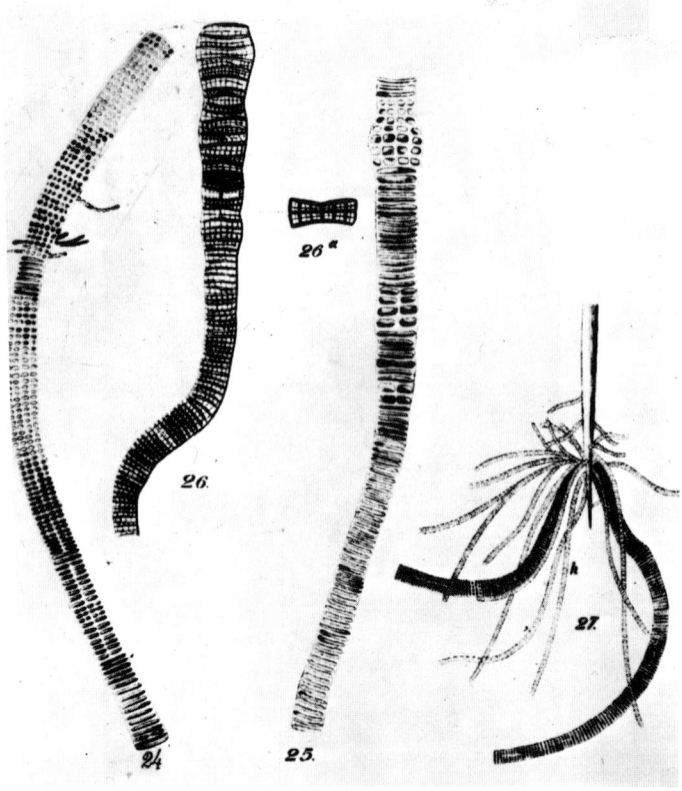

Figure 14.5 *24,* "*Phragmidiothrix multiseptata*" filament showing arrangement and septation of disk-shaped cells; *25,* part of filaments showing enlarged end and individual disk-shaped cells of uneven diameter; *26,* part of filament with much cross-septation and longitudinal septation; *26a,* part of filament at higher magnification; *27,* a bristle of *Gammarus locusta* with two attached filaments of "*Phragmidiothrix*" (wide filaments) and several narrow filaments of young *Beggiatoa alba.* (Approximately ×330) (Reproduced from A. Engler, Über die Pilzvegetation des Weissen oder Todten Grundes in der Kieler Bucht. Vierter Ber. d. Kommission z. wiss. Untersuchg. der Meere, Abt. I, 187-193, 1883.)

colorless sheath that is **not encrusted** with iron or manganese compounds. Cell walls are distinctive and of even thickness throughout the filament. Multiplication is by cross-septation and, in certain regions of the filament, by both cross-septation and longitudinal septation, forming *Sarcina*-like aggregates of small, cuboidal propagation cells. Septating cells have greater diameters, giving rise to irregular filament widths (Figure 14.5). This genus has not been studied in pure culture.

Type (and only) species: *"Phragmidiothrix multiseptata."*

Characteristics of the species: As described for the genus.

Genus **Sphaerotilus**

Straight rods, 1.2–2.5 × 2–10 μm, usually arranged **in single chains within sheaths** of uniform width, **which may be attached by** means of **holdfasts** to walls of containers, submerged plants, stones, and other surfaces. Single or paired cells released from the sheaths are **motile by** means of **a bundle of subpolar flagella,** sometimes so intertwined as to give the appearance of a single large "unit flagellum." **Sheaths are usually thin without encrustation by ferric and manganic oxides.** They cannot always be easily recognized when completely filled with cells. If parts of the sheaths are vacated by cells, recognition of the organism cannot be misinterpreted (Figure 14.6). Resting stages are not known. Gram negative and may store poly-β-hydroxybutyrate (PHB) granules. Chemoorganotrophs; **metabolism is respiratory,** never fermentative. Grow at very low concentrations of dissolved oxygen (below 0.1 mg/L). Temperature range is 10–37°C; optimum is 20–30°C. Optimum pH is 6.5–7.6. **Alcohols, several organic acids, and sugars are used as sources of carbon and energy.** Ammonium salts and nitrates may serve as the nitrogen source in the presence of vitamin B_{12} or methionine, but growth is better with peptone, casamino acids, mixtures of aspartic and glutamic acids, and vitamin B_{12}.

Type (and only) species: *Sphaerotilus natans.*

Characteristics of the species: As described for the genus.

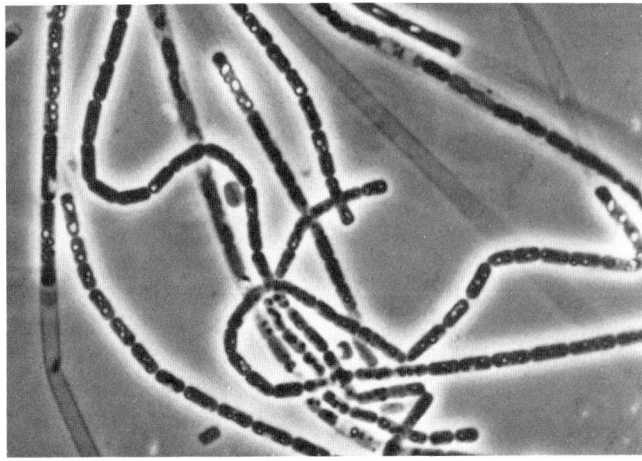

Figure 14.6 Filaments of *Sphaerotilus natans* as observed by phase microscopy. Note the empty sheaths that appear as transparent tubes. The bright areas in some cells are PHB granules. (×1118) (Reproduced with permission from E. G. Mulder and W. L. van Veen, Antonie van Leeuwenhoek J. Microbiol. Serol. *29:* 121-153, 1963)

Important Notes for Users of This Manual
Unless otherwise indicated in footnotes to tables, the meanings of symbols are as follows:

    + 90% or more of strains are positive
    − 90% or more of strains are negative
    d 11-89% of strains are positive
    v strain instability (not equivalent to "d")
    D Different reactions in different taxa (species of a genus or genera of a family)

All other symbols are defined in footnotes to tables.

Table 14.1 Differentiation among genera of heterotrophic sheathed bacteria [a]

Characteristics	"Clonothrix"	Crenothrix	Haliscomenobacter	Leptothrix	"Lieskeella"	"Phragmidiothrix"	Sphaerotilus
Nontapering filaments	-	-	+	+	+	-	+
Tapering filaments	+	+	-	-	-	+	-
Flagellated cells	-	-	-	+	-	-	+
False branching of filaments [b]	+	+	-	D	-	-	+
True branching of filaments	-	-	+	-	-	-	-
Filament diameter (μm)	3–7	0.6–5.0	0.4–0.5	0.6–1.4	0.6	3–6	1.2–2.5
Metal oxide enumeration:							
Fe(III)	D	D	-	+	-	-	D
Mn(IV)	D	D	-	+	-	-	-
Disk-shaped cells	-	-	-	-	-	+	-
Longitudinal cell septation	-	+	-	-	-	+	-
Mol % G + C	?	?	49	70–71	?	?	70

[a] Symbols: see standard definitions; ?, unknown.
[b] False branches occur when cells attach to an existing sheathed filament and develop a new filament, which appears as a branch. True branches are formed as an outgrowth of a filament.

Table 14.2 Differentiation among species of Leptothrix [a]

Characteristics	L. cholodnii	"L. discophora"	L. lopholea	L. ochracea	"L. pseudo-ochracea"
Cells size (μm):					
Width	0.17–1.3	0.6–0.8	1.0–1.4	1.0	0.8–1.3
Length	2–7	1–4	3–7	2–4	5–12
Flagella:					
Polar monotrichous	+	+	-	+	+
Subpolar polytrichous	-	-	+	-	-
Holdfasts	-	-	+	-	-
Response to high nutrient concentration	+	-	-	?	+/-
False branches	-	+/-	+	-	-
Pure cultures available	+	+	+	-	+

[a] Symbols: see standard definitions; ?, unknown.

NONPHOTOSYNTHETIC, NONFRUITING GLIDING BACTERIA

Organisms in this group have the ability to **move upon a solid surface without any visible means of locomotion (gliding motility).** All are **Gram negative.** They are **morphologically diverse** and may be cocci, spirals, long flexible rods, filaments that may or may not be bound by a sheath, flexible gliding trichomes, and rosette formers. Some may contain gas vesicles, and others contain sulfur inclusions. With the possible exception of some strains of *Beggiatoa* which have been grown chemoautotrophically, all organisms in this group that have been obtained in pure culture have been grown chemoheterotrophically; however, some may be capable of mixotrophic growth and, perhaps, others of chemoautotrophic growth.

Key to the subgroups and other genera of **Group 15**

I. Cells occur singly.

    A. Cells are spherical, or nearly so. Some are large (> 5 μm in diameter) and elongate.

        1. Cells deposit sulfur globules when in the presence of hydrogen sulfide.

            Genus *Achromatium* (see Subgroup 3)

        2. Cells do not deposit sulfur globules when in the presence of hydrogen sulfide. Twitching, gliding motility.

            Genus *Agitococcus*

    B. Rod-shaped cells that grow singly and may range from short (i.e., < 10 μm) to long (i.e., 10 to > 100 μm) and filamentous. Rarely multicellular.

        Subgroup 1. Single-celled gliding bacteria

II. Cells that occur as multicellular filaments.

    A. Multicellular filaments (trichomes) that are flattened, with the long axis of the individual cells perpendicular to the long axis of the filament. From the oral cavity of warm-blooded vertebrates.

        Subgroup 2. Flattened filamentous gliding bacteria

B. Multicellular filamentous bacteria (one genus is single celled) that produce intracellular sulfur globules when in the presence of hydrogen sulfide.

Subgroup 3. Gliding sulfur-oxidizing bacteria

C. Multicellular filamentous gliding bacteria that do not produce sulfur globules when in the presence of hydrogen sulfide.

1. Filaments are helically or spirally coiled. Do not produce a sheath.

Genus *Saprospira*

2. Rod-shaped cells that form long filaments enclosed within a sheath. Not coiled.

Genus *Herpetosiphon*

3. Filaments are composed of spherical cells.

Genus *"Isosphaera"*

D. Rod-shaped cells that form long, flexible gliding filaments. Strictly anaerobic, using sulfate and other oxidized sulfur compounds as electron acceptors. Produce hydrogen sulfide.

Genus *Desulfonema* (see also Group 7)

E. Rod-shaped cells that form long, U-shaped multicellular trichomes. The rounded, central portion of the trichome glides forward, occasionally rising off of the surface temporarily and redepositing at another site. The ends of the trichome rotate, causing parallel tracks to be made on the substrate. The polymer remaining in the tracks left by ends and the central portion may eventually become encrusted with iron.

Genus *Toxothrix*

F. Rod-shaped cells that grow in motile multicellular filaments. No pigments are produced. In the presence of hydrogen sulfide these organisms do not deposit sulfur inclusions.

Genus *Vitreoscilla*

G. Rod-shaped cells that form filaments that may be single or aggregated into bands or bundles. May contain gas vesicles but do not contain sulfur inclusions. Probably anaerobic.

Subgroup 4. The pelonemas

H. Rod-shaped cells that form nonmotile filaments. Gonidia are produced and are motile by gliding. Attach to a solid surface via a holdfast. Rosettes are produced. Usually of marine origin.

Genus *Leucothrix*

SUBGROUP 1: SINGLE CELLED, ROD-SHAPED GLIDING BACTERIA

Very **short to elongate Gram-negative rods,** which are often pleomorphic, **motile by gliding or nonmotile,** and usually yellow, orange, or red due to cell-bound pigments. **Often show a color shift** from yellow to shades of red, brown, or purple when the colonies are flooded **with alkali. Chemoorganotrophs,** aerobic, and facultatively or obligately anaerobic. Often **degrade** one or several **biomacromolecules** such as pro-teins (e.g., casein and gelatin), starch, dextran, yeast cell-wall glucans, cellulose, pectin, agar, or chitin. Free living in terrestrial, freshwater, and marine environments and in sewage plants, where they are very common and occur in large numbers. Some colonize fish and may be pathogenic. Others may colonize or invade humans and other animals and may be pathogenic under certain circumstances.

Differentiation of the genera in **Subgroup 1:** The genera in this group can be differentiated by the following key.

Key to the genera of **Subgroup 1**

I. Rod-shaped cells that grow singly and may range from short (i.e., < 10 µm) to long (i.e., 10 to > 100 µm) and filamentous. Rarely multicellular. Motile by gliding or nonmotile. Colonies are usually yellow, orange, or red due to the production of nondiffusible pigments. In the presence of sulfide these organisms do not deposit sulfur. The mol% G+C of the DNA ranges from about 30–50.

   A. Morphologically differentiated resting cells are not produced.

      1. Short to elongated (i.e., typically < 15 µm in length) gliding rods.

         a. Strictly aerobic and facultatively anaerobic free-living species of terrestrial, freshwater, and marine habitats, with some colonizing fish.

            Genus *Cytophaga*

         b. Capnophilic, do not grow under a normal atmosphere; known only from the oral cavity.

            Genus *Capnocytophaga*

      2. Very long (i.e., up to 5–100 µm or longer) gliding or nonmotile filaments.

         a. Freshwater or soil organisms.

            i. Thermophilic. Grow at 60°C.

               Genus *Thermonema*

            ii. Not thermophilic. Do not grow at 60°C.

               Genus *Flexibacter*

         b. Marine organisms

            i. Sheaths not produced.

               Genus *Microscilla*

485

    ii. Sheaths may be produced.

        Genus *Flexithrix*

  B. Resting cells (microcysts) are produced

    1. Cellulose degraded

        Genus *Sporocytophaga*

    2. Chitin degraded

        Genus *Chitinophaga*

II. Rod-shaped cells that grow singly and may become very long (up to 70 µm). Colonies are cream, pink, or yellow-brown. May produce a brown water-soluble pigment. Degrade chitin. Lyse a wide variety of microorganisms. The mol% G+C of the DNA ranges from 65–71.

    Genus *Lysobacter*

Genus **Capnocytophaga**

Short to elongate flexible rods or filaments 0.4–0.6 µm in diameter and 2.5–5.7 µm in length. Ends of cells are usually round to tapered. Cells can be pleomorphic. Capsules and sheaths are not formed. Resting stages are not known. **Motile by gliding.** Gram negative. **Facultatively anaerobic. Growth** occurs in air supplemented **with 5.0% CO$_2$.** Optimum temperature is 35–37°C. Some strains are reported to grow aerobically without CO$_2$. **Primary isolation** and initial in vitro growth **requires CO$_2$.** Colonies are yellow-orange. Chemoorganotrophic, with **fermentative** type **metabolism.** This organism utilizes variable carbohydrates as fermentable substrates and energy source. Fermentation of glucose yields chiefly **acetate and succinate** as **major acidic end products;** there are trace amounts of isovalerate. Polysaccharides such as dextran, glycogen, inulin, or starch may be **fermented.** *O*-nitrophenyl-β-D-galactoside (ONPG)-positive. Found in association with animal and human hosts. May be pathogenic; they are **frequent isolates from oral sites** and are also recovered from pulmonary lesions, blood, and abscesses, as well as from healthy oral and nonoral sites in their hosts.

Type species: *Capnocytophaga ochracea.*

Differentiation of the species of the genus **Capnocytophaga:** See Table 15.1.

Editorial note: C. canimorsus and *C. cynodegmi* were not included in *Bergey's Manual of Systematic Bacteriology.*

These species were described by Brenner et al. (Int. J. Syst. Bacteriol. *40:* 105–106, 1990; effective publication: J. Clin. Microbiol. *27:* 231–235, 1989).

Genus **Chitinophaga**

Flexible rods with rounded ends, 0.5–0.8 µm by ca. 40 µm when fully developed. The rods occur singly. **A resting stage (microcyst), 0.8–0.9 µm in diameter is formed,** but is not highly refractile. **Macroscopic fruiting bodies are not formed. Motile by gliding and flexing.** Gram negative. Aerobic. Optimum temperatures are: 23–24°C; maximum: 37–40°C; and minimum: 10–12°C. Optimum pH for growth is 7; maximum: 8–10; minimum: 4. **Chemoorganotrophic,** oxidative, or fermentative. Acid but no gas produced from some carbohydrates. **Chitin is hydrolyzed, but cellulose and agar are not.** Congo red is absorbed. Cell masses are yellow.

Type (and only) species: *Chitinophaga pinensis.*

Characteristics of the species: As described for the genus.

Genus **Cytophaga**

Very short to moderately long rods, 0.3–0.8 × 1.5–15 µm, only rarely longer with rounded or slightly tapered ends. The longer rods are flexible. **Motile by gliding.**

Gram negative. On solid media with a low nutrient content (e.g., below 0.1% peptone), **the colonies are spreading swarms,** sometimes penetrating the agar, often very delicate and occasionally with a reddish or greenish iridescence. On substrates with a higher nutrient content (e.g., above 0.3% peptone), the colonies usually become compact, often convex, with a smooth or wavy edge, sometimes sunken into the agar. **Cell mass is usually** more or less intensely pigmented, **yellow, orange, or red,** due to cell-bound carotenoids, flexirubin type pigments, or both. When covered with alkali (e.g., 20% KOH solution: flexirubin reaction), colonies may reversibly change their color from yellow to purple- or red-brown. Unpigmented species or strains also occur. **Strict aerobes** or **facultative anaerobes.** Some may use NO_3^- as terminal electron acceptor; **Chemoorganotrophs. Metabolism is respiratory or fermentative.** In the latter case, acetate, propionate, and succinate may be produced. Organic acids may, however, also arise during growth of strictly aerobic stains, particularly on sugar-containing media. **All decompose** one or several kinds of **organic macromolecules,** mainly various proteins and polysaccharides, including cellulose, agar, chitin, pectin, and starch. Optimum temperature is 20–35°C. Optimum pH is around 7. Common in soil, decomposing organic matter, freshwater, and marine habitats.

Type species: *Cytophaga hutchinsonii.*

Differentiation of the species of the genus **Cytophaga:** See Table 15.2.

Genus **Flexibacter**

Rod-shaped cells of variable length, typically 10–50 μm long, slender, 0.2–0.6 μm wide, flexible, and with tapering or rounded ends (Figure 15.1). **Gliding,** but nonmotile stages may occur. **In some species** there is a **(cyclic) change in cell morphology.** In young liquid cultures or on plates at the edge of spreading colonies or swarms, the cells are thread-like, 20–50 μm and more in length, and very agile, actively twisting and bending (Figure 15.2). These thread-shaped cells have no cross-walls or, at best, very few cross-walls. In aging cultures or behind the swarm edge, the cells become shorter and shorter until only short, stout, nonmotile and phase-dark rods are left, clearly fatter than the thread cells, sometimes almost coccoid. **Cell mass** usually **yellow to orange,** sometimes very pale or even colorless. The **pigments** are cell-bound and often **carotenoids and/or of the flexirubin type.** In the latter case, the colonies

show an immediate and **reversible color change** to **red-brown or purple when covered with alkali solution** (e.g., 20% KOH; reversed by 1% HCl). Gram negative. **Chemoorganotrophic.** Strictly aerobic or facultatively anaerobic. Peptones and amino acid mixtures serve as nitrogen sources. Various sugars are utilized. Some grow on simple defined media, e.g., with glucose and NH_4^+. Cellulose and agar are not attacked, but chitin and starch are often degraded. Found widely distributed and common in soil and freshwater.

Type species: *Flexibacter flexilis.*

Differentiation of the species of the genus **Flexibacter:** See Table 15.3.

Editorial note: The species *F. maritimus* was not included in *Bergey's Manual of Systematic Bacteriology.* It was described by Wakabayashi et al., (Int. J. Syst. Bacteriol. *36:* 396–398, 1986).

Genus **Flexithrix**

Very long **nonmotile, multicellular uniseriate and sheathed and may release, as a second form, flexible gliding cells** (Figure 15.3). Gram negative. Depending on the culture conditions, the organism may grow alternatively in the one or the other form. For example, liquid cultures in relatively rich media contain exclusively flexible rods that give the suspension a silky appearance when shaken. In such cultures, the organism may be mistaken for a *Microscilla* or a *Flexibacter.* **Chemoorganotrophic** with respiratory metabolism. Found in marine habitats.

Type (and only) species: *Flexithrix dorotheae.*

Characteristics of the species: As described for the genus.

Genus **Lysobacter**

Thin rods, 0.2–0.5 × 1.0–15 μm (sometimes up to 70 μm). Gram negative. Gliding motility. **Flexible. Colonies are highly mucoid, cream-colored, pink, or yellow-brown;** many strains also produce a brown, water-soluble **pigment.** Growth in broth culture is silky. Most strains are resistant to actinomycin D. **Degrade chitin** and often other polysaccharides, but do not degrade filter-paper cellulose, and infrequently degrade agar. They are strongly proteolytic and characteristically **lyse a variety of microorganisms** (Gram-negative and

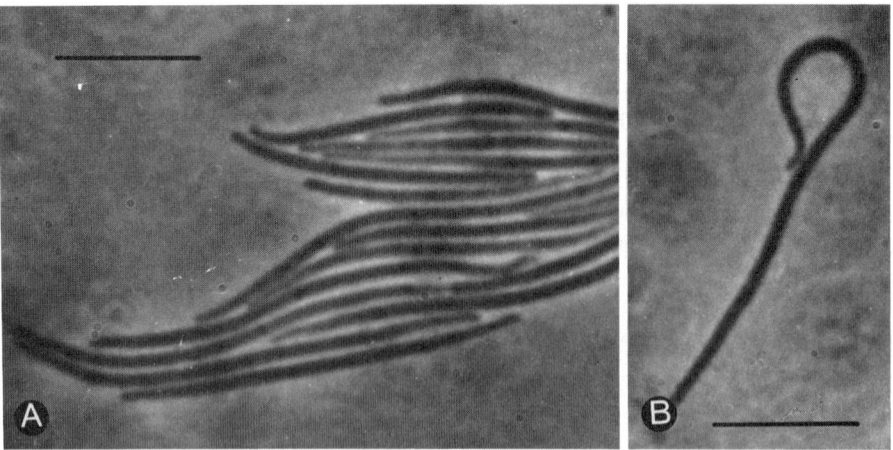

Figure 15.1. *Flexibacter flexilis* type strain cells on starch-casitone agar. Culture 2 days old. Phase contrast. *A*, characteristic arrangement on agar surfaces. *Bar*, 6 μm. *B*, high flexibility typical of the organism. *Bar*, 6 μm.

Gram-positive bacteria including actinomycetes, blue-green and green algae, yeasts, and filamentous fungi) and also nematodes. Found in soil and fresh water.

Type species: *Lysobacter enzymogenes.*

Differentiation of the species of the genus **Lysobacter:** See Table 15.4.

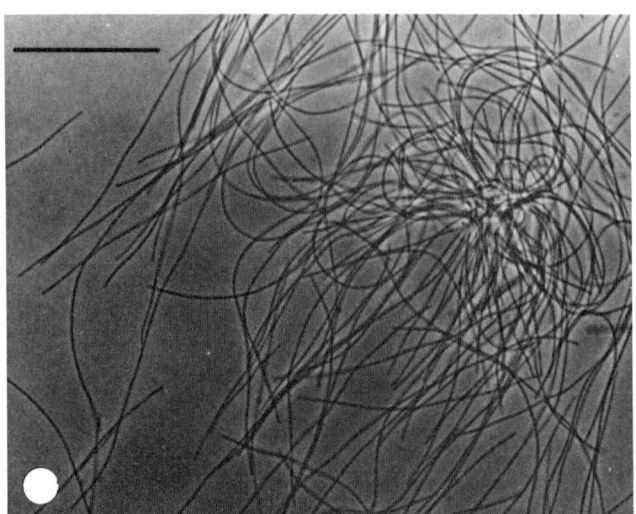

Figure 15.2. *Flexibacter elegans* type strain threads from glutamate-yeast extract-glucose medium. Phase contrast in liquid. *Bar*, 24 μm.

Genus **Microscilla**

Long, thin, flexible, thread-like rods usually measuring from 10 to > 100 μm. **Motile by gliding. Cell mass** is more or less intensely **orange or yellow.**

Strictly aerobic chemoheterotrophs. All **grow on peptones** as sole source of nitrogen. Chitin and cellulose are not attacked, but other polysaccharides including carboxymethyl cellulose (CMC) may be decomposed. **Marine,** from coastal habitats. Do not grow below half-strength sea water.

Type species: *Microscilla marina.*

Differentiation of the species of the genus **Microscilla:** See Table 15.5.

Genus **Sporocytophaga**

Flexible rods with rounded ends 0.3–0.5 × 5–8 μm, occurring singly. Gram negative. Sphaeroplasts and distorted cells occur in older cultures. A resting stage, the **microcyst,** is formed. Motile by **gliding.** Chemoorganotrophs. Metabolism is respiratory with molecular oxygen used as terminal electron acceptor. **Cellulose, cellobiose, glucose,** and (for some strains) **mannose are the only known sources of carbon and energy.** Agar and chitin are not known to be attacked. Either ammonium or nitrate ions, or peptone, urea, or yeast extract can serve as sole nitrogen sources. Amino acids, peptones, yeast extract, or nutrient agar (Difco) cannot serve as sole carbon and energy sources. No organic growth factor requirements are known. Catalase positive. Strict aerobe. Optimum temperature is about 30°C.

Type (and only) species: *Sporocytophaga myxococcoides.*

Editorial note: This genus is not included in *Bergey's Manual of Systematic Bacteriology.* It was described in 1989 by Hudson et al. (Int. J. Syst. Bacteriol. *39:* 485–487).

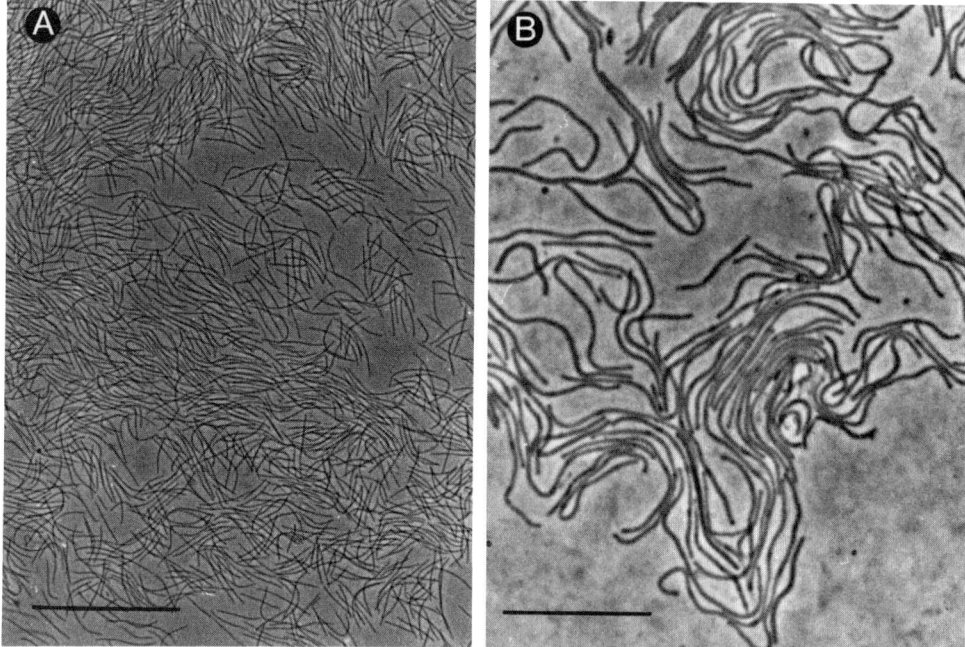

Figure 15.3. *Flexithrix dorotheae* gliding cells from agar cultures. Slide mounts, phase contrast. *A*, survey. *Bar*, 50 μm. *B, bar*, 12 μm.

Genus **Thermonema**

Unicellular filaments, usually about 0.25–0.3 μm in diameter and **60 μm** up **to several hundred μm in length.** Gram negative. Aerobic. Chemoheterotrophic. Produce an orange carotenoid pigment and do not produce flexirubin. Motile by gliding. Aminopeptidase positive. Oxidase and catalase positive. **Thermophilic, growing optimally at 60°C** and at temperatures up to 70°C. Alpha- and beta-galactosidase negative and DNAse positive. Proteolytic; **do not hydrolyze cellulose or starch.** Isolated from hot springs.

Type (and only) species: *Thermonema lapsum.*

Characteristics of the species: As described for the genus.

SUBGROUP 2: FLATTENED, FILAMENTOUS GLIDING BACTERIA

Multicellular filaments that are **flat rather than cylindrical. The width of an individual cell is greater than its length. Gliding motility** when the flat side of the filament is in contact with a surface. Chemoorganotrophic. Aerobic. Some may produce acid aerobically from carbohydrates. Optimum temperature is 37°C. **Found in the oral cavity of warm-blooded vertebrates.** Organisms in this group are recognizable on their morphology alone.

Differentiation of the genera in **Subgroup 2:** The genera in this group can be differentiated with the following key:

Key to the genera of **Subgroup 2**

A. Filament ends are square. Cells are paired within the filament.

Genus *Alysiella*

B. Filament has rounded ends and is often segmented into groups of eight cells.

Genus *Simonsiella*

Genus **Alysiella**

Organisms that exist in characteristic **flat, ribbon-like multicellular filaments. The long axis of the individual cells is perpendicular to the long axis of the filament. The cells within the filament are paired,** and in axenic culture the filament often breaks up into groups of two or four cells. The width of an individual cell (the width of a filament) is about 0.6 μm. The length of the filament is quite variable. The filament does not show either a dorsal-ventral differentiation or a convex-concave curvature in transverse cross-section. The **ends of the individual filaments are square** (Figure 15.4). Gram negative. **Motile by gliding** of the entire filament in the direction of the long axis. Aerobic. Chemoorganotrophic. Found in the oral cavity of warm-blooded vertebrates and is not known to be pathogenic.

Type (and only) species: *Alysiella filiformis.*

Characteristics of the species: As described for the genus.

Genus **Simonsiella**

Organisms that exist in characteristic **multicellular filaments** and are often segmented into groups of eight cells. **The long axis of an individual cell is perpendicular to the long axis of the filament** (Figure 15.5). The diameter of the filaments (the width of the individual cells) can be 2.0–8.0 μm, and the length of filaments may vary from about 10 to over 50 μm. Individual cells within the filaments may be 0.5–1.3 μm long. In thin sections cut perpendicular to the long axis of the filament, the **cells are flattened and curved to yield a crescent-shaped, convex-concave (dorsal-ventral) asymmetry. The ends of the individual filaments are**

rounded. Gram negative. **Motile by gliding** of the entire filament in the direction of the long axis. Aerobic. Chemoorganotrophic. Found in the oral cavity of warm-blooded vertebrates.

Type species: *Simonsiella muelleri.*

Differentiation of the species of the genus **Simonsiella:** See Table 15.6.

SUBGROUP 3: SULFUR-OXIDIZING, GLIDING BACTERIA

Cells of widely varying sizes, and with the exception of one genus, they **occur mostly in filaments.** Filaments usually demonstrate flexing or bending and can usually be considered as distinct multicellular organisms. **Motile by gliding. Sulfur is deposited internally in the presence of hydrogen sulfide, and often from thiosulfate.** Aerobic to microaerophilic. **Chemoorganotrophic** and chemolithotrophic nutrition are known; **mixotrophic nutrition postulated** for several. Metabolism is **respiratory.** Found in freshwater and marine environments.

Differentiation of the genera in **Subgroup 3:** The genera in this group can be differentiated with the following key.

Key to the genera in **Subgroup 3**

A. Cells occur singly. Cells are spherical or nearly so. Some are large (> 5 μm in diameter) and elongate.

    Genus *Achromatium*

B. Cells occur in multicellular filaments that are motile by gliding.

    1. Filaments are helically or spirally coiled.

        Genus *Thiospirillopsis*

    2. Filaments may be bent or straight but are not permanently coiled. Filaments are not enclosed in a sheath.

        Genus *Beggiatoa*

    3. Filaments may be bent or straight but are not permanently coiled. Filaments are found singly or multiply within sheaths.

        Genus *Thioploca*

C. Cells occur in multicellular filaments that do not glide. Filaments are enclosed within a sheath. Gonidia are produced from the open end of the sheath are motile by gliding.

Genus *Thiothrix*

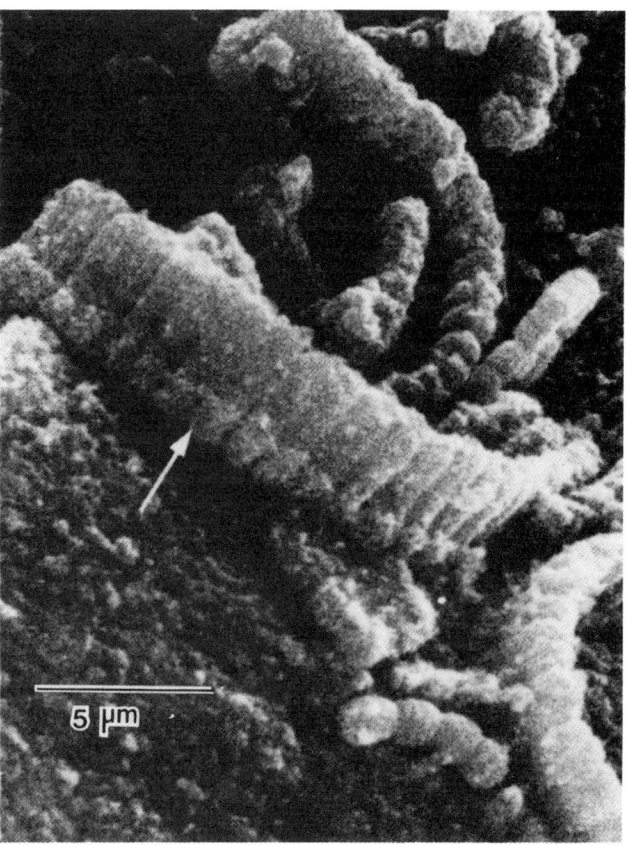

Figure 15.4. Scanning electron micrograph of an *Alysiella* filament attached to the epithelium of the bovine tongue. The *arrow* indicates the fringe of fibers that attach the bacterial filament by its side to the substrate. The palisade organization of the filament is characteristic of *Alysiella*. (Reproduced with permission from R.P. McCowen, K.-J. Cheng and J.W. Costerton, Applied and Environmental Microbiology *37:* 1224-1229, 1979, © American Society for Microbiology.)

Genus **Achromatium**

Cells are **spherical to ovoid or cylindrical** with **hemispherical ends,** 5–33 × 15–100 μm (Figure 15.6). Stain Gram negative. **Sulfur droplets and large spherules of calcium carbonate are typical inclusions.** Cell division is by constriction in the middle. **Movement** is of a **slow, rolling jerky type on a solid surface.** Locomotion in the type species may be caused by peritrichous filaments within the slime layer that surrounds the cells. **Aerobic.** Pure cultures have not been obtained. Apparently requires H$_2$S. Not pigmented. Catalase negative. Found in or on mud or fresh or saline waters.

Type species: *Achromatium oxaliferum.*

Differentiation of the species of the genus **Achromatium:** See Table 15.7.

Genus **Beggiatoa**

Colorless cells, about 1.0 to nearly 200 μm in diameter and about 2–10 μm in length, occur in **filaments** with diameters of a constant width. **May exist as single cells or in filaments** containing up to 50 or more cells. Cells in filaments of the thinner strains (< 7 μm in diameter)

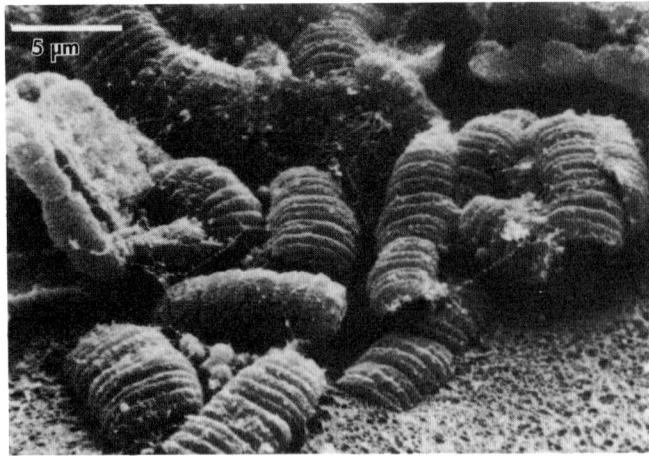

Figure 15.5. Scanning electron micrograph of the edge of a colony of *Simonsiella* obtained from a cat. (Micrograph taken by J. Pangborn; reproduced with permission from J. Pangborn, D.A. Kuhn and J.R. Woods, Archives of Microbiology *113:* 197-204, 1977, ©Springer-Verlag.)

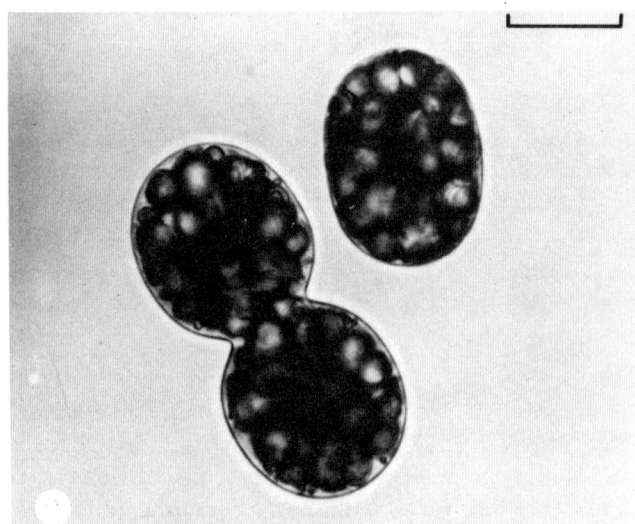

Figure 15.6. *Achromatium oxaliferum* cells. Bar, 20 μm. (Reproduced with permission from J.W.M. la Rivière and K. Schmidt, *In* Starr, Stolp, Trüper, Balows and Schlegel (Eds.), *The Prokaryotes. A Handbook on Habitats, Isolation, and Identification of Bacteria,* 1981, ©Springer-Verlag, Berlin.)

are cylindrical. In wider strains, cells are usually disk-shaped and typically are wider than they are long. **Filaments occur singly or in cottony masses.** In marine environments they may form a carpet at the sediment-water interface. Filament dispersion is by sacrificial cell death (necridial cells) and filament breakage or via simple disintegration. Disintegration may result in the production of single or double cells (hormogonia), which grow to produce a new filament. **Cells contain sulfur inclusions when they are grown in the presence of hydrogen sulfide** and, with some strains, thiosulfate. Intracellular inclusions of poly-β-

hydroxybutyric acid (PHB) or polyphosphate may be present. **Sheaths are not present.** A holdfast may be present when the filaments are attached to submerged objects in flowing water. Gram negative. **Hormogonia and filaments are motile by gliding.** Aerobic or microaerophilic. Chemoorganotrophic and facultatively autotrophic. Some strains may also grow mixotrophically. One marine strain is autotrophic. **Metabolism is respiratory,** with molecular oxygen used as the terminal electron acceptor. **Beggiatoas are gradient organisms existing in horizontal layers in sediments at the interface between the underlying anoxic sulfide-liberating zone and the overlying oxic zone.** Growth may occur between 0 and 40°C.

Type (and only) species: *Beggiatoa alba.*

Characteristics of the species: As described for the genus.

Editorial note: In the older literature there have been many species of *Beggiatoa* described solely on the width of the trichomes. Because of a lack of cultures, all species except *B. alba* were dropped from the Approved Lists. Recent work, particularly around the deep-ocean hydrothermal vents and hydrocarbon seeps, has demonstrated filaments of *Beggiatoa*-like sulfide-oxidizing gliding organisms with diameters ranging from 1 μm to nearly 200 μm.

Genus **Thioploca**

Flexible, uniseriate filaments made up **of numerous cells,** generally with numerous **sulfur inclusions,** occur in parallel or braided **fascicles, enclosed** by **a common**

sheath of variable width. The **number of filaments within a sheath is variable.** Within one sheath, filaments may be of fairly uniform or of greatly differing diameters. The sheath is frequently encrusted with detritus. Individual **filaments show gliding movement,** and they may emerge from the end of or from breaks in the sheath. The long filaments are of **uniform diameter;** their **terminal segments** may be **tapered or rounded.** Not isolated in pure culture.

Type species: *Thioploca schmidlei.*

Differentiation of the species of the genus **Thioploca:** See Table 15.8.

Genus "Thiospirillopsis"

Colorless sulfur bacteria occurring in spirally wound filaments. Continuous cell envelope; cells may or may not be visible in filaments. **Exhibit a creeping motility combined with rotation,** so that the trichomes move forward with a corkscrew motion. The tips may oscillate. Not grown in pure culture.

Type (and only) species: *"Thiospirillopsis floridana"*

Characteristics of the species: As described for the genus.

Genus **Thiothrix**

Rods, about 1.0–1.5 μm in diameter, which exist in **multicellular** rigid **filaments** of uniform diameter **within a sheath** (Figure 15.7). Rosettes may be produced. The closed end of the sheath may have a holdfast. Capsules are not produced. Resting stages not known. Gram negative. The **gonidia** are **motile by gliding.** No flagella are present, but a tuft of fimbria may be present at one end of the gonidium. Aerobic. Optimum temperature is 25–30°C; maximum is about 32–34°C; minimum is about 6–8°C. Isolates to date are mixotrophic, requiring any of several small organic compounds as well as a reduced inorganic sulfur source. Cells contain sulfur inclusions when grown in the presence of a reduced inorganic sulfur compound. **By light microscopy the globules appear to be internal.** Cytochrome oxidase positive and catalase negative. Found in sulfide-containing flowing water and in activated sludge sewage systems.

Type (and only) species: *Thiothrix nivea.*

Characteristics of the species: As described for the genus.

Editorial note: For a description of several classified isolates of *Thiothrix*-like organisms see Williams and Unz, Appl. Env. Microbiol. *49:* 887–898, 1985.

SUBGROUP 4: THE PELONEMAS

Cells cylindrical or short rods, colorless, and arranged in **unbranched filaments,** usually visible without staining. Filaments straight, flexuous, or spirally coiled, with or without a polymer sheath, **occurring singly or aggregated in bands or bundles.** Single filaments of two genera may show a gliding movement combined with rotation around the longitudinal axis. Cells **may contain gas vesicles,** but the cells are always unpigmented and lack sulfur globules. Reproduction is by binary fission of the cells and by fragmentation of the filaments or even of the aggregates. Some species in one genus appear to form arthrospore-like resting cells.

Differentiation of the genera in **Subgroup 4:** See Table 15.9.

Editorial note: None of the genera and species in this subgroup have been cultivated or validly published. There are a number of species described in the four genera and the reader is referred to *Bergey's Manual of Systematic Bacteriology,* Volume 3, for further information and partial keys to their identification.

Genus "Achroonema"

Cells range from 0.3–0.5 to 5–6.8 μm in width; they are arranged in multicellular, colorless, unbranched filaments that are always **more or less motile** and may be up to 5 mm long. Protoplasm is homogeneous, granular, or differentiated into lighter peripheral and darker central portions. **Filaments are straight** or nearly so **or in very loose spirals;** in some species, filaments are **constricted at cross-walls.** Three species appear to form **arthrospore-like resting cells; one species contains gas vesicles** in older cells. Multiply by fragmentation of filaments. Aquatic and have not been cultivated.

Type species: *"Achroonema spiroideum."*

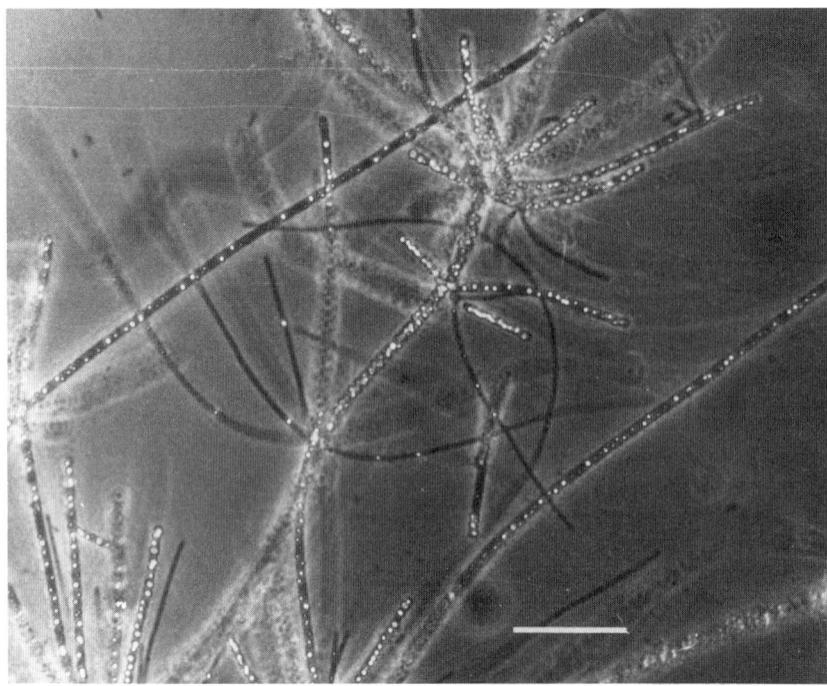

Figure 15.7. Phase micrograph of *Thiothrix nivea* from a freshwater sulfide-containing spring. *Bar,* 10 μm.

Genus "Desmanthos"

Colorless cells in **unbranched,** more or less straight **filaments** that are **thicker at the base than at the apex; filaments are in bundles, the base of which is enclosed in a common hyaline, gelatinous sheath of variable thickness;** the filaments at the top are free and divergent (Figure 15.8). Bundles are attached by a hold-fast or are partially buried in bottom mud. **Propagation is by fragmentation of the filaments** and probably by longitudinal separation of the bundle-filaments. Motility is not observed. Not cultured.

Type (and only) species: *"Desmanthos thiokrenophilum."*

Characteristics of the species: As described for the genus.

Genus "Pelonema"

Cells colorless, more or less cylindrical, **usually with gas vesicles** in the center. Cells arranged in **multicellular, unbranched straight** or loosely flexuous filaments of uniform thickness, which may or may not be constricted at the cross-walls (Figure 15.9). Filaments of **some species** occasionally show a **slow gliding motility.** Propagation is by fragmentation of filaments.

Resting cells unknown. Anaerobic and presumably heterotrophic. **Found in the hypolimnion and on the sediment surface of deeper lakes or stratified ponds.** Pure cultures are not available.

Type species: *"Pelonema tenue."*

Differentiation of the species of the genus **Pelonema:** See Table 15.9.

Genus "Peloploca"

Cells rod-shaped, 0.3–1.0 × 0.6–10 μm, arranged in filaments of variable length, and **with or without a common sheath.** Cells of **some species contain gas vesicles.** There may or may not be constrictions at cross-walls (Figure 15.10). **Filaments bound together laterally to form rigid bundles or flat ribbons,** which may be straight, undulate in one plane, or spirally wound; the ends may appear more or less fibrous. Nonmotile. Gram reaction is not recorded. Propagation is by fragmentation of bundles or ribbons. Aquatic, nonmotile but free floating, or in sediment surface layers. Have not been cultivated.

Type species: *"Peloploca undulata."*

494

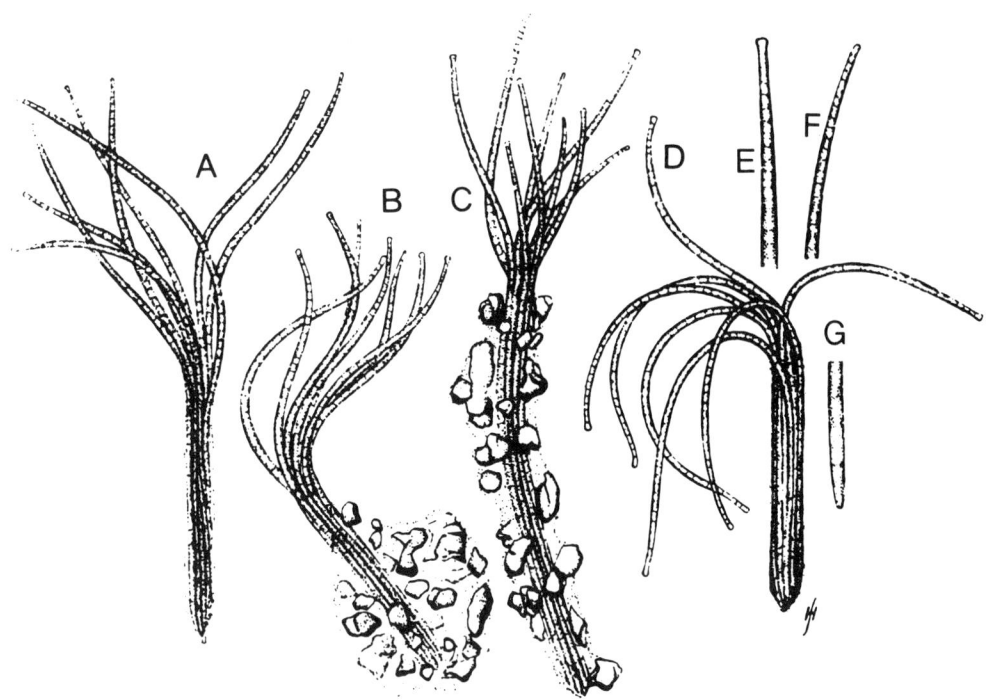

Figure 15.8. *"Desmanthos thiokrenophilum."* A-D show filament bundles with varying amounts of terminal polymer which, in some cases, allowed the incorporation of small mineral particles. E-G show ends of filaments. Reproduced from H. Skuja, Sven. Boto. Tidskr. *52:* 437–444, 1958.

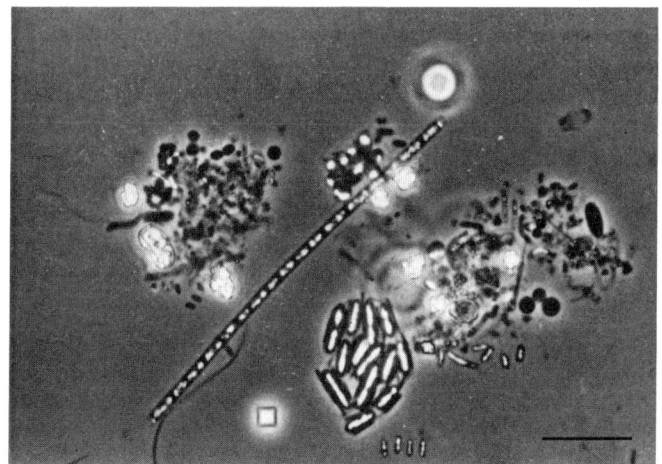

Figure 15.9. *"Pelonema pseudovacuolatum"* sensu Skuja 1956. The sample came from Wintergreen Lake (Michigan, U.S.A.) on August 26 from a 5-m depth (anaerobic hypolimnion). *Bar,* 1 μm.

OTHER GENERA

Genus **Agitococcus**

Spherical cells 1–2 μm in diameter (Figure 15.11). Stain Gram negative. May distend with sudanophilic granules in old cultures. Capsules are produced. **Twitching, gliding motility.** No flagella or fimbria are present. **Aerobic.** Grow between 15–30°C and pH 6–10. **Chemoorganotrophic. Do not oxidize or ferment glucose. Actively metabolize Tween compounds.** Oxidase positive, catalase negative. Found in fresh water.

Type (and only) species: *Agitococcus lubricus.*

Characteristics of the species: As described for the genus.

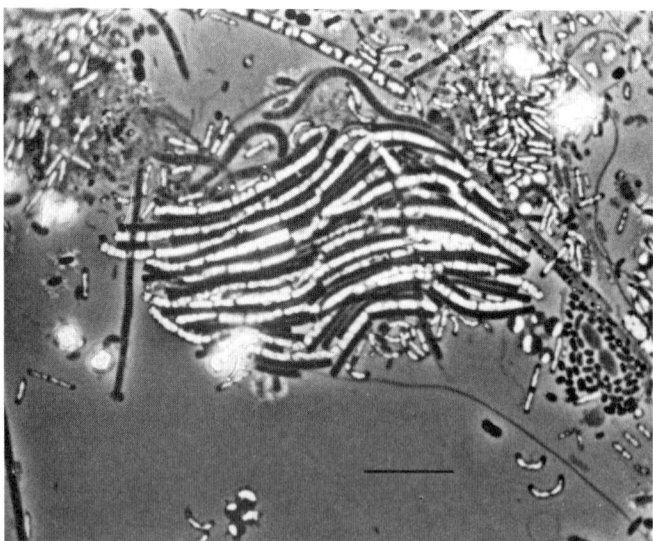

Figure 15.10. *"Peloploca undulata"* from the anaerobic hypolimnion of Cassidy Lake (Michigan, U.S.A.) from a 7-m depth. The sample was collected in August. Gas-vacuolated cell filaments alternate with one to two filaments without gas vesicles. *"P. undulata"* is a consortium of two different organisms. *Bar,* 1 μm.

ing), but electron microscopy of thin sections exhibit a characteristic **Gram-negative** cell wall. Poly-β-hydroxybutyrate is often stored. Cells attach to surfaces where they exhibit **gliding motility. Strictly anaerobic.** Marine forms usually require brackish water or seawater concentrations of NaCl, $MgCl_2$, and sometimes $CaCl_2$. Growth occurs between 10–36°C. Chemoorganotrophs or chemolithotrophs; metabolism is respiratory. **Sulfate and other oxidized sulfur compounds serve as electron acceptors and are reduced to hydrogen sulfide. Fatty acids and other organic acids are used as electron donors** and carbon sources; oxidation of electron donors is complete and results in the production of carbon dioxide. **Anoxic media** containing a reductant and vitamins are necessary for growth. Found in sulfate-rich freshwater or marine habitats.

Type species: *Desulfonema limicola.*

Differentiation of the species of the genus **Desulfonema:** See Table 15.10.

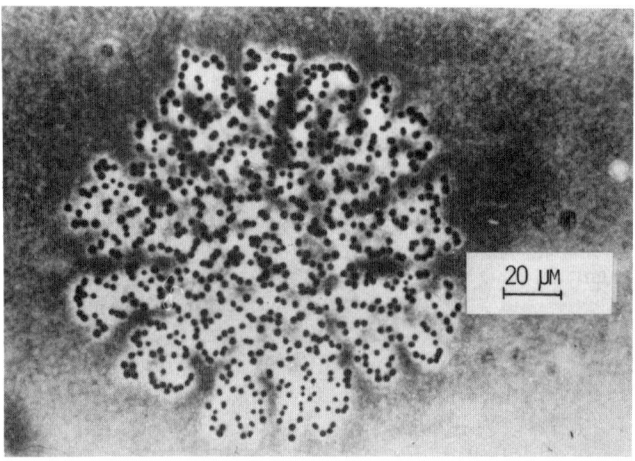

Figure 15.11. *Agitococcus lubricus* strain on an isolation plate of lake water agar after incubation for 21 h at 22°C. (Reproduced with permission from P.D. Franzmann and V.B.D. Skerman, Int. J. Syst. Bacteriol. *31:* 177-183 (Fig. 1), 1981. © International Union of Microbiological Societies.)

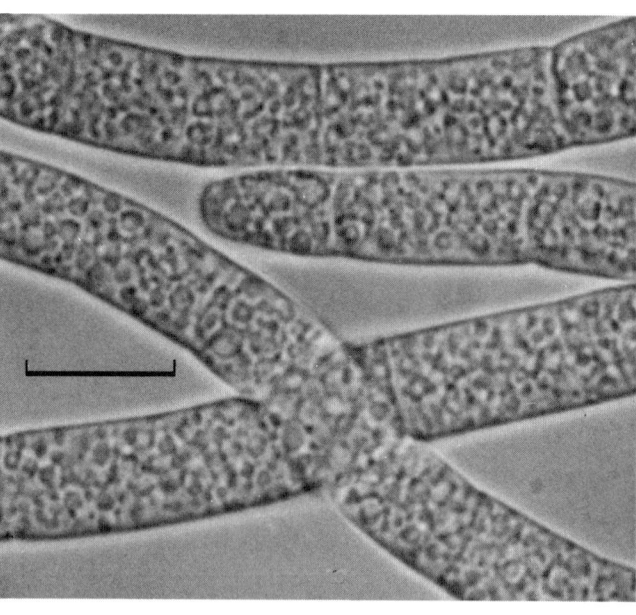

Figure 15.12. Brightfield photomicrograph of *Desulfonema magnum* isolated from marine mud and grown on benzoate as electron donor. *Bar,* 10 μm.

Genus **Desulfonema** *(See also Group 7)*

Cells arranged in **uniseriately multicellular, flexible filaments.** Filaments are 3–8 μm in diameter and up to 1 mm in length. Individual cells are 2.5–13 μm long (Figure 15.12). May stain Gram positive (uneven stain-

Editorial note: Filaments of *D. limicola* do not glide regularly; the most conspicuous motility characteristic of this species is a twitching or jerky swinging of the filaments when they creep out of sediment particles. The gliding movement of *D. magnum* is more regular.

Genus Herpetosiphon

Unbranched, flexible rods or filaments, 0.5–1.5 × 5–150 μm or longer, consisting of individual cells 2–3 μm long. Often **ensheathed,** although sheaths may be difficult to visualize. Stains Gram negative. **Motile by gliding.** Strictly aerobic. On dilute media, colonies may show a rough swirled texture, and heavy growth may result in large fruiting body-like structures that do not contain resting stages. May be mesophilic or thermophilic, and some species require sea water. Chemoorganotrophic. **Metabolism is respiratory, with molecular oxygen used as terminal electron acceptor. Produce yellow, orange, or red** carotenoid **pigments.** Cellulose and chitin may be degraded; carboxymethyl cellulose may be depolymerized; or starch may be hydrolyzed. They may lyse and digest living bacterial and yeast cells. Gelatin is liquified. Indole, ammonia or H_2S are not produced. Found in fresh waters, marine shores, soil, dung, decaying plant material, and hot springs.

Type species: *Herpetosiphon aurantiacus.*

Differentiation of the species of the genus **Herpetosiphon:** See Table 15.11.

Genus "Isosphaera"

Cells are **spherical,** 2.5–3.0 μm in diameter, **forming chains of indefinite length,** but often having over 100 cells. Cell **growth and division is by** the production of **intercalary buds** that occur in the chain axis. There is no branching of chains. Colonies produce a pink carotenoid pigment. Gas vesicles are often present in newly isolated cultures. Cells exhibit **gliding motility** when medium is made **with 0.4% agarose or 0.5% Gelrite®,** but **not with 1.5% agar.** Obligate aerobe. Chemoheterotrophic. Utilize glucose and lactate as sole carbon sources at concentration of 0.025%. Grows between 34°C and 55°C, with an optimum at 41°C.

Type (and only) species: *"Isopheara pallida."*

Characteristics of the species: As described for the genus.

Genus Leucothrix

Long filaments composed of short cylindrical or ovoid cells (gonidia), cross-walls clearly visible, **colorless, unbranched:** typically uniform filaments may taper from base to apex, under some conditions showing an apical beady chain of gonidia connected end to end. In nature, filaments **usually are attached to solid substrates by means of inconspicuous holdfasts; stalks absent. Filaments do not glide** but may wave sporadically from side to side. **Dispersal is by means of gonidia** (single cells arising from cells of the filaments by rounding up, often released primarily from apices, but they may also be formed in an intercalary fashion); **gonidia often,** but not always, **show jerky gliding motion on solid substrates. Rosette formation** is a key diagnostic characteristic of the genus but is found rarely in nature, although it is frequently seen in a laboratory culture. The rosettes may be formed of gonidia or, after gonidial growth of several or more filaments attached at their bases (Figure 15.13). **Strictly aerobic, heterotrophic. Aquatic, usually marine,** although one freshwater strain has been isolated. Most strains require NaCl for growth; optimum concentration is about 1.5% NaCl; grows at concentrations of 0.3–6.0% NaCl. Most strains do not require growth factors. Optimum temperature is 25°C; maximum; 30–35°C; minimum; 0°C, forming visible colonies within 1–2 weeks. Strains from tropical waters are more stenothermal, not growing below 15°C. Catalase positive.

Type (and only) species: *Leucothrix mucor.*

Characteristics of the species: As described for the genus.

Genus Saprospira

Helical filaments, multicellular, unbranched and without sheaths, 10–500 μm long, filaments 0.5–3.0 μm wide (Figure 15.14). Length of cells is 1.5–5.5 μm. Gram negative. **Gliding** is in longitudinal direction; simultaneously the screws rotate around their long axes. Resting stages are not known. Colonies often have a regular pattern of stripes. **Strictly aerobic organotrophs** requiring amino acid mixtures, peptones, or proteins as nitrogen source and often, perhaps always, as carbon and energy sources. Most strains are pigmented: pink, yellow, orange, or brick red. Aquatic organisms living in bottom sediments in marine and freshwater environments.

Type species: *Saprospira grandis.*

Differentiation of the species of the genus **Saprospira:** See Table 15.12.

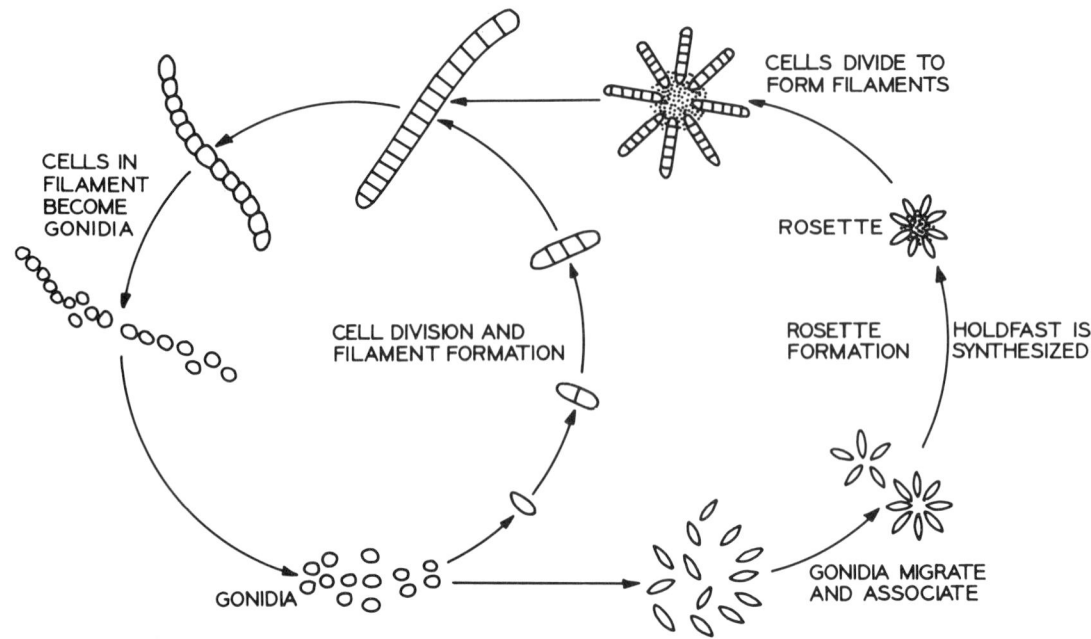

Figure 15.13. A simplified life cycle of *Leucothrix mucor*.

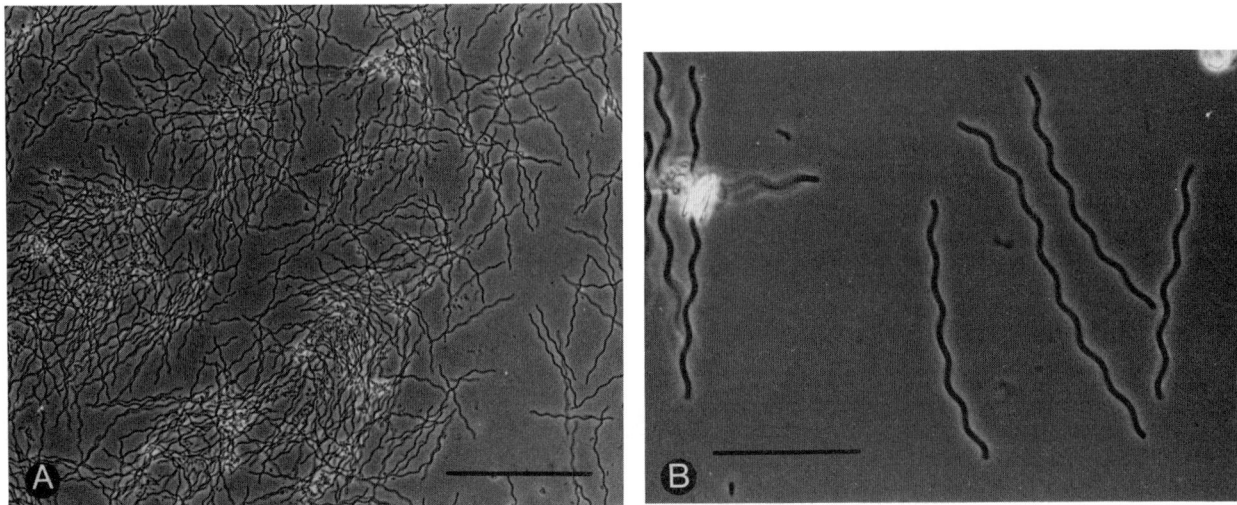

Figure 15.14. *Saprospira grandis* filaments. *A*, filaments from a liquid culture, survey at low magnification. Although the pitch and, particularly, the length of the screws are variable, the width and the diameter of the filaments are essentially constant. Phase contrast. *Bar*, 80 μm. *B*, filaments at higher magnification. Phase contrast. *Bar*, 25 μm.

Genus **Toxothrix**

Cells are cylindrical, colorless, 0.5–0.75 × 3–6 μm, and **in filaments** (trichomes) up to 400 μm long (Figure 15.15). Cross-walls in the filament might not be seen by phase contrast microscopy. A dense body (polyphosphate?) is often located at either end of the cell. **Filaments are often U-shaped and rotating**

while slowly moving forward with the rounded part in the lead; a mucoid substance excreted from several sites on the trailing ends, **is deposited as a double track** ("railroad track") **of twisted strings** each 0.2 μm wide. **Fan-shaped structures may be deposited later-ally** along the tracks, as the arms of the U move from side to side, and between the tracks, as a result of the middle section being lifted and then touched down

498

again. **Oxidized iron may be deposited on the mucoid threads,** rendering them yellowish brown and brittle and giving them a diameter of 2.5 μm. Have not been obtained in pure culture, but chemoorganotrophic and psychrophilic cultures have been maintained from long periods at 5 and 10°C. **Filaments are extremely fragile during laboratory examination, and explosive disintegration of filaments has been observed after short periods under the microscope.** Grow attached to surfaces and develop best at reduced oxygen tensions and a pH range of 5.1–7.7. Widely distributed in cold iron springs, brooks, forest ponds, and lakes containing ferrous iron and with reduced oxygen tension.

Type (and only) species: *Toxothrix trichogenes.*

Characteristics of the species: As described for the genus.

Genus **Vitreoscilla**

Organisms exist in **colorless filaments,** which contain cells with diameters of 1 to about 3 μm; **cells may be clearly delimited and barrel-shaped or may be undelimited and cylindrical (similar to *Beggiatoa*)** (Figures 15.16 and 15.17). Filaments may contain from 1 to > 40 cells. Cell division is by transverse binary fission; dispersion is by fragmentation of filaments or by sacrificial cell death and necridia formation in species similar to the beggiatoas. Resting stages are not known. **Sheaths or holdfasts are not produced. Gram negative. Motile by gliding;** no locomotor organelles known. **Aerobic or microaerophilic. Chemoorganotrophs. Metabolism is respiratory** with molecular oxygen as sole known terminal electron acceptor. **Sulfur inclusions are not formed** from hydrogen sulfide or thiosulfate. Nutritional requirements vary among species, with the simplest organic requirement being

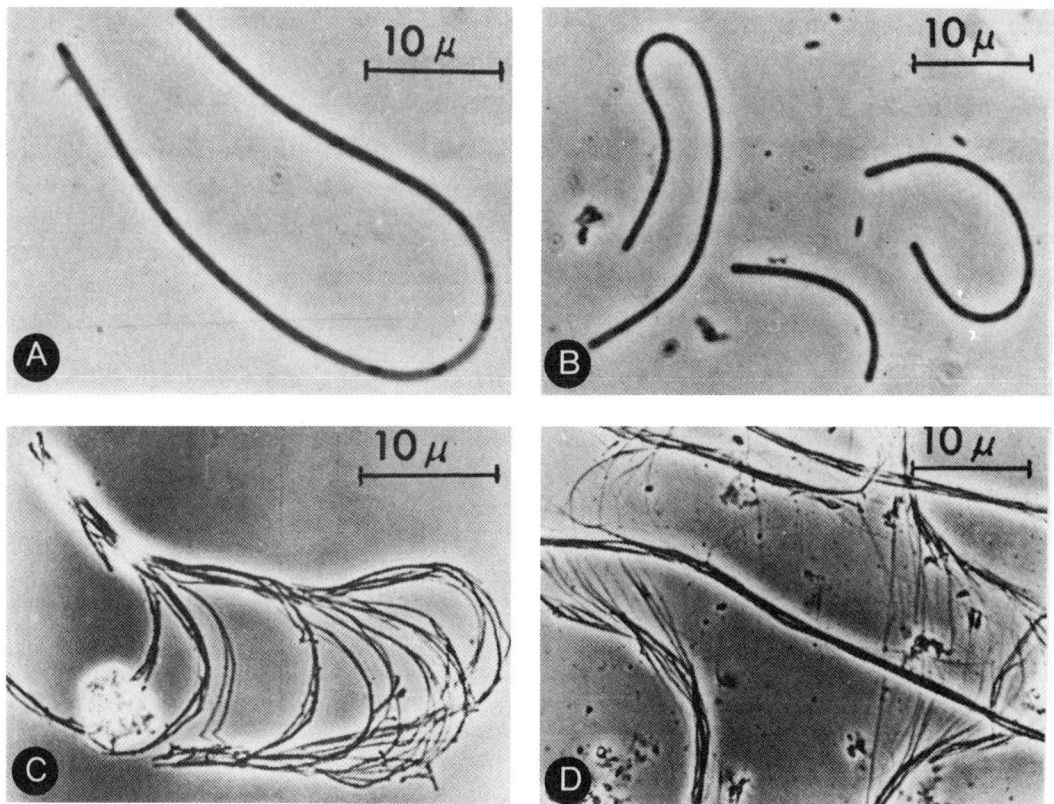

Figure 15.15. *Toxothrix trichogenes* observed in a small iron spring catch basin. *A* and *B*, laboratory wet mounts of living trichomes during the first minute. Phase-contrast micrographs. *C* and *D*, excreted polymer coated with iron oxides, from which, through the peculiar type of motion, arose fan-shaped structures *(C)* or double tracks *(D)*. (*A-D* are reproduced with permission from J.M. Krul, P. Hirsch, and J.T. Staley, Antonie van Leeuwenhoek J. Microbiol. *36:* 409-420, 1970, ©Martinus Nijhoff Publishers. Reprinted by permission of Kluwer Academic Publishers.)

acetate as sole combined carbon and energy source and with the more complex requirements being for groups of amino acids. The larger *Vitreoscilla* species show similarities to certain *Beggiatoa* strains, with the exception of their inability to deposit sulfur when exposed to hydrogen sulfide.

Type species: *Vitreoscilla beggiatoides.*

Differentiation of the species of the genus **Vitreoscilla:** See Table 15.13.

Editorial note: This genus consists of two distinct types of organisms that are differentiated by both morphology and physiology. One type, which includes the type species, is morphologically similar to *Beggiatoa* and can grow on acetate alone. The other type (*V. stercoraria*) is a streptobacillus and requires amino acids.

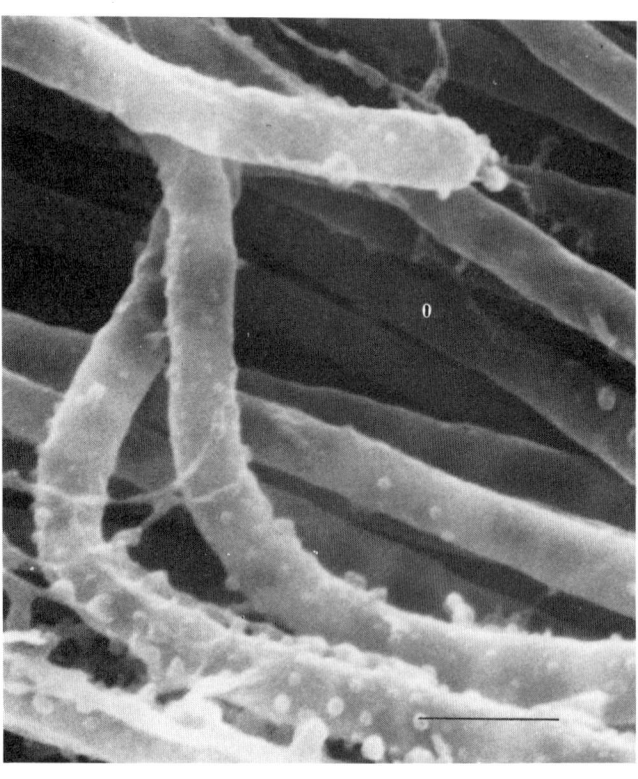

Figure 15.16. Scanning electron micrograph of *Vitreoscilla beggiatoides* strain B23SS, with continuous cell envelope and slimelike matrix shown. *Bar,* 5 μm.

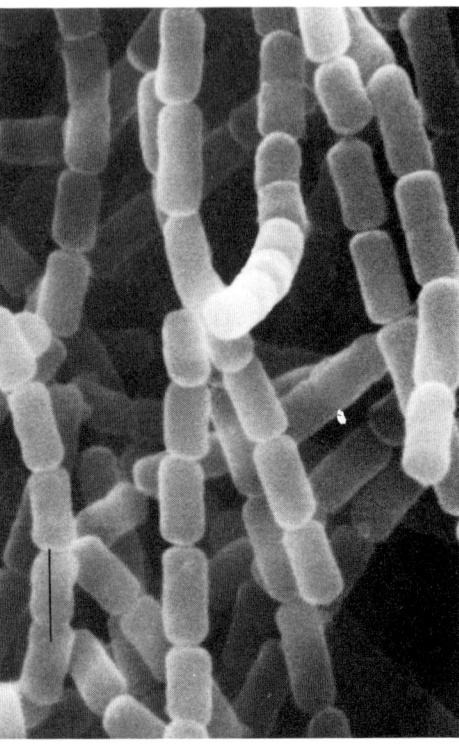

Figure 15.17. Scanning electron micrograph of *Vitreoscilla stercoraria* strain VT-1, with the discontinuous cell walls and the connecting material between the cells shown. *Bar,* 1 μm.

Table 15.1   Differential characteristics of the species of the genus *Capnocytophaga*[a]

Characteristic	C. canimorsus	C. cynodegmi	C. gingivalis	C. ochracea	C. sputigena
Hydrolysis of:					
Esculin	d	d	+	+	+
Starch	ND	ND	−	+	−
Dextran	ND	ND	−	+	−
Gelatin	−	−	−	−	+
ONPG	+	+	−	+	+
Acid production from:					
Lactose	+	+	−	+	d
Melibiose	−	+	−	−	−
Raffinose	−	+	+	+	+
Oxidase	+	+	−	−	−
Catalase	+	+	−	−	−

[a] Symbols: +, positive; −, negative; d, results vary among strains; ND, not determined.

Important Notes for Users of This Manual

Unless otherwise indicated in footnotes to tables, the meanings of symbols are as follows:

+ 90% or more of strains are positive

− 90% or more of strains are negative

d 11-89% of strains are positive

v strain instability (not equivalent to "d")

D Different reactions in different taxa (species of a genus or genera of a family)

All other symbols are defined in footnotes to tables.

Table 15.2 Differential characteristics of the species of the genus *Cytophaga*[a]

Characteristic	C. agarovorans	C. allerginae	C. aprica	C. aquatilis	C. arvensicola	C. aurantiaca
Length of cells (μm)	12–30 (8–50)	3–9	6–30 (–50)	2–6 (–50)	0.6–2.0 (–4.0)	3–5
Width of cells (μm)	0.3–0.4	0.3	0.5–0.7	0.5–0.7	0.4	0.3
Cells ends tapering	–	–	+	+/–	–	+/–
Color of cell mass	Pink to salmon	Bright yellow	Orange	Yellow	Yellow-orange	Bright orange
Flexirubin reaction [b]	ND	+	–	+	+	–
Carotenoid reaction	ND	ND	Saproxanthin	ND	ND	ND
Suitable as nitrogen source:						
Peptones	+	ND	+	+	+	+
Casamino acids	+	ND	+	+	ND	ND
Single amino acids	ND	ND	Glu	Glu	ND	ND
Urea	ND	ND	ND	ND	ND	ND
NH_4^+	+	ND	ND	+	+	+
NO_3^-	+	ND	–	+	ND	+
Grows on peptone alone	+	ND	+	+	+	–
NH_3 released from peptones	ND	ND	+	+	ND	ND
Glucose sole carbon and energy source	+	ND	–	+	ND	+
Acid from carbohydrates (Aerobically)	ND	–	+	+	ND	ND
Degradation of:						
Gelatin	+	+	+	+	–	ND
Casein	ND	+	–	+	ND	–
Starch	+	+	+	+	–	–
Cellulose (filter paper)	ND	–	–	–	ND	+
CMC	ND	+	+	+	ND	+
Agar	+	–	+	–	ND	–
Carrageenan	ND	ND	ND	ND	ND	ND
Alginate	ND	–	+	–	ND	ND
Pectin	ND	ND	ND	+	ND	ND
Chitin	–	+	–	+	ND	ND
Inulin	ND	ND	ND	ND	ND	ND
Autoclaved yeast cells [e]	–	ND	ND	+	ND	ND
Hemolysis (kind of erythrocyte)	ND	ND	ND	–	ND	ND
Lipases	ND	ND	ND	B/T	ND	ND
DNase	ND	–	ND	+	+	ND
Phosphatase	ND	ND	ND	ND	ND	ND
Urease	ND	–	ND	–	–	ND
Indole produced	ND	–	ND	–	–	ND
H_2S produced	ND	–	+	–	–	ND
NO_3^- reduced [f]	ND	–	–[c]	NH_3	+	ND
Strict aerobe (a)/facultative anaerobe (f)	f	a	a	f	a	a
Catalase	+	+	–	+	+	–
Oxidase	ND	+	+	–	+	+
Optimum temperature (°C)	28–37	ND	25–30	20–25	ND	20–25
Maximum temperature (°C)	ND	>37	<40	30	>37	ND
Highest NaCl concentration tolerated (%)	3	1.5	6	2	1	ND
Growth on seawater media	+	ND	+	–	ND	–
Habitat	Marine	Fresh water	Marine	Fresh water	Soil	Soil
Mol% G + C of DNA of type strain (range of species)	41	35	35 (35–37)	32/34	43–46	42

Footnotes are at end of table

Table 15.2 *(continued)*

Characteristic	C. columnaris	C. diffluens	C. fermentans	C. flevensis	C. heparina	C. hutchinsonii
Length of cells (µm)	2–12	4–10 (–30)	2–10 (–30)	2–5	1–9	2–5
Width of cells (µm)	0.4	0.5	0.5–0.7	0.5–0.7	0.3	0.4
Cells ends tapering	–	+	–	–	+/–	+/–
Color of cell mass	Golden yellow	Orange	Bright yellow	Yellow	Yellow- gray	Bright yellow
Flexirubin reaction [b]	+	–	ND	–	–	+
Carotenoid reaction	ND	Saproxanthin	ND	ND	ND	ND
Suitable as nitrogen source:						
Peptones	+	+	+	+	+	+
Casamino acids	+	+	ND	+	+	+
Single amino acids	ND	Glu	Glu, Asn	Glu, Asn	Asp	Glu, Asp
Urea	ND	ND	–	+/–	+	ND
NH_4^+	–	ND	+	+	+	+
NO_3^-	–	–[c]	–	+	–	+
Grows on peptone alone	+	+	ND	+	+	–
NH_3 released from peptones	+	+	ND	+	+	+
Glucose sole carbon and energy source	–	–[c]	+	+	+	+
Acid from carbohydrates (Aerobically)	–	–[c]	+	+	+	+
Degradation of:						
Gelatin	+	+	ND	–	–	ND
Casein	+	–[c]	ND	–	ND	ND
Starch	–	+	+	–	ND	–
Cellulose (filter paper)	–	–[c]	–	ND	–	+
CMC	ND	+	ND	+	–	+
Agar	–	+	+	–	–	–
Carrageenan	ND	ND	ND	–	ND	ND
Alginate	ND	+	–	+	ND	ND
Pectin	ND	ND	ND	–	ND	–
Chitin	–	–	–	+	–	–
Inulin	ND	ND	ND	–	ND	ND
Autoclaved yeast cells [e]	+	–	ND	–	–	ND
Hemolysis (kind of erythrocyte)	ND	ND	ND	–(Sheep)	+(Sheep)	ND
Lipases	ND	ND	ND	B	ND	ND
DNase	ND	ND	ND	ND	ND	ND
Phosphatase	ND	ND	ND	+	+	ND
Urease	ND	ND	ND	–	–	ND
Indole produced	–	ND	ND	–	–	ND
H_2S produced	+	–[d]	ND	–	ND	ND
NO_3^- reduced [f]	–	–[c]	–	+	–	ND
Strict aerobe (a)/facultative anaerobe (f)	a	a	f	a	a	a
Catalase	+	–	–	–	+	–
Oxidase	+	+	–	+	+	+
Optimum temperature (°C)	25–30	25–30	30	25	20–25	30
Maximum temperature (°C)	<37	<45[c]	ND	<35	30	ND
Highest NaCl concentration tolerated (%)	0.5	6	ND	ND	3	ND
Growth on seawater media	–	+	+	+	ND	–
Habitat	Fresh water	Marine	Marine	Fresh water	Soil	Soil
Mol% G + C of DNA of type strain (range of species)	35 (35–37)	42 (40–42)	42 (40–42)	42	42	39

Footnotes are at end of table

Table 15.2 *(continued)*

Characteristic	C. johnsonii	C. latercula	C. lytica	C. marina	C. marinoflava	C. pectinovora
Length of cells (μm)	2–5 (–25)	1–5 (10–40)	1.5–3.5	2–5	1–3	1–5
Width of cells (μm)	0.3–0.4	0.3–0.4	0.4	0.4	0.5–0.6	0.4–0.5
Cells ends tapering	+/–	+/–	–	ND	+/–	+/–
Color of cell mass	Yellow	Orange- red	Bright yellow	Yellow-orange	Yellow	Yellow
Flexirubin reaction [b]	+	–	–	ND	–	+
Carotenoid reaction	Zeaxanthin	ND	Zeaxanthin	ND	ND	ND
Suitable as nitrogen source:						
Peptones	+	+	+	+	+	+
Casamino acids	+	+	+	+	ND	+
Single amino acids	Glu, Asn	Glu	Glu	–	ND	Glu, Asn
Urea	+	ND	ND	–	ND	+
NH_4^+	+	ND	ND	–	ND	+
NO_3^-	+/–	+	–	–	ND	+
Grows on peptone alone	+	+	+	+	+	+
NH_3 released from peptones	+	+	+	+	+	+
Glucose sole carbon and energy source	+	+	ND	–	ND	+
Acid from carbohydrates (Aerobically)	+	+	ND	–	+	+
Degradation of:						
Gelatin	ND/+	+	+	+	ND	+
Casein	+	–	–	+	ND	+
Starch	+	–	+	–	+	+
Cellulose (filter paper)	–	–	–	–	–	–
CMC	+	+	+	–	ND	+
Agar	–	+	+	–	–	–
Carrageenan	ND	ND	ND	ND	ND	ND
Alginate	+	+	+	ND	ND	+
Pectin	+	–	ND	ND	ND	+
Chitin	+	+	–	–	ND	+
Inulin	+	ND	ND	ND	ND	+
Autoclaved yeast cells [e]	+	+/–	–	ND	–	+
Hemolysis (kind of erythrocyte)	ND	ND	ND	ND	ND	– (Horse)/+ (Sheep)
Lipases	ND	ND	ND	T	T	B
DNase	+	ND	ND	ND	ND	+
Phosphatase	+	ND	ND	ND	ND	+
Urease	+	ND	ND	+	ND	–
Indole produced	–	ND	ND	–	–	–
H_2S produced	–	+	+	–	–	–
NO_3^- reduced [f]	–	–	–	+	+	+
Strict aerobe (a)/facultative anaerobe (f)	a	a	a	a	a	a
Catalase	+	–	+	+	+	+
Oxidase	+	+	+	+	+	+
Optimum temperature (°C)	25–30	20–25	22–30	30	20–30	20–25
Maximum temperature (°C)	<37	<35	35–40	<37	<37	<37
Highest NaCl concentration tolerated (%)	1	3.5	6		ND	1
Growth on seawater media	–	+	+	+	+	–
Habitat	Soil, fresh water	Marine	Marine	Marine	Marine	Soil
Mol% G + C of DNA of type strain (range of species)	42	34	33 (32–34)	42	37	34

Footnotes are at end of table

Table 15.2 *(continued)*

Characteristic	C. psychrophila	C. saccharophila	C. salmonicolor	C. succinicans	C. uliginosa
Length of cells (μm)	1.5–3.5 (–7.5)	2.5–6	4–6 (2–30)	4–6	1.2–4
Width of cells (μm)	0.4–0.5	0.5–0.6	0.3–0.7	0.5	0.4
Cells ends tapering	–	+	–	–	+/–
Color of cell mass	Yellow	Yellow	Yellow to salmon	Bright yellow	Golden yellow
Flexirubin reaction [b]	+	+	ND	–	+
Carotenoid reaction	ND	–	ND	ND	ND
Suitable as nitrogen source:					
Peptones	+	+	+	+	+
Casamino acids	+	ND	+	+	ND
Single amino acids	ND	ND	ND	Glu	ND
Urea	ND	ND	ND	ND	ND
NH_4^+	ND	+	+	+	ND
NO_3^-	ND	ND	+	+	ND
Grows on peptone alone	+	+	+	+	+
NH_3 released from peptones	+	+	+	+	ND
Glucose sole carbon and energy source	–	+	+	+	ND
Acid from carbohydrates (Aerobically)	–	ND	ND	+	+
Degradation of:					
Gelatin	+	+	+	+	+
Casein	+	ND	ND	+	+
Starch	–	+	+	+	+
Cellulose (filter paper)	–	–	ND	–	–
CMC	ND	+	ND	ND	ND
Agar	–	+	–	–	+
Carrageenan	ND	+	ND	ND	ND
Alginate	ND	ND	ND	ND	ND
Pectin	–	+	ND	ND	ND
Chitin	–	–	–	–	+
Inulin	ND	ND	ND	ND	ND
Autoclaved yeast cells [e]	–	+	–	+	–
Hemolysis (kind of erythrocyte)	ND	ND	ND	ND	ND
Lipases	B	ND	ND	ND	–
DNase	ND	ND	ND	ND	ND
Phosphatase	ND	+	ND	ND	ND
Urease	ND	ND	ND	ND	–
Indole produced	–	–	ND	ND	–
H_2S produced	–	+	ND	ND	–
NO_3^- reduced [f]	–	+	ND	+ [c,d]	+
Strict aerobe (a)/facultative anaerobe (f)	a	a	f	f	a
Catalase	–	–	+	+/–	+
Oxidase	+	–	ND	ND	+
Optimum temperature (°C)	20	25–30	28–30	25	20–30
Maximum temperature (°C)	<25	<37	ND	>37	ND
Highest NaCl concentration tolerated (%)	0.8	2	3	ND	ND
Growth on seawater media	–	–	+	–	+
Habitat	Fresh water	Fresh water	Marine	Fresh water	Marine
Mol% G + C of DNA of type strain (range of species)	35 (35–37)	32 (32–36)	37	38 (38)	42

Footnotes are at end of table

Table 15.2 *(continued)*

[a] Symbols: +, positive; –, negative; +/–, characteristic is not clearly positive or negative or varies with the culture conditions; d, varies among isolates; ND, not determined; T, active against Tweens; and B, active against tributyrin.

[b] Color change from yellow to red, purple or red-brown when 20% KOH is added, and back to yellow again when the latter is replaced with 10% HCl.

[c] Characteristic variable: the type strain is negative, whereas other strains may be positive.

[d] Characteristic variable with different strains: the type strain is positive, whereas other strains may be negative.

[e] On yeast agar, e.g., VY/2 agar, if required with seawater.

[f] NO_3^-, reduced to NO_2^-; NO_2^-, reduced to NH_3.

Important Notes for Users of This Manual

Unless otherwise indicated in footnotes to tables, the meanings of symbols are as follows:

    + 90% or more of strains are positive

    – 90% or more of strains are negative

    d 11-89% of strains are positive

    v strain instability (not equivalent to "d")

    D Different reactions in different taxa (species of a genus or genera of a family)

All other symbols are defined in footnotes to tables.

Table 15.3 Differential characteristics of the species of the genus *Flexibacter*[a]

Characteristic	F. aurantiacus	F. canadensis	F. chinensis	F. elegans	F. filiformis	F. flexilis
Length of threads (μm)	5–20	3–60	4–16 (>60)	1.0 – >80	30–80	10– >50
Width of threads (μm)	ND	0.3	0.4–0.5	0.4	0.4–0.5	0.5
Cyclic shape change	ND	ND	ND	–	+	–
Color of cell mass	Yellow	White	Pink/orange/colorless[c]	Bright orange	Golden yellow	Orange[d]
Flexirubin reaction[e]	ND	ND	ND	–	+	+
Carotenoids present	ND	ND	ND	Saproxanthin	Zeaxanthin	Saproxanthin
Relation to oxygen	Aerobe	Aerobe	Aerobe[f]	Aerobe	Aerobe	Aerobe
Growth on peptones alone	ND	ND	ND	–	+	+
Suitable as sole nitrogen source:						
Peptones	ND	ND	+	+	+	+
Casamino acids	+	+	+	+	+	+
Glutamic acid	+	+	+	–	+	–
NH_4^+	ND	+	ND	–	+	–
NO_3^-	+	–	+	–	+	–
Sugars metabolized	d	+	+	+	+	+
Acid from glucose	d	+	+	+	+	+
Degradation of:						
Gelatin	+	+	–	+	+	+
Casein	ND	ND	ND	ND	+	ND
Starch	+	+	+	–	+	+
Chitin	ND	–	–	–	+	–
Yeast cells[g]	ND	ND	ND	–	+	–
Indole produced	ND	–	ND	–	–	–
H_2S produced	–	+	ND	–	–	+
NO_3^- reduced	–	+	+	–	–	–
Catalase	+	+	ND	–	–	–
Oxidase	ND	+	ND	+	+	+
DNase	ND	+	–	ND	–	ND
Growth on seawater medium	+	ND	ND	+	–	–
Maximum NaCl concentration (%)	ND	2	ND	2.4	0.3	ND
Optimum temperature (°C)	ND	18–30	30	40–45	40–45	40–45
Maximum growth temperature (°C)	30	40	ND	40–45	40–45	40–45
Optimum pH	ND	6–8	ND	7	7	7
Habitat	Soil	Soil	Fresh water	Fresh water	Soil/fresh water	Fresh water
Mol% G + C of DNA	31.5–32	37	34	ND	46–47	40–43
Mol% G + C of DNA of type strain	32	37	34	48	47	41

Footnotes are at end of table

Table 15.3 *(continued)*

Characteristic	*F. litoralis*[b]	*F. maritimus*	*F. polymorphus*[b]	*F. roseolus*	*F. ruber*	*F. sancti*
Length of threads (μm)	>180	2–30	>200	>50	>50	15 (>50)
Width of threads (μm)	0.5–0.7	0.5	1.1	ND	ND	0.5
Cyclic shape change	ND	ND	ND	ND	ND	ND
Color of cell mass	Brick red	Light yellow	Peach	Red	Red	Yellow
Flexirubin reaction [e]	ND	–	ND	ND	ND	+
Carotenoids present	Flexixanthin	+	Saproxanthin	+	+	+
Relation to oxygen	Aerobe	Aerobe	Aerobe	Aerobe	Aerobe	Aerobe
Growth on peptones alone	ND	ND	ND	ND	ND	ND
Suitable as sole nitrogen source:						
Peptones	+	ND	+	+	+	+
Casamino acids	ND	+	+	+	+	+
Glutamic acid	–	ND	+	–	+	–
NH_4^+	ND	ND	–	ND	ND	ND
NO_3^-	–	ND	–	–	+	+
Sugars metabolized	–	–	ND	–	+	+
Acid from glucose	–	–	ND	–	+	+
Degradation of:						
Gelatin	+	+	+	+	+	+
Casein	ND	+	ND	ND	ND	ND
Starch	+	–	–	–	–	+
Chitin	ND	–	ND	ND	ND	ND
Yeast cells [g]	ND	ND	ND	ND	ND	+
Indole produced	ND	–	ND	ND	ND	ND
H_2S produced	–	–	–	–	–	d
NO_3^- reduced	–	+	ND	–	+	d
Catalase	–	+	–	–	–	–
Oxidase	ND	+	ND	ND	ND	ND
DNase	ND	ND	ND	ND	ND	ND
Growth on seawater medium	+	+	+	+	–	–
Maximum NaCl concentration (%)	ND	ND	ND	ND	ND	ND
Optimum temperature (°C)	ND	30	ND	ND	ND	ND
Maximum growth temperature (°C)	30–35	34	~30	40	40–45	35
Optimum pH	ND	ND	>32	ND	ND	ND
Habitat	Marine	Marine fish	Marine	Hot springs	Hot springs	Soil
Mol% G + C of DNA	ND	31.3–32.5	ND	34.5–38	37	46–47
Mol% G + C of DNA of type strain	31	32	29	38	37	47

[a] Symbols: +, 90% or more of strains are positive; –, 90% or more of strains are negative; ND, not determined.

[b] These species are listed as *Species Incertae Sedis* with the genus *Microscilla* in *Bergey's Manual of Systematic Bacteriology*. *F. polymorphus* is more nearly like *Microscilla* but *F. litoralis* differs from *Flexibacter* and *Microscilla*.

[c] Colonies initially pink, then orange, and colorless on the third day.

[d] On many media very pale.

[e] Color change from yellow to red, purple to red-brown when alkali is added.

[f] Can grow anaerobically.

[g] In yeast agar, e.g., VY/2 agar.

Table 15.4 Differential characteristics of the species of the genus *Lysobacter*[a,b]

Characteristic	*L. antibioticus*	*L. brunescens*	*L. enzymogenes*	*L. gummosus*
Broth culture viscous	+	−	+	+ heavy
Colonies:				
Type of growth	Sloppy, mucoid	Spreading, thin	Sloppy, mucoid	Pulvinate, gummy
Surface smooth	+	+ or −	+ or −	+
Opacity	Tl or Op	Tp or Tl	Tp, Tl, or Op	Tl or Op
Color	Pi to Br–Pi or O–Br	Y to Ch	Dark C to deep Y–C	Pale Y–G to Y–G
Brown, water soluble pigment	Weak to heavy	Weak to heavy	None to moderate	−
Myxin crystals in old cultures	Usually +	−	−	−
Urea as nitrogen source	−	−	+ or −	+
Acid from:				
Cellobiose	+	−	+	+
Sucrose	−	−	+	+
Lactose	d	−	+	+
Lipase-Tween 20	d	−	+	+
Hydrolysis of:				
Alginate	d	−	+ or −	−
CMC	+	−	+	+
Pectate	−[c]	+	+	+
SYS or NBS starch	−[d]	+	−	−
Potato starch	−[e]	+	+ or −	−
Sheep erythrocytes	α or β	γ or −	α, β or γ	α
H$_2$S produced	−	+	d	−
Citrate as sole carbon source	+	−	+	+
Growth on:				
MacConkey's	d, colorless	−	+, colorless or −	−
EMB	+, Pi	−	+, Pi	+, Pi
Completely inhibited by 0.1% SLS	+	+	+ or −	−
Susceptible to:				
Chloramphenicol (30 μg)	−	+	d	−
Penicillin G (10 U)	−	d	−	−
Actinomycin D	−	+	d	d
Habitat	Soil	Fresh water	Soil	Soil

[a] Symbols: +, 90% or more strains are positive; −, 90% or more of strains are negative; d, 11–89% of strains are positive.

[b] Abbreviations: Br, brown(ish); C, cream-colored; Ch, chocolate; EMB, eosin methylene blue; G, gray; NBS, nutrient broth-starch; O, orange; Op, opaque; Pi, pink(ish); SLS, sodium lauryl sulfate; SYS, salts-yeast extract-starch; Tl, translucent; Tp, transparent; and Y, yellowish.

[c] Growth but no liquifaction.

[d] Only one strain recorded on SYS.

[e] Only one strain recorded.

Table 15.5 Differential characteristics of the species of the genus *Microscilla*[a]

Characteristic	"M. aggregans"	M. arenaria	"M. furvescens"	M. marina	"M. sericea"	"M. tractuosa"
Length of threads (μm)	20– >100	>20	10–50	>150	30– >100	5– >50
Color of cell mass	Yellow	Orange	Orange	Orange	Orange	Orange
Main carotenoid [b]	Zeaxanthin	Saproxanthin	Saproxanthin	Saproxanthin	Saproxanthin	Saproxanthin
Salinity range (S [c])	1/2–2	1/2–2	ND	1–2	1/2–2	0–2 [d]
Maximum temperature (°C)	35–45	30–35	ND	30–35	30–35	30–45
Suitable as sole nitrogen source:						
Peptones	+	+	+	+	+	+
Casamino acids	+	+	+	+/–	–	+
Glutamate	+	ND	+	–	–	+
NO_3^-	+/–	ND	+	–	–	–
Utilization of glucose	+	ND	+	–	+	+
Degradation of:						
Gelatin	–	–	+	+	+	+/–
Starch	–	+	+	–	+	+/–
Alginate	–	+	+	–	+	–
CMC	+/–	+	+	–	–	–
Agar	–	–	+	–	–	–
NO_3^- reduced	–	–	–	–	–	–/+
Mol% G + C of DNA	37–42	ND	ND	ND	38–39	35–40
Mol% G + C of DNA of type strain	37	32.5	44	42	39	37

[a] Symbols: ND, no information; +, 90% or more of strains are positive; +/–, most strains are positive; –, 90% or more of strains are negative; and –/+, most strains are negative.

[b] Type of carotenoid inferred from electron spectra of crude extracts and chromatographic data.

[c] Expressed as manifolds of seawater (S); 0 = freshwater.

[d] Some strains do not grow below 1/2 S.

Table 15.6 Differential characteristics of the species of the genus *Simonsiella*[a]

Characteristic	S. crassa	S. muelleri	S. steedae
Acid from:			
Glucose	+	+	−
Maltose	+	+	−
Trehalose	+	−	−
Ribose	+	−	−
Fructose	+	−	−
Sucrose	+	−	−
Mannitol	+	−	−
Growth at:			
27°C	+	+	−
43°C	+	−	−
pH 6.0	+	+	−
pH 8.0	+	−	−
Source	Sheep	Humans	Dogs

[a] Symbols: +, 90% or more of strains are positive; −, 90% or more of strains are negative.

Table 15.7 Differential characteristics of the species of the genus *Achromatium*[a]

Characteristic	A. oxaliferum	"A. volutans"
Calcium carbonate inclusions	+	−
Sulfur inclusions	+	+
Found in fresh water	+	−
Found in saline waters	+	+

[a] Symbols: +, positive; −, negative.

Table 15.8 Differential characteristics of the species of the genus *Thioploca*[a]

Characteristic	T. araucae	T. chileae	T. ingrica	"T. marina"	"T. minima"	T. schmidlei
Habitat and sediments:						
Fresh water	−	−	+	−	+	+
Brackish water	−	−	+	−	−	+
Marine water	+	+	−	+	−	−
Size (μm):						
Filament diameter	30–43	12–20	2.0–4.5	2.5–5.0	0.8–1.5	5.0–9.0

[a] Symbols: +, positive; −, negative.

Table 15.9 Differential characteristics of Subgroup 4, the Pelonemas [a]

Characteristics	"Achroonema"	"Desmanthos"	"Pelonema"	"Peloploca"
Filaments:				
Single	+	−	+	−
In bands or bundles	−	+	−	+
Straight or wavy	D	+	+	+
Coiled (but not wavy)	D	−	−	D
Constricted at cross-walls	D	−	D	D
With arthrospores (resting cells)	D	−	−	−
With gas vesicles	D	−	+	D
Gliding	+	−	+/−[b]	−
Holdfasts	−	+	−	−
Sheaths	D	+	−	D

[a] Symbols: +, 90% or more of strains are positive; −, 90% or more of strains are negative; and D, different reactions in different species.

[b] Three of six species gliding.

Table 15.10 Differential characteristics of the species of the genus *Desulfonema* [a]

Characteristic	*D. limicola*	*D. magnum*
Diameter of filament (μm)	3	6–8
Length of individual cell (μm)	2.5–3.5	9–13
Serve as electron donor and carbon source:		
$H_2 + CO_2$	+	−
Formate	+	−
Malate	−	+
Benzoate	−	+
Serve as electron acceptors:		
$S_2O_3^{2-}$	+	−
SO_3^{2-}	+	−

[a] Symbols: +, 90% or more of strains are positive; −, 90% or more of strains are negative.

Table 15.11 Differential characteristics of the species of the genus *Herpetosiphon*[a]

Characteristic	H. aurantiacus	H. cohaerens	H. geysericola	H. nigricans	H. persicus
Color of cell mass	Orange	Orange	Orange	Yellow	Orange
Cellulose hydrolysis	−	−	+	−	−
Starch hydrolysis	+	−	+	−	−
Tyrosine degraded	+	−	ND	+	−
Catalase	+	−	+	−	−
Seawater required	−	+	−	+	+
Thermophilic	−	−	−	−	−
NO_3^- as sole nitrogen source	ND	−	ND	+	+
Mol% G + C of DNA	48.1	44.9	48.5	53.1	52.6

[a] Symbols: +, 90% or more of strains are positive; −, 90% or more of strains are negative; ND, not determined.

Table 15.12 Differential characteristics of the species of the genus *Saprospira*[a,b]

Characteristic	"S. albida"	"S. flammula"	"S. gigantea"	S. grandis	"S. nana"	"S. thermalis"	"S. toviformis"
Length of cells (μm)	2–3	2–3	ND	1–5.5	1.5–3	2–5	1–2.5
Diameter of filaments (μm)	0.8–1.2	1	1.5–3	0.8–1.2	0.5	1	0.8
Width of helix (μm)	1.5–2	1.5	5–9	1.4–2	ND	1.5–2.5	1.5
Pitch of helix (μm)	3–7	3–4	25–40	4–10	2.3–3	7–17	4–9
Length of filaments (μm)	10–500	10–500	400	6–500	36	10–500	10–500
Growth factors required	+	−	ND	+	ND	−	−
Starch degraded	ND	ND	ND	−	ND	+	−
Optimal temperature (°C)	30	30 (ND)	ND	30–37	ND	35	25
Color of cell mass	Pinkish tinge	Bright orange	ND	Orange-red	ND	Pink	Orange-yellow
Mol% G + C of DNA	42–47	48	ND	46–48	ND	35–37	38
Habitat	Fresh water	Fresh water	Marine	Marine	Marine	Fresh water	Marine

[a] Symbols: +, 90% or more of strains are positive; −, 90% or more of strains are negative; and ND, not determined.

[b] There are no cultures available of species in quotation marks, and several have never been studied in pure culture.

Table 15.13 Differential characteristics of the species of the genus *Vitreoscilla*[a]

Characteristic	*V. beggiatoides*	*V. filiformis*	*V. stercoraria*
Colony type	L, C	L	L
Filament type[b]:			
Continuous cell wall	+	+	−
Discontinuous cell wall	−	−	+
Filament diameter (μm)	2.5–3.0	1.0–1.5	1.0
Mol% G + C of DNA	42	59–63	50–51
Habitat:			
Freshwater sediments	+	+	−
Cow dung	−	−	+
Obligately requires amino acid mixtures	−	−	+
Growth on:			
Nutrient agar	−	−	+
Acetate plus salts	+	+	−
Use as sole carbon and energy sources:			
Glucose	−	+	−
Citrate	−	+	−
Lactate	−	+	−
Glutamate	−	+	−
Succinate	+	+	−
Acetate	+	+	−
Utilization of nitrate as sole nitrogen source	+	+	−

[a] Symbols: L, linguiformis type colonies; C, circuitans type colonies; +, 90% or more of the strains are positive; and −, 90% or more of the strains are negative.

[b] Cells within filaments having continuous cell walls share the outer cell wall layers, and only shallow constrictions are noticed between cells. Strains containing discontinuous filament cell walls consist of chains of individual cells held together by "connective material."

Important Notes for Users of This Manual

Unless otherwise indicated in footnotes to tables, the meanings of symbols are as follows:

+ 90% or more of strains are positive

− 90% or more of strains are negative

d 11-89% of strains are positive

v strain instability (not equivalent to "d")

D Different reactions in different taxa (species of a genus or genera of a family)

All other symbols are defined in footnotes to tables.

GROUP 16

THE FRUITING, GLIDING BACTERIA: THE MYXOBACTERIA

Rod-shaped Gram-negative bacteria that move by gliding and produce fruiting bodies and dessication-resistant myxospores. Cells are relatively large, measuring 0.6–1.2 μm in width and 2–10 μm in length. They occur in two morphological types, viz. (a) **slender, flexible rods with more or less tapering ends and** (b) **relatively stout, cylindrical rods with rounded ends** (Fig. 16.1). The fatty acid patterns are dominated by branched fatty acids and by C16 and C16:1. Menaquinone MK-8 is the only respiratory quinone in all species. Myxobacterial cells are usually yellow, orange, or red, pigmented by carotenoids; the main pigments are almost always fatty acid esters of carotenoid glucosides (the only known exception is *Nannocystis,* with aromatic carotenoids).

The myxobacteria **move by gliding** along interfaces, such as agar or glass surfaces. They also may penetrate the substrate and move within, for example, 1.2–1.5% agar gels. The gliding cells always leave slime trails. As a consequence of the gliding movements, the colonies of the myxobacteria **spread over the substrate surface** and therefore are called swarms. Gliding and spreading depend critically on the composition of the substrate. As a rule, spreading is observed on lean media; on rich ones (e.g., with 1–2 % peptone), the myxobacterial colonies remain small and convex, with an entire edge. Normally the cells are unevenly distributed within the swarm but concentrate in radial veins and sometimes in massive ridges along the periphery of the swarm. The characteristic morphology of their colonies usually allows one to distinguish myxobacteria in the vegetative state at a glance from other gliding bacteria.

Under starvation conditions the **cells accumulate or aggregate** at certain sites within the swarm, forming large globular or ridge-shaped masses containing approximately 10^4–10^6 individual cells. These masses further **differentiate into structures called fruiting bodies** that vary substantially in shape and organization with different species. The least elaborate types consist of slime and dormant cells and have no special boundary layer, and the slime may be soft and even deliquescent or hard and cartilaginous. Dormant cells may be encased in a morphologically distinct, brightly colored, tough wall that remains as an empty husk after germination, is spherical or ovoid, and is known as a sporangiole (sometimes the term sporangium is used). Fruiting bodies may be either simple or composite (i.e., consisting of several units). Those that are devoid of sporangioles can always be regarded as simple, although sometimes several fruiting bodies are clustered together and the boundaries of the individual fruiting bodies are not always clearly recognizable. Accessory structures may be produced, such as a slime stalk in the case of *Myxococcus stipitatus.* The composite fruiting bodies consist of a variable number of sporangioles, which either form a more or less tight package or a cluster on top of a stalk. As the composite fruiting body arises from a single aggregate mass, the whole fruiting body, and not the individual sporangiole, is homologous to a simple fruiting body, for example, of the *Myxococcus* type. During cultivation the typical morphology of the fruiting bodies often quickly gets lost. Fruiting bodies range in size between (10) 100 and 600 μm and are conspicuous by bright colors and shining surfaces. The size of individual fruiting bodies in a single culture may vary by a factor of 5–10 in the linear dimension. Sometimes the fruiting bodies are arranged in concentric rings, or they concentrate along the radial veins. Inside the maturing fruiting body the vegetative cells **convert into dormant myxospores** (formerly known as microcysts). Myxospores are desiccation resistant and show a moderate temperature tolerance, surviving 58–60°C for 10–60 min suspended in liquids.

Strictly aerobic, require a neutral pH, and are adapted to the mineral load normally found in soil and fresh water. Most of them are mesophiles, with a temperature optimum of 30–35°C. They are **chemoorganotrophs, capable of decomposing many different biomacromolecules.**

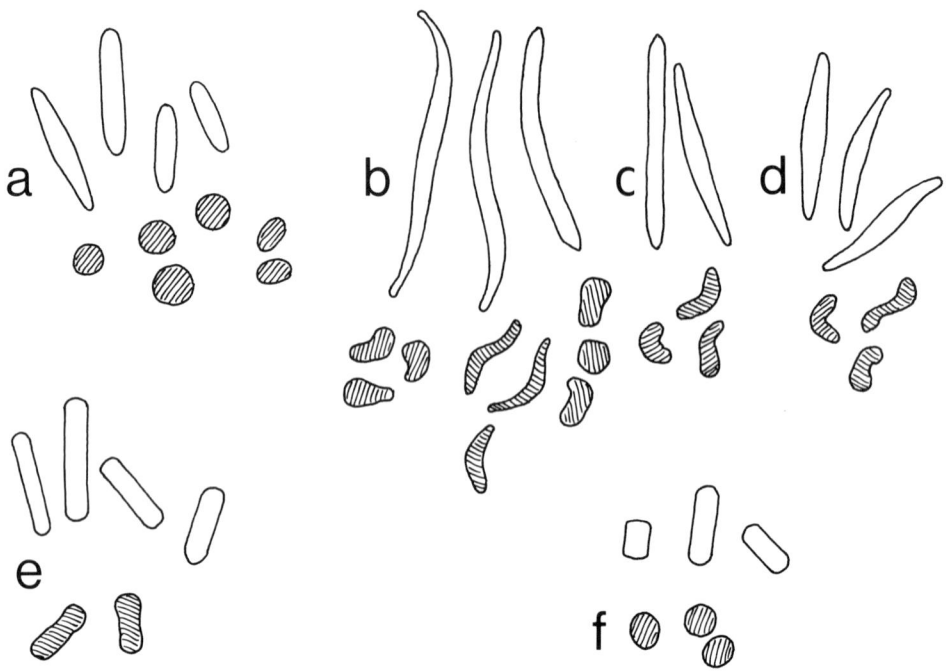

Figure 16.1 Cell types of myxobacteria (vegetative cells above, corresponding myxospores below): **a.** *Myxococcus, "Corallococcus," Angiococcus.* **b.** *Archangium, Cystobacter.* **c.** *Melittangium.* **d.** *Stigmatella.* **e.** *Polyangium, "Sorangium," Chondromyces.* **f.** *Nannocystis.*

Species are bacteriolytic, efficiently degrading whole cells of bacteria, yeasts, and other microorganisms and their various constituents. One group is able to decompose crystalline cellulose (e.g., filter paper). Several myxobacteria (e.g., *Nannocystis*) synthesize sterols in considerable quantities.

Myxobacteria are found worldwide. They seem to be particularly abundant in warm, semidry, and dry environments, such as steppes and semideserts of the subtropical and temperate zones. Typical habitats of myxobacteria are soils in the neutral pH and normal salt range, decaying organic material, including dung of herbivorous animals, and rotting wood, bark of living and dead trees, and also fresh water.

The taxonomy of the myxobacteria rests almost entirely on morphological characteristics, which may create serious problems for the determinative microbiologist. There are about 40 species, and the delimitation of quite a few is not particularly sound and would need a reexamination. In recent years, 16S rRNA studies have shown that the group as a whole is phylogenetically coherent and belongs to the *Proteobacteria,* delta branch. In the past, other gliding bacteria, especially the cytophagas and *Cytophaga*-like bacteria and *Lysobacter,* were often confused with the myxobacteria and have also been labeled so. As we know today, those bacteria are not related to the myxobacteria at all, and it would be highly desirable that use of the term *myxobacteria* be restricted to the true myxobacteria.

Key to the genera of the **Myxobacteria**

I. Vegetative cells are slender and flexible, with their ends more or less tapering (Fig. 16.1); myxospores are much different in shape, mostly short, fat rods or spheres (Fig. 16.1); swarm colonies are essentially on top of the agar surface, with a pattern of radial veins and with flame-like extensions at the edge; fruiting bodies are with or without sporangioles, simple or composite; slime sheet of swarm stains with 0.01% aqueous Congo red solution; 67–70 mol% G + C.

A. Myxospores are spherical or ovoid. Vegetative cells are moderately long (3–8 µm), cigar- or boat-shaped; swarm colonies are relatively unstructured, but in the border zone often characteristic undulating radial veins are developed; swarm sheet is soft, slimy; fruiting bodies are mostly simple, without sporangioles (but sporangioles are produced by *Angiococcus*).

   1. Fruiting bodies that are simple, without sporangioles

      a. Fruiting bodies are soft slimy humps or smooth knobs, often with a constriction at their base or even a slime stalk

Genus *Myxococcus*

      b. Fruiting bodies are hard, slimy, of cartilaginous consistency, pustules and ridges, often with finger- or staghorn-like protrusions, or of bizarre coralloid shape.

Genus *"Corallococcus"*

   2. Fruiting bodies that are composite, with small, more or less spherical brown-orange to red-brown sporangioles

Genus *Angiococcus*

B. Myxospores are short, fat rods, sometimes almost spherical, but then irregular in outline (not smooth as in A). Vegetative cells are long and slender (5–12 µm), with tapering ends, sometimes boat-shaped; swarm colonies are conspicuously structured, with long, straight or slightly curved, more or less branched radial veins; swarm sheet are often very tough; fruiting bodies mostly with sporangioles, either composite or with a single sporangiole on an individual stalk, but simple fruiting bodies without sporangioles do also occur (*Archangium*).

   1. Fruiting bodies are simple, without sporangioles, variable in shape: usually meandering ridges or cushion-like masses consisting of convoluted rolls of hardened slime and myxospores; myxospores are short, fat rods; vegetative cells are long, slender, and needle-shaped.

Genus *Archangium*

   2. Fruiting bodies consist of sporangioles.

      a. Vegetative cells are long (6–12 µm), slender, needle-shaped.

         i. Fruiting bodies are clusters of sporangioles, which sometimes are embedded in a conspicuous, common slime capsule (matrix), sitting directly on the substrate.

Genus *Cystobacter*

         ii. Fruiting bodies are single, tiny sporangioles, each on a delicate stalk.

Genus *Melittangium*

      b. Vegetative cells are moderately long (5–8 µm), more or less boat-shaped; fruiting bodies consist of a stalk bearing one or several sporangioles.

Genus *Stigmatella*

517

II. Vegetative cells are cylindrical, rather stout, with rounded ends (Fig. 16.1); myxospores are usually very similar in shape and size to vegetative cells although optically refractile (exception: *Nannocystis* with spherical myxospores); swarm colonies are often subsurface, at least partly so, veins may or may not be present (occasionally very pronounced), edge often with broad lobes or fan-like sectors; fruiting bodies always with sporangioles and almost always composite (exceptions: *Nannocystis, "Haploangium"*); slime does not stain with Congo red solution; no 2-hydroxy fatty acids; G + C 70–72 mol%.

A. Myxospores are cylindrical rod-shaped cells, optically more or less refractile but otherwise morphologically very similar to vegetative cells. The fruiting bodies always consist of sporangioles and are almost always composite (exception: *"Haploangium"*).

1. Fruiting bodies are single sporangioles.

   a. Fruiting body is a large (60–200 μm), globular, sessile sporangiole, often with a wrinkled surface, found on bark and rotting wood.

   Genus *"Haploangium"*

   b. Fruiting bodies are tiny (typically 6–40 μm), ovoid or spherical sporangioles on and within the substrate, solitary, scattered, or sometimes clustered together; myxospores are spherical or nearly so and optically refractile.

   The vegetative cells are cylindrical with blunt, sometimes almost square ends; they may become very short, nearly cube-shaped. The fruiting bodies are sporangioles and are single, scattered, or assembled in clusters and packages; the swarm colonies on agar plates typically show a much corroded surface.

   Genus *Nannocystis*

2. Fruiting bodies consist of clusters of packages of sporangioles that are located directly on or within the substrate.

   a. Sporangioles of several species are very large (60–400 μm); in others, they are moderately sized, sometimes embedded in a conspicuous, common slime envelope (matrix); of the bacteriolytic nutritional type.

   Genus *Polyangium*

   b. Sporangioles usually are rather small (20–30 μm), produced in tremendous numbers; cellulose decomposers.

   Genus *"Sorangium"*

   c. Fruiting bodies are clusters of orange sporangioles borne on long, branched or unbranched, white slime stalks, very large (300–600 μm) and impressive.

   Genus *Chondromyces*

Genus **Angiococcus**

Vegetative cells are moderately long, 4–6 μm, and cigar- to slightly boat-shaped. Fruiting bodies are bright orange-brown to red-brown, **spherical or ovoid sporangioles, 50–70 μm in diameter,** arranged in clusters and, often, in more or less meandering strings. Myxospores are spherical, optically refractile, and

1.5–1.9 μm in diameter. Swarm colonies show wavy veins. Some strains vigorously decompose chitin.

Type (and only) species: *Angiococcus disciformis.*

Characteristics of the species: As described for the genus.

Genus Archangium

Vegetative cells are slender, **tapered,** flexible **rods,** 0.6–0.8 × 6–12 μm. **Fruiting bodies are a mass of solid slime and myxospores and are not differentiated into sporangioles.** Their color varies from pale orange to brown and even violet, and they may become rather large. Their shape is variable: cushion-like pads, meandering ridges, or, most typical, intestine-like twisting, convoluted rolls, with a bulging, knotted surface, sometimes extended into finger-like projections. **Myxospores** are **very short rods, ellipsoids or spheres,** 1.0–2.0 × 1.5–2.8 μm, and are refractile or optically dense. The thin, film-like swarm colonies form a tough slime sheet and often show a pattern of densely packed, long, fine, branched radial veins. Cells do not etch or erode agar media. Congo red is adsorbed. *Archangium* is difficult to identify reliably because under cultivation many myxobacteria soon produce degenerate fruiting bodies that can hardly be distinguished from those of *Archangium.* This is the case with *Cystobacter, Melittangium, Stigmatella,* and *Polyangium* strains. In some cases the problem is further aggravated because the vegetative cells and myxospores are also nearly identical (*Melittangium,* several *Cystobacter* species), so that doubts arise concerning whether *Archangium* really exists as a taxon of its own. Because typical *Archangium* fruiting bodies are also often found on natural substrates (e.g., rabbit dung), it seems justified to retain the genus for the moment.

Type (and only) species: *Archangium gephyra.*

Characteristics of the species: As described for the genus.

Genus Chondromyces

Vegetative cells are **cylindrical, untapered rods** with blunt rounded ends, 1.0–1.2 (1.4) × 3–8 (14) μm. Fruiting bodies consist of **clusters of bright orange sporangioles borne on long, white slime stalks,** the latter often with a disk-shaped base. Fruiting bodies are **larger** than those of most other myxobacteria (300–700 μm high) and are easily seen with the naked eye. Swarm colonies grow on and below the agar surface, producing large, shallow depressions in the agar with a narrow orange band along the swarm edge where the vegetative cells concentrate. Vegetative slime **does not adsorb Congo red dye.** Temperature range is 18–37°C (optimum, 28–30°C). The mol% G + C of the DNA is 69–70 (T_m).

Type species: *Chondromyces crocatus.*

Key to the species of the genus Chondromyces

I. Slime stalk is normally branched, often repeatedly so (see also B), with sporangioles in dense clusters at the ends of the branches, ovoid, relatively small, 10–30 μm in diameter, sessile or on very short peduncles; fruiting body may become very large, 700 μm and more in height; pure cultures have a characteristic, pyridine-like odor.

C. crocatus

II. Slime stalk is unbranched.

A. Sporangioles are turnip-shaped, each one ending with a 15–35-μm-long (white) tail; rather large, 20–50 μm wide and 40–100 μm long (without the tail), usually sessile or with a short peduncle; the sporangioles often occur, perhaps 100 together, in large dense tufts; fruiting bodies are large and often stout, up to 700 μm and more high; they are often found on rotting wood.

C. apiculatus

B. Stalk bears few (usually 2–12) but large sporangioles, 70–170 μm in diameter; the sporangioles are bowl-shaped to narrowly cylindrical, with a flattened outer end from which many 10–30-μm-long, hair-like, white tails arise; the sporangioles usually sit on short peduncles that may be relatively stout and rather look like branches of the stalk. Branches are restricted to the uppermost part of the stalk.

C. lanuginosus

C. Sporangioles are bell-shaped, with a flattened outerend sometimes bearing a short tip in its center; in culture the sporangioles often lose their typical shape and become ovoid; they are relatively small, 25–40 × 40–70 μm, and are usually borne on thin and sometimes rather long peduncles (e.g., 5–7 × 35–40 μm) so that they may bend over and hang more or less downward; the main stalk often is very slender (e.g., 25 × 350 μm); the fruiting bodies may become large, around 700 μm in height.

C. pediculatus

D. These are rather small oval sporangioles (18 × 20–50 μm) that are arranged in chains up to 300 μm long and sometimes branched; the individual sporangioles are separated within the chain by shorter or longer pieces of shriveled slime, and the whole structure resembles a string of beads; the chains originate as a tuft from the end of a 180–360-μm-long stalk; the whole fruiting body may reach 650 μm in height.

C. catenulatus

Genus *"Corallococcus"*

Vegetative cells are moderately long (3–6 μm) and cigar-shaped. The fruiting body consists of solid slime and **spherical myxospores.** Fruiting bodies are extremely variable in shape; they may be **simple pustules** or **straight to meandering ridges,** often with tapering tails of myxospores at their ends or sides, sometimes **branched or star-shaped;** these pads often bear **finger-or hand-shaped projections** of variable size which end either in tips or in tiny globules; sometimes the whole fruiting body is a **columnar, repeatedly branched, coral-like mass.** They decompose starch, and some strains utilize carbohydrate if added as oligo- or a degradable polysaccharide.

Type species: *"Corallococcus macrosporus."*

Differentiation of the species of the genus **"Corallococcus":** See Table 16.1.

Genus **Cystobacter**

Vegetative cells are slender, **tapered,** flexible rods, 0.6–0.8 × 8–15 μm. **Sporangioles are usually sessile,** occurring singly or in clusters; are rounded, elongated, or coiled; and are surrounded by a definite slime envelope or wall. They are either free or embedded in a second slimy layer. **Myxospores are rod-shaped, phase-dense, or refractile** and are rigid. Swarm colonies are thin and film-like with a tough, sometimes extremely tenacious slime sheath and with long, fine, branched radial veins. Vegetative colonies do not etch or erode agar media. **Congo red is adsorbed.** The minimum nutritional requirements are not known, but all species are easily cultivated on media containing enzymatically hydrolyzed protein, salts, and starch or glycogen. The latter is not utilized. **Cellulose is not digested.**

Type species: *Cystobacter fuscus.*

Key to the species of the genus **Cystobacter**

I. Myxospores are long and slender, 0.9–1.6 × 3–7 μm, and somewhat irregular in outline, with slightly tapering or rounded ends and are usually S- or C-like curved and optically refractile; sporangioles are large, 80–200 μm in diameter; they are chestnut brown and efficient chitin-decomposers.

A. Sporangioles are smoothly spherical or ovoid, of a bright chestnut brown, glistening, assembled in large, more or less dense, normally single-layered clusters, and typically embedded in a thick, transparent common slime matrix.

C. fuscus

B. Sporangioles are oval to elongated, occasionally incompletely septated and rather irregular in shape; they tend to vary substantially in size even within one cluster but may become very large; sporangioles are arranged in dense, often elongated string-like clusters; they may be piled on top of one another and even point finger-like upward.

C. ferrugineus

II. Myxospores are shorter and stouter, usually very short, fat rods, sometimes almost irregularly spherical, 1.3–2.4 × 1.4–5 μm.

A. Sporangioles are spherical, oval, or elongated, usually somewhat asymmetric, large, (30) 50–140 (160) μm in diameter, dull red-brown to (rarely) violet, arranged in dense clusters, often string-like and piled on top of one another, sometimes with finger-like projections; swarm colony may become deep violet, but the pigment is usually produced only in confined areas and often not at all; it does not decompose chitin.

"C. violaceus"

B. Sporangioles are oval, somewhat irregular in shape, resembling potato tubers, sometimes incompletely septated, rather large, (35) 50–80 (100) μm in diameter; they are yellow-brown or orange-brown to fawn; they are arranged in dense clusters, often string-like and sometimes piled on top of one another; the whole fruiting body is covered by a delicate, pleated slime sheet.

"C. velatus"

C. Sporangioles are spherical to oval, pale brownish yellow to bright orange and orange-brown, smaller than those of the other *Cystobacter* species, 20–50 (70) μm in diameter; they are arranged in dense clusters and strings; often arising within the agar (which seems never to happen with the other *Cystobacter* species) and then usually tightly packed together, so that the sporangioles become polygonal and the fruiting body resembles rather that of a *Polyangium* or *Sorangium;* vegetative cells are slightly shorter than those of the other species, 0.7–0.9 × 4.5–11 μm, needle- or boat-shaped; the swarm colonies sometimes are yellow or orange.

C. minus

Genus "Haploangium"

Fruiting bodies are conspicuous, more or less globular, and solitary sporangioles, 60–140 (200) μm in diameter, bright orange to orange-red or orange-brown, often with a wrinkled surface. Vegetative cell shape is not known. Myxospores are cylindrical, 0.7–0.8 × 2.5–3.5 μm. They are found on tree bark and rotting wood and have not been cultivated.

Type species: None designated.

Differentiation of the species of the genus **"Haploangium":** See Table 16.2. The two species *"H. rugiseptum"* and *"H. minor"* were included as valid species in the genus *Polyangium* in *Bergey's Manual of Systematic Bacteriology,* Volume 3, pp. 2159–2162, but they are considered sufficiently different from other polyangia to be placed in a separate genus here.

Genus **Melittangium**

Vegetative cells are long, slender, tapered rods, 0.6–0.8 × 5–12 μm. **Fruiting bodies consist of a tiny sporangiole located on a delicate, short, white stalk.** Sporangioles may be spherical, ovoid, bean-shaped, or flat like a mushroom cap, 20–50 μm in diameter, and pale yellow to red-brown. Myxospores are **short to moderately long, relatively slender, irregular rods, often C- or S-like curved** and with slightly tapering or rounded ends, 0.9–1.2 × 1.5–4.0 μm. Under cultivation the fruiting bodies degenerate very soon to *Archangium*-like structures of a more or less bright orange color; in that case, the shape of the myxospores is a distinguishing although not very dependable characteristic. Swarm colonies show a pattern of long, branching, radial veins and do not etch or erode agar media. Congo red is adsorbed. Minimum nutritional requirements are unknown, but this organism is cultivable on media containing enzymatically hydrolyzed protein.

Type species: *Melittangium boletus.*

Differentiation of the species of the genus **Melittangium:** See Table 16.3.

Genus **Myxococcus**

Vegetative cells are slender rods with tapering ends, 0.4–0.7 × 2.0–10.0 μm. Fruiting bodies are relatively large, globular masses containing **refractile, spherical, or ellipsoidal microcysts** up to 2.3 μm in the largest dimension. All species may produce a short pedicle, and one species (*M. stipitatus*) produces a long stalk. The **microcysts** are **not** enclosed **in a sporangiole.** They are chemoorganotrophs. Metabolism is respiratory. Those that have been studied require several amino acids and are capable of hydrolyzing protein, starch, nucleic acids, and various fatty acid esters. They do not utilize carbohydrates. **They are known to lyse bacteria, yeast, and certain filamentous fungi.** They are noncellulolytic and strict aerobes. They fail to grow at 40°C or above. **Colonies adsorb Congo red.** They are resistant to 10-unit disks of penicillin and sensitive to 10-μg disks of neomycin and tetracycline, to 5-μg disks of erythromycin, and to 250 μg/ml kanamycin.

The mol% G + C of the DNA of the examined species is 68–71.

Type species: *Myxococcus fulvus.*

Differentiation of the species of the genus **Myxococcus:** See Table 16.4.

Genus **Nannocystis**

Vegetative cells are stout rods with blunt ends, 1.1–2.0 × 1.5–5.0 μm. Myxospores are ovoid or nearly spherical, 0.75–1.5 μm. Fruiting bodies are tiny solitary sporangioles that are spherical or oval, 6–30 μm in diameter, and usually embedded in the agar; on the surface, often much larger, irregularly shaped sporangioles, 50–150 μm in diameter are produced. Swarm colony deeply corrodes the agar surface. Cells are aerobic organotrophic organisms. The mol% G + C of the DNA is 70–72 (Bd, T_m).

Type (and only) species: *Nannocystis exedens.*

Characteristics of the species: As described for the genus.

Genus **Polyangium**

Vegetative cells are cylindrical and of uniform diameter, **with blunt, rounded ends,** 0.6–1.2 × 3–8 μm. Fruiting bodies consist of **spherical to oval, sometimes elongated, sporangioles that occur in groups, sometimes closely packed together,** and often, especially when embedded in the agar, packed so tightly that they become polygonal. The sporangiole cluster is surrounded by a communal slime envelope or matrix, and sometimes is elevated on a stout slime stalk. Sporangioles may pile on top of one another and form finger-like projectons. **Myxospores resemble vegetative cells** and are not refractile or phase-dense, **lacking a slime capsule** in the examined species. Colonies of all examined species etch, erode, and penetrate the agar. **Congo red is not adsorbed** by the vegetative slime. Cellulose may or may not be hydrolyzed.

Type species: *Polyangium vitellinum.*

Key to the species of the genus **Polyangium**

I. Sporangioles are large, 40–200 μm in diameter, spherical, oval, or elongated, and bright yellow to orange.

A. Sporangioles are large to very large, 75–200 μm across, spherical or oval, golden yellow to orange, in groups of up to 20 in a common transparent slime envelope; they are found particularly on wet rotting wood and bark.

P. vitellinum

B. Sporangioles are large to very large, 80–180 μm in diameter, more or less spherical, bright yellow, solitary or in groups of two or, rarely, three or four in a common, colorless to faintly yellow slime envelope; the sporangioles are surrounded by a thin, unpigmented wall and are easy to crush; they are known to come from soil and rabbit dung.

P. luteum

C. Sporangioles are of variable size, measuring 60–150 μm, are more or less elongated, often irregular in shape, and subdivided by incomplete septa; they are bright yellow-orange to orange-brown. The sporangioles usually occur in large groups of 20–100 on top of or within the agar. They are often arranged in convoluted chains that may rest on a conspicuous, stout, cushion-like stalk; if the convoluted bacterial mass on top of the stalk is not differentiated into sporangioles, which often happens, then the whole structure corresponds exactly with the *"Archangium thaxteri"* of Jahn (1924). It is found in soil.

"P. thaxteri"

II. Sporangioles are smaller, 10–60 μm across, and more or less brown or orange.

A. Sporangioles are very small, 10–15 μm across, polygonal to spherical, yellow-orange to orange-red. The sporangioles are packed together in large numbers to form flat, cushion-like fruiting bodies about 500 μm diameter; fruiting bodies are covered by a thin, colorless slime layer.

P. sorediatum

B. Sporangioles are small to moderately large, 10–70 μm across, colorless, smokey gray or brownish, often within the agar.

1. Sporangioles are small, 20–35 μm in diameter, more or less oval to slightly polygonal, colorless to brownish, usually in tight packages of variable size.

P. spumosum

2. Sporangioles are medium-sized, 20–60 μm in diameter, spherical, oval, or slightly polygonal; the sporangioles occur in flat groups of variable size, sometimes rather large ones with dozens of sporangioles, often relatively loosely packed and in the shape of a ring.

P. fumosum

523

3. Sporangioles are medium sized, 30–60 µm across; they are red-brown.

    a. Sporangioles are more or less spherical, usually 30–40 µm in diameter, red-brown, in groups of 2–15 in a characteristic orange-yellow; they have a common slime envelope; they are found in soil.

P. aureum

    b. Sporangioles are more or less spherical, 25–40 µm in diameter, red-brown, usually in groups of two to eight in a colorless communal slime envelope, found in the interior of *Cladophora* (green algae) cells in fresh water; the organism has not been reported again since its discovery in 1925; conceivably it is not an obligate parasite of algae, or it may even be restricted to fresh water.

P. parasiticum

Genus "Sorangium"

Vegetative cells are cylindrical rods, 0.9–1.2 × 2.5–6 (8) µm, often with bright polar granules. Myxospores are optically refractile, cylindrical rods with rounded ends, 0.8–1.2 × 1.5–3.0 µm. Fruiting bodies consist of **tiny spherical or polygonal sporangioles tightly packed together to form small, dense parcels.** The parcels are clustered together into large masses and are produced in enormous numbers, especially in filter paper cultures, forming a dense continuous layer. On filter paper the sporangioles often are arranged in strings and ribbons along the cellulose fibers. Fruiting body color is bright and varies among yellow, orange, orange-brown, light to dark red-brown, deep brown, and black. Swarm colonies remain on the surface of the agar, forming a system of long radial veins. They **decompose cellulose.** Many strains decompose chitin. They grow on simple defined media with glucose, nitrate, and the usual basic mineral salts.

Type (and only) species: *"Sorangium cellulosum."*

Characteristics of the species: As described for the genus.

Genus Stigmatella

Vegetative cells are straight rods with tapered ends, 0.6–0.8 × 4–10 µm. Myxospores are short, fat, and somewhat crooked, 1.0–1.5 × 1.5–4 µm, phase-dense or refractile. Fruiting bodies consist of **sporangioles on stalks. Sporangioles are spherical, ovoid, or sometimes club-shaped, bright orange to dark red-brown,** and (25) 40–80 (100) µm in diameter. The stalk is usually white and in its upper part is often yellow to brown, 25–45 µm wide and 40–140 µm long. The whole fruiting body measures between 80 and 200 µm in height. Vegetative swarms on agar are yellow-orange to red-brown, do not penetrate or etch agar, and adsorb Congo red. This organism can be cultivated on media containing enzymatically hydrolyzed protein. Nitrate is not reduced. This organism is catalase positive and oxidase negative and hydrolyzes starch, Tween 80, indoxyl acetate, RNA, DNA, gelatin, casein, and esculin. Urea is usually hydrolyzed. It is aerobic. The temperature range is 18–37°C (optimum, 30°C). The mol% G + C of the DNA is 68.5–68.7 (T_m).

Type species: *Stigmatella aurantiaca.*

Differentiation of the species of the genus **Stigmatella:** See Table 16.5.

Table 16.1   Differential characteristics of the species of "Corallococcus"[a]

Characteristics	"C. coralloides"	"C. exiguus"	"C. macrosporus"
Diameter of myxospores (µm)	1.3–1.9	1.5–1.8	1.8–2.5
Color of fruiting bodies	White, pink to brick red	Brown	Sulfur yellow
Fruiting bodies with projections	+	−	+
Fruiting body population on yeast agar	Loose to moderately dense	Very dense	Loose to moderately dense

[a] Symbols: see standard definitions.

Table 16.2   Differential characteristics of the species of "Haploangium"[a]

Characteristics	"H. minor"	"H. rugiseptum"
Color of sporangiole	Orange-brown	Bright orange
Wall of sporangiole with an outer layer	−	+
Surface of sporangiole heavily wrinkled	−	+
Myxospores packed in spherical masses	+	−

[a] Symbols: see standard definitions.

Table 16.3   Differential characteristics of the species of Melittangium[a]

Characteristics	M. boletus	M. lichenicola
Sporangiole with a flat underside, like a mushroom cap	+	−
Sporangiole spherical to ovoid	−	+
Preferentially on bark and rotting wood	−	+
Mainly in soil and on rabbit dung	+	−

[a] Symbols: see standard definitions.

Table 16.4   Differential characteristics of the species of Myxococcus

Characteristics	M. fulvus	M. stipitatus	M. virescens	M. xanthus
Diameter of myxospores (µm)	1.3–1.6	1.1–1.3	2.0–2.5	1.8–2.2
Shape of myxospores	Spherical	Ovoid	Spherical	Spherical
Fruiting body with a stalk	−	+	−	−
Color of fruiting bodies [a]	White, pink to red	White to fawn	Greenish yellow	Orange
Swarm with a yellow fluorescence (366 nm)	−	+	−	−
Swarm bright greenish yellow	−	−	+	− (+)

[a] Pigmentation is often light-induced; therefore typical colors may be seen only in illuminated cultures.

Table 16.5   Differential characteristics of the species of Stigmatella[a]

Characteristics	S. aurantiaca	S. erecta
Fruiting body a single sporangiole on a stalk	−	+
Fruiting body a cluster of sporangioles on a stalk	+	−
Mainly found on rotting wood	+	−
Mainly found in soil and on dung	−	+

[a] Symbols: see standard definitions.

GROUP 17

GRAM-POSITIVE COCCI

This group consists of genera that are quite diverse but are placed together for convenience because of the **spherical shape** of their cells and their **Gram-positive staining.** They **do not form endospores.** (*Sporosarcina,* Group 18, is coccoid but forms endospores.) In the great majority, cells are **spherical or only slightly oval,** and the Gram stain almost always shows a definite positive reaction. **Motility is uncommon.** The genera fall into reasonably distinct groupings of **aerobic, facultatively anaerobic, and strictly anaerobic** genera. The **arrangement of cells** and the **occurrence of catalase** are other features convenient for separating genera.

The major differential features of the genera are presented in Tables 17.1 (aerobic genera), 17.2 (facultatively anaerobic genera), and 17.3 (strictly anaerobic genera).

Genus **Aerococcus**

Cells spherical, 1.0–2.0 μm in diameter, **forming tetrads in liquid media.** Gram-positive, nonmotile, facultative anaerobes, but they **grow best at reduced oxygen tension and grow poorly in air or anaerobically.** Produce H_2O_2 during aerobic growth, and there is **marked greening on blood agar.** Chemoorganotrophic, with a respiratory metabolism. Produce acid without gas from various carbohydrates catalase **negative or weak.** Gelatin is not liquefied, and nitrate is not reduced. The optimum temperature is 30°C; grows at 10°C but not at 45°C; **grows at pH 9.6, in 10% NaCl, and in 40% bile.** Common as an airborne organism in hospitals and has been associated with disease of lobsters.

Type (and only) species: *Aerococcus viridans.*

Characteristics of the species: As described for the genus.

Genus **Coprococcus**

Cells spherical, 0.8–1.5 μm in diameter, and sometimes ovoid. Arranged in pairs or short chains. **Gram-positive,** nonmotile. **Strictly anaerobic** cells produce small, whitish colonies on blood agar, sometimes with weak α-hemolysis. Chemoorganotrophs, with a **fermentative metabolism; carbohydrates are required or highly stimulatory for growth. Fermentation products include butyrate, acetate, and gas. Catalase negative.** The optimum temperature is 37°C. Found in human intestine and feces.

Type species: *Coprococcus eutactus.*

Differentiation of the species of the genus **Coprococcus:** See Table 17.4.

Genus **Deinobacter**

Editorial note: The genus *Deinobacter* was not included in *Bergey's Manual of Systematic Bacteriology.* It was established by Oyaizu et al. (Int. J. Syst. Bacteriol. *37:* 62–67, 1987) for an organism that is highly resistant to γ-rays and that, except for cell shape, closely resembles the radiation-resistant genus *Deinococcus* (*q.v.*), to which it seems close on ribosomal RNA evidence.

Cells rod-shaped and 0.6–1.2 × 1.5–4.0 μm. **Gram negative,** but the cell wall has a complex layered structure, somewhat similar to that of Gram-positive organisms and very similar to that of *Deinococcus.* Nonmotile. **Aerobic. Colonies pink or red.** Chemoorganotrophic, metabolism respiratory, and **inactive toward carbohydrates. Catalase positive, oxidase negative,** and indole negative; nitrate is not reduced. Gelatin is usually hydrolyzed. The optimum temperature is 30–35°C. **Highly resistant to γ-radiation.** Probably widely distributed in the environment.

Type (and only) species: *Deinobacter grandis.*

Characteristics of the species: As described for the genus.

Genus **Deinococcus**

Cells spherical, 0.5–3.5 μm in diameter, **in pairs or tetrads,** and **appear large** in relation to other cocci. **Gram-positive** cells have a complex cell wall profile **with several distinct layers.** Nonmotile, nonsporing. **Aerobic,** colonies are usually **red or orange.** Chemoorganotrophic, with a **respiratory metabolism, inactive toward carbohydrates. Catalase positive. Highly resistant to γ-radiation.** Lipids contain large amounts of **palmitoleate.** The optimum temperature is 30–37°C. Probably widely distributed in the environment.

Type species: *Deinococcus radiodurans.*

Differentiation of the species of the genus **Deinococcus:** See Table 17.5.

Genus **Enterococcus**

Editorial note: The genus *Enterococcus* was not included in *Bergey's Manual of Systematic Bacteriology,* where some of its species were included in the genus *Streptococcus.* A proposal was made by Schleifer and Kilpper-Bälz (Int. J. Sysy. Bacteriol. *34:* 31–34, 1984) to transfer streptococci that had long been informally referred to as enterococci to a separate genus *Enterococcus.* In that publication *Streptococcus faecalis* and *S. faecium* were transferred to the new genus as *Enterococcus faecalis* (type species) and *E. faecium,* respectively. Subsequently *Streptococcus avium* and *S. gallinarum* were transferred by Collins et al. (Int. J. Syst. Bacteriol. *34:* 220–223, 1984), and new species of *Enterococcus* were established for other earlier, streptococcal species and subspecies (i.e., *E. durans, E. casseliflavus,* and *E. malodoratus*). A number of newly described species have since been added.

Cells spherical or ovoid, 0.6–2.0 × 0.6–2.5 μm, **occurring in pairs or short chains** in liquid media. **Endospores are not formed. Gram positive.** Sometimes motile by scanty flagella. Lack obvious capsules. **Facultative anaerobes,** chemoorganotrophs with fermentative metabolism; a wide range of carbohydrates are fermented with the production of **mainly L(+)-lactic acid** but no gas and a final pH of 4.2–4.6. **Nutritional requirements are complex. Catalase negative. Usually grow at both 10° and 45°C** (optimum 37°C), **at pH 9.6, with 6.5% NaCl, and with 40% bile.** Seldom reduce nitrate. Usually ferment lactose. Usually of

Lancefield serological Group D. Occur widely in the environment, particularly in feces of vertebrates; sometimes cause pyogenic infections.

Type species: *Enterococcus faecalis.*

Differentiation of the species of the genus **Enterococcus:** See Table 17.6.

Genus **Gemella**

Cells spherical or elongate, 0.5–0.8 × 0.5–1.4 μm, **often in pairs with adjacent sides flattened, or in pairs of cells of unequal size.** Sometimes in short chains. **Gram positive or Gram variable,** cells easily decolorize, although the cell wall is of the Gram-positive type. Nonmotile, nonsporing. **Facultatively anaerobic** but may not grow in air on first isolation. Colonies on rabbit or horse blood agar are small, often with α- or β-hemolysis. Chemoorganotrophic, requiring nutritionally rich media containing protein. **Metabolism fermentative,** producing **mainly L(+)-lactic acid** but no gas from glucose and a few other carbohydrates. **Catalase negative** and oxidase negative. Gelatin is not hydrolyzed. The optimum temperature is 37°C, with no growth at 10°C or 45°C. H_2O_2 may be produced aerobically. Obligate parasites in the human mouth, intestine, and respiratory tract and may be confused with species of *Neisseria, Veillonella,* or *Streptococcus.*

Type species: *Gemella haemolysans.*

Differentiation of the species of the genus **Gemella:** See Table 17.7.

Genus **Lactococcus**

Editorial note: The genus *Lactococcus* was not included in *Bergey's Manual of Systematic Bacteriology.* It was established by Schleifer et al. (Int. J. Syst. Bacteriol. *36:* 354–356, 1986; effective publication: Syst. Appl. Microbiol. *6:* 183–195, 1985) for the lactic streptococci *Streptococcus lactis* and *Streptococcus raffinolactis* and certain lactobacilli (*Lactobacillus hordniae* and *Lactobacillus xylosus*), together with some newly described species.

Cells spherical or ovoid, 0.5–1.2 × 0.5–1.5 μm, **occurring in pairs and short chains** in liquid media. **Endospores not formed. Gram-positive.** Nonmotile and without capsules. **Facultative anaerobes.** Chemoorganotrophs **with fermentative metabolism;** a number of carbohydrates are fermented with the production of **mainly L(+)-lactic acid** but no gas. **Nutritional requirements complex. Catalase negative** and oxidase negative. The optimum temperature is 30°C. **Grow at 10°C but not at 45°C, nor with 0.5% NaCl. Usually of Lancefield serological Group N.** Found in dairy and plant products.

Type species: *Lactococcus lactis.*

Differentiation of the species of the genus **Lactococcus:** See Table 17.8.

Genus **Leuconostoc**

Cells spherical or **somewhat longer than broad when in the usual pairs or chains,** 0.5–0.7 × 0.7–1.2 μm. Sometimes short rods with rounded ends in long chains. **Gram positive, nonmotile, and nonsporing.** Grow rather slowly, producing small colonies that may be slimy on media containing sucrose. **Facultative anaerobes,** chemoorganotrophs with **obligate requirement for a fermentable carbohydrate, requiring nutritionally rich media.** The optimum growth temperature is 20–30°C. Glucose is fermented with the **production of acid and usually gas; major fermentation products are ethanol and D(−)-lactate.** Fermentation is mainly restricted to mono- and disaccharides. **Catalase negative, arginine not hydrolyzed,** indole-negative, nonhemolytic, nitrates not reduced. The final pH in glucose broth is 4.4–5.0. Widely distributed on plants and in dairy and other food products. Nonpathogenic to plants and animals, but *L. citreum* and *L. pseudomesenteroides* have been isolated from human sources.

Type species: *Leuconostoc mesenteroides.*

Differentiation of the species of the genus **Leuconostoc:** See Table 17.9.

Genus **Marinococcus**

Editorial note: The genus *Marinococcus* was not included in *Bergey's Manual of Systematic Bacteriology.* It was established for the organism *Planococcus halophilus*

by Hao et al. (Int. J. Syst. Bacteriol. *35:* 535, 1985; effective publication: Hao et al., J. Gen. Appl. Microbiol. *30:* 449–459, 1984), and two new species were added later.

Cells spherical, 1.0–2.0 μm in diameter, occurring **singly, in pairs, or occasionally in tetrads. Gram positive to Gram variable. Often motile** by a few flagella. Nonsporing. **Aerobic.** Chemoorganotrophs with respiratory metabolism, giving smooth convex colonies often with yellow-orange nondiffusing pigment. **Catalase positive.** The optimum temperature is 30–37°C. **Moderate halophiles, requiring 0.5–20% NaCl in media.** Found in marine habitats and saline soil.

Type species: *Marinococcus halophilus.*

Differentiation of the species of the genus **Marinococcus:** See Table 17.10.

Genus **Melissococcus**

Editorial note: The genus *Melissococcus* was mentioned briefly in *Bergey's Manual of Systematic Bacteriology* as a *genus incertae sedis.* It was established for the organism that causes European foulbrood of honey bees (Bailey and Collins, Int. J. Syst. Bacteriol. *33:* 672–674, 1983, where the name is spelled *Melisococcus* because of a typographical error; effective publication: J. Appl. Bacteriol. *53:* 215–217, 1982).

Cells ovoid to lanceolate, 0.8–1.0 × 0.8–1.5 μm, **in chains,** sometimes pleomorphic or short rods. **Gram positive** but are easily destained. Nonmotile, nonsporing, **facultatively anaerobic but requiring added CO₂.** Chemoorganotrophic, **requiring nutritionally rich media for growth and added cysteine or cystine,** and produce small colonies up to 1 mm in diameter. Ferment glucose or fructose but seldom other carbohydrates and **produce weak acidity** (final pH 5.3); the major product is lactic acid. A Na:K ratio of 1 is required. The optimum temperature is 35°C. Reacts with Lancefield Group D antiserum. Causative agent of European foulbrood of honey bees.

Type (and only) species: *Melissococcus pluton.*

Characteristics of the species: As described for the genus.

Genus **Micrococcus**

Cells spherical, 0.5–2.0 μm in diameter, occurring in pairs, **tetrads, or irregular clusters, not in chains. Gram positive.** Seldom motile, nonsporing. **Strictly aerobic. Colonies usually pigmented in shades of yellow or red.** Chemoorganotrophs, with a **respiratory metabolism,** often producing little or no acid from carbohydrates. Usually grow on simple media. **Catalase positive** and often oxidase positive, although weakly. Usually halotolerant, grow with 5% NaCl. Contain cytochromes and are **resistant to lysostaphin** (Schleifer and Kloos, J. Clin. Microbiol. *1:* 337–338, 1975). The optimum temperature is 25–37°C. Occur primarily on mammalian skin and in soil but commonly are isolated from food products and the air.

Type species: *Micrococcus luteus.*

Differentiation of the species of the genus **Micrococcus:** See Table 17.11.

Genus **Pediococcus**

Cells spherical, never elongated, 1.0–2.0 μm in diameter. **Division occurs alternately in two planes at right angles to form tetrads** under favorable conditions, although sometimes only pairs of cells are seen. **Single cells are rare, and chains are not formed. Gram-positive, nonmotile, nonsporing. Facultative anaerobes,** though some strains are inhibited on incubation in air. Chemoorganotrophic, the cells **require nutritionally rich media and fermentable carbohydrate** (mainly mono- and disaccharides). Glucose is fermented with the **production of acid but no gas; the major product is DL or L(+)-lactate. Catalase negative,** cytochromes absent. Nitrate is not reduced. The optimum growth temperature is 25–40°C. Occur in vegetable material and food products; nonpathogenic to plants or animals.

Type species: *Pediococcus damnosus.*

Differentiation of species of the genus **Pediococcus:** See Table 17.12.

Genus **Peptococcus**

Cells spherical and 0.3–1.2 μm in diameter, **arrangement is variable: in pairs,** tetrads, clumps, or short chains. **Gram-positive,** nonmotile, nonsporing. **Anaerobic.** Chemoorganotrophic, fermentative, and **require nutritionally rich media. Carbohydrates are not attacked, and H$_2$ is produced from peptone.** Usually catalase negative; sometimes give a weak reaction. Indole negative, and nitrate is not reduced. The optimum temperature is 37°C. The genus is differentiated from *Peptostreptococcus* mainly by its higher mol% G + C content of the DNA (50–51), but the single species forms black colonies on blood agar. Obligate parasites of human mucous membranes also from clinical specimens.

Type (and only) species: *Peptococcus niger.*

Characteristics of the species: As described for the genus.

Genus **Peptostreptococcus**

Cells spherical, 0.5–1.2 μm in diameter, and sometimes ovoid; **arrangement is variable: in pairs, tetrads, clumps, or chains. Gram-positive,** nonmotile, nonsporing. **Anaerobic.** Chemoorganotrophic and fermentative, **requiring nutritionally rich media.** Metabolize peptone to mainly acetic acid; their attack on carbohydrates is usually weak or absent. **Usually catalase negative,** but weak or pseudocatalase reactions may occur. Some members produce indole and reduce nitrate. The optimum temperature is 37°C. The genus is differentiated from *Peptococcus* mainly by its lower mol% G + C content of the DNA (27–45). Obligate parasites of the mouth, mucous membranes, and intestinal tract of mammals, and may play a part in purulent infections.

Types species: *Peptostreptococcus anaerobius.*

Differentiation of the species of the genus **Peptostreptococcus:** Differentiation of these organisms requires special techniques for anaerobic cultivation and products of metabolism (see *Bergey's Manual of Systematic Bacteriology,* Volume 2, pp. 1083–1092).

Genus **Planococcus**

Cells spherical, 1.0–1.2 μm in diameter, occurring singly or in pairs and sometimes in tetrads. **Gram-positive to Gram-variable. Motile** by one or two flagella per cell. **Endospores are not formed. Aerobic. Colonies are yellow-orange.** Chemoorganotrophic with a respiratory metabolism; **carbohydrates are seldom metabolized. Catalase positive,** oxidase negative, and nitrate is not reduced. Gelatin is usually hydrolyzed. **Halotolerant;** show poor growth with less than 1% or more than 15% NaCl. The optimum temperature is 27–37°C. They are widely distributed in marine habitats.

Type species: *Planococcus citreus.*

Differentiation of the species of the genus **Planococcus:** See Table 17.13.

Genus **Ruminococcus**

Cells spherical or slightly elongated, may have pointed ends, 0.3–1.5 × 0.7–1.8 μm, and arranged **in pairs and chains.** May be motile with 1–3 flagella per cell. Nonsporing. **Stain weakly. Gram positive or Gram negative,** though cell wall structure is of the Gram-positive type. **Strict anaerobes** requiring special methods for study. Chemoorganotrophs with a fermentative metabolism, utilizing carbohydrates with the **production of mixed acids, ethanol, CO_2, and H_2. Catalase negative;** nitrate is not reduced, and ammonia is not produced from amino acids. Growth occurs at a temperature of 20–45°C (optimum 40°C). Inhabit the rumen, large bowel, and cecum of mammals.

Type species: *Ruminococcus flavefaciens.*

Differentiation of the species of the genus **Ruminococcus:** These bacteria require special anaerobic techniques for cultivation, and the species are difficult to distinguish (see *Bergey's Manual of Systematic Bacteriology,* Volume 2, pp. 1093–1097).

Genus **Saccharococcus**

Editorial note: The genus *Saccharococcus* was not included in *Bergey's Manual of Systematic Bacteriology.* It was established by Nystrand (Int. J. Syst. Bacteriol. *34:* 503–505, 1984; effective publication: Syst. Appl. Microbiol. *5:* 204–219, 1984) for a thermophilic organism isolated from sugar solutions in a sugar refinery.

Cells spherical, 1.0–1.5 μm in diameter, and are **arranged in clusters.** Nonmotile, and no endospores have been reported. **Gram positive. Probably facultatively anaerobic.** Chemoorganotrophic, **producing acid, mainly L(+)-lactate,** but no gas from glucose, with the final pH below 4.6. **Catalase positive,** oxidase positive, methyl red positive cells. **Thermophilic,** with an optimal temperature of 68°C and a maximum of 78°C. Found in a sugar refinery; natural habitat unknown.

Type (and only) species: *Saccharococcus thermophilus.*

Characteristics of the species: As described for the genus.

Genus **Salinicoccus**

Editorial note: The genus *Salinicoccus* was not included in *Bergey's Manual of Systematic Bacteriology.* It was established by Ventosa et al. (Int. J. Syst. Bacteriol. *40:* 320–321, 1990; effective publication: Syst. Appl. Microbiol. *13:* 29–33, 1990) for a moderately halophilic organism, *S. roseus.*

Cells spherical, 1.0–1.5 μm in diameter, **arranged in pairs or tetrads. Gram-positive, nonmotile,** nonsporing. Form **pink or red colonies. Aerobic** and chemoorganotrophic; produce no acid from carbohydrates. **Catalase positive** and oxidase positive. Grow at temperatures of 15–40°C (optimum 37°C). A **moderate halophile,** requiring 0.9–25% NaCl (optimum 10%). They are not close to *Marinococcus* on evidence of DNA-DNA pairing. Isolated from marine sources.

Type species: *Salinicoccus roseus.*

Characteristics of the species: As described for the genus.

Genus **Sarcina**

Cells spherical or almost so, 1.8–3 μm in diameter, and occur in cuboidal packets of eight or more. **Division occurs in three perpendicular planes.** Some cells occur singly or in pairs and tetrads; cells are usually flattened in areas of contact with adjacent cells. **Gram positive** and nonmotile. Spore formation has been reported but is not usually seen. **Anaerobic.** Chemoorganotrophic, requiring nutritionally rich media and carbohydrates. **Metabolism is fermentative with carbohydrates as substrates,** producing **acetic acid, H_2, CO_2,** and other compounds. **Catalase negative.** The optimum temperature is 30–37°C. Widely distributed in the environment, commonly isolated from mammalian intestinal tract and from cereal seeds.

Type species: *Sarcina ventriculi.*

Differentiation of the species of the genus **Sarcina:** See Table 17.14.

Genus **Staphylococcus**

Cells spherical, 0.5–1.5 µm in diameter, **occuring singly,** in pairs, and in irregular clusters. **Gram positive,** nonmotile, nonsporing. **Facultative anaerobes.** Chemoorganotrophic, with **both respiratory and fermentative metabolism.** Colonies are usually opaque and may be white or cream and sometimes yellow to orange. **Usually catalase positive,** cytochromes present but usually oxidase negative. Nitrate is often reduced to nitrite. **Susceptible to lysis by lysostaphin but not by lysozyme** (Schleifer and Kloos, J. Clin. Microbiol. *1:* 337–338, 1975). Usually grow with 10% NaCl. The optimum temperature is 30–37°C. Mainly associated with the skin and mucous membranes of warm-blooded vertebrates but are often isolated from food products, dust, and water. Some species are opportunistic pathogens of humans and animals or produce extracellular toxins.

Type species: *Staphylococcus aureus.*

Differentiation of the species and subspecies of the genus **Staphylococcus:** See Table 17.15.

Genus **Stomatococcus**

Cells spherical, 0.9–1.3 µm in diameter, **usually in clusters,** within which pairs and tetrads are commonly seen. **Gram positive, nonmotile,** nonsporing, **encapsulated. Facultatively anaerobic.** Colonies are round, convex, **mucoid,** and **adherent to the agar.** Chemoorganotrophic, requiring nutritionally rich media. They have a fermentative metabolism, giving acid but not gas from carbohydrates. **Unable to grow on nutrient agar with 5% NaCl. Weakly positive or negative for catalase.** Oxidase negative, contain cytochromes, and are

Voges-Proskauer positive. Nitrate is reduced to nitrite. The optimum temperature is 37°C. Commensal in the human mouth and upper respiratory tract and may be implicated in infectious processes.

Type (and only) species: *Stomatococcus mucilaginosus.*

Characteristics of the species: As given for the genus.

Genus **Streptococcus**

Editorial note: The genus *Streptococcus* in *Bergey's Manual of Systematic Bacteriology* covered the organisms that are here treated in the genera *Enterococcus, Lactococcus,* and *Streptococcus,* which broadly comprise, respectively, the enterococci, the lactic streptococci, and the pyogenic, oral, and anaerobic streptococci. The genus *Melissococcus (q.v.)* contains an organism previously known as "*Streptococcus pluton.*" It is likely that most of the anaerobic streptococci (whose speciation is currently confused) will be transferred to genera such as *Peptostreptococcus,* but they are retained here for the present.

Cells spherical or ovoid, 0.5–2.0 µm in diameter, **occurring in pairs or chains** when grown in liquid media; they are sometimes elongated in the axis of the chain to a lanceolate shape. **Nonmotile,** nonsporing, and **Gram positive.** Some species are encapsulated. **Facultatively anaerobic.** Chemoorganotrophs, **requiring nutritionally rich media for growth** and sometimes 5% CO_2. **Metabolism fermentative, producing mainly lactate** but no gas. **Catalase negative. Commonly attack red blood cells,** with either greenish discoloration (α-hemolysis) or complete clearing (β-hemolysis). Growth is usually restricted to a temperature of 25–45°C (optimum 37°C). Parasites of vertebrates, mainly inhabiting the mouth and upper respiratory tract; some species are pathogenic for humans and animals. Various antigens associated with Lancefield serological groups are characteristic of some of the species and are required for accurate identification.

Type species: *Streptococcus pyogenes.*

Differentiation of the species of the genus **Streptococcus:** Differential tables for streptococci have commonly assumed that the user knew whether a strain belonged to the pyogenic, oral, or anaerobic groups. However, there is difficulty in differentiating these groups, so that a combined table is given as Table 17.16. Many species, particularly in the oral group, are undergoing active study, with consequent rearrangement of taxonomy and continuing emendation of descriptions, so that a number of areas are still not very clear (see Killian et al., Int. J. Syst. Bacteriol. *39:* 471–484, 1989; Facklam, Eur. J. Microbiol. *3:* 91–93, 1984).

Genus **Trichococcus**

Editorial note: The genus *Trichococcus* was not included in *Bergey's Manual of Systematic Bacteriology.* It was established by Scheff et al. (Int. J. Syst. Bacteriol. *34:* 355–357, 1984; effective publication: Appl. Microbiol. Biotechnol. *19:* 114–119, 1984) for a Gram-positive coccus from sewage sludge that forms very long chains.

Cells spherical to ovoid, 1.0–1.5 × 1.0–2.5 μm, **forming long chains** containing 50–200 cells, **often showing knots** in shaken liquid media. **Gram-positive,** nonmotile, nonsporing. Form flocs in liquid media and **sinuous chains on solid media. Facultatively anaerobic.** Chemoorganotrophic and fermentative, they grow slowly. Acid is formed from various carbohydrates. **Catalase positive,** oxidase negative, indole negative, nitrate not reduced, methyl red positive. The **optimum temperature** is 25–30°C. Found in sewage sludge.

Type (and only) species: *Trichococcus flocculiformis.*

Characteristics of the species: As given for the genus.

Genus **Vagococcus**

Editorial note: The genus *Vagococcus* was not described in *Bergey's Manual of Systematic Bacteriology.* It was established for motile Gram-positive cocci from water, *Vagococcus fluvialis,* by Collins et al. (Int. J. Syst. Bacteriol. *40:* 212, 1990; effective publication: J. Appl. Bacteriol. *67:* 453–460, 1989). A second species, *V. salmoninarum* from fish, was described by Wallbanks et al. (Int. J. Syst. Bacteriol. *40:* 224–230, 1990).

Cells are spheres, ovals, or short rods, 0.5–1.2 × 0.5–2.0 μm, occurring **singly, in pairs, or in short chains. Gram positive** and nonsporing. Some are motile by peritrichous flagella. **Facultative anaerobes.** Chemoorganotrophic, with a fermentative metabolism. Produce acid but no gas from a number of carbohydrates. Glucose fermentation yields mainly L(+)-lactate. **Catalase negative.** Nitrates are not reduced. The optimum temperature is 25–35°C. Isolated from water or from salmonid fish, but their pathogenicity is not clear. Some are of Lancefield serological Group N.

Type species: *Vagococcus fluvialis.*

Differentiation of the species of the genus **Vagococcus:** See Table 17.17.

Important Notes for Users of This Manual

Unless otherwise indicated in footnotes to tables, the meanings of symbols are as follows:

+ 90% or more of strains are positive
− 90% or more of strains are negative
d 11-89% of strains are positive
v strain instability (not equivalent to "d")
D Different reactions in different taxa (species of a genus or genera of a family)

All other symbols are defined in footnotes to tables.

Table 17.1 Differential properties of strictly aerobic genera of Gram-positive cocci [a]

Characteristics	*Deinobacter*	*Deinococcus*	*Marinococcus*	*Micrococcus* [b]	*Planococcus*	*Salinicoccus*
Predominant arrangement of cells (other than single cells), most common appearance listed first	Pairs of rods	Pairs, tetrads	Pairs, tetrads	Clusters, tetrads	Pairs, tetrads	Pairs, tetrads, clumps
Motility	–	–	+	–	+	–
Obligate halophiles, requiring 7.5% NaCl for growth	–	–	+	–	–	+
Catalase reaction	+	+	+	+	+	+
Cytochromes present	ND	+	ND	+	+	ND
Acidity from utilizable carbohydrates	No acid	Weak or no acid	No acid	Often acid	No acid	No acid
Peptidoglycan: [c]						
Position 1	Ala	Ala	ND	Ala	Ala	ND
Position 2	Orn	Orn	mDpm	Lys	Lys	Lys
Main interpeptide bridge	Gly_2	Gly_2	ND	Ala_{1-4}, Asp, Thr-Ala_3, or Ser_2-D-Glu	D-Glu	Gly_5
Teichoic acid in cell wall	ND	–	ND	–	–	ND
Major menaquinones [d]	MK-8	MK-8	MK-6, MK-7	MK-7, MK-8, MK-9, often hydrogenated	MK-8	none
Mol% G + C of DNA	69	62–70	44–47	64–75	39–52	51

[a] Symbols: +, 90% or more of strains are positive; –, 90% or more of strains are negative; ND, not determined.
[b] One species sometimes facultatively anaerobic.
[c] Symbolism of Schleifer and Kandler (Bacteriol. Rev. *36:* 407–477, 1972).
[d] Symbolism of Collins and Jones (Microbiol. Rev. *45:* 316–354, 1981).

Important Notes for Users of This Manual

Unless otherwise indicated in footnotes to tables, the meanings of symbols are as follows:

+ 90% or more of strains are positive
– 90% or more of strains are negative
d 11-89% of strains are positive
v strain instability (not equivalent to "d")
D Different reactions in different taxa (species of a genus or genera of a family)

All other symbols are defined in footnotes to tables.

Table 17.2 Differential properties of facultatively anaerobic genera of Gram-positive cocci [a]

Characteristics	Aerococcus[b]	Enterococcus	Gemella[c]	Lactococcus	Leuconostoc	Melissococcus[d]	Pediococcus
Predominant arrangement of cells (other than single cells), most common appearance first	Tetrads, pairs	Pairs, chains	Pairs, short chains	Pairs, short chains	Pairs, chains	Chains of lanceolate cocci	Tetrads, some pairs
Motility	−	d	−	−	−	−	−
Growth at:							
10°C	+	+	−	+	+	−	ND
45°C	−	+	−	−	−	−	D
pH 9.6	+	+	ND	−	ND	−	D
Growth with:							
6.5% NaCl	+	+	ND	−	d	−	D
40% bile	−[h]	+	ND	D	ND	ND	D
Catalase reaction	−	−[i]	−	−	−	ND	−
Cytochromes present	−	ND	−[k]	ND	−	ND	−
Major fermentation products from utilizable carbohydrates anaerobically (if not known, acidity recorded)	ND (acid)	Lactate	Lactate	Lactate	Lactate, ethanol	Lactate	Lactate
Peptidoglycan:[l]							
Position 1	Ala	Ala	Ala	Ala	Ala	Ala	Ala
Position 2	Lys	Lys	Lys	Lys	Lys	Lys	Lys
Main interpeptide bridge	None	Ala3, D-Asp	Ala2–3	D-Asp, Gly-Ala, Ser-Ala, Thr-Ala	Ala2, Ser-Ala2 or Ser2	ND	Ala-D-Asp or Ala
Teichoic acid in cell wall	+	+	ND	+	ND	ND	ND
Major menaquinones[m]	ND	D	ND	D	None	None	None
Mol% G + C of DNA	35–40	34–42	30–35	38–40	38–44	29–30	34–42

Footnotes are at end of table.

Table 17.2 (continued)

Characteristics	Saccharococcus[e]	Staphylococcus[f]	Stomatococcus	Streptococcus[g]	Trichococcus	Vagococcus
Predominant arrangement of cells (other than single cells), most common appearance first	Clusters	Clusters, pairs	Clusters, some pairs, usually capsulated	Chains, pairs	Very long chains of coccoid or oval cells	Pairs, short rods, short chains
Motility	-	-	-	-	-	d
Growth at:						
10°C	-	D	-	D	+	+
45°C	d	D	-	D	ND	-
pH 9.6	ND	ND	-	D	ND	-
Growth with:						
6.5% NaCl	ND	+	-	D	ND	-
40% bile	ND	D	-	D	ND	ND
Catalase reaction	+	+	+ weak	-	-	-
Cytochromes present	ND	+	+	-[k]	-	ND
Major fermentation products from utilizable carbohydrates anaerobically (if not known, acidity recorded)	L(+)-Lactate	Lactate	ND (acid)	Lactate	ND (acid)	Lactate
Peptidoglycan:[l]						
Position 1	ND	Ala	Ala	Ala	ND	ND
Position 2	ND	Lys	Lys	Lys	Lys	Lys
Main interpeptide bridge	mDpm	Gly$_{5-6}$, Ala-Gly$_4$ or Gly$_{3-5}$, Ser$_{1-2}$	Ala, Gly or Ser	Ala$_{1-4}$, Thr-Ala, Thr-Gly	D-Asp	ND
Teichoic acid in cell wall	-	+	-	D	-	+
Major menaquinones[m]	ND	MK-6, MK-7, MK-8	ND	D	ND	ND
Mol% G + C of DNA	48	30–39	56–61	34–46	47	33–37

[a] Symbols: +, 90% or more of strains are positive; −, 90% or more of strains are negative; d, 11–89% of strains are positive; D, substantial proportion of species differ; ND, not determined.

[b] Grows best at reduced oxygen concentration; sometimes does not grow anaerobically.

[c] Weak anaerobic growth; may not grow in air on first isolation.

[d] Requires 5% CO_2 and cystine or cysteine for growth.

[e] Thermophilic; optimum growth temperature is 68°C.

[f] Rarely strictly aerobic.

[g] Rarely strictly anaerobic.

[h] Sometimes weak catalase or pseudocatalase reaction.

[i] May produce a pseudocatalase.

[k] A few strains synthesize cytochrome on aerobic media supplemented with haemin.

[l] Symbolism of Schleifer and Kandler (Bacteriol. Rev. 36: 407–447, 1972).

[m] Symbolism of Collins and Jones (Microbiol. Rev. 45: 316–354, 1981).

Table 17.3 Differential properties of the strictly anaerobic genera of Gram-positive cocci [a,b]

Characteristics	*Coprococcus*	*Peptococcus*	*Peptostreptococcus*	*Ruminococcus*	*Sarcina* [c]
Predominant arrangement of cells (other than single cells), most common appearance listed first	Pairs, chains of pairs	Pairs, tetrads	Pairs, chains, tetrads	Pairs, chains	Cuboidal packets
Catalase reaction	–	–[d]	–[d]	–	–
Cytochromes present	ND	–	–	ND	ND
Major fermentation products from utilizable carbohydrates anaerobically (if not fermentative, acidity recorded)	Butyrate, acetate, lactate, formate or propionate, H_2	NA (no acid)	NA (rarely acid)	Acetate, lactate, succinate, formate, ethanol, CO_2, H_2	Acetate, ethanol, H_2, CO_2 or butyrate
Peptidoglycan: [e]					
Position 1	Ala	Ala	Ala or Gly	Ala	Ala
Position 2	m-A_2pm	Lys	Lys or Orn	m-A_2pm	LL-A_2pm
Main interpeptide bridge	None	D-Asp	D-Asp, D-Glu, Gly, Lys-D-Glu, or Gly-D-Asp	None	Gly
Mol% G + C of DNA	39–42	50–51	27–45	39–46	28–31

[a] Symbols: +, 90% or more of strains positive; –, 90% or more of strains are negative; ND, not determined; NA, not applicable.
[b] Teichoic acid and menaquinones not reported for these genera.
[c] Sometimes aerotolerant.
[d] Sometimes weak catalase or pseudocatalase reaction.
[e] Symbolism of Schleifer and Kandler (Bacteriol. Rev. *36*: 407–477, 1972).

Table 17.4 Characteristics differentiating the species of *Coprococcus* [a]

Characteristics	*C. catus*	*C. comes*	*C. eutactus*
Acid produced from:			
Sucrose	–	+	+
Melezitose	–	–	+
Xylose	–	+	–
Major fermentation products	Butyrate, propionate	Lactate	Formate
Consistently occurring minor products	Acetate	Butyrate, acetate	Butyrate, lactate, acetate

[a] Symbols: +, 90% or more of strains are positive; –, 90% or more of strains are negative.

Table 17.5 Characteristics differentiating the species of *Deinococcus* [a]

Characteristics	*D. proteolyticus*	*D. radiodurans*	*D. radiophilus*	*D. radiopugnans*
Growth with 5% NaCl	–	d	+	–
Nitrate reduction	–	d	–	+
Esculin hydrolysis	+	–	–	–
ONPG hydrolysis [b]	–	–	–	+
Colony color	Orange-red	Red	Orange-red	Orange-red

[a] Symbols: +, 90% or more of strains are positive; –, 90% or more of strains are negative; d, 11–89% of strains are positive.
[b] ONPG, orthonitrophenyl-β-galactoside.

Bergey's Manual of Determinative Bacteriology-9

Table 17.6 Characteristics differentiating the species of *Enterococcus*[a]

Characteristics	E. avium[b,c]	E. casseliflavus[b,c]	E. cecorum[b,d]	E. dispar[e]	E. durans[c]	E. faecalis[b]	E. faecium[b]	E. gallinarum[b,c]	E. hirae[f]
Motility	−	+	−	−	−	(−)	(−)	−	−
Growth at:									
45°C	+	+	+	−	+	+	+	+	+
50°C	−	−	ND	−	−	(−)	(+)	−	−
Growth in:									
6.5% NaCl	+	+	−	+	+	+	+	+	+
0.04% tellurite	−	+	ND	ND	−	+	−	(+)	−
0.01% tetrazolium	ND	ND	ND	ND	−	+	−	+	ND
0.1% methylene blue milk	d	ND	d	ND	+	+	ND	d	ND
Yellow pigment produced	−	+	−	−	−	−	−	−	ND
Hemolysis	α	ND	α	ND	α, β	(β)	(α)	α, β	−
H$_2$S production	+	−	ND	−	−	ND	ND	−	ND
Ammonia from arginine	ND	ND	−	−	ND	ND	+	+	ND
Arginine hydrolysis	−	(+)	−	+	+	+	+	+	+
Hippurate hydrolysis	d	−	−	d	d	(+)	+	+	−
Voges-Proskauer reaction	−	+	+	ND	ND	ND	ND	ND	+
Acid from:									
D-Xylose	−	+	−	ND	−	−	−	+	−
L-Rhamnose	+	(+)	−	ND	−	d	−	−	−
Sucrose	+	+	+	+	−	+	d	+	+
Lactose	+	+	+	+	+	+	+	+	+
Melibiose	−	+	+	+	−	−	d	+	+
Raffinose	−	+	+	+	−	−	−	+	d
Melezitose	+	−	+	ND	−	(+)	−	−	(−)
Glycerol	+	−	−	+	−	+	+	+	(−)
Adonitol	+	−	−	ND	−	−	−	−	−
Sorbitol	+	−	−	−	−	(+)	−	−	−
Mannitol	+	+	−	−	(−)	+	(+)	+	−
Lancefield serological group	Q(D)	D	Not D	Not D	D	D	D	D	D

[a] Symbols: +, 90% or more of strains are positive; (+), 80–89% of strains are positive; d, 21–79% of strains are positive; (−), 11–20% of strains are positive; −, 90% or more of strains are negative; α, usually α-hemolytic; (α), sometimes α-hemolytic; β, usually β-hemolytic; (β), sometimes β-hemolytic; α,β, may be α- or β-hemolytic; Q(D), group Q but may cross-react with Group D; ND, not determined.

[b] Included in *Streptococcus* in *Bergey's Manual of Systematic Bacteriology*.

[c] Collins et al. (Int. J. Syst. Bacteriol. *34:* 220–223, 1984).

[d] Williams et al. (Int. J. Syst. Bacteriol. *39:* 495–497, 1989; effective publication: Lett. Appl. Microbiol. *8:* 185–189, 1989; see also Devriese et al., Int. J. Syst. Bacteriol. *3:* 772–776, 1983).

Table 17.6 (continued)

Characteristics	E. malodoratus [c]	E. mundtii [g]	E. pseudoavium [h]	E. raffinosus [h]	E. saccharolyticus [b,j]	E. seriolicida [j]	E. solitarius [h]
Motility	–	–	–	–	–	–	–
Growth at:							
45°C	–	+	+	+	+	+	+
50°C	–	–	ND	ND	–	–	ND
Growth in:							
6.5% NaCl	+	+	–	+	+	+	+
0.04% tellurite	–	–	ND	ND	ND	–	ND
0.01% tetrazolium	ND	ND	ND	ND	ND	+	ND
0.1% methylene blue milk	ND	ND	ND	ND	ND	+	ND
Yellow pigment produced	–	+	ND	ND	–	–	–
Hemolysis	ND	–	α	ND	–	α	ND
H_2S production	+	–	ND	ND	ND	–	ND
Ammonia from arginine	ND	ND	ND	ND	ND	–	ND
Arginine hydrolysis	–	+	–	–	–	+	+
Hippurate hydrolysis	d	–	+	–	–	–	ND
Voges-Proskauer reaction	–	+	+	+	–	+	d
Acid from:							
D-Xylose	d	+	ND	ND	–	–	–
L-Rhamnose	+	(+)	ND	ND	–	–	–
Sucrose	+	+	–	+	+	–	+
Lactose	+	+	+	+	+	–	–
Melibiose	+	+	–	+	+	–	ND
Raffinose	+	(+)	–	+	+	–	ND
Melezitose	–	–	–	ND	+	–	+
Glycerol	d	d	–	+	–	–	ND
Adonitol	+	–	ND	ND	–	–	–
Sorbitol	+	d	+	+	+	+	+
Mannitol	+	+	+	+	+	+	+
Lancefield serological group	D	D	ND	(D)	Not D	Not D	D

[e] Collins et al. (Int. J. Syst. Bacteriol. 41: 456–458, 1991; effective publication: Lett. Appl. Microbiol. 12: 95–98, 1991).

[f] Farrow and Collins (Int. J. Syst. Bacteriol. 35: 73–75, 1985).

[g] Collins et al. (Int. J. Syst. Bacteriol. 36: 8–12, 1986).

[h] Collins et al. (Int. J. Syst. Bacteriol. 39: 371, 1989; effective publication: FEMS Microbiol. Lett. 57: 283–288, 1989).

[i] Rodrigues and Collins (Int. J. Syst. Bacteriol. 41: 178–179, 1991; effective publication: FEMS Microbiol. Lett. 71: 231–234, 1990; see also Farrow et al. (Syst. Appl. Microbiol. 5: 467–482, 1984).

[j] Kasuda et al. (Int. J. Syst. Bacteriol. 41: 406–409, 1991).

Table 17.7 Characteristics differentiating the species of Gemella[a]

Characteristics	G. haemolysans	G. morbillorum[b]
Type of hemolysis on rabbit or horse blood agar	β	α or none
Acid from:		
Fructose	+	−
Glucose	+	+
Maltose	+	w
Sucrose	+	w
Reduce nitrite	d	−

[a] Symbols: +, 90% or more of strains are positive; −, 90% or more of strains are negative; w, 90% or more of strains are positive but weak; d, 11–89% of strains positive.

[b] Transferred from the genus *Streptococcus* (*Bergey's Manual of Systematic Bacteriology*, Volume 2, p. 1067) by Kilpper-Bälz and Schleifer (Int. J. Syst. Bacteriol. *38*: 442–443, 1988).

Table 17.8 Characteristics differentiating species and subspecies of Lactococcus[a]

Characteristics	L. garvieae[b,c]	L. lactis			L. piscium[c]	L. plantarum[b,c]	L. raffinolactis[b,d]
		subsp. cremoris[b,d,e]	subsp. hordniae[b,f]	subsp. lactis[b,d,e,g]			
Growth at 40°C	+	−	−	(+)	−	−	−
Growth with 4% NaCl	+	−	−	+	ND	+	−
Arginine hydrolysis	+	−	+	+	−	−	(−)
Acid from lactose	+	+	−	+	+	−	+
Acid from mannitol	(+)	−	−	(−)	+	+	d
Acid from raffinose	−	−	−	−	+	−	+
Pyrrolidonylarylamidase	+	−	−	(−)	ND	−	−

[a] Symbols: +, 90% or more of strains are positive; (+), 80–89% of strains are positive; d, 21–79% of strains are positive; (−), 11–20% of strains are positive; −, 90% or more of strains are negative.

[b] Transferred from other genera by Schleifer et al. (Int. J. Syst. Bacteriol. *36*: 354–356, 1986; effective publication: Syst. Appl. Microbiol. *6*: 183–195, 1985).

[c] First described as species of *Streptococcus* (Collins et al., Int. J. Syst. Bacteriol. *34*: 270–271, 1984; effective publication: J. Gen. Microbiol. *129*: 3427–3431, 1983).

[d] Included in *Streptococcus* in *Bergey's Manual of Systematic Bacteriology*.

[e] No subspecies of *Streptococcus lactis* were recognized in *Bergey's Manual of Systematic Bacteriology*, but Garvie and Farrow (Int. J. Syst. Bacteriol. *32*: 453–455, 1982) considered these justified as subspecies, and Schleifer et al. concur.

[f] First described as *Lactobacillus hordniae* (Lattore-Guzmann et al., Int. J. Syst. Bacteriol. *27*: 362–370, 1977).

[g] Schleifer et al. (Int. J. Syst. Bacteriol. *36*: 354–356, 1986; effective publication: Syst. Appl. Microbiol. *6*: 183–195, 1985) consider that *Streptococcus lactis* subsp. *diacetilactis* of Garvie and Farrow (Int. J. Syst. Bacteriol. *32*: 453–455, 1982) and *Lactobacillus xylosus* (Kitahara, J. Agric. Chem. Soc. Japan *14*: 1449–1465, 1938) are synonymous with *Lactococcus lactis* subsp. *lactis*.

[h] New species of *Lactococcus* described by Williams et al. (Int. J. Syst. Bacteriol. *40*: 320–321, 1990; effective publication: FEMS Microbiol. Lett. *68*: 109–114, 1990).

Table 17.9 Characteristics differentiating the species and subspecies of *Leuconostoc*[a]

Characteristics	*L. amelibiosum*[b]	*L. carnosum*[c]	*L. citreum*[d]	*L. gelidum*[c]	*L. lactis*
Lemon-yellow pigment	–	–	+	–	–
Acid from:					
L-Arabinose	+	–	+	+	–
Fructose	+	+	+	+	+
Maltose	+	–	+	d	+
Melibiose	–	(–)	–	+	d
Salicin	+	–	+	+	d
Sucrose	+	+	+	+	+
Trehalose	+	+	+	+	–
Dextran formed from sucrose	+	+	ND	(+)	–
Esculin hydrolyzed	ND	d	+	+	–
Growth at pH 4.8	ND	ND	ND	ND	–
Growth with 10% ethanol	ND	ND	ND	ND	–
Growth at 37°C	+	(–)	d	(–)	+

Table 17.9 *(continued)*

Characteristics	*L. mesenteroides* subsp. *cremoris*	*L. mesenteroides* subsp. *dextranicum*	*L. mesenteroides* subsp. *mesenteroides*	*L. oenos*	*L. paramesenteroides*	*L. pseudomesenteroides*[d]
Lemon-yellow pigment	(–)	–	–	–	–	–
Acid from:						
L-Arabinose	–	–	+	d	d	d
Fructose	–	+	+	+	+	+
Maltose	d	+	+	–	+	+
Melibiose	d	+	+	d	+	(+)
Salicin	–	d	(–)	d	–	d
Sucrose	–	+	+	–	+	d
Trehalose	–	+	+	+	+	+
Dextran formed from sucrose	–	+	+	–	–	ND
Esculin hydrolyzed	–	d	d	+	d	d
Growth at pH 4.8	–	–	–	+	d	ND
Growth with 10% ethanol	–	–	–	+	–	ND
Growth at 37°C	–	+	d	d	d	+

[a] Symbols: +, 90% or more of strains are positive; (+), 80–89% of strains are positive; d, 21–79% of strains are positive; (–), 11–20% of strains are positive; –, 90% or more of strains are negative; ND, not determined.

[b] Schillinger et al. (Int. J. Syst. Bacteriol. *39:* 495–497, 1989; effective publication: Syst. Appl. Microbiol. *12:* 48–55, 1989).

[c] Shaw and Harding (Int. J. Syst. Bacteriol. *39:* 217–223, 1989).

[d] Farrow et al. (Int. J. Syst. Bacteriol. *39:* 279–283, 1989).

Table 17.10   Characteristics differentiating among species of *Marinococcus*[a]

Characteristics	*M. albus*[b]	*M. halophilus*[b]	*M. hispanicus*[c]
Motility	+	+	−
Pigment	None	Yellow-orange	Reddish orange
Acid from:			
Fructose	−	−	+
Glucose	−	+	+
Mannitol	−	+	+
Trehalose	−	+	−
D-Xylose	−	+	−
Oxidase	+	−	+
Reduction of nitrate	+	−	+

[a] Symbols: +, 90% or more of strains are positive; −, 90% or more of strains are negative.

[b] Hao et al. (Int. J. Syst. Bacteriol. *35:* 535, 1985; effective publication: J. Gen. Appl. Bacteriol. *30:* 449–459, 1984).

[c] Marquez et al. (Int. J. Syst. Bacteriol. *40:* 165–169, 1990).

Table 17.11   Characteristics differentiating the species of *Micrococcus*[a]

Characteristics	*M. agilis*	*M. halobius*[b]	*M. kristinae*	*M. luteus*	*M. lylae*	*M. nishinomiyaensis*	*M. roseus*	*M. sedentarius*	*M. varians*
Color of colonies	Red	Unpigmented	Pale orange	Yellow	Cream-white	Orange	Pink or orange-red	Cream-white or yellow	Yellow
Motility	+	−	−	−	−	−	+	−	−
Acid from:									
Glucose	−	+	+	−	−	d	+	−	+
Glycerol	−	+	+	−	−	−	−	−	−
Mannose	−	−	+	−	−	−	−	−	−
Lactose	−	+	−	−	−	−	−	−	−
Hydrolysis of:									
Esculin	+	ND	+	−	−	−	−	−	−
Gelatin	+	−	−	+	+	+	−	+	+
Nitrate reduced to nitrite	−	−	−	−	−	d	+	−	+
Arginine dihydrolase	−	−	−	−	−	−	−	+	−
Oxidase	−	+	+	+	+	+	−	−	−
Growth on:									
Inorganic nitrogen agar[c]	−	ND	−	+	−	−	−	−	−
Simmons citrate agar	−	−	−	−	−	−	−	−	+
Nutrient agar with 7.5% NaCl	−	+	+	+	+	−	+	+	+
Growth at 37°C	−	+	+	+	+	+	+	+	+

[a] Symbols: +, 90% or more of strains are positive; d, 11–89% of strains are positive; −, 90% or more of strains are negative; ND, not determined.

[b] In media supplemented with 5% NaCl.

[c] Method of Kloos et al. (Int. J. Syst. Bacteriol. *24:* 79–101, 1974).

Table 17.12 Characteristics differentiating the species of *Pediococcus*[a]

Characteristics	P. acidilactici	P. damnosus	P. dextrinicus	P. halophilus	P. inopinatus	P. parvulus	P. pentosaceus	P. urinaeequi
Growth at:								
35°C	+	−	+	+	+	+	+	+
40°C	+	−	+	−	−	−	+	+
50°C	+	−	−	−	−	−	−	−
Growth at pH 4.2	+	+	−	−	−	+	+	−
Growth at pH 7.5	+	−	+	+	d	+	+	+
Growth at pH 8.5	d	−	−	+	−	−	d	+
Growth with 4% NaCl	+	−	+	d	+	+	+	+
Growth with 6.5% NaCl	+	−	−	+	d	+	+	+
Growth with 18% NaCl	−	−	−	+	−	−	−	−
Arginine hydrolysis	+	−	−	−	−	−	+	−
Acid from dextrin	−	−	+	−	d	−	−	+
Acid from starch	−	−	+	−	−	−	−	−

[a] Symbols: +, 90% or more of strains are positive; d, 11–89% of strains are positive; −, 90% or more of strains are negative.

Table 17.13 Characteristics differentiating the species of *Planococcus*[a,b]

Characteristics	P. citreus	P. kocurii[c]
Growth on nutrient agar containing 15% NaCl	+	−

[a] Symbols: +, 90% or more of strains are positive; −, 90% or more of strains are negative.

[b] The organism described in *Bergey's Manual of Systematic Bacteriology* as *Planococcus halophilus* has been transferred to the new genus *Marinococcus* (*q.v.*).

[c] Hao and Komagata (Int. J. Syst. Bacteriol. *36:* 573–576, 1986; effective publication: J. Gen. Appl. Microbiol. *31:* 441–455, 1985).

Table 17.14 Characteristics differentiating species of *Sarcina*[a]

Characteristics	S. maxima	S. ventriculi
Cell diameter (μm)	2–3	1.8–2.4
Ethanol production	−	+
Butyrate production	+	−
Fermentation of D-xylose	+	−
Fermentation of melibiose	−	+
Formation of cellulose[b]	−	+

[a] Symbols: +, 90% or more of strains are positive; −, 90% or more of strains are negative.

[b] Methods for detection of cellulose are described in *Bergey's Manual of Systematic Bacteriology*, Volume 2, pp. 1101–1102.

Table 17.15 Characteristics differentiating the species and subspecies of the genus Staphylococcus[a]

Characteristics	S. arlettae[b]	S. aureus subsp. anaerobius[c]	subsp. aureus	S. auricularis	S. capitis subsp. capitis[d]	subsp. ureolyticus[d]
Colony diameter > 5 mm[j]	+	−	+	−	−	d
Colony pigment (carotenoid)	+	−	+w	−	−	d
Aerobic growth	+	−[k]	+	+	+	+
Anaerobic growth (thioglycolate)	−	+	+	−w	(+)	(+)
Growth on NaCl agar:						
10% (w/v)	ND	+	+	+	+	ND
15% (w/v)	ND	d	w	w	−w	ND
Growth at:						
15°C	ND	ND	+	−	−	ND
45°C	ND	−	+	+	+	ND
Cytochrome c (oxidase test)	−	−	−		−	−
Lactic acid production:						
L(+)−Isomer	w	+	+	w	+	ND
D(−)−Isomer	w	−	+	−	−w	ND
Acetoin production	ND	−	+	d	d	d
FDP-aldolase:						
Class I	ND	+	+	+	+	ND
Class II	ND	−	−	−	−	ND
Acid (aerobically) from:						
D-Xylose	+	−	−	−	−	−
L-Arabinose	+	−	−	−	−	−
D-Cellobiose	−	−	−	−	−	−
D-Fucose	w	ND	−	−	−	ND
Raffinose	+	−	−	−	−	−
Salicin	ND	−	−	−	−	ND
Sucrose	+	+	+	d	(+)	+
Maltose	+	+	+	(+)	−	+
D-Mannitol	+	−	+	−	+	+
D-Mannose	+w	−	+	−	+	+
D-Trehalose	+	−	+	(+)	−	−
α-Lactose	+	−	+	−	−	d
D-Galactose	d	−	+	−	−	ND
β-D-Fructose	+	+	+	+	+	ND
D-Melezitose	+	−	−	−	−	ND
D-Turanose	+	ND	+w	(d)	−	d
D-Ribose	+	−	+	−	−	ND
Xylitol	−	−	−	−	−	ND
Hyaluronidase	−	+	+	ND	ND	ND
Growth on (NH₄)₂SO₄ (nitrogen source)	ND	ND	−	ND	−	ND
Nitrate reduction	−	−	+	(d)	d	d
Alkaline phosphatase	+	+	+	−	−	−

Footnotes are at end of table

Table 17.15 *(continued)*

Characteristics	S. arlettae[b]	S. aureus subsp. anaerobius[c]	S. aureus subsp. aureus	S. auricularis	S. capitis subsp. capitis[d]	S. capitis subsp. ureolyticus[d]
Arginine dihydrolase	–	ND	+w	d	d	ND
Urease	–	ND	+w	–	–	+
Coagulase (rabbit plasma)	–	+	+	–	–	–
Clumping factor	–	–	+	–	–	–
Fibrinolysin	–	ND	de	ND	ND	ND
Hemolysis	–	+	+	–	–w	(d)
Deoxyribonuclease (DNase agar)	–	+	+	–w	w	ND
Heat-stable nuclease	–	+	+	–	–	–
β-Glucosidase	ND	–	+	–	–	–
β-Glucuronidase	ND	–	–	–	–	–
β-Galactosidase	ND	–	–	(d)	–	–
Novobiocin resistance (MIC ≥ 1.6 μg/ml)	+	–	–	–	–	–

Table 17.15 *(continued)*

Characteristics	S. caprae	S. carnosus	S. caseolyticus	S. chromogenes[c]	S. cohnii subsp. cohnii[e]	S. cohnii subsp. urealyticus[e]
Colony diameter > 5 mm[j]	d	+	–	+	d	+
Colony pigment (carotenoid)	–	–	d	+	–	d
Aerobic growth	+	+	+	+	+	+
Anaerobic growth (thioglycolate)	+	+	–w	+	d	(+)
Growth on NaCl agar:						
10% (w/v)	ND	+	ND	+	+	+
15% (w/v)	ND	+	ND	–w	d	d
Growth at:						
15°C	ND	+	ND	+	+	+
45°C	+	+	ND	–w	d	d
Cytochrome c (oxidase test)	–	–	+	–	–	–
Lactic acid production:						
L(+)–Isomer	+	+	w	+	w	w
D(–)–Isomer	–	+	–	–	–	w
Acetoin production	+	+	–	–	d	d
FDP-aldolase:						
Class I	+	+	–	ND	+	ND
Class II	–	–	+	ND	–	ND

Footnotes are at end of table

545

Table 17.15 *(continued)*

Characteristics	S. caprae	S. carnosus	S. caseolyticus	S. chromogenes [c]	S. cohnii subsp. cohnii [e]	S. cohnii subsp. urealyticus [e]
Acid (aerobically) from:						
D-Xylose	–	–	–	–	–	–
L-Arabinose	–	–	ND	–	–	–
D-Cellobiose	–	–	ND	–	–	–
D-Fucose	–	–	ND	–	–	–
Raffinose	–	–	ND	–	–	–
Salicin	–	–	ND	–	–	–
Sucrose	–	–	d	+	–	–
Maltose	d	–	+	d	(d)	(+)
D-Mannitol	d	+	–	d	d	d
D-Mannose	+	+	–	+	(d)	+
D-Trehalose	+	d	d	+	+	+
α-Lactose	+	d	+	+	–	+
D-Galactose	+	d	+	+	–	d
β-D-Fructose	–	+	+	+	+	+
D-Melezitose	–	–	ND	–	–	–
D-Turanose	–	–	–	d	–	–
D-Ribose	–	ND	+	+	–	–
Xylitol	–	–	–	–	(d)	(d)
Hyaluronidase	–	ND	ND	–	ND	ND
Growth on (NH₄)₂SO₄ (nitrogen source)	ND	ND	ND	ND	d	d
Nitrate reduction	+	+	+	+	–	–
Alkaline phosphatase	+	+	ND	+	–	+w
Arginine dihydrolase	+	+	ND	+	–	–w
Urease	+	–	ND	d	–	+
Coagulase (rabbit plasma)	–	–	–	–	–	–
Clumping factor	–	–	–	–	–	–
Fibrinolysin	–	ND	ND	–	ND	ND
Hemolysis	(+)	–	–	–	(d)	(d)
Deoxyribonuclease (DNase agar)	+	w	ND	w	–w	–w
Heat-stable nuclease	–	d	ND	–w	–	–
β-Glucosidase	–	–	ND	d	–	–
β-Glucuronidase	–	–	ND	–	–	+
β-Galactosidase	–	+	ND	–	–	+
Novobiocin resistance (MIC ≥ 1.6 µg/ml)	–	–	–	–	+	+

Footnotes are at end of table

Table 17.15 *(continued)*

Characteristics	S. delphini[f]	S. epidermidis	S. equorum[b]	S. felis[g]	S. gallinarum	S. haemolyticus	S. hominis
Colony diameter > 5 mm[j]	+	−	+	+	+	+	−
Colony pigment (carotenoid)	−	−	−	−	d	d	d
Aerobic growth	+	+	+	+	+	+	+
Anaerobic growth (thioglycolate)	+	+	−	+	+	(+)	−w
Growth on NaCl agar:							
10% (w/v)	+	w	ND	+	ND	+	w
15% (w/v)	+	−	ND	+	ND	d	−
Growth at:							
15°C	ND	−w	ND	+w	ND	−w	−w
45°C	+	+	−	+	+w	+	+
Cytochrome c (oxidase test)	ND	−	−	−	−	−	−
Lactic acid production:							
L(+)–Isomer	ND	+	w	ND	+	−	d
D(−)–Isomer	ND	−	w	ND	−	+	+
Acetoin production	−	+	ND	−	−	d	d
FDP-aldolase:							
Class I	ND	+	ND	ND	+	+	+
Class II	ND	−	ND	ND	−	−	−
Acid (aerobically) from:							
D-Xylose	−	−	+	−	+	−	−
L-Arabinose	−	−	+	−	+	−	−
D-Cellobiose	ND	−	−w	−	+	−	−
D-Fucose	ND	−	−	ND	w	−	−
Raffinose	ND	−	−	−	+	−	−
Salicin	ND	−	ND	−	+	−	−
Sucrose	+	+	+	d	+	+	(+)
Maltose	ND	+	+	−	+	+	+
D-Mannitol	+	−	+	d	+	d	−
D-Mannose	+	(+)	+	+	+	−	−
D-Trehalose	−	−	+	+	+	+	d
α-Lactose	+	d	d	+	d	d	d
D-Galactose	ND	d	+	d	+	d	d
β-D-Fructose	+	+	+	+	+	d	+
D-Melezitose	ND	(d)	+	−	+	−	d
D-Turanose	ND	d	d	ND	+	d	d
D-Ribose	ND	d	+	−w	+	d	−
Xylitol	−	−	−	−	d	−	−
Hyaluronidase	ND	d	−	−	−	ND	ND
Growth on (NH4)2SO4 (nitrogen source)	ND	−	ND	ND	ND	−	−
Nitrate reduction	+	+w	+	+	+	d	d
Alkaline phosphatase	+	+	+	+	+	−	−
Arginine dihydrolase	+	+w	−	+	−	+	d

Footnotes are at end of table

Table 17.15 (continued)

Characteristics	S. delphini[f]	S. epidermidis	S. equorum[b]	S. felis[g]	S. gallinarum	S. haemolyticus	S. hominis
Urease	+	+	+	+	+	−	+
Coagulase (rabbit plasma)	+	−	−	−	−	−	−
Clumping factor	−	−	−	−	−	−	−
Fibrinolysin	−	d	−	ND	−	ND	ND
Hemolysis	+	−w	d	−w	w	(+)	−w
Deoxyribonuclease (DNase agar)	w	−w	−	ND	ND	ds	−w
Heat-stable nuclease	−	−w	−	−w	−	−	−
β-Glucosidase	ND	(d)	ND	−	+	d	−
β-Glucuronidase	ND	−	ND	−	−	d	−
β-Galactosidase	ND	−	ND	+	−w	−	−
Novobiocin resistance (MIC ≥ 1.6 µg/ml)	−	−	+	−	+	−	−

Table 17.15 (continued)

Characteristics	S. hyicus[c]	S. intermedius	S. kloosii[b]	S. lentus	S. lugdunensis[h]	S. saccharolyticus
Colony diameter > 5 mm[j]	+	+	+	−	−	−
Colony pigment (carotenoid)	−	−	d	d	d	−
Aerobic growth	+	+	+	+	+	−w
Anaerobic growth (thioglycolate)	+	(+)	−	−w	+	+
Growth on NaCl agar:						
10% (w/v)	+	+	ND	+	+	ND
15% (w/v)	−w	d	ND	−w	+	ND
Growth at:						
15°C	+	+	ND	−w	ND	−w
45°C	−w	+	ND	−	+	+
Cytochrome c (oxidase test)	−	−	−	+	−	−
Lactic acid production:						
L(+)–Isomer	+	+	w	+	w	w
D(−)–Isomer	−	−	w	−	+	−
Acetoin production	−	−	ND	−	+	ND
FDP-aldolase:						
Class I	+	+	ND	+	ND	+
Class II	+	+	ND	−	ND	−
Acid (aerobically) from:						
D-Xylose	−	−	−	−w	−	−
L-Arabinose	−	−	d	d	−	−
D-Cellobiose	−	−	−	+	−	−
D-Fucose	−	−	−	d	ND	ND
Raffinose	−	−	−w	+	−	ND
Salicin	−	−	ND	d	ND	ND
Sucrose	+	+	−	+	+	−

Footnotes are at end of table

Table 17.15 *(continued)*

Characteristics	S. hyicus[c]	S. intermedius	S. kloosii[b]	S. lentus	S. lugdunensis[h]	S. saccharolyticus
Maltose	−	(w)	+	d	+	−
D-Mannitol	−	(d)	+	+	−	−
D-Mannose	+	+	−	(+)	+	(+)
D-Trehalose	+	+	+	+	+	−
α-Lactose	+	d	d	d	+	−
D-Galactose	+	+	+w	d	ND	ND
β-D-Fructose	+	+	+	(+)	+	(+)
D-Melezitose	−	−	+w	−	ND	ND
D-Turanose	−	d	w	−	d	ND
D-Ribose	+	+	+	+	−	ND
Xylitol	−	−	−w	−	−	−
Hyaluronidase	+	ND	−	ND	ND	ND
Growth on (NH₄)₂SO₄ (nitrogen source)	ND	−	ND	+	ND	ND
Nitrate reduction	+	+	−	+	+	+
Alkaline phosphatase	+	+	+	w	−	ND
Arginine dihydrolase	+	ds	−	−	−	+
Urease	d	+	d	−	d	ND
Coagulase (rabbit plasma)	d	+	−	−	−	−
Clumping factor	−	d	−	−	+	−
Fibrinolysin	d	−	−	ND	−	ND
Hemolysis	−	d	d	−	w	−
Deoxyribonuclease (DNase agar)	+	+	−	+w	ND	ND
Heat-stable nuclease	+	+	−	ND	−	ND
β-Glucosidase	d	d	ND	+	ND	ND
β-Glucuronidase	d	−	ND	−	−	ND
β-Galactosidase	−	ds	ND	−	−	ND
Novobiocin resistance (MIC ≥ 1.6 µg/ml)	−	−	+	+	−	−

Table 17.15 *(continued)*

Characteristics	S. saprophyticus	S. schleiferi subsp. coagulans[i]	S. schleiferi subsp. schleiferi[h,i]	S. sciuri	S. simulans	S. warneri	S. xylosus
Colony diameter > 5 mm[j]	+	+	−	+	+	d	+
Colony pigment (carotenoid)	d	−	−	d	−	d	d
Aerobic growth	+	+	+	+	+	+	+
Anaerobic growth (thioglycolate)	(+)	+	+	(+)	+	+	d
Growth on NaCl agar:							
10% (w/v)	+	ND	+	+	+	+	+
15% (w/v)	d	ND	+	d	w	w	d

Footnotes are at end of table

Table 17.15 *(continued)*

Characteristics	S. saprophyticus	S. schleiferi subsp. coagulans[i]	S. schleiferi subsp. schleiferi[h,i]	S. sciuri	S. simulans	S. warneri	S. xylosus
Growth at:							
15°C	+	ND	ND	+	+	d	+
45°C	d	ND	+	−w	+	+	−w
Cytochrome c (oxidase test)	−	−	−	+	−	−	−
Lactic acid production:							
L(+)–Isomer	w	ND	+	+	+	+	w
D(−)–Isomer	w	ND	w	−	d	+	−w
Acetoin production	+	+	+	−	−w	+	d
FDP-aldolase:							
Class I	+	ND	ND	+	+	+	+
Class II	−	ND	ND	−	−	−	−
Acid (aerobically) from:							
D-Xylose	−	−	−	−	−	−	+
L-Arabinose	−	−	−	d	−	−	+
D-Cellobiose	−	−	−	+	−	−	−
D-Fucose	−	ND	ND	+	−	−	−
Raffinose	−	−	−	−	−	−	−
Salicin	−	ND	ND	+	−	−	d
Sucrose	+	d	−	+	+	+	+
Maltose	+	−	−	(d)	−w	(+)	+
D-Mannitol	d	d	−	+	+	d	d
D-Mannose	−	+	+	(d)	d	−	+
D-Trehalose	+	−	d	+	d	+	+
α-Lactose	d	d	−	−w	+	ds	d
D-Galactose	−	+	ND	(+)	−w	d	d
β-D-Fructose	+	+	w	+	+	+	+
D-Melezitose	−	−	ND	d	−	ds	−
D-Turanose	+	ND	−	−	−	d	d
D-Ribose	−	+	−	+	d	d	d
Xylitol	d	−	−	−	−	−	−w
Hyaluronidase	ND	−	ND	ND	ND	ND	ND
Growth on (NH₄)₂SO₄ (nitrogen source)	d	ND	ND	+	−	−	+
Nitrate reduction	−	+	+	+	+	−w	d
Alkaline phosphatase	−	+	+	+w	w	−	d
Arginine dihydrolase	−w	+	+	−	+	d	−
Urease	+	+	−	−	+	+	+
Coagulase (rabbit plasma)	−	+	−	−	−	−	−
Clumping factor	−	−	+	−	−	−	−
Fibrinolysin	ND	ND	−	ND	ND	ND	ND
Hemolysis	−	+	−w	−	−w	(ds)	−w

Footnotes are at end of table

Table 17.15 *(continued)*

| Characteristics | S. saprophyticus | S. schleiferi | | S. sciuri | S. simulans | S. warneri | S. xylosus |
		subsp. coagulans[i]	subsp. schleiferi[h,i]				
Deoxyribonuclease (DNase agar)	–	ND	ND	+w	w	ds	–w
Heat-stable nuclease	–	+	+	ND	–w	–	–
β-Glucosidase	d	ND	ND	+	–	+	+
β-Glucuronidase	–	ND	–	–	d	d	d
β-Galactosidase	d	ND	d	–	+	–	+
Novobiocin resistance (MIC ≥ 1.6 μg/ml)	+	–	–	+	–	–	+

[a] Symbols: +, 90% or more strains are positive; –, 90% or more strains are negative; d, 11–89% strains are positive; ( ), delayed reaction; w, weak reaction; –w, negative to weak reaction; +w, positive to weak reaction; ds, test differentiates subspecies (not separated out above in heading); de, test differentiates ecotypes; ND, test not determined.

[b] Schleifer et al. (Int. J. Syst. Bacteriol. *35:* 223–225, 1985; Syst. Appl. Microbiol. *5:* 501–509, 1984).

[c] *S. hyicus* subsp. *hyicus* and *S. hyicus* subsp. *chromogenes* in *Bergey's Manual of Systematic Bacteriology;* raised to species status by Hájek et al. (Int. J. Syst. Bacteriol. *37:* 179–180, 1987; effective publication: Syst. Appl. Microbiol. *8:* 169–173, 1986).

[d] Bannerman and Kloos (Int. J. Syt. Bacteriol. *41:* 144–147, 1991); the spelling of the second subspecific epithet is as in the original, *ureolyticus;* however, the preferred form (Appendix 9 of the Bacteriological Code) is not clear.

[e] Kloos and Wolfshohl (Int. J. Syst. Bacteriol. *41:* 284–289, 1991) formally named the subspecies *S. cohnii* subsp. *cohnii* and *S. cohnii* subsp. *urealyticus* (originally spelled *urealyticum* through a typographic error), which in *Bergey's Manual of Systematic Bacteriology* were referred to as *S. cohnii* subsp. 1 and *S. cohnii* subsp. 2, respectively. The subspecific epithet *urealyticus* is, however, illegitimate because of the prior subspecific epithet *ureolyticus* in *S. capitis* subsp. *ureolyticus* (Rules 12b and 13c of the Bacteriological Code), of which *urealyticus* is a later homonym as an orthographic variant. The epithet *urealyticus* in *S. cohnii* subsp. *urealyticus* will therefore have to be replaced.

[f] Varaldo et al. (Int. J. Syst. Bacteriol. *38:* 436–439, 1988).

[g] Igimi et al. (Int. J. Syst. Bacteriol. *39:* 373–377, 1989).

[h] Freney et al. (Int. J. Syst. Bacteriol. *38:* 168–172, 1988).

[i] Igimi et al. (Int. J. Syst. Bacteriol. *40:* 409–411, 1990).

[j] Colony diameter is determined after incubation on P agar (Kloos et al., Int. J. Syst. Bacteriol. *24:* 79–101, 1974; Kloos and Schleifer, J. Clin. Microbiol. *1:* 82–88, 1975) at 34–35°C for 3 days and at room temperature (≅ 25°C) for an additional 2 days.

[k] May grow on repeated subculture.

Important Notes for Users of This Manual

Unless otherwise indicated in footnotes to tables, the meanings of symbols are as follows:

+ 90% or more of strains are positive

– 90% or more of strains are negative

d 11-89% of strains are positive

v strain instability (not equivalent to "d")

D Different reactions in different taxa (species of a genus or genera of a family)

All other symbols are defined in footnotes to tables.

Table 17.16 Characteristics differentiating the species of *Streptococcus*[a,b]

Characteristics	Pyogenic streptococci						
				S. equi			
	S. agalactiae	*S. canis*[c]	*S. dysgalactiae*	subsp. *equi*	"*S. equi* subsp. *equisimilis*"[d]	subsp. *zooepidemicus*[e]	*S. iniae*
Growth in air	+	+	+	+	+	+	+
Growth in air plus 5% CO_2	+	+	+	+	+	+	+
Growth anaerobically	+	+	+	+	+	+	+
Growth at:							
10°C	d	ND	−	−	−	−	+
45°C	−	ND	−	−	−	−	−
pH 9.6	−	ND	−	−	−	−	−
Growth with:							
6.5% NaCl	d	−	−	−	−	−	−
40% bile	d	−	−	−	−	−	−
0.25% optocin	+	ND	+	+	+	+	+
Alpha-hemolysis	−	−	+	−	−	−	+[f]
Beta-hemolysis	d	+	−	+	+	+	+
Hydrolysis of:							
Arginine	+	+	+	+	+	+	ND
Hippurate	+	−	−	−	−	−	−
Esculin	−	+	−	(−)	−	(−)	+
Acid from[s]:							
Inulin	−	−	−	−	−	−	−
Lactose	d	−	+	−	d	+	−
Mannitol	−	−	−	−	−	−	+
Raffinose	−	−	−	−	−	−	−
Ribose	+	+	ND	−	+	(+)	+
Salicin	d	+	d	+	ND	ND	+
Sorbitol	−	−	d	−	−	+	−
Trehalose	+	(−)	+	−	+	−	+
Production of[s]:							
Alkaline phosphatase	+	+	+	+	+	+	ND
α-Galactosidase	−	d	−	−	−	−	ND
β-Glucuronidase	d	(−)	+	+	+	+	ND
β-Galactosidase	−	(+)	−	−	−	−	ND
Pyrrolidonearylamidase	−	−	−	−	−	−	−
Voges-Proskauer test	+	−	−	−	−	−	ND
Lancefield serological group	B	G	C	C	C	C	Uncertain

Footnotes are at end of table

Table 17.16 (continued)

Characteristics	Pyogenic streptococci (continued)			Oral streptococci			
	S. porcinus[f]	S. pyogenes	S. suis[g]	S. adjacens[h]	S. cricetus	S. defectivus[h]	S. downei[i]
Growth in air	+	+	+	+	d	+	+
Growth in air plus 5% CO₂	+	+	+	+	d	+	+
Growth anaerobically	+	+	+	+	+	+	+
Growth at:							
10°C	ND	–	ND	ND	–	ND	ND
45°C	–	–	ND	ND	d	ND	–
pH 9.6	ND	–	ND	ND	–	ND	–
Growth with:							
6.5% NaCl	d	–	ND	ND	d	ND	–
40% bile	d	–	+	ND	d	ND	d
0.25% optocin	+	+	ND	+	+	+	+
Alpha-hemolysis	–	–	+	+	–	+	ND
Beta-hemolysis	(+)	+	d	–	–	–	ND
Hydrolysis of:							
Arginine	+	+	ND	–	–	–	–
Hippurate	–	–	–	–	–	–	–
Esculin	(+)	(–)	+	ND	d	ND	–
Acid from [s]:							
Inulin	–	–	+	d	d	–	+
Lactose	d	+	+	–	+	d	+
Mannitol	+	–	–	–	+	–	+
Raffinose	–	–	–	–	+	d	–
Ribose	+	–	ND	–	ND	–	ND
Salicin	+	+	+	ND	+	ND	+
Sorbitol	+	–	–	–	+	–	–
Trehalose	+	+	+	–	+	+	+
Production of [s]:							
Alkaline phosphatase	+	+	–	–	ND	–	ND
α-Galactosidase	–	–	+	–	ND	+	ND
β-Glucuronidase	+	(–)	+	d	ND	–	ND
β-Galactosidase	–	–	–	–	ND	+	ND
Pyrrolidonearylamidase	–	+	–	+	ND	+	ND
Voges-Proskauer test	+	–	ND	ND	+	ND	+
Lancefield serological group	E, P, U or V	A	D, R or S	None	None	Sometimes H[t]	ND

Footnotes are at end of table

553

Table 17.16 *(continued)*

Characteristics	Oral streptococci *(continued)*							
	S. ferus	*S. gordonii*[j]	*S. macacae*[k]	*"S. milleri"*[l]	*"S. mitior"*	*S. mitis*	*S. mutans*	*S. oralis*
Growth in air	d	+	w	d	+	+	d	+
Growth in air plus 5% CO_2	d	+	+	d	+	+	d	+
Growth anaerobically	+	+	+	+	+	+	+	+
Growth at:								
10°C	–	ND	ND	–	–	+	–	(–)
45°C	–	ND	–	d	d	d	d	d
pH 9.6	–	ND	–	–	–	ND	–	–
Growth with:								
6.5% NaCl	–	–	–	–	–	–	–	–
40% bile	ND	ND	+	d	–	d	d	–
0.25% optocin	+	+	+	ND	+	ND	+	+
Alpha-hemolysis	–	+	+	–	+	(+)	–	+
Beta-hemolysis	–	–	–	d	–	–	–	–
Hydrolysis of:								
Arginine	–	+	–	d	–	d	(–)	(–)
Hippurate	–	–	ND	–	–	–	–	–
Esculin	+	+	+	d	–	d	+	(–)
Acid from [s]:								
Inulin	+	d	–	d	(–)	d	+	–
Lactose	ND	+	ND	(+)	d	+	+	+
Mannitol	+	–	+	(–)	–	–	+	–
Raffinose	–	d	+	d	d	d	+	d
Ribose	ND	ND	–	ND	ND	ND	ND	ND
Salicin	+	+	ND	(+)	ND	d	+	–
Sorbitol	+	–	+	(–)	–	(–)	+	–
Trehalose	ND	+	+	(+)	d	d	+	+
Production of [s]:								
Alkaline phosphatase	ND	+	ND	+	d	d	–	d
α-Galactosidase	ND	d	ND	d	–	(+)	d	d
β-Glucuronidase	ND	–	ND	–	–	–	–	–
β-Galactosidase	ND	(+)	ND	–	d	d	–	+
Pyrrolidonearylamidase	ND	ND	ND	–	–	–	–	ND
Voges-Proskauer test	d	–	ND	d	d	–	+	d
Lancefield serological group	None	None	ND	Sometimes F or G[t]	Sometimes H, K or O[t]	None	None	None

Footnotes are at end of table

Table 17.16 (continued)

Characteristics	Oral streptococci (continued)					
	S. pneumoniae	S. rattus	S. salivarius	S. sanguis	S. sobrinus	S. vestibularis [m]
Growth in air	+	d	+	+	d	+
Growth in air plus 5% CO_2	+	d	+	+	d	+
Growth anaerobically	+	+	+	+	+	+
Growth at:						
10°C	−	−	−	−	−	−
45°C	−	d	(+)	d	d	−
pH 9.6	−	−	−	−	−	ND
Growth with:						
6.5% NaCl	−	−	−	−	d	ND
40% bile	−	d	d	d	d	−
0.25% optocin	−	+	+	+	+	+
Alpha-hemolysis	+	−	d	+	−	+
Beta-hemolysis	(−)	−	−	−	−	−
Hydrolysis of:						
Arginine	(+)	+	−	+	−	−
Hippurate	−	−	−	−	−	ND
Esculin	d	+	+	d	d	(+)
Acid from [s]:						
Inulin	d	+	+	(+)	d	−
Lactose	+	+	(+)	+	+	+
Mannitol	(−)	+	−	(−)	+	−
Raffinose	+	+	+	d	d	−
Ribose	−	ND	ND	ND	ND	−
Salicin	−	+	+	+	−	+
Sorbitol	−	+	−	(−)	d	−
Trehalose	+	+	(+)	(+)	d	d
Production of [s]:						
Alkaline phosphatase	−	ND	d	d	ND	ND
α-Galactosidase	(+)	ND	−	d	ND	ND
β-Glucuronidase	−	ND	d	−	ND	ND
β-Galactosidase	d	ND	−	d	ND	ND
Pyrrolidonearylamidase	d	ND	−	−	ND	ND
Voges-Proskauer test	−	+	(+)	−	+	(+)
Lancefield serological group	None	None	Often K	H	None	ND

Footnotes are at end of table

Table 17.16 (continued)

Characteristics	Anaerobic streptococci				Other streptococci	
	S. hansenii	S. morbillorum	S. parvulus	S. pleomorphus	S. acidominimus	S. alactolyticus [n]
Growth in air	−	−	−	−	+	+
Growth in air plus 5% CO_2	−	−	−	−	+	+
Growth anaerobically	+	+	+	+	+	+
Growth at:						
10°C	−	−	−	−	−	ND
45°C	+	ND	+	+	−	+
pH 9.6	ND	ND	ND	ND	−	ND
Growth with:						
6.5% NaCl	−	ND	−	ND	ND	+
40% bile	ND	ND	−	ND	ND	ND
0.25% optocin	ND	+	ND	ND	ND	ND
Alpha-hemolysis	−	d	−	−	d	d
Beta-hemolysis	−	−	−	−	−	−
Hydrolysis of:						
Arginine	−	−	−	ND	(−)	−
Hippurate	−	−	+	−	dw	−
Esculin	d	−	+	ND	−	+
Acid from [s]:						
Inulin	−	ND	+	ND	−	−
Lactose	+	−	+	−	+	−
Mannitol	ND	−	−	−	+	(+)
Raffinose	+	ND	−	ND	−	+
Ribose	ND	ND	−	ND	−	−
Salicin	−	−	+	−	+	(+)
Sorbitol	ND	ND	−	ND	+	−
Trehalose	ND	ND	+	ND	(+)	d
Production of [s]:						
Alkaline phosphatase	ND	ND	ND	ND	(+)	−
α-Galactosidase	ND	ND	ND	ND	−	+
β-Glucuronidase	ND	ND	ND	ND	d	−
β-Galactosidase	ND	ND	ND	ND	d	−
Pyrrolidonearylamidase	ND	ND	ND	ND	−	−
Voges-Proskauer test	d	ND	−	ND	+	+
Lancefield serological group	ND	ND	ND	ND	ND	ND

Footnotes are at end of table

Table 17.16 (continued)

Characteristics	Other streptococci (continued)					
	S. bovis	S. equinus	S. hyointestinalis[o]	S. intestinalis[p]	S. thermophilus[q]	S. uberis
Growth in air	+	+	+	+	+	+
Growth in air plus 5% CO_2	+	+	+	+	+	+
Growth anaerobically	+	+	+	+	+	+
Growth at:						
10°C	–	–	ND	–	–	+
45°C	d	+	ND	+	+	–
pH 9.6	d	–	ND	–	–	–
Growth with:						
6.5% NaCl	–	–	–	–	–	–
40% bile	+	+	–	–	–	d
0.25% optocin	+	ND	ND	ND	ND	ND
Alpha-hemolysis	dw	dw	+	–	d	d
Beta-hemolysis	–	–	–	+	–	–
Hydrolysis of:						
Arginine	–	–	–	–	d	+
Hippurate	–	–	–	–	–	+
Esculin	+	+	+	+	–	+
Acid from [s]:						
Inulin	d	(–)	–	ND	–	+
Lactose	+	–	+	–	+	+
Mannitol	d	–	–	–	–	+
Raffinose	ND	(–)	d	–	ND	(–)
Ribose	–	–	–	–	ND	+
Salicin	+	(+)	+	+	–	+
Sorbitol	d	–	–	–	–	+
Trehalose	d	ND	+	–	ND	+
Production of [s]:						
Alkaline phosphatase	–	–	+	ND	ND	d
α-Galactosidase	d	–	d	ND	ND	–
β-Glucuronidase	(–)	–	–	ND	ND	+
β-Galactosidase	(–)	–	–	ND	ND	–
Pyrrolidonearylamidase	–	–	–	ND	ND	–
Voges-Proskauer test	ND	+	+	ND	ND	ND
Lancefield serological group	D	D	None	Sometimes G[t]	Uncertain	Sometimes E[t]

Footnotes are at end of table

Table 17.16 *(continued)*

a Symbols: +, 90% or more of strains are positive; (+), 80–89% of strains are positive; d, 21–79% of strains are positive; (−), 11–20% of strains are positive; −, 90% or more of strains are negative; w, weak growth or reaction; ND, not determined.

b Taken mainly from *Bergey's Manual of Systematic Bacteriology,* Schleifer and Kilpper-Bälz (Syst. Appl. Microbiol. *10:* 1–19, 1987), Colman et al. (*In* Board et al., *Identification Methods in Applied and Environmental Microbiology,* Academic Press, London, 1992), and Bridge and Sneath (J. Gen. Microbiol. *129:* 565–597, 1983). Species are arranged largely according to Schleifer and Kilpper-Bälz.

c Not included in *Bergey's Manual of Systematic Bacteriology;* proposed by Devriese et al. (Int. J. Syst. Bacteriol. *36:* 422–425, 1986).

d Not included in *Bergey's Manual of Systematic Bacteriology;* description mainly from Colman et al. (in Board et al., *Identification Methods in Applied and Environmental Microbiology,* in press, Academic Press, London, 1992).

e Not included in *Bergey's Manual of Systematic Bacteriology;* proposed by Farrow and Collins (Int. J. Syst. Bacteriol. *35:* 223–225, 1985; effective publication: Syst. Appl. Microbiol. *5:* 483–493, 1985).

f Not described in *Bergey's Manual of Systematic Bacteriology;* proposed by Collins et al. (Int. J. Syst. Bacteriol. *35:* 223–225, 1985; effective publication: Syst. Appl. Microbiol. *5:* 402–413, 1984).

g Proposed by Kilpper-Bälz and Schleifer (Int. J. Syst. Bacteriol. *37:* 160–162, 1987).

h Not included in *Bergey's Manual of Systematic Bacteriology;* proposed by Bouvet et al. (Int. J. Syst. Bacteriol. *39:* 290–294, 1989).

i Not described in *Bergey's Manual of Systematic Bacteriology;* proposed by Whiley et al. (Int. J. Syst. Bacteriol. *38:* 25–29, 1988).

j Not included in *Bergey's Manual of Systematic Bacteriology;* proposed by Killian et al. (Int. J. Syst. Bacteriol. *39:* 471–484, 1989).

k Not described in *Bergey's Manual of Systematic Bacteriology;* proposed by Beighton et al. (Int. J. Syst. Bacteriol. *34:* 332–335, 1984).

l A complex including strains named *Streptococcus anginosus, S. constellatus,* and *S. intermedius;* see *Bergey's Manual of Systematic Bacteriology* and Schleifer and Kilpper-Bälz (Syst. Appl. Microbiol. *10:* 1–19, 1987). Coykendall et al. (Int. J. Syst. Bacteriol. *37:* 222–228, 1987) note that if these are considered conspecific, the correct name is *S. anginosus.* The epithet *milleri* is retained here for diagnostic convenience pending further study (see also Facklam, Eur. J. Microbiol. *3:* 91–93, 1984).

m Not included in *Bergey's Manual of Systematic Bacteriology;* proposed by Whiley and Hardie (Int. J. Syst. Bacteriol. *38:* 335–339, 1988).

n Not included in *Bergey's Manual of Systematic Bacteriology;* proposed by Farrow et al. (Int. J. Syst. Bacteriol. *35:* 223–225, 1985; effective publication: Syst. Appl. Microbiol. *5:* 467–482, 1985).

o Not included in *Bergey's Manual of Systematic Bacteriology;* proposed by Devriese et al. (Int. J. Syst. Bacteriol. *38:* 440–441, 1988).

p Not included in *Bergey's Manual of Systematic Bacteriology;* proposed by Robinson et al. (Int. J. Syst. Bacteriol. *38:* 245–248, 1988). It is unusual in being urease positive.

q Has been considered to be a subspecies of *Streptococcus salivarius* by Farrow and Collins (Int. J. Syst. Bacteriol. *34:* 355–357, 1984; effective publication: J. Gen. Microbiol. *130:* 357–362, 1984), but see Schleifer et al. (Syst. Appl. Microbiol. *14:* 386–388, 1991).

r With outer ring of α-hemolysis.

s Additional information from API Company, API 20 STREP Instruction Manual (API Co., La Balme les Grottes, France).

t Only the most common serological groups are listed.

Table 17.17 Characteristics differentiating the species of *Vagococcus*[a]

Characteristics	*V. fluvialis*	*V. salmoninarum*
Acid from:		
Glycerol	d	−
Sorbitol	+	−
D-Tagatose	−	+
H$_2$S produced	−	+
Growth at 40°C	+	−

a Symbols: +, 90% or more of strains are positive; −, 90% or more of strains are negative; d, 11–89% of strains are positive.

ENDOSPORE-FORMING GRAM-POSITIVE RODS AND COCCI

This is a group of convenience whose taxonomic affinities are currently uncertain. Rods or cocci, and sometimes filaments, are 0.3–2 µm in diameter (except *Oscillospira*). **Most stain Gram positive** and possess a thick cell wall of the Gram-positive type. Cells are **usually motile** by peritrichous flagella and **form heat-resistant endospores, which are highly refractile** and possess an impermeable spore coat; they **stain poorly** with the usual stains but may be stained by special methods. Endospores may be confused with lipid inclusions. They may be scanty and are best sought in old cultures. It is wise to confirm their nature by testing that cultures survive heating at 70–80°C for 10 min followed by cultivation under suitable conditions. Metabolism and ecology are very varied. Endospore-forming genera that are treated elsewhere are *Thermoactinomyces* and *Sporomusa*. Many strains of *Clostridium* sporulate very poorly and are weakly Gram positive; they may therefore seem to belong to genera such as *Eubacterium, Bifidobacterium, Bacteroides,* or *Fusobacterium.*

Differentiation of the genera in **Group 18:** See Table 18.1.

Genus **Amphibacillus**

Editorial note: The genus *Amphibacillus* was not included in *Bergey's Manual of Systematic Bacteriology.* The genus was created in 1990 by Niimura et al. (Int. J. Syst. Bacteriol. *40:* 297–301). It includes only one species, *A. xylanus.*

Straight rods are 0.3–0.5 × 0.9–1.9 µm. **Gram positive** in early stages of growth. **Endospores oval and central,** somewhat swelling the cell, but are soon liberated by cell lysis. Motile by peritrichous flagella. **Facultatively anaerobic,** amphibacilli **grow well and sporulate under both aerobic and anaerobic conditions** on glucose peptone yeast extract media at pH 10. Chemoorganotrophic cells **producing ethanol, acetic acid, and formic acid from glucose under anaerobic**

conditions and acetic acid under aerobic conditions. Oxygen consumption does not occur through the cytochrome system, but NADH oxidase plays a role. **Catalase negative and oxidase negative.** Grow at a temperature of 25–45°C. Probably widely distributed in decaying plant material.

Type (and only) species: *Amphibacillus xylanus.*

Characteristics of the species: **Attacks xylan both aerobically and anaerobically,** as well as a number of other carbohydrates. Isolated from compost of animal manure, grass, and rice straw and probably are widely distributed in composts of manure with grass.

Genus **Bacillus**

Cells are rod-shaped and straight, 0.5–2.5 × 1.2–10 µm, and often are arranged in pairs or chains, with rounded or squared ends. Cells **stain Gram positive** and are **motile by peritrichous flagella. Endospores are oval** or sometimes round or cylindrical and are very resistant to many adverse conditions. There is **not more than one spore per cell, and sporulation is not repressed by exposure to air. Aerobic or facultatively anaerobic,** With wide diversity of physiological ability with respect to heat, pH, and salinity. **Chemoorganotrophs,** with a fermentative or respiratory metabolism. **Usually catalase positive.** Found in a wide range of habitats; a few species are pathogenic to vertebrates or invertebrates.

Type species: *Bacillus subtilis.*

Differentiation of the species of the genus **Bacillus:** Species differentiation is difficult because of their large number and the often incomplete descriptions of a number of newly reported species. Reference should be made to *Bergey's Manual of Systematic Bacteriology* for properties that distinguish *Bacillus anthracis,* a pathogen of humans and animals, from the closely allied *B. cereus.*

Genus **Clostridium**

Cells are rod-shaped, 0.3–2.0 × 1.5–20.0 μm, and are often arranged in pairs or short chains, with rounded or sometimes pointed ends. **Commonly pleomorphic. Usually stain Gram positive in young cultures, usually motile** by peritrichous flagella. **Form oval or spherical endospores** that usually distend the cell. Most species are chemoorganotrophic; some are chemoautotrophic or chemolithotrophic as well. **May be saccharolytic, proteolytic, neither, or both.** Usually they produce mixtures of organic acids and alcohols from carbohydrates or peptones. **Do not carry out a dissimilatory sulfate reduction. Usually catalase negative** and obligately anaerobic; if growth occurs in air, it is scanty and sporulation is inhibited. Metabolically they are very diverse, with optimum temperatures of 10–65°C. Widespread in the environment. Many species produce potent exotoxins, and some are pathogenic for animals because of either wound infections or the absorption of toxins.

Type species: *Clostridium butyricum.*

Differentiation of the species of the genus **Clostridium:** Differentiation is best carried out by specialist laboratories because of the exacting growth conditions and test procedures that are required. Many new species have been described recently; the genus now contains about 100 species.

Genus **Desulfotomaculum**

Straight or curved rods, 0.3–1.5 × 3–9 μm, are usually single. **Stain Gram negative,** but the cell wall is of the Gram-positive type. Cells are **motile,** with oval or round endospores that are terminal to subterminal, causing slight swelling of the cells. **Strict anaerobes and chemoorganotrophic,** the cells have a respiratory metabolism. **Sulfates, sulfites, and reducible sulfur compounds act as electron acceptors and are reduced to H$_2$S. Contain cytochrome of the protohaem class but not cytochrome c_3. Catalase negative.** Growth many occur at temperatures of 20–70°C (optimum 30–55°C). Widely distributed in soil, the rumen, and elsewhere.

Type species: *Desulfotomaculum nigrificans.*

Differentiation of the species of the genus **Desulfotomaculum:** See Table 18.2.

Genus **Oscillospira**

Large rods or filaments, 3–6 μm in diameter, are divided by closely spaced cross-walls into numerous disk-shaped cells. Reproduction is by transverse fission. Cells **stain Gram negative and are motile by numerous lateral flagella.** Endospores may be formed. Oscillospirae are not yet cultivated. Exposure to air abolishes motility, thus suggesting that the organism is an anaerobe. It occurs in the alimentary tract of herbivorous animals. For further details see *Bergey's Manual of Systematic Bacteriology.*

Type (and only) species: *Oscillospira guilliermondii.*

Characteristics of the species: As described for the genus.

Genus **Sporohalobacter**

Editorial note: The genus *Sporohalobacter* was not included in *Bergey's Manual of Systematic Bacteriology.* It was proposed by Oren et al. (Int. J. Syst. Bacteriol. *38:* 136–137, 1988; effective publication: Syst. Appl. Microbiol. *9:* 239–246, 1987) for the organism *Clostridium lortetii* because it has a wall of Gram-negative morphology. It contains two species, *S. lortetii* and *S. marismortui.*

Rods, 0.6–0.8 × 2.5–13 μm, **stain Gram negative.** Motile by peritrichous flagella. **Endospores are round and terminal, swelling the rod** to give a **drumstick appearance. Obligately anaerobic and halophilic, require 0.5–2 M NaCl for growth.** Chemoorganotrophic, with a **fermentative metabolism,** producing acetate, ethanol, CO$_2$, and H$_2$ from carbohydrates. **Catalase negative** and oxidase negative. The optimum temperature is 35–45°C. Found in saline habitats.

Type species: *Sporohalobacter lortetii.*

Differentiation of the species of the genus **Sporohalobacter:** See Table 18.3.

Genus **Sporolactobacillus**

Straight rods, 0.7–0.8 × 3–5 μm, occur singly, in pairs, or rarely in short chains. Cells **stain Gram positive** and are motile by peritrichous flagella. Endospores are scantily formed, ellipsoidal, and terminal, swelling the cell. **Facultative anaerobes but show scanty growth in air. Produces homolactic fermentation of hexoses to D(–)-lactic acid. Without catalase or cytochromes; nitrate is not reduced, and indole is not formed.** The optimum temperature is 35°C. Isolated from chicken feed and soil and are probably widespread in the environment.

Type (and only) species: *Sporolactobacillus inulinus.*

Characteristics of the species: As described for the genus.

Genus **Sporosarcina**

Cells are spherical or oval, 1–2 × 2–3 μm, and are arranged mainly as diplococci and tetrads but sometimes as cubical packets. They **stain Gram positive and are motile by a few flagella on each cell. Endospores are round** (0.5–1.5 μm in diameter). Chemoorganotrophic and **strictly aerobic,** Growing well on nutrient agar, forming cream to orange colonies; *S. halophila* requires the addition of 3% NaCl and 0.5% MgCl$_2$. Cells grow at a temperature of 15–37°C. Widely distributed in soils, including salt marshes.

Type species: *Sporosarcina ureae.*

Differentiation of the species of the genus **Sporosarcina:** See Table 18.4.

Genus **Sulfidobacillus**

Editorial note: The genus *Sulfidobacillus* was not included in *Bergey's Manual of Systematic Bacteriology,* although it was noted briefly. The genus was created by Golovachova and Karavaiko (Int. J. Syst. Bacteriol. *41:* 178–179, 1991; effective publication: Microbioloiya *47:* 815–822, 1978) for a sulfur-oxidizing endospore-forming rod.

Rods, 0.6–0.8 × 1.0–6.0 μm, occur singly, in pairs, or in short chains. Cells **stain Gram positive and are nonmotile. Endospores are round or slightly oval,** subterminal or terminal, and swell the cell. **Strictly aerobic,** facultatively chemolithotrophic. **Sulfur, iron, and pyrite are oxidized** but can be adapted to grow with glucose. **The optimum pH is 1.9–2.4,** and the optimum temperature is 50°C. Isolated from pyrite deposits.

Type (and only) species: *Sulfidobacillus thermosulfooxidans.*

Characteristics of the species: As described for the genus.

Genus **Syntrophospora**

Editorial note: The genus *Syntrophospora* was not included in *Bergey's Manual of Systematic Bacteriology.* It was established as a new genus to contain the species *Clostridium bryantii* because it differs considerably from other species of *Clostridium* in its 16S rRNA sequence (Zhao et al., Int. J. Syst. Bacteriol. *40:* 40–44, 1990).

Cells are rod-shaped; Gram stain is variable, but the cell wall has a Gram-positive ultrastructure. **Endospores are oval and terminal** and swell the cell. **Strictly anaerobic and chemoorganotrophic,** use saturated fatty acids, butyrate and longer chained, for growth, and are β-oxidized to acetate and H$_2$ or, together with odd-numbered straight-chain fatty acids, to acetate, propionate, and H$_2$ in syntropic association with H$_2$-scavenging anaerobes. Some strains can be adapted to grow on crotonate. The optimum temperature is 30°C. **Sulfate and nitrate are not reduced.** Found in aquatic and marine sediments.

Type (and only) species: *Syntrophospora bryantii.*

Characteristics of the species: As described for the genus.

Table 18.1 Differential characteristics of endospore-forming bacteria and similar genera [a]

Characteristics	Genera with endospores										Genera without endospores				
	Amphibacillus	Bacillus	Clostridium	Desulfotomaculum	Oscillospira	Sporohalobacter	Sporolactobacillus	Sporosarcina	Sulfidobacillus	Syntrophospora	Kurthia	Lactobacillus	Marinococcus	Planococcus	Salinicoccus
Rod-shaped in young cultures	+	+	+	+	+	+	+	−	+	+	+	+	−	−	−
Diameter over 2.5 μm	ND	−	D	−	+	−	−	−	−	−	d	−	−	−	−
Filaments	−	−	−	−	+[b]	−	−	−	−	−	−	−[e]	−	−	−
Rods or filaments curved	−	−	D	D	+	−	−	NA	−	ND	−	−	NA	NA	NA
Cocci in tetrads or packets	−	−	−	−	−	−	−	+	−	−	−	−	d	d	d
Endospores produced	+	+	+	+	+	+	+	+	+	+	−	−	−	−	−
Motile	+	+	+	+	+	+	+	+	−	ND	+	−	+	+	−
Stain Gram positive at least in young cultures	+	+	+[c]	d	−	+	+	+	+	+	+	+	+	+	+
Strict aerobes	−	D	−	−	ND	−	−	+	+	−	+	−	+	+	+
Facultative anaerobes or microaerophiles	+	D	−	−	ND	−	+	−	−	−	−	+	−	−	−
Strict anaerobes	−	−	+[d]	+	+	+	−	−	−	+	−	−	−	−	−
Product of carbohydrate fermentation is almost all lactate	ND	D	−	−	ND	−	+	−	ND	−	−	D	−	−	−
Sulfate actively reduced to sulfide	−	−	−	+	ND	ND	−	−	−	−	−	−	−	−	−
Catalase	−	+	−	−	ND	−	−	+	ND	ND	+	−	+	+	+
Oxidase	−	D	−	ND	ND	−	ND	+	ND	ND	−	−	+	−	−
Marked acidity from glucose	+	+	D	−	ND	+	+	−	ND	ND	−	+	+	−	−
Nitrate reduced to nitrite	−	D	D	ND	ND	ND	−	D	ND	−	−	−	D	−	ND
Requires 3–12% NaCl for growth	−	D	−	−	−	+	−	D	−	−	−	−	−	−	−

[a] Symbols: +, 90% or more of strains are positive; −, 90% or more of strains are negative; d, 11–89% of strains are positive; D, substantial proportion of species differ; NA, not applicable; ND, not determined.

[b] Morphologically similar to Caryophanon; broad discs of cells form filament.

[c] Rarely stain Gram negative.

[d] Rarely aerotolerant.

[e] Long chains common, but not filaments.

Table 18.2 Differential characteristics of species of *Desulfotomaculum*[a]

Characteristics	D. acetoxidans	D. antarcticum	D. geothermicum[b]	D. guttoideum[c]	D. kuznetsovii[d]	D. nigrificans	D. orientis	D. ruminis	D. sapomandens[e]
Rods markedly curved	–	–	–	–	–	–	+	–	–
Flagella	Polar (thick)	Peritrichous	ND	Peritrichous	Peritrichous	Peritrichous	Peritrichous	Peritrichous	ND
Gas vacuoles within cells	+	ND	+	ND	–	–	–	–	+
Growth with sulfate plus:									
Lactate	–	+	+	ND	+	+	+	+	–
Formate	ND	–	+	–	+	–	–	+	+
Acetate	+	–	+	–	+	–	–	–	+
Glucose	–	+	ND	ND	–	–	–	–	–
Growth with pyruvate without sulfate	–	ND	ND	+	ND	+	–	+	+
Gelatinase	–	+	ND	ND	ND	–	–	–	ND
Thermophilic (poor growth below 40°C)	–	–	+	–	+	+	–	–	–

[a] Symbols: +, 90% or more of strains are positive; –, 90% or more of strains are negative; ND, not determined.

[b] Daumas et al. (Int. J. Syst. Bacteriol. *40:* 105–106, 1990; effective publication: Antonie van Leewenhoek *54:* 165–178, 1988).

[c] Gogotova and Vainstein (Int. J. Syst. Bacteriol. *36:* 573–576, 1986; effective publication: Mikrobiologiya *52:* 789–792, 1983).

[d] Nazina et al. (Int. J. Syst. Bacteriol. *40:* 470–471, 1990; effective publication: Mikrobiologiya *57:* 823–827, 1988).

[e] Cord-Ruwish and Garcia (Int. J. Syst. Bacteriol. *40:* 105–106, 1990; effective publication: FEMS Microbiol. Lett. *29:* 325–330, 1985); produces gas vacuoles within cells.

Important Notes for Users of This Manual

Unless otherwise indicated in footnotes to tables, the meanings of symbols are as follows:

- + 90% or more of strains are positive
- – 90% or more of strains are negative
- d 11-89% of strains are positive
- v strain instability (not equivalent to "d")
- D Different reactions in different taxa (species of a genus or genera of a family)

All other symbols are defined in footnotes to tables.

Table 18.3 Differential characteristics of species of *Sporohalobacter*[a]

Characteristics	S. lortetii	S. marismortui
Cells contain gas vacuoles	+	−
Readily utilizes carbohydrates as main energy source	−	+
Grows in media containing 0.5 M NaCl	−	+
Indole production	+	−

[a] Symbols: +, 90% or more of strains are positive; −, 90% or more of strains are negative.

Table 18.4 Differential characteristics of species of *Sporosarcina*[a]

Characteristics	S. halophila	S. ureae
Growth in:		
Nutrient broth	−	+
Nutrient broth plus 10% NaCl and 0.5% MgCl$_2$	+	−
Hydrolysis of:		
Casein	+	−
Gelatin	+	−
Tyrosine	−	+
Starch	+	−
Pullulan	+	−
Urea	−	+
Nitrate reduced to nitrite	−	+

[a] Symbols: +, 90% or more of strains are positive; −, 90% or more of strains are negative.

GROUP 19

REGULAR, NONSPORING GRAM-POSITIVE RODS

This group comprises genera that have in common only a few morphological and physiological characteristics but that form a convenient grouping for purposes of identification. **Cross-reference, however, should always be made to Group 20,** the irregular, nonsporing Gram-positive rods, because the **major difference** is in the **regularity of cell shape,** and this is **often difficult to decide.** Genera of Group 19 are **rod-shaped** (varying from short, almost coccal, rods to elongated rods, filaments, or trichomes) but exhibit a **regular shape** with **little pleomorphism.** They are **Gram positive, nonsporing,** seldom pigmented, mesophilic, and **chemoorganotrophic,** and they **grow only in complex media.** They are usually associated with plants or animals or with decaying organic matter; several genera contain pathogens.

Differentiation of the genera in **Group 19:** See Table 19.1.

Genus **Brochothrix**

Regular, unbranched rods are 0.6–0.7 × 1–2 µm and occur singly, in chains, or **in long filamentous chains that fold into knotted masses.** Coccoid forms appear in old cultures. **Gram positive, the cells are not encapsulated, are nonmotile, and are nonsporing.** *Brochothrix* species are facultative anaerobes and are nonpigmented. They **grow at a temperature of 0–30°C** (optimum 20–25°C). The major fermentation product from glucose is L(+)-lactate. Cells are **catalase positive** and contain cytochromes. They stain **methyl red and Voges-Proskauer positive.** *Brochothrix* species occur mainly in meat products but are widely distributed in the environment.

Type species: *Brochothrix thermosphacta.*

Differentiation of the species of the genus **Brochothrix:** See Table 19.2.

Genus **Carnobacterium**

Editorial note: The genus *Carnobacterium* was not included in *Bergey's Manual of Systematic Bacteriology.* It was created by Collins et al. (Int. J. Syst. Bacteriol. *37:* 310–316, 1987) for certain members of the genus *Lactobacillus* that show major differences from most species in that genus and for also some newly described bacterial species.

Straight, slender rods, 0.5–0.7 × 1.0–2.0 µm, occur singly or in pairs and sometimes in short chains. Cells are **Gram positive and may or may not be motile.** They are nonsporing chemoorganotrophs that are **heterofermentative, producing mainly L(+)-lactate from glucose.** They **grow at 10°C but not at 45°C** (optimum 30°C). Carnobacteria are **catalase negative,** and nitrates are not reduced. The major $C_{18:1}$ straight-chain fatty acid is oleic. Cells are found in meat products and fish; one species (*C. piscicola*) is pathogenic for salmonid fish.

Type species: *Carnobacterium divergens.*

Differentiation of the species of the genus **Carnobacterium:** See Table 19.3.

Genus **Caryophanon**

Straight or slightly curved **multicellular rods** (i.e., trichomes) are 1.4–3.2 × 10–20 µm; each cell is 1.0–2.0 µm long. Ends are rounded or tapered. **Gram-positive cells are motile by peritrichous flagella.** Nonsporing and **strictly aerobic,** *Caryophanon* species have a very variable morphology in culture unless special media are used (see *Bergey's Manual of Systematic Bacteriology,* Volume 2, p. 1259).

565

Cells are chemoorganotrophic. **Acetate is the only major utilizable carbon source,** and cells are **catalase positive.** The optimum growth temperature is 25–30°C. These bacteria are associated with cattle dung.

Type species: *Caryophanon latum.*

Differentiation of the species of the genus **Caryophanon:** See Table 19.4.

Genus **Erysipelothrix**

Straight or slightly curved, slender rods, 0.2–0.4 × 0.8–2.5 µm, have a tendency to form **long filaments, often 60 µm or more long. Gram-positive, non-motile, nonsporing cells are without capsules.** They are **not acid-fast** and are chemoorganotrophic. Cells are **facultatively anaerobic and catalase negative.** The optimum temperature is 30–37°C. **Fermentative activity is weak,** with acid but no gas from glucose and a few other carbohydrates. *Erysipelothrix* species are widely distributed in nature and are usually parasitic on mammals, birds, and fish; some strains are pathogenic for mammals and birds.

Type species: *Erysipelothrix rhusiopathiae.*

Differentiation of the species of the genus **Erysipelothrix:** A second species of the genus, *Erysipelothrix tonsillarum,* was described by Takahashi et al. (Int. J. Syst. Bacteriol. *37:* 166–168, 1987); this species can be differentiated from *E. rhusiopathiae* only by DNA-DNA homology. However, the strains belong to one serovar, serovar 7, and show little evidence of virulence for swine.

Genus **Kurthia**

Regular, unbranched rods with rounded ends **in young cultures,** 0.8–1.2 × 2–4 µm, occur in **long chains. Older cultures** (over 2 days) are usually **composed of coccoid cells. Gram-positive** cells are usually **motile by peritrichous flagella.** They do not form endospores and are not acid-fast. Cells are **strictly aerobic.** Growth on yeast nutrient agar shows **rhizoid colonies,** with loops and whorls of chains of rods at the edge (**"Medusa-head appearance"**), and on nutrient gelatin slants the growth has a **"bird's feather"** appearance. Cells are chemoorganotrophs, with a respiratory, not fermentative, metabolism, with **weak acidity from**

glucose. The optimum growth temperature is 25–30°C. Nonpathogenic, *Kurthia* species are widely distributed in the environment and are common in animal feces and meat products.

Type species: *Kurthia zopfii.*

Differentiation of the species of the genus **Kurthia:** See Table 19.5.

Genus **Lactobacillus**

Cells are rod-shaped and usually regular, 0.5–1.2 × 1.0–10.0 µm. They are **usually long rods** but sometimes are almost coccoid, **commonly in short chains. Gram-positive, nonsporing cells are rarely motile** by peritrichous flagella. **Facultative anaerobes,** sometimes microaerophilic, grow poorly in air but better under reduced oxygen tension; some are anaerobes on isolation. Growth is generally enhanced by 5% CO_2. Colonies on agar media are usually 2–5 mm, convex, entire, opaque, and without pigment. Chemoorganotrophs, these cells **require rich, complex media;** their metabolism is **fermentative and saccharoclastic; at least half of the end-product carbon is lactate. Nitrates are not reduced, gelatin is not liquefied, and cells are catalase and cytochrome negative.** The major $C_{18:1}$ straight-chain fatty acid is *cis*-vaccenic. The optimum growth temperature is 30–40°C. Lactobacilli are widely distributed in the environment, especially in animal and vegetable food products; they normally inhabit the gastrointestinal tract of birds and mammals and the mammalian vagina. They are rarely pathogenic.

Type species: *Lactobacillus delbrueckii.*

Differentiation of the species of the genus **Lactobacillus:** This large genus requires special expertise to identify the species; many test reactions are weak and dependent on the composition of the media and the exact cultural conditions.

Genus **Listeria**

Regular, short rods, 0.4–0.5 × 0.5–2 µm with rounded ends, are sometimes almost coccoid, **occurring singly or in short chains** and less often in long filaments. Cells are **Gram positive, nonsporing, not acid-fast, and not encapsulated.** They are **motile** by a few peritrichous flagella **when grown at 20–25°C.** Cells are **facultative anaerobes.** Colonies on nutrient agar have a low convex profile and are translucent and entire, bluish gray by normal illumination and with a **characteristic blue-green sheen under oblique illumination.** Chemoorganotrophs, the cells have a **fermentative metabolism of glucose that yields mainly L(+)-lactate.** Cells are **catalase positive and oxidase negative. Cytochromes are produced.** The optimum temperature for growth (but not for motility, see above) is 30–37°C. Listerias are widely distributed in the environment; some species are pathogenic for humans and animals.

Type species: *Listeria monocytogenes.*

Differentiation of the species of the genus **Listeria:** See Table 19.6.

Genus **Renibacterium**

Short regular rods, 0.3–1.0 × 1.0–1.5 μm, occur often in pairs and sometimes in short chains. **Strongly Gram-positive cells** are nonencapsulated, nonmotile, nonsporing, and **aerobic.** This bacterium is **slow growing. The optimum growth temperature is 15–18°C; there is no growth at 37°C. Cysteine is required for growth,** and growth is enhanced by blood or serum. Cells are chemoorganotrophic. There is **no acid production from sugars.** Cells are **catalase positive** and oxidase negative, and gelatin is not liquefied. Renibacteria are obligate pathogens of salmonid fish.

Type (and only) species: *Renibacterium salmoninarum.*

Characteristics of the species: As described for the genus.

Important Notes for Users of This Manual

Unless otherwise indicated in footnotes to tables, the meanings of symbols are as follows:

+ 90% or more of strains are positive

− 90% or more of strains are negative

d 11-89% of strains are positive

v strain instability (not equivalent to "d")

D Different reactions in different taxa (species of a genus or genera of a family)

All other symbols are defined in footnotes to tables.

Table 19.1 Differential characteristics of genera of regular, nonsporing Gram-positive rods[a]

Characteristics	Brochothrix	Carnobacterium	Caryophanon	Erysipelothrix	Kurthia	Lactobacillus	Listeria	Renibacterium
Cell morphology	Slender rods, often filaments	Slender straight rods	Short rods in chains	Slender rods, often filaments	Regular rods in chains, cocci in old cultures	Rods, usually straight, sometimes cocco-bacilli	Short rods, often short chains and filaments	Short rods, often in pairs
Multicellular rods (trichomes)	-	-	+	-	-	-	-	-
Diameter of rods (µm)	0.6-0.8	0.5-0.7	1.4-3.2	0.2-0.5	0.7-0.9	0.5-1.6	0.4-0.5	0.3-1.0
Motility (if motile, peritrichate flagella)	-	D	+[b]	-	+[b]	-[c]	-[d]	-
Strictly aerobic	-	-	+	-	+[e]	-	-	+
Facultatively anaerobic or microaerophilic	+	+	-	+	-	+	+	-
Catalase reaction	+	-	+	-	+	-[f]	+	+
Major fermentation products from utilizable carbohydrates anaerobically (if NA, acidity from glucose is noted)	Mainly lactate	Mainly L(+)-lactate	NA (no acid)	Lactate	NA (no acid)	Mainly lactate; may give some acetate, ethanol, CO_2	Lactate	NA (no acid)
Peptidoglycan: Group[g]	A	ND	A	B	A	A	A	A
Diamino acid	m-DAP	m-DAP	Lys	Lys	Lys	Lys, m-DAP, Orn	m-DAP	Lys
Major fatty acids[h]	S, A, I	S, U, sometimes C	ND	S, A, I, U	S, A, I	S, U, sometimes C	S, A, I	A
Major menaquinone[i]	MK-7	ND	MK-6	None	MK-7	None[j]	MK-7	MK-9
Habitat	Meat products, nonpathogenic	Food products; one species is a pathogen of fish	Cow dung, nonpathogenic	Widespread, may be pathogenic in vertebrates	Animal feces, meat products, nonpathogenic	Widespread in fermentable materials, rarely pathogenic	Widespread in decaying matter, may be vertebrate pathogen	Pathogen in salmonid fish

[a] Symbols: +, 90% or more strains are positive; -, 90% or more of strains are negative; D, substantial proportion of species differ; NA, not applicble; ND, not determined.

[b] Numerous flagella.

[c] Rarely motile.

[d] Motile at 20-25°C, poorly motile at 37°C.

[e] Rhizoid colonies.

[f] Some strains weakly positive.

[g] Symbolism of Schleifer and Kandler (Bacteriol. Rev. 36: 407-477, 1972).

[h] S, straight-chain saturated; U, monounsaturated; A, anteiso-methyl branched; I, iso-methyl branched; C, cyclopropane ring fatty acids.

[i] Symbolism of Collins and Jones (Microbiol. Rev. 45: 316-354, 1981).

[j] L. mali contains MK-8 and MK-9; a menaquinone is also found in L. casei subsp. rhamnosus.

Table 19.2 Differential characteristics of species of the genus *Brochothrix*[a]

Characteristics	B. campestris[b]	B. thermosphacta
Growth with 8% NaCl in 2 days	–	+
Growth with and reduction of 0.05% potassium tellurite	–	+
Hippurate hydrolysis	+	–
Acid from rhamnose	+	–

[a] Symbols: +, 90% or more of strains are positive; –, 90% or more of strains are negative.

[b] Talon et al. (Int. J. Syst. Bacteriol. *38:* 99–102, 1988).

Table 19.3 Differential characteristics of species of *Carnobacterium*[a]

Characteristics	C. divergens[b]	C. gallinarum[c]	C. mobile[c]	C. piscicola[d]
Motility	–	–	+	–
Acid from amygdalin	+	+	–	+
Acid from inulin	–	–	+	+
Acid from mannitol	–	–	–	+
Acid from α-methyl-D-glucoside	–	+	–	+
Acid from D-xylose	–	+	–	–
Voges-Proskauer test	+	+	(–)	+

[a] Symbols: +, 90% or more of strains are positive; (–), 11–20% of strains are positive; –, 90% or more of strains are negative.

[b] Synonym of *Lactobacillus divergens.*

[c] Collins et al. (Int. J. Syst. Bacteriol. *37:* 310–316, 1987).

[d] Synonym of *Lactobacillus piscicola;* the species *Lactobacillus carnis* described by Shaw and Harding (Int. J. Syt. Bacteriol. *36:* 354–356, 1986; Syst. App. Microbiol. *6:* 291–297, 1985) is considered a synonym of *C. piscicola* by Collins et al.

Table 19.4 Differential characteristics of the genus *Caryophanon*

Characteristics	C. latum	C. tenue
Trichome width (µm)	2.8–3.2	1.4–2.0
Number of visible cross-septa per cell	2–3	1

Table 19.5 Differential characteristics of the species of the genus *Kurthia*[a]

Characteristics	K. gibsonii	K. sibirica[b]	K. zopfii
Growth at 45°C	+	–	–
Growth with 7% NaCl	+	–	+
Colonies cream or yellow	+	+	–
Phosphatase	+	+	–
Acid from ethanol in 0.1% yeast extract medium [c]	–	–	+

[a] Symbols: +, 90% or more of strains are positive; –, 90% or more of strains are negative.

[b] Described by Belikova et al. (Int. J. Syst. Bacteriol. *38:* 220-222, 1988; Mikrobiologiya *55:* 831–836, 1986).

[c] See *Bergey's Manual of Systematic Bacteriology,* Volume 2, p. 1256 and Shaw and Keddie (Syst. Appl. Microbiol. *4:* 253–276, 1983).

Table 19.6 Differential characteristics of the species of the genus *Listeria*[a,b]

Characteristics	L. grayi	L. innocua	L. ivanovii	L. monocytogenes	L. murrayi	L. seeligeri	L. welshimeri
β-Hemolysis	−	−	+[c]	+[d]	−	+	−
CAMP test (*Staphylococcus aureus*)[e]	−	−	−	+	−	+	−
CAMP test (*Rhodococcus equi*)[e]	−	−	+	−	−	−	−
Acid production from:							
Mannitol	+	−	−	−	+	−	−
α-Methyl-D-mannoside	ND	+	−	+	ND	−	+
L-Rhamnose	−	d	−	+	d	−	d
Soluble starch	+	−	−	−	+	ND	ND
D-Xylose	−	−	+	−	−	+	+
Hippurate hydrolysis	−	+	+	+	−	ND	ND
Reduction of nitrate	−	−	−	−	+	ND	ND
Pathogenicity for mice	−	−	+	+	−	−	−

[a] Symbols: +, 90% or more of strains are positive; −, 90% or more of strains are negative; d, 11–89% of strains are positive; ND, not determined.

[b] The species *Listeria denitrificans,* which was in this genus in *Bergey's Manual of Systematic Bacteriology,* has been transferred to a separate genus *Jonesia* in Group 20.

[c] Usually a wide zone or multiple zones.

[d] A few strains negative.

[e] See *Bergey's Manual of Systematic Bacteriology,* Volume 2, p. 1236.

Important Notes for Users of This Manual

Unless otherwise indicated in footnotes to tables, the meanings of symbols are as follows:

+ 90% or more of strains are positive

− 90% or more of strains are negative

d 11-89% of strains are positive

v strain instability (not equivalent to "d")

D Different reactions in different taxa (species of a genus or genera of a family)

All other symbols are defined in footnotes to tables.

IRREGULAR, NONSPORING GRAM-POSITIVE RODS

A diverse collection of genera grouped together for practical purposes only. The majority comprise rods with a visibly irregular shape, which stain Gram positive, grow in the presence of air, and do not form endospores; few of these are motile. Some genera, however, are strict anaerobes or stain Gram negative, and a few are regularly motile. The user is reminded to consult other groups for genera that are readily confused with those in Group 20: in particular Group 19, the regular Gram-positive rods, and Group 22, the nocardioforms. Also, species of *Clostridium* are easily confused with the anaerobic genera in this group if endospores are not observed.

As for the Gram-positive cocci (Group 17), the genera are conveniently considered under the strictly aerobic genera, the facultative anaerobes, and the strict anaerobes, as arranged in Tables 20.1–20.3. The cell morphology and the occurrence of catalase are also particularly helpful in separating the genera.

Differentiation of the genera in **Group 20:** See Tables 20.1–20.3.

Genus **Acetobacterium**

Cells ovoid or short rods, 0.8–1.2 × 1.5–2.5 μm, **occurring singly, in pairs and occasionally in short chains;** the cells taper to blunt ends, and cell irregularity is not marked. **Gram-positive. Motile with one or two subterminal flagella.** Nonsporing, **strictly anaerobic,** and chemolithotrophic, **oxidizing hydrogen and reducing carbon dioxide to yield acetic acid.** Also capable of **chemoorganotrophic fermentation of fructose and other substrates to yield almost exclusively acetic acid. Acetic acid is the organic end product of metabolism. Catalase negative.** The optimum growth temperature is 30°C. Widely distributed in aquatic anaerobic environments.

Type species: *Acetobacterium woodii.*

Differentiation of the species of the genus **Acetobacterium:** See Table 20.4.

Genus **Acetogenium**

Editorial note: The genus *Acetogenium* was not included in *Bergey's Manual of Systematic Bacteriology.* It was established by Leigh and Wolfe (Int. J. Syst. Bacteriol. *33:* 886, 1983; Leigh et al., Arch. Microbiol. *129:* 275–280, 1981).

Straight rods, 0.8–1.2 × 1.5–3.0 μm, mostly arranged singly or in pairs. **Stains Gram negative,** but the cell wall is of the Gram-positive type. **Nonmotile, nonsporing, strictly anaerobic.** Chemolithotrophic, **oxidize hydrogen, and reduce CO_2 to yield acetic acid.** Have an obligate growth requirement for cysteine or sulfide. Also capable of **chemoorganotrophic fermentation of carbohydrates and other substrates to yield principally acetic acid. Catalase negative. Thermophilic, grow at a temperature of 50–72°C (optimum 66°C).** Found in the mud of a tropical lake.

Type (and only) species: *Acetogenium kivui.*

Characteristics of the species: As described for the genus.

Genus **Actinomyces**

Slender, straight or slightly curved rods 0.2–1.0 × 2–5 μm, **and filaments** 10–50 μm long, **with true branching. Short rods, often with clubbed ends,** may occur **singly, in pairs, in V and Y arrangements, and in palisades** (also sometimes in short chains or small clusters). **Branched rods are common;** filaments may branch, **with swollen or clubbed ends. Gram positive** but often stain irregularly, giving rise to a beaded appearance. Nonmotile, nonsporing, not acid-fast. Conidia are not produced. **Facultative anaerobes,**

requiring added CO_2 for good growth both in air and anaerobically. Some species grow poorly in air even with added CO_2. Some species produce filamentous microcolonies. Mature colonies (7–14 days) are commonly rough and crumbly in texture, occasionally with red pigment. Chemoorganotrophic and fermentative, producing acid but no gas from carbohydrates and yielding formic, acetic, lactic, and succinic acids but not propionic acid. Catalase production and nitrate reduction are variable. Indole negative. The optimum growth temperature is 35–37°C. Occur mainly in the oral cavity and on mucous membranes of warm-blooded vertebrates; commonly cause pyogenic infections in association with other concomitant bacteria.

Type species: *Actinomyces bovis.*

Differentiation of the species of the genus **Actinomyces:** Differentiation is difficult because of variable test reactions and because some species are best differentiated by protein gel electrophoresis (see *Bergey's Manual of Systematic Bacteriology;* Dent and Williams, Int. J. Syst. Bacteriol. *36:* 392–395, 1986; Johnson et al., Int. J. Syst. Bacteriol. *40:* 273–286, 1990).

Genus **Aeromicrobium**

Editorial note: The genus *Aeromicrobium* was not included in *Bergey's Manual of Systematic Bacteriology.* It was established by Miller et al. (Int. J. Syst. Bacteriol. *41:* 363–368, 1991) for an organism that differs from *Arthrobacter globiformis* mainly in 16S rRNA sequence and in cell wall peptidoglycan and fatty acid composition.

Irregular small short rods and occasional cocci, 0.5 × 0.5–1.2 μm, predominantly arranged singly, without V arrangement, mycelium, or branching. Cells do not become mainly coccoid in old cultures. Gram-positive, nonmotile, nonsporing. Aerobic. Colonies on tryptone yeast extract agar that are smooth, convex, and beige. Chemoorganotrophic, grow on simple media if vitamins are added. Metabolism is respiratory, producing weak acid from fructose and some other carbohydrates. Catalase positive. Nitrate is not reduced. The optimum growth temperature is about 30°C. Isolated from tropical soil and produce the antibiotic erythromycin.

Type (and only) species: *Aeromicrobium erythreum.*

Characteristics of the species: As described for the genus.

Genus **Agromyces**

Cells are composed of branched, slender filamentous elements 0.3–1.0 μm in diameter in the early phases of growth, fragmenting to yield coccoid and irregular cells in older cultures. Gram positive, with Gram-negative segments in older cultures. Nonmotile, endospores not formed, cells not acid-fast; neither aerial mycelia nor conidia are formed. Aerobic or microaerophilic with poor or no growth anaerobically, do not require CO_2 for growth. Chemoorganotrophic, with a respiratory metabolism. Acid without gas is produced from fructose and other carbohydrates but very slowly from glucose. Cells are difficult to adapt to culture but, when adapted, grow well on heart infusion agar. Catalase negative, oxidase negative, do not reduce nitrate, indole negative. The optimum growth temperature is 30°C. Widely distributed in soil, where it is very numerous.

Type (and only) species: *Agromyces ramosus.*

Characteristics of the species: As described for the genus.

Genus **Arachnia**

Editorial note: The genus *Arachnia* was included in *Bergey's Manual of Systematic Bacteriology* with a single species, *A. propionica.* It has been suggested (Charfreitag et al., Int. J. Syst. Bacteriol. *38:* 354–357, 1988) that this species should be transferred to the genus *Propionibacterium,* but it is retained here for determinative reasons because of its characteristic morphology.

Short, irregular rods, 0.3–0.8 × 3.0–5.0 μm, some showing branching, and slender branching filaments 5–20 μm or more long, especially in young cultures in liquid media, commonly break up in older cultures into short rods. Rods are of variable length, often with clubbed ends, and are commonly arranged in pairs, in Y or V configuration, or in parallel rows forming palisades. Swollen spherical cells up to 5.0 μm in diameter may be seen. Gram-positive but may stain unevenly, giving a beaded appearance. Nonmotile, non-acid-fast, do not form endospores or conidia and are without capsules. Facultative anaerobes. Young microcolonies are composed of tangled filaments; older colonies commonly are heaped or convoluted, white, and opaque, with an undulant edge; **no aerial filaments.** Chemoorganotrophic, requiring nutritionally rich media; fermentative, yielding acid but no gas from many carbohydrates. The **major end products** of glucose fermentation are **acetic and propionic acids. Catalase negative**, indole negative, reduce nitrate to nitrite. The optimum growth temperature is 35–37°C. Inhabit the human oral cavity and are occasionally involved in infective processes.

Type (and only) species: *Arachnia propionica.*

Characteristics of the species: As described for the genus.

Genus **Arcanobacterium**

Slender, irregular rods, 0.3–0.8 × 1.0–5.0 μm, in young cultures; cells may show clubbed ends **sometimes arranged in V formation but there are no filaments.** In older cultures, organisms **segment into short, irregular rods and cocci. Gram positive, nonmotile,** not acid-fast, without endospores. **Facultatively anaerobic.** Grow slowly on nutrient agar; growth is better on horse blood agar, giving small, convex, translucent colonies surrounded by a **zone of complete hemolysis** after 2 days at 37°C. Growth is enhanced by the addition of CO_2. Chemoorganotrophic, requiring nutritionally rich media. **Metabolism is fermentative,** yielding acid but no gas from glucose and a few other carbohydrates, with the production of mainly acetic, lactic, and succinic acids. Usually **catalase negative;** some strains show weak activity. Indole negative, nitrate is reduced to nitrite. The optimum growth temperature is 37°C. Obligate parasites of the pharynx of humans and farm animals; occasionally, they cause pharyngeal or skin lesions.

Type (and only) species: *Arcanobacterium haemolyticum.*

Characteristics of the species: As described for the genus.

Genus **Arthrobacter**

Cells in young cultures are irregular rods, 0.8–1.2 × 1.0–8.0 μm, **often V-shaped** and with clubbed ends, **but there are no filaments. As growth proceeds the rods segment into small cocci,** 0.6–1.0 μm in diameter, **arranged singly, in pairs, and in irregular clumps. This marked rod-coccus growth cycle** is characteristic of *Arthrobacter* and *Pimelobacter;* stationary phase cultures consist almost entirely of cocci. **Gram positive** but easily decolorized. The rods of some species are motile. Nonsporing, not acid-fast. **Aerobic.** Chemoorganotrophic, usually grow on simple media plus biotin, with an **oxidative metabolism.** Little or no acid and no gas is produced from glucose and other carbohydrates. **Catalase positive.** The optimum growth temperature is 25–30°C. Widely distributed in the environment, principally in soils.

Type species: *Arthrobacter globiformis.*

Differentiation of the species of the genus **Arthrobacter:** The differentiation of species is difficult because many are still poorly studied and comparative data are scanty.

Genus **Aureobacterium**

Irregular short rods, 0.4–0.6 × 0.6–3 μm, occur singly or in irregular groups; **many cells are arranged in V forms.** In older cultures rods become shorter but a **marked rod-coccus cycle does not occur.** Branching is uncommon; **no mycelium is produced.** Cells are **Gram positive** in young cultures. Some species are motile. Nonsporing, not acid-fast. **Aerobic. Colonies are pigmented in shades of yellow,** and the pigment is nondiffusing. Chemoorganotrophic, require nutritionally rich media. Metabolism is respiratory, yielding weak acid but no gas from glucose and other carbohydrates. The optimum temperature is 25–30°C. **Catalase positive.** Some species require siderophores such as terregens factor (found in soil extract). Found in soil and dairy products but are probably widely distributed in the environment.

Type species: *Aureobacterium liquefaciens.*

Differentiation of the species of the genus **Aureobacterium:** See Table 20.5.

Genus **Bifidobacterium**

Rods of very varied shapes, 0.5–1.3 × 1.5–8 μm, usually **somewhat curved and clubbed and are often branched.** Arranged singly, in pairs, **in V arrangements,** sometimes in chains, in palisades of parallel cells, or in rosettes. Occasionally exhibit swollen coccoid forms. **Gram positive, often stain irregularly. Nonmotile,** nonsporing, **non-acid-fast. Anaerobic.** A few species can grow in air enriched with 10% CO_2. There is no growth below pH 4.5 or above 8.5. Chemoorganotrophs, **actively ferment carbohydrates,** with the production mainly of **acetic and lactic acids in the molar ratio of 3:2; CO_2 is not produced.** Butyric and propionic acids are not produced. **Catalase negative** (rarely positive when grown in air with added CO_2). Usually require various vitamins. The optimum growth temperature is 37–41°C. Found in the mouth and intestinal tract of warm-blooded vertebrates, in insects, and in sewage; have been implicated in human infective processes but usually are considered nonpathogenic.

Type species: *Bifidobacterium bifidum.*

Differentiation of the species of the genus **Bifidobacterium:** This differentiation requires specialized techniques for strict anaerobiosis and metabolic studies.

Genus **Brachybacterium**

Editorial note: The genus *Brachybacterium* was not included in *Bergey's Manual of Systematic Bacteriology.* It was established by Collins et al. (Int. J. Syst. Bacteriol. *38:* 45–48, 1988) for the new species *B. faecium*, which is similar to *Brevibacterium* but with a distinctive pattern of lipids and menaquinones and the ability to acidify glucose in peptone media.

In young cultures, slender rods, 0.5–0.75 × 1.5–2.5 µm, **are irregular in shape.** Some cells are arranged in V formation, and these **segment into cocci in old, stationary-phase cultures. Gram positive,** nonmotile, nonsporing, not acid-fast. **Aerobic,** colonies white or pale yellow. Chemoorganotrophic, **metabolism respiratory,** yielding acid from glucose and some other carbohydrates. **Catalase positive,** oxidase negative, and may reduce nitrate. The optimum temperature is 25–30°C. Isolated from poultry litter.

Type (and only) species: *Brachybacterium faecium.*

Characteristics of the species: As described for the genus.

Genus **Brevibacterium**

Cells in young cultures are irregular rods, 0.6–1.2 × 1.5–6 µm, arranged singly or in pairs and often at an angle to give V formations. **Branching may occur, but a mycelium is not formed. In older cultures the rods

segment into small cocci. Gram positive** but easily decolorized. Nonmotile, nonsporing, not acid-fast. **Strict aerobes;** colonies may show yellow-orange or purple pigmentation. Chemoorganotrophs, metabolism respiratory. **Little or no acid is produced from glucose or other carbohydrates. Catalase positive,** gelatin and casein are usually hydrolyzed, often produce methanethiol from L-methionine. The optimum growth temperature is 20–35°C. Widely distributed in dairy products and are found on human skin.

Type species: *Brevibacterium linens.*

Differentiation of the species of the genus **Brevibacterium:** See Table 20.6.

Editorial note: Four definite species of the genus were included in *Bergey's Manual of Systematic Bacteriology,* and these are shown in Table 20.6. Additional species that have been poorly studied or are of uncertain generic position were also treated, and some are now included under the genera *Arthrobacter, Corynebacterium,* and *Microbacterium.*

Genus **Butyrivibrio**

Editorial note: The genus *Butyrivibrio* has been regarded as Gram positive or Gram negative, and for this reason it is noted in Group 20 as well as in Group 6.

Curved rods, 0.3–0.8 × 1.0–5.0 µm, are arranged singly or in chains or filaments, which may be helical. **Stain Gram negative,** but the cell wall is of the Gram-positive type. Cells are **motile by a few polar or subpolar flagella;** motility is rapid and vibratory, though often only a few cells in a culture are motile. **Strictly anaerobic.** Growth is slow below 30°C; there is no growth at 50°C (optimum 37°C). Chemoorganotrophic **metabolism fermentative;** glucose is fermented **with butyrate as a major product** and sometimes lactate. There is little growth in the absence of carbohydrates, but cellulose, starch, and other polysaccharides are often attacked. **Catalase negative,** may reduce nitrate. Occur in the rumen of ruminants and occasionally in mammalian feces; they are nonpathogenic.

Type species: *Butyrivibrio fibrisolvens.*

Differentiation of the species of the genus **Butyrivibrio:** See Table 20.7.

Genus **Caseobacter**

Editorial note: The genus *Caseobacter* was described in *Bergey's Manual of Systematic Bacteriology* with the single species *C. polymorphus.* Collins et al. (Int. J. Syst. Bacteriol. *39:* 7–9, 1989) suggested that it should be transferred to the genus *Corynebacterium,* but it is retained here to assist with determination.

In young cultures, cells are **irregular rods** that are **club-shaped or tapered,** 0.8–1.2 × 2.0–4.0 µm, and rarely branched, not forming filaments. **Cells occur singly, in pairs, in V configurations, and in palisade formations** of parallel rods. **In old cultures the cells become ovoid or coccoid. Gram-positive,** nonmotile, nonsporing, without capsules. **Strictly aerobic.** Colonies on yeast extract glucose agar are convex, are **gray or slightly red,** and have a **dry appearance.** Chemoorganotrophic, usually require organic nitrogen compounds. **Metabolism is respiratory,** giving acid from glucose. Usually halotolerant, grow with 8% NaCl. The optimum temperature is 25–30°C. **Catalase positive, oxidase negative,** and indole negative. Occur on the rind of cheeses.

Type (and only) species: *Caseobacter polymorphus.*

Characteristics of the species: As described for the genus.

Genus **Cellulomonas**

Editorial note: The genus *Cellulomonas* was included in *Bergey's Manual of Systematic Bacteriology* with eight species. One of these, *C. cellulans,* is often considered a synonym of *Oerskovia xanthineolytica* (*q.v.*), and it is possible that *Oerskovia* would best be combined with *Cellulomonas* by transferring *O. turbata* to become *Cellulomonas turbata.* However, the arrangement in the manual is retained here pending further study.

In young cultures, slender, irregular rods, 0.5–0.6 × 2.0–5.0 µm, are straight or slightly curved; some rods are in pairs at an angle to each other **giving V formations;** rods occasionally show branching, **but no mycelium is formed.** In old cultures, the rods are usually short, and a few cocci may occur. Stain **Gram positive but are easily decolorized. Often motile** by one or a few flagella. Nonsporing, non-acid-fast. **Facultative anaerobes,** but some grow very poorly anaerobically. Growth on peptone-yeast extract agar gives usually **con-**vex, **yellow colonies.** Chemoorganotrophic. The **metabolism is respiratory and also fermentative,** giving acid from glucose and various other carbohydrates, both aerobically and anaerobically. **Catalase positive** and **cellulolytic.** Nitrate is reduced to nitrite. The optimum temperature is 30°C. Widely distributed in soils and decaying vegetable matter.

Type species: *Cellulomonas flavigena.*

Differentiation of the species of the genus **Cellulomonas:** See Table 20.8.

Genus **Clavibacter**

Editorial note: The genus *Clavibacter* was not a separate genus from *Corynebacterium* in *Bergey's Manual of Systematic Bacteriology* because it was established just before publication of the manual. It was proposed for the aerobic phytopathogenic species whose cell walls contain 2,4-diaminobutyric acid instead of *meso*-diaminopimelic acid as in the other species of *Corynebacterium* (Davis et al., Int. J. Syst. Bacteriol. *34:* 107–117, 1984). The species of *Clavibacter* were distinguished in the tables in the manual and are here treated under the new genus name. Certain other phytopathogens have recently been transferred from *Corynebacterium* to *Curtobacterium* (*q.v.*).

Straight or slightly curved slender rods, 0.4–0.75 × 0.8–2.5 µm, **irregular** and often wedge- or club-shaped; **predominantly arranged singly but often are arranged in pairs in a V configuration** and sometimes in palisades. **Old cultures are not predominantly coccoid. Gram positive,** nonmotile, nonsporing, not acid-fast. **Obligate aerobes, require nutritionally rich media on which they grow slowly.** Chemoorganotrophic, **metabolism respiratory,** producing weak acidity from glucose and some other carbohydrates. The optimum temperature is 20–29°C; **seldom grow above 35°C.** Some strains produce yellow or blue pigments. **Catalase positive,** oxidase negative, nitrate not reduced, indole negative. Obligate parasites of various flowering plants, in which they are pathogenic.

Type species: *Clavibacter michiganensis.*

Differentiation of the species and subspecies of the genus **Clavibacter:** See Table 20.9.

Genus **Coriobacterium**

Editorial note: The genus *Coriobacterium* was not included in *Bergey's Manual of Systematic Bacteriology.* It was proposed by Haas and König (Int. J. Syst. Bacteriol. *38:* 382–384, 1988) for an anaerobic organism similar to *Bifidobacterium* but differing in producing hydrogen and lacking fructose-6-phosphate phosphoketolase.

Irregular rods, 0.4–1.2 × 0.5–2.0 μm, **mainly pear-shaped in long chains. Gram-positive,** nonmotile, nonsporing. **Strictly anaerobic,** yielding colonies composed of long chains in loops. Chemoorganotrophic, **metabolism fermentative,** yielding from glucose mainly acetate, L(+)-lactate, ethanol, CO_2, and H_2. They ferment a range of carbohydrates. The optimum growth temperature is 25–30°C. Occur in the intestine of red soldier bugs (*Pyrrhocoris apterus*) and are probably symbiotic.

Type (and only) species: *Coriobacterium glomerans.*

Characteristics of the species: As described for the genus.

Genus **Corynebacterium**

Editorial note: It was noted in *Bergey's Manual of Systematic Bacteriology* that the traditional genus *Corynebacterium* was heterogeneous and that there were proposals to divide it into several genera. A number of such proposals have now received general assent, and in particular the aerobic plant pathogens containing 2,4-diaminobutyric acid in the cell wall are treated under the new genus *Clavibacter* (*q.v.*).

Straight or slightly curved, slender rods have tapered or sometimes clubbed ends and are 0.3–0.8 × 1.5–8.0 μm; one species (*C. matruchotii*) has a whip-handle shape. Cells are usually arranged singly or in pairs, **often in a V formation or in palisades of several parallel cells. Gram positive,** though some cells stain unevenly, giving a beaded appearance. **Metachromatic granules** of polymetaphosphate are commonly formed within the cells. **Nonmotile,** nonsporing, **not acid-fast. Facultative anaerobes,** commonly requiring nutritionally rich media such as serum or blood media, on which colonies are usually convex and semiopaque, with a mat surface. Chemoorganotrophs with **fermentative metabolism,** most species produce acid without gas from glucose and some other carbohydrates. **Catalase positive,** often reduce nitrate and tellurite. Rarely acid-

ify lactose or raffinose or liquefy gelatin. Primarily obligate parasites of mucous membranes or skin of mammals, but occasionally they are found in other sources; some species are pathogenic for mammals.

Type species: *Corynebacterium diphtheriae.*

Differentiation of the species of the genus **Corynebacterium:** See Table 20.10.

Genus **Curtobacterium**

Small, short, irregular rods in young cultures, 0.4–0.6 × 0.6–3.0 μm, **become coccoid in old cultures.** Arranged singly or sometimes in pairs, **often in a V formation;** no branching is found. **Gram positive, but cells from old cultures are easily decolorized. Generally motile** by peritrichous flagella. Nonsporing, not acid-fast. Metachromatic granules are absent. **Obligately aerobic,** they yield smooth, convex colonies on nutrient agar; the colonies are **usually yellow or orange.** Chemoorganotrophic, not especially exacting nutritionally. **Metabolism is respiratory, yielding small amounts of acid from glucose** and some other carbohydrates. **Catalase positive.** The optimum growth temperature is 25–30°C. Occur on plants, in soil, and in oil brine; *C. flaccumfaciens* is a plant pathogen.

Type species: *Curtobacterium citreum.*

Differentiation of the species of the genus **Curtobacterium:** See Table 20.11.

Genus **Dermabacter**

Editorial note: The genus *Dermabacter* was not included in *Bergey's Manual of Systematic Bacteriology.* It was established for a new species *Dermabacter hominus* by Jones and Collins (Int. J. Syst. Bacteriol. *39:* 93–94, 1989; effective publication: FEMS Microbiol. Lett. *51:* 51–56, 1988), which is distinguished from other genera by containing *meso*-diaminopimelic acid in the cell wall, containing the menaquinone MK-9, and having a mol% G + C content of 62.

Short rods, 0.5–0.6 × 1.0–2.0 μm (D. Jones, personal communication), **Gram positive, nonmotile,** nonsporing, and not acid-fast. **Facultatively anaerobic,** yielding small, creamy white, low convex colonies on nutrient agar. Chemoorganotrophic, **metabolism fermentative,** yielding acid from glucose and some other carbohydrates. **Catalase positive,** oxidase negative. Hydrolyze casein, starch, and esculin but not hippurate. **Gelatin is liquefied.** Nitrate is not reduced. Voges-Proskauer negative. They grow at 37°C but not at 10°C, and some strains grow at 45°C. Grow with 5% NaCl. Isolated from human skin.

Type (and only) species: *Dermabacter hominus.*

Characteristics of the species: As described for the genus.

Genus **Eubacterium**

Rods are usually irregular in size, varying considerably among species (0.2–2.0 × 0.3–10 μm), and rarely form filaments. Species **vary in shape from cocci to long rods.** Cells are usually **irregular,** often with swollen or tapered ends, and are sometimes curved. They are **usually arranged singly, in pairs, or in chains. Gram positive in young cultures. Motility is variable,** nonsporing. **Strict anaerobes;** many require special anaerobic techniques for growth and require nutritionally rich media, on which colonies are usually low convex or flat. Chemoorganotrophic, **metabolism fermentative;** some attack carbohydrates. The product of metabolism from glucose or peptone is usually a **mixture of acids including** large amounts of **butyric, acetic,** or **formic** acids, with visible H₂ gas. **Catalase negative** and usually indole negative. Nitrate may be reduced, and gelatin may be liquefied. Found in cavities of animals, feces, animal and plant products, and soil; some species are opportunistic pathogens of vertebrates.

Type species: *Eubacterium limosum.*

Differentiation of the species of the genus **Eubacterium:** Special techniques are required for species differentiation, and *Bergey's Manual of Systematic Bacteriology* and the *Anaerobe Laboratory Manual* of Holdeman et al., eds (4th ed., Anaerobe Laboratory, Virginia Polytechnic Institute and State University, Blacksburg, Virginia, 1977), should be consulted. *Eubacterium* is a genus of convenience, and poorly sporing strains of *Clostridium* are readily confused with species of *Eubacterium.*

Genus **Exiguobacterium**

Editorial note: The genus *Exiguobacterium* was not included in *Bergey's Manual of Systematic Bacteriology.* It was established by Collins et al. (Int. J. Syst. Bacteriol. *34:* 91–92, 1984; effective publication: J. Gen. Microbiol. *129:* 2037–2042, 1983) for an alkalophilic bacterium, *E. aurantiacum,* that had a peptidoglycan of type lysine-glycine and contained mainly MK-7 menaquinones.

Rods in young cultures become almost coccoid in old cultures (1.1–1.2 × 1.4–3.2 μm). **Gram positive,** non-acid-fast, nonsporing. **Motile** by **peritrichous flagella.** **Facultatively anaerobic,** forming flat, **pale orange** colonies on nutrient agar; the pigment does not diffuse. **Alkalophilic,** cells grow at pH 6.5–11.5. Chemoorganotrophic, **metabolism fermentative,** giving acid from glucose, sucrose, galactose, and some other sugars; the main products are **lactate, acetate, and formate. Catalase positive and oxidase negative. Nitrate is reduced.** Starch, casein, and gelatin are hydrolyzed. The optimum growth temperature is 37°C. Isolated from potato-processing effluent.

Type (and only) species: *Exiguobacterium aurantiacum.*

Characteristics of the species: As described for the genus.

Genus **Falcivibrio**

Editorial note: The genus *Falcivibrio* was not included in *Bergey's Manual of Systematic Bacteriology.* It was proposed by Hammann et al. (Int. J. Syst. Bacteriol. *34:* 355–357, 1984; effective publication: Syst. Appl. Microbiol. *5:* 81–96, 1984) and contains two species, *F. grandis* and *F. vaginalis.* This genus may be a later synonym of *Mobiluncus* (q.v.).

Slender, curved rods, are variable in size and shape (0.4–0.6 × 1.2–6.0 μm). **Arranged singly or in pairs; pairs show a sickle configuration.** The cells may show clubbed or tapering ends. **Gram positive in young cultures;** older cultures stain Gram negative with Gram-positive granules. **Motile** by polar or subpolar flagella or by lateral flagella from the concave side of the cell. Nonsporing. **Anaerobic,** these cells may become tolerant to 5% O₂ on culture. On blood agar, colonies are small, white, 1 mm in diameter, and sometimes show weak greening or weak β-hemolysis. Chemoorganotrophic, requiring nutritionally rich media; **growth is**

stimulated by fumarate, which is reduced to succinate. Metabolism is fermentative and may produce acid from glucose and some other carbohydrates. Catalase negative, oxidase negative, may reduce nitrate, indole negative. Optimum growth temperature is 37°C. Isolated from the human vagina.

Type species: *Falcivibrio grandis.*

Differentiation of the species of the genus **Falcivibrio:** See Table 20.12.

Genus **Gardnerella**

Editorial note: The genus *Gardnerella* has been regarded as Gram positive or Gram negative, and for this reason it is noted in Group 20 as well as in Group 5.

Pleomorphic rods about 0.5 μm in diameter and 1.5–2.5 μm in length form no capsules or endospores. Gram negative to Gram variable. Nonmotile. Facultatively anaerobic. Fastidious in growth requirements but do not need X factor or V factor. Catalase and oxidase negative. Chemoorganotrophic, having a fermentative type of metabolism. The optimum growth temperature is 35–37°C. Acid but no gas is produced from glucose and some other carbohydrates. Do not reduce nitrates. Hippurate is hydrolyzed. Human blood but not sheep blood is hemolyzed. Found in the human genital/urinary tract and are considered to be a major cause of bacterial "nonspecific" vaginitis.

Type (and only) species: *Gardnerella vaginalis.*

Differentiation of the genus **Gardnerella** *from other genera:* Characteristics that differentiate *Gardnerella* from other genera in Group 20 are given in Table 20.2 and from genera in Subgroup 4 of Group 5 are given in Table 5.1. Differentiation of *Gardnerella* from other genera is given in Table 5.60.

Characteristics of **Gardnerella vaginalis:** Little or no growth on nutrient agar or on most common selective media. On Vaginalis agar, colonies are 0.5 mm in diameter after 48 hours of incubation in a candle extinction jar or in a CO_2 incubator. Special procedures are used to test for acid production from carbohydrates, oxidase, catalase, hemolysis, and hippurate hydrolysis (Greenwood and Pickett, *Bergey's Manual of Systematic Bacteriology,* Volume 1, pp. 587–591, 1984). Negative

reactions in tests for arginine dihydrolase, esculin hydrolysis, gelatin hydrolysis, H_2S, indole, lysine decarboxylase, phenylalanine deaminase, ornithine decarboxylase, Tween 80 hydrolysis, urease, and acetoin (Voges-Proskauer). No acid production from arbutin, cellobiose, glycerol, inositol, mannitol, melibiose, raffinose, rhamnose, and salicin. Variable reactions in tests for lipase, o-nitrophenyl-β-D-galactopyranoside, and acid production from L-arabinose, fructose, galactose, inulin, lactose, mannose, sucrose, and xylose. Positive reactions in tests for H_2O_2 inhibition, methyl red, starch hydrolysis, and acid production from dextrin, maltose, ribose, and starch.

Genus **Jonesia**

Editorial note: The genus *Jonesia* was not included in *Bergey's Manual of Systematic Bacteriology.* It was created by Rocourt et al. (Int. J. Syst. Bacteriol. *37:* 266–270, 1987) for the organism previously known as *Listeria denitrificans.*

Irregular, slender rods, 0.3–0.5 × 2–3 μm, mainly arranged singly; branched Y and club-like forms occur. Filaments and coccoid cells may be present in old cultures. Gram positive; many cells in old cultures are readily decolorized, but coccoid forms always stain Gram positive. Motile by peritrichous flagella at both 25°C and 37°C. Nonsporing, not acid-fast. Facultatively anaerobic, giving on nutrient agar small, convex, smooth colonies that are grayish and translucent to opaque; they become yellowish in 10–20 days. Colonies do not appear blue or blue-green by oblique transmitted light. Growth occurs at a temperature of 10–40°C (optimum 30°C); grow with 5% but not 10% NaCl. Chemoorganotrophic, metabolism fermentative, producing acid but not gas from glucose and a number of other carbohydrates, including cellobiose, starch, and salicin but not ribose, rhamnose, mannitol, or raffinose. Methyl red positive, Voges-Proskauer-negative, Catalase positive, oxidase negative, and nitrate reduced. Hydrolyze cellulose and starch but not gelatin or casein. Hippurate hydrolysis is weak or negative. Isolated from cooked ox blood, but their natural habitat is not known.

Type (and only) species: *Jonesia denitrificans.*

Characteristics of the species: As described for the genus.

Genus **Lachnospira**

Editorial note: The genus *Lachnospira* has been regarded as Gram positive or Gram negative, and for this reason it is noted in Group 20 as well as in Group 6.

Curved rods with bluntly pointed ends, 0.4–0.6 × 2.0–4.0 μm, cells may be helical with one or more turns. Cells are arranged singly, in pairs, or in chains or filaments; cells in chains or filaments may be only slightly curved. **Weakly Gram positive;** the Gram stain may be negative except in very young cultures. **Motile by a single lateral or subpolar flagellum.** Nonsporing. **Strictly anaerobic. Colonies filamentous and woolly in appearance.** Chemoorganotrophic, **metabolism fermentative;** the products of glucose fermentation are **formate, lactate, acetate, ethanol, CO_2, and H_2. Pectin is fermented,** giving the same products plus **methanol.** Succinate, butyrate, and propionate are not produced. Ferment a range of carbohydrates but not amino acids. The optimum growth temperature is 40°C. Catalase negative, nitrate not reduced, indole negative, and gelatin not liquefied. Found in the rumen of bovines, where it is important in the digestion of pectin.

Type (and only) species: *Lachnospira multiparus.*

Characteristics of the species: As described for the genus.

Genus **Microbacterium**

Slender, irregular rods in young cultures, 0.4–0.8 × 1.0–4.0 μm, arranged singly or **in pairs, when some are arranged at an angle to give V formations.** Primary branching is uncommon, and mycelia are not produced. In old cultures, rods are shorter and cocci may occur, but **there is no marked rod-coccus cycle. Gram positive,** non-acid-fast, nonsporing. Nonmotile or motile by one to three flagella. **Aerobic;** weak anaerobic growth may occur. On yeast extract-peptone-glucose agar, colonies are opaque and glistening, often with yellowish pigmentation. Chemoorganotrophic, **metabolism primarily respiratory** but may be weakly fermentative. Acid is produced from glucose and some other carbohydrates. Nutritional requirements are complex. **Catalase positive.** The optimum growth temperature is 30°C. Found in dairy products, sewage, and insects.

Type species: *Microbacterium lacticum.*

Differentiation of the species of the genus **Microbacterium:** See Table 20.13.

Genus **Mobiluncus**

Editorial note: The genus *Mobiluncus* was not included in *Bergey's Manual of Systematic Bacteriology.* It was established by Spiegel and Roberts (Int. J. Syst. Bacteriol. *34:* 177–184, 1984) and contains two species, *M. curtisii* (with two subspecies) and *M. mulieris.* The genus may be a synonym of *Falcivibrio (q.v.).* If so, *Mobiluncus* has priority, and it seems probable that *F. grandis* is a later synonym of *M. mulieris* and that *F. vaginalis* is a later synonym of *M. curtisii,* but comparative studies are needed to clarify the position.

Slender, curved rods, 0.4–0.6 × 1.2–4.0 μm, have tapered ends and are variable in shape and size. Arranged singly and **sometimes in pairs with a gullwing appearance. Gram variable or Gram negative,** but the cell wall is of the Gram-positive type. **Motile by multiple lateral or subpolar flagella.** Nonsporing. **Anaerobic,** give colorless, translucent, smooth, convex colonies. Chemoorganotrophic, **metabolism fermentative,** fermenting glucose, often weakly, to yield acetate, succinate, and sometimes lactate. **Growth is not stimulated by a mixture of formate and fumarate.** The optimum growth temperature is 37°C. **Catalase negative,** oxidase negative, and indole negative. Nitrate may be reduced. Isolated from the human vagina and may be implicated in vaginitis.

Type species: *Mobiluncus curtisii.*

Differentiation of the species and subspecies of the genus **Mobiluncus:** See Table 20.14.

Genus **Pimelobacter**

Editorial note: The genus *Pimelobacter* was not described in *Bergey's Manual of Systematic Bacteriology* as a genus distinct from *Arthrobacter.* It was proposed by Suzuki and Komagata (Int. J. Syst. Bacteriol. *33:* 673–675, 1983; effective publication: J. Gen. Appl. Bacteriol. *29:* 59–71, 1983) to accommodate arthrobacters containing LL-diaminopimelic acid in their cell walls. They transferred two species of *Arthrobacter,* *A. simplex* and *A. tumescens,* to which they added a new species, *Pimelobacter jensenii.* O'Donnell et al. (Int. J. Syst. Bacteriol. *33:* 896–897, 1983; effective publication: Arch. Microbiol. *133:* 323–329, 1982) suggested moving *Arthrobacter simplex* to the genus *Nocardioides,* thus making *Pimelobacter* redundant. Collins et al. (Int. J. Syst. Bacteriol. *39:* 1–6, 1989) suggested also moving *P. jensenii* to *Nocardioides.* However, *Nocardioides* has a very different morphology, and *Pimelobacter* is treated here as by Suzuki and Komagata to assist in determination of these organisms. However, a new genus, *Terrabacter,* has also been proposed for *A. tumescens,* so this organism is here included in both *Pimelobacter* and *Terrabacter.*

Cells in young cultures are irregular rods, 0.5–1.2 × 1.0–7.0 μm, **mostly arranged singly and often showing branching. Old cultures** contain **mainly coccoid cells.** This **marked rod-coccus cycle** is characteristic of *Pimelobacter* and *Arthrobacter.* **Gram positive;** old cultures may be easily decolorized. Cells may be motile by peritrichous flagella. Nonsporing, non-acid-fast. Aerobic cells yield smooth white or yellowish white colonies. Chemoorganotrophic and **nutritionally nonexacting,** except for thiamine in *P. tumescens.* **Metabolism is respiratory. Acid is not produced from carbohydrates in peptone-based media,** although glucose is assimilated as the sole carbon source. The optimum growth temperature is 25–30°C. **Catalase positive.** Gelatin is usually hydrolyzed, and nitrates may be reduced. Widely distributed in soil.

Type species: *Pimelobacter simplex.*

Differentiation of the species of the genus **Pimelobacter:** See Table 20.15.

Genus **Propionibacterium**

Pleomorphic rods, 0.5–0.8 × 1–5 μm, **are often club-shaped with one end rounded and the other tapered;** some cells may be **coccoid, bifid, or branched,** but they are not filamentous. Cells occur singly, in pairs or short chains, in V or Y configurations, or in clumps with a "Chinese character" arrangement. **Gram positive,** nonmotile, and nonsporing. **Facultative anaerobes but have variable aerotolerance;** most grow somewhat in air but better anaerobically, giving on blood agar colonies that are usually convex, semiopaque, glistening, and often pigmented in shades of cream to reddish. Chemoorganotrophic with complex nutritional requirements, have a **metabolism fermentative,** producing from glucose and some other carbohydrates **large amounts of propionic and acetic acids and**

often small amounts of gas. The optimum growth temperature is 30–37°C. **Usually catalase positive.** They are found mainly in cheese and dairy products and on human skin. Readily confused with some species of *Corynebacterium* or *Clostridium.*

Type species: *Propionibacterium freudenreichii.*

Differentiation of the species of the genus **Propionibacterium:** See Table 20.16.

Genus **Rarobacter**

Editorial note: The genus *Rarobacter* was not included in *Bergey's Manual of Systematic Bacteriology.* It was established by Yamamoto et al. (Int. J. Syst. Bacteriol. *38:* 7–11, 1988) for a bacterium that lyses cells of yeasts.

Small, irregular rods, 0.2–0.3 × 0.8–1.0 μm, occur mainly singly **but show V arrangements.** Rods show no branching. **Gram positive in young cultures** and Gram variable in old cultures. Motile by multitrichous flagella. Nonsporing, non acid-fast. **Facultative anaerobes** but require catalase, hemoglobin, or hemin for growth in air; they require carbon dioxide for anaerobic growth. Chemoorganotrophic, they require thiamine and biotin, yielding **pale yellow, opaque, convex, smooth colonies. Metabolism fermentative,** producing acid without gas from glucose and some other carbohydrates. **No assimilation was observed with a wide range of organic acids.** Reported as catalase positive, **oxidase positive,** and indole negative; nitrate is not reduced. Starch, casein, and gelatin are hydrolyzed, but not cellulose. The optimum growth temperature is 30°C. Isolated from wastewaters of alcoholic beverage factories; adhere to and lyse living yeast cells (*Saccharomyces, Hansenula*).

Type (and only) species: *Rarobacter faecitabidus.*

Characteristics of the species: As described for the genus.

Genus **Rothia**

Irregular rods, 0.8–1.2 × 1.0–5.0 μm, **have irregular swellings and clubbed ends** up to 5.0 μm in diameter; usually, a **mixture of cocci, rods, and filaments** is present. **Gram-positive,** non-acid-fast, nonmotile, nonsporing cells are without visible capsules. **Facultatively anaerobic,** yield convex or convoluted, glistening colonies aerobically and frequently spider-like, filamentous colonies anaerobically. Chemoorganotrophic, requiring nutritionally rich media. **Metabolism fermentative,** producing **mainly lactic** and some acetic acid from glucose but **no propionic acid.** Ferments a range of carbohydrates. **Catalase positive,** indole negative, and nitrate is reduced. The optimum growth temperature is 35–37°C. Normally inhabit the human mouth and throat; may become an opportunistic pathogen.

Type (and only) species: *Rothia dentocariosa.*

Characteristics of the species: As described for the genus.

Genus **Rubrobacter**

Editorial note: The genus *Rubrobacter* was not included in *Bergey's Manual of Systematic Bacteriology.* It was established by Suzuki et al. (Int. J. Syst. Bacteriol. *39:* 93–94, 1989; effective publication: FEMS Microbiol. Lett. *52:* 33–40, 1988) for a highly radiotolerant organism previously known as *Arthrobacter radiotolerans,* because of a characteristic cellular fatty acid composition with 12-methyl-hexadecanoic acid as the major component. This fatty acid distinguishes *Rubrobacter* from the genera *Deinococcus* and *Deinobacter,* which are also highly resistant to γ-radiation.

Irregular rods, 0.8–1.0 × 1.0–4.0 μm, are shorter in old cultures; there is **no marked rod-coccus cycle.** No branching or aerial mycelium is observed. **Gram-positive,** nonmotile, nonsporing, non-acid-fast. **Aerobic,** give smooth, opaque, **reddish pink colonies** on nutrient agar, **developing slowly** (1 mm in diameter after two weeks). Chemoorganotrophic, **with simple nutritional requirements** and a **respiratory metabolism.** Utilize glucose and a few other carbohydrates for growth but **produce no acid on the usual media. Catalase positive** and indole negative; nitrate is reduced, and gelatin is not liquefied. **Ammonium sulfate is utilized as the sole nitrogen source. The optimum growth temperature is 46–48°C.** Isolated from a radioactive hot spring in Japan.

Type (and only) species: *Rubrobacter radiotolerans.*

Characteristics of the species: As described for the genus.

Genus **Sphaerobacter**

Editorial note: The genus *Sphaerobacter* was not included in *Bergey's Manual of Systematic Bacteriology.* It was set up by Demharter et al. (Int. J. Syst. Bacteriol. *39:* 495–497, 1989; effective publication: Syst. Appl. Microbiol. *11:* 261–266, 1989) for a new, moderately thermophilic species of the actinomycete subdivision of Gram-positive bacteria.

Irregular rods, 0.4–1.2 × 1.5–5 μm, are variable in shape, **with ovoid, swollen, and club-shaped forms. Arranged singly or in pairs,** do not branch, and sometimes show V configurations. **Gram-positive,** nonmotile, nonsporing cells. **Aerobic, optimum temperature is 55°C.** On nutrient agar these cells form circular, opaque colonies. Chemoorganotrophic, with a respiratory metabolism. No acid is produced from glucose. **Catalase positive and oxidase positive;** starch is hydrolyzed but not gelatin, casein, or cellulose. Isolated from heated sewage sludge.

Type (and only) species: *Sphaerobacter thermophilus.*

Characteristics of the species: As described for the genus.

Genus **Terrabacter**

Editorial note: The genus *Terrabacter* was not included in *Bergey's Manual of Systematic Bacteriology.* It was established by Collins et al. (Int. J. Syst. Bacteriol. *39:* 1–6, 1989) for the organism previously known as *Arthrobacter tumescens* or *Pimelobacter tumescens* on the basis of ribosomal ribonucleic acid sequences and fatty acid and polar lipid composition.

In young cultures, cells are long, irregular rods, 0.6–1.2 × 2.0–6.0 μm (sometimes much longer), which then show branching. **Mainly coccoid in old cultures; there is a pronounced rod-coccus cycle of growth.** No aerial mycelium. **Gram positive** in both young and old cultures. Usually nonmotile, but motile strains occur. Nonsporing, non acid-fast. **Aerobic,** give glossy, gray or white colonies. Chemoorganotrophic with **simple nutritional requirements but require thiamine** when growth occurs on ammonium salts or nitrate together with glucose in mineral salts medium. **Metabolism is respiratory,** no acid is produced from glucose or other carbohydrates in peptone media. Can utilize a wide range of organic compounds for growth, including glucose, mannitol, acetate, and succinate but not rhamnose or citrate. **Catalase positive** and oxidase negative. Nitrate is reduced, and gelatin is hydrolyzed, but not cellulose. The optimum growth temperature is 25–30°C. Found in soil.

Type (and only) species: *Terrabacter tumescens.*

Characteristics of the species: As described for the genus.

Genus **Thermoanaerobacter**

In young cultures, rods, 0.5–0.8 × 4–8 μm, are arranged singly, in pairs, or in chains, **becoming pleomorphic in old cultures** and showing **chains of rods interspersed with coccoid cells; the cocci may be large** (0.8–1.5 μm in diameter). Long filaments may occur. **Gram positive only in very young cultures,** but the cell wall is of the Gram-positive type. **Motile by peritrichous flagella,** nonsporing. **Strictly anaerobic,** and **thermophilic,** growing at a temperature of 35–78°C (optimum 68°C). Colonies are smooth and white. Chemoorganotrophic, and **metabolism is fermentative; main products from carbohydrates are ethanol and CO$_2$.** Nutritionally exacting, require yeast extract and ferment a range of carbohydrates. **Catalase negative,** nitrates are not reduced, indole negative, and gelatin is not hydrolyzed. Starch is hydrolyzed but not cellulose. Isolated from geothermal springs. They resemble in many ways the poorly studied organism *Thermoanaerobium* (*q.v.*).

Type (and only) species: *Thermoanaerobacter ethanolicus.*

Characteristics of the species: As described for the genus.

Genus **Thermoanaerobium**

Editorial note: The genus *Thermoanaerobium* was not described in *Bergey's Manual of Systematic Bacteriology* because of restricted information. It is briefly described here, but further studies are clearly needed. The genus was described by Zeikus et al. (Int. J. Syst. Bacteriol. *33:* 672–674, 1983; effective publication: Arch. Microbiol. *122:* 41–48, 1979). It is very similar to *Thermoanaerobacter* (*q.v.*).

Irregular rods, 0.8–1.0 × 2–20 μm, are arranged singly, in pairs, and in short chains in young cultures, but **older cultures show long chains of rods interspersed with short rods or cocci. Gram positive, nonmotile,** nonsporing. **Anaerobic, thermophilic,** grow at a temperature of 45–75°C (optimum 70°C), yielding round, mucoid, flat, nonpigmented colonies. Chemoorganotrophic, with a **fermentative metabolism,** producing from glucose **lactate, ethanol, acetate, H$_2$, and CO$_2$,** but not formate, succinate, or butanediol. Nutritionally exacting, require yeast extract and fermentable carbohydrate; ferment a range of carbohydrates. **Catalase negative.** Isolated from geothermal springs.

Type (and only) species: *Thermoanaerobium brockii.*

Characteristics of the species: As described for the genus.

Important Notes for Users of This Manual

Unless otherwise indicated in footnotes to tables, the meanings of symbols are as follows:

- \+ 90% or more of strains are positive
- − 90% or more of strains are negative
- d 11-89% of strains are positive
- v strain instability (not equivalent to "d")
- D Different reactions in different taxa (species of a genus or genera of a family)

All other symbols are defined in footnotes to tables.

Table 20.1 Differential characteristics of aerobic, irregular, nonsporing Gram-positive rods [a]

Characteristics	Aeromicrobium	Arthrobacter	Aureobacterium	Brachybacterium	Brevibacterium	Caseobacter	Clavibacter
Cell morphology	Irregular short rods	Marked rod-coccus cycle	Irregular rods, many arranged in V-forms	Slender irregular rods with rod-coccus cycle	Rod-coccus cycle	Irregular rods, some coccoid forms	Irregular rods
Gram stain	+	+[b]	+	+	+	+	+
Motility	−	D	D	−	−	−	−
Catalase	+	+	+	+	+	+	+
Peptidoglycan:							
Group [c]	ND	A	B	ND	A	A	B
Diamino acid	LL-DAP	Lys	D-Orn	meso-DAP	meso-DAP	meso-DAP	DAB
N-Glycolyl residue	ND	−	d	ND	−	−	−
Wall arabino-galactan polymer	ND	−	−	ND	−	+	−
Mycolic acids	ND	−	−	−	−	+	−
Major fatty acid types [d]	T,U,S,	S,A,I	S,A,I	S,I,A	S,A,I	S,U,T	S,A,I
Major menaquinones [e]	ND	MK-9(H₂), MK-9, MK-8	MK-11, MK-12	MK-7	MK-8(H₂), MK-7(H₂)	MK-9(H₂), MK-8(H₂)	MK-10, MK-9
Habitat and pathogenicity	Soil	Soil	Dairy, sewage, soil, and insect sources	Poultry litter	Cheese, skin	Cheese	Plant material; pathogenic for various plants

Footnotes are at end of table

583

Table 20.1 (continued)

Characteristics	Curtobacterium	Microbacterium	Pimelobacter	Rubrobacter	Sphaerobacter	Terrabacter
Cell morphology	Irregular rods	Irregular rods, some coccoid forms	Marked rod-coccus cycle	Irregular rods	Irregular rods, often clubbed and in V formation	Marked rod-coccus cycle
Gram stain	+[b]	+	+[b]	+	+	+[b]
Motility	D	D	+	−	−	−
Catalase	+	+	+	+	+	+
Peptidoglycan:						
Group[c]	B	B	ND	ND	ND	ND
Diamino acid	D-Orn	Lys	LL-DAP	Lys	L-Orn	LL-DAP
N-Glycolyl residue	−	+	ND	ND	ND	ND
Wall arabino-galactan polymer	−	−	ND	ND	ND	ND
Mycolic acids	−	−	−	−	ND	−
Major fatty acid types[d]	S,A,I(H)	S,A,I	S,A,I	12-H,A,2-OH-FA	ND	I,S,A
Major menaquinones[e]	MK-9	MK-12, MK-11, MK-10	MK-8(H$_4$)	MK-8	MK-8	MK-8(H$_4$)
Habitat and pathogenicity	Plants, soil, oil brines; C. flaccumfaciens is a plant pathogen	Dairy, sewage, and insect sources	Soil	Geothermal springs	Sewage sludge	Soil

[a] Symbols: +, 90% or more of strains are positive; −, 90% or more of strains are negative; d, 11–89% of strains are positive; D, substantial proportion of species differ; ND, not determined.

[b] Often readily decolorized.

[c] Designation as Schleifer and Kandler (Bacteriol. Rev. 36: 407–477, 1972).

[d] S, straight-chain saturated; U, monounsaturated; A, anteiso-methyl-branched; I, iso-methyl branched; T, 10-methyl branched (tuberculostearic acid); A, iso-methyl branched; sometimes present; 12-H, 12-methyl-hexadecanoic acid; 2-OH-FA, 2-hydroxylated fatty acids.

[e] Symbolism of Collins and Jones (Microbiol. Rev. 45: 316–354, 1981).

Table 20.2 Differential characteristics of the facultatively anaerobic, irregular, nonsporing Gram-positive rods[a]

Characteristics	Actinomyces	Agromyces	Arachnia	Arcanobacterium	Cellulomonas[b]	Corynebacterium	Dermabacter
Cell morphology	Rods and filaments with some branching	Branched, filamentous elements	Rods, filaments with some branching	Short irregular rods; coccoid forms may predominate	Irregular rods, with some coccoid forms	Rods, clubbed forms[d]	Short rods
Gram stain	+	+	+	+	+[e]	+	+
Motility	–	–	–	–	D	–	–
Catalase	–	–	–	–	+	+	+
Peptidoglycan:							
Group[g]	A	B	A	A	A	A	ND
Diamino acid	Lys, Orn	DAB	LL-DAP	Lys	L-Orn	meso-DAP	meso-DAP
N-Glycolyl residue	–	–	ND	–	ND	–	ND
Wall arabino-galactan polymer	–	–	–	–	–	+	ND
Mycolic acids	–	–	–	–	–	+	–
Major fatty acid types[h]	S,U,(C)	S,A,I	A,(S),(I),(U)	S,U	S,A,I	S,U,(T)	I,A
Major menaquinones[i]	MK-10(H4)	MK-11, MK-12	MK-9(H4)	MK-9(H4)	MK-9(H4)	MK-9(H2), MK-8(H2)	MK-9, MK-8, MK-7
Habitat and pathogenicity	Mammalian oral and other cavities; many are pathogens	Soil	Human oral cavity; can be human pathogen	Human and animal sources; pathogenic	Soil and rotting vegetation	Humans, animals; some species are pathogenic	Human skin

Footnotes are at end of table

585

Table 20.2 *(continued)*

Characteristics	Exiguobacterium	Gardnerella	Jonesia	Propionibacterium[c]	Rarobacter	Rothia
Cell morphology	Rods, with cocci in old cultures	Irregular small rods	Irregular slender rods with some clubbed and Y forms	Irregular rods, branched forms, and cocci	Small irregular rods with some V forms	Irregular rods, filamentous and coccoid forms
Gram stain	+	−[f]	+	+	+	+
Motility	+	−	+	−	+	−
Catalase	+	−	+	+	+	+
Peptidoglycan: Group[g]	ND	A	ND	A	A	A
Diamino acid	Lys	Lys	Lys	LL-DAP, *meso*-DAP	L-Orn	Lys
N-Glycolyl residue	ND	ND	ND	ND	ND	ND
Wall arabino-galactan polymer	−	−	−	−	ND	−
Mycolic acids	−	−	−	−	ND	−
Major fatty acid types[h]	ND	S,U	S,A,I	A,I,(S)	A	S,A,I
Major menaquinones[i]	MK-7	−	MK-9	MK-9(H₄)	MK-9	MK-7
Habitat and pathogenicity	Potato-processing effluent	Human genital and urinary tract; may cause vaginitis	Unknown	Dairy products, human skin; some are pathogenic for humans	Alcoholic beverage factories	Human oral cavity; opportunistic pathogen

[a] Symbols: +, 90% or more of strains are positive; −, 90% or more of strains are negative; D, substantial proportion of species differ; ND, not determined.

[b] Very similar except for morphology to *Oerskovia* (q.v.).

[c] May be confused with *Clostridium* if spores not observed.

[d] *Corynebacterium matruchotii* has a characteristic whip-handle morphology.

[e] Easily decolorized.

[f] Gram reaction variable.

[g] Designation as Schleifer and Kandler (Bacteriol. Rev. *36:* 407–477, 1972).

[h] S, straight-chain unsaturated; U, monounsaturated; A, *anteiso*-methyl branched; I, *iso*-methyl branched; C, cyclopropane ring; T, 10-methyl branched (tuberculostearic acid); H, cyclohexyl fatty acids; parentheses indicate sometimes present or in small amounts.

[i] Symbolism of Collins and Jones (Microbiol. Rev. *45:* 316–354, 1981).

Table 20.3 Differential characteristics of anaerobic, irregular, nonsporing Gram-positive rods[a]

Characteristics	Acetobacterium	Acetogenium	Bifidobacterium[b]	Butyrivibrio	Coriobacterium	Eubacterium[b]
Cell morphology	Oval shaped short rods	Rods	Very irregular rods with branching	Curved rods, may be helical	Irregular pear-shaped rods in chains	Irregular rods often in pairs or chains
Gram stain	+	–	+	–	+	+
Motility	+	–	–	+	–	D
Catalase	–	–	–	–	ND	–
Thermophilic, poor growth below 40°C	–	+	–	–	–	–
Peptidoglycan:						
Group[e]	B[f]	ND	A	ND	ND	D
Diamino acid	Orn	ND	Lys, Orn	ND	Lys	D
N-Glycolyl residues	ND	ND	ND	ND	ND	ND
Wall arabino-galactan polymer	–	ND	–	–	ND	–
Mycolic acids	–	ND	–	–	ND	–
Major fatty acid types[g]	ND	ND	S,U	various (S,U,A,I,DCFA)	ND	ND
Major menaquinones[h]	ND	ND	–	ND	ND	ND
Habitat and pathogenicity	Anaerobic fresh water and marine sediments, and sewage	Tropical lake mud	Intestines of humans and animals; sewage; pathogenicity doubtful	Rumen	Intestine of insects	Intestinal tract of humans and animals, plants, soil; some are pathogenic for humans

Footnotes are at end of table

587

Table 20.3 (continued)

Characteristics	Falcivibrio	Lachnospira	Mobiluncus	Thermoanaerobacter	Thermoanaerobium
Cell morphology	Slender curved rods, often in pairs	Curved rods or filaments	Slender curved rods, often in pairs	Rods, irregular in older cultures, some filamentous or coccoid forms	Irregular rods; in old cultures chains of rods interspersed with cocci
Gram stain	+	−[c]	−[c]	+[d]	+
Motility	+	+	+	+	−
Catalase	−	−	−	−	−
Thermophilic, poor growth below 40°C	−	−	−	+	+
Peptidoglycan:					
Group [e]	ND	ND	ND	ND	ND
Diamino acid	ND	ND	ND	meso-DAP	ND
N-Glycolyl residues	ND	ND	ND	ND	ND
Wall arabino-galactan polymer	ND	−	ND	−	ND
Mycolic acids	ND	−	ND	−	ND
Major fatty acid types [g]	ND	S,U,3-OH-FA	ND	ND	ND
Major menaquinones [h]	ND	ND	ND	ND	ND
Habitat and pathogenicity	Human vagina	Rumen	Human vagina	Hot springs	Hot springs

[a] Symbols: +, 90% or more of strains are positive; −, 90% or more of strains are negative; D, substantial proportion of species differ; ND, not determined.

[b] May be confused with Clostridium if spores not observed.

[c] Sometimes weakly Gram positive.

[d] Easily decolorized.

[e] Designation as Schleifer and Kandler (Bacteriol. Rev. 36: 407–477, 1972).

[f] This Group B type of peptidoglycan differs from that of the Group B genera in Tables 20.1 and 20.2 in having a seryl residue in Position 1 of the peptide subunit.

[g] S, straight-chain saturated; U, monounsaturated; A, anteiso-methyl branched; I, iso-methyl branched; 3-OH-FA, 3-hydroxylated long chain fatty acids; DCFA, dicarboxylic fatty acids with vicinal dimethyl branching; parentheses indicate sometimes present or in small amounts.

[h] Symbolism of Collins and Jones (Microbiol. Rev. 45: 316–354, 1981).

Table 20.4 Characteristics differentiating the species of *Acetobacterium*[a]

Characteristics	A. carbinolicum [b]	A. wieringae	A. woodii
Fermentation of:			
Glucose	d	–	+
Glycerate	ND	–	+
Glycerol	+	+w	–
Fructose	d	+	+
Ethanol	+	–	–

[a] Symbols: +, 90% or more of strains are positive; –, 90% or more of strains are negative; d, 11–89% of strains are positive; w, weak reaction; ND, not determined.

[b] Eichler and Schink (Int. J. Syst. Bacteriol. *35:* 375–376, 1985; effective publication: Arch. Microbiol. *140:* 147–157, 1984).

Table 20.5 Characteristics differentiating the species of *Aureobacterium*[a]

Characteristics	A. barkeri	A. flavescens	A. liquefaciens	A. saperdae	A. terregens	A. testaceum
Pigment of colonies [b]	Yellow	Yellow	Bright yellow	Yellow	Yellowish brown	Yellow to orange-red
Diameter of colonies (mm)	1–3	0.2–0.5	1–3	1–3	1–2	1–3
Motility	+	–	–	+	–	+
Terregens factor needed	ND	+	–	ND	+	ND
Gelatin hydrolysis	ND	+	+	–	–	+
Tellurite (0.05% w/v) reduced	+	ND	–	–	ND	+

[a] Symbols: +, 90% or more of strains are positive; –, 90% or more of strains are negative; ND, not determined.

[b] On nutrient agar supplemented with soil extract.

Table 20.6 Characteristics differentiating the species of *Brevibacterium*[a]

Characteristics	B. casei [b]	B. epidermidis [b]	B. iodinum	B. linens
Color of colonies	Grey	Grey	Grey [c]	Yellow-orange [d]
Pink-red coloration when growth treated with 5 M KOH	–	–	–	+
Crystals of iodinin formed	–	–	+	–
Oxidase	–	–	+	–
Survival at 60°C	+	+	–	–
Major menaquinones	MK-8(H₂), MK-7 (H₂)	MK-8(H₂)	MK-8(H₂)	MK-8(H₂)

[a] Symbols: +, 90% or more of strains are positive; –, 90% or more of strains are negative.

[b] Distinguishable also by DNA-DNA pairing and mol% G + C ratios.

[c] May show crystals of iodinin, giving a shimmering purple coloration.

[d] When incubated in light.

Table 20.7 Characteristics differentiating the species of *Butyrivibrio*[a]

Characteristics	B. crossotus	B. fibrisolvens
Single polar flagellum	–	+
Several polar or subpolar flagella	+	–
Produce H_2 from glucose or maltose	–	+
Fermentation of glucose	+w	+
Fermentation of sucrose	–	+

[a] Symbols: +, 90% or more of strains are positive; –, 90% or more of strains are negative; w, weak reaction.

Table 20.8 Characteristics differentiating the species of *Cellulomonas*[a]

Characteristics	C. biazotea	C. cellasea	C. cellulans[b]	C. fimi	C. flavigena	C. gelida	C. uda
Motility	+	–	ND	+	–	+	–
Utilization as sole carbon source:							
D-Ribose	–	–	+	–	+	–	–
Raffinose	+	–	–	–	–	–	–
L(+)-Lactate	+[c]	+	+	+[c]	–	–	–
Proline	–	+	+	–	–	–	–
Acid from dextrin	–	–	ND	+	+	+	+

[a] Symbols: +, 90% or more of strains are positive; –, 90% or more of strains are negative; ND, not determined.

[b] *C. cellulans* has been considered synonymous with a number of other little studied cellulose-digesting bacteria (see *Bergey's Manual of Systematic Bacteriology*, Volume 2, p. 1329).

[c] Conflicting reports in the literature.

Important Notes for Users of This Manual

Unless otherwise indicated in footnotes to tables, the meanings of symbols are as follows:

+ 90% or more of strains are positive

– 90% or more of strains are negative

d 11-89% of strains are positive

v strain instability (not equivalent to "d")

D Different reactions in different taxa (species of a genus or genera of a family)

All other symbols are defined in footnotes to tables.

Table 20.9 Characteristics differentiating the species and subspecies of *Clavibacter*[a,b]

Characteristics	C. iranicus	C. michiganensis subsp. insidiosus	subsp. michiganensis	subsp. nebraskensis	subsp. sepedonicus	subsp. tesselarius	C. rathayi	C. tritici	C. xyli subsp. cynodontis	subsp. xyli
Colonies with:										
Yellow or orange pigment	+	+	+	+	−	+	+	+	+	−
Blue or grey pigment	−	d	d	−	−	−	−	+	−	−
Acid produced from:										
Inulin	−	−	−	−	−	−	−	+	−	−
Mannitol	−	−	−	−	+	+	+	+	+	+
Mannose	+	+	+	+	d	ND	−	+	+	+w
Melezitose	+	−	−	−	−	−	−	−	−	−
Sorbitol	−	−	−	+	+	+	−	−	−	−
Utilization of:										
Acetate	−	−	+	+	+	+	−	+	−	−
Citrate	+	+	+	+	+	+	+	+	+	−
Lactate	−	−	d	+	−	ND	−	−	−	−
Propionate	−	−	−	+w	−	−	−	−	−	−
Succinate	+	−	+	+	+	ND	+	+	−	−
Hydrolysis of:										
Gelatin	−	−	+w	−	−	−	+	−	−	−
Soluble starch	−	−	d	d	d	+	d	−	+	−
Methyl red	−	+	−	d	−	−	d	−	−	−
H$_2$S from peptone	+	−	+	d	−	−	+	+	−	−

[a] Symbols: +, 90% or more of strains are positive; −, 90% or more of strains are negative; d, 11–89% of strains are positive; w, weak reaction; ND, not determined.

[b] From *Bergey's Manual of Systematic Bacteriology* and Davis et al. (Int. J. Syst. Bacteriol. *34*: 107–117, 1984).

Table 20.10 Characteristics differentiating the species of *Corynebacterium*[a]

Characteristics	C. amycolatum[b]	C. bovis	C. callunae	C. cystitidis	C. diphtheriae	C. flavescens	C. glutamicum	C. jeikeium[c]	C. kutscheri	C. matruchotii[d]
Acid produced from:										
Glucose	+	+	+	+	+	+	+	+	+	+
Arabinose	–	d	–	–	+	–	–	–	–	–
Xylose	–	–	–	+	–	–	–	ND	–	–
Rhamnose	–	–	–	–	+	–	–	–	–	–
Fructose	+	+	+	+	+	+	+	–	+	+
Galactose	–	+	–	–	+	+	–	+	–	–
Mannose	+	–	+	–	+	+	+	–	+	+
Lactose	–	d	–	–	–	–	–	–	–	–
Maltose	(+)	+	+	+	+	–	+	d	+	+
Sucrose	d	–	+	–	–	–	+	–	+	+
Trehalose	d	d	+	+	–	–	+	–	d	–
Raffinose	–	–	–	–	d	–	–	–	–	d
Salicin	–	–	+	–	+	–	–	–	+	+
Dextrin	ND	d	–	+	+	–	–	–	+	+
Starch	–	–	–	+	d	–	–	–	+	–
Hydrolysis of:										
Esculin	–	–	–	–	–	ND	–	–	–	+
Hippurate	d	+	+	+	–	–	+	ND	+	+
Urea	d	–	+	+	–	–	+	–	+	d
Tyrosine	–	–	ND	–	–	ND	ND	–	–	ND
Casein	–	–	–	–	–	ND	–	–	–	ND
Phosphatase	ND	+	ND	–	–	–	ND	–	–	–
Pyrazinamidase	ND	+	ND	+	–	–	ND	ND	+	+
Methyl red	+	–	+	–	+	+	+	–	–	–
Nitrate reduced to nitrite	ND	–	–	–	+	–	+	–	+	+

Footnotes are at end of table

Table 20.10 *(continued)*

Characteristics	*C. minutissimum*	*C. mycetoides*	*C. paurometabolum*	*C. pilosum*	*C. pseudodiphtheriticum*	*C. pseudotuberculosis*	*C. renale*	*C. striatum*	*C. vitarumen*	*C. xerosis*
Acid produced from:										
Glucose	+	+	–	+	–	+	+	+	+	+
Arabinose	ND	–	–	–	–	d	–	–	ND	–
Xylose	–	–	–	–	–	–	–	–	ND	–
Rhamnose	ND	ND	–	–	–	–	–	–	ND	–
Fructose	+	ND	–	+	–	+	+	+	+	+
Galactose	ND	–	–	–	–	+	–	d	+	+
Mannose	d	–	–	+	–	+	+	+	+	+
Lactose	–	–	–	–	–	–	–	d	–	–
Maltose	+	–	–	+	–	+	d	+	+	–
Sucrose	+	ND	–	–	–	d	–	–	+	+
Trehalose	–	d	–	+	–	–	d	d	+	–
Raffinose	–	–	–	–	–	–	–	–	ND	–
Salicin	ND	ND	–	–	–	–	–	–	+	+
Dextrin	ND	–	–	+	–	d	+	+	ND	–
Starch	–	ND	–	+	–	–	–	+	–	–
Hydrolysis of:										
Esculin	ND	–	+	–	–	–	–	–	+	–
Hippurate	+	ND	–	+	+	–	+	+	–	+
Urea	–	–	–	+	+	+	+	–	+	–
Tyrosine	ND	ND	–	–	–	–	–	+	ND	–
Casein	–	ND	–	–	–	–	+	–	ND	–
Phosphatase	+	+	+	–	–	–	–	+	–	–
Pyrazinamidase	+	ND	+	+	+	–	+	ND	+	+
Methyl red	–	–	–	–	–	+	–	+	+	–
Nitrate reduced to nitrite	–	–	–	+	+	d	–	–	+	+

[a] Symbols: +, 90% or more of strains are positive; (+), 80–89% of strains are positive; d, 21–79% of strains are positive; (–), 11–20% of strains are positive; –, 90% or more of strains are negative; ND, not determined.

[b] Collins et al. (Int. J. Syst. Bacteriol. *38:* 449, 1988; effective publication: FEMS Microbiol. Lett. *49:* 349–352, 1988).

[c] Jackman et al. (Int. J. Syst. Bacteriol. *38:* 136–137, 1988; effective publication: Syst. Appl. Microbiol. *9:* 83–90, 1987).

[d] *C. matruchotii* has a whip-handle morphology.

Table 20.11 Characteristics differentiating the species of *Curtobacterium*[a]

Characteristics	C. albidum	C. citreum	C. flaccumfaciens[b]	C. luteum	C. plantarum[c]	C. pusillum
Colony color	Ivory	Yellow	Yellow, orange or pink	Golden yellow	Yellow	Pale yellow
Motility	–	+	+	+	+	+
Hydrolysis of:						
Starch	+	+	ND	–	ND	+
Casein	+	–	+	+	–	+
Acid from:						
L-Arabinose	–	+w	–	+w	d	+w
Mannose	–	+w	dw	+	+	+w
Sorbose	–	+w	–	–	–	–
Sucrose	–	–	–	–	–	+w
Assimilation of:						
L(+)-Lactic acid	+	+	+	–	ND	+
D(–)-Lactic acid	+	+	+	–	ND	+
Malic acid	–	+	+	+	d	–
Gluconic acid	+	+	+	+	+	–
Propionic acid	–	–	–	–	+	d

[a] Symbols: +, 90% or more of strains are positive; –, 90% or more of strains are negative; d, 11–89% of strains are positive; w, weak reaction; ND, not determined.

[b] For purposes of plant quarantine *C. flaccumfaciens* is divided into four pathovars; pv. *flaccumfaciens* causing vascular wilt of bean (*Phaseolus vulgaris*); pv. *betae* causing vascular wilt and leaf spot of red beet (*Beta vulgaris*); pv. *oortii* causing vascular disease and leaf and bulb spot of tulips (*Tulipa* spp.); and pv. *poinsettiae* causing stem canker and leaf spot of poinsettia (*Euphorbia pulcherrima*).

[c] Dunleavy (Int. J. Syst. Bacteriol. *39:* 240–249, 1989).

Table 20.12 Characteristics differentiating the species of *Falcivibrio*[a]

Characteristics	F. grandis	F. vaginalis
Length of cells (μm)	2.5–5.0	1.0–5.0
Degree of cell curvature	Marked	Slight
Origin of flagella	Polar or subpolar	Concave side of cell
Acid from glucose, maltose, starch, fructose, galactose, ribose, D-xylose	Fermented, often weakly	Not fermented
Acid from cellobiose, lactose, mannitol, salicin	–	–
Hydrolysis of esculin	–	–
Reduction of nitrate	–	(+)

[a] Symbols: (+), 80% or more of strains positive; –, 90% or more of strains are negative.

Table 20.13 Characteristics differentiating the species of *Microbacterium*[a]

Characteristics	M. arborescens[b]	M. imperiale	M. lacticum	M. laevaniformans
Color of colonies	Dirty orange	Red-orange	Yellow-white	Yellow
Motility	–	+	–	–
Survive heating at 63°C for 30 min	ND	–	+	+
Acid produced from:				
L-Arabinose	+w	+	–	–
D-Xylose	+w	d	–	–
Sucrose	+w	+	–	+
Nitrate reduced	–	–	+	–
H₂S produced	+	–	–	+

[a] Symbols: +, 90% or more of strains are positive; –, 90% or more of strains are negative; d, 11–89% of strains are positive; w, weak reaction; ND, not determined.

[b] Imai et al. (Int. J. Syst. Bacteriol. *35:* 535, 1985; effective publication: Curr. Microbiol. *11:* 281–284, 1984), transferred from the genus *Flavobacterium* on account of its Gram-positivity.

Table 20.14 Characteristics differentiating the species and subspecies of *Mobiluncus*[a]

Characteristics	M. curtisii		M. mulieris
	subsp. curtisii	subsp. holmesii	
Mean length of cells (μm)	1.7	1.7	2.9
Origin of flagella	lateral	lateral or subpolar	lateral
Acid from:			
Glucose	+w	(–)vw	d
Maltose	+w	(–)vw	+
Fructose	dvw	dvw	d
Ribose	–	–	dw
Starch	–	dvw	d
Lactose	dvw	–	(–)vw
Mannitol	–	–	–
Migrates through soft agar	+	–	–
Hydrolysis of:			
Starch	+	+	+
Esculin	–	–	–
Hippurate	+	+	–
Reduction of nitrate	–	+	–
NH₄ from arginine	+	+	–

[a] Symbols: +, 90% or more of strains are positive; –, 90% or more of strains are negative; (–), 11–20% of strains are positive; d, 21–89% of strains are positive; w, weak reaction; vw, very weak reaction.

Table 20.15 Characteristics that differentiate the species of *Pimelobacter*[a]

Characteristics	*P. jensenii*	*P. simplex*	*P. tumescens*[b]
DNAase	+w	+	+
Urease	+	−	−
Assimilation of:			
Glucose	+	+	+
Mannitol	−	−	+
Rhamnose	+	−	−
Acetate	+	+	+
Succinate	−	+	+
Citrate	−	+w	−

[a] Symbols: +, 90% or more of strains are positive; −, 90% or more of strains are negative; w, weak reaction.

[b] See also *Terrabacter tumescens*.

Table 20.16 Characteristics differentiating the species of *Propionibacterium*[a,b]

Characteristics	*P. acidipropionici*	*P. acnes*	*P. avidum*	*P. freudenreichii*	*P. granulosum*	*P. jensenii*	*P. lymphophilum*	*P. thoenii*
Usual pigmentation of colonies	White	White to gray	White to cream	May be tan or pink	White to gray	White to pink	White	Orange to red-brown
β-Hemolysis	−	d	(+)	−	(−)	−	−	+
Acid produced from:								
Maltose	+	−	+	−	+	+	+	+
Sucrose	+	−	+	−	+	+	d	+
L-Arabinose	+	−	d	+	−	−	−	−
Cellobiose	+	−	−	−	−	d	−	−
Glycerol	+	d	+	+	+	+	−	+
Hydrolysis of:								
Esculin	+	−	+	+	−	+	−	+
Gelatin	−	+	+	−	d	−	d	−

[a] Symbols: +, 90% or more of strains are positive; (+), 80–89% of strains are positive; (−), 11–20% of strains are positive; −, 90% or more of strains are negative; d, 21–79% of strains are positive.

[b] It has been suggested that *Arachnia propionica* should be transferred to *Propionibacterium* but for determinative purposes it is retained as *Arachnia*.

GROUP 21

THE MYCOBACTERIA

The group contains a single genus, *Mycobacterium.* These are slender, rod-shaped bacteria, characteristically acid-fast, aerobic, slow-growing, and free-living or pathogens of vertebrates. Species of *Mycobacterium* may be confused with other related genera (see Table 21.1). The property of acid-fastness, due to waxy materials in the cell walls, is particularly important for recognizing mycobacteria. These bacteria are commonly described as acid-alcohol-fast, implying that after staining they resist decolorization with acidified alcohol as well as with strong mineral acids. Some other bacteria are partially acid-fast and are readily decolorized by alcohol, though they may resist decolorization by weak acid. The staining technique must be carefully performed: it is easy to decolorize stained preparations, and endospores and waxy material may stain acid-fast.

Genus Mycobacterium

Straight or slightly curved rods, 0.2–0.7 × 1.0–10 μm, **sometimes branching;** filamentous or mycelium-like growth may occur but is readily fragmented into rods or cocci. **Acid-alcohol-fast** at some stage of growth. **Not readily stained** by Gram's method, **usually weakly Gram positive. No aerial hyphae are grossly visible. Nonmotile, nonsporing;** without conidia or capsules. **Aerobic** and chemoorganotrophic. **Growth is slow** or **very slow;** visible colonies appear in 2–60 days at optimum temperature. Colonies are often pink, orange, or yellow, especially when exposed to light, **pigment is not diffusing,** surface commonly dull or rough. Some species are fastidious, requiring special supplements (e.g., *M. paratuberculosis*), or are noncultivable (*M. leprae*).

Catalase positive, arylsulfatase positive, lysozyme resistant. Widely distributed in soil and water; some species are obligate parasites and pathogens of vertebrates.

Type species: *Mycobacterium tuberculosis.*

Differentiation of the species of the genus **Mycobacterium:** Mycobacteria require special methods for study because many grow very slowly and need special media, and differentiation of species is best carried out in a specialist laboratory. Many of the rapidly growing species can, however, be identified reasonably easily, but those that grow slowly show several complexes of very similar species, which are hard to distinguish. Therefore, two tables are presented: Table 21.2 for selected slow-growing mycobacteria and Table 21.3 for the rapidly growing species. The distinction between slow and rapid growers is determined as follows: slow growers require over 7 days at optimum temperature on nutritionally rich media to yield easily visible single colonies from very dilute inocula; rapid growers show visible colonies in 7 days or less. Certain species are intermediate in their growth rates, and these are indicated in the tables. *M. leprae* is not cultivable, and several other species are only cultivated with great difficulty (see *Bergey's Manual of Systematic Bacteriology*).

Mycobacteria should be handled in a safety cabinet to prevent dissemination in case the human pathogen *Mycobacterium tuberculosis* should occur among the cultures, and infected material should be disinfected by heat because mycobacteria are relatively resistant to chemical disinfectants.

Table 21.1 Differentiation of *Mycobacterium* from other genera

Characteristics	*Mycobacterium*	*Corynebacterium*	*Nocardia*	*Rhodococcus*
Morphology	Rods, occasionally branched filaments; rarely aerial mycelium	Pleomorphic rods, often club-shaped; commonly in angular and palisade arrangement	Mycelium, later fragmenting into rods and cocci; usually some aerial mycelium	Scanty mycelium, fragmenting into irregular rods and cocci; no aerial mycelium
Acid-fast	Usually strongly acid-fast	Sometimes weakly acid-fast	Often partially acid-fast	Often partially acid-fast
Acid-alcohol-fast	At least some cells in young cultures are acid-alcohol-fast	Negative	Negative	Negative
Degree of Gram staining	Weak	Strong	Usually strong	Usually strong
Rate of growth: time to visible colonies	2–60 days	1–2 days	1–5 days	1–3 days
Reaction to penicillin	Usually resistant	Sensitive	Resistant	Sensitive
Production of arylsulfatase	Positive, sometimes slow	Negative	Uncommon	Negative

---

Important Notes for Users of This Manual

Unless otherwise indicated in footnotes to tables, the meanings of symbols are as follows:

+ 90% or more of strains are positive

− 90% or more of strains are negative

d 11-89% of strains are positive

v strain instability (not equivalent to "d")

D Different reactions in different taxa (species of a genus or genera of a family)

All other symbols are defined in footnotes to tables.

Table 21.2 Characteristics differentiating selected slowly growing species of Mycobacterium [a]

Characteristics	"M. avium complex"	"M. bovis complex" [b]	M. cookii [c]	M. flavescens [d]	M. gastri	M. gordonae	M. kansasii	M. malmoense	M. marinum	M. scrofulaceum	M. simiae	M. szulgai	"M. terrae complex" [b]	M. tuberculosis	M. xenopi
Urease	−	+	−	+	+	−	+	d	+	+	+	d	−	+	−
Pyrazinamidase in agar	+	−	−	ND	d	d	d	d	+	+	d	+	d	+	d
Nitrate reduction	−	−	−	+	−	−	+	−	−	d	−	+	d	+	−
Acid phosphatase	−	+	+	d	+	d	+	−	+	−	−	+	+	+	−
β-Galactosidase	−	ND	ND	ND	−	−	−	−	−	−	−	−	+	ND	−
Catalase, > 45 mm foam	d	−	ND	+	−	+	+	−	d	+	+	+	+	−	−
Tween hydrolysis, 10 days	−	−	−	+	+	+	+	+	+	−	−	d	+	d	−
α-Esterase	+	+	ND	d	−	+	−	d	−	d	d	d	d	+	+
Niacin accumulates	−	−	ND	−	−	−	−	−	−	−	d	−	−	+	−
Photochromogenic	−	−	−	−	−	−	+	−	+	−	+	−[e]	−	−	−
Scotochromogenic	(−)	−	+	+	−	+	−	−	−	+	−	+[e]	−	−	d
Grows at 25°C	d	−	+	+	+	+	+	d	+	+	+	+	+	+	−
Grows at 45°C	d	−	−	d	−	−	−	−	−	−	−	−	−	−	+
Resists inhibition by:															
Picric acid, 2 mg/ml	−	−	−	+	−	−	−	−	−	−	d	−	ND	−	−
p-Nitrobenzoic acid, 0.5 mg/ml	(+)	−	ND	+	−	d	d	+	d	d	+	+	d	−	d
Hydroxylamine HCl, 0.5 mg/ml	d	−	ND	−	−	d	d	d	+	d	+	−	+	−	−
Isoniazid, 1 µg/ml	+	−	+	−	−	d	d	+	d	+	+	+	ND	+	−
Isoniazid, 10 µg/ml	d	−	+	−	−	−	d	−	d	d	d	−	+	−	−
Thiacetazone, 10 µg/ml	+	−	ND	+	−	+	−	+	d	d	+	+	+	−	+
Thiophene-2-carboxylic acid hydrazide, 1 µg/ml	+	−	ND	+	+	+	+	+	+	+	+	+	+	+	+
NaCl, 5%	−	−	ND	+	−	−	−	−	d	−	−	−	d	−	−

[a] Symbols: +, 90% or more strains positive; (+), 80-90% of strains positive; d, 21-79% of strains positive; (−) 11-20% of strains positive; −, 90% or more of strains negative; ND, not determined.

[b] "M. bovis complex" includes M. bovis, BCG, and M. africanum. "M. terrae complex" includes M. terrae, M. nonchromogenicum, and M. triviale. "M. avium complex" includes M. avium and M. intracellulare. A new subspecies, not described in Bergey's Manual of Systematic Bacteriology, M. avium subsp. silvaticum has been proposed by Thorel et al. (Int. J. Syst. Bacteriol. 40: 254–260, 1990), who also propose that M. paratuberculosis should be another subspecies, M. avium subsp. paratuberculosis, but current data are not adequate for easy differentiation of the subspecies.

[c] Not described in Bergey's Manual of Systematic Bacteriology; proposed by Kazda et al. (Int. J. Syst. Bacteriol. 40: 217–223, 1990).

[d] M. flavescens is a rapid grower that can easily be confused with M. szulgai.

[e] M. szulgai may be photochromogenic when grown at 25°C.

Table 21.3 Characteristics differentiating the rapidly growing species and subspecies of *Mycobacterium*[a]

Characteristics	*M. agri*	*M. aichiense*	*M. aurum*	*M. austroafricanum*[b]	*M. chelonae* subsp. *abscessus*	*M. chelonae* subsp. *chelonae*	*M. chitae*	*M. chubuense*	*M. diernhoferi*[b]	*M. duvalii*	*M. fallax*[c,d]
Degradation of *p*-aminosalicylic acid	–	–	–	–	+	+	–	–	–	–	ND
Growth on MacConkey without crystal violet (28°C)	ND	ND	–	ND	+	+	–	ND	ND	ND	ND
Arylsulfatase, 3 day	d	+	d	+	+	+	–	–	–	–	–
Growth with NH₂OH·HCl, 500 µg/ml	–	–	d	–	+	+	–	–	–	–	ND
Colonies pigmented	–	+	+	+	–	–	+	+	–	+	–
Colonies photochromogenic	–	–	–	+[j]	–	–	–	–	–	–	–
Grows at 45°C	+	–	–	–	–	–	–	–	–	–	–
Grows at 52°C	–	–	–	–	–	–	–	–	–	–	+
Nitrate reduction	+	–	d	+	–	(–)	+	+	+	+	+
Iron uptake (28°C)	ND	ND	+	ND	–	–	d	ND	ND	ND	ND
NaCl tolerance (5%, 28°C)	ND	ND	d	+	+	–	d	ND	+	ND	–
Citrate utilized (28°C)	d	–	+	+	–	+	–	–	d	–	–
Mannitol utilized (28°C)	–	+	+	+	–	–	–	(–)	+	+	–
Acid from arabinose	ND	ND	+	ND	–	–	–	ND	ND	–	ND
Acid from xylose	ND	ND	+	ND	–	–	–	ND	ND	–	ND
Acid from dulcitol	ND	ND	–	ND	–	–	–	ND	ND	–	ND
Growth with malachite green, 0.01%	ND	ND	–	ND	+	+	–	ND	ND	ND	ND
Growth with pyronin B, 0.01%	ND	ND	–	ND	+	+	–	ND	ND	ND	ND
Oxalate utilized	ND	ND	–	ND	–	–	–	ND	ND	ND	ND
Allantoinamidase	+	–	d	–	–	–	–	–	(+)	–	ND
Benzamidase	–	–	d	–	–	–	–	–	–	–	ND
Isonicotinamidase	–	–	–	–	–	–	–	–	–	–	ND
Succinamidase	–	–	(–)	–	–	–	–	–	–	–	ND
Acid phosphatase	–	+	–	–	+	+	+	–	d	–	ND
Growth with picrate, 0.2%	+	+	+	+	+	(–)	+	+	+	ND	ND

Footnotes are at end of table

Table 21.3 (continued)

Characteristics	M. flavescens [c]	M. fortuitum subsp. acetamido-lyticum [c,e]	M. fortuitum subsp. fortuitum	M. fortuitum subsp. peregrinum	M. gadium	M. gilvum	M. komossense	M. moriokaense [f]	M. neoaurum	M. obuense
Degradation of p-aminosalicylic acid	–	–	+	+	ND	–	d	–	–	+
Growth on MacConkey without crystal violet (28°C)	–	ND	+	+	ND	ND	–	ND	–	ND
Arylsulfatase, 3 day	d	+	+	+	–	–	–	+	–	–
Growth with NH$_2$OH·HCl, 500 µg/ml	–	+	+	+	ND	–	–	–	d	–
Colonies pigmented	+	–	–	–	+	+	+	–	+	+
Colonies photochromogenic	d	–	–	–	–	–	–	–	–	–
Grows at 45°C	d	–	–	–	–	–	–	–	–	–
Grows at 52°C	–	–	–	–	+	–	–	–	–	–
Nitrate reduction	+	+	+	+	+	+	–	+	(+)	–
Iron uptake (28°C)	–	ND	+	+	–	ND	+	ND	ND	ND
NaCl tolerance (5%, 28°C)	+	–	+	+	ND	ND	–	d	ND	ND
Citrate utilized (28°C)	–	d	d	d	–	–	+	–	+	–
Mannitol utilized (28°C)	d	–	–	+	ND	+	+	+	+	+
Acid from arabinose	–	ND	–	–	–	–	–	ND	(+)	ND
Acid from xylose	–	ND	–	–	–	–	–	ND	(+)	ND
Acid from dulcitol	–	ND	–	–	–	–	–	ND	–	ND
Growth with malachite green, 0.01%	–	ND	+	+	ND	ND	ND	ND	ND	ND
Growth with pyronin B, 0.01%	–	ND	+	+	ND	ND	ND	ND	–	ND
Oxalate utilized	–	ND	–	–	–	ND	+	ND	–	ND
Allantoinamidase	–	–	+	+	–	d	–	d	+	+
Benzamidase	–	–	–	–	–	–	–	–	–	–
Isonicotinamidase	–	–	–	–	–	–	–	–	–	–
Succinamidase	–	–	–	–	–	–	+	–	–	–
Acid phosphatase	–	–	+	+	ND	d	+	–	–	–
Growth with picrate, 0.2%	+	–	+	+	–	ND	ND	+	ND	+

Footnotes are at end of table

Table 21.3 (continued)

Characteristics	M. parafortuitum	M. phlei	M. porcinum [g]	M. poriferae [h]	M. pulveris [c,i]	M. rhodesiae	M. senegalense	M. smegmatis	M. sphagni	M. thermoresistibile	M. tokaiense	M. vaccae
Degradation of p-aminosalicylic acid	–	–	+	–	–	–	ND	–	d	–	–	–
Growth on MacConkey without crystal violet (28°C)	–	–	ND	ND	ND	ND	ND	–	–	–	ND	–
Arylsulfatase, 3 day	d	–	+	–	–	+	+	–	+	–	+	(–)
Growth with NH$_2$OH·HCl, 500 µg/ml	–	–	+	–	–	–	ND	–	+	–	–	–
Colonies pigmented	+	+	–	+	–	+	–	–	+	+	+	+
Colonies photochromogenic	+	–	–	–	–	–	–	–	–	–	–	+
Grows at 45°C	–	+	–	–	d	–	–	+	–	+	–	–
Grows at 52°C	–	+	–	–	–	–	–	+	–	+	–	–
Nitrate reduction	d	+	–	–	+	–	–	+	+	+	–	+
Iron uptake (28°C)	+	+	ND	+	ND	ND	–	+	–	–	ND	+
NaCl tolerance (5%, 28°C)	d	+	+	+	+	ND	+	+	–	+	ND	d
Citrate utilized (28°C)	+	+	+	+	–	–	+	+	d	–	–	+
Mannitol utilized (28°C)	+	+	+	+	–	+	ND	+	+	–	+	+
Acid from arabinose	+	+	ND	–	ND	ND	ND	+	–	–	ND	+
Acid from xylose	+	+	ND	+	ND	ND	ND	d	–	–	ND	+
Acid from dulcitol	–	–	ND	ND	ND	ND	ND	+	–	–	ND	–
Growth with malachite green, 0.01%	–	+	ND	ND	ND	ND	ND	+	ND	–	ND	–
Growth with pyronin B, 0.01%	–	–	ND	ND	ND	ND	ND	+	ND	–	ND	–
Oxalate utilized	–	–	ND	ND	ND	ND	ND	+	(+)	–	ND	–
Allantoinamidase	–	–	+	–	–	–	+	+	–	–	+	+
Benzamidase	–	–	–	+	–	–	+	d	–	–	+	+
Isonicotinamidase	–	–	–	–	–	–	+	d	–	–	+	+
Succinamidase	–	–	+	–	–	–	ND	+	–	–	+	+
Acid phosphatase	–	+	–	ND	+	+	ND	–	+	–	–	–
Growth with picrate, 0.2%	+	+	+	+	–[k]	+	ND	+	ND	+	+	+

Footnotes are at end of table

602

Table 21.3 *(continued)*

[a] Symbols: +, 90% or more of strains positive; (+), 80–89% of strains positive; d, 21–79% of strains positive; −, 90% or more strains negative; ND, not determined.

[b] Supplementary data from Tsukamura et al. (Int. J. Syst. Bacteriol. *33*: 460–469, 1983).

[c] Growth rate is not as rapid as that of most; may be confused with slowly growing mycobacteria.

[d] Supplementary data from Lévy-Frébault et al. (Int. J. Syst. Bacteriol. *33*: 336–343, 1983).

[e] Not described in *Bergey's Manual of Systematic Bacteriology*; proposed by Tsukamura et al. (Int. J. Syst. Bacteriol. *36*: 489, 1986; effective publication: Microbiol. Immunol. *30*: 97–110, 1986).

[f] Not described in *Bergey's Manual of Systematic Bacteriology*; proposed by Tsukamura et al. (Int. J. Syst. Bacteriol. *36*: 333–338, 1986).

[g] Supplementary data from Tsukamura et al. (Int. J. Syst. Bacteriol. *33*: 162–165, 1983).

[h] Not described in *Bergey's Manual of Systematic Bacteriology*; proposed by Padgitt and Moshier (Int. J. Syst. Bacteriol. *37*: 186–191, 1987).

[i] Supplementary data from Tsukamura et al. (Int. J. Syst. Bacteriol. *33*: 811–815, 1983).

[j] Uncertain; see Tsukamura et al. (Int. J. Syst. Bacteriol. *33*: 460–469, 1983, Table 2 and text description).

[k] May grow after repeated cultivation on laboratory media.

THE ACTINOMYCETES

GROUPS 22–29

GROWTH AND EXAMINATION OF ACTINOMYCETES—SOME GUIDELINES

(This is an amended reprint of a chapter by Tom Cross in Volume 4 of
Bergey's Manual of Systematic Bacteriology, pp. 2340–2343.)

Many bacteriologists encounter difficulties when first faced with the task of growing, subculturing, and examining actinomycetes. Familiarity and some years of practice with cultures of *Escherichia coli* or *Bacillus subtilis* do not prepare the laboratory worker for some of the problems that will be encountered, and a mycologist can often find the task slightly easier. The aim here is therefore to suggest some **simple methods and guidelines that will enable the bacteriologist to begin the study of an actinomycete.** It is hoped that experience and confidence will be gained, along with perhaps the breathing space to consult papers and articles for the specific methods and media required.

The term "actinomycete" now encompasses a wide range of bacteria. Some workers would include genera such as *Micrococcus, Arthrobacter,* and *Cellulomonas,* which can be handled in the same way as most other oxidative, Gram-positive bacteria. However, one must remember that the cells of two of these genera, *Arthrobacter* and *Cellulomonas,* can show **characteristic morphological variation** during exponential and stationary-phase growth in broth culture. *Cellulomonas* and *Oerskovia* strains appear as short, motile rods in broth cultures but grow as feathery colonies composed of radiating, branching filaments when cultured on dilute, nutrient-deficient media solidified with agar. One must examine possible actinomycetes growing in broth after incubation for various times by phase-contrast microscopy of wet mounts as well as high-magnification observation of stained smears, and one must examine the appearance of colonies growing on agar media.

Some actinomycete genera are **anaerobic** (e.g., *Actinomyces*) or require very **specialized growth media and incubation conditions** (e.g., *Frankia*). The investigator must consult the relevant sections in Volume 4 of *Bergey's Manual of Systematic Bacteriology* and the specialist papers quoted for specific information on these organisms.

It is the common sporoactinomycetes, bacteria with branching hyphae and specialized spore-bearing structures, and the nocardioform actinomycetes that are now attracting considerable interest from biotechnologists, geneticists, and ecologists. Strains of the genera *Streptomyces* and *Thermomonospora* (sporoactinomycetes; Figure 1), *Nocardia* and *Rhodococcus* (nocardioforms, Figure 2), and *Actinoplanes* and *Dactylosporangium* (actinoplanetes, Figure 3), for example, have been studied intensively by a relatively small group of microbiologists specifically interested in their taxonomy and ability to produce secondary metabolites. There is an extensive but very scattered literature on methods that should be used, but unfortunately it will not be found in the commonly available methods manuals. When discussing methods in this text, the above genera will be used as examples of actinomycetes exhibiting differentiation.

Many actinomycetes will grow on the common bacteriological media used in the laboratory, such as nutrient agar, Trypticase soy agar, blood agar, and even brain-heart infusion agar. Indeed, suspected clinical isolates of *Nocardia asteroides* and *Streptomyces somaliensis* should be first cultivated on a rich nutritive medium, but then one should expect to obtain what must be regarded as atypical growth. Tough, leathery colonies will appear, but they usually lack any aerial mycelium so typical of

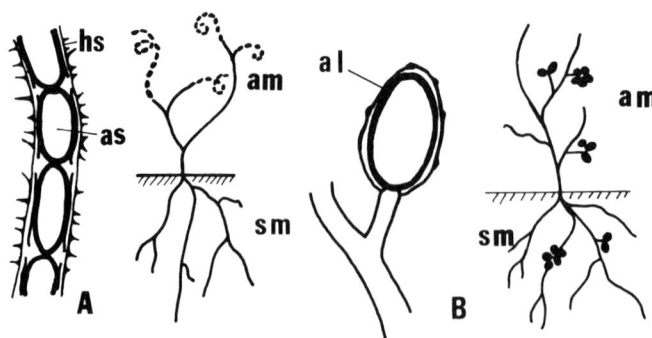

Figure 1. Sporoactinomycetes. *A, Streptomyces* species have chains of spores on the aerial mycelium (*am*), which are normally absent from the substrate mycelium (*sm*). These spores are arthrospores (*as*), regular segments of hyphae with a thickened spore wall surrounded by a hydrophobic sheath (*hs*) that may bear spines or hairs. *B, Thermomonospora* species can have clusters of single spores, aleuriospores (*al*), on both aerial and substrate mycelium.

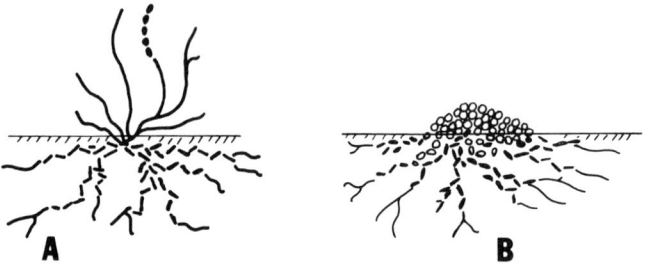

Figure 2. Nocardioform actinomycetes. *A, Nocardia* species with a fragmenting substrate mycelium and limited aerial mycelium that can bear chains of arthrospores. *B, Rhodococcus* species have a rapidly fragmenting substrate mycelium; the segments become rounded, and the colony usually consists of a mass of coccoid elements.

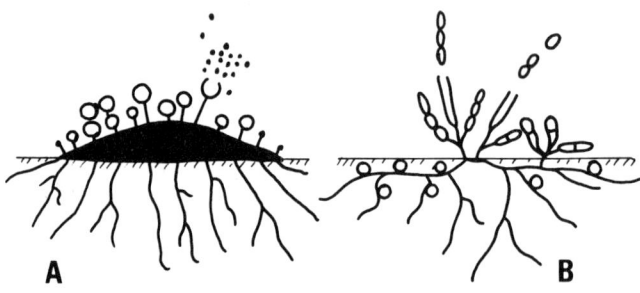

Figure 3. Actinoplanetes. *A, Actinoplanes* species form globose sporangia on the surface of the colony; these sporangia liberate motile zoospores. *B, Dactylosporangium* species have tubular sporangia containing few potentially motile spores together with aleuriospores (globose bodies) on the substrate mycelium.

Streptomyces strains, and it will be impossible to see the spore clusters embedded in the colony that are characteristic of micromonosporae.

Sporoactinomycetes require special media to allow differentiation and the development of characteristic spores and pigments. Some of these media are not available from commercial suppliers and must be pre-

pared in the laboratory. Those containing colloidal chitin, soil extract, or decoctions of plant materials require lengthy preparation times but are extremely useful. For example, the pale, shiny, hard colonies of a *Streptomyces* species on nutrient agar can be transformed into bright yellow colonies with a powdery white aerial mycelium and spirals of arthrospores when the organism is subcultured onto a more suitable growth medium, such as oatmeal or inorganic salts starch agars. Other species grow thinly but sporulate profusely on tap water agar, where they use the polysaccharide as a carbon source, or on water agar containing trace amounts of yeast extract or peptone.

Outgrowths from a spore or fragments of mycelium (colony-forming units, CFUs) develop into hyphae that penetrate the agar (**substrate mycelium**) and hyphae that branch repeatedly and become cemented together on the surface of the agar to form a **tough, leathery colony.** The density and consistency of the colony will depend on the composition of the medium. Nocardioform actinomycetes exhibit **fragmentation;** the hyphae break up into rods and cocci, thus leading to soft or friable colonies. In strains of certain genera (e.g., *Streptomyces*), the colony becomes covered with **aerial mycelium:** free, erect hyphae surrounded by a hydrophobic sheath that grow into the air away from the colony. These hyphae are initially white but assume a range of colors when spore formation begins. Colonies then appear powdery or velvety and can then be readily distinguished from the more typical bacterial colonies.

Growth can be slow. A branching mycelium growing at the surface of transparent agar can be seen with the aid of a microscope after 24 hours, and visible colonies may appear in 3–4 days, but mature aerial mycelium with spores may take 7–14 days to develop, and some very slow-growing strains may require up to 1 month's incubation. Lengthy incubation times can result in the evaporation of the medium, so thick agar plates are required. Thermophilic species incubated at 45–55°C require a humid incubator.

One requires patience when working with actinomycetes, and one must have the ability to plan and run several experiments concurrently to avoid wasting time.

The **morphology** of an actinomycete growing on agar can provide useful and rapid clues to its identity, but viewing isolated colonies can give little worthwhile information (Figure 4). Examine the organism streaked in a cross-hatched pattern on the surface of the agar (Figure 5), first with a stereomicroscope and then with a transmitted light microscope with a × 40 long-working-distance objective to avoid water condensation on the front lens. The appearance of hyphae within the agar (presence of spores, fragmentation, etc.) and the nature of the spores on the aerial hyphae are important but can only be observed when growth is thin and the medium promotes differentiation.

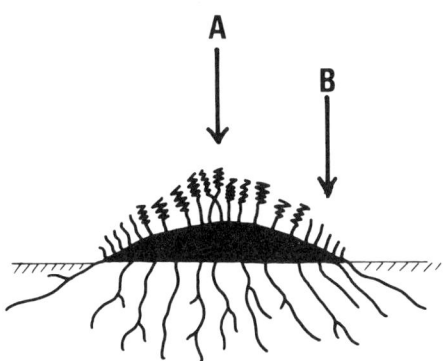

Figure 4. The mature, spore-bearing hyphae in the center of the colony (*A*) cannot be seen with a transmitted light microscope because of the density of the colony. Only young, immature aerial hyphae occur at the margins of the colony (*B*), where adequate light levels are possible. Actinomycete hyphae are too small to be seen with a stereomicroscope.

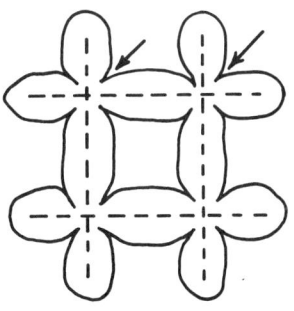

Figure 5. Cross-hatched streak plate. View directly under a microscope fitted with a long working distance objective (×25 and ×40). One should look for mature hyphae with spores in the angles of the streaks.

For detailed light microscope studies on sporulating structures and spore arrangements in some genera such as *Microbispora, Microtetraspora,* and *Saccharomonospora,* it is advisable to use **slide cultures** (Figure 6) incubated in a moist chamber. The thin squares of nutritive agar or agarose beneath a coverslip can be examined under high magnification and transmitted light will not be obscured by heavy vegetative growth. Mycelium adhering to coverslips placed at an angle in growing cultures (Figure 7) can be transferred to a slide and also examined at high magnifications. For **scanning electron microscopy,** an agar block carrying a sporulating colony should be fixed in osmium tetroxide, gradually dehydrated in alcohol before being subjected to critical point drying, and coated with gold. Sections for **transmission electron microscopy** can also be prepared from agar blocks or from growth scraped off the surface of cellophane strips laid on the surface agar.

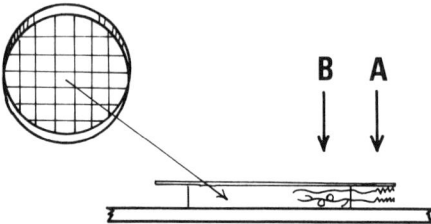

Figure 6. Slide culture. A thin agar block, cut from poured plate, is placed on a sterile microscope slide and inoculated, and a sterile coverslip is applied. After incubation in a moist chamber, view the slide culture directly on the microscope stage, when it should be possible to see the aerial mycelium (*A*) and the substrate mycelium (*B*) within the agar.

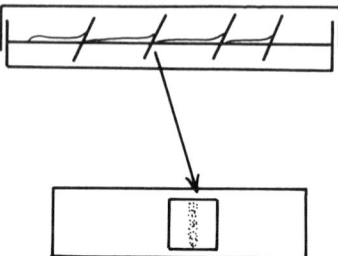

Figure 7. Inclined coverslips for observing actinomycete morphology. Inoculate the agar plate with coverslips inserted at an angle. After incubation, withdraw the coverslips and mount them, upper surface down, in water containing a wetting agent.

The saprophytic actinomycetes are oxidative and may grow poorly when the air supply is restricted. The screw caps of slope cultures must be loose to allow adequate aeration, and some strains show better growth in tubes with cotton-wool plugs or spring caps that allow better gas exchange. Actinomycetes will grow at a lower a_w than most Gram-negative bacteria, enabling one to use 1.5–2.0% (weight/volume) agar in media and up to 3.0% (w/v) for strains of some genera. This may sound extravagant but can be most useful when one wishes to remove the very hard growth of a strain reluctant to produce aerial mycelium and spores. A rough, stiff loop is then an essential tool for abrading the colony and collecting sufficient mycelial fragments for an efficient transfer.

Actinomycetes will grow in broth but need to be cultivated under specialized conditions. The growth of a streptomycete in a stationary broth tube is usually restricted to a surface pellicle and perhaps a cottony sediment, leaving the broth quite clear. Liquid cultures require considerable **aeration and agitation** to give the even, suspended growth required for most chemotaxo-

nomic studies. Tubes and flasks must be incubated on a shaker at high speeds (e.g., 200–250 rpm) to give the supply of oxygen and mixing necessary for maximum growth. Even, diffuse mycelial growth may require the higher agitation and mixing rates achieved by internal baffles or springs within the flask, but in turn this will be attained only with an **adequate inoculum.** Unlike a typical dividing bacterium that will give a smooth suspension of cells on continued incubation, an actinomycete spore or piece of mycelium will germinate, elongate, intertwine, and grow as a pellet. **High numbers of growth sources** (spores or hyphal fragments) must be introduced into a flask or fermenter to achieve heavy, dispersed growth. Hence, there is an initial need for a good sporulation medium or, failing that, a method that will allow homogenization of vegetative mycelium to give many growth sources.

Spore suspensions can be prepared by detaching the spores from aerial hyphae with a loop or scraper and placing them in a suspending medium containing a wetting agent. The arthrospores of streptomycetes are hydrophobic because of the enveloping sheath, and the wetting agent aids their even suspension in the diluent. Free spores may be removed from lawns of aerial mycelium by rolling glass beads or agar cylinders over the surface. Cultures of actinoplanetes bearing sporangia should be flooded with diluent and allowed to stand at room temperature for at least 30 minutes to permit spore hydration and maturation before the surface is scraped with a loop.

Pure cultures are of vital importance for all taxonomic purposes. **Checks for purity** should be undertaken frequently. The filamentous growth in broths contained in shake flasks or tubes may be a **fungal contaminant** that would be revealed by examining a sample dispersed in lactophenol-cotton blue. Nonactinomycete, **contaminating bacteria** may be hidden within the sporulating mycelial growth on agar, only to outgrow the actinomycete in shake flasks. The actinomycete colonies on plate or slope cultures should be examined carefully with a hand lens, and subcultures should be spread on rich agar media or transferred to a broth more suitable for the growth of nonactinomycete contaminants.

Actinomycete contaminants are also a problem. The dry, hydrophobic spores of streptomycetes can be dispersed within a laboratory or sterile room, thus having the potential for contaminating other actinomycetes—an association that is more difficult to detect. One must then streak out to obtain isolated colonies and examine carefully for differences in colony color and size. Finally, in laboratories where soil and litter are being sampled, one must always be on the lookout for **mites** that can wander from plate to plate and among incubators, causing havoc and ruining many experiments.

Morphological characters are still widely used for characterizing genera, for example, the presence or absence of spores on the substrate mycelium or the formation of zoospores in specialized spore vesicles or sporangia. The spores of *Streptosporangium* species are formed within a sporangium but are nonmotile, whereas those of *Actinoplanes* and *Spirillospora* developing within similar-shaped sporangia are motile. They may not exhibit movement when first released, and it may take 20–60 minutes of suspension in water, an amino acid solution, or dilute broth plus a wetting agent to induce motility. The ability to produce **motile spores** is more widespread in the actinomycetes than was previously suspected.

Preservation of both sporing and nonsporulating actinomycetes can be achieved by freeze-drying or storage in liquid nitrogen. Freezing suspensions in 20% (v/v) glycerol at –20° to –40°C has proved to be a very useful method in a busy laboratory, and one can even preserve a vegetative inoculum for flasks by freezing a glycerol suspension.

In summary, the following points should be emphasized:

1. Grow actinomycetes on suitable media.

2. Provide sufficient aeration for growth of oxidative species in broth cultures—this means vigorous agitation.

3. All subcultures and particularly broth cultures must receive an adequate inoculum.

4. Morphological examination by the traditional bacteriological stained smear method will invariably give misleading or incomplete information.

5. Examine organisms growing on agar.

6. Be patient!

Most of the methods briefly mentioned in this guide will have to be supplemented with details from specialist papers, but the general books, chapters, and papers listed below should prove useful.

Further Reading

Dietz, A. and D. W. Thayer (Editors). 1980. *Actinomycete Taxonomy*. Society for Industrial Microbiology, Arlington, Virginia, Special Publication No. 6.

Goodfellow, M. and D. E. Minnikin (Editors). 1984. *Chemical Techniques for Bacterial Systematics*. Academic Press, London.

Starr, M. P., H. Stolp, H. G. Trüper, A. Balows, and H. G. Schlegel (Editors). 1981. Section U, The Actinomycetes. In *The Prokaryotes, a Handbook on Habitats, Isolation and Identification of Bacteria*, Volume 2. Springer-Verlag, Berlin, pp. 1915–2123.

Waksman, S. A. 1961. *The Actinomycetes*, Volume 2. *Classification, Identification and Description of Genera and Species*. Williams & Wilkins, Baltimore.

Williams, S. T. and T. Cross. 1971. Actinomycetes. In Booth (Editor), *Methods in Microbiology*, Volume 4. Academic Press, London, pp. 295–334.

A PRACTICAL GUIDE TO GENERIC IDENTIFICATION OF ACTINOMYCETES

(This is an amended reprint of a chapter by Hubert Lechevalier in Volume 4 of
Bergey's Manual of Systematic Bacteriology, pp. 2344–2347.)

Groups 22 through 29 contain the organisms that are the most filamentous of all bacteria. Some approach, and in some cases may equal, the morphological complexity of the imperfect fungi with which they have been confused in the past. They are aerobic Gram-positive bacteria that form branching filaments or hyphae that may persist as a stable mycelium or may break up into rod-shaped or coccoid elements. Motility, when present, is due to flagellation. The genera in this section (Groups 22–29) are divided into eight groups. The purpose of this introduction is to guide the user to the most appropriate group(s) and genera.

Morphology. Nothing beats morphology for simplicity, but even with considerable experience generic identification based on morphology alone is rarely secure. The tools of the actinomycetic morphologist are a brightfield microscope equipped with a long-working-distance condenser and objectives. These permit one to observe undisturbed cultures grown in petri dishes. The best media for the observation of the morphology of actinomycetes are lean media such as water agar with traces of nutrients such as casein hydrolysate, yeast extract, or starch. For certain features, such as the determination of the nature of the surfaces of spores, the use of an electron microscope is essential. Morphological features include the following:

1. **Mycelium,** which may be stable or fugacious. If it breaks down, one should observe the shape of the elements and their motility (*Oerskovia* spp. release flagellate elements—Group 22). One should note if both a substrate and an aerial mycelium are formed, or only a substrate mycelium (very common), or only aerial hyphae (extremely rare—*Sporichthya*, Group 25). The mycelium may bear intercalary vesicles that do not contain spores (*Intrasporangium*, Group 25) or contain many spores (*Frankia*, Group 23).

2. **Conidia,** a general term used to refer to any asexual spores that are not intercalary chlamydospores or sporangiospores. Actinomycetes form conidia in a variety of ways:

a. *Single conidia* are found in several genera. In the genus *Thermoactinomyces* (Group 28), these are bacterial endospores that can be recognized by their thermostability.

Nonthermostable single conidia are found in the genera *Saccharomonospora* and *Promicromonospora* (Group 22), *Micromonospora* (Group 24), and *Thermomonospora* (Group 27). In addition, members of the genera *Frankia, Dactylosporangium,* and *Intrasporangium* may form terminal vesicles that may be confused with spores. Also, many different organisms, such as actinomadurae, will form single vesicles when grown under adverse conditions.

b. *Pairs of conidia.* Longitudinal pairs of conidia are the key characteristic of the genus *Microbispora* (Group 26) when formed only on the aerial mycelium.

c. *Short chains of conidia.* Although it is difficult to say how long a short chain of conidia can be, chains of up to 20 spores are usually considered to be "short." Representatives of the following genera may form such chains: *Nocardia, Pseudonocardia,* and *Saccharomonospora* (Group 22); *Streptoverticillium* and *Sporichthya* (Group 25); *Actinomadura* and *Microtetraspora* (Group 26); *Streptoalloteichus* (Group 27); and *Glycomyces* (Group 29). This type of morphology may also be encountered in the genera *Amycolata* and *Amycolatopsis* (Group 22) and *Catellatospora* (Group 24). Some of the streptomycetes of the *Microellobosporia* type form extremely short chains of spores, which are surrounded by an envelope that can be seen even by light microscopy.

d. *Long chains of conidia* are formed by strains belonging to various genera. These may include *Nocardia, Nocardioides, Pseudonocardia, Saccharopolyspora, Actinopolyspora,* and *Amycolatopsis* (Group 22); *Streptomyces* and *Streptoverticillium* (Group 25); *Actinosynnema, Nocardiopsis,* and *Streptoalloteichus* (Group 27); and *Kibdelosporangium, Kitasatosporia, Glycomyces,* and *Saccharothrix* (Group 29).

e. The conidia-bearing hyphae may be united into synnemata releasing motile spores (*Actinosynnema,* Group 27).

3. **Sporangia** are bags that contain spores. These may be borne: (a) on well-developed aerial hyphae or on the surface of colonies with little or no aerial hyphae (*Actinoplanes, Ampullariella, Pilimelia, Dactylosporangium* [Group 24]); *Planobispora, Planomonospora, Spirillospora, Streptosporangium* (Group 26); or (b) mainly within the agar (*Kineosporia,* Group 25).

4. **Other structures.** Some actinomycetes form unusual structures. Already mentioned are the spore-bearing synnemata found in the genus *Actinosynnema.* Organisms in Group 23 form masses of spores that are the result of division in several planes, rather than division perpendicular to the axis of the hyphae. These spore-bearing structures are called multilocular sporangia.

Many actinomycetes will form spherical structures on their aerial hyphae. These may be nothing more than drops of condensed water that enclose a curled chain of spores, or these structures may contain hyphae embedded in an amorphous matrix (*Kibdelosporangium,* Group 29).

Sclerotia are globose structures formed by some of the streptomycetes. These sclerotia contain not spores but cells filled with lipids. They germinate as a whole, as do the pseudosporangia of kibdelosporangia.

The morphology of aerial and subsurface growth of most of the genera in Groups 22–29 is presented schematically in Figure 8. Morphology is useful for the identification of some genera but not for all.

Cell Chemistry. Lechevalier and Lechevalier (Ann. Inst. Pasteur *108:* 662–673, 1965) pointed out that the actinomycetes can be separated into broad groups on the basis of morphological and chemical criteria. This approach is still the simplest and best for generic identification.

For a reader to reach the pages of this manual that are most likely to be of interest in the generic identification of a specific actinomycete, two pieces of chemical information are most useful: (a) which dibasic amino acid is present in the cell wall of the unknown, and (b) which diagnostic sugars are present in its whole cell hydrolysate. In both cases, necessary basic information can be obtained using one-way paper chromatography of whole cell hydrolysates (Lechevalier and Lechevalier, In Dietz and Thayer (ed.), *Actinomycete Taxonomy,* Spec. Publ. 6, Soc. Ind. Microbiol., Arlington, VA, pp. 227–291, 1980). Such a simple method will reveal whether the unknown contains no diaminopimelic acid (DAP) or the *meso-* or the L- form of this dibasic acid.

If no DAP is present, two main generic possibilities exist among the organisms covered in this manual: *Oerskovia* and *Promicromonospora* (Group 22). If the unknown does not fit in these two genera and is not an anomalous *Actinoplanes* or *Actinomadura,* an examination of organisms as described in *Bergey's Manual of Systematic Bacteriology,* Volume 2, is recommended.

If the L isomer of DAP (cell wall type I) is present as a major constituent, the number of generic possibilities among the organisms treated in this manual is still limited: *Nocardioides* (Group 22) and *Intrasporangium* (Group 25), all of the genera treated in Group 25, and *Kitasatosporia* (Group 29). Also to keep in mind are members of some genera included in Group 20 (*Arachnia*).

Although the number of genera characterized by the presence of cell walls of type I is small, the number of species may be large. Unfortunately, species are still poorly defined in the genus *Streptomyces,* and even with numerical taxonomy identification to species may become a nightmare.

If the organism to be identified contains major amounts of *meso-*DAP, many generic possibilities exist, and a knowledge of its whole cell sugar pattern (WCSP) is essential.

The presence of xylose and arabinose (WCSP of type D) will usually indicate a type II cell wall (*meso-*DAP + glycine), and the organism, if described in this volume, should be in Group 24 or in Group 29 (*Glycomyces*). Also to be considered is *Catellatospora* (Group 24). The

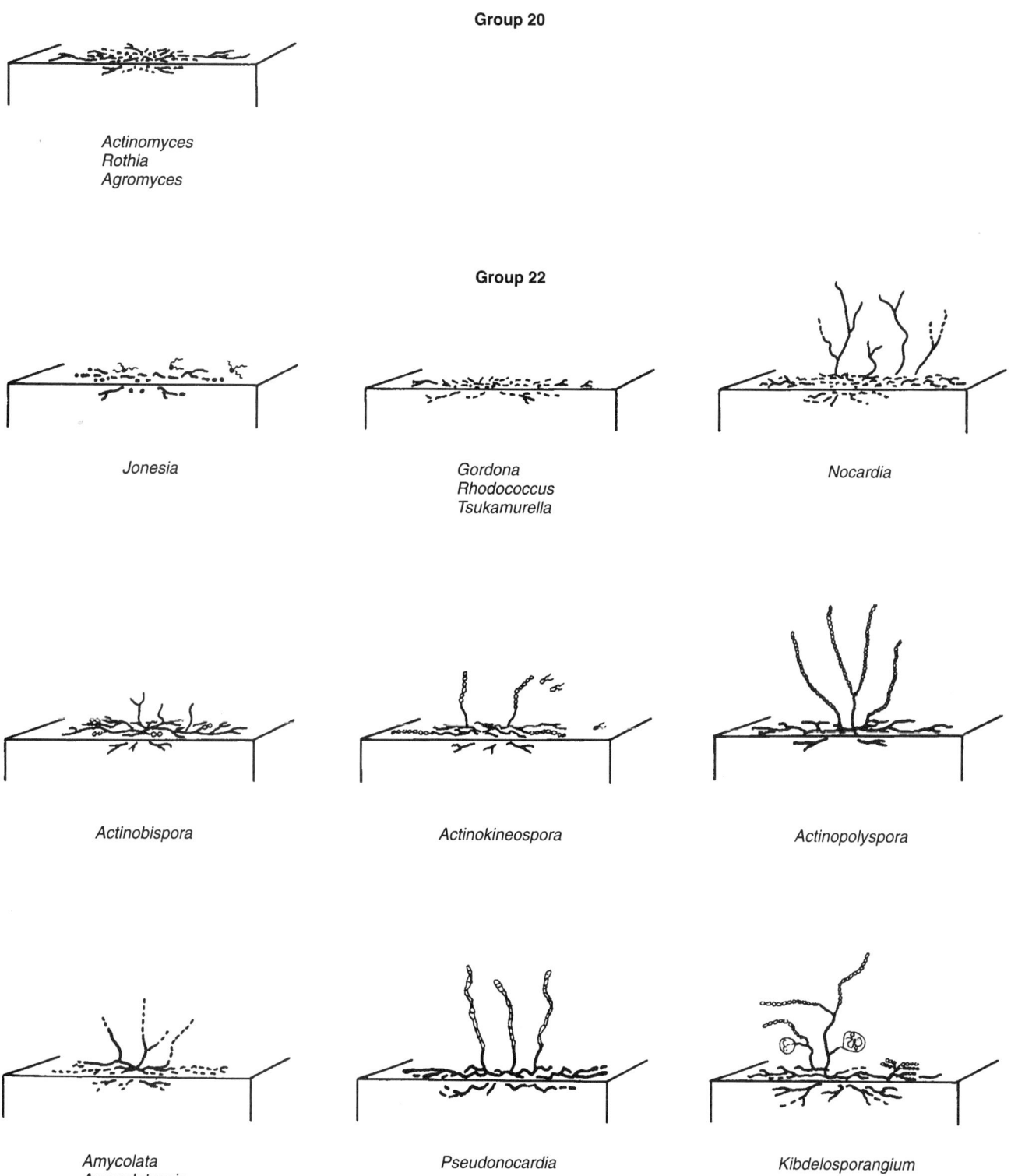

Group 20

Actinomyces
Rothia
Agromyces

Group 22

Jonesia

Gordona
Rhodococcus
Tsukamurella

Nocardia

Actinobispora

Actinokineospora

Actinopolyspora

Amycolata
Amycolatopsis
Pseudomycolata

Pseudonocardia

Kibdelosporangium

Figure 8. Schematic diagrams of substrate and aerial growth of genera of *Actinomycetales* and other genera with similar morphology.

Group 22 (continued)

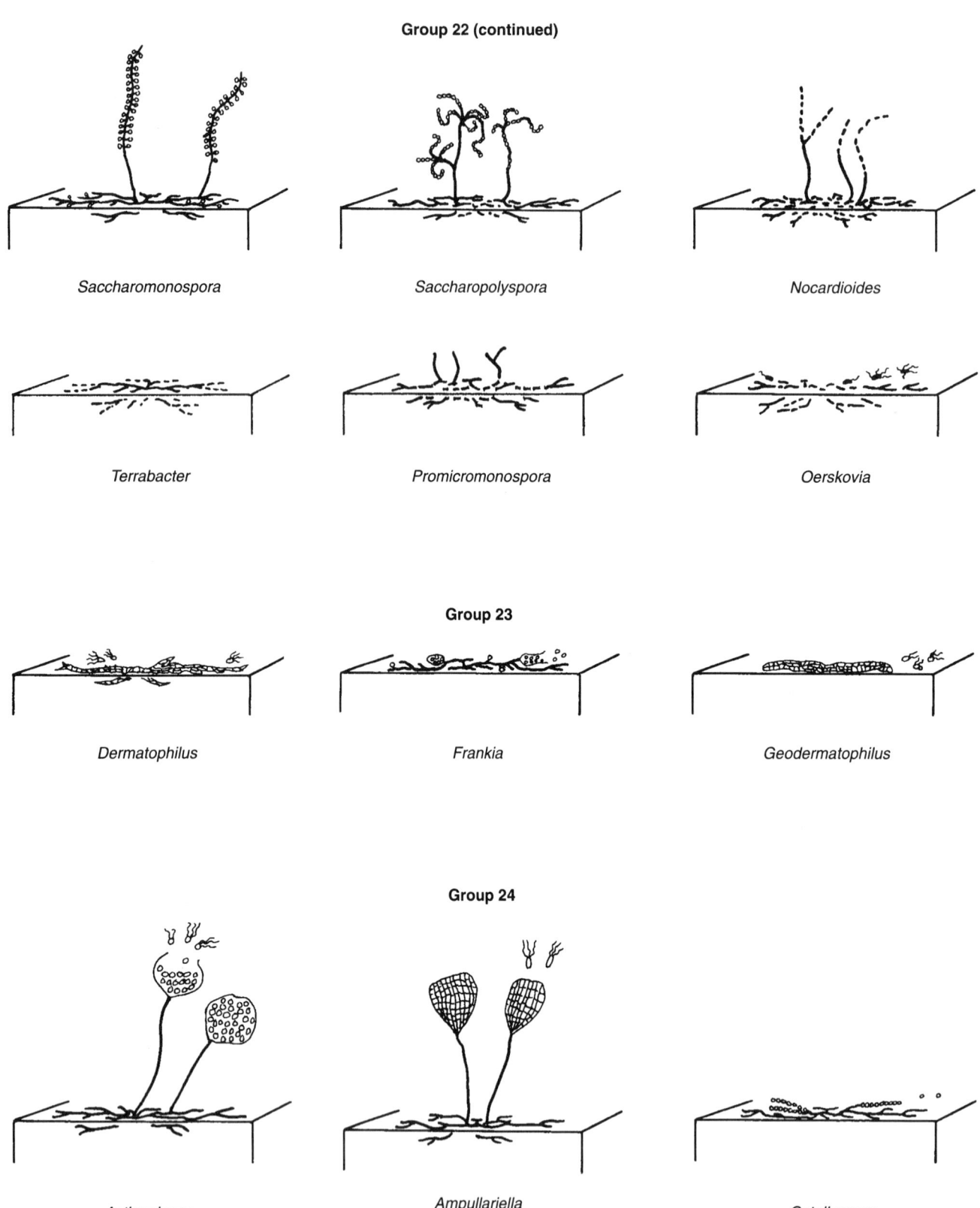

Saccharomonospora

Saccharopolyspora

Nocardioides

Terrabacter

Promicromonospora

Oerskovia

Group 23

Dermatophilus

Frankia

Geodermatophilus

Group 24

Actinoplanes

Ampullariella
Pilimelia

Catellospora

Figure 8. (continued)

Group 24 (continued)

Dactylosporangium

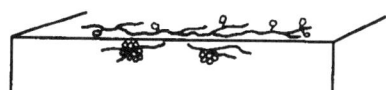

Micromonospora

Group 25

Intrasporangium

Kineosporia

Sporichthya

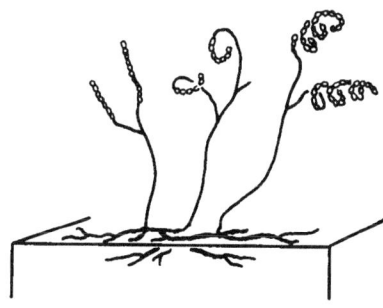

Streptomyces

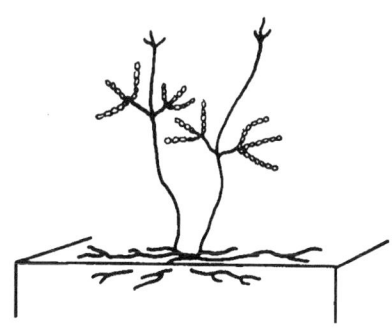

Streptoverticillium

Group 26

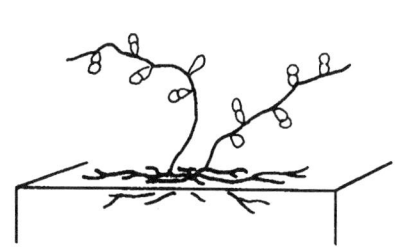

Microbispora

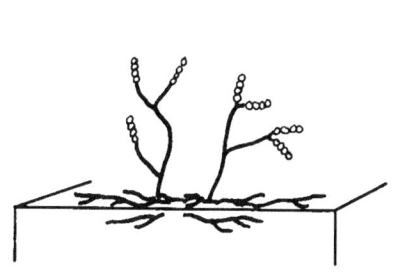

Microtetraspora

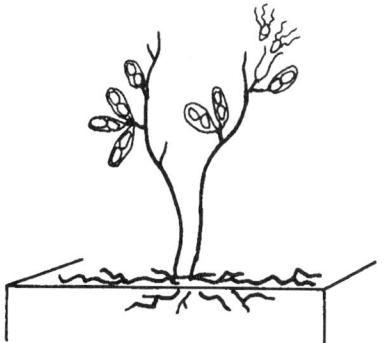

Planobispora

Figure 8. (continued)

Group 26 (continued)

Planomonospora

Spirillospora

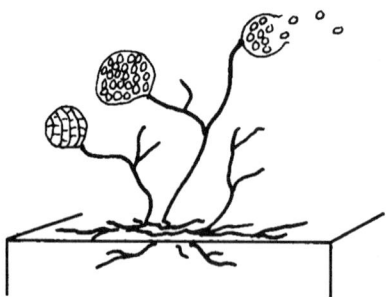

Streptosporangium

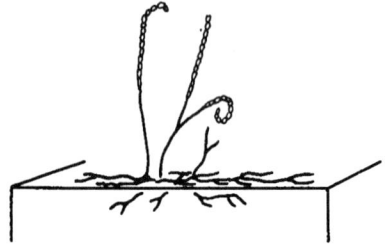

Actinomadura

Group 27

Actinosynnema

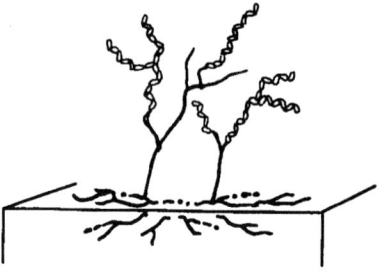

Nocardiopsis

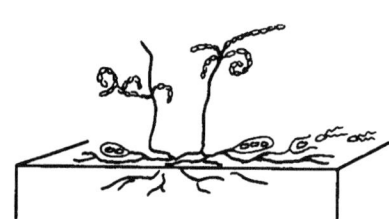

Streptoalloteichus

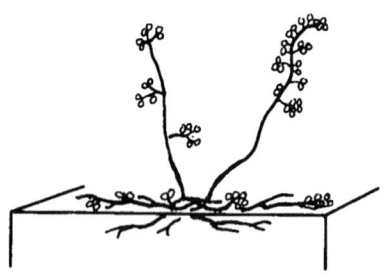

Thermomonospora

Figure 8. (continued)

Group 28

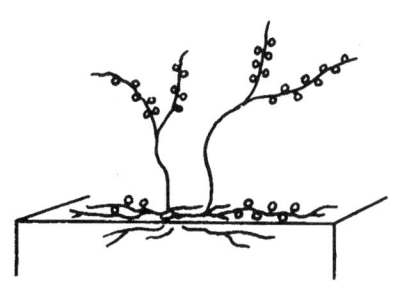

Thermoactinomyces

Group 29

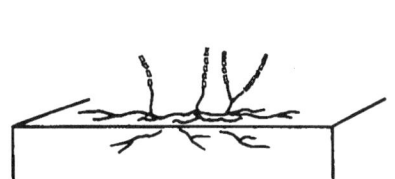

Glycomyces

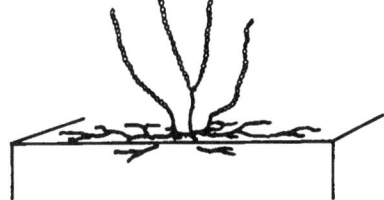

Kitasatosporia

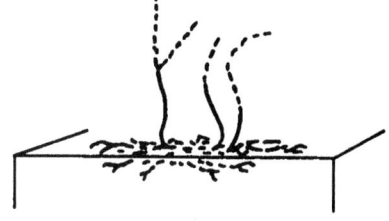

Saccharothrix

Figure 8. (continued)

sole exception to this rule is *Frankia,* where a WCSP of type D corresponds to a cell wall composition of type III.

The presence of madurose (3-*O*-methyl-D-galactose, WCSP of type B) will indicate a cell wall of type III of the *Actinomadura* variety and will lead to Group 26 or, if multilocular sporangia are formed, to some strains of *Frankia* and to the genus *Dermatophilus* (Group 23). The presence of rhamnose without other diagnostic sugars indicates a possible *Saccharothrix.*

Organisms containing fucose in their WCSP may belong to the genus *Frankia* or to *Actinoplanes.* Some plant pathogenic corynebacteria also contain fucose.

Arabinose and galactose present in the whole cell hydrolysates (WCSP of type A, cell wall type IV) will point to many of the organisms discussed in Group 22 or to the mycobacteria (Group 21) and the corynebacteria (Group 20). In addition, species of *Kibdelosporangium* (Group 29), *Amycolata,* and *Amycolatopsis* (Group 22) also have cell walls of type IV.

The large group of organisms with type IV cell walls can be separated into (a) those with mycolic acids such as the corynebacteria, the mycobacteria, and the nocardiae and (b) those without mycolic acids (*Pseudonocardia, Saccharomonospora, Saccharopolyspora, Actinopolyspora, Amycolata, Amycolatopsis,* and *Kibdelosporangium*).

On the basis of their molecular weights, the mycolates can be separated into the small corynomycolates, the medium nocardomycolates, and the large mycolates of mycobacteria.

The absence of any of the above-mentioned sugars in the whole cell hydrolysates is an indication of cell walls of type III without madurose (WCSP of type C). Organisms with this type of chemistry are very diverse and include some with multilocular sporangia (*Geodermatophilus,* Group 23), the organisms covered in Groups 27 and 28, and two of the four genera treated in Group 29 (*Saccharothrix* and *Kitasatosporia*). Strains of *Kitasatosporia* are anomalous in the sense that their mycelium has cell walls of type III but the cell walls of their spores are of type I.

The information discussed above is summarized in Table 1.

Important Notes for Users of This Manual

Unless otherwise indicated in footnotes to tables, the meanings of symbols are as follows:

- \+ 90% or more of strains are positive
- – 90% or more of strains are negative
- d 11-89% of strains are positive
- v strain instability (not equivalent to "d")
- D Different reactions in different taxa (species of a genus or genera of a family)

All other symbols are defined in footnotes to tables.

A BRIEF GUIDE TO GENERIC GROUPS

Gram-positive. Usually form **branching filaments** (0.5–1.0 μm in diameter). Filaments may fragment into **irregularly sized elements** or remain stable and produce **arthrospores.** Spores are produced singly, in chains of various lengths, or in sporangia. Spores are usually **nonmotile,** but some genera produce **flagellate spores.** Some genera that do not produce branching filaments are phylogenetically related to this group. Genera are distinguished on the basis of their **morphology** and **marker chemical constituents** of the cell wall, membranes, and whole cell hydrolysates. Mainly aerobic, but some genera are facultatively or obligately anaerobic. **Chemoheterotrophic,** using a wide variety of energy sources, including complex polymers. Mainly **free living** in a wide range of habitats. Some are **pathogens** of humans, animals, or plants. **Symbiotic** nitrogen-fixing associations are formed with plants.

A brief guide to generic groups:

Group 22. Nocardioform actinomycetes

This is a heterogeneous group, many of which form filaments which fragment into shorter elements. Aerial growth is formed by some genera and may produce chains of spores. Genera are distinguished primarily by wall chemotypes, the presence or absence of mycolic acids, and other chemical characters.

Subgroup 1 Mycolic acid-containing bacteria

Subgroup 2 *Pseudonocardia* and related genera

Subgroup 3 *Nocardioides* and *Terrabacter*

Subgroup 4 *Promicromonospora* and related genera

Group 23. Genera with multilocular sporangia

These genera form filaments that divide by longitudinal and transverse septa. This produces large numbers of coccoid-like elements, which may be motile (*Dermatophilus, Geodermatophilus*) or nonmotile (*Frankia*).

Group 24. Actinoplanetes

Stable filaments are formed, with little or no aerial growth. Motile spores are produced in sporangia (*Actinoplanes, Ampullariella, Dactylosporangium, Pilimelia*), or nonmotile spores are produced singly (*Micromonospora*) or in chains (*Catellatospora*). Cell walls contain *meso*-DAP and glycine, and arabinose and xylose are found in whole cell hydrolysates.

Group 25. Streptomycetes and related genera

A heterogeneous group, all of which have cell walls containing L-DAP and glycine. Stable filaments are formed and may produce extensive aerial growth with long spore chains (*Streptomyces, Streptoverticillium*). Other genera (*Intrasporangium, Kineosporia, Sporichthya*) produce little or no aerial growth and have a variety of spore forms.

Group 26. Maduromycetes

Stable filaments are formed and produce varying amounts of aerial growth, which bears spores. Short chains of nonmotile arthrospores are produced by *Microbispora* (two spores), *Microtetraspora* (four spores), and *Actinomadura* (varying number). Other genera produce spores in sporangia which are motile (*Planobispora, Planomonospora, Spirillospora*) or nonmotile (*Streptosporangium*). The cell walls contain *meso*-DAP, and cell hydrolysates contain madurose.

Subgroup 1 *Streptosporangium* and related genera

Subgroup 2 *Actinomadura*

Group 27. Thermomonospora and related genera

Stable filaments are formed and produce aerial growth bearing spores that are single (*Thermomonospora*), in chains (*Actinosynnema, Nocardiopsis*), or in sporangia-like structures (*Streptoalloteichus*). The cell walls contain *meso*-DAP, but no characteristic amino acids or sugars in whole cell hydrolysates.

Group 28. Thermoactinomycetes

This comprises only one genus, *Thermoactinomyces*. The stable filaments produce aerial growth. Single spores (which are endospores) are formed on both aerial and vegetative filaments. All species are thermophilic. The cell walls contain *meso*-DAP but no characteristic amino acids or sugars.

Group 29. Other genera

This group comprises three genera that cannot at present be assigned to other groups. They all produce aerial growth bearing chains of spores.

Important Notes for Users of This Manual

Unless otherwise indicated in footnotes to tables, the meanings of symbols are as follows:

- \+ 90% or more of strains are positive
- − 90% or more of strains are negative
- d 11-89% of strains are positive
- v strain instability (not equivalent to "d")
- D Different reactions in different taxa (species of a genus or genera of a family)

All other symbols are defined in footnotes to tables.

Table 1   Guide to the chemical and morphological properties of the genera of actinomycetes

Diagnostic amino acid	Diagnostic sugar	Typical key morphological features and additional chemical properties	Possible generic assignment (group number)
No DAP [a]	NA [a]	Only substrate mycelium formed; breaks into motile elements	Oerskovia (22)
		Only substrate mycelium formed; breaks into motile coccoid elements in older cultures	Jonesia (22)
		Sterile aerial mycelium formed; substrate mycelium breaking up into nonmotile elements	Promicromonospora (22)
	Xylose	Sporangia with motile spores	Actinoplanes (24)
	Madurose	Short chains of conidia on the aerial mycelium	Actinomadura (26)
L-DAP	NA	Both aerial and substrate mycelia breaking up into fragments	Nocardioides (22)
		Only substrate mycelium formed; breaks into rods and cocci	Terrabacter (22)
		Only substrate mycelium formed bearing terminal or subterminal vesicles	Intrasporangium (25)
		Aerial mycelium with long chains of spores	Streptomyces (25) Kitasatosporia (29)
		Sclerotia formed (Chainia type)	Streptomyces (25)
		Very short chains of large conidia formed (Microellobosporia type) on aerial and vegetative mycelium	Streptomyces (25)
		Whorls of straight chains of conidia formed	Streptoverticillium (25)
		No aerial mycelium; club-shaped sporangia formed terminally on the vegetative mycelium	Kineosporia (25)
		Aerial mycelium only, motile elements formed	Sporichthya (25)
meso-DAP [b]	Xylose and arabinose	No sporangia; single conidia formed on substrate mycelia, often in large black mucoid masses	Micromonospora (24)
		No sporangia; short chains of conidia formed protruding from the surface of the colonies	Catellatospora (24)
		Chains of conidia on aerial mycelium	Glycomyces (29)
		Dactyloid oligosporic sporangia protruding from the surface of the colonies; spores motile	Dactylosporangium (24)
		Sporangia containing spherical motile spores formed on the surface of colonies	Actinoplanes (24)
		Same; rod-shaped sporangiospores motile by polar flagella	Ampullariella (24)
		Same; sporangiospores with lateral flagella	Pilimelia (24)
		Multilocular sporangia formed; spores nonmotile	Frankia (23)

621

Table 1   *(continued)*

Diagnostic amino acid	Diagnostic sugar	Typical key morphological features and additional chemical properties	Possible generic assignment (group number)
meso-DAP	Madurose	Short chains of conidia on aerial mycelium, often curled into a crozier	*Actinomadura* (26)
		Chains of conidia with only two spores	*Microbispora* (26)
		Chains of conidia mainly with four (2–6) spores	*Microtetraspora* (26)
		Sporangia formed with two motile spores	*Planobispora* (26)
		Sporangia formed with only one motile spore	*Planomonospora* (26)
		Spherical sporangia formed on aerial mycelium containing many motile rod-shaped spores	*Spirillospora* (26)
		Spherical sporangia formed on aerial mycelium containing many aplanospores	*Streptosporangium* (26)
		Multilocular sporangia formed	*Dermatophilus* (23) *Frankia* (23)
meso-DAP	Fucose	Multilocular sporangia formed	*Frankia* (23)
		Sporangia with motile spores	*Actinoplanes* (24)
	Rhamnose and galactose	Both aerial and substrate hyphae fragment into nonmotile elements	*Saccharothrix* (29)
	Rhamnose, galactose, and mannose	*Streptomyces* type of morphology	*Streptoalloteichus* (27)
	Galactose	*Streptomyces* type of morphology	*Kitasatosporia* (29)
	Arabinose and galactose	NMA [a] present; morphology ranging from fugaceous substrate mycelium only to *Streptomyces*-like	*Nocardia* (22)
		NMA present; soft, salmon to pink organisms	*Rhodococcus* (22)
		NMA present; only substrate mycelium formed, which breaks into rods and cocci	*Gordona* (22)
		NMA present; only substrate mycelium formed, which breaks into single, paired, or massed rods	*Tsukamurella* (22)
		NMA present; paired spores formed on substrate mycelium; aerial mycelium sparse	*Actinobispora* (22)
		NMA absent; long cylindrical spores on the aerial mycelium; spores formed by budding	*Pseudonocardia* (22)
		NMA absent; single spores formed mainly on the aerial hyphae	*Saccharomonospora* (22)
		NMA absent; very long chains of conidia on the aerial mycelium	*Saccharopolyspora* (22)
		NMA absent; long chains of conidia on the aerial mycelium; halophile	*Actinopolyspora* (22)

Table 1 *(continued)*

Diagnostic amino acid	Diagnostic sugar	Typical key morphological features and additional chemical properties	Possible generic assignment (group number)
		NMA absent; substrate mycelium tends to break into nonmotile elements; aerial hyphae may be formed and may also segment	*Amycolata* (22) *Amycolatopsis* (22)
		NMA absent; aerial mycelium bearing curled hyphae embedded in an amorphous matrix	*Kibdelosporangium* (29)
		NMA absent; long chain of spores formed on aerial and substrate mycelium; spores become motile in an aqueous environment	*Actinokineosporia* (22)
		NMA absent; aerial mycelium tends to fragment into rods and cocci; short chains of spores also formed	*Pseudoamycolata* (22)
	No diagnostic sugar	Single conidia formed; these are heat-resistant bacterial endospores	*Thermoactinomyces* (28)
		Same as above but the spores are not heat-resistant	*Thermomonospora* (27)
		Long chains of spores formed by the aerial hyphae	*Nocardiopsis* (27)
		Aerial hyphae, often united into synnemata releasing motile spores	*Actinosynnema* (27)
		Multilocular sporangia releasing motile spores	*Geodermatophilus* (27)

[a] Abbreviations: DAP, diaminopimelic acid; NA, not applicable; NMA, nocardomycolic acid.
[b] May also contain hydroxy forms of DAP that may even replace *meso*-DAP.

GROUP 22

NOCARDIOFORM ACTINOMYCETES

This group is divided into four subgroups: 1, mycolic acid-containing bacteria; 2, *Pseudonocardia* and related genera; 3, *Nocardioides* and *Terrabacter;* 4, *Promicromonospora* and related genera.

SUBGROUP 1: MYCOLIC ACID-CONTAINING BACTERIA

Nonmotile actinomycetes that show considerable morphological diversity. Corynebacteria, tsukamurellae, many rhodococci and most mycobacteria form **straight to curved rods or irregular elements,** whereas nocardiae, some rhodococci, and a few mycobacteria typically produce **a branched mycelium that sooner or later fragments into bacillary and coccoid elements that are markedly pleomorphic.** Nocardiae and some mycobacteria have aerial hyphae that range from being sparse and invisible to the naked eye to completely covering the substrate mycelium with a white down. Neither endospores nor sclerotia are present. **Stain Gram positive or Gram variable** and are usually acid-fast at some stage of growth. **Mainly aerobic, but some corynebacteria are facultatively anaerobic. Chemoorganotrophic.** Diverse compounds are used as the sole carbon and energy sources. **Catalase positive. Cell wall peptidoglycan is based upon *meso*-diaminopimelic acid and is of the Alγ type. The glycan moiety of the cell wall contains *N*-glycolyl residues, except for corynebacteria, which have muramic acid in the *N*-acetylated form. Major cell wall sugars are arabinose and galactose. The wall envelope contains α-branched β-hydroxylated fatty acids and the mycolic acids, with 22 to 90 carbon atoms and major proportions of straight-chain saturated, monounsaturated, and, apart from most corynebacterial species, 10-methyl (tuberculostearic) branched-chain acids. Contain diphosphatidylglycerol, phosphatidylinositol, phosphatidylinositol mannosides, and, with the exception of most corynebacteria, phosphatidylethanolamine;** neither phosphatidylcholine nor phospholipids containing glucosamine are present. Some strains also produce trehalose dimycolates and other glycolipids. Menaquinones (vitamin K₂) are the sole respiratory quinones. Widely distributed, notably in soil, though some species live in association with animals. Certain strains are important pathogens of animals, including human beings.

Differentiation of the genera in **Subgroup 1:** See Table 22.1.

Examination of whole-organism methanolysates for the presence of mycolic acids is the first stage in the chemical procedure. Qualitative evaluation of mycolic acids can be easily and quickly achieved using the thin-layer chromatographic system mentioned in the table. When the presence of mycolic acids has been detected, their esters can be isolated and studied further by pyrolysis gas chromatography. Nocardiae, apart from *N. amarae,* rhodococci, and tsukamurellae can be separated solely on the basis of their predominant menaquinone composition (see Table 22.1).

Genus **Gordona**

Editorial note: The genus *Gordona* was not included in *Bergey's Manual of Systematic Bacteriology.* The genus was reintroduced in 1988 by Stackebrandt et al. (J. Gen. Appl. Microbiol. *34:* 341–348) for four species classified in the genus *Rhodococcus.* A further species, *Rhodococcus obuensis* (Tsukamura, Microbiol. Immunol. *26:* 1101–1119, 1982) was subsequently considered to be a subjective synonym of *Gordona* (*Rhodococcus*) *sputi* (Zakrzewska-Czerwinska et al., J. Gen. Microbiol. *134:* 2807–2813, 1988).

Short rods and cocci. Rough brownish, pink, or orange to red colonies are formed on glucose yeast extract agar, Sauton's agar, and egg media. **Stain Gram positive or Gram variable.** Contain mycobactins. **Usually partially acid-fast. Aerobic. The glycan moiety of the cell wall has *N*-glycolyl residues. The wall envelope contains mycolic acids with 48 to 66 carbon atoms and up to four double bonds and major proportions of straight-chain saturated, monounsaturated, and 10-methyl (tuberculostearic)-branched fatty acids. Fatty acid esters released on pyrolysis gas chromatography of mycolic esters have**

16 to 18 carbon atoms. **Cells contain diphospha-tidylglycerol, phosphatidylethanolamine, and phosphatidylinositol dimannosides as major phospholipids. Dihydrogenated menaquinone with nine isoprene units (MK-9(H$_2$)) is the predominant isoprenolog.** Isolated from soil and from sputa of patients with bronchiectasis and cavitary pulmonary tuberculosis.

Type species: *Gordona bronchialis.*

Differentiation of the species of the genus **Gordona:** See Table 22.2.

Genus **Nocardia**

Rudimentary to extensively branched vegetative hyphae, 0.5–1.2 μm in diameter, growing on the surface of and penetrating agar media; these hyphae often fragment in situ or on mechanical disruption into **bacteroid, rod-shaped to coccoid elements. Aerial hyphae,** at times visible only microscopically, **are almost always produced.** Short-to-long chains of well to poorly formed conidia may occasionally be found on the aerial hyphae and, more rarely, on both aerial and substrate hyphae. **Stain Gram positive to Gram variable. Usually partially acid-fast. Aerobic.** Colonies have a chalky, mat or velvety appearance and may be brown, tan, pink, orange, red, purple, gray, or white; smooth or granular; and irregular, wrinkled, or heaped. Soluble brown or yellow pigments may be produced. **The glycan moiety of the cell wall has** *N*-**glycolyl residues. The wall envelope contains mycolic acids with 46 to 60 carbons and up to 3 double bonds and major proportions of straight-chain, unsaturated, and 10-methyl (tuberculostearic)-branched fatty acids. Fatty acid esters released on pyrolysis gas chromatography of mycolic esters have 12 to 18 carbon atoms and may be saturated or unsaturated.** These cells contain diphosphatidylglycerol, phosphatidylethanolamine, phospha-tidylinositol, and phosphatidylinositol mannosides as major phospholipids. **The predominant menaquinone in most nocardiae corresponds to a hexahydrogenated menaquinone with eight isoprene units in which the two end units are cyclized** (i.e., II,III-tetrahydro-ω[2,6,6-trimethylcyclohex-2-enylmethyl]menaquinone-6); *N. amarae* **contains dihydrogenated menaquinone with nine isoprene units (MK-9(H$_2$)) as the major isoprenolog.** Widely distributed and abundant in soil. Some strains are pathogenic opportunists for humans and ani-

mals; they include causal agents of actinomycete mycetoma and nocardiosis.

Type species: *Nocardia asteroides.*

Differentiation of the species of the genus **Nocardia:** See Table 22.3.

Genus **Rhodococcus**

Rods to extensively branched vegetative mycelium may be formed. In all strains the morphogenetic cycle is initiated with the coccus or short rod stage, with different organisms showing a succession of more or less complex morphological stages by which the completion of the growth cycle is achieved. Thus, **cocci may merely germinate into short rods, form filaments with side projections, show elementary branching, or, in the most differentiated forms, produce extensively branched hyphae.** The next generation of cocci or short rods is formed by **fragmentation of the rods, filaments, and hyphae.** Some strains produce feeble, microscopically visible aerial hyphae, which may be branched, or aerial synnemata consisting of unbranched filaments that coalesce and project upward. Colonies may be rough, smooth, or mucoid and pigmented buff, cream, yellow, orange, or red, although colorless variants do occur. **Stain Gram positive. Usually partially acid-fast. Aerobic.** Do not contain mycobactins. **Sensitive to lysozyme. The glycan moiety of the cell wall has** *N*-**glycolyl residues. The wall envelope contains mycolic acids with 34 to 52 carbon atoms and up to 3 double bonds and major proportions of straight-chain saturated, unsaturated, and 10-methyl (tuberculostearic)-branched fatty acids. Fatty acid esters released on pyrolysis gas chromatography of mycolic esters have 12 to 18 carbon atoms. Cells contain diphosphatidylglycerol, phosphatidylethanolamine, and phosphatidylinositol dimannosides as major phospholipids. Dihydrogenated menaquinone with eight isoprene units (MK-8(H$_2$)) form the predominant isoprenolog.** Widely distributed but particularly abundant in soil and herbivore dung. Some strains are pathogenic for animals, including human beings.

Type species: *Rhodococcus rhodochrous.*

Differentiation of the species of the genus **Rhodococcus:** See Table 22.4.

Genus Tsukamurella

Editorial note: The genus *Tsukamurella* was not included in *Bergey's Manual of Systematic Bacteriology*. The genus was created in 1988 by Collins et al. (Int. J. Syst. Bacteriol. *38:* 385–391) when organisms previously classified as *Corynebacterium paurometabolum* and *Rhodococcus aurantiacus* were reduced to a single species, *T. paurometabolum*. A second species, *T. wratislaviensis*, was subsequently described (Goodfellow et al., Zbl. Bakt. *275:* 162–178, 1991).

Straight to slightly curved rods (0.5–0.8 × 1.0–5 µm) occur singly, in pairs, or in masses. Very short rods are also present. White/creamy to orange small colonies (0.5–2.0 mm in diameter) with convex elevation and entire edges (sometimes rhizoidal) are dryish but easily emulsified. **Stain Gram positive. Weakly to strongly acid-fast. Obligate aerobes. The glycan moiety of the cell wall has *N*-glycolyl residues. The cell envelope contains mycolic acids with 62 to 78 carbon atoms and up to six double bonds and major proportions of straight-chain saturated, unsaturated, and 10-methyl (tuberculostearic)-branched fatty acids. Fatty acid esters on pyrolysis gas chromatography of mycolic esters have 20 to 22 carbon atoms. Cells contain diphosphatidylglycerol, phosphatidylethanolamine, phosphatidylinositol, and phosphatidylinositol mannosides as major phospholipids. Unsaturated menaquinones with nine isoprene units form the predominant isoprenolog.** Isolated from soil, human sputum, and the mycetomes and ovaries of bedbugs (*Cimex lectularis*). Some strains have been reported to cause infection of the lung, lethal meningitis, and necrotizing tenosynovitis.

Type species: *Tsukamurella paurometabolum*.

Differentiation of the species of the genus **Tsukamurella:** See Table 22.5.

SUBGROUP 2: PSEUDONOCARDIA AND RELATED GENERA

Morphologically and physiologically diverse actinomycetes that form extensively branched vegetative and aerial hyphae. Smooth, spiny, or hairy spores are produced singly, in pairs, or in chains of variable length on aerial hyphae. Some species may also produce spores on vegetative hyphae. Kibdelosporangiae also produce sporangia-like structures on aerial hyphae. **Fragmentation of the mycelium can occur but is generally not pronounced.** Neither endospores nor sclerotia are present. **Nonmotile** apart from *Actinokineospora*. **Stain Gram positive. Non-acid fast. Aerobic. Generally chemoorganotrophic, though some are facultatively autotrophic.** Diverse compounds are used as the sole carbon and energy sources. **Catalase positive. The wall peptidoglycan is of the A1γ type; it contains alanine, glutamic acid, glucosamine, *meso*-diaminopimelic acid, and muramic acid with the sugars arabinose and galactose (wall chemotype IV). The glycan moiety of the cell wall has *N*-acetylated residues. The cell envelope does not contain mycolic acids but has major amounts of mainly *meso*-methyl components belonging to the *iso*- and *anteiso*-fatty acid series. Phospholipid composition is variable. Menaquinones (vitamin K₂) are the sole respiratory quinones.** Isolated from a wide range of habitats, notably soil and vegetable matter; some species cause hypersensitivity diseases.

Differentiation of the genera in **Subgroup 2:** See Table 22.6.

Genus Actinobispora

Editorial note: The genus *Actinobispora* was not included in *Bergey's Manual of Systematic Bacteriology*. The genus was created in 1991 by Jiang et al. (Int. J. Syst. Bacteriol. *41:* 526–528). It includes only one species, *A. yunnanensis*.

Long, irregularly branched, vegetative mycelium (0.3–0.6 µm in diameter) that does not fragment. Many paired, smooth-walled spores (0.7–0.9 µm in diameter, 1.0–1.4 µm long) are borne in longitudinal pairs on vegetative hyphae. Sparse and irregularly branched aerial hyphae (0.4–0.6 µm in diameter) carry single or longitudinal pairs of smooth-walled spores (0.6–0.9 µm in diameter). Mesophilic. Whole-organism hydrolysates have arabinose, galactose, and xylose as characteristic sugars. **Contain phosphatidylethanolamine and glucosamine-containing phospholipids. Dihydrogenated menaquinones with seven or nine isoprene units are the major isoprenologs.** Isolated from soil.

Type (and only) species: *Actinobispora yunnanensis*.

Characteristics of the species: As described for the genus.

Genus **Actinokineospora**

Editorial note: The genus *Actinokineospora* was not included in *Bergey's Manual of Systematic Bacteriology*. The genus was created in 1988 by Hasegawa (Actinomycetology *2:* 31–45). It includes only one species, *A. riparia.*

Colorless to brownish, moderately branched vegetative hyphae about 0.5 μm in diameter break up into rods during the preparation of smears but do not fragment in plate culture. White aerial mycelium are formed but not abundantly. **Aerial hyphae bear long chains of smooth-walled conidia that can differentiate into zoospores in an aqueous environment. The rod-shaped zoospores (0.5–0.6 x 0.7–1.0 μm) bear multi-trichous flagella and eventually form several germ tubes. Spores are covered with multilayered integuments with cores separated from cortex-like coats by a boundary region.** Chains of spores are also produced on vegetative hyphae. Growth occurs at 23–41°C and pH 5.0–9.0. **It does not occur in lysozyme broth or in the presence of 5% NaCl.** Rich in branched-chain fatty acids. **Unsaturated menaquinone with ten isoprene units (MK-10) is the major isoprenolog.** Whole-organism hydrolysates contain arabinose, galactose, glucose, mannose, and rhamnose. **Contain phosphatidylethanolamine, diphosphatidylglycerol, and phosphatidylinositol as major phospholipids.** The only strain isolated was from a soil sample collected adjacent to a river in Japan.

Type (and only) species: *Actinokineospora riparia.*

Characteristics of the species: As described for the genus.

Genus **Actinopolyspora**

Branching vegetative hyphae (0.4–2.0 μm in diameter) are formed. The vegetative mycelium is mostly unfragmented; fragmentation is occasionally observed near the colony center. **Long chains of smooth-walled spores are formed basipetally on aerial hyphae.** Spores are oval to cylindrical or coccoid to short rod-shaped. Spores are not observed on vegetative hyphae. In ultrathin sections, there are small cells between the mature spores. **Moderately to extremely halophilic. The wall envelope contains saturated, unsaturated, and *iso-* and *anteiso*-branched fatty acids; only traces of 10-methyl-branched components are evident. Cells contain diphosphatidylglycerol, phospha-** tidylcholine, phosphatidylglycerol, phosphatidylinositol, and phosphatidylmethylethanolamine. **Tetrahydrogenated menaquinones with either nine or ten isoprene units (MK-9(H$_4$), MK-10(H$_4$)) predominate.** Isolated from soil.

Type species: *Actinopolyspora halophila.*

Differentiation of the species of the genus **Actinopolyspora:** See Table 22.7.

The two species are only represented by the type strains, though a mutant of *A. halophila* ATCC 27976 is available.

Genus **Amycolata**

Editorial note: The genus *Amycolata* was not included in *Bergey's Manual of Systematic Bacteriology*. The genus was created in 1986 by Lechevalier et al. (Int. J. Syst. Bacteriol. *36:* 29–37) for organisms previously classified as *Nocardia autotrophica, Nocardia hydrocarbonoxydans,* and *Nocardia saturnea.* It now includes four species.

Branching vegetative hyphae (0.5–2.0 μm in diameter) tend to fragment into squarish elements. Aerial mycelium may or may not be present. **When formed, aerial hyphae may be stable or may differentiate into long chains of smooth-walled ellipsoidal to cylindrical spores. Spore chains are also produced on vegetative hyphae. The hyphal wall of both aerial and vegetative mycelium is covered by an electron-dense layer. Multiple divisions in intercalary or terminal portions of hyphae occur. Mesophilic.** Some strains are facultatively autotrophic. The wall envelope is rich in fatty acids of the *iso-/anteiso*-branched series. **Contain diphosphatidylglycerol, phosphatidylcholine, and phosphatidylmethylethanolamine as major phospholipids; phosphatidylinositol mannosides variably present, and phosphatidylethanolamine is absent. Di- and tetrahydrogenated menaquinones with eight isoprene units (MK-8(H$_2$, H$_4$)) predominate.** Isolated from diverse habitats including soil, decomposing vegetable matter, *Alnus* rhizosphere and root nodules, and clinical material.

Type species: *Amycolata autotrophica.*

Differentiation of the species of the genus **Amycolata:** See Table 22.8.

Genus **Amycolatopsis**

Editorial note: The genus *Amycolatopsis* was not included in *Bergey's Manual of Systematic Bacteriology.* The genus was created in 1986 by Lechevalier et al. (Int. J. Syst. Bacteriol. *36:* 29–37) for organisms previously classified as *Nocardia orientalis, Nocardia mediterranei, Nocardia rugosa,* and *Nocardia sulphurea.* The genus contains seven species.

Branching substrate hyphae (0.5–2.0 μm in diameter) fragment into squarish elements. Aerial mycelium may or may not be present. **Aerial hyphae, when formed, may be sterile or may differentiate into long chains of smooth-walled, squarish to ellipsoidal spore-like structures. Spore chains are also produced on vegetative hyphae. Mesophilic.** Some strains are facultatively autotrophic. The wall envelope is rich in fatty acids of the *iso-/anteiso*-branched series. **Contain phosphatidylethanolamine and phosphatidylglycerol with diphosphatidylglycerol, phosphatidylinositol, and phosphatidylinositol mannosides variably present; phosphatidylcholine and glucosamine-containing phospholipids are absent. The predominant isoprenologs are di-, tetra- and hexahydrogenated menaquinones with nine isoprene units (MK-9(H$_2$, H$_4$, H$_6$)).** Isolated from soil, vegetable matter, and clinical specimens.

Type species: *Amycolatopsis orientalis.*

Differentiation of the species of the genus **Amycolatopsis:** See Table 22.9.

Genus **Kibdelosporangium**

Filamentous, with well-developed, moderately branched vegetative hyphae (0.3–1.0 μm in diameter) that penetrate the agar and form a compact layer on top of the agar. **Vegetative hyphae may exhibit varying degrees of fragmentation and frequently bear specialized structures that appear to be dichotomously branched, septate hyphae radiating from a common stalk. Aerial hyphae bear long chains of smooth-walled, rod-shaped (0.4 μm wide x 0.8–2.8 μm long), and sporangium-like structures (9–22 μm in diameter) that are usually round but may be irregular. These structures are surrounded by a well-defined wall but contain hyphae embedded in an amorphous matrix rather than spores. They germinate directly, forming one or more germ tubes. Mesophilic.**

Whole-organism hydrolysates contain arabinose and galactose; traces of madurose may also be present. Fatty acids are mainly of the *iso-/anteiso*-branched series. **Contain diphosphatidylglycerol, phosphatidylethanolamine, and phosphatidylinositol mannosides; phosphatidylmethylethanolamine may also be present.** Widely distributed, but with low frequency, in soil.

Type species: *Kibdelosporangium aridum.*

Differentiation of the species of the genus **Kibdelosporangium:** See Table 22.10.

Genus **Pseudoamycolata**

Editorial note: The genus *Pseudoamycolata* was not included in *Bergey's Manual of Systematic Bacteriology.* The genus was created in 1989 by Akimov et al. (Int. J. Syst. Bacteriol. *39:* 457–461). It includes only one species, *P. halophobica.*

Yellowish orange vegetative and white to cream aerial hyphae tend to fragment into rod-shaped and oval elements. Short chains of spores may be produced on aerial hyphae. Swollen hyphal segments (up to 3 μm long) with transverse and longitudinal septa may be present. **Mesophilic.** Hydrolyze arbutin, esculin, gelatin, urea, and xanthine but not allantoin or starch. **Do not grow in the presence of 3% NaCl.** Saturated branched-chain fatty acids of the *iso-* and *anteiso-* series predominate; straight-chain saturated, unsaturated, 2-hydroxy-, and 10-methyl-branched components, including tuberculostearic acid (trace), are found as minor components. **Tetrahydrogenated menaquinone with eight isoprene units (MK-8(H$_4$)) is the major isoprenolog. Contain diphosphatidylglycerol, phosphatidylethanolamine, phospha-tidylinositol mannosides, and small amounts of phospha-tidylmethylethanolamine.** Isolated from soil.

Type (and only) species: *Pseudoamycolata halophobica.*

Characteristics of the species: As described for the genus.

Genus **Pseudonocardia**

Vegetative and aerial mycelium-bearing spores in chains; hyphae are segmented and **often zig-zag shaped,** with a tendency to form apical or intercalary swellings. Spores are smooth-walled or spiny, varying greatly in size, usually 0.5 to 1.0 μm wide by 1.5 to 3.5 μm long but varying in length from 1.0 to 4.5 μm. Hyphal elongation is by acropetal budding. Segments act directly as spores or are secondarily divided in the sporulation process. The hyphal wall has two layers, cross-walls are interspace septa, and spores are without intersporal pads. **Mesophilic or thermophilic.** The cell envelope contains *iso-, anteiso-,* and 10-methyl-branched saturated, *iso*-branched monounsaturated, and straight-chain saturated and monounsaturated components. **Contain hydroxyphosphatidylethanolamine, diphosphatidylglycerol, phosphatidylcholine, phosphatidylinositol, and phosphatidylmethylethanolamine as major phospholipids. Tetrahydrogenated menaquinone with nine isoprene units (MK-9(H$_4$)) is the major isoprenolog.** Isolated from rotting plant remains and from soil.

Type species: *Pseudonocardia thermophila.*

Differentiation of the species of the genus **Pseudonocardia:** See Table 22.11.

Genus Saccharomonospora

Produce predominantly single spores on aerial hyphae. The spores are heat-sensitive, nonmotile aleuriospores that are either sessile or formed at the tips of simple unbranched sporophores of variable length. On agar media, a branched vegetative mycelium forms leathery colonies, usually covered in aerial mycelium in which spores are densely packed along the hyphae. The aerial mycelium is initially white, becoming gray-green to dark green or bluish; green pigmentation may also occur on the vegetative mycelium and diffuse into the surrounding medium. Variants that are nonpigmented or lilac have also been observed. Spores on the vegetative hyphae and in pairs or short chains on the aerial hyphae are occasionally present. Amino acid and vitamin supplements (e.g., yeast extract) are required for growth. Growth is optimal in the pH range of 7.0–10.0. **Mesophilic and thermophilic. Growth is not inhibited by 3% NaCl.** Deaminase and phosphatase are produced. Casein, gelatin, starch, xylan, and **tyrosine are degraded. There is no activity against cellulose.** The cell envelope contains a mixture of *iso-/anteiso*-branched and straight-chain unsaturated components. **Contain diphosphatidylglycerol, phosphatidylethanolamine, phosphatidylglycerol, phosphatidylinositol, and phosphatidylmethylethanolamine**

as major phospholipids. Tetrahydrogenated menaquinone with nine isoprene units (MK-9(H$_4$)) is the major isoprenolog.** Common in composts, manure, and overheated fodders but are also found in soil, lake sediments, and peat.

Type species: *Saccharomonospora viridis.*

Differentiation of the species of the genus **Saccharomonospora:** See Table 22.12.

Genus Saccharopolyspora

Vegetative mycelium is well developed, branched, septate 0.4–0.8 μm in diameter. **Fragments into rod-shaped elements** about 1.0 x 0.5 μm, more often in older parts of the colony and seldom near the growing margins. **Aerial mycelium** is 0.5–1.2 μm in diameter, straight or in spirals, and characteristically is **segmented into beadlike chains of spores,** 0.7–1.5 x 0.5–0.7 μm, **usually separated by lengths of "empty" hypha and retained in a sheath. Mesophilic and thermophilic. Colonies are thin, raised or convex, slightly wrinkled, and mucoid or gelatinous** in appearance, with **sparse aerial mycelium, often produced in tufts** and mostly in the older parts. **Resistant to many antibiotics but susceptible to lysozyme.** The wall envelope is rich in *iso-* and *anteiso*-branched-chain fatty acids. **Contain diphosphatidylglycerol, phosphatidylcholine, phosphatidylglycerol, and phosphatidylmethylethanolamine.** Tetrahydrogenated menaquinone with nine isoprene units (MK-9(H$_4$)) is the predominant isoprenolog. Common in fodder, grain, and sugar cane bagasse and are also found in soil. Spores of some strains can cause extensive allergic alveolitis in sensitized subjects.

Type species: *Saccharopolyspora hirsuta.*

Differentiation of the species of the genus **Saccharopolyspora:** See Table 22.13.

SUBGROUP 3: NOCARDIOIDES AND TERRABACTER

Editorial note: The genus *Nocardioides* was classified with the nocardioform actinomycetes in *Bergey's Manual of Systematic Bacteriology.* The related genus *Pimelobacter,* which was not recognized in the manual, was proposed by Suzuki and Komagata (J. Gen. Appl.

Microbiol. *29:* 59–71) for *Arthrobacter* species with LL-diaminopimelic acid in the peptidoglycan (see *Arthrobacter*, Addendum II in *Bergey's Manual of Systematic Bacteriology*). Initially, three *Pimelobacter* species were recognized, namely *P. jensenii, P. (Arthrobacter) simplex,* and *P. (Arthrobacter) tumescens*. Nesterenko et al. (Microbiol. Zhurnal *47:* 3–12, 1985) created the family *Nocardioidaceae* to accommodate the genera *Nocardioides* and *Pimelobacter*. O'Donnell et al. (Arch. Microbiol. *133:* 323–329, 1982) proposed that *A. (Pimelobacter) simplex* be assigned to the genus *Nocardioides* as *N. simplex*. Subsequently, Collins et al. (Int. J. Syst. Bacteriol. *39:* 1–6, 1989) transferred *P. jensenii* to the genus *Nocardioides* as *N. jensenii* and raised *P. tumescens* to generic status as *Terrabacter tumescens*. An additional *Nocardioides* species, *N. fastidiosa,* was created in 1989 by Collins and Stackebrandt (Microbiol. Lett. *57:* 289–294).

The inclusion of the three *Pimelobacter* species in the genus *Nocardioides* makes the latter heterogeneous with respect to morphology, though the extent of the morphological diversity is debatable. Prauser (Int. J. Syst. Bacteriol. *26:* 58–65, 1976) noted that strain IMET 7801 lost the ability to form aerial mycelium on continued subculture. Furthermore, the original *Nocardioides* species exhibited a developmental cycle in which aging vegetative mycelia fragmented into irregular rod-shaped to coccoid elements similar to the "coryneform morphology" shown by subsequent additions to the genus.

Pleomorphic elements to branched vegetative and aerial mycelium are formed. When produced, aerial and vegetative mycelia break up into fragments that may be **irregular, rodlike, or coccoid.** Motile and nonmotile strains occur. **Stain Gram positive.** Non-acid-fast. **Strictly aerobic.** Chemoorganotrophic, with oxidative metabolism. **Catalase positive. The peptidoglycan contains LL-diaminopimelic acid, glutamic acid, and glycine and is of the A3γ type.** Complex fatty acids are primarily of the *iso-, anteiso-,* and 10-methyl-branched types. Mycolic acids are absent. **Tetrahydrogenated menaquinones with eight or nine isoprene units (MK-8(H₄), 9(H₄)) predominate.** Occur in soil.

Differentiation of the genera in **Subgroup 3:** See Table 22.14.

Genus **Nocardioides**

Pleomorphic elements to branched vegetative mycelium are formed. The vegetative mycelium shows abundantly branching hyphae growing on the surface and penetrating into agar media; they break up into fragments that may be irregular, rodlike, or coccoid. When formed, aerial mycelium consists of irregular, sparsely and irregularly branching, or unbranched hyphae that **break up into short to elongated rodlike fragments.** The fragments of both the primary and the aerial mycelium give rise to new mycelia. Motile and nonmotile strains occur. **Susceptible to specific phages.** Fatty acid profiles contain *iso-, anteiso-,*straight chain, unsaturated, and tuberculostearic (10-methyloctadecanoic) acid types with 14-methylpentadecanoic acid predominating. **Contain diphosphatidylglycerol, phosphatidylglycerol, and two incompletely characterized lipids;** phosphatidylethanolamine and other nitrogenous phospholipids are lacking. **Tetrahydrogenated menaquinones with either eight or nine isoprene units (MK-8(H₄), 9(H₄)) predominate.** Found worldwide in soil and also in herbage.

Type species: *Nocardioides albus.*

Differentiation of the species of the genus **Nocardioides:** See Table 22.15.

Genus **Terrabacter**

Editorial note: The genus *Terrabacter* was not included in *Bergey's Manual of Systematic Bacteriology*. The genus was created in 1989 by Collins et al. (Int. J. Syst. Bacteriol. *39:* 1–6) for organisms previously classified as *Arthrobacter tumescens* or *Pimelobacter tumescens*.

The rod-coccus growth cycle occurs during growth on complex media. Cells from older cultures are coccoid (0.5–0.8 μm in diameter). After transfer to fresh complex media, long, irregular rods (0.6–1.2 x 2.0–6.0 μm) are formed. **The long rods show primary branching. Aerial mycelium is not produced.** Generally nonmotile (occasionally motile strains occur). Growth occurs at 10–35°C. The optimum temperature for growth is approximately 25–30°C. Does not survive heating at 63°C for 30 min. **Thiamine is required for growth.** Acid is not produced from glucose and other sugars in peptone-based media. Grows in the presence of 5% NaCl. Oxidase negative. Cellulose is not hydrolyzed. Nitrate is reduced to nitrite. Fatty acids are mainly of the *iso*-branched types; straight-chain saturated, *anteiso*-methyl-branched, and monounsaturated *iso*-methyl-branched acids are also present. **Contain diphosphatidylglycerol, phosphatidylethanolamine, phosphatidylinositol, and some unknown amino-containing phosphoglycolipids. Tetrahydrogenated menaquinone with eight isoprene units saturated at Sites II and III (MK-8(II, III-H$_4$)) is the predominant isoprenolog.** Occurs in soil.

Type (and only) species: *Terrabacter tumescens.*

Characteristics of the species: As described for the genus.

SUBGROUP 4: PROMICROMONOSPORA AND RELATED GENERA

Editorial note: The genera *Promicromonospora* and *Oerskovia* were classified with the nocardioform actinomycetes in *Bergey's Manual of Systematic Bacteriology* primarily on the basis of their ability to form a mycelium that fragmented. These genera have little in common with other nocardioform actinomycetes but share many properties with one another and with the genus *Jonesia.*

Irregular rods to extensively branched vegetative and aerial mycelia are formed. Endospores are not produced. Stain Gram positive. Non-acid-fast. **Aerobic to facultatively anaerobic. Mesophilic.** The peptidoglycan contains **L-lysine and is of the A4α type.** Long chain fatty acids consist predominantly of straight-chain saturated *anteiso-* and *iso-* methyl branched-chain acids. **Mycolic acids are absent.** Major polar lipids include **phosphatidylglycerol and glucosamine-containing phospholipids** (phospholipid type 5). Menaquinones (vitamin K$_2$) are the sole respiratory quinones. Isolated from soil and vegetable matter.

Differentiation of the genera in **Subgroup 4:** See Table 22.16.

Genus **Jonesia**

Editorial note: The genus *Jonesia* was not included in *Bergey's Manual of Determinative Bacteriology.* The genus was created in 1987 by Rocourt et al. (Int. J. Syst. Bacteriol. *37:* 266–270) for an organism previously clas-

sified in the manual as *Listeria denitrificans.* It includes only one species, *Jonesia denitrificans.*

Irregular rods, 0.3–0.5 μm in diameter and 2–3 μm in length, show branched Y and clublike forms. Filamentous and coccoid cells may occur in older cultures. **Gram positive,** but many cells, especially in older cultures, are readily decolorized. Coccoid forms that develop in older cultures always stain Gram positive. The coccoid elements give rise to rod forms on transfer to a fresh medium. **Motile by means of peritrichous flagella at both 25° and 37°C. Facultatively anaerobic.** Colonies on nutrient agar are 0.5–1.5 mm in diameter after 24–48 hours, convex, smooth, edge entire, grayish, and translucent to opaque, becoming yellowish in 10–20 days. Grows at 10–40°C; the optimum growth temperature is around 30°C. Catalase positive and oxidase negative. **Teichoic acids are of the poly (ribitol phosphate) type** (ribitol-galactosamine-galactoses). Lipoteichoic acid is absent. **12-Methyltetradecanoic acid (*anteiso*-C$_{15:0}$) and 14-methylhexadecanoic acid (*anteiso*-C$_{17:0}$)** are also present. Contain diphosphatidylglycerol, phosphatidylinositol, two phosphoglycolipids, and small amounts of two unidentified phospholipids as major components. **Unsaturated menaquinone with nine isoprene units (MK-9) is the predominant isoprenolog.** The single isolate known originated from cooked ox blood.

Type (and only) species: *Jonesia denitrificans.*

Characteristics of the species: As described for the genus.

Genus **Oerskovia**

Extensively branching vegetative hyphae, about 0.5 μm in diameter, grow on the surface of and penetrate into agar media, **breaking up** into rod-shaped, **motile,** flagellate elements. Rods are monotrichous when small but peritrichous when long. Nonmotile strains also occur. **No aerial mycelium is formed.** Stain Gram positive, with part of the thallus becoming Gram negative with age. Growth appears coryneform to bacteroid in smears. **Facultatively anaerobic** on trypticase-soy medium and catalase negative when grown anaerobically. Glucose is metabolized both oxidatively and fermentatively. Catalase positive when grown aerobically. Contain major amounts of 12-methyltetradecanoic acid (*anteiso*-C$_{15:0}$) **with lesser amounts of hexadecanoic (*iso*-C$_{16:0}$) and 14-methylhexadecanoic (*anteiso*-C$_{17:0}$)** acids. Diphosphatidylglycerol, phosphatidylglycerol, and phospholipids of unknown structure containing glucosamine are the predominant polar lipids. **Tetrahydrogenated menaquinone with nine isoprene units (MK-9(H₄)) is the major isoprenolog.** Found in soil, decaying plant materials, brewery sewage, aluminum hydroxide gels, and clinical specimens, including blood samples.

Type species: *Oerskovia turbata.*

Differentiation of the species of the genus **Oerskovia:** See Table 22.17.

Genus Promicromonospora

Branching septate **hyphae** (0.5–1.0 μm in diameter) grow on the surface of and penetrate into the agar before **breaking into fragments of various sizes and shapes.** Fragmentation finally results in nonmotile, Y- or V-shaped, rodlike, coccoid, chlamydospore-like, and other spore-shaped elements. All of them may give rise to new mycelia. Aerial hyphae in different strains may vary in abundance (sometimes discernible only microscopically). The aerial hyphae are straight to curved and sometimes sparsely branched. They usually fragment into rodlike or elongated coccoid elements. **Growth is pasty to leathery. Aerobic.** Chemoorganotrophic. Glucose metabolism is oxidative and rarely fermentative. **Susceptible to taxon-specific phages.** Wall teichoic acids are absent. **12- and 13-Methyltetradecanoic** (*anteiso-* and *iso-*C₁₅:₀) **acids are the predominant fatty acids.** Diagnostic phospholipids are represented by phosphatidylglycerol and an unidentified glucosamine-containing phospholipid. Tetrahydrogenated menaquinone with nine isoprene units (MK-9(H₄)) is the predominant isoprenolog. Mainly found in soils.

Type species: *Promicromonospora citrea.*

Differentiation of the species of the genus **Promicromonospora:** See Table 22.18.

Important Notes for Users of This Manual

Unless otherwise indicated in footnotes to tables, the meanings of symbols are as follows:

+ 90% or more of strains are positive

− 90% or more of strains are negative

d 11-89% of strains are positive

v strain instability (not equivalent to "d")

D Different reactions in different taxa (species of a genus or genera of a family)

All other symbols are defined in footnotes to tables.

Table 22.1 Differential characteristics of the genera that encompass mycolic acid-containing bacteria [a]

Characteristics	Corynebacterium	Gordona	Mycobacterium	Nocardia	Rhodococcus	Tsukamurella
Morphological properties:						
Cell morphology	Straight to slightly curved rods with tapering ends which reproduce by snapping division, so that V-forms and palisades are seen. Club-shaped forms may also be observed.	Rods and cocci	Slightly curved or straight rods; sometimes branching; filaments or mycelial type growth may occur but on slight disturbance usually becomes fragmented into rods and coccoid elements.	Substrate mycelium; latter fragments into rods and cocci.	Rods to extensively branched substrate mycelium; latter fragments into irregular rods and cocci.	Straight to slightly curved rods occur singly, in pairs, or in masses.
Aerial mycelium	Absent	Absent	Usually absent	Sparse to moderate, nearly always formed; short to long chains of conidia may occasionally be found on the aerial hyphae and more rarely on both aerial and substrate hyphae.	Absent	Absent
Time for visible colonies	1–2 days	1–3 days	2–40 days	1–5 days	1–3 days	1–3 days
Degree of acid-fastness (not necessarily also alcohol-fastness)	Sometimes weakly acid-fast	Usually partially acid-fast	Usually strongly acid-fast	Often partially acid-fast	Often partially acid-fast	Weak to strongly acid-fast
Chemical properties:						
Arylsulphatase produced	–	–	+	–	–	–
Acyl group of muramic acid [b]	N-acetylated	N-glycolated	N-glycolated	N-glycolated	N-glycolated	N-glycolated
Mycolic acids:						
Overall size (number of carbons) [c]	22–38	48–66	60–90	46–90	34–52	64–78
Number of double bonds [c]	0–2	1–4	1–3	0–3	0–3	1–6
Fatty acid esters released on pyrolysis [d] (number of carbons)	8–18	16–18	22–26	12–18	12–16	20–22
Pattern derived from thin-layer chromatography [e]	Single spot	Single spot	Multispot	Single spot	Single spot	Single spot

Footnotes are at end of table

634

Table 22.1 *(continued)*

Characteristics	Corynebacterium	Gordona	Mycobacterium	Nocardia	Rhodococcus	Tsukamurella
Other fatty acids:						
Tuberculostearic acid[f]	Absent	Present	Present	Present	Present	Present
Phospholipids:						
Phosphatidylethanolamine[g]	Absent	Present	Present	Present	Present	Present
Predominant menaquinone[h]	MK-8(H$_2$) or MK-9(H$_2$)	MK-9(H$_2$)	MK-8(H$_2$)	MK-8(H$_{4\ 10}$-cycl[i])	MK-8(H$_2$)	MK-9

[a] Symbols: see standard definitions.

[b] Acyl group was detected using a simple glycolate test (Uchida and Aida, J. Gen. Appl. Microbiol. 25: 169–183, 1979).

[c] Detected by mass spectrometry (Alshamaony et al., J. Gen. Microbiol. 92: 188–199, 1976; Collins et al., J. Gen. Microbiol. 128: 129–149, 1982). In mycobacterial mycolic acids, double bonds may be converted to cyclopropane rings; methyl branches and oxygen functions may be present (Dobson et al., In Goodfellow and Minnikin (Eds.), Chemical Methods in Bacterial Systematics, Academic Press, London, 1985, pp. 237–265; Minnikin et al., Arch. Microbiol. 139: 225–231, 1984).

[d] Goodfellow et al., J. Gen. Microbiol. 109: 57–68, 1978; Collins et al., J. Gen. Microbiol. 128: 129–149, 1982.

[e] Detected by thin-layer chromatography (Minnikin et al., J. Gen. Microbiol. 88: 200–204, 1978; J. Chromatogr. 188: 221–233, 1980).

[f] Corynebacterium bovis, Corynebacterium minutissimum, and Corynebacterium variabilis contain tuberculostearic acid (Lechevalier et al., Biochem. Syst. Ecol. 5: 249–260, 1977; Collins et al., J. Gen. Microbiol. 128: 2503–2509, 1982; Kroppenstedt, In Goodfellow and Minnikin (Eds.), Chemical Methods in Bacterial Systematics, Academic Press, London, 1985, pp. 173–199); Mycobacterium gordonae lacks substantial amounts of tuberculostearic acid (Tisdall et al., J. Clin. Microbiol. 10: 506–514, 1979; Minnikin et al., J. Gen. Microbiol. 131: 2013–2021, 1985).

[g] Detected by thin-layer chromatography (Lechevalier et al., Biochem. System. Ecol., 5: 249–260, 1977; Minnikin et al., Int. J. Syst. Bacteriol. 27: 104–117, 1977).

[h] Menaquinones are detected by chromatographic or physicochemical analysis (Collins et al., J. Gen. Microbiol. 100: 221–230, 1977; Collins, In Goodfellow and Minnikin (Eds.), Chemical Methods in Bacterial Systematics, Academic Press, London, 1985, pp. 267–287; Kroppenstedt, 1985, as above). Abbreviations exemplified by MK-9(H$_2$), as in menaquinone having two of the eight isoprene units hydrogenated.

[i] Nocardiae were originally reported to have predominant amounts of MK-8(H$_4$). However, the major component was shown to correspond to a hexahydrogenated menaquinone with eight isoprene units in which the end two units of the multiprenyl side chain were cyclized (Collins et al., FEMS Microbiol. Lett. 41: 35–39, 1987; Howarth et al., Biochem. Biophys. Res. Commun. 140: 916–923, 1986). The predominant isoprenolog in N. amarae is MK-9(H$_2$) (Goodfellow et al., J. Gen. Microbiol. 128: 1283–1297, 1982).

Table 22.2 Differential characteristics of the species of the genus *Gordona*[a,b]

Characteristics	G. bronchialis	G. rubropertincta	G. sputi	G. terrae
Cleavage of 7-amino-4-methylcoumarin (-7AMC) substrates:				
Endopeptidase:				
Z-Glycine-glycine-proline-7AMC [c]	−	+	−	+
Exopeptidases:				
D-Alanine-7AMC	−	+	−	+
iso-Leucine-7AMC	−	−	+	−
Pyroglutamate-7AMC	−	−	−	+
Serine-7AMC	+	+	−	+
Valine-7AMC	−	+	+	+
Cleavage of 4-methylumbelliferone (4MU-) substrates:				
Glycosides:				
4MU-β-D-galactopyranoside	+	−	+	−
4MU-β-D-mannopyranoside	+		+	+
Inorganic ester:				
4MU-pyrophosphate	+	+	+	−
Growth on sole carbon sources (%, w/v):				
D-Cellobiose (1.0)	+	−	+	+
D-Galactose (1.0)	−	+	+	+
meso-Inositol (1.0)	+	−	−	−
Raffinose (1.0)	−	−	+	+
L-Rhamnose (1.0)	−	−	−	+
Butan-2,3-diol (1.0)	−	−	+	+
Butyric acid (Na salt) (0.1)	−	+	−	−
Citric acid (Na salt) (0.1)	−	+	−	+
Propan-1-ol (0.1)	+	−	+	+
L-Serine (0.1)	−	+	−	+
Spermine (0.1)	+	−	+	+
Growth on sole carbon and nitrogen sources (0.1%, w/v):				
Acetamide	+	+	−	−
D-Alanine	+	−	−	−
L-Alanine	−	−	−	+

[a] Symbols: see standard definitions.

[b] Test procedures: enzyme tests (see Goodfellow et al., FEMS Microbiol. Lett. *43:* 39–44, 1987; FEMS Microbiol. Lett. *44:* 349–355, 1987); growth on sole carbon sources (see Goodfellow, J. Gen. Microbiol. *69:* 33–80, 1971); use of nitrogenous compounds as simultaneous nitrogen and carbon sources (see Tsukamura, Tubercle *48:* 311–328, 1967).

[c] Z, the *N*-terminal carbobenzoxy group of endopeptidase substrates.

Table 22.3 Differential characteristics of the species of the genus *Nocardia*[a,b]

Characteristics	N. amarae	N. asteroides	N. brasiliensis	N. brevicatena	N. carnea	N. farcinica	N. nova[c]	N. otitidis-caviarum	N. pinensis[c]	N. seriolae[c]	N. transvalensis	N. vaccinii
Decomposition of:												
Casein	−	−	+	−	−	−	ND	−	−	−	−	−
Elastin	−	−	+	−	−	−	ND	−	ND	ND	−	−
Esculin	+	+	+	+	+	+	ND	+	−	+	+	+
Hypoxanthine	−	−	+	−	−	−	ND	+	−	−	+	−
Testosterone	−	+	+	+	+	+	ND	−	ND	ND	−	−
Tyrosine	−	−	+	−	−	−	ND	−	−	−	−	−
Xanthine	−	−	−	−	−	−	ND	+	−	−	d	−
Growth on sole carbon sources (%, w/v):												
Adonitol (1.0)	−	−	−	+	−	−	ND	−	ND	ND	+	−
L-Arabinose (1.0)	−	−	−	−	+	−	−	−	ND	ND	−	d
meso-Inositol (1.0)	+	−	+	+	−	−	−	+	ND	ND	−	+
D-Mannitol (1.0)	+	−	+	+	−	−	−	+	ND	−	−	+
L-Rhamnose (1.0)	+	−ᵈ	+	−	−	+ᵈ	−	−	ND	−	−	+
Adipic acid (0.1)	+	−	−	−	d	−	+	−	ND	ND	+	−
iso-Butanol	−	−	d	ND	+	+	+	−	ND	−	ND	−
Pimelic acid (0.1)	+	−	−	−	−	−	ND	−	ND	ND	−	+
Sebacic acid (0.1)	+	+	−	−	d	d	ND	+	ND	ND	−	−
Production of:												
Acid phosphatase	−	+	+	ND	+	−	−	+	ND	−	ND	+
α-Esterase	−	−	−	ND	−	−	+	−	ND	−	ND	+
β-Esterase	−	−	+	ND	+	+	+	d	ND	−	ND	+
Nitrate reductase	+	+	+	−	+	+	+	+	−	ND	+	+
Urease	+	+	+	−	−	+	+	+	+	−	+	+
Resistance to lysozyme	−	+	+	+	+	+	ND	+	−	+	+	+

[a] Symbols: see standard definitions.

[b] For details of test procedures see Goodfellow (J. Gen. Microbiol. 69: 33–80, 1971), Lacey and Goodfellow (J. Gen. Microbiol. 88: 75–85, 1975), Goodfellow and Alderson (J. Gen. Microbiol. 100: 99–122, 1977), Goodfellow and Pirouz (J. Gen. Microbiol. 128: 503–527, 1982), and Käppler (Beitr. Klin. Tuberk. 130: 1–4, 1965).

[c] N. nova, N. pinensis and N. seriolae were not included in Bergey's Manual of Systematic Bacteriology. The species N. nova was created in 1990 by Yano et al. (Int. J. Syst. Bacteriol. 40: 170–174). N. pinensis was created in 1989 by Blackall et al. (J. Gen. Microbiol. 135: 1547–1588), and N. seriolae was created in 1988 by Kudo et al. (Int. J. Syst. Bacteriol. 38: 173–178).

[d] N. asteroides and N. farcinica, the most frequent agents of nocardiosis, can also be distinguished by the ability of the latter to use iso-amyl alcohol, 2,3-butylene glycol, and 1,3-propylene glycol as sole carbon sources (see Schaal, In A. von Graevenitz (Eds.), CRC Handbook Series in Clinical Laboratory Sciences, Section E, Clinical Microbiology, Volume 1, CRC Press, Cleveland, pp. 131–158.

Table 22.4 Differential characteristics of the species of the genus *Rhodococcus*[a,b]

Characteristics	*R. aichiensis*[c]	*R. chlorophenolicus*	*R. coprophilus*	*R. equi*	*R. erythropolis*	*R. fascians*	*R. globerulus*
Morphogenetic sequence[d]	R–C	R–C	H–R–C	R–C	EB–R–C	H–R–C	EB–R–C
Cleavage of 7-amino-4-methylcoumarin (7-AMC substrates):							
Endopeptidase:							
Z-phenylalanine-arginine-7AMC[e]	+	+	+	−	−	+	+
Exopeptidase:							
β-Alanine	−	−	+	+	+	−	d
Pyroglutamate	+	+	+	−	+	+	−
Threonine-7AMC	+	+	+	−	+	+	−
Cleavage of 4-methylumbelliferone (4MU-) substrates:							
Glucosides:							
4MU-β-D-cellobiopyranoside	d	−	−	−	+	−	+
4MU-β-D-galactopyranoside	+	−	−	+	−	−	+
4MU-β-D-glucopyranoside	+	+	+	+	+	+	+
4MU-β-D-xylopyranoside	+	+	−	+	+	−	+
Inorganic ester:							
4MU-phosphate	+	+	+	+	+	−	+
Degradation of tyrosine	ND	ND	−	−	d	+	−
Growth on sole carbon sources (%, w/v):							
Lactose (1.0)	+	−	−	−	−	−	−
Maltose (1.0)	ND	−	+	−	+	−	+
Mannose (1.0)	+	+	+	d	−	+	−
Androsterone (0.1)	−	−	+	d	+	−	−
Benzoic acid (Na salt) (0.1)	+	+	+	d	+	−	+
Butan-2,3-diol (0.1)	+	−	−	−	+	−	−
Citraconic acid (0.1)	−	+	−	−	+	+	+
Citric acid (Na salt) (0.1)	+	−	−	−	+	−	−
D-Mandelic acid (0.1)	−	−	−	−	+	+	−
DL-Norleucine (0.1)	−	+	+	d	+	−	+
Pimelic acid (0.1)	+	−	−	d	+	−	+
Spermine (0.1)	+	+	−	d	−	+	−
Testosterone (0.1)	−	+	+	d	+	−	−
Growth on sole carbon and nitrogen sources (0.1%, w/v):							
Acetamide	+	−	−	d	−	−	−
L-Asparagine	+	−	−	−	+	+	−

Footnotes are at end of table

Table 22.4 *(continued)*

Characteristics	R. luteus	R. marinonascens	R. maris	R. rhodnii	R. rhodochrous	R. roseus	R. ruber
Morphogenetic sequence[d]	EB–R–C	H–R–C	R–C	EB–R–C	EB–R–C	R–C	H–R–C
Cleavage of 7-amino-4-methylcoumarin (7-AMC substrates):							
Endopeptidase:							
Z-phenylalanine-arginine-7AMC[e]	+	+	+	+	–	ND	–
Exopeptidase:							
β-Alanine	+	+	+	+	+	ND	+
Pyroglutamate	+	+	–	+	+	ND	+
Threonine-7AMC	+	+	–	+	+	ND	+
Cleavage of 4-methylumbelliferone (4MU-) substrates:							
Glucosides:							
4MU-β-D-cellobiopyranoside	–	–	–	–	–	ND	+
4MU-β-D-galactopyranoside	–	+	–	–	–	ND	–
4MU-β-D-glucopyranoside	–	–	–	–	+	ND	+
4MU-β-D-xylopyranoside	–	–	–	–	+	ND	+
Inorganic ester:							
4MU-phosphate	+	–	–	–	+	ND	+
Degradation of tyrosine	–	d	–	+	+	ND	+
Growth on sole carbon sources (%, w/v):							
Lactose (1.0)	+	–	–	–	+	ND	+
Maltose (1.0)	–	–	–	–	+	–	+
Mannose (1.0)	+	+	–	–	+	+	+
Androsterone (0.1)	–	–	+	–	+	ND	+
Benzoic acid (Na salt) (0.1)	–	+	+	+	+	+	+
Butan-2,3-diol (0.1)	+	–	+	–	–	+	+
Citraconic acid (0.1)	+	–	+	–	–	ND	–
Citric acid (Na salt) (0.1)	+	–	–	–	+	–	+
D-Mandelic acid (0.1)	+	–	+	–	–	ND	–
DL-Norleucine (0.1)	+	–	–	–	+	ND	+
Pimelic acid (0.1)	–	–	–	–	+	ND	+
Spermine (0.1)	+	–	–	–	–	ND	–
Testosterone (0.1)	+	–	+	+	+	ND	+
Growth on sole carbon and nitrogen sources (0.1%, w/v):							
Acetamide	–	–	–	+	+	+	+
L-Asparagine	+	+	+	+	+	–	–

Footnotes are at end of table

Table 22.4 (continued)

^{a} Symbols: see standard definitions.

^{b} Test procedures: enzyme tests (see Goodfellow et al., FEMS Microbiol. Lett. 43: 39–44, 1987; FEMS Microbiol. Lett. 44: 349–355, 1987), use of nitrogenous compounds as simultaneous nitrogen and carbon sources (see Tsukamura, Tubercle 48: 311–338, 1967), tyrosine degradation and growth on sole carbon sources (see Goodfellow, J. Gen. Microbiol. 69: 33–80, 1971).

^{c} R. roseus was not included in Bergey's Manual of Systematic Bacteriology. The species R. roseus was reintroduced in 1991 by Tsukamura et al. (Int. J. Syst. Bacteriol. 41: 385–389). R. aichiensis is now known to be a good taxospecies (Goodfellow, Zbl. Bakt. 274: 299–315, 1990) but was listed as species incertae sedis in Bergey's Manual of Systematic Bacteriology.

^{d} EB–R–C, elementary branching-rod-coccus growth cycle; R–C, rod-coccus growth cycle; H–R–C, hyphae-rod-coccus growth cycle.

^{e} Z, the N-terminal carbobenzoxy group of endopeptidase substrates.

Table 22.5 Differential characteristics of the species of the genus Tsukamurella^{a,b}

Characteristics	T. paurometabolum	T. wratislaviensis
Cleavage of 7-amino-4-methylcoumarin (-7AMC) substrates:		
Endopeptidase: Z-glycine-glycine-proline-7AMC^c	+	–
Exopeptidase: D-Alanine-7AMC	–	+
Cleavage of 4-methylumbelliferone (4MU-) substrates:		
Glycosides:		
4MU-α-L-arabinofuranoside	+	–
4MU-α-L-arabinopyranoside	–	+
4MU-α-D-mannopyranoside	+	–
4MU-β-D-mannopyranoside	+	–
4MU-β-D-xylopyranoside	+	–
Growth on sole carbon sources (%, w/v):		
Cellobiose (1.0)	+	–
Galactose (1.0)	+	–
Androsterone (0.1)	–	+
Citraconic acid (0.1)	–	+
m-Hydroxybenzoic acid (0.1)	–	+
p-Hydroxybenzoic acid (0.1)	–	+
D-Mandelic acid (0.1)	–	+
Tartaric acid (Na salt) (0.1)	+	–
Testosterone (0.1)	–	+

^{a} Symbols: see standard definitions.

^{b} Test procedures: enzyme tests (see Goodfellow et al., FEMS Microbiol. Lett. 43: 39–44, 1987; FEMS Microbiol. Lett. 44: 349–355, 1987); growth on sole carbon sources (see Goodfellow, J. Gen. Microbiol. 69: 33–80, 1971).

^{c} Z, the N-terminal carbobenzoxy group of endopeptidase substrates.

Table 22.6 Differential characteristics of *Pseudonocardia* and related genera [a]

Characteristics	*Actinobispora*	*Actinokineospora*	*Actinopolyspora*	*Amycolata*	*Amycolatopsis*	*Kibdelosporangium*	*Pseudoamycolata*	*Pseudonocardia*	*Saccharomonospora*	*Saccharopolyspora*
Morphological properties:										
Acropetal budding	−	−	−	−	−	−	−	+	−	−
Sporangia-like structures	−	−	−	−	−	+	−	−	−	−
Spores produced on substrate hyphae	+	+	−	+	+	−	+	+	−	+
Spores borne:										
Singly	+	−	−	−	−	−	−	−	+	−
In pairs	+	−	−	−	−	−	−	−	−	−
In chains	−	+	+	+	+	+	+	+	−	+
Spore surface ornamentation	Smooth	Smooth	Smooth	Smooth	Smooth	Smooth	ND	Smooth, spiny	Smooth, warty	Hairy, smooth
Spores motile	−	+	−	−	−	−	−	−	−	−
Substrate mycelium with an outer sheath	ND	ND	−	+	−	ND	ND	+	−	−
Aerial mycelium with an outer sheath	ND	ND	+	+	−	+	ND	+	+	+
Chemical properties:										
Phospholipid [b]	4	2	3	3	2	2	2	3	2	3
Predominant menaquinone(s) [c]	MK-7(H$_2$) and MK-9(H$_2$)	MK-10	MK-9(H$_4$) and MK-10(H$_4$)	MK-8(H$_2$, H$_4$)	MK-9(H$_2$, H$_4$, H$_6$)	ND	MK-8(H$_4$)	MK-8(H$_4$)	MK-9(H$_4$)	MK-9(H$_4$)
Moderately extreme halophile	−	−	+	−	−	−	−	−	−	−

[a] Symbols: see standard definitions.

[b] Characteristic phospholipid types in addition to phosphatidylinositol (which is always present): 2, only phosphatidylethanolamine; 3, phosphatidylcholine (with phosphatidylethanolamine, phosphatidylmethylethanolamine, and phosphatidylglycerol variable); and 4, phospholipids containing glucosamine (with phosphatidylethanolamine and phosphatidylmethylethanolamine variable). The phospholipid patterns are detected by thin-layer chromatography (see Lechevalier et al., Biochem. Syst. Ecol. 5: 249–260, 1977; Minnikin et al., Int. J. Syst. Bacteriol. 27: 104–117, 1977).

[c] Menaquinones detected by chromatographic or physicochemical analysis (Collins et al., J. Gen. Microbiol. 100: 221–230, 1977; Collins, In Goodfellow and Minnikin (Eds.), Chemical Methods in Bacterial Systematics, Academic Press, London, 1985, pp. 267–287; Kroppenstedt, ibid, pp. 173–199. Abbreviations exemplified by MK-8(H$_4$), menaquinone having four of the eight isoprene units hydrogenated.

Table 22.7 Differential characteristics of the species of the genus *Actinopolyspora*[a,b]

Characteristics	A. halophila ATCC 27976 [T]	A. mortivallis JCM 7550 [T]
Acid produced from:		
Glucose	–	+
Rhamnose	+	–
Sucrose	–	+
Xylose	–	+
Growth on sole carbon sources (%, w/v):		
L-Arabinose	+	–
meso-Inositol	+	–
Mannitol	+	–
Rhamnose	+	–
Milk peptonization	–	+
Nitrate reduction	–	+
Xanthine degradation	–	+
Optimum temperature	37°C	45°C
Optimum growth at sodium chloride concentration (%, w/v)	20	15–20

[a] Symbols: +, positive; –, negative.

[b] *A. mortivallis* was not included in *Bergey's Manual of Systematic Bacteriology*. It was created in 1991 by Yoshida et al. (Int. J. Syst. Bacteriol. *41:* 15–20). This reference includes details of the procedures required to carry out the diagnostic tests.

Important Notes for Users of This Manual

Unless otherwise indicated in footnotes to tables, the meanings of symbols are as follows:

+ 90% or more of strains are positive

– 90% or more of strains are negative

d 11-89% of strains are positive

v strain instability (not equivalent to "d")

D Different reactions in different taxa (species of a genus or genera of a family)

All other symbols are defined in footnotes to tables.

Table 22.8 Differential characteristics of the species of the genus *Amycolata*[a,b]

Characteristics	A. alni[c]	A. autotrophica	A. hydrocarbonoxydans NRRL B16171[T]	A. saturnea NRRL B16172[T]
Color of substrate mycelium	Brown to orange	Yellow to orange	Orange	Orange
Acid produced from:				
Adonitol	+	+	−	−
Cellobiose	−	+	+	+
meso-Erythritol	+	+	+	−
Galactose	+	+	+	−
meso-Inositol	−	+	+	+
Lactose	−	−	+	−
Mannitol	+	+	−	+
α-Methyl-D-glucoside	d	d	−	+
Salicin	+	−	+	−
Sorbitol	+	−	−	−
Decarboxylation of:				
Benzoate	+	d	−	+
Citrate	+	+	−	−
Decomposition of:				
Adenine	+	+	−	−
Hypoxanthine	+	+	−	+
Tyrosine	+	+	−	+
Xanthine	d	d	−	+
Growth on 5% NaCl	+	+	−	−
Growth on sole carbon sources (1%, w/v):				
D-Gluconate	+	−	+	+
α-Methyl-D-glucoside	+	−	+	+
meso-Inositol	−	+	+	+
Resistance to carbenicillin (10 µg/ml)	+	−	−	−
Growth on sole nitrogen sources:				
L-Histidine	+	−	−	−
L-Proline	+	d	−	+

[a] Symbols: see standard definitions.

[b] Test procedures: see Goodfellow (J. Gen. Microbiol. *69:* 33–80, 1971), Gordon et al. (J. Gen. Microbiol. *109:* 69–78, 1978), Lechevalier (Int. J. Syst. Bacteriol. *22:* 260–264, 1972), Mishra et al. (J. Clin. Microbiol. *11:* 728–736, 1980), and Shirling and Gottlieb (Int. J. Syst. Bacteriol. *16:* 313–340, 1966).

[c] *A. alni* was created in 1989 by Evtushenko et al. (Int. J. Syst. Bacteriol. *39:* 72–77).

Table 22.9 Differential characteristics of the species of the genus Amycolatopsis[a,b]

Characteristics	A. azurea[c]	A. fastidiosa[c]	A. orientalis subsp. lurida NRRL 2430[T]	A. orientalis subsp. orientalis	A. mediterranei ATCC 13685[T]	A. methanolica[c] NCIB 11946[T]	A. rugosa ATCC 43014[T]	A. sulphurea ATCC 27624[T]
Color of aerial mycelium:								
Blue	+	–	–	–	–	–	–	–
Pink	+	+	–	–	–	–	–	–
White	+	+	+	+	+	+	–	+
Yellowish green	–	–	!	–	–	–	–	+
Production of soluble pigments	+	+	–	–	–	–	+	–
Color of soluble pigments:								
Blue	+	–	–	–	–	–	–	–
Brown	–	+	–	–	–	–	+	–
Yellow	–	+	–	–	–	–	–	–
Acid produced from:								
Adonitol	+	–	+	+	–	ND	+	–
L-Arabinose	+	–	+	+	+	ND	+	+
Cellobiose	+	–	+	+	+	ND	–	–
Erythritol	+	–	+	+	–	ND	+	–
Galactose	+	–	+	+	+	ND	+	+
meso-Inositol	+	–	+	+	+	ND	–	+
Lactose	+	–	+	+	+	ND	–	–
Mannitol	+	–	+	+	+	ND	+	+
Raffinose	+	–	–	d	+	ND	–	–
Decarboxylation of:								
Benzoate	–	–	–	d	–	ND	+	–
Citrate	+	+	+	+	+	ND	–	+
Decomposition of:								
Hypoxanthine	–	–	+	+	+	+	+	+
Xanthine	–	–	+	d	–	–	+	–
Growth of 5% NaCl	+	+	d	ND	–	+	+	+
Production of:								
Amylase	–	–	–	+	–	–	–	–
Nitrate reductase	+	+	–	+	–	+	–	–
Urease	+	+	+	+	+	–	+	–

[a] Symbols: see standard definitions.

[b] Test procedures: see Gordon et al. (J. Gen. Microbiol. 109: 69–78, 1978) and Lechevalier (Int. J. Syst. Bacteriol. 22: 260–264, 1972).

[c] Pseudonocardia azurea and Pseudonocardia fastidiosa were transferred to the genus Amycolatopsis as Amycolatopsis azurea and Amycolatopsis fastidiosa by Henssen et al. (Int. J. Syst. Bacteriol. 37: 292–295, 1987). Amycolatopsis methanolica was created in 1990 by De Boer et al. (Int. J. Syst. Bacteriol. 40: 194–204).

Table 22.10 Differential characteristics of the species of the genus *Kibdelosporangium*[a,b]

Characteristics	K. aridum	K. aridum subsp. largum[c] ATCC 39922[T]	K. philippinense[c] NRRL 18198[T]
Formation of aerial mycelia	Poor	Moderate	Good
Formation of sporangium-like structures	Sparse	Many	Profuse
Size of sporangia (μm)	9–22	12–32	1–8
Size of aerial spores (μm)	0.4 × 2.8	0.5 × 0.8–3.2	0.4 × 1.6
Reverse color	Off-white	Off-white to grayish brown	Orange yellow
Whole organism wall components:			
Madurose	+	+	−
L-Rhamnose	+	−	−
Decomposition of:			
Allantoin	+	+	−
Guanine	+	+	−
Growth on sole carbon sources (%, w/v):			
L-Arabinose	+	+	−
Dextrin	+	+	−
Glycogen	+	+	−
α-Methyl-D-glucoside	+	+	−
D-Melezitose	−	ND	+
Raffinose	+	+	−
Salicin	ND	+	−
Peptonization of milk	+	+	−
Production of nitrase reductase	−	−	+
Sodium chloride tolerance (%)	5–7	7	2
Resistance to:			
Bacitracin (1[10U])	+	ND	−
Gentamicin (10 μg)	+	ND	−
Oleandomycin (15 μg)	−	ND	+
Streptomycin (10 μg)	+	ND	−
Temperature range (°C)	5–45	15–42	20–40

[a] Symbols: see standard definitions.

[b] Test procedures: see Gordon (J. Gen. Microbiol. *45:* 355–364, 1966), Gordon and Mihn (Ann. NY Acad. Sci. *98:* 628–636, 1962), Mertz and Yao (Int. J. Syst. Bacteriol. *38:* 282–286, 1988), and Shearer et al. (Int. J. Syst. Bacteriol. *36:* 47–54, 1986).

[c] *K. philippinense* and *K. aridum* subsp. *largum* were not included in *Bergey's Manual of Systematic Bacteriology.* The species *K. philippinense* was created by Mertz and Yao (see above). The subspecies *K. aridum* subsp. *largum* was created in 1986 by Shearer et al. (J. Antibiot. *39:* 1386–1394).

Table 22.11 Differential characteristics of the species of the genus *Pseudonocardia*[a]

Characteristics	P. compacta	P. spinosa	P. thermophila
Morphology:			
Zig-zag hyphae frequent	−	+	+
Many hyphal swellings	+	+	−
Spore ornamentation	Smooth	Spiny	Smooth
Growth on cellulose agar	−	+	+
Optimum temperature (°C)	20–30	20–30	40–50

[a] Symbols: see standard definitions.

Table 22.12 Differential characteristics of the genus *Saccharomonospora*[a,b]

Characteristics	S. azurea[c] S1A 86128[T]	S. caesia[c]	S. cyanea[c] ATCC 43724[T]	S. glauca[c]	S. viridis[d]
Aerial mycelium color	Azure	Green	Dark blue	Light to bluish green	Green
Spore ornamentation	Smooth	Warty	Warty	Warty	Warty
Growth on sole carbon sources (1%, w/v):					
D-Arabinose	ND	+	ND	+	−
L-Arabinose	−	+	−	+	−
Galactose	−	+	+	ND	−
Glucose	+	d	−	+	d
Mannitol	−	+	−	+	d
Mannose	+	+	+	ND	−
Melibiose	+	−	−	ND	ND
Rhamnose	+	−	+	ND	−
Ribose	+	−	+	ND	ND
Sucrose	+	−	+	−	+
Xylose	+	d	+	d	−
Growth temperature range (°C)	24–40	28–50	24–40	37–60	28–60

[a] Symbols: see standard definitions.
[b] Test procedures: see Goodfellow and Pirouz (J. Gen. Microbiol. *128:* 503–517, 1982) and Greiner-Mai et al. (Int. J. Syst. Bacteriol. *38:* 398–405, 1988).
[c] *S. azurea, S. caesia, S. cyanea,* and *S. glauca* were not included in *Bergey's Manual of Systematic Bacteriology. S. azurea* was created by Runmao (Int. J. Syst. Bacteriol. *37:* 60–61, 1987), *S. cyanea* by Runmao et al. (Int. J. Syst. Bacteriol. *38:* 444–446, 1988), and *S. glauca* by Greiner-Mai et al. (Int. J. Syst. Bacteriol. *38:* 398–405, 1988).
[d] *S. internatus,* listed as a species *incertae sedis* in *Bergey's Manual of Systematic Bacteriology,* was reduced to a synonym of *S. viridis* by Greiner-Mai et al. (as above).

Table 22.13 Differential characteristics of the species of the genus *Saccharopolyspora*[a,b]

Characteristics	S. erythraea[d] NRRL 2338[T]	S. gregorii[d]	S. hirsuta	S. hordei[d]	S. rectivirgula[d]	S. spinosa[d] NRRL 18395[T]	S. taberi[d] NRRL 16173[T]
Aerial mycelium produced	+	−[c]	+	−[c]	+	+	−
Spores on vegetative mycelium	−	−	−	−	+	−	−
Spore ornamentation	Spiny	Smooth	Hairy	Smooth	Smooth or irregularly rough	Spiny	ND
Acid production from:							
D-Arabinose	+	ND	−	ND	ND	−	−
Erythritol	+	ND	−	ND	ND	−	+
Galactose	+	ND	+	ND	ND	−	+
Melibiose	+	ND	−	ND	ND	−	d
α-Methyl-D-glucoside	−	ND	+	ND	ND	−	+
Rhamnose	+	ND	+	ND	ND	−	+
Xylose	+	ND	+	ND	ND	−	d
Decarboxylation of:							
Benzoate	−	ND	+	ND	ND	−	−
Mucate	−	ND	+	ND	ND	−	−
Tartrate	−	ND	+	ND	ND	−	−
Decomposition of:							
Adenine	+	−	+	d	−	−	+
Allantoin	−	−	−	−	−	+	−
Elastin	+	+	+	−	−	+	ND
Esculin	+	+	+	+	+	−	+
Starch	+	+	+	+	−	−	+
Urea	−	−	+	−	d	+	−
Xanthine	+	+	+	+	+	−	+
NaCl tolerance (%)	ND	13	15	10	10	11	12
Nitrate reduction	+	d	−	d	d	+	+
Phosphatase	−	ND	+	ND	+	+	+
Temperature range (°C)	22–42	10–35	25–50	20–60	30–63	15–37	20–42

[a] Symbols: see standard definitions.

[b] Test procedures: see Athalye et al. (Int. J. Syst. Bacteriol. *35:* 86–98, 1985), Gordon et al. (Int. J. Syst. Bacteriol. *24:* 54–63, 1974), and Shirling and Gottlieb (Int. J. Syst. Bacteriol. *16:* 313–340, 1966).

[c] Some strains are occasionally positive.

[d] *S. erythraea, S. gregorii, S. hordei, S. rectivirgula, S. spinosa,* and *S. taberi* were not included in *Bergey's Manual of Systematic Bacteriology. S. erythraea* was created in 1987 by Labeda (Int. J. Syst. Bacteriol. *37:* 19–22) for a strain previously classified as *Streptomyces erythraeus,* and *S. rectivirgula* and *S. taberi* in 1989 by Korn-Wendisch et al. (Int. J. Syst. Bacteriol. *39:* 430–441) for organisms previously classified as *Faenia rectivirgula* and *Saccharopolyspora hirsuta* subsp. *taberi,* respectively. *S. gregorii* and *S. hordei* were created in 1989 by Goodfellow et al. (J. Gen. Microbiol. *135:* 2125–2139), and *S. spinosa* in 1980 by Mertz and Yao (Int. J. Syst. Bacteriol. *40:* 34–39).

Table 22.14 Differential properties separating the genera *Nocardioides* and *Terrabacter*[a]

Characteristics	Nocardioides	Terrabacter
Diagnostic fatty acids:		
13-Methyltetradecanoic acid	−	+
14-Methylpentadecanoic acid	+	−
Thiamine required for growth	−	+
Susceptibility to *Nocardioides* phages	+	−

[a] For details of test procedures, see Prauser (Int. J. Syst. Bacteriol. *26:* 58–65, 1976) and O'Donnell et al. (Arch. Microbiol. *133:* 323–329, 1982).

Table 22.15 Differential characteristics of the species of the genus *Nocardioides*[a,b]

Characteristics	N. albus	N. fastidiosa	N. jensenii	N. luteus	N. simplex
Well-developed vegetative mycelium	Present	Absent	Absent	Present	Absent
Aerial mycelium color	White	−	−	Cream	−
Motility	−	+	−	−	+
Nutritionally exacting	−	+	−	−	−
Acid from:					
Rhamnose	+	ND	−	−	−
Sucrose	+	ND	−	−	−
Predominant menaquinone	MK-8(H$_4$)	MK-9(H$_4$)	MK-8(H$_4$)	MK-8(H$_4$)	MK-8(H$_4$)

[a] Symbols: see standard definitions.
[b] For details of test procedures see Prauser (Int. J. Syst. Bacteriol. *26:* 58–65, 1976) and O'Donnell et al. (Arch. Microbiol. *133:* 323–329, 1982).

Table 22.16 Differential properties separating the genus *Promicromonospora* from related genera [a,b]

Characteristics	Jonesia	Oerskovia	Promicromonospora
Aerial mycelium produced	−	−	+[c]
Motile elements formed	+	+[d]	−
Strictly aerobic	−	−	+
Facultatively anaerobic	+	+	−[e]
Major menaquinone	MK-9	MK-9(H$_4$)	MK-9(H$_4$)

[a] Symbols: see standard definitions.
[b] For details of test procedures see Lechevalier et al. (Int. J. Syst. Bacteriol. *22:* 260–264, 1972) and Collins et al. (J. Gen. Microbiol. *100:* 221–230, 1977).
[c] Lacking in some strains.
[d] Some strains are nonmotile (nonmotile oerskoviae [NMO]).
[e] Some are strains positive.

Table 22.17 Differential characteristics of the genus *Oerskovia* and two types of nonmotile *Oerskovia*-like strains (NMOs) [a,b]

Characteristics	O. turbata	O. xanthineolytica	NMO	
			Type A	Type B
Acid from:				
Melibiose	–	+	d	–
Raffinose	–	+	d	d
Cell wall	Type VI plus galactose	Type VI plus galactose	Type VI plus galactose	Type VI
Cytochrome a_1	+	–	ND	ND
Degradation of:				
Casein	+	+	d	–
Hypoxanthine	–	+	+	–
Urea	–	–	–	+
Xanthine	–	+	+	–
Growth at 42°C	–	+	d	–
Motility	+	+	–	–
Production of:				
Cellulose	–	–	+	+
Phosphatase (24h)	–	+	+	–

[a] Symbols: see standard definitions.

[b] For details of test procedures see Becker et al. (Appl. Microbiol. *13:* 236–242, 1965) and Lechevalier (J. Lab. Clin. Med. *71:* 934–944, 1968; Int. J. Syst. Bacteriol. *22:* 260–264, 1972).

Important Notes for Users of This Manual

Unless otherwise indicated in footnotes to tables, the meanings of symbols are as follows:

+ 90% or more of strains are positive

– 90% or more of strains are negative

d 11-89% of strains are positive

v strain instability (not equivalent to "d")

D Different reactions in different taxa (species of a genus or genera of a family)

All other symbols are defined in footnotes to tables.

Table 22.18 Differential characteristics of the species of the genus *Promicromonospora*[a,b]

Characteristics	P. citrea ATCC 15908[T]	P. enterophila	P. sukumoe[c]
Anaerobic growth on tryptic soy agar	−	+	−
Degradation of:			
Gelatin	+	−	+
RNA	+	−	ND
Tween 80	+	−	ND
Tyrosine	+	−	+
Growth at:			
5°C	−	+	−
42°C	+	−	−
pH 3	+	−	−
Nitrate to nitrite reduction	−	+	+
Tolerance to 9% NaCl	+	−	−
Utilization of:			
Mannitol	+	−	ND
Raffinose	+	−	ND
Rhamnose	+	−	ND
Citrate	+	−	+
Malate	+	−	+

[a] Symbols: see standard definitions.

[b] For test details see Jáger et al. (Int. J. Syst. Bacteriol. *33:* 525–531, 1983).

[c] *P. sukumoe* was not included in *Bergey's Manual of Systematic Bacteriology.* The species *P. sukumoe* was created in 1983 by Jáger et al. (as above).

GENERA WITH MULTILOCULAR SPORANGIA

Genus Dermatophilus

Aerial mycelium develops in atmospheres containing added CO_2. **Substrate mycelium** consists of **long tapering filaments, branching laterally at right angles; septa formed in transverse and in horizontal and vertical longitudinal planes give rise to up to eight parallel rows of coccoid cells (spores), each of which becomes motile by a tuft of flagella.** Gram positive. Walls contain *meso*-diaminopimelic acid (*meso*-DAP); madurose is present in whole-cell hydrolysates. Polar lipids include phosphatidylglycerol, diphosphatidylglycerol, and phosphatidylinositol. Aerobic and facultatively anaerobic, chemoorganotropic, and **nonfermentative, but acid is produced from certain carbohydrates. Catalase positive** and are not acid-fast. Growth is reported only on complex media; minimum nutritional requirements are unknown. Optimum temperature is ~37°C. **Parasitic on mammals,** especially the domestic herbivores; pathology is usually limited to exudative dermatitis, which may, however, be severe and life-threatening; on rare occasions *Dermatophilus* causes subcutaneous abscesses and lymph node granulomas.

Type (and only) species: *Dermatophilus congolensis.*

Characteristics of the species: As described for the genus.

Genus Frankia

Vegetative hyphae with limited to extensive branching are 0.5–2.0 μm in diameter and occasionally wider. **No aerial mycelium are formed. Round to irregularly shaped multilocular sporangia are borne terminally, laterally, or in an intercalary position on the vegetative hyphae.** Lateral sporangia are usually borne on sporangiophores; some may be sessile. Sporangia up to 100 μm in length are formed by septation in three planes of the cytoplasm of preexisting thin-walled swell-ings. **Sporangiospores are nonmotile,** of irregular (often somewhat polygonal) shape, 1–5 μm in size, and colorless to black, showing multilaminar outer membrane-like layers in thin section. Spores are not thermally resistant. Sporangiospores do not usually develop and mature simultaneously so that developing sporangia may contain spores of various ages and sizes. **Terminal or laterally borne "vesicles" may be formed.** These are thick-walled swellings that show various irregular septations in thin section and are probably the site of nitrogenase activity in strains fixing atmospheric nitrogen. Vesicles may also be formed under conditions where no nitrogen fixation takes place. All of these morphological characteristics may be expressed both in vitro and in planta, although nitrogen-fixing nodules without vesicles are known. Intra- and extracellular pigments are common. Cells are Gram positive to Gram variable. **Cell walls contain *meso*-diaminopimelic acid (*meso*-DAP), glutamic acid, alanine, muramic acid, and glucosamine. Whole-cell sugar patterns show xylose (without arabinose), madurose, or fucose, or cells may contain only glucose or galactose,** sugars not previously found to have taxonomic significance in the Actinomycetales. Many strains contain 2-*O*-methyl-D-mannose, and most contain rhamnose. **Phospholipids comprise phosphatidylinositol mannosides, phosphatidylinositol, and diphosphatidylglycerol.** Fatty acids are normal, branched chain, and monounsaturated. **No mycolates are present.** Aerobic to microaerophilic and will not grow under anaerobic conditions. Catalase positive, chemoorganotrophic, and mesophilic. **Usually very slow-growing (doubling time of 1–7 days). Most strains are capable of fixing atmospheric nitrogen in vitro and in planta. Most strains are symbiotic with certain angiospermous plants, inducing nodules on the roots of suitable hosts.** They may be found free in the soil.

Type (and only) species: *Frankia alni.*

Characteristics of the species: As described for the genus.

Editorial note: To date, no clear-cut speciation seems possible inside the genus.

Genus Geodermatophilus

Cells produce a **muriform, tuber-shaped, noncapsulated, holocarpic multilocular thallus** containing masses of cuboid cells 0.5–2.0 μm in diameter. The thallus, under favorable environmental conditions, breaks up, releasing cuboid and coccoid nonmotile cells. Some of these cells may develop into **elliptical to lanceolate zoospores** that are propelled by a terminal tuft of long flagella. **Mycelium is rudimentary, and aerial mycelium is not produced.** Cells are Gram positive. Cell wall contains *meso*-diaminopimelic acid (*meso*-DAP), together with glutamic acid, alanine, glucosamine, and muramic acid. Whole-cell hydrolysates do not contain madurose. Phospholipids include phosphatidylethanolamine, phosphatidylinositol mannosides, phosphatidylinositol, phosphatidylglycerol, and diphosphatidylglycerol. Fatty acids are of the branched-chain type. Cells are aerobic, chemoorganotrophic, and mesophilic. They inhabit soil.

Type (and only) species: *Geodermatophilus obscurus.*

Characteristics of the species: As described for the genus.

Editorial note: Five subspecies of *Geodermatophilus obscurus* are listed in *Bergey's Manual of Systematic Bacteriology.* The carbohydrate utilization and acid production of four of them are given in Table 23.1.

Table 23.1 Carbohydrate utilization and acid production in *Geodermatophilus obscurus* subspecies [a]

Carbohydrate	*obscurus*	*"amargosae"*	*"utahensis"*	*"dictyosporus"*
D-Arabinose	0a	0	2a	0a
L-Arabinose	3A	3A	1	3A
D-Galactose	3	3A	3	3
D-Glucose	3a	3A	3a	3a
Glycerol	3a	3A	1	3A
Inositol	3	2	2	0
β-Lactose	0	1	0	3
D-Fructose	3a	3A	2	3A
D-Mannitol	3	3	3a	3
Melezitose	0	1	1	3
L-Rhamnose	1a	2A	1	1A
D-Ribose	0A	1A	1a	1a
Sucrose	3	3	3	3
D-Xylose	3a	3A	3	3A

[a] 0, no growth; 1, poor growth; 2, fair growth; 3, good growth. A, acid production relatively consistent; a, sporadic or transient production of acid. Dulcitol, α-melibiose, and raffinose are not utilized.

GROUP 24

ACTINOPLANETES

Genus **Actinoplanes**

Produce a fine, nonfragmenting, branching mycelium. Aerial mycelium is scanty or absent. Under certain conditions, many strains have **hyphae arranged in palisade formation. Highly colored.** Diffusible pigments of various colors may be produced. Spores are produced within **sporangia** (spore vesicles) that are **spherical or subspherical to very irregular,** 3–20 × 6–30 μm, borne on short sporangiophores, or sessile, occasionally within the agar. When palisade formation is present, the sporangia are mainly produced at the tip of the palisade hyphae. **Spores are spherical, subspherical, or short rods,** variously arranged within the sporangia, where they are formed by fragmentation of the internal hypha either directly or after one or more ramifications. Upon immersion in water, **motile spores** are released from the sporangia, but in some instances motility begins some time after spore release. **Polar flagella** are present in motile spores. Gram positive, although part of the vegetative growth may be Gram negative. Non-acid-fast. The cell wall contains *meso*-diaminopimelic acid (*meso*-DAP) and glycine. D-Xylose and L-arabinose are present in whole-cell hydrolysates. Among the phospholipids, phosphatidylethanolamine is present, while phosphatidylmethylethanolamine, phosphatidylcholine, glucosamine-containing phospholipids, and phosphatidylglycerol are absent. Aerobic. The spores have microaerophilic behavior. Chemoorganotrophic. Most strains **do not require organic growth factors.** Mesophilic or moderately thermophilic. For most, the growth temperature range is 15–35°C. The habitat is soil and decaying plant material.

Type species: *Actinoplanes philippinensis.*

Differentiation of the species of the genus **Actinoplanes:** See Tables 24.1, 24.2, and 24.3.

Genus **Ampullariella**

Hyphae of **substrate mycelium** are 0.2–1.2 μm in diameter, branched, and septate. True aerial mycelium is not developed. **Sporangia** are produced above the surface of the substrate. They are **irregular, cylindrical, lobate, bottle-shaped, flask-shaped, or digitate,** 5.0–20.0 μm wide and 8.0–30.0 μm long. Numerous **spores** are produced within the sporangium, **arranged in parallel chains.** Spores are **rod-shaped** (0.5–1.0 × 2.0–4.0 μm) and motile. Colonies on various complex agar media are elevated and convoluted. The **color of substrate mycelium is usually orange, red, brown, or black,** sometimes with white areas. Gram positive. The peptidoglycan of the cell wall contains *meso*-diaminopimelic acid (*meso*-DAP) and glycine as distinguishing components. **Xylose and arabinose** are characteristic whole-cell sugars. Phosphatidylethanolamine characterizes the phospholipid pattern. *Iso-* and *anteiso-* branched and saturated or unsaturated fatty acids. MK-9(H$_4$) is the main component of the menaquinones. Aerobic, chemoorganotrophic, and mesophilic; optimum growth temperature is 25°C. The habitat is soil and fresh water.

Type species: *Ampullariella regularis.*

Differentiation of the species of the genus **Ampullariella:** See Table 24.4.

Genus **Catellatospora**

Editorial note: The genus *Catellatospora* was not included in *Bergey's Manual of Systematic Bacteriology.* The genus was created in 1986 by Asano and Kawamoto (Int. J. Syst. Bacteriol. *36:* 512–517) and revised in 1989 by Asano et al. (Int. J. Syst. Bacteriol. *39:* 309–313).

Vegetative hyphae are **branched but not fragmented. No true aerial mycelium is produced.** Short chains of **nonmotile spores** arise singly or in tufts from the vegetative hyphae on the surface of agar media. Gram positive. Cell walls contain *meso*-**diaminopimelic (*meso*-DAP) and 3-hydroxydiaminopimelic acids, glycine, and glycolic acid. Xylose and arabinose are present in cell hydrolysates.** Characteristic phospholipids osphatidylethanolamine, phosphatidylinositol, phosphatidylinositol mannosides. The major menaquinones are MK-9 (*C. citrea, C. citrea* subsp. *methionotrophica, C. tsunoense*) and MK-10 (*C. ferruginea* and *C. tsunoense*). Aerobic, chemoorganotrophic, and mesophilic. Inhabit soil.

Type species: *Catellatospora citrea.*

Differentiation of the species of the genus **Catellatospora:** See Table 24.5.

Genus **Dactylosporangium**

Finger-shaped to claviform sporangia (0.6–1.4 × 2.5–6.0 μm) are **formed on short sporangiophores on the substrate mycelium.** They are developed singly or in clusters above the surface of the substrate. **Each sporangium contains a single row of normally three to four spores. The spores are oblong, ellipsoidal, ovoid, or slightly pyriform** (0.4–1.3 × 0.5–1.8 μm) and **motile** by means of a polarly inserted tuft of flagella. True aerial mycelium is not formed. Hyphae of the substrate mycelium are 0.5–1.0 μm in diameter, branched, and rarely septate. Large globose spores (1.7–2.8 μm in diameter) are formed on short branches on substrate mycelium. Colonies grow on various agar media. They are compact, somewhat tough and leathery, and mostly flat or sometimes elevated with a smooth to slightly wrinkled surface. **The substrate mycelium is pale to deep orange, rose or wine-colored to brown.** Organisms are Gram positive and not acid-fast. The peptidoglycan of the cell walls contains *meso*-diaminopimelic acid (*meso*-DAP) and glycine, with xylose and arabinose as characteristic sugars of whole-cell hydrolysates. Characteristic phospholipids are phosphatidylinositol mannosides, phosphatidylinositol, phosphatidylethanolamine, and diphosphatidylglycerol. Branched *iso-* and *anteiso-* fatty acids are main components. Aerobic and chemoorganotrophic, with an optimum growth temperature of 25–37°C and a pH optimum of 6.0–7.0. Found in soil, plant debris, and lake sediments.

Type species: *Dactylosporangium aurantiacum.*

Differentiation of the species of the genus **Dactylosporangium:** See Table 24.6.

Genus **Micromonospora**

Well-developed, branched, septate mycelium averages 0.5 μm in diameter. **Nonmotile spores** are borne singly, sessile, or on short or long sporophores that often occur in branched clusters. Sporophore development is monopodial or in some cases sympodial. **Aerial mycelium is absent** or in some cultures appears irregularly as a restricted white or grayish bloom. Gram positive and not acid-fast. **Walls contain *meso*-diaminopimelic acid (*meso*-DAP) and/or its 3-hydroxy derivative and glycine. Xylose and arabinose are present in cell hydrolysates.** Characteristic phospholipids are phosphatidylethanolamine, phosphatidylinositol, and phosphatidylinositol mannosides. Major menaquinones are MK-9(H$_4$), MK-10(H$_4$), MK-10(H$_6$), or MK-12(H$_6$). *Iso-* and *anteiso-* branched fatty acids predominant. Aerobic to microaerobic. Chemoorganotrophic, and are sensitive to pH values below 6.0. Growth occurs normally at 20–40°C but not above 50°C. Inhabit soil, water, marine environments, and sediments. A few strains, anaerobic and of uncertain taxonomy, have been reported in the intestinal tract of termites and the rumen of sheep.

Type species: *Micromonospora chalcea.*

Differentiation of the species of the genus **Micromonospora:** See Table 24.7.

Genus **Pilimelia**

Sporangia are produced **on the surface of the substrate** on sporangiophores. Sporangia are **spherical, ovoid, pyriform, campanulate,** or **cylindrical** and are approximately 10–15 μm in size. Sporangia contain numerous **spores in chains that are arranged in parallel or irregularly swirl-like rows.** Spores (zoospores) are rod-shaped (0.4 × 1.2 μm) and motile by means of a laterally inserted tuft of flagella. Nonmotile spores are developed in free chains arranged similarly to the zoospores. **Colonies grow only on complex media. They are small, compact, soft pasty, or solid. Substrate mycelia are pale lemon yellow, golden yellow, orange, or pale brown, turning brown to dark with age.** Hyphae of substrate mycelium are 0.2–0.8 μm in diameter, branched, and septate. **True aerial mycelium is not developed.** The organisms are Gram positive. The peptidoglycan of the cell walls contains *meso*-diaminopimelic acid (*meso*-DAP) and glycine, with xylose and arabinose as characteristic sugars of whole-cell hydrolysates. Phospholipids are represented by phosphatidylethanolamine. There are *iso-* branched saturated and unsaturated fatty acids with odd numbers of carbon atoms and no *anteiso-* fatty acids. Aerobic and chemoorganotrophic, with optimal growth at pH 6.5–7.5 and at 20–30°C (minimum 10°C, maximum 38°C). Inhabit soil. Strains decompose keratinic substances (hair of mammals).

Type species: *Pilimelia terevasa.*

Differentiation of the species of the genus **Pilimelia:** See Tables 24.8 and 24.9.

Important Notes for Users of This Manual

Unless otherwise indicated in footnotes to tables, the meanings of symbols are as follows:

+ 90% or more of strains are positive
− 90% or more of strains are negative
d 11-89% of strains are positive
v strain instability (not equivalent to "d")
D Different reactions in different taxa (species of a genus or genera of a family)

All other symbols are defined in footnotes to tables.

Table 24.1 Morphological and physiological characteristics of the species of the genus *Actinoplanes*[a]

Species	Sporangial shape and size (µm)	Spore arrangement shape and size (µm)	Aerial mycelium	Mycelial color	Soluble pigment	Physiological characteristics[b]
A. philippinensis (ATCC 12247)	Globose to oval (8–25)	Coils; globose (1–1.2)	Absent	Yellow to orange-brown	Brown	A,b,C,D,E,F,g,h,I,J
A. utahensis (ATCC 14539)	Irregular (5–18)	Coils; subglobose (1–2)	Absent	Orange to brown-orange	Absent	A,B,C,D,E,F,g,h,I,J
A. missouriensis (ATCC 14538)	Globose, subglobose irregular (6–14)	Coils; globose (1–1.2)	Absent	Orange	Absent	a,b,C,D,E,F,g,h,I,j
A. brasiliensis (ATCC 25844)	Irregular to umbelliform (3.5–11.5)	Coils; subglobose (1.2 × 1.7–2.3)	Absent	Orange	Absent	a,b,c,D,E,F,g,h,I,J
A. italicus (ATCC 27366)	Globose to oval (6–11)	Coils; globose to oval (1–2)	Absent	Cherry red	Cherry red	A,B,C,D,e,F,g,H,I,J
A. rectilineatus (ATCC 29234)	Cylindrical (6–14 × 10–15)	Long rows (1.5–2)	Short sterile	Orange	Absent	B,C,D,f,G,H,I,J
A. deccanensis (ATCC 21983)	Globose (4–7)	Coils; globose (1–1.5)	Absent	Orange	Absent	a,B,d,E,F,g,h,I,J
A. ferrugineus (ATCC 29868)	Globose to irregular	Coils; globose (0.9–1)	Short sterile	Rusty brown	Brown	a,B,C,D,E,F,I,J
A. auranticolor (ATCC 15330)	Very irregular, lobed (6–25 × 8–15)	Irregular; rods (0.5–0.7 × 1–1.5)	Absent	Orange	Absent yellowish or amber	B,c,D,f,I,J
A. globisporus (ATCC 23056)	Irregular (3–5 × 4–7)	Coils; globose (0.8–1)	Rudimentary chlamydospores	Cream to light orange	Absent	a,b,D,e,f,g,h,I,j
"*A. ianthinogenes*" (ATCC 21884)	Globose (8–15)	Coils; subglobose (1.4–1.8)	Absent	Violet	Absent	a,b,C,D,E,F,g,h,I,J
"*A. garbadinensis*" (ATCC 31049)	Globose, lobate, irregular (7–12)	Coils; subglobose (1–1.5)	Rudimentary sterile	Orange	Brown	A,B,C,d,e,F,g,h,I,J
"*A. liguriae*" (ATCC 31048)	Globose to oval (15–25)	Coils; globose (1.5–2)	Absent	Orange	Yellow-amber	a,b,c,D,e,f,g,h,I,j
"*A. teichomyceticus*" (ATCC 31121)	Globose to oval (15–25)	Coils; globose to oval (1.5–2)	Long sterile	Orange	Absent	A,B,c,D,e,F,g,H,I,J

Footnotes are at end of table

656

Table 24.1 *(continued)*

Species	Sporangial shape and size (μm)	Spore arrangement shape and size (μm)	Aerial mycelium	Mycelial color	Soluble pigment	Physiological characteristics [b]
"*A. caeruleus*" (ATCC 33937)	Globose to irregular (6–16)	Globose to oval (1.3–2)	Absent	Tan to blue	Yellowish brown	b,C,D,F,J
A. consettensis [c] (NCIB 20027)	Globose	Irregular	Absent	Light to yellow-brown	Absent	D,f,I
"*A. derwentensis*" [c] (NCIB 12875)	Spherical	Irregular	Absent	Orange to dark orange	Absent	D,I
A. durhamensis [c] (NCIB 20041)	Globose	Irregular	Absent	Light to dark orange	Absent	a,D,I
A. humidus [c] (NCIB 20000)	Spherical	Irregular	Absent	Light yellow-orange to brown	Dark	D,F,I
A. palleronii [c] (NCIB 20021)	Spherical	Irregular	Absent	Light to yellow-brown	Absent	a,B,D,I

[a] Modified from Parenti and Coronelli (*Ann. Rev. Microbiol. 33*: 389–412, 1973).

[b] A, H$_2$S formation; B, melanin production; C, tyrosine degradation; D, casein hydrolysis; E, calcium malate degradation; F, nitrate reduction; G, litmus milk coagulation; H, litmus milk peptonization; I, starch hydrolysis; J, gelatin liquefaction. Positive and negative properties are indicated by capital and small letters, respectively. Letters are missing when the results have not been reported.

[c] After Goodfellow et al. (*J. Gen. Microbiol. 136*: 19–36, 1990).

657

Important Notes for Users of This Manual

Unless otherwise indicated in footnotes to tables, the meanings of symbols are as follows:

+ 90% or more of strains are positive

− 90% or more of strains are negative

d 11–89% of strains are positive

v strain instability (not equivalent to "d")

D Different reactions in different taxa (species of a genus or genera of a family)

All other symbols are defined in footnotes to tables.

Table 24.2 Carbon assimilation by members of various _Actinoplanes_ species[a]

Carbon sources	_A. auranticolor_ (ATCC 15330)	_A. brasiliensis_ (ATCC 25844)	_A. caeruleus_ (ATCC 33937)	_A. consettensis_[b] (NCIB 20027)	_A. deccanensis_ (ATCC 21983)	_A. derwentensis_[b] (NCIB 12875)	_A. durhamensis_[b] (NCIB 20041)	_A. ferrugineus_[c] (ATCC 29868)
D-Xylose	+	+	−	+	+	+	+	+
L-Arabinose	+	+	−	+	+	+	+	+
D-Glucose	+	+	+	+	+	+	+	+
D-Fructose	+	+	+	+	−	+	+	±
D-Mannose	ND	ND	+	+	+	+	+	+
L-Rhamnose	+	+	+	+	+	ND	+	−
m-Inositol	+	+	+	−	−	+	+	−
D-Mannitol	+	+	+	+	+	+	ND	+
Sucrose	+	+	+	ND	+	ND	+	−
Lactose	+	ND	+	+	−	+	ND	+
Salicin	+	ND	+	ND	−	ND	ND	+
Raffinose	+	−	−	ND	−	+	ND	−
Cellulose	−	+	ND	ND	−	ND	ND	−

Table 24.2 _(continued)_

Carbon sources	"_A. garbadinensis_" (ATCC 31049)	_A. globisporus_ (ATCC 23056)	_A. humidus_[b] (NCIB 20000)	"_A. ianthinogenes_" (ATCC 21884)	_A. italicus_ (ATCC 27366)	"_A. liguriae_" (ATCC 31048)	_A. missouriensis_ (ATCC 14538)
D-Xylose	+	+	+	+	+	+	+
L-Arabinose	+	+	+	+	+	+	+
D-Glucose	+	+	+	+	+	+	+
D-Fructose	+	+	+	+	ND	+	ND
D-Mannose	+	+	+	+	ND	+	ND
L-Rhamnose	+	+	ND	+	+	+	+
m-Inositol	−	+	−	−	+	+	−
D-Mannitol	+	+	+	+	+	−	+
Sucrose	+	+	ND	+	+	−	+
Lactose	+	+	+	−	ND	−	ND
Salicin	+	ND	ND	+	ND	−	ND
Raffinose	−	−	ND	−	−	−	ND
Cellulose	−	−	ND	−	−	−	−

Footnotes are at end of table

Table 24.2 (continued)

Carbon sources	*A. palleronii*[b] (NCIB 20021)	*A. philippinensis* (ATCC 12427)	*A. rectilineatus*[d] (ATCC 29234)	*"A. teichomyceticus"* (ATCC 31121)	*A. utahensis* (ATCC 14539)
D-Xylose	+	+	+	+	+
L-Arabinose	−	+	+	+	+
D-Glucose	+	+	+	+	+
D-Fructose	+	+	+	+	+
D-Mannose	+	ND	+	+	ND
L-Rhamnose	ND	+	+	−	+
m-Inositol	ND	+	+	−	−
D-Mannitol	ND	+	+	+	+
Sucrose	ND	+	−	+	+
Lactose	ND	ND	+	±	ND
Salicin	ND	ND	+	±	ND
Raffinose	ND	+	−	−	−
Cellulose	ND	+	−	−	−

[a] Symbols: +, positive; −, negative; ±, weak or slow.

[b] After Goodfellow et al. (J. Gen. Microbiol. *136*: 19–36, 1990).

[c] Growth at the expense of many more organic compounds is reported by Palleroni (Int. J. Syst. Bacteriol. *29*: 51–55, 1979).

[d] Only acid formation is recorded here. Additional results are given by Lechevalier and Lechevalier (Int. J. Syst. Bacteriol. *25*: 371–376, 1975).

Table 24.3 Substrates of taxonomic value for the differentiation of some *Actinoplanes* species[a,b]

Compounds	*A. brasiliensis*	*A. ferrugineus*	*A. italicus*	*A. missouriensis* (ATCC 14538)	*A. missouriensis* (IMRU 824)	*A. philippinensis*	*A. utahensis*
D-Ribose	–	–	+	–	–	+	+
D-Arabinose	+	+	–	+	–	–	–
L-Fucose	–	+	–	+	–	–	+
L-Rhamnose	+	–	+	+	+	+	+
Sucrose	+	–	+	+	+	+	+
Melibiose	+	–	–	–	–	–	–
Melezitose	–	–	+	–	+	+	–
Adipate, pimelate, or suberate	–	–	+	–	–	+	+
D-Malate	–	–	±	–	–	–	+
β-Hydroxybutyrate	+	–	+	+	+	+	+
D,L-Lactate	+	–	+	+	+	+	–
Citrate	–	–	+	+	–	–	–
α-Ketoglutarate	–	–	+	+	–	–	–
Arabitol	+	+	–	–	+	–	–
Xylitol	+	–	–	–	–	–	–
Sorbitol	+	–	±	+	±	±	±
Phenylacetate	–	±	–	–	+	–	–
Quinate	–	–	+	–	–	+	+
β-Alanine	–	+	+	–	–	+	+
L-Asparagine	+	±	–	–	–	±	±
L-Glutamine	+	–	+	+	+	–	+
L-Arginine	–	–	+	–	–	+	+
L-Citrulline	–	–	±	+	–	±	+
L-Ornithine	+	–	±	–	–	–	–
γ-Aminobutyrate	+	–	+	–	+	+	–
L-Histidine	–	–	+	+	–	+	–
L-Proline	+	+	+	–	+	+	+
L-Tyrosine	±	–	+	–	–	–	+
Spermine	–	+	–	–	–	–	–

[a] N. J. Palleroni, unpublished data.

[b] Symbols: +, good growth; ±, poor growth; –, no growth.

Table 24.4 Characteristics differentiating the species of the genus *Ampullariella*[a]

Characteristics	A. campanulata	A. digitata	A. lobata	A. regularis
Sporangial shape: [b,c]				
Bottle	−	+	−	+
Cylindrical	+	−	+	+
Subcylindrical	−	+	−	−
Bell	+	−	−	−
Lobed	+	+	+	−
Irregular	+	−	+	−
Pyriform	+	−	−	−
Digitate	−	+	−	−
Sporangial size: [b,c]				
5–14 × 8–30 μm	−	−	−	+
5–15 × 6–12 μm	+	−	−	−
4–20 × 12–23 μm	−	−	+	−
3–9 × 6–12 μm	−	+	−	−
Color of colonies on Czapek agar: [b,d]				
Orange	−	−	+	+
Red	+	−	+	+
Brown	+	+	−	−
Black	+	−	−	−
Color of colonies on Peptone-Czapek agar: [b,d]				
Orange	+	−	+	+
Ocher	−	−	−	+
Red	+	−	+	−
Utilization of: [d,e]				
L-Arabinose	+	−[d]; +[e]	+	+
D-Fructose	+	+	+	+
D-Glucose	+	+	+	+
Inositol	−	+	−	−
D-Mannitol	+	−	+	−
Raffinose	−	+	+	−
Sucrose	+	+	+	+
D-Ribose	+	+	+	+
Rhamnose	+	+	+	+
D-Xylose	+	+	+	+
Glycerol	+	+	+	+
Hydrolysis of:				
Starch [e]	+	+	+	+
Casein [b,e]	+	+[b]; −[e]	+	+
Tyrosine [e]	−	−	−	−
Gelatin [e]	−	v	+	+
Reduction of nitrate [e]	+	+	+	+
Skim milk [e]**:**				
Peptonization	+	+	−	−
Coagulation	+	+	+	+
Production of melanoid pigments [e]	−	+	−	−

[a] Symbols: see standard definitions; v, strain instability.

[b] Data from Couch (J. Elisha Mitchell Sci. Soc. *79:* 53–70, 1963).

[c] Data from Couch and Bland (*In* Buchanan and Gibbons (Eds.), *Bergey's Manual of Determinative Bacteriology*, 8th ed., Williams & Wilkins Co., Baltimore, 1974, pp. 717–718).

[d] Data from Schäfer (Beiträge zur Klassifizierung und Taxonomie der Actinoplanceen, Dissertation, Marburg, 1973).

[e] Data from Nonomura et al. (J. Ferment. Technol. *57:* 79–85, 1979).

Table 24.5 Differential characteristics of the species of the genus *Catellatospora*[a]

Characteristics	*C. citrea*	*C. citrea* subsp. *methionotrophica*	*C. ferruginea*	*C. matsumotoense*	*C. tsunoense*
MK-9 major menaquinone	+	+	−	−	+
MK-10 major menaquinone	−	−	+	+	−
3-*O*-Methylrhamnose in cell walls	−	−	+	+	−
Resistance to novobiocin	−	−	+	+	−
Growth requirements:					
Thiamine required for growth	−	−	+	+	+
Methionine required for growth	−	+	−	−	−
Utilization of:					
D-Fructose	−	−	+	+	+
Raffinose	−	−	+	+	+
L-Rhamnose	+	+	+	−	+
D-Mannitol	−	−	+	−	−
α-Methyl-D-glucoside	−	−	+	−	−

[a] +/-: 100% strains positive/negative, Asano et al. (Int. J. Syst. Bacteriol. *39:* 309–313, 1989).

Important Notes for Users of This Manual

Unless otherwise indicated in footnotes to tables, the meanings of symbols are as follows:

+   90% or more of strains are positive

−   90% or more of strains are negative

d   11-89% of strains are positive

v   strain instability (not equivalent to "d")

D   Different reactions in different taxa (species of a genus or genera of a family)

All other symbols are defined in footnotes to tables.

Table 24.6 Physiological characteristics of the species of the genus *Dactylosporangium*[a]

Characteristics	*D. aurantiacum*	*D. matsuzakiense*[b,c]	*D. roseum*[d]	*D. thailandense*	*D. vinaceum*[b]
Utilization of:					
D-Fructose	+	+	+	+	+
Sucrose	+	+	+	+	+
D-Glucose	+	+	+	+	+
D-Xylose	+	+	v	+	+
D-Mannitol	+	+	−	+	+
L-Arabinose	+	+	v	+	+
L-Rhamnose	+	+	−	+	+
Raffinose	+[e]; −[b,c]	−	−	+[e]; −[b,c]	−
D-Mannose[e]	+	ND	ND	+	ND
Lactose[e]	+	ND	ND	+	ND
Maltose[e]	+	ND	ND	+	ND
Dextrin[e]	+	ND	ND	+	ND
Inulin[e]	+	ND	ND	+	ND
D-Galactose[e]	+	ND	ND	+	ND
Glycerol	−	−	ND	−	−
D-Sorbitol[e]	−	ND	ND	−	ND
D-Dulcitol[e]	−	ND	ND	−	ND
Inositol	−	−	−	−	−
D-Ribose[e]	−	ND	ND	+	ND
Sorbose[e]	−	ND	ND	ND	ND
D-Melibiose	+	−	ND	−	ND
Hydrolysis of:					
Starch	+	+	−	+	+
Cellulose[e]	−	ND	ND	−	ND
Tyrosine	−[e]	ND	ND	+[f]	ND
Gelatin	−[e]; +[b]	−	+	+[e]; −[b]	+
Casein	+[e]; −[b]	−	−	+[e]; −[b]	+
Coagulation of milk	−	−	−	−	+
Digestion of calcium malate[e]	−	ND	ND	−	ND
Reduction of nitrate	+	−	−	−	−
Production of melanoid pigments	−	−	−	−	−
Production of H$_2$S[e,f]	+	ND	ND	+	ND
Optimum for growth:					
pH[e]	6.0–7.0	ND	ND	6.0–7.0	ND
Temperature (°C)	28–37[e]	25–37	28–37	28–37[e]	25–37
NaCl tolerance:					
Up to 1.5% (w/v)	+[b]	−	+	+[b]	+
Up to 3.0% (w/v)	+[b]	−	−	−[b]	+

[a] Symbols: see standard definitions; v, strain instability; ND, not determined.

[b] Data from Shomura et al. (Int. J. Syst. Bacteriol. *33:* 309–313, 1983).

[c] Data from Shomura et al. (J. Antibiot. *33:* 924–930, 1980).

[d] Data from Shomura et al. (Int. J. Syst. Bacteriol. *35:* 1–4, 1985).

[e] Data from Thiemann et al. (Arch. Mikrobiol. *58:* 42–52, 1967).

[f] Data from Thiemann (*In* Buchanan and Gibbons (Eds.), *Bergey's Manual of Determinative Bacteriology*, 8th ed., Williams & Wilkins Co., Baltimore, 1974, pp. 721–722).

Table 24.7 Differential characteristics of the species of the genus *Micromonospora*[a]

Characteristics	M. carbonacea		M. chalcea	M. coerulea	M. echinospora		
	subsp. aurantiaca	subsp. carbonacea			subsp. echinospora	subsp. ferruginea	subsp. pallida
Diagnostic mycelial pigment[b]	ND	ND	ND	B-G	Pu	Pu	ND
Diffusible pigment[b]	(pYF)	ND	Y	ND	ND	ND	(RF)
Growth on:							
Czapek-sucrose agar	-	-	-	-	+	+	+
Potato slice	+	+	+	+	-	-	-
Carbohydrate utilization:							
α-Melibiose	+	+	+	+	-	-	-
Raffinose	v	v	+	+	-	-	-
D-Mannitol	-	-	-	+	-	-	-
L-Rhamnose	-	-	-	-	+	+	+
Glycerol	-	-	-	-	-	-	-
Inositol	-	-	-	-	-	-	-
D-Ribose	-	-	-	-	-	+	d
Glycosidase activity:							
α-Galactosidase	+	+	+	+	-	-	-
β-Xylosidase	+	+	+	-	d	d	+
α-Mannosidase	-	-	-	+	-	-	-
Nitrate reduction	-	+	v	-	v	-	+
Maximum NaCl tolerance (%, w/v)	3	3	5	1.5	3	3	3
Cell component: 3-Hydroxydiaminopimelate	+	+	-	-	+	+	+
Main menaquinone	MK-9	MK-9	MK-10	MK-10	MK-10	MK-10	MK-12

Footnotes are at end of table

Table 24.7(*continued*)

Characteristics	*M. halophytica*		*M. inositola*	*M. olivasterospora*	*M. purpureochromogenes*
	subsp. *halophytica*	subsp. *nigra*			
Diagnostic mycelial pigment [b]	ND	ND	ND	ND	ND
Diffusible pigment [b]	RBr	ND	ND	O–G	dBr
Growth on:					
Czapek-sucrose agar	+	+	–	+	v
Potato slice	–	–	+	–	v
Carbohydrate utilization:					
α-Melibiose	+	+	+	–	+
Raffinose	+	+	+	–	+
D-Mannitol	–	–	+	–	–
L-Rhamnose	–	–	–	–	–
Glycerol	–	–	–	–	+
Inositol	–	–	v	–	–
D-Ribose	–	–	–	+	–
Glycosidase activity:					
α-Galactosidase	+	+	+	–	+
β-Xylosidase	+	+	+	–	–
α-Mannosidase	–	–	+	–	–
Nitrate reduction	+	+	–	–	–
Maximum NaCl tolerance (%, w/v)	4	4	1.5	3	1.5
Cell component: 3-Hydroxydiaminopimelate	+	+	+	+	–
Main menaquinone	MK-9	MK-9	MK-10	MK-10	MK-10

[a] Symbols: see standard definitions.
[b] Pu, purple; B, blue; G, green; R, red; F, fluorescent; O, olive; d, dark; Br, brown; Y, yellow; p, pale.

665

Table 24.8 Differential characteristics of the species of the genus *Pilimelia*[a]

Characteristics	*P. anulata*	*P. columellifera*	*P. terevasa*
Sporangial shape:			
Spherical	−	+	+
Pyriform	−	+	−
Campanulate	−	−	+
Flabelliform	−	−	+
Cylindrical	+	−	−
Sporangiophore:			
Septate	+	−	+
Annulate	+	−	−
Extended as columella	−	+	−
Arrangement of spore chains:			
Parallel rows	+	−	+
Swirl-like	−	+	−
Consistency of colonies:			
Solid	−	+	−
Soft	+	−	+
Color of colonies:			
Lemon yellow, yellow-gray	+	−	+
Golden yellow, orange	−	+	−

[a] Symbols: see standard definitions.

Table 24.9 Other characteristics of the species of the genus *Pilimelia*[a,b]

Characteristics	*P. anulata*	*P. columellifera* subsp. *columellifera*	*P. columellifera* subsp. *pallida*	*P. terevasa*
Hydrolysis of starch	−	−	−	−
Degradation of tyrosine	+	−	−	+
Production of melanoid pigments [c]	v	v	+	v
Liquefaction of gelatin	−	+	+	−
Peptonization of casein	+	+	+	+
Reduction of nitrate	−	+	−	−
pH growth range	6.5–7.8	6.5–7.6	5.0–7.5	6.5–7.6
Temperature growth range (°C)	15–35	15–35	10–30	10–35

[a] Data from studies of type strains (Vobis et al., Syst. Appl. Microbiol. *8:* 67–74, 1986).

[b] Symbols: see standard definitions; v, strain instability.

[c] Data from Kane Hanton (*In* Buchanan and Gibbons (Eds.), *Bergey's Manual of Determinative Bacteriology,* 8th ed., Williams & Wilkins Co., Baltimore, 1974, pp. 718–719), Schäfer (Beiträge zur Klassifizierung und Taxonomie der Actinoplanaceen, Dissertation, Marburg, 1973), and Vobis et al. (Syst. Appl. Microbiol. *8:* 67–74, 1986).

STREPTOMYCETES AND RELATED GENERA

Genus **Intrasporangium**

Branching mycelium, about 1.0 μm in diameter, **has a tendency to break into fragments of various sizes and shapes. Aerial mycelium is never observed. Oval and lemon-shaped vesicles (5–15 μm in diameter) are formed intercalary and/or at the hyphal apices.** In some of the vesicles (termed *sporangia* in the original description), several round or oval bodies (1.2–1.5 μm in diameter) may be observed in older cultures. These are nonmotile but may undergo a brownian movement within the mature vesicles. When released into fresh medium, the sporelike cells germinate, giving rise to a branching mycelium. Stain Gram positive, are non-acid-fast. Aerobic. **Chemoorganotrophic with an oxidative type of metabolism and possess catalase activity.** They grow best at 28–37°C; no growth occurs at 45°C. **Prefer complex media,** especially **containing peptone and yeast extract.** There is no growth on the majority of mineral synthetic media routinely used for actinomycetes. **The peptidoglycan contains LL-diaminopimelic acid and is of the A3γ type. Straight-chain saturated and unsaturated fatty acids predominate. Cells contain an unknown, diagnostic glucosamine-containing phospholipid (phospholipid type 4). Unsaturated menaquinone with eight isoprene units (MK-8) is the major isoprenolog.** The original strain was isolated under nonselective conditions on plates of meat-peptone agar exposed to the atmosphere of a school dining room.

Type (and only) species: *Intrasporangium calvum.*

Characteristics of the species: As described for the genus. The formation of an extensive mycelium and intramycelial vesicles, the fatty acid spectra, and the phospholipid composition separate *Intrasporangium* from other aerobic actinomycetes that have DNA rich in guanine plus cytosine and LL-diaminopimelic acid in the wall peptidoglycan.

Genus **Kineosporia**

Editorial note: The description of the genus *Kineosporia* was amended in 1989 by Ito et al. (Int. J. Syst. Bacteriol. *39:* 168–173) after chemotaxonomic and morphological studies of the type strain of the type species.

Colonies on agar medium **lack aerial mycelia,** form **central projections with radiating vegetative hyphae,** and are occasionally accompanied by bunches of spore clusters in the agar. Mature colonies have a gelatinous matrix that confers a **glossy appearance.** Spores, which are spherical to ovoid or pyriform with a long axis of 1–2 μm, are catenated around the central projection or are located singly or in aggregate at the tips of hyphae. The spores are **motile** with **polar tufts of flagella.** Gram positive. The peptidoglycan of the cell wall contains **both L- and *meso*-diaminopimelic acid (*meso*-DAP), suggesting that the former is present in the mycelium and the latter in the spores.** There are **no characteristic sugar patterns.** A diagnostic phospholipid is phosphatidylcholine. MK-9(H$_4$) is present as the main menaquinone component. There are **no *iso-/anteiso*-branched fatty acids and no mycolic acids.** Aerobic. Chemoorganotrophic, using a range of simple sugars. Optimum temperature for growth and sporulation is 20–30°C, with no growth at or above 37°C. The habitat is soil.

Type (and only) species: *Kineosporia aurantiaca.*

Characteristics of the species: As described for the genus.

Genus **Sporichthya**

Very short, sparse aerial mycelium is composed of hyphae (0.5–1.0 μm in diameter) that grow on the surface of solid media. **The aerial hyphae are main-**

tained upright at the surface of the medium by holdfasts, which are outgrowths of the wall of the basal cell. **Primary (substrate) mycelium is not formed. The sparingly branched aerial mycelium divides into rod-shaped to coccoid spores, which,** in the presence of water, **may become polarly flagellate and motile.** On nitrogen-rich media the spores swell to give rise to swollen elements of various shapes, including fish-shaped structures, which may be motile. **Gram variable.** Young cells tend to be Gram negative; older cells are mainly Gram positive. Cell sections examined with an electron microscope reveal a **Gram-positive type of cell wall.** Cell wall preparations contain major amounts of L-diaminopimelic acid (L-DAP), glycine, alanine, glutamic acid, glucosamine, and muramic acid. Because of the difficulties of obtaining enough cells for analysis, there are no data on phospholipid or menaquinone composition. **Facultatively anaerobic,** chemoorganotrophic, and mesophilic. Sporichthyae are rare; all of those isolated to date were from cultivated soil.

Type (and only) species: *Sporichthya polymorpha.*

Characteristics of the species: As described for the genus.

Genus Streptomyces

Vegetative hyphae (0.5–2.0 μm in diameter) produce an extensively branched mycelium that rarely fragments. **The aerial mycelium at maturity forms chains of three to many spores.** A few species bear short chains of spores on the substrate mycelium. Sclerotia-, pycnidial-, sporangia-, and synnemata-like structures may be formed by some species. Spores are nonmotile. **Form discrete and lichenoid, leathery or butyrous colonies.** Initially, colonies are relatively smooth surfaced, but later they develop a weft of aerial mycelium that may appear floccose, granular, powdery, or velvety. **Produce a wide variety of pigments responsible for the color of the vegetative and aerial mycelia. Colored diffusible pigments may also be formed.** Many strains produce one or more antibiotics. Gram positive but not acid-alcohol-fast. The cell wall peptidoglycan contains major amounts of L-diaminopimelic acid (L-DAP). Lack mycolic acids but contain major amounts of **saturated, *iso*-, and *anteiso*- fatty acids;** possess either **hexa- or octahydrogenated menaquinones with nine isoprene units** as the predominant isoprenolog; and have complex polar lipid patterns that typically contain **diphosphatidylglycerol, phosphatidylethanolamine, phosphatidylinositol, and phosphatidylinositol man-**

nosides. Aerobes. Chemoorganotrophic, having an oxidative type of metabolism. Catalase positive and generally reduce nitrates to nitrites and degrade adenine, esculin, casein, gelatin, hypoxanthine, starch, and L-tyrosine. **Use a wide range of organic compounds as sole sources of carbon for energy and growth.** The temperature optimum is 25–35°C; some species grow at temperatures within the psychrophilic and thermophilic range; the optimum pH range for growth is 6.5–8.0. **Widely distributed and abundant in soil,** including composts. A few species are pathogenic for animals and humans; others are phytopathogens.

Type species: *Streptomyces albus.*

Differentiation of the species of the genus **Streptomyces:** The differentiation of species in this genus remains difficult. For a percentage positive probability matrix, see Table 25.1.

Genus Streptoverticillium

The substrate mycelium, 0.8–1.2 μm in diameter, is branching. **The aerial mycelium consists of long, straight filaments bearing at more or less regular intervals branches (three to six) arranged in whorls (verticils). The appearance at ~100 × magnification is of "barbed wire." Each branch of the verticil produces at its apex an umbel that consists of two to many chains of spherical to ellipsoidal spores.** Spore chains may be straight or flexuous or may terminate in hooks. **Spiral spore chains have not been observed.** Reproduction occurs either from fragments of substrate and/or aerial mycelium or from germination of spores. Spores are smooth-surfaced to slightly rough. **Neither spiny nor hairy spores have been observed.** On primary isolation, colonies are small and discrete, developing a weft of aerial mycelium that typically appears cottony. **Media low in available carbohydrates are particularly suitable for spore production.** Most species produce soluble pigments and colored substrate and aerial mycelium. They are resistant to lysozyme and to neomycin; they produce compounds that exhibit antifungal, antibacterial, antiprotozoal, and antitumor activity and are sensitive to antibacterial agents and actinophages. Gram positive. Cell walls contain L-diaminopimelic acid (L-DAP). Cells contain major amounts of saturated, *iso*-, and *anteiso*-fatty acids, MK-9(H$_6$) and MK-9(H$_8$) menaquinones, and the phospholipids (diphosphatidylglycerol, phosphatidylethanolamine, phosphatidylinositol, and phosphatidylinositol mannosides). Aerobic, substrate growth may develop under reduced oxygen tensions or increased carbon dioxide concentrations. Chemoorganotrophic. Mesophilic; optimum growth occurs at 26–32°C and at pH 6.5–8.0. These are mostly saprophytes in soil.

Type species: *Streptoverticillium baldaccii.*

Differentiation of the species of the genus **Streptoverticillium:** For a percentage positive probability matrix, see Table 25.2.

Important Notes for Users of This Manual

Unless otherwise indicated in footnotes to tables, the meanings of symbols are as follows:
 	
   + 90% or more of strains are positive
   − 90% or more of strains are negative
   d 11-89% of strains are positive
   v strain instability (not equivalent to "d")
   D Different reactions in different taxa (species of a genus or genera of a family)
 	
All other symbols are defined in footnotes to tables.

Table 25.1 A percentage positive probability matrix for streptomycete species defined by the major clusters of Williams et al. (J. Gen. Microbiol. 129: 1815–1830, 1983)

Characteristics	S. albidoflavus	S. albus	S. antibioticus	S. anulatus	S. chromofuscus	S. cyaneus	S. diastaticus	S. exfoliatus	S. fulvissimus	S. griseoflavus	S. griseoruber	S. griseoviridis	S. halstedii	S. lavendulae	S. lydicus	S. microflavus	S. olivaceoviridis	S. phaeochromogenes	S. purpureus	S. rimosus	S. rochei	S. violaceus	S. violaceusniger
Spore chain Rectiflexibles	70	1	20	90	22	5	42	99	67	1	1	1	77	92	1	20	1	1	99	14	4	38	1
Spore chains Spirales	1	99	60	38	78	82	58	1	22	1	99	99	23	17	99	40	86	67	1	57	64	50	99
Spore mass red	5	1	1	5	1	31	16	39	66	17	12	99	1	83	9	1	14	1	25	1	7	13	1
Spore mass grey	1	1	99	3	34	31	48	38	11	1	78	1	99	8	91	80	71	1	75	1	77	13	99
Mycelial pigment red-orange	1	1	1	1	1	21	16	11	89	1	67	1	1	1	1	20	1	1	1	1	8	1	1
Diffusible pigment produced	20	1	1	11	11	33	10	39	33	1	99	99	8	1	1	20	29	50	99	1	4	1	1
Diffusible pigment yellow-brown	15	1	1	11	11	13	5	28	1	1	1	99	1	1	1	1	29	33	99	1	4	1	1
Melanin on peptone-yeast-iron agar	5	1	80	13	33	97	47	61	67	17	22	1	1	99	1	80	14	17	75	1	4	88	1
Melanin on tyrosine agar	10	1	60	24	22	85	47	61	1	17	22	17	1	92	1	80	14	33	1	1	1	99	1
Antibiosis against:																							
Bacillus subtilis NCIB 3610	5	17	40	50	11	44	21	56	78	17	1	99	8	92	73	80	1	17	50	99	35	50	67
Micrococcus luteus NCIB 196	5	17	20	71	11	33	21	44	89	17	1	99	8	83	99	99	1	17	50	86	35	38	99
Candida albicans CBS 562	85	1	99	21	1	3	1	1	11	1	1	33	8	67	27	1	1	1	75	57	19	1	17
Saccharomyces cerevisiae CBS 1171	80	1	99	13	1	5	1	1	22	1	1	33	8	67	27	1	29	17	1	86	15	1	50
Streptomyces murinus ISP 5091	10	17	80	61	22	62	5	39	99	67	22	80	15	99	99	99	1	1	75	99	39	88	50
Aspergillus niger LIV 131	55	17	80	32	1	10	11	6	22	1	1	33	1	75	99	1	1	1	50	99	27	25	33
Lecithinase activity	15	1	40	3	11	10	1	50	44	1	1	50	15	99	64	60	1	1	75	86	4	75	1
Lipolysis	99	99	1	99	89	49	26	94	99	99	89	83	99	33	18	40	86	83	25	99	73	63	99
Pectin hydrolysis	5	1	40	53	22	56	68	61	1	99	99	67	77	8	1	99	14	50	1	14	42	13	50
Nitrate reduction	20	1	80	79	22	36	47	83	1	83	89	33	23	50	9	1	14	17	99	86	27	99	83
H$_2$S production	85	83	1	90	89	90	79	89	67	99	99	99	92	42	1	99	99	83	75	14	92	63	99
Hippurate hydrolysis	1	1	40	13	13	3	28	44	44	67	56	67	77	1	30	40	50	83	1	71	9	88	67
Elastin degradation	90	50	1	87	67	41	37	89	67	17	44	50	85	92	36	60	57	99	99	99	50	88	83
Xanthine degradation	95	83	80	95	22	80	53	94	99	1	78	83	99	83	82	60	29	99	99	86	96	88	1
Arbutin degradation	99	99	99	99	99	54	53	99	67	50	99	99	99	92	99	99	1	67	75	71	96	99	99

Footnotes are at end of table

Table 25.1 *(continued)*

Characteristics	S. albidoflavus	S. albus	S. antibioticus	S. anulatus	S. chromofuscus	S. cyaneus	S. diastaticus	S. exfoliatus	S. fulvissimus	S. griseoflavus	S. griseoruber	S. griseoviridis	S. halstedii	S. lavendulae	S. lydicus	S. microflavus	S. olivaceoviridis	S. phaeochromogenes	S. purpureus	S. rimosus	S. rochei	S. violaceus	S. violaceusniger
Resistance to:																							
Neomycin (50 µg/ml)	1	1	1	1	1	1	1	11	89	1	1	1	1	50	18	1	1	1	75	99	8	25	1
Rifampicin (50 µg/ml)	40	99	20	66	33	46	68	11	99	50	78	83	31	33	9	99	86	50	99	99	89	99	83
Oleandomycin (100 µg/ml)	70	83	1	84	1	13	26	44	99	83	1	17	99	8	9	1	1	33	75	71	46	13	50
Penicillin G (10 i.u.)	90	99	1	74	44	64	47	44	99	83	78	99	92	58	91	20	14	33	25	99	92	99	17
Growth at 45°C	1	99	80	5	67	41	16	17	22	1	1	67	23	17	1	60	86	67	1	43	77	1	50
Growth with (%, w/v):																							
NaCl (7.0)	85	99	20	74	44	18	32	22	22	33	78	83	92	1	55	1	29	83	25	99	92	38	1
Sodium azide (0.01)	75	99	60	32	56	15	5	23	1	33	11	99	85	1	18	60	57	67	1	71	62	63	50
Phenol (0.1)	90	17	80	92	22	64	95	72	44	99	89	83	85	58	9	99	86	99	1	71	96	99	1
Potassium tellurite (0.001)	85	50	60	87	67	46	74	83	56	99	78	99	99	42	55	80	57	99	1	71	73	99	17
Thallous acetate (0.001)	90	1	40	87	54	13	21	67	22	50	33	50	92	17	46	1	14	33	1	14	54	63	1
Utilization of:																							
DL-α-Amino-*n*-butyric acid	65	1	20	37	67	31	32	61	89	17	33	1	54	42	9	60	57	67	1	1	12	88	99
L-Cysteine	60	1	80	61	67	72	79	50	44	17	78	83	69	33	46	60	99	17	1	29	50	38	33
L-Valine	35	17	60	37	33	69	74	50	99	1	56	17	62	17	27	40	86	99	1	57	15	50	33
L-Phenylalanine	70	17	40	61	11	67	16	83	89	17	33	33	77	42	99	99	71	83	1	86	46	99	67
L-Histidine	40	99	99	74	78	85	68	78	99	17	99	83	69	8	36	40	99	83	1	99	77	25	99
L-Hydroxyproline	1	67	40	37	1	28	21	89	78	17	22	17	23	42	55	20	1	67	1	29	8	88	83
Sucrose	45	33	80	26	33	92	74	28	34	83	44	17	23	50	73	99	86	83	25	1	81	38	33
meso-Inositol	45	33	80	32	89	92	84	6	99	83	99	67	69	25	91	40	57	99	75	99	96	63	67
Mannitol	90	99	80	99	99	95	90	1	99	99	99	99	69	8	91	99	99	99	25	99	99	38	99
L-Rhamnose	20	17	60	82	67	92	95	61	22	83	99	83	69	17	18	80	99	99	1	1	96	38	67
Raffinose	5	33	60	18	22	99	84	33	89	33	99	50	31	8	82	99	99	99	1	86	69	50	83
D-Melezitose	55	67	80	71	22	72	26	22	44	67	56	33	46	33	82	40	29	99	99	57	81	13	83
Adonitol	50	99	20	66	22	82	16	1	89	1	22	67	8	8	82	1	14	99	1	99	35	50	83
D-Melibiose	25	50	99	32	44	98	95	1	89	67	78	33	77	17	82	40	99	83	25	99	96	75	83
Dextran	20	1	1	76	78	59	16	6	22	50	44	17	69	1	1	1	14	50	1	14	89	13	17
Xylitol	5	1	1	21	1	21	16	1	11	1	1	1	1	1	55	40	29	99	25	86	46	1	17

Table 25.2 A percentage positive probability matrix for *Streptoverticillium* clusters

Cluster[a]	Aerial mycelium cottony	Spores yellow	Spores white	Melanin produced	Utilization of:						
					Mannitol	D-Melibiose	Raffinose	Sorbitol	Coumarin	L-Methionine	L-Proline
S. baldaccii (17)	99	35	15	55	10	10	20	1	30	10	99
S. cinnamoneum (14)	94	67	22	28	6	6	17	1	39	44	94
S. griseocarneum (13)	93	13	67	40	13	1	40	27	1	7	87
S. hachijoense (2)	99	1	99	1	1	50	99	50	50	1	50
S. salmonis (3)	99	1	1	99	1	1	1	33	1	1	99
S. ladakanum (2)	50	1	50	1	1	1	1	1	50	99	50
S. mobaraense (3)	75	1	1	1	1	1	50	1	1	25	99
"*S. morookaense*" (2)	99	1	1	1	99	50	50	1	50	99	99
S. abikoense (5)	99	80	20	80	80	1	1	60	60	80	99
"*S. olivoreticulum* subsp. *cellulophilum*" (2)	99	1	99	50	99	99	99	99	1	99	99
S. albireticuli (1)	99	99	1	99	99	1	1	1	99	1	99
"*S. alboverticillatum*" (1)	99	1	99	1	1	1	1	1	99	99	99
S. album (1)	99	99	1	1	1	1	1	1	1	99	99
S. kashmirense (1)	99	1	1	99	1	99	99	99	1	1	1
S. kishiwadense (1)	99	1	99	1	1	1	1	1	99	1	99
S. liacinum (1)	99	1	1	99	1	1	1	1	99	1	1
S. orinoci (1)	1	99	1	1	1	1	99	1	1	1	1
S. rectiverticillatum (1)	99	1	1	99	99	99	1	1	1	1	99
"*S. reticulum* subsp. *protomycicum*" (1)	1	1	1	1	1	1	1	1	1	1	1
"*S. sapporonense*" (1)	99	1	1	1	1	1	1	1	1	99	1
S. thioluteum (1)	1	99	1	1	1	99	1	99	99	99	99
"*S. verticillum* subsp. *quintum*" (1)	1	1	1	1	1	99	99	99	99	99	99
"*S. verticillum* subsp. *tsukushiense*" (1)	99	1	1	1	99	99	99	99	99	1	99
"*S. viridoflavum*" (1)	1	1	99	1	1	1	1	1	99	99	99

Footnotes are at end of table

Table 25.2 *(continued)*

Cluster[a]	Utilization of *(continued)*		Acid production from:					Degradation of:			
	Shikimic acid	DL-α-Amino-butyric acid	D-Galactose	*meso*-Inositol	D-Fructose	D-Ribose	D-Trehalose	Esculin	Citrate	DNA	Hypoxanthine
S. baldaccii (17)	15	35	25	80	25	85	99	1	99	60	40
S. cinnamoneum (14)	11	50	22	99	11	94	72	28	99	83	1
S. griseocarneum (13)	13	47	13	93	20	80	87	1	80	99	33
S. hachijoense (2)	1	1	1	50	1	1	99	1	99	99	1
S. salmonis (3)	1	67	1	99	1	99	99	1	99	99	99
S. ladakanum (2)	1	50	1	1	99	99	1	99	1	99	1
S. mobaraense (3)	25	50	25	50	1	99	25	99	25	75	1
"*S. morookaense*" (2)	99	99	99	99	99	99	99	99	99	1	1
S. abikoense (5)	60	80	40	99	80	99	1	60	99	99	60
"*S. olivoreticulum* subsp. *cellulophilum*" (2)	1	1	1	99	1	50	50	1	99	99	1
S. albireticuli (1)	99	1	1	99	99	99	99	1	99	1	1
"*S. alboverticillatum*" (1)	1	99	99	99	1	1	99	1	99	99	1
S. album (1)	99	1	1	1	1	99	99	99	1	99	1
S. kashmirense (1)	1	1	1	99	1	1	99	1	99	99	1
S. kishiwadense (1)	1	1	1	99	99	99	1	1	99	99	99
S. lilacinum (1)	1	1	1	1	99	1	99	1	99	1	1
S. orinoci (1)	99	99	99	1	1	99	99	99	1	1	1
"*S. rectiverticillatum*" (1)	1	1	99	99	99	99	99	1	99	99	99
"*S. reticulum* subsp. *protomycicum*" (1)	1	99	1	99	1	99	99	1	1	1	1
"*S. sapporonense*" (1)	99	99	99	99	1	1	1	99	99	1	1
S. thioluteum (1)	1	1	1	1	1	99	99	99	1	1	1
"*S. verticillum* subsp. *quintum*" (1)	1	99	99	1	1	1	1	99	99	1	1
"*S. verticillum* subsp. *tsukushiense*" (1)	1	99	99	99	1	99	1	99	99	1	99
"*S. viridoflavum*" (1)	99	1	1	99	1	99	99	1	99	99	99

Footnotes are at end of table

673

Table 25.2 (continued)

Cluster[a]	Degradation of (continued)				Growth at 12°C	Growth with:				
	L-Tyrosine	Tween 20	NO₃⁻ reduction	H₂S production		NaCl (5.0%, w/v)	1-Phenylethanol (0.3%, w/v)	Potassium tellurite (0.01%, w/v)	Crystal violet (0.01%, w/v)	Malachite green (0.01%, w/v)
S. baldaccii (17)	70	65	25	75	95	85	45	80	20	10
S. cinnamoneum (14)	61	67	17	83	61	39	33	6	28	22
S. griseocarneum (13)	53	1	13	93	67	80	73	67	53	20
S. hachijoense (2)	99	99	1	99	99	1	1	50	1	1
S. salmonis (3)	99	67	99	33	99	99	33	99	1	1
S. ladakanum (2)	50	50	50	99	99	1	1	50	1	50
S. mobaraense (3)	25	75	99	75	50	99	99	1	25	25
"S. morookaense" (2)	50	99	99	99	1	99	99	50	99	1
S. abikoense (5)	99	40	1	60	20	99	20	40	99	99
"S. olivoreticulum subsp. cellulophilum" (2)	99	99	1	50	99	50	1	1	99	50
S. albireticuli (1)	1	1	1	99	99	99	1	99	1	1
"S. alboverticillatum" (1)	99	99	1	99	99	99	1	1	1	1
S. album (1)	99	99	1	1	1	1	99	1	1	1
S. kashmirense (1)	1	99	1	1	1	1	99	99	99	99
S. kishiwadense (1)	1	1	1	99	1	1	99	1	99	99
S. lilacinum (1)	1	1	1	1	1	99	99	99	99	1
S. orinoci (1)	1	1	1	99	1	99	99	1	1	1
S. rectiverticillatum (1)	99	99	99	99	99	99	1	1	1	1
"S. reticulum subsp. protomycicum" (1)	1	99	99	99	99	1	1	1	99	99
S. sapporonense (1)	1	99	1	99	99	99	1	1	1	1
S. thioluteum (1)	99	99	1	1	99	1	99	1	1	1
"S. verticillum subsp. quintum" (1)	99	1	1	99	1	99	1	1	99	1
"S. verticillum subsp. tsukushiense" (1)	1	1	1	1	1	99	1	1	1	99
"S. viridoflavum" (1)	1	99	1	99	99	1	1	1	1	1

Footnotes are at end of table

Table 25.2 *(continued)*

Cluster[a]	Resistance to:						Antibiosis to:		
	Azlocillin (30 µg/ml)	Carbenicillin (100 µg/ml)	Cephaloridine (30 µg/ml)	Cephalothin (30 µg/ml)	Cephamandole (30 µg/ml)	Colistin (30 µg/ml)	Aspergillus niger	Bacillus subtilis	Candida albicans
S. baldaccii (17)	1	20	10	1	1	5	20	15	25
S. cinnamoneum (14)	1	39	17	17	17	6	89	44	67
S. griseocarneum (13)	33	60	73	73	73	40	73	27	73
S. hachijoense (2)	50	50	1	50	50	1	99	99	99
S. salmonis (3)	33	67	33	1	1	67	99	67	33
S. ladakanum (2)	99	99	99	99	99	1	1	50	1
S. mobaraense (3)	50	99	50	1	50	25	99	25	25
"S. morookaense" (2)	50	50	1	1	1	50	99	1	1
S. abikoense (5)	1	99	80	99	99	20	80	1	80
"S. olivoreticulum subsp. cellulophilum" (2)	50	99	99	99	99	1	99	50	99
S. albireticuli (1)	1	1	1	1	1	1	99	1	99
"S. alboverticillatum" (1)	99	99	1	1	99	99	1	99	1
S. album (1)	1	99	1	1	1	1	1	1	1
S. kashmirense (1)	1	99	99	1	99	99	1	1	1
S. kishiwadense (1)	1	1	1	1	1	1	99	99	99
S. liacinum (1)	99	99	1	1	1	1	1	1	1
S. orinoci (1)	99	99	99	99	1	99	99	99	1
S. rectiverticillatum (1)	1	99	1	1	1	1	99	99	1
"S. reticulum subsp. protomycicum" (1)	1	1	1	99	1	1	1	1	1
"S. sapporonense" (1)	1	99	99	99	1	99	99	99	1
S. thioluteum (1)	1	1	99	1	1	1	99	1	99
"S. verticillum subsp. quintum" (1)	99	99	1	1	99	99	1	99	1
"S. verticillum subsp. tsukushiense" (1)	1	1	99	1	1	1	99	1	1
"S. viridoflavum" (1)	1	1	1	99	99	99	99	1	99

[a] The number of strains in each cluster is given in parentheses.

MADUROMYCETES

This group is divided into two subgroups: 1, *Streptosporangium* and related taxa; 2, *Actinomadura*.

Differentiation of the genera in **Subgroup 1:** See Table 26.1.

SUBGROUP 1: STREPTOSPORANGIUM AND RELATED TAXA

Editorial note: This subgroup encompasses six of the seven taxa classified as maduromycetes in *Bergey's Manual of Systematic Bacteriology*. The revised taxon includes the genera *Microbispora, Microtetraspora, Planobispora, Planomonospora, Spirillospora,* and *Streptosporangium* but excludes the genus *Actinomadura,* which is related to the genus *Thermomonospora* (see Kroppenstedt et al., Syst. Appl. Microbiol. *13:* 148–160, 1990).

Actinomycetes that form a branched, nonfragmenting vegetative mycelium that bears aerial hyphae that can differentiate either into two or more arthrospores or into spore vesicles (sporangia) encompassing one to many spores. Nonmotile and motile spores are formed. Neither endospores nor sclerotia are produced. Stain Gram positive. Non-acid-fast. **Aerobic. Chemoorganotrophic. Mesophilic and thermophilic. The wall peptidoglycan is of the A1γ type.** It contains alanine, glutamic acid, glucosamine, *meso*-diaminopimelic acid, and muramic acid but lacks characteristic sugars (wall chemotype III). **The glycan moiety of the wall contains *N*-acetylated muramic acid.** The cell envelope is rich in saturated, unsaturated, *iso-, anteiso-* (variable), and methyl-branched fatty acids. **Cells contain major amounts of glucosamine-containing polar lipids** (phosphatidylethanolamine and phosphatidylmethylethanolamine variable) apart from *Spirillospora,* which has either phosphatidylethanolamine or phosphatidylglycerol as the major phospholipid. **Partially hydrogenated menaquinones with nine isoprene units (MK-9(H_2, H_4, H_6)) are the predominant isoprenologs. Whole-organism hydrolysates yield madurose (3-*O*-methyl-D-galactose).** Isolated from soil.

Genus **Microbispora**

A stable, branched mycelium carries **spores in characteristic longitudinal pairs on aerial mycelium; spores are not usually formed on the substrate mycelium.** Spores are sessile or on short sporophores, spherical to oval (usually 1.2 to 1.6 μm in diameter), and nonmotile. The spore surface is smooth. **Mesophilic and thermophilic.** Most strains require B vitamins, particularly thiamine, for growth on synthetic media. **A tetrahydrogenated menaquinone with nine isoprene units (MK-9(H_4)) is the predominant isoprenolog.** Common in soil.

Type species: *Microbispora rosea.*

Differentiation of the species of the genus **Microbispora:** See Table 26.2.

Editorial note: There is evidence that the genus *Microbispora* is overspeciated (see Miyadoh et al., J. Gen. Microbiol. *136:* 1905–1913, 1990). These workers have proposed that *M. rosea, M. amethystogenes, M. chromogenes, M. diastatica, M. indica, M. karnatakensis,* and *M. parva* should be combined into the species *M. rosea* subsp. *rosea* and that *M. aerata, M. thermodiastatica,* and *M. thermorosea* should be combined and transferred to the new subspecies *M. rosea* subsp. *aerata.*

Genus **Microtetraspora**

Editorial note: The genus *Microtetraspora* was included in *Bergey's Manual of Systematic Bacteriology,* but the description of this taxon has been emended to accommodate species previously classified as members of the *Actinomadura pusilla* group. The revised genus encompasses 18 species.

677

Stable, extensively branched vegetative mycelium is formed. When produced, aerial hyphae may carry chains of up to 30 nonmotile spores. Spore chains may be straight, hooked (open spirals), or tightly closed spirals, and spore surfaces may be folded, irregular, smooth, or warty. Aerial mycelium may be blue-gray, cream, gray, pink, violet, yellow, or white. Some species require B vitamins for growth. The temperature range is usually 20–45°C, although strains of some species grow at 55°C. The wall envelope contains major proportions of 14-methylpentadecanoic and 10-methylheptadecanoic fatty acids and small amounts of α-hydroxylated *iso*-hexadecanoic acid. Cells contain diphosphatidylglycerol, hydroxylated phosphatidylethanolamine, uncharacterized glycolipids, and a glucosamine-containing phospholipid. Tetrahydrogenated menaquinone with nine isoprene units saturated at Sites III and VIII (MK-9(H$_4$; III, VIII)) is the predominant isoprenolog. Whole-organism hydrolysates contain glucose, ribose, and madurose, although the latter may be difficult to detect. Widely distributed in soil.

Type species: *Microtetraspora glauca.*

Differentiation of the species of the genus **Microtetraspora:** See Table 26.3.

Genus Planobispora

Substrate and aerial mycelia are developed on agar media. Substrate hyphae (0.5–2.0 μm in diameter) are irregularly branched, occasionally septate, and nonfragmenting. Aerial hyphae (1.0 μm in diameter) are sparsely branched and rarely septate. Cylindrical to clavate spore vesicles/sporangia (1.0–1.2 μm wide × 6.0–8.0 μm long), each containing a longitudinal pair of spores, are formed singly or in bundles on short ramifications of the aerial hyphae. The vesicular envelope is smooth and contains fibrillar elements. A transverse septum or diaphragm connected to the vesicular envelope divides the two spores. The spores (zoospores) are straight or slightly curved with rounded ends and are motile by means of peritrichous flagella. They become motile only after being dispersed for some time and usually germinate with one or two polar germ tubes. Colonies grown on agar media are flat or occasionally elevated. The substrate mycelium either is without distinctive color or is rose-colored. The aerial mycelium, which is developed only on certain agar media, is white or with a light rose tinge. Grows well at 28–40°C and pH 6.0–9.0. The cell envelope contains straight-chain acids, *iso*- and 10-methyl-branched acids, and unsaturated fatty acids; *anteiso*-branched acids are not present. Contain diphosphatidylglycerol, phosphatidylethanolamine, phosphatidylinositol, and unknown glucosamine-containing phospholipids. Di- and tetrahydrogenated menaquinones are the major isoprenologs. Madurose is the characteristic sugar or whole-organism hydrolysate. Isolated from soil, albeit very rarely.

Type species: *Planobispora longispora.*

Differentiation of the species of the genus **Planobispora:** See Table 26.4.

Genus Planomonospora

Substrate and aerial mycelia develop on various agar media. Substrate hyphae (0.6–1.0 μm in diameter) are irregularly branched, occasionally septate, and nonfragmenting. Aerial hyphae (0.5–1.0 μm in diameter) are sparsely branched and rarely septate. Cylindrical to clavate spore vesicles/sporangia (1.0–1.5 μm wide × 3.5–5.5 μm long), each containing a single spore, are formed only on the aerial mycelium. The spores (zoospores) are fusiform and motile by peritrichous flagella. Colonies grown on complex agar media are raised or flat with a rugose or smooth surface. The substrate mycelium is either rose to light orange or brown-violet to light brown. The aerial mycelium is white with a rose tinge or grayish white. Grows well at 28–37°C and at pH 7.0–8.0. The cell envelope contains straight-chain acids, *iso*- and 10-methyl-branched acids, and unsaturated fatty acids; *anteiso*-branched acids are not present. Contain diphosphatidylglycerol, lyso-diphosphatidylglycerol, phosphatidylethanolamine, phosphatidylinositol, and unknown glucosamine-containing phospholipids. Madurose is the characteristic sugar of whole-organism hydrolysates. Isolated from diverse soils.

Type species: *Planomonospora parontospora.*

Differentiation of the species of the genus **Planomonospora:** See Table 26.5.

Genus **Spirillospora**

Substrate and aerial hyphae are 0.2–1.0 μm thick, branched, and septate. **Spherical to vermiform spore vesicles/sporangia** (5.0–24 μm in diameter) **are produced on the aerial mycelium.** The sporangial envelope encloses numerous spores that are arranged in **coiled and branched spore chains. The spores are rod-shaped or curved** (0.5–0.7 × 2.0–6.0 μm) **and motile by means of one to seven subpolarly inserted flagella.** Free, exposed spores in regular or irregular coils may be found among the aerial hyphae. When flooded with water, the coils break up into rod-shaped to curved spores that subsequently become motile. **The color of the substrate mycelium is white to pale yellow or pale buffy pink to red; the aerial mycelium is usually white.** Mesophilic and grow at 18–35°C. Tetra- and hexahydrogenated menaquinones with nine isoprene units (MK-9(H_4, H_6)) are the predominant isoprenologs. **Madurose is the characteristic sugar of whole-organism hydrolysates.** Infrequently isolated from soil.

Type species: *Spirillospora albida.*

Differentiation of the species of the genus **Spirillospora:** See Table 26.6.

Genus **Streptosporangium**

Stable, branched, mycelium-producing **globose spore vesicles/sporangia** (usually 10 μm in diameter) are formed **on aerial mycelium. Arthrospores are formed by septation of a coiled, unbranched hypha within the spore vesicle:** they are spherical, oval, or rod-shaped, 0.2–1.3 × 3.5 μm (usually 1.5 × 1.5 μm), and **nonmotile.** Mesophilic, a few species are thermotolerant. Some species require B vitamins for growth. Di- and tetrahydrogenated menaquinones with nine isoprene units (MK-9(H_2, H_4)) are the predominant isoprenolog. **Madurose is the characteristic sugar of whole-organism hydrolysates.** Isolated from soil, dung, and leaf litter.

Type species: *Streptosporangium roseum.*

Differentiation of the species of the genus **Streptosporangium:** See Table 26.7.

SUBGROUP 2: ACTINOMADURA

Genus **Actinomadura**

Editorial note: The genus *Actinomadura* has undergone marked taxonomic revision since the publication of *Bergey's Manual of Systematic Bacteriology* (see Kroppenstedt et al., Syst. Appl. Microbiol. *13:* 148–160, 1990). The transfer of the *A. pusilla* group to the genus *Microtetraspora* leaves the emended genus *Actinomadura* as a relatively homogeneous taxon that encompasses 25 species. The genus *Excellospora* was listed as *incertae sedis* in *Bergey's Manual of Systematic Bacteriology.* However, *E. viridinigra* was recognized as a synonym of *E. rubrobrunea* and transferred to the emended genus *Actinomadura* as *A. rubrobrunea* (see Kroppenstedt et al., as above).

Extensively branching vegetative hyphae form a dense **nonfragmenting substrate mycelium; aerial mycelium is moderately developed** or absent. At maturity, **the aerial mycelium forms short or occasionally long chains of arthrospores. Spore chains are straight, hooked** (open loops), or **irregular spirals** (1–4 turns). **Spore surface is folded, irregular, rugose, smooth, spiny, or warty.** Aerial mycelium may be blue, brown, cream, gray, green, pink, red, white, or yellow. Colonies have a leathery or cartilaginous appearance when aerial mycelium is absent. **Aerobic. Chemoorganotrophic.** Temperature range is 10–60°C. **Stain Gram positive** and are non-acid-fast. **The peptidoglycan contains *meso*-diaminopimelic acid and N-acetylated muramic acid and is of the A1γ type.** Whole-organism hydrolysates contain galactose, glucose, mannose, ribose, and madurose, the latter sometimes in trace amounts. *Actinomadura* species have a complex mixture of fatty acids with **hexadecanoic, 14-methylpentadecanoic, and 10-methyloctadecanoic acids predominating. Mycolic acids are not produced. Cells contain diphosphatidylglycerol and phosphatidylinositol as major phospholipids. Hexahydrogenated menaquinone with nine isoprene units saturated at Sites II, III and VIII (MK-9(H_6, II, III, VIII))** are the major isoprenolog. Widely distributed in soil. Some strains are pathogenic for animals, including human beings.

Type species: *Actinomadura madurae.*

Differentiation of the species of the genus **Actinomadura:** See Table 26.8.

Table 26.1 Differential characteristics of *Streptosporangium* and related genera[a]

Characteristics	Microbispora	Microtetraspora	Planobispora	Planomonospora	Spirillospora	Streptosporangium
Morphological properties:						
Aerial mycelium and spores:						
Absent or in chains	+	+	−	−	−	−
Spore vesicle containing spores	−	−	+	+	+	+
Spores per chain/spore vesicle	Two	Two to many	Two	One	Many	Many
Spore motility	−	−	+	+	+	−
Temperature range	Mesophilic and thermophilic	Mesophilic and thermophilic	Mesophilic	Mesophilic	Mesophilic	Mesophilic
Chemical properties:						
Fatty acid profile[b]	3c	3c	3c	3c	3a	3c
Predominant menaquinone[c]	MK-9(H$_4$)	MK-9(H$_4$)	MK-9(H$_2$,H$_4$)	MK-9(H$_2$,H$_4$,H$_6$)	MK-9(H$_4$,H$_6$)	MK-9(H$_2$,H$_4$)
Phospholipid pattern[d]	4[e]	4	4	4	1/2	4

[a] Symbols: see standard definitions.

[b] Saturated, unsaturated, *iso-*, *anteiso-* (variable), and methyl-branched fatty acids (see Kroppenstedt, *In* Goodfellow and Minnikin (Eds.), *Chemical Methods in Bacterial Systematics*, Academic Press, London, 1985, pp. 173–199).

[c] Menaquinones detected by chromatographic or physicochemical analysis (Collins et al., J. Gen. Microbiol. *100*: 221–230, 1977; Collins, *In* Goodfellow and Minnikin (Eds.), *Chemical Methods in Bacterial Systematics*, Academic Press, London, 1985, pp. 267–287; Kroppenstedt, 1985, as above). Abbreviations exemplified by MK-9(H$_4$); menaquinone having four of the nine isoprene units hydrogenated.

[d] Detected by thin-layer chromatography (Lechevalier et al., Biochem. Syst. Ecol. *5*: 249–260; Minnikin et al., Int. J. Syst. Bacteriol. *27*: 104–117, 1977).

[e] Characteristic phospholipids: 1, phosphatidylglycerol (variable); 2, only phosphatidylethanolamine; 4, phospholipids containing glucosamine (with phosphatidylethanolamine and phosphatidylmethylethanolamine variable). All preparations contain phosphatidylinositol (see Lechevalier et al., as above).

Table 26.2 Differential characteristics of the species of the genus *Microbispora*[a,b]

Characteristics	*M. aerata*	*M. amethystogenes*	*M. bispora*	*M. chromogenes*	*M. diastatica*	*M. indica*	*M. kamatakensis*	*M. parva*	*M. rosea*	*M. thermodiastatica*	*M. thermorosea*
Color of aerial mycelium	Pink	Pink	White	Pink	Pink	Pinkish white	White	Pink	Pale pink	Pink	Pink
Color of vegetative mycelium	Yellowish brown	Light brown	Yellowish brown	Orange	Yellowish brown	Violet orange	Yellowish pink	Light brown	Orange	Yellowish brown	Yellowish brown
Growth at:											
25°C	−	+	−	+	+	+	+	+	+	−	−
35°C	+	+	−	+	+	+	+	+	+	+	+
50°C	+	−	+	−	−	+	+	+	−	+	+
55°C	+	−	+	−	−	−	−	−	−	+	+
60°C	−	−	+	−	−	−	−	−	−	−	−
Iodinin production	+	+	−	−	−	−	−	+	−	−	−
Nitrate reduction	+	+	−	+	−	+	+	−	+	−	−
Starch hydrolysis	+	−	−	+	+	−	+	−	−	+	−
Soluble pigments[c]	−	−	−	Light yellowish pink	−	Deep orange-yellow	Deep orange-yellow	−	−	−	−
Utilization of:											
Arabinose	+	+	−	+	+	+	−	+	+	+	+
Glycerol	+	+	−	+	+	−	−	+	+	+	+
Inositol	−	+	+	+	−	−	+	−	−	−	−
Rhamnose	−	−	+	−	+	+	+	+	+	−	−

[a] Symbols: see standard definitions.

[b] For details of test procedures, see Nonomura and Ohara (J. Ferment. Technol. *49*: 887–894, 1971), Rao et al. (Int. J. Syst. Bacteriol. *37*: 181–185, 1987), and Shirling and Gottlieb (Int. J. Syst. Bacteriol. *16*: 313–340, 1966).

[c] Other than pale yellow-brown.

Table 26.3 Differential characteristics of the species of the genus *Microtetraspora*[a,b]

Characteristics	M. africana[c]	M. angiospora[d]	M. fastidiosa[e]	M. ferruginea[e]	M. flexuosa[e]	M. fusca
Spore chain morphology:						
Hooks, curled	−	−	−	+	+	−
Pseudosporangia	−	−	−	−	−	−
Spirals:						
1–2 turns	−	−	+	+	+	−
3–5 turns	−	+	+	−	−	−
Straight:						
Two spores	−	−	−	−	−	−
Many spores	+	−	−	−	−	−
Spore surface morphology:						
Folded	−	−	−	+	−	−
Irregular, uneven	−	−	+	−	−	−
Ridged	−	+	−	−	−	−
Smooth	+	−	+	+	−	+
Warty	−	−	−	−	+	−
Wrinkled	−	−	−	−	−	−
Yeast extract-malt extract (ISP Medium 2):						
Aerial mycelium	ND	ND	Trace	Orange-pink	−	ND
Substrate mycelium	ND	ND	Brown	Orange	−	ND
Oatmeal agar (ISP Medium 3):						
Aerial mycelium	Greyish blue	White	White-pink	White-pink	White-yellow	Trace
Substrate mycelium	Yellow	White-ocher	Colorless	Pink	Brown	Colorless
Soluble pigment	Yellowish brown	−	−	−	−	−
Inorganic salts-starch agar (ISP Medium 4):						
Aerial mycelium	Greyish blue	ND	Colorless, pink	−	−	ND
Substrate mycelium	Red, brownish violet	ND	Colorless	Colorless, brown	−	ND
Degradation of:						
Casein	ND	+	+	+	+	ND
DNA	ND	+	+	+	+	ND
Elastin	ND	+	ND	+	ND	ND
Esculin	ND	+	+	−	+	+
Gelatin	ND	+	+	−	+	−
Hypoxanthine	ND	+	+	+	−	−
Starch	ND	−	−	+	+	−
Testosterone	ND	+	−	+	+	−
Tyrosine	ND	+	−	+	+	+
Xanthine	ND	−	−	−	−	−
Reduction of nitrate	ND	−	+	+	+	−

Footnotes are at end of table

Table 26.3 *(continued)*

Characteristics	M. glauca	M. helvata[e]	M. niveoalba	M. polychroma[e]	M. pusilla[e]	M. recticatena[e]
Spore chain morphology:						
Hooks, curled	–	+	–	–	–	–
Pseudosporangia	–	+	–	–	+	–
Spirals:						
1–2 turns	–	–	–	–	–	–
3–5 turns	–	–	–	–	–	–
Straight:						
Two spores	–	–	–	–	–	–
Many spores	–	–	–	–	–	+
Spore surface morphology:						
Folded	–	–	–	–	–	–
Irregular, uneven	–	–	–	–	–	–
Ridged	–	–	–	–	–	–
Smooth	+	+	+	–	+	–
Warty	–	–	–	–	–	–
Wrinkled	–	–	–	–	+	–
Yeast extract-malt extract (ISP Medium 2):						
Aerial mycelium	ND	White-yellow	ND	ND	Trace	ND
Substrate mycelium	ND	Yellow-brown	ND	ND	Brown-red	ND
Oatmeal agar (ISP Medium 3):						
Aerial mycelium	Blue-gray	Trace, white	ND	Trace	White-cream	White-cream
Substrate mycelium	Blue-green	Yellow-brown	ND	Colorless, brown	Gray-brown	Dark yellow-brown
Soluble pigment	–	–	ND	–	–	–
Inorganic salts-starch agar (ISP Medium 4):						
Aerial mycelium	ND	White-cream	ND	–	Trace	–
Substrate mycelium	ND	Yellow-brown	ND	Colorless, yellow-brown	Colorless, brown	Yellow-pink
Degradation of:						
Casein	+	–	+	–	–	ND
DNA	+	–	+	–	+	ND
Elastin	–	ND	+	ND	–	ND
Esculin	+	+	+	+	+	ND
Gelatin	+	–	+	+	+	ND
Hypoxanthine	+	–	+	+	+	ND
Starch	+	–	+	–	–	ND
Testosterone	+	+	+	–	+	ND
Tyrosine	+	–	+	–	+	ND
Xanthine	+	–	+	–	–	ND
Reduction of nitrate	+	+	+		+	ND

Footnotes are at end of table

Table 26.3 *(continued)*

Characteristics	M. roseola[e]	M. roseoviolacea[e]	M. rubra[e]	M. salmonea[e]	M. spiralis[e]	M. turkmeniaca[e]
Spore chain morphology:						
Hooks, curled	−	−	+	+	+	−
Pseudosporangia	−	+	−	−	−	−
Spirals:						
1–2 turns	−	−	+	+	+	+
3–5 turns	+	−	+	−	−	−
Straight:						
Two spores	−	−	−	−	−	−
Many spores	−	−	−	−	−	+
Spore surface morphology:						
Folded	+	−	−	−	−	−
Irregular, uneven	−	−	+	−	−	−
Ridged	−	−	−	−	−	−
Smooth	−	+	+	−	−	+
Warty	−	−	−	+	+	−
Wrinkled	−	−	−	−	−	−
Yeast extract-malt extract (ISP Medium 2):						
Aerial mycelium	Pink	White-pink	Trace	Cream-pink	White-yellow	ND
Substrate mycelium	Red-brown	Purple-red	Red-brown	Brown	Yellow-brown	ND
Oatmeal agar (ISP Medium 3):						
Aerial mycelium	Pink	Pink-violet	Trace	Pink	White-yellow	Trace
Substrate mycelium	Brown-red	Violet	Orange, red	Red	Yellow-brown	Violet-red
Soluble pigment	−	Violet	Red	−	−	Pink-violet
Inorganic salts-starch agar (ISP Medium 4):						
Aerial mycelium	−	White	−	Trace	−	−
Substrate mycelium	−	White, pink	−	Brown	Yellow-brown	Gray-violet
Degradation of:						
Casein	−	−	−	+	−	+
DNA	−	+	−	+	+	−
Elastin	ND	−	+	+	ND	ND
Esculin	+	+	−	+	+	+
Gelatin	+	+	+	+	−	ND
Hypoxanthine	+	+	+	+	−	+
Starch	−	−	+	−	−	−
Testosterone	+	−	+	+	+	−
Tyrosine	+	−	+	+	+	−
Xanthine	−	−	−	−	−	−
Reduction of nitrate	+	+	+	+	+	−

Footnotes are at end of table

Table 26.3 *(continued)*

[a] Symbols: see standard definitions.

[b] For details of test procedures, see Athalye et al. (Int. J. Syst. Bacteriol. *35:* 86–98, 1985), Greiner-Mai et al. (Syst. Appl. Microbiol. *9:* 97–109, 1987), and Meyer (Z. Allg. Mikrobiol. *19:* 37–44, 1979).

[c,d,e] Species transferred to the genus *Microtetraspora* from the genera *Nocardiopsis, Micropolyspora,* and *Actinomadura,* respectively (see Kroppenstedt et al., Syst. Appl. Microbiol. *13:* 148–160, 1990).

Table 26.4 **Differential characteristics of the species of the genus *Planobispora*** [a,b]

Characteristics	*P. longispora* DSM 43041 [T]	*P. rosea* DSM 43051 [T]
Degradation of:		
Chitin	–	+
Esculin	–	+
Hypoxanthine	+	–
Growth on sole carbon sources (%, w/v):		
Galactose	–	+
Melezitose	–	+
Rhamnose	+	–
Coagulation of litmus milk	+	–
Peptonization of litmus milk	+	–

[a] +, positive; –, negative.

[b] For details of test procedures, see Goodfellow and Pirouz (J. Gen. Microbiol. *128:* 503–527, 1982) and Thiemann (*In* Prauser (Ed.), *The Actinomycetales,* VEB Gustav Fischer Verlag, Jena, 1970, pp. 245–257).

Important Notes for Users of This Manual

Unless otherwise indicated in footnotes to tables, the meanings of symbols are as follows:

- \+ 90% or more of strains are positive
- – 90% or more of strains are negative
- d 11-89% of strains are positive
- v strain instability (not equivalent to "d")
- D Different reactions in different taxa (species of a genus or genera of a family)

All other symbols are defined in footnotes to tables.

Table 26.5 Differential characteristics of the species of the genus *Planomonospora*[a,b]

| Characteristics | P. parontospora | | P. venezuelensis |
	subsp. antibiotica	subsp. parontospora	
Vegetative mycelium	Light rose	Rose to light orange	Violet-brown
Aerial mycelium	Hyaline to white	White with a rose tinge	White to grayish white
Spore vesicles sessile	Yes	Yes	No
Arrangement of spore vesicles	Double parallel rows	Parallel rows on a curved sporangiophore	Singly or in groups on short lateral branches
Melanin produced	+	−	ND
Growth on melezitose as a sole carbon source	ND	+	−
Gelatin liquefaction	+	−	+
Peptonization of milk	−	+	−
Sensitivity to demethylchlortetracycline	ND	−	+

[a] Symbols: see standard definitions.

[b] For test methods see Thiemann et al. (Arch. Microbiol. *58:* 42-52, 1967).

Table 26.6 Differential characteristics of the species of the genus *Spirillospora*[a,b]

Characteristics	S. albida	S. rubra
Share of spore vesicle	Spherical to vermiform	Spherical
Color of vegetative mycelium	White to pale yellow or buffy pink	Red to reddish brown
Production of soluble pigment on tyrosine agar	Clay-colored pigment	None
Degradation of tyrosine	+	−

[a] Symbols: see standard definitions.

[b] For details of test procedures, see Couch (J. Elisha Mitchell Sci. Soc. *79:* 53–70, 1963) and Goodfellow and Pirouz (J. Gen. Microbiol. *128:* 503–527, 1982).

Table 26.7 Differential characteristics of the species of the genus *Streptosporangium*[a,b]

Characteristics	S. albidum	S. album	S. amethystogenes	S. carneum[c]	S. corrugatum
Color of vegetative mycelium:					
Brown-black	–	–	–	–	–
Red or orange	–	–	–	–	–
Yellowish brown to brown	+	+	+	+	+
Color of spore mass:					
Flesh	–	–	–	+	–
Greenish gray	–	–	–	–	–
Pink	–	–	+	+	–
White	+	+	–	–	+
Spore vesicle size:					
1–5 μm	–	–	–	–	+
6–10 μm	–	+	+	+	–
11–20 μm	+	–	–	–	–
21–30 μm	+	–	–	–	–
31–50 μm	–	–	–	–	–
Sporangiophore size:					
Short (10 μm)	–	+	+	+	+
Long (50 μm)	+	–	–	–	–
Spore shape:					
Spherical-oval	+	+	+	+	+
Rod-like	–	–	–	–	–
Soluble pigments	–	–	–	+	–
B vitamins required	–	+	+	ND	–
Growth at:					
42°C	–	–	–	–	–
50°C	–	–	–	–	–
Gelatin liquefaction	–	+	–	–	ND
Iodinin production	–	–	+	ND	–
Nitrate reduction	+	–	+	–	–
Starch hydrolysis	–	–	+	–	–
Utilization of:					
Adonitol	ND	+	+	ND	+
Arabinose	ND	+	ND	ND	+
Galactose	ND	+	–	ND	–
Glycerol	ND	–	ND	ND	–
Inositol	ND	–	+	ND	–
Mannitol	ND	+	ND	ND	–
Rhamnose	ND	–	+	ND	–
Turanose	ND	+	ND	ND	–

Footnotes are at end of table

Table 26.7 *(continued)*

Characteristics	*S. fragile*	*S. longisporum*	*S. nondiastaticum*	*S. pseudovulgare*	*S. roseum*
Color of vegetative mycelium:					
Brown-black	+	−	−	−	−
Red or orange	−	+	+	+	+
Yellowish brown to brown	−	−	+	+	+
Color of spore mass:					
Flesh	−	−	−	−	−
Greenish gray	−	−	−	−	−
Pink	+	+	+	+	+
White	−	−	−	−	−
Spore vesicle size:					
1–5 μm	−	−	−	−	−
6–10 μm	+	+	−	+	+
11–20 μm	+	+	+	−	+
21–30 μm	−	−	−	−	−
31–50 μm	−	−	−	−	−
Sporangiophore size:					
Short (10 μm)	+	+	+	+	+
Long (50 μm)	−	−	−	−	−
Spore shape:					
Spherical-oval	+	−	+	+	+
Rod-like	−	+	−	−	−
Soluble pigments	+	−	−	−	+
B vitamins required	−	−	+	+	+
Growth at:					
42°C	+	−	+	+	−
50°C	−	−	−	−	−
Gelatin liquefaction	−	ND	+	+	+
Iodinin production	−	−	−	−	−
Nitrate reduction	+	+	+	+	+
Starch hydrolysis	+	+	−	+	+
Utilization of:					
Adonitol	−	+	+	+	+
Arabinose	+	+	+	+	+
Galactose	+	−	+	−	+
Glycerol	−	−	−	+	+
Inositol	−	−	−	−	+
Mannitol	+	−	+	+	−
Rhamnose	+	−	−	−	+
Turanose	+	−	+	+	+

Footnotes are at end of table

Table 26.7 *(continued)*

Characteristics	S. violaceochromogenes	S. viridialbum	S. viridogriseum subsp. kofuense	S. viridogriseum subsp. viridogriseum	S. vulgare
Color of vegetative mycelium:					
Brown-black	−	−	−	−	−
Red or orange	−	−	−	−	+
Yellowish brown to brown	+	+	+	+	+
Color of spore mass:					
Flesh	−	−	−	−	−
Greenish gray	−	+	+	+	−
Pink	+	−	−	−	+
White	−	−	−	−	−
Spore vesicle size:					
1–5 μm	−	−	−	−	−
6–10 μm	+	+	−	−	+
11–20 μm	−	−	+	−	−
21–30 μm	−	−	−	+	−
31–50 μm	−	−	−	+	−
Sporangiophore size:					
Short (10 μm)	+	+	+	−	+
Long (50 μm)	−	−	+	+	−
Spore shape:					
Spherical-oval	+	+	−	+	+
Rod-like	−	−	+	−	−
Soluble pigments	+	−	−	−	−
B vitamins required	−	+	−	−	+
Growth at:					
42°C	−	−	+	+	−
50°C	−	−	+	+	−
Gelatin liquefaction	+	d	+	+	d
Iodinin production	−	−	−	−	−
Nitrate reduction	+	d	−	−	−
Starch hydrolysis	+	+	+	+	+
Utilization of:					
Adonitol	ND	−	−	−	+
Arabinose	ND	−	−	−	+
Galactose	ND	−	+	+	+
Glycerol	ND	−	+	+	+
Inositol	+	+	+	+	+
Mannitol	ND	+	+	+	−
Rhamnose	+	+	+	+	+
Turanose	ND	+	+	−	+

[a] Symbols: see standard definitions.

[b] For details of test procedures, see Goodfellow and Pirouz (J. Gen. Microbiol. *128:* 503–527, 1982).

[c] *S. carneum* was not included in *Bergey's Manual of Systematic Bacteriology*. The species *S. carneum* was created in 1990 by Mertz and Yao (Int. J. Syst. Bacteriol. *40:* 247–253).

Table 26.8 Differential characteristics of the species of the genus *Actinomadura*[a,b]

Characteristics	A. atramentaria[c]	A. aurantiaca	A. citrea	A. coerulea	A. cremea	A. echinospora[d]	A. fibrosa[c]	A. fulvescens	A. hibisca[c]	A. kijaniata	A. libanotica	A. livida	A. luteofluorescens
Morphology:													
Spore chains[e]	b,str	hs	h	h,s	h,s	b	a	s	str	sp	h	h,s	h
Spore surface	s	w	u	w	w	sp	ND	s	s	s	f	u	w
Degradation of:													
Arbutin	+	ND	+	+	+	+	ND	+	ND	+	+	+	d
Casein	+	-	+	+	+	+	+	+	+	+	-	+	+
DNA	-	d	-	-	d	+	ND	+	ND	+	-	+	+
Elastin	+	ND	+	+	+	+	ND	+	ND	+	+	-	-
Esculin	+	-	+	-	+	+	+	+	+	+	+	+	-
Gelatin	-	+	+	+	+	+	+	+	+	+	-	+	+
Guanine	-	-	d	+	-	-	-	-	ND	-	-	+	+
Hypoxanthine	-	ND	+	+	-	-	+	+	+	+	-	-	+
Starch	-	-	+	-	-	-	+	+	-	+	+	+	+
Testosterone	+	+	+	+	+	+	+	+	ND	-	+	+	+
Tyrosine	+	-	+	-	-	+	+	-	ND	+	-	+	-
Urea	ND	ND	-	-	+	ND	+	ND	-	+	ND	-	+
Xanthine	-	-	-	-	-	-	-	-	-	+	-	-	-
Nitrate reduction	+	ND	+	+	+	-	ND	ND	+	ND	+	+	d

Footnotes are at end of table

690

Table 26.8 *(continued)*

Characteristics	A. macra	A. madurae	A. oligospora [c]	A. pelletieri	A. rubrobrunea	A. rugatobispora [d]	A. spadix	A. umbrina [c]	A. verrucosospora	A. vinacea	A. viridis [d]	A. yumaensis
Morphology:												
Spore chains [e]	h,s	h,s	h	h,s	h,s	b	psp	h,s,str	h,s	str	b,str	h
Spore surface	s	w	s	w	sp	r	s	s	w	u	rd	s
Degradation of:												
Arbutin	–	+	ND	–	ND	ND	+	ND	+	ND	ND	ND
Casein	+	+	+	+	ND	ND	+	ND	+	+	ND	+
DNA	–	d	+	–	ND	ND	–	ND	+	–	ND	ND
Elastin	+	+	–	+	ND	ND	–	ND	+	ND	ND	ND
Esculin	–	+	+	–	ND	ND	+	ND	+	+	ND	+
Gelatin	+	+	+	+	ND	+	–	ND	+	+	+	+
Guanine	–	d	–	–	ND	ND	–	ND	d	ND	ND	–
Hypoxanthine	+	+	–	d	ND	ND	–	ND	+	–	ND	d
Starch	–	d	–	–	ND	+	+	ND	+	–	+	+
Testosterone	–	+	–	–	ND	ND	–	ND	+	–	ND	ND
Tyrosine	+	+	–	+	ND	ND	–	ND	+	–	ND	+
Urea	–	–	+	–	ND	ND	–	ND	–	ND	ND	ND
Xanthine	–	–	–	–	ND	ND	–	ND	–	–	ND	+
Nitrate reduction	+	+	–	+	ND	–	+	ND	d	ND	ND	+

[a] Symbols: see standard definitions.

[b] For details of test procedures, see Goodfellow and Pirouz (J. Gen. Microbiol. *128*: 503–507, 1982) and Miyadoh et al. (J. Gen. Microbiol. *136*: 1905–1913, 1990).

[c] Species not recognized in *Bergey's Manual of Systematic Bacteriology*. *A. atramentaria* was created by Miyadoh et al. (Int. J. Syst. Bacteriol. *37*: 242–246, 1987), *A. fibrosa* by Mertz and Yao (Int. J. Syst. Bacteriol. *40*: 28–33, 1990), *A. hibisca* by Tomita et al. (J. Antibiot. *43*: 755–762, 1990), *A. oligospora* by Mertz and Yao (Int. J. Syst. Bacteriol. *36*: 179–182, 1986), *A. fulvescens* by Terekhova et al. (Antibiotiki *27*: 87–92, 1982), and *A. umbrina* by Galatenko et al. (Antibiotiki *26*: 803–807, 1981).

[d] *A. echinospora* was transferred from the genus *Microbispora* (*M. echinospora*) by Kroppenstedt et al. (Syst. Appl. Bacteriol. *13*: 148–160, 1990), *A. rubrobrunea* from the genus *Excellospora* (*E. rubrobrunea*) also by Kroppenstedt et al., and *A. rugatobispora* from the genus *Microtetraspora* (*M. viridis*) by Miyadoh et al. (J. Gen. Microbiol. *136*: 1905–1913, 1990), who also reduced *A. malachitica* to a synonym of *A. viridis*.

[e] Abbreviations: a, asporogenous; b, two spores (bispores); h, hooks, curled; psp, pseudosporangia; s, spirals of 1 to 2 turns; sp, spirals of 2 to 4 turns; and str, straight.

[f] Abbreviations: f, folded; r, rugose; rd, ridged; s, smooth; sp, spiny; u, irregular, uneven; and w, warty.

THERMOMONOSPORA AND RELATED GENERA

These **aerobic spore-forming actinomycetes** produce a branched vegetative mycelium bearing aerial hyphae. Chemoorganotrophs. **The cell wall contains *meso*-diaminopimelic acid** but no characteristic sugars or other amino acids. Mycolic acids are absent. **Menaquinones typically contain 9 or 10 isoprene units (MK-9, MK-10).** The spore arrangement and morphology are distinctive for each genus and provide the means of differentiating genera within this group.

Differentiation of the genera in **Group 27:** See Table 27.1.

Genus Actinosynnema

Substrate hyphae penetrate the agar medium and form **synnemata, dome-like bodies, or flat colonies on the agar surface.** Aerial hyphae arise from these structures and bear **chains of spores that can form flagella** in an aqueous environment. Mesophilic. A range of sugars can be used as sources of carbon and energy. Melanoid pigments cannot be formed from tyrosine. The main habitat seems to be the surface of fresh plant tissue at riversides.

Characteristics useful in the differentiation of *Actinosynnema* from some of the other actinomycete genera that produce motile spores are presented in Table 27.2.

Type species: *Actinosynnema mirum.*

Differentiation of the species of the genus **Actinosynnema:** See Table 27.3.

Genus Nocardiopsis

Substrate mycelium is well developed, and hyphae are long and densely branched; fragmentation into coccoid and bacillary elements may occur. Aerial mycelium is usually well developed and abundant; **aerial hyphae completely fragment into spores of various lengths.** Growth temperature is 10–45°C. A range of compounds can be used as sources of carbon and energy, and there is a diversity of degradative activity in the genus. Can be isolated from soil and vegetable matter, and strains have been recovered from clinical material of animal and human origin.

Editorial note: Since the publication of *Bergey's Manual of Systematic Bacteriology,* Volume 4, many species have been transferred to the genus *Saccharothrix* on chemotaxonomic grounds (Grund and Kroppenstedt, Syst. Appl. Microbiol. *12:* 267-274, 1989), and the genus *Nocardiopsis* was consolidated (Grund and Kroppenstedt, Int. J. Syst. Bacteriol. *40:* 5-11, 1990). The differentiation of the genera *Nocardiopsis* and *Saccharothrix* is summarized in Table 27.4.

Type species: *Nocardiopsis dassonvillei.*

Differentiation of the species of the genus **Nocardiopsis:** See Table 27.5.

Genus Streptoalloteichus

Slender, well-branched hyphae. **The aerial mycelium bears chains of 5–50 spores,** 0.5–1.2 μm diameter. The **vegetative hyphae bear oval or spherical sporangium-like vessels, which envelop one spore or a single row of two to four spores. The spores are motile with a single polar flagellum.** Whole-cell hydrolysates contain galactose, mannose, and rhamnose but not diagnostic sugars such as madurose, arabinose, or xylose. Phospholipids include phosphatidylethanolamine but not phosphatidylglucosamine or phosphatidylcholine. Isolated from soil.

Type (and only) species: *Streptoalloteichus hindustanus.*

Characteristics of the species: The morphological and chemotaxonomic characteristics are as given for the genus. Aerial mycelium is usually abundant and white, turning pale yellow on sporulation. Hexoses are utilized as carbon sources, but pentoses and sugar alcohols are not; **gelatin, starch, and casein are degraded. The temperature range for growth is 20–54°C** (optimum, 45°C). *Streptoalloteichus* can grow in the presence of NaCl (5%, w/v) and exhibits resistance to a range of antibiotics including kanamycin, ampicillin, tetracycline, and novobiocin.

Genus **Thermomonospora**

Produce **single, heat-sensitive, nonmotile aleuriospores** on aerial hyphae. On agar media, a branched, nonfragmenting vegetative mycelium forms leathery colonies usually covered with aerial mycelium. Spores may be sessile but are more often formed at the tips of unbranched or branched sporophores; in many strains, repeated sporophore branching leads to the formation of spore clusters. Spores may also be produced on the substrate hyphae. Aerial mycelium production and sporulation are often optimal at pH > 8.0. A wide range of compounds, including polymeric substrates, can be used as sources of carbon and energy. **Spores are killed by treatment at 90°C for 30 min** in aqueous suspension, and **all strains are sensitive to novobiocin (50 μg·ml^{-1}). Thermophilic strains are common in manures, composts, and overheated fodders;** mesophiles can be isolated from soil.

Characteristics that enable the differentiation of *Thermomonospora* from other monosporic actinomycete genera are presented in Table 27.6.

Type species: *Thermomonospora curvata.*

Differentiation of the species of the genus **Thermomonospora:** See Table 27.7.

Important Notes for Users of This Manual

Unless otherwise indicated in footnotes to tables, the meanings of symbols are as follows:

+ 90% or more of strains are positive
− 90% or more of strains are negative
d 11-89% of strains are positive
v strain instability (not equivalent to "d")
D Different reactions in different taxa (species of a genus or genera of a family)

All other symbols are defined in footnotes to tables.

Table 27.1 Differential characteristics of the genera in Group 27

Characteristics	Actinosynnema	Nocardiopsis	Streptoalloteichus	Thermomonospora
Single spores	−	−	−	+
Chains of arthrospores	+	+	+	−
Sporangia-like structures	−	−	+	−
Synnemata	+	−	−	−
Motile spores	+	−	+	−

Table 27.2 Characteristics that differentiate *Actinosynnema* from morphologically and/or chemotaxonomically similar actinomycete genera

Genus	DAP[a] isomer in peptidoglycan	Madurose in whole-cell hydrolysates	Aerial mycelium	Motile spores on aerial mycelium	Sporangia with motile spores
Actinosynnema	meso-DAP	−	+	+	−
Dermatophilus	meso-DAP	+	−	−	+
Geodermatophilus	meso-DAP	−	−	−	+
Planobispora	meso-DAP	+	+	−	+
Planomonospora	meso-DAP	+	+	−	+
Spirillospora	meso-DAP	+	+	−	+
Sporichthya	L-DAP	−	+	+	−

[a] DAP, diaminopimelic acid.

Table 27.3 Differentiation of the species of the genus *Actinosynnema*

Characteristics	A. mirum	A. pretiosum[a]
Growth at:		
10°C	+	−
38°C	−	+
Utilization of:		
Melibiose	−	+
Raffinose	−	+
Fragmentation of substrate hyphae in liquid media	−	+
Antibiotics produced	Nocardicins	Ansamitocins, Dnacins, Macbecins, Nocardicins, Tomaymycin

[a] The subspecies of *A. pretiosum* subsp. *pretiosum* and subsp. *auranticum* seem to differ only in the darkness and texture of the colonies.

Table 27.4 Differentiation of the genera *Nocardiopsis* and *Saccharothrix*

Characteristics	Nocardiopsis	Saccharothrix
Morphology of sporulating aerial mycelium	Long chains	Long chains
Rhamnose and galactose in whole-cell hydrolysates	−	+
Predominant menaquinones	MK-10	MK-9
Diagnostic phospholipid: Phosphatidylcholine	+	−

Table 27.5 Differentiation of the species of the genus *Nocardiopsis*

Characteristics	N. africana	N. alborubidus	N. albus[a]	N. dassonvillei	N. listeri
Color of substrate mycelium [b]	O, B	C, B	C, Y	Y, O, B	C
Color of aerial mycelium [b]	Bl	W	W	W,Cr,Y,Gr	(W)
Degradation of:					
Hypoxanthine	ND	−	+	+	−
Xanthine	ND	−	+	+	+
Utilization of:					
Xylose	ND	+	−	+	+
Melezitose	ND	+	−	−	−
Cellobiose	ND	−	+	+	+

[a] *N. albus* subsp. *albus* and *N. albus* subsp. *prasina* are differentiated by a few physiological characteristics.

[b] Abbreviations: O, orange; B, brown; Bl, blue; C, colorless; W, white; Y, yellow; Cr, Cream; Gr, grey; ( ), aerial mycelium often absent.

Table 27.6 Characteristics that differentiate the genus *Thermomonospora* from other monosporic actinomycete genera

Characteristics	Micromonospora	Saccharomonospora	Thermoactinomyces	Thermomonospora
Aerial mycelium	−	+	+	+
Heat-resistant endospores	−	−	+	−
Glycine in cell wall	+	−	−	−
Sugars in whole-cell hydrolysates:				
Arabinose	+	+	−	−
Galactose	−	+	−	−
Xylose	+	−	−	−

Table 27.7 Characteristics differentiating the species of the genus *Thermomonospora*

Characteristics	T. alba	T. chromogena	T. curvata	T. formosensis[b]	T. fusca	T. mesophila
Colony reverse color	Pale yellow	Brown	Yellow/ orange	Pink/ light orange	Pale yellow	Brown
Aerial mycelium spores in clusters	d	+	−	−	+	−
Growth at:						
30°C	d	−	d	+	−	+
53°C	d	+	d	−	+	+
Growth in:						
Crystal violet (0.00002%, w/v)	−	+	+	ND	+	+
Tetrazolium chloride (0.002%, w/v)	−	d	d	ND	+	−
Novobiocin (10 µg/ml)	−	−	−	ND	−	+
Kanamycin (25 µg/ml)	−	+	−	ND	−	−
Degradation of:						
Tyrosine, xanthine, hypoxanthine	−	+	−	ND	−	+
Starch	+	−	+	−	+	+
Pectin	+	d	−	ND	+	+
Elastin	d	+	−	ND	+	+
Cellulose powder	+	−	+	ND	+	−
Growth on [a]:						
D-Galactose	d	+	−	+	+	+
D-Ribose	−	d	+	ND	−	−
Sucrose	+	−	+	+	+	
Lactose	d	−	−	−	+	−
L-Arabinose	−	−	−	−	−	+
Nitrate reduction	d	+	+	−	−	+
Galactose and madurose in whole-cell hydrolysates	−	−	−	+	−	−

[a] Utilization as carbon source at 1.0% (w/v).

[b] *T. formosensis* was cited in *Bergey's Manual of Systematic Bacteriology*, Volume 4, as species *incertae sedis*.

THERMOACTINOMYCETES

Well developed substrate and aerial mycelium **morphology is typical of actinomycetes. Single spores** are borne on both aerial and vegetative hyphae. The spores are **true bacterial endospores;** they contain dipicolinic acid, exhibit the characteristic fine structure of endospores, and are heat resistant. The wall structure is Gram positive; wall peptidoglycan contains *meso*-diaminopimelic acid but no characteristic sugars or other amino acids. Aerobic and saprophytic chemoorganotrophs utilize a range of sugars as carbon and energy sources and are able to degrade various polymeric substrates.

Genus Thermoactinomyces

This is the only genus in the group, the definition of which can be extended as follows. Substrate mycelium is branched, septate, and 0.4–0.8 μm in diameter. Aerial mycelium is 0.5–1.0 μm in diameter, variable in amount, and sometimes transient, lysing to leave a layer of spores. The single spores are sessile or on unbranched or branched sporophores. Spores are globose, often ridged, and 0.5–1.5 μm in diameter. With the exception of one species, *T. peptonophilus*, all are thermophilic (temperature optimum, 50–60°C) and fast growing, giving flat or ridged colonies on nutrient media. The **thermophilic species are found widely in nature** but are most numerous in molding hay and cereal grains and in decaying vegetable material and composts that have heated spontaneously to temperatures of 50°C or more. Thermophilic thermoactinomycetes are morphologically similar to *Thermomonospora* spp., which share many of their habitats; thermoactinomycetes can be readily differentiated from other thermophilic actinomycetes by the **heat resistance properties of their spores (90°C for 30 min) and resistance to novobiocin (25 μg·ml⁻¹).** (Cellulolytic activity is absent in *Thermoactinomyces* but common in *Thermomonospora.*) *Thermoactinomyces* spores are extremely long-lived and can be isolated from soil and sediment cores. They have been **implicated in various forms of hypersensitivity pneumonitis** (extrinsic allergic alveolitis), especially farmer's lung disease. The single mesophilic species, *T. peptonophilus,* has rarely been isolated and only from soil; it is not well characterized. Characteristics useful in distinguishing *Thermoactinomyces* from other monosporic actinomycete genera are given in Table 27.2.

Type species: *Thermoactinomyces vulgaris.*

Differentiation of the species of the genus **Thermoactinomyces:** See Table 28.1.

Important Notes for Users of This Manual

Unless otherwise indicated in footnotes to tables, the meanings of symbols are as follows:

  + 90% or more of strains are positive
  − 90% or more of strains are negative
  d 11-89% of strains are positive
  v strain instability (not equivalent to "d")
  D Different reactions in different taxa (species of a genus or genera of a family)

All other symbols are defined in footnotes to tables.

Table 28.1 Characteristics differentiating the species of the genus *Thermoactinomyces*[a]

Characteristics	T. dichotomicus	T. intermedius	T. peptonophilus	T. putidus	T. sacchari	T. thalpophilus	T. vulgaris
Aerial mycelium:							
Abundant	+	+	d	+	−	+	+
Transient	−	−	−	+	+	−	−
White	−	+	+	+	+	+	+
Yellow	+	−	−	−	−	−	−
Soluble pigment:							
Pink-red	−	−	−	−	−	+	−
Yellowish gray	−	−	−	+	−	−	−
Growth at:							
30°C	−	−	+	d	−	d	−
55°C	+	+	−	d	+	+	+
Spores sessile:	−	d	+	−	−	+	−
On unbranched sporophores	−	d	−	+	+	−	+
On dichotomously branched sporophores	+	−	−	−	−	−	−
Growth in the presence of novobiocin, 25 μg·ml^{-1}	+	+	−	+	+	+	+
Melanin production on CYC agar with 0.5% (w/v) L-tyrosine	−	+	NT	+	−	+	−
Degradation of:							
Arbutin	−	+	NT	d	d	−	+
Chitin	NT	−	NT	d	−	+	−
Esculin	−	+	NT	d	+	−	
Hypoxanthine	+	−	NT	NT	−	−	−
Starch	−	−	NT	+	+	+	−
Tyrosine	−	+	NT	+	−	+	−
Xanthine	+	−	NT	NT	−	−	−
Utilization as carbon source:							
Mannitol	+	NT	NT	−	+	+	d
Sucrose	+	NT	NT	+	−	d	d
Trehalose	NT	NT	NT	−	d	+	d

[a] Symbols: see standard definitions; NT, not tested

OTHER GENERA

This group comprises three genera whose combination of morphological and chemotaxonomic characteristics do not permit their simple assignment to an established group of actinomycete genera. They do, however, have two properties in common: the production of **long chains of spores on the aerial mycelium and the absence of mycolic acids** in the cell envelope. Cells are aerobic chemoorganotrophs and are isolated from soil.

Differentiation of the genera in **Group 29:** See Table 29.1.

Genus **Glycomyces**

The vegetative mycelium does not fragment, but **short chains of conidia** develop on the aerial hyphae or substrate mycelium in species lacking an aerial mycelium. Cell wall contains major amounts of *meso*-diaminopimelic acid (*meso*-DAP) and glycine. **Whole-cell hydrolysates contain xylose and arabinose. Nitrogen-containing phospholipids are absent;** significant amounts of phosphatidyl mannosides are present. This unique combination of spore chain morphology, cell wall chemistry, and phospholipid pattern is very diagnostic for this genus.

Type species: *Glycomyces harbinensis.*

Differentiation of the species of the genus **Glycomyces:** See Table 29.2.

Genus **Kitasatosporia**

Aerial hyphae bear long chains of spores. The vegetative mycelium does not fragment. No sporangia, synnemata, sclerotia, or zoospores are formed. **Whole-cell hydrolysates contain both L- and *meso*-isomers of diaminopimelic acid (DAP);** *meso*-DAP seems to originate from the mycelium, whereas spores, whether from surface or submerged cultures, contain L-DAP. **Whole-cell hydrolysates also contain galactose. Phosphatidyl ethanolamine is the diagnostic phospholipid.**

The detection of both L- and *meso*-isomers of DAP in whole-cell hydrolysates is the key characteristic enabling distinction of *Kitasatosporia* strains from morphologically similar actinomycetes.

Type species: *Kitasatosporia setae.*

Differentiation of the species of the genus **Kitasatosporia:** See Table 29.3.

Genus **Saccharothrix**

Vegetative and aerial hyphae fragment into ovoid to bacillary nonmotile elements. The aerial hyphae exhibit a characteristic "zig-zag" morphology during this fragmentation/sporulation. Cell wall contains *meso*-diaminopimelic acid but not glycine. **Whole-cell hydrolysates contain galactose and rhamnose in major amounts as diagnostic components. The predominant menaquinones found are those with nine isoprene units (MK-9). The diagnostic phospholipid present is phosphatidyl ethanolamine.**

Subsequent to the publication of *Bergey's Manual of Systematic Bacteriology*, Volume 4, a number of new *Saccharothrix* species have been described. Furthermore, the detection of rhamnose and galactose in whole-cell hydrolysates of a number of *Nocardiopsis* species led to their transfer to *Saccharothrix* (Grund and Kroppenstedt, Syst. Appl. Microbiol. *12:* 267–274, 1989) supported by data on phospholipid, fatty acid, and menaquinone composition. The differentiation of *Saccharothrix* and *Nocardiopsis* is outlined in Table 29.1. The genus description has also been amended to permit the inclusion of actinomycetes in which glucosamine-containing phospholipids can be found in addition to phosphatidyl ethanolamine (Labeda and Lechevalier, Int. J. Syst. Bacteriol. *39:* 420–423, 1989).

Type species: *Saccharothrix australiensis.*

Differentiation of the species of the genus **Saccharothrix:** See Table 29.4.

Table 29.1 Comparison of Group 29 genera with selected actinomycete genera

Characteristics	Actinomadura	Glycomyces	Kibdelosporangium	Kitasatosporia	Nocardiopsis	Saccharopolyspora	Saccharothrix	Streptomyces
Aerial spore chains	+	+	+	+	+	+	+	+
Mycolic acids	–	–	–	–	–	–	–	–
Wall DAP[a]:								
L-DAP	–	–	–	+	–	–	–	+
meso-DAP	+	+	+	+	+	+	+	–
Diagnostic: whole-cell sugars[b]	mad	xyl, ara	ara, gal	gal	None	ara, gal	rha, gal	None

[a] DAP, diaminopimelic acid.

[b] Abbreviations: xyl, xylose; ara, arabinose; gal, galactose; mad, madurose; rha, rhamnose.

Table 29.2 Differentiation of the species of the genus *Glycomyces*[a]

Characteristics	G. harbinensis	G. rutgersensis	G. tenuis
Aerial mycelium	Sparse	Abundant	Absent
Substrate mycelium color	Light cream	White, yellow	White, light pink
Major menaquinones	MK-10(H$_{4,6}$); MK-10 (H$_{2,6}$)	MK-10(H$_{2,6}$)	MK-9(H$_6$), MK-10(H$_6$), MK-11(H$_6$)
Utilization of:			
Citrate	–	+	–
Lactate	+	–	+
Acid from:			
Mannitol	+	–	–
Sucrose	–	–	+
Growth at 42°C	–	+	–
Growth in the presence of sodium azide (0.1%)	+	+	–

[a] Abridged from Evtushenko et al. (Int. J. Syst. Bacteriol. *41:* 154-157, 1991).

Table 29.3   Differentiation of the species of the genus *Kitasatosporia*[a]

Characteristics	K. griseola	K. mediocidica	K. phosalacinea	K. setae
Soluble pigment	Pink	Yellow to brown	Light tan	Yellowish brown
Nitrate reduction	–	–	+	–
Utilization of:				
Raffinose	+	–	+	–
Rhamnose	–	–	+	–
Sucrose	–	+	+	–
Antibiotic produced	Setamycin	Mediocidicin	Phosalacine	Setamycin

[a] Abridged from Labeda (Int. J. Syst. Bacteriol. *38:* 287-290, 1988).

Table 29.4   Differentiation of the species of the genus *Saccharothrix*[a,b]

Characteristics	S. aerocolonigenes	S. australiensis	S. coeruleofusca	S. cryophilis	S. espanaensis	S. flava	S. longispora	S. mutabilis	S. syringae	S. texasensis	S. waywayandensis
Utilization of:											
Citrate	+	–	–	–	V	–	+	+	–	–	+
Nitrate reduction	+	+	–	+	W	–	+	+	–	+	–
Degradation of:											
Adenine	–	–	–	–	–	–	+	–	–	+	–
Casein	+	+	+	–	+	–	+	+	+	+	+
Hypoxanthine	+	–	–	–	+	+	–	+	–	+	+
Starch	+	–	+	–	–	ND	+	+	+	+	+
Tyrosine	+	+	–	+	–	+	+	+	+	+	+

[a] See standard abbreviations; V, variable response; W, weak response; ND, not determined.

[b] Data taken from Meyer (Genus *Nocardiopsis*, In *Bergey's Manual of Systematic Bacteriology,* Volume 4); Labeda and Lechevalier (Int. J. Syst. Bacteriol. *39:* 420–423, 1989); Grund and Kroppenstedt (Syst. Appl. Microbiol. *12:* 267–274, 1989).

Important Notes for Users of This Manual

Unless otherwise indicated in footnotes to tables, the meanings of symbols are as follows:

   + 90% or more of strains are positive

   – 90% or more of strains are negative

   d 11-89% of strains are positive

   v strain instability (not equivalent to "d")

   D Different reactions in different taxa (species of a genus or genera of a family)

All other symbols are defined in footnotes to tables.

GROUP 30

THE MYCOPLASMAS (OR MOLLICUTES): CELL WALL-LESS BACTERIA

This group consists of very small procaryotes totally **devoid of cell walls,** bounded by a plasma membrane only and incapable of synthesis of peptidoglycan and its precursors. Consequently, resistant to penicillin and its analogues and sensitive to lysis by osmotic shock, detergents, alcohols, and specific antibody plus complement. **Pleomorphic,** varying in shape from spherical or pear-shaped structures (0.3–0.8 μm in diameter) to **branched or helical filaments,** some with organized attachment structures. Genome replication precedes but is not necessarily synchronized with cell division. Budding forms and chains of beads may thus be observed, as well as classic binary fission. **Usually nonmotile,** some species, however, show **gliding motility** on liquid-covered surfaces. Other species that occur as helical filaments show **rotary, flexional, and translational motility.** No resting stages are known. **Gram negative.**

The species recognized thus far can be grown on artificial cell-free media of diverse complexity. Most species **require sterols** and fatty acids for growth. However, certain strains may grow poorly in artificial media and may be more readily isolated by cell-culture procedures. Most species are **facultatively anaerobic,** but some are **obligate anaerobes** that are killed by exposure to minute quantities of oxygen. Colonies on solid media are minute, usually much smaller than 1 mm in diameter. There is a tendency for the organisms to penetrate and grow inside the medium. Under suitable conditions, most species **form colonies that have a characteristic "fried egg" appearance.** One genus of cell wall-less bacteria, *Thermoplasma,* is an archaeobacterium and is discussed in Group 34.

All mollicutes are **parasites, commensals, or saprophytes,** and many are **pathogens of humans, animals, plants, and insects.** Genome size varies from about 5×10^8 to 1×10^9 daltons, among the smallest recorded in procaryotes. The mol% G + C of the DNA is low, ranging from ~23–40% (Bd, T_m).

Key to the genera of **mycoplasmas (mollicutes)**

I. Facultatively anaerobic or microaerophilic

    A. Sterol required

        1. Cells helical during logarithmic growth

            Genus *Spiroplasma*

        2. Cells not helical

           a. Urease positive

               Genus *Ureaplasma*

           b. Urease negative

               Genus *Mycoplasma*

    B. Sterol not required

        Genus *Acholeplasma*

II. Obligately anaerobic

    A. Sterol required

        Genus *Anaeroplasma*

    B. Sterol not required

        Genus *Asteroleplasma*

Genus **Acholeplasma**

Cells **spherical,** with a minimum diameter of ~300 nm, **and filamentous,** usually ~2–5 μm in length. Modes of reproduction are as for the genus *Mycoplasma.* **Cells are bounded by a plasma membrane only.** Generally more susceptible to lysis by osmotic shock at 37°C than other mycoplasmas. Gram negative. Nonmotile. Facultative anaerobes, most having a fermentative type of metabolism. **Colonies on solid media containing animal serum usually show a "fried egg" appearance** and may reach 2–3 mm in diameter. The temperature range for growth is 20–40°C. Chemoorganotrophic. Carbohydrates serve as the fermentable substrates. Carbohydrate transport occurs through an active carrier-mediated process different from the phosphoenolpyruvate-dependent phosphotransferase system (PEP-PTS) found in some *Mycoplasma* species. **Arginine and urea are not hydrolyzed.** Phosphatase activity is weak or negative. Reduced nicotinamide adenine dinucleotide ($NADH_2$) activity is located in the plasma membrane. Cells possess lactic dehydrogenases specifically activated by fructose-1,6-diphosphate. **Serum or cholesterol is not required for growth.** Agar colonies do not show hemadsorption of red blood cells from a variety of hosts. All species are resistant, or only very slightly sensitive, to 1.5% digitonin. They are absolutely resistant to penicillin, not being inhibited by penicillin G, ampicillin, cloxacillin, or methacillin in concentrations of at least 4000 µg/ml. Apparently, parasites of a wide range of vertebrate hosts. Pathogenicity has not been clearly established. May also be a part of plant and insect flora. The mol% G + C of the DNA is ~26–35.7 (T_m, Bd). The genome size is ~1.0×10^9 daltons. The species utilize UGA codon as a stop signal. Other characteristics are as for the genus *Mycoplasma*.

Type species: *Acholeplasma laidlawii.*

Differentiation of the species of the genus **Acholeplasma:** See Tables 30.1 and 30.2.

Genus **Anaeroplasma**

Cells of young (16- to 18-hour-old) cultures are **coccoid,** 0.5–2.0 µm in diameter. Older cells have a variety of pleomorphic forms. **Gram negative. Nonmotile. Obligately anaerobic;** the inhibitory effect of oxygen on growth is not alleviated during repeated subcultures. Optimum temperature is 37°C; there is no growth at 26°C and 47°C. Optimum pH is 6.5–7.0. **Sterols are required for growth.** Surface colonies have a dense center with a translucent periphery, a "fried egg" appearance. Subsurface colonies are golden, irregular, and often multilobed. Strains vary in their ability to ferment various carbohydrates. The products of carbohydrate fermentation include acids (generally acetic, formic, propionic, lactic, and succinic), ethanol, and gases

(primarily CO_2, but some strains also produce H_2). Bacteriolytic and nonbacteriolytic strains of anaerobic mycoplasmas are described. **Occur in the bovine and ovine rumen.** The genome size is ~1.0×10^9 daltons. The mol% G + C of the DNA is 29–34 (T_m, Bd).

Type species: *Anaeroplasma abactoclasticum.*

Differentiation of the species of the genus **Anaeroplasma:** See Table 30.3.

Genus **Asteroleplasma**

Editorial note: The genus *Asteroleplasma* was not included in *Bergey's Manual of Systematic Bacteriology.* The genus was created in 1987 by Robinson and Freundt (Int. J. Syst. Bacteriol. *37:* 78–81). It contains one species, *A. anaerobium.*

Characteristics of this genus are as described for *Anaeroplasma* except these cells are not lytic for bacteria, do not require sterols, and have a mol% G + C of DNA of 40.3–40.5. They are also serologically distinct from species of *Anaeroplasma.*

Type (and only) species: *Asteroleplasma anaerobium.*

Characteristics of the species: As described for the genus.

Genus **Mycoplasma**

Pleomorphic, varying in shape from spherical, slightly ovoid, or pear-shaped (0.3–0.8 µm in diameter) to **slender branched filaments** of uniform diameter, ranging in length from a few to 150 µm. Cells **lack a cell wall** and are bounded by a plasma membrane only. Gram negative. **Usually nonmotile, but gliding motility has been described in some species. Facultatively anaerobic,** possessing a truncated flavin-terminated electron transport chain **devoid of quinones and cytochromes.** Colonies are very small (usually less than 1 mm in diameter). **The typical colony,** under adequate growth conditions, **has a "fried egg" appearance.** Catalase negative. **Chemoorganotrophic,** using either sugars or arginine as the major energy source. **Require cholesterol** or related sterols for growth. **Parasites and pathogens** of a wide range of mammalian and avian hosts. Some species occur on plant surfaces and in insects. The mol% G + C of the DNA ranges from 23 to 40 (T_m and Bd), and the genome size of the species examined varies from ~5.0–9.0 × 10⁸ daltons. UGA codon is read as tryptophan signal.

Type species: *Mycoplasma mycoides.*

Differentiation of the species of the genus **Mycoplasma:** See Table 30.4. Also, differentiation of many of the species will require serological determination.

Genus **Spiroplasma**

Cells are pleomorphic, varying in size and shape from **helical** and branched nonhelical filaments to spherical or ovoid. The helical forms, usually 100–200 nm in diameter and 3–5 µm in length, typically occur during the logarithmic phase of growth and in some species persist during the stationary phase. Spherical cells ~300 nm in diameter and nonhelical filaments are frequently seen in the stationary phase or in all growth phases in suboptimal growth media. **Helical filaments are motile, with flexional and twitching movements, and often show an apparent rotatory motility. Flagella, periplasmic fibrils, or other organelles of locomotion are not present, but intracellular fibrils have been demonstrated.** Cells divide by binary fission. Facultatively anaerobic. Temperature range for growth is 20–41°C. **Colonies are frequently diffuse,** reflecting the motility of the cells during active growth. The colony type is strongly dependent on the agar concentration. Colony sizes vary from 0.1–4.0 mm. Colonies formed on solid media by nonmotile variants or by cultures growing on inadequate media typically attain diameters of 200 µm or less and exhibit a typical umbonate appearance. Colonies of motile, fast-growing spiroplasmas are diffuse, often with satellite colonies developing from foci adjacent to the initial site of colony development. Light to heavy turbidity is produced in liquid cultures. Chemoorganotrophic. Possess a phosphoenolpyruvate phosphotransferase system for glucose. Reduced nicotinamide adenine dinucleotide (NADH₂) oxidase activity is located only in the cytoplasm. Unable to synthesize fatty acids from acetate. Acid is produced from glucose. Variable fermentation of other carbohydrates occurs. Most strains hydrolyze arginine. There is no hydrolysis of urea, arbutin, or esculin. No liquefaction of coagulated horse serum occurs. Adsorption of guinea pig red blood cells is variable. **Cholesterol** (or possibly other sterols) **is required for growth.** Some species require an optimum osmolality of the culture medium for primary growth, usually in the range of 300–800 mOsm. Media containing mycoplasma broth base, serum, and other supplements are required for primary growth, but on adaptation growth may occur in less complex media. Defined media are available for some strains. Resistant to 10,000 units of penicillin per ml and is sensitive to erythromycin and tetracycline. Sensitive to 1.5% digitonin in the disk test. **Isolated from ticks, from the hemolymph and guts of insects, from vascular plant fluids and insects that feed on the fluids, and from the surfaces of flowers and other plant parts.** The type species is pathogenic for citrus (grapefruit and orange), producing "stubborn" disease; experimental or natural infections are also established in corn, horseradish, periwinkle, radish, broad bean, etc. Some species are pathogenic under experimental conditions for insects, chicken embryos, and for a variety of suckling rodents (rats, mice, hamsters, and rabbits). The mol% G + C of the DNA is 25–31 (*T_m*, Bd), and genome size ranges from 6 × 10⁸ to 1.4 × 10⁹ daltons.

Type species: *Spiroplasma citri.*

Differentiation of the species of the genus **Spiroplasma:** See Tables 30.5 and 30.6.

Genus **Ureaplasma**

Cells from 18- to 24-hour-old cultures are round or coccobacillary, ~330 nm in diameter. A variety of **pleomorphic forms** may be seen, depending on the strain, the age of the culture, and the method of examination. Gram negative. Nonmotile. Microaerophilic. Optimum temperature is 37°C; poor growth occurs at 22°C, and no growth occurs at 42°C. **Optimum pH is ~6.0. Colonies generally are small, ~15–60 µm in diameter,** and may not have zones of surface growth (and therefore may lack the "fried egg" appearance of colonies of most other members of the mollicutes). Growth is retarded by thallous acetate (0.05%), 5-iodo-2′-deoxyuridine (125 µg/ml), hydroxyurea (500 µg/ml), acetohydroxamic acid (1 mM), the tetracyclines, erythromycin, streptomycin, chloramphenicol, gentamicin, and kanamycin but not by the penicillins. **All strains hydrolyze urea with the production of ammonia.** Arginine and the usual carbohydrates are not metabolized. **Occurs predominantly in the mouth, respiratory tract, and urogenital tract of humans and various animal species.** The mol% G + C of the DNA of ureaplasmas is 26.9–30.2. Genome size varies from 5–7 × 10⁸ daltons.

Differentiation of the species of the genus **Ureaplasma:** There are currently five species recognized in the genus, and they are separated on the basis of host, serology, and DNA base ratio. Human strains are *U. urealyticum;* bovine strains are *U. diversum;* certain avian strains are *U. gallorale* (Koshimizu et al., Int. J. Syst. Bacteriol.*37:* 333–338, 1987); *U. cati* and *U. felinum* occur in cats (Harasawa et al., Int. J. Syst. Bacteriol. *40:* 45–51, 1990).

Important Notes for Users of This Manual

Unless otherwise indicated in footnotes to tables, the meanings of symbols are as follows:

  + 90% or more of strains are positive

  − 90% or more of strains are negative

  d 11-89% of strains are positive

  v strain instability (not equivalent to "d")

  D Different reactions in different taxa (species of a genus or genera of a family)

All other symbols are defined in footnotes to tables.

Table 30.1 Differential characteristics of the species of the genus *Acholeplasma*[a]

Characteristics	*A. axanthum*	*A. entomophilum*[b]	*A. equifetale*	*A. florum*[c]	*A. granularum*	*A. hippikon*	*A. laidlawii*	*A. modicum*	*A. morum*	*A. oculi*	*A. parvum*	*A. seiffertii*[d]
Fermentation of mannose	−	−	d	−	−	+	(−)	−	−	−		+
Esculin hydrolysis	+		d	−	−	−	(+)	−	+	+	−	
Arbutin hydrolysis	4+	−	−	−	−	−	(±)	−	1+	2+		−
Carotenoid pigments produced[e]	−	−	−	−	+	−	+	−	−	+	−	
Film and spots	−	−	+	+	−	+	−[f]	−	−	−		
Mol % G + C of DNA	31	30	30.5	27.3	30–32	33.1	31–36	29	34	26–27	29.1	30

[a] Symbols: see standard definitions.

[b] The species *A. entomophilum* was not included in *Bergey's Manual of Systematic Bacteriology*. See Tully et al. (Int. J. Syst. Bacteriol. *38:* 164–167, 1988).

[c] The species *A. florum* was not included in *Bergey's Manual of Systematic Bacteriology*. See McCoy et al. (Int. J. Syst. Bacteriol. *34:* 11–15, 1984).

[d] The species *A. seiffertii* was not included in *Bergey's Manual of Systematic Bacteriology*. See Bonnet et al. (Int. J. Syst. Bacteriol. *41:* 45–49, 1991).

[e] Presence of carotenoids is based upon light absorption at 438 nm and when the cell yield from 100–500 ml of culture medium is tested.

[f] Positive film and spot reactions occur when the organisms are grown in serum fraction medium supplemented with Tween 80 and $CaCl_2$.

Important Notes for Users of This Manual

Unless otherwise indicated in footnotes to tables, the meanings of symbols are as follows:

+ 90% or more of strains are positive

− 90% or more of strains are negative

d 11-89% of strains are positive

v strain instability (not equivalent to "d")

D Different reactions in different taxa (species of a genus or genera of a family)

All other symbols are defined in footnotes to tables.

Table 30.2 Other characteristics of the species of the genus *Acholeplasma* [a]

Characteristics	A. axanthum	A. entomophilum	A. equifetale	A. florum	A. granularum	A. hippikon	A. laidlawii	A. modicum	A. morum	A. oculi	A. parvum	A. seiffertii
Acid production from:												
Cellobiose	+						+			+		
Dextrin	+						+					
Dulcitol	−						−					
Fructose	−		+			+				+		
Galactose	+		+			+				+		
Glucose	+	+	+	+	+	+	+	+	+	+	−	+
Glycerol	+						−			−		
Glycogen	+						+					
Lactose	−						−					
Maltose	+		+			+	+					
Mannitol	−						−			−		
Salicin	+						−			−	−	
Sorbitol	−						−			−		
Starch	+						+					
Sucrose	−		+			+	−			−		
Xylose	−						−			+		
Gelatin liquefaction			−		−	−	d					
Digestion of:												
Casein			−		−	−	−					
Coagulated blood serum	−		−	−	−	−	−	−	−			
Adsorption of red blood cells:												
Guinea pig	−	+	−	−	−	−	−	−	−			+
Human	−											
Ox							−					
Bovine			−				−					
Ovine			−				−					
Canine			−				−					
Rabbit			−				−					
Chicken			−				−					
Equine			−				−					
Reduction of tetrazolium:												
Aerobically	+		−[b]		weak	−[b]	weak	+		+		
Anaerobically	+		−[b]		+	−[b]	+	+	+	+		
Phosphatase activity	−		−		−	−	−	−	−			

[a] Symbols: see standard definitions.

[b] Variable reduction of tetrazolium occurs in liquid media, and no reduction occurs on solid media.

Table 30.3 Differential characteristics of the species of the genus *Anaeroplasma*[a]

Characteristics	A. abactoclasticum	A. bactoclasticum	A. intermedium[b]	A. varium[b]
Bacteriolytic	−	+	+	+
Digest casein	−	+	+	+
Products of fermentation:				
Acetate, formate, lactate, ethanol, carbon dioxide	+	+	+	+
Succinate	+	−	−	−
Hydrogen, propionate	−	+	+	+
Mol % G + C of DNA	29.3–29.5	33.7	33.4	32.5

[a] Symbols: see standard definitions.

[b] Differentiation of these two species from *A. bactoclasticum* and each other is based on DNA/DNA homology and serology.

Important Notes for Users of This Manual

Unless otherwise indicated in footnotes to tables, the meanings of symbols are as follows:

    + 90% or more of strains are positive

    − 90% or more of strains are negative

    d 11-89% of strains are positive

    v strain instability (not equivalent to "d")

    D Different reactions in different taxa (species of a genus or genera of a family)

All other symbols are defined in footnotes to tables.

Table 30.4 Differential characteristics of the species of the genus *Mycoplasma* [a]

Characteristic	Glucose catabolism	Mannose catabolism	Arginine hydrolysis	Phos-phatase	Film and spots	Tetrazolium reduction (Ae/An) [b]	Gelatin hydrolysis	Coagulated serum digestion	Casein digestion	Hemad-sorption [c]	Mol% G+C of DNA
M. agalactiae	−	−	−	+	d	+/+	−	−	−	+	30.5–34.2
M. alkalescens	−		+	+	−	−/−		−		−	25.9
M. alvi	+		+	−	−	−/+					26.4
M. anatis	+	d	−	+	+	−/+				−	26.6
M. anseris [d]	−		+	−	−	−/−					27.4–26.0
M. arginini	−	−	+	−	−	−/+				−	27.6–28.6
M. arthritidis	−	−	+	+	−	−/−	+		−	−	30.0–32.6
M. bovigenitalium	−	−	−	+	+	−/+	−		−	d	28.1–30.4
M. bovirhinis	+	−	−	d	−	+/+	−	d	+	d	24.5–27.3
M. bovis	−	−	d	+	d	+/+	−	−	−	x	27.8–32.9
M. bovoculi	+		d	d	+	+/+	−			−	29.0
M. buccale	−	−	+	+	−	−/+	−	−	−	−	25.0–26.4
M. californicum	−		−	+	−	−/d					31.9
M. canadense	−	−	+	W	−	−/+				−	29.0
M. canis	+	−	−	−	−	−/+	d	−		+	28.4–29.1
M. capricolum	+	+	d	+	−	+/+		+		−	24.1–25.5
M. caviae	+		v	+	+	−/[e]				−	
M. cavipharyngis [f]	+	+	−	−	−	W/W			+	+	30.0
M. citelli	+	+	−	+	+	W/W		−		−	27.4
M. cloacale [g]	−		+	−	−	−/+		−	−	−	26.0
M. collis [h]	+	+	−	−	−	+/+					28.0
M. columbinasale	−	−	+	+	+	−/−			+	−[i]	32.0
M. columbinum	−	−	+	−	+	−/+	−		−	−[i]	27.3
M. columborale	+	−	−	−	−	−/+	−	−	−	−[i]	29.2
M. conjunctivae	+	+	−	−	−	W/+				−	
M. cricetuli	+	+	+	+	+	+/+					
M. cynos	+	+	−	+	+	W/+			−	+	25.8
M. dispar	+	+	−	−	−	+/+				−	28.5–29.3
M. edwardii	+	−	−	+	+	−/+	−	−	−	d	29.2
M. ellychniae [j]	+		−	−	−					−	27.5
M. equigenitalium	+	+	−	+	+	−/−			−	−	31.5
M. equirhinis	−	−	+	−	+	−/−				−	
M. fastidiosum	+		−	−	+	−/−				+	32.3
M. faucium	−	−	+	−	−	−/−				+[f]	

Footnotes are at end of table

Table 30.4 (continued)

Characteristic	Glucose catabolism	Mannose catabolism	Arginine hydrolysis	Phosphatase	Film and spots	Tetrazolium reduction (Ae/An) [b]	Gelatin hydrolysis	Coagulated serum digestion	Casein digestion	Hemadsorption [c]	Mol% G+C of DNA
M. felifaucium [l]	–	–	+	+	+	–/–		–	–	+	31.0
M. feliminutum	+	x	–	–	d	–/+	–	–	–	–	29.1
M. felis	+	–	–	+	+	–/+	–	–	–	–	25.2
M. fermentans	+	–	+	d	+	–/+	–	–	–	–	27.5–28.7
M. flocculare	x	x	–	–	w	–/w		–		–	
M. gallinaceum	+		–	–	–	–/–				–[i]	28.0
M. gallinarum	–	–	+	–	+	+/+	–	–		–	26.5–28.0
M. gallisepticum	+	+	–	–	–	+/+	–	–	–	+	31.8–35.7
M. gallopavonis	+		–	–	–	d/d				–	27.0
M. gateae	–	–	+	–	–	–/w	–	–	–		28.5
M. genitalium	+		–	–	–	w/+		–		+	32.4
M. glycophilum [m]	+		–	w	–	–/w				+[n]	27.5
M. hominis	–	–	+	–	–	–/–	–		–	–	27.3–33.7
M. hyopharyngis [o]	–		+	+	+	–/ND	–	–	–	–	24.0
M. hyopneumoniae	x	x	–	–	w	–/w				–	27.5
M. hyorhinis	+	–	–	+	–	+/+	–	–	–	–	27.3–27.8
M. hyosynoviae	–	–	+	–	+	–/–		–		–	28.0
M. iners	–	–	+	–	+	–/–	–	–	–	–	29.1–29.6
M. iowae	+		+	–	–	+/+				+	25.0
M. lactucae [p]	+		–	–	–					+	30.0
M. lipofaciens	+		+	–	+	–/+		–			24.5
M. lipophilum	–	–	+		+					–	29.7
M. lucivorax [q]	+		–	+	+						27.4
M. luminosum [r]	+		+		+					+	28.8
M. maculosum	–	–	+	+	+	–/+	–	–	–	–	26.7–29.6
M. melaleucae [s]	+	–	–		–					–	27.0
M. meleagridis	–	–	+	+	–	–/+				d	27.0–28.6
M. moatsii	+		+		+	–/–					25.7
M. mobile [t]	+	+	–	±	+	+/+				+	23.5
M. molare	+	+	–	–	+	+/+	–	–	–		26.0
M. muris	–		+	–	+	–/+				+	24.9
M. mustelae	+		–	+	+	–/+	–	–			28.2

Footnotes are at end of table

713

Table 30.4 (continued)

Characteristic	Glucose catabolism	Mannose catabolism	Arginine hydrolysis	Phosphatase	Film and spots	Tetrazolium reduction (Ae/An)[b]	Gelatin hydrolysis	Coagulated serum digestion	Casein digestion	Hemadsorption[c]	Mol% G+C of DNA
M. mycoides subsp. mycoides	+	+	–	–	–	+/+	+[u]	+ or W[v]	+ or W[v]	–	26.1–27.1
M. mycoides subsp. capri	+	+	–	–	–	+/+	+	+	+	–	24.0–26.0
M. neurolyticum	+	+	–	–	–	–/+	–	–	–	–	22.8–26.2
M. opalescens	–	–	+	+	+	–/–	–	–	–	–[k]	29.2
M. orale	–	–	+	–	–	–/–	–	–	–	+[k]	24.0–28.2
M. ovipneumoniae	+		–	–	–	W/+	–	–	–	–	29.0
M. oxoniensis[w]	+	+	–	+	+	–/–				+	30.5
M. penetrans[x]	+		+	+	–	+/+				–	25.9
M. phocacerebrale[y]	–	+	+	+	+	+/–		–		–	26.5
M. phocarhinis[z]	–		–	+	+	+/–				–	27.8
M. phocidae[aa]	–		+	+	+	–/–	–	–		–	25.5
M. pirum[bb]	+		+	–	–		–			–	
M. pneumoniae	+	+	–	–	–	+/+	–	–	–	+	38.6–40.8
M. primatum	–	–	+	+	–	–/–	–	–	–	–[i]	28.6
M. pullorum	+		–	–	–	–/–	–			–[j]	29.0
M. pulmonis	+	+	–	–	+	–/+	–	–	–	d	27.5–29.2
M. putrefaciens	+	+	–	+	+	W/+		–		+	28.9
M. salivarium	–	–	+	–	+	–/W	–	–	–	–	27.3–31.4
M. somnilux[cc]	+		–		–					–	27.4
M. spermatophilum[dd]	–	–	+	W	–	–/–		–	–	–	32.0
M. spumans	–	–	+	+	–	–/–				+	28.4–29.1
M. sualvi	+		+	–	–	–/+					23.7
M. subdolum	–	–	+	v	d	–/–	–	–	–	–	28.8
M. synoviae	+		–	–	+	–/W				d	34.2
M. testudinis[ee]	+	+	–	–	–	–/–					35.0
M. verecundum	–	–	–	+	+	–/d		–	–	–	27.0–29.2

Footnotes are at end of table

Table 30.4 (continued)

a Symbols: see standard definitions; also x, not definitely settled.

b Ae, aerobically; An, anaerobically.

c Hemadsorption: unless otherwise indicated, the test is performed with red blood cells (RBCs) from different animal species (often with RBCs from the animal species from which the mycoplasma originated, plus guinea pig RBCs; sometimes with a variety of RBCs).

d The species *M. anseris* was not included in *Bergey's Manual of Systematic Bacteriology*. See Bradbury et al. (Int. J. Syst. Bacteriol. *38*: 74–76, 1988).

e Not tested anaerobically.

f The species *M. cavipharyngis* was not included in *Bergey's Manual of Systematic Bacteriology*. See Hill (J. Gen. Microbiol. *130*: 3183–3188, 1984; Int. J. Syst. Bacteriol. *39*: 371, 1989).

g The species *M. cloacale* was not included in *Bergey's Manual of Systematic Bacteriology*. See Bradbury and Forrest (Int. J. Syst. Bacteriol. *34*: 389–392, 1984).

h The species *M. collis* was not included in *Bergey's Manual of Systematic Bacteriology*. See Hill (Int. J. Syst. Bacteriol. *33*: 847–851, 1983).

i Only chicken RBCs were tested.

j The species *M. ellychniae* was not included in *Bergey's Manual of Systematic Bacteriology*. See Tully et al. (Int. J. Syst. Bacteriol. *39*: 284–289, 1989).

k Adsorbs chicken but not human, monkey, rat, or guinea pig RBCs.

l The species *M. felifaucium* was not included in *Bergey's Manual of Systematic Bacteriology*. See Hill (J. Gen. Microbiol. *132*: 1923–1928, 1986; Int. J. Syst. Bacteriol. *38*: 449, 1988).

m The species *M. glycophilum* was not included in *Bergey's Manual of Systematic Bacteriology*. See Forrest and Bradbury (J. Gen. Microbiol. *130*: 597–603, 1984; Int. J. Syst. Bacteriol. *34*: 355, 1984).

n Adsorbs turkey but not chicken RBCs.

o The species *M. hyopharyngis* was not included in *Bergey's Manual of Systematic Bacteriology*. See Erickson et al. (Int. J. Syst. Bacteriol. *36*: 55–59, 1986).

p The species *M. lactucae* was not included in *Bergey's Manual of Systematic Bacteriology*. See Rose et al. (Int. J. Syst. Bacteriol. *40*: 138–142, 1990).

q The species *M. lucivorax* was not included in *Bergey's Manual of Systematic Bacteriology*. See Williamson et al. (Int. J. Syst. Bacteriol. *40*: 160–164, 1990).

r The species *M. luminosum* was not included in *Bergey's Manual of Systematic Bacteriology*. See Williamson et al. (Int. J. Syst. Bacteriol. *40*: 160–164, 1990).

s The species *M. melaleucae* was not included in *Bergey's Manual of Systematic Bacteriology*. See Tully et al. (Int. J. Syst. Bacteriol. *40*: 143–147, 1990).

t The species *M. mobile* was not included in *Bergey's Manual of Systematic Bacteriology*. See Kirchhoff et al. (Int. J. Syst. Bacteriol. *37*: 192–197, 1987).

u Hydrolysis of gelatin not specified for large colony (LC) and small colony (SC) strains.

v LC strains digest coagulated serum and casein more vigorously than do SC strains.

w The species *M. oxoniensis* was not included in *Bergey's Manual of Systematic Bacteriology*. See Hill (Int. J. Syst. Bacteriol. *41*: 21–25, 1991).

x The species *M. penetrans* was not included in *Bergey's Manual of Systematic Bacteriology*. See Lo et al. (Int. J. Syst. Bacteriol. *42*: 357–364, 1992).

y The species *M. phocacerebrale* was not included in *Bergey's Manual of Systematic Bacteriology*. See Giebel et al. (Int. J. Syst. Bacteriol. *41*: 39–44, 1991).

z The species *M. phocarhinis* was not included in *Bergey's Manual of Systematic Bacteriology*. See Giebel et al. (Int. J. Syst. Bacteriol. *41*: 39–44, 1991).

aa The species *M. phocidae* was not included in *Bergey's Manual of Systematic Bacteriology*. See Ruhnke and Madoff (Int. J. Syst. Bacteriol. *42*: 211–214, 1992).

bb The species *M. pirum* was not included in *Bergey's Manual of Systematic Bacteriology*. See Del Giudice et al. (Int. J. Syst. Bacteriol. *35*: 285–291, 1985).

cc The species *M. somnilux* was not included in *Bergey's Manual of Systematic Bacteriology*. See Williamson et al. (Int. J. Syst. Bacteriol. *40*: 160–164, 1990).

dd The species *M. spermatophilum* was not included in *Bergey's Manual of Systematic Bacteriology*. See Hill (Int. J. Syst. Bacteriol. *41*: 229–233, 1991).

ee The species *M. testudinis* was not included in *Bergey's Manual of Systematic Bacteriology*. See Hill (Int. J. Syst. Bacteriol. *35*: 489–496, 1985).

715

Table 30.5 Hosts, pathogenicity, and relatedness of the species, serogroups, and subgroups of the genus *Spiroplasma*[a]

Binomial and/or common name	Group[a] or subgroup	Strain[b]	G + C content (mol%)	Principal host	Disease incited
S. citri	I-1	Maroc-R8A2[T] (27556) C189 (27665), Israel	26	Dicots, leafhoppers	Citrus stubborn
S. melliferum[c]	I-2	BC-3[T] (33219), AS 576 (29416)	26	Bees	Honeybee spiroplasmosis
S. kunkelii[d]	I-3	E275[T] (29320), I-747 (29051), B655 (33289)	26	Maize, leafhoppers	Corn stunt
277F spiroplasma	I-4	277F (29761)	26	Rabbit tick	None known
Green leaf bug spiroplasma	I-5	LB-12 (33649)	26	Green-leaf bug	None known
Maryland flower spiroplasma	I-6	M55 (33502), ET-1	28	Flowers, *Eristalis* fly	None known
Cocos spiroplasma	I-7	N525 (33287), N628	26	Coconut palm	None known
S. phoeniceum[e]; Vinca spiroplasma	I-8	P40[T] (43115)	26	Catharanthus roseus	Periwinkle disease
Sex ratio spiroplasmas	II	DW-1 (43153)	26	Drosophila	Sex ratio trait
S. floricola	III	23-6[T] (29989), BNR1 (33220), OBMG (33221)	26	Insects, flowers	None known
S. apis	IV	B31[T] (33834), SR3 (33095), PPS1 (33450)	30	Bees, flowers	"May disease"
S. mirum	V	SMCA[T] (29335), GT-48 (29334), TP-2 (33503)	30	Rabbit ticks	Suckling mouse cataract disease
Ixodes spiroplasma	VI	Y32 (33835)	25	Ixodes pacificus ticks	None known
Monobia spiroplasma	VII	MQ-1 (33825)	28	Monobia wasp	None known
Syrphid spiroplasma	VIII	EA-1 (33826)	30	Eristalis arbustorum fly	None known
Cotinus spiroplasma	IX	CN-5 (33827)	29	Cotinus beetle	None known
S. culicicola[f]	X	AES-1[T] (35112)	26	Aedes mosquito	None known
Monobia spiroplasma	XI	MQ-4 (35262)	26	Monobia wasp	None known
Cucumber beetle spiroplasma	XII	DU-1 (43210)	25	Diabrotica undecimpunctata beetle	None known
S. sabaudiense[g]	XIII	Ar-1343[T] (43303)	30	Aedes mosquito	None known
Ellychnia spiroplasma	XIV	EC-1 (43212)	26	Ellychnia corrusca beetle	None known
Leafhopper spiroplasma	XV	I-25 (43262)	26	Cicadulina leafhopper	
Cantharis spiroplasma	XVI	CC-1 (43207), Ar-1357, MQ-6	26	Cantharis beetle, mosquito, wasp	None known
Deer fly spiroplasma	XVII	DF-1 (43209)	29	Chrysops fly	None known
Tabanid spiroplasma	XVIII	TN-1 (43211)	25	Tabanus nigrovittatus	None known
Firefly spiroplasma	XIX	PUP-1 (43206)	26	Photuris pennsylvanicus beetle	None known
Colorado potato beetle spiroplasma	XX	LD-1 (43213)	25	Leptinotarsa decemlineata	None known
Flower spiroplasma	XXI	W115 (43260)	24	Prunus flower	None known
S. taiwanense[h]	XXII	CT-1[T] (43302)	25	Culex tritaeniorhynchus	None known
Tabanid spiroplasma	XXIII	TG-1 (43525)	26	Tabanus gladiator	None known
S. chinense[i]	XXIV	CCH (43960)	29	Calystegia hederaceae flowers	None known

Footnotes are at end of table

GROUP 30 THE MYCOPLASMAS (OR MOLLICUTES): CELL WALL-LESS BACTERIA

Table 30.5 *(continued)*

[a] Groups assigned on the basis of failure to cross-react in growth inhibition, metabolism inhibition, and deformation tests.

[b] Accession numbers (in parentheses) are those of the American Type Culture Collection (ATCC). ND, not done.

[c] The species *S. melliferum* was not included in *Bergey's Manual of Systematic Bacteriology.* See Clark et al. (Int. J. Syst. Bacteriol. *35:* 296–308, 1985).

[d] The species *S. kunkelii* was not included in *Bergey's Manual of Systematic Bacteriology.* See Whitcomb et al. (Int. J. Syst. Bacteriol. *36:* 170–178, 1986).

[e] The species *S. phoeniceum* was not included in *Bergey's Manual of Systematic Bacteriology.* See Saillard et al. (Int. J. Syst. Bacteriol. *37:* 106–115, 1987).

[f] The species *S. culicicola* was not included in *Bergey's Manual of Systematic Bacteriology.* See Hung et al. (Int. J. Syst. Bacteriol. *37:* 365–370, 1987).

[g] The species *S. sabaudiense* was not included in *Bergey's Manual of Systematic Bacteriology.* See Abalain-Colloc et al. (Int. J. Syst. Bacteriol. *37:* 260–265, 1987).

[h] The species *S. taiwanense* was not included in *Bergey's Manual of Systematic Bacteriology.* See Abalain-Colloc et al. (Int. J. Syst. Bacteriol. *38:* 103–107, 1987).

[i] The species *S. chinense* was not included in *Bergey's Manual of Systematic Bacteriology.* See Guo et al. (Int. J. Syst. Bacteriol. *40:* 421–425, 1990).

Table 30.6 **Physiological characteristics of the species of the genus *Spiroplasma*** [a]

Characteristics	*S. apis*	*S. chinense*	*S. citri*	*S. culicicola*	*S. floricola*	*S. kunkelii*	*S. melliferum*	*S. mirum*	*S. phoeniceum*	*S. sabaudiense*	*S. taiwanense*
Acid production from:											
Glucose	+	+	+	+	+	+	+	+[b]	+	+	+
Mannose		−	d	+	+		+	d			
Fructose		+	+	+	+			d			
Sucrose, trehalose, glycerol		+						d			
Lactose, mannitol, xylose, cellobiose, galactose, salicin								−			
Arginine hydrolysis	+	+	+	−	+	+	+	+	+	+	−
Tetrazolium, anaerobic reduction	+	+				d	d	+			
Arbutin hydrolysis								−			
Phosphatase activity	−	+	−	−	−	−		+	+		
Film and spot reaction	+	d	+			+		+	+	−	−
Adsorption of guinea pig red blood cells	−	d	−			−	−	−	+	−	−
Liquefaction of coagulated serum			−			−	−	−			
Optimum temperature, °C	32	25–37	32	31	34	30–32	32–35	30–37	28–32	30	25–30

[a] Symbols: see standard definitions.
[b] Fermentation is very slow in broth without amino acid supplement.

717

GROUP 31

THE METHANOGENS

Rods, cocci, pseudosarcinae, spirals, and multicellular aggregates, motile or nonmotile. Stain Gram negative or Gram positive, but cells have neither murein nor outer membrane. Membranes are composed of ether-linked isoprenoids. **Very strictly anaerobic. Chemoautotrophic or chemoheterotrophic, with methane always the product of catabolic metabolism.** $H_2 + CO_2$, formate, acetate, methyl compounds (methanol, methylamines, methylsulfides, and perhaps methylselenides), methanol $+ H_2$, or alcohols $+ CO_2$ serve as carbon and energy sources. Many strains are obligately or facultatively autotrophic. Ammonia serves as the nitrogen source (although some strains can also use amino acids or fix molecular nitrogen), and sulfide or sulfur serves as the sulfur source. (Sulfur is reduced to sulfide, but this reaction is apparently not coupled to growth (Stetter and Gaag, Nature *503:* 309–311, 1983).) Cells are free-living, occupying the terminal position in the anaerobic food web when CO_2 is the major electron acceptor (O_2, sulfate, and nitrate are absent) by reducing CO_2 to methane, by splitting acetate to methane and CO_2, or by degrading methylamines and methylsulfides in anoxic, saline, sulfate-containing environments.

The differentiation of methanogens from other organisms is very unambiguous: all methanogens are microbes that produce methane as a major catabolic product. No other organisms but methanogens fall into this category. Methanogens can be divided into three taxonomic subgroups:

Differentiation of the genera included in **Group 31:**

A. Rod-shaped, lancet-shaped, or coccoid methanogens that catabolize $H_2 + CO_2$, formate, or $H_2 +$ methanol; cell walls contain pseudomurein.

Subgroup 1: *Methanobacterium, Methanobrevibacter, Methanosphaera,* and *Methanothermus;* see Table 31.1 for differentiation of genera.

B. Coccoid-, rod-, spiral-, or plate-shaped methanogens grow on $H_2 + CO_2$, formate, or alcohols $+ CO_2$; pseudomurein is absent and cells are lysed by detergent, but spiral cells have a resistant sheath.

Subgroup 2: *Methanococcus, Methanocorpusculum, Methanoculleus, Methanogenium, Methanolacinia, Methanomicrobium, Methanoplanus,* and *Methanospirillum;* see Table 31.2 for differentiation of genera.

C. Pseudosarcinae, cocci, or sheathed rods able to grow on trimethylamine or acetate.

Subgroup 3: *Methanococcoides, Methanohalobium, Methanohalophilus, Methanolobus, Methanosarcina,* and *Methanothrix;* see Table 31.3 for differentiation of genera.

The genera of Subgroup 1 may be assigned to genera on the basis of morphology, growth temperature, and catabolic substrate, as shown in Table 31.1.

Subgroup 2 comprises a genetically diverse group of organisms that are often difficult to assign to genera. Classification of these organisms may require specialized phylogenetic tests such as interstrain DNA hybridization, but often a strain may be classified after a simple comparison of its whole-cell proteins with those of related type strains (by denaturing gel electrophoresis). *Methanolacinia, Methanospirillum,* and *Methanomicrobium* can be differentiated from the other species of this group by their morphologies, but the other organisms are difficult or impossible to distinguish except by specialized laboratories.

Subgroup 3 contains acetitrophic sheathed rods, methylotrophic cocci and pseudosarcinae, and acetitrophic cocci and pseudosarcinae. The sheathed rods (*Methanothrix*) can be easily differentiated from other members of this subgroup by their distinctive morphology. *Methanosarcina* can be differentiated from the other five genera by its ability to use either acetate or $H_2 + CO_2$ as substrate; most strains can use both. The remaining four genera in this subgroup are halophiles and are restricted to methylotrophic substrates. These latter four genera can be distinguished from each other by their degree of halophily. *Methanohalobium* is a genus of extreme halophiles, with optimal Na⁺ concentration above 2 M. Strains of *Methanohalophilus* generally grow fastest at Na⁺ of 0.5–2 M, although some alkaliphilic species assigned to this genus grow well at slightly lower salinities. The other two genera in Subgroup 3, *Methanolobus* and *Methanococcoides*, grow fastest with Na⁺ at 0.1–0.6 M. These latter two genera may be difficult to differentiate.

Editorial note: Halomethanococcus was proposed as a genus of halophilic methanogens (see Yu and Kawamura [J. Gen. Appl. Microbiol. *33:* 303–310, 1987, and Int. J. Syst. Bacteriol. *38:* 328, 1988]), but the type strain of the only proposed species *(H. doii)* is unavailable, and the genus appears to be synonymous with *Methanohalophilus*. Thus, the genus *Halomethanococcus* and the species *H. doii* are not included in this taxonomy.

The genera *Methanothrix* and *Methanosaeta* are subjectively synonymous. The former name has priority, but its legitimacy is not clear because the purity of the culture on which the description was based has been questioned (Boone, Int. J. Syst. Bacteriol. *41:* 588–589, 1991).

The genus *"Methanopyrus"* was proposed without a type species (Huber et al., Nature *342:* 833–834, 1989), but later the species *"Methanopyrus kandleri"* was proposed; neither name has been validly published.

SUBGROUP 1

Genus **Methanobacterium**

Curved, crooked to straight rods, long to filamentous, about 0.5–1.0 µm in width. **Endospores are not formed.** Gram variable; cell walls are composed of pseudomurein. **Nonmotile.** Cells produce fimbriae (based on examination of 2 species). **Very strictly anaerobic.** Optimum temperatures are 37–45°C for mesophilic species and 55°C or greater for thermophiles. Energy metabolism occurs by reduction of CO_2 to CH_4; electron donors are limited to H_2, formate, secondary alcohols, and CO; methylamines and acetate are not catabolized. Ammonia may serve as sole nitrogen source, and sulfide may serve as sulfur source. Nitrogen may be fixed (in each of two species tested

[Belay et al., Biochim. Biophys. Acta *971:* 233-245, 1988]), and sulfur may be reduced to sulfide. May be found in flooded soils, sediments, or other anoxic environments where inorganic electron acceptors (such as nitrate or sulfate) other than CO_2 are limiting.

Type species: *Methanobacterium formicicum.*

Differentiation of the species of the genus **Methanobacterium:** See Table 31.4.

Genus **Methanobrevibacter**

Oval rods or cocci to short rods, usually occurring in pairs or chains; about 0.6 µm wide and 0.8–2 µm long. **Endospores are not formed. Stain Gram positive. Cell walls are composed of pseudomurein. Nonmotile and very strictly anaerobic.** Optimum temperatures are 37–40°C. Energy metabolism reduction of CO_2 to CH_4; electron donors are limited to H_2, formate, and CO; acetate, methylamines, and secondary alcohols are not catabolized. Ammonia or N_2 (Fardeau et al., Arch. Microbiol. *148:* 128–131, 1987; Belay et al., Biochim. Biophys. Acta *971:* 233–245, 1988) are the nitrogen source, and sulfide or sulfur may serve as sulfur source. One or more B vitamins are required. May be found in gastrointestinal tracts, anaerobic digestors, or other anoxic environments where inorganic electron acceptors (such as nitrate or sulfate) other than CO_2 are limiting.

Type species: *Methanobrevibacter ruminantium.*

Differentiation of the species of the genus **Methanobrevibacter:** See Table 31.5.

Genus **Methanosphaera**

Irregular cocci, 0.6–1.2 µm, occurring singly or in pairs. **Endospores are not formed. Stain Gram positive. Nonmotile. Very strictly anaerobic.** Optimum growth is at 35–40°C and at pH 6.8. Requires acetate, CO_2, isoleucine, and thiamin as organic growth factors. Energy metabolism occurs by reduction of methanol to CH_4 with H_2 as the electron donor. Both H_2 and methanol are required, and no other catabolic substrates support growth. Ammonia serves as nitrogen source, and sulfide or S⁰ serves as sulfur source. May be found in gastrointestinal tracts.

Type species: *Methanosphaera stadtmanae.*

Differentiation of the species of the genus **Methanosphaera:** *M. stadtmanae* has been isolated from gastrointestinal tracts. The characteristics are as described for the genus. The only other species in this genus, *M. cuniculi,* cannot be distinguished from *M. stadtmanae* by known physiological tests, although it can be distinguished antigenically, by DNA hybridization, or by electrophoretic analysis of proteins; its characteristics are also as described for the genus.

Editorial note: M. cuniculi was not included in *Bergey's Manual of Systematic Bacteriology.* See Biavati et al. (Appl. Environ. Microbiol. *54:* 768–771, 1988, and Int. J. Syst. Bacteriol. *40:* 470–471, 1990).

Genus **Methanothermus**

Straight to slightly curved rods, 0.3–0.4 μm wide and 1–3 μm long, occurring singly or in short chains. The cell wall contains pseudomurein and an outer S-layer. **Endospores are not formed. Stain Gram positive.** Appear to be motile, although electron microscopy fails to reveal flagella. **Very strictly anaerobic.** Optimal growth is at 83–88°C and pH 6.5. **Grow autotrophically by conversion of H_2 + CO_2 to methane;** acetate, formate, methylamines, and methanol are not catabolized. Ammonia serves as nitrogen source, and sulfide or S^0 serves as sulfur source. May be found in thermal sulfataric fields.

Type species: *Methanothermus fervidus.*

Differentiation of species of the genus **Methanothermus:** See Table 31.6.

SUBGROUP 2

Genus **Methanococcus**

Irregular cocci, 1–2 μm in diameter. **Endospores are not formed. Stain Gram negative.** Susceptible to lysis by detergents or hypotonic shock. **Nonmotile. Very strictly anaerobic.** Optimum NaCl concentration is 0.1–0.8 M. Mesophilic (optimum 35–40°C), thermophilic (optimum 65°C), or extremely thermophilic (optimum 85–91°C). H_2 + CO_2 and usually formate are catabolic substrates for CH_4 production; acetate or methylamines are not catabolized. Ammonia and some-

times N_2, or alanine are nitrogen sources, and sulfide or sulfur is sulfur source. May be found in anoxic salt-marsh, marine, or estuarine environments.

Type species: *Methanococcus vannielii.*

Differentiation of species of the genus **Methanococcus:** See Table 31.7.

Editorial note: The species *M. halophilus,* described in *Bergey's Manual of Systematic Bacteriology,* is transferred to the genus *Methanohalophilus,* and the species *M. frisius* has been transferred to the genus *Methanosarcina* as *M. frisia* (Blotevogel et al., Int. J. Syst. Bacteriol. *39:* 91–92, 1989).

Genus **Methanocorpusculum**

Small, irregular cocci, usually < 1 μm in diameter. **Endospores are not formed. Stain Gram negative.** Susceptible to lysis by detergents or hypotonic shock. **Nonmotile. Very strictly anaerobic.** Optimum NaCl concentration is 0.1–0.25 M. Temperature optimum is 30–40°C. H_2 + CO_2, formate, and sometimes secondary alcohols + CO_2 are catabolic substrates for CH_4 production; acetate or methylamines are not catabolized. Acetate and either yeast extract or peptones are required as anabolic substrates. Ammonia serves as nitrogen source, and sulfide serves as sulfur source. May be found in anaerobic digestors or anoxic lake sediments.

Type species: *Methanocorpusculum parvum.*

Differentiation of species of the genus **Methanocorpusculum:** See Table 31.8.

Genus **Methanoculleus**

Editorial note: The genus *Methanoculleus* was not included in *Bergey's Manual of Systematic Bacteriology.* The genus was created in 1989 by Maestrojuán et al. (Int. J. Syst. Bacteriol. *40:* 117–122, 1990).

Irregular cocci, 1–2 μm in diameter. **Endospores are not formed. Stain Gram negative.** Susceptible to lysis by detergents or hypotonic shock. **Nonmotile. Very strictly anaerobic.** Optimum NaCl concentration is 0.1–0.25 M. Temperature optimum is 35–40°C. H_2 + CO_2, formate, and sometimes secondary alcohols + CO_2

are catabolic substrates for CH_4 production; acetate or methylamines are not catabolized. Ammonia serves as nitrogen source, and sulfide serves as sulfur source. May be found in anaerobic digestors or anoxic lake sediments.

Type species: *Methanoculleus bourgense.*

Differentiation of species of the genus **Methanoculleus:** See Table 31.9.

Genus **Methanogenium**

Irregular cocci, 1–3 μm in diameter. **Endospores are not formed. Stain Gram negative.** Susceptible to lysis by detergents or hypotonic shock. **Nonmotile,** although flagella or fimbriae may be present. **Very strictly anaerobic.** Optimum NaCl concentration is 0.1–0.5 M, and optimum pH is near neutral. Temperature optimum is 20–40°C, with some strains growing poorly above 30°C. $H_2 + CO_2$, formate, and sometimes ethanol or secondary alcohols + CO_2 are catabolic substrates for CH_4 production; acetate or methylamines are not catabolized. Ammonia serves as nitrogen source, and sulfide serves as sulfur source. Tungsten may be required. Found in anoxic marine sediments or lake sediments.

Type species: *Methanogenium cariaci.*

Differentiation of species of the genus **Methanogenium:** See Table 31.10.

Editorial note: M. aggregans was transferred to the genus *Methanocorpusculum* (Xun et al., Int. J. Syst. Bacteriol. *39*: 109–111, 1989), and *M. bourgense, M. marisnigri, M. olentangyi,* and *M. thermophilicum* were transferred to the genus *Methanoculleus* (Maestrojuán et al., Int. J. Syst. Bacteriol. *40*: 117–122, 1990).

Genus **Methanolacinia**

Editorial note: The genus *Methanolacinia* was not included in *Bergey's Manual of Systematic Bacteriology.* The genus was created in 1989 by Zellner et al. (J. Gen. Appl. Microbiol. *35*: 185–202; Int. J. Syst. Bacteriol. *40*: 470–471, 1990).

Short, irregular rods, 0.6 μm in diameter and 1.5–2.5 μm long. **Endospores are not formed. Stain Gram**

negative. Susceptible to lysis by detergent. **Nonmotile or very weakly motile:** flagella may be present. **Very strictly anaerobic.** Optimum temperature is 40°C, optimum pH 6.6–7.2. Energy metabolism is by reduction of CO_2 to CH_4 with H_2 (or secondary alcohols) as electron donor. Acetate, formate, methylamines, methanol, and methanol + H_2 are not catabolized. Ammonia may serve as the sole nitrogen source, and sulfide may serve as sulfur source; acetate is required. May be found in marine sediments.

Type (and only) species: *Methanolacinia paynteri.*

Characteristics of the species: M. paynteri has been isolated from marine sediments of a mangrove swamp. The characteristics are as described for the genus.

Genus **Methanomicrobium**

Short, curved rods, 0.7 μm in diameter and 1.5–2 μm long. **Endospores are not formed. Stain Gram negative. Motile. Very strictly anaerobic.** Optimum temperature is 40°C, optimum pH 6.1–6.9. Energy metabolism is by reduction of CO_2 to CH_4 with H_2 or formate as electron donor. Acetate, methylamines, and methanol are not catabolized. Ammonia serves as sole nitrogen source, and sulfide or S^0 serves as sulfur source. Growth factors found in rumen fluid are required (these include acetate, isobutyrate, isovalerate, 2-methylbutyrate, tryptophan, pyridoxine, thiamine, biotin, and vitamin B_{12} [Tanner and Wolfe, Appl. Environ. Microbiol. *54*: 625–628, 1988]). May be found in the bovine rumen.

Type (and only) species: *Methanomicrobium mobile.*

Characteristics of the species: M. mobile has been isolated from a bovine rumen. The characteristics are as described for the genus.

Editorial note: The species M. paynteri has been transferred to the genus *Methanolacinia* (Zellner et al., J. Gen. Appl. Microbiol. *35*: 185–202, 1989, and Int. J. Syst. Bacteriol. *40*: 470–471, 1990).

Genus **Methanoplanus**

Very irregular cocci or plate-shaped, 0.1–0.3 μm thick and 1.5–3 μm wide. **Endospores are not formed. Stain Gram negative.** Susceptible to lysis by detergents or hypotonic shock. **Flagella or pili are present. Very strictly anaerobic.** Temperature optimum is 30–40°C, pH optimum near neutral. H$_2$ + CO$_2$ or formate are catabolic substrates for CH$_4$ production; acetate or methylamines are not catabolized. Ammonia serves as nitrogen source, and sulfide or S^0 serves as sulfur source. Acetate, peptones, and yeast extract may be required as growth factors. May be found in anoxic sediments or as endosymbionts in sapropelic marine ciliates.

Type species: *Methanoplanus limicola.*

Differentiation of species of the genus **Methanoplanus:** See Table 31.11.

Genus **Methanospirillum**

Symmetrically curved rods, forming an α-helix, 0.4–0.5 μm wide and 7–10 μm long, occurring singly or in chains up to several hundred μm in length. The cells are enclosed within a sheath. **Endospores are not formed. Stain Gram negative. Motile, with polar, tufted flagella. Very strictly anaerobic.** Optimum growth is at 35–40°C and pH 7–7.5. Cysteine and acetate are required by some strains, and acetate and vitamins are stimulatory. Energy is produced by conversion of H$_2$ + CO$_2$, formate, or sometimes secondary alcohols + CO$_2$ to methane. Acetate, methylamines, methanol, and other alcohols are not catabolized. Ammonia or N$_2$ (Belay et al., Biochim. Biophys. Acta *971:* 233–245, 1988) serves as nitrogen source, and sulfide serves as sulfur source. May be found in the bovine rumen or in anaerobic digestors.

Type (and only) species: *Methanospirillum hungatei.*

Characteristics of the species: M. hungatei has been isolated from the bovine rumen, anaerobic digestors, and from mud of a ditch. The characteristics are as described for the genus.

SUBGROUP 3

Genus **Methanococcoides**

Extremely irregular cocci, occurring singly or in pairs; about 1 μm in diameter. **Endospores are not formed. Stain Gram negative.** Susceptible to lysis by detergent or hypotonic shock. **Nonmotile. Very strictly anaerobic.** Good growth occurs with NaCl concentration of

0.2–0.6 M, Mg^{2+} concentration of 25–200 mM, temperature 30–35°C, and pH 7–7.5. Energy metabolism occurs by dismutation of methylamines to CH$_4$, CO$_2$, and NH$_3$; acetate, formate, and H$_2$ are not catabolized, even when provided in combination with trimethylamine. Ammonia (or methylamines) is the sole nitrogen source, and sulfide may serve as sulfur source; biotin is required. Found in anaerobic marine sediments.

Type (and only) species: *Methanococcoides methylutens.*

Characteristics of the species: M. methylutens was isolated from deep organic sediments of a marine trench. The characteristics are as described for the genus.

Genus **Methanohalobium**

Editorial note: The genus *Methanohalobium* was not included in *Bergey's Manual of Systematic Bacteriology.* The genus was created in 1987 by Zhilina and Zavarzin (Dokl. Akad. Nauk SSSR *293:* 464–468; Int. J. Syst. Bacteriol. *38:* 136–137, 1988). It includes only one species, *M. evestigatus.*

Very irregular cocci, 0.5–1.5 μm in diameter. **Endospores are not formed. Stain Gram negative. Nonmotile. Very strictly anaerobic.** Optimum NaCl concentration is > 3 M. Temperature optimum is 50–55°C; pH optimum is 6.5–7.5. Methylamines and methanol (Menaia and Boone, unpublished data) are the sole catabolic substrates, with H$_2$ + CO$_2$, formate, acetate, or alcohols + CO$_2$ not catabolized. Ammonia serves as nitrogen source, and sulfide serves as sulfur source. Requires B-vitamins or yeast extract. May be found in saline ponds and lagoons.

Type (and only) species: *Methanohalobium evestigatus.*

Characteristics of the species: M. evestigatus has been isolated from a salt lagoon near Arabata, Sivasch, Ukraine. The characteristics are as described for the genus.

Genus **Methanohalophilus**

Editorial note: The genus *Methanohalophilus* was not included in *Bergey's Manual of Systematic Bacteriology.* The genus was created in 1988 by Paterek and Smith (Int. J. Syst. Bacteriol. *38:* 122–123). The species *Methanococcus halophilus,* which was listed in *Bergey's*

Manual of Systematic Bacteriology as a *species incertae sedis,* has since been transferred to *Methanohalophilus* (Wilharm et al., Int. J. Syst. Bacteriol. *41:* 558–562, 1991).

Extremely irregular cocci, occurring singly or in pairs; 0.5–3 μm in diameter. **Endospores are not formed. Stain Gram negative.** Susceptible to lysis by detergent or hypotonic shock. **Nonmotile. Very strictly anaerobic.** Optimum NaCl concentration is 0.5–2 M. Optimum temperatures are 35–45°C. Energy metabolism occurs by dismutation of methylamines to CH_4, CO_2, and NH_3 or sometimes methylsulfides to CH_4, CO_2, and H_2S; acetate, formate, and H_2 are not catabolized, even when provided in combination with trimethylamine. Ammonia (or methylamines) is the sole nitrogen source, and sulfide (or methylsulfides) may serve as sulfur source; biotin or *p*-aminobenzoate are sometimes required. May be found in natural or artificial saline lakes or saline ground water.

Type species: *Methanohalophilus mahii.*

Differentiation of species of the genus **Methanohalophilus:** See Table 31.12.

Genus **Methanolobus**

Irrregular cocci, about 1 μm in diameter, sometimes forming loose aggregates. **Endospores are not formed. Stain Gram negative.** Susceptible to lysis by detergents or hypotonic shock. **Sometimes motile by monotrichous flagella. Very strictly anaerobic.** Optimum NaCl concentration is 0.1–0.6 M. Optimum temperature is 35–40°C and optimum pH is 6.5–6.8. No organic growth factors are required. Energy metabolism occurs by dismutation of methyl amines or methanol to CO_2 and CH_4 (and NH_3 in the case of methyl amines). Some strains dismutate methyl sulfides. $H_2 + CO_2$, formate, acetate, and alcohols other than methanol are not catabolized. Ammonia (or methylamines) serves as nitrogen source, and sulfide or S^0 (or sometimes methyl sulfides) serves as sulfur source. May be found in anoxic marine sediments or lake sediments.

Type species: *Methanolobus tindarius.*

Differentiation of species of the genus **Methanolobus:** See Table 31.13.

Genus **Methanosarcina**

Irregular spheroid bodies 1–1000 μm or more in diameter, **occurring alone or typically in aggregates of cells.** Small aggregates typically appear as sarcina ("pseudosarcina"), except that the division planes are not perpendicular. **Sometimes the bodies occur as large cysts** with a common outer wall surrounding individual coccoid cells. **Endospores are not formed. Gram stain results are variable.** Individual coccoid cells may be susceptible to lysis by detergents or hypotonic shock. **Nonmotile. Very strictly anaerobic.** Optimum NaCl concentration is 0.1–0.5 M. Temperature optimum is 30–40°C for mesophiles and 50–55°C for thermophiles. Organic growth factors generally are not required. Energy metabolism occurs by dismutation of methylamines or methanol to CO_2 and CH_4 (and NH_3 in the case of methylamines); methanol and methylamines are reduced to CH_4 when grown with H_2. Other catabolic substrates include $H_2 + CO_2$ or acetate, but never formate. Ammonia, methylamines, and often N_2 serve as nitrogen source, and sulfide or S^0 serves as sulfur source. May be found in anoxic marine sediments, lake sediments, or anaerobic digestors.

Type species: *Methanosarcina barkeri.*

Differentiation of species of the genus **Methanosarcina:** See Table 31.14.

Genus **Methanothrix**

Large, sheathed rods, 0.8–1.2 μm in diameter and 2–3 μm long. The sheaths may enclose long chains of these rods, which sometimes appear to be continuous when viewed by phase-contrast microscopy. **Endospores are not formed. Stain Gram negative. Some strains contain gas vacuoles. Nonmotile. Very strictly anaerobic.** Optimum temperature is 35–40°C for mesophiles and 65°C for thermophiles. Optimum pH is 7.1–7.8. Optimum NaCl concentration is < 0.8 M (unpublished data, Menaia and Boone). Energy metabolism occurs by aceticlastic degradation of acetate to CO_2 and CH_4. $H_2 + CO_2$, formate, methylamines, and methanol are not catabolized. Ammonia is the sole nitrogen source, and sulfide may serve as sulfur source. Some mesophilic strains are reported to require yeast extract, and other reports indicate that yeast extract is inhibitory and that there are no organic growth factors; thermophiles may require biotin. Found in anaerobic digestors.

Type species: *Methanothrix soehngenii.*

Differentiation of species of the genus **Methanothrix:** See Table 31.15.

Editorial note: The nomenclature of the genus *Methanothrix* is in controversy, partly because of the designation of an impure type culture for the type species *M. soehngenii.* Various authors have interpreted Rules 30 and 31 of the Code of Bacteriological Nomenclature to deem the genus invalid on the basis of an impure type culture (e.g., Patel and Sprott, Int. J. Syst. Bacteriol. *40:* 79–82, 1990). These authors renamed the genus *Methanosaeta* with the type species *M. concilii,* based on a pure type culture. Boone (Int. J. Syst. Bacteriol. *41:* 588–589, 1991) has proposed a neotype culture (strain GP6) for *Methanothrix soehngenii* with retention of the genus name and has asked for an opinion from the Judicial Commission on the matter. Patel (Int. J. Syst. Bacteriol. *42:* 324–326, 1992) disputes Boone's request and argues for the retention of *Methanosaeta.* The question is now before the Judicial Commission to decide. The thermophilic strains of *Methanothrix* (*Methanosaeta*) were put into a species *Methanosaeta thermoacetophila* by Patel and Sprott (ibid.), but the type strain was also impure. Recently, Kamagata et al. (Int. J. Syst. Bacteriol. *42:* 463–468, 1992) have named a species *Methanothrix thermophila* to include the thermophilic strains, with their pure strain PT as the type.

Important Notes for Users of This Manual

Unless otherwise indicated in footnotes to tables, the meanings of symbols are as follows:

- \+ 90% or more of strains are positive
- − 90% or more of strains are negative
- d 11-89% of strains are positive
- v strain instability (not equivalent to "d")
- D Different reactions in different taxa (species of a genus or genera of a family)

All other symbols are defined in footnotes to tables.

Table 31.1 Differential characteristics of the genera of Group 31 Subgroup 1 [a]

Characteristic	*Methanobacterium*	*Methanobrevibacter*	*Methanosphaera*	*Methanothermus*
Shape:				
Filaments	d	–	–	–
Rods	+	–	–	+
Short rods or lancet-shaped	–	+	–	–
Cocci	–	+	+	–
Growth temperature	<70°C	<45°C	<45°C	>60°C
Catabolic substrates:				
$H_2 + CO_2$	+	+	–	+
Formate	d	d	–	d
Acetate	–	–	–	–
Methyl compounds	–	–	–	–
Methanol + H_2	–	–	+	–
Organic growth factors:				
Required:				
Acetate	–	+	+	–
Cysteine	–	–	–	–
Vitamins	–	+	+	–
Coenzyme M	–	d	–	–
2-Methyl butyrate or isoleucine	–	d	+	–
Stimulatory:				
Acetate	+	+	+	–
Cysteine	+	d	–	–
Vitamins	–	+	+	–
Peptone, yeast extract, or rumen fluid	–	d	–	–

[a] Symbols: see standard definitions.

Important Notes for Users of This Manual

Unless otherwise indicated in footnotes to tables, the meanings of symbols are as follows:

+ 90% or more of strains are positive

– 90% or more of strains are negative

d 11-89% of strains are positive

v strain instability (not equivalent to "d")

D Different reactions in different taxa (species of a genus or genera of a family)

All other symbols are defined in footnotes to tables.

Table 31.2 Differential characteristics of the genera of Group 31 Subgroup 2[a]

Characteristic	Methano-coccus	Methano-corpusculum	Methano-culleus	Methano-genium	Methano-lacinia	Methano-microbium	Methano-planus	Methano-spirillum
Size and shape:								
Small coccoid (<2 μm)	+	+	+	−	−	−	−	−
Large coccoid (>1.5 μm)	−	−	−	+	d	−	−	−
Plate shaped (0.1–0.5 × 1.5–3 μm)	−	+	+	−	−	−	+	−
Irregular rod (0.6 × 1.5–2.5 μm)	−	−	−	−	d	−	−	−
Curved rod (0.7 × 1.5–2 μm)	−	−	−	−	−	+	−	−
α-Helical spiral (0.4 × 7–10 μm)	−	−	−	−	−	−	−	+
Gram stain reaction [b]	−	−	−	−	−	−	−	−
Motility	+	+	−[c]	−[c]	−	+	d[c]	+
Catabolic substrates:								
H₂ + CO₂	+	+	+	+	+	+	+	+
Formate	+	+	+	+	−	+	+	+
Organic growth factors:								
Required:								
Acetate	−	+	+	+	+	−	+	d
Yeast extract or vitamins	−	+[d]	d	+	−	−	−	−
Amino acids or peptones	−	+[d]	d	d	−	−	−	−
Rumen fluid	−	−	−	−	−	+	−	−
Stimulatory:								
Acetate	+	+	+	+	+	−	+	d
Yeast extract or vitamins	−	+	d	+	d	−	+	+
Amino acids or peptones	+	+	d	d	d	−	+	d
Rumen fluid	−	d	−	−	−	+	−	−

[a] Symbols: see standard definitions.
[b] Some strains, which have S-layers as the only cell wall, disintegrate upon drying; these are indicated to have a negative Gram-stain reaction, fixed.
[c] Flagella observed by electron microscopy, but motility not documented.
[d] Either yeast extract or peptone is required.

Table 31.3 Differential characteristics of the genera of Group 31 Subgroup 3 [a]

Characteristic	Methano-coccoides	Methano-halobium	Methano-halophilus	Methano-lobus	Methano-sarcina	Methano-thrix
Shape:						
Irregular coccus (1–3 µm)	+	+	+	+	+	−
Pseudosarcina (1–3 µm)	−	−	−	−	+	−
Large aggregate or "cyst" (cells 1–3 µm, aggregates up to 1 cm)	−	−	−	−	+	−
Sheathed rod (1 × 1–3 µm)	−	−	−	−	−	+
Gram stain reaction	−	−	−	−	d,D,v [b]	−
Catabolic substrates:						
$H_2 + CO_2$	−	−	−	−	d	−
Formate	−	−	−	−	−	−
Acetate	−	−	−	−	+	+
Methanol	+	+	+	+	+	−
Mono-, di-, or trimethylamine	+	+	+	+	+	−
Dimethylsulfide	ND	ND	D	D	−	ND
Methanethiol	ND	ND	D	D	−	ND
Organic growth factors:						
Biotin	+	ND [c]	d	−	−	+
p-Aminobenzoate	−	ND [c]	d	−	D	ND [c]
Optimal Na$^+$ concentration (M):						
<0.6	+	−	−	+	+	+
0.5–2	−	−	+	−	−	−
>2	−	+	−	−	−	−

[a] Symbols: see standard definitions.

[b] For some strains, Gram stain reaction depends on growth conditions and history.

[c] Requires yeast extract or B vitamin mixture.

Important Notes for Users of This Manual

Unless otherwise indicated in footnotes to tables, the meanings of symbols are as follows:

+ 90% or more of strains are positive

− 90% or more of strains are negative

d 11-89% of strains are positive

v strain instability (not equivalent to "d")

D Different reactions in different taxa (species of a genus or genera of a family)

All other symbols are defined in footnotes to tables.

Table 31.4 Differentiation of the species of the genus *Methanobacterium*

Characteristic	M. alcaliphilum	M. bryantii	M. espanolae [a]	M. formicicum	M. ivanovii [b]	M. palustre [c]	M. thermoaggregans [d]	M. thermoalcaliphilum [e]	M. thermoautotrophicum	M. thermoformicicum [f]	M. thermophilum [g]	M. uliginosum	M. wolfei
Morphology:													
Long rods	+	+	+	+	+	+	+	+	+	+	+	+	+
Filaments	+	+	−	+	+	+	+	+	+	d	−	−	−
Catabolic substrate:													
H$_2$ + CO$_2$	+	+	+	+	+	+	+	+	+	+	+	+	+
Formate	−	−	−	+	−	+	−[h]	−	−	+	−	−	−
Alcohols + CO$_2$	+[i]	+	−[j]	−	+[k]	−[h]			−	−[l]	−[h]		−
Growth conditions:													
Temperature:													
Growth at 68°C	−	−	−	−	−	−	+	+	+	−	+	−	+
Growth at 55°C	−	−	−	−[m]	−	−	+	+	+	+	+	−	+
Fastest growth at <45°C	+	+	+	+	+	+	−	−	−	−	−	+	−
pH:													
5.8–6.3	−	−	+	−	−	−	−	−	−	−	−	−	−
7.0–7.5	−	+	−	+	+	+	+	+	+	+	−	+	+
8.1–9.1	+	−	−	−	−	−	−	−	−	−	+	−	−
Fastest growth in mineral medium	−	−	−	−	−		−	−	+	d[l,n]	−		+

[a] The species *M. espanolae* was not included in *Bergey's Manual of Systematic Bacteriology*. See Patel et al. (Int. J. Syst. Bacteriol. *40:* 12–18, 1990).

[b] The species *M. ivanovii* was not included in *Bergey's Manual of Systematic Bacteriology*. See Belayev et al. (Appl. Environ. Microbiol. *45:* 691–697, 1983), Jain et al. (Syst. Appl. Microbiol. *9:* 77–82, 1987, and Int. J. Syst. Bacteriol. *38:* 136, 1988).

[c] The species *M. palustre* was not included in *Bergey's Manual of Systematic Bacteriology*. See Zellner et al. (Arch. Microbiol. *151:* 1–9, 1989, and Int. J. Syst. Bacteriol. *40:* 470, 1990).

[d] The species *M. thermoaggregans* was not included in *Bergey's Manual of Systematic Bacteriology*. See Blotevogel and Fischer (Arch. Microbiol. *142:* 218–222, 1985, and Int. J. Syst. Bacteriol. *38:* 221, 1988).

[e] The species *M. thermoalcaliphilum* was not included in *Bergey's Manual of Systematic Bacteriology*. See Blotevogel et al. (Arch. Microbiol. *142:* 211–217, 1985, and Int. J. Syst. Bacteriol. *38:* 220, 1988).

[f] See Schauer and Ferry (J. Bacteriol. *142:* 800–807, 1980), Zhilina and Illiarnov (Microbiology [USSR] *53:* 647–651, 1985), Touzel et al. (Arch. Microbiol. *149:* 291–296, 1988), and Zhao et al. (Appl. Environ. Microbiol. *52:*1227–1229, 1986).

[g] The species *M. thermophilum* was not included in *Bergey's Manual of Systematic Bacteriology*. See Laurinavichus et al. (Mikrobiologiya *57:* 1035–1041, 1987, and Int. J. Syst. Bacteriol. *40:* 320, 1990).

[h] Reported as using only H$_2$ + CO$_2$, but because compounds tested as possible substrates were not reported it is not clear whether alcohols were tested.

[i] Can oxidize propanol while reducing CO$_2$ to CH$_4$; this reaction may not be coupled to growth.

[j] Zellner and Winter (FEMS Microbiol. Lett. *44:* 323–328, 1987).

[k] Can grow by oxidizing 2–propanol while reducing CO$_2$ to CH$_4$; can also oxidize 2–butanol but without measurable growth.

[l] Touzel et al. (Arch. Microbiol. *149:* 291–296, 1988).

[m] Strain JF-1, identified as *M. formicicum* before the species *M. thermoformicicum* was described, may belong to the latter species; this strain grows optimally at about 55°C.

[n] Zhilina and Illiarnov (Microbiology [USSR] *53:* 647–651, 1985); and Zhao et al., Appl. Environ. Microbiol. *52:* 1227–1229, 1986.

Table 31.5   Differentiation of the species of the genus *Methanobrevibacter* [a]

Characteristic	M. arboriphilicus	M. ruminantium	M. smithii
Morphology:			
Width *ca.* 0.6 μm	+	+	+
Length <1.5 μm	−	+	+
Length 1–3 μm	+	−	−
Motility	d [b,c]	− [d]	+ [d]
Catabolic substrates:			
$H_2 + CO_2$	+	+	+
Formate	d [b]	+	+
Organic growth factors:			
Acetate	−	+	+
B vitamins	+	+	+
Coenzyme M	−	+	−
2-Methylbutyrate	−	+	−
Amino acids	+ [e]	+	−
pH optimum:			
6–7	−	+ [d]	−
7–7.5	−	−	+ [d]
7.5–8	+ [b]	−	−

[a] Symbols: see standard definitions.
[b] Morii et al. (Agric. Biol. Chem. *47:* 2781–2789, 1983).
[c] Doddema et al. (FEMS Microbiol. Lett. *5:* 135–138, 1979).
[d] Balch et al. (Microbiol. Rev. *43:* 260–296, 1979).
[e] Cysteine is required (Zehnder and Wuhrmann, Arch. Microbiol. *111:* 199–205, 1977).

Table 31.6   Differentiation of the species of the genus *Methanothermus* [a]

Characteristic	M. fervidus	M. sociabilis
Optimum growth temperature:		
83°C	+	−
88°C	−	+
Growth in large clusters (1–3 mm)	−	+
Nitrogen fixation		− [b]

[a] Symbols: see standard definitions.
[b] Lauerer et al. (Syst. Appl. Microbiol. *8:* 100–105, 1986).

Table 31.7 Differentiation of the species of the genus *Methanococcus*[a]

Characteristic	M. deltae[b]	M. igneus[c]	M. jannaschii	M. maripaludis	M. thermolithotrophicus	M. vannielii	M. voltae[d]
Motility	−[e]	−	+	+	+	+	+
Catabolic substrates:							
$H_2 + CO_2$	+	+	+	+	+	+	+
Formate	+	−	−	+	+	+	+
Secondary alcohols			−[f]	−[f]	−[g]	−[g]	−[f]
Organic growth factors:							
None	−	+	+	+	+	+	−
Acetate, leucine, isoleucine	+	−	−	−	−	−	+
Nitrogen source:							
NH_3	+	+	+	+	+	+	+
N_2	+			+	+	−	−
Alanine	+			+		−	−
Inorganic requirements:							
W			+[h]			+[i]	
Se		−	+[h]	−[j]	−[k]	+[i]	−[j]
Growth temperature:							
20°C	+	−	−	+	−	+	+
30°C	+	−	−	+	+	+	+
40°C	+	−	−	+	+	+	+
45°C	+	+	−	+	+	−	+
50°C	−	+	+	−	+	−	−
70°C	−	+	+	−	+	−	−
85°C	−	+	+	−	−	−	−
90°C	−	+	+	−	−	−	−
95°C	−	−	−	−	−	−	−
pH range:							
5–7		+	+	−	−	−	−
6.5–8	+[l]	−	−	+	+	−	+
7–9	+[l]	−	−	−	−	+	−

[a] Symbols: see standard definitions.

[b] The species *M. deltae* was validated after publication of *Bergey's Manual of Systematic Bacteriology.* See Corder et al. (Int. J. Syst. Bacteriol. *38:* 221, 1988).

[c] The species *M. igneus* was not included in *Bergey's Manual of Systematic Bacteriology.* See Burggraf et al. (Syst. Appl. Microbiol. *13:* 263–269, 1990, and Int. J. Syst. Bacteriol. *40:* 470, 1990).

[d] For emended description of *M. voltae*, see Ward et al. (Int. J. Syst. Bacteriol. *39:* 493–494, 1989).

[e] Nonmotile, but flagella are present.

[f] Widdel et al. (Arch. Microbiol. *150:* 477–481, 1988).

[g] Zellner and Winter (FEMS Microbiol. Lett. *44:* 323–328, 1987).

[h] Zhao et al. (Arch. Microbiol. *150:* 178–183, 1988).

[i] Pecher and Böck (FEMS Microbiol. Lett. *10:* 295–297, 1981).

[j] Stimulatory but not required.

[k] Required only for growth on formate.

[l] Grows at pH 7.2–7.4, but the growth range is unknown.

Table 31.8 Differentiation of the species of the genus *Methanocorpusculum*[a]

Characteristic	M. aggregans[b]	M. bavaricum[c]	M. labreanum[d]	M. parvum	M. sinense[c]
Catabolic substrates:					
$H_2 + CO_2$	+	+	+	+	+
Formate	+	+	+	+	+
Secondary alcohols + CO_2		+		+	−
Optimum temperature range:					
30–35°C	−	−	−	−	+
35–40°C	+	+	+	+	−

[a] Symbols: see standard definitions.

[b] The species *M. aggregans* was transferred to the genus *Methanocorpusculum* after publication of *Bergey's Manual of Systematic Bacteriology*. See Xun et al. (Int. J. Syst. Bacteriol. *39:* 109–111, 1989).

[c] The species *M. bavaricum* and *M. sinense* were not included in *Bergey's Manual of Systematic Bacteriology*. See Zellner et al. (Arch Microbiol. *151:* 381–390, 1989, and Int. J. Syst. Bacteriol. *39:* 371, 1989).

[d] The species *M. labreanum* was not included in *Bergey's Manual of Systematic Bacteriology*. See Zhao et al. (Int. J. Syst. Bacteriol. *39:* 10–13, 1989).

Table 31.9 Differentiation of the species of the genus *Methanoculleus*[a]

Characteristic	M. bourgense[b]	M. marisnigri[c]	M. olentangyi[d]	M. thermophilicum[e]
Catabolic substrates:				
$H_2 + CO_2$	+	+	+	+
Formate	+	+	+	+
Secondary alcohols + CO_2	d[f]	+[f,g]	−[f]	d[g]
Organic growth factors:				
None	−	−	−	−
Acetate	+	d[f]	+	+
Peptones	−	d[f,h]	−	+[h]
Yeast extract or vitamins	−	d[f,h]	−	+[h]
Optimum temperature range:				
<40°C	+	+	+	−
55–60°C	−	−	−	+

[a] Symbols: see standard definitions.

[b] The species *Methanogenium bourgense* was transferred to the genus *Methanoculleus* after publication of *Bergey's Manual of Systematic Bacteriology*. See Maestrojuán et al. (Int. J. Syst. Bacteriol. *40:* 117–122, 1990).

[c] The species *Methanogenium marisnigri* was transferred to the genus *Methanoculleus* after publication of *Bergey's Manual of Systematic Bacteriology*. See Maestrojuán et al. (Int. J. Syst. Bacteriol. *40:* 117–122, 1990).

[d] The species *Methanogenium olentangyi* was transferred to the genus *Methanoculleus* after publication of *Bergey's Manual of Systematic Bacteriology*. See Maestrojuán et al. (Int. J. Syst. Bacteriol. *40:* 117–122, 1990).

[e] The species *Methanogenium thermophilicum* was transferred to the genus *Methanoculleus* after publication of *Bergey's Manual of Systematic Bacteriology*. See Maestrojuán et al. (Int. J. Syst. Bacteriol. *40:* 117–122, 1990).

[f] Maestrojuán et al. (Int. J. Syst. Bacteriol. *40:* 117–122, 1990).

[g] Widdel et al. (Arch. Microbiol. *150:* 477–481, 1988).

[h] Either peptones, yeast extract, or vitamins are required.

Table 31.10   Differentiation of the species of the genus *Methanogenium*[a]

Characteristic	M. cariaci	M. liminatans[b]	M. organophilum[c]	M. tationis
Catabolic substrates:				
$H_2 + CO_2$	+	+	+	+
Formate	+	+	+	+
Ethanol or secondary alcohols + CO_2	–[d]	+	+[d]	+[e]
Organic growth factors:				
None	–	–	–	–
Acetate	+	+	+	+
Peptones	–	–	–	+[f]
Yeast extract or vitamins	+	–	+	+[f]
Growth temperature:				
Growth at 15°C	+	–	+	–
Optimum 20–30°C	+[g]	–	–	–
Optimum 30–37°C	+[g]	–	+	–
Optimum 35–40°C	+[g]	+	–	+
Salt requirements:				
Growth with 0.1 M NaCl	–	+	–	+
Optimum <0.3 M NaCl	–	+	–	+
Optimum 0.3–0.6 M NaCl	+	+	+	–

[a] Symbols: see standard definitions.

[b] The species *M. liminatans* was not included in *Bergey's Manual of Systematic Bacteriology*. See Burggraf et al. (Syst. Appl. Microbiol. *13:* 263–269, 1990, and Int. J. Syst. Bacteriol. *40:* 470, 1980).

[c] The species *M. organophilum* was not included in *Bergey's Manual of Systematic Bacteriology*. See Widdel et al. (Arch. Microbiol. *150:* 477–481, 1988, and Int. J. Syst. Bacteriol. *39:* 93, 1989).

[d] Zellner and Winter (FEMS Microbiol. Lett. *44:* 323–328, 1987).

[e] Maestrojuán et al. (Int. J. Syst. Bacteriol. *40:* 117–142, 1990).

[f] Growth requires the presence of either peptones or yeast extract.

[g] Good growth occurs at the lower temperature range, but cultures may adapt to the higher range. See Zabel et al. (Arch. Microbiol. *137:* 308–315, 1984) and Maestrojuán et al. (Int. J. Syst. Bacteriol. *40:* 117–142, 1990).

Table 31.11   Differentiation of the species of the genus *Methanoplanus*[a]

Characteristic	M. endosymbiosus	M. limicola
Catabolic substrates:		
$H_2 + CO_2$	+	+
Formate	+	+
Ethanol or secondary alcohols + CO_2		–
Temperature optimum:		
Optimum 30°C	+	–
Optimum 40°C	–	+

[a] Symbols: see standard definitions.

Table 31.12 Differentiation of the species of the genus *Methanohalophilus*[a]

Characteristic	M. halophilus[b]	M. mahii[c]	M. oregonensis[d]	M. zhilinae[e]
Catabolic substrates:				
Methylamines	+	+	+	+
Methanol	+	+	+	+
Methylsulfides	−[f]		+	+
Organic growth factors:				
Poor growth with no additions	+	+[g]	+	−
Good growth with no additions	−	−[g]	−	+
Stimulation by biotin	+	+[g]	+[g]	−
Stimulation by *p*-aminobenzoate	−	−		−
Optimal growth temperature:				
35–40°C	+	+	+	−
35–45°C	−	−	−	+
Optimal NaCl concentration:				
0.5–1.5 M	−	−	+	+
1–2.5 M	+	+	−	−
Optimal pH:				
7.5	+	+	−	−
8.5–9.5	−	−	+	+

[a] Symbols: see standard definitions.

[b] The species *Methanococcus halophilus*, described in *Bergey's Manual of Systematic Bacteriology*, has since been transferred to the genus *Methanohalophilus* as *M. halophilus* (Wilharm et al., Int. J. Syst. Bacteriol. *41*: 588–562, 1991).

[c] The species *M. mahii* was not included in *Bergey's Manual of Systematic Bacteriology*. See Paterek and Smith (Int. J. Syst. Bacteriol. *38*: 122–123, 1988).

[d] The species *M. oregonensis* was not included in *Bergey's Manual of Systematic Bacteriology*. See Liu et al. (Int. J. Syst. Bacteriol. *40*: 111–116, 1990).

[e] The species *M. zhilinae* was not included in *Bergey's Manual of Systematic Bacteriology*. See Mathrani et al. (Int. J. Syst. Bacteriol. *38*: 139–142, 1988).

[f] Ni and Boone, 10th Int. Symp. Environ. Biogeochem., San Francisco, 1991).

[g] Liu and Boone (unpublished).

Table 31.13 Differentiation of the species of the genus _Methanolobus_[a]

Characteristic	_M. siciliae_[b]	_M. tindarius_	_M. vulcani_[c]
Catabolic substrates:			
Methylamines or methanol	+	+	+
Methylsulfides	+		−[d]
Growth temperature:			
Optimum at 37°C	+	+	+
Growth at 40°C	+	+	+
Growth at 45°C	+	−	+
Growth at 48°C	+	−	−

[a] Symbols: see standard definitions.

[b] The species _M. siciliae_ was proposed in _Bergey's Manual of Systematic Bacteriology_ although it had not been characterized; later this was done (Ni and Boone, Int J. Syst. Bacteriol. _41:_ 410–416, 1991).

[c] The species _M. vulcani_ was proposed in _Bergey's Manual of Systematic Bacteriology_ although it was characterized beyond the characters indicated here.

[d] Ni and Boone (10th Int. Symp. Environ. Biogeochem., San Francisco, 1991).

Table 31.14 Differentiation of the species of the genus _Methanosarcina_[a]

Characteristic	_M. acetivorans_	_M. barkeri_	_M. frisia_[b]	_M. mazei_	_M. thermophila_	_M. vacuolata_[c]
Catabolic substrates:						
Methylamines or methanol	+	+	+	+	+	+
$H_2 + CO_2$	−	+	+	+,v	+,v	+
Acetate	+	+	−	d	+	+
Organic growth factors:						
None	+	+	+	d[d]	−	+
_p_–Aminobenzoate	−	−	−	−	+	−
Nitrogen fixation	+[e]				−[f]	
Optimal growth temperature:						
35–40°C	+	+	+	+	−	+
50–55°C	−	−	−	−	+	−
Optimal pH	6.5[g]	6[g]	6.5–7.2	7[g]	7	6[g]
Optimum NaCl concentration:						
<0.15 M	−	+[g]	−	+[g]	+[h]	+[g]
0.15–0.6 M	+	−[g]	+	−[g]	+[h]	−[g]

[a] Symbols: see standard definitions.

[b] The species _M. frisus_ was classified in _Bergey's Manual of Systematic Bacteriology_ as a species of _Methanococcus_. It was later transferred to the genus _Methanosarcina_ as _M. frisia_ (Blotevogel and Fischer, Int. J. Syst. Bacteriol. _39:_ 91–92, 1989).

[c] The species _M. vacuolata_ was not included in _Bergey's Manual of Systematic Bacteriology_. See Zhilina and Zavarzin (Int. J. Syst. Bacteriol. _37:_ 281–283, 1987).

[d] One strain requires yeast extract (Liu et al., Appl. Environ. Microbiol. _49:_ 608–613, 1985).

[e] Murray and Zinder (Nature _312:_ 284–286, 1984), Bomar et al. (FEMS Microbiol. Lett. _31:_ 47–55, 1985), and Lobo and Zinder (Appl. Environ. Microbiol. _54:_ 1656–1661, 1988).

[f] Murray and Zinder (Appl. Environ. Microbiol. _50:_ 49–55, 1985).

[g] Maestrojuán and Boone (Int. J. Syst. Bacteriol. _41:_ 267–274, 1991).

[h] Cells can adapt and grow well at Na[+] concentrations from near zero to 1.2 M (Sowers and Gunsalus, J. Bacteriol. _170:_ 998–1002, 1988).

Table 31.15 Differentiation of the species of the genus *Methanothrix*[a]

Characteristic	M. soehngenii	M. thermophila
Optimum growth temperature:		
35–40°C	+	–
55–65°C	–	+
Vitamin requirements:		
Biotin	+	ND
Mixture	–	+

[a] Symbols: see standard definitions.

Important Notes for Users of This Manual

Unless otherwise indicated in footnotes to tables, the meanings of symbols are as follows:

+ 90% or more of strains are positive

– 90% or more of strains are negative

d 11-89% of strains are positive

v strain instability (not equivalent to "d")

D Different reactions in different taxa (species of a genus or genera of a family)

All other symbols are defined in footnotes to tables.

GROUP 32

ARCHAEAL SULFATE REDUCERS

These are irregular coccoid cells, often triangular, 0.4–1.3 μm in diameter, that occur singly or in pairs. Flagella may be present or absent. Cells stain **Gram negative. Blue-greenish fluorescence occurs at 420 nm.** Cells form greenish black, smooth colonies with a diameter of 1–2 mm. Cells are **strictly anaerobic.** They show chemolithotrophic, chemoorganotrophic, or chemomixotrophic growth. **Autotrophic growth occurs with thiosulfate and H$_2$, but with sulfate there is very little growth.** Under heterotrophic conditions formate, lactate, glucose, starch, and proteins are used as electron donors and sulfate, sulfite, or thiosulfate can function as electron acceptors. H$_2$S is formed. S^0 can be reduced, but no growth is obtained. S^0 inhibits growth in the presence of sulfate, sulfite, and thiosulfate. Temperature range is 60–95°C, with the optimum around 83°C, and pH range is 4.5–7.5, with the optimum around 6. Salt range is (NaCl) 0.9–3.6%. The species are isolated from shallow (near Vulcano, Italy) and abyssal marine hydrothermal systems (Guaymas hot vent area, Gulf of California, Mexico).

Genus **Archaeoglobus**

The characteristics are as described for the group.

Type species: *Archaeoglobus fulgidus.*

Differentiation of the species of the genus **Archaeoglobus:**

1. Chemolithotrophic growth in the presence of H$_2$, CO$_2$, and thiosulfate. Monopolar polytrichous flagellation. Low amounts of methane (up to 0.1 μm/ml) formed.

   A.   *A. fulgidus*

2. Obligate mixotrophic growth requiring H$_2$ and an organic carbon source (e.g., acetate). Flagella absent. No methane formed.

   A.   *A. profundus*

Editorial note: The species of *Archaeoglobus* were not included in *Bergey's Manual of Systematic Bacteriology.* The species were created in 1988 (*A. fulgidus*) by Stetter (Int. J. Syst. Bacteriol. *38:* 328-329, 1988; effective publication: Syst. Appl. Microbiol. *10:* 172-173) and in 1990 (*A. profundus*) by Burggraf et al. (Int. J. Syst. Bacteriol. *40:* 320-321, 1990; effective publication: Syst. Appl. Microbiol. *13:* 24-28).

Important Notes for Users of This Manual

Unless otherwise indicated in footnotes to tables, the meanings of symbols are as follows:

- + 90% or more of strains are positive
- − 90% or more of strains are negative
- d 11-89% of strains are positive
- v strain instability (not equivalent to "d")
- D Different reactions in different taxa (species of a genus or genera of a family)

All other symbols are defined in footnotes to tables.

EXTREMELY HALOPHILIC, AEROBIC ARCHAEOBACTERIA (HALOBACTERIA)

Coccoid or irregular rod-shaped bacteria, 0.8–2.0 μm for coccoid forms, 0.3–1.2 x 1.0–15.0 μm for rod-shaped forms. Motile by tufts of polar flagella, or they are nonmotile. Stain Gram negative (rods) or Gram variable (cocci). Cocci occur singly or in pairs, tetrads, or irregular refractile clusters where the outlines of individual cells are indistinct (Figure 33.1). The majority of rod-shaped forms have a characteristic flat cell morphology and exhibit a multitude of pleomorphic forms from regular rod or ribbon-like cells (Figure 33.2) to disks, irregular triangles, or rectangles (Figure 33.3). The cells of the rod-shaped forms lyse when suspended in distilled water and may exhibit spherical morphology in agar-grown culture or under adverse conditions. Gas vacuoles may be present. **Colonies are various shades of red because of the presence of carotenoid pigments** and may become pink or white if gas vacuoles are produced.

Aerobic; some are able to grow anaerobically in the presence of nitrate. **Chemoheterotrophic.** Carbohy-drates, alcohols, carboxylic acids, or amino acids serve as carbon and energy sources.

Require at least 1.5 M NaCl for growth, most growing optimally at 2–4 M NaCl. **Some members are alkaliphilic,** growing only at pH > 8.5. They occur in nature when the salt concentration is high (i.e., in salt lakes, soda lakes, salterns, and saline soils). One type occurs in proteinaceous products heavily salted with solar salt.

Major differential features of the genera in **Group 33:** See Table 33.1

Current genera are largely defined by chemotaxonomic criteria, notably polar lipid composition. The lipids of all isolates to date contain diphytanyl (C_{20},C_{20}) or phytanyl sesterterpanyl (C_{20},C_{25}) derivatives of phosphatidyl glycerol (PG) and phosphatidyl glycerol phosphate (PGP). The presence or absence of phosphatidyl glycerol sulfate (PGS) and of particular representatives of a family of glycolipids based on a mannosylglucosyl

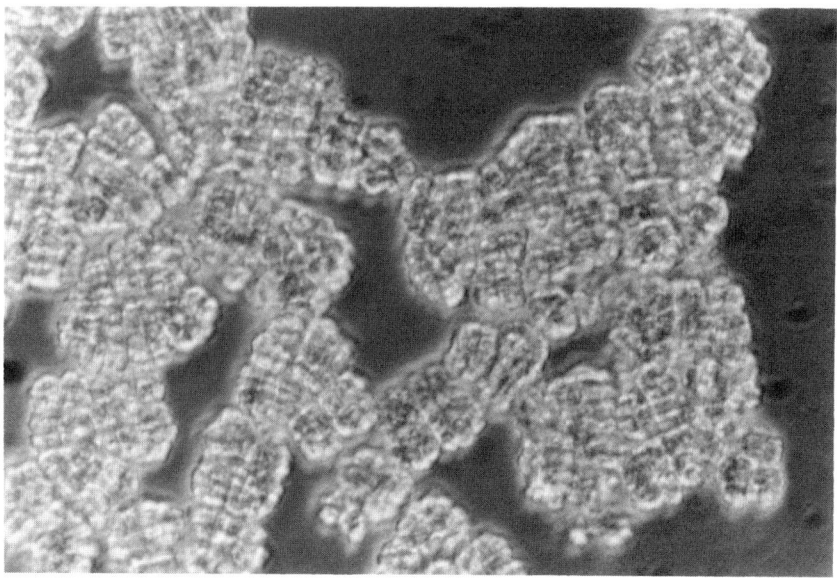

Figure 33.1. *Halococcus morrhuae.* Phase.

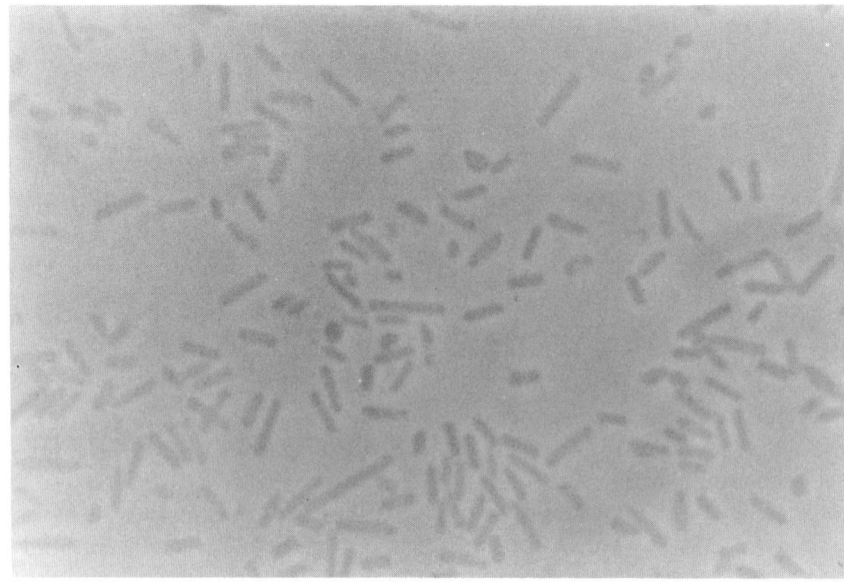

Figure 33.2. *Halobacterium salinarium.* Phase. (Courtesy of A. Ventosa and M. C. Gutierrez)

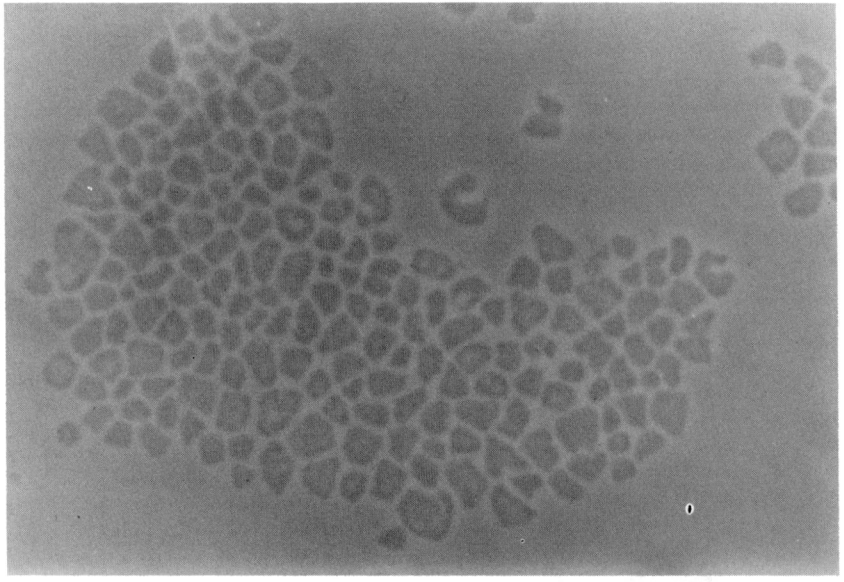

Figure 33.3. *Haloarcula* sp. Phase. (Courtesy of A. Ventosa and M. C. Gutierrez)

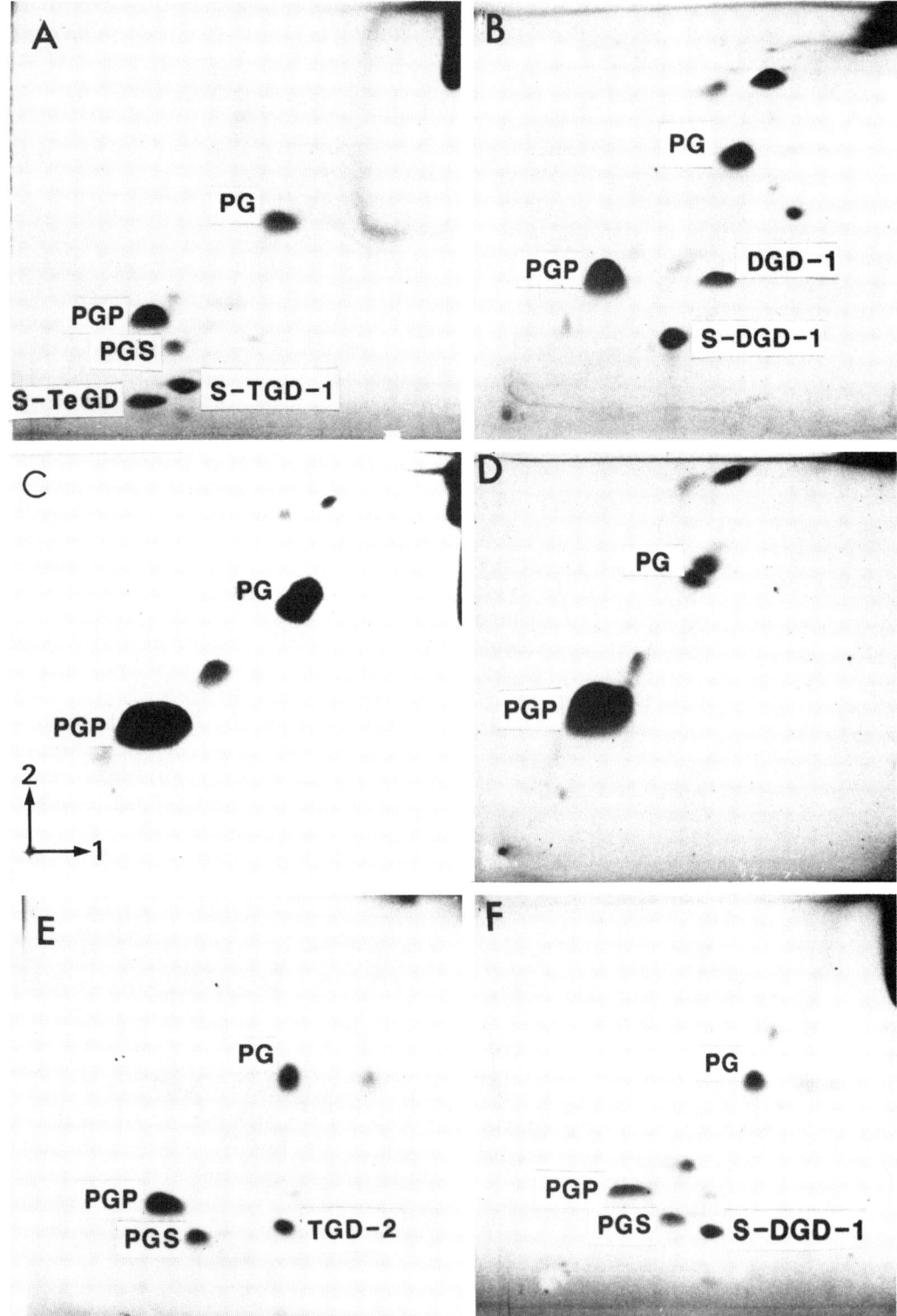

Figure 33.4. Polar lipid patterns of halobacteria. A. *Halobacterium salinarium;* **B.** *Haloferax volcanii;* **C.** *Natronococcus occultus;* **D.** *Natronobacterium gregoryi;* **E.** *Haloarcula vallismortis;* **F.** *Halobacterium saccharovorum.* **S-DGD-1,** sulfated mannosyl glucosyl glycolipid; **S-TGD-1,** sulfated galactosyl mannosyl glucosyl glycolipid; **TGD-2,** glucosyl mannosyl glucosyl glycolipid; **S-TeGD,** sulfated digalactosyl mannosyl glucosyl glycolipid; **PG,** phosphatidyl glycerol; **PGP,** phosphatidyl glycerol phosphate; **PGS,** phosphatidyl glycerol sulfate. For structures and methods see *Bergey's Manual of Systematic Bacteriology.* C_{20},C_{20} and C_{20},C_{25} forms of PG are resolved for those organisms possessing both core lipids (Figure 33.4C,D).

diphytanyl glycerol is used to readily assign isolates to genera. Figure 33.4 illustrates polar lipid patterns characteristic of particular genera or groups.

Identification within genera is often not an easy task. There has been a lack of standardization in differential test methods, in particular over growth enhancement/acid production in the presence of carbon and energy sources. DNA/DNA hybridization with appropriate type strains is usually necessary to establish relationships unequivocally.

Genus **Haloarcula**

Irregular rods (0.3–1.0 × 1.0–6.0 μm) or flat triangles or rectangles 1.0–3.0 μm across. Some strains are motile. Stain Gram negative. Aerobic. **Chemoheterotrophic utilizing carbohydrates,** carboxylic acids, alcohols, and amino acids as carbon and energy sources. **Acids are produced from sugars.** The polar lipids are characterized by C_{20},C_{20} derivatives of TGD-2. Found in neutral saline environments such as salt lakes, marine salterns, and saline soils.

Type species: *Haloarcula vallismortis.*

Differentiation of the species of the genus **Haloarcula:** See Table 33.2.

Editorial note: The species *Haloarcula japonica* was not included in *Bergey's Manual of Systematic Bacteriology.* The species was described by Takashina et al. (Int. J. Syst. Bacteriol. *41:* 178-179, 1991; effective publication: Syst. Appl. Microbiol. *13:* 177-181, 1990). *Haloarcula marismortui* is the organism previously referred to as *"Halobacterium marismortui"* in *Bergey's Manual of Systematic Bacteriology.* Reclassification and deposition of a type culture was carried out by Oren et al. in 1990 (Int. J. Syst. Bacteriol. *40:* 209-210).

Genus **Halobacterium**

Irregular rod-shaped cells 0.5–1.2 x 1.0–6.0 μm. Motile. Some isolates have gas vacuoles. Stain Gram negative. Aerobic. Chemoheterotrophic. **Amino acids are required for growth. Carbohydrates are not utilized. Strongly proteolytic.** The polar lipids are characterized by C_{20},C_{20} derivatives of S-TGD-1 and S-TeGD. Found in neutral, highly saline environments

such as salt lakes and marine salterns and associated with the spoilage of salted fish and hides.

Type species: *Halobacterium salinarium.*

Differentiation of **H. salinarium** *from* **species incertae sedis:** See Table 33.3.

Editorial note: Halobacterium as redefined in *Bergey's Manual of Systematic Bacteriology* contains *Halobacterium salinarium* as the only species. The taxonomic position of the *species incertae sedis* listed under *Halobacterium* in *Bergey's Manual of Systematic Bacteriology* remains to be unequivocally established. *Halobacterium saccharovorum* has a characteristic polar lipid pattern (Lanzotti et al., FEMS Microbiol. Lett. *55:* 223-228, 1988) that it shares with the type strain of *Halobacterium trapanicum* (Figure 33.4F). However, care should be exercised over *H. trapanicum* strains since at least one nontype strain exists that has quite different properties.

Recent 16S rRNA sequence comparisons have indicated that *Halobacterium saccharovorum* is distinct from other halobacterial groups (Lodwick et al., Syst. Appl. Microbiol. *14:* 352-357, 1991). *Halobacterium sodomense,* although in the same rRNA hybridization group as *Halobacterium trapanicum* and *Halobacterium saccharovorum,* has a sulfated diglycosyl lipid based on a mannosyl glycosyl diether glycerol but with different linkages (Trincone et al., J. Gen. Microbiol. *136:* 2327-2331, 1990).

The species *Halobacterium lacusprofundi* was not included in *Bergey's Manual of Systematic Bacteriology.* The species was described by Franzmann et al. (Int. J. Syst. Bacteriol. *39:* 205-206, 1988; effective publication: Syst. Appl. Microbiol. *11:* 20-27, 1988). However, lipid analyses (Tindall, FEMS Microbiol. Lett. *66:* 199-201, 1990) and 16S rRNA analyses (Holmes et al., Nucleic Acid Res. *7:* 4607, 1990) indicate that the organism is not a member of the genus *Halobacterium sensu strictu* and has more affinity with the *Halobacterium saccharovorum* group. The species *Halobacterium denitrificans* described in *Bergey's Manual of Systematic Bacteriology* has now been transferred to the genus *Haloferax* (Tindall et. al., Int. J. Syst. Bacteriol. *39:* 359-360, 1989). The species *Halobacterium distributus,* published in 1988 (Zvyagintseva and Tarasov, Mikrobiologiya *56:* 839-844; Int. J. Syst. Bacteriol. *39:* 495-497, 1989) has an insufficiently detailed description to allow unequivocal assignment to a particular genus.

Genus Halococcus

Cocci, 0.8–1.5 μm in diameter, occur singly or as pairs, tetrads, or irregular clusters. **Nonmotile.** Stain Gram variable. **Cells are stable when suspended in distilled water. Aerobic.** Chemoheterotrophic. The polar lipids are characterized by C_{20},C_{20} and C_{20},C_{25} derivatives. PGS is absent. Found in neutral salt lakes, marine salterns, and saline soils. Have also been isolated from seawater.

Type species: *Halococcus morrhuae.*

Differentiation of the species of the genus **Halococcus:** See Table 33.4.

Editorial note: The species *H. saccharolyticus* was not included in *Bergey's Manual of Systematic Bacteriology.* The species was described by Montero et al. (Int. J. Syst. Bacteriol. *40:* 105-106, 1990; effective publication: Syst. Appl. Microbiol. *12:* 167-171, 1990). The species *H. turkmenicus,* published in 1988 (Zvygintseva and Tarasov, Mikrobiologiya, *56:* 839-844; Int. J. Syst. Bacteriol. *39:* 495-497, 1989), has an insufficiently detailed description to allow a clear distinction between this organism and *H. saccharolyticus.*

Genus Haloferax

Irregular rods, disks, or cups (1.0–3.0 x 2.0–3.0 μm). Motile. One isolate has gas vacuoles. Stain Gram negative. Aerobic. **Chemoheterotrophic, utilizing carbohydrates,** carboxylic acids, alcohols, and amino acids as carbon and energy sources. **Acid is produced from sugars. Polyhydroxybutyrate is accumulated under certain conditions.** The polar lipids are characterized by C_{20},C_{20} derivatives of S-DGD-1. PGS is absent. Found in neutral saline environments such as salt lakes and marine salterns.

Type species: *Haloferax volcanii.*

Differentiation of the species of the genus **Haloferax:** See Table 33.5.

Editorial note: The species *Haloferax denitrificans,* originally classified in the genus *Halobacterium* as *Halobacterium denitrificans,* was reclassified in 1989 by Tindall et al. (Int. J. Syst. Bacteriol. *39:* 359-360).

Genus Natronobacterium

Rods in liquid culture, 0.5–1.0 x 2–15 μm. Some strains are motile. Aerobic. Stain Gram negative. **Alkaliphilic,** with no growth occurring below pH 8.5. Chemoheterotrophic. The polar lipids are characterized by C_{20},C_{20} and C_{20},C_{25} derivatives. Glycolipids and PGS are absent. Found in soda lakes, alkaline salterns, and soda soils.

Type species: *Natronobacterium gregoryi.*

Differentiation of the species of the genus **Natronobacterium:** See Table 33.6.

Genus Natronococcus

Cocci, 1.0–2.0 μm in diameter, occur singly or in pairs, tetrads, or irregular clusters. **Nonmotile.** Stain Gram variable. Cells suspended in distilled water leak cell contents but appear intact. Aerobic. **Alkaliphilic,** with no growth occurring below pH 8.5. Chemoorganotrophic. The polar lipids are characterized by C_{20},C_{20} and C_{20},C_{25} derivatives. PGS and glycolipids are absent. Found in soda lakes, alkaline salterns, and soda soils.

Type (and only) species: *Natronococcus occultus.*

Characteristics of the species: As described for the genus.

Table 33.1 Differentiation of the genera of extremely halophilic archaeobacteria [a]

Characteristic	Haloarcula	Halobacterium	Halococcus	Haloferax	Natronobacterium	Natronococcus
Cell shape	Irregular rods, triangles, rectangles	Irregular rods	Cocci	Irregular rods, disks	Irregular rods	Cocci
Cell dimensions (μm)	0.3–4.0 x 0.5–5.0	0.5–1.2 x 1.0–6.0	0.8–1.5	0.4–3.0 x 0.6–3.0	0.5–1.0 x 2.0–15.0	1.0–2.0
Gram stain	Negative	Negative	Variable	Negative	Negative	Variable
Motile	D	+	−	D	D	−
pH range for growth	5.0–8.0	5.0–8.0	5.0–8.0	5.0–8.0	8.5–11.0	8.5–11.0
Carbohydrates used as carbon and energy sources	+	−	D	+	D	−
Glycerol ether core	C_{20},C_{20}	C_{20},C_{20}	C_{20},C_{20}^{+}	C_{20},C_{20}	C_{20},C_{20}^{+}	C_{20},C_{20}^{+}
Lipid(s)			C_{20},C_{25}		C_{20},C_{25}	C_{20},C_{25}
PGS present	+	+	−	−	−	−
Glycolipid(s) present	TGD-2	S-TGD-1 S-TeGD	S-DGD-1	S-DGD-1	−	−

[a] Symbols: see standard definitions.

Table 33.2 Differential characteristics of the species of the genus Haloarcula [a]

Characteristic	H. hispanica	H. japonica	H. marismortui	H. vallismortis
Cell shape	Pleomorphic rods	Irregular flat triangles, rectangles	Irregular flat disks, rectangles	Pleomorphic rods
Cell dimensions (μm)	0.3–0.4 x 0.5–1.0	1.5–2.5 x 4.0–5.0	1.5–2.0 x 2.0–4.0	0.6–1.0 x 3.0–5.0
Motile	+	+	−	+
Starch hydrolysis	+	−	+*	+*
Gelatin hydrolysis	+	−	−	−
Indole	d	+	−	+
Acid produced from:				
Mannose	d	−	−	−
Mannitol	+	+	+	−
Arabinose	+	+	−[b]	−[b]

[a] Symbols: see standard definitions.
[b] Conflicting reports.

Table 33.3 Differential characteristics of *Halobacterium salinarium* and *species incertae sedis*[a]

Characteristic	H. salinarium	H. lacusprofundi	H. saccharovorum	H. sodomense	H. trapanicum
Cell shape	Pleomorphic rods	Pleomorphic rods	Rods	Rods	Pleomorphic rods
Cell dimensions (μm)	0.5–1.0 x 1.0–6.0	0.7–1.0 x 2.0–12.0	0.6–1.2 x 2.0–2.5	0.5–0.7 x 2.5–5.0	0.7–1.0 x 1.5–3.0
Motile	+	d	+	+	−
Presence of gas vacuoles	d	−	−	−	−
Growth at 10°C	−	+	−	−	−
Starch hydrolysis	−	−	−	+	−
Gelatin hydrolysis	+	−	−	−	−
Nitrate reduction	−	+	−[b]	+	+
Acid produced from:					
Glucose	−	−	+	+	−
Galactose	−	−	+	−	−
Lactose	−	−	+	−	−
Xylose	−	−	+	+	−
Mannitol	−	−	+	−	−

[a] Symbols: see standard definitions.
[b] Conflicting reports.

Table 33.4 Differential characteristics of the species of the genus *Halococcus*[a]

Characteristic	H. morrhuae	H. saccharolyticus
Tween 80 hydrolysis	+	−
Acid production from glucose	−	+
Utilization of organic compounds as carbon and energy source:		
Glucose	−	+
Lactose	−	+
Ethanol	−	+
Propanol	−	+

[a] Symbols: see standard definitions.

Table 33.5 Differential characteristics of the species of the genus _Haloferax_[a]

Characteristic	H. denitrificans	H. gibbonsii	H. mediterranei	H. volcanii
Cell shape	Pleomorphic rods, disks	Pleomorphic rods	Pleomorphic rods	Flattened disks, cups
Cell dimensions (μm)	0.8–1.5 x 2.0–3.0	0.4–0.6 x 2.0–2.5	0.5–1.0 x 2.0–3.0	1.0–3.0 x 2.0–3.0
Motile	−	−	+	−
Presence of gas vacuoles	−	−	+	−
Starch hydrolysis	−	−	+	−
Gelatin hydrolysis	+	−	+	−
Tween 80 hydrolysis	−	+	+	−
Nitrate reduction	+	−	+	−[b]
Indole	−	+	+	+
Acid produced from galactose	+	+	d	−

[a] Symbols: see standard definitions.
[b] Conflicting reports.

Table 33.6 Differential characteristics of the species of the genus _Natronobacterium_[a]

Characteristic	N. gregoryi	N. magadii	N. pharaonis
Cell shape	Rods	Pleomorphic rods	Pleomorphic rods
Cell dimensions (μm)	0.5–0.7 x 10.0–15.0	0.8–1.0 x 1.0–3.0	0.7–0.9 x 2.0–4.0
Motile	−	+	+
Nitrate reduction	−	−	+[b]
Gelatin hydrolysis	−	−	+
Growth stimulated by:			
Acetate	+	+	−[b]
Ribose	+	−	−
Fructose	+	−	−
Mannitol	+	−	−
Sucrose	+	−	−
Sensitivity to:			
Erythromycin	+	+	−
Flavomycin	+	+	−

[a] Symbols: see standard definitions.
[b] Conflicting reports.

CELL WALL-LESS ARCHAEOBACTERIA

Genus **Thermoplasma**

Pleomorphic cells range from spheres (0.1–5 μm in diameter) to filaments. Cells **lack a cell wall** and are bound by the cell membrane, ~7 nm thick. The cell **membrane contains ether lipids** with 40-carbon isoprenoid-branched diglycerol tetraethers. Cells are Gram negative and **may be motile and flagellated.**

Obligately thermophilic, *Thermoplasma* cells grow at 33–67°C; **obligately acidophilic,** they grow at pH 0.5–4. Cells lyse at neutral pH and grow in salt solutions up to half-strength marine water.

Cells are **facultatively anaerobic.** Anaerobic growth is enhanced with elemental sulfur, which is reduced to H$_2$S. Cells are **chemoorganotrophic, requiring yeast extract** for growth. *Thermoplasma* species are resistant to the antibiotics ampicillin, streptomycin, bacitracin, vancomycin, chloramphenicol, and rifampicin. Elonga-

tion factor G is ADP-ribosylated by diphtheria toxin. Colonies on agar media are small (0.3-mm diameter), brown, flat, and granular and may show a typical "fried egg" appearance at pH 2. *Thermoplasma* species are isolated from self-heating coal refuse piles and acidic solfatara fields.

Type species: *Thermoplasma acidophilum.*

Differentiation of species in the genus **Thermoplasma:** See Table 34.1.

Table 34.1   Differentiation between species of the genus *Thermoplasma*

Characteristic	*T. acidophilum*	*T. volcanicum*
Temperature range for growth	45–63°C	33–67°C
Mol% G + C	~46	~38

Important Notes for Users of This Manual

Unless otherwise indicated in footnotes to tables, the meanings of symbols are as follows:

+ 90% or more of strains are positive

− 90% or more of strains are negative

d 11-89% of strains are positive

v strain instability (not equivalent to "d")

D Different reactions in different taxa (species of a genus or genera of a family)

All other symbols are defined in footnotes to tables.

EXTREMELY THERMOPHILIC AND HYPERTHERMOPHILIC S⁰-METABOLIZERS

These **rods, filaments, cocci, or disk-shaped cells** show no evidence of spores or resting stages. All species stain Gram negative. **Motile or nonmotile** cells are **aerobic, strictly anaerobic, or facultatively anaerobic** and show chemoautotrophic or chemoheterotrophic growth. **Under anaerobic conditions S⁰ is reduced to H₂S; under aerobic conditions H₂S or S⁰ is oxidized to H₂SO₄. H₂ or organic compounds serve as electron donors. The growth temperature is 45–110°C,** with optimum growth at 70–105°C. No mesophilic species are known. Habitats are continental solfatara fields or marine hydrothermal systems.

Differentiation of the subgroups and genera in **Group 35:** See Table 35.1.

SUBGROUP 1

Subgroup 1 consists of four genera (see Table 35.1) and includes the thermoacidophilic aerobic or facultatively aerobic coccoid organisms, which grow optimally below pH 4.0.

Genus **Acidianus**

Irregularly lobed cocci occur almost singly. They are 0.5–2 μm in diameter and are Gram negative. Flagella are absent. The **facultatively anaerobic** cells show lithotrophic or chemotrophic growth. **Lithotrophic aerobic growth is via S⁰ oxidation. Lithotrophic anaerobic growth is by means of S⁰ reduction with H₂ as the electron donor.** Growth occurs at a **pH of 1–6, a temperature of 45–96°C,** and a salt range (NaCl) of 0.1–4%.

Type species: *Acidianus infernus.*

Differentiation of the two species of the genus **Acidianus:**

1. Optimal temperature for growth is 88°C with a maximum at 96°C. Aerobic growth occurs strictly lithotrophically by sulfur oxidation.

   A. infernus

2. Optimal temperature for growth is around 70°C with a maximum at 75°C. Aerobic growth occurs lithotrophically by sulfur oxidation or chemotrophically by oxidation of organic material (e.g., yeast extract).

   A. brierleyi

Genus **Desulfurolobus**

Editorial note: The genus *Desulfurolobus* was not included in *Bergey's Manual of Systematic Bacteriology*. The genus was created in 1986 by Zillig et al. (Syst. Appl. Microbiol. *8:* 197–203).

Type (and only) species: *D. ambivalens.*

Characteristics of the species: As described for the genus *Acidianus*. The genus *Desulfurolobus* cannot be readily distinguished from the genus *Acidianus*.

Differentiation of **D. ambivalens** *and the two species of* **Acidianus:** *D. ambivalens* may be distinguished from *A. infernus* and *A. brierleyi* by its optimal growth temperature of 80°C and its maximal growth temperature of 87°C.

Genus **Metallosphaera**

Editorial note: The genus *Metallosphaera* was not included in *Bergey's Manual of Systematic Bacteriology*.

The genus was created in 1989 by Huber et al. (Syst. Appl. Microbiol. *12:* 38–47).

Regular to slightly irregular coccoid cells are 0.8–1.2 µm in width. Cells are Gram negative. Flagella are absent, but pilus-like structures are found. **Strict aerobic growth** occurs at a pH of 1.0–4.5. **The growth temperature is 50–80°C, with an optimum around 75°C.** There is both chemolithoautotrophic and organotrophic growth. **Chemolithoautotrophic growth occurs on sulfidic ores** such as pyrite, sphalerite, chalcopyrite, or elemental sulfur. Metallosphaeras extract metal ions from sulfidic ores and produce sulfuric acid. Organotrophic growth occurs on yeast extract, beef extract, peptone, tryptone, and casamino acids. Sugars are not used. The organisms have been isolated from solfataric fields in Italy.

Type (and only) species: *Metallosphaera sedula.*

Characteristics of the species: As described for the genus.

Genus **Sulfolobus**

Highly irregular lobed cocci usually occur singly. Cells are 0.8–2 µm in diameter. **Pilus-like and pseudopodium-like structures are often found.** Flagella are absent. Growth is **strictly aerobic** and at low ionic strength. Under microaerophilic conditions Fe^{3+} and MoO_4^{2-} can be used as electron acceptors. The **pH range is 1–6,** and the **temperature range is 55–87°C.** Lithotrophic and organotrophic growth both occur. **Lithotrophic growth occurs via oxidation of sulfide or tetrathionate.** Produce sulfuric acid. S^0 is oxidized or not oxidized. Organotrophic growth occurs by oxidation of complex organic material like yeast extract, sugars, or amino acids. *Sulfolobus* species were isolated worldwide from continental solfatara fields.

Type species: *Sulfolobus acidocaldarius.*

Differentiation of the species of the genus **Sulfolobus:**

1. Optimal growth temperature around 80°C.

    S. acidocaldarius

2. Optimal growth temperature at 87°C.

    S. solfataricus

3. *Sulfolobus shibatae.*

   The species *Sulfolobus shibatae* cannot be readily distinguished from the other two species of the genus by chemical methods.

Editorial note: The species *Sulfolobus shibatae* was not included in *Bergey's Manual of Systematic Bacteriology.* The species was created in 1990 by Grogan et al. (Arch. Microbiol. *154:* 594–599).

SUBGROUP 2

Subgroup 2 consists of three genera (see Table 35.1). The rod-shaped or filamentous organisms are strict anaerobes, which grow optimally above pH 4.

Genus **Pyrobaculum**

Editorial note: The genus *Pyrobaculum* was not included in *Bergey's Manual of Systematic Bacteriology.* The genus was created in 1987 by Huber et al. (Arch. Microbiol. *149:* 95–101).

Rod-shaped cells with rectangular ends occur singly and in V-, X-, and raft-shaped aggregates. Terminal spheres appear during the exponential growth phase. Cells are 1.5–8 µm in length and about 0.5 µm in width and are motile due to peritrichous or bipolar polytrichous flagellation. Gram-negative cells form gray or greenish black colonies. The **temperature range is 74–103°C, with an optimum at 100°C; the pH range is 5–7;** the salt range (NaCl) is 0–0.8%. **Strict anaerobes,** these cells show facultative chemolithoautotrophic or strict heterotrophic growth. **Molecular hydrogen or complex organic substances are used as electron donors.** Heterotrophic growth occurs on yeast extract, peptone, extract of meat, and archaeal and bacterial cell homogenates by respiration of sulfur, L-cystine, or oxidized glutathione or, depending on the strains, on sulfite and thiosulfate. **Lithoautotrophic growth occurs on H_2, CO_2, and elemental sulfur by H_2/S-autotrophy.** Elemental sulfur is strictly required as the electron acceptor for autotrophic growth. The organisms live in boiling neutral to alkaline solfataric waters in the Azores, Iceland, and Italy.

Type species: *Pyrobaculum islandicum.*

Differentiation of the two species of the genus **Pyrobaculum:**

1. Facultative autotrophic growth; polar polytrichous flagellation; growth at 74°C and 0.8% NaCl; S⁰, sulfite, thiosulfate, L-cystine, and oxidized glutathione as electron acceptor.

    P. islandicum

2. Obligately heterotrophic growth; peritrichous flagellation; no growth at 74°C or in the presence of 0.8% NaCl; S⁰, L-cystine, and oxidized glutathione as electron acceptor.

    P. organotrophum

Genus **Thermofilum**

Thin rods are 0.15–0.35 µm in diameter and 1 to >100 µm in length. The median size is 5–10 µm. Cells **often have terminal pili and terminal spherical protrusions.** They are Gram negative and **strictly anaerobic. Heterotrophic growth occurs on yeast extract. Utilize peptides by sulfur respiration.** One species requires a cell component (polar lipid) from *Thermoproteus tenax.* **The pH range is 4.0–6.7, with an optimum at 5. The temperature range is 70–95°C, with an optimum at 85–90°C.** Organisms live in solfataric hot springs and water holes.

Type species: *Thermofilum pendens.*

Differentiation of the two species of the genus **Thermofilum:**

1. Requirement for a cell component (polar lipid) of *Thermoproteus tenax* or cell extracts of other archaea for growth.

    T. pendens

2. Cell components of *Thermoproteus tenax* not required for growth.

    T. librum

Editorial note: The species *Thermofilum librum* was not included in *Bergey's Manual of Systematic Bacteriology.*

The species was cited in 1986 by Stetter (In *Thermophiles: General, Molecular, and Applied Microbiology,* T. D. Brock, ed, J. Wiley & Sons, Chichester, England, pp. 39–74).

Genus **Thermoproteus**

Rigid rods are about 0.4 µm in diameter and from 1 to almost 100 µm in length. No cell septa are found. Cell division is by constriction, branching, and bud formation. Gram-negative cells do not have flagella. Lateral and/or terminal pili are present. Spheres protrude terminally. The **temperature range is 70–97°C, and the pH range is 2.5–6.** The best growth is at pH 5–6.5 and around 88°C. Chemolithoautotrophic or chemoheterotrophic growth occurs. **Chemolithoautotrophic growth is by H₂/S⁰ autotrophy with CO₂ as the carbon source.** H₂S is formed. **Chemoheterotrophic growth is by sulfur respiration of various organic substrates** including glucose, starch, glycogen, fumarate, and amino acids. Malate can replace sulfur as the electron acceptor. The organisms occur worldwide in solfataric hot waters and mud holes.

Type species: *Thermoproteus tenax.*

Differentiation of the two species of the genus **Thermoproteus:**

1. Chemolithotrophic growth with CO₂ as carbon source; optimal growth at pH 5.

    T. tenax

2. Facultatively chemolithotrophic growth with CO₂ as carbon source; acetate can be utilized instead of CO₂ as carbon source; optimal growth at pH 6.5.

    T. neutrophilus

SUBGROUP 3

Subgroup 3 consists of seven genera (see Table 35.1). The coccoid or disk-shaped organisms are strict anaerobes, which grow optimally above pH 4.

Genus **Desulfurococcus**

Coccoid cells are 0.5–1 µm in diameter and are Gram negative. Flagella may be present or absent. **Strictly anaerobic and show heterotrophic growth on proteins, peptides, or carbohydrates by sulfur respiration or fermentation. The temperature range is 70–95°C, with an optimum around 85°C. The pH range is 4.5–7, with an optimum around 6.** Organisms live in solfataric hot springs.

Type species: *Desulfurococcus mucosus.*

Differentiation of the three species of the genus **Desulfurococcus:**

1. Flagella absent; slime produced. In the presence of 1% yeast extract, giant cells up to 10 µm in diameter are formed. Sugars are not used.

    D. mucosus

2. Monopolar polytrichous flagellated; slime not produced; sugars not used.

    D. mobilis

3. Sugars used (e.g., 0.5% glucose in the presence of 0.02% yeast extract).

    D. saccharovorans

Editorial note: The species *D. saccharovorans* was not included in *Bergey's Manual of Systematic Bacteriology.* The species was created in 1986 by Stetter (In *Thermophiles: General, Molecular, and Applied Microbiology,* T. D. Brock, ed, J. Wiley & Sons, Chichester, England, pp. 39–74).

Genus **Hyperthermus**

Editorial note: The genus *Hyperthermus* was not included in *Bergey's Manual of Systematic Bacteriology.* The genus was created in 1990 by Zillig et al. (J. Bacteriol. *172:* 3959–3965).

Irregular coccoid spheres occur singly, in pairs, or in clumps of several hundred cells. Vacuoles are present. Cells are around 1.5 µm in diameter; flagella are absent, and pili are present. **Strictly anaerobic cells show heterotrophic growth by fermentation of peptide mixtures** (tryptone, peptone). **During fermentation, sulfur is reduced to H_2S with H_2 as an additional means for energy generation.** Fermentation products are CO_2, 1-butanol, acetic acid, phenylacetic acid, and hydroxyphenylacetic acid. **Optimal growth occurs at 95–106°C with 17 g NaCl/l and at pH 7.** The temperature maximum is 108°C. The natural habitats are hot solfataric sites on the sea floor.

Type (and only) species: *Hyperthermus butylicus.*

Characteristics of the species: As described for the genus.

Genus **Pyrococcus**

Slightly irregular cocci occur singly or in pairs and are 0.8–2.5 µm in width. Cells are monopolar polytrichous flagellated. **Strictly anaerobic,** they show **heterotrophic growth by fermentation or sulfur respiration** on peptone, tryptone, yeast extract, meat extract, extracts of bacteria and archaea, casein, starch, maltose, and casamino acids. During fermentation CO_2 and H_2 are formed, and in the presence of elemental sulfur H_2 is removed by H_2S formation (detoxification of H_2 by H_2S formation). **The temperature range is 70–103°C, with an optimum at 100°C. The pH range is 5–9,** with an optimum around 7. The salt range (NaCl) is 0.5–5%, with an optimum around 2%. The organisms live in geothermally heated marine sediments worldwide.

Type species: *Pyrococcus furiosus.*

Differentiation of the two species of the genus **Pyrococcus:**

1. Fermentation of peptones, yeast extract, and polysaccharides.

    P. furiosus

2. No fermentation of peptones, yeast extract, and polysaccharides; sulfur respiration; H_2S formed; utilization of elemental sulfur as an electron acceptor for these substrates.

    P. woesei

Genus **Pyrodictium**

Disk- to dish-shaped cells are 0.3–2.5 μm in diameter and about 0.2 μm thick. **They produce networks of hollow fibers,** with a diameter of 0.04–0.08 μm, which connect the cells. Cultures may grow in flocs. Cells are Gram negative with no flagella or motility and are **strictly anaerobic. The temperature range is 80–110°C, with an optimum at 105°C.** The pH range is 5–7, with an optimum around 5.5. The salt range (NaCl) is 0.1%–12%, with an optimum around 1.5%. Chemolithotrophic, mixotrophic, or fermentative growth may occur. **Chemolithotrophic growth occurs by hydrogen-sulfur autotrophy.** H_2S is formed from H_2 and S^0. Mixotrophic growth occurs on H_2 and thiosulfate in the presence of bacterial cell extracts or yeast extract. The organisms were isolated from submarine hydrothermal vents on the sea floor off Vulcano, Italy, and from hot deep sea vents near Guaymas, Mexico.

Type species: *Pyrodictium occultum.*

Differentiation of the three species of the genus **Pyrodictium:**

1. Hydrogen-sulfur autotrophy.

    P. occultum or *P. brockii*

The two species *P. occultum* and *P. brockii* cannot readily be distinguished.

2. Fermentation.

    P. abyssi

Editorial note: The species *P. abyssi* was not included in *Bergey's Manual of Systematic Bacteriology.* The species was created in 1990 by Stetter et al. (FEMS Microbiol. Rev. *75:* 117–124).

Genus **Staphylothermus**

Slightly irregular cocci occur singly, in pairs, as short chains, and as aggregates of up to 100 cells, each 0.5–15 μm in width. Cells are Gram negative and without flagella. They are **strictly anaerobic. Heterotrophic growth** occurs on peptone, tryptone, yeast extract, meat extract, and extracts of bacteria and archaea **in the presence of sulfur. H_2S, CO_2, acetate, and isovalerate are formed as metabolic products. The temperature range is 65–98°C, with an optimum at 92°C** in the presence of 0.1% yeast extract and 0.5% peptone. The organisms have been isolated from an anaerobic marine sediment near Vulcano, Italy, and from anaerobic samples of "black smokers" at the East Pacific Rise.

Type (and only) species: *Staphylothermus marinus.*

Characteristics of the species: As described for the genus.

Genus **Thermococcus**

Cocci occur singly and in pairs. Cells are 0.5–3 μm in diameter. There is monopolar polytrichous flagellation, or flagella are absent. Cells are **strictly anaerobic,** with **sulfur respiration or fermentation. In the presence of S^0, H_2S is formed.** Peptides, carbohydrates, yeast extract, tryptone, meat extract, and casein can function as carbon and energy sources. **The temperature range is 50–98°C.** The pH range is 4–8. The salt range (NaCl) is 1.8–6.5%. The natural biotopes are shallow solfataric marine water holes or hot sediments, worldwide.

Type species: *Thermococcus celer.*

Differentiation of the three species of the genus **Thermococcus:**

1. Flagella present; temperature maximum 93°C.

    T. celer

2. Flagella absent; temperature maximum 98°C; no growth on sugars or casamino acids.

    T. litoralis

3. Growth on peptides and polysaccharides; obligate sulfur dependent; no growth on amino acids and sugars.

    T. stetteri

Editorial note: The species *T. litoralis* and *T. stetteri* were not included in *Bergey's Manual of Systematic Bacteriology.* The first species (*T. litoralis*) was created in

1990 by Neuner et al. (Arch. Microbiol. *153:* 205–207), and the second species (*T. stetteri*) was created in 1989 by Miroshnichenko et al. (Syst. Appl. Microbiol. *12:* 257–262).

Genus **Thermodiscus**

Editorial note: The genus *Thermodiscus* was not included in *Bergey's Manual of Systematic Bacteriology.* The genus was created in 1986 by Stetter et al. (In *Thermophiles: General, Molecular, and Applied Microbiology,* T. D. Brock, ed, J. Wiley & Sons, Chichester, England, pp. 39–74).

Highly irregular dish- to disk-shaped cells are 0.3–3 µm in diameter and 0.1–0.2 µm thick. Pilus-like structures about 0.01 µm in diameter and up to 15 µm in length are present. Cells are Gram negative. Flagella and motility are absent. **Strictly anaerobic, these cells show heterotrophic growth on yeast extract in the presence (sulfur respiration) or absence of sulfur. The temperature range is 75–98°C, with an optimum around 88°C. The pH range is 5–7,** with an optimum of 5.5. The salt range is 1–4%, with an optimum around 2%. Organisms occur in submarine solfatara fields close to Vulcano, Italy.

Type (and only) species: *Thermodiscus maritimus.*

Characteristics of the species: As described for the genus.

Important Notes for Users of This Manual

Unless otherwise indicated in footnotes to tables, the meanings of symbols are as follows:

+ 90% or more of strains are positive

− 90% or more of strains are negative

d 11-89% of strains are positive

v strain instability (not equivalent to "d")

D Different reactions in different taxa (species of a genus or genera of a family)

All other symbols are defined in footnotes to tables.

Table 35.1 Differentiation of the subgroups and genera of extremely thermophilic and hyperthermophilic S⁰-metabolizers

Characteristic	Subgroup 1				Subgroup 2			Subgroup 3						
	Acidianus	*Desulfurolobus*	*Metallosphaera*	*Sulfolobus*	*Pyrobaculum*	*Thermofilum*	*Thermoproteus*	*Desulfurococcus*	*Hyperthermus*	*Pyrococcus*	*Pyrodictium*	*Staphylothermus*	*Thermococcus*	*Thermodiscus*
1. Morphology	Coccoid	Coccoid	Coccoid	Coccoid	Rod	Filament	Rod	Coccoid	Coccoid	Coccoid	Disk-shaped	Coccoid	Coccoid	Disk-shaped
2. Flagella	–	–	–	–	+	–	–	+/–	–	+	–	–	+/–	–
3. Optimal growth below pH 4	+	+	+	+	–	–	–	–	–	–	–	–	–	–
4. S⁰ reduction	+	+	–	–	+	+	+	+	+	+	+	+	+	+
5. S⁰ oxidation	+	+	+	+/–	–	–	–	–	–	–	–	–	–	–
6. H₂S oxidation	+	+	+	+	–	–	–	–	–	–	–	–	–	–
7. Ore leaching	+/–	?	+	–	–	–	–	–	–	–	–	–	–	–
8. Aerobic growth	+	+	+	+	–	–	–	–	–	–	–	–	–	–
9. Anaerobic growth	+	+	–	–	+	+	+	+	+	+	+	+	+	+
10. Temperature range (°C)	45–96	?–87	50–80	55–87	74–103	70–95	70–97	70–95	?–108	70–103	80–110	65–98	50–98	75–98
11. pH range	1–6	?	1–4.5	1–6	5–7	4–6.7	2.5–6	4.5–7	?	5–9	5–7	?	4–8	5–7
12. NaCl range (%)	0.1–4	?	?	?	0–0.8	?	?	?	?	?	0.1–12	?	1.8–6.5	1–4
13. Energy source: H₂	+	+	–	–	+	+	+	–	+	–	+/–	–	–	–
Carbohydrates	+	+	+	+	–	–	+	+/–	–	+	+/–	+	+	?
14. Fermentation	–	–	–	–	–	–	?	+/–	+	+/–	+/–	?	+/–	?

Symbols: +, present or positive; –, absent or negative; +/–, present or absent, positive or negative; ?, not described.
Subgroup 1 is defined by Characteristics 3, 6, and 8.
Subgroup 2 is defined by Characteristic 1.
Subgroup 3 is defined by Characteristics 1, 3, and 9.

755

INDEX OF SCIENTIFIC NAMES OF BACTERIA

Key to the Index

Nomenclature: Every bacterial name mentioned in this manual is listed in the index. Specific epithets are listed individually (followed by the genus in parentheses) and also under the genus as a subentry.

Lower case: Genera, species, and subspecies.

CAPITALS: Taxa higher than genus (tribe, family, order, class, division, kingdom).

Pagination

Lightface: Pages on which taxa are mentioned.

Boldface: Page on which the description of a taxon is given.

Index

INDEX OF SCIENTIFIC NAMES OF BACTERIA

coli (Campylobacter), 58
coli (Escherichia), 178, 179, 180, 181, 182, 188, 209, 224-225, 233-234, 428, 605
collis (Mycoplasma), 712, 715
columbinasale (Mycoplasma), 712
columbinum (Mycoplasma), 712
columborale (Mycoplasma), 712
columellifera (Pilimelia), 666
columnaris (Cytophaga), 503
Colwellia, 190, 193, 264-265
    hadaliensis, 193, 264-265
    psychroerythrus, 193, 194
colwelliana (Alteromonas), 121, 122, 132, 133
Comamonas, 55, **80**, 106, 138, 157
    acidovorans, 80, 138, 157
    terrigena, 55, 80, 138
    testosteroni, 80, 138, 157
comes (Coprococcus), 537
commune (Thermodesulfobacterium), 337
communis (Alteromonas), 43, 87, 133, 144
communis (Marinomonas), 39, 43, 48, 51, 87, 121, 122, 144
compacta (Pseudonocardia), 646
compransoris (Pseudomonas), 125, 160, 167
concilii (Methanosaeta), 725
concisus (Campylobacter), 58
congolensis (Dermatophilus), 651
conjunctivae (Mycoplasma), 712
consettensis (Actinoplanes), 657-658
consociatum (Prosthecomicrobium), 474
constellatus (Streptococcus), 558
cookii (Mycobacterium), 599
Coprococcus, **527**, 537
    catus, 537
    comes, 537
    eutactus, 527, 537
coprophilus (Rhodococcus), 638
Corallococcus, 516, 517, **520**, 525
    coralloides, 525
    exiguus, 525
    macrosporus, 520
coralloides (Corallococcus), 525
coriaceae (Borrelia), 33
Coriobacterium, **576**, 587
    glomerans, 576
coronata (Siderocapsa), 453
corporis (Bacteroides), 321
corrodens (Eikenella), 199, 286
corrugata (Pseudomonas), 159
corrugatum (Streptosporangium), 687
Corynebacterium, 15, 574, 575, **576**, 580, 585, 592-593, 598, 634-635
    amycolatum, 592
    bovis, 592, 635
    callunae, 592
    cystitidis, 592
    diphtheriae, 576, 592
    flavescens, 592
    glutamicum, 592
    jeikeium, 592

kutscheri, 592
matruchotii, 297, 576, 586, 592, 593
minutissimum, 593, 635
mycetoides, 593
paurometabolum, 593, 627
pilosum, 593
pseudodiphtheriticum, 593
pseudotuberculosis, 593
renale, 593
striatum, 593
variabilis, 635
vitarumen, 593
xerosis, 593
costicola (Vibrio), 253, 262-263
Coxiella, 15
crassa (Simonsiella), 511
cremea (Actinomadura), 691
Crenothrix, 477, **478**, 482
    polyspora, 477, 478
crescentus (Caulobacter), 472
criceti (Veillonella), 350
cricetuli (Mycoplasma), 712
cricetus (Streptococcus), 553
Crinalium, 382, 392, **393**, 399
    epipsammum, 382
Cristispira, **28**, 32
    pectinis, 28
crocatus (Chondromyces), 519
crocidurae (Borrelia), 33
crossotus (Butyrivibrio), 323, 590
crunogena (Thiomicrospira), 436, 451
cryaerophila (Campylobacter), 51, 58, 61
cryaerophilus (Arcobacter), 61
cryocrescens (Kluyvera), 182, 212, 223, 235
cryophilis (Saccharothrix), 703
cryotolerans (Nitrosomonas), 455
cryptum (Acidiphilium), 72, 127, 428, 439, 452
culicicola (Spiroplasma), 716-717
cuniculi (Branhamella), 147, 148, 149
cuniculi (Methanosphaera), 721
cuniculi (Moraxella), 147, 148, 149
cuniculi (Neisseria), 149
cupida (Deleya), 81, 139
cupidus (Alcaligenes), 81, 131, 139
Cupriavidus, **80-81**, 106
    necator, 80-81
cuprinus (Thiobacillus), 435, 452
curtisii (Mobiluncus), 579, 595
Curtobacterium, 575, **576**, 584, 594
    albidum, 594
    citreum, 576, 594
    flaccumfaciens, 576, 584, 594
    luteum, 594
    plantarum, 594
    pusillum, 594
curva (Wolinella), **45-46**, 51, 60, 61, 305, 334
curvata (Thermomonospora), 694, 697
curvatus (Desulfobacter), 344
curvus (Campylobacter), 61
cyanea (Saccharomonospora), 646
cyaneus (Streptomyces), 670-671

Cyanobacterium, 385, 422
Cyanobacterium-*cluster* (Synechococcus-*culture group*), 385, 422
Cyanobium, 385, 422
Cyanobium-*cluster* (Synechococcus-*culture group*), 385, 422
Cyanobotrys, 412
Cyanospira, 401, 402, 404
Cyanothece, 380, **382-383**, 385, 386, 389, 421
Cyanothece-*culture group*, 381, 383, 384
Cyclobacterium, **66**, 68, 69
    marinum, 66, 69
    marinus, 66
Cylindrospermum, 399, 401, 402, **404**
cynodegmi (Capnocytophaga), 486, 501
cynos (Mycoplasma), 712
cypripedii (Erwinia), 208, 230-232, 248
cystitidis (Corynebacterium), 592
Cystobacter, 516, 517, 519, **520**
    ferrugineus, 521
    fuscus, 520
    minus, 521
    velatus, 521
    violaceus, 521
Cytophaga, 485, **486-487**, 502-506, 516
    agarovorans, 502
    allerginae, 502
    aprica, 502
    aquatilis, 502
    arvensicola, 502
    aurantiaca, 502
    columnaris, 503
    diffluens, 503
    fermentans, 503
    flevensis, 503
    heparina, 503
    hutchinsonii, 487, 503
    johnsonii (*sic*), 504
    latercula, 504
    lytica, 504
    marina, 504
    marinoflava, 504
    pectinovora, 504
    psychrophila, 505
    saccharophila, 505
    salmonicolor, 505
    succinicans, 505
    uliginosa, 505

Dactylococcopsis, **384**
    salina, 380, 384
Dactylosporangium, 605, 606, 611, 612, 615, 619, 621, **654**, 663
    aurantiacum, 654, 663
    matsuzakiense, 663
    roseum, 663
    thailandense, 663
    vinaceum, 663
dagmatis (Pasteurella), 196, 280-281
damnosus (Pediococcus), 530, 543
damsela (Listonella), 193, 262-263
damsela (Vibrio), 193, 262-263, 273-274

764